Advanced Engineering Mathematics

Math 105A

Author
CSU Sacramento

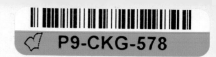

ISBN 9781119918387

Printed in the United States of America SKY10029973_092221

List of Titles

Advanced Engineering Mathematics, 10th edition
by Erwin Kreyszig
Copyright © 2012, ISBN: 978-0-470-45836-5

Table of Contents

Systems of Units. Some Important Conversion Factors

The most important systems of units are shown in the table below. The mks system is also known as the *International System of Units* (abbreviated *SI*), and the abbreviations sec (instead of s), gm (instead of g), and nt (instead of N) are also used.

System of units	Length	Mass	Time	Force
cgs system	centimeter (cm)	gram (g)	second (s)	dyne
mks system	meter (m)	kilogram (kg)	second (s)	newton (nt)
Engineering system	foot (ft)	slug	second (s)	pound (lb)

1 inch (in.) = 2.540000 cm

1 foot (ft) = 12 in. = 30.480000 cm

1 yard (yd) = 3 ft = 91.440000 cm

1 statute mile (mi) = 5280 ft = 1.609344 km

1 nautical mile = 6080 ft = 1.853184 km

1 acre = 4840 yd^2 = 4046.8564 m^2

1 mi^2 = 640 acres = 2.5899881 km^2

1 fluid ounce = 1/128 U.S. gallon = 231/128 $in.^3$ = 29.573730 cm^3

1 U.S. gallon = 4 quarts (liq) = 8 pints (liq) = 128 fl oz = 3785.4118 cm^3

1 British Imperial and Canadian gallon = 1.200949 U.S. gallons = 4546.087 cm^3

1 slug = 14.59390 kg

1 pound (lb) = 4.448444 nt

1 newton (nt) = 10^5 dynes

1 British thermal unit (Btu) = 1054.35 joules

1 joule = 10^7 ergs

1 calorie (cal) = 4.1840 joules

1 kilowatt-hour (kWh) = 3414.4 Btu = $3.6 \cdot 10^6$ joules

1 horsepower (hp) = 2542.48 Btu/h = 178.298 cal/sec = 0.74570 kW

1 kilowatt (kW) = 1000 watts = 3414.43 Btu/h = 238.662 cal/s

$°F = °C \cdot 1.8 + 32$

$1° = 60' = 3600'' = 0.017453293$ radian

For further details see, for example, D. Halliday, R. Resnick, and J. Walker, *Fundamentals of Physics.* 9th ed., Hoboken, N. J: Wiley, 2011. See also AN American National Standard, ASTM/IEEE Standard Metric Practice, Institute of Electrical and Electronics Engineers, Inc. (IEEE), 445 Hoes Lane, Piscataway, N. J. 08854, website at www.ieee.org.

Differentiation

$(cu)' = cu' \qquad (c \text{ constant})$

$(u + v)' = u' + v'$

$(uv)' = u'v + uv'$

$\left(\dfrac{u}{v}\right)' = \dfrac{u'v - uv'}{v^2}$

$\dfrac{du}{dx} = \dfrac{du}{dy} \cdot \dfrac{dy}{dx} \qquad$ (Chain rule)

$(x^n)' = nx^{n-1}$

$(e^x)' = e^x$

$(e^{ax})' = ae^{ax}$

$(a^x)' = a^x \ln a$

$(\sin x)' = \cos x$

$(\cos x)' = -\sin x$

$(\tan x)' = \sec^2 x$

$(\cot x)' = -\csc^2 x$

$(\sinh x)' = \cosh x$

$(\cosh x)' = \sinh x$

$(\ln x)' = \dfrac{1}{x}$

$(\log_a x)' = \dfrac{\log_a e}{x}$

$(\arcsin x)' = \dfrac{1}{\sqrt{1 - x^2}}$

$(\arccos x)' = -\dfrac{1}{\sqrt{1 - x^2}}$

$(\arctan x)' = \dfrac{1}{1 + x^2}$

$(\operatorname{arccot} x)' = -\dfrac{1}{1 + x^2}$

Integration

$\int uv' \, dx = uv - \int u'v \, dx$ (by parts)

$\int x^n \, dx = \dfrac{x^{n+1}}{n+1} + c \qquad (n \neq -1)$

$\int \dfrac{1}{x} \, dx = \ln |x| + c$

$\int e^{ax} \, dx = \dfrac{1}{a} e^{ax} + c$

$\int \sin x \, dx = -\cos x + c$

$\int \cos x \, dx = \sin x + c$

$\int \tan x \, dx = -\ln |\cos x| + c$

$\int \cot x \, dx = \ln |\sin x| + c$

$\int \sec x \, dx = \ln |\sec x + \tan x| + c$

$\int \csc x \, dx = \ln |\csc x - \cot x| + c$

$\int \dfrac{dx}{x^2 + a^2} = \dfrac{1}{a} \arctan \dfrac{x}{a} + c$

$\int \dfrac{dx}{\sqrt{a^2 - x^2}} = \arcsin \dfrac{x}{a} + c$

$\int \dfrac{dx}{\sqrt{x^2 + a^2}} = \operatorname{arcsinh} \dfrac{x}{a} + c$

$\int \dfrac{dx}{\sqrt{x^2 - a^2}} = \operatorname{arccosh} \dfrac{x}{a} + c$

$\int \sin^2 x \, dx = \tfrac{1}{2}x - \tfrac{1}{4} \sin 2x + c$

$\int \cos^2 x \, dx = \tfrac{1}{2}x + \tfrac{1}{4} \sin 2x + c$

$\int \tan^2 x \, dx = \tan x - x + c$

$\int \cot^2 x \, dx = -\cot x - x + c$

$\int \ln x \, dx = x \ln x - x + c$

$\int e^{ax} \sin bx \, dx$

$= \dfrac{e^{ax}}{a^2 + b^2} (a \sin bx - b \cos bx) + c$

$\int e^{ax} \cos bx \, dx$

$= \dfrac{e^{ax}}{a^2 + b^2} (a \cos bx + b \sin bx) + c$

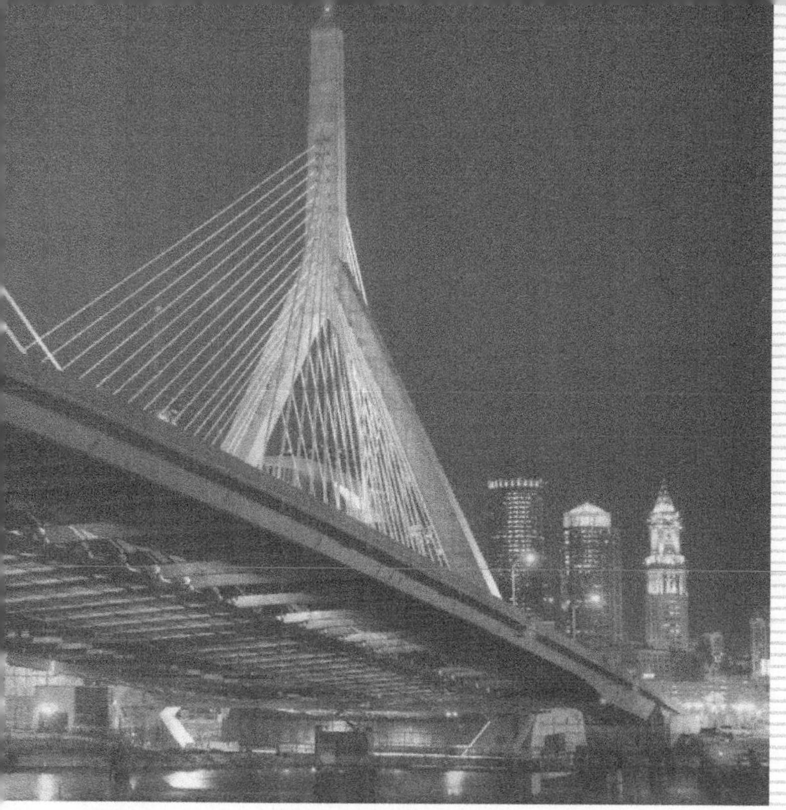

PART A

Ordinary Differential Equations (ODEs)

Many physical laws and relations can be expressed mathematically in the form of differential equations. Thus it is natural that this book opens with the study of differential equations and their solutions. Indeed, many engineering problems appear as differential equations.

The main objectives of Part A are twofold: the study of ordinary differential equations and their most important methods for solving them and the study of modeling.

Ordinary differential equations **(ODEs)** are differential equations that depend on a single variable. The more difficult study of partial differential equations (PDEs), that is, differential equations that depend on several variables, is covered in Part C.

Modeling is a crucial general process in engineering, physics, computer science, biology, medicine, environmental science, chemistry, economics, and other fields that translates a physical situation or some other observations into a "mathematical model." Numerous examples from engineering (e.g., mixing problem), physics (e.g., Newton's law of cooling), biology (e.g., Gompertz model), chemistry (e.g., radiocarbon dating), environmental science (e.g., population control), etc. shall be given, whereby this process is explained in detail, that is, how to set up the problems correctly in terms of differential equations.

For those interested in solving ODEs numerically on the computer, look at Secs. 21.1–21.3 of Chapter 21 of Part F, that is, ***numeric methods for ODEs***. These sections are kept independent by design of the other sections on numerics. ***This allows for the study of numerics for ODEs directly after Chap. 1 or 2.***

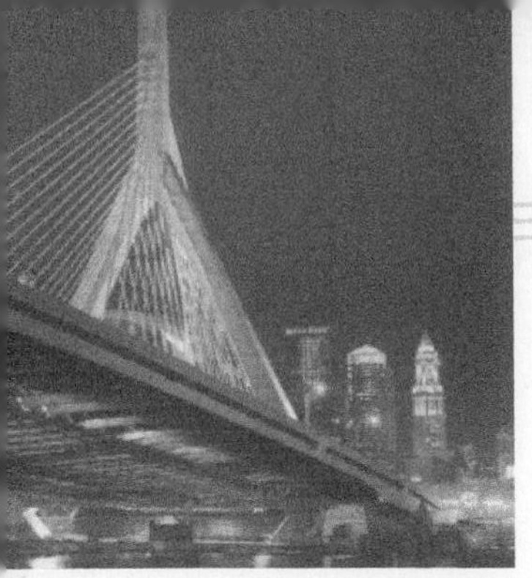

CHAPTER 1

First-Order ODEs

Chapter 1 begins the study of ordinary differential equations (ODEs) by deriving them from physical or other problems (***modeling***), solving them by standard mathematical methods, and interpreting solutions and their graphs in terms of a given problem. The simplest ODEs to be discussed are ODEs ***of the first order*** because they involve only the first derivative of the unknown function and no higher derivatives. These unknown functions will usually be denoted by $y(x)$ or $y(t)$ when the independent variable denotes time t. The chapter ends with a study of the existence and uniqueness of solutions of ODEs in Sec. 1.7.

Understanding the basics of ODEs requires solving problems by hand (paper and pencil, or typing on your computer, but first without the aid of a CAS). In doing so, you will gain an important conceptual understanding and feel for the basic terms, such as ODEs, direction field, and initial value problem. If you wish, you can use your **Computer Algebra System (CAS)** for checking solutions.

COMMENT. ***Numerics for first-order ODEs can be studied immediately after this chapter.*** See Secs. 21.1–21.2, which are independent of other sections on numerics.

Prerequisite: Integral calculus.
Sections that may be omitted in a shorter course: 1.6, 1.7.
References and Answers to Problems: App. 1 Part A, and App. 2.

1.1 Basic Concepts. Modeling

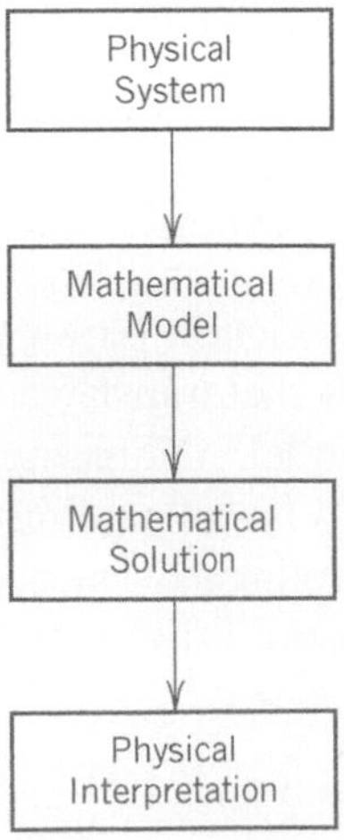

Fig. 1. Modeling, solving, interpreting

If we want to solve an engineering problem (usually of a physical nature), we first have to formulate the problem as a mathematical expression in terms of variables, functions, and equations. Such an expression is known as a mathematical **model** of the given problem. The process of setting up a model, solving it mathematically, and interpreting the result in physical or other terms is called *mathematical modeling* or, briefly, **modeling**.

Modeling needs experience, which we shall gain by discussing various examples and problems. (Your computer may often help you in *solving* but rarely in *setting up* models.)

Now many physical concepts, such as velocity and acceleration, are derivatives. Hence a model is very often an equation containing derivatives of an unknown function. Such a model is called a **differential equation**. Of course, we then want to find a solution (a function that satisfies the equation), explore its properties, graph it, find values of it, and interpret it in physical terms so that we can understand the behavior of the physical system in our given problem. However, before we can turn to methods of solution, we must first define some basic concepts needed throughout this chapter.

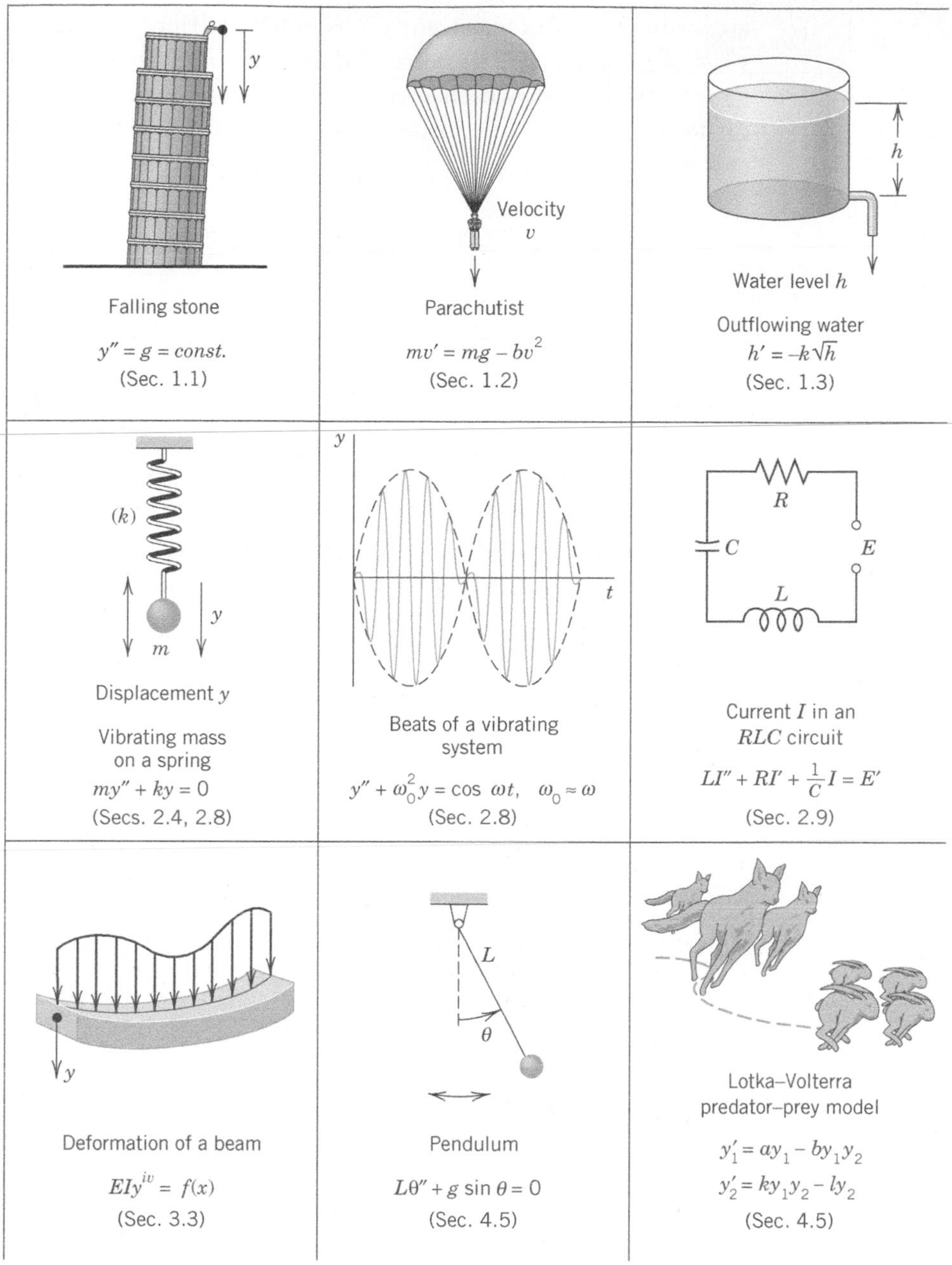

Fig. 2. Some applications of differential equations

An **ordinary differential equation (ODE)** is an equation that contains one or several derivatives of an unknown function, which we usually call $y(x)$ (or sometimes $y(t)$ if the independent variable is time t). The equation may also contain y itself, known functions of x (or t), and constants. For example,

$$y' = \cos x \tag{1}$$

$$y'' + 9y = e^{-2x} \tag{2}$$

$$y'y''' - \tfrac{3}{2}y'^2 = 0 \tag{3}$$

are ordinary differential equations (ODEs). Here, as in calculus, y' denotes dy/dx, $y'' = d^2y/dx^2$, etc. The term *ordinary* distinguishes them from *partial differential equations* (PDEs), which involve partial derivatives of an unknown function of *two or more* variables. For instance, a PDE with unknown function u of two variables x and y is

$$\frac{\partial^2 u}{\partial x^2} + \frac{\partial^2 u}{\partial y^2} = 0.$$

PDEs have important engineering applications, but they are more complicated than ODEs; they will be considered in Chap. 12.

An ODE is said to be of **order** n if the nth derivative of the unknown function y is the highest derivative of y in the equation. The concept of order gives a useful classification into ODEs of first order, second order, and so on. Thus, (1) is of first order, (2) of second order, and (3) of third order.

In this chapter we shall consider **first-order ODEs**. Such equations contain only the first derivative y' and may contain y and any given functions of x. Hence we can write them as

$$F(x, y, y') = 0 \tag{4}$$

or often in the form

$$y' = f(x, y).$$

This is called the *explicit form*, in contrast to the *implicit form* (4). For instance, the implicit ODE $x^{-3}y' - 4y^2 = 0$ (where $x \neq 0$) can be written explicitly as $y' = 4x^3y^2$.

Concept of Solution

A function

$$y = h(x)$$

is called a **solution** of a given ODE (4) on some open interval $a < x < b$ if $h(x)$ is defined and differentiable throughout the interval and is such that the equation becomes an identity if y and y' are replaced with h and h', respectively. The curve (the graph) of h is called a **solution curve**.

Here, **open interval** $a < x < b$ means that the endpoints a and b are not regarded as points belonging to the interval. Also, $a < x < b$ includes *infinite intervals* $-\infty < x < b$, $a < x < \infty$, $-\infty < x < \infty$ (the real line) as special cases.

EXAMPLE 1 Verification of Solution

Verify that $y = c/x$ (c an arbitrary constant) is a solution of the ODE $xy' = -y$ for all $x \neq 0$. Indeed, differentiate $y = c/x$ to get $y' = -c/x^2$. Multiply this by x, obtaining $xy' = -c/x$; thus, $xy' = -y$, the given ODE. ■

EXAMPLE 2 Solution by Calculus. Solution Curves

The ODE $y' = dy/dx = \cos x$ can be solved directly by integration on both sides. Indeed, using calculus, we obtain $y = \int \cos x \, dx = \sin x + c$, where c is an arbitrary constant. This is a ***family of solutions***. Each value of c, for instance, 2.75 or 0 or -8, gives one of these curves. Figure 3 shows some of them, for $c = -3, -2, -1, 0, 1, 2, 3, 4$. ■

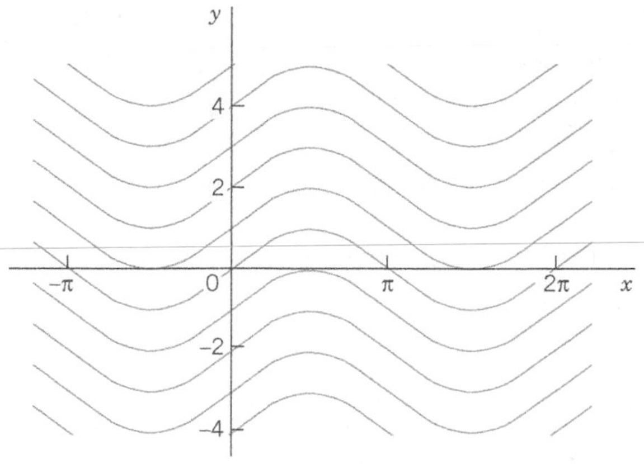

Fig. 3. Solutions $y = \sin x + c$ of the ODE $y' = \cos x$

EXAMPLE 3 (A) Exponential Growth. (B) Exponential Decay

From calculus we know that $y = ce^{0.2t}$ has the derivative

$$y' = \frac{dy}{dt} = 0.2e^{0.2t} = 0.2y.$$

Hence y is a solution of $y' = 0.2y$ (Fig. 4A). This ODE is of the form $y' = ky$. With positive-constant k it can model exponential growth, for instance, of colonies of bacteria or populations of animals. It also applies to humans for small populations in a large country (e.g., the United States in early times) and is then known as **Malthus's law**.[1] We shall say more about this topic in Sec. 1.5.

(B) Similarly, $y' = -0.2$ (with a minus on the right) has the solution $y = ce^{-0.2t}$, (Fig. 4B) modeling **exponential decay**, as, for instance, of a radioactive substance (see Example 5). ■

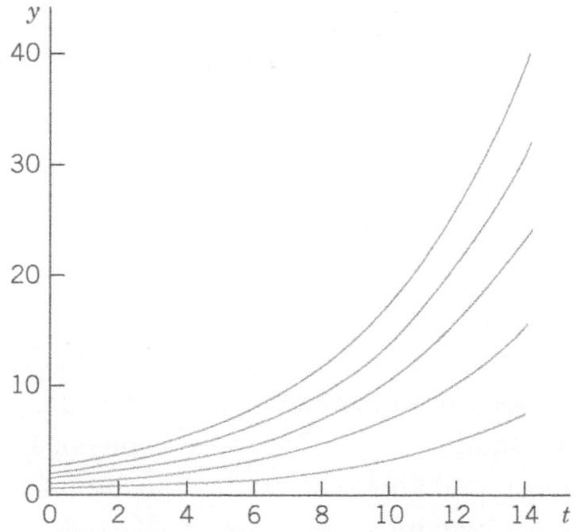

Fig. 4A. Solutions of $y' = 0.2y$ in Example 3 **(exponential growth)**

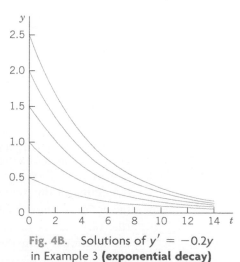

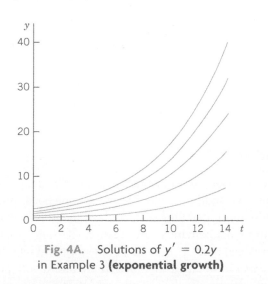

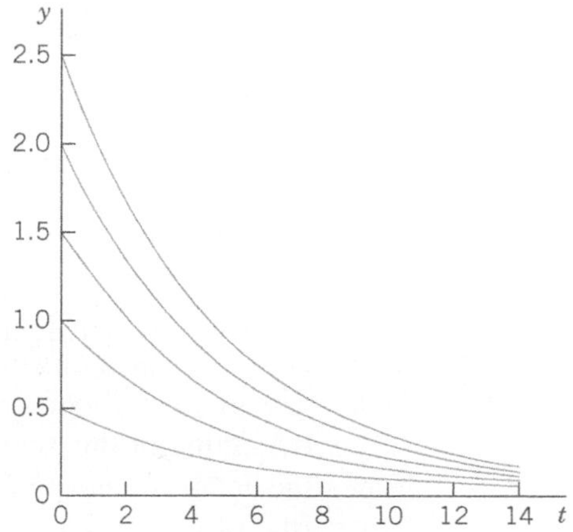

Fig. 4B. Solutions of $y' = -0.2y$ in Example 3 **(exponential decay)**

[1]Named after the English pioneer in classic economics, THOMAS ROBERT MALTHUS (1766–1834).

We see that each ODE in these examples has a solution that contains an arbitrary constant c. Such a solution containing an arbitrary constant c is called a **general solution** of the ODE.

(We shall see that c is sometimes not completely arbitrary but must be restricted to some interval to avoid complex expressions in the solution.)

We shall develop methods that will give general solutions *uniquely* (perhaps except for notation). Hence we shall say ***the*** general solution of a given ODE (instead of ***a*** general solution).

Geometrically, the general solution of an ODE is a family of infinitely many solution curves, one for each value of the constant c. If we choose a specific c (e.g., $c = 6.45$ or 0 or -2.01) we obtain what is called a **particular solution** of the ODE. A particular solution does not contain any arbitrary constants.

In most cases, general solutions exist, and every solution not containing an arbitrary constant is obtained as a particular solution by assigning a suitable value to c. Exceptions to these rules occur but are of minor interest in applications; see Prob. 16 in Problem Set 1.1.

Initial Value Problem

In most cases the unique solution of a given problem, hence a particular solution, is obtained from a general solution by an **initial condition** $y(x_0) = y_0$, with given values x_0 and y_0, that is used to determine a value of the arbitrary constant c. Geometrically this condition means that the solution curve should pass through the point (x_0, y_0) in the xy-plane. An ODE, together with an initial condition, is called an **initial value problem**. Thus, if the ODE is explicit, $y' = f(x, y)$, the initial value problem is of the form

$$y' = f(x, y), \qquad y(x_0) = y_0. \tag{5}$$

EXAMPLE 4 Initial Value Problem

Solve the initial value problem

$$y' = \frac{dy}{dx} = 3y, \qquad y(0) = 5.7.$$

Solution. The general solution is $y(x) = ce^{3x}$; see Example 3. From this solution and the initial condition we obtain $y(0) = ce^0 = c = 5.7$. Hence the initial value problem has the solution $y(x) = 5.7e^{3x}$. This is a particular solution. ■

More on Modeling

The general importance of modeling to the engineer and physicist was emphasized at the beginning of this section. We shall now consider a basic physical problem that will show the details of the typical steps of modeling. Step 1: the transition from the physical situation (the physical system) to its mathematical formulation (its mathematical model); Step 2: the solution by a mathematical method; and Step 3: the physical interpretation of the result. This may be the easiest way to obtain a first idea of the nature and purpose of differential equations and their applications. Realize at the outset that your ***computer*** (your ***CAS***) may perhaps give you a hand in Step 2, but Steps 1 and 3 are basically your work.

And Step 2 requires a solid knowledge and good understanding of solution methods available to you—*you* have to choose the method for your work by hand or by the computer. Keep this in mind, and always check computer results for errors (which may arise, for instance, from false inputs).

EXAMPLE 5 Radioactivity. Exponential Decay

Given an amount of a radioactive substance, say, 0.5 g (gram), find the amount present at any later time.

Physical Information. Experiments show that at each instant a radioactive substance decomposes—and is thus decaying in time—proportional to the amount of substance present.

Step 1. Setting up a mathematical model of the physical process. Denote by $y(t)$ the amount of substance still present at any time t. By the physical law, the time rate of change $y'(t) = dy/dt$ is proportional to $y(t)$. This gives the ***first-order ODE***

$$\frac{dy}{dt} = -ky \tag{6}$$

where the constant k is positive, so that, because of the minus, we do get decay (as in [B] of Example 3). The value of k is known from experiments for various radioactive substances (e.g., $k = 1.4 \cdot 10^{-11}\ \text{sec}^{-1}$, approximately, for radium ${}^{226}_{88}\text{Ra}$).

Now the given initial amount is 0.5 g, and we can call the corresponding instant $t = 0$. Then we have the ***initial condition*** $y(0) = 0.5$. This is the instant at which our observation of the process begins. It motivates the term *initial condition* (which, however, is also used when the independent variable is not time or when we choose a t other than $t = 0$). Hence the mathematical model of the physical process is the ***initial value problem***

$$\frac{dy}{dt} = -ky, \qquad y(0) = 0.5. \tag{7}$$

Step 2. Mathematical solution. As in (B) of Example 3 we conclude that the ODE (6) models exponential decay and has the general solution (with arbitrary constant c but definite given k)

$$y(t) = ce^{-kt}. \tag{8}$$

We now determine c by using the initial condition. Since $y(0) = c$ from (8), this gives $y(0) = c = 0.5$. Hence the particular solution governing our process is (cf. Fig. 5)

$$y(t) = 0.5e^{-kt} \qquad (k > 0). \tag{9}$$

Always check your result—it may involve human or computer errors! Verify by differentiation (chain rule!) that your solution (9) satisfies (7) as well as $y(0) = 0.5$:

$$\frac{dy}{dt} = -0.5ke^{-kt} = -k \cdot 0.5e^{-kt} = -ky, \qquad y(0) = 0.5e^{0} = 0.5.$$

Step 3. Interpretation of result. Formula (9) gives the amount of radioactive substance at time t. It starts from the correct initial amount and decreases with time because k is positive. The limit of y as $t \to \infty$ is zero. ■

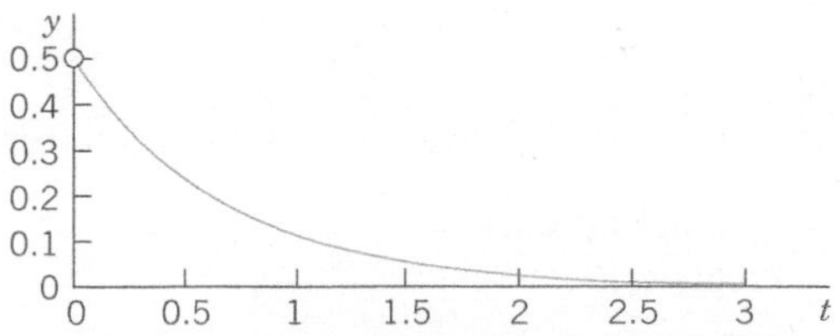

Fig. 5. Radioactivity (Exponential decay, $y = 0.5e^{-kt}$, with $k = 1.5$ as an example)

PROBLEM SET 1.1

1–8 CALCULUS

Solve the ODE by integration or by remembering a differentiation formula.

1. $y' + 2 \sin 2\pi x = 0$
2. $y' + xe^{-x^2/2} = 0$
3. $y' = y$
4. $y' = -1.5y$
5. $y' = 4e^{-x} \cos x$
6. $y'' = -y$
7. $y' = \cosh 5.13x$
8. $y''' = e^{-0.2x}$

9–15 VERIFICATION. INITIAL VALUE PROBLEM (IVP)

(a) Verify that y is a solution of the ODE. **(b)** Determine from y the particular solution of the IVP. **(c)** Graph the solution of the IVP.

9. $y' + 4y = 1.4, \quad y = ce^{-4x} + 0.35, \quad y(0) = 2$
10. $y' + 5xy = 0, \quad y = ce^{-2.5x^2}, \quad y(0) = \pi$
11. $y' = y + e^x, \quad y = (x + c)e^x, \quad y(0) = \frac{1}{2}$
12. $yy' = 4x, \quad y^2 - 4x^2 = c \ (y > 0), \quad y(1) = 4$
13. $y' = y - y^2, \quad y = \dfrac{1}{1 + ce^{-x}}, \quad y(0) = 0.25$
14. $y' \tan x = 2y - 8, \quad y = c \sin^2 x + 4, \quad y(\frac{1}{2}\pi) = 0$
15. Find two constant solutions of the ODE in Prob. 13 by inspection.
16. **Singular solution.** An ODE may sometimes have an additional solution that cannot be obtained from the general solution and is then called a *singular solution.* The ODE $y'^2 - xy' + y = 0$ is of this kind. Show by differentiation and substitution that it has the general solution $y = cx - c^2$ and the singular solution $y = x^2/4$. Explain Fig. 6.

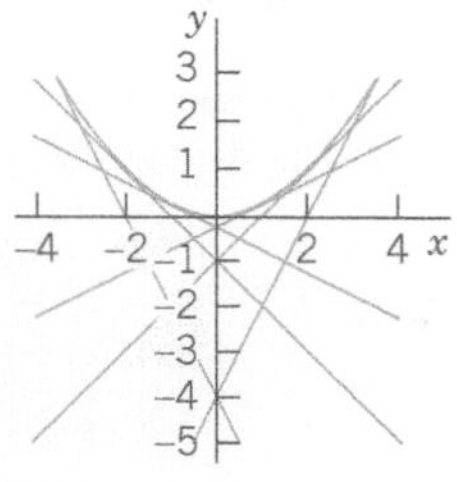

Fig. 6. Particular solutions and singular solution in Problem 16

17–20 MODELING, APPLICATIONS

These problems will give you a first impression of modeling. Many more problems on modeling follow throughout this chapter.

17. **Half-life.** The *half-life* measures exponential decay. It is the time in which half of the given amount of radioactive substance will disappear. What is the half-life of $^{226}_{88}\mathrm{Ra}$ (in years) in Example 5?
18. **Half-life.** Radium $^{224}_{88}\mathrm{Ra}$ has a half-life of about 3.6 days.

    (a) Given 1 gram, how much will still be present after 1 day?

    (b) After 1 year?
19. **Free fall.** In dropping a stone or an iron ball, air resistance is practically negligible. Experiments show that the acceleration of the motion is constant (equal to $g = 9.80 \text{ m/sec}^2 = 32 \text{ ft/sec}^2$, called the **acceleration of gravity**). Model this as an ODE for $y(t)$, the distance fallen as a function of time t. If the motion starts at time $t = 0$ from rest (i.e., with velocity $v = y' = 0$), show that you obtain the familiar law of free fall
$$y = \tfrac{1}{2}gt^2.$$
20. **Exponential decay. Subsonic flight.** The efficiency of the engines of subsonic airplanes depends on air pressure and is usually maximum near 35,000 ft. Find the air pressure $y(x)$ at this height. *Physical information.* The rate of change $y'(x)$ is proportional to the pressure. At 18,000 ft it is half its value $y_0 = y(0)$ at sea level. *Hint.* Remember from calculus that if $y = e^{kx}$, then $y' = ke^{kx} = ky$. Can you see without calculation that the answer should be close to $y_0/4$?

1.2 Geometric Meaning of $y' = f(x, y)$. Direction Fields, Euler's Method

A first-order ODE

$$y' = f(x, y) \tag{1}$$

has a simple geometric interpretation. From calculus you know that the derivative $y'(x)$ of $y(x)$ is the slope of $y(x)$. Hence a solution curve of (1) that passes through a point (x_0, y_0) must have, at that point, the slope $y'(x_0)$ equal to the value of f at that point; that is,

$$y'(x_0) = f(x_0, y_0).$$

Using this fact, we can develop graphic or numeric methods for obtaining approximate solutions of ODEs (1). This will lead to a better conceptual understanding of an ODE (1). Moreover, such methods are of practical importance since many ODEs have complicated solution formulas or no solution formulas at all, whereby numeric methods are needed.

Graphic Method of Direction Fields. Practical Example Illustrated in Fig. 7. We can show directions of solution curves of a given ODE (1) by drawing short straight-line segments (lineal elements) in the xy-plane. This gives a **direction field** (or *slope field*) into which you can then fit (approximate) solution curves. This may reveal typical properties of the whole family of solutions.

Figure 7 shows a direction field for the ODE

$$y' = y + x \tag{2}$$

obtained by a CAS (Computer Algebra System) and some approximate solution curves fitted in.

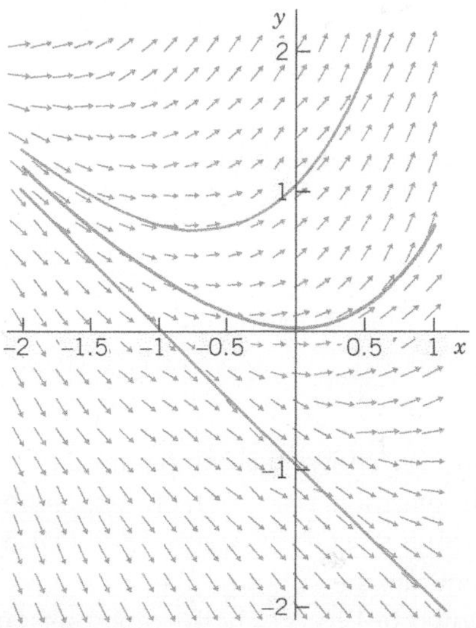

Fig. 7. Direction field of $y' = y + x$, with three approximate solution curves passing through (0, 1), (0, 0), (0, −1), respectively

If you have no CAS, first draw a few *level curves* $f(x, y) = \text{const}$ of $f(x, y)$, then parallel lineal elements along each such curve (which is also called an **isocline**, meaning a curve of equal inclination), and finally draw approximation curves fit to the lineal elements.

We shall now illustrate how numeric methods work by applying the simplest numeric method, that is Euler's method, to an initial value problem involving ODE (2). First we give a brief description of Euler's method.

Numeric Method by Euler

Given an ODE (1) and an initial value $y(x_0) = y_0$, **Euler's method** yields approximate solution values at equidistant x-values $x_0, x_1 = x_0 + h, x_2 = x_0 + 2h, \cdots$, namely,

$$y_1 = y_0 + hf(x_0, y_0) \qquad \text{(Fig. 8)}$$

$$y_2 = y_1 + hf(x_1, y_1), \qquad \text{etc.}$$

In general,

$$y_n = y_{n-1} + hf(x_{n-1}, y_{n-1})$$

where the step h equals, e.g., 0.1 or 0.2 (as in Table 1.1) or a smaller value for greater accuracy.

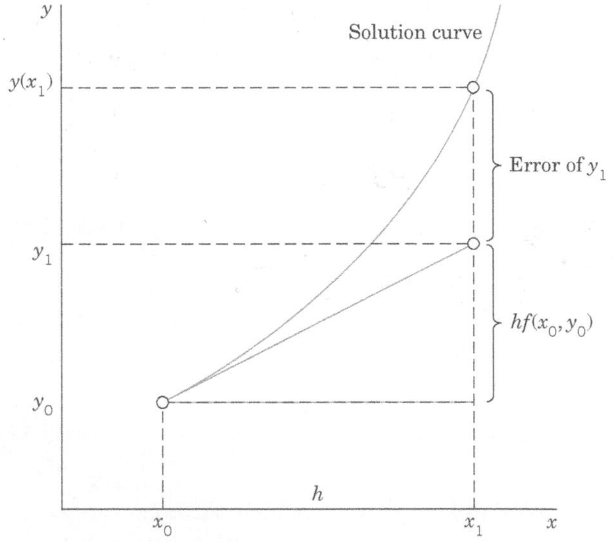

Fig. 8. First Euler step, showing a solution curve, its tangent at (x_0, y_0), step h and increment $hf(x_0, y_0)$ in the formula for y_1

Table 1.1 shows the computation of $n = 5$ steps with step $h = 0.2$ for the ODE (2) and initial condition $y(0) = 0$, corresponding to the middle curve in the direction field. We shall solve the ODE exactly in Sec. 1.5. For the time being, verify that the initial value problem has the solution $y = e^x - x - 1$. The solution curve and the values in Table 1.1 are shown in Fig. 9. These values are rather inaccurate. The errors $y(x_n) - y_n$ are shown in Table 1.1 as well as in Fig. 9. Decreasing h would improve the values, but would soon require an impractical amount of computation. Much better methods of a similar nature will be discussed in Sec. 21.1.

Table 1.1. Euler method for $y' = y + x, y(0) = 0$ for $x = 0, \cdots, 1.0$ with step $h = 0.2$

n	x_n	y_n	$y(x_n)$	Error
0	0.0	0.000	0.000	0.000
1	0.2	0.000	0.021	0.021
2	0.4	0.04	0.092	0.052
3	0.6	0.128	0.222	0.094
4	0.8	0.274	0.426	0.152
5	1.0	0.488	0.718	0.230

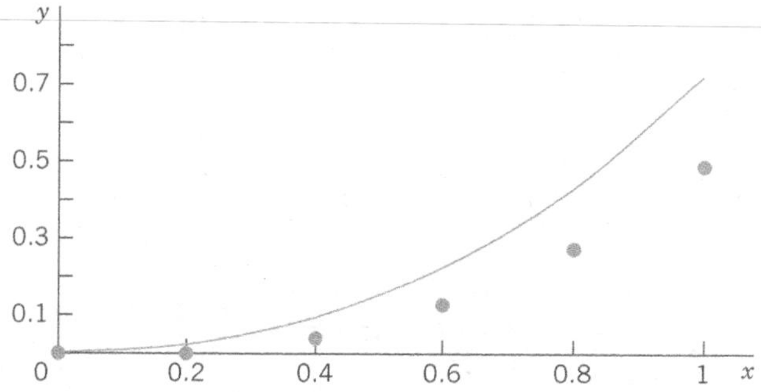

Fig. 9. Euler method: Approximate values in Table 1.1 and solution curve

PROBLEM SET 1.2

1–8 DIRECTION FIELDS, SOLUTION CURVES

Graph a direction field (by a CAS or by hand). In the field graph several solution curves by hand, particularly those passing through the given points (x, y).

1. $y' = 1 + y^2$, $(\frac{1}{4}\pi, 1)$
2. $yy' + 4x = 0$, $(1, 1), (0, 2)$
3. $y' = 1 - y^2$, $(0, 0), (2, \frac{1}{2})$
4. $y' = 2y - y^2$, $(0, 0), (0, 1), (0, 2), (0, 3)$
5. $y' = x - 1/y$, $(1, \frac{1}{2})$
6. $y' = \sin^2 y$, $(0, -0.4), (0, 1)$
7. $y' = e^{y/x}$, $(2, 2), (3, 3)$
8. $y' = -2xy$, $(0, \frac{1}{2}), (0, 1), (0, 2)$

9–10 ACCURACY OF DIRECTION FIELDS

Direction fields are very useful because they can give you an impression of all solutions without solving the ODE, which may be difficult or even impossible. To get a feel for the accuracy of the method, graph a field, sketch solution curves in it, and compare them with the exact solutions.

9. $y' = \cos \pi x$
10. $y' = -5y^{1/2}$ (Sol. $\sqrt{y} + \frac{5}{2}x = c$)
11. **Autonomous ODE.** This means an ODE not showing x (the independent variable) *explicitly*. (The ODEs in Probs. 6 and 10 are autonomous.) What will the level curves $f(x, y) = \text{const}$ (also called *isoclines* = curves of equal inclination) of an autonomous ODE look like? Give reason.

12–15 MOTIONS

Model the motion of a body B on a straight line with velocity as given, $y(t)$ being the distance of B from a point $y = 0$ at time t. Graph a direction field of the model (the ODE). In the field sketch the solution curve satisfying the given initial condition.

12. Product of velocity times distance constant, equal to 2, $y(0) = 2$.
13. Distance = Velocity × Time, $y(1) = 1$
14. Square of the distance plus square of the velocity equal to 1, initial distance $1/\sqrt{2}$
15. **Parachutist.** Two forces act on a parachutist, the attraction by the earth mg (m = mass of person plus equipment, $g = 9.8 \text{ m/sec}^2$ the acceleration of gravity) and the air resistance, assumed to be proportional to the square of the velocity $v(t)$. Using **Newton's second law** of motion (mass × acceleration = resultant of the forces), set up a model (an ODE for $v(t)$). Graph a direction field (choosing m and the constant of proportionality equal to 1). Assume that the parachute opens when $v = 10 \text{ m/sec}$. Graph the corresponding solution in the field. What is the limiting velocity? Would the parachute still be sufficient if the air resistance were only proportional to $v(t)$?

16. CAS PROJECT. Direction Fields. Discuss direction fields as follows.

(a) Graph portions of the direction field of the ODE (2) (see Fig. 7), for instance, $-5 \leqq x \leqq 2$, $-1 \leqq y \leqq 5$. Explain what you have gained by this enlargement of the portion of the field.

(b) Using implicit differentiation, find an ODE with the general solution $x^2 + 9y^2 = c$ $(y > 0)$. Graph its direction field. Does the field give the impression that the solution curves may be semi-ellipses? Can you do similar work for circles? Hyperbolas? Parabolas? Other curves?

(c) Make a conjecture about the solutions of $y' = -x/y$ from the direction field.

(d) Graph the direction field of $y' = -\frac{1}{2}y$ and some solutions of your choice. How do they behave? Why do they decrease for $y > 0$?

17–20 EULER'S METHOD

This is the simplest method to explain numerically solving an ODE, more precisely, an initial value problem (IVP). (More accurate methods based on the same principle are explained in Sec. 21.1.) Using the method, to get a feel for numerics as well as for the nature of IVPs, solve the IVP numerically with a PC or a calculator, 10 steps. Graph the computed values and the solution curve on the same coordinate axes.

17. $y' = y, \quad y(0) = 1, \quad h = 0.1$

18. $y' = y, \quad y(0) = 1, \quad h = 0.01$

19. $y' = (y - x)^2, \quad y(0) = 0, \quad h = 0.1$
Sol. $y = x - \tanh x$

20. $y' = -5x^4y^2, \quad y(0) = 1, \quad h = 0.2$
Sol. $y = 1/(1 + x)^5$

1.3 Separable ODEs. Modeling

Many practically useful ODEs can be reduced to the form

$$g(y)\, y' = f(x) \tag{1}$$

by purely algebraic manipulations. Then we can integrate on both sides with respect to x, obtaining

$$\int g(y)\, y'\, dx = \int f(x)\, dx + c. \tag{2}$$

On the left we can switch to y as the variable of integration. By calculus, $y'dx = dy$, so that

$$\int g(y)\, dy = \int f(x)\, dx + c. \tag{3}$$

If f and g are continuous functions, the integrals in (3) exist, and by evaluating them we obtain a general solution of (1). This method of solving ODEs is called the **method of separating variables**, and (1) is called a **separable equation**, because in (3) the variables are now separated: x appears only on the right and y only on the left.

EXAMPLE 1 Separable ODE

The ODE $y' = 1 + y^2$ is separable because it can be written

$$\frac{dy}{1 + y^2} = dx. \qquad \text{By integration,} \qquad \arctan y = x + c \qquad \text{or} \qquad y = \tan(x + c).$$

It is very important to introduce the constant of integration immediately when the integration is performed. If we wrote $\arctan y = x$, then $y = \tan x$, and *then* introduced c, we would have obtained $y = \tan x + c$, which is not a solution (when $c \neq 0$). Verify this. ■

EXAMPLE 2 Separable ODE

The ODE $y' = (x + 1)e^{-x}y^2$ is separable; we obtain $y^{-2}\,dy = (x + 1)e^{-x}\,dx$.

By integration, $\quad -y^{-1} = -(x + 2)e^{-x} + c, \qquad y = \dfrac{1}{(x + 2)e^{-x} - c}$. ■

EXAMPLE 3 Initial Value Problem (IVP). Bell-Shaped Curve

Solve $y' = -2xy$, $y(0) = 1.8$.

Solution. By separation and integration,

$$\frac{dy}{y} = -2x\,dx, \qquad \ln y = -x^2 + \tilde{c}, \qquad y = ce^{-x^2}.$$

This is the general solution. From it and the initial condition, $y(0) = ce^0 = c = 1.8$. Hence the IVP has the solution $y = 1.8e^{-x^2}$. This is a particular solution, representing a bell-shaped curve (Fig. 10). ■

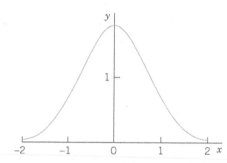

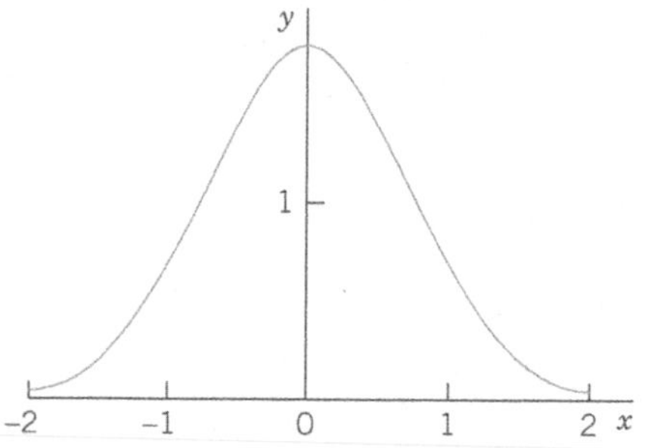

Fig. 10. Solution in Example 3 (bell-shaped curve)

Modeling

The importance of modeling was emphasized in Sec. 1.1, and separable equations yield various useful models. Let us discuss this in terms of some typical examples.

EXAMPLE 4 Radiocarbon Dating[2]

In September 1991 the famous Iceman (Oetzi), a mummy from the Neolithic period of the Stone Age found in the ice of the Oetztal Alps (hence the name "Oetzi") in Southern Tyrolia near the Austrian–Italian border, caused a scientific sensation. When did Oetzi approximately live and die if the ratio of carbon ${}^{14}_{6}C$ to carbon ${}^{12}_{6}C$ in this mummy is 52.5% of that of a living organism?

Physical Information. In the atmosphere and in living organisms, the ratio of radioactive carbon ${}^{14}_{6}C$ (made radioactive by cosmic rays) to ordinary carbon ${}^{12}_{6}C$ is constant. When an organism dies, its absorption of ${}^{14}_{6}C$ by breathing and eating terminates. Hence one can estimate the age of a fossil by comparing the radioactive carbon ratio in the fossil with that in the atmosphere. To do this, one needs to know the half-life of ${}^{14}_{6}C$, which is 5715 years (*CRC Handbook of Chemistry and Physics,* 83rd ed., Boca Raton: CRC Press, 2002, page 11–52, line 9).

Solution. ***Modeling.*** Radioactive decay is governed by the ODE $y' = ky$ (see Sec. 1.1, Example 5). By separation and integration (where t is time and y_0 is the initial ratio of ${}^{14}_{6}C$ to ${}^{12}_{6}C$)

$$\frac{dy}{y} = k\,dt, \qquad \ln|y| = kt + c, \qquad y = y_0 e^{kt} \qquad (y_0 = e^c).$$

[2]Method by WILLARD FRANK LIBBY (1908–1980), American chemist, who was awarded for this work the 1960 Nobel Prize in chemistry.

Next we use the half-life $H = 5715$ to determine k. When $t = H$, half of the original substance is still present. Thus,

$$y_0 e^{kH} = 0.5 y_0, \qquad e^{kH} = 0.5, \qquad k = \frac{\ln 0.5}{H} = -\frac{0.693}{5715} = -0.0001\,213.$$

Finally, we use the ratio 52.5% for determining the time t when Oetzi died (actually, was killed),

$$e^{kt} = e^{-0.0001213t} = 0.525, \qquad t = \frac{\ln 0.525}{-0.0001\,213} = 5312. \qquad \textit{Answer:} \qquad \text{About 5300 years ago.}$$

Other methods show that radiocarbon dating values are usually too small. According to recent research, this is due to a variation in that carbon ratio because of industrial pollution and other factors, such as nuclear testing. ■

EXAMPLE 5 Mixing Problem

Mixing problems occur quite frequently in chemical industry. We explain here how to solve the basic model involving a single tank. The tank in Fig. 11 contains 1000 gal of water in which initially 100 lb of salt is dissolved. Brine runs in at a rate of 10 gal/min, and each gallon contains 5 lb of dissoved salt. The mixture in the tank is kept uniform by stirring. Brine runs out at 10 gal/min. Find the amount of salt in the tank at any time t.

Solution. ***Step 1. Setting up a model.*** Let $y(t)$ denote the amount of salt in the tank at time t. Its time rate of change is

$$y' = \text{Salt inflow rate} - \text{Salt outflow rate} \qquad \textbf{Balance law.}$$

5 lb times 10 gal gives an inflow of 50 lb of salt. Now, the outflow is 10 gal of brine. This is $10/1000 = 0.01$ ($= 1\%$) of the total brine content in the tank, hence 0.01 of the salt content $y(t)$, that is, $0.01\ y(t)$. Thus the model is the ODE

$$y' = 50 - 0.01y = -0.01(y - 5000). \tag{4}$$

Step 2. Solution of the model. The ODE (4) is separable. Separation, integration, and taking exponents on both sides gives

$$\frac{dy}{y - 5000} = -0.01\,dt, \qquad \ln|y - 5000| = -0.01t + c^*, \qquad y - 5000 = ce^{-0.01t}.$$

Initially the tank contains 100 lb of salt. Hence $y(0) = 100$ is the initial condition that will give the unique solution. Substituting $y = 100$ and $t = 0$ in the last equation gives $100 - 5000 = ce^0 = c$. Hence $c = -4900$. Hence the amount of salt in the tank at time t is

$$y(t) = 5000 - 4900e^{-0.01t}. \tag{5}$$

This function shows an exponential approach to the limit 5000 lb; see Fig. 11. Can you explain physically that $y(t)$ should increase with time? That its limit is 5000 lb? Can you see the limit directly from the ODE?

The model discussed becomes more realistic in problems on pollutants in lakes (see Problem Set 1.5, Prob. 35) or drugs in organs. These types of problems are more difficult because the mixing may be imperfect and the flow rates (in and out) may be different and known only very roughly. ■

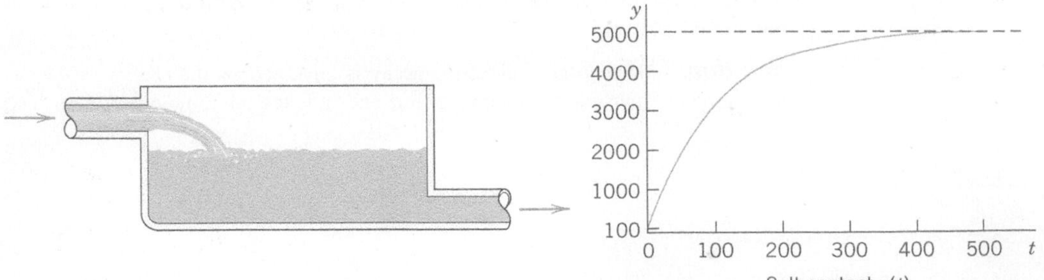

Fig. 11. Mixing problem in Example 5

EXAMPLE 6 Heating an Office Building (Newton's Law of Cooling[3])

Suppose that in winter the daytime temperature in a certain office building is maintained at 70°F. The heating is shut off at 10 P.M. and turned on again at 6 A.M. On a certain day the temperature inside the building at 2 A.M. was found to be 65°F. The outside temperature was 50°F at 10 P.M. and had dropped to 40°F by 6 A.M. What was the temperature inside the building when the heat was turned on at 6 A.M.?

Physical information. Experiments show that the time rate of change of the temperature T of a body B (which conducts heat well, for example, as a copper ball does) is proportional to the difference between T and the temperature of the surrounding medium **(Newton's law of cooling).**

Solution. ***Step 1. Setting up a model.*** Let $T(t)$ be the temperature inside the building and T_A the outside temperature (assumed to be constant in Newton's law). Then by Newton's law,

$$\frac{dT}{dt} = k(T - T_A). \tag{6}$$

Such experimental laws are derived under idealized assumptions that rarely hold exactly. However, even if a model seems to fit the reality only poorly (as in the present case), it may still give valuable qualitative information. To see how good a model is, the engineer will collect experimental data and compare them with calculations from the model.

Step 2. General solution. We cannot solve (6) because we do not know T_A, just that it varied between 50°F and 40°F, so we follow the ***Golden Rule:*** *If you cannot solve your problem, try to solve a simpler one.* We solve (6) with the unknown function T_A replaced with the average of the two known values, or 45°F. For physical reasons we may expect that this will give us a reasonable approximate value of T in the building at 6 A.M.

For constant $T_A = 45$ (or any other *constant* value) the ODE (6) is separable. Separation, integration, and taking exponents gives the general solution

$$\frac{dT}{T - 45} = k\,dt, \qquad \ln|T - 45| = kt + c^*, \qquad T(t) = 45 + ce^{kt} \qquad (c = e^{c^*}).$$

Step 3. Particular solution. We choose 10 P.M. to be $t = 0$. Then the given initial condition is $T(0) = 70$ and yields a particular solution, call it T_p. By substitution,

$$T(0) = 45 + ce^0 = 70, \qquad c = 70 - 45 = 25, \qquad T_p(t) = 45 + 25e^{kt}.$$

Step 4. Determination of k. We use $T(4) = 65$, where $t = 4$ is 2 A.M. Solving algebraically for k and inserting k into $T_p(t)$ gives (Fig. 12)

$$T_p(4) = 45 + 25e^{4k} = 65, \qquad e^{4k} = 0.8, \qquad k = \tfrac{1}{4}\ln 0.8 = -0.056, \qquad T_p(t) = 45 + 25e^{-0.056t}.$$

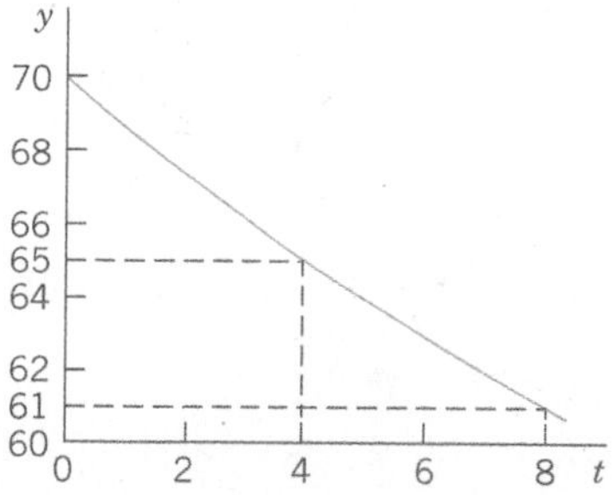

Fig. 12. Particular solution (temperature) in Example 6

[3]Sir ISAAC NEWTON (1642–1727), great English physicist and mathematician, became a professor at Cambridge in 1669 and Master of the Mint in 1699. He and the German mathematician and philosopher GOTTFRIED WILHELM LEIBNIZ (1646–1716) invented (independently) the differential and integral calculus. Newton discovered many basic physical laws and created the method of investigating physical problems by means of calculus. His *Philosophiae naturalis principia mathematica* (*Mathematical Principles of Natural Philosophy*, 1687) contains the development of classical mechanics. His work is of greatest importance to both mathematics and physics.

Step 5. Answer and interpretation. 6 A.M. is $t = 8$ (namely, 8 hours after 10 P.M.), and

$$T_p(8) = 45 + 25e^{-0.056 \cdot 8} = 61[^\circ\text{F}].$$

Hence the temperature in the building dropped 9°F, a result that looks reasonable. ■

EXAMPLE 7 Leaking Tank. Outflow of Water Through a Hole (Torricelli's Law)

This is another prototype engineering problem that leads to an ODE. It concerns the outflow of water from a cylindrical tank with a hole at the bottom (Fig. 13). You are asked to find the height of the water in the tank at any time if the tank has diameter 2 m, the hole has diameter 1 cm, and the initial height of the water when the hole is opened is 2.25 m. When will the tank be empty?

Physical information. Under the influence of gravity the outflowing water has velocity

$$v(t) = 0.600\sqrt{2gh(t)} \qquad \text{(7)} \quad \textbf{(Torricelli's law[4]),}$$

where $h(t)$ is the height of the water above the hole at time t, and $g = 980 \text{ cm/sec}^2 = 32.17 \text{ ft/sec}^2$ is the acceleration of gravity at the surface of the earth.

Solution. ***Step 1. Setting up the model.*** To get an equation, we relate the decrease in water level $h(t)$ to the outflow. The volume ΔV of the outflow during a short time Δt is

$$\Delta V = Av\,\Delta t \qquad (A = \text{Area of hole}).$$

ΔV must equal the change ΔV^* of the volume of the water in the tank. Now

$$\Delta V^* = -B\,\Delta h \qquad (B = \text{Cross-sectional area of tank})$$

where $\Delta h\ (> 0)$ is the decrease of the height $h(t)$ of the water. The minus sign appears because the volume of the water in the tank decreases. Equating ΔV and ΔV^* gives

$$-B\,\Delta h = Av\,\Delta t.$$

We now express v according to Torricelli's law and then let Δt (the length of the time interval considered) approach 0—this is a ***standard way*** of obtaining an ODE as a model. That is, we have

$$\frac{\Delta h}{\Delta t} = -\frac{A}{B}v = -\frac{A}{B}0.600\sqrt{2gh(t)}$$

and by letting $\Delta t \to 0$ we obtain the ODE

$$\frac{dh}{dt} = -26.56\frac{A}{B}\sqrt{h},$$

where $26.56 = 0.600\sqrt{2 \cdot 980}$. This is our model, a first-order ODE.

Step 2. General solution. Our ODE is separable. A/B is constant. Separation and integration gives

$$\frac{dh}{\sqrt{h}} = -26.56\frac{A}{B}dt \qquad \text{and} \qquad 2\sqrt{h} = c^* - 26.56\frac{A}{B}t.$$

Dividing by 2 and squaring gives $h = (c - 13.28At/B)^2$. Inserting $13.28A/B = 13.28 \cdot 0.5^2\pi/100^2\pi = 0.000332$ yields the general solution

$$h(t) = (c - 0.000332t)^2.$$

[4]EVANGELISTA TORRICELLI (1608–1647), Italian physicist, pupil and successor of GALILEO GALILEI (1564–1642) at Florence. The "contraction factor" 0.600 was introduced by J. C. BORDA in 1766 because the stream has a smaller cross section than the area of the hole.

Step 3. Particular solution. The initial height (the initial condition) is $h(0) = 225$ cm. Substitution of $t = 0$ and $h = 225$ gives from the general solution $c^2 = 225$, $c = 15.00$ and thus the particular solution (Fig. 13)

$$h_p(t) = (15.00 - 0.000332t)^2.$$

Step 4. Tank empty. $h_p(t) = 0$ if $t = 15.00/0.000332 = 45{,}181$ [sec] $= 12.6$ [hours].

Here you see distinctly the ***importance of the choice of units***—we have been working with the cgs system, in which time is measured in seconds! We used $g = 980\ \text{cm/sec}^2$.

Step 5. Checking. Check the result. ■

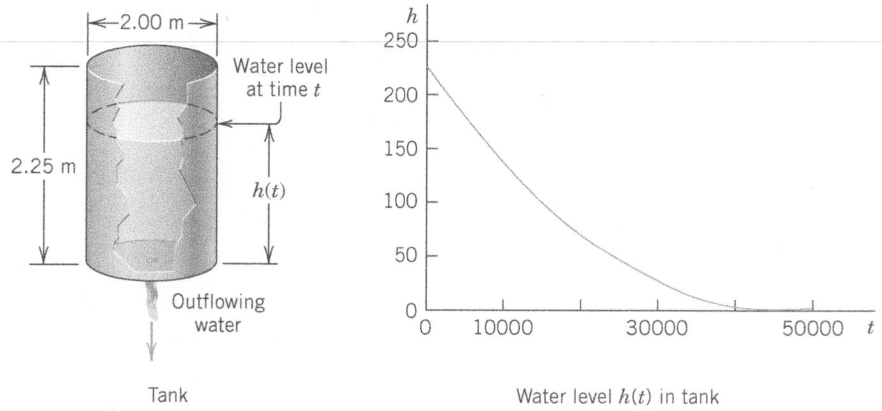

Fig. 13. Example 7. Outflow from a cylindrical tank ("leaking tank"). Torricelli's law

Extended Method: Reduction to Separable Form

Certain nonseparable ODEs can be made separable by transformations that introduce for y a new unknown function. We discuss this technique for a class of ODEs of practical importance, namely, for equations

$$y' = f\left(\frac{y}{x}\right). \tag{8}$$

Here, f is any (differentiable) function of y/x, such as $\sin(y/x)$, $(y/x)^4$, and so on. (Such an ODE is sometimes called a *homogeneous ODE*, a term we shall not use but reserve for a more important purpose in Sec. 1.5.)

The form of such an ODE suggests that we set $y/x = u$; thus,

$$y = ux \qquad \text{and by product differentiation} \qquad y' = u'x + u. \tag{9}$$

Substitution into $y' = f(y/x)$ then gives $u'x + u = f(u)$ or $u'x = f(u) - u$. We see that if $f(u) - u \neq 0$, this can be separated:

$$\frac{du}{f(u) - u} = \frac{dx}{x}. \tag{10}$$

EXAMPLE 8 Reduction to Separable Form

Solve

$$2xyy' = y^2 - x^2.$$

Solution. To get the usual explicit form, divide the given equation by $2xy$,

$$y' = \frac{y^2 - x^2}{2xy} = \frac{y}{2x} - \frac{x}{2y}.$$

Now substitute y and y' from (9) and then simplify by subtracting u on both sides,

$$u'x + u = \frac{u}{2} - \frac{1}{2u}, \qquad u'x = -\frac{u}{2} - \frac{1}{2u} = \frac{-u^2 - 1}{2u}.$$

You see that in the last equation you can now separate the variables,

$$\frac{2u\,du}{1 + u^2} = -\frac{dx}{x}. \qquad \text{By integration,} \qquad \ln(1 + u^2) = -\ln|x| + c^* = \ln\left|\frac{1}{x}\right| + c^*.$$

Take exponents on both sides to get $1 + u^2 = c/x$ or $1 + (y/x)^2 = c/x$. Multiply the last equation by x^2 to obtain (Fig. 14)

$$x^2 + y^2 = cx. \qquad \text{Thus} \qquad \left(x - \frac{c}{2}\right)^2 + y^2 = \frac{c^2}{4}.$$

This general solution represents a family of circles passing through the origin with centers on the x-axis. ■

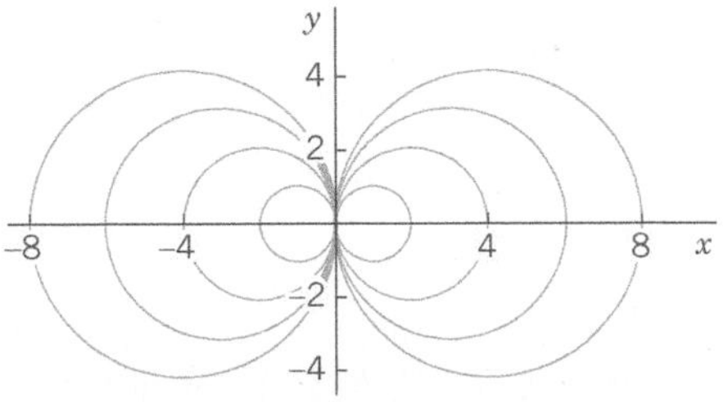

Fig. 14. General solution (family of circles) in Example 8

PROBLEM SET 1.3

1. CAUTION! **Constant of integration.** Why is it important to introduce the constant of integration immediately when you integrate?

2–10 GENERAL SOLUTION

Find a general solution. Show the steps of derivation. Check your answer by substitution.

2. $y^3y' + x^3 = 0$
3. $y' = \sec^2 y$
4. $y' \sin 2\pi x = \pi y \cos 2\pi x$
5. $yy' + 36x = 0$
6. $y' = e^{2x-1}y^2$
7. $xy' = y + 2x^3 \sin^2 \frac{y}{x}$ (Set $y/x = u$)
8. $y' = (y + 4x)^2$ (Set $y + 4x = v$)
9. $xy' = y^2 + y$ (Set $y/x = u$)
10. $xy' = x + y$ (Set $y/x = u$)

11–17 INITIAL VALUE PROBLEMS (IVPs)

Solve the IVP. Show the steps of derivation, beginning with the general solution.

11. $xy' + y = 0, \quad y(4) = 6$
12. $y' = 1 + 4y^2, \quad y(1) = 0$
13. $y' \cosh^2 x = \sin^2 y, \quad y(0) = \frac{1}{2}\pi$
14. $dr/dt = -2tr, \quad r(0) = r_0$
15. $y' = -4x/y, \quad y(2) = 3$
16. $y' = (x + y - 2)^2, \quad y(0) = 2$ (Set $v = x + y - 2$)
17. $xy' = y + 3x^4 \cos^2(y/x), \quad y(1) = 0$ (Set $y/x = u$)
18. **Particular solution.** Introduce limits of integration in (3) such that y obtained from (3) satisfies the initial condition $y(x_0) = y_0$.

19–36 MODELING, APPLICATIONS

19. Exponential growth. If the growth rate of the number of bacteria at any time t is proportional to the number present at t and doubles in 1 week, how many bacteria can be expected after 2 weeks? After 4 weeks?

20. Another population model.

(a) If the birth rate and death rate of the number of bacteria are proportional to the number of bacteria present, what is the population as a function of time.

(b) What is the limiting situation for increasing time? Interpret it.

21. Radiocarbon dating. What should be the ${}^{14}_{6}C$ content (in percent of y_0) of a fossilized tree that is claimed to be 3000 years old? (See Example 4.)

22. Linear accelerators are used in physics for accelerating charged particles. Suppose that an alpha particle enters an accelerator and undergoes a constant acceleration that increases the speed of the particle from 10^3 m/sec to 10^4 m/sec in 10^{-3} sec. Find the acceleration a and the distance traveled during that period of 10^{-3} sec.

23. Boyle–Mariotte's law for ideal gases.[5] Experiments show for a gas at low pressure p (and constant temperature) the rate of change of the volume $V(p)$ equals $-V/p$. Solve the model.

24. Mixing problem. A tank contains 400 gal of brine in which 100 lb of salt are dissolved. Fresh water runs into the tank at a rate of 2 gal/min.The mixture, kept practically uniform by stirring, runs out at the same rate. How much salt will there be in the tank at the end of 1 hour?

25. Newton's law of cooling. A thermometer, reading 5°C, is brought into a room whose temperature is 22°C. One minute later the thermometer reading is 12°C. How long does it take until the reading is practically 22°C, say, 21.9°C?

26. Gompertz growth in tumors. The Gompertz model is $y' = -Ay \ln y$ $(A > 0)$, where $y(t)$ is the mass of tumor cells at time t. The model agrees well with clinical observations. The declining growth rate with increasing $y > 1$ corresponds to the fact that cells in the interior of a tumor may die because of insufficient oxygen and nutrients. Use the ODE to discuss the growth and decline of solutions (tumors) and to find constant solutions. Then solve the ODE.

27. Dryer. If a wet sheet in a dryer loses its moisture at a rate proportional to its moisture content, and if it loses half of its moisture during the first 10 min of drying, when will it be practically dry, say, when will it have lost 99% of its moisture? First guess, then calculate.

28. Estimation. Could you see, practically without calculation, that the answer in Prob. 27 must lie between 60 and 70 min? Explain.

29. Alibi? Jack, arrested when leaving a bar, claims that he has been inside for at least half an hour (which would provide him with an alibi). The police check the water temperature of his car (parked near the entrance of the bar) at the instant of arrest and again 30 min later, obtaining the values 190°F and 110°F, respectively. Do these results give Jack an alibi? (Solve by inspection.)

30. Rocket. A rocket is shot straight up from the earth, with a net acceleration (= acceleration by the rocket engine minus gravitational pullback) of $7t\,\text{m/sec}^2$ during the initial stage of flight until the engine cut out at $t = 10$ sec. How high will it go, air resistance neglected?

31. Solution curves of $y' = g(y/x)$. Show that any (nonvertical) straight line through the origin of the xy-plane intersects all these curves of a given ODE at the same angle.

32. Friction. If a body slides on a surface, it experiences friction F (a force against the direction of motion). Experiments show that $|F| = \mu|N|$ (*Coulomb's*[6] *law of kinetic friction without lubrication*), where N is the normal force (force that holds the two surfaces together; see Fig. 15) and the constant of proportionality μ is called the *coefficient of kinetic friction*. In Fig. 15 assume that the body weighs 45 nt (about 10 lb; see front cover for conversion). $\mu = 0.20$ (corresponding to steel on steel), $\alpha = 30°$, the slide is 10 m long, the initial velocity is zero, and air resistance is negligible. Find the velocity of the body at the end of the slide.

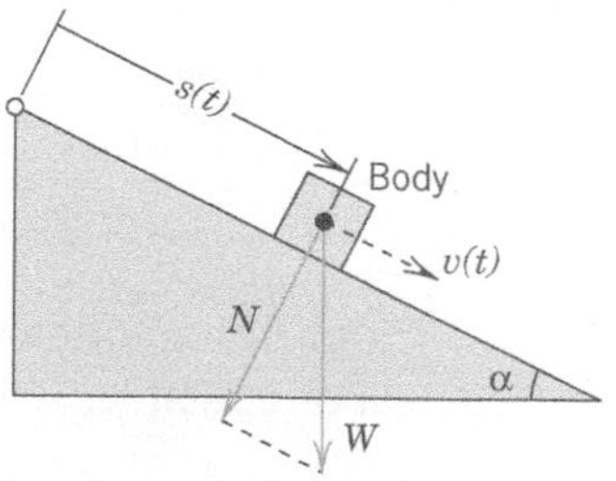

Fig. 15. Problem 32

[5]ROBERT BOYLE (1627–1691), English physicist and chemist, one of the founders of the Royal Society. EDME MARIOTTE (about 1620–1684), French physicist and prior of a monastry near Dijon. They found the law experimentally in 1662 and 1676, respectively.

[6]CHARLES AUGUSTIN DE COULOMB (1736–1806), French physicist and engineer.

33. Rope. To tie a boat in a harbor, how many times must a rope be wound around a bollard (a vertical rough cylindrical post fixed on the ground) so that a man holding one end of the rope can resist a force exerted by the boat 1000 times greater than the man can exert? First guess. Experiments show that the change ΔS of the force S in a small portion of the rope is proportional to S and to the small angle $\Delta\phi$ in Fig. 16. Take the proportionality constant 0.15. The result should surprise you!

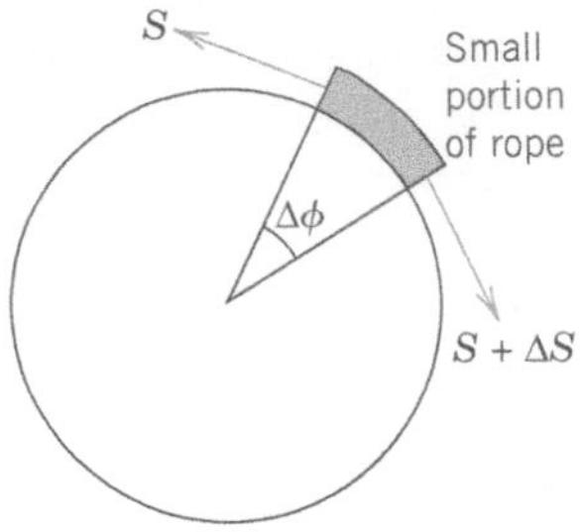

Fig. 16. Problem 33

34. TEAM PROJECT. Family of Curves. A family of curves can often be characterized as the general solution of $y' = f(x, y)$.

(a) Show that for the circles with center at the origin we get $y' = -x/y$.

(b) Graph some of the hyperbolas $xy = c$. Find an ODE for them.

(c) Find an ODE for the straight lines through the origin.

(d) You will see that the product of the right sides of the ODEs in (a) and (c) equals -1. Do you recognize this as the condition for the two families to be orthogonal (i.e., to intersect at right angles)? Do your graphs confirm this?

(e) Sketch families of curves of your own choice and find their ODEs. Can every family of curves be given by an ODE?

35. CAS PROJECT. Graphing Solutions. A CAS can usually graph solutions, even if they are integrals that cannot be evaluated by the usual analytical methods of calculus.

(a) Show this for the five initial value problems $y' = e^{-x^2}$, $y(0) = 0, \pm 1, \pm 2$ graphing all five curves on the same axes.

(b) Graph approximate solution curves, using the first few terms of the Maclaurin series (obtained by termwise integration of that of y') and compare with the exact curves.

(c) Repeat the work in (a) for another ODE and initial conditions of your own choice, leading to an integral that cannot be evaluated as indicated.

36. TEAM PROJECT. Torricelli's Law. Suppose that the tank in Example 7 is hemispherical, of radius R, initially full of water, and has an outlet of 5 cm^2 cross-sectional area at the bottom. (Make a sketch.) Set up the model for outflow. Indicate what portion of your work in Example 7 you can use (so that it can become part of the general method independent of the shape of the tank). Find the time t to empty the tank (a) for any R, (b) for $R = 1$ m. Plot t as function of R. Find the time when $h = R/2$ (a) for any R, (b) for $R = 1$ m.

1.4 Exact ODEs. Integrating Factors

We recall from calculus that if a function $u(x, y)$ has continuous partial derivatives, its **differential** (also called its *total differential*) is

$$du = \frac{\partial u}{\partial x}\,dx + \frac{\partial u}{\partial y}\,dy.$$

From this it follows that if $u(x, y) = c = \text{const}$, then $du = 0$.

For example, if $u = x + x^2y^3 = c$, then

$$du = (1 + 2xy^3)\,dx + 3x^2y^2\,dy = 0$$

or

$$y' = \frac{dy}{dx} = -\frac{1 + 2xy^3}{3x^2y^2},$$

an ODE that we can solve by going backward. This idea leads to a powerful solution method as follows.

A first-order ODE $M(x, y) + N(x, y)y' = 0$, written as (use $dy = y'dx$ as in Sec. 1.3)

$$M(x, y)\,dx + N(x, y)\,dy = 0 \tag{1}$$

is called an **exact differential equation** if the **differential form** $M(x, y)\,dx + N(x, y)\,dy$ is **exact**, that is, this form is the differential

$$du = \frac{\partial u}{\partial x}\,dx + \frac{\partial u}{\partial y}\,dy \tag{2}$$

of some function $u(x, y)$. Then (1) can be written

$$du = 0.$$

By integration we immediately obtain the general solution of (1) in the form

$$u(x, y) = c. \tag{3}$$

This is called an **implicit solution**, in contrast to a solution $y = h(x)$ as defined in Sec. 1.1, which is also called an *explicit solution*, for distinction. Sometimes an implicit solution can be converted to explicit form. (Do this for $x^2 + y^2 = 1$.) If this is not possible, your CAS may graph a figure of the **contour lines** (3) of the function $u(x, y)$ and help you in understanding the solution.

Comparing (1) and (2), we see that (1) is an exact differential equation if there is some function $u(x, y)$ such that

$$\text{(a)}\quad \frac{\partial u}{\partial x} = M, \qquad \text{(b)}\quad \frac{\partial u}{\partial y} = N. \tag{4}$$

From this we can derive a formula for checking whether (1) is exact or not, as follows.

Let M and N be continuous and have continuous first partial derivatives in a region in the xy-plane whose boundary is a closed curve without self-intersections. Then by partial differentiation of (4) (see App. 3.2 for notation),

$$\frac{\partial M}{\partial y} = \frac{\partial^2 u}{\partial y\,\partial x},$$

$$\frac{\partial N}{\partial x} = \frac{\partial^2 u}{\partial x\,\partial y}.$$

By the assumption of continuity the two second partial derivaties are equal. Thus

$$\frac{\partial M}{\partial y} = \frac{\partial N}{\partial x}. \tag{5}$$

This condition is not only necessary but also sufficient for (1) to be an exact differential equation. (We shall prove this in Sec. 10.2 in another context. Some calculus books, for instance, [GenRef 12], also contain a proof.)

If (1) is exact, the function $u(x, y)$ can be found by inspection or in the following systematic way. From (4a) we have by integration with respect to x

$$u = \int M\,dx + k(y); \tag{6}$$

in this integration, y is to be regarded as a constant, and $k(y)$ plays the role of a "constant" of integration. To determine $k(y)$, we derive $\partial u/\partial y$ from (6), use (4b) to get dk/dy, and integrate dk/dy to get k. (See Example 1, below.)

Formula (6) was obtained from (4a). Instead of (4a) we may equally well use (4b). Then, instead of (6), we first have by integration with respect to y

$$u = \int N\,dy + l(x). \tag{6*}$$

To determine $l(x)$, we derive $\partial u/\partial x$ from (6*), use (4a) to get dl/dx, and integrate. We illustrate all this by the following typical examples.

EXAMPLE 1 An Exact ODE

Solve

$$\cos(x + y)\,dx + (3y^2 + 2y + \cos(x + y))\,dy = 0. \tag{7}$$

Solution. ***Step 1. Test for exactness.*** Our equation is of the form (1) with

$$M = \cos(x + y),$$

$$N = 3y^2 + 2y + \cos(x + y).$$

Thus

$$\frac{\partial M}{\partial y} = -\sin(x + y),$$

$$\frac{\partial N}{\partial x} = -\sin(x + y).$$

From this and (5) we see that (7) is exact.

Step 2. Implicit general solution. From (6) we obtain by integration

$$u = \int M\,dx + k(y) = \int \cos(x + y)\,dx + k(y) = \sin(x + y) + k(y). \tag{8}$$

To find $k(y)$, we differentiate this formula with respect to y and use formula (4b), obtaining

$$\frac{\partial u}{\partial y} = \cos(x + y) + \frac{dk}{dy} = N = 3y^2 + 2y + \cos(x + y).$$

Hence $dk/dy = 3y^2 + 2y$. By integration, $k = y^3 + y^2 + c^*$. Inserting this result into (8) and observing (3), we obtain the *answer*

$$u(x, y) = \sin(x + y) + y^3 + y^2 = c.$$

Step 3. Checking an implicit solution. We can check by differentiating the implicit solution $u(x, y) = c$ implicitly and see whether this leads to the given ODE (7):

$$du = \frac{\partial u}{\partial x}\,dx + \frac{\partial u}{\partial y}\,dy = \cos(x+y)\,dx + (\cos(x+y) + 3y^2 + 2y)\,dy = 0. \tag{9}$$

This completes the check. ■

EXAMPLE 2 An Initial Value Problem

Solve the initial value problem

$$(\cos y \sinh x + 1)\,dx - \sin y \cosh x\,dy = 0, \qquad y(1) = 2. \tag{10}$$

Solution. You may verify that the given ODE is exact. We find u. For a change, let us use (6*),

$$u = -\int \sin y \cosh x\,dy + l(x) = \cos y \cosh x + l(x).$$

From this, $\partial u/\partial x = \cos y \sinh x + dl/dx = M = \cos y \sinh x + 1$. Hence $dl/dx = 1$. By integration, $l(x) = x + c^*$. This gives the general solution $u(x, y) = \cos y \cosh x + x = c$. From the initial condition, $\cos 2 \cosh 1 + 1 = 0.358 = c$. Hence the answer is $\cos y \cosh x + x = 0.358$. Figure 17 shows the particular solutions for $c = 0, 0.358$ (thicker curve), 1, 2, 3. Check that the answer satisfies the ODE. (Proceed as in Example 1.) Also check that the initial condition is satisfied. ■

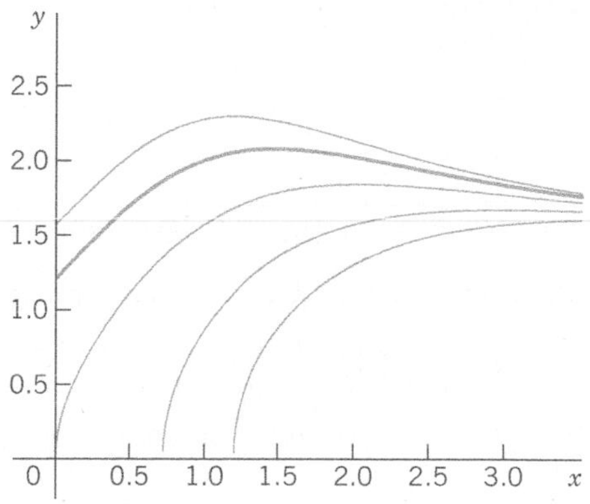

Fig. 17. Particular solutions in Example 2

EXAMPLE 3 WARNING! Breakdown in the Case of Nonexactness

The equation $-y\,dx + x\,dy = 0$ is not exact because $M = -y$ and $N = x$, so that in (5), $\partial M/\partial y = -1$ but $\partial N/\partial x = 1$. Let us show that in such a case the present method does not work. From (6),

$$u = \int M\,dx + k(y) = -xy + k(y), \qquad \text{hence} \qquad \frac{\partial u}{\partial y} = -x + \frac{dk}{dy}.$$

Now, $\partial u/\partial y$ should equal $N = x$, by (4b). However, this is impossible because $k(y)$ can depend only on y. Try (6*); it will also fail. Solve the equation by another method that we have discussed. ■

Reduction to Exact Form. Integrating Factors

The ODE in Example 3 is $-y\,dx + x\,dy = 0$. It is not exact. However, if we multiply it by $1/x^2$, we get an exact equation [check exactness by (5)!],

$$\frac{-y\,dx + x\,dy}{x^2} = -\frac{y}{x^2}\,dx + \frac{1}{x}\,dy = d\left(\frac{y}{x}\right) = 0. \tag{11}$$

Integration of (11) then gives the general solution $y/x = c = \text{const}$.

This example gives the idea. All we did was to multiply a given nonexact equation, say,

$$P(x, y)\,dx + Q(x, y)\,dy = 0, \tag{12}$$

by a function F that, in general, will be a function of both x and y. The result was an equation

$$FP\,dx + FQ\,dy = 0 \tag{13}$$

that is exact, so we can solve it as just discussed. Such a function $F(x, y)$ is then called an **integrating factor** of (12).

EXAMPLE 4 Integrating Factor

The integrating factor in (11) is $F = 1/x^2$. Hence in this case the exact equation (13) is

$$FP\,dx + FQ\,dy = \frac{-y\,dx + x\,dy}{x^2} = d\left(\frac{y}{x}\right) = 0. \qquad \text{Solution} \qquad \frac{y}{x} = c.$$

These are straight lines $y = cx$ through the origin. (Note that $x = 0$ is also a solution of $-y\,dx + x\,dy = 0$.)

It is remarkable that we can readily find other integrating factors for the equation $-y\,dx + x\,dy = 0$, namely, $1/y^2$, $1/(xy)$, and $1/(x^2 + y^2)$, because

$$\frac{-y\,dx + x\,dy}{y^2} = d\left(\frac{x}{y}\right), \quad \frac{-y\,dx + x\,dy}{xy} = -d\left(\ln\frac{x}{y}\right), \quad \frac{-y\,dx + x\,dy}{x^2 + y^2} = d\left(\arctan\frac{y}{x}\right). \tag{14}$$ ■

How to Find Integrating Factors

In simpler cases we may find integrating factors by inspection or perhaps after some trials, keeping (14) in mind. In the general case, the idea is the following.

For $M\,dx + N\,dy = 0$ the exactness condition (5) is $\partial M/\partial y = \partial N/\partial x$. Hence for (13), $FP\,dx + FQ\,dy = 0$, the exactness condition is

$$\frac{\partial}{\partial y}(FP) = \frac{\partial}{\partial x}(FQ). \tag{15}$$

By the product rule, with subscripts denoting partial derivatives, this gives

$$F_yP + FP_y = F_xQ + FQ_x.$$

In the general case, this would be complicated and useless. So we follow the ***Golden Rule:*** *If you cannot solve your problem, try to solve a simpler one*—the result may be useful (and may also help you later on). Hence we look for an integrating factor depending only on ***one*** variable: fortunately, in many practical cases, there are such factors, as we shall see. Thus, let $F = F(x)$. Then $F_y = 0$, and $F_x = F' = dF/dx$, so that (15) becomes

$$FP_y = F'Q + FQ_x.$$

Dividing by FQ and reshuffling terms, we have

$$\frac{1}{F}\frac{dF}{dx} = R, \qquad \text{where} \qquad R = \frac{1}{Q}\left(\frac{\partial P}{\partial y} - \frac{\partial Q}{\partial x}\right). \tag{16}$$

This proves the following theorem.

THEOREM 1

Integrating Factor $F(x)$

If (12) is such that the right side R of (16) depends only on x, then (12) has an integrating factor $F = F(x)$, which is obtained by integrating (16) and taking exponents on both sides.

$$F(x) = \exp \int R(x)\, dx. \tag{17}$$

Similarly, if $F^* = F^*(y)$, then instead of (16) we get

$$\frac{1}{F^*}\frac{dF^*}{dy} = R^*, \qquad \text{where} \qquad R^* = \frac{1}{P}\left(\frac{\partial Q}{\partial x} - \frac{\partial P}{\partial y}\right) \tag{18}$$

and we have the companion

THEOREM 2

Integrating Factor $F^*(y)$

If (12) is such that the right side R^* of (18) depends only on y, then (12) has an integrating factor $F^* = F^*(y)$, which is obtained from (18) in the form

$$F^*(y) = \exp \int R^*(y)\, dy. \tag{19}$$

EXAMPLE 5 **Application of Theorems 1 and 2. Initial Value Problem**

Using Theorem 1 or 2, find an integrating factor and solve the initial value problem

$$(e^{x+y} + ye^y)\, dx + (xe^y - 1)\, dy = 0, \quad y(0) = -1 \tag{20}$$

Solution. ***Step 1. Nonexactness.*** The exactness check fails:

$$\frac{\partial P}{\partial y} = \frac{\partial}{\partial y}(e^{x+y} + ye^y) = e^{x+y} + e^y + ye^y \qquad \text{but} \qquad \frac{\partial Q}{\partial x} = \frac{\partial}{\partial x}(xe^y - 1) = e^y.$$

Step 2. Integrating factor. General solution. Theorem 1 fails because R [the right side of (16)] depends on both x and y.

$$R = \frac{1}{Q}\left(\frac{\partial P}{\partial y} - \frac{\partial Q}{\partial x}\right) = \frac{1}{xe^y - 1}(e^{x+y} + e^y + ye^y - e^y).$$

Try Theorem 2. The right side of (18) is

$$R^* = \frac{1}{P}\left(\frac{\partial Q}{\partial x} - \frac{\partial P}{\partial y}\right) = \frac{1}{e^{x+y} + ye^y}(e^y - e^{x+y} - e^y - ye^y) = -1.$$

Hence (19) gives the integrating factor $F^*(y) = e^{-y}$. From this result and (20) you get the exact equation

$$(e^x + y)\, dx + (x - e^{-y})\, dy = 0.$$

Test for exactness; you will get 1 on both sides of the exactness condition. By integration, using (4a),

$$u = \int (e^x + y)\,dx = e^x + xy + k(y).$$

Differentiate this with respect to y and use (4b) to get

$$\frac{\partial u}{\partial y} = x + \frac{dk}{dy} = N = x - e^{-y}, \qquad \frac{dk}{dy} = -e^{-y}, \qquad k = e^{-y} + c^*.$$

Hence the general solution is

$$u(x, y) = e^x + xy + e^{-y} = c.$$

Setp 3. Particular solution. The initial condition $y(0) = -1$ gives $u(0, -1) = 1 + 0 + e = 3.72$. Hence the answer is $e^x + xy + e^{-y} = 1 + e = 3.72$. Figure 18 shows several particular solutions obtained as level curves of $u(x, y) = c$, obtained by a CAS, a convenient way in cases in which it is impossible or difficult to cast a solution into explicit form. Note the curve that (nearly) satisfies the initial condition.

Step 4. Checking. Check by substitution that the answer satisfies the given equation as well as the initial condition. ■

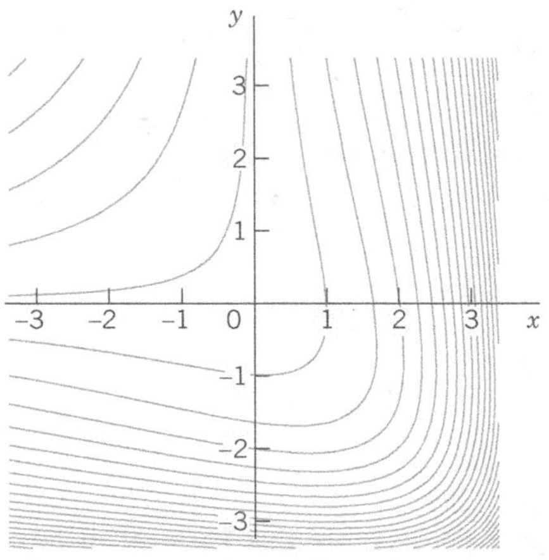

Fig. 18. Particular solutions in Example 5

PROBLEM SET 1.4

1–14 ODEs. INTEGRATING FACTORS

Test for exactness. If exact, solve. If not, use an integrating factor as given or obtained by inspection or by the theorems in the text. Also, if an initial condition is given, find the corresponding particular solution.

1. $2xy\,dx + x^2\,dy = 0$
2. $x^3 dx + y^3 dy = 0$
3. $\sin x \cos y\,dx + \cos x \sin y\,dy = 0$
4. $e^{3\theta}(dr + 3r\,d\theta) = 0$
5. $(x^2 + y^2)\,dx - 2xy\,dy = 0$
6. $3(y + 1)\,dx = 2x\,dy, \quad (y + 1)x^{-4}$
7. $2x \tan y\,dx + \sec^2 y\,dy = 0$
8. $e^x(\cos y\,dx - \sin y\,dy) = 0$
9. $e^{2x}(2\cos y\,dx - \sin y\,dy) = 0, \quad y(0) = 0$
10. $y\,dx + [y + \tan(x + y)]\,dy = 0, \quad \cos(x + y)$
11. $2\cosh x \cos y\,dx = \sinh x \sin y\,dy$
12. $(2xy\,dx + dy)e^{x^2} = 0, \quad y(0) = 2$
13. $e^{-y}\,dx + e^{-x}(-e^{-y} + 1)\,dy = 0, \quad F = e^{x+y}$
14. $(a + 1)y\,dx + (b + 1)x\,dy = 0, \quad y(1) = 1,$ $F = x^a y^b$
15. **Exactness.** Under what conditions for the constants a, b, k, l is $(ax + by)\,dx + (kx + ly)\,dy = 0$ exact? Solve the exact ODE.

16. TEAM PROJECT. Solution by Several Methods. Show this as indicated. Compare the amount of work.

(a) $e^y(\sinh x\,dx + \cosh x\,dy) = 0$ as an exact ODE and by separation.

(b) $(1 + 2x)\cos y\,dx + dy/\cos y = 0$ by Theorem 2 and by separation.

(c) $(x^2 + y^2)\,dx - 2xy\,dy = 0$ by Theorem 1 or 2 and by separation with $v = y/x$.

(d) $3x^2 y\,dx + 4x^3\,dy = 0$ by Theorems 1 and 2 and by separation.

(e) Search the text and the problems for further ODEs that can be solved by more than one of the methods discussed so far. Make a list of these ODEs. Find further cases of your own.

17. WRITING PROJECT. Working Backward. Working backward from the solution to the problem is useful in many areas. Euler, Lagrange, and other great masters did it. To get additional insight into the idea of integrating factors, start from a $u(x, y)$ of your choice, find $du = 0$, destroy exactness by division by some $F(x, y)$, and see what ODE's solvable by integrating factors you can get. Can you proceed systematically, beginning with the simplest $F(x, y)$?

18. CAS PROJECT. Graphing Particular Solutions. Graph particular solutions of the following ODE, proceeding as explained.

$$dy - y^2 \sin x\,dx = 0. \qquad (21)$$

(a) Show that (21) is not exact. Find an integrating factor using either Theorem 1 or 2. Solve (21).

(b) Solve (21) by separating variables. Is this simpler than (a)?

(c) Graph the seven particular solutions satisfying the following initial conditions $y(0) = 1$, $y(\pi/2) = \pm\frac{1}{2}$, $\pm\frac{2}{3}$, ± 1 (see figure below).

(d) Which solution of (21) do we not get in (a) or (b)?

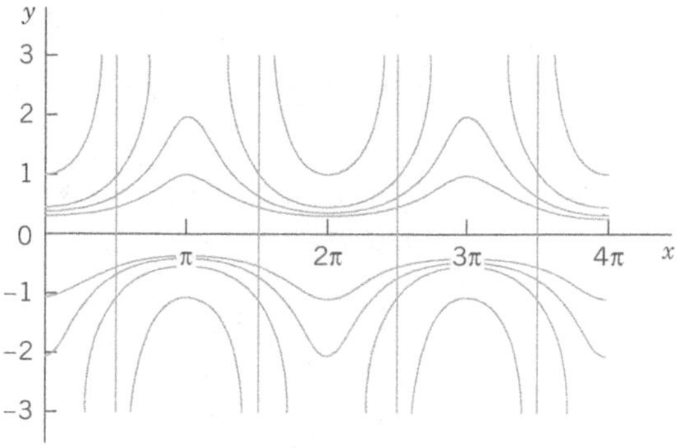

Particular solutions in CAS Project 18

1.5 Linear ODEs. Bernoulli Equation. Population Dynamics

Linear ODEs or ODEs that can be transformed to linear form are models of various phenomena, for instance, in physics, biology, population dynamics, and ecology, as we shall see. A first-order ODE is said to be **linear** if it can be brought into the form

$$y' + p(x)y = r(x), \qquad (1)$$

by algebra, and **nonlinear** if it cannot be brought into this form.

The defining feature of the linear ODE (1) is that it is linear in both the unknown function y and its derivative $y' = dy/dx$, whereas p and r may be ***any*** given functions of x. If in an application the independent variable is time, we write t instead of x.

If the first term is $f(x)y'$ (instead of y'), divide the equation by $f(x)$ to get the **standard form** (1), with y' as the first term, which is practical.

For instance, $y' \cos x + y \sin x = x$ is a linear ODE, and its standard form is $y' + y \tan x = x \sec x$.

The function $r(x)$ on the right may be a force, and the solution $y(x)$ a displacement in a motion or an electrical current or some other physical quantity. In engineering, $r(x)$ is frequently called the **input**, and $y(x)$ is called the **output** or the *response* to the input (and, if given, to the initial condition).

Homogeneous Linear ODE. We want to solve (1) in some interval $a < x < b$, call it J, and we begin with the simpler special case that $r(x)$ is zero for all x in J. (This is sometimes written $r(x) \equiv 0$.) Then the ODE (1) becomes

$$y' + p(x)y = 0 \tag{2}$$

and is called **homogeneous**. By separating variables and integrating we then obtain

$$\frac{dy}{y} = -p(x)\,dx, \qquad \text{thus} \qquad \ln|y| = -\int p(x)\,dx + c^*.$$

Taking exponents on both sides, we obtain the general solution of the homogeneous ODE (2),

$$y(x) = ce^{-\int p(x)dx} \qquad (c = \pm e^{c^*} \quad \text{when} \quad y \gtrless 0); \tag{3}$$

here we may also choose $c = 0$ and obtain the **trivial solution** $y(x) = 0$ for all x in that interval.

Nonhomogeneous Linear ODE. We now solve (1) in the case that $r(x)$ in (1) is not everywhere zero in the interval J considered. Then the ODE (1) is called **nonhomogeneous**. It turns out that in this case, (1) has a pleasant property; namely, it has an integrating factor depending only on x. We can find this factor $F(x)$ by Theorem 1 in the previous section or we can proceed directly, as follows. We multiply (1) by $F(x)$, obtaining

$$Fy' + pFy = rF. \tag{1*}$$

The left side is the derivative $(Fy)' = F'y + Fy'$ of the product Fy if

$$pFy = F'y, \qquad \text{thus} \qquad pF = F'.$$

By separating variables, $dF/F = p\,dx$. By integration, writing $h = \int p\,dx$,

$$\ln|F| = h = \int p\,dx, \qquad \text{thus} \qquad F = e^h.$$

With this F and $h' = p$, Eq. (1*) becomes

$$e^h y' + h' e^h y = e^h y' + (e^h)' y = (e^h y)' = re^h.$$

By integration,

$$e^h y = \int e^h r\,dx + c.$$

Dividing by e^h, we obtain the desired solution formula

$$y(x) = e^{-h}\left(\int e^h r\,dx + c\right), \qquad h = \int p(x)\,dx. \tag{4}$$

This reduces solving (1) to the generally simpler task of evaluating integrals. For ODEs for which this is still difficult, you may have to use a numeric method for integrals from Sec. 19.5 or for the ODE itself from Sec. 21.1. We mention that h has nothing to do with $h(x)$ in Sec. 1.1 and that the constant of integration in h does not matter; see Prob. 2.

The structure of (4) is interesting. The only quantity depending on a given initial condition is c. Accordingly, writing (4) as a sum of two terms,

$$y(x) = e^{-h}\int e^{h} r\,dx + ce^{-h}, \tag{4*}$$

we see the following:

$$\text{Total Output} = \text{Response to the Input } r + \text{Response to the Initial Data.} \tag{5}$$

EXAMPLE 1 First-Order ODE, General Solution, Initial Value Problem

Solve the initial value problem

$$y' + y \tan x = \sin 2x, \qquad y(0) = 1.$$

Solution. Here $p = \tan x$, $r = \sin 2x = 2 \sin x \cos x$, and

$$h = \int p\,dx = \int \tan x\,dx = \ln |\sec x|.$$

From this we see that in (4),

$$e^h = \sec x, \qquad e^{-h} = \cos x, \qquad e^h r = (\sec x)(2 \sin x \cos x) = 2 \sin x,$$

and the general solution of our equation is

$$y(x) = \cos x \left(2 \int \sin x\,dx + c\right) = c \cos x - 2 \cos^2 x.$$

From this and the initial condition, $1 = c \cdot 1 - 2 \cdot 1^2$; thus $c = 3$ and the solution of our initial value problem is $y = 3 \cos x - 2 \cos^2 x$. Here $3 \cos x$ is the response to the initial data, and $-2 \cos^2 x$ is the response to the input $\sin 2x$. ■

EXAMPLE 2 Electric Circuit

Model the ***RL*-circuit** in Fig. 19 and solve the resulting ODE for the current $I(t)$ A (amperes), where t is time. Assume that the circuit contains as an **EMF** $E(t)$ (electromotive force) a battery of $E = 48$ V (volts), which is constant, a ***resistor*** of $R = 11\ \Omega$ (ohms), and an ***inductor*** of $L = 0.1$ H (henrys), and that the current is initially zero.

Physical Laws. A current I in the circuit causes a **voltage drop** RI across the resistor **(Ohm's law)** and a voltage drop $LI' = L\,dI/dt$ across the conductor, and the sum of these two voltage drops equals the EMF **(Kirchhoff's Voltage Law, KVL)**.

Remark. In general, KVL states that "The voltage (the electromotive force EMF) impressed on a closed loop is equal to the sum of the voltage drops across all the other elements of the loop." For Kirchoff's Current Law (KCL) and historical information, see footnote 7 in Sec. 2.9.

Solution. According to these laws the model of the RL-circuit is $LI' + RI = E(t)$, in standard form

$$I' + \frac{R}{L} I = \frac{E(t)}{L}. \tag{6}$$

We can solve this linear ODE by (4) with $x = t$, $y = I$, $p = R/L$, $h = (R/L)t$, obtaining the general solution

$$I = e^{-(R/L)t} \left(\int e^{(R/L)t} \frac{E(t)}{L} dt + c \right).$$

By integration,

(7)
$$I = e^{-(R/L)t} \left(\frac{E}{L} \frac{e^{(R/L)t}}{R/L} + c \right) = \frac{E}{R} + ce^{-(R/L)t}.$$

In our case, $R/L = 11/0.1 = 110$ and $E(t) = 48/0.1 = 480 = $ const; thus,

$$I = \tfrac{48}{11} + ce^{-110t}.$$

In modeling, one often gets better insight into the nature of a solution (and smaller roundoff errors) by inserting given numeric data only near the end. Here, the general solution (7) shows that the current approaches the limit $E/R = 48/11$ faster the larger R/L is, in our case, $R/L = 11/0.1 = 110$, and the approach is very fast, from below if $I(0) < 48/11$ or from above if $I(0) > 48/11$. If $I(0) = 48/11$, the solution is constant (48/11 A). See Fig. 19.

The initial value $I(0) = 0$ gives $I(0) = E/R + c = 0$, $c = -E/R$ and the particular solution

(8)
$$I = \frac{E}{R}(1 - e^{-(R/L)t}), \qquad \text{thus} \qquad I = \frac{48}{11}(1 - e^{-110t}). \qquad \blacksquare$$

$R = 11\ \Omega$

$E = 48$ V

$L = 0.1$ H
Circuit

Current $I(t)$

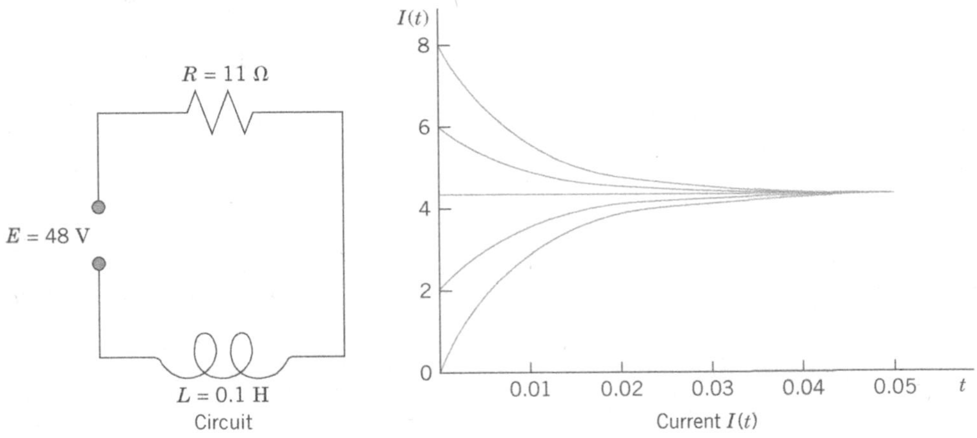

Fig. 19.   RL-circuit

EXAMPLE 3   **Hormone Level**

Assume that the level of a certain hormone in the blood of a patient varies with time. Suppose that the time rate of change is the difference between a sinusoidal input of a 24-hour period from the thyroid gland and a continuous removal rate proportional to the level present. Set up a model for the hormone level in the blood and find its general solution. Find the particular solution satisfying a suitable initial condition.

Solution.   *Step 1. Setting up a model.* Let $y(t)$ be the hormone level at time t. Then the removal rate is $Ky(t)$. The input rate is $A + B \cos \omega t$, where $\omega = 2\pi/24 = \pi/12$ and A is the average input rate; here $A \geqq B$ to make the input rate nonnegative. The constants A, B, K can be determined from measurements. Hence the model is the linear ODE

$$y'(t) = \text{In} - \text{Out} = A + B \cos \omega t - Ky(t), \qquad \text{thus} \qquad y' + Ky = A + B \cos \omega t.$$

The initial condition for a particular solution y_{part} is $y_{\text{part}}(0) = y_0$ with $t = 0$ suitably chosen, for example, 6:00 A.M.

Step 2. General solution. In (4) we have $p = K = $ const, $h = Kt$, and $r = A + B \cos \omega t$. Hence (4) gives the general solution (evaluate $\int e^{Kt} \cos \omega t\ dt$ by integration by parts)

$$y(t) = e^{-Kt}\int e^{Kt}\left(A + B\cos\omega t\right)dt + ce^{-Kt}$$

$$= e^{-Kt}e^{Kt}\left[\frac{A}{K} + \frac{B}{K^2 + \omega^2}\left(K\cos\omega t + \omega\sin\omega t\right)\right] + ce^{-Kt}$$

$$= \frac{A}{K} + \frac{B}{K^2 + (\pi/12)^2}\left(K\cos\frac{\pi t}{12} + \frac{\pi}{12}\sin\frac{\pi t}{12}\right) + ce^{-Kt}.$$

The last term decreases to 0 as t increases, practically after a short time and regardless of c (that is, of the initial condition). The other part of $y(t)$ is called the **steady-state solution** because it consists of constant and periodic terms. The entire solution is called the **transient-state solution** because it models the transition from rest to the steady state. These terms are used quite generally for physical and other systems whose behavior depends on time.

Step 3. Particular solution. Setting $t = 0$ in $y(t)$ and choosing $y_0 = 0$, we have

$$y(0) = \frac{A}{K} + \frac{B}{K^2 + (\pi/12)^2}\,\frac{u}{\pi}K + c = 0, \qquad \text{thus} \qquad c = -\frac{A}{K} - \frac{KB}{K^2 + (\pi/12)^2}.$$

Inserting this result into $y(t)$, we obtain the particular solution

$$y_{\text{part}}(t) = \frac{A}{K} + \frac{B}{K^2 + (\pi/12)^2}\left(K\cos\frac{\pi t}{12} + \frac{\pi}{12}\sin\frac{\pi t}{12}\right) - \left(\frac{A}{K} + \frac{KB}{K^2 + (\pi/12)^2}\right)e^{-K}$$

with the steady-state part as before. To plot y_{part} we must specify values for the constants, say, $A = B = 1$ and $K = 0.05$. Figure 20 shows this solution. Notice that the transition period is relatively short (although K is small), and the curve soon looks sinusoidal; this is the response to the input $A + B\cos\left(\frac{1}{12}\pi t\right) = 1 + \cos\left(\frac{1}{12}\pi t\right)$. ■

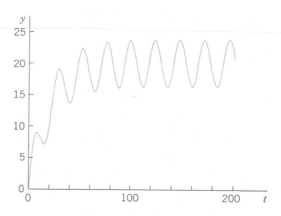

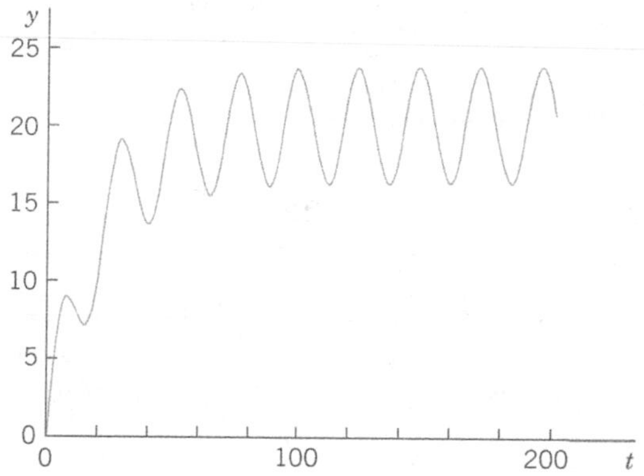

Fig. 20. Particular solution in Example 3

Reduction to Linear Form. Bernoulli Equation

Numerous applications can be modeled by ODEs that are nonlinear but can be transformed to linear ODEs. One of the most useful ones of these is the **Bernoulli equation**[7]

$$y' + p(x)y = g(x)y^a \qquad (a \text{ any real number}). \tag{9}$$

[7] JAKOB BERNOULLI (1654–1705), Swiss mathematician, professor at Basel, also known for his contribution to elasticity theory and mathematical probability. The method for solving Bernoulli's equation was discovered by Leibniz in 1696. Jakob Bernoulli's students included his nephew NIKLAUS BERNOULLI (1687–1759), who contributed to probability theory and infinite series, and his youngest brother JOHANN BERNOULLI (1667–1748), who had profound influence on the development of calculus, became Jakob's successor at Basel, and had among his students GABRIEL CRAMER (see Sec. 7.7) and LEONHARD EULER (see Sec. 2.5). His son DANIEL BERNOULLI (1700–1782) is known for his basic work in fluid flow and the kinetic theory of gases.

If $a = 0$ or $a = 1$, Equation (9) is linear. Otherwise it is nonlinear. Then we set

$$u(x) = [y(x)]^{1-a}.$$

We differentiate this and substitute y' from (9), obtaining

$$u' = (1 - a)y^{-a}y' = (1 - a)y^{-a}(gy^a - py).$$

Simplification gives

$$u' = (1 - a)(g - py^{1-a}),$$

where $y^{1-a} = u$ on the right, so that we get the linear ODE

$$u' + (1 - a)pu = (1 - a)g. \tag{10}$$

For further ODEs reducible to linear form, see Ince's classic [A11] listed in App. 1. See also Team Project 30 in Problem Set 1.5.

EXAMPLE 4 **Logistic Equation**

Solve the following Bernoulli equation, known as the **logistic equation** (or **Verhulst equation**[8]):

$$y' = Ay - By^2 \tag{11}$$

Solution. Write (11) in the form (9), that is,

$$y' - Ay = -By^2$$

to see that $a = 2$, so that $u = y^{1-a} = y^{-1}$. Differentiate this u and substitute y' from (11),

$$u' = -y^{-2}y' = -y^{-2}(Ay - By^2) = B - Ay^{-1}.$$

The last term is $-Ay^{-1} = -Au$. Hence we have obtained the linear ODE

$$u' + Au = B.$$

The general solution is [by (4)]

$$u = ce^{-At} + B/A.$$

Since $u = 1/y$, this gives the general solution of (11),

$$y = \frac{1}{u} = \frac{1}{ce^{-At} + B/A} \tag{12}$$ (Fig. 21)

Directly from (11) we see that $y \equiv 0$ ($y(t) = 0$ for all t) is also a solution. ■

---

[8] PIERRE-FRANÇOIS VERHULST, Belgian statistician, who introduced Eq. (8) as a model for human population growth in 1838.

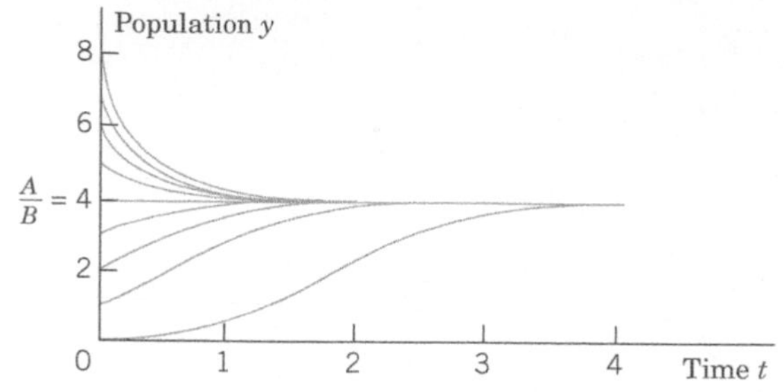

Fig. 21. Logistic population model. Curves (9) in Example 4 with $A/B = 4$

Population Dynamics

The logistic equation (11) plays an important role in **population dynamics**, a field that models the evolution of populations of plants, animals, or humans over time t. If $B = 0$, then (11) is $y' = dy/dt = Ay$. In this case its solution (12) is $y = (1/c)e^{At}$ and gives exponential growth, as for a small population in a large country (the United States in early times!). This is called ***Malthus's law***. (See also Example 3 in Sec. 1.1.)

The term $-By^2$ in (11) is a "braking term" that prevents the population from growing without bound. Indeed, if we write $y' = Ay[1 - (B/A)y]$, we see that if $y < A/B$, then $y' > 0$, so that an initially small population keeps growing as long as $y < A/B$. But if $y > A/B$, then $y' < 0$ and the population is decreasing as long as $y > A/B$. The limit is the same in both cases, namely, A/B. See Fig. 21.

We see that in the logistic equation (11) the independent variable t does not occur explicitly. An ODE $y' = f(t, y)$ in which t does not occur explicitly is of the form

$$y' = f(y) \tag{13}$$

and is called an **autonomous ODE**. Thus the logistic equation (11) is autonomous.

Equation (13) has constant solutions, called **equilibrium solutions** or **equilibrium points**. These are determined by the zeros of $f(y)$, because $f(y) = 0$ gives $y' = 0$ by (13); hence $y = \text{const}$. These zeros are known as **critical points** of (13). An equilibrium solution is called **stable** if solutions close to it for some t remain close to it for all further t. It is called **unstable** if solutions initially close to it do not remain close to it as t increases. For instance, $y = 0$ in Fig. 21 is an unstable equilibrium solution, and $y = 4$ is a stable one. Note that (11) has the critical points $y = 0$ and $y = A/B$.

EXAMPLE 5 **Stable and Unstable Equilibrium Solutions. "Phase Line Plot"**

The ODE $y' = (y - 1)(y - 2)$ has the stable equilibrium solution $y_1 = 1$ and the unstable $y_2 = 2$, as the direction field in Fig. 22 suggests. The values y_1 and y_2 are the zeros of the parabola $f(y) = (y - 1)(y - 2)$ in the figure. Now, since the ODE is autonomous, we can "condense" the direction field to a "phase line plot" giving y_1 and y_2, and the direction (upward or downward) of the arrows in the field, and thus giving information about the stability or instability of the equilibrium solutions. ■

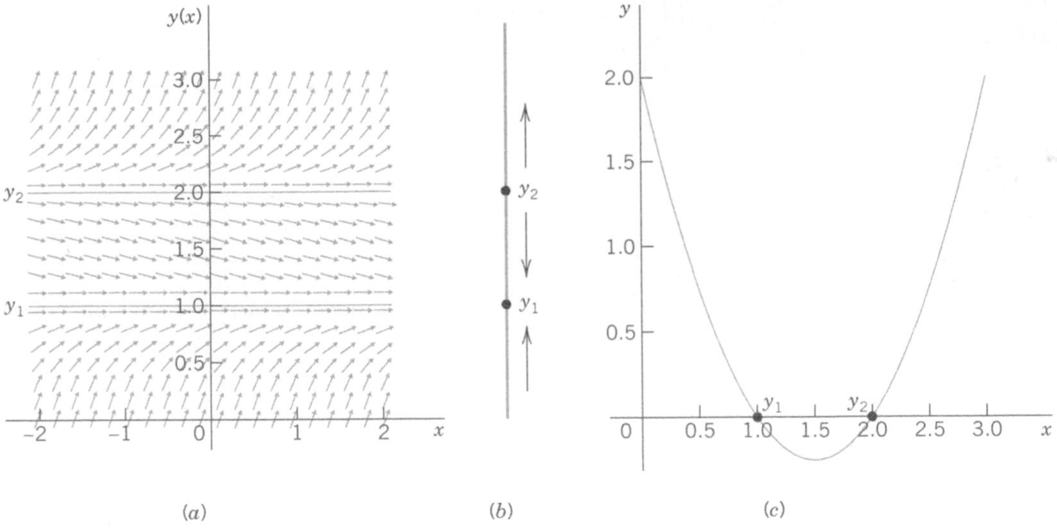

Fig. 22. Example 5. (A) Direction field. (B) "Phase line". (C) Parabola $f(y)$

A few further population models will be discussed in the problem set. For some more details of population dynamics, see C. W. Clark. *Mathematical Bioeconomics: The Mathematics of Conservation* 3rd ed. Hoboken, NJ, Wiley, 2010.

Further applications of linear ODEs follow in the next section.

PROBLEM SET 1.5

1. **CAUTION!** Show that $e^{-\ln x} = 1/x$ (not $-x$) and $e^{-\ln(\sec x)} = \cos x$.
2. **Integration constant.** Give a reason why in (4) you may choose the constant of integration in $\int p\,dx$ to be zero.

3–13 GENERAL SOLUTION. INITIAL VALUE PROBLEMS

Find the general solution. If an initial condition is given, find also the corresponding particular solution and graph or sketch it. (Show the details of your work.)

3. $y' - y = 5.2$
4. $y' = 2y - 4x$
5. $y' + ky = e^{-kx}$
6. $y' + 2y = 4\cos 2x, \quad y(\tfrac{1}{4}\pi) = 3$
7. $xy' = 2y + x^3 e^x$
8. $y' + y\tan x = e^{-0.01x}\cos x, \quad y(0) = 0$
9. $y' + y\sin x = e^{\cos x}, \quad y(0) = -2.5$
10. $y'\cos x + (3y - 1)\sec x = 0, \quad y(\tfrac{1}{4}\pi) = 4/3$
11. $y' = (y - 2)\cot x$
12. $xy' + 4y = 8x^4, \quad y(1) = 2$
13. $y' = 6(y - 2.5)\tanh 1.5x$

14. **CAS EXPERIMENT.** **(a)** Solve the ODE $y' - y/x = -x^{-1}\cos(1/x)$. Find an initial condition for which the arbitrary constant becomes zero. Graph the resulting particular solution, experimenting to obtain a good figure near $x = 0$.

    (b) Generalizing (a) from $n = 1$ to arbitrary n, solve the ODE $y' - ny/x = -x^{n-2}\cos(1/x)$. Find an initial condition as in (a) and experiment with the graph.

15–20 GENERAL PROPERTIES OF LINEAR ODEs

These properties are of practical and theoretical importance because they enable us to obtain new solutions from given ones. Thus in modeling, whenever possible, we prefer linear ODEs over nonlinear ones, which have no similar properties.

Show that nonhomogeneous linear ODEs (1) and homogeneous linear ODEs (2) have the following properties. Illustrate each property by a calculation for two or three equations of your choice. Give proofs.

15. The sum $y_1 + y_2$ of two solutions y_1 and y_2 of the homogeneous equation (2) is a solution of (2), and so is a scalar multiple ay_1 for any constant a. These properties are not true for (1)!

16. $y = 0$ (that is, $y(x) = 0$ for all x, also written $y(x) \equiv 0$) is a solution of (2) [not of (1) if $r(x) \neq 0$!], called the **trivial solution**.

17. The sum of a solution of (1) and a solution of (2) is a solution of (1).

18. The difference of two solutions of (1) is a solution of (2).

19. If y_1 is a solution of (1), what can you say about cy_1?

20. If y_1 and y_2 are solutions of $y_1' + py_1 = r_1$ and $y_2' + py_2 = r_2$, respectively (with the same p!), what can you say about the sum $y_1 + y_2$?

21. Variation of parameter. Another method of obtaining (4) results from the following idea. Write (3) as cy^*, where y^* is the exponential function, which is a solution of the homogeneous linear ODE $y^{*\prime} + py^* = 0$. Replace the arbitrary constant c in (3) with a function u to be determined so that the resulting function $y = uy^*$ is a solution of the nonhomogeneous linear ODE $y' + py = r$.

22–28 NONLINEAR ODEs

Using a method of this section or separating variables, find the general solution. If an initial condition is given, find also the particular solution and sketch or graph it.

22. $y' + y = y^2, \quad y(0) = -\frac{1}{3}$

23. $y' + xy = xy^{-1}, \quad y(0) = 3$

24. $y' + y = -x/y$

25. $y' = 3.2y - 10y^2$

26. $y' = (\tan y)/(x - 1), \quad y(0) = \frac{1}{2}\pi$

27. $y' = 1/(6e^y - 2x)$

28. $2xyy' + (x - 1)y^2 = x^2e^x \quad (\text{Set } y^2 = z)$

29. REPORT PROJECT. Transformation of ODEs. We have transformed ODEs to separable form, to exact form, and to linear form. The purpose of such transformations is an extension of solution methods to larger classes of ODEs. Describe the key idea of each of these transformations and give three typical examples of your choice for each transformation. Show each step (not just the transformed ODE).

30. TEAM PROJECT. Riccati Equation. Clairaut Equation. Singular Solution.
A **Riccati equation** is of the form

$$y' + p(x)y = g(x)y^2 + h(x). \tag{14}$$

A **Clairaut equation** is of the form

$$y = xy' + g(y'). \tag{15}$$

(a) Apply the transformation $y = Y + 1/u$ to the Riccati equation (14), where Y is a solution of (14), and obtain for u the linear ODE $u' + (2Yg - p)u = -g$. Explain the effect of the transformation by writing it as $y = Y + v$, $v = 1/u$.

(b) Show that $y = Y = x$ is a solution of the ODE $y' - (2x^3 + 1)\, y = -x^2y^2 - x^4 - x + 1$ and solve this Riccati equation, showing the details.

(c) Solve the Clairaut equation $y'^2 - xy' + y = 0$ as follows. Differentiate it with respect to x, obtaining $y''(2y' - x) = 0$. Then solve (A) $y'' = 0$ and (B) $2y' - x = 0$ separately and substitute the two solutions (a) and (b) of (A) and (B) into the given ODE. Thus obtain (a) a general solution (straight lines) and (b) a parabola for which those lines (a) are tangents (Fig. 6 in Prob. Set 1.1); so (b) is the envelope of (a). Such a solution (b) that cannot be obtained from a general solution is called a **singular solution**.

(d) Show that the Clairaut equation (15) has as solutions a family of straight lines $y = cx + g(c)$ and a singular solution determined by $g'(s) = -x$, where $s = y'$, that forms the envelope of that family.

31–40 MODELING. FURTHER APPLICATIONS

31. Newton's law of cooling. If the temperature of a cake is 300°F when it leaves the oven and is 200°F ten minutes later, when will it be practically equal to the room temperature of 60°F, say, when will it be 61°F?

32. Heating and cooling of a building. Heating and cooling of a building can be modeled by the ODE

$$T' = k_1(T - T_a) + k_2(T - T_\omega) + P,$$

where $T = T(t)$ is the temperature in the building at time t, T_a the outside temperature, T_w the temperature wanted in the building, and P the rate of increase of T due to machines and people in the building, and k_1 and k_2 are (negative) constants. Solve this ODE, assuming $P = \text{const}$, $T_w = \text{const}$, and T_a varying sinusoidally over 24 hours, say, $T_a = A - C\cos(2\pi/24)t$. Discuss the effect of each term of the equation on the solution.

33. Drug injection. Find and solve the model for drug injection into the bloodstream if, beginning at $t = 0$, a constant amount A g/min is injected and the drug is simultaneously removed at a rate proportional to the amount of the drug present at time t.

34. Epidemics. A model for the spread of contagious diseases is obtained by assuming that the rate of spread is proportional to the number of contacts between infected and noninfected persons, who are assumed to move freely among each other. Set up the model. Find the equilibrium solutions and indicate their stability or instability. Solve the ODE. Find the limit of the proportion of infected persons as $t \to \infty$ and explain what it means.

35. Lake Erie. Lake Erie has a water volume of about 450 km^3 and a flow rate (in and out) of about 175 km^2

per year. If at some instant the lake has pollution concentration $p = 0.04\%$, how long, approximately, will it take to decrease it to $p/2$, assuming that the inflow is much cleaner, say, it has pollution concentration $p/4$, and the mixture is uniform (an assumption that is only imperfectly true)? First guess.

36. Harvesting renewable resources. Fishing. Suppose that the population $y(t)$ of a certain kind of fish is given by the logistic equation (11), and fish are caught at a rate Hy proportional to y. Solve this so-called *Schaefer model.* Find the equilibrium solutions y_1 and $y_2 (> 0)$ when $H < A$. The expression $Y = Hy_2$ is called the **equilibrium harvest** or **sustainable yield** *corresponding to H.* Why?

37. Harvesting. In Prob. 36 find and graph the solution satisfying $y(0) = 2$ when (for simplicity) $A = B = 1$ and $H = 0.2$. What is the limit? What does it mean? What if there were no fishing?

38. Intermittent harvesting. In Prob. 36 assume that you fish for 3 years, then fishing is banned for the next 3 years. Thereafter you start again. And so on. This is called *intermittent harvesting.* Describe qualitatively how the population will develop if intermitting is continued periodically. Find and graph the solution for the first 9 years, assuming that $A = B = 1$, $H = 0.2$, and $y(0) = 2$.

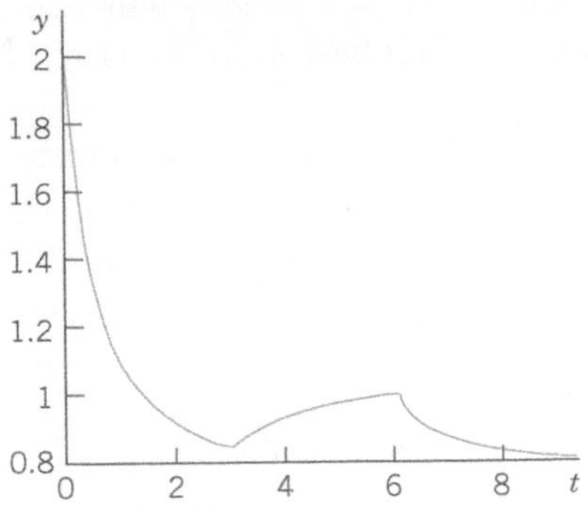

Fig. 23. Fish population in Problem 38

39. Extinction vs. unlimited growth. If in a population $y(t)$ the death rate is proportional to the population, and the birth rate is proportional to the chance encounters of meeting mates for reproduction, what will the model be? Without solving, find out what will eventually happen to a small initial population. To a large one. Then solve the model.

40. Air circulation. In a room containing 20,000 ft^3 of air, 600 ft^3 of fresh air flows in per minute, and the mixture (made practically uniform by circulating fans) is exhausted at a rate of 600 cubic feet per minute (cfm). What is the amount of fresh air $y(t)$ at any time if $y(0) = 0$? After what time will 90% of the air be fresh?

1.6 Orthogonal Trajectories. *Optional*

An important type of problem in physics or geometry is to find a family of curves that intersects a given family of curves at right angles. The new curves are called **orthogonal trajectories** of the given curves (and conversely). Examples are curves of equal temperature **(isotherms)** and curves of heat flow, curves of equal altitude **(contour lines)** on a map and curves of steepest descent on that map, curves of equal potential (**equipotential curves**, curves of equal voltage—the ellipses in Fig. 24) and curves of electric force (the parabolas in Fig. 24).

Here the **angle of intersection** between two curves is defined to be the angle between the tangents of the curves at the intersection point. *Orthogonal* is another word for *perpendicular.*

In many cases orthogonal trajectories can be found using ODEs. In general, if we consider $G(x, y, c) = 0$ to be a given family of curves in the xy-plane, then each value of c gives a particular curve. Since c is one parameter, such a family is called a **one-parameter family of curves.**

In detail, let us explain this method by a family of ellipses

$$\frac{1}{2}x^2 + y^2 = c \qquad (c > 0) \tag{1}$$

and illustrated in Fig. 24. We assume that this family of ellipses represents electric equipotential curves between the two black ellipses (equipotential surfaces between two elliptic cylinders in space, of which Fig. 24 shows a cross-section). We seek the orthogonal trajectories, the curves of electric force. Equation (1) is a *one-parameter family* with *parameter c*. Each value of c (> 0) corresponds to one of these ellipses.

Step 1. Find an ODE for which the given family is a general solution. Of course, this ODE must no longer contain the parameter c. Differentiating (1), we have $x + 2yy' = 0$. Hence the ODE of the given curves is

$$y' = f(x, y) = -\frac{x}{2y}. \tag{2}$$

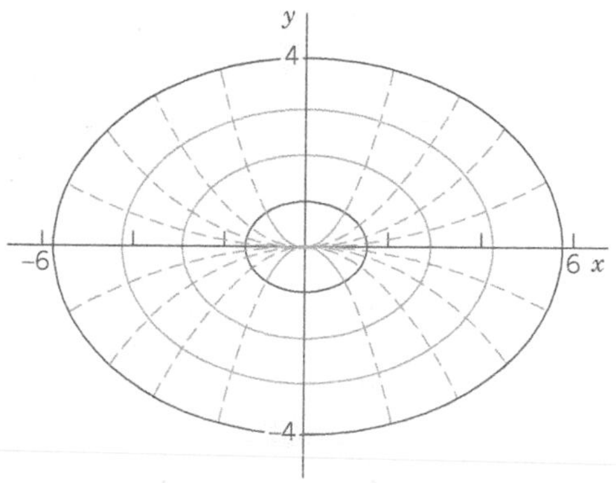

Fig. 24. Electrostatic field between two ellipses (elliptic cylinders in space): Elliptic equipotential curves (equipotential surfaces) and orthogonal trajectories (parabolas)

Step 2. Find an ODE for the orthogonal trajectories $\tilde{y} = \tilde{y}(x)$. This ODE is

$$\tilde{y}' = -\frac{1}{f(x, \tilde{y})} = +\frac{2\tilde{y}}{x} \tag{3}$$

with the same f as in (2). Why? Well, a given curve passing through a point (x_0, y_0) has slope $f(x_0, y_0)$ at that point, by (2). The trajectory through (x_0, y_0) has slope $-1/f(x_0, y_0)$ by (3). The product of these slopes is -1, as we see. From calculus it is known that this is the condition for orthogonality (perpendicularity) of two straight lines (the tangents at (x_0, y_0)), hence of the curve and its orthogonal trajectory at (x_0, y_0).

Step 3. Solve (3) by separating variables, integrating, and taking exponents:

$$\frac{d\tilde{y}}{\tilde{y}} = 2\frac{dx}{x}, \qquad \ln|\tilde{y}| = 2\ln x + c, \qquad \tilde{y} = c^* x^2.$$

This is the family of orthogonal trajectories, the quadratic parabolas along which electrons or other charged particles (of very small mass) would move in the electric field between the black ellipses (elliptic cylinders).

PROBLEM SET 1.6

1–3 FAMILIES OF CURVES

Represent the given family of curves in the form $G(x, y; c) = 0$ and sketch some of the curves.

1. All ellipses with foci -3 and 3 on the x-axis.

2. All circles with centers on the cubic parabola $y = x^3$ and passing through the origin (0, 0).

3. The catenaries obtained by translating the catenary $y = \cosh x$ in the direction of the straight line $y = x$.

4–10 ORTHOGONAL TRAJECTORIES (OTs)

Sketch or graph some of the given curves. Guess what their OTs may look like. Find these OTs.

4. $y = x^2 + c$ **5.** $y = cx$

6. $xy = c$ **7.** $y = c/x^2$

8. $y = \sqrt{x + c}$ **9.** $y = ce^{-x^2}$

10. $x^2 + (y - c)^2 = c^2$

11–16 APPLICATIONS, EXTENSIONS

11. Electric field. Let the electric **equipotential lines** (curves of constant potential) between two concentric cylinders with the z-axis in space be given by $u(x, y) = x^2 + y^2 = c$ (these are circular cylinders in the xyz-space). Using the method in the text, find their orthogonal trajectories (the curves of electric force).

12. Electric field. The lines of electric force of two opposite charges of the same strength at $(-1, 0)$ and $(1, 0)$ are the circles through $(-1, 0)$ and $(1, 0)$. Show that these circles are given by $x^2 + (y - c)^2 = 1 + c^2$. Show that the **equipotential lines** (which are orthogonal trajectories of those circles) are the circles given by $(x + c^*)^2 + \tilde{y}^2 = c^{*2} - 1$ (dashed in Fig. 25).

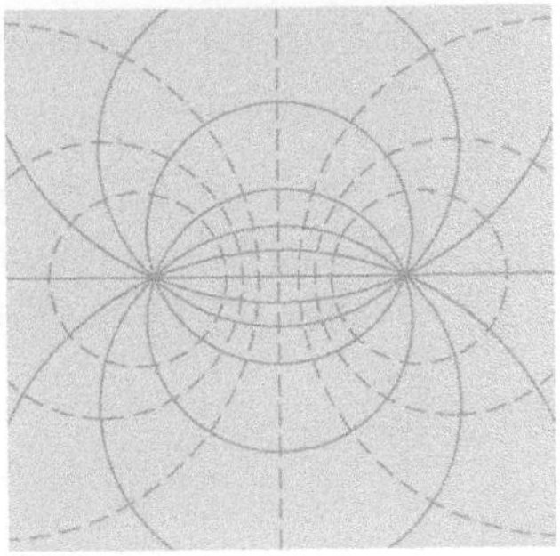

Fig. 25. Electric field in Problem 12

13. Temperature field. Let the **isotherms** (curves of constant temperature) in a body in the upper half-plane $y > 0$ be given by $4x^2 + 9y^2 = c$. Find the orthogonal trajectories (the curves along which heat will flow in regions filled with heat-conducting material and free of heat sources or heat sinks).

14. Conic sections. Find the conditions under which the orthogonal trajectories of families of ellipses $x^2/a^2 + y^2/b^2 = c$ are again conic sections. Illustrate your result graphically by sketches or by using your CAS. What happens if $a \to 0$? If $b \to 0$?

15. Cauchy–Riemann equations. Show that for a family $u(x, y) = c =$ const the orthogonal trajectories $v(x, y) = c^* =$ const can be obtained from the following *Cauchy–Riemann equations* (which are basic in complex analysis in Chap. 13) and use them to find the orthogonal trajectories of $e^x \sin y =$ const. (Here, subscripts denote partial derivatives.)

$$u_x = v_y, \qquad u_y = -v_x$$

16. Congruent OTs. If $y' = f(x)$ with f independent of y, show that the curves of the corresponding family are congruent, and so are their OTs.

1.7 Existence and Uniqueness of Solutions for Initial Value Problems

The initial value problem

$$|y'| + |y| = 0, \qquad y(0) = 1$$

has no solution because $y = 0$ (that is, $y(x) = 0$ for all x) is the only solution of the ODE. The initial value problem

$$y' = 2x, \qquad y(0) = 1$$

has precisely one solution, namely, $y = x^2 + 1$. The initial value problem

$$xy' = y - 1, \qquad y(0) = 1$$

has infinitely many solutions, namely, $y = 1 + cx$, where c is an arbitrary constant because $y(0) = 1$ for all c.

From these examples we see that an **initial value problem**

$$y' = f(x, y), \qquad y(x_0) = y_0 \tag{1}$$

may have no solution, precisely one solution, or more than one solution. This fact leads to the following two fundamental questions.

Problem of Existence

Under what conditions does an initial value problem of the form (1) *have at least one solution (hence one or several solutions)?*

Problem of Uniqueness

Under what conditions does that problem have at most one solution (hence excluding the case that is has more than one solution)?

Theorems that state such conditions are called **existence theorems** and **uniqueness theorems**, respectively.

Of course, for our simple examples, we need no theorems because we can solve these examples by inspection; however, for complicated ODEs such theorems may be of considerable practical importance. Even when you are sure that your physical or other system behaves uniquely, occasionally your model may be oversimplified and may not give a faithful picture of reality.

THEOREM 1

Existence Theorem

Let the right side $f(x, y)$ of the ODE in the initial value problem

$$y' = f(x, y), \qquad y(x_0) = y_0 \tag{1}$$

be continuous at all points (x, y) in some rectangle

$$R: |x - x_0| < a, \qquad |y - y_0| < b \qquad \text{(Fig. 26)}$$

and **bounded** *in R; that is, there is a number K such that*

$$|f(x, y)| \leqq K \qquad \textit{for all } (x, y) \textit{ in } R. \tag{2}$$

Then the initial value problem (1) *has at least one solution $y(x)$. This solution exists at least for all x in the subinterval $|x - x_0| < \alpha$ of the interval $|x - x_0| < a$; here, α is the smaller of the two numbers a and b/K.*

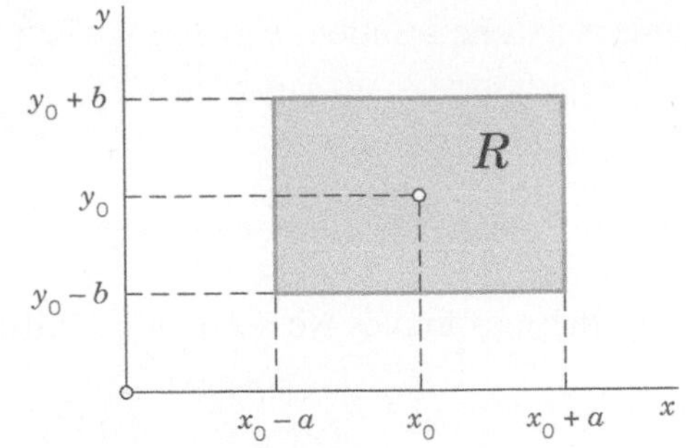

Fig. 26. Rectangle R in the existence and uniqueness theorems

(*Example of Boundedness.* The function $f(x, y) = x^2 + y^2$ is bounded (with $K = 2$) in the square $|x| < 1, |y| < 1$. The function $f(x, y) = \tan(x + y)$ is not bounded for $|x + y| < \pi/2$. Explain!)

THEOREM 2

Uniqueness Theorem

Let f and its partial derivative $f_y = \partial f/\partial y$ be continuous for all (x, y) in the rectangle R (Fig. 26) and bounded, say,

$$\text{(3)} \qquad \text{(a)}\ |f(x, y)| \leqq K, \qquad \text{(b)}\ |f_y(x, y)| \leqq M \qquad \textit{for all } (x, y) \textit{ in } R.$$

Then the initial value problem (1) has at most one solution $y(x)$. Thus, by Theorem 1, the problem has precisely one solution. This solution exists at least for all x in that subinterval $|x - x_0| < \alpha$.

Understanding These Theorems

These two theorems take care of almost all practical cases. Theorem 1 says that if $f(x, y)$ is continuous in some region in the xy-plane containing the point (x_0, y_0), then the initial value problem (1) has at least one solution.

Theorem 2 says that if, moreover, the partial derivative $\partial f/\partial y$ of f with respect to y exists and is continuous in that region, then (1) can have at most one solution; hence, by Theorem 1, it has precisely one solution.

Read again what you have just read—these are entirely new ideas in our discussion.

Proofs of these theorems are beyond the level of this book (see Ref. [A11] in App. 1); however, the following remarks and examples may help you to a good understanding of the theorems.

Since $y' = f(x, y)$, the condition (2) implies that $|y'| \leqq K$; that is, the slope of any solution curve $y(x)$ in R is at least $-K$ and at most K. Hence a solution curve that passes through the point (x_0, y_0) must lie in the colored region in Fig. 27 bounded by the lines l_1 and l_2 whose slopes are $-K$ and K, respectively. Depending on the form of R, two different cases may arise. In the first case, shown in Fig. 27a, we have $b/K \geqq a$ and therefore $\alpha = a$ in the existence theorem, which then asserts that the solution exists for all x between $x_0 - a$ and $x_0 + a$. In the second case, shown in Fig. 27b, we have $b/K < a$. Therefore, $\alpha = b/K < a$, and all we can conclude from the theorems is that the solution exists for all x between $x_0 - b/K$ and $x_0 + b/K$. For larger or smaller x's the solution curve may leave the rectangle R, and since nothing is assumed about f outside R, nothing can be concluded about the solution for those larger or amaller x's; that is, for such x's the solution may or may not exist—we don't know.

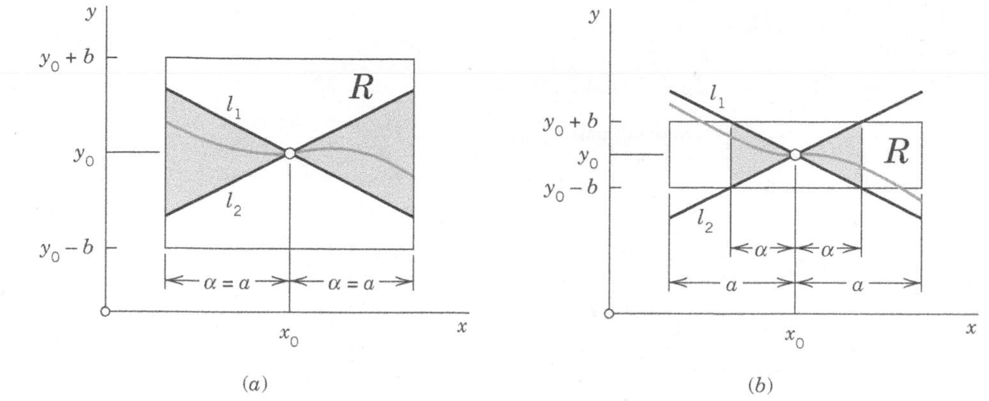

Fig. 27. The condition (2) of the existence theorem. (a) First case. (b) Second case

Let us illustrate our discussion with a simple example. We shall see that our choice of a rectangle R with a large base (a long x-interval) will lead to the case in Fig. 27b.

EXAMPLE 1 **Choice of a Rectangle**

Consider the initial value problem

$$y' = 1 + y^2, \qquad y(0) = 0$$

and take the rectangle R; $|x| < 5$, $|y| < 3$. Then $a = 5$, $b = 3$, and

$$|f(x, y)| = |1 + y^2| \leqq K = 10,$$

$$\left|\frac{\partial f}{\partial y}\right| = 2|y| \leqq M = 6,$$

$$\alpha = \frac{b}{K} = 0.3 < a.$$

Indeed, the solution of the problem is $y = \tan x$ (see Sec. 1.3, Example 1). This solution is discontinuous at $\pm\pi/2$, and there is no *continuous* solution valid in the entire interval $|x| < 5$ from which we started. ■

The conditions in the two theorems are sufficient conditions rather than necessary ones, and can be lessened. In particular, by the mean value theorem of differential calculus we have

$$f(x, y_2) - f(x, y_1) = (y_2 - y_1) \left.\frac{\partial f}{\partial y}\right|_{y=\tilde{y}}$$

where (x, y_1) and (x, y_2) are assumed to be in R, and $\tilde{y}$ is a suitable value between y_1 and y_2. From this and (3b) it follows that

$$|f(x, y_2) - f(x, y_1)| \leqq M|y_2 - y_1|. \tag{4}$$

It can be shown that (3b) may be replaced by the weaker condition (4), which is known as a **Lipschitz condition.**[9] However, continuity of $f(x, y)$ is not enough to guarantee the *uniqueness* of the solution. This may be illustrated by the following example.

EXAMPLE 2 Nonuniqueness

The initial value problem

$$y' = \sqrt{|y|}. \qquad y(0) = 0$$

has the two solutions

$$y = 0 \qquad \text{and} \qquad y^* = \begin{cases} x^2/4 & \text{if } x \geqq 0 \\ -x^2/4 & \text{if } x < 0 \end{cases}$$

although $f(x, y) = \sqrt{|y|}$ is continuous for all y. The Lipschitz condition (4) is violated in any region that includes the line $y = 0$, because for $y_1 = 0$ and positive y_2 we have

$$\text{(5)} \qquad \frac{|f(x, y_2) - f(x, y_1)|}{|y_2 - y_1|} = \frac{\sqrt{y_2}}{y_2} = \frac{1}{\sqrt{y_2}}, \qquad (\sqrt{y_2} > 0)$$

and this can be made as large as we please by choosing y_2 sufficiently small, whereas (4) requires that the quotient on the left side of (5) should not exceed a fixed constant M. ■

PROBLEM SET 1.7

1. **Linear ODE.** If p and r in $y' + p(x)y = r(x)$ are continuous for all x in an interval $|x - x_0| \leq a$, show that $f(x, y)$ in this ODE satisfies the conditions of our present theorems, so that a corresponding initial value problem has a unique solution. Do you actually need these theorems for this ODE?
2. **Existence?** Does the initial value problem $(x - 2)y' = y$, $y(2) = 1$ have a solution? Does your result contradict our present theorems?
3. **Vertical strip.** If the assumptions of Theorems 1 and 2 are satisfied not merely in a rectangle but in a vertical infinite strip $|x - x_0| < a$, in what interval will the solution of (1) exist?
4. **Change of initial condition.** What happens in Prob. 2 if you replace $y(2) = 1$ with $y(2) = k$?
5. **Length of x-interval.** In most cases the solution of an initial value problem (1) exists in an x-interval larger than that guaranteed by the present theorems. Show this fact for $y' = 2y^2$, $y(1) = 1$ by finding the best possible a (choosing b optimally) and comparing the result with the actual solution.
6. **CAS PROJECT. Picard Iteration.** **(a)** Show that by integrating the ODE in (1) and observing the initial condition you obtain

$$\text{(6)} \qquad y(x) = y_0 + \int_{x_0}^{x} f(t, y(t))\,dt.$$

This form (6) of (1) suggests **Picard's Iteration Method**[10] which is defined by

$$\text{(7)} \quad y_n(x) = y_0 + \int_{x_0}^{x} f(t, y_{n-1}(t))\,dt, \quad n = 1, 2, \cdots.$$

It gives approximations $y_1, y_2, y_3, \ldots$ of the unknown solution y of (1). Indeed, you obtain y_1 by substituting $y = y_0$ on the right and integrating—this is the first step—then y_2 by substituting $y = y_1$ on the right and integrating—this is the second step—and so on. Write

[9] RUDOLF LIPSCHITZ (1832–1903), German mathematician. Lipschitz and similar conditions are important in modern theories, for instance, in partial differential equations.

[10] EMILE PICARD (1856–1941). French mathematician, also known for his important contributions to complex analysis (see Sec. 16.2 for his famous theorem). Picard used his method to prove Theorems 1 and 2 as well as the convergence of the sequence (7) to the solution of (1). In precomputer times, the iteration was of little *practical* value because of the integrations.

a program of the iteration that gives a printout of the first approximations $y_0, y_1, \ldots, y_N$ as well as their graphs on common axes. Try your program on two initial value problems of your own choice.

(b) Apply the iteration to $y' = x + y$, $y(0) = 0$. Also solve the problem exactly.

(c) Apply the iteration to $y' = 2y^2$, $y(0) = 1$. Also solve the problem exactly.

(d) Find all solutions of $y' = 2\sqrt{y}$, $y(1) = 0$. Which of them does Picard's iteration approximate?

(e) Experiment with the conjecture that Picard's iteration converges to the solution of the problem for any initial choice of y in the integrand in (7) (leaving y_0 outside the integral as it is). Begin with a simple ODE and see what happens. When you are reasonably sure, take a slightly more complicated ODE and give it a try.

7. **Maximum α.** What is the largest possible α in Example 1 in the text?

8. **Lipschitz condition.** Show that for a linear ODE $y' + p(x)y = r(x)$ with continuous p and r in $|x - x_0| \leqq a$ a Lipschitz condition holds. This is remarkable because it means that for a *linear* ODE the continuity of $f(x, y)$ guarantees not only the existence but also the uniqueness of the solution of an initial value problem. (Of course, this also follows directly from (4) in Sec. 1.5.)

9. **Common points.** Can two solution curves of the same ODE have a common point in a rectangle in which the assumptions of the present theorems are satisfied?

10. **Three possible cases.** Find all initial conditions such that $(x^2 - x)y' = (2x - 1)y$ has no solution, precisely one solution, and more than one solution.

CHAPTER 1 REVIEW QUESTIONS AND PROBLEMS

1. Explain the basic concepts ordinary and partial differential equations (ODEs, PDEs), order, general and particular solutions, initial value problems (IVPs). Give examples.
2. What is a linear ODE? Why is it easier to solve than a nonlinear ODE?
3. Does every first-order ODE have a solution? A solution formula? Give examples.
4. What is a direction field? A numeric method for first-order ODEs?
5. What is an exact ODE? Is $f(x)\,dx + g(y)\,dy = 0$ always exact?
6. Explain the idea of an integrating factor. Give two examples.
7. What other solution methods did we consider in this chapter?
8. Can an ODE sometimes be solved by several methods? Give three examples.
9. What does modeling mean? Can a CAS solve a model given by a first-order ODE? Can a CAS set up a model?
10. Give problems from mechanics, heat conduction, and population dynamics that can be modeled by first-order ODEs.

11–16 DIRECTION FIELD: NUMERIC SOLUTION

Graph a direction field (by a CAS or by hand) and sketch some solution curves. Solve the ODE exactly and compare. In Prob. 16 use Euler's method.

11. $y' + 2y = 0$
12. $y' = 1 - y^2$
13. $y' = y - 4y^2$
14. $xy' = y + x^2$
15. $y' + y = 1.01 \cos 10x$
16. Solve $y' = y - y^2$, $y(0) = 0.2$ by Euler's method (10 steps, $h = 0.1$). Solve exactly and compute the error.

17–21 GENERAL SOLUTION

Find the general solution. Indicate which method in this chapter you are using. Show the details of your work.

17. $y' + 2.5y = 1.6x$
18. $y' - 0.4y = 29 \sin x$
19. $25yy' - 4x = 0$
20. $y' = ay + by^2 \quad (a \neq 0)$
21. $(3xe^y + 2y)\,dx + (x^2e^y + x)\,dy = 0$

22–26 INITIAL VALUE PROBLEM (IVP)

Solve the IVP. Indicate the method used. Show the details of your work.

22. $y' + 4xy = e^{-2x^2}, \quad y(0) = -4.3$
23. $y' = \sqrt{1 - y^2}, \quad y(0) = 1/\sqrt{2}$
24. $y' + \frac{1}{2}y = y^3, \quad y(0) = \frac{1}{3}$
25. $3 \sec y\,dx + \frac{1}{3} \sec x\,dy = 0, \quad y(0) = 0$
26. $x \sinh y\,dy = \cosh y\,dx, \quad y(3) = 0$

27–30 MODELING, APPLICATIONS

27. **Exponential growth.** If the growth rate of a culture of bacteria is proportional to the number of bacteria present and after 1 day is 1.25 times the original number, within what interval of time will the number of bacteria (a) double, (b) triple?

28. Mixing problem. The tank in Fig. 28 contains 80 lb of salt dissolved in 500 gal of water. The inflow per minute is 20 lb of salt dissolved in 20 gal of water. The outflow is 20 gal/min of the uniform mixture. Find the time when the salt content $y(t)$ in the tank reaches 95% of its limiting value (as $t \to \infty$).

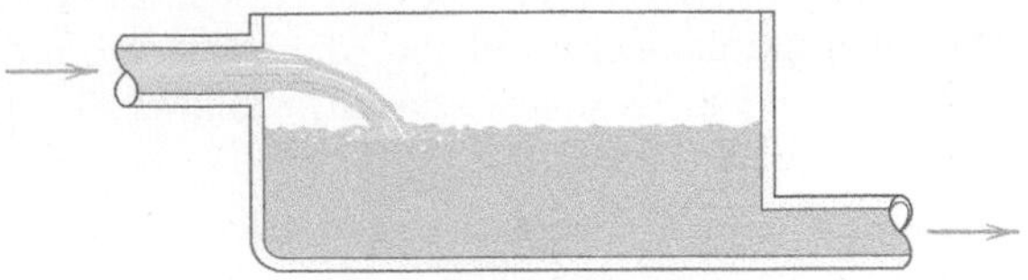

Fig. 28. Tank in Problem 28

29. Half-life. If in a reactor, uranium $^{237}_{97}U$ loses 10% of its weight within one day, what is its half-life? How long would it take for 99% of the original amount to disappear?

30. Newton's law of cooling. A metal bar whose temperature is 20°C is placed in boiling water. How long does it take to heat the bar to practically 100°C, say, to 99.9°C, if the temperature of the bar after 1 min of heating is 51.5°C? First guess, then calculate.

SUMMARY OF CHAPTER 1
First-Order ODEs

This chapter concerns **ordinary differential equations (ODEs) of first order** and their applications. These are equations of the form

$$(1) \qquad F(x, y, y') = 0 \qquad \text{or in explicit form} \qquad y' = f(x, y)$$

involving the derivative $y' = dy/dx$ of an unknown function y, given functions of x, and, perhaps, y itself. If the independent variable x is time, we denote it by t.

In Sec. 1.1 we explained the basic concepts and the process of **modeling**, that is, of expressing a physical or other problem in some mathematical form and solving it. Then we discussed the method of direction fields (Sec. 1.2), solution methods and models (Secs. 1.3–1.6), and, finally, ideas on existence and uniqueness of solutions (Sec. 1.7).

A first-order ODE usually has a **general solution**, that is, a solution involving an arbitrary constant, which we denote by c. In applications we usually have to find a unique solution by determining a value of c from an **initial condition** $y(x_0) = y_0$. Together with the ODE this is called an **initial value problem**

$$(2) \qquad y' = f(x, y), \qquad y(x_0) = y_0 \qquad (x_0, y_0 \text{ given numbers})$$

and its solution is a **particular solution** of the ODE. Geometrically, a general solution represents a family of curves, which can be graphed by using **direction fields** (Sec. 1.2). And each particular solution corresponds to one of these curves.

A **separable ODE** is one that we can put into the form

$$(3) \qquad g(y)\,dy = f(x)\,dx \qquad \text{(Sec. 1.3)}$$

by algebraic manipulations (possibly combined with transformations, such as $y/x = u$) and solve by integrating on both sides.

An **exact ODE** is of the form

(4) $$M(x, y)\,dx + N(x, y)\,dy = 0 \qquad \text{(Sec. 1.4)}$$

where $M\,dx + N\,dy$ is the **differential**

$$du = u_x\,dx + u_y\,dy$$

of a function $u(x, y)$, so that from $du = 0$ we immediately get the implicit general solution $u(x, y) = c$. This method extends to nonexact ODEs that can be made exact by multiplying them by some function $F(x, y,)$, called an **integrating factor** (Sec. 1.4).

Linear ODEs

(5) $$y' + p(x)y = r(x)$$

are very important. Their solutions are given by the integral formula (4), Sec. 1.5. Certain nonlinear ODEs can be transformed to linear form in terms of new variables. This holds for the **Bernoulli equation**

$$y' + p(x)y = g(x)y^a \qquad \text{(Sec. 1.5).}$$

Applications and ***modeling*** are discussed throughout the chapter, in particular in Secs. 1.1, 1.3, 1.5 (***population dynamics***, etc.), and 1.6 (***trajectories***).

Picard's ***existence*** and ***uniqueness theorems*** are explained in Sec. 1.7 (and *Picard's iteration* in Problem Set 1.7).

Numeric methods for first-order ODEs can be studied in Secs. 21.1 and 21.2 immediately after this chapter, as indicated in the chapter opening.

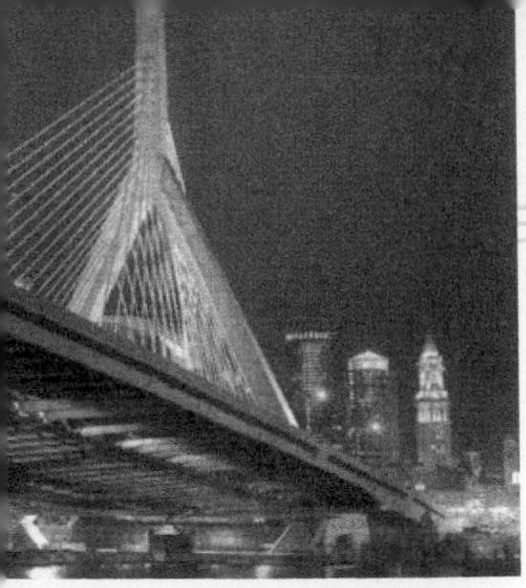

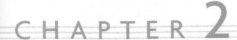

CHAPTER 2

Second-Order Linear ODEs

Many important applications in mechanical and electrical engineering, as shown in Secs. 2.4, 2.8, and 2.9, are modeled by linear ordinary differential equations (linear ODEs) of the second order. Their theory is representative of all linear ODEs as is seen when compared to linear ODEs of third and higher order, respectively. However, the solution formulas for second-order linear ODEs are simpler than those of higher order, so it is a natural progression to study ODEs of second order first in this chapter and then of higher order in Chap. 3.

Although ordinary differential equations (ODEs) can be grouped into linear and nonlinear ODEs, nonlinear ODEs are difficult to solve in contrast to linear ODEs for which many beautiful standard methods exist.

Chapter 2 includes the derivation of general and particular solutions, the latter in connection with initial value problems.

For those interested in solution methods for Legendre's, Bessel's, and the hypergeometric equations consult Chap. 5 and for Sturm–Liouville problems Chap. 11.

COMMENT. ***Numerics for second-order ODEs can be studied immediately after this chapter.*** See Sec. 21.3, which is independent of other sections in Chaps. 19–21.

Prerequisite: Chap. 1, in particular, Sec. 1.5.
Sections that may be omitted in a shorter course: 2.3, 2.9, 2.10.
References and Answers to Problems: App. 1 Part A, and App. 2.

2.1 Homogeneous Linear ODEs of Second Order

We have already considered first-order linear ODEs (Sec. 1.5) and shall now define and discuss linear ODEs of second order. These equations have important engineering applications, especially in connection with mechanical and electrical vibrations (Secs. 2.4, 2.8, 2.9) as well as in wave motion, heat conduction, and other parts of physics, as we shall see in Chap. 12.

A second-order ODE is called **linear** if it can be written

$$y'' + p(x)y' + q(x)y = r(x) \tag{1}$$

and **nonlinear** if it cannot be written in this form.

The distinctive feature of this equation is that it is *linear in y and its derivatives,* whereas the functions p, q, and r on the right may be any given functions of x. If the equation begins with, say, $f(x)y''$, then divide by $f(x)$ to have the **standard form** (1) with y'' as the first term.

The definitions of homogeneous and nonhomogenous second-order linear ODEs are very similar to those of first-order ODEs discussed in Sec. 1.5. Indeed, if $r(x) \equiv 0$ (that is, $r(x) = 0$ for all x considered; read "$r(x)$ is identically zero"), then (1) reduces to

$$y'' + p(x)y' + q(x)y = 0 \tag{2}$$

and is called **homogeneous**. If $r(x) \not\equiv 0$, then (1) is called **nonhomogeneous**. This is similar to Sec. 1.5.

An example of a nonhomogeneous linear ODE is

$$y'' + 25y = e^{-x}\cos x,$$

and a homogeneous linear ODE is

$$xy'' + y' + xy = 0, \qquad \text{written in standard form} \qquad y'' + \frac{1}{x}y' + y = 0.$$

Finally, an example of a nonlinear ODE is

$$y''y + y'^2 = 0.$$

The functions p and q in (1) and (2) are called the **coefficients** of the ODEs.

Solutions are defined similarly as for first-order ODEs in Chap. 1. A function

$$y = h(x)$$

is called a *solution* of a (linear or nonlinear) second-order ODE on some open interval I if h is defined and twice differentiable throughout that interval and is such that the ODE becomes an identity if we replace the unknown y by h, the derivative y' by h', and the second derivative y'' by h''. Examples are given below.

Homogeneous Linear ODEs: Superposition Principle

Sections 2.1–2.6 will be devoted to **homogeneous** linear ODEs (2) and the remaining sections of the chapter to nonhomogeneous linear ODEs.

Linear ODEs have a rich solution structure. For the homogeneous equation the backbone of this structure is the *superposition principle* or *linearity principle,* which says that we can obtain further solutions from given ones by adding them or by multiplying them with any constants. Of course, this is a great advantage of homogeneous linear ODEs. Let us first discuss an example.

EXAMPLE 1 Homogeneous Linear ODEs: Superposition of Solutions

The functions $y = \cos x$ and $y = \sin x$ are solutions of the homogeneous linear ODE

$$y'' + y = 0$$

for all x. We verify this by differentiation and substitution. We obtain $(\cos x)'' = -\cos x$; hence

$$y'' + y = (\cos x)'' + \cos x = -\cos x + \cos x = 0.$$

Similarly for $y = \sin x$ (verify!). We can go an important step further. We multiply $\cos x$ by any constant, for instance, 4.7, and $\sin x$ by, say, -2, and take the sum of the results, claiming that it is a solution. Indeed, differentiation and substitution gives

$$(4.7 \cos x - 2 \sin x)'' + (4.7 \cos x - 2 \sin x) = -4.7 \cos x + 2 \sin x + 4.7 \cos x - 2 \sin x = 0. \quad \blacksquare$$

In this example we have obtained from $y_1 (= \cos x)$ and $y_2 (= \sin x)$ a function of the form

(3) $$y = c_1 y_1 + c_2 y_2 \qquad (c_1, c_2 \text{ arbitrary constants}).$$

This is called a **linear combination** of y_1 and y_2. In terms of this concept we can now formulate the result suggested by our example, often called the **superposition principle** or **linearity principle**.

THEOREM 1

> **Fundamental Theorem for the Homogeneous Linear ODE (2)**
>
> *For a homogeneous linear ODE (2), any linear combination of two solutions on an open interval I is again a solution of (2) on I. In particular, for such an equation, sums and constant multiples of solutions are again solutions.*

PROOF   Let y_1 and y_2 be solutions of (2) on I. Then by substituting $y = c_1 y_1 + c_2 y_2$ and its derivatives into (2), and using the familiar rule $(c_1 y_1 + c_2 y_2)' = c_1 y_1' + c_2 y_2'$, etc., we get

$$y'' + py' + qy = (c_1 y_1 + c_2 y_2)'' + p(c_1 y_1 + c_2 y_2)' + q(c_1 y_1 + c_2 y_2)$$

$$= c_1 y_1'' + c_2 y_2'' + p(c_1 y_1' + c_2 y_2') + q(c_1 y_1 + c_2 y_2)$$

$$= c_1(y_1'' + py_1' + qy_1) + c_2(y_2'' + py_2' + qy_2) = 0,$$

since in the last line, $(\cdots) = 0$ because y_1 and y_2 are solutions, by assumption. This shows that y is a solution of (2) on I. $\blacksquare$

CAUTION!   Don't forget that this highly important theorem holds for *homogeneous linear* ODEs only but **does not hold** for nonhomogeneous linear or nonlinear ODEs, as the following two examples illustrate.

EXAMPLE 2   **A Nonhomogeneous Linear ODE**

Verify by substitution that the functions $y = 1 + \cos x$ and $y = 1 + \sin x$ are solutions of the nonhomogeneous linear ODE

$$y'' + y = 1,$$

but their sum is not a solution. Neither is, for instance, $2(1 + \cos x)$ or $5(1 + \sin x)$. $\blacksquare$

EXAMPLE 3   **A Nonlinear ODE**

Verify by substitution that the functions $y = x^2$ and $y = 1$ are solutions of the nonlinear ODE

$$y''y - xy' = 0,$$

but their sum is not a solution. Neither is $-x^2$, so you cannot even multiply by -1! $\blacksquare$

Initial Value Problem. Basis. General Solution

Recall from Chap. 1 that for a first-order ODE, an *initial value problem* consists of the ODE and one *initial condition* $y(x_0) = y_0$. The initial condition is used to determine the *arbitrary constant* c in the *general solution* of the ODE. This results in a unique solution, as we need it in most applications. That solution is called a *particular solution* of the ODE. These ideas extend to second-order ODEs as follows.

For a second-order homogeneous linear ODE (2) an **initial value problem** consists of (2) and two **initial conditions**

$$y(x_0) = K_0, \qquad y'(x_0) = K_1. \tag{4}$$

These conditions prescribe given values K_0 and K_1 of the solution and its first derivative (the slope of its curve) at the same given $x = x_0$ in the open interval considered.

The conditions (4) are used to determine the two arbitrary constants c_1 and c_2 in a **general solution**

$$y = c_1 y_1 + c_2 y_2 \tag{5}$$

of the ODE; here, y_1 and y_2 are suitable solutions of the ODE, with "suitable" to be explained after the next example. This results in a unique solution, passing through the point (x_0, K_0) with K_1 as the tangent direction (the slope) at that point. That solution is called a **particular solution** of the ODE (2).

EXAMPLE 4 **Initial Value Problem**

Solve the initial value problem

$$y'' + y = 0, \qquad y(0) = 3.0, \qquad y'(0) = -0.5.$$

Solution. ***Step 1. General solution.*** The functions $\cos x$ and $\sin x$ are solutions of the ODE (by Example 1), and we take

$$y = c_1 \cos x + c_2 \sin x.$$

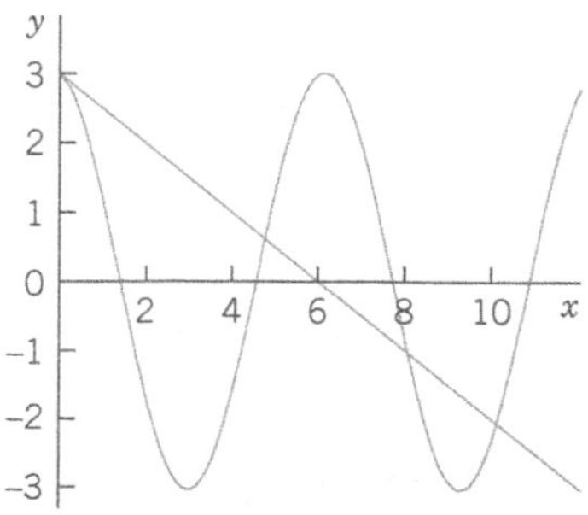

Fig. 29. Particular solution and initial tangent in Example 4

This will turn out to be a general solution as defined below.

Step 2. Particular solution. We need the derivative $y' = -c_1 \sin x + c_2 \cos x$. From this and the initial values we obtain, since $\cos 0 = 1$ and $\sin 0 = 0$,

$$y(0) = c_1 = 3.0 \qquad \text{and} \qquad y'(0) = c_2 = -0.5.$$

This gives as the solution of our initial value problem the particular solution

$$y = 3.0 \cos x - 0.5 \sin x.$$

Figure 29 shows that at $x = 0$ it has the value 3.0 and the slope -0.5, so that its tangent intersects the x-axis at $x = 3.0/0.5 = 6.0$. (The scales on the axes differ!) ■

Observation. Our choice of y_1 and y_2 was general enough to satisfy both initial conditions. Now let us take instead two proportional solutions $y_1 = \cos x$ and $y_2 = k \cos x$, so that $y_1/y_2 = 1/k = \text{const}$. Then we can write $y = c_1 y_1 + c_2 y_2$ in the form

$$y = c_1 \cos x + c_2(k \cos x) = C \cos x \qquad \text{where} \qquad C = c_1 + c_2 k.$$

Hence we are no longer able to satisfy two initial conditions with only one arbitrary constant C. Consequently, in defining the concept of a general solution, we must exclude proportionality. And we see at the same time why the concept of a general solution is of importance in connection with initial value problems.

DEFINITION

General Solution, Basis, Particular Solution

A **general solution** of an ODE (2) on an open interval I is a solution (5) in which y_1 and y_2 are solutions of (2) on I that are not proportional, and c_1 and c_2 are arbitrary constants. These y_1, y_2 are called a **basis** (or a **fundamental system**) of solutions of (2) on I.

A **particular solution** of (2) on I is obtained if we assign specific values to c_1 and c_2 in (5).

For the definition of an *interval* see Sec. 1.1. Furthermore, as usual, y_1 and y_2 are called *proportional* on I if for all x on I,

$$\text{(6)} \qquad \text{(a)} \quad y_1 = ky_2 \qquad \text{or} \qquad \text{(b)} \quad y_2 = ly_1$$

where k and l are numbers, zero or not. (Note that (a) implies (b) if and only if $k \neq 0$).

Actually, we can reformulate our definition of a basis by using a concept of general importance. Namely, two functions y_1 and y_2 are called **linearly independent** on an interval I where they are defined if

$$\text{(7)} \quad k_1y_1(x) + k_2y_2(x) = 0 \qquad \text{everywhere on } I \text{ implies} \qquad k_1 = 0 \text{ and } k_2 = 0.$$

And y_1 and y_2 are called **linearly dependent** on I if (7) also holds for some constants k_1, k_2 not both zero. Then, if $k_1 \neq 0$ or $k_2 \neq 0$, we can divide and see that y_1 and y_2 are proportional,

$$y_1 = -\frac{k_2}{k_1}y_2 \qquad \text{or} \qquad y_2 = -\frac{k_1}{k_2}y_1.$$

In contrast, in the case of linear *independence* these functions are not proportional because then we cannot divide in (7). This gives the following

DEFINITION

Basis (Reformulated)

A **basis** of solutions of (2) on an open interval I is a pair of linearly independent solutions of (2) on I.

If the coefficients p and q of (2) are continuous on some open interval I, then (2) has a general solution. It yields the unique solution of any initial value problem (2), (4). It includes all solutions of (2) on I; hence (2) has no *singular solutions* (solutions not obtainable from of a general solution; see also Problem Set 1.1). All this will be shown in Sec. 2.6.

EXAMPLE 5 Basis, General Solution, Particular Solution

$\cos x$ and $\sin x$ in Example 4 form a basis of solutions of the ODE $y'' + y = 0$ for all x because their quotient is $\cot x \neq \text{const}$ (or $\tan x \neq \text{const}$). Hence $y = c_1 \cos x + c_2 \sin x$ is a general solution. The solution $y = 3.0 \cos x - 0.5 \sin x$ of the initial value problem is a particular solution. ■

EXAMPLE 6 Basis, General Solution, Particular Solution

Verify by substitution that $y_1 = e^x$ and $y_2 = e^{-x}$ are solutions of the ODE $y'' - y = 0$. Then solve the initial value problem

$$y'' - y = 0, \qquad y(0) = 6, \qquad y'(0) = -2.$$

Solution. $(e^x)'' - e^x = 0$ and $(e^{-x})'' - e^{-x} = 0$ show that e^x and e^{-x} are solutions. They are not proportional, $e^x/e^{-x} = e^{2x} \neq \text{const}$. Hence e^x, e^{-x} form a basis for all x. We now write down the corresponding general solution and its derivative and equate their values at 0 to the given initial conditions,

$$y = c_1 e^x + c_2 e^{-x}, \qquad y' = c_1 e^x - c_2 e^{-x}, \qquad y(0) = c_1 + c_2 = 6, \qquad y'(0) = c_1 - c_2 = -2.$$

By addition and subtraction, $c_1 = 2$, $c_2 = 4$, so that the *answer* is $y = 2e^x + 4e^{-x}$. This is the particular solution satisfying the two initial conditions. ■

Find a Basis if One Solution Is Known. Reduction of Order

It happens quite often that one solution can be found by inspection or in some other way. Then a second linearly independent solution can be obtained by solving a first-order ODE. This is called the method of **reduction of order**.[1] We first show how this method works in an example and then in general.

EXAMPLE 7 Reduction of Order if a Solution Is Known. Basis

Find a basis of solutions of the ODE

$$(x^2 - x)y'' - xy' + y = 0.$$

Solution. Inspection shows that $y_1 = x$ is a solution because $y_1' = 1$ and $y_1'' = 0$, so that the first term vanishes identically and the second and third terms cancel. The idea of the method is to substitute

$$y = uy_1 = ux, \qquad y' = u'x + u, \qquad y'' = u''x + 2u'$$

into the ODE. This gives

$$(x^2 - x)(u''x + 2u') - x(u'x + u) + ux = 0.$$

ux and $-xu$ cancel and we are left with the following ODE, which we divide by x, order, and simplify,

$$(x^2 - x)(u''x + 2u') - x^2u' = 0, \qquad (x^2 - x)u'' + (x - 2)u' = 0.$$

This ODE is of first order in $v = u'$, namely, $(x^2 - x)v' + (x - 2)v = 0$. Separation of variables and integration gives

$$\frac{dv}{v} = -\frac{x - 2}{x^2 - x}\,dx = \left(\frac{1}{x - 1} - \frac{2}{x}\right)dx, \qquad \ln|v| = \ln|x - 1| - 2\ln|x| = \ln\frac{|x - 1|}{x^2}.$$

---

[1]Credited to the great mathematician JOSEPH LOUIS LAGRANGE (1736–1813), who was born in Turin, of French extraction, got his first professorship when he was 19 (at the Military Academy of Turin), became director of the mathematical section of the Berlin Academy in 1766, and moved to Paris in 1787. His important major work was in the calculus of variations, celestial mechanics, general mechanics (*Mécanique analytique*, Paris, 1788), differential equations, approximation theory, algebra, and number theory.

We need no constant of integration because we want to obtain a particular solution; similarly in the next integration. Taking exponents and integrating again, we obtain

$$v = \frac{x-1}{x^2} = \frac{1}{x} - \frac{1}{x^2}, \qquad u = \int v\,dx = \ln|x| + \frac{1}{x}, \qquad \text{hence} \qquad y_2 = ux = x\ln|x| + 1.$$

Since $y_1 = x$ and $y_2 = x \ln |x| + 1$ are linearly independent (their quotient is not constant), we have obtained a basis of solutions, valid for all positive x. ■

In this example we applied **reduction of order** to a homogeneous linear ODE [see (2)]

$$y'' + p(x)y' + q(x)y = 0.$$

Note that we now take the ODE in standard form, with y'', not $f(x)y''$—this is essential in applying our subsequent formulas. We assume a solution y_1 of (2), on an open interval I, to be known and want to find a basis. For this we need a second linearly independent solution y_2 of (2) on I. To get y_2, we substitute

$$y = y_2 = uy_1, \qquad y' = y_2' = u'y_1 + uy_1', \qquad y'' = y_2'' = u''y_1 + 2u'y_1' + uy_1''$$

into (2). This gives

$$u''y_1 + 2u'y_1' + uy_1'' + p(u'y_1 + uy_1') + quy_1 = 0. \tag{8}$$

Collecting terms in u'', u', and u, we have

$$u''y_1 + u'(2y_1' + py_1) + u(y_1'' + py_1' + qy_1) = 0.$$

Now comes the main point. Since y_1 is a solution of (2), the expression in the last parentheses is zero. Hence u is gone, and we are left with an ODE in u' and u''. We divide this remaining ODE by y_1 and set $u' = U$, $u'' = U'$,

$$u'' + u'\frac{2y_1' + py_1}{y_1} = 0, \qquad \text{thus} \qquad U' + \left(\frac{2y_1'}{y_1} + p\right)U = 0.$$

This is the desired first-order ODE, the reduced ODE. Separation of variables and integration gives

$$\frac{dU}{U} = -\left(\frac{2y_1'}{y_1} + p\right)dx \qquad \text{and} \qquad \ln|U| = -2\ln|y_1| - \int p\,dx.$$

By taking exponents we finally obtain

$$U = \frac{1}{y_1^2}e^{-\int p\,dx}. \tag{9}$$

Here $U = u'$, so that $u = \int U\,dx$. Hence the desired second solution is

$$y_2 = y_1 u = y_1 \int U\,dx.$$

The quotient $y_2/y_1 = u = \int U\,dx$ cannot be constant (since $U > 0$), so that y_1 and y_2 form a basis of solutions.

PROBLEM SET 2.1

REDUCTION OF ORDER is important because it gives a simpler ODE. A general second-order ODE $F(x, y, y', y'') = 0$, linear or not, can be reduced to first order if y does not occur explicitly (Prob. 1) or if x does not occur explicitly (Prob. 2) or if the ODE is homogeneous linear and we know a solution (see the text).

1. **Reduction.** Show that $F(x, y', y'') = 0$ can be reduced to first order in $z = y'$ (from which y follows by integration). Give two examples of your own.
2. **Reduction.** Show that $F(y, y', y'') = 0$ can be reduced to a first-order ODE with y as the independent variable and $y'' = (dz/dy)z$, where $z = y'$; derive this by the chain rule. Give two examples.

3–10 REDUCTION OF ORDER

Reduce to first order and solve, showing each step in detail.

3. $y'' + y' = 0$
4. $2xy'' = 3y'$
5. $yy'' = 3y'^2$
6. $xy'' + 2y' + xy = 0, \quad y_1 = (\cos x)/x$
7. $y'' + y'^3 \sin y = 0$
8. $y'' = 1 + y'^2$
9. $x^2y'' - 5xy' + 9y = 0, \quad y_1 = x^3$
10. $y'' + (1 + 1/y)y'^2 = 0$

11–14 APPLICATIONS OF REDUCIBLE ODEs

11. **Curve.** Find the curve through the origin in the xy-plane which satisfies $y'' = 2y'$ and whose tangent at the origin has slope 1.
12. **Hanging cable.** It can be shown that the curve $y(x)$ of an inextensible flexible homogeneous cable hanging between two fixed points is obtained by solving $y'' = k\sqrt{1 + y'^2}$, where the constant k depends on the weight. This curve is called *catenary* (from Latin *catena* = the chain). Find and graph $y(x)$, assuming that $k = 1$ and those fixed points are $(-1, 0)$ and $(1, 0)$ in a vertical xy-plane.
13. **Motion.** If, in the motion of a small body on a straight line, the sum of velocity and acceleration equals a positive constant, how will the distance $y(t)$ depend on the initial velocity and position?
14. **Motion.** In a straight-line motion, let the velocity be the reciprocal of the acceleration. Find the distance $y(t)$ for arbitrary initial position and velocity.

15–19 GENERAL SOLUTION. INITIAL VALUE PROBLEM (IVP)

(More in the next set.) **(a)** Verify that the given functions are linearly independent and form a basis of solutions of the given ODE. **(b)** Solve the IVP. Graph or sketch the solution.

15. $4y'' + 25y = 0, \quad y(0) = 3.0, \quad y'(0) = -2.5,$
    $\cos 2.5x, \sin 2.5x$
16. $y'' + 0.6y' + 0.09y = 0, \quad y(0) = 2.2, \ y'(0) = 0.14,$
    $e^{-0.3x}, xe^{-0.3x}$
17. $4x^2y'' - 3y = 0, \quad y(1) = -3, \quad y'(1) = 0,$
    $x^{3/2}, x^{-1/2}$
18. $x^2y'' - xy' + y = 0, \quad y(1) = 4.3, \quad y'(1) = 0.5,$
    $x, x \ln x$
19. $y'' + 2y' + 2y = 0, \quad y(0) = 0, \quad y'(0) = 15,$
    $e^{-x} \cos x, e^{-x} \sin x$
20. **CAS PROJECT. Linear Independence.** Write a program for testing linear independence and dependence. Try it out on some of the problems in this and the next problem set and on examples of your own.

2.2 Homogeneous Linear ODEs with Constant Coefficients

We shall now consider second-order homogeneous linear ODEs whose coefficients a and b are constant,

$$y'' + ay' + by = 0. \tag{1}$$

These equations have important applications in mechanical and electrical vibrations, as we shall see in Secs. 2.4, 2.8, and 2.9.

To solve (1), we recall from Sec. 1.5 that the solution of the first-order linear ODE with a constant coefficient k

$$y' + ky = 0$$

is an exponential function $y = ce^{-kx}$. This gives us the idea to try as a solution of (1) the function

$$y = e^{\lambda x}. \tag{2}$$

Substituting (2) and its derivatives

$$y' = \lambda e^{\lambda x} \qquad \text{and} \qquad y'' = \lambda^2 e^{\lambda x}$$

into our equation (1), we obtain

$$(\lambda^2 + a\lambda + b)e^{\lambda x} = 0.$$

Hence if λ is a solution of the important **characteristic equation** (or *auxiliary equation*)

$$\lambda^2 + a\lambda + b = 0 \tag{3}$$

then the exponential function (2) is a solution of the ODE (1). Now from algebra we recall that the roots of this quadratic equation (3) are

$$\lambda_1 = \tfrac{1}{2}\left(-a + \sqrt{a^2 - 4b}\right), \qquad \lambda_2 = \tfrac{1}{2}\left(-a - \sqrt{a^2 - 4b}\right). \tag{4}$$

(3) and (4) will be basic because our derivation shows that the functions

$$y_1 = e^{\lambda_1 x} \qquad \text{and} \qquad y_2 = e^{\lambda_2 x} \tag{5}$$

are solutions of (1). Verify this by substituting (5) into (1).

From algebra we further know that the quadratic equation (3) may have three kinds of roots, depending on the sign of the discriminant $a^2 - 4b$, namely,

(Case I) *Two real roots if* $a^2 - 4b > 0$,
(Case II) *A real double root if* $a^2 - 4b = 0$,
(Case III) *Complex conjugate roots if* $a^2 - 4b < 0$.

Case I. Two Distinct Real-Roots λ_1 and λ_2

In this case, a basis of solutions of (1) on any interval is

$$y_1 = e^{\lambda_1 x} \qquad \text{and} \qquad y_2 = e^{\lambda_2 x}$$

because y_1 and y_2 are defined (and real) for all x and their quotient is not constant. The corresponding general solution is

$$y = c_1 e^{\lambda_1 x} + c_2 e^{\lambda_2 x}. \tag{6}$$

EXAMPLE 1 General Solution in the Case of Distinct Real Roots

We can now solve $y'' - y = 0$ in Example 6 of Sec. 2.1 systematically. The characteristic equation is $\lambda^2 - 1 = 0$. Its roots are $\lambda_1 = 1$ and $\lambda_2 = -1$. Hence a basis of solutions is e^x and e^{-x} and gives the same general solution as before,

$$y = c_1 e^x + c_2 e^{-x}.$$ ■

EXAMPLE 2 Initial Value Problem in the Case of Distinct Real Roots

Solve the initial value problem

$$y'' + y' - 2y = 0, \qquad y(0) = 4, \qquad y'(0) = -5.$$

Solution. *Step 1.* ***General solution.*** The characteristic equation is

$$\lambda^2 + \lambda - 2 = 0.$$

Its roots are

$$\lambda_1 = \tfrac{1}{2}(-1 + \sqrt{9}) = 1 \qquad \text{and} \qquad \lambda_2 = \tfrac{1}{2}(-1 - \sqrt{9}) = -2$$

so that we obtain the general solution

$$y = c_1 e^x + c_2 e^{-2x}.$$

Step 2. Particular solution. Since $y'(x) = c_1 e^x - 2c_2 e^{-2x}$, we obtain from the general solution and the initial conditions

$$y(0) = c_1 + c_2 = 4,$$
$$y'(0) = c_1 - 2c_2 = -5.$$

Hence $c_1 = 1$ and $c_2 = 3$. This gives the *answer* $y = e^x + 3e^{-2x}$. Figure 30 shows that the curve begins at $y = 4$ with a negative slope (-5, but note that the axes have different scales!), in agreement with the initial conditions. ■

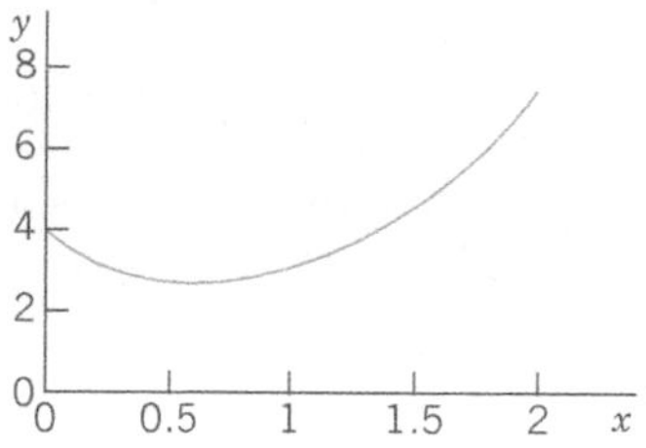

Fig. 30. Solution in Example 2

Case II. Real Double Root $\lambda = -a/2$

If the discriminant $a^2 - 4b$ is zero, we see directly from (4) that we get only one root, $\lambda = \lambda_1 = \lambda_2 = -a/2$, hence only one solution,

$$y_1 = e^{-(a/2)x}.$$

To obtain a second independent solution y_2 (needed for a basis), we use the method of reduction of order discussed in the last section, setting $y_2 = uy_1$. Substituting this and its derivatives $y_2' = u'y_1 + uy_1'$ and y_2'' into (1), we first have

$$(u''y_1 + 2u'y_1' + uy_1'') + a(u'y_1 + uy_1') + buy_1 = 0.$$

Collecting terms in u'', u', and u, as in the last section, we obtain

$$u''y_1 + u'(2y_1' + ay_1) + u(y_1'' + ay_1' + by_1) = 0.$$

The expression in the last parentheses is zero, since y_1 is a solution of (1). The expression in the first parentheses is zero, too, since

$$2y_1' = -ae^{-ax/2} = -ay_1.$$

We are thus left with $u''y_1 = 0$. Hence $u'' = 0$. By two integrations, $u = c_1x + c_2$. To get a second independent solution $y_2 = uy_1$, we can simply choose $c_1 = 1$, $c_2 = 0$ and take $u = x$. Then $y_2 = xy_1$. Since these solutions are not proportional, they form a basis. Hence in the case of a double root of (3) a basis of solutions of (1) on any interval is

$$e^{-ax/2}, \qquad xe^{-ax/2}.$$

The corresponding general solution is

$$y = (c_1 + c_2x)e^{-ax/2}. \tag{7}$$

WARNING! If λ is a *simple* root of (4), then $(c_1 + c_2x)e^{\lambda x}$ with $c_2 \neq 0$ is ***not*** a solution of (1).

EXAMPLE 3 General Solution in the Case of a Double Root

The characteristic equation of the ODE $y'' + 6y' + 9y = 0$ is $\lambda^2 + 6\lambda + 9 = (\lambda + 3)^2 = 0$. It has the double root $\lambda = -3$. Hence a basis is e^{-3x} and xe^{-3x}. The corresponding general solution is $y = (c_1 + c_2x)e^{-3x}$. ■

EXAMPLE 4 Initial Value Problem in the Case of a Double Root

Solve the initial value problem

$$y'' + y' + 0.25y = 0, \qquad y(0) = 3.0, \quad y'(0) = -3.5.$$

Solution. The characteristic equation is $\lambda^2 + \lambda + 0.25 = (\lambda + 0.5)^2 = 0$. It has the double root $\lambda = -0.5$. This gives the general solution

$$y = (c_1 + c_2x)e^{-0.5x}.$$

We need its derivative

$$y' = c_2e^{-0.5x} - 0.5(c_1 + c_2x)e^{-0.5x}.$$

From this and the initial conditions we obtain

$$y(0) = c_1 = 3.0, \qquad y'(0) = c_2 - 0.5c_1 = 3.5; \qquad \text{hence} \qquad c_2 = -2.$$

The particular solution of the initial value problem is $y = (3 - 2x)e^{-0.5x}$. See Fig. 31. ■

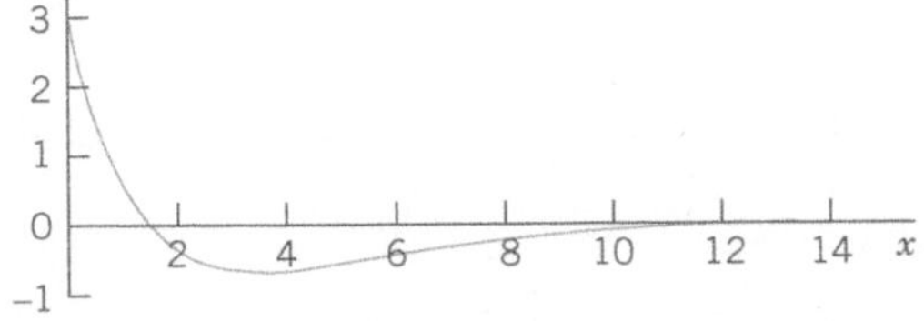

Fig. 31. Solution in Example 4

Case III. Complex Roots $-\frac{1}{2}a + i\omega$ and $-\frac{1}{2}a - i\omega$

This case occurs if the discriminant $a^2 - 4b$ of the characteristic equation (3) is negative. In this case, the roots of (3) are the complex $\lambda = -\frac{1}{2}a \pm i\omega$ that give the complex solutions of the ODE (1). However, we will show that we can obtain a basis of *real* solutions

$$(8) \qquad y_1 = e^{-ax/2} \cos \omega x, \qquad y_2 = e^{-ax/2} \sin \omega x \qquad (\omega > 0)$$

where $\omega^2 = b - \frac{1}{4}a^2$. It can be verified by substitution that these are solutions in the present case. We shall derive them systematically after the two examples by using the complex exponential function. They form a basis on any interval since their quotient $\cot \omega x$ is not constant. Hence a real general solution in Case III is

$$(9) \qquad y = e^{-ax/2}(A \cos \omega x + B \sin \omega x) \qquad (A, B \text{ arbitrary}).$$

EXAMPLE 5 Complex Roots. Initial Value Problem

Solve the initial value problem

$$y'' + 0.4y' + 9.04y = 0, \qquad y(0) = 0, \qquad y'(0) = 3.$$

Solution. ***Step 1. General solution.*** The characteristic equation is $\lambda^2 + 0.4\lambda + 9.04 = 0$. It has the roots $-0.2 \pm 3i$. Hence $\omega = 3$, and a general solution (9) is

$$y = e^{-0.2x}(A \cos 3x + B \sin 3x).$$

Step 2. Particular solution. The first initial condition gives $y(0) = A = 0$. The remaining expression is $y = Be^{-0.2x} \sin 3x$. We need the derivative (chain rule!)

$$y' = B(-0.2e^{-0.2x} \sin 3x + 3e^{-0.2x} \cos 3x).$$

From this and the second initial condition we obtain $y'(0) = 3B = 3$. Hence $B = 1$. Our solution is

$$y = e^{-0.2x} \sin 3x.$$

Figure 32 shows y and the curves of $e^{-0.2x}$ and $-e^{-0.2x}$ (dashed), between which the curve of y oscillates. Such "damped vibrations" (with $x = t$ being time) have important mechanical and electrical applications, as we shall soon see (in Sec. 2.4). ■

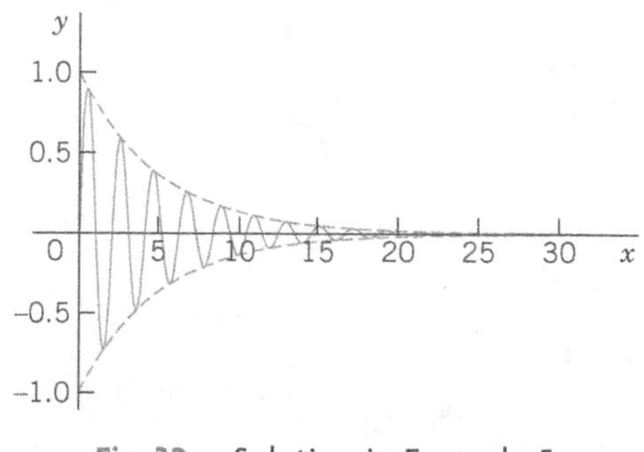

Fig. 32. Solution in Example 5

EXAMPLE 6 Complex Roots

A general solution of the ODE

$$y'' + \omega^2 y = 0 \qquad (\omega \text{ constant, not zero})$$

is

$$y = A \cos \omega x + B \sin \omega x.$$

With $\omega = 1$ this confirms Example 4 in Sec. 2.1. ■

Summary of Cases I–III

Case	Roots of (2)	Basis of (1)	General Solution of (1)
I	Distinct real λ_1, λ_2	$e^{\lambda_1 x}, e^{\lambda_2 x}$	$y = c_1 e^{\lambda_1 x} + c_2 e^{\lambda_2 x}$
II	Real double root $\lambda = -\frac{1}{2}a$	$e^{-ax/2}, xe^{-ax/2}$	$y = (c_1 + c_2 x)e^{-ax/2}$
III	Complex conjugate $\lambda_1 = -\frac{1}{2}a + i\omega$, $\lambda_2 = -\frac{1}{2}a - i\omega$	$e^{-ax/2} \cos \omega x$ $e^{-ax/2} \sin \omega x$	$y = e^{-ax/2}(A \cos \omega x + B \sin \omega x)$

It is very interesting that in applications to mechanical systems or electrical circuits, these three cases correspond to three different forms of motion or flows of current, respectively. We shall discuss this basic relation between theory and practice in detail in Sec. 2.4 (and again in Sec. 2.8).

Derivation in Case III. Complex Exponential Function

If verification of the solutions in (8) satisfies you, skip the systematic derivation of these real solutions from the complex solutions by means of the complex exponential function e^z of a complex variable $z = r + it$. We write $r + it$, not $x + iy$ because x and y occur in the ODE. The definition of e^z in terms of the real functions e^r, $\cos t$, and $\sin t$ is

$$e^z = e^{r+it} = e^r e^{it} = e^r(\cos t + i \sin t). \tag{10}$$

This is motivated as follows. For real $z = r$, hence $t = 0$, $\cos 0 = 1$, $\sin 0 = 0$, we get the real exponential function e^r. It can be shown that $e^{z_1+z_2} = e^{z_1}e^{z_2}$, just as in real. (Proof in Sec. 13.5.) Finally, if we use the Maclaurin series of e^z with $z = it$ as well as $i^2 = -1$, $i^3 = -i$, $i^4 = 1$, etc., and reorder the terms as shown (this is permissible, as can be proved), we obtain the series

$$\begin{aligned} e^{it} &= 1 + it + \frac{(it)^2}{2!} + \frac{(it)^3}{3!} + \frac{(it)^4}{4!} + \frac{(it)^5}{5!} + \cdots \\ &= 1 - \frac{t^2}{2!} + \frac{t^4}{4!} - + \cdots + i\left(t - \frac{t^3}{3!} + \frac{t^5}{5!} - + \cdots\right) \\ &= \cos t + i \sin t. \end{aligned}$$

(Look up these real series in your calculus book if necessary.) We see that we have obtained the formula

$$e^{it} = \cos t + i \sin t, \tag{11}$$

called the **Euler formula**. Multiplication by e^r gives (10).

For later use we note that $e^{-it} = \cos(-t) + i\sin(-t) = \cos t - i \sin t$, so that by addition and subtraction of this and (11),

$$\cos t = \tfrac{1}{2}(e^{it} + e^{-it}), \qquad \sin t = \frac{1}{2i}(e^{it} - e^{-it}). \tag{12}$$

After these comments on the definition (10), let us now turn to Case III.

In Case III the radicand $a^2 - 4b$ in (4) is negative. Hence $4b - a^2$ is positive and, using $\sqrt{-1} = i$, we obtain in (4)

$$\tfrac{1}{2}\sqrt{a^2 - 4b} = \tfrac{1}{2}\sqrt{-(4b - a^2)} = \sqrt{-(b - \tfrac{1}{4}a^2)} = i\sqrt{b - \tfrac{1}{4}a^2} = i\omega$$

with ω defined as in (8). Hence in (4),

$$\lambda_1 = \tfrac{1}{2}a + i\omega \qquad \text{and, similarly,} \qquad \lambda_2 = \tfrac{1}{2}a - i\omega.$$

Using (10) with $r = -\frac{1}{2}ax$ and $t = \omega x$, we thus obtain

$$e^{\lambda_1 x} = e^{-(a/2)x + i\omega x} = e^{-(a/2)x}(\cos \omega x + i \sin \omega x)$$

$$e^{\lambda_2 x} = e^{-(a/2)x - i\omega x} = e^{-(a/2)x}(\cos \omega x - i \sin \omega x).$$

We now add these two lines and multiply the result by $\frac{1}{2}$. This gives y_1 as in (8). Then we subtract the second line from the first and multiply the result by $1/(2i)$. This gives y_2 as in (8). These results obtained by addition and multiplication by constants are again solutions, as follows from the superposition principle in Sec. 2.1. This concludes the derivation of these real solutions in Case III.

PROBLEM SET 2.2

1–15 GENERAL SOLUTION

Find a general solution. Check your answer by substitution. ODEs of this kind have important applications to be discussed in Secs. 2.4, 2.7, and 2.9.

1. $4y'' - 25y = 0$
2. $y'' + 36y = 0$
3. $y'' + 6y' + 8.96y = 0$
4. $y'' + 4y' + (\pi^2 + 4)y = 0$
5. $y'' + 2\pi y' + \pi^2 y = 0$
6. $10y'' - 32y' + 25.6y = 0$
7. $y'' + 4.5y' = 0$
8. $y'' + y' + 3.25y = 0$
9. $y'' + 1.8y' - 2.08y = 0$
10. $100y'' + 240y' + (196\pi^2 + 144)y = 0$
11. $4y'' - 4y' - 3y = 0$
12. $y'' + 9y' + 20y = 0$
13. $9y'' - 30y' + 25y = 0$
14. $y'' + 2k^2y' + k^4y = 0$
15. $y'' + 0.54y' + (0.0729 + \pi)y = 0$

16–20 FIND AN ODE

$y'' + ay' + by = 0$ for the given basis.

16. $e^{2.6x}$, $e^{-4.3x}$
17. $e^{-\sqrt{5}x}$, $xe^{-\sqrt{5}x}$
18. $\cos 2\pi x$, $\sin 2\pi x$
19. $e^{(-2+i)x}$, $e^{(-2-i)x}$
20. $e^{-3.1x}\cos 2.1x$, $e^{-3.1x}\sin 2.1x$

21–30 INITIAL VALUES PROBLEMS

Solve the IVP. Check that your answer satisfies the ODE as well as the initial conditions. Show the details of your work.

21. $y'' + 25y = 0$, $y(0) = 4.6$, $y'(0) = -1.2$
22. The ODE in Prob. 4, $y(\frac{1}{2}) = 1$, $y'(\frac{1}{2}) = -2$
23. $y'' + y' - 6y = 0$, $y(0) = 10$, $y'(0) = 0$
24. $4y'' - 4y' - 3y = 0$, $y(-2) = e$, $y'(-2) = -e/2$
25. $y'' - y = 0$, $y(0) = 2$, $y'(0) = -2$
26. $y'' - k^2y = 0\ (k \neq 0)$, $y(0) = 1$, $y'(0) = 1$

27. The ODE in Prob. 5,
$y(0) = 4.5, \quad y'(0) = -4.5\pi - 1 = 13.137$

28. $8y'' - 2y' - y = 0, \quad y(0) = -0.2, \quad y'(0) = -0.325$

29. The ODE in Prob. 15, $y(0) = 0, \quad y'(0) = 1$

30. $9y'' - 30y' + 25y = 0, \quad y(0) = 3.3, \quad y'(0) = 10.0$

31–36 **LINEAR INDEPENDENCE** is of basic importance, in this chapter, in connection with general solutions, as explained in the text. Are the following functions linearly independent on the given interval? Show the details of your work.

31. e^{kx}, xe^{kx}, any interval

32. $e^{ax}, e^{-ax}, \quad x > 0$

33. $x^2, x^2 \ln x, \quad x > 1$

34. $\ln x, \ln(x^3), \quad x > 1$

35. $\sin 2x, \cos x \sin x, \quad x < 0$

36. $e^{-x} \cos \frac{1}{2}x, 0, \quad -1 \leqq x \leqq 1$

37. Instability. Solve $y'' - y = 0$ for the initial conditions $y(0) = 1, y'(0) = -1$. Then change the initial conditions to $y(0) = 1.001, y'(0) = -0.999$ and explain why this small change of 0.001 at $t = 0$ causes a large change later, e.g., 22 at $t = 10$. This is instability: a small initial difference in setting a quantity (a current, for instance) becomes larger and larger with time t. This is undesirable.

38. TEAM PROJECT. General Properties of Solutions

(a) Coefficient formulas. Show how a and b in (1) can be expressed in terms of λ_1 and λ_2. Explain how these formulas can be used in constructing equations for given bases.

(b) Root zero. Solve $y'' + 4y' = 0$ (i) by the present method, and (ii) by reduction to first order. Can you explain why the result must be the same in both cases? Can you do the same for a general ODE $y'' + ay' = 0$?

(c) Double root. Verify directly that $xe^{\lambda x}$ with $\lambda = -a/2$ is a solution of (1) in the case of a double root. Verify and explain why $y = e^{-2x}$ is a solution of $y'' - y' - 6y = 0$ but xe^{-2x} is not.

(d) Limits. Double roots should be limiting cases of distinct roots λ_1, λ_2 as, say, $\lambda_2 \to \lambda_1$. Experiment with this idea. (Remember l'Hôpital's rule from calculus.) Can you arrive at $xe^{\lambda_1 x}$? Give it a try.

2.3 Differential Operators. *Optional*

This short section can be omitted without interrupting the flow of ideas. It will not be used subsequently, except for the notations Dy, D^2y, etc. to stand for y', y'', etc.

Operational calculus means the technique and application of operators. Here, an **operator** is a transformation that transforms a function into another function. Hence differential calculus involves an operator, the **differential operator** D, which transforms a (differentiable) function into its derivative. In operator notation we write $D = \frac{d}{dx}$ and

$$Dy = y' = \frac{dy}{dx}. \tag{1}$$

Similarly, for the higher derivatives we write $D^2y = D(Dy) = y''$, and so on. For example, $D \sin = \cos$, $D^2 \sin = -\sin$, etc.

For a homogeneous linear ODE $y'' + ay' + by = 0$ with constant coefficients we can now introduce the **second-order differential operator**

$$L = P(D) = D^2 + aD + bI,$$

where I is the **identity operator** defined by $Iy = y$. Then we can write that ODE as

$$Ly = P(D)y = (D^2 + aD + bI)y = 0. \tag{2}$$

P suggests "polynomial." L is a **linear operator**. By definition this means that if Ly and Lw exist (this is the case if y and w are twice differentiable), then $L(cy + kw)$ exists for any constants c and k, and

$$L(cy + kw) = cLy + kLw.$$

Let us show that from (2) we reach agreement with the results in Sec. 2.2. Since $(De^{\lambda})(x) = \lambda e^{\lambda x}$ and $(D^2 e^{\lambda})(x) = \lambda^2 e^{\lambda x}$, we obtain

$$(3) \qquad \begin{aligned} Le^{\lambda}(x) &= P(D)e^{\lambda}(x) = (D^2 + aD + bI)e^{\lambda}(x) \\ &= (\lambda^2 + a\lambda + b)e^{\lambda x} = P(\lambda)e^{\lambda x} = 0. \end{aligned}$$

This confirms our result of Sec. 2.2 that $e^{\lambda x}$ *is a solution of the ODE* (2) *if and only if* λ *is a solution of the characteristic equation* $P(\lambda) = 0$.

$P(\lambda)$ is a polynomial in the usual sense of algebra. If we replace λ by the operator D, we obtain the "operator polynomial" $P(D)$. *The point of this operational calculus is that* $P(D)$ *can be treated just like an algebraic quantity.* In particular, we can factor it.

EXAMPLE 1 Factorization, Solution of an ODE

Factor $P(D) = D^2 - 3D - 40I$ and solve $P(D)y = 0$.

Solution. $D^2 - 3D - 40I = (D - 8I)(D + 5I)$ because $I^2 = I$. Now $(D - 8I)y = y' - 8y = 0$ has the solution $y_1 = e^{8x}$. Similarly, the solution of $(D + 5I)y = 0$ is $y_2 = e^{-5x}$. This is a basis of $P(D)y = 0$ on any interval. From the factorization we obtain the ODE, as expected,

$$\begin{aligned} (D - 8I)(D + 5I)y &= (D - 8I)(y' + 5y) = D(y' + 5y) - 8(y' + 5y) \\ &= y'' + 5y' - 8y' - 40y = y'' - 3' - 40y = 0. \end{aligned}$$

Verify that this agrees with the result of our method in Sec. 2.2. This is not unexpected because we factored $P(D)$ in the same way as the characteristic polynomial $P(\lambda) = \lambda^2 - 3\lambda - 40$. ■

It was essential that L in (2) had *constant* coefficients. Extension of operator methods to variable-coefficient ODEs is more difficult and will not be considered here.

If operational methods were limited to the simple situations illustrated in this section, it would perhaps not be worth mentioning. Actually, the power of the operator approach appears in more complicated engineering problems, as we shall see in Chap. 6.

PROBLEM SET 2.3

1–5 APPLICATION OF DIFFERENTIAL OPERATORS

Apply the given operator to the given functions. Show all steps in detail.

1. $D^2 + 2D$; $\cosh 2x$, $e^{-x} + e^{2x}$, $\cos x$
2. $D - 3I$; $3x^2 + 3x$, $3e^{3x}$, $\cos 4x - \sin 4x$
3. $(D - 2I)^2$; e^{2x}, xe^{2x}, e^{-2x}
4. $(D + 6I)^2$; $6x + \sin 6x$, xe^{-6x}
5. $(D - 2I)(D + 3I)$; e^{2x}, xe^{2x}, e^{-3x}

6–12 GENERAL SOLUTION

Factor as in the text and solve.

6. $(D^2 + 4.00D + 3.36I)y = 0$
7. $(4D^2 - I)y = 0$
8. $(D^2 + 3I)y = 0$
9. $(D^2 - 4.20D + 4.41I)y = 0$
10. $(D^2 + 4.80D + 5.76I)y = 0$
11. $(D^2 - 4.00D + 3.84I)y = 0$
12. $(D^2 + 3.0D + 2.5I)y = 0$

13. Linear operator. Illustrate the linearity of L in (2) by taking $c = 4, k = -6, y = e^{2x}$, and $w = \cos 2x$. Prove that L is linear.

14. Double root. If $D^2 + aD + bI$ has distinct roots μ and λ, show that a particular solution is $y = (e^{\mu x} - e^{\lambda x})/(\mu - \lambda)$. Obtain from this a solution $xe^{\lambda x}$ by letting $\mu \to \lambda$ and applying l'Hôpital's rule.

15. Definition of linearity. Show that the definition of linearity in the text is equivalent to the following. If $L[y]$ and $L[w]$ exist, then $L[y + w]$ exists and $L[cy]$ and $L[kw]$ exist for all constants c and k, and $L[y + w] = L[y] + L[w]$ as well as $L[cy] = cL[y]$ and $L[kw] = kL[w]$.

2.4 Modeling of Free Oscillations of a Mass–Spring System

Linear ODEs with constant coefficients have important applications in mechanics, as we show in this section as well as in Sec. 2.8, and in electrical circuits as we show in Sec. 2.9. In this section we model and solve a basic mechanical system consisting of a mass on an elastic spring (a so-called "mass–spring system," Fig. 33), which moves up and down.

Setting Up the Model

We take an ordinary coil spring that resists extension as well as compression. We suspend it vertically from a fixed support and attach a body at its lower end, for instance, an iron ball, as shown in Fig. 33. We let $y = 0$ denote the position of the ball when the system is at rest (Fig. 33b). Furthermore, we choose ***the downward direction as positive***, thus regarding downward forces as *positive* and upward forces as *negative*.

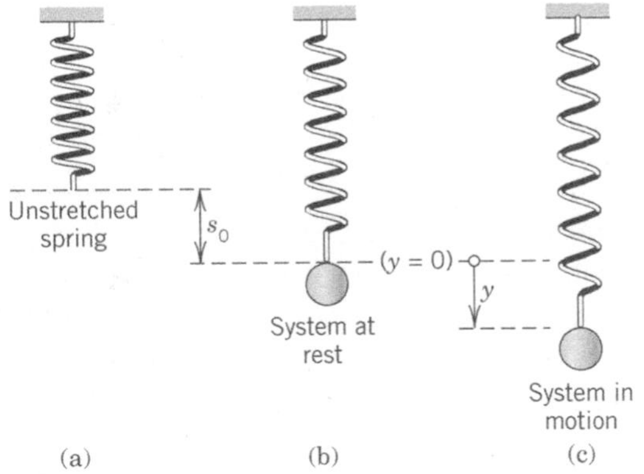

Fig. 33. Mechanical mass–spring system

We now let the ball move, as follows. We pull it down by an amount $y > 0$ (Fig. 33c). This causes a spring force

$$F_1 = -ky \quad \text{(1)} \qquad \textbf{(Hooke's law}^2\textbf{)}$$

proportional to the stretch y, with $k\,(>0)$ called the **spring constant**. The minus sign indicates that F_1 points upward, against the displacement. It is a *restoring force*: It wants to restore the system, that is, to pull it back to $y = 0$. Stiff springs have large k.

[2]ROBERT HOOKE (1635–1703), English physicist, a forerunner of Newton with respect to the law of gravitation.

Note that an additional force $-F_0$ is present in the spring, caused by stretching it in fastening the ball, but F_0 has no effect on the motion because it is in equilibrium with the weight W of the ball, $-F_0 = W = mg$, where $g = 980 \text{ cm/sec}^2 = 9.8 \text{ m/sec}^2 = 32.17 \text{ ft/sec}^2$ is the **constant of gravity at the Earth's surface** (not to be confused with the *universal gravitational constant* $G = gR^2/M = 6.67 \cdot 10^{-11} \text{ nt m}^2/\text{kg}^2$, which we shall not need; here $R = 6.37 \cdot 10^6$ m and $M = 5.98 \cdot 10^{24}$ kg are the Earth's radius and mass, respectively).

The motion of our mass–spring system is determined by **Newton's second law**

$$\text{Mass} \times \text{Acceleration} = my'' = \text{Force} \tag{2}$$

where $y'' = d^2y/dt^2$ and "Force" is the resultant of all the forces acting on the ball. (For systems of units, see the inside of the front cover.)

ODE of the Undamped System

Every system has damping. Otherwise it would keep moving forever. But if the damping is small and the motion of the system is considered over a relatively short time, we may disregard damping. Then Newton's law with $F = -F_1$ gives the model $my'' = -F_1 = -ky$; thus

$$my'' + ky = 0. \tag{3}$$

This is a homogeneous linear ODE with constant coefficients. A general solution is obtained as in Sec. 2.2, namely (see Example 6 in Sec. 2.2)

$$y(t) = A \cos \omega_0 t + B \sin \omega_0 t \qquad \omega_0 = \sqrt{\frac{k}{m}}. \tag{4}$$

This motion is called a **harmonic oscillation** (Fig. 34). Its *frequency* is $f = \omega_0/2\pi$ Hertz[3] (= cycles/sec) because cos and sin in (4) have the period $2\pi/\omega_0$. The frequency f is called the **natural frequency** of the system. (We write ω_0 to reserve ω for Sec. 2.8.)

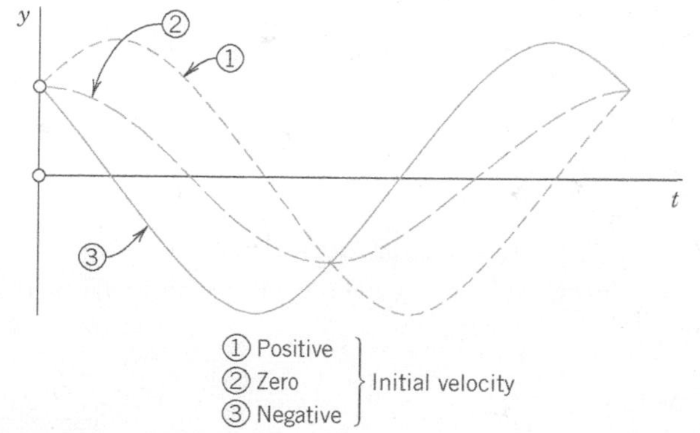

Fig. 34. Typical harmonic oscillations (4) and (4*) with the same $y(0) = A$ and different initial velocities $y'(0) = \omega_0 B$, positive ①, zero ②, negative ③

[3]**HEINRICH HERTZ (1857–1894)**, German physicist, who discovered electromagnetic waves, as the basis of wireless communication developed by GUGLIELMO MARCONI (1874–1937), Italian physicist (Nobel prize in 1909).

An alternative representation of (4), which shows the physical characteristics of amplitude and phase shift of (4), is

$$y(t) = C \cos(\omega_0 t - \delta) \tag{4*}$$

with $C = \sqrt{A^2 + B^2}$ and phase angle δ, where $\tan \delta = B/A$. This follows from the addition formula (6) in App. 3.1.

EXAMPLE 1 Harmonic Oscillation of an Undamped Mass–Spring System

If a mass–spring system with an iron ball of weight $W = 98$ nt (about 22 lb) can be regarded as undamped, and the spring is such that the ball stretches it 1.09 m (about 43 in.), how many cycles per minute will the system execute? What will its motion be if we pull the ball down from rest by 16 cm (about 6 in.) and let it start with zero initial velocity?

Solution. Hooke's law (1) with W as the force and 1.09 meter as the stretch gives $W = 1.09k$; thus $k = W/1.09 = 98/1.09 = 90$ [kg/sec^2] $= 90$ [nt/meter]. The mass is $m = W/g = 98/9.8 = 10$ [kg]. This gives the frequency $\omega_0/(2\pi) = \sqrt{k/m}/(2\pi) = 3/(2\pi) = 0.48$ [Hz] $= 29$ [cycles/min].

From (4) and the initial conditions, $y(0) = A = 0.16$ [meter] and $y'(0) = \omega_0 B = 0$. Hence the motion is

$$y(t) = 0.16 \cos 3t \text{ [meter]} \qquad \text{or} \qquad 0.52 \cos 3t \text{ [ft]} \qquad \text{(Fig. 35).}$$

If you have a chance of experimenting with a mass–spring system, don't miss it. You will be surprised about the good agreement between theory and experiment, usually within a fraction of one percent if you measure carefully. ■

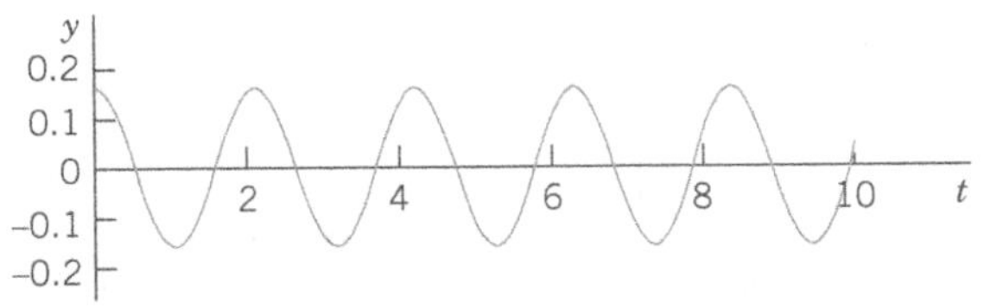

Fig. 35. Harmonic oscillation in Example 1

ODE of the Damped System

To our model $my'' = -ky$ we now add a damping force

$$F_2 = -cy',$$

obtaining $my'' = -ky - cy'$; thus the ODE of the damped mass–spring system is

$$my'' + cy' + ky = 0. \tag{5}$$ (Fig. 36)

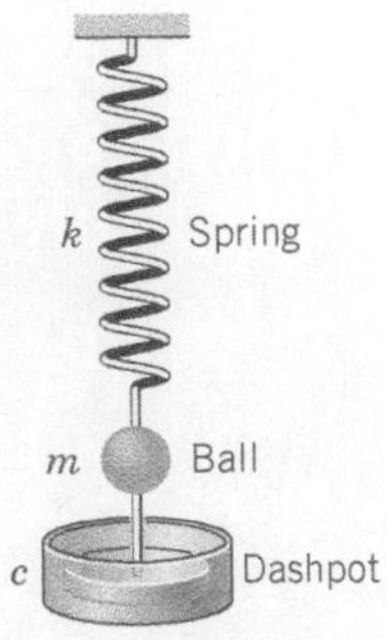

Fig. 36. Damped system

Physically this can be done by connecting the ball to a dashpot; see Fig. 36. We assume this damping force to be proportional to the velocity $y' = dy/dt$. This is generally a good approximation for small velocities.

The constant c is called the *damping constant.* Let us show that c is positive. Indeed, the damping force $F_2 = -cy'$ acts *against* the motion; hence for a downward motion we have $y' > 0$ which for positive c makes F negative (an upward force), as it should be. Similarly, for an upward motion we have $y' < 0$ which, for $c > 0$ makes F_2 positive (a downward force).

The ODE (5) is homogeneous linear and has constant coefficients. Hence we can solve it by the method in Sec. 2.2. The characteristic equation is (divide (5) by m)

$$\lambda^2 + \frac{c}{m}\lambda + \frac{k}{m} = 0.$$

By the usual formula for the roots of a quadratic equation we obtain, as in Sec. 2.2,

$$\text{(6)} \quad \lambda_1 = -\alpha + \beta, \quad \lambda_2 = -\alpha - \beta, \quad \text{where} \quad \alpha = \frac{c}{2m} \quad \text{and} \quad \beta = \frac{1}{2m}\sqrt{c^2 - 4mk}.$$

It is now interesting that depending on the amount of damping present—whether a lot of damping, a medium amount of damping or little damping—three types of motions occur, respectively:

Case I.	$c^2 > 4mk.$	*Distinct real roots* λ_1, λ_2.	**(Overdamping)**
Case II.	$c^2 = 4mk.$	*A real double root.*	**(Critical damping)**
Case III.	$c^2 < 4mk.$	*Complex conjugate roots.*	**(Underdamping)**

They correspond to the three Cases I, II, III in Sec. 2.2.

Discussion of the Three Cases

Case I. Overdamping

If the damping constant c is so large that $c^2 > 4mk$, then λ_1 and λ_2 are distinct real roots. In this case the corresponding general solution of (5) is

$$\textbf{(7)} \qquad y(t) = c_1 e^{-(\alpha-\beta)t} + c_2 e^{-(\alpha+\beta)t}.$$

We see that in this case, damping takes out energy so quickly that the body does not oscillate. For $t > 0$ both exponents in (7) are negative because $\alpha > 0, \beta > 0$, and $\beta^2 = \alpha^2 - k/m < \alpha^2$. Hence both terms in (7) approach zero as $t \to \infty$. Practically speaking, after a sufficiently long time the mass will be at rest at the *static equilibrium position* ($y = 0$). Figure 37 shows (7) for some typical initial conditions.

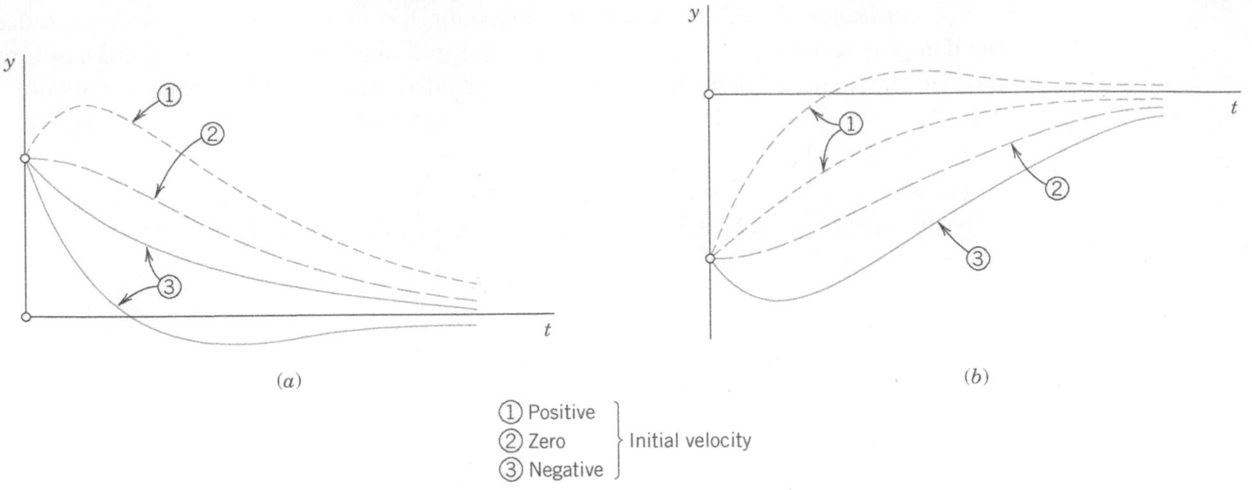

Fig. 37. Typical motions (7) in the overdamped case
(a) Positive initial displacement
(b) Negative initial displacement

Case II. Critical Damping

Critical damping is the border case between nonoscillatory motions (Case I) and oscillations (Case III). It occurs if the characteristic equation has a double root, that is, if $c^2 = 4mk$, so that $\beta = 0$, $\lambda_1 = \lambda_2 = -\alpha$. Then the corresponding general solution of (5) is

$$y(t) = (c_1 + c_2 t)e^{-\alpha t}. \tag{8}$$

This solution can pass through the equilibrium position $y = 0$ at most once because $e^{-\alpha t}$ is never zero and $c_1 + c_2 t$ can have at most one positive zero. If both c_1 and c_2 are positive (or both negative), it has no positive zero, so that y does not pass through 0 at all. Figure 38 shows typical forms of (8). Note that they look almost like those in the previous figure.

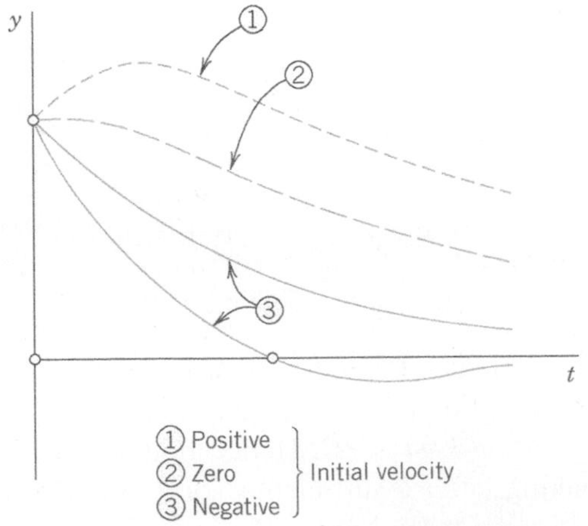

Fig. 38. Critical damping [see (8)]

Case III. Underdamping

This is the most interesting case. It occurs if the damping constant c is so small that $c^2 < 4mk$. Then β in (6) is no longer real but pure imaginary, say,

$$\beta = i\omega^* \quad \text{where} \quad \omega^* = \frac{1}{2m}\sqrt{4mk - c^2} = \sqrt{\frac{k}{m} - \frac{c^2}{4m^2}} \quad (>0). \tag{9}$$

(We now write ω^* to reserve ω for driving and electromotive forces in Secs. 2.8 and 2.9.) The roots of the characteristic equation are now complex conjugates,

$$\lambda_1 = -\alpha + i\omega^*, \qquad \lambda_2 = -\alpha - i\omega^*$$

with $\alpha = c/(2m)$, as given in (6). Hence the corresponding general solution is

$$y(t) = e^{-\alpha t}(A \cos \omega^* t + B \sin \omega^* t) = Ce^{-\alpha t} \cos (\omega^* t - \delta) \tag{10}$$

where $C^2 = A^2 + B^2$ and $\tan \delta = B/A$, as in (4*).

This represents **damped oscillations**. Their curve lies between the dashed curves $y = Ce^{-\alpha t}$ and $y = -Ce^{-\alpha t}$ in Fig. 39, touching them when $\omega^* t - \delta$ is an integer multiple of π because these are the points at which $\cos (\omega^* t - \delta)$ equals 1 or -1.

The frequency is $\omega^*/(2\pi)$ Hz (hertz, cycles/sec). From (9) we see that the smaller c (>0) is, the larger is ω^* and the more rapid the oscillations become. If c approaches 0, then ω^* approaches $\omega_0 = \sqrt{k/m}$, giving the harmonic oscillation (4), whose frequency $\omega_0/(2\pi)$ is the natural frequency of the system.

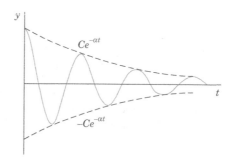

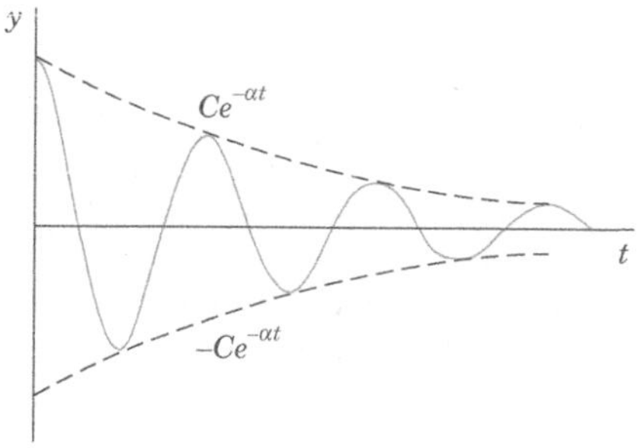

Fig. 39. Damped oscillation in Case III [see (10)]

EXAMPLE 2 The Three Cases of Damped Motion

How does the motion in Example 1 change if we change the damping constant c from one to another of the following three values, with $y(0) = 0.16$ and $y'(0) = 0$ as before?

$$\text{(I) } c = 100 \text{ kg/sec}, \qquad \text{(II) } c = 60 \text{ kg/sec}, \qquad \text{(III) } c = 10 \text{ kg/sec}.$$

Solution. It is interesting to see how the behavior of the system changes due to the effect of the damping, which takes energy from the system, so that the oscillations decrease in amplitude (Case III) or even disappear (Cases II and I).

(I) With $m = 10$ and $k = 90$, as in Example 1, the model is the initial value problem

$$10y'' + 100y' + 90y = 0, \qquad y(0) = 0.16 \text{ [meter]}, \qquad y'(0) = 0.$$

The characteristic equation is $10\lambda^2 + 100\lambda + 90 = 10(\lambda + 9)(\lambda + 1) = 0$. It has the roots -9 and -1. This gives the general solution

$$y = c_1 e^{-9t} + c_2 e^{-t}. \qquad \text{We also need} \qquad y' = -9c_1 e^{-9t} - c_2 e^{-t}.$$

The initial conditions give $c_1 + c_2 = 0.16$, $-9c_1 - c_2 = 0$. The solution is $c_1 = -0.02$, $c_2 = 0.18$. Hence in the overdamped case the solution is

$$y = -0.02e^{-9t} + 0.18e^{-t}.$$

It approaches 0 as $t \to \infty$. The approach is rapid; after a few seconds the solution is practically 0, that is, the iron ball is at rest.

(II) The model is as before, with $c = 60$ instead of 100. The characteristic equation now has the form $10\lambda^2 + 60\lambda + 90 = 10(\lambda + 3)^2 = 0$. It has the double root -3. Hence the corresponding general solution is

$$y = (c_1 + c_2 t)e^{-3t}. \qquad \text{We also need} \qquad y' = (c_2 - 3c_1 - 3c_2 t)e^{-3t}.$$

The initial conditions give $y(0) = c_1 = 0.16$, $y'(0) = c_2 - 3c_1 = 0$, $c_2 = 0.48$. Hence in the critical case the solution is

$$y = (0.16 + 0.48t)e^{-3t}.$$

It is always positive and decreases to 0 in a monotone fashion.

(III) The model now is $10y'' + 10y' + 90y = 0$. Since $c = 10$ is smaller than the critical c, we shall get oscillations. The characteristic equation is $10\lambda^2 + 10\lambda + 90 = 10[(\lambda + \frac{1}{2})^2 + 9 - \frac{1}{4}] = 0$. It has the complex roots [see (4) in Sec. 2.2 with $a = 1$ and $b = 9$]

$$\lambda = -0.5 \pm \sqrt{0.5^2 - 9} = -0.5 \pm 2.96i.$$

This gives the general solution

$$y = e^{-0.5t}(A \cos 2.96t + B \sin 2.96t).$$

Thus $y(0) = A = 0.16$. We also need the derivative

$$y' = e^{-0.5t}(-0.5A \cos 2.96t - 0.5B \sin 2.96t - 2.96A \sin 2.96t + 2.96B \cos 2.96t).$$

Hence $y'(0) = -0.5A + 2.96B = 0$, $B = 0.5A/2.96 = 0.027$. This gives the solution

$$y = e^{-0.5t}(0.16 \cos 2.96t + 0.027 \sin 2.96t) = 0.162e^{-0.5t} \cos (2.96t - 0.17).$$

We see that these damped oscillations have a smaller frequency than the harmonic oscillations in Example 1 by about 1% (since 2.96 is smaller than 3.00 by about 1%). Their amplitude goes to zero. See Fig. 40. ■

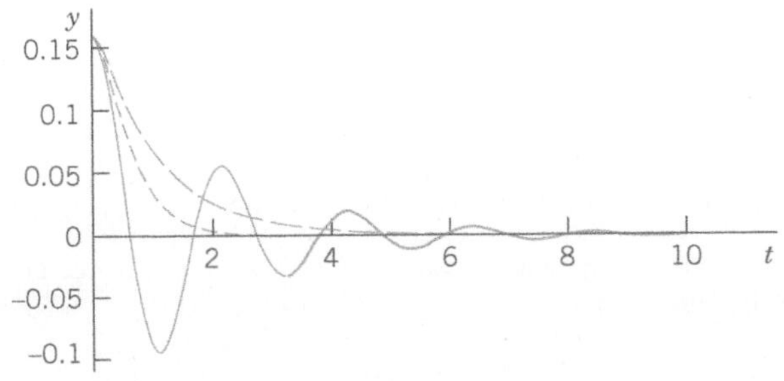

Fig. 40. The three solutions in Example 2

This section concerned ***free motions*** of mass–spring systems. Their models are *homogeneous* linear ODEs. *Nonhomogeneous* linear ODEs will arise as models of **forced motions**, that is, motions under the influence of a "driving force." We shall study them in Sec. 2.8, after we have learned how to solve those ODEs.

PROBLEM SET 2.4

1–10 HARMONIC OSCILLATIONS (UNDAMPED MOTION)

1. **Initial value problem.** Find the harmonic motion (4) that starts from y_0 with initial velocity v_0. Graph or sketch the solutions for $\omega_0 = \pi$, $y_0 = 1$, and various v_0 of your choice on common axes. At what t-values do all these curves intersect? Why?
2. **Frequency.** If a weight of 20 nt (about 4.5 lb) stretches a certain spring by 2 cm, what will the frequency of the corresponding harmonic oscillation be? The period?
3. **Frequency.** How does the frequency of the harmonic oscillation change if we (i) double the mass, (ii) take a spring of twice the modulus? First find qualitative answers by physics, then look at formulas.
4. **Initial velocity.** Could you make a harmonic oscillation move faster by giving the body a greater initial push?
5. **Springs in parallel.** What are the frequencies of vibration of a body of mass $m = 5$ kg (i) on a spring of modulus $k_1 = 20$ nt/m, (ii) on a spring of modulus $k_2 = 45$ nt/m, (iii) on the two springs in parallel? See Fig. 41.

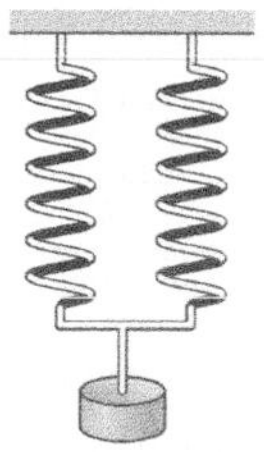

Fig. 41. Parallel springs (Problem 5)

6. **Spring in series.** If a body hangs on a spring s_1 of modulus $k_1 = 8$, which in turn hangs on a spring s_2 of modulus $k_2 = 12$, what is the modulus k of this combination of springs?
7. **Pendulum.** Find the frequency of oscillation of a pendulum of length L (Fig. 42), neglecting air resistance and the weight of the rod, and assuming θ to be so small that $\sin\theta$ practically equals θ.

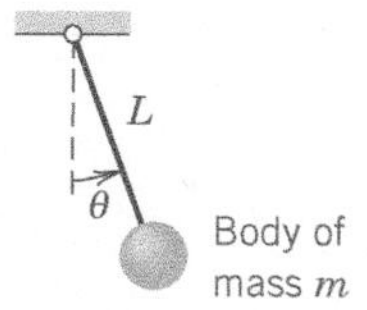

Fig. 42. Pendulum (Problem 7)

8. **Archimedian principle.** This principle states that the buoyancy force equals the weight of the water displaced by the body (partly or totally submerged). The cylindrical buoy of diameter 60 cm in Fig. 43 is floating in water with its axis vertical. When depressed downward in the water and released, it vibrates with period 2 sec. What is its weight?

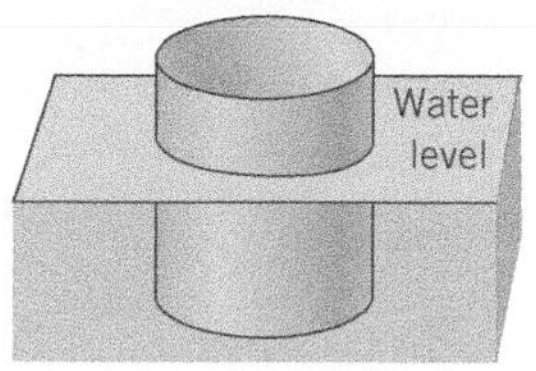

Fig. 43. Buoy (Problem 8)

9. **Vibration of water in a tube.** If 1 liter of water (about 1.06 US quart) is vibrating up and down under the influence of gravitation in a U-shaped tube of diameter 2 cm (Fig. 44), what is the frequency? Neglect friction. First guess.

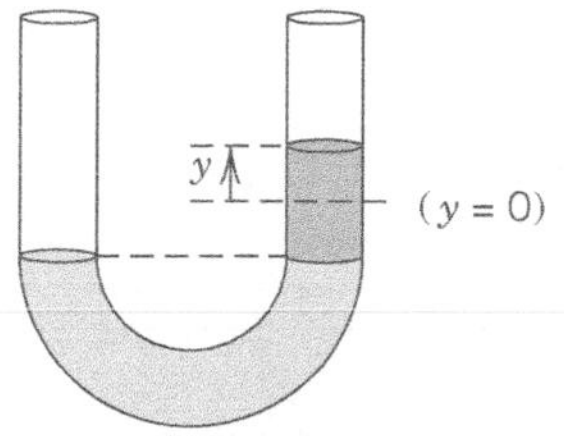

Fig. 44. Tube (Problem 9)

10. **TEAM PROJECT. Harmonic Motions of Similar Models.** The ***unifying power of mathematical methods*** results to a large extent from the fact that different physical (or other) systems may have the same or very similar models. Illustrate this for the following three systems

    (a) Pendulum clock. A clock has a 1-meter pendulum. The clock ticks once for each time the pendulum completes a full swing, returning to its original position. How many times a minute does the clock tick?

    (b) Flat spring (Fig. 45). The harmonic oscillations of a flat spring with a body attached at one end and horizontally clamped at the other are also governed by (3). Find its motions, assuming that the body weighs 8 nt (about 1.8 lb), the system has its static equilibrium 1 cm below the horizontal line, and we let it start from this position with initial velocity 10 cm/sec.

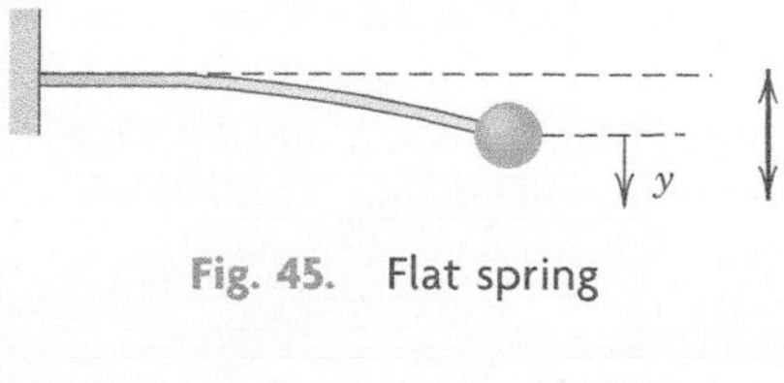

Fig. 45. Flat spring

(c) Torsional vibrations (Fig. 46). Undamped torsional vibrations (rotations back and forth) of a wheel attached to an elastic thin rod or wire are governed by the equation $I_0\theta'' + K\theta = 0$, where θ is the angle measured from the state of equilibrium. Solve this equation for $K/I_0 = 13.69 \text{ sec}^{-2}$, initial angle $30°(= 0.5235 \text{ rad})$ and initial angular velocity $20° \text{ sec}^{-1} (= 0.349 \text{ rad} \cdot \text{sec}^{-1})$.

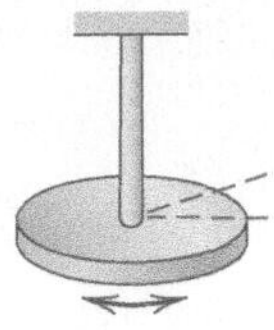

Fig. 46. Torsional vibrations

11–20 DAMPED MOTION

11. Overdamping. Show that for (7) to satisfy initial conditions $y(0) = y_0$ and $v(0) = v_0$ we must have $c_1 = [(1 + \alpha/\beta)y_0 + v_0/\beta]/2$ and $c_2 = [(1 - \alpha/\beta)y_0 - v_0/\beta]/2$.

12. Overdamping. Show that in the overdamped case, the body can pass through $y = 0$ at most once (Fig. 37).

13. Initial value problem. Find the critical motion (8) that starts from y_0 with initial velocity v_0. Graph solution curves for $\alpha = 1$, $y_0 = 1$ and several v_0 such that (i) the curve does not intersect the t-axis, (ii) it intersects it at $t = 1, 2, \dots, 5$, respectively.

14. Shock absorber. What is the smallest value of the damping constant of a shock absorber in the suspension of a wheel of a car (consisting of a spring and an absorber) that will provide (theoretically) an oscillation-free ride if the mass of the car is 2000 kg and the spring constant equals 4500 kg/sec^2?

15. Frequency. Find an approximation formula for ω^* in terms of ω_0 by applying the binomial theorem in (9) and retaining only the first two terms. How good is the approximation in Example 2, III?

16. Maxima. Show that the maxima of an underdamped motion occur at equidistant t-values and find the distance.

17. Underdamping. Determine the values of t corresponding to the maxima and minima of the oscillation $y(t) = e^{-t} \sin t$. Check your result by graphing $y(t)$.

18. Logarithmic decrement. Show that the ratio of two consecutive maximum amplitudes of a damped oscillation (10) is constant, and the natural logarithm of this ratio called the *logarithmic decrement*, equals $\Delta = 2\pi\alpha/\omega^*$. Find Δ for the solutions of $y'' + 2y' + 5y = 0$.

19. Damping constant. Consider an underdamped motion of a body of mass $m = 0.5$ kg. If the time between two consecutive maxima is 3 sec and the maximum amplitude decreases to $\frac{1}{2}$ its initial value after 10 cycles, what is the damping constant of the system?

20. CAS PROJECT. Transition Between Cases I, II, III. Study this transition in terms of graphs of typical solutions. (Cf. Fig. 47.)

(a) ***Avoiding unnecessary generality is part of good modeling.*** Show that the initial value problems (A) and (B),

$$\text{(A)} \quad y'' + cy' + y = 0, \quad y(0) = 1, \quad y'(0) = 0$$

(B) the same with different c and $y'(0) = -2$ (instead of 0), will give practically as much information as a problem with other m, k, $y(0)$, $y'(0)$.

(b) ***Consider*** **(A).** Choose suitable values of c, perhaps better ones than in Fig. 47, for the transition from Case III to II and I. Guess c for the curves in the figure.

(c) ***Time to go to rest.*** Theoretically, this time is infinite (why?). Practically, the system is at rest when its motion has become very small, say, less than 0.1% of the initial displacement (this choice being up to us), that is in our case,

$$(11) \quad |y(t)| < 0.001 \quad \text{for all } t \text{ greater than some } t_1.$$

In engineering constructions, damping can often be varied without too much trouble. Experimenting with your graphs, find empirically a relation between t_1 and c.

(d) ***Solve*** **(A)** ***analytically.*** Give a reason why the solution c of $y(t_2) = -0.001$, with t_2 the solution of $y'(t) = 0$, will give you the best possible c satisfying (11).

(e) Consider (B) empirically as in (a) and (b). What is the main difference between (B) and (A)?

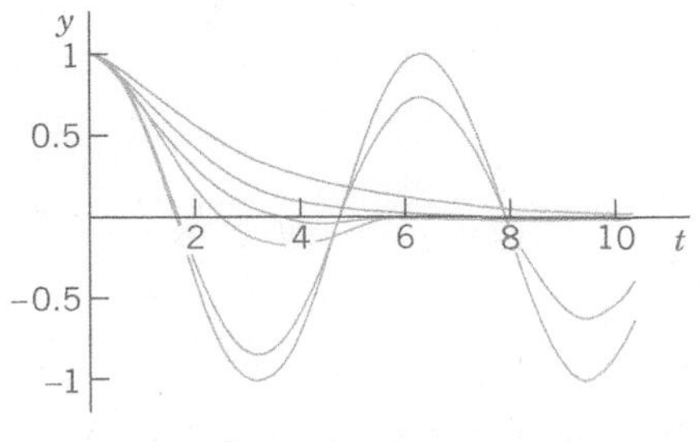

Fig. 47. CAS Project 20

2.5 Euler–Cauchy Equations

Euler–Cauchy equations[4] are ODEs of the form

$$x^2y'' + axy' + by = 0 \tag{1}$$

with given constants a and b and unknown function $y(x)$. We substitute

$$y = x^m, \qquad y' = mx^{m-1}, \qquad y'' = m(m-1)x^{m-2}$$

into (1). This gives

$$x^2m(m-1)x^{m-2} + axmx^{m-1} + bx^m = 0$$

and we now see that $y = x^m$ was a rather natural choice because we have obtained a common factor x^m. Dropping it, we have the auxiliary equation $m(m-1) + am + b = 0$ or

$$m^2 + (a-1)m + b = 0. \qquad \text{(Note: } a - 1, \text{ not } a.) \tag{2}$$

Hence $y = x^m$ is a solution of (1) if and only if m is a root of (2). The roots of (2) are

$$m_1 = \tfrac{1}{2}(1-a) + \sqrt{\tfrac{1}{4}(1-a)^2 - b}, \qquad m_2 = \tfrac{1}{2}(1-a) - \sqrt{\tfrac{1}{4}(1-a)^2 - b}. \tag{3}$$

Case I. Real different roots m_1 and m_2 give two real solutions

$$y_1(x) = x^{m_1} \qquad \text{and} \qquad y_2(x) = x^{m_2}.$$

These are linearly independent since their quotient is not constant. Hence they constitute a basis of solutions of (1) for all x for which they are real. The corresponding general solution for all these x is

$$y = c_1x^{m_1} + c_2x^{m_2} \qquad (c_1, c_2 \text{ arbitrary}). \tag{4}$$

EXAMPLE 1 General Solution in the Case of Different Real Roots

The Euler–Cauchy equation $x^2y'' + 1.5xy' - 0.5y = 0$ has the auxiliary equation $m^2 + 0.5m - 0.5 = 0$. The roots are 0.5 and -1. Hence a basis of solutions for all positive x is $y_1 = x^{0.5}$ and $y_2 = 1/x$ and gives the general solution

$$y = c_1\sqrt{x} + \frac{c_2}{x} \qquad (x > 0). \qquad ■$$

---

[4] **LEONHARD EULER (1707–1783)** was an enormously creative Swiss mathematician. He made fundamental contributions to almost all branches of mathematics and its application to physics. His important books on algebra and calculus contain numerous basic results of his own research. The great French mathematician AUGUSTIN LOUIS CAUCHY (1789–1857) is the father of modern analysis. He is the creator of complex analysis and had great influence on ODEs, PDEs, infinite series, elasticity theory, and optics.

Case II. A real double root $m_1 = \frac{1}{2}(1 - a)$ occurs if and only if $b = \frac{1}{4}(a - 1)^2$ because then (2) becomes $[m + \frac{1}{2}(a - 1)]^2$, as can be readily verified. Then a solution is $y_1 = x^{(1-a)/2}$, and (1) is of the form

$$(5) \qquad x^2y'' + axy' + \tfrac{1}{4}(1 - a)^2 y = 0 \qquad \text{or} \qquad y'' + \frac{a}{x}y' + \frac{(1-a)^2}{4x^2}y = 0.$$

A second linearly independent solution can be obtained by the method of reduction of order from Sec. 2.1, as follows. Starting from $y_2 = uy_1$, we obtain for u the expression (9) Sec. 2.1, namely,

$$u = \int U\,dx \qquad \text{where} \qquad U = \frac{1}{y_1^2}\exp\left(-\int p\,dx\right).$$

From (5) in standard form (second ODE) we see that $p = a/x$ (not ax; this is essential!). Hence $\exp \int(-p\,dx) = \exp(-a \ln x) = \exp(\ln x^{-a}) = 1/x^a$. Division by $y_1^2 = x^{1-a}$ gives $U = 1/x$, so that $u = \ln x$ by integration. Thus, $y_2 = uy_1 = y_1 \ln x$, and y_1 and y_2 are linearly independent since their quotient is not constant. The general solution corresponding to this basis is

$$(6) \qquad y = (c_1 + c_2 \ln x)\, x^m, \qquad m = \tfrac{1}{2}(1 - a).$$

EXAMPLE 2 **General Solution in the Case of a Double Root**

The Euler–Cauchy equation $x^2y'' - 5xy' + 9y = 0$ has the auxiliary equation $m^2 - 6m + 9 = 0$. It has the double root $m = 3$, so that a general solution for all positive x is

$$y = (c_1 + c_2 \ln x)\, x^3.$$

■

Case III. Complex conjugate roots are of minor practical importance, and we discuss the derivation of real solutions from complex ones just in terms of a typical example.

EXAMPLE 3 **Real General Solution in the Case of Complex Roots**

The Euler–Cauchy equation $x^2y'' + 0.6xy' + 16.04y = 0$ has the auxiliary equation $m^2 - 0.4m + 16.04 = 0$. The roots are complex conjugate, $m_1 = 0.2 + 4i$ and $m_2 = 0.2 - 4i$, where $i = \sqrt{-1}$. We now use the trick of writing $x = e^{\ln x}$ and obtain

$$x^{m_1} = x^{0.2+4i} = x^{0.2}(e^{\ln x})^{4i} = x^{0.2}e^{(4 \ln x)i},$$
$$x^{m_2} = x^{0.2-4i} = x^{0.2}(e^{\ln x})^{-4i} = x^{0.2}e^{-(4 \ln x)i}.$$

Next we apply Euler's formula (11) in Sec. 2.2 with $t = 4 \ln x$ to these two formulas. This gives

$$x^{m_1} = x^{0.2}[\cos(4 \ln x) + i \sin(4 \ln x)],$$
$$x^{m_2} = x^{0.2}[\cos(4 \ln x) - i \sin(4 \ln x)].$$

We now add these two formulas, so that the sine drops out, and divide the result by 2. Then we subtract the second formula from the first, so that the cosine drops out, and divide the result by $2i$. This yields

$$x^{0.2}\cos(4 \ln x) \qquad \text{and} \qquad x^{0.2}\sin(4 \ln x)$$

respectively. By the superposition principle in Sec. 2.2 these are solutions of the Euler–Cauchy equation (1). Since their quotient $\cot(4 \ln x)$ is not constant, they are linearly independent. Hence they form a basis of solutions, and the corresponding real general solution for all positive x is

$$(8) \qquad y = x^{0.2}[A \cos(4 \ln x) + B \sin(4 \ln x)].$$

Figure 48 shows typical solution curves in the three cases discussed, in particular the real basis functions in Examples 1 and 3. ■

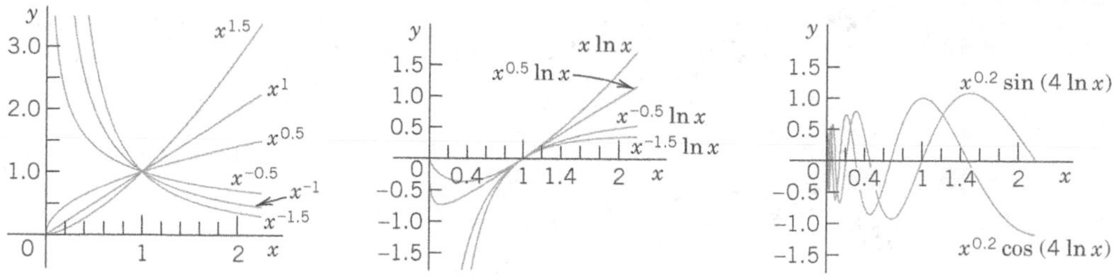

Case I: Real roots Case II: Double root Case III: Complex roots

Fig. 48. Euler–Cauchy equations

EXAMPLE 4 Boundary Value Problem. Electric Potential Field Between Two Concentric Spheres

Find the electrostatic potential $v = v(r)$ between two concentric spheres of radii $r_1 = 5$ cm and $r_2 = 10$ cm kept at potentials $v_1 = 110$ V and $v_2 = 0$, respectively.

Physical Information. $v(r)$ is a solution of the Euler–Cauchy equation $rv'' + 2v' = 0$, where $v' = dv/dr$.

Solution. The auxiliary equation is $m^2 + m = 0$. It has the roots 0 and -1. This gives the general solution $v(r) = c_1 + c_2/r$. From the "boundary conditions" (the potentials on the spheres) we obtain

$$v(5) = c_1 + \frac{c_2}{5} = 110. \qquad v(10) = c_1 + \frac{c_2}{10} = 0.$$

By subtraction, $c_2/10 = 110$, $c_2 = 1100$. From the second equation, $c_1 = -c_2/10 = -110$. *Answer:* $v(r) = -110 + 1100/r$ V. Figure 49 shows that the potential is not a straight line, as it would be for a potential between two parallel plates. For example, on the sphere of radius 7.5 cm it is not $110/2 = 55$ V, but considerably less. (What is it?) ■

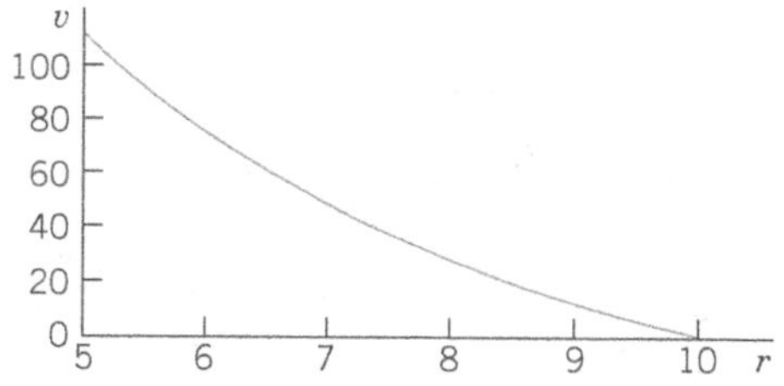

Fig. 49. Potential $v(r)$ in Example 4

PROBLEM SET 2.5

1. **Double root.** Verify directly by substitution that $x^{(1-a)/2} \ln x$ is a solution of (1) if (2) has a double root, but $x^{m_1} \ln x$ and $x^{m_2} \ln x$ are not solutions of (1) if the roots m_1 and m_2 of (2) are different.

2–11 GENERAL SOLUTION

Find a real general solution. Show the details of your work.

2. $x^2y'' - 20y = 0$
3. $5x^2y'' + 23xy' + 16.2y = 0$
4. $xy'' + 2y' = 0$
5. $4x^2y'' + 5y = 0$
6. $x^2y'' + 0.7xy' - 0.1y = 0$
7. $(x^2D^2 - 4xD + 6I)y = C$
8. $(x^2D^2 - 3xD + 4I)y = 0$
9. $(x^2D^2 - 0.2xD + 0.36I)y = 0$
10. $(x^2D^2 - xD + 5I)y = 0$
11. $(x^2D^2 - 3xD + 10I)y = 0$

12–19 INITIAL VALUE PROBLEM

Solve and graph the solution. Show the details of your work.

12. $x^2y'' - 4xy' + 6y = 0, \quad y(1) = 0.4, \quad y'(1) = 0$

13. $x^2y'' + 3xy' + 0.75y = 0, \quad y(1) = 1,$ $y'(1) = -1.5$

14. $x^2y'' + xy' + 9y = 0, \quad y(1) = 0, \quad y'(1) = 2.5$

15. $x^2y'' + 3xy' + y = 0, \quad y(1) = 3.6, \quad y'(1) = 0.4$

16. $(x^2D^2 - 3xD + 4I)y = 0, \quad y(1) = -\pi, \quad y'(1) = 2\pi$

17. $(x^2D^2 + xD + I)y = 0, \quad y(1) = 1, \quad y'(1) = 1$

18. $(9x^2D^2 + 3xD + I)y = 0, \quad y(1) = 1, \quad y'(1) = 0$

19. $(x^2D^2 - xD - 15I)y = 0, \quad y(1) = 0.1,$ $y'(1) = -4.5$

20. TEAM PROJECT. Double Root

(a) Derive a second linearly independent solution of (1) by reduction of order; but instead of using (9), Sec. 2.1, perform all steps directly for the present ODE (1).

(b) Obtain $x^m \ln x$ by considering the solutions x^m and x^{m+s} of a suitable Euler–Cauchy equation and letting $s \to 0$.

(c) Verify by substitution that $x^m \ln x$, $m = (1 - a)/2$, is a solution in the critical case.

(d) Transform the Euler–Cauchy equation (1) into an ODE with constant coefficients by setting $x = e^t$ $(x > 0)$.

(e) Obtain a second linearly independent solution of the Euler–Cauchy equation in the "critical case" from that of a constant-coefficient ODE.

2.6 Existence and Uniqueness of Solutions. Wronskian

In this section we shall discuss the general theory of homogeneous linear ODEs

$$y'' + p(x)y' + q(x)y = 0 \tag{1}$$

with continuous, but otherwise arbitrary, ***variable coefficients*** p and q. This will concern the existence and form of a general solution of (1) as well as the uniqueness of the solution of initial value problems consisting of such an ODE and two initial conditions

$$y(x_0) = K_0, \qquad y'(x_0) = K_1 \tag{2}$$

with given x_0, K_0, and K_1.

The two main results will be Theorem 1, stating that such an initial value problem always has a solution which is unique, and Theorem 4, stating that a general solution

$$y = c_1y_1 + c_2y_2 \qquad (c_1, c_2 \text{ arbitrary}) \tag{3}$$

includes all solutions. Hence *linear* ODEs with continuous coefficients have no *"singular solutions"* (solutions not obtainable from a general solution).

Clearly, no such theory was needed for constant-coefficient or Euler–Cauchy equations because everything resulted explicitly from our calculations.

Central to our present discussion is the following theorem.

THEOREM 1

Existence and Uniqueness Theorem for Initial Value Problems

If $p(x)$ and $q(x)$ are continuous functions on some open interval I (see Sec. 1.1) *and x_0 is in I, then the initial value problem consisting of* (1) *and* (2) *has a unique solution $y(x)$ on the interval I.*

The proof of existence uses the same prerequisites as the existence proof in Sec. 1.7 and will not be presented here; it can be found in Ref. [A11] listed in App. 1. Uniqueness proofs are usually simpler than existence proofs. But for Theorem 1, even the uniqueness proof is long, and we give it as an additional proof in App. 4.

Linear Independence of Solutions

Remember from Sec. 2.1 that a general solution on an open interval I is made up from a **basis** y_1, y_2 on I, that is, from a pair of linearly independent solutions on I. Here we call y_1, y_2 **linearly independent** on I if the equation

$$k_1y_1(x) + k_2y_2(x) = 0 \quad \text{on } I \qquad \text{implies} \qquad k_1 = 0, \quad k_2 = 0. \tag{4}$$

We call y_1, y_2 **linearly dependent** on I if this equation also holds for constants k_1, k_2 not both 0. In this case, and only in this case, y_1 and y_2 are proportional on I, that is (see Sec. 2.1),

$$\text{(a)} \quad y_1 = ky_2 \qquad \text{or} \qquad \text{(b)} \quad y_2 = ly_1 \qquad \qquad \text{for all on } I. \tag{5}$$

For our discussion the following criterion of linear independence and dependence of solutions will be helpful.

THEOREM 2

Linear Dependence and Independence of Solutions

Let the ODE (1) *have continuous coefficients $p(x)$ and $q(x)$ on an open interval I. Then two solutions y_1 and y_2 of* (1) *on I are linearly dependent on I if and only if their* "**Wronskian**"

$$W(y_1, y_2) = y_1y_2' - y_2y_1' \tag{6}$$

is 0 *at some x_0 in I. Furthermore, if $W = 0$ at an $x = x_0$ in I, then $W = 0$ on I; hence, if there is an x_1 in I at which W is not* 0, *then y_1, y_2 are linearly independent on I.*

PROOF **(a)** Let y_1 and y_2 be linearly dependent on I. Then (5a) or (5b) holds on I. If (5a) holds, then

$$W(y_1, y_2) = y_1y_2' - y_2y_1' = ky_2y_2' - y_2ky_2' = 0.$$

Similarly if (5b) holds.

(b) Conversely, we let $W(y_1, y_2) = 0$ for some $x = x_0$ and show that this implies linear dependence of y_1 and y_2 on I. We consider the linear system of equations in the unknowns k_1, k_2

$$\begin{aligned} k_1y_1(x_0) + k_2y_2(x_0) &= 0 \\ k_1y_1'(x_0) + k_2y_2'(x_0) &= 0. \end{aligned} \tag{7}$$

To eliminate k_2, multiply the first equation by y_2' and the second by $-y_2$ and add the resulting equations. This gives

$$k_1 y_1(x_0) y_2'(x_0) - k_1 y_1'(x_0) y_2(x_0) = k_1 W(y_1(x_0), y_2(x_0)) = 0.$$

Similarly, to eliminate k_1, multiply the first equation by $-y_1'$ and the second by y_1 and add the resulting equations. This gives

$$k_2 W(y_1(x_0), y_2(x_0)) = 0.$$

If W were not 0 at x_0, we could divide by W and conclude that $k_1 = k_2 = 0$. Since W is 0, division is not possible, and the system has a solution for which k_1 and k_2 are not both 0. Using *these numbers* k_1, k_2, we introduce the function

$$y(x) = k_1 y_1(x) + k_2 y_2(x).$$

Since (1) is homogeneous linear, Fundamental Theorem 1 in Sec. 2.1 (the superposition principle) implies that this function is a solution of (1) on I. From (7) we see that it satisfies the initial conditions $y(x_0) = 0$, $y'(x_0) = 0$. Now another solution of (1) satisfying the same initial conditions is $y^* \equiv 0$. Since the coefficients p and q of (1) are continuous, Theorem 1 applies and gives uniqueness, that is, $y \equiv y^*$, written out

$$k_1 y_1 + k_2 y_2 \equiv 0 \qquad \text{on } I.$$

Now since k_1 and k_2 are not both zero, this means linear dependence of y_1, y_2 on I.

(c) We prove the last statement of the theorem. If $W(x_0) = 0$ at an x_0 in I, we have linear dependence of y_1, y_2 on I by part (b), hence $W \equiv 0$ by part (a) of this proof. Hence in the case of linear dependence it cannot happen that $W(x_1) \neq 0$ at an x_1 in I. If it does happen, it thus implies linear independence as claimed. ■

For calculations, the following formulas are often simpler than (6).

$$\textbf{(6*)} \quad W(y_1, y_2) = (a) \quad \left(\frac{y_2}{y_1}\right)' y_1^2 \quad (y_1 \neq 0) \qquad \text{or} \qquad (b) \quad -\left(\frac{y_1}{y_2}\right)' y_2^2 \quad (y_2 \neq 0).$$

These formulas follow from the quotient rule of differentiation.

Remark. Determinants. Students familiar with second-order determinants may have noticed that

$$W(y_1, y_2) = \begin{vmatrix} y_1 & y_2 \\ y_1' & y_2' \end{vmatrix} = y_1 y_2' - y_2 y_1'.$$

This determinant is called the *Wronski determinant*[5] or, briefly, the **Wronskian**, of two solutions y_1 and y_2 of (1), as has already been mentioned in (6). Note that its four entries occupy the same positions as in the linear system (7).

[5]Introduced by WRONSKI (JOSEF MARIA HÖNE, 1776–1853), Polish mathematician.

EXAMPLE 1 **Illustration of Theorem 2**

The functions $y_1 = \cos \omega x$ and $y_2 = \sin \omega x$ are solutions of $y'' + \omega^2 y = 0$. Their Wronskian is

$$W(\cos \omega x, \sin \omega x) = \begin{vmatrix} \cos \omega x & \sin \omega x \\ -\omega \sin \omega x & \omega \cos \omega x \end{vmatrix} = y_1 y_2' - y_2 y_1' = \omega \cos^2 \omega x + \omega \sin^2 \omega x = \omega.$$

Theorem 2 shows that these solutions are linearly independent if and only if $\omega \neq 0$. Of course, we can see this directly from the quotient $y_2/y_1 = \tan \omega x$. For $\omega = 0$ we have $y_2 = 0$, which implies linear dependence (why?). ■

EXAMPLE 2 **Illustration of Theorem 2 for a Double Root**

A general solution of $y'' - 2y' + y = 0$ on any interval is $y = (c_1 + c_2 x)e^x$. (Verify!). The corresponding Wronskian is not 0, which shows linear independence of e^x and xe^x on any interval. Namely,

$$W(x, xe^x) = \begin{vmatrix} e^x & xe^x \\ e^x & (x+1)e^x \end{vmatrix} = (x+1)e^{2x} - xe^{2x} = e^{2x} \neq 0.$$ ■

A General Solution of (1) Includes All Solutions

This will be our second main result, as announced at the beginning. Let us start with existence.

THEOREM 3 **Existence of a General Solution**

If $p(x)$ and $q(x)$ are continuous on an open interval I, then (1) *has a general solution on I.*

PROOF By Theorem 1, the ODE (1) has a solution $y_1(x)$ on I satisfying the initial conditions

$$y_1(x_0) = 1, \qquad y_1'(x_0) = 0$$

and a solution $y_2(x)$ on I satisfying the initial conditions

$$y_2(x_0) = 0, \qquad y_2'(x_0) = 1.$$

The Wronskian of these two solutions has at $x = x_0$ the value

$$W(y_1(0), y_2(0)) = y_1(x_0)y_2'(x_0) - y_2(x_0)y_1'(x_0) = 1.$$

Hence, by Theorem 2, these solutions are linearly independent on I. They form a basis of solutions of (1) on I, and $y = c_1 y_1 + c_2 y_2$ with arbitrary c_1, c_2 is a general solution of (1) on I, whose existence we wanted to prove. ■

We finally show that a general solution is as general as it can possibly be.

THEOREM 4

A General Solution Includes All Solutions

If the ODE (1) *has continuous coefficients* $p(x)$ *and* $q(x)$ *on some open interval* I, *then every solution* $y = Y(x)$ *of* (1) *on* I *is of the form*

$$Y(x) = C_1 y_1(x) + C_2 y_2(x) \tag{8}$$

where y_1, y_2 *is any basis of solutions of* (1) *on* I *and* C_1, C_2 *are suitable constants.*

Hence (1) *does not have* **singular solutions** *(that is, solutions not obtainable from a general solution).*

PROOF Let $y = Y(x)$ be any solution of (1) on I. Now, by Theorem 3 the ODE (1) has a general solution

$$y(x) = c_1 y_1(x) + c_2 y_2(x) \tag{9}$$

on I. We have to find suitable values of c_1, c_2 such that $y(x) = Y(x)$ on I. We choose any x_0 in I and show first that we can find values of c_1, c_2 such that we reach agreement at x_0, that is, $y(x_0) = Y(x_0)$ and $y'(x_0) = Y'(x_0)$. Written out in terms of (9), this becomes

$$\begin{aligned} &\text{(a)} \quad c_1 y_1(x_0) + c_2 y_2(x_0) = Y(x_0) \\ &\text{(b)} \quad c_1 y_1'(x_0) + c_2 y_2'(x_0) = Y'(x_0). \end{aligned} \tag{10}$$

We determine the unknowns c_1 and c_2. To eliminate c_2, we multiply (10a) by $y_2'(x_0)$ and (10b) by $-y_2(x_0)$ and add the resulting equations. This gives an equation for c_1. Then we multiply (10a) by $-y_1'(x_0)$ and (10b) by $y_1(x_0)$ and add the resulting equations. This gives an equation for c_2. These new equations are as follows, where we take the values of $y_1, y_1', y_2, y_2', Y, Y'$ at x_0.

$$c_1(y_1 y_2' - y_2 y_1') = c_1 W(y_1, y_2) = Y y_2' - y_2 Y'$$

$$c_2(y_1 y_2' - y_2 y_1') = c_2 W(y_1, y_2) = y_1 Y' - Y y_1'.$$

Since y_1, y_2 is a basis, the Wronskian W in these equations is not 0, and we can solve for c_1 and c_2. We call the (unique) solution $c_1 = C_1, c_2 = C_2$. By substituting it into (9) we obtain from (9) the particular solution

$$y^*(x) = C_1 y_1(x) + C_2 y_2(x).$$

Now since C_1, C_2 is a solution of (10), we see from (10) that

$$y^*(x_0) = Y(x_0), \qquad y^{*\prime}(x_0) = Y'(x_0).$$

From the uniqueness stated in Theorem 1 this implies that y^* and Y must be equal everywhere on I, and the proof is complete. ■

Reflecting on this section, we note that homogeneous linear ODEs with continuous variable coefficients have a conceptually and structurally rather transparent existence and uniqueness theory of solutions. Important in itself, this theory will also provide the foundation for our study of nonhomogeneous linear ODEs, whose theory and engineering applications form the content of the remaining four sections of this chapter.

PROBLEM SET 2.6

1. Derive (6*) from (6).

2–8 BASIS OF SOLUTIONS. WRONSKIAN

Find the Wronskian. Show linear independence by using quotients and confirm it by Theorem 2.

2. $e^{4.0x}, e^{-1.5x}$

3. $e^{-0.4x}, e^{-2.6x}$

4. $x, 1/x$

5. x^3, x^2

6. $e^{-x} \cos \omega x, e^{-x} \sin \omega x$

7. $\cosh ax, \sinh ax$

8. $x^k \cos (\ln x), x^k \sin (\ln x)$

9–15 ODE FOR GIVEN BASIS. WRONSKIAN. IVP

(a) Find a second-order homogeneous linear ODE for which the given functions are solutions. **(b)** Show linear independence by the Wronskian. **(c)** Solve the initial value problem.

9. $\cos 5x, \sin 5x, \quad y(0) = 3, \quad y'(0) = -5$

10. $x^{m_1}, x^{m_2}, \quad y(1) = -2, \quad y'(1) = 2m_1 - 4m_2$

11. $e^{-2.5x} \cos 0.3x, e^{-2.5x} \sin 0.3x, \quad y(0) = 3,$ $y'(0) = -7.5$

12. $x^2, x^2 \ln x, \quad y(1) = 4, \quad y'(1) = 6$

13. $1, e^{-2x}, \quad y(0) = 1, \quad y'(0) = -1$

14. $e^{-kx} \cos \pi x, e^{-kx} \sin \pi x, \quad y(0) = 1,$ $y'(0) = -k - \pi$

15. $\cosh 1.8x, \sinh 1.8x, \quad y(0) = 14.20, \quad y'(0) = 16.38$

16. TEAM PROJECT. Consequences of the Present Theory. This concerns some noteworthy general properties of solutions. Assume that the coefficients p and q of the ODE (1) are continuous on some open interval I, to which the subsequent statements refer.

(a) Solve $y'' - y = 0$ (a) by exponential functions, (b) by hyperbolic functions. How are the constants in the corresponding general solutions related?

(b) Prove that the solutions of a basis cannot be 0 at the same point.

(c) Prove that the solutions of a basis cannot have a maximum or minimum at the same point.

(d) Why is it likely that formulas of the form (6*) should exist?

(e) Sketch $y_1(x) = x^3$ if $x \geqq 0$ and 0 if $x < 0$, $y_2(x) = 0$ if $x \geqq 0$ and x^3 if $x < 0$. Show linear independence on $-1 < x < 1$. What is their Wronskian? What Euler–Cauchy equation do y_1, y_2 satisfy? Is there a contradiction to Theorem 2?

(f) Prove **Abel's formula**[6]

$$W(y_1(x), y_2(x)) = c \exp \left[-\int_{x_0}^{x} p(t)\, dt \right]$$

where $c = W(y_1(x_0), y_2(x_0))$. Apply it to Prob. 6. *Hint:* Write (1) for y_1 and for y_2. Eliminate q algebraically from these two ODEs, obtaining a first-order linear ODE. Solve it.

2.7 Nonhomogeneous ODEs

We now advance from homogeneous to nonhomogeneous linear ODEs.

Consider the second-order nonhomogeneous linear ODE

$$y'' + p(x)y' + q(x)y = r(x) \tag{1}$$

where $r(x) \not\equiv 0$. We shall see that a "general solution" of (1) is the sum of a general solution of the corresponding homogeneous ODE

[6]**NIELS HENRIK ABEL** (1802–1829), Norwegian mathematician.

$$y'' + p(x)y' + q(x)y = 0 \tag{2}$$

and a "particular solution" of (1). These two new terms "general solution of (1)" and "particular solution of (1)" are defined as follows.

DEFINITION

General Solution, Particular Solution

A **general solution** of the nonhomogeneous ODE (1) on an open interval I is a solution of the form

$$y(x) = y_h(x) + y_p(x); \tag{3}$$

here, $y_h = c_1y_1 + c_2y_2$ is a general solution of the homogeneous ODE (2) on I and y_p is any solution of (1) on I containing no arbitrary constants.

A **particular solution** of (1) on I is a solution obtained from (3) by assigning specific values to the arbitrary constants c_1 and c_2 in y_h.

Our task is now twofold, first to justify these definitions and then to develop a method for finding a solution y_p of (1).

Accordingly, we first show that a general solution as just defined satisfies (1) and that the solutions of (1) and (2) are related in a very simple way.

THEOREM 1

Relations of Solutions of (1) to Those of (2)

(a) *The sum of a solution y of* (1) *on some open interval I and a solution $\tilde{y}$ of* (2) *on I is a solution of* (1) *on I. In particular,* (3) *is a solution of* (1) *on I.*

(b) *The difference of two solutions of* (1) *on I is a solution of* (2) *on I.*

PROOF **(a)** Let $L[y]$ denote the left side of (1). Then for any solutions y of (1) and $\tilde{y}$ of (2) on I,

$$L[y + \tilde{y}] = L[y] + L[\tilde{y}] = r + 0 = r.$$

(b) For any solutions y and y^* of (1) on I we have $L[y - y^*] = L[y] - L[y^*] = r - r = 0$. ■

Now for *homogeneous ODEs* (2) we know that general solutions include all solutions. We show that the same is true for nonhomogeneous ODEs (1).

THEOREM 2

A General Solution of a Nonhomogeneous ODE Includes All Solutions

If the coefficients $p(x)$, $q(x)$, and the function $r(x)$ in (1) *are continuous on some open interval I, then every solution of* (1) *on I is obtained by assigning suitable values to the arbitrary constants c_1 and c_2 in a general solution* (3) *of* (1) *on I.*

PROOF Let y^* be any solution of (1) on I and x_0 any x in I. Let (3) be any general solution of (1) on I. This solution exists. Indeed, $y_h = c_1y_1 + c_2y_2$ exists by Theorem 3 in Sec. 2.6

because of the continuity assumption, and y_p exists according to a construction to be shown in Sec. 2.10. Now, by Theorem 1(b) just proved, the difference $Y = y^* - y_p$ is a solution of (2) on I. At x_0 we have

$$Y(x_0) = y^*(x_0) - y_p(x_0). \qquad Y'(x_0) = y^{*\prime}(x_0) - y_p'(x_0).$$

Theorem 1 in Sec. 2.6 implies that for these conditions, as for any other initial conditions in I, there exists a unique particular solution of (2) obtained by assigning suitable values to c_1, c_2 in y_h. From this and $y^* = Y + y_p$ the statement follows. ∎

Method of Undetermined Coefficients

Our discussion suggests the following. *To solve the nonhomogeneous ODE* (1) *or an initial value problem for* (1), *we have to solve the homogeneous ODE* (2) *and find any solution* y_p *of* (1), so that we obtain a general solution (3) of (1).

How can we find a solution y_p of (1)? One method is the so-called **method of undetermined coefficients**. It is much simpler than another, more general, method (given in Sec. 2.10). Since it applies to models of vibrational systems and electric circuits to be shown in the next two sections, it is frequently used in engineering.

More precisely, the method of undetermined coefficients is suitable for linear ODEs with ***constant coefficients a and b***

$$y'' + ay' + by = r(x) \tag{4}$$

when $r(x)$ is an exponential function, a power of x, a cosine or sine, or sums or products of such functions. These functions have derivatives similar to $r(x)$ itself. This gives the idea. We choose a form for y_p similar to $r(x)$, but with unknown coefficients to be determined by substituting that y_p and its derivatives into the ODE. Table 2.1 on p. 82 shows the choice of y_p for practically important forms of $r(x)$. Corresponding rules are as follows.

Choice Rules for the Method of Undetermined Coefficients

(a) Basic Rule. *If* $r(x)$ *in* (4) *is one of the functions in the first column in Table* 2.1, *choose* y_p *in the same line and determine its undetermined coefficients by substituting* y_p *and its derivatives into* (4).

(b) Modification Rule. *If a term in your choice for* y_p *happens to be a solution of the homogeneous ODE corresponding to* (4), *multiply this term by* x (*or by* x^2 *if this solution corresponds to a double root of the characteristic equation of the homogeneous ODE*).

(c) Sum Rule. *If* $r(x)$ *is a sum of functions in the first column of Table* 2.1, *choose for* y_p *the sum of the functions in the corresponding lines of the second column.*

The Basic Rule applies when $r(x)$ is a single term. The Modification Rule helps in the indicated case, and to recognize such a case, we have to solve the homogeneous ODE first. The Sum Rule follows by noting that the sum of two solutions of (1) with $r = r_1$ and $r = r_2$ (and the same left side!) is a solution of (1) with $r = r_1 + r_2$. (Verify!)

The method is self-correcting. A false choice for y_p or one with too few terms will lead to a contradiction. A choice with too many terms will give a correct result, with superfluous coefficients coming out zero.

Let us illustrate Rules (a)–(c) by the typical Examples 1–3.

Table 2.1 Method of Undetermined Coefficients

Term in $r(x)$	Choice for $y_p(x)$
$ke^{\gamma x}$	$Ce^{\gamma x}$
$kx^n \ (n = 0, 1, \cdots)$	$K_n x^n + K_{n-1}x^{n-1} + \cdots + K_1 x + K_0$
$k \cos \omega x$ $k \sin \omega x$	$K \cos \omega x + M \sin \omega x$
$ke^{\alpha x} \cos \omega x$ $ke^{\alpha x} \sin \omega x$	$e^{\alpha x}(K \cos \omega x + M \sin \omega x)$

EXAMPLE 1 Application of the Basic Rule (a)

Solve the initial value problem

$$y'' + y = 0.001x^2, \qquad y(0) = 0, \qquad y'(0) = 1.5. \tag{5}$$

Solution. ***Step 1. General solution of the homogeneous ODE.*** The ODE $y'' + y = 0$ has the general solution

$$y_h = A \cos x + B \sin x.$$

Step 2. Solution y_p of the nonhomogeneous ODE. We first try $y_p = Kx^2$. Then $y_p'' = 2K$. By substitution, $2K + Kx^2 = 0.001x^2$. For this to hold for all x, the coefficient of each power of x (x^2 and x^0) must be the same on both sides; thus $K = 0.001$ and $2K = 0$, a contradiction.

The second line in Table 2.1 suggests the choice

$$y_p = K_2 x^2 + K_1 x + K_0. \qquad \text{Then} \qquad y_p'' + y_p = 2K_2 + K_2 x^2 + K_1 x + K_0 = 0.001x^2.$$

Equating the coefficients of x^2, x, x^0 on both sides, we have $K_2 = 0.001, K_1 = 0, 2K_2 + K_0 = 0$. Hence $K_0 = -2K_2 = -0.002$. This gives $y_p = 0.001x^2 - 0.002$, and

$$y = y_h + y_p = A \cos x + B \sin x + 0.001x^2 - 0.002.$$

Step 3. Solution of the initial value problem. Setting $x = 0$ and using the first initial condition gives $y(0) = A - 0.002 = 0$, hence $A = 0.002$. By differentiation and from the second initial condition,

$$y' = y_h' + y_p' = -A \sin x + B \cos x + 0.002x \qquad \text{and} \qquad y'(0) = B = 1.5.$$

This gives the answer (Fig. 50)

$$y = 0.002 \cos x + 1.5 \sin x + 0.001x^2 - 0.002.$$

Figure 50 shows y as well as the quadratic parabola y_p about which y is oscillating, practically like a sine curve since the cosine term is smaller by a factor of about $1/1000$. ■

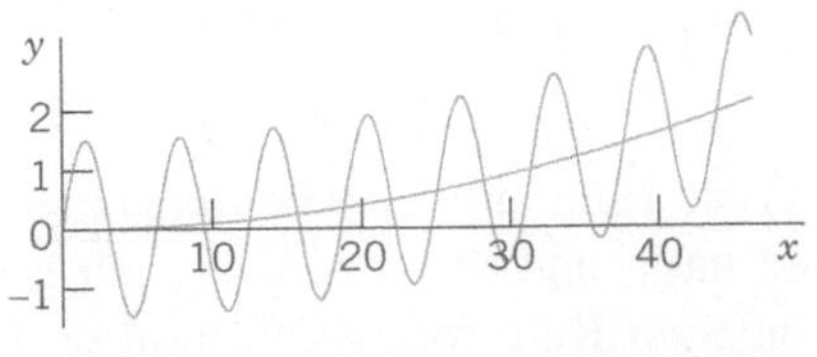

Fig. 50. Solution in Example 1

EXAMPLE 2 Application of the Modification Rule (b)

Solve the initial value problem

$$y'' + 3y' + 2.25y = -10e^{-1.5x}, \qquad y(0) = 1, \qquad y'(0) = 0. \tag{6}$$

Solution. ***Step 1. General solution of the homogeneous ODE.*** The characteristic equation of the homogeneous ODE is $\lambda^2 + 3\lambda + 2.25 = (\lambda + 1.5)^2 = 0$. Hence the homogeneous ODE has the general solution

$$y_h = (c_1 + c_2x)e^{-1.5x}.$$

Step 2. Solution y_p of the nonhomogeneous ODE. The function $e^{-1.5x}$ on the right would normally require the choice $Ce^{-1.5x}$. But we see from y_h that this function is a solution of the homogeneous ODE, which corresponds to a *double root* of the characteristic equation. Hence, according to the Modification Rule we have to multiply our choice function by x^2. That is, we choose

$$y_p = Cx^2e^{-1.5x}. \quad \text{Then} \quad y_p' = C(2x - 1.5x^2)e^{-1.5x}, \quad y_p'' = C(2 - 3x - 3x + 2.25x^2)e^{-1.5x}.$$

We substitute these expressions into the given ODE and omit the factor $e^{-1.5x}$. This yields

$$C(2 - 6x + 2.25x^2) + 3C(2x - 1.5x^2) + 2.25Cx^2 = -10.$$

Comparing the coefficients of x^2, x, x^0 gives $0 = 0, 0 = 0, 2C = -10$, hence $C = -5$. This gives the solution $y_p = -5x^2e^{-1.5x}$. Hence the given ODE has the general solution

$$y = y_h + y_p = (c_1 + c_2x)e^{-1.5x} - 5x^2e^{-1.5x}.$$

Step 3. Solution of the initial value problem. Setting $x = 0$ in y and using the first initial condition, we obtain $y(0) = c_1 = 1$. Differentiation of y gives

$$y' = (c_2 - 1.5c_1 - 1.5c_2x)e^{-1.5x} - 10xe^{-1.5x} + 7.5x^2e^{-1.5x}.$$

From this and the second initial condition we have $y'(0) = c_2 - 1.5c_1 = 0$. Hence $c_2 = 1.5c_1 = 1.5$. This gives the answer (Fig. 51)

$$y = (1 + 1.5x)e^{-1.5x} - 5x^2e^{-1.5x} = (1 + 1.5x - 5x^2)e^{-1.5x}.$$

The curve begins with a horizontal tangent, crosses the x-axis at $x = 0.6217$ (where $1 + 1.5x - 5x^2 = 0$) and approaches the axis from below as x increases. ∎

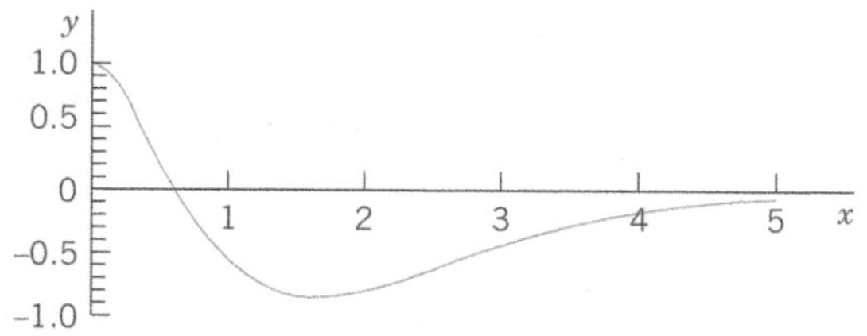

Fig. 51. Solution in Example 2

EXAMPLE 3 Application of the Sum Rule (c)

Solve the initial value problem

$$y'' + 2y' + 0.75y = 2\cos x - 0.25\sin x + 0.09x, \qquad y(0) = 2.78, \qquad y'(0) = -0.43. \tag{7}$$

Solution. ***Step 1. General solution of the homogeneous ODE.*** The characteristic equation of the homogeneous ODE is

$$\lambda^2 + 2\lambda + 0.75 = (\lambda + \tfrac{1}{2})(\lambda + \tfrac{3}{2}) = 0$$

which gives the general solution $y_h = c_1e^{-x/2} + c_2e^{-3x/2}$.

Step 2. Particular solution of the nonhomogeneous ODE. We write $y_p = y_{p1} + y_{p2}$ and, following Table 2.1, (C) and (B),

$$y_{p1} = K\cos x + M\sin x \quad \text{and} \quad y_{p2} = K_1 x + K_0.$$

Differentiation gives $y'_{p1} = -K\sin x + M\cos x$, $y''_{p1} = -K\cos x - M\sin x$ and $y'_{p2} = 1$, $y''_{p2} = 0$. Substitution of y_{p1} into the ODE in (7) gives, by comparing the cosine and sine terms,

$$-K + 2M + 0.75K = 2, \qquad -M - 2K + 0.75M = -0.25,$$

hence $K = 0$ and $M = 1$. Substituting y_{p2} into the ODE in (7) and comparing the x- and x^0-terms gives

$$0.75K_1 = 0.09, \quad 2K_1 + 0.75K_0 = 0, \qquad \text{thus} \qquad K_1 = 0.12, \quad K_0 = -0.32.$$

Hence a general solution of the ODE in (7) is

$$y = c_1 e^{-x/2} + c_2 e^{-3x/2} + \sin x + 0.12x - 0.32.$$

Step 3. Solution of the initial value problem. From y, y' and the initial conditions we obtain

$$y(0) = c_1 + c_2 - 0.32 = 2.78, \qquad y'(0) = -\tfrac{1}{2}c_1 - \tfrac{3}{2}c_2 + 1 + 0.12 = -0.4.$$

Hence $c_1 = 3.1$, $c_2 = 0$. This gives the solution of the IVP (Fig. 52)

$$y = 3.1e^{-x/2} + \sin x + 0.12x - 0.32. \qquad \blacksquare$$

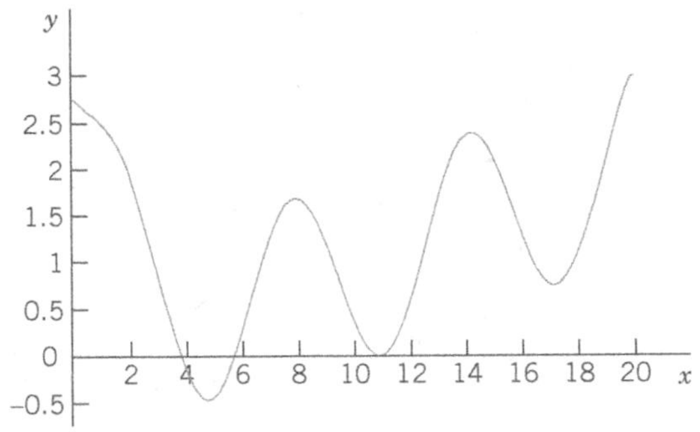

Fig. 52. Solution in Example 3

Stability. The following is important. If (and only if) all the roots of the characteristic equation of the homogeneous ODE $y'' + ay' + by = 0$ in (4) are negative, or have a negative real part, then a general solution y_h of this ODE goes to 0 as $x \to \infty$, so that the "**transient solution**" $y = y_h + y_p$ of (4) approaches the "**steady-state solution**" y_p. In this case the nonhomogeneous ODE and the physical or other system modeled by the ODE are called **stable**; otherwise they are called **unstable.** For instance, the ODE in Example 1 is unstable.

Applications follow in the next two sections.

PROBLEM SET 2.7

1–10 NONHOMOGENEOUS LINEAR ODEs: GENERAL SOLUTION

Find a (real) general solution. State which rule you are using. Show each step of your work.

1. $y'' + 5y' + 4y = 10e^{-3x}$
2. $10y'' + 50y' + 57.6y = \cos x$
3. $y'' + 3y' + 2y = 12x^2$
4. $y'' - 9y = 18\cos \pi x$
5. $y'' + 4y' + 4y = e^{-x}\cos x$
6. $y'' + y' + (\pi^2 + \tfrac{1}{4})y = e^{-x/2}\sin \pi x$

7. $(D^2 + 2D + \frac{3}{4}I)y = 3e^x + \frac{9}{2}x$
8. $(3D^2 + 27I)y = 3 \cos x + \cos 3x$
9. $(D^2 - 16I)y = 9.6e^{4x} + 30e^x$
10. $(D^2 + 2D + I)y = 2x \sin x$

11–18 NONHOMOGENEOUS LINEAR ODEs: IVPs

Solve the initial value problem. State which rule you are using. Show each step of your calculation in detail.

11. $y'' + 3y = 18x^2, \quad y(0) = -3, \quad y'(0) = 0$
12. $y'' + 4y = -12 \sin 2x, \quad y(0) = 1.8, \quad y'(0) = 5.0$
13. $8y'' - 6y' + y = 6 \cosh x, \quad y(0) = 0.2,$ $y'(0) = 0.05$
14. $y'' + 4y' + 4y = e^{-2x} \sin 2x, \quad y(0) = 1,$ $y'(0) = -1.5$
15. $(x^2D^2 - 3xD + 3I)y = 3 \ln x - 4,$ $y(1) = 0, \quad y'(1) = 1; \quad y_p = \ln x$
16. $(D^2 - 2D)y = 6e^{2x} - 4e^{-2x}, \quad y(0) = -1, \quad y'(0) = 6$
17. $(D^2 + 0.2D + 0.26I)y = 1.22e^{0.5x}, \quad y(0) = 3.5,$ $y'(0) = 0.35$
18. $(D^2 + 2D + 10I)y = 17 \sin x - 37 \sin 3x,$ $y(0) = 6.6, \quad y'(0) = -2.2$
19. **CAS PROJECT. Structure of Solutions of Initial Value Problems.** Using the present method, find, graph, and discuss the solutions y of initial value problems of your own choice. Explore effects on solutions caused by changes of initial conditions. Graph $y_p, y, y - y_p$ separately, to see the separate effects. Find a problem in which **(a)** the part of y resulting from y_h decreases to zero, **(b)** increases, **(c)** is not present in the answer y. Study a problem with $y(0) = 0, y'(0) = 0$. Consider a problem in which you need the Modification Rule (a) for a simple root, (b) for a double root. Make sure that your problems cover all three Cases I, II, III (see Sec. 2.2).
20. **TEAM PROJECT. Extensions of the Method of Undetermined Coefficients. (a)** Extend the method to products of the function in Table 2.1, **(b)** Extend the method to Euler–Cauchy equations. Comment on the practical significance of such extensions.

2.8 Modeling: Forced Oscillations. Resonance

In Sec. 2.4 we considered vertical motions of a mass–spring system (vibration of a mass m on an elastic spring, as in Figs. 33 and 53) and modeled it by the ***homogeneous*** linear ODE

$$my'' + cy' + ky = 0. \tag{1}$$

Here $y(t)$ as a function of time t is the displacement of the body of mass m from rest.

The mass–spring system of Sec. 2.4 exhibited only free motion. This means no external forces (outside forces) but only internal forces controlled the motion. The internal forces are forces within the system. They are the force of inertia my'', the damping force cy' (if $c > 0$), and the spring force ky, a restoring force.

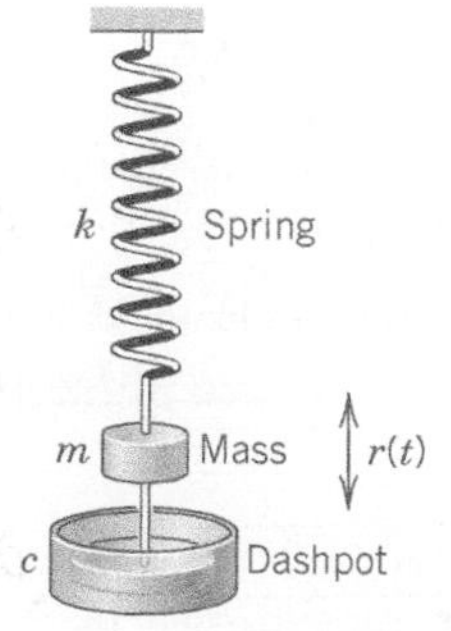

Fig. 53. Mass on a spring

We now extend our model by including an additional force, that is, the external force $r(t)$, on the right. Then we have

$$my'' + cy' + ky = r(t). \tag{2*}$$

Mechanically this means that at each instant t the resultant of the internal forces is in equilibrium with $r(t)$. The resulting motion is called a **forced motion** with **forcing function** $r(t)$, which is also known as **input** or **driving force**, and the solution $y(t)$ to be obtained is called the **output** or the **response** *of the system to the driving force.*

Of special interest are periodic external forces, and we shall consider a driving force of the form

$$r(t) = F_0 \cos \omega t \qquad (F_0 > 0, \omega > 0).$$

Then we have the nonhomogeneous ODE

$$my'' + cy' + ky = F_0 \cos \omega t. \tag{2}$$

Its solution will reveal facts that are fundamental in engineering mathematics and allow us to model resonance.

Solving the Nonhomogeneous ODE (2)

From Sec. 2.7 we know that a general solution of (2) is the sum of a general solution y_h of the homogeneous ODE (1) plus any solution y_p of (2). To find y_p, we use the method of undetermined coefficients (Sec. 2.7), starting from

$$y_p(t) = a \cos \omega t + b \sin \omega t. \tag{3}$$

By differentiating this function (chain rule!) we obtain

$$\begin{aligned} y_p' &= -\omega a \sin \omega t + \omega b \cos \omega t, \\ y_p'' &= -\omega^2 a \cos \omega t - \omega^2 b \sin \omega t. \end{aligned}$$

Substituting y_p, y_p', and y_p'' into (2) and collecting the cosine and the sine terms, we get

$$[(k - m\omega^2)a + \omega cb] \cos \omega t + [-\omega ca + (k - m\omega^2)b] \sin \omega t = F_0 \cos \omega t.$$

The cosine terms on both sides must be equal, and the coefficient of the sine term on the left must be zero since there is no sine term on the right. This gives the two equations

$$\begin{aligned} (k - m\omega^2)a + \quad \omega cb \quad &= F_0 \\ -\omega ca \quad + (k - m\omega^2)b &= 0 \end{aligned} \tag{4}$$

for determining the unknown coefficients a and b. This is a linear system. We can solve it by elimination. To eliminate b, multiply the first equation by $k - m\omega^2$ and the second by $-\omega c$ and add the results, obtaining

$$(k - m\omega^2)^2 a + \omega^2 c^2 a = F_0(k - m\omega^2).$$

Similarly, to eliminate a, multiply (the first equation by ωc and the second by $k - m\omega^2$ and add to get

$$\omega^2 c^2 b + (k - m\omega^2)^2 b = F_0 \omega c.$$

If the factor $(k - m\omega^2)^2 + \omega^2 c^2$ is not zero, we can divide by this factor and solve for a and b,

$$a = F_0 \frac{k - m\omega^2}{(k - m\omega^2)^2 + \omega^2 c^2}, \qquad b = F_0 \frac{\omega c}{(k - m\omega^2)^2 + \omega^2 c^2}.$$

If we set $\sqrt{k/m} = \omega_0 \,(> 0)$ as in Sec. 2.4, then $k = m\omega_0^2$ and we obtain

$$\textbf{(5)} \qquad a = F_0 \frac{m(\omega_0^2 - \omega^2)}{m^2(\omega_0^2 - \omega^2)^2 + \omega^2 c^2}, \qquad b = F_0 \frac{\omega c}{m^2(\omega_0^2 - \omega^2)^2 + \omega^2 c^2}.$$

We thus obtain the general solution of the nonhomogeneous ODE (2) in the form

$$(6) \qquad y(t) = y_h(t) + y_p(t).$$

Here y_h is a general solution of the homogeneous ODE (1) and y_p is given by (3) with coefficients (5).

We shall now discuss the behavior of the mechanical system, distinguishing between the two cases $c = 0$ (no damping) and $c > 0$ (damping). These cases will correspond to two basically different types of output.

Case 1. Undamped Forced Oscillations. Resonance

If the damping of the physical system is so small that its effect can be neglected over the time interval considered, we can set $c = 0$. Then (5) reduces to $a = F_0/[m(\omega_0^2 - \omega^2)]$ and $b = 0$. Hence (3) becomes (use $\omega_0{}^2 = k/m$)

$$(7) \qquad y_p(t) = \frac{F_0}{m(\omega_0^2 - \omega^2)} \cos \omega t = \frac{F_0}{k[1 - (\omega/\omega_0)^2]} \cos \omega t.$$

Here we must assume that $\omega^2 \neq \omega_0{}^2$; physically, the frequency $\omega/(2\pi)$ [cycles/sec] of the driving force is different from the *natural frequency* $\omega_0/(2\pi)$ of the system, which is the frequency of the free undamped motion [see (4) in Sec. 2.4]. From (7) and from (4*) in Sec. 2.4 we have the general solution of the "undamped system"

$$\textbf{(8)} \qquad y(t) = C \cos(\omega_0 t - \delta) + \frac{F_0}{m(\omega_0^2 - \omega^2)} \cos \omega t.$$

We see that this output is a ***superposition of two harmonic oscillations*** *of the frequencies just mentioned.*

Resonance. We discuss (7). We see that the maximum amplitude of y_p is (put $\cos \omega t = 1$)

$$a_0 = \frac{F_0}{k} \rho \quad \text{where} \quad \rho = \frac{1}{1 - (\omega/\omega_0)^2}. \tag{9}$$

a_0 depends on ω and ω_0. If $\omega \to \omega_0$, then ρ and a_0 tend to infinity. This excitation of large oscillations by matching input and natural frequencies ($\omega = \omega_0$) is called **resonance.** ρ is called the **resonance factor** (Fig. 54), and from (9) we see that $\rho/k = a_0/F_0$ is the ratio of the amplitudes of the particular solution y_p and of the input $F_0 \cos \omega t$. We shall see later in this section that resonance is of basic importance in the study of vibrating systems.

In the case of resonance the nonhomogeneous ODE (2) becomes

$$y'' + \omega_0^2 y = \frac{F_0}{m} \cos \omega_0 t. \tag{10}$$

Then (7) is no longer valid, and, from the Modification Rule in Sec. 2.7, we conclude that a particular solution of (10) is of the form

$$y_p(t) = t(a \cos \omega_0 t + b \sin \omega_0 t).$$

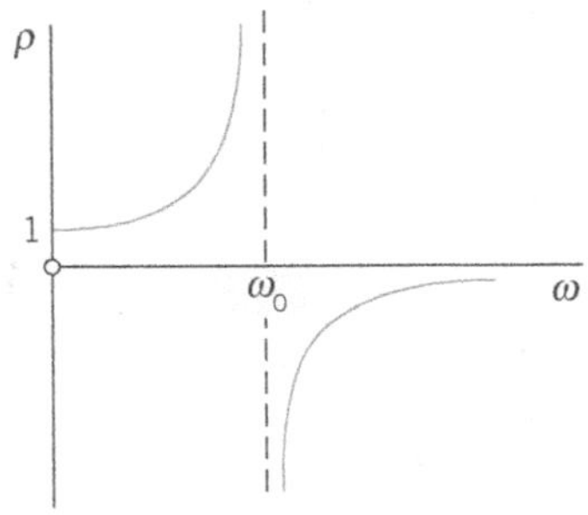

Fig. 54. Resonance factor $\rho(\omega)$

By substituting this into (10) we find $a = 0$ and $b = F_0/(2m\omega_0)$. Hence (Fig. 55)

$$y_p(t) = \frac{F_0}{2m\omega_0} t \sin \omega_0 t. \tag{11}$$

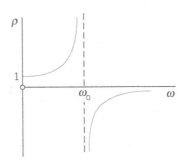

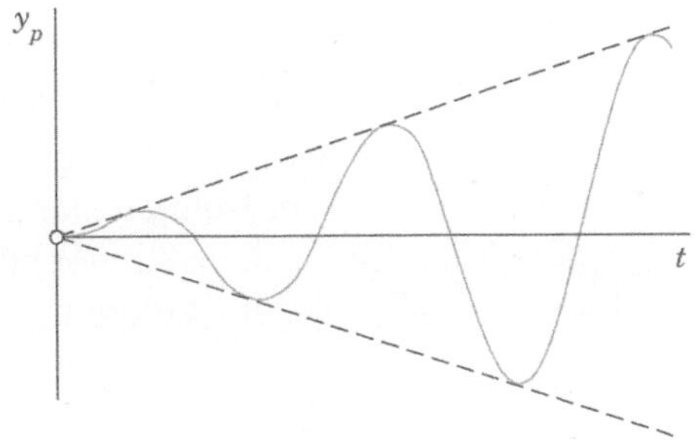

Fig. 55. Particular solution in the case of resonance

We see that, because of the factor t, the amplitude of the vibration becomes larger and larger. Practically speaking, systems with very little damping may undergo large vibrations that can destroy the system. We shall return to this practical aspect of resonance later in this section.

Beats. Another interesting and highly important type of oscillation is obtained if ω is close to ω_0. Take, for example, the particular solution [see (8)]

$$y(t) = \frac{F_0}{m(\omega_0^2 - \omega^2)}(\cos \omega t - \cos \omega_0 t) \qquad (\omega \neq \omega_0). \tag{12}$$

Using (12) in App. 3.1, we may write this as

$$y(t) = \frac{2F_0}{m(\omega_0^2 - \omega^2)} \sin\left(\frac{\omega_0 + \omega}{2}t\right) \sin\left(\frac{\omega_0 - \omega}{2}t\right).$$

Since ω is close to ω_0, the difference $\omega_0 - \omega$ is small. Hence the period of the last sine function is large, and we obtain an oscillation of the type shown in Fig. 56, the dashed curve resulting from the first sine factor. This is what musicians are listening to when they ***tune*** their instruments.

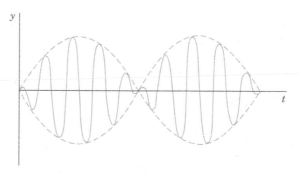

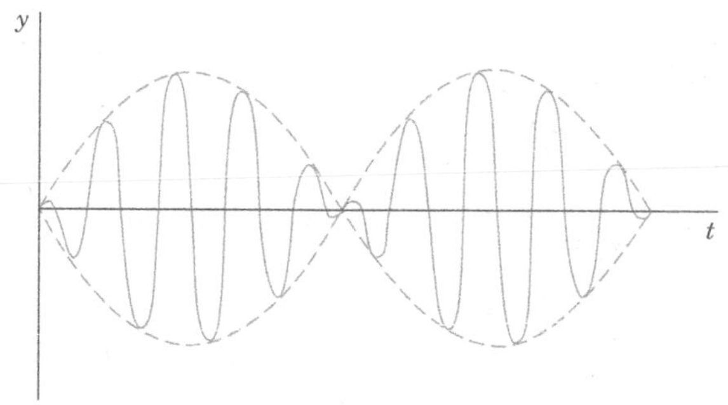

Fig. 56. Forced undamped oscillation when the difference of the input and natural frequencies is small (**"beats"**)

Case 2. Damped Forced Oscillations

If the damping of the mass–spring system is not negligibly small, we have $c > 0$ and a damping term cy' in (1) and (2). Then the general solution y_h of the homogeneous ODE (1) approaches zero as t goes to infinity, as we know from Sec. 2.4. Practically, it is zero after a sufficiently long time. Hence the "**transient solution**" (6) of (2), given by $y = y_h + y_p$, approaches the "**steady-state solution**" y_p. This proves the following.

THEOREM 1 **Steady-State Solution**

After a sufficiently long time the output of a damped vibrating system under a purely sinusoidal driving force [see (2)] will practically be a harmonic oscillation whose frequency is that of the input.

Amplitude of the Steady-State Solution. Practical Resonance

Whereas in the undamped case the amplitude of y_p approaches infinity as ω approaches ω_0, this will not happen in the damped case. In this case the amplitude will always be finite. But it may have a maximum for some ω depending on the damping constant c. This may be called **practical resonance**. It is of great importance because if c is not too large, then some input may excite oscillations large enough to damage or even destroy the system. Such cases happened, in particular in earlier times when less was known about resonance. Machines, cars, ships, airplanes, bridges, and high-rising buildings are vibrating mechanical systems, and it is sometimes rather difficult to find constructions that are completely free of undesired resonance effects, caused, for instance, by an engine or by strong winds.

To study the amplitude of y_p as a function of ω, we write (3) in the form

$$y_p(t) = C^* \cos(\omega t - \eta). \tag{13}$$

C^* is called the **amplitude** of y_p and η the **phase angle** or **phase lag** because it measures the lag of the output behind the input. According to (5), these quantities are

$$\begin{aligned} C^*(\omega) &= \sqrt{a^2 + b^2} = \frac{F_0}{\sqrt{m^2(\omega_0^2 - \omega^2)^2 + \omega^2 c^2}}, \\ \tan \eta(\omega) &= \frac{b}{a} = \frac{\omega c}{m(\omega_0^2 - \omega^2)}. \end{aligned} \tag{14}$$

Let us see whether $C^*(\omega)$ has a maximum and, if so, find its location and then its size. We denote the radicand in the second root in C^* by R. Equating the derivative of C^* to zero, we obtain

$$\frac{dC^*}{d\omega} = F_0\left(-\frac{1}{2}R^{-3/2}\right)[2m^2(\omega_0^2 - \omega^2)(-2\omega) + 2\omega c^2].$$

The expression in the brackets [. . .] is zero if

$$c^2 = 2m^2(\omega_0^2 - \omega^2) \qquad (\omega_0^2 = k/m). \tag{15}$$

By reshuffling terms we have

$$2m^2\omega^2 = 2m^2{\omega_0}^2 - c^2 = 2mk - c^2.$$

The right side of this equation becomes negative if $c^2 > 2mk$, so that then (15) has no real solution and C^* decreases monotone as ω increases, as the lowest curve in Fig. 57 shows. If c is smaller, $c^2 < 2mk$, then (15) has a real solution $\omega = \omega_{max}$, where

$$\omega_{max}^2 = \omega_0^2 - \frac{c^2}{2m^2}. \tag{15*}$$

From (15*) we see that this solution increases as c decreases and approaches ω_0 as c approaches zero. See also Fig. 57.

The size of $C^*(\omega_{\max})$ is obtained from (14), with $\omega^2 = \omega_{\max}^2$ given by (15*). For this ω^2 we obtain in the second radicand in (14) from (15*)

$$m^2(\omega_0^2 - \omega_{\max}^2)^2 = \frac{c^4}{4m^2} \quad \text{and} \quad \omega_{\max}^2 c^2 = \left(\omega_0^2 - \frac{c^2}{2m^2}\right)c^2.$$

The sum of the right sides of these two formulas is

$$(c^4 + 4m^2\omega_0^2 c^2 - 2c^4)/(4m^2) = c^2(4m^2\omega_0^2 - c^2)/(4m^2).$$

Substitution into (14) gives

$$C^*(\omega_{\max}) = \frac{2mF_0}{c\sqrt{4m^2\omega_0^2 - c^2}}. \tag{16}$$

We see that $C^*(\omega_{\max})$ is always finite when $c > 0$. Furthermore, since the expression

$$c^2 4m^2\omega_0^2 - c^4 = c^2(4mk - c^2)$$

in the denominator of (16) decreases monotone to zero as c^2 ($<2mk$) goes to zero, the maximum amplitude (16) increases monotone to infinity, in agreement with our result in Case 1. Figure 57 shows the **amplification** C^*/F_0 (ratio of the amplitudes of output and input) as a function of ω for $m = 1$, $k = 1$, hence $\omega_0 = 1$, and various values of the damping constant c.

Figure 58 shows the phase angle (the lag of the output behind the input), which is less than $\pi/2$ when $\omega < \omega_0$, and greater than $\pi/2$ for $\omega > \omega_0$.

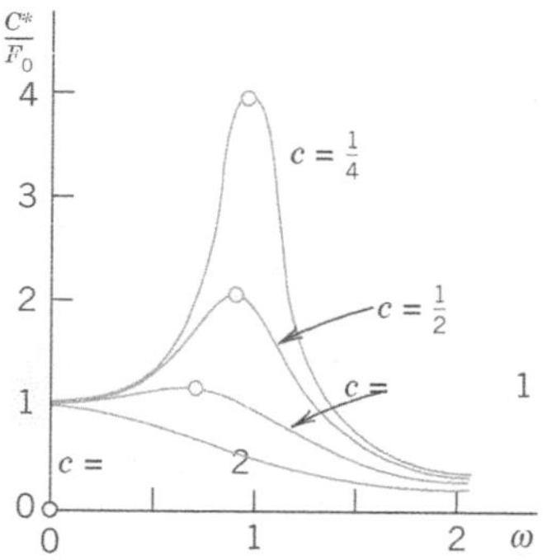

Fig. 57. Amplification C^*/F_0 as a function of ω for $m = 1$, $k = 1$, and various values of the damping constant c

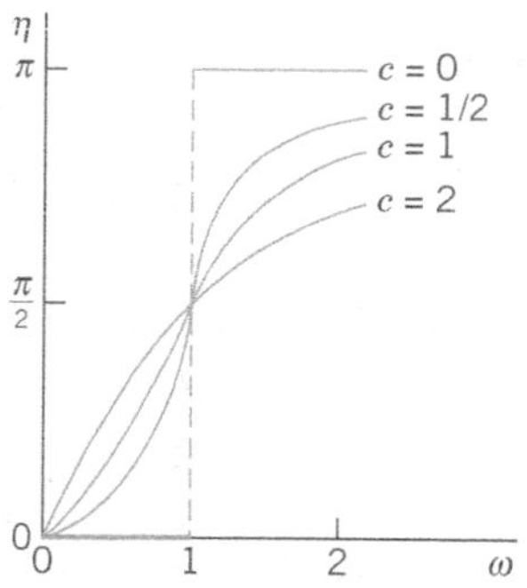

Fig. 58. Phase lag η as a function of ω for $m = 1$, $k = 1$, thus $\omega_0 = 1$, and various values of the damping constant c

PROBLEM SET 2.8

1. **WRITING REPORT. Free and Forced Vibrations.** Write a condensed report of 2–3 pages on the most important similarities and differences of free and forced vibrations, with examples of your own. No proofs.

2. **Which of Probs.** 1–18 in Sec. 2.7 (with x = time t) can be models of mass–spring systems with a harmonic oscillation as steady-state solution?

3–7 STEADY-STATE SOLUTIONS

Find the steady-state motion of the mass–spring system modeled by the ODE. Show the details of your work.

3. $y'' + 6y' + 8y = 42.5 \cos 2t$

4. $y'' + 2.5y' + 10y = -13.6 \sin 4t$

5. $(D^2 + D + 4.25I)y = 22.1 \cos 4.5t$

6. $(D^2 + 4D + 3I)y = \cos t + \frac{1}{3}\cos 3t$

7. $(4D^2 + 12D + 9I)y = 225 - 75 \sin 3t$

8–15 TRANSIENT SOLUTIONS

Find the transient motion of the mass–spring system modeled by the ODE. Show the details of your work.

8. $2y'' + 4y' + 6.5y = 4 \sin 1.5t$

9. $y'' + 3y' + 3.25y = 3 \cos t - 1.5 \sin t$

10. $y'' + 16y = 56 \cos 4t$

11. $(D^2 + 2I)y = \cos \sqrt{2}t + \sin\sqrt{2}t$

12. $(D^2 + 2D + 5I)y = 4 \cos t + 8 \sin t$

13. $(D^2 + I)y = \cos \omega t, \omega^2 \neq 1$

14. $(D^2 + I)y = 5e^{-t} \cos t$

15. $(D^2 + 4D + 8I)y = 2 \cos 2t + \sin 2t$

16–20 INITIAL VALUE PROBLEMS

Find the motion of the mass–spring system modeled by the ODE and the initial conditions. Sketch or graph the solution curve. In addition, sketch or graph the curve of $y - y_p$ to see when the system practically reaches the steady state.

16. $y'' + 25y = 24 \sin t, \quad y(0) = 1, \quad y'(0) = 1$

17. $(D^2 + 4I)y = \sin t + \frac{1}{3} \sin 3t + \frac{1}{5} \sin 5t,$
$y(0) = 0, \quad y'(0) = \frac{3}{35}$

18. $(D^2 + 8D + 17I)y = 474.5 \sin 0.5t, \quad y(0) = -5.4,$
$y'(0) = 9.4$

19. $(D^2 + 2D + 2I)y = e^{-t/2} \sin \frac{1}{2}t, \quad y(0) = 0,$
$y'(0) = 1$

20. $(D^2 + 5I)y = \cos \pi t - \sin \pi t, \quad y(0) = 0, \quad y'(0) = 0$

21. **Beats.** Derive the formula after (12) from (12). Can we have beats in a damped system?

22. **Beats.** Solve $y'' + 25y = 99 \cos 4.9t, \quad y(0) = 2,$ $y'(0) = 0$. How does the graph of the solution change if you change **(a)** $y(0)$, **(b)** the frequency of the driving force?

23. **TEAM EXPERIMENT. Practical Resonance.**
(a) Derive, in detail, the crucial formula (16).
(b) By considering dC^*/dc show that $C^*(\omega_{max})$ increases as c ($\leqq \sqrt{2mk}$) decreases.
(c) Illustrate practical resonance with an ODE of your own in which you vary c, and sketch or graph corresponding curves as in Fig. 57.
(d) Take your ODE with c fixed and an input of two terms, one with frequency close to the practical resonance frequency and the other not. Discuss and sketch or graph the output.
(e) Give other applications (not in the book) in which resonance is important.

24. **Gun barrel.** Solve $y'' + y = 1 - t^2/\pi^2$ if $0 \leqq t \leqq \pi$ and 0 if $t \to \infty$; here, $y(0) = 0, y'(0) = 0$. This models an undamped system on which a force F acts during some interval of time (see Fig. 59), for instance, the force on a gun barrel when a shell is fired, the barrel being braked by heavy springs (and then damped by a dashpot, which we disregard for simplicity). *Hint:* At π both y and y' must be continuous.

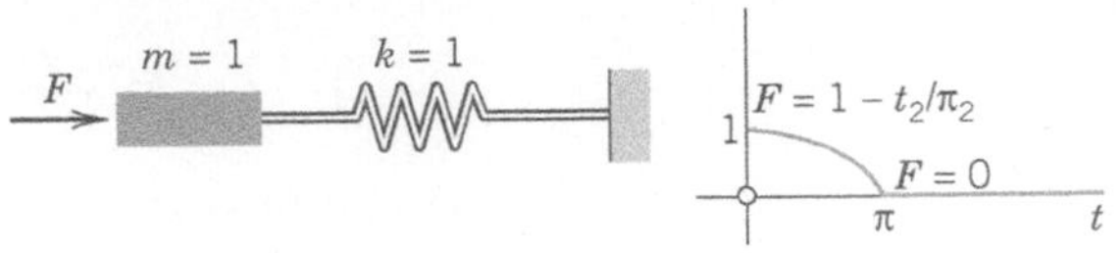

Fig. 59. Problem 24

25. **CAS EXPERIMENT. Undamped Vibrations.**
(a) Solve the initial value problem $y'' + y = \cos \omega t$, $\omega^2 \neq 1, y(0) = 0, y'(0) = 0$. Show that the solution can be written

$$y(t) = \frac{2}{1 - \omega^2} \sin [\tfrac{1}{2}(1 + \omega)t] \sin [\tfrac{1}{2}(1 - \omega)t].$$

(b) Experiment with the solution by changing ω to see the change of the curves from those for small ω (>0) to beats, to resonance, and to large values of ω (see Fig. 60).

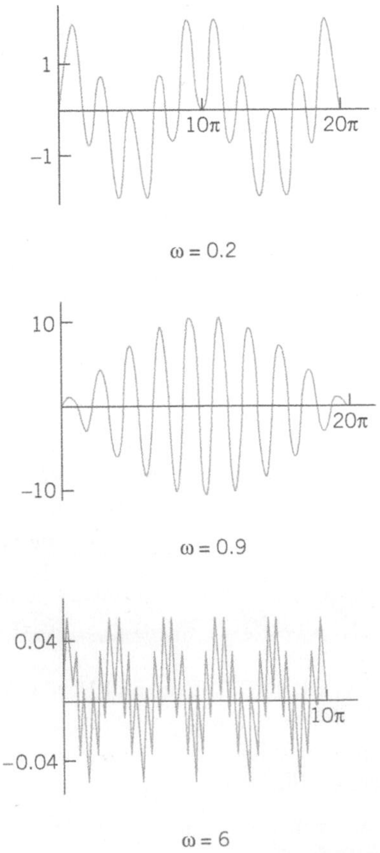

Fig. 60. Typical solution curves in CAS Experiment 25

2.9 Modeling: Electric Circuits

Designing good models is a task the computer cannot do. Hence setting up models has become an important task in modern applied mathematics. The best way to gain experience in successful modeling is to carefully examine the modeling process in various fields and applications. Accordingly, modeling electric circuits will be *profitable for all students*, not just for electrical engineers and computer scientists.

Figure 61 shows an ***RLC*-circuit**, as it occurs as a basic building block of large electric networks in computers and elsewhere. An *RLC*-circuit is obtained from an *RL*-circuit by adding a capacitor. Recall Example 2 on the *RL*-circuit in Sec. 1.5: The model of the *RL*-circuit is $LI' + RI = E(t)$. It was obtained by **KVL** (Kirchhoff's Voltage Law)[7] by equating the voltage drops across the resistor and the inductor to the EMF (electromotive force). Hence we obtain the model of the *RLC*-circuit simply by adding the voltage drop Q/C across the capacitor. Here, C F (farads) is the capacitance of the capacitor. Q coulombs is the charge on the capacitor, related to the current by

$$I(t) = \frac{dQ}{dt}, \qquad \text{equivalently} \qquad Q(t) = \int I(t)\, dt.$$

See also Fig. 62. Assuming a sinusoidal EMF as in Fig. 61, we thus have the model of the *RLC*-circuit

$$E(t) = E_0 \sin \omega t$$

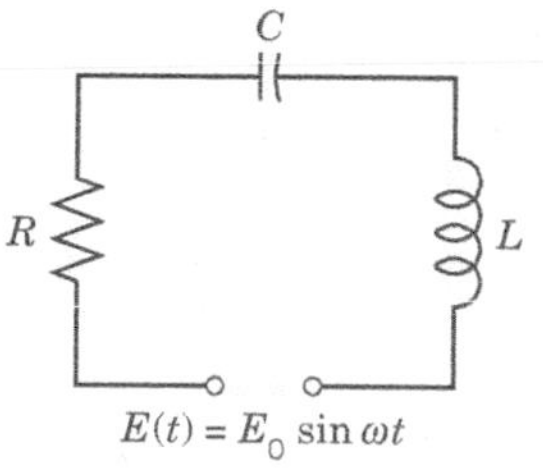

Fig. 61.   *RLC*-circuit

Name	Symbol	Notation		Unit	Voltage Drop
Ohm's Resistor	—\/\/\/—	R	Ohm's Resistance	ohms (Ω)	RI
Inductor	—⌢⌢⌢⌢—	L	Inductance	henrys (H)	$L\dfrac{dI}{dt}$
Capacitor	—⊣⊢—	C	Capacitance	farads (F)	Q/C

Fig. 62.   Elements in an *RLC*-circuit

---

[7]GUSTAV ROBERT KIRCHHOFF (1824–1887), German physicist. Later we shall also need **Kirchhoff's Current Law (KCL):**

At any point of a circuit, the sum of the inflowing currents is equal to the sum of the outflowing currents.

The units of measurement of electrical quantities are named after ANDRÉ MARIE AMPÈRE (1775–1836), French physicist, CHARLES AUGUSTIN DE COULOMB (1736–1806), French physicist and engineer, MICHAEL FARADAY (1791–1867), English physicist, JOSEPH HENRY (1797–1878), American physicist, GEORG SIMON OHM (1789–1854), German physicist, and ALESSANDRO VOLTA (1745–1827), Italian physicist.

$$LI' + RI + \frac{1}{C}\int I\,dt = E(t) = E_0 \sin \omega t. \tag{1'}$$

This is an "integro-differential equation." To get rid of the integral, we differentiate (1′) with respect to t, obtaining

$$LI'' + RI' + \frac{1}{C}I = E'(t) = E_0\omega \cos \omega t. \tag{1}$$

This shows that the current in an RLC-circuit is obtained as the solution of this nonhomogeneous second-order ODE (1) with constant coefficients.

In connection with initial value problems, we shall occasionally use

$$LQ'' + RQ'' + \frac{1}{C}Q = E(t), \tag{1''}$$

obtained from (1′) and $I = Q'$.

Solving the ODE (1) for the Current in an *RLC*-Circuit

A general solution of (1) is the sum $I = I_h + I_p$, where I_h is a general solution of the homogeneous ODE corresponding to (1) and I_p is a particular solution of (1). We first determine I_p by the method of undetermined coefficients, proceeding as in the previous section. We substitute

$$\begin{aligned} I_p &= a \cos \omega t + b \sin \omega t \\ I_p' &= \omega(-a \sin \omega t + b \cos \omega t) \\ I_p'' &= \omega^2(-a \cos \omega t - b \sin \omega t) \end{aligned} \tag{2}$$

into (1). Then we collect the cosine terms and equate them to $E_0\omega \cos \omega t$ on the right, and we equate the sine terms to zero because there is no sine term on the right,

$$L\omega^2(-a) + R\omega b + a/C = E_0\omega \qquad \text{(Cosine terms)}$$

$$L\omega^2(-b) + R\omega(-a) + b/C = 0 \qquad \text{(Sine terms).}$$

Before solving this system for a and b, we first introduce a combination of L and C, called the **reactance**

$$S = \omega L - \frac{1}{\omega C}. \tag{3}$$

Dividing the previous two equations by ω, ordering them, and substituting S gives

$$\begin{aligned} -Sa + Rb &= E_0 \\ -Ra - Sb &= 0. \end{aligned}$$

We now eliminate b by multiplying the first equation by S and the second by R, and adding. Then we eliminate a by multiplying the first equation by R and the second by $-S$, and adding. This gives

$$-(S^2 + R^2)a = E_0 S, \qquad (R^2 + S^2)b = E_0 R.$$

We can solve for a and b,

$$\text{(4)} \qquad a = \frac{-E_0 S}{R^2 + S^2}, \qquad b = \frac{E_0 R}{R^2 + S^2}.$$

Equation (2) with coefficients a and b given by (4) is the desired particular solution I_p of the nonhomogeneous ODE (1) governing the current I in an RLC-circuit with sinusoidal electromotive force.

Using (4), we can write I_p in terms of "physically visible" quantities, namely, amplitude I_0 and phase lag θ of the current behind the EMF, that is,

$$\text{(5)} \qquad I_p(t) = I_0 \sin(\omega t - \theta)$$

where [see (14) in App. A3.1]

$$I_0 = \sqrt{a^2 + b^2} = \frac{E_0}{\sqrt{R^2 + S^2}}, \qquad \tan\theta = -\frac{a}{b} = \frac{S}{R}.$$

The quantity $\sqrt{R^2 + S^2}$ is called the **impedance**. Our formula shows that the impedance equals the ratio E_0/I_0. This is somewhat analogous to $E/I = R$ (Ohm's law) and, because of this analogy, the impedance is also known as the **apparent resistance**.

A general solution of the homogeneous equation corresponding to (1) is

$$I_h = c_1 e^{\lambda_1 t} + c_2 e^{\lambda_2 t}$$

where λ_1 and λ_2 are the roots of the characteristic equation

$$\lambda^2 + \frac{R}{L}\lambda + \frac{1}{LC} = 0.$$

We can write these roots in the form $\lambda_1 = -\alpha + \beta$ and $\lambda_2 = -\alpha - \beta$, where

$$\alpha = \frac{R}{2L}, \qquad \beta = \sqrt{\frac{R^2}{4L^2} - \frac{1}{LC}} = \frac{1}{2L}\sqrt{R^2 - \frac{4L}{C}}.$$

Now in an actual circuit, R is never zero (hence $R > 0$). From this it follows that I_h approaches zero, theoretically as $t \to \infty$, but practically after a relatively short time. Hence the transient current $I = I_h + I_p$ tends to the steady-state current I_p, and after some time the output will practically be a harmonic oscillation, which is given by (5) and whose frequency is that of the input (of the electromotive force).

EXAMPLE 1 ***RLC*-Circuit**

Find the current $I(t)$ in an RLC-circuit with $R = 11\ \Omega$ (ohms), $L = 0.1$ H (henry), $C = 10^{-2}$ F (farad), which is connected to a source of EMF $E(t) = 110 \sin (60 \cdot 2\pi t) = 110 \sin 377\, t$ (hence 60 Hz = 60 cycles/sec, the usual in the U.S. and Canada; in Europe it would be 220 V and 50 Hz). Assume that current and capacitor charge are 0 when $t = 0$.

Solution. ***Step 1. General solution of the homogeneous ODE.*** Substituting R, L, C and the derivative $E'(t)$ into (1), we obtain

$$0.1I'' + 11I' + 100I = 110 \cdot 377 \cos 377t.$$

Hence the homogeneous ODE is $0.1I'' + 11I' + 100I = 0$. Its characteristic equation is

$$0.1\lambda^2 + 11\lambda + 100 = 0.$$

The roots are $\lambda_1 = -10$ and $\lambda_2 = -100$. The corresponding general solution of the homogeneous ODE is

$$I_h(t) = c_1 e^{-10t} + c_2 e^{-100t}.$$

Step 2. Particular solution I_p of (1). We calculate the reactance $S = 37.7 - 0.3 = 37.4$ and the steady-state current

$$I_p(t) = a \cos 377t + b \sin 377t$$

with coefficients obtained from (4) (and rounded)

$$a = \frac{-110 \cdot 37.4}{11^2 + 37.4^2} = -2.71, \qquad b = \frac{110 \cdot 11}{11^2 + 37.4^2} = 0.796.$$

Hence in our present case, a general solution of the nonhomogeneous ODE (1) is

$$(6) \qquad I(t) = c_1 e^{-10t} + c_2 e^{-100t} - 2.71 \cos 377t + 0.796 \sin 377t.$$

Step 3. Particular solution satisfying the initial conditions. **How to use $Q(0) = 0$?** We finally determine c_1 and c_2 from the in initial conditions $I(0) = 0$ and $Q(0) = 0$. From the first condition and (6) we have

$$(7) \qquad I(0) = c_1 + c_2 - 2.71 = 0, \qquad \text{hence} \qquad c_2 = 2.71 - c_1.$$

We turn to $Q(0) = 0$. The integral in (1′) equals $\int I\, dt = Q(t)$; see near the beginning of this section. Hence for $t = 0$, Eq. (1′) becomes

$$LI'(0) + R \cdot 0 = 0, \qquad \text{so that} \qquad I'(0) = 0.$$

Differentiating (6) and setting $t = 0$, we thus obtain

$$I'(0) = -10c_1 - 100c_2 + 0 + 0.796 \cdot 377 = 0, \qquad \text{hence by (7),} \qquad -10c_1 = 100(2.71 - c_1) - 300.1.$$

The solution of this and (7) is $c_1 = -0.323$, $c_2 = 3.033$. Hence the answer is

$$I(t) = -0.323e^{-10t} + 3.033e^{-100t} - 2.71 \cos 377t + 0.796 \sin 377t.$$

You may get slightly different values depending on the rounding. Figure 63 shows $I(t)$ as well as $I_p(t)$, which practically coincide, except for a very short time near $t = 0$ because the exponential terms go to zero very rapidly. Thus after a very short time the current will practically execute harmonic oscillations of the input frequency 60 Hz = 60 cycles/sec. Its maximum amplitude and phase lag can be seen from (5), which here takes the form

$$I_p(t) = 2.824 \sin (377t - 1.29).$$ ■

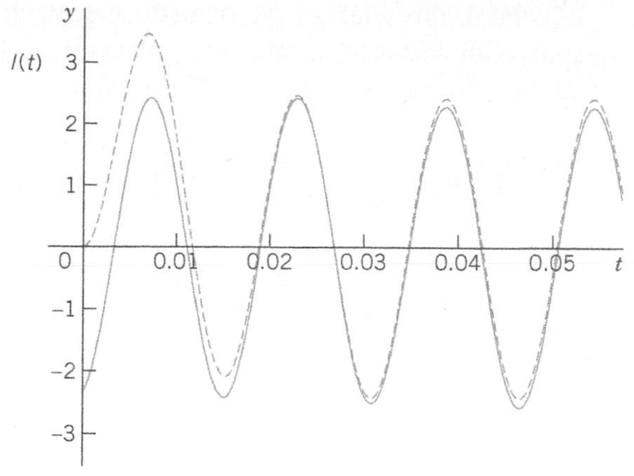

Fig. 63. Transient (upper curve) and steady-state currents in Example 1

Analogy of Electrical and Mechanical Quantities

Entirely different physical or other systems may have the same mathematical model. For instance, we have seen this from the various applications of the ODE $y' = ky$ in Chap. 1. Another impressive demonstration of this ***unifying power of mathematics*** is given by the ODE (1) for an electric *RLC*-circuit and the ODE (2) in the last section for a mass–spring system. Both equations

$$LI'' + RI' + \frac{1}{C}I = E_0\omega \cos \omega t \qquad \text{and} \qquad my'' + cy' + ky = F_0 \cos \omega t$$

are of the same form. Table 2.2 shows the analogy between the various quantities involved. The inductance L corresponds to the mass m and, indeed, an inductor opposes a change in current, having an "inertia effect" similar to that of a mass. The resistance R corresponds to the damping constant c, and a resistor causes loss of energy, just as a damping dashpot does. And so on.

This analogy is ***strictly quantitative*** in the sense that to a given mechanical system we can construct an electric circuit whose current will give the exact values of the displacement in the mechanical system when suitable scale factors are introduced.

The ***practical importance*** of this analogy is almost obvious. The analogy may be used for constructing an "electrical model" of a given mechanical model, resulting in substantial savings of time and money because electric circuits are easy to assemble, and electric quantities can be measured much more quickly and accurately than mechanical ones.

Table 2.2 Analogy of Electrical and Mechanical Quantities

Electrical System	Mechanical System
Inductance L	Mass m
Resistance R	Damping constant c
Reciprocal $1/C$ of capacitance	Spring modulus k
Derivative $E_0\omega \cos \omega t$ of electromotive force	Driving force $F_0 \cos \omega t$
Current $I(t)$	Displacement $y(t)$

Related to this analogy are **transducers**, devices that convert changes in a mechanical quantity (for instance, in a displacement) into changes in an electrical quantity that can be monitored; see Ref. [GenRef11] in App. 1.

PROBLEM SET 2.9

1–6 *RLC*-CIRCUITS: SPECIAL CASES

1. *RC*-Circuit. Model the *RC*-circuit in Fig. 64. Find the current due to a constant E.

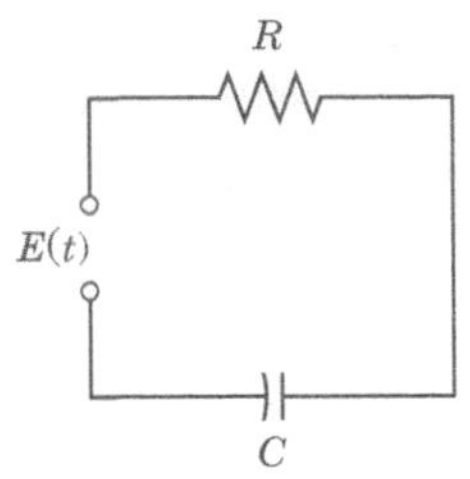

Fig. 64. *RC*-circuit

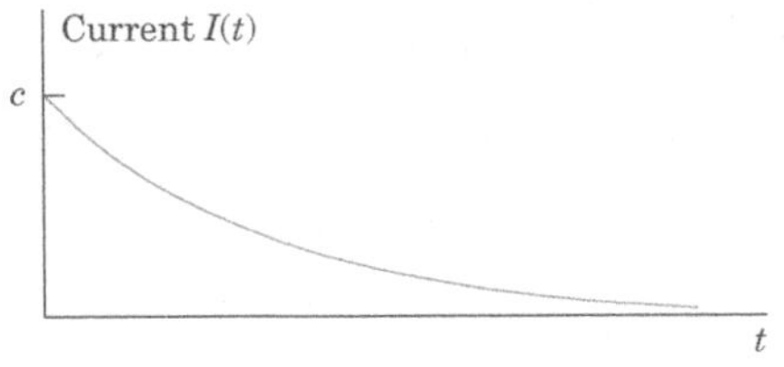

Fig. 65. Current 1 in Problem 1

2. *RC*-Circuit. Solve Prob. 1 when $E = E_0 \sin \omega t$ and R, C, E_0, and ω are arbitrary.

3. *RL*-Circuit. Model the *RL*-circuit in Fig. 66. Find a general solution when R, L, E are any constants. Graph or sketch solutions when $L = 0.25$ H, $R = 10\ \Omega$, and $E = 48$ V.

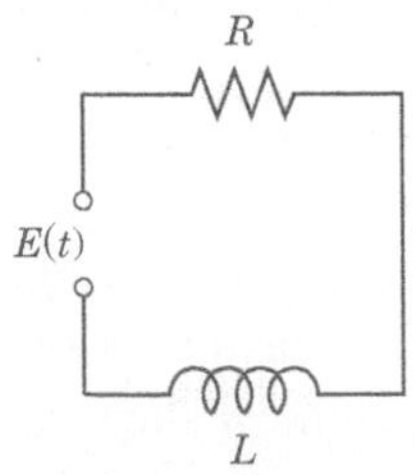

Fig. 66. *RL*-circuit

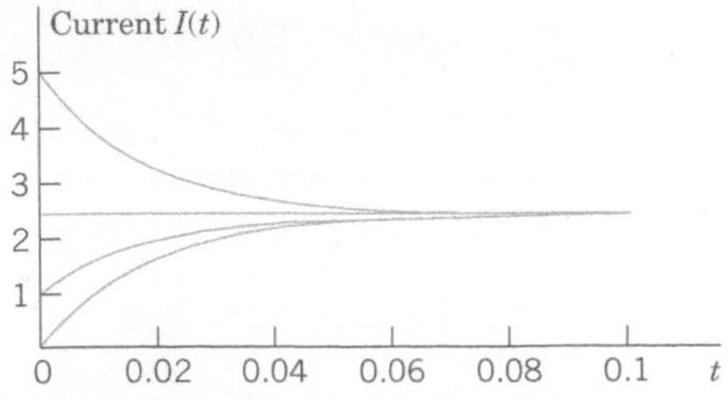

Fig. 67. Currents in Problem 3

4. *RL*-Circuit. Solve Prob. 3 when $E = E_0 \sin \omega t$ and R, L, E_0, and are arbitrary. Sketch a typical solution.

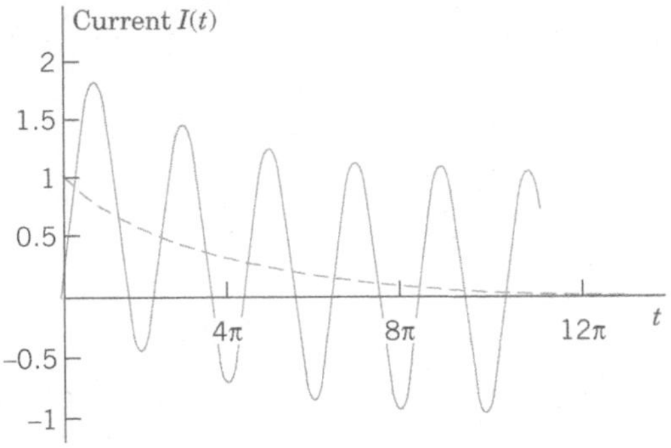

Fig. 68. Typical current $I = e^{-0.1t} + \sin(t - \frac{1}{4}\pi)$ in Problem 4

5. *LC*-Circuit. This is an *RLC*-circuit with negligibly small R (analog of an undamped mass–spring system). Find the current when $L = 0.5$ H, $C = 0.005$ F, and $E = \sin t$ V, assuming zero initial current and charge.

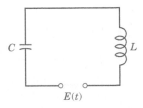

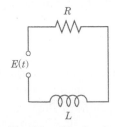

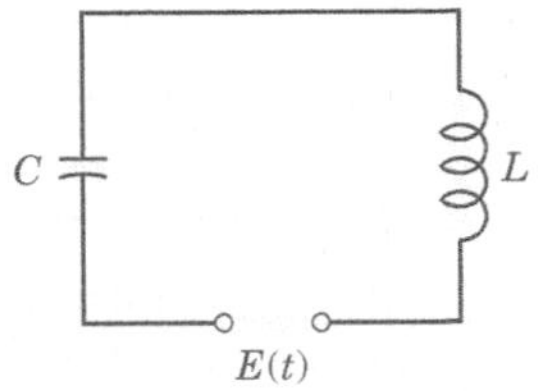

Fig. 69. *LC*-circuit

6. *LC*-Circuit. Find the current when $L = 0.5$ H, $C = 0.005$ F, $E = 2t^2$ V, and initial current and charge zero.

7–18 GENERAL *RLC*-CIRCUITS

7. Tuning. In tuning a stereo system to a radio station, we adjust the tuning control (turn a knob) that changes C (or perhaps L) in an *RLC*-circuit so that the amplitude of the steady-state current (5) becomes maximum. For what C will this happen?

8–14 Find the **steady-state current** in the *RLC*-circuit in Fig. 61 for the given data. Show the details of your work.

8. $R = 4\ \Omega$, $L = 0.5$ H, $C = 0.1$ F, $E = 500 \sin 2t$ V

9. $R = 4\ \Omega$, $L = 0.1$ H, $C = 0.05$ F, $E = 110$ V

10. $R = 2\ \Omega$, $L = 1$ H, $C = \frac{1}{20}$ F, $E = 157 \sin 3t$ V

11. $R = 12\ \Omega$, $L = 0.4$ H, $C = \frac{1}{80}$ F, $E = 220 \sin 10t$ V

12. $R = 0.2\ \Omega$, $L = 0.1$ H, $C = 2$ F, $E = 220 \sin 314t$ V

13. $R = 12$, $L = 1.2$ H, $C = \frac{20}{3} \cdot 10^{-3}$ F, $E = 12{,}000 \sin 25t$ V

14. Prove the claim in the text that if $R \neq 0$ (hence $R > 0$), then the transient current approaches I_p as $t \to \infty$.

15. Cases of damping. What are the conditions for an *RLC*-circuit to be (I) overdamped, (II) critically damped, (III) underdamped? What is the critical resistance R_{crit} (the analog of the critical damping constant $2\sqrt{mk}$)?

16–18 Solve the **initial value problem** for the *RLC*-circuit in Fig. 61 with the given data, assuming zero initial current and charge. Graph or sketch the solution. Show the details of your work.

16. $R = 8\ \Omega$, $L = 0.2$ H, $C = 12.5 \cdot 10^{-3}$ F, $E = 100 \sin 10t$ V

17. $R = 6\ \Omega$, $L = 1$ H, $C = 0.04$ F, $E = 600 (\cos t + 4 \sin t)$ V

18. $R = 18\ \Omega$, $L = 1$ H, $C = 12.5 \cdot 10^{-3}$ F, $E = 820 \cos 10t$ V

19. WRITING REPORT. Mechanic-Electric Analogy. Explain Table 2.2 in a 1–2 page report with examples, e.g., the analog (with $L = 1$ H) of a mass–spring system of mass 5 kg, damping constant 10 kg/sec, spring constant 60 kg/sec^2, and driving force $220 \cos 10t$ kg/sec.

20. Complex Solution Method. Solve $L\tilde{I}'' + R\tilde{I}' + \tilde{I}/C = E_0 e^{i\omega t}$, $i = \sqrt{-1}$, by substituting $I_p = Ke^{i\omega t}$ (K unknown) and its derivatives and taking the real part I_p of the solution $\tilde{I}_p$. Show agreement with (2), (4). *Hint:* Use (11) $e^{i\omega t} = \cos \omega t + i \sin \omega t$; cf. Sec. 2.2, and $i^2 = -1$.

2.10 Solution by Variation of Parameters

We continue our discussion of nonhomogeneous linear ODEs, that is

$$y'' + p(x)y' + q(x)y = r(x). \tag{1}$$

In Sec. 2.6 we have seen that a general solution of (1) is the sum of a general solution y_h of the corresponding homogeneous ODE and any particular solution y_p of (1). To obtain y_p when $r(x)$ is not too complicated, we can often use the *method of undetermined coefficients,* as we have shown in Sec. 2.7 and applied to basic engineering models in Secs. 2.8 and 2.9.

However, since this method is restricted to functions $r(x)$ whose derivatives are of a form similar to $r(x)$ itself (powers, exponential functions, etc.), it is desirable to have a method valid for more general ODEs (1), which we shall now develop. It is called the **method of variation of parameters** and is credited to Lagrange (Sec. 2.1). Here p, q, r in (1) may be variable (given functions of x), but we assume that they are continuous on some open interval I.

Lagrange's method gives a particular solution y_p of (1) on I in the form

$$y_p(x) = -y_1 \int \frac{y_2 r}{W}\, dx + y_2 \int \frac{y_1 r}{W}\, dx \tag{2}$$

where y_1, y_2 form a basis of solutions of the corresponding homogeneous ODE

$$y'' + p(x)y' + q(x)y = 0 \tag{3}$$

on I, and W is the Wronskian of y_1, y_2,

$$W = y_1 y_2' - y_2 y_1' \qquad \text{(see Sec. 2.6).} \tag{4}$$

CAUTION! The solution formula (2) is obtained under the assumption that the ODE is written in standard form, with y'' as the first term as shown in (1). If it starts with $f(x)y''$, divide first by $f(x)$.

The integration in (2) may often cause difficulties, and so may the determination of y_1, y_2 if (1) has variable coefficients. If you have a choice, use the previous method. It is simpler. Before deriving (2) let us work an example for which you *do need* the new method. (Try otherwise.)

EXAMPLE 1 Method of Variation of Parameters

Solve the nonhomogeneous ODE

$$y'' + y = \sec x = \frac{1}{\cos x}.$$

Solution. A basis of solutions of the homogeneous ODE on any interval is $y_1 = \cos x$, $y_2 = \sin x$. This gives the Wronskian

$$W(y_1, y_2) = \cos x \cos x - \sin x\,(-\sin x) = 1.$$

From (2), choosing zero constants of integration, we get the particular solution of the given ODE

$$\begin{aligned} y_p &= -\cos x \int \sin x \sec x\,dx + \sin x \int \cos x \sec x\,dx \\ &= \cos x \ln|\cos x| + x \sin x \qquad \text{(Fig. 70)} \end{aligned}$$

Figure 70 shows y_p and its first term, which is small, so that $x \sin x$ essentially determines the shape of the curve of y_p. (Recall from Sec. 2.8 that we have seen $x \sin x$ in connection with resonance, except for notation.) From y_p and the general solution $y_h = c_1y_1 + c_2y_2$ of the homogeneous ODE we obtain the *answer*

$$y = y_h + y_p = (c_1 + \ln|\cos x|) \cos x + (c_2 + x) \sin x.$$

Had we included integration constants $-c_1, c_2$ in (2), then (2) would have given the additional $c_1 \cos x + c_2 \sin x = c_1y_1 + c_2y_2$, that is, a general solution of the given ODE directly from (2). This will always be the case. ■

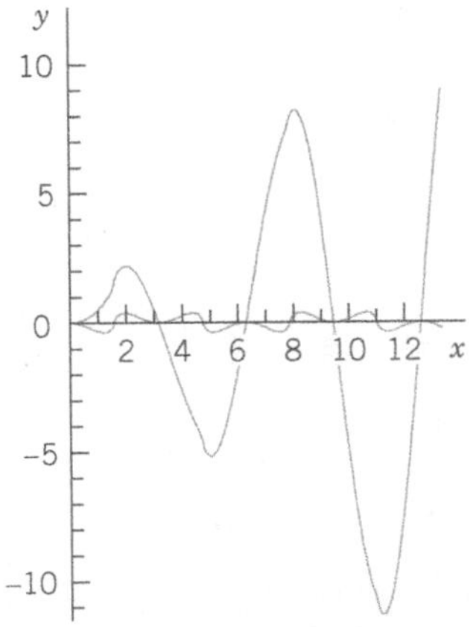

Fig. 70. Particular solution y_p and its first term in Example 1

Idea of the Method. Derivation of (2)

What idea did Lagrange have? What gave the method the name? Where do we use the continuity assumptions?

The idea is to start from a general solution

$$y_h(x) = c_1y_1(x) + c_2y_2(x)$$

of the homogeneous ODE (3) on an open interval I and to replace the constants ("the parameters") c_1 and c_2 by functions $u(x)$ and $v(x)$; this suggests the name of the method. We shall determine u and v so that the resulting function

$$y_p(x) = u(x)y_1(x) + v(x)y_2(x) \tag{5}$$

is a particular solution of the nonhomogeneous ODE (1). Note that y_h exists by Theorem 3 in Sec. 2.6 because of the continuity of p and q on I. (The continuity of r will be used later.)

We determine u and v by substituting (5) and its derivatives into (1). Differentiating (5), we obtain

$$y_p' = u'y_1 + uy_1' + v'y_2 + vy_2'.$$

Now y_p must satisfy (1). This is *one* condition for *two* functions u and v. It seems plausible that we may impose a *second* condition. Indeed, our calculation will show that we can determine u and v such that y_p satisfies (1) and u and v satisfy as a second condition the equation

$$u'y_1 + v'y_2 = 0. \tag{6}$$

This reduces the first derivative y_p' to the simpler form

$$y_p' = uy_1' + vy_2'. \tag{7}$$

Differentiating (7), we obtain

$$y_p'' = u'y_1' + uy_1'' + v'y_2' + vy_2''. \tag{8}$$

We now substitute y_p and its derivatives according to (5), (7), (8) into (1). Collecting terms in u and terms in v, we obtain

$$u(y_1'' + py_1' + qy_1) + v(y_2'' + py_2' + qy_2) + u'y_1' + v'y_2' = r.$$

Since y_1 and y_2 are solutions of the homogeneous ODE (3), this reduces to

$$u'y_1' + v'y_2' = r. \tag{9a}$$

Equation (6) is

$$u'y_1 + v'y_2 = 0. \tag{9b}$$

This is a linear system of two algebraic equations for the unknown functions u' and v'. We can solve it by elimination as follows (or by Cramer's rule in Sec. 7.6). To eliminate v', we multiply (9a) by $-y_2$ and (9b) by y_2' and add, obtaining

$$u'(y_1y_2' - y_2y_1') = -y_2r, \qquad \text{thus} \qquad u'W = -y_2r.$$

Here, W is the Wronskian (4) of y_1, y_2. To eliminate u' we multiply (9a) by y_1, and (9b) by $-y_1'$ and add, obtaining

$$v'(y_1y_2' - y_2y_1') = -y_1r, \qquad \text{thus} \qquad v'W = y_1r.$$

Since y_1, y_2 form a basis, we have $W \neq 0$ (by Theorem 2 in Sec. 2.6) and can divide by W,

$$u' = -\frac{y_2r}{W}, \qquad v' = \frac{y_1r}{W}. \tag{10}$$

By integration,

$$u = -\int \frac{y_2r}{W}\,dx, \qquad v = \int \frac{y_1r}{W}\,dx.$$

These integrals exist because $r(x)$ is continuous. Inserting them into (5) gives (2) and completes the derivation. ■

PROBLEM SET 2.10

1–13 GENERAL SOLUTION

Solve the given nonhomogeneous linear ODE by variation of parameters or undetermined coefficients. Show the details of your work.

1. $y'' + 9y = \sec 3x$
2. $y'' + 9y = \csc 3x$
3. $x^2y'' - 2xy' + 2y = x^3 \sin x$
4. $y'' - 4y' + 5y = e^{2x} \csc x$
5. $y'' + y = \cos x - \sin x$
6. $(D^2 + 6D + 9I)y = 16e^{-3x}/(x^2 + 1)$
7. $(D^2 - 4D + 4I)y = 6e^{2x}/x^4$
8. $(D^2 + 4I)y = \cosh 2x$
9. $(D^2 - 2D + I)y = 35x^{3/2}e^x$
10. $(D^2 + 2D + 2I)y = 4e^{-x} \sec^3 x$
11. $(x^2D^2 - 4xD + 6I)y = 21x^{-4}$
12. $(D^2 - I)y = 1/\cosh x$
13. $(x^2D^2 + xD - 9I)y = 48x^5$
14. **TEAM PROJECT. Comparison of Methods. Invention.** The undetermined-coefficient method should be used whenever possible because it is simpler. Compare it with the present method as follows.

    (a) Solve $y'' + 4y' + 3y = 65 \cos 2x$ by both methods, showing all details, and compare.

    (b) Solve $y'' - 2y' + y = r_1 + r_2$, $r_1 = 35x^{3/2}e^x$ $r_2 = x^2$ by applying each method to a suitable function on the right.

    (c) Experiment to invent an undetermined-coefficient method for nonhomogeneous Euler–Cauchy equations.

CHAPTER 2 REVIEW QUESTIONS AND PROBLEMS

1. Why are linear ODEs preferable to nonlinear ones in modeling?
2. What does an initial value problem of a second-order ODE look like? Why must you have a general solution to solve it?
3. By what methods can you get a general solution of a nonhomogeneous ODE from a general solution of a homogeneous one?
4. Describe applications of ODEs in mechanical systems. What are the electrical analogs of the latter?
5. What is resonance? How can you remove undesirable resonance of a construction, such as a bridge, a ship, or a machine?
6. What do you know about existence and uniqueness of solutions of linear second-order ODEs?

7–18 GENERAL SOLUTION

Find a general solution. Show the details of your calculation.

7. $4y'' + 32y' + 63y = 0$
8. $y'' + y' - 12y = 0$
9. $y'' + 6y' + 34y = 0$
10. $y'' + 0.20y' + 0.17y = 0$
11. $(100D^2 - 160D + 64I)y = 0$
12. $(D^2 + 4\pi D + 4\pi^2 I)y = 0$
13. $(x^2D^2 + 2xD - 12I)y = 0$
14. $(x^2D^2 + xD - 9I)y = 0$
15. $(2D^2 - 3D - 2I)y = 13 - 2x^2$
16. $(D^2 + 2D + 2I)y = 3e^{-x} \cos 2x$
17. $(4D^2 - 12D + 9I)y = 2e^{1.5x}$
18. $yy'' = 2y'^2$

19–22 INITIAL VALUE PROBLEMS

Solve the problem, showing the details of your work. Sketch or graph the solution.

19. $y'' + 16y = 17e^x, \quad y(0) = 6, \quad y'(0) = -2$

20. $y'' - 3y' + 2y = 10 \sin x, \quad y(0) = 1, \quad y'(0) = -6$

21. $(x^2D^2 + xD - I)y = 16x^3, \quad y(1) = -1, \quad y'(1) = 1$

22. $(x^2D^2 + 15xD + 49I)y = 0, \quad y(1) = 2,$ $y'(1) = -11$

23–30 APPLICATIONS

23. Find the steady-state current in the *RLC*-circuit in Fig. 71 when $R = 2\text{ k}\Omega\ (2000\ \Omega), L = 1\text{ H}, C = 4 \cdot 10^{-3}\text{ F}$, and $E = 110 \sin 415t$ V (66 cycles/sec).

24. Find a general solution of the homogeneous linear ODE corresponding to the ODE in Prob. 23.

25. Find the steady-state current in the *RLC*-circuit in Fig. 71 when $R = 50\ \Omega, L = 30\text{ H}, C = 0.025\text{ F}$, $E = 200 \sin 4t$ V.

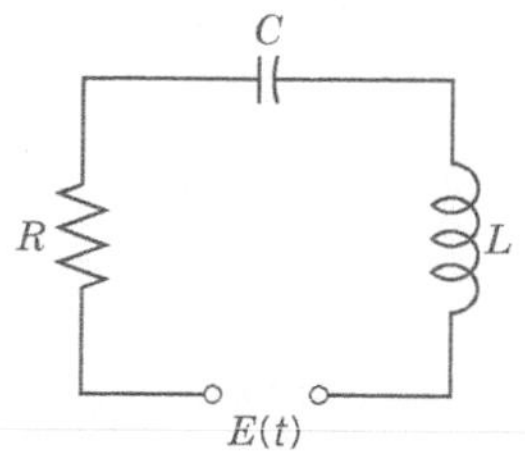

Fig. 71. *RLC*-circuit

26. Find the current in the *RLC*-circuit in Fig. 71 when $R = 40\ \Omega, \quad L = 0.4\text{ H}, \quad C = 10^{-4}\text{ F}, \quad E =$ $220 \sin 314t$ V (50 cycles/sec).

27. Find an electrical analog of the mass–spring system with mass 4 kg, spring constant 10 kg/sec^2, damping constant 20 kg/sec, and driving force $100 \sin 4t$ nt.

28. Find the motion of the mass–spring system in Fig. 72 with mass 0.125 kg, damping 0, spring constant 1.125 kg/sec^2, and driving force $\cos t - 4 \sin t$ nt, assuming zero initial displacement and velocity. For what frequency of the driving force would you get resonance?

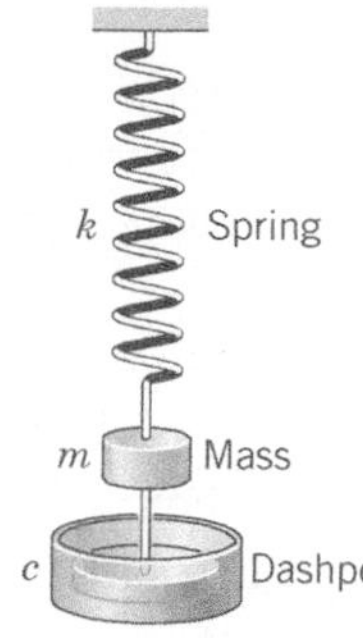

Fig. 72. Mass–spring system

29. Show that the system in Fig. 72 with $m = 4, c = 0,$ $k = 36$, and driving force $61 \cos 3.1t$ exhibits beats. *Hint:* Choose zero initial conditions.

30. In Fig. 72, let $m = 1$ kg, $c = 4$ kg/sec, $k = 24$ kg/sec^2, and $r(t) = 10 \cos \omega t$ nt. Determine w such that you get the steady-state vibration of maximum possible amplitude. Determine this amplitude. Then find the general solution with this ω and check whether the results are in agreement.

SUMMARY OF CHAPTER 2
Second-Order Linear ODEs

Second-order linear ODEs are particularly important in applications, for instance, in mechanics (Secs. 2.4, 2.8) and electrical engineering (Sec. 2.9). A second-order ODE is called **linear** if it can be written

$$y'' + p(x)y' + q(x)y = r(x) \qquad \text{(Sec. 2.1).} \tag{1}$$

(If the first term is, say, $f(x)y''$, divide by $f(x)$ to get the "**standard form**" (1) with y'' as the first term.) Equation (1) is called **homogeneous** if $r(x)$ is zero for all x considered, usually in some open interval; this is written $r(x) \equiv 0$. Then

$$y'' + p(x)y' + q(x)y = 0. \tag{2}$$

Equation (1) is called **nonhomogeneous** if $r(x) \not\equiv 0$ (meaning $r(x)$ is not zero for some x considered).

For the homogeneous ODE (2) we have the important **superposition principle** (Sec. 2.1) that a linear combination $y = ky_1 + ly_2$ of two solutions y_1, y_2 is again a solution.

Two *linearly independent* solutions y_1, y_2 of (2) on an open interval I form a **basis** (or **fundamental system**) of solutions on I. and $y = c_1y_1 + c_2y_2$ with arbitrary constants c_1, c_2 a **general solution** of (2) on I. From it we obtain a **particular solution** if we specify numeric values (numbers) for c_1 and c_2, usually by prescribing two **initial conditions**

$$y(x_0) = K_0, \qquad y'(x_0) = K_1 \qquad (x_0, K_0, K_1 \text{ given numbers; Sec. 2.1}). \tag{3}$$

(2) and (3) together form an **initial value problem**. Similarly for (1) and (3).

For a nonhomogeneous ODE (1) a **general solution** is of the form

$$y = y_h + y_p \qquad \text{(Sec. 2.7).} \tag{4}$$

Here y_h is a general solution of (2) and y_p is a particular solution of (1). Such a y_p can be determined by a general method (***variation of parameters***, Sec. 2.10) or in many practical cases by the ***method of undetermined coefficients***. The latter applies when (1) has constant coefficients p and q, and $r(x)$ is a power of x, sine, cosine, etc. (Sec. 2.7). Then we write (1) as

$$y'' + ay' + by = r(x) \qquad \text{(Sec. 2.7).} \tag{5}$$

The corresponding homogeneous ODE $y' + ay' + by = 0$ has solutions $y = e^{\lambda x}$, where λ is a root of

$$\lambda^2 + a\lambda + b = 0. \tag{6}$$

Hence there are three cases (Sec. 2.2):

Case	Type of Roots	General Solution
I	Distinct real λ_1, λ_2	$y = c_1e^{\lambda_1 x} + c_2e^{\lambda_2 x}$
II	Double $-\frac{1}{2}a$	$y = (c_1 + c_2x)e^{-ax/2}$
III	Complex $-\frac{1}{2}a \pm i\omega^*$	$y = e^{-ax/2}(A\cos\omega^* x + B\sin\omega^* x)$

Here ω^* is used since ω is needed in driving forces.

Important applications of (5) in mechanical and electrical engineering in connection with ***vibrations*** and ***resonance*** are discussed in Secs. 2.4, 2.7, and 2.8.

Another large class of ODEs solvable "algebraically" consists of the **Euler–Cauchy equations**

$$x^2y'' + axy' + by = 0 \qquad \text{(Sec. 2.5).} \tag{7}$$

These have solutions of the form $y = x^m$, where m is a solution of the auxiliary equation

$$m^2 + (a - 1)m + b = 0. \tag{8}$$

Existence and uniqueness of solutions of (1) and (2) is discussed in Secs. 2.6 and 2.7, and ***reduction of order*** in Sec. 2.1.

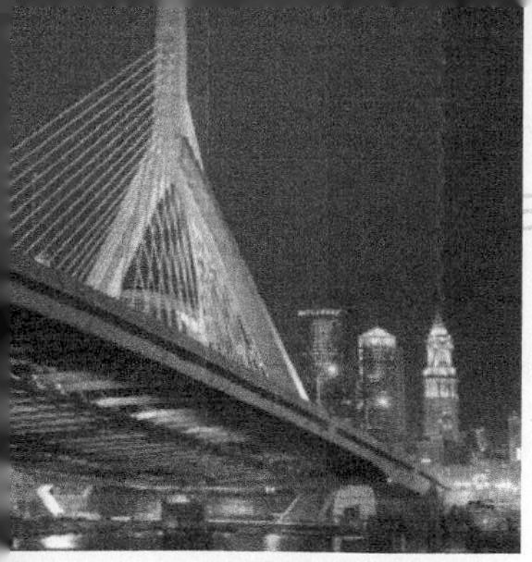

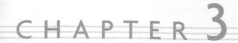

CHAPTER 3

Higher Order Linear ODEs

The concepts and methods of solving linear ODEs of order $n = 2$ extend nicely to linear ODEs of higher order n, that is, $n = 3, 4$, etc. This shows that the theory explained in Chap. 2 for second-order linear ODEs is attractive, since it can be extended in a straightforward way to arbitrary n. We do so in this chapter and notice that the formulas become more involved, the variety of roots of the characteristic equation (in Sec. 3.2) becomes much larger with increasing n, and the Wronskian plays a more prominent role.

The concepts and methods of solving second-order linear ODEs extend readily to linear ODEs of higher order.

This chapter follows Chap. 2 naturally, since the results of Chap. 2 can be readily extended to that of Chap. 3.

Prerequisite: Secs. 2.1, 2.2, 2.6, 2.7, 2.10.
References and Answers to Problems: App. 1 Part A, and App. 2.

3.1 Homogeneous Linear ODEs

Recall from Sec. 1.1 that an ODE is of ***n*th order** if the nth derivative $y^{(n)} = d^n y/dx^n$ of the unknown function $y(x)$ is the highest occurring derivative. Thus the ODE is of the form

$$F(x, y, y', \cdots, y^{(n)}) = 0$$

where lower order derivatives and y itself may or may not occur. Such an ODE is called **linear** if it can be written

$$y^{(n)} + p_{n-1}(x)y^{(n-1)} + \cdots + p_1(x)y' + p_0(x)y = r(x). \tag{1}$$

(For $n = 2$ this is (1) in Sec. 2.1 with $p_1 = p$ and $p_0 = q$.) The **coefficients** $p_0, \cdots, p_{n-1}$ and the function r on the right are any given functions of x, and y is unknown. $y^{(n)}$ has coefficient 1. We call this the **standard form**. (If you have $p_n(x)y^{(n)}$, divide by $p_n(x)$ to get this form.) An nth-order ODE that cannot be written in the form (1) is called **nonlinear**.

If $r(x)$ is identically zero, $r(x) \equiv 0$ (zero for all x considered, usually in some open interval I), then (1) becomes

$$y^{(n)} + p_{n-1}(x)y^{(n-1)} + \cdots + p_1(x)y' + p_0(x)y = 0 \tag{2}$$

and is called **homogeneous**. If $r(x)$ is not identically zero, then the ODE is called **nonhomogeneous**. This is as in Sec. 2.1.

A **solution** of an nth-order (linear or nonlinear) ODE on some open interval I is a function $y = h(x)$ that is defined and n times differentiable on I and is such that the ODE becomes an identity if we replace the unknown function y and its derivatives by h and its corresponding derivatives.

Sections 3.1–3.2 will be devoted to homogeneous linear ODEs and Section 3.3 to nonhomogeneous linear ODEs.

Homogeneous Linear ODE: Superposition Principle, General Solution

The basic **superposition or linearity principle** of Sec. 2.1 extends to nth order homogeneous linear ODEs as follows.

THEOREM 1

Fundamental Theorem for the Homogeneous Linear ODE (2)

For a homogeneous linear ODE (2), *sums and constant multiples of solutions on some open interval I are again solutions on I.* (This does ***not*** hold for a nonhomogeneous or nonlinear ODE!)

The proof is a simple generalization of that in Sec. 2.1 and we leave it to the student.

Our further discussion parallels and extends that for second-order ODEs in Sec. 2.1. So we next define a general solution of (2), which will require an extension of linear independence from 2 to n functions.

DEFINITION

General Solution, Basis, Particular Solution

A **general solution** of (2) on an open interval I is a solution of (2) on I of the form

$$y(x) = c_1 y_1(x) + \cdots + c_n y_n(x) \qquad (c_1, \cdots, c_n \text{ arbitrary}) \tag{3}$$

where $y_1, \cdots, y_n$ is a **basis** (or **fundamental system**) of solutions of (2) on I; that is, these solutions are linearly independent on I, as defined below.

A **particular solution** of (2) on I is obtained if we assign specific values to the n constants $c_1, \cdots, c_n$ in (3).

DEFINITION

Linear Independence and Dependence

Consider n functions $y_1(x), \cdots, y_n(x)$ defined *on some interval I.*

These functions are called **linearly independent** on I if the equation

$$k_1 y_1(x) + \cdots + k_n y_n(x) = 0 \qquad \text{on } I \tag{4}$$

implies that all $k_1, \cdots, k_n$ are zero. These functions are called **linearly dependent** on I if this equation also holds on I for some $k_1, \cdots, k_n$ not all zero.

If and only if $y_1, \cdots, y_n$ are linearly dependent on I, we can express (at least) one of these functions on I as a "**linear combination**" of the other $n - 1$ functions, that is, as a sum of those functions, each multiplied by a constant (zero or not). This motivates the term "linearly dependent." For instance, if (4) holds with $k_1 \neq 0$, we can divide by k_1 and express y_1 as the linear combination

$$y_1 = -\frac{1}{k_1}(k_2 y_2 + \cdots + k_n y_n).$$

Note that when $n = 2$, these concepts reduce to those defined in Sec. 2.1.

EXAMPLE 1 Linear Dependence

Show that the functions $y_1 = x^2$, $y_2 = 5x$, $y_3 = 2x$ are linearly dependent on any interval.

Solution. $y_2 = 0y_1 + 2.5y_3$. This proves linear dependence on any interval. ■

EXAMPLE 2 Linear Independence

Show that $y_1 = x$, $y_2 = x^2$, $y_3 = x^3$ are linearly independent on any interval, for instance, on $-1 \leqq x \leqq 2$.

Solution. Equation (4) is $k_1 x + k_2 x^2 + k_3 x^3 = 0$. Taking (a) $x = -1$, (b) $x = 1$, (c) $x = 2$, we get

$$\text{(a) } -k_1 + k_2 - k_3 = 0, \qquad \text{(b) } k_1 + k_2 + k_3 = 0, \qquad \text{(c) } 2k_1 + 4k_2 + 8k_3 = 0.$$

$k_2 = 0$ from (a) + (b). Then $k_3 = 0$ from (c) -2(b). Then $k_1 = 0$ from (b). This proves linear independence.

A better method for testing linear independence of solutions of ODEs will soon be explained. ■

EXAMPLE 3 General Solution. Basis

Solve the fourth-order ODE

$$y^{\text{iv}} - 5y'' + 4y = 0 \qquad (\text{where } y^{\text{iv}} = d^4y/dx^4).$$

Solution. As in Sec. 2.2 we substitute $y = e^{\lambda x}$. Omitting the common factor $e^{\lambda x}$, we obtain the characteristic equation

$$\lambda^4 - 5\lambda^2 + 4 = 0.$$

This is a quadratic equation in $\mu = \lambda^2$, namely,

$$\mu^2 - 5\mu + 4 = (\mu - 1)(\mu - 4) = 0.$$

The roots are $\mu = 1$ and 4. Hence $\lambda = -2, -1, 1, 2$. This gives four solutions. A general solution on any interval is

$$y = c_1 e^{-2x} + c_2 e^{-x} + c_3 e^{x} + c_4 e^{2x}$$

provided those four solutions are linearly independent. This is true but will be shown later. ■

Initial Value Problem. Existence and Uniqueness

An **initial value problem** for the ODE (2) consists of (2) and n **initial conditions**

$$\text{(5)} \qquad y(x_0) = K_0, \qquad y'(x_0) = K_1, \qquad \cdots, \qquad y^{(n-1)}(x_0) = K_{n-1}$$

with given x_0 in the open interval I considered, and given $K_0, \cdots, K_{n-1}$.

In extension of the existence and uniqueness theorem in Sec. 2.6 we now have the following.

THEOREM 2

Existence and Uniqueness Theorem for Initial Value Problems

If the coefficients $p_0(x), \cdots, p_{n-1}(x)$ of (2) are continuous on some open interval I and x_0 is in I, then the initial value problem (2), (5) has a unique solution $y(x)$ on I.

Existence is proved in Ref. [A11] in App. 1. Uniqueness can be proved by a slight generalization of the uniqueness proof at the beginning of App. 4.

EXAMPLE 4 **Initial Value Problem for a Third-Order Euler–Cauchy Equation**

Solve the following initial value problem on any open interval I on the positive x-axis containing $x = 1$.

$$x^3y''' - 3x^2y'' + 6xy' - 6y = 0, \qquad y(1) = 2, \qquad y'(1) = 1, \qquad y''(1) = -4.$$

Solution. ***Step 1. General solution.*** As in Sec. 2.5 we try $y = x^m$. By differentiation and substitution,

$$m(m-1)(m-2)x^m - 3m(m-1)x^m + 6mx^m - 6x^m = 0.$$

Dropping x^m and ordering gives $m^3 - 6m^2 + 11m - 6 = 0$. If we can guess the root $m = 1$. We can divide by $m - 1$ and find the other roots 2 and 3, thus obtaining the solutions x, x^2, x^3, which are linearly independent on I (see Example 2). [In general one shall need a root-finding method, such as Newton's (Sec. 19.2), also available in a CAS (Computer Algebra System).] Hence a general solution is

$$y = c_1x + c_2x^2 + c_3x^3$$

valid on any interval I, even when it includes $x = 0$ where the coefficients of the ODE divided by x^3 (to have the standard form) are not continuous.

Step 2. Particular solution. The derivatives are $y' = c_1 + 2c_2x + 3c_3x^2$ and $y'' = 2c_2 + 6c_3x$. From this, and y and the initial conditions, we get by setting $x = 1$

$$\begin{aligned}
&\text{(a)}\ y(1) = c_1 + c_2 + c_3 = 2\\
&\text{(b)}\ y'(1) = c_1 + 2c_2 + 3c_3 = 1\\
&\text{(c)}\ y''(1) = 2c_2 + 6c_3 = -4.
\end{aligned}$$

This is solved by Cramer's rule (Sec. 7.6), or by elimination, which is simple, as follows. (b) − (a) gives (d) $c_2 + 2c_3 = -1$. Then (c) − 2(d) gives $c_3 = -1$. Then (c) gives $c_2 = 1$. Finally $c_1 = 2$ from (a). *Answer:* $y = 2x + x^2 - x^3$. ■

Linear Independence of Solutions. Wronskian

Linear independence of solutions is crucial for obtaining general solutions. Although it can often be seen by inspection, it would be good to have a criterion for it. Now Theorem 2 in Sec. 2.6 extends from order $n = 2$ to any n. This extended criterion uses the **Wronskian** W of n solutions $y_1, \cdots, y_n$ defined as the nth-order determinant

$$W(y_1, \cdots, y_n) = \begin{vmatrix} y_1 & y_2 & \cdots & y_n \\ y_1' & y_2' & \cdots & y_n' \\ \cdot & \cdot & \cdots & \cdot \\ y_1^{(n-1)} & y_2^{(n-1)} & \cdots & y_n^{(n-1)} \end{vmatrix}. \tag{6}$$

Note that W depends on x since $y_1, \cdots, y_n$ do. The criterion states that these solutions form a basis if and only if W is not zero; more precisely:

THEOREM 3

Linear Dependence and Independence of Solutions

Let the ODE (2) have continuous coefficients $p_0(x), \cdots, p_{n-1}(x)$ on an open interval I. Then n solutions $y_1, \cdots, y_n$ of (2) on I are linearly dependent on I if and only if their Wronskian is zero for some $x = x_0$ in I. Furthermore, if W is zero for $x = x_0$, then W is identically zero on I. Hence if there is an x_1 in I at which W is not zero, then $y_1, \cdots, y_n$ are linearly independent on I, so that they form a basis of solutions of (2) on I.

PROOF **(a)** Let $y_1, \cdots, y_n$ be linearly dependent solutions of (2) on I. Then, by definition, there are constants $k_1, \cdots, k_n$ *not all zero*, such that for all x in I,

$$k_1 y_1 + \cdots + k_n y_n = 0. \tag{7}$$

By $n - 1$ differentiations of (7) we obtain for all x in I

$$\begin{aligned} k_1 y_1' + \cdots + k_n y_n' &= 0 \\ &\vdots \\ k_1 y_1^{(n-1)} + \cdots + k_n y_n^{(n-1)} &= 0. \end{aligned} \tag{8}$$

(7), (8) is a homogeneous linear system of algebraic equations with a nontrivial solution $k_1, \cdots, k_n$. Hence its coefficient determinant must be zero for every x on I, by Cramer's theorem (Sec. 7.7). But that determinant is the Wronskian W, as we see from (6). Hence W is zero for every x on I.

(b) Conversely, if W is zero at an x_0 in I, then the system (7), (8) with $x = x_0$ has a solution $k_1^*, \cdots, k_n^*$, not all zero, by the same theorem. With these constants we define the solution $y^* = k_1^* y_1 + \cdots + k_n^* y_n$ of (2) on I. By (7), (8) this solution satisfies the initial conditions $y^*(x_0) = 0, \cdots, y^{*(n-1)}(x_0) = 0$. But another solution satisfying the same conditions is $y \equiv 0$. Hence $y^* \equiv y$ by Theorem 2, which applies since the coefficients of (2) are continuous. Together, $y^* = k_1^* y_1 + \cdots + k_n^* y_n \equiv 0$ on I. This means linear dependence of $y_1, \cdots, y_n$ on I.

(c) If W is zero at an x_0 in I, we have linear dependence by (b) and then $W \equiv 0$ by (a). Hence if W is not zero at an x_1 in I, the solutions $y_1, \cdots, y_n$ must be linearly independent on I. ■

EXAMPLE 5 **Basis, Wronskian**

We can now prove that in Example 3 we do have a basis. In evaluating W, pull out the exponential functions columnwise. In the result, subtract Column 1 from Columns 2, 3, 4 (without changing Column 1). Then expand by Row 1. In the resulting third-order determinant, subtract Column 1 from Column 2 and expand the result by Row 2:

$$W = \begin{vmatrix} e^{-2x} & e^{-x} & e^{x} & e^{2x} \\ -2e^{-2x} & -e^{-x} & e^{x} & 2e^{2x} \\ 4e^{-2x} & e^{-x} & e^{x} & 4e^{2x} \\ -8e^{-2x} & -e^{-x} & e^{x} & 8e^{2x} \end{vmatrix} = \begin{vmatrix} 1 & 1 & 1 & 1 \\ -2 & -1 & 1 & 2 \\ 4 & 1 & 1 & 4 \\ -8 & -1 & 1 & 8 \end{vmatrix} = \begin{vmatrix} 1 & 3 & 4 \\ -3 & -3 & 0 \\ 7 & 9 & 16 \end{vmatrix} = 72. \quad ■$$

A General Solution of (2) Includes All Solutions

Let us first show that general solutions always exist. Indeed, Theorem 3 in Sec. 2.6 extends as follows.

THEOREM 4

> **Existence of a General Solution**
>
> *If the coefficients $p_0(x), \cdots, p_{n-1}(x)$ of (2) are continuous on some open interval I, then (2) has a general solution on I.*

PROOF We choose any fixed x_0 in I. By Theorem 2 the ODE (2) has n solutions $y_1, \cdots, y_n$, where y_j satisfies initial conditions (5) with $K_{j-1} = 1$ and all other K's equal to zero. Their Wronskian at x_0 equals 1. For instance, when $n = 3$, then $y_1(x_0) = 1$, $y_2'(x_0) = 1$, $y_3''(x_0) = 1$, and the other initial values are zero. Thus, as claimed,

$$W(y_1(x_0), y_2(x_0), y_3(x_0)) = \begin{vmatrix} y_1(x_0) & y_2(x_0) & y_3(x_0) \\ y_1'(x_0) & y_2'(x_0) & y_3'(x_0) \\ y_1''(x_0) & y_2''(x_0) & y_3''(x_0) \end{vmatrix} = \begin{vmatrix} 1 & 0 & 0 \\ 0 & 1 & 0 \\ 0 & 0 & 1 \end{vmatrix} = 1.$$

Hence for any n those solutions $y_1, \cdots, y_n$ are linearly independent on I, by Theorem 3. They form a basis on I, and $y = c_1 y_1 + \cdots + c_n y_n$ is a general solution of (2) on I. ∎

We can now prove the basic property that, from a general solution of (2), every solution of (2) can be obtained by choosing suitable values of the arbitrary constants. Hence an nth-order *linear* ODE has no **singular solutions**, that is, solutions that cannot be obtained from a general solution.

THEOREM 5

> **General Solution Includes All Solutions**
>
> *If the ODE (2) has continuous coefficients $p_0(x), \cdots, p_{n-1}(x)$ on some open interval I, then every solution $y = Y(x)$ of (2) on I is of the form*
>
> (9) $$Y(x) = C_1 y_1(x) + \cdots + C_n y_n(x)$$
>
> *where $y_1, \cdots, y_n$ is a basis of solutions of (2) on I and $C_1, \cdots, C_n$ are suitable constants.*

PROOF Let Y be a given solution and $y = c_1 y_1 + \cdots + c_n y_n$ a general solution of (2) on I. We choose any fixed x_0 in I and show that we can find constants $c_1, \cdots, c_n$ for which y and its first $n - 1$ derivatives agree with Y and its corresponding derivatives at x_0. That is, we should have at $x = x_0$

(10)
$$
\begin{aligned}
c_1 y_1 + \cdots + c_n y_n &= Y \\
c_1 y_1' + \cdots + c_n y_n' &= Y' \\
&\;\;\vdots \\
c_1 y_1^{(n-1)} + \cdots + c_n y_n^{(n-1)} &= Y^{(n-1)}.
\end{aligned}
$$

But this is a linear system of equations in the unknowns $c_1, \cdots, c_n$. Its coefficient determinant is the Wronskian W of $y_1, \cdots, y_n$ at x_0. Since $y_1, \cdots, y_n$ form a basis, they are linearly independent, so that W is not zero by Theorem 3. Hence (10) has a unique solution $c_1 = C_1, \cdots, c_n = C_n$ (by Cramer's theorem in Sec. 7.7). With these values we obtain the particular solution

$$y^*(x) = C_1 y_1(x) + \cdots + C_n y_n(x)$$

on I. Equation (10) shows that y^* and its first $n - 1$ derivatives agree at x_0 with Y and its corresponding derivatives. That is, y^* and Y satisfy, at x_0, the same initial conditions. The uniqueness theorem (Theorem 2) now implies that $y^* \equiv Y$ on I. This proves the theorem. ■

This completes our theory of the homogeneous linear ODE (2). Note that for $n = 2$ it is identical with that in Sec. 2.6. This had to be expected.

PROBLEM SET 3.1

1–6 BASES: TYPICAL EXAMPLES

To get a feel for higher order ODEs, show that the given functions are solutions and form a basis on any interval. Use Wronskians. In Prob. 6, $x > 0$,

1. $1, x, x^2, x^3, \quad y^{\text{iv}} = 0$
2. $e^x, e^{-x}, e^{2x}, \quad y''' - 2y'' - y' + 2y = 0$
3. $\cos x, \sin x, x \cos x, x \sin x, \quad y^{\text{iv}} + 2y'' + y = 0$
4. $e^{-4x}, xe^{-4x}, x^2e^{-4x}, \quad y''' + 12y'' + 48y' + 64y = 0$
5. $1, e^{-x} \cos 2x, e^{-x} \sin 2x, \quad y''' + 2y'' + 5y' = 0$
6. $1, x^2, x^4, \quad x^2y''' - 3xy'' + 3y' = 0$
7. **TEAM PROJECT. General Properties of Solutions of Linear ODEs.** These properties are important in obtaining new solutions from given ones. Therefore extend Team Project 38 in Sec. 2.2 to nth-order ODEs. Explore statements on sums and multiples of solutions of (1) and (2) systematically and with proofs. Recognize clearly that no new ideas are needed in this extension from $n = 2$ to general n.

8–15 LINEAR INDEPENDENCE

Are the given functions linearly independent or dependent on the half-axis $x \geq 0$? Give reason.

8. $x^2, 1/x^2, 0$
9. $\tan x, \cot x, 1$
10. $e^{2x}, xe^{2x}, x^2e^{2x}$
11. $e^x \cos x, e^x \sin x, e^x$
12. $\sin^2 x, \cos^2 x, \cos 2x$
13. $\sin x, \cos x, \sin 2x$
14. $\cos^2 x, \sin^2 x, 2\pi$
15. $\cosh 2x, \sinh 2x, e^{2x}$
16. **TEAM PROJECT. Linear Independence and Dependence.** (a) Investigate the given question about a set S of functions on an interval I. Give an example. Prove your answer.

    (1) If S contains the zero function, can S be linearly independent?

    (2) If S is linearly independent on a subinterval J of I, is it linearly independent on I?

    (3) If S is linearly dependent on a subinterval J of I, is it linearly dependent on I?

    (4) If S is linearly independent on I, is it linearly independent on a subinterval J?

    (5) If S is linearly dependent on I, is it linearly independent on a subinterval J?

    (6) If S is linearly dependent on I, and if T contains S, is T linearly dependent on I?

    (b) In what cases can you use the Wronskian for testing linear independence? By what other means can you perform such a test?

3.2 Homogeneous Linear ODEs with Constant Coefficients

We proceed along the lines of Sec. 2.2, and generalize the results from $n = 2$ to arbitrary n. We want to solve an nth-order homogeneous linear ODE with constant coefficients, written as

$$y^{(n)} + a_{n-1}y^{(n-1)} + \cdots + a_1y' + a_0y = 0 \tag{1}$$

where $y^{(n)} = d^n y/dx^n$, etc. As in Sec. 2.2, we substitute $y = e^{\lambda x}$ to obtain the characteristic equation

$$\lambda^{(n)} + a_{n-1}\lambda^{(n-1)} + \cdots + a_1\lambda + a_0 y = 0 \tag{2}$$

of (1). If λ is a root of (2), then $y = e^{\lambda x}$ is a solution of (1). To find these roots, you may need a numeric method, such as Newton's in Sec. 19.2, also available on the usual CASs. For general n there are more cases than for $n = 2$. We can have distinct real roots, simple complex roots, multiple roots, and multiple complex roots, respectively. This will be shown next and illustrated by examples.

Distinct Real Roots

If all the n roots $\lambda_1, \cdots, \lambda_n$ of (2) are real and different, then the n solutions

$$y_1 = e^{\lambda_1 x}, \quad \cdots, \quad y_n = e^{\lambda_n x}. \tag{3}$$

constitute a basis for all x. The corresponding general solution of (1) is

$$y = c_1 e^{\lambda_1 x} + \cdots + c_n e^{\lambda_n x}. \tag{4}$$

Indeed, the solutions in (3) are linearly independent, as we shall see after the example.

EXAMPLE 1 Distinct Real Roots

Solve the ODE $y''' - 2y'' - y' + 2y = 0$.

Solution. The characteristic equation is $\lambda^3 - 2\lambda^2 - \lambda + 2 = 0$. It has the roots $-1, 1, 2$; if you find one of them by inspection, you can obtain the other two roots by solving a quadratic equation (explain!). The corresponding general solution (4) is $y = c_1 e^{-x} + c_2 e^{x} + c_3 e^{2x}$. ■

Linear Independence of (3). Students familiar with nth-order determinants may verify that, by pulling out all exponential functions from the columns and denoting their product by $E = \exp[\lambda_1 + \cdots + \lambda_n)x]$, the Wronskian of the solutions in (3) becomes

$$\begin{aligned} W &= \begin{vmatrix} e^{\lambda_1 x} & e^{\lambda_2 x} & \cdots & e^{\lambda_n x} \\ \lambda_1 e^{\lambda_1 x} & \lambda_2 e^{\lambda_2 x} & \cdots & \lambda_n e^{\lambda_n x} \\ \lambda_1^2 e^{\lambda_1 x} & \lambda_2^2 e^{\lambda_2 x} & \cdots & \lambda_n^2 e^{\lambda_n x} \\ \cdot & \cdot & \cdots & \cdot \\ \lambda_1^{n-1} e^{\lambda_1 x} & \lambda_2^{n-1} e^{\lambda_2 x} & \cdots & \lambda_n^{n-1} e^{\lambda_n x} \end{vmatrix} \\ &= E \begin{vmatrix} 1 & 1 & \cdots & 1 \\ \lambda_1 & \lambda_2 & \cdots & \lambda_n \\ \lambda_1^2 & \lambda_2^2 & \cdots & \lambda_n^2 \\ \cdot & \cdot & \cdots & \cdot \\ \lambda_1^{n-1} & \lambda_2^{n-1} & \cdots & \lambda_n^{n-1} \end{vmatrix}. \end{aligned} \tag{5}$$

The exponential function E is never zero. Hence $W = 0$ if and only if the determinant on the right is zero. This is a so-called **Vandermonde** or **Cauchy determinant.**[1] It can be shown that it equals

$$(-1)^{n(n-1)/2}V \tag{6}$$

where V is the product of all factors $\lambda_j - \lambda_k$ with $j < k\ (\leqq n)$; for instance, when $n = 3$ we get $-V = -(\lambda_1 - \lambda_2)(\lambda_1 - \lambda_3)(\lambda_2 - \lambda_3)$. This shows that the Wronskian is not zero if and only if all the n roots of (2) are different and thus gives the following.

THEOREM 1

Basis

Solutions $y_1 = e^{\lambda_1 x}, \cdots, y_n = e^{\lambda_n x}$ of (1) (with any real or complex λ_j's) form a basis of solutions of (1) on any open interval if and only if all n roots of (2) are different.

Actually, Theorem 1 is an important special case of our more general result obtained from (5) and (6):

THEOREM 2

Linear Independence

Any number of solutions of (1) of the form $e^{\lambda x}$ are linearly independent on an open interval I if and only if the corresponding λ are all different.

Simple Complex Roots

If complex roots occur, they must occur in conjugate pairs since the coefficients of (1) are real. Thus, if $\lambda = \gamma + i\omega$ is a simple root of (2), so is the conjugate $\bar{\lambda} = \gamma - i\omega$, and two corresponding linearly independent solutions are (as in Sec. 2.2, except for notation)

$$y_1 = e^{\gamma x} \cos \omega x, \qquad y_2 = e^{\gamma x} \sin \omega x.$$

EXAMPLE 2 **Simple Complex Roots. Initial Value Problem**

Solve the initial value problem

$$y''' - y'' + 100y' - 100y = 0, \qquad y(0) = 4, \qquad y'(0) = 11, \qquad y''(0) = -299.$$

Solution. The characteristic equation is $\lambda^3 - \lambda^2 + 100\lambda - 100 = 0$. It has the root 1, as can perhaps be seen by inspection. Then division by $\lambda - 1$ shows that the other roots are $\pm 10i$. Hence a general solution and its derivatives (obtained by differentiation) are

$$y = c_1 e^x + A \cos 10x + B \sin 10x,$$

$$y' = c_1 e^x - 10A \sin 10x + 10B \cos 10x,$$

$$y'' = c_1 e^x - 100A \cos 10x - 100B \sin 10x.$$

[1]ALEXANDRE THÉOPHILE VANDERMONDE (1735–1796), French mathematician, who worked on solution of equations by determinants. For CAUCHY see footnote 4, in Sec. 2.5.

From this and the initial conditions we obtain, by setting $x = 0$,

$$\text{(a)}\ c_1 + A = 4, \qquad \text{(b)}\ c_1 + 10B = 11, \qquad \text{(c)}\ c_1 - 100A = -299.$$

We solve this system for the unknowns A, B, c_1. Equation (a) minus Equation (c) gives $101A = 303$, $A = 3$. Then $c_1 = 1$ from (a) and $B = 1$ from (b). The solution is (Fig. 73)

$$y = e^x + 3\cos 10x + \sin 10x.$$

This gives the solution curve, which oscillates about e^x (dashed in Fig. 73). ■

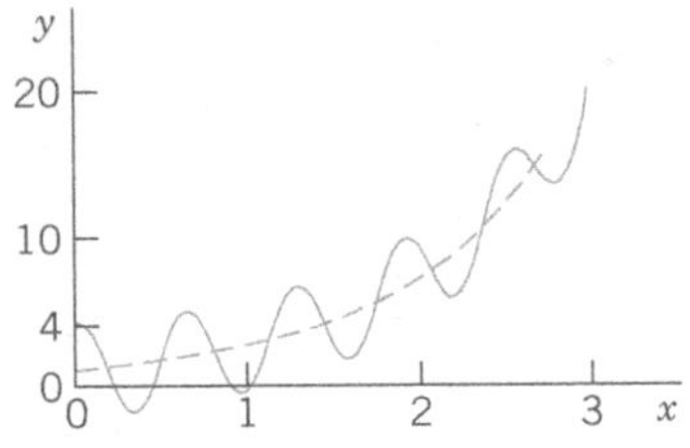

Fig. 73. Solution in Example 2

Multiple Real Roots

If a real double root occurs, say, $\lambda_1 = \lambda_2$, then $y_1 = y_2$ in (3), and we take y_1 and xy_1 as corresponding linearly independent solutions. This is as in Sec. 2.2.

More generally, if λ is a real root of order m, then m corresponding linearly independent solutions are

$$e^{\lambda x}, \quad xe^{\lambda x}, \quad x^2e^{\lambda x}, \quad \cdots, \quad x^{m-1}e^{\lambda x}. \tag{7}$$

We derive these solutions after the next example and indicate how to prove their linear independence.

EXAMPLE 3 Real Double and Triple Roots

Solve the ODE $y^{v} - 3y^{iv} + 3y''' - y'' = 0$.

Solution. The characteristic equation $\lambda^5 - 3\lambda^4 + 3\lambda^3 - \lambda^2 = 0$ has the roots $\lambda_1 = \lambda_2 = 0$, and $\lambda_3 = \lambda_4 = \lambda_5 = 1$, and the answer is

$$y = c_1 + c_2x + (c_3 + c_4x + c_5x^2)e^x. \tag{8}$$ ■

***Derivation of* (7).** We write the left side of (1) as

$$L[y] = y^{(n)} + a_{n-1}y^{(n-1)} + \cdots + a_0y.$$

Let $y = e^{\lambda x}$. Then by performing the differentiations we have

$$L[e^{\lambda x}] = (\lambda^n + a_{n-1}\lambda^{n-1} + \cdots + a_0)e^{\lambda x}.$$

Now let λ_1 be a root of mth order of the polynomial on the right, where $m \leqq n$. For $m < n$ let $\lambda_{m+1}, \cdots, \lambda_n$ be the other roots, all different from λ_1. Writing the polynomial in product form, we then have

$$L[e^{\lambda x}] = (\lambda - \lambda_1)^m h(\lambda) e^{\lambda x}$$

with $h(\lambda) = 1$ if $m = n$, and $h(\lambda) = (\lambda - \lambda_{m+1}) \cdots (\lambda - \lambda_n)$ if $m < n$. Now comes the key idea: We differentiate on both sides with respect to λ,

$$\text{(9)} \qquad \frac{\partial}{\partial \lambda} L[e^{\lambda x}] = m(\lambda - \lambda_1)^{m-1} h(\lambda) e^{\lambda x} + (\lambda - \lambda_1)^m \frac{\partial}{\partial \lambda} [h(\lambda) e^{\lambda x}].$$

The differentiations with respect to x and λ are independent and the resulting derivatives are continuous, so that we can interchange their order on the left:

$$\text{(10)} \qquad \frac{\partial}{\partial \lambda} L[e^{\lambda x}] = L\left[\frac{\partial}{\partial \lambda} e^{\lambda x}\right] = L[xe^{\lambda x}].$$

The right side of (9) is zero for $\lambda = \lambda_1$ because of the factors $\lambda - \lambda_1$ (and $m \geqq 2$ since we have a multiple root!). Hence $L[xe^{\lambda_1 x}] = 0$ by (9) and (10). This proves that $xe^{\lambda_1 x}$ is a solution of (1).

We can repeat this step and produce $x^2 e^{\lambda_1 x}, \cdots, x^{m-1} e^{\lambda_1 x}$ by another $m - 2$ such differentiations with respect to λ. Going one step further would no longer give zero on the right because the lowest power of $\lambda - \lambda_1$ would then be $(\lambda - \lambda_1)^0$, multiplied by $m! h(\lambda)$ and $h(\lambda_1) \neq 0$ because $h(\lambda)$ has no factors $\lambda - \lambda_1$; so we get *precisely* the solutions in (7).

We finally show that the solutions (7) are linearly independent. For a specific n this can be seen by calculating their Wronskian, which turns out to be nonzero. For arbitrary m we can pull out the exponential functions from the Wronskian. This gives $(e^{\lambda x})^m = e^{\lambda m x}$ times a determinant which by "row operations" can be reduced to the Wronskian of $1, x, \cdots, x^{m-1}$. The latter is constant and different from zero (equal to $1!2! \cdots (m-1)!$). These functions are solutions of the ODE $y^{(m)} = 0$, so that linear independence follows from Theroem 3 in Sec. 3.1.

Multiple Complex Roots

In this case, real solutions are obtained as for complex simple roots above. Consequently, if $\lambda = \gamma + i\omega$ is a **complex double root**, so is the conjugate $\bar{\lambda} = \gamma - i\omega$. Corresponding linearly independent solutions are

$$\textbf{(11)} \qquad e^{\gamma x} \cos \omega x, \quad e^{\gamma x} \sin \omega x, \quad xe^{\gamma x} \cos \omega x, \quad xe^{\gamma x} \sin \omega x.$$

The first two of these result from $e^{\lambda x}$ and $e^{\bar{\lambda} x}$ as before, and the second two from $xe^{\lambda x}$ and $xe^{\bar{\lambda} x}$ in the same fashion. Obviously, the corresponding general solution is

$$\text{(12)} \qquad y = e^{\gamma x}[(A_1 + A_2 x) \cos \omega x + (B_1 + B_2 x) \sin \omega x].$$

For *complex triple roots* (which hardly ever occur in applications), one would obtain two more solutions $x^2 e^{\gamma x} \cos \omega x$, $x^2 e^{\gamma x} \sin \omega x$, and so on.

PROBLEM SET 3.2

1–6 GENERAL SOLUTION

Solve the given ODE. Show the details of your work.

1. $y''' + 25y' = 0$
2. $y^{iv} + 2y'' + y = 0$
3. $y^{iv} + 4y'' = 0$
4. $(D^3 - D^2 - D + I)y = 0$
5. $(D^4 + 10D^2 + 9I)y = 0$
6. $(D^5 + 8D^3 + 16D)y = 0$

7–13 INITIAL VALUE PROBLEM

Solve the IVP by a CAS, giving a general solution and the particular solution and its graph.

7. $y''' + 3.2y'' + 4.81y' = 0,\quad y(0) = 3.4,\quad y'(0) = -4.6,$ $y''(0) = 9.91$
8. $y''' + 7.5y'' + 14.25y' - 9.125y = 0,\quad y(0) = 10.05,$ $y'(0) = -54.975,\quad y''(0) = 257.5125$
9. $4y''' + 8y'' + 41y' + 37y = 0,\quad y(0) = 9,$ $y'(0) = -6.5,\quad y''(0) = -39.75$
10. $y^{iv} + 4y = 0,\quad y(0) = \frac{1}{2},\quad y'(0) = -\frac{3}{2},\quad y''(0) = \frac{5}{2},$ $y'''(0) = -\frac{7}{2}$
11. $y^{iv} - 9y'' - 400y = 0,\quad y(0) = 0,\quad y'(0) = 0,$ $y''(0) = 41,\quad y'''(0) = 0$
12. $y^{v} - 5y''' + 4y' = 0,\quad y(0) = 3,\quad y'(0) = -5,$ $y''(0) = 11,\quad y'''(0) = -23,\quad y^{iv}(0) = 47$
13. $y^{iv} + 0.45y''' - 0.165y'' + 0.0045y' - 0.00175y = 0,$ $y(0) = 17.4,\quad y'(0) = -2.82,\quad y''(0) = 2.0485,$ $y'''(0) = -1.458675$
14. **PROJECT. Reduction of Order.** This is of practical interest since a single solution of an ODE can often be guessed. For second order, see Example 7 in Sec. 2.1.

    (a) How could you reduce the order of a linear constant-coefficient ODE if a solution is known?

    (b) Extend the method to a variable-coefficient ODE

    $$y''' + p_2(x)y'' + p_1(x)y' + p_0(x)y = 0.$$

    Assuming a solution y_1 to be known, show that another solution is $y_2(x) = u(x)y_1(x)$ with $u(x) = \int z(x)\,dx$ and z obtained by solving

    $$y_1z'' + (3y_1' + p_2y_1)z' + (3y_1'' + 2p_2y_1' + p_1y_1)z = 0.$$

    (c) Reduce

    $$x^3y''' - 3x^2y'' + (6 - x^2)xy' - (6 - x^2)y = 0,$$

    using $y_1 = x$ (perhaps obtainable by inspection).
15. **CAS EXPERIMENT. Reduction of Order.** Starting with a basis, find third-order linear ODEs with variable coefficients for which the reduction to second order turns out to be relatively simple.

3.3 Nonhomogeneous Linear ODEs

We now turn from homogeneous to nonhomogeneous linear ODEs of nth order. We write them in standard form

$$y^{(n)} + p_{n-1}(x)y^{(n-1)} + \cdots + p_1(x)y' + p_0(x)y = r(x) \tag{1}$$

with $y^{(n)} = d^ny/dx^n$ as the first term, and $r(x) \not\equiv 0$. As for second-order ODEs, a general solution of (1) on an open interval I of the x-axis is of the form

$$y(x) = y_h(x) + y_p(x). \tag{2}$$

Here $y_h(x) = c_1y_1(x) + \cdots + c_ny_n(x)$ is a general solution of the corresponding homogeneous ODE

$$y^{(n)} + p_{n-1}(x)y^{(n-1)} + \cdots + p_1(x)y' + p_0(x)y = 0 \tag{3}$$

on I. Also, y_p is any solution of (1) on I containing no arbitrary constants. If (1) has continuous coefficients and a continuous $r(x)$ on I, then a general solution of (1) exists and includes all solutions. Thus (1) has no singular solutions.

An **initial value problem** for (1) consists of (1) and n **initial conditions**

$$y(x_0) = K_0, \quad y'(x_0) = K_1, \quad \cdots, \quad y^{(n-1)}(x_0) = K_{n-1} \tag{4}$$

with x_0 in I. Under those continuity assumptions it has a unique solution. The ideas of proof are the same as those for $n = 2$ in Sec. 2.7.

Method of Undetermined Coefficients

Equation (2) shows that for solving (1) we have to determine a particular solution of (1). For a constant-coefficient equation

$$y^{(n)} + a_{n-1}y^{(n-1)} + \cdots + a_1 y' + a_0 y = r(x) \tag{5}$$

($a_0, \cdots, a_{n-1}$ constant) and special $r(x)$ as in Sec. 2.7, such a $y_p(x)$ can be determined by the **method of undetermined coefficients,** as in Sec. 2.7, using the following rules.

(A) Basic Rule as in Sec. 2.7.

(B) Modification Rule. *If a term in your choice for $y_p(x)$ is a solution of the homogeneous equation* (3), *then multiply this term by x^k, where k is the smallest positive integer such that this term times x^k is not a solution of* (3).

(C) Sum Rule as in Sec. 2.7.

The practical application of the method is the same as that in Sec. 2.7. It suffices to illustrate the typical steps of solving an initial value problem and, in particular, the new Modification Rule, which includes the old Modification Rule as a particular case (with $k = 1$ or 2). We shall see that the technicalities are the same as for $n = 2$, except perhaps for the more involved determination of the constants.

EXAMPLE 1 **Initial Value Problem. Modification Rule**

Solve the initial value problem

$$y''' + 3y'' + 3y' + y = 30e^{-x}, \quad y(0) = 3, \quad y'(0) = -3, \quad y''(0) = -47. \tag{6}$$

Solution. *Step 1.* The characteristic equation is $\lambda^3 + 3\lambda^2 + 3\lambda + 1 = (\lambda + 1)^3 = 0$. It has the triple root $\lambda = -1$. Hence a general solution of the homogeneous ODE is

$$\begin{aligned} y_h &= c_1 e^{-x} + c_2 x e^{-x} + c_3 x^2 e^{-x} \\ &= (c_1 + c_2 x + c_3 x^2)e^{-x}. \end{aligned}$$

Step 2. If we try $y_p = Ce^{-x}$, we get $-C + 3C - 3C + C = 30$, which has no solution. Try Cxe^{-x} and Cx^2e^{-x}. The Modification Rule calls for

$$y_p = Cx^3e^{-x}.$$

Then

$$\begin{aligned} y_p' &= C(3x^2 - x^3)e^{-x}, \\ y_p'' &= C(6x - 6x^2 + x^3)e^{-x}, \\ y_p''' &= C(6 - 18x + 9x^2 - x^3)e^{-x}. \end{aligned}$$

Substitution of these expressions into (6) and omission of the common factor e^{-x} gives

$$C(6 - 18x + 9x^2 - x^3) + 3C(6x - 6x^2 + x^3) + 3C(3x^2 - x^3) + Cx^3 = 30.$$

The linear, quadratic, and cubic terms drop out, and $6C = 30$. Hence $C = 5$. This gives $y_p = 5x^3e^{-x}$.

Step 3. We now write down $y = y_h + y_p$, the general solution of the given ODE. From it we find c_1 by the first initial condition. We insert the value, differentiate, and determine c_2 from the second initial condition, insert the value, and finally determine c_3 from $y''(0)$ and the third initial condition:

$$y = y_h + y_p = (c_1 + c_2x + c_3x^2)e^{-x} + 5x^3e^{-x}, \qquad y(0) = c_1 = 3$$

$$y' = [-3 + c_2 + (-c_2 + 2c_3)x + (15 - c_3)x^2 - 5x^3]e^{-x}, \qquad y'(0) = -3 + c_2 = -3, \qquad c_2 = 0$$

$$y'' = [3 + 2c_3 + (30 - 4c_3)x + (-30 + c_3)x^2 + 5x^3]e^{-x}, \qquad y''(0) = 3 + 2c_3 = -47, \qquad c_3 = -25.$$

Hence the *answer* to our problem is (Fig. 73)

$$y = (3 - 25x^2)e^{-x} + 5x^3e^{-x}.$$

The curve of y begins at (0, 3) with a negative slope, as expected from the initial values, and approaches zero as $x \to \infty$. The dashed curve in Fig. 74 is y_p. ■

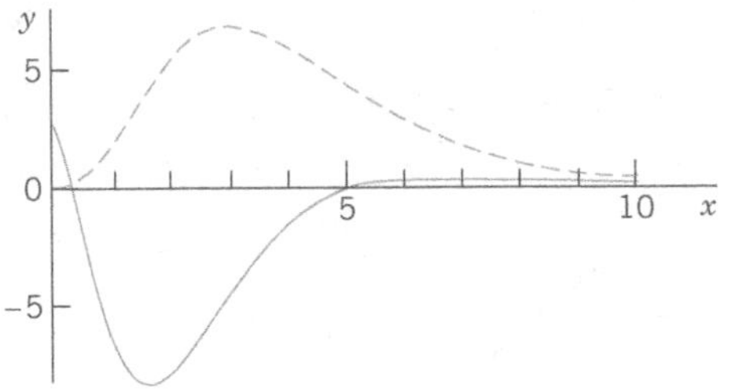

Fig. 74. y and y_p (dashed) in Example 1

Method of Variation of Parameters

The method of variation of parameters (see Sec. 2.10) also extends to arbitrary order n. It gives a particular solution y_p for the nonhomogeneous equation (1) (in standard form with $y^{(n)}$ as the first term!) by the formula

$$y_p(x) = \sum_{k=1}^{n} y_k(x) \int \frac{W_k(x)}{W(x)} r(x)\, dx \tag{7}$$

$$= y_1(x) \int \frac{W_1(x)}{W(x)} r(x)\, dx + \cdots + y_n(x) \int \frac{W_n(x)}{W(x)} r(x)\, dx$$

on an open interval I on which the coefficients of (1) and $r(x)$ are continuous. In (7) the functions $y_1, \cdots, y_n$ form a basis of the homogeneous ODE (3), with Wronskian W, and W_j $(j = 1, \cdots, n)$ is obtained from W by replacing the jth column of W by the column $[0 \quad 0 \quad \cdots \quad 0 \quad 1]^\mathsf{T}$. Thus, when $n = 2$, this becomes identical with (2) in Sec. 2.10,

$$W = \begin{vmatrix} y_1 & y_2 \\ y_1' & y_2' \end{vmatrix}, \qquad W_1 = \begin{vmatrix} 0 & y_2 \\ 1 & y_2' \end{vmatrix} = -y_2, \qquad W_2 = \begin{vmatrix} y_1 & 0 \\ y_1' & 1 \end{vmatrix} = y_1.$$

The proof of (7) uses an extension of the idea of the proof of (2) in Sec. 2.10 and can be found in Ref [A11] listed in App. 1.

EXAMPLE 2 Variation of Parameters. Nonhomogeneous Euler–Cauchy Equation

Solve the nonhomogeneous Euler–Cauchy equation

$$x^3y''' - 3x^2y'' + 6xy' - 6y = x^4 \ln x \qquad (x > 0).$$

Solution. *Step 1. General solution of the homogeneous ODE.* Substitution of $y = x^m$ and the derivatives into the homogeneous ODE and deletion of the factor x^m gives

$$m(m-1)(m-2) - 3m(m-1) + 6m - 6 = 0.$$

The roots are 1, 2, 3 and give as a basis

$$y_1 = x, \qquad y_2 = x^2, \qquad y_3 = x^3.$$

Hence the corresponding general solution of the homogeneous ODE is

$$y_h = c_1x + c_2x^2 + c_3x^3.$$

Step 2. Determinants needed in (7). These are

$$W = \begin{vmatrix} x & x^2 & x^3 \\ 1 & 2x & 3x^2 \\ 0 & 2 & 6x \end{vmatrix} = 2x^3$$

$$W_1 = \begin{vmatrix} 0 & x^2 & x^3 \\ 0 & 2x & 3x^2 \\ 1 & 2 & 6x \end{vmatrix} = x^4$$

$$W_2 = \begin{vmatrix} x & 0 & x^3 \\ 1 & 0 & 3x^2 \\ 0 & 1 & 6x \end{vmatrix} = -2x^3$$

$$W_3 = \begin{vmatrix} x & x^2 & 0 \\ 1 & 2x & 0 \\ 0 & 2 & 1 \end{vmatrix} = x^2.$$

Step 3. Integration. In (7) we also need the right side $r(x)$ of our ODE in standard form, obtained by division of the given equation by the coefficient x^3 of y'''; thus, $r(x) = (x^4 \ln x)/x^3 = x \ln x$. In (7) we have the simple quotients $W_1/W = x/2$, $W_2/W = -1$, $W_3/W = 1/(2x)$. Hence (7) becomes

$$\begin{aligned} y_p &= x \int \frac{x}{2} x \ln x \, dx - x^2 \int x \ln x \, dx + x^3 \int \frac{1}{2x} x \ln x \, dx \\ &= \frac{x}{2}\left(\frac{x^3}{3} \ln x - \frac{x^3}{9}\right) - x^2\left(\frac{x^2}{2} \ln x - \frac{x^2}{4}\right) + \frac{x^3}{2}(x \ln x - x). \end{aligned}$$

Simplification gives $y_p = \frac{1}{6}x^4 \left(\ln x - \frac{11}{6}\right)$. Hence the answer is

$$y = y_h + y_p = c_1x + c_2x^2 + c_3x^3 + \tfrac{1}{6}x^4\left(\ln x - \tfrac{11}{6}\right).$$

Figure 75 shows y_p. Can you explain the shape of this curve? Its behavior near $x = 0$? The occurrence of a minimum? Its rapid increase? Why would the method of undetermined coefficients not have given the solution? ■

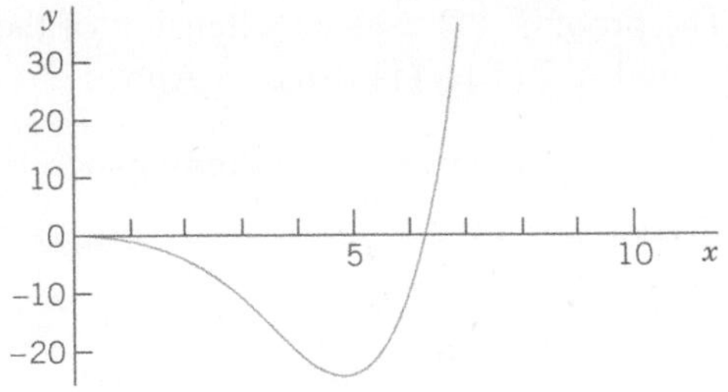

Fig. 75. Particular solution y_p of the nonhomogeneous Euler–Cauchy equation in Example 2

Application: Elastic Beams

Whereas second-order ODEs have various applications, of which we have discussed some of the more important ones, higher order ODEs have much fewer engineering applications. An important fourth-order ODE governs the bending of elastic beams, such as wooden or iron girders in a building or a bridge.

A related application of vibration of beams does not fit in here since it leads to PDEs and will therefore be discussed in Sec. 12.3.

EXAMPLE 3 Bending of an Elastic Beam under a Load

We consider a beam B of length L and constant (e.g., **rectangular**) cross section and homogeneous elastic material (e.g., steel); see Fig. 76. We assume that under its own weight the beam is bent so little that it is practically straight. If we apply a load to B in a vertical plane through the axis of symmetry (the x-axis in Fig. 76), B is bent. Its axis is curved into the so-called **elastic curve** C (or **deflection curve**). It is shown in elasticity theory that the bending moment $M(x)$ is proportional to the curvature $k(x)$ of C. We assume the bending to be small, so that the deflection $y(x)$ and its derivative $y'(x)$ (determining the tangent direction of C) are small. Then, by calculus, $k = y''/(1 + y'^2)^{3/2} \approx y''$. Hence

$$M(x) = EIy''(x).$$

EI is the constant of proportionality. E is *Young's modulus of elasticity* of the material of the beam. I is the moment of inertia of the cross section about the (horizontal) z-axis in Fig. 76.

Elasticity theory shows further that $M''(x) = f(x)$, where $f(x)$ is the load per unit length. Together,

$$EIy^{\text{iv}} = f(x). \qquad (8)$$

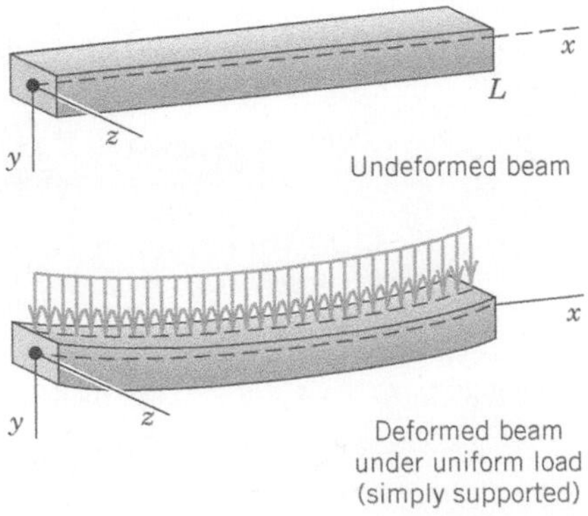

Fig. 76. Elastic beam

In applications the most important supports and corresponding boundary conditions are as follows and shown in Fig. 77.

(A) Simply supported $y = y'' = 0$ at $x = 0$ and L

(B) Clamped at both ends $y = y' = 0$ at $x = 0$ and L

(C) Clamped at $x = 0$, free at $x = L$ $y(0) = y'(0) = 0, y''(L) = y'''(L) = 0.$

The boundary condition $y = 0$ means no displacement at that point, $y' = 0$ means a horizontal tangent, $y'' = 0$ means no bending moment, and $y''' = 0$ means no shear force.

Let us apply this to the uniformly loaded simply supported beam in Fig. 76. The load is $f(x) \equiv f_0 = \text{const}$. Then (8) is

$$y^{\text{iv}} = k, \qquad k = \frac{f_0}{EI}. \tag{9}$$

This can be solved simply by calculus. Two integrations give

$$y'' = \frac{k}{2}x^2 + c_1 x + c_2.$$

$y''(0) = 0$ gives $c_2 = 0$. Then $y''(L) = L(\frac{1}{2}kL + c_1) = 0, c_1 = -kL/2$ (since $L \neq 0$). Hence

$$y'' = \frac{k}{2}(x^2 - Lx).$$

Integrating this twice, we obtain

$$y = \frac{k}{2}\left(\frac{1}{12}x^4 - \frac{L}{6}x^3 + c_3 x + c_4\right)$$

with $c_4 = 0$ from $y(0) = 0$. Then

$$y(L) = \frac{kL}{2}\left(\frac{L^3}{12} - \frac{L^3}{6} + c_3\right) = 0, \qquad c_3 = \frac{L^3}{12}.$$

Inserting the expression for k, we obtain as our solution

$$y = \frac{f_0}{24EI}(x^4 - 2Lx^3 + L^3 x).$$

Since the boundary conditions at both ends are the same, we expect the deflection $y(x)$ to be "symmetric" with respect to $L/2$, that is, $y(x) = y(L - x)$. Verify this directly or set $x = u + L/2$ and show that y becomes an even function of u,

$$y = \frac{f_0}{24EI}\left(u^2 - \frac{1}{4}L^2\right)\left(u^2 - \frac{5}{4}L^2\right).$$

From this we can see that the maximum deflection in the middle at $u = 0 (x = L/2)$ is $5f_0L^4/(16 \cdot 24EI)$. Recall that the positive direction points downward. ■

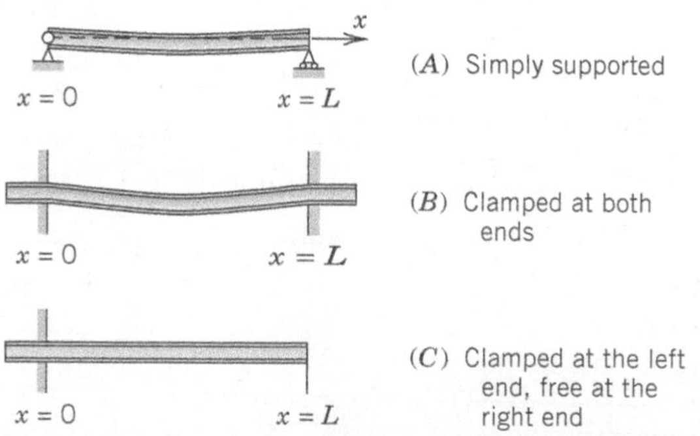

Fig. 77. Supports of a beam

PROBLEM SET 3.3

1–7 GENERAL SOLUTION

Solve the following ODEs, showing the details of your work.

1. $y''' + 3y'' + 3y' + y = e^x - x - 1$
2. $y''' + 2y'' - y' - 2y = 1 - 4x^3$
3. $(D^4 + 10D^2 + 9I)y = 6.5 \sinh 2x$
4. $(D^3 + 3D^2 - 5D - 39I)y = -300 \cos x$
5. $(x^3D^3 + x^2D^2 - 2xD + 2I)y = x^{-2}$
6. $(D^3 + 4D)y = \sin x$
7. $(D^3 - 9D^2 + 27D - 27I)y = 27 \sin 3x$

8–13 INITIAL VALUE PROBLEM

Solve the given IVP, showing the details of your work.

8. $y^{iv} - 5y'' + 4y = 10e^{-3x}$, $y(0) = 1$, $y'(0) = 0$, $y''(0) = 0$, $y'''(0) = 0$
9. $y^{iv} + 5y'' + 4y = 90 \sin 4x$, $y(0) = 1$, $y'(0) = 2$, $y''(0) = -1$, $y'''(0) = -32$
10. $x^3y''' + xy' - y = x^2$, $y(1) = 1$, $y'(1) = 3$, $y''(1) = 14$
11. $(D^3 - 2D^2 - 3D)y = 74e^{-3x} \sin x$, $y(0) = -1.4$, $y'(0) = 3.2$, $y''(0) = -5.2$
12. $(D^3 - 2D^2 - 9D + 18I)y = e^{2x}$, $y(0) = 4.5$, $y'(0) = 8.8$, $y''(0) = 17.2$
13. $(D^3 - 4D)y = 10 \cos x + 5 \sin x$, $y(0) = 3$, $y'(0) = -2$, $y''(0) = -1$

14. **CAS EXPERIMENT. Undetermined Coefficients.** Since variation of parameters is generally complicated, it seems worthwhile to try to extend the other method. Find out experimentally for what ODEs this is possible and for what not. *Hint:* Work backward, solving ODEs with a CAS and then looking whether the solution could be obtained by undetermined coefficients. For example, consider

$$y''' - 3y'' + 3y' - y = x^{1/2}e^x$$

and

$$x^3y''' + x^2y'' - 2xy' + 2y = x^3 \ln x.$$

15. **WRITING REPORT. Comparison of Methods.** Write a report on the method of undetermined coefficients and the method of variation of parameters, discussing and comparing the advantages and disadvantages of each method. Illustrate your findings with typical examples. Try to show that the method of undetermined coefficients, say, for a third-order ODE with constant coefficients and an exponential function on the right, can be derived from the method of variation of parameters.

CHAPTER 3 REVIEW QUESTIONS AND PROBLEMS

1. What is the superposition or linearity principle? For what nth-order ODEs does it hold?
2. List some other basic theorems that extend from second-order to nth-order ODEs.
3. If you know a general solution of a homogeneous linear ODE, what do you need to obtain from it a general solution of a corresponding nonhomogeneous linear ODE?
4. What form does an initial value problem for an nth-order linear ODE have?
5. What is the Wronskian? What is it used for?

6–15 GENERAL SOLUTION

Solve the given ODE. Show the details of your work.

6. $y^{iv} - 3y'' - 4y = 0$
7. $y''' + 4y'' + 13y' = 0$
8. $y''' - 4y'' - y' + 4y = 30e^{2x}$
9. $(D^4 - 16I)y = -15 \cosh x$
10. $x^2y''' + 3xy'' - 2y' = 0$
11. $y''' + 4.5y'' + 6.75y' + 3.375y = 0$
12. $(D^3 - D)y = \sinh 0.8x$
13. $(D^3 + 6D^2 + 12D + 8I)y = 8x^2$
14. $(D^4 - 13D^2 + 36I)y = 12e^x$
15. $4x^3y''' + 3xy' - 3y = 10$

16–20 INITIAL VALUE PROBLEM

Solve the IVP. Show the details of your work.

16. $(D^3 - D^2 - D + I)y = 0$, $y(0) = 0$, $Dy(0) = 1$, $D^2y(0) = 0$
17. $y''' + 5y'' + 24y' + 20y = x$, $y(0) = 1.94$, $y'(0) = -3.95$, $y'' = -24$
18. $(D^4 - 26D^2 + 25I)y = 50(x + 1)^2$, $y(0) = 12.16$, $Dy(0) = -6$, $D^2y(0) = 34$, $D^3y(0) = -130$
19. $(D^3 + 9D^2 + 23D + 15I)y = 12\exp(-4x)$, $y(0) = 9$, $Dy(0) = -41$, $D^2y(0) = 189$
20. $(D^3 + 3D^2 + 3D + I)y = 8 \sin x$, $y(0) = -1$, $y'(0) = -3$, $y''(0) = 5$

SUMMARY OF CHAPTER 3

Higher Order Linear ODEs

Compare with the similar Summary of Chap. 2 (the case n = 2).
Chapter 3 extends Chap. 2 from order $n = 2$ to arbitrary order n. An ***n*th-order linear ODE** is an ODE that can be written

$$y^{(n)} + p_{n-1}(x)y^{(n-1)} + \cdots + p_1(x)y' + p_0(x)y = r(x) \tag{1}$$

with $y^{(n)} = d^n y/dx^n$ as the first term; we again call this the **standard form**. Equation (1) is called **homogeneous** if $r(x) \equiv 0$ on a given open interval I considered, **nonhomogeneous** if $r(x) \not\equiv 0$ on I. For the homogeneous ODE

$$y^{(n)} + p_{n-1}(x)y^{(n-1)} + \cdots + p_1(x)y' + p_0(x)y = 0 \tag{2}$$

the ***superposition principle*** (Sec. 3.1) holds, just as in the case $n = 2$. A **basis** or **fundamental system** of solutions of (2) on I consists of n linearly independent solutions $y_1, \cdots, y_n$ of (2) on I. A **general solution** of (2) on I is a linear combination of these,

$$y = c_1 y_1 + \cdots + c_n y_n \qquad (c_1, \cdots, c_n \text{ arbitrary constants}). \tag{3}$$

A **general solution** of the nonhomogeneous ODE (1) on I is of the form

$$y = y_h + y_p \qquad \text{(Sec. 3.3).} \tag{4}$$

Here, y_p is a particular solution of (1) and is obtained by two methods (***undetermined coefficients*** or ***variation of parameters***) explained in Sec. 3.3.

An **initial value problem** for (1) or (2) consists of one of these ODEs and n initial conditions (Secs. 3.1, 3.3)

$$y(x_0) = K_0, \quad y'(x_0) = K_1, \quad \cdots, \quad y^{(n-1)}(x_0) = K_{n-1} \tag{5}$$

with given x_0 in I and given $K_0, \cdots, K_{n-1}$. If $p_0, \cdots, p_{n-1}, r$ are continuous on I, then general solutions of (1) and (2) on I exist, and initial value problems (1), (5) or (2), (5) have a unique solution.

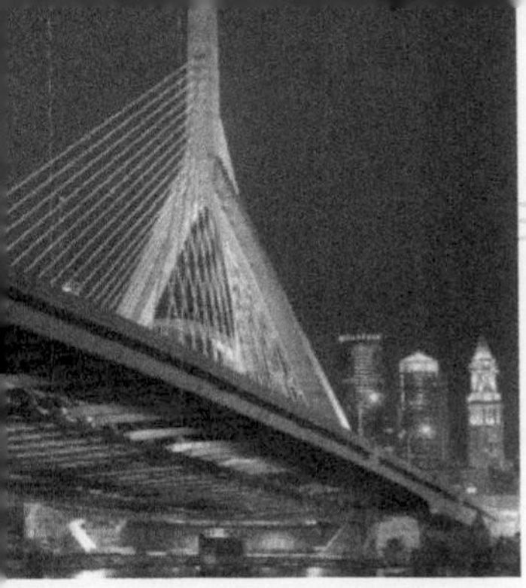

CHAPTER 4

Systems of ODEs. Phase Plane. Qualitative Methods

Tying in with Chap. 3, we present another method of solving higher order ODEs in Sec. 4.1. This converts any nth-order ODE into a system of n first-order ODEs. We also show some applications. Moreover, in the same section we solve systems of first-order ODEs that occur directly in applications, that is, not derived from an nth-order ODE but dictated by the application such as two tanks in mixing problems and two circuits in electrical networks. (The elementary aspects of vectors and matrices needed in this chapter are reviewed in Sec. 4.0 and are probably familiar to most students.)

In Sec. 4.3 we introduce a totally different way of looking at systems of ODEs. The method consists of examining the general behavior of whole families of solutions of ODEs in the ***phase plane***, and aptly is called the phase plane method. It gives information on the **stability** of solutions. (*Stability of a physical system* is desirable and means roughly that a small change at some instant causes only a small change in the behavior of the system at later times.) This approach to systems of ODEs is a **qualitative method** because it depends only on the nature of the ODEs and does not require the actual solutions. This can be very useful because it is often difficult or even impossible to solve systems of ODEs. In contrast, the approach of actually solving a system is known as a *quantitative method.*

The phase plane method has many applications in control theory, circuit theory, population dynamics and so on. Its use in linear systems is discussed in Secs. 4.3, 4.4, and 4.6 and its even more important use in nonlinear systems is discussed in Sec. 4.5 with applications to the pendulum equation and the Lokta–Volterra population model. The chapter closes with a discussion of nonhomogeneous linear systems of ODEs.

NOTATION. We continue to denote unknown functions by y; thus, $y_1(t), y_2(t)$—analogous to Chaps. 1–3. (Note that some authors use x for functions, $x_1(t), x_2(t)$ when dealing with systems of ODEs.)

Prerequisite: Chap. 2.
References and Answers to Problems: App. 1 Part A, and App. 2.

4.0 For Reference: Basics of Matrices and Vectors

For clarity and simplicity of notation, we use matrices and vectors in our discussion of linear systems of ODEs. We need only a few elementary facts (and not the bulk of the material of Chaps. 7 and 8). Most students will very likely be already familiar with these facts. Thus ***this section is for reference only***. *Begin with Sec. 4.1 and consult 4.0 as needed.*

Most of our linear systems will consist of two linear ODEs in two unknown functions $y_1(t)$, $y_2(t)$,

$$(1) \qquad \begin{aligned} y_1' &= a_{11}y_1 + a_{12}y_2, \\ y_2' &= a_{21}y_1 + a_{22}y_2, \end{aligned} \qquad \text{for example,} \qquad \begin{aligned} y_1' &= -5y_1 + 2y_2 \\ y_2' &= \ \ 13y_1 + \tfrac{1}{2}y_2 \end{aligned}$$

(perhaps with additional *given* functions $g_1(t)$, $g_2(t)$ on the right in the two ODEs).

Similarly, a linear system of n first-order ODEs in n unknown functions $y_1(t), \cdots, y_n(t)$ is of the form

$$(2) \qquad \begin{aligned} y_1' &= a_{11}y_1 + a_{12}y_2 + \cdots + a_{1n}y_n \\ y_2' &= a_{21}y_1 + a_{22}y_2 + \cdots + a_{2n}y_n \\ &\cdots\cdots\cdots\cdots\cdots\cdots\cdots\cdots \\ y_n' &= a_{n1}y_1 + a_{n2}y_2 + \cdots + a_{nn}y_n \end{aligned}$$

(perhaps with an additional given function on the right in each ODE).

Some Definitions and Terms

Matrices. In (1) the (constant or variable) coefficients form a $\mathbf{2 \times 2}$ **matrix A**, that is, an array

$$(3) \qquad \mathbf{A} = [a_{jk}] = \begin{bmatrix} a_{11} & a_{12} \\ a_{21} & a_{22} \end{bmatrix}, \qquad \text{for example,} \qquad \mathbf{A} = \begin{bmatrix} -5 & 2 \\ 13 & \frac{1}{2} \end{bmatrix}.$$

Similarly, the coefficients in (2) form an $\boldsymbol{n \times n}$ **matrix**

$$(4) \qquad \mathbf{A} = [a_{jk}] = \begin{bmatrix} a_{11} & a_{12} & \cdots & a_{1n} \\ a_{21} & a_{22} & \cdots & a_{2n} \\ \cdot & \cdot & \cdots & \cdot \\ a_{n1} & a_{n2} & \cdots & a_{nn} \end{bmatrix}.$$

The $a_{11}, a_{12}, \cdots$ are called **entries**, the horizontal lines **rows**, and the vertical lines **columns**. Thus, in (3) the first row is $[a_{11} \quad a_{12}]$, the second row is $[a_{21} \quad a_{22}]$, and the first and second columns are

$$\begin{bmatrix} a_{11} \\ a_{21} \end{bmatrix} \quad \text{and} \quad \begin{bmatrix} a_{12} \\ a_{22} \end{bmatrix}.$$

In the "***double subscript notation***" for entries, the first subscript denotes the *row* and the second the *column* in which the entry stands. Similarly in (4). The **main diagonal** is the diagonal $a_{11} \quad a_{22} \quad \cdots \quad a_{nn}$ in (4), hence $a_{11} \quad a_{22}$ in (3).

We shall need only **square matrices**, that is, matrices with the same number of rows and columns, as in (3) and (4).

Vectors. A **column vector x** with n **components** $x_1, \cdots, x_n$ is of the form

$$\mathbf{x} = \begin{bmatrix} x_1 \\ x_2 \\ \vdots \\ x_n \end{bmatrix}, \qquad \text{thus if } n = 2, \qquad \mathbf{x} = \begin{bmatrix} x_1 \\ x_2 \end{bmatrix}.$$

Similarly, a **row vector v** is of the form

$$\mathbf{v} = [v_1 \quad \cdots \quad v_n], \qquad \text{thus if } n = 2, \text{ then} \qquad \mathbf{v} = [v_1 \quad v_2].$$

Calculations with Matrices and Vectors

Equality. Two $n \times n$ matrices are *equal* if and only if corresponding entries are equal. Thus for $n = 2$, let

$$\mathbf{A} = \begin{bmatrix} a_{11} & a_{12} \\ a_{21} & a_{22} \end{bmatrix} \qquad \text{and} \qquad \mathbf{B} = \begin{bmatrix} b_{11} & b_{12} \\ b_{21} & b_{22} \end{bmatrix}.$$

Then $\mathbf{A} = \mathbf{B}$ if and only if

$$a_{11} = b_{11}, \qquad a_{12} = b_{12}$$
$$a_{21} = b_{21}, \qquad a_{22} = b_{22}.$$

Two column vectors (or two row vectors) are *equal* if and only if they both have n components and corresponding components are equal. Thus, let

$$\mathbf{v} = \begin{bmatrix} v_1 \\ v_2 \end{bmatrix} \quad \text{and} \quad \mathbf{x} = \begin{bmatrix} x_1 \\ x_2 \end{bmatrix}. \qquad \text{Then} \qquad \mathbf{v} = \mathbf{x} \quad \text{if and only if} \quad \begin{aligned} v_1 &= x_1 \\ v_2 &= x_2. \end{aligned}$$

Addition is performed by adding corresponding entries (or components); here, matrices must both be $n \times n$, and vectors must both have the same number of components. Thus for $n = 2$,

$$(5) \qquad \mathbf{A} + \mathbf{B} = \begin{bmatrix} a_{11} + b_{11} & a_{12} + b_{12} \\ a_{21} + b_{21} & a_{22} + b_{22} \end{bmatrix}, \qquad \mathbf{v} + \mathbf{x} = \begin{bmatrix} v_1 + x_1 \\ v_2 + x_2 \end{bmatrix}.$$

Scalar multiplication (multiplication by a number c) is performed by multiplying each entry (or component) by c. For example, if

$$\mathbf{A} = \begin{bmatrix} 9 & 3 \\ -2 & 0 \end{bmatrix}, \qquad \text{then} \qquad -7\mathbf{A} = \begin{bmatrix} -63 & -21 \\ 14 & 0 \end{bmatrix}.$$

If

$$\mathbf{v} = \begin{bmatrix} 0.4 \\ -13 \end{bmatrix}, \quad \text{then} \quad 10\mathbf{v} = \begin{bmatrix} 4 \\ -130 \end{bmatrix}.$$

Matrix Multiplication. The product $\mathbf{C} = \mathbf{AB}$ (in this order) of two $n \times n$ matrices $\mathbf{A} = [a_{jk}]$ and $\mathbf{B} = [b_{jk}]$ is the $n \times n$ matrix $\mathbf{C} = [c_{jk}]$ with entries

$$c_{jk} = \sum_{m=1}^{n} a_{jm} b_{mk} \qquad \begin{array}{l} j = 1, \cdots, n \\ k = 1, \cdots, n, \end{array} \tag{6}$$

that is, multiply each entry in the jth *row* of $\mathbf{A}$ by the corresponding entry in the kth *column* of $\mathbf{B}$ and then add these n products. One says briefly that this is a "multiplication of rows into columns." For example,

$$\begin{bmatrix} 9 & 3 \\ -2 & 0 \end{bmatrix} \begin{bmatrix} 1 & -4 \\ 2 & 5 \end{bmatrix} = \begin{bmatrix} 9 \cdot 1 + 3 \cdot 2 & 9 \cdot (-4) + 3 \cdot 5 \\ -2 \cdot 1 + 0 \cdot 2 & (-2) \cdot (-4) + 0 \cdot 5 \end{bmatrix},$$

$$= \begin{bmatrix} 15 & -21 \\ -2 & 8 \end{bmatrix}.$$

CAUTION! Matrix multiplication is ***not commutative,*** $\mathbf{AB} \neq \mathbf{BA}$ in general. In our example,

$$\begin{bmatrix} 1 & -4 \\ 2 & 5 \end{bmatrix} \begin{bmatrix} 9 & 3 \\ -2 & 0 \end{bmatrix} = \begin{bmatrix} 1 \cdot 9 + (-4) \cdot (-2) & 1 \cdot 3 + (-4) \cdot 0 \\ 2 \cdot 9 + 5 \cdot (-2) & 2 \cdot 3 + 5 \cdot 0 \end{bmatrix}$$

$$= \begin{bmatrix} 17 & 3 \\ 8 & 6 \end{bmatrix}.$$

Multiplication of an $n \times n$ matrix $\mathbf{A}$ by a vector $\mathbf{x}$ with n components is defined by the same rule: $\mathbf{v} = \mathbf{Ax}$ is the vector with the n components

$$v_j = \sum_{m=1}^{n} a_{jm} x_m \qquad j = 1, \cdots, n.$$

For example,

$$\begin{bmatrix} 12 & 7 \\ -8 & 3 \end{bmatrix} \begin{bmatrix} x_1 \\ x_2 \end{bmatrix} = \begin{bmatrix} 12x_1 + 7x_2 \\ -8x_1 + 3x_2 \end{bmatrix}.$$

Systems of ODEs as Vector Equations

Differentiation. The *derivative* of a matrix (or vector) with variable entries (or components) is obtained by differentiating each entry (or component). Thus, if

$$\mathbf{y}(t) = \begin{bmatrix} y_1(t) \\ y_2(t) \end{bmatrix} = \begin{bmatrix} e^{-2t} \\ \sin t \end{bmatrix}, \quad \text{then} \quad \mathbf{y}'(t) = \begin{bmatrix} y_1'(t) \\ y_2'(t) \end{bmatrix} = \begin{bmatrix} -2e^{-2t} \\ \cos t \end{bmatrix}.$$

Using matrix multiplication and differentiation, we can now write (1) as

$$\text{(7)} \qquad \mathbf{y}' = \begin{bmatrix} y_1' \\ y_2' \end{bmatrix} = \mathbf{A}\mathbf{y} = \begin{bmatrix} a_{11} & a_{12} \\ a_{21} & a_{22} \end{bmatrix} \begin{bmatrix} y_1 \\ y_2 \end{bmatrix}, \quad \text{e.g.,} \quad \mathbf{y}' = \begin{bmatrix} -5 & 2 \\ 13 & \frac{1}{2} \end{bmatrix} \begin{bmatrix} y_1 \\ y_2 \end{bmatrix}.$$

Similarly for (2) by means of an $n \times n$ matrix $\mathbf{A}$ and a column vector $\mathbf{y}$ with n components, namely, $\mathbf{y}' = \mathbf{A}\mathbf{y}$. The vector equation (7) is equivalent to two equations for the components, and these are precisely the two ODEs in (1).

Some Further Operations and Terms

Transposition is the operation of writing columns as rows and conversely and is indicated by T. Thus the transpose $\mathbf{A}^{\mathsf{T}}$ of the 2×2 matrix

$$\mathbf{A} = \begin{bmatrix} a_{11} & a_{12} \\ a_{21} & a_{22} \end{bmatrix} = \begin{bmatrix} -5 & 2 \\ 13 & \frac{1}{2} \end{bmatrix} \quad \text{is} \quad \mathbf{A}^{\mathsf{T}} = \begin{bmatrix} a_{11} & a_{21} \\ a_{12} & a_{22} \end{bmatrix} = \begin{bmatrix} -5 & 13 \\ 2 & \frac{1}{2} \end{bmatrix}.$$

The transpose of a column vector, say,

$$\mathbf{v} = \begin{bmatrix} v_1 \\ v_2 \end{bmatrix}, \quad \text{is a row vector,} \quad \mathbf{v}^{\mathsf{T}} = [v_1 \quad v_2],$$

and conversely.

Inverse of a Matrix. The $n \times n$ **unit matrix I** is the $n \times n$ matrix with main diagonal $1, 1, \cdots, 1$ and all other entries zero. If, for a given $n \times n$ matrix $\mathbf{A}$, there is an $n \times n$ matrix $\mathbf{B}$ such that $\mathbf{AB} = \mathbf{BA} = \mathbf{I}$, then $\mathbf{A}$ is called **nonsingular** and $\mathbf{B}$ is called the **inverse** of $\mathbf{A}$ and is denoted by $\mathbf{A}^{-1}$; thus

$$\text{(8)} \qquad \mathbf{A}\mathbf{A}^{-1} = \mathbf{A}^{-1}\mathbf{A} = \mathbf{I}.$$

The inverse exists if the determinant det $\mathbf{A}$ of $\mathbf{A}$ is not zero.

If $\mathbf{A}$ has no inverse, it is called **singular**. For $n = 2$,

$$\text{(9)} \qquad \mathbf{A}^{-1} = \frac{1}{\det \mathbf{A}} \begin{bmatrix} a_{22} & -a_{12} \\ -a_{21} & a_{11} \end{bmatrix},$$

where the **determinant** of $\mathbf{A}$ is

$$\text{(10)} \qquad \det \mathbf{A} = \begin{vmatrix} a_{11} & a_{12} \\ a_{21} & a_{22} \end{vmatrix} = a_{11}a_{22} - a_{12}a_{21}.$$

(For general n, see Sec. 7.7, but this will not be needed in this chapter.)

Linear Independence. r given vectors $\mathbf{v}^{(1)}, \cdots, \mathbf{v}^{(r)}$ with n components are called a *linearly independent set* or, more briefly, **linearly independent**, if

$$\text{(11)} \qquad c_1\mathbf{v}^{(1)} + \cdots + c_r\mathbf{v}^{(r)} = \mathbf{0}$$

implies that all scalars $c_1, \cdots, c_r$ must be zero; here, $\mathbf{0}$ denotes the **zero vector**, whose n components are all zero. If (11) also holds for scalars not all zero (so that at least one of these scalars is not zero), then these vectors are called a *linearly dependent set* or, briefly, **linearly dependent**, because then at least one of them can be expressed as a **linear combination** of the others; that is, if, for instance, $c_1 \neq 0$ in (11), then we can obtain

$$\mathbf{v}^{(1)} = -\frac{1}{c_1}(c_2\mathbf{v}^{(2)} + \cdots + c_r\mathbf{v}^{(r)}).$$

Eigenvalues, Eigenvectors

Eigenvalues and eigenvectors will be very important in this chapter (and, as a matter of fact, throughout mathematics).

Let $\mathbf{A} = [a_{jk}]$ be an $n \times n$ matrix. Consider the equation

$$\mathbf{Ax} = \lambda\mathbf{x} \tag{12}$$

where λ is a scalar (a real or complex number) to be determined and $\mathbf{x}$ is a vector to be determined. Now, for every λ, a solution is $\mathbf{x} = \mathbf{0}$. A scalar λ such that (12) holds for some vector $\mathbf{x} \neq \mathbf{0}$ is called an **eigenvalue** of $\mathbf{A}$, and this vector is called an **eigenvector** of $\mathbf{A}$ corresponding to this eigenvalue λ.

We can write (12) as $\mathbf{Ax} - \lambda\mathbf{x} = \mathbf{0}$ or

$$(\mathbf{A} - \lambda\mathbf{I})\mathbf{x} = \mathbf{0}. \tag{13}$$

These are n linear algebraic equations in the n unknowns $x_1, \cdots, x_n$ (the components of $\mathbf{x}$). For these equations to have a solution $\mathbf{x} \neq \mathbf{0}$, the determinant of the coefficient matrix $\mathbf{A} - \lambda\mathbf{I}$ must be zero. This is proved as a basic fact in linear algebra (Theorem 4 in Sec. 7.7). In this chapter we need this only for $n = 2$. Then (13) is

$$\begin{bmatrix} a_{11} - \lambda & a_{12} \\ a_{21} & a_{22} - \lambda \end{bmatrix} \begin{bmatrix} x_1 \\ x_2 \end{bmatrix} = \begin{bmatrix} 0 \\ 0 \end{bmatrix}; \tag{14}$$

in components,

$$\begin{aligned} (a_{11} - \lambda)x_1 + \quad a_{12}x_2 \quad &= 0 \\ a_{21}x_1 \quad + (a_{22} - \lambda)x_2 &= 0. \end{aligned} \tag{14*}$$

Now $\mathbf{A} - \lambda\mathbf{I}$ is singular if and only if its determinant $\det(\mathbf{A} - \lambda\mathbf{I})$, called the **characteristic determinant** of $\mathbf{A}$ (also for general n), is zero. This gives

$$\begin{aligned} \det(\mathbf{A} - \lambda\mathbf{I}) &= \begin{vmatrix} a_{11} - \lambda & a_{12} \\ a_{21} & a_{22} - \lambda \end{vmatrix} \\ &= (a_{11} - \lambda)(a_{22} - \lambda) - a_{12}a_{21} \\ &= \lambda^2 - (a_{11} + a_{22})\lambda + a_{11}a_{22} - a_{12}a_{21} = 0. \end{aligned} \tag{15}$$

This quadratic equation in λ is called the **characteristic equation** of **A**. Its solutions are the eigenvalues λ_1 and λ_2 of **A**. First determine these. Then use (14*) with $\lambda = \lambda_1$ to determine an eigenvector $\mathbf{x}^{(1)}$ of **A** corresponding to λ_1. Finally use (14*) with $\lambda = \lambda_2$ to find an eigenvector $\mathbf{x}^{(2)}$ of **A** corresponding to λ_2. Note that if **x** is an eigenvector of **A**, so is $k\mathbf{x}$ with any $k \neq 0$.

EXAMPLE 1 **Eigenvalue Problem**

Find the eigenvalues and eigenvectors of the matrix

$$\mathbf{A} = \begin{bmatrix} -4.0 & 4.0 \\ -1.6 & 1.2 \end{bmatrix} \qquad (16)$$

Solution. The characteristic equation is the quadratic equation

$$\det |\mathbf{A} - \lambda\mathbf{I}| = \begin{vmatrix} -4-\lambda & 4 \\ -1.6 & 1.2-\lambda \end{vmatrix} = \lambda^2 + 2.8\lambda + 1.6 = 0.$$

It has the solutions $\lambda_1 = -2$ and $\lambda_2 = -0.8$. These are the eigenvalues of **A**.

Eigenvectors are obtained from (14*). For $\lambda = \lambda_1 = -2$ we have from (14*)

$$\begin{aligned} (-4.0 + 2.0)x_1 + \quad 4.0x_2 \quad &= 0 \\ -1.6x_1 \quad + (1.2 + 2.0)x_2 &= 0. \end{aligned}$$

A solution of the first equation is $x_1 = 2, x_2 = 1$. This also satisfies the second equation. (Why?) Hence an eigenvector of **A** corresponding to $\lambda_1 = -2.0$ is

$$\mathbf{x}^{(1)} = \begin{bmatrix} 2 \\ 1 \end{bmatrix}. \qquad \text{Similarly,} \qquad \mathbf{x}^{(2)} = \begin{bmatrix} 1 \\ 0.8 \end{bmatrix} \qquad (17)$$

is an eigenvector of **A** corresponding to $\lambda_2 = -0.8$, as obtained from (14*) with $\lambda = \lambda_2$. Verify this. ■

4.1 Systems of ODEs as Models in Engineering Applications

We show how systems of ODEs are of practical importance as follows. We first illustrate how systems of ODEs can serve as models in various applications. Then we show how a higher order ODE (with the highest derivative standing alone on one side) can be reduced to a first-order system.

EXAMPLE 1 **Mixing Problem Involving Two Tanks**

A mixing problem involving a single tank is modeled by a single ODE, and you may first review the corresponding Example 3 in Sec. 1.3 because the principle of modeling will be the same for two tanks. The model will be a system of two first-order ODEs.

Tank T_1 and T_2 in Fig. 78 contain initially 100 gal of water each. In T_1 the water is pure, whereas 150 lb of fertilizer are dissolved in T_2. By circulating liquid at a rate of 2 gal/min and stirring (to keep the mixture uniform) the amounts of fertilizer $y_1(t)$ in T_1 and $y_2(t)$ in T_2 change with time t. How long should we let the liquid circulate so that T_1 will contain at least half as much fertilizer as there will be left in T_2?

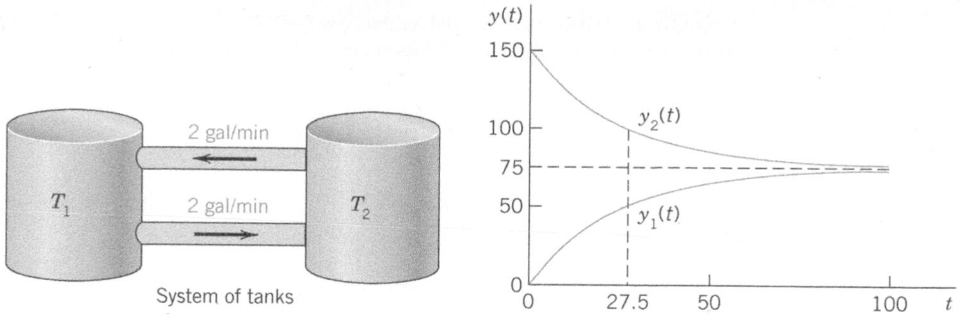

Fig. 78. Fertilizer content in Tanks T_1 (lower curve) and T_2

Solution. *Step 1.* **Setting up the model.** As for a single tank, the time rate of change $y_1'(t)$ of $y_1(t)$ equals inflow minus outflow. Similarly for tank T_2. From Fig. 78 we see that

$$y_1' = \text{Inflow/min} - \text{Outflow/min} = \frac{2}{100}y_2 - \frac{2}{100}y_1 \qquad \text{(Tank } T_1\text{)}$$

$$y_2' = \text{Inflow/min} - \text{Outflow/min} = \frac{2}{100}y_1 - \frac{2}{100}y_2 \qquad \text{(Tank } T_2\text{).}$$

Hence the mathematical model of our mixture problem is the system of first-order ODEs

$$\begin{aligned} y_1' &= -0.02y_1 + 0.02y_2 && \text{(Tank } T_1\text{)} \\ y_2' &= 0.02y_1 - 0.02y_2 && \text{(Tank } T_2\text{).} \end{aligned}$$

As a vector equation with column vector $\mathbf{y} = \begin{bmatrix} y_1 \\ y_2 \end{bmatrix}$ and matrix $\mathbf{A}$ this becomes

$$\mathbf{y}' = \mathbf{A}\mathbf{y}, \qquad \text{where} \qquad \mathbf{A} = \begin{bmatrix} -0.02 & 0.02 \\ 0.02 & -0.02 \end{bmatrix}.$$

Step 2. **General solution.** As for a single equation, we try an exponential function of t,

$$\text{(1)} \qquad \mathbf{y} = \mathbf{x}e^{\lambda t}. \qquad \text{Then} \qquad \mathbf{y}' = \lambda \mathbf{x} e^{\lambda t} = \mathbf{A}\mathbf{x}e^{\lambda t}.$$

Dividing the last equation $\lambda \mathbf{x} e^{\lambda t} = \mathbf{A}\mathbf{x}e^{\lambda t}$ by $e^{\lambda t}$ and interchanging the left and right sides, we obtain

$$\mathbf{A}\mathbf{x} = \lambda\mathbf{x}.$$

We need nontrivial solutions (solutions that are not identically zero). Hence we have to look for eigenvalues and eigenvectors of $\mathbf{A}$. The eigenvalues are the solutions of the characteristic equation

$$\text{(2)} \qquad \det(\mathbf{A} - \lambda\mathbf{I}) = \begin{vmatrix} -0.02 - \lambda & 0.02 \\ 0.02 & -0.02 - \lambda \end{vmatrix} = (-0.02 - \lambda)^2 - 0.02^2 = \lambda(\lambda + 0.04) = 0.$$

We see that $\lambda_1 = 0$ (which can very well happen—don't get mixed up—it is eigen*vectors* that must not be zero) and $\lambda_2 = -0.04$. Eigenvectors are obtained from (14*) in Sec. 4.0 with $\lambda = 0$ and $\lambda = -0.04$. For our present $\mathbf{A}$ this gives [we need only the first equation in (14*)]

$$-0.02x_1 + 0.02x_2 = 0 \qquad \text{and} \qquad (-0.02 + 0.04)x_1 + 0.02x_2 = 0,$$

respectively. Hence $x_1 = x_2$ and $x_1 = -x_2$, respectively, and we can take $x_1 = x_2 = 1$ and $x_1 = -x_2 = 1$. This gives two eigenvectors corresponding to $\lambda_1 = 0$ and $\lambda_2 = -0.04$, respectively, namely,

$$\mathbf{x}^{(1)} = \begin{bmatrix} 1 \\ 1 \end{bmatrix} \quad \text{and} \quad \mathbf{x}^{(2)} = \begin{bmatrix} 1 \\ -1 \end{bmatrix}.$$

From (1) and the superposition principle (which continues to hold for systems of homogeneous linear ODEs) we thus obtain a solution

$$\mathbf{y} = c_1\mathbf{x}^{(1)}e^{\lambda_1 t} + c_2\mathbf{x}^{(2)}e^{\lambda_2 t} = c_1\begin{bmatrix} 1 \\ 1 \end{bmatrix} + c_2\begin{bmatrix} 1 \\ -1 \end{bmatrix}e^{-0.04t} \tag{3}$$

where c_1 and c_2 are arbitrary constants. Later we shall call this a **general solution.**

***Step 3.* Use of initial conditions.** The initial conditions are $y_1(0) = 0$ (no fertilizer in tank T_1) and $y_2(0) = 150$. From this and (3) with $t = 0$ we obtain

$$\mathbf{y}(0) = c_1\begin{bmatrix} 1 \\ 1 \end{bmatrix} + c_2\begin{bmatrix} 1 \\ -1 \end{bmatrix} = \begin{bmatrix} c_1 + c_2 \\ c_1 - c_2 \end{bmatrix} = \begin{bmatrix} 0 \\ 150 \end{bmatrix}.$$

In components this is $c_1 + c_2 = 0$, $c_1 - c_2 = 150$. The solution is $c_1 = 75$, $c_2 = -75$. This gives the answer

$$\mathbf{y} = 75\mathbf{x}^{(1)} - 75\mathbf{x}^{(2)}e^{-0.04t} = 75\begin{bmatrix} 1 \\ 1 \end{bmatrix} - 75\begin{bmatrix} 1 \\ -1 \end{bmatrix}e^{-0.04t}.$$

In components,

$$y_1 = 75 - 75e^{-0.04t} \qquad \text{(Tank } T_1\text{, lower curve)}$$

$$y_2 = 75 + 75e^{-0.04t} \qquad \text{(Tank } T_2\text{, upper curve).}$$

Figure 78 shows the exponential increase of y_1 and the exponential decrease of y_2 to the common limit 75 lb. Did you expect this for physical reasons? Can you physically explain why the curves look "symmetric"? Would the limit change if T_1 initially contained 100 lb of fertilizer and T_2 contained 50 lb?

***Step 4.* Answer.** T_1 contains half the fertilizer amount of T_2 if it contains 1/3 of the total amount, that is, 50 lb. Thus

$$y_1 = 75 - 75e^{-0.04t} = 50, \qquad e^{-0.04t} = \tfrac{1}{3}, \qquad t = (\ln 3)/0.04 = 27.5.$$

Hence the fluid should circulate for at least about half an hour. ■

EXAMPLE 2 Electrical Network

Find the currents $I_1(t)$ and $I_2(t)$ in the network in Fig. 79. Assume all currents and charges to be zero at $t = 0$, the instant when the switch is closed.

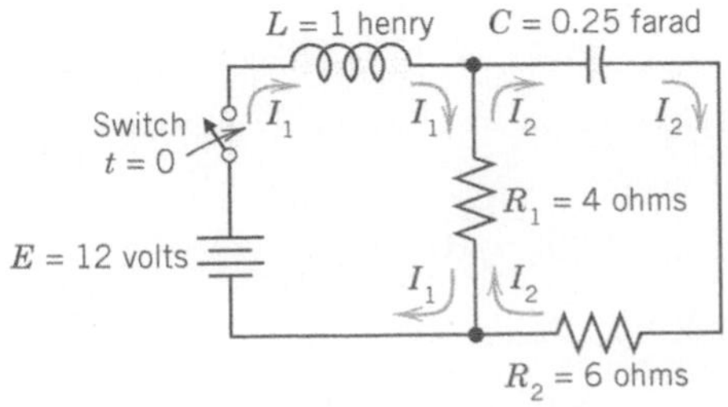

Fig. 79. Electrical network in Example 2

Solution. ***Step 1.* Setting up the mathematical model.** The model of this network is obtained from Kirchhoff's Voltage Law, as in Sec. 2.9 (where we considered single circuits). Let $I_1(t)$ and $I_2(t)$ be the currents in the left and right loops, respectively. In the left loop, the voltage drops are $LI_1' = I_1'$ [V] over the inductor and $R_1(I_1 - I_2) = 4(I_1 - I_2)$ [V] over the resistor, the difference because I_1 and I_2 flow through the resistor in opposite directions. By Kirchhoff's Voltage Law the sum of these drops equals the voltage of the battery; that is, $I_1' + 4(I_1 - I_2) = 12$, hence

$$I_1' = -4I_1 + 4I_2 + 12. \tag{4a}$$

In the right loop, the voltage drops are $R_2I_2 = 6I_2$ [V] and $R_1(I_2 - I_1) = 4(I_2 - I_1)$ [V] over the resistors and $(I/C)\int I_2\,dt = 4\int I_2\,dt$ [V] over the capacitor, and their sum is zero,

$$6I_2 + 4(I_2 - I_1) + 4\int I_2\,dt = 0 \qquad \text{or} \qquad 10I_2 - 4I_1 + 4\int I_2\,dt = 0.$$

Division by 10 and differentiation gives $I_2' - 0.4I_1' + 0.4I_2 = 0$.

To simplify the solution process, we first get rid of $0.4I_1'$, which by (4a) equals $0.4(-4I_1 + 4I_2 + 12)$. Substitution into the present ODE gives

$$I_2' = 0.4I_1' - 0.4I_2 = 0.4(-4I_1 + 4I_2 + 12) - 0.4I_2$$

and by simplification

$$I_2' = -1.6I_1 + 1.2I_2 + 4.8. \tag{4b}$$

In matrix form, (4) is (we write **J** since **I** is the unit matrix)

$$\mathbf{J}' = \mathbf{AJ} + \mathbf{g}, \quad \text{where} \quad \mathbf{J} = \begin{bmatrix} I_1 \\ I_2 \end{bmatrix}, \quad \mathbf{A} = \begin{bmatrix} -4.0 & 4.0 \\ -1.6 & 1.2 \end{bmatrix}, \quad \mathbf{g} = \begin{bmatrix} 12.0 \\ 4.8 \end{bmatrix}. \tag{5}$$

***Step 2.* Solving (5).** Because of the vector **g** this is a *nonhomogeneous* system, and we try to proceed as for a single ODE, solving first the *homogeneous* system $\mathbf{J}' = \mathbf{AJ}$ (thus $\mathbf{J}' - \mathbf{AJ} = \mathbf{0}$) by substituting $\mathbf{J} = \mathbf{x}e^{\lambda t}$. This gives

$$\mathbf{J}' = \lambda\mathbf{x}e^{\lambda t} = \mathbf{A}\mathbf{x}e^{\lambda t}, \qquad \text{hence} \qquad \mathbf{Ax} = \lambda\mathbf{x}.$$

Hence, to obtain a nontrivial solution, we again need the eigenvalues and eigenvectors. For the present matrix **A** they are derived in Example 1 in Sec. 4.0:

$$\lambda_1 = -2, \qquad \mathbf{x}^{(1)} = \begin{bmatrix} 2 \\ 1 \end{bmatrix}; \qquad \lambda_2 = -0.8, \qquad \mathbf{x}^{(2)} = \begin{bmatrix} 1 \\ 0.8 \end{bmatrix}.$$

Hence a "general solution" of the homogeneous system is

$$\mathbf{J}_h = c_1\mathbf{x}^{(1)}e^{-2t} + c_2\mathbf{x}^{(2)}e^{-0.8t}.$$

For a particular solution of the nonhomogeneous system (5), since **g** is constant, we try a constant column vector $\mathbf{J}_p = \mathbf{a}$ with components a_1, a_2. Then $\mathbf{J}_p' = \mathbf{0}$, and substitution into (5) gives $\mathbf{Aa} + \mathbf{g} = \mathbf{0}$; in components,

$$\begin{aligned} -4.0a_1 + 4.0a_2 + 12.0 &= 0 \\ -1.6a_1 + 1.2a_2 + 4.8 &= 0. \end{aligned}$$

The solution is $a_1 = 3$, $a_2 = 0$; thus $\mathbf{a} = \begin{bmatrix} 3 \\ 0 \end{bmatrix}$. Hence

$$\mathbf{J} = \mathbf{J}_h + \mathbf{J}_p = c_1\mathbf{x}^{(1)}e^{-2t} + c_2\mathbf{x}^{(2)}e^{-0.8t} + \mathbf{a}; \tag{6}$$

in components,

$$\begin{aligned} I_1 &= 2c_1e^{-2t} + c_2e^{-0.8t} + 3 \\ I_2 &= c_1e^{-2t} + 0.8c_2e^{-0.8t}. \end{aligned}$$

The initial conditions give

$$I_1(0) = 2c_1 + c_2 + 3 = 0$$
$$I_2(0) = c_1 + 0.8c_2 = 0.$$

Hence $c_1 = -4$ and $c_2 = 5$. As the solution of our problem we thus obtain

$$\mathbf{J} = -4\mathbf{x}^{(1)}e^{-2t} + 5\mathbf{x}^{(2)}e^{-0.8t} + \mathbf{a}. \tag{7}$$

In components (Fig. 80b),

$$I_1 = -8e^{-2t} + 5e^{-0.8t} + 3$$
$$I_2 = -4e^{-2t} + 4e^{-0.8t}.$$

Now comes an important idea, on which we shall elaborate further, beginning in Sec. 4.3. Figure 80a shows $I_1(t)$ and $I_2(t)$ as two separate curves. Figure 80b shows these two currents as a single curve $[I_1(t), I_2(t)]$ in the I_1I_2-plane. This is a parametric representation with time t as the parameter. It is often important to know in which sense such a curve is traced. This can be indicated by an arrow in the sense of increasing t, as is shown. The I_1I_2-plane is called the **phase plane** of our system (5), and the curve in Fig. 80b is called a **trajectory**. We shall see that such "**phase plane representations**" are far more important than graphs as in Fig. 80a because they will give a much better qualitative overall impression of the general behavior of whole families of solutions, not merely of one solution as in the present case. ■

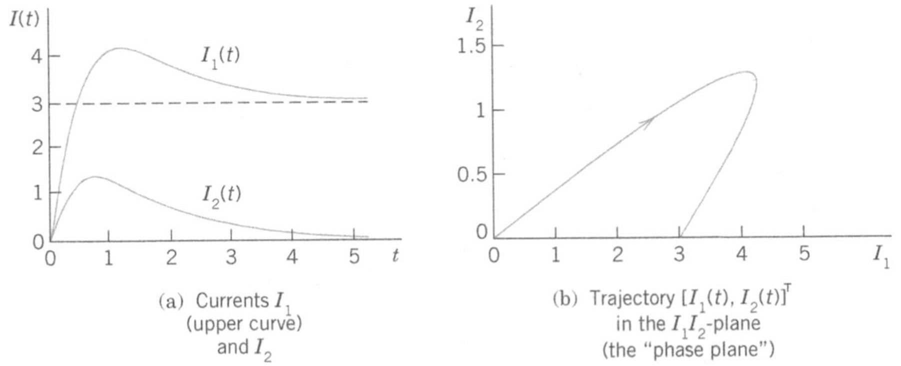

Fig. 80. Currents in Example 2

Remark. In both examples, by growing the dimension of the problem (from one tank to two tanks or one circuit to two circuits) we also increased the number of ODEs (from one ODE to two ODEs). This "growth" in the problem being reflected by an "increase" in the mathematical model is attractive and affirms the quality of our mathematical modeling and theory.

Conversion of an nth-Order ODE to a System

We show that an nth-order ODE of the general form (8) (see Theorem 1) can be converted to a system of n first-order ODEs. This is practically and theoretically important—practically because it permits the study and solution of single ODEs by methods for systems, and theoretically because it opens a way of including the theory of higher order ODEs into that of first-order systems. This conversion is another reason for the importance of systems, in addition to their use as models in various basic applications. The idea of the conversion is simple and straightforward, as follows.

THEOREM 1

Conversion of an ODE

An nth-order ODE

$$y^{(n)} = F(t, y, y', \cdots, y^{(n-1)}) \tag{8}$$

can be converted to a system of n first-order ODEs by setting

$$y_1 = y, \quad y_2 = y', \quad y_3 = y'', \cdots, y_n = y^{(n-1)}. \tag{9}$$

This system is of the form

$$\begin{aligned} y_1' &= y_2 \\ y_2' &= y_3 \\ &\vdots \\ y_{n-1}' &= y_n \\ y_n' &= F(t, y_1, y_2, \cdots, y_n). \end{aligned} \tag{10}$$

PROOF The first $n - 1$ of these n ODEs follows immediately from (9) by differentiation. Also, $y_n' = y^{(n)}$ by (9), so that the last equation in (10) results from the given ODE (8). ∎

EXAMPLE 3 **Mass on a Spring**

To gain confidence in the conversion method, let us apply it to an old friend of ours, modeling the free motions of a mass on a spring (see Sec. 2.4)

$$my'' + cy' + ky = 0 \qquad \text{or} \qquad y'' = -\frac{c}{m}y' - \frac{k}{m}y.$$

For this ODE (8) the system (10) is linear and homogeneous,

$$\begin{aligned} y_1' &= y_2 \\ y_2' &= -\frac{k}{m}y_1 - \frac{c}{m}y_2. \end{aligned}$$

Setting $\mathbf{y} = \begin{bmatrix} y_1 \\ y_2 \end{bmatrix}$, we get in matrix form

$$\mathbf{y}' = \mathbf{A}\mathbf{y} = \begin{bmatrix} 0 & 1 \\ -\dfrac{k}{m} & -\dfrac{c}{m} \end{bmatrix} \begin{bmatrix} y_1 \\ y_2 \end{bmatrix}.$$

The characteristic equation is

$$\det(\mathbf{A} - \lambda\mathbf{I}) = \begin{vmatrix} -\lambda & 1 \\ -\dfrac{k}{m} & -\dfrac{c}{m} - \lambda \end{vmatrix} = \lambda^2 + \frac{c}{m}\lambda + \frac{k}{m} = 0.$$

It agrees with that in Sec. 2.4. For an illustrative computation, let $m = 1, c = 2$, and $k = 0.75$. Then

$$\lambda^2 + 2\lambda + 0.75 = (\lambda + 0.5)(\lambda + 1.5) = 0.$$

This gives the eigenvalues $\lambda_1 = -0.5$ and $\lambda_2 = -1.5$. Eigenvectors follow from the first equation in $\mathbf{A} - \lambda\mathbf{I} = \mathbf{0}$, which is $-\lambda x_1 + x_2 = 0$. For λ_1 this gives $0.5x_1 + x_2 = 0$, say, $x_1 = 2$, $x_2 = -1$. For $\lambda_2 = -1.5$ it gives $1.5x_1 + x_2 = 0$, say, $x_1 = 1$, $x_2 = -1.5$. These eigenvectors

$$\mathbf{x}^{(1)} = \begin{bmatrix} 2 \\ -1 \end{bmatrix}, \quad \mathbf{x}^{(2)} = \begin{bmatrix} 1 \\ -1.5 \end{bmatrix} \quad \text{give} \quad \mathbf{y} = c_1 \begin{bmatrix} 2 \\ -1 \end{bmatrix} e^{-0.5t} + c_2 \begin{bmatrix} 1 \\ -1.5 \end{bmatrix} e^{-1.5t}.$$

This vector solution has the first component

$$y = y_1 = 2c_1 e^{-0.5t} + c_2 e^{-1.5t}$$

which is the expected solution. The second component is its derivative

$$y_2 = y_1' = y' = -c_1 e^{-0.5t} - 1.5c_2 e^{-1.5t}.$$ ■

PROBLEM SET 4.1

1–6 MIXING PROBLEMS

1. Find out, without calculation, whether doubling the flow rate in Example 1 has the same effect as halfing the tank sizes. (Give a reason.)
2. What happens in Example 1 if we replace T_1 by a tank containing 200 gal of water and 150 lb of fertilizer dissolved in it?
3. Derive the eigenvectors in Example 1 without consulting this book.
4. In Example 1 find a "general solution" for any ratio a = (flow rate)/(tank size), tank sizes being equal. Comment on the result.
5. If you extend Example 1 by a tank T_3 of the same size as the others and connected to T_2 by two tubes with flow rates as between T_1 and T_2, what system of ODEs will you get?
6. Find a "general solution" of the system in Prob. 5.

7–9 ELECTRICAL NETWORK

In Example 2 find the currents:

7. If the initial currents are 0 A and −3 A (minus meaning that $I_2(0)$ flows against the direction of the arrow).
8. If the capacitance is changed to $C = 5/27$ F. (General solution only.)
9. If the initial currents in Example 2 are 28 A and 14 A.

10–13 CONVERSION TO SYSTEMS

Find a general solution of the given ODE **(a)** by first converting it to a system, **(b)**, as given. Show the details of your work.

10. $y'' + 3y' + 2y = 0$ **11.** $4y'' - 15y' - 4y = 0$

12. $y''' + 2y'' - y' - 2y = 0$

13. $y'' + 2y' - 24y = 0$

14. TEAM PROJECT. Two Masses on Springs. (a) Set up the model for the (undamped) system in Fig. 81. **(b)** Solve the system of ODEs obtained. *Hint.* Try $\mathbf{y} = \mathbf{x}e^{\omega t}$ and set $\omega^2 = \lambda$. Proceed as in Example 1 or 2. **(c)** Describe the influence of initial conditions on the possible kind of motions.

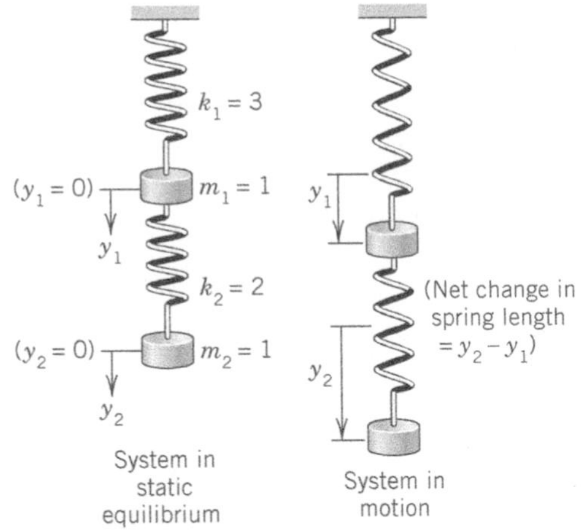

Fig. 81. Mechanical system in Team Project

15. CAS EXPERIMENT. Electrical Network. (a) In Example 2 choose a sequence of values of C that increases beyond bound, and compare the corresponding sequences of eigenvalues of $\mathbf{A}$. What limits of these sequences do your numeric values (approximately) suggest?

(b) Find these limits analytically.

(c) Explain your result physically.

(d) Below what value (approximately) must you decrease C to get vibrations?

4.2 Basic Theory of Systems of ODEs. Wronskian

In this section we discuss some basic concepts and facts about system of ODEs that are quite similar to those for single ODEs.

The first-order systems in the last section were special cases of the more general system

$$
\begin{aligned}
y_1' &= f_1(t, y_1, \cdots, y_n) \\
y_2' &= f_2(t, y_1, \cdots, y_n) \\
&\cdots \\
y_n' &= f_n(t, y_1, \cdots, y_n).
\end{aligned} \tag{1}
$$

We can write the system (1) as a vector equation by introducing the column vectors $\mathbf{y} = [y_1 \ \cdots \ y_n]^\mathsf{T}$ and $\mathbf{f} = [f_1 \ \cdots \ f_n]^\mathsf{T}$ (where T means *transposition* and saves us the space that would be needed for writing $\mathbf{y}$ and $\mathbf{f}$ as columns). This gives

$$\mathbf{y}' = \mathbf{f}(t, \mathbf{y}). \tag{1}$$

This system (1) includes almost all cases of practical interest. For $n = 1$ it becomes $y_1' = f_1(t, y_1)$ or, simply, $y' = f(t, y)$, well known to us from Chap. 1.

A **solution** of (1) on some interval $a < t < b$ is a set of n differentiable functions

$$y_1 = h_1(t), \quad \cdots, \quad y_n = h_n(t)$$

on $a < t < b$ that satisfy (1) throughout this interval. In vector from, introducing the "*solution vector*" $\mathbf{h} = [h_1 \ \cdots \ h_n]^\mathsf{T}$ (a column vector!) we can write

$$\mathbf{y} = \mathbf{h}(t).$$

An **initial value problem** for (1) consists of (1) and n given **initial conditions**

$$y_1(t_0) = K_1, \qquad y_2(t_0) = K_2, \qquad \cdots, \qquad y_n(t_0) = K_n, \tag{2}$$

in vector form, $\mathbf{y}(t_0) = \mathbf{K}$, where t_0 is a specified value of t in the interval considered and the components of $\mathbf{K} = [K_1 \ \cdots \ K_n]^\mathsf{T}$ are given numbers. Sufficient conditions for the existence and uniqueness of a solution of an initial value problem (1), (2) are stated in the following theorem, which extends the theorems in Sec. 1.7 for a single equation. (For a proof, see Ref. [A7].)

THEOREM 1

Existence and Uniqueness Theorem

Let $f_1, \cdots, f_n$ in (1) be continuous functions having continuous partial derivatives $\partial f_1/\partial y_1, \cdots, \partial f_1/\partial y_n, \cdots, \partial f_n/\partial y_n$ in some domain R of $ty_1y_2\cdots y_n$-space containing the point $(t_0, K_1, \cdots, K_n)$. Then (1) has a solution on some interval $t_0 - \alpha < t < t_0 + \alpha$ satisfying (2), and this solution is unique.

Linear Systems

Extending the notion of a *linear* ODE, we call (1) a **linear system** if it is linear in $y_1, \cdots, y_n$; that is, if it can be written

$$\begin{aligned} y_1' &= a_{11}(t)y_1 + \cdots + a_{1n}(t)y_n + g_1(t) \\ &\vdots \\ y_n' &= a_{n1}(t)y_1 + \cdots + a_{nn}(t)y_n + g_n(t). \end{aligned} \tag{3}$$

As a vector equation this becomes

$$\mathbf{y}' = \mathbf{A}\mathbf{y} + \mathbf{g} \tag{3}$$

where $\mathbf{A} = \begin{bmatrix} a_{11} & \cdots & a_{1n} \\ \cdot & \cdots & \cdot \\ a_{n1} & \cdots & a_{nn} \end{bmatrix}, \quad \mathbf{y} = \begin{bmatrix} y_1 \\ \vdots \\ y_n \end{bmatrix}, \quad \mathbf{g} = \begin{bmatrix} g_1 \\ \vdots \\ g_n \end{bmatrix}.$

This system is called **homogeneous** if $\mathbf{g} = \mathbf{0}$, so that it is

$$\mathbf{y}' = \mathbf{A}\mathbf{y}. \tag{4}$$

If $\mathbf{g} \neq \mathbf{0}$, then (3) is called **nonhomogeneous**. For example, the systems in Examples 1 and 3 of Sec. 4.1 are homogeneous. The system in Example 2 of that section is nonhomogeneous.

For a linear system (3) we have $\partial f_1/\partial y_1 = a_{11}(t), \cdots, \partial f_n/\partial y_n = a_{nn}(t)$ in Theorem 1. Hence for a linear system we simply obtain the following.

THEOREM 2

Existence and Uniqueness in the Linear Case

Let the a_{jk}'s and g_j's in (3) *be continuous functions of t on an open interval $\alpha < t < \beta$ containing the point $t = t_0$. Then* (3) *has a solution $\mathbf{y}(t)$ on this interval satisfying* (2), *and this solution is unique.*

As for a single homogeneous linear ODE we have

THEOREM 3

Superposition Principle or Linearity Principle

If $\mathbf{y}^{(1)}$ and $\mathbf{y}^{(2)}$ are solutions of the **homogeneous linear** system (4) *on some interval, so is any linear combination $\mathbf{y} = c_1\mathbf{y}^{(1)} + c_1\mathbf{y}^{(2)}$.*

PROOF Differentiating and using (4), we obtain

$$\begin{aligned} \mathbf{y}' &= [c_1\mathbf{y}^{(1)} + c_1\mathbf{y}^{(2)}]' \\ &= c_1\mathbf{y}^{(1)\prime} + c_2\mathbf{y}^{(2)\prime} \\ &= c_1\mathbf{A}\mathbf{y}^{(1)} + c_2\mathbf{A}\mathbf{y}^{(2)} \\ &= \mathbf{A}(c_1\mathbf{y}^{(1)} + c_2\mathbf{y}^{(2)}) = \mathbf{A}\mathbf{y}. \end{aligned}$$ ■

The general theory of linear systems of ODEs is quite similar to that of a single linear ODE in Secs. 2.6 and 2.7. To see this, we explain the most basic concepts and facts. For proofs we refer to more advanced texts, such as [A7].

Basis. General Solution. Wronskian

By a **basis** or a **fundamental system** of solutions of the homogeneous system (4) on some interval J we mean a linearly independent set of n solutions $\mathbf{y}^{(1)}, \cdots, \mathbf{y}^{(n)}$ of (4) on that interval. (We write J because we need $\mathbf{I}$ to denote the unit matrix.) We call a corresponding linear combination

$$\mathbf{y} = c_1\mathbf{y}^{(1)} \cdots + c_n\mathbf{y}^{(n)} \qquad (c_1, \cdots, c_n \text{ arbitrary}) \tag{5}$$

a **general solution** of (4) on J. It can be shown that if the $a_{jk}(t)$ in (4) are continuous on J, then (4) has a basis of solutions on J, hence a general solution, which includes every solution of (4) on J.

We can write n solutions $\mathbf{y}^{(1)}, \cdots, \mathbf{y}^{(n)}$ of (4) on some interval J as columns of an $n \times n$ matrix

$$\mathbf{Y} = [\mathbf{y}^{(1)} \quad \cdots \quad \mathbf{y}^{(n)}]. \tag{6}$$

The determinant of $\mathbf{Y}$ is called the **Wronskian** of $\mathbf{y}^{(1)}, \cdots, \mathbf{y}^{(n)}$, written

$$W(\mathbf{y}^{(1)}, \cdots, \mathbf{y}^{(n)}) = \begin{vmatrix} y_1^{(1)} & y_1^{(2)} & \cdots & y_1^{(n)} \\ y_2^{(1)} & y_2^{(2)} & \cdots & y_2^{(n)} \\ \cdot & \cdot & \cdots & \cdot \\ y_n^{(1)} & y_n^{(2)} & \cdots & y_n^{(n)} \end{vmatrix}. \tag{7}$$

The columns are these solutions, each in terms of components. These solutions form a basis on J if and only if W is not zero at any t_1 in this interval. W is either identically zero or nowhere zero in J. (This is similar to Secs. 2.6 and 3.1.)

If the solutions $\mathbf{y}^{(1)}, \cdots, \mathbf{y}^{(n)}$ in (5) form a basis (a fundamental system), then (6) is often called a **fundamental matrix**. Introducing a column vector $\mathbf{c} = [c_1 \quad c_2 \quad \cdots \quad c_n]^\mathsf{T}$, we can now write (5) simply as

$$\mathbf{y} = \mathbf{Y}\mathbf{c}. \tag{8}$$

Furthermore, we can relate (7) to Sec. 2.6, as follows. If y and z are solutions of a second-order homogeneous linear ODE, their Wronskian is

$$W(y, z) = \begin{vmatrix} y & z \\ y' & z' \end{vmatrix}.$$

To write this ODE as a system, we have to set $y = y_1, y' = y_1' = y_2$ and similarly for z (see Sec. 4.1). But then $W(y, z)$ becomes (7), except for notation.

4.3 Constant-Coefficient Systems. Phase Plane Method

Continuing, we now assume that our homogeneous linear system

$$\mathbf{y}' = \mathbf{A}\mathbf{y} \tag{1}$$

under discussion has ***constant coefficients***, so that the $n \times n$ matrix $\mathbf{A} = [a_{jk}]$ has entries not depending on t. We want to solve (1). Now a single ODE $y' = ky$ has the solution $y = Ce^{kt}$. So let us try

$$\mathbf{y} = \mathbf{x}e^{\lambda t}. \tag{2}$$

Substitution into (1) gives $\mathbf{y}' = \lambda\mathbf{x}e^{\lambda t} = \mathbf{A}\mathbf{y} = \mathbf{A}\mathbf{x}e^{\lambda t}$. Dividing by $e^{\lambda t}$, we obtain the **eigenvalue problem**

$$\mathbf{A}\mathbf{x} = \lambda\mathbf{x}. \tag{3}$$

Thus the nontrivial solutions of (1) (solutions that are not zero vectors) are of the form (2), where λ is an eigenvalue of $\mathbf{A}$ and $\mathbf{x}$ is a corresponding eigenvector.

We assume that $\mathbf{A}$ has a linearly independent set of n eigenvectors. This holds in most applications, in particular if $\mathbf{A}$ is symmetric ($a_{kj} = a_{jk}$) or skew-symmetric ($a_{kj} = -a_{jk}$) or has n *different* eigenvalues.

Let those eigenvectors be $\mathbf{x}^{(1)}, \cdots, \mathbf{x}^{(n)}$ and let them correspond to eigenvalues $\lambda_1, \cdots, \lambda_n$ (which may be all different, or some—or even all—may be equal). Then the corresponding solutions (2) are

$$\mathbf{y}^{(4)} = \mathbf{x}^{(1)}e^{\lambda_1 t}, \quad \cdots, \quad \mathbf{y}^{(n)} = \mathbf{x}^{(n)}e^{\lambda_n t}. \tag{4}$$

Their Wronskian $W = W(\mathbf{y}^{(1)}, \cdots, \mathbf{y}^{(n)})$ [(7) in Sec. 4.2] is given by

$$W = (\mathbf{y}^{(1)}, \cdots, \mathbf{y}^{(n)}) = \begin{vmatrix} x_1^{(1)}e^{\lambda_1 t} & \cdots & x_1^{(n)}e^{\lambda_n t} \\ x_2^{(1)}e^{\lambda_1 t} & \cdots & x_2^{(n)}e^{\lambda_n t} \\ \cdot & \cdots & \cdot \\ x_n^{(1)}e^{\lambda_1 t} & \cdots & x_n^{(n)}e^{\lambda_n t} \end{vmatrix} = e^{\lambda_1 t + \cdots + \lambda_n t} \begin{vmatrix} x_1^{(1)} & \cdots & x_1^{(n)} \\ x_2^{(1)} & \cdots & x_2^{(n)} \\ \cdot & \cdots & \cdot \\ x_n^{(1)} & \cdots & x_n^{(n)} \end{vmatrix}.$$

On the right, the exponential function is never zero, and the determinant is not zero either because its columns are the n linearly independent eigenvectors. This proves the following theorem, whose assumption is true if the matrix $\mathbf{A}$ is symmetric or skew-symmetric, or if the n eigenvalues of $\mathbf{A}$ are all different.

THEOREM 1

General Solution

If the constant matrix **A** *in the system* (1) *has a linearly independent set of n eigenvectors, then the corresponding solutions* $\mathbf{y}^{(1)}, \cdots, \mathbf{y}^{(n)}$ *in* (4) *form a basis of solutions of* (1), *and the corresponding general solution is*

$$\mathbf{y} = c_1\mathbf{x}^{(1)}e^{\lambda_1 t} + \cdots + c_n\mathbf{x}^{(n)}e^{\lambda_n t}. \tag{5}$$

How to Graph Solutions in the Phase Plane

We shall now concentrate on systems (1) with constant coefficients consisting of two ODEs

$$\mathbf{y}' = \mathbf{A}\mathbf{y}; \qquad \text{in components,} \qquad \begin{aligned} y_1' &= a_{11}y_1 + a_{12}y_2 \\ y_2' &= a_{21}y_1 + a_{22}y_2. \end{aligned} \tag{6}$$

Of course, we can graph solutions of (6),

$$\mathbf{y}(t) = \begin{bmatrix} y_1(t) \\ y_2(t) \end{bmatrix}, \tag{7}$$

as two curves over the t-axis, one for each component of $\mathbf{y}(t)$. (Figure 80a in Sec. 4.1 shows an example.) But we can also graph (7) as a single curve in the y_1y_2-plane. This is a *parametric representation* (*parametric equation*) with parameter t. (See Fig. 80b for an example. Many more follow. Parametric equations also occur in calculus.) Such a curve is called a **trajectory** (or sometimes an *orbit* or *path*) of (6). The y_1y_2-plane is called the **phase plane**.[1] If we fill the phase plane with trajectories of (6), we obtain the so-called **phase portrait** of (6).

Studies of solutions in the phase plane have become quite important, along with advances in computer graphics, because a phase portrait gives a good general qualitative impression of the entire family of solutions. Consider the following example, in which we develop such a phase portrait.

EXAMPLE 1

Trajectories in the Phase Plane (Phase Portrait)

Find and graph solutions of the system.

In order to see what is going on, let us find and graph solutions of the system

$$\mathbf{y}' = \mathbf{A}\mathbf{y} = \begin{bmatrix} -3 & 1 \\ 1 & -3 \end{bmatrix}\mathbf{y}, \qquad \text{thus} \qquad \begin{aligned} y_1' &= -3y_1 + y_2 \\ y_2' &= y_1 - 3y_2. \end{aligned} \tag{8}$$

[1] A name that comes from physics, where it is the y-(mv)-plane, used to plot a motion in terms of position y and velocity $y' = v$ (m = mass); but the name is now used quite generally for the y_1y_2-plane.

The use of the phase plane is a **qualitative method**, a method of obtaining general qualitative information on solutions without actually solving an ODE or a system. This method was created by HENRI POINCARÉ (1854–1912), a great French mathematician, whose work was also fundamental in complex analysis, divergent series, topology, and astronomy.

Solution. By substituting $\mathbf{y} = \mathbf{x}e^{\lambda t}$ and $\mathbf{y}' = \lambda \mathbf{x}e^{\lambda t}$ and dropping the exponential function we get $\mathbf{Ax} = \lambda\mathbf{x}$. The characteristic equation is

$$\det(\mathbf{A} - \lambda\mathbf{I}) = \begin{vmatrix} -3-\lambda & 1 \\ 1 & -3-\lambda \end{vmatrix} = \lambda^2 + 6\lambda + 8 = 0.$$

This gives the eigenvalues $\lambda_1 = -2$ and $\lambda_2 = -4$. Eigenvectors are then obtained from

$$(-3-\lambda)x_1 + x_2 = 0.$$

For $\lambda_1 = -2$ this is $-x_1 + x_2 = 0$. Hence we can take $\mathbf{x}^{(1)} = [1 \quad 1]^\mathsf{T}$. For $\lambda_2 = -4$ this becomes $x_1 + x_2 = 0$, and an eigenvector is $\mathbf{x}^{(2)} = [1 \quad -1]^\mathsf{T}$. This gives the general solution

$$\mathbf{y} = \begin{bmatrix} y_1 \\ y_2 \end{bmatrix} = c_1\mathbf{y}^{(1)} + c_2\mathbf{y}^{(2)} = c_1\begin{bmatrix} 1 \\ 1 \end{bmatrix}e^{-2t} + c_2\begin{bmatrix} 1 \\ -1 \end{bmatrix}e^{-4t}.$$

Figure 82 shows a phase portrait of some of the trajectories (to which more trajectories could be added if so desired). The two straight trajectories correspond to $c_1 = 0$ and $c_2 = 0$ and the others to other choices of c_1, c_2. ■

The method of the phase plane is particularly valuable in the frequent cases when solving an ODE or a system is inconvenient of impossible.

Critical Points of the System (6)

The point $\mathbf{y} = \mathbf{0}$ in Fig. 82 seems to be a common point of all trajectories, and we want to explore the reason for this remarkable observation. The answer will follow by calculus. Indeed, from (6) we obtain

$$\frac{dy_2}{dy_1} = \frac{y_2'\,dt}{y_1'\,dt} = \frac{y_2'}{y_1'} = \frac{a_{21}y_1 + a_{22}y_2}{a_{11}y_1 + a_{12}y_2}. \tag{9}$$

This associates with every point P: (y_1, y_2) a unique tangent direction dy_2/dy_1 of the trajectory passing through P, except for the point $P = P_0$: $(0, 0)$, where the right side of (9) becomes $0/0$. This point P_0, at which dy_2/dy_1 becomes undetermined, is called a **critical point** of (6).

Five Types of Critical Points

There are five types of critical points depending on the geometric shape of the trajectories near them. They are called **improper nodes, proper nodes, saddle points, centers**, and **spiral points**. We define and illustrate them in Examples 1–5.

EXAMPLE 1 **(*Continued*) Improper Node (Fig. 82)**

An **improper node** is a critical point P_0 at which all the trajectories, except for two of them, have the same limiting direction of the tangent. The two exceptional trajectories also have a limiting direction of the tangent at P_0 which, however, is different.

The system (8) has an improper node at $\mathbf{0}$, as its phase portrait Fig. 82 shows. The common limiting direction at $\mathbf{0}$ is that of the eigenvector $\mathbf{x}^{(1)} = [1 \quad 1]^\mathsf{T}$ because e^{-4t} goes to zero faster than e^{-2t} as t increases. The two exceptional limiting tangent directions are those of $\mathbf{x}^{(2)} = [1 \quad -1]^\mathsf{T}$ and $-\mathbf{x}^{(2)} = [-1 \quad 1]^\mathsf{T}$. ■

EXAMPLE 2 Proper Node (Fig. 83)

A **proper node** is a critical point P_0 at which every trajectory has a definite limiting direction and for any given direction **d** at P_0 there is a trajectory having **d** as its limiting direction.

The system

$$\text{(10)} \qquad \mathbf{y}' = \begin{bmatrix} 1 & 0 \\ 0 & 1 \end{bmatrix} \mathbf{y}, \qquad \text{thus} \qquad \begin{aligned} y_1' &= y_1 \\ y_2' &= y_2 \end{aligned}$$

has a proper node at the origin (see Fig. 83). Indeed, the matrix is the unit matrix. Its characteristic equation $(1 - \lambda)^2 = 0$ has the root $\lambda = 1$. Any $\mathbf{x} \neq \mathbf{0}$ is an eigenvector, and we can take $[1 \quad 0]^\mathsf{T}$ and $[0 \quad 1]^\mathsf{T}$. Hence a general solution is

$$\mathbf{y} = c_1 \begin{bmatrix} 1 \\ 0 \end{bmatrix} e^t + c_2 \begin{bmatrix} 0 \\ 1 \end{bmatrix} e^t \qquad \text{or} \qquad \begin{aligned} y_1 &= c_1 e^t \\ y_2 &= c_2 e^t \end{aligned} \qquad \text{or} \qquad c_1 y_2 = c_2 y_1. \qquad ■$$

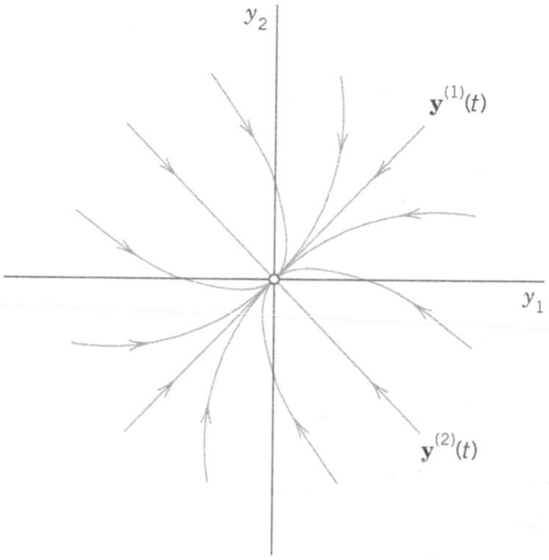

Fig. 82. Trajectories of the system (8) (Improper node)

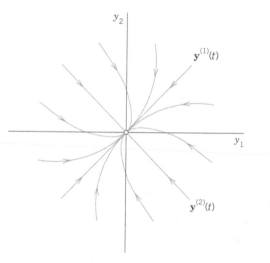

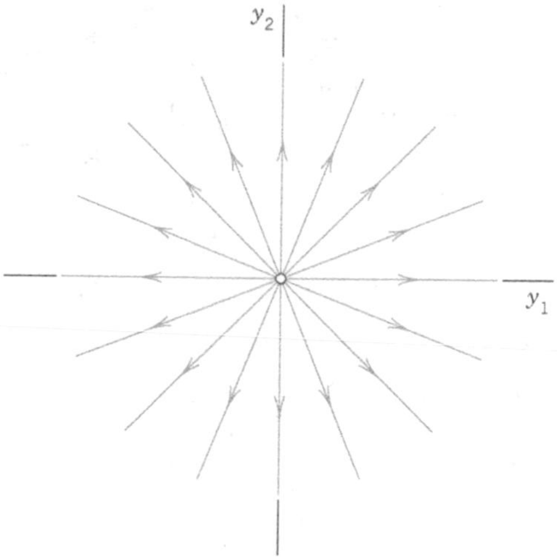

Fig. 83. Trajectories of the system (10) (Proper node)

EXAMPLE 3 Saddle Point (Fig. 84)

A **saddle point** is a critical point P_0 at which there are two incoming trajectories, two outgoing trajectories, and all the other trajectories in a neighborhood of P_0 bypass P_0.

The system

$$\text{(11)} \qquad \mathbf{y}' = \begin{bmatrix} 1 & 0 \\ 0 & -1 \end{bmatrix} \mathbf{y}, \qquad \text{thus} \qquad \begin{aligned} y_1' &= \ \ y_1 \\ y_1' &= -y_2 \end{aligned}$$

has a saddle point at the origin. Its characteristic equation $(1 - \lambda)(-1 - \lambda) = 0$ has the roots $\lambda_1 = 1$ and $\lambda_2 = -1$. For $\lambda = 1$ an eigenvector $[1 \quad 0]^\mathsf{T}$ is obtained from the second row of $(\mathbf{A} - \lambda\mathbf{I})\mathbf{x} = \mathbf{0}$, that is, $0x_1 + (-1 - 1)x_2 = 0$. For $\lambda_2 = -1$ the first row gives $[0 \quad 1]^\mathsf{T}$. Hence a general solution is

$$\mathbf{y} = c_1 \begin{bmatrix} 1 \\ 0 \end{bmatrix} e^t + c_2 \begin{bmatrix} 0 \\ 1 \end{bmatrix} e^{-t} \qquad \text{or} \qquad \begin{aligned} y_1 &= c_1 e^t \\ y_2 &= c_2 e^{-t} \end{aligned} \qquad \text{or} \qquad y_1 y_2 = \text{const.}$$

This is a family of hyperbolas (and the coordinate axes); see Fig. 84. ■

EXAMPLE 4 Center (Fig. 85)

A **center** is a critical point that is enclosed by infinitely many closed trajectories.

The system

$$\mathbf{y}' = \begin{bmatrix} 0 & 1 \\ -4 & 0 \end{bmatrix}\mathbf{y}, \qquad \text{thus} \qquad \begin{aligned} &\text{(a)}\quad y_1' = y_2 \\ &\text{(b)}\quad y_2' = -4y_1 \end{aligned} \tag{12}$$

has a center at the origin. The characteristic equation $\lambda^2 + 4 = 0$ gives the eigenvalues $2i$ and $-2i$. For $2i$ an eigenvector follows from the first equation $-2ix_1 + x_2 = 0$ of $(\mathbf{A} - \lambda\mathbf{I})\mathbf{x} = \mathbf{0}$, say, $[1 \quad 2i]^\mathsf{T}$. For $\lambda = -2i$ that equation is $-(-2i)x_1 + x_2 = 0$ and gives, say, $[1 \quad -2i]^\mathsf{T}$. Hence a complex general solution is

$$\mathbf{y} = c_1 \begin{bmatrix} 1 \\ 2i \end{bmatrix} e^{2it} + c_2 \begin{bmatrix} 1 \\ -2i \end{bmatrix} e^{-2it}, \qquad \text{thus} \qquad \begin{aligned} y_1 &= c_1e^{2it} + c_2e^{-2it} \\ y_2 &= 2ic_1e^{2it} - 2ic_2e^{-2it}. \end{aligned} \tag{12*}$$

A real solution is obtained from (12*) by the Euler formula or directly from (12) by a trick. (Remember the trick and call it a method when you apply it again.) Namely, the left side of (a) times the right side of (b) is $-4y_1y_1'$. This must equal the left side of (b) times the right side of (a). Thus,

$$-4y_1y_1' = y_2y_2'. \qquad \text{By integration,} \qquad 2y_1^2 + \tfrac{1}{2}y_2^2 = \text{const.}$$

This is a family of ellipses (see Fig. 85) enclosing the center at the origin. ■

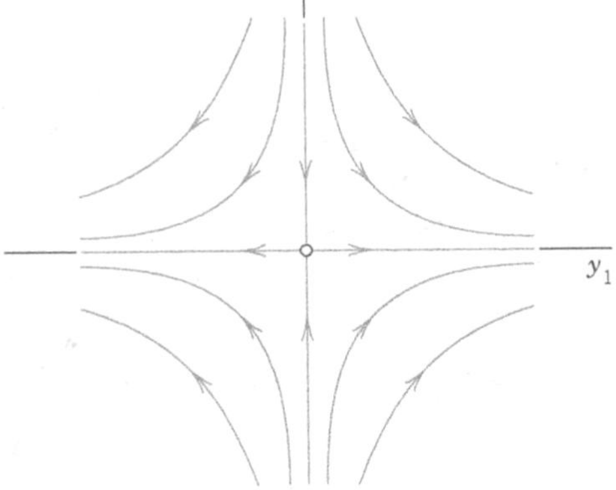

Fig. 84. Trajectories of the system (11) (Saddle point)

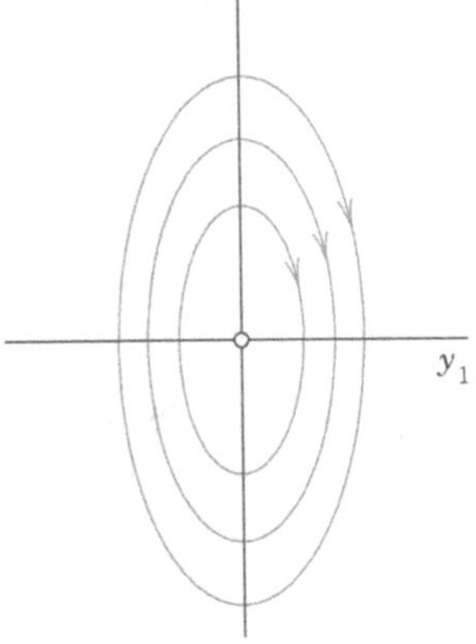

Fig. 85. Trajectories of the system (12) (Center)

EXAMPLE 5 Spiral Point (Fig. 86)

A **spiral point** is a critical point P_0 about which the trajectories spiral, approaching P_0 as $t \to \infty$ (or tracing these spirals in the opposite sense, away from P_0).

The system

$$\mathbf{y}' = \begin{bmatrix} -1 & 1 \\ -1 & -1 \end{bmatrix}\mathbf{y}, \qquad \text{thus} \qquad \begin{aligned} y_1' &= -y_1 + y_2 \\ y_2' &= -y_1 - y_2 \end{aligned} \tag{13}$$

has a spiral point at the origin, as we shall see. The characteristic equation is $\lambda^2 + 2\lambda + 2 = 0$. It gives the eigenvalues $-1 + i$ and $-1 - i$. Corresponding eigenvectors are obtained from $(-1 - \lambda)x_1 + x_2 = 0$. For

$\lambda = -1 + i$ this becomes $-ix_1 + x_2 = 0$ and we can take $[1 \quad i]^\mathsf{T}$ as an eigenvector. Similarly, an eigenvector corresponding to $-1 - i$ is $[1 \quad -i]^\mathsf{T}$. This gives the complex general solution

$$\mathbf{y} = c_1 \begin{bmatrix} 1 \\ i \end{bmatrix} e^{(-1+i)t} + c_2 \begin{bmatrix} 1 \\ -i \end{bmatrix} e^{(-1-i)t}.$$

The next step would be the transformation of this complex solution to a real general solution by the Euler formula. But, as in the last example, we just wanted to see what eigenvalues to expect in the case of a spiral point. Accordingly, we start again from the beginning and instead of that rather lengthy systematic calculation we use a shortcut. We multiply the first equation in (13) by y_1, the second by y_2, and add, obtaining

$$y_1 y_1' + y_2 y_2' = -(y_1^2 + y_2^2).$$

We now introduce polar coordinates r, t, where $r^2 = y_1^2 + y_2^2$. Differentiating this with respect to t gives $2rr' = 2y_1 y_1' + 2y_2 y_2'$. Hence the previous equation can be written

$$rr' = -r^2, \qquad \text{Thus,} \qquad r' = -r, \qquad dr/r = -dt, \qquad \ln |r| = -t + c^*, \qquad r = ce^{-t}.$$

For each real c this is a spiral, as claimed (see Fig. 86). ■

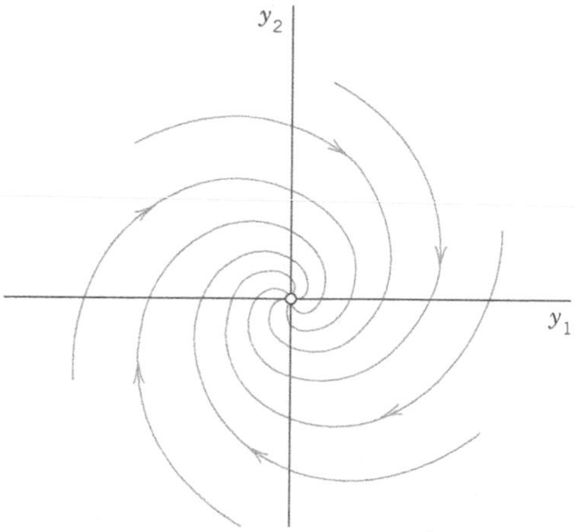

Fig. 86. Trajectories of the system (13) (Spiral point)

EXAMPLE 6 No Basis of Eigenvectors Available. Degenerate Node (Fig. 87)

This cannot happen if $\mathbf{A}$ in (1) is symmetric ($a_{kj} = a_{jk}$, as in Examples 1–3) or skew-symmetric ($a_{kj} = -a_{jk}$, thus $a_{jj} = 0$). And it does not happen in many other cases (see Examples 4 and 5). Hence it suffices to explain the method to be used by an example.

Find and graph a general solution of

$$\text{(14)} \qquad \mathbf{y}' = \mathbf{A}\mathbf{y} = \begin{bmatrix} 4 & 1 \\ -1 & 2 \end{bmatrix} \mathbf{y}.$$

Solution. $\mathbf{A}$ is not skew-symmetric! Its characteristic equation is

$$\det(\mathbf{A} - \lambda\mathbf{I}) = \begin{vmatrix} 4-\lambda & 1 \\ -1 & 2-\lambda \end{vmatrix} = \lambda^2 - 6\lambda + 9 = (\lambda - 3)^2 = 0.$$

It has a double root $\lambda = 3$. Hence eigenvectors are obtained from $(4 - \lambda)x_1 + x_2 = 0$, thus from $x_1 + x_2 = 0$, say, $\mathbf{x}^{(1)} = [1 \quad -1]^\mathsf{T}$ and nonzero multiples of it (which do not help). The method now is to substitute

$$\mathbf{y}^{(2)} = \mathbf{x}te^{\lambda t} + \mathbf{u}e^{\lambda t}$$

with constant $\mathbf{u} = [u_1 \quad u_2]^\mathsf{T}$ into (14). (The $\mathbf{x}t$-term alone, the analog of what we did in Sec. 2.2 in the case of a double root, would not be enough. Try it.) This gives

$$\mathbf{y}^{(2)\prime} = \mathbf{x}e^{\lambda t} + \lambda\mathbf{x}te^{\lambda t} + \lambda\mathbf{u}e^{\lambda t} = \mathbf{A}\mathbf{y}^{(2)} = \mathbf{A}\mathbf{x}te^{\lambda t} + \mathbf{A}\mathbf{u}e^{\lambda t}.$$

On the right, $\mathbf{A}\mathbf{x} = \lambda\mathbf{x}$. Hence the terms $\lambda\mathbf{x}te^{\lambda t}$ cancel, and then division by $e^{\lambda t}$ gives

$$\mathbf{x} + \lambda\mathbf{u} = \mathbf{A}\mathbf{u}, \qquad \text{thus} \qquad (\mathbf{A} - \lambda\mathbf{I})\mathbf{u} = \mathbf{x}.$$

Here $\lambda = 3$ and $\mathbf{x} = [1 \quad -1]^\mathsf{T}$, so that

$$(\mathbf{A} - 3\mathbf{I})\mathbf{u} = \begin{bmatrix} 4-3 & 1 \\ -1 & 2-3 \end{bmatrix}\mathbf{u} = \begin{bmatrix} 1 \\ -1 \end{bmatrix}, \qquad \text{thus} \qquad \begin{aligned} u_1 + u_2 &= 1 \\ -u_1 - u_2 &= -1. \end{aligned}$$

A solution, linearly independent of $\mathbf{x} = [1 \quad -1]^\mathsf{T}$, is $\mathbf{u} = [0 \quad 1]^\mathsf{T}$. This yields the answer (Fig. 87)

$$\mathbf{y} = c_1\mathbf{y}^{(1)} + c_2\mathbf{y}^{(2)} = c_1\begin{bmatrix} 1 \\ -1 \end{bmatrix}e^{3t} + c_2\left(\begin{bmatrix} 1 \\ -1 \end{bmatrix}t + \begin{bmatrix} 0 \\ 1 \end{bmatrix}\right)e^{3t}.$$

The critical point at the origin is often called a **degenerate node**. $c_1\mathbf{y}^{(1)}$ gives the heavy straight line, with $c_1 > 0$ the lower part and $c_1 < 0$ the upper part of it. $\mathbf{y}^{(2)}$ gives the right part of the heavy curve from 0 through the second, first, and—finally—fourth quadrants. $-\mathbf{y}^{(2)}$ gives the other part of that curve. ■

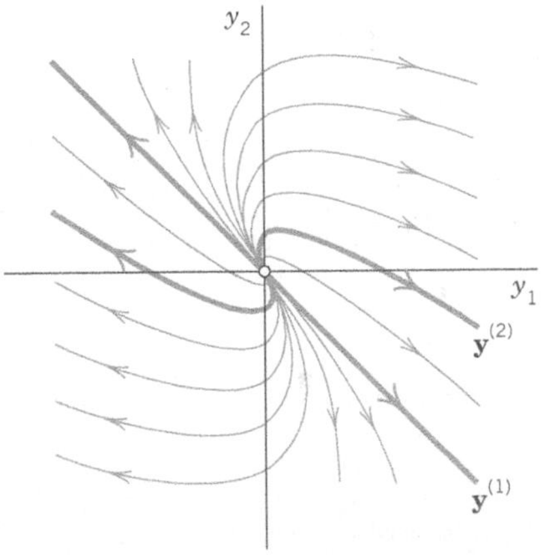

Fig. 87. Degenerate node in Example 6

We mention that for a system (1) with three or more equations and a triple eigenvalue with only one linearly independent eigenvector, one will get two solutions, as just discussed, and a third linearly independent one from

$$\mathbf{y}^{(3)} = \tfrac{1}{2}\mathbf{x}t^2e^{\lambda t} + \mathbf{u}te^{\lambda t} + \mathbf{v}e^{\lambda t} \qquad \text{with } \mathbf{v} \text{ from} \qquad \mathbf{u} + \lambda\mathbf{v} = \mathbf{A}\mathbf{v}.$$

PROBLEM SET 4.3

1–9 GENERAL SOLUTION

Find a real general solution of the following systems. Show the details.

1. $y_1' = y_1 + y_2$
$y_2' = 3y_1 - y_2$

2. $y_1' = 6y_1 + 9y_2$
$y_2' = y_1 + 6y_2$

3. $y_1' = y_1 + 2y_2$
$y_2' = \frac{1}{2}y_1 + y_2$

4. $y_1' = -8y_1 - 2y_2$
$y_2' = 2y_1 - 4y_2$

5. $y_1' = 2y_1 + 5y_2$
$y_2' = 5y_1 + 12.5y_2$

6. $y_1' = 2y_1 - 2y_2$
$y_2' = 2y_1 + 2y_2$

7. $y_1' = y_2$
$y_2' = -y_1 + y_3$
$y_3' = -y_2$

8. $y_1' = 8y_1 - y_2$
$y_2' = y_1 + 10y_2$

9. $y_1' = 10y_1 - 10y_2 - 4y_3$
$y_2' = -10y_1 + y_2 - 14y_3$
$y_3' = -4y_1 - 14y_2 - 2y_3$

10–15 IVPs

Solve the following initial value problems.

10. $y_1' = 2y_1 + 2y_2$
$y_2' = 5y_1 - y_2$
$y_1(0) = 0, \quad y_2(0) = 7$

11. $y_1' = 2y_1 + 5y_2$
$y_2' = -\frac{1}{2}y_1 - \frac{3}{2}y_2$
$y_1(0) = -12, \quad y_2(0) = 0$

12. $y_1' = y_1 + 3y_2$
$y_2' = \frac{1}{3}y_1 + y_2$
$y_1(0) = 12, \quad y_2(0) = 2$

13. $y_1' = y_2$
$y_2' = y_1$
$y_1(0) = 0, \quad y_2(0) = 2$

14. $y_1' = -y_1 - y_2$
$y_2' = y_1 - y_2$
$y_1(0) = 1, \quad y_2(0) = 0$

15. $y_1' = 3y_1 + 2y_2$
$y_2' = 2y_1 + 3y_2$
$y_1(0) = 0.5, \quad y_2(0) = -0.5$

16–17 CONVERSION

Find a general solution by conversion to a single ODE.

16. The system in Prob. 8.

17. The system in Example 5 of the text.

18. Mixing problem, Fig. 88. Each of the two tanks contains 200 gal of water, in which initially 100 lb (Tank T_1) and 200 lb (Tank T_2) of fertilizer are dissolved. The inflow, circulation, and outflow are shown in Fig. 88. The mixture is kept uniform by stirring. Find the fertilizer contents $y_1(t)$ in T_1 and $y_2(t)$ in T_2.

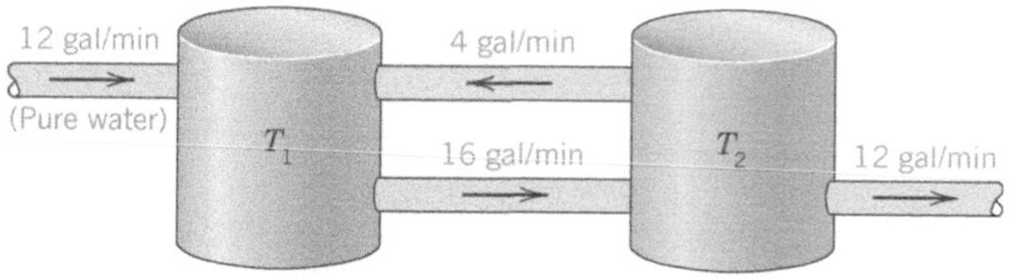

Fig. 88. Tanks in Problem 18

19. Network. Show that a model for the currents $I_1(t)$ and $I_2(t)$ in Fig. 89 is

$$\frac{1}{C}\int I_1\,dt + R(I_1 - I_2) = 0, \quad LI_2' + R(I_2 - I_1) = 0.$$

Find a general solution, assuming that $R = 3\ \Omega$, $L = 4$ H, $C = 1/12$ F.

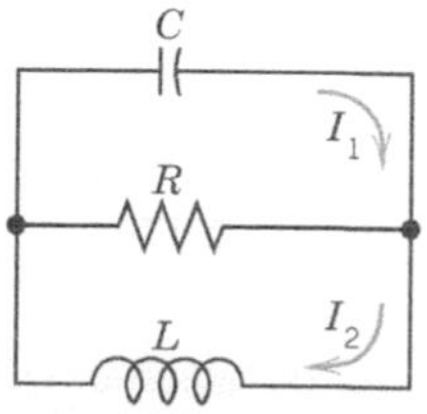

Fig. 89. Network in Problem 19

20. CAS PROJECT. Phase Portraits. Graph some of the figures in this section, in particular Fig. 87 on the degenerate node, in which the vector $\mathbf{y}^{(2)}$ depends on t. In each figure highlight a trajectory that satisfies an initial condition of your choice.

4.4 Criteria for Critical Points. Stability

We continue our discussion of homogeneous linear systems with constant coefficients (1). Let us review where we are. From Sec. 4.3 we have

$$(1) \qquad \mathbf{y}' = \mathbf{A}\mathbf{y} = \begin{bmatrix} a_{11} & a_{12} \\ a_{21} & a_{22} \end{bmatrix} \mathbf{y}, \qquad \text{in components,} \qquad \begin{aligned} y_1' &= a_{11}y_1 + a_{12}y_2 \\ y_2' &= a_{21}y_1 + a_{22}y_2. \end{aligned}$$

From the examples in the last section, we have seen that we can obtain an overview of families of solution curves if we represent them parametrically as $\mathbf{y}(t) = [y_1(t) \quad y_2(t)]^{\mathsf{T}}$ and graph them as curves in the y_1y_2-plane, called the **phase plane**. Such a curve is called a **trajectory** of (1), and their totality is known as the **phase portrait** of (1).

Now we have seen that solutions are of the form

$$\mathbf{y}(t) = \mathbf{x}e^{\lambda t}. \qquad \text{Substitution into (1) gives} \qquad \mathbf{y}'(t) = \lambda\mathbf{x}e^{\lambda t} = \mathbf{A}\mathbf{y} = \mathbf{A}\mathbf{x}e^{\lambda t}.$$

Dropping the common factor $e^{\lambda t}$, we have

$$(2) \qquad \mathbf{A}\mathbf{x} = \lambda\mathbf{x}.$$

Hence $\mathbf{y}(t)$ is a (nonzero) solution of (1) if λ is an eigenvalue of $\mathbf{A}$ and $\mathbf{x}$ a corresponding eigenvector.

Our examples in the last section show that the general form of the phase portrait is determined to a large extent by the type of **critical point** of the system (1) defined as a point at which dy_2/dy_1 becomes undetermined, 0/0; here [see (9) in Sec. 4.3]

$$(3) \qquad \frac{dy_2}{dy_1} = \frac{y_2'\,dt}{y_1'\,dt} = \frac{a_{21}y_1 + a_{22}y_2}{a_{11}y_1 + a_{12}y_2}.$$

We also recall from Sec. 4.3 that there are various types of critical points.

What is now new, is that we shall see how these types of critical points are related to the eigenvalues. The latter are solutions $\lambda = \lambda_1$ and λ_2 of the characteristic equation

$$(4) \qquad \det(\mathbf{A} - \lambda\mathbf{I}) = \begin{vmatrix} a_{11} - \lambda & a_{12} \\ a_{21} & a_{22} - \lambda \end{vmatrix} = \lambda^2 - (a_{11} + a_{22})\lambda + \det\mathbf{A} = 0.$$

This is a quadratic equation $\lambda^2 - p\lambda + q = 0$ with coefficients p, q and discriminant Δ given by

$$(5) \qquad p = a_{11} + a_{22}, \qquad q = \det\mathbf{A} = a_{11}a_{22} - a_{12}a_{21}, \qquad \Delta = p^2 - 4q.$$

From algebra we know that the solutions of this equation are

$$(6) \qquad \lambda_1 = \tfrac{1}{2}(p + \sqrt{\Delta}), \qquad \lambda_2 = \tfrac{1}{2}(p - \sqrt{\Delta}).$$

Furthermore, the product representation of the equation gives

$$\lambda^2 - p\lambda + q = (\lambda - \lambda_1)(\lambda - \lambda_2) = \lambda^2 - (\lambda_1 + \lambda_2)\lambda + \lambda_1\lambda_2.$$

Hence p is the sum and q the product of the eigenvalues. Also $\lambda_1 - \lambda_2 = \sqrt{\Delta}$ from (6). Together,

$$p = \lambda_1 + \lambda_2, \qquad q = \lambda_1\lambda_2, \qquad \Delta = (\lambda_1 - \lambda_2)^2. \tag{7}$$

This gives the criteria in Table 4.1 for classifying critical points. A derivation will be indicated later in this section.

Table 4.1 Eigenvalue Criteria for Critical Points (Derivation after Table 4.2)

Name	$p = \lambda_1 + \lambda_2$	$q = \lambda_1\lambda_2$	$\Delta = (\lambda_1 - \lambda_2)^2$	Comments on λ_1, λ_2
(a) Node		$q > 0$	$\Delta \geqq 0$	Real, same sign
(b) Saddle point		$q < 0$		Real, opposite signs
(c) Center	$p = 0$	$q > 0$		Pure imaginary
(d) Spiral point	$p \neq 0$		$\Delta < 0$	Complex, not pure imaginary

Stability

Critical points may also be classified in terms of their stability. Stability concepts are basic in engineering and other applications. They are suggested by physics, where **stability** means, roughly speaking, that a small change (a small disturbance) of a physical system at some instant changes the behavior of the system only slightly at all future times t. For critical points, the following concepts are appropriate.

DEFINITIONS

Stable, Unstable, Stable and Attractive

A critical point P_0 of (1) is called **stable**[2] if, roughly, all trajectories of (1) that at some instant are close to P_0 remain close to P_0 at all future times; precisely: if for every disk D_ϵ of radius $\epsilon > 0$ with center P_0 there is a disk D_δ of radius $\delta > 0$ with center P_0 such that every trajectory of (1) that has a point P_1 (corresponding to $t = t_1$, say) in D_δ has all its points corresponding to $t \geqq t_1$ in D_ϵ. See Fig. 90.

P_0 is called **unstable** if P_0 is not stable.

P_0 is called **stable and attractive** (or *asymptotically stable*) if P_0 is stable and every trajectory that has a point in D_δ approaches P_0 as $t \to \infty$. See Fig. 91.

Classification criteria for critical points in terms of stability are given in Table 4.2. Both tables are summarized in the **stability chart** in Fig. 92. In this chart region of instability is dark blue.

[2]In the sense of the Russian mathematician ALEXANDER MICHAILOVICH LJAPUNOV (1857–1918), whose work was fundamental in stability theory for ODEs. This is perhaps the most appropriate definition of stability (and the only we shall use), but there are others, too.

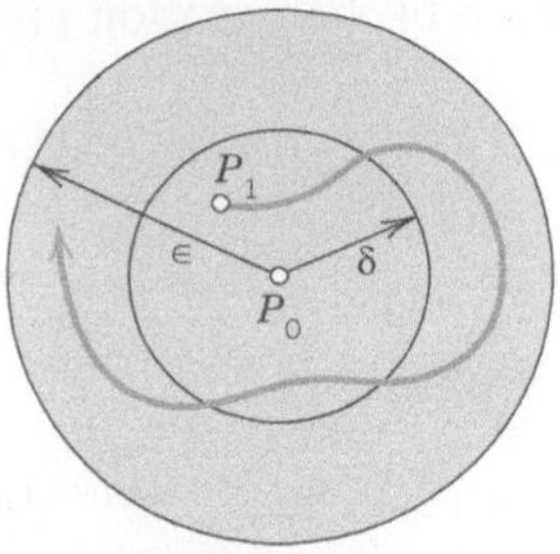

Fig. 90. Stable critical point P_0 of (1) (The trajectory initiating at P_1 stays in the disk of radius ϵ.)

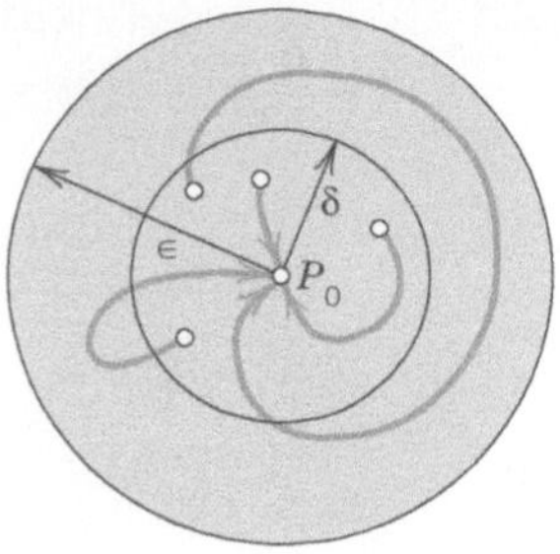

Fig. 91. Stable and attractive critical point P_0 of (1)

Table 4.2 Stability Criteria for Critical Points

Type of Stability	$p = \lambda_1 + \lambda_2$		$q = \lambda_1\lambda_2$
(a) Stable and attractive	$p < 0$		$q > 0$
(b) Stable	$p \leqq 0$		$q > 0$
(c) Unstable	$p > 0$	OR	$q < 0$

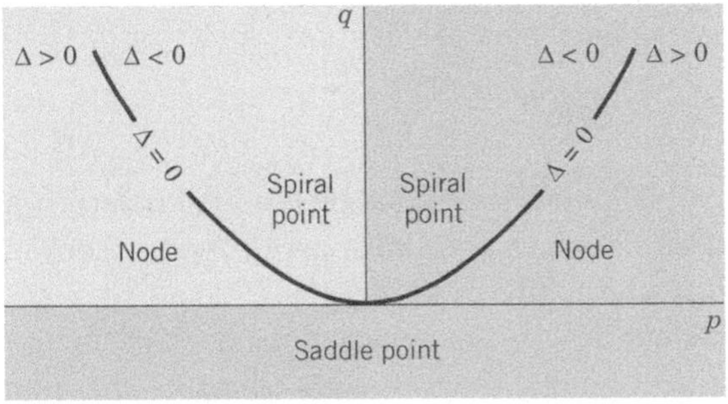

Fig. 92. Stability chart of the system (1) with p, q, Δ defined in (5). Stable and attractive: The second quadrant without the q-axis. Stability also on the positive q-axis (which corresponds to centers). Unstable: Dark blue region

We indicate how the criteria in Tables 4.1 and 4.2 are obtained. If $q = \lambda_1\lambda_2 > 0$, both of the eigenvalues are positive or both are negative or complex conjugates. If also $p = \lambda_1 + \lambda_2 < 0$, both are negative or have a negative real part. Hence P_0 is stable and attractive. The reasoning for the other two lines in Table 4.2 is similar.

If $\Delta < 0$, the eigenvalues are complex conjugates, say, $\lambda_1 = \alpha + i\beta$ and $\lambda_2 = \alpha - i\beta$. If also $p = \lambda_1 + \lambda_2 = 2\alpha < 0$, this gives a spiral point that is stable and attractive. If $p = 2\alpha > 0$, this gives an unstable spiral point.

If $p = 0$, then $\lambda_2 = -\lambda_1$ and $q = \lambda_1\lambda_2 = -\lambda_1^2$. If also $q > 0$, then $\lambda_1^2 = -q < 0$, so that λ_1, and thus λ_2, must be pure imaginary. This gives periodic solutions, their trajectories being closed curves around P_0, which is a center.

EXAMPLE 1 Application of the Criteria in Tables 4.1 and 4.2

In Example 1, Sec 4.3, we have $\mathbf{y}' = \begin{bmatrix} -3 & 1 \\ 1 & -3 \end{bmatrix}\mathbf{y}$, $p = -6$, $q = 8$, $\Delta = 4$, a node by Table 4.1(a), which is stable and attractive by Table 4.2(a). ■

EXAMPLE 2 Free Motions of a Mass on a Spring

What kind of critical point does $my'' + cy' + ky = 0$ in Sec. 2.4 have?

Solution. Division by m gives $y'' = -(k/m)y - (c/m)y'$. To get a system, set $y_1 = y$, $y_2 = y'$ (see Sec. 4.1). Then $y_2' = y'' = -(k/m)y_1 - (c/m)y_2$. Hence

$$\mathbf{y}' = \begin{bmatrix} 0 & 1 \\ -k/m & -c/m \end{bmatrix} \mathbf{y}, \qquad \det(\mathbf{A} - \lambda\mathbf{I}) = \begin{vmatrix} -\lambda & 1 \\ -k/m & -c/m - \lambda \end{vmatrix} = \lambda^2 + \frac{c}{m}\lambda + \frac{k}{m} = 0.$$

We see that $p = -c/m$, $q = k/m$, $\Delta = (c/m)^2 - 4k/m$. From this and Tables 4.1 and 4.2 we obtain the following results. Note that in the last three cases the discriminant Δ plays an essential role.

No damping. $c = 0, p = 0, q > 0$, a center.
Underdamping. $c^2 < 4mk, p < 0, q > 0, \Delta < 0$, a stable and attractive spiral point.
Critical damping. $c^2 = 4mk, p < 0, q > 0, \Delta = 0$, a stable and attractive node.
Overdamping. $c^2 > 4mk, p < 0, q > 0, \Delta > 0$, a stable and attractive node. ■

PROBLEM SET 4.4

1–10 TYPE AND STABILITY OF CRITICAL POINT

Determine the type and stability of the critical point. Then find a real general solution and sketch or graph some of the trajectories in the phase plane. Show the details of your work.

1. $y_1' = y_1$, $y_2' = 2y_2$

2. $y_1' = -4y_1$, $y_2' = -3y_2$

3. $y_1' = y_2$, $y_2' = -9y_1$

4. $y_1' = 2y_1 + y_2$, $y_2' = 5y_1 - 2y_2$

5. $y_1' = -2y_1 + 2y_2$, $y_2' = -2y_1 - 2y_2$

6. $y_1' = -6y_1 - y_2$, $y_2' = -9y_1 - 6y_2$

7. $y_1' = y_1 + 2y_2$, $y_2' = 2y_1 + y_2$

8. $y_1' = -y_1 + 4y_2$, $y_2' = 3y_1 - 2y_2$

9. $y_1' = 4y_1 + y_2$, $y_2' = 4y_1 + 4y_2$

10. $y_1' = y_2$, $y_2' = -5y_1 - 2y_2$

11–18 TRAJECTORIES OF SYSTEMS AND SECOND-ORDER ODEs. CRITICAL POINTS

11. Damped oscillations. Solve $y'' + 2y' + 2y = 0$. What kind of curves are the trajectories?

12. Harmonic oscillations. Solve $y'' + \frac{1}{9}y = 0$. Find the trajectories. Sketch or graph some of them.

13. Types of critical points. Discuss the critical points in (10)–(13) of Sec. 4.3 by using Tables 4.1 and 4.2.

14. Transformation of parameter. What happens to the critical point in Example 1 if you introduce $\tau = -t$ as a new independent variable?

15. Perturbation of center. What happens in Example 4 of Sec. 4.3 if you change $\mathbf{A}$ to $\mathbf{A} + 0.1\mathbf{I}$, where $\mathbf{I}$ is the unit matrix?

16. Perturbation of center. If a system has a center as its critical point, what happens if you replace the matrix $\mathbf{A}$ by $\tilde{\mathbf{A}} = \mathbf{A} + k\mathbf{I}$ with any real number $k \neq 0$ (representing measurement errors in the diagonal entries)?

17. Perturbation. The system in Example 4 in Sec. 4.3 has a center as its critical point. Replace each a_{jk} in Example 4, Sec. 4.3, by $a_{jk} + b$. Find values of b such that you get **(a)** a saddle point, **(b)** a stable and attractive node, **(c)** a stable and attractive spiral, **(d)** an unstable spiral, **(e)** an unstable node.

18. CAS EXPERIMENT. Phase Portraits. Graph phase portraits for the systems in Prob. 17 with the values of b suggested in the answer. Try to illustrate how the phase portrait changes "continuously" under a continuous change of b.

19. WRITING PROBLEM. Stability. Stability concepts are basic in physics and engineering. Write a two-part report of 3 pages each (A) on general applications in which stability plays a role (be as precise as you can), and (B) on material related to stability in this section. Use your own formulations and examples; do not copy.

20. Stability chart. Locate the critical points of the systems (10)–(14) in Sec. 4.3 and of Probs. 1, 3, 5 in this problem set on the stability chart.

4.5 Qualitative Methods for Nonlinear Systems

Qualitative methods are methods of obtaining qualitative information on solutions without actually solving a system. These methods are particularly valuable for systems whose solution by analytic methods is difficult or impossible. This is the case for many practically important **nonlinear systems**

$$(1) \qquad \mathbf{y}' = \mathbf{f}(\mathbf{y}), \qquad \text{thus} \qquad \begin{aligned} y_1' &= f_1(y_1, y_2) \\ y_2' &= f_2(y_1, y_2). \end{aligned}$$

In this section we extend **phase plane methods**, as just discussed, from linear systems to nonlinear systems (1). We assume that (1) is **autonomous**, that is, the independent variable t does not occur explicitly. (All examples in the last section are autonomous.) We shall again exhibit entire families of solutions. This is an advantage over numeric methods, which give only one (approximate) solution at a time.

Concepts needed from the last section are the **phase plane** (the y_1y_2-plane), **trajectories** (solution curves of (1) in the phase plane), the **phase portrait** of (1) (the totality of these trajectories), and **critical points** of (1) (points (y_1, y_2) at which both $f_1(y_1, y_2)$ and $f_2(y_1, y_2)$ are zero).

Now (1) may have several critical points. Our approach shall be to discuss one critical point after another. If a critical point P_0 is not at the origin, then, for technical convenience, we shall move this point to the origin before analyzing the point. More formally, if P_0: (a, b) is a critical point with (a, b) not at the origin $(0, 0)$, then we apply the translation

$$\tilde{y}_1 = y_1 - a, \qquad \tilde{y}_2 = y_2 - b$$

which moves P_0 to $(0, 0)$ as desired. Thus we can assume P_0 to be the origin $(0, 0)$, and for simplicity we continue to write y_1, y_2 (instead of $\tilde{y}_1, \tilde{y}_2$). We also assume that P_0 is **isolated**, that is, it is the only critical point of (1) within a (sufficiently small) disk with center at the origin. If (1) has only finitely many critical points, that is automatically true. (Explain!)

Linearization of Nonlinear Systems

How can we determine the kind and stability property of a critical point P_0: $(0, 0)$ of (1)? In most cases this can be done by **linearization** of (1) near P_0, writing (1) as $\mathbf{y}' = \mathbf{f}(\mathbf{y}) = \mathbf{A}\mathbf{y} + \mathbf{h}(\mathbf{y})$ and dropping $\mathbf{h}(\mathbf{y})$, as follows.

Since P_0 is critical, $f_1(0, 0) = 0$, $f_2(0, 0) = 0$, so that f_1 and f_2 have no constant terms and we can write

$$(2) \qquad \mathbf{y}' = \mathbf{A}\mathbf{y} + \mathbf{h}(\mathbf{y}), \qquad \text{thus} \qquad \begin{aligned} y_1' &= a_{11}y_1 + a_{12}y_2 + h_1(y_1, y_2) \\ y_2' &= a_{21}y_1 + a_{22}y_2 + h_2(y_1, y_2). \end{aligned}$$

A is constant (independent of t) since (1) is autonomous. One can prove the following (proof in Ref. [A7], pp. 375–388, listed in App. 1).

THEOREM 1

Linearization

If f_1 and f_2 in (1) *are continuous and have continuous partial derivatives in a neighborhood of the critical point P_0:* (0, 0), *and if* det $\mathbf{A} \neq 0$ *in* (2), *then the kind and stability of the critical point of* (1) *are the same as those of the* **linearized system**

$$\text{(3)} \qquad \mathbf{y}' = \mathbf{A}\mathbf{y}, \qquad \text{thus} \qquad \begin{aligned} y_1' &= a_{11}y_1 + a_{12}y_2 \\ y_2' &= a_{21}y_1 + a_{22}y_2. \end{aligned}$$

Exceptions occur if $\mathbf{A}$ *has equal or pure imaginary eigenvalues; then* (1) *may have the same kind of critical point as* (3) *or a spiral point.*

EXAMPLE 1

Free Undamped Pendulum. Linearization

Figure 93a shows a pendulum consisting of a body of mass m (the bob) and a rod of length L. Determine the locations and types of the critical points. Assume that the mass of the rod and air resistance are negligible.

Solution. *Step 1. Setting up the mathematical model.* Let θ denote the angular displacement, measured counterclockwise from the equilibrium position. The weight of the bob is mg (g the acceleration of gravity). It causes a restoring force $mg \sin\theta$ tangent to the curve of motion (circular arc) of the bob. By Newton's second law, at each instant this force is balanced by the force of acceleration $mL\theta''$, where $L\theta''$ is the acceleration; hence the resultant of these two forces is zero, and we obtain as the mathematical model

$$mL\theta'' + mg\sin\theta = 0.$$

Dividing this by mL, we have

$$\text{(4)} \qquad \theta'' + k\sin\theta = 0 \qquad \left(k = \frac{g}{L}\right).$$

When θ is very small, we can approximate $\sin\theta$ rather accurately by θ and obtain as an *approximate* solution $A\cos\sqrt{k}t + B\sin\sqrt{k}t$, but the *exact* solution for any θ is not an elementary function.

Step 2. Critical points* $(0, 0), (\pm 2\pi, 0), (\pm 4\pi, 0), \cdots$*, Linearization. To obtain a system of ODEs, we set $\theta = y_1, \theta' = y_2$. Then from (4) we obtain a nonlinear system (1) of the form

$$\text{(4*)} \qquad \begin{aligned} y_1' &= f_1(y_1, y_2) = y_2 \\ y_2' &= f_2(y_1, y_2) = -k\sin y_1. \end{aligned}$$

The right sides are both zero when $y_2 = 0$ and $\sin y_1 = 0$. This gives infinitely many critical points $(n\pi, 0)$, where $n = 0, \pm 1, \pm 2, \cdots$. We consider (0, 0). Since the Maclaurin series is

$$\sin y_1 = y_1 - \tfrac{1}{6}y_1^3 + - \cdots \approx y_1,$$

the linearized system at (0, 0) is

$$\mathbf{y}' = \mathbf{A}\mathbf{y} = \begin{bmatrix} 0 & 1 \\ -k & 0 \end{bmatrix}\mathbf{y}, \qquad \text{thus} \qquad \begin{aligned} y_1' &= y_2 \\ y_2' &= -ky_1. \end{aligned}$$

To apply our criteria in Sec. 4.4 we calculate $p = a_{11} + a_{22} = 0$, $q = \det\mathbf{A} = k = g/L\,(>0)$, and $\Delta = p^2 - 4q = -4k$. From this and Table 4.1(c) in Sec. 4.4 we conclude that (0, 0) is a center, which is always stable. Since $\sin\theta = \sin y_1$ is periodic with period 2π, the critical points $(n\pi, 0)$, $n = \pm 2, \pm 4, \cdots$, are all centers.

Step 3. Critical points* $(\pm\pi, 0), (\pm 3\pi, 0), (\pm 5\pi, 0), \cdots$*, Linearization. We now consider the critical point $(\pi, 0)$, setting $\theta - \pi = y_1$ and $(\theta - \pi)' = \theta' = y_2$. Then in (4),

$$\sin\theta = \sin(y_1 + \pi) = -\sin y_1 = -y_1 + \tfrac{1}{6}y_1^3 - + \cdots \approx -y_1$$

and the linearized system at $(\pi, 0)$ is now

$$\mathbf{y}' = \mathbf{A}\mathbf{y} = \begin{bmatrix} 0 & 1 \\ k & 0 \end{bmatrix}\mathbf{y}, \qquad \text{thus} \qquad \begin{aligned} y_1' &= y_2 \\ y_2' &= ky_1. \end{aligned}$$

We see that $p = 0$, $q = -k\ (<0)$, and $\Delta = -4q = 4k$. Hence, by Table 4.1(b), this gives a saddle point, which is always unstable. Because of periodicity, the critical points $(n\pi, 0)$, $n = \pm 1, \pm 3, \cdots$, are all saddle points. These results agree with the impression we get from Fig. 93b. ■

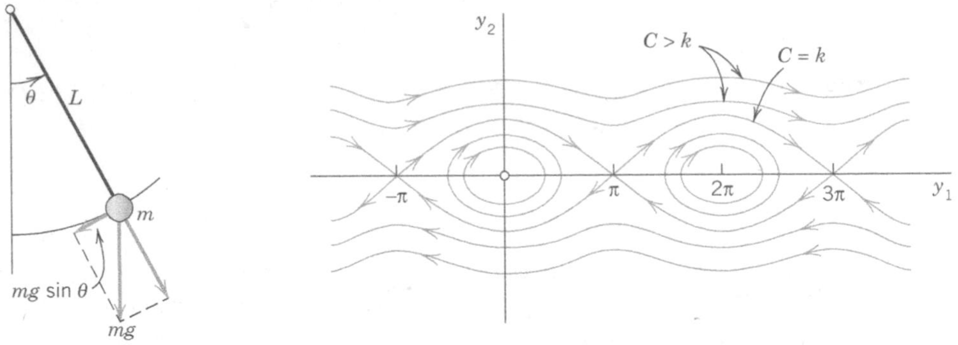

(a) Pendulum (b) Solution curves $y_2(y_1)$ of (4) in the phase plane

Fig. 93. Example 1 (*C* will be explained in Example 4.)

EXAMPLE 2 **Linearization of the Damped Pendulum Equation**

To gain further experience in investigating critical points, as another practically important case, let us see how Example 1 changes when we add a damping term $c\theta'$ (damping proportional to the angular velocity) to equation (4), so that it becomes

$$\theta'' + c\theta' + k \sin\theta = 0 \tag{5}$$

where $k > 0$ and $c \geqq 0$ (which includes our previous case of no damping, $c = 0$). Setting $\theta = y_1$, $\theta' = y_2$, as before, we obtain the nonlinear system (use $\theta'' = y_2'$)

$$\begin{aligned} y_1' &= y_2 \\ y_2' &= -k \sin y_1 - cy_2. \end{aligned}$$

We see that the critical points have the same locations as before, namely, $(0, 0)$, $(\pm\pi, 0)$, $(\pm 2\pi, 0)$, $\cdots$. We consider $(0, 0)$. Linearizing $\sin y_1 \approx y_1$ as in Example 1, we get the linearized system at $(0, 0)$

$$\mathbf{y}' = \mathbf{A}\mathbf{y} = \begin{bmatrix} 0 & 1 \\ -k & -c \end{bmatrix}\mathbf{y}, \qquad \text{thus} \qquad \begin{aligned} y_1' &= y_2 \\ y_2' &= -ky_1 - cy_2. \end{aligned} \tag{6}$$

This is identical with the system in Example 2 of Sec. 4.4, except for the (positive!) factor m (and except for the physical meaning of y_1). Hence for $c = 0$ (no damping) we have a center (see Fig. 93b), for small damping we have a spiral point (see Fig. 94), and so on.

We now consider the critical point $(\pi, 0)$. We set $\theta - \pi = y_1$, $(\theta - \pi)' = \theta' = y_2$ and linearize

$$\sin\theta = \sin(y_1 + \pi) = -\sin y_1 \approx -y_1.$$

This gives the new linearized system at $(\pi, 0)$

$$\mathbf{y}' = \mathbf{A}\mathbf{y} = \begin{bmatrix} 0 & 1 \\ k & -c \end{bmatrix}\mathbf{y}, \qquad \text{thus} \qquad \begin{aligned} y_1' &= y_2 \\ y_2' &= ky_1 - cy_2. \end{aligned} \tag{6*}$$

For our criteria in Sec. 4.4 we calculate $p = a_{11} + a_{22} = -c$, $q = \det \mathbf{A} = -k$, and $\Delta = p^2 - 4q = c^2 + 4k$. This gives the following results for the critical point at $(\pi, 0)$.

No damping. $c = 0, p = 0, q < 0, \Delta > 0$, a saddle point. See Fig. 93b.
Damping. $c > 0, p < 0, q < 0, \Delta > 0$, a saddle point. See Fig. 94.

Since $\sin y_1$ is periodic with period 2π, the critical points $(\pm 2\pi, 0), (\pm 4\pi, 0), \cdots$ are of the same type as $(0, 0)$, and the critical points $(-\pi, 0), (\pm 3\pi, 0), \cdots$ are of the same type as $(\pi, 0)$, so that our task is finished.

Figure 94 shows the trajectories in the case of damping. What we see agrees with our physical intuition. Indeed, damping means loss of energy. Hence instead of the closed trajectories of periodic solutions in Fig. 93b we now have trajectories spiraling around one of the critical points $(0, 0), (\pm 2\pi, 0), \cdots$. Even the wavy trajectories corresponding to whirly motions eventually spiral around one of these points. Furthermore, there are no more trajectories that connect critical points (as there were in the undamped case for the saddle points). ■

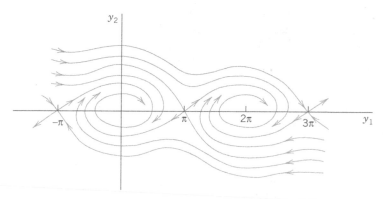

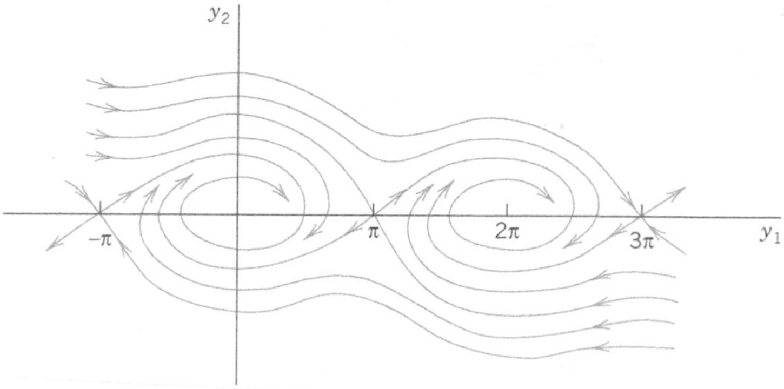

Fig. 94. Trajectories in the phase plane for the damped pendulum in Example 2

Lotka–Volterra Population Model

EXAMPLE 3 Predator–Prey Population Model[3]

This model concerns two species, say, rabbits and foxes, and the foxes prey on the rabbits.

Step 1. Setting up the model. We assume the following.

1. Rabbits have unlimited food supply. Hence, if there were no foxes, their number $y_1(t)$ would grow exponentially, $y_1' = ay_1$.
2. Actually, y_1 is decreased because of the kill by foxes, say, at a rate proportional to $y_1 y_2$, where $y_2(t)$ is the number of foxes. Hence $y_1' = ay_1 - by_1y_2$, where $a > 0$ and $b > 0$.
3. If there were no rabbits, then $y_2(t)$ would exponentially decrease to zero, $y_2' = -ly_2$. However, y_2 is increased by a rate proportional to the number of encounters between predator and prey; together we have $y_2' = -ly_2 + ky_1y_2$, where $k > 0$ and $l > 0$.

This gives the (nonlinear!) Lotka–Volterra system

$$(7) \qquad \begin{aligned} y_1' &= f_1(y_1, y_2) = ay_1 - by_1y_2 \\ y_2' &= f_2(y_1, y_2) = ky_1y_2 - ly_2. \end{aligned}$$

[3]Introduced by ALFRED J. LOTKA (1880–1949), American biophysicist, and VITO VOLTERRA (1860–1940), Italian mathematician, the initiator of functional analysis (see [GR7] in App. 1).

Step 2. Critical point* (0, 0), *Linearization. We see from (7) that the critical points are the solutions of

$$(7^*) \qquad f_1(y_1, y_2) = y_1(a - by_2) = 0, \qquad f_2(y_1, y_2) = y_2(ky_1 - l) = 0.$$

The solutions are $(y_1, y_2) = (0, 0)$ and $\left(\frac{l}{k}, \frac{a}{b}\right)$. We consider (0, 0). Dropping $-by_1y_2$ and ky_1y_2 from (7) gives the linearized system

$$\mathbf{y}' = \begin{bmatrix} a & 0 \\ 0 & -l \end{bmatrix} \mathbf{y}.$$

Its eigenvalues are $\lambda_1 = a > 0$ and $\lambda_2 = -l < 0$. They have opposite signs, so that we get a saddle point.

Step. 3. Critical point (l/k, a/b), Linearization. We set $y_1 = \tilde{y}_1 + l/k$, $y_2 = \tilde{y}_2 + a/b$. Then the critical point $(l/k, a/b)$ corresponds to $(\tilde{y}_1, \tilde{y}_2) = (0, 0)$. Since $\tilde{y}_1' = y_1', \tilde{y}_2' = y_2'$, we obtain from (7) [factorized as in (7*)]

$$\tilde{y}_1' = \left(\tilde{y}_1 + \frac{l}{k}\right)\left[a - b\left(\tilde{y}_2 + \frac{a}{b}\right)\right] = \left(\tilde{y}_1 + \frac{l}{k}\right)(-b\tilde{y}_2)$$

$$\tilde{y}_2' = \left(\tilde{y}_2 + \frac{a}{b}\right)\left[k\left(\tilde{y}_1 + \frac{l}{k}\right) - l\right] = \left(\tilde{y}_2 + \frac{a}{b}\right)k\tilde{y}_1.$$

Dropping the two nonlinear terms $-b\tilde{y}_1\tilde{y}_2$ and $k\tilde{y}_1\tilde{y}_2$, we have the linearized system

$$(7^{**}) \qquad \begin{aligned} &\text{(a)} \quad \tilde{y}_1' = -\frac{lb}{k}\tilde{y}_2 \\ &\text{(b)} \quad \tilde{y}_2' = \frac{ak}{b}\tilde{y}_1. \end{aligned}$$

The left side of (a) times the right side of (b) must equal the right side of (a) times the left side of (b),

$$\frac{ak}{b}\tilde{y}_1\tilde{y}_1' = -\frac{lb}{k}\tilde{y}_2\tilde{y}_2'. \qquad \text{By integration,} \qquad \frac{ak}{b}\tilde{y}_1^2 + \frac{lb}{k}\tilde{y}_2^2 = \text{const.}$$

This is a family of ellipses, so that the critical point $(l/k, a/b)$ of the linearized system (7**) is a center (Fig. 95). It can be shown, by a complicated analysis, that the nonlinear system (7) also has a center (rather than a spiral point) at $(l/k, a/b)$ surrounded by closed trajectories (not ellipses).

We see that the predators and prey have a cyclic variation about the critical point. Let us move counterclockwise around the ellipse, beginning at the right vertex, where the rabbits have a maximum number. Foxes are sharply increasing in number until they reach a maximum at the upper vertex, and the number of rabbits is then sharply decreasing until it reaches a minimum at the left vertex, and so on. Cyclic variations of this kind have been observed in nature, for example, for lynx and snowshoe hare near the Hudson Bay, with a cycle of about 10 years.

For models of more complicated situations and a systematic discussion, see C. W. Clark, *Mathematical Bioeconomics: The Mathematics of Conservation,* 3rd ed. Hoboken, NJ, Wiley, 2010. ■

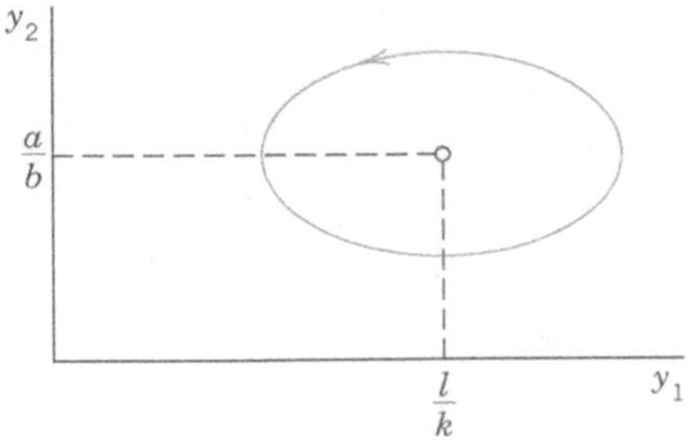

Fig. 95. Ecological equilibrium point and trajectory of the linearized Lotka–Volterra system (7**)

Transformation to a First-Order Equation in the Phase Plane

Another phase plane method is based on the idea of transforming a second-order **autonomous ODE** (an ODE in which t does not occur explicitly)

$$F(y, y', y'') = 0$$

to first order by taking $y = y_1$ as the independent variable, setting $y' = y_2$ and transforming y'' by the chain rule,

$$y'' = y_2' = \frac{dy_2}{dt} = \frac{dy_2}{dy_1}\frac{dy_1}{dt} = \frac{dy_2}{dy_1}y_2.$$

Then the ODE becomes of first order,

$$F\left(y_1, y_2, \frac{dy_2}{dy_1}y_2\right) = 0 \tag{8}$$

and can sometimes be solved or treated by direction fields. We illustrate this for the equation in Example 1 and shall gain much more insight into the behavior of solutions.

EXAMPLE 4 An ODE (8) for the Free Undamped Pendulum

If in (4) $\theta'' + k \sin\theta = 0$ we set $\theta = y_1$, $\theta' = y_2$ (the angular velocity) and use

$$\theta'' = \frac{dy_2}{dt} = \frac{dy_2}{dy_1}\frac{dy_1}{dt} = \frac{dy_2}{dy_1}y_2, \qquad \text{we get} \qquad \frac{dy_2}{dy_1}y_2 = -k\sin y_1.$$

Separation of variables gives $y_2\,dy_2 = -k \sin y_1\,dy_1$. By integration,

$$\tfrac{1}{2}y_2^2 = k\cos y_1 + C \qquad (C \text{ constant}). \tag{9}$$

Multiplying this by mL^2, we get

$$\tfrac{1}{2}m(Ly_2)^2 - mL^2k\cos y_1 = mL^2C.$$

We see that these three terms are **energies**. Indeed, y_2 is the angular velocity, so that Ly_2 is the velocity and the first term is the kinetic energy. The second term (including the minus sign) is the potential energy of the pendulum, and mL^2C is its total energy, which is constant, as expected from the law of conservation of energy, because there is no damping (no loss of energy). The type of motion depends on the total energy, hence on C, as follows.

Figure 93b shows trajectories for various values of C. These graphs continue periodically with period 2π to the left and to the right. We see that some of them are ellipse-like and closed, others are wavy, and there are two trajectories (passing through the saddle points $(n\pi, 0)$, $n = \pm 1, \pm 3, \cdots$) that separate those two types of trajectories. From (9) we see that the smallest possible C is $C = -k$; then $y_2 = 0$, and $\cos y_1 = 1$, so that the pendulum is at rest. The pendulum will change its direction of motion if there are points at which $y_2 = \theta' = 0$. Then $k \cos y_1 + C = 0$ by (9). If $y_1 = \pi$, then $\cos y_1 = -1$ and $C = k$. Hence if $-k < C < k$, then the pendulum reverses its direction for a $|y_1| = |\theta| < \pi$, and for these values of C with $|C| < k$ the pendulum oscillates. This corresponds to the closed trajectories in the figure. However, if $C > k$, then $y_2 = 0$ is impossible and the pendulum makes a whirly motion that appears as a wavy trajectory in the y_1y_2-plane. Finally, the value $C = k$ corresponds to the two "separating trajectories" in Fig. 93b connecting the saddle points. ■

The phase plane method of deriving a single first-order equation (8) may be of practical interest not only when (8) can be solved (as in Example 4) but also when a solution is not possible and we have to utilize fields (Sec. 1.2). We illustrate this with a very famous example:

EXAMPLE 5 Self-Sustained Oscillations. Van der Pol Equation

There are physical systems such that for small oscillations, energy is fed into the system, whereas for large oscillations, energy is taken from the system. In other words, large oscillations will be damped, whereas for small oscillations there is "negative damping" (feeding of energy into the system). For physical reasons we expect such a system to approach a periodic behavior, which will thus appear as a closed trajectory in the phase plane, called a **limit cycle**. A differential equation describing such vibrations is the famous **van der Pol equation**[4]

$$y'' - \mu(1 - y^2)y' + y = 0 \qquad (\mu > 0, \text{ constant}). \tag{10}$$

It first occurred in the study of electrical circuits containing vacuum tubes. For $\mu = 0$ this equation becomes $y'' + y = 0$ and we obtain harmonic oscillations. Let $\mu > 0$. The damping term has the factor $-\mu(1 - y^2)$. This is negative for small oscillations, when $y^2 < 1$, so that we have "negative damping," is zero for $y^2 = 1$ (no damping), and is positive if $y^2 > 1$ (positive damping, loss of energy). If μ is small, we expect a limit cycle that is almost a circle because then our equation differs but little from $y'' + y = 0$. If μ is large, the limit cycle will probably look different.

Setting $y = y_1, y' = y_2$ and using $y'' = (dy_2/dy_1)y_2$ as in (8), we have from (10)

$$\frac{dy_2}{dy_1}y_2 - \mu(1 - y_1^2)y_2 + y_1 = 0. \tag{11}$$

The isoclines in the y_1y_2-plane (the phase plane) are the curves $dy_2/dy_1 = K = \text{const}$, that is,

$$\frac{dy_2}{dy_1} = \mu(1 - y_1^2) - \frac{y_1}{y_2} = K.$$

Solving algebraically for y_2, we see that the isoclines are given by

$$y_2 = \frac{y_1}{\mu(1 - y_1^2) - K} \qquad \text{(Figs. 96, 97).}$$

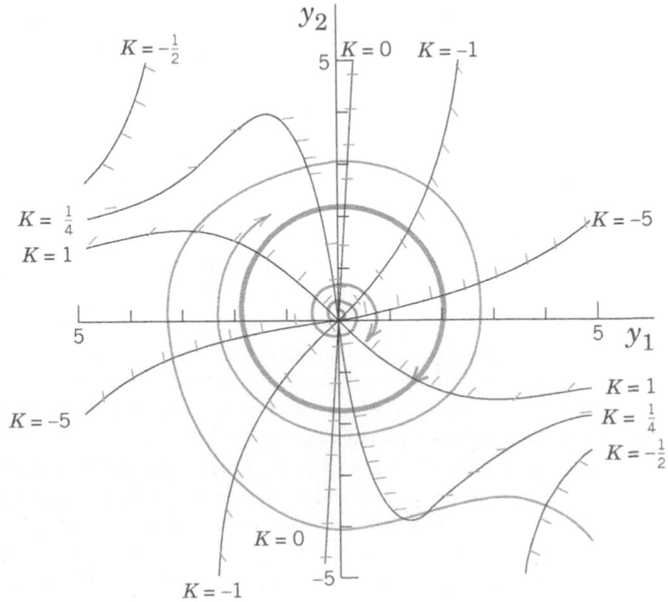

Fig. 96. Direction field for the van der Pol equation with $\mu = 0.1$ in the phase plane, showing also the limit cycle and two trajectories. See also Fig. 8 in Sec. 1.2

[4]BALTHASAR VAN DER POL (1889–1959), Dutch physicist and engineer.

Figure 96 shows some isoclines when μ is small, $\mu = 0.1$, the limit cycle (almost a circle), and two (blue) trajectories approaching it, one from the outside and the other from the inside, of which only the initial portion, a small spiral, is shown. Due to this approach by trajectories, a limit cycle differs conceptually from a closed curve (a trajectory) surrounding a center, which is not approached by trajectories. For larger μ the limit cycle no longer resembles a circle, and the trajectories approach it more rapidly than for smaller μ. Figure 97 illustrates this for $\mu = 1$. ■

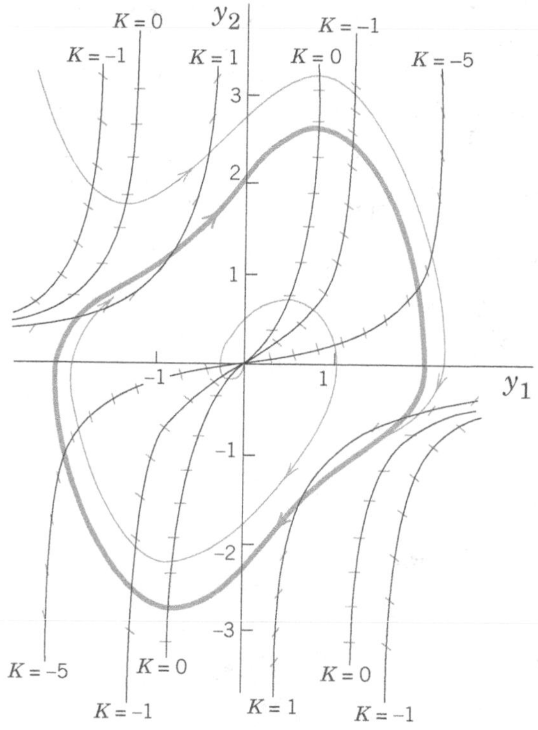

Fig. 97. Direction field for the van der Pol equation with $\mu = 1$ in the phase plane, showing also the limit cycle and two trajectories approaching it

PROBLEM SET 4.5

1. **Pendulum.** To what state (position, speed, direction of motion) do the four points of intersection of a closed trajectory with the axes in Fig. 93b correspond? The point of intersection of a wavy curve with the y_2-axis?

2. **Limit cycle.** What is the essential difference between a limit cycle and a closed trajectory surrounding a center?

3. **CAS EXPERIMENT. Deformation of Limit Cycle.** Convert the van der Pol equation to a system. Graph the limit cycle and some approaching trajectories for $\mu = 0.2, 0.4, 0.6, 0.8, 1.0, 1.5, 2.0$. Try to observe how the limit cycle changes its form continuously if you vary μ continuously. Describe in words how the limit cycle is deformed with growing μ.

4–8 CRITICAL POINTS. LINEARIZATION

Find the location and type of all critical points by linearization. Show the details of your work.

4. $y_1' = 4y_1 - y_1^2$, $y_2' = y_2$
5. $y_1' = y_2$, $y_2' = -y_1 + \frac{1}{2}y_1^2$
6. $y_1' = y_2$, $y_2' = -y_1 - y_1^2$
7. $y_1' = -y_1 + y_2 - y_2^2$, $y_2' = -y_1 - y_2$
8. $y_1' = y_2 - y_2^2$, $y_2' = y_1 - y_1^2$

9–13 CRITICAL POINTS OF ODEs

Find the location and type of all critical points by first converting the ODE to a system and then linearizing it.

9. $y'' - 9y + y^3 = 0$
10. $y'' + y - y^3 = 0$
11. $y'' + \cos y = 0$
12. $y'' + 9y + y^2 = 0$

13. $y'' + \sin y = 0$

14. TEAM PROJECT. Self-sustained oscillations.
(a) Van der Pol equation. Determine the type of the critical point at $(0, 0)$ when $\mu > 0, \mu = 0, \mu < 0$.
(b) Rayleigh equation. Show that the Rayleigh equation[5]

$$Y'' - \mu(1 - \tfrac{1}{3}Y'^2)Y' + Y = 0 \quad (\mu > 0)$$

also describes self-sustained oscillations and that by differentiating it and setting $y = Y'$ one obtains the van der Pol equation.
(c) Duffing equation. The Duffing equation is

$$y'' + \omega_0^2 y + \beta y^3 = 0$$

where usually $|\beta|$ is small, thus characterizing a small deviation of the restoring force from linearity. $\beta > 0$ and $\beta < 0$ are called the cases of a *hard spring* and a *soft spring*, respectively. Find the equation of the trajectories in the phase plane. (Note that for $\beta > 0$ all these curves are closed.)

15. Trajectories. Write the ODE $y'' - 4y + y^3 = 0$ as a system, solve it for y_2 as a function of y_1, and sketch or graph some of the trajectories in the phase plane.

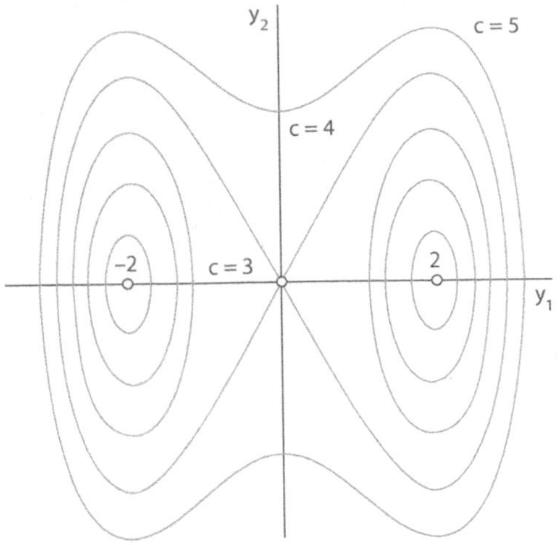

Fig. 98. Trajectories in Problem 15

4.6 Nonhomogeneous Linear Systems of ODEs

In this section, the last one of Chap. 4, we discuss methods for solving nonhomogeneous linear systems of ODEs

$$\text{(1)} \qquad \mathbf{y}' = \mathbf{A}\mathbf{y} + \mathbf{g} \qquad \text{(see Sec. 4.2)}$$

where the vector $\mathbf{g}(t)$ is not identically zero. We assume $\mathbf{g}(t)$ and the entries of the $n \times n$ matrix $\mathbf{A}(t)$ to be continuous on some interval J of the t-axis. From a general solution $\mathbf{y}^{(h)}(t)$ of the homogeneous system $\mathbf{y}' = \mathbf{A}\mathbf{y}$ on J and a **particular solution** $\mathbf{y}^{(p)}(t)$ of (1) on J [i.e., a solution of (1) containing no arbitrary constants], we get a solution of (1),

$$\text{(2)} \qquad \mathbf{y} = \mathbf{y}^{(h)} + \mathbf{y}^{(p)}.$$

$\mathbf{y}$ is called a **general solution** of (1) on J because it includes every solution of (1) on J. This follows from Theorem 2 in Sec. 4.2 (see Prob. 1 of this section).

Having studied homogeneous linear systems in Secs. 4.1–4.4, our present task will be to explain methods for obtaining particular solutions of (1). We discuss the method of

[5] LORD RAYLEIGH (JOHN WILLIAM STRUTT) (1842–1919), English physicist and mathematician, professor at Cambridge and London, known by his important contributions to the theory of waves, elasticity theory, hydrodynamics, and various other branches of applied mathematics and theoretical physics. In 1904 he was awarded the Nobel Prize in physics.

undetermined coefficients and the method of the variation of parameters; these have counterparts for a single ODE, as we know from Secs. 2.7 and 2.10.

Method of Undetermined Coefficients

Just as for a single ODE, this method is suitable if the entries of $\mathbf{A}$ are constants and the components of $\mathbf{g}$ are constants, positive integer powers of t, exponential functions, or cosines and sines. In such a case a particular solution $\mathbf{y}^{(p)}$ is assumed in a form similar to $\mathbf{g}$; for instance, $\mathbf{y}^{(p)} = \mathbf{u} + \mathbf{v}t + \mathbf{w}t^2$ if $\mathbf{g}$ has components quadratic in t, with $\mathbf{u}$, $\mathbf{v}$, $\mathbf{w}$ to be determined by substitution into (1). This is similar to Sec. 2.7, except for the Modification Rule. It suffices to show this by an example.

EXAMPLE 1 **Method of Undetermined Coefficients. Modification Rule**

Find a general solution of

$$(3) \qquad \mathbf{y}' = \mathbf{A}\mathbf{y} + \mathbf{g} = \begin{bmatrix} -3 & 1 \\ 1 & -3 \end{bmatrix} \mathbf{y} + \begin{bmatrix} -6 \\ 2 \end{bmatrix} e^{-2t}.$$

Solution. A general equation of the homogeneous system is (see Example 1 in Sec. 4.3)

$$(4) \qquad \mathbf{y}^{(h)} = c_1 \begin{bmatrix} 1 \\ 1 \end{bmatrix} e^{-2t} + c_2 \begin{bmatrix} 1 \\ -1 \end{bmatrix} e^{-4t}.$$

Since $\lambda = -2$ is an eigenvalue of $\mathbf{A}$, the function e^{-2t} on the right side also appears in $\mathbf{y}^{(h)}$, and we must apply the Modification Rule by setting

$$\mathbf{y}^{(p)} = \mathbf{u}te^{-2t} + \mathbf{v}e^{-2t} \qquad\qquad \text{(rather than } \mathbf{u}e^{-2t}).$$

Note that the first of these two terms is the analog of the modification in Sec. 2.7, but it would not be sufficient here. (Try it.) By substitution,

$$\mathbf{y}^{(p)\prime} = \mathbf{u}e^{-2t} - 2\mathbf{u}te^{-2t} - 2\mathbf{v}e^{-2t} = \mathbf{A}\mathbf{u}te^{-2t} + \mathbf{A}\mathbf{v}e^{-2t} + \mathbf{g}.$$

Equating the te^{-2t}-terms on both sides, we have $-2\mathbf{u} = \mathbf{A}\mathbf{u}$. Hence $\mathbf{u}$ is an eigenvector of $\mathbf{A}$ corresponding to $\lambda = -2$; thus [see (5)] $\mathbf{u} = a[1 \quad 1]^{\mathsf{T}}$ with any $a \neq 0$. Equating the other terms gives

$$\mathbf{u} - 2\mathbf{v} = \mathbf{A}\mathbf{v} + \begin{bmatrix} -6 \\ 2 \end{bmatrix} \qquad \text{thus} \qquad \begin{bmatrix} a \\ a \end{bmatrix} - \begin{bmatrix} 2v_1 \\ 2v_2 \end{bmatrix} = \begin{bmatrix} -3v_1 + v_2 \\ v_1 - 3v_2 \end{bmatrix} + \begin{bmatrix} -6 \\ 2 \end{bmatrix}.$$

Collecting terms and reshuffling gives

$$v_1 - v_2 = -a - 6$$
$$-v_1 + v_2 = -a + 2.$$

By addition, $0 = -2a - 4$, $a = -2$, and then $v_2 = v_1 + 4$, say, $v_1 = k$, $v_2 = k + 4$, thus, $\mathbf{v} = [k \quad k + 4]^{\mathsf{T}}$. We can simply choose $k = 0$. This gives the *answer*

$$(5) \qquad \mathbf{y} = \mathbf{y}^{(h)} + \mathbf{y}^{(p)} = c_1 \begin{bmatrix} 1 \\ 1 \end{bmatrix} e^{-2t} + c_2 \begin{bmatrix} 1 \\ -1 \end{bmatrix} e^{-4t} - 2 \begin{bmatrix} 1 \\ 1 \end{bmatrix} te^{-2t} + \begin{bmatrix} 0 \\ 4 \end{bmatrix} e^{-2t}.$$

For other k we get other $\mathbf{v}$; for instance, $k = -2$ gives $\mathbf{v} = [-2 \quad 2]^{\mathsf{T}}$, so that the *answer* becomes

$$(5^*) \qquad \mathbf{y} = c_1 \begin{bmatrix} 1 \\ 1 \end{bmatrix} e^{-2t} + c_2 \begin{bmatrix} 1 \\ -1 \end{bmatrix} e^{-4t} - 2 \begin{bmatrix} 1 \\ 1 \end{bmatrix} te^{-2t} + \begin{bmatrix} -2 \\ 2 \end{bmatrix} e^{-2t}, \qquad \text{etc.} \qquad \blacksquare$$

Method of Variation of Parameters

This method can be applied to nonhomogeneous linear systems

$$\mathbf{y}' = \mathbf{A}(t)\mathbf{y} + \mathbf{g}(t) \tag{6}$$

with variable $\mathbf{A} = \mathbf{A}(t)$ and general $\mathbf{g}(t)$. It yields a particular solution $\mathbf{y}^{(p)}$ of (6) on some open interval J on the t-axis if a general solution of the homogeneous system $\mathbf{y}' = \mathbf{A}(t)\mathbf{y}$ on J is known. We explain the method in terms of the previous example.

EXAMPLE 2 Solution by the Method of Variation of Parameters

Solve (3) in Example 1.

Solution. A basis of solutions of the homogeneous system is $[e^{-2t} \quad e^{-2t}]^{\mathsf{T}}$ and $[e^{-4t} \quad -e^{-4t}]^{\mathsf{T}}$. Hence the general solution (4) of the homogeneous system may be written

$$\mathbf{y}^{(h)} = \begin{bmatrix} e^{-2t} & e^{-4t} \\ e^{-2t} & -e^{-4t} \end{bmatrix} \begin{bmatrix} c_1 \\ c_2 \end{bmatrix} = \mathbf{Y}(t)\mathbf{c}. \tag{7}$$

Here, $\mathbf{Y}(t) = [\mathbf{y}^{(1)} \quad \mathbf{y}^{(2)}]^{\mathsf{T}}$ is the fundamental matrix (see Sec. 4.2). As in Sec. 2.10 we replace the constant vector $\mathbf{c}$ by a variable vector $\mathbf{u}(t)$ to obtain a particular solution

$$\mathbf{y}^{(p)} = \mathbf{Y}(t)\mathbf{u}(t).$$

Substitution into (3) $\mathbf{y}' = \mathbf{A}\mathbf{y} + \mathbf{g}$ gives

$$\mathbf{Y}'\mathbf{u} + \mathbf{Y}\mathbf{u}' = \mathbf{A}\mathbf{Y}\mathbf{u} + \mathbf{g}. \tag{8}$$

Now since $\mathbf{y}^{(1)}$ and $\mathbf{y}^{(2)}$ are solutions of the homogeneous system, we have

$$\mathbf{y}^{(1)\prime} = \mathbf{A}\mathbf{y}^{(1)}, \qquad \mathbf{y}^{(2)\prime} = \mathbf{A}\mathbf{y}^{(2)}, \qquad \text{thus} \qquad \mathbf{Y}' = \mathbf{A}\mathbf{Y}.$$

Hence $\mathbf{Y}'\mathbf{u} = \mathbf{A}\mathbf{Y}\mathbf{u}$, so that (8) reduces to

$$\mathbf{Y}\mathbf{u}' = \mathbf{g}. \qquad \text{The solution is} \qquad \mathbf{u}' = \mathbf{Y}^{-1}\mathbf{g};$$

here we use that the inverse $\mathbf{Y}^{-1}$ of $\mathbf{Y}$ (Sec. 4.0) exists because the determinant of $\mathbf{Y}$ is the Wronskian W, which is not zero for a basis. Equation (9) in Sec. 4.0 gives the form of $\mathbf{Y}^{-1}$,

$$\mathbf{Y}^{-1} = \frac{1}{-2e^{-6t}} \begin{bmatrix} -e^{-4t} & -e^{-4t} \\ -e^{-2t} & e^{-2t} \end{bmatrix} = \frac{1}{2} \begin{bmatrix} e^{2t} & e^{2t} \\ e^{4t} & -e^{4t} \end{bmatrix}.$$

We multiply this by $\mathbf{g}$, obtaining

$$\mathbf{u}' = \mathbf{Y}^{-1}\mathbf{g} = \frac{1}{2} \begin{bmatrix} e^{2t} & e^{2t} \\ e^{4t} & -e^{4t} \end{bmatrix} \begin{bmatrix} -6e^{-2t} \\ 2e^{-2t} \end{bmatrix} = \frac{1}{2} \begin{bmatrix} -4 \\ -8e^{2t} \end{bmatrix} = \begin{bmatrix} -2 \\ -4e^{2t} \end{bmatrix}.$$

Integration is done componentwise (just as differentiation) and gives

$$\mathbf{u}(t) = \int_0^t \begin{bmatrix} -2 \\ -4e^{2\tilde{t}} \end{bmatrix} d\tilde{t} = \begin{bmatrix} -2t \\ -2e^{2t} + 2 \end{bmatrix}$$

(where + 2 comes from the lower limit of integration). From this and $\mathbf{Y}$ in (7) we obtain

$$\mathbf{Y}\mathbf{u} = \begin{bmatrix} e^{-2t} & e^{-4t} \\ e^{-2t} & -e^{-4t} \end{bmatrix} \begin{bmatrix} -2t \\ -2e^{2t} + 2 \end{bmatrix} = \begin{bmatrix} -2te^{-2t} - 2e^{-2t} + 2e^{-4t} \\ -2te^{-2t} + 2e^{-2t} - 2e^{-4t} \end{bmatrix} = \begin{bmatrix} -2t - 2 \\ -2t + 2 \end{bmatrix} e^{-2t} + \begin{bmatrix} 2 \\ -2 \end{bmatrix} e^{-4t}.$$

The last term on the right is a solution of the homogeneous system. Hence we can absorb it into $\mathbf{y}^{(h)}$. We thus obtain as a general solution of the system (3), in agreement with (5*).

$$\mathbf{y} = c_1 \begin{bmatrix} 1 \\ 1 \end{bmatrix} e^{-2t} + c_2 \begin{bmatrix} 1 \\ -1 \end{bmatrix} e^{-4t} - 2 \begin{bmatrix} 1 \\ 1 \end{bmatrix} te^{-2t} + \begin{bmatrix} -2 \\ 2 \end{bmatrix} e^{-2t}. \tag{9}$$ ■

PROBLEM SET 4.6

1. Prove that (2) includes every solution of (1).

2–7 GENERAL SOLUTION

Find a general solution. Show the details of your work.

2. $y_1' = y_1 + y_2 + 10 \cos t$
$y_2' = 3y_1 - y_2 - 10 \sin t$

3. $y_1' = y_2 + e^{3t}$
$y_2' = y_1 - 3e^{3t}$

4. $y_1' = 4y_1 - 8y_2 + 2 \cosh t$
$y_2' = 2y_1 - 6y_2 + \cosh t + 2 \sinh t$

5. $y_1' = 4y_1 + y_2 + 0.6t$
$y_2' = 2y_1 + 3y_2 - 2.5t$

6. $y_1' = 4y_2$
$y_2' = 4y_1 - 16t^2 + 2$

7. $y_1' = -3y_1 - 4y_2 + 11t + 15$
$y_2' = 5y_1 + 6y_2 + 3e^{-t} - 15t - 20$

8. CAS EXPERIMENT. Undetermined Coefficients. Find out experimentally how general you must choose $\mathbf{y}^{(p)}$, in particular when the components of $\mathbf{g}$ have a different form (e.g., as in Prob. 7). Write a short report, covering also the situation in the case of the modification rule.

9. Undetermined Coefficients. Explain why, in Example 1 of the text, we have some freedom in choosing the vector $\mathbf{v}$.

10–15 INITIAL VALUE PROBLEM

Solve, showing details:

10. $y_1' = -3y_1 - 4y_2 + 5e^t$
$y_2' = 5y_1 + 6y_2 - 6e^t$
$y_1(0) = 19, \quad y_2(0) = -23$

11. $y_1' = y_2 + 6e^{2t}$
$y_2' = y_1 - e^{2t}$
$y_1(0) = 1, \quad y_2(0) = 0$

12. $y_1' = y_1 + 4y_2 - t^2 + 6t$
$y_2' = y_1 + y_2 - t^2 + t - 1$
$y_1(0) = 2, \quad y_2(0) = -1$

13. $y_1' = y_2 - 5 \sin t$
$y_2' = -4y_1 + 17 \cos t$
$y_1(0) = 5, \quad y_2(0) = 2$

14. $y_1' = 4y_2 + 5e^t$
$y_2' = -y_1 - 20e^{-t}$
$y_1(0) = 1, \quad y_2(0) = 0$

15. $y_1' = y_1 + 2y_2 + e^{2t} - 2t$
$y_2' = -y_2 + 1 + t$
$y_1(0) = 1, \quad y_2(0) = -4$

16. WRITING PROJECT. Undetermined Coefficients. Write a short report in which you compare the application of the method of undetermined coefficients to a single ODE and to a system of ODEs, using ODEs and systems of your choice.

17–20 NETWORK

Find the currents in Fig. 99 (Probs. 17–19) and Fig. 100 (Prob. 20) for the following data, showing the details of your work.

17. $R_1 = 2\ \Omega, R_2 = 8\ \Omega, L = 1$ H, $C = 0.5$ F, $E = 200$ V

18. Solve Prob. 17 with $E = 440 \sin t$ V and the other data as before.

19. In Prob. 17 find the particular solution when currents and charge at $t = 0$ are zero.

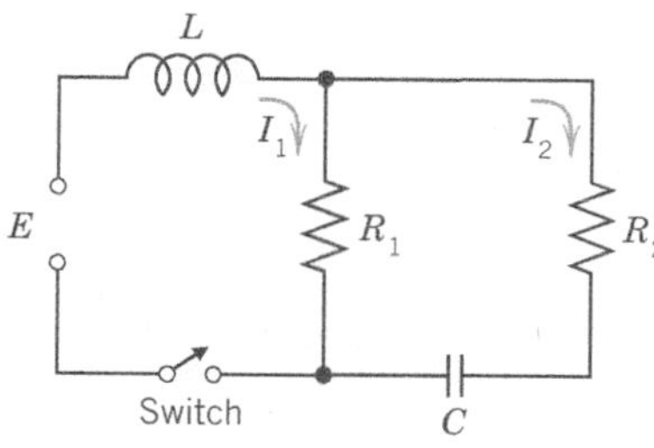

Fig. 99. Problems 17–19

20. $R_1 = 1\ \Omega$, $R_2 = 1.4\ \Omega$, $L_1 = 0.8$ H, $L_2 = 1$ H, $E = 100$ V, $I_1(0) = I_2(0) = 0$

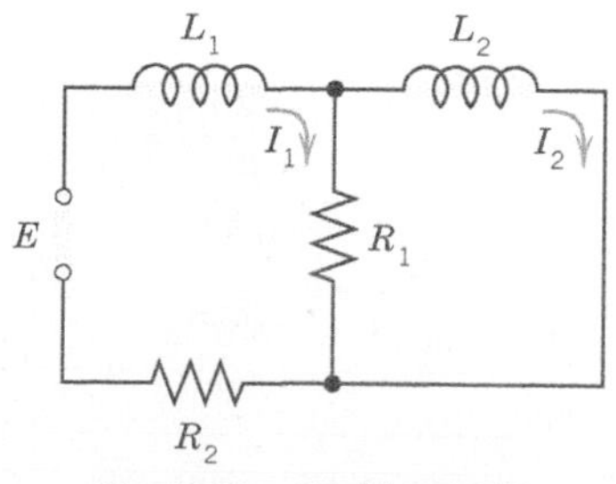

Fig. 100. Problem 20

CHAPTER 4 REVIEW QUESTIONS AND PROBLEMS

1. State some applications that can be modeled by systems of ODEs.
2. What is population dynamics? Give examples.
3. How can you transform an ODE into a system of ODEs?
4. What are qualitative methods for systems? Why are they important?
5. What is the phase plane? The phase plane method? A trajectory? The phase portrait of a system of ODEs?
6. What are critical points of a system of ODEs? How did we classify them? Why are they important?
7. What are eigenvalues? What role did they play in this chapter?
8. What does stability mean in general? In connection with critical points? Why is stability important in engineering?
9. What does linearization of a system mean?
10. Review the pendulum equations and their linearizations.

11–17 GENERAL SOLUTION. CRITICAL POINTS

Find a general solution. Determine the kind and stability of the critical point.

11. $y_1' = 2y_2$, $y_2' = 8y_1$

12. $y_1' = 5y_1$, $y_2' = y_2$

13. $y_1' = -2y_1 + 5y_2$, $y_2' = -y_1 - 6y_2$

14. $y_1' = 3y_1 + 4y_2$, $y_2' = 3y_1 + 2y_2$

15. $y_1' = -3y_1 - 2y_2$, $y_2' = -2y_1 - 3y_2$

16. $y_1' = 4y_2$, $y_2' = -4y_1$

17. $y_1' = -y_1 + 2y_2$, $y_2' = -2y_1 - y_2$

18–19 CRITICAL POINT

What kind of critical point does $\mathbf{y}' = \mathbf{A}\mathbf{y}$ have if $\mathbf{A}$ has the eigenvalues

18. -4 and 2

19. $2 + 3i, 2 - 3i$

20–23 NONHOMOGENEOUS SYSTEMS

Find a general solution. Show the details of your work.

20. $y_1' = 2y_1 + 2y_2 + e^t$, $y_2' = -2y_1 - 3y_2 + e^t$

21. $y_1' = 4y_2$, $y_2' = 4y_1 + 32t^2$

22. $y_1' = y_1 + y_2 + \sin t$, $y_2' = 4y_1 + y_2$

23. $y_1' = y_1 + 4y_2 - 2\cos t$, $y_2' = y_1 + y_2 - \cos t + \sin t$

24. Mixing problem. Tank T_1 in Fig. 101 initially contains 200 gal of water in which 160 lb of salt are dissolved. Tank T_2 initially contains 100 gal of pure water. Liquid is pumped through the system as indicated, and the mixtures are kept uniform by stirring. Find the amounts of salt $y_1(t)$ and $y_2(t)$ in T_1 and T_2, respectively.

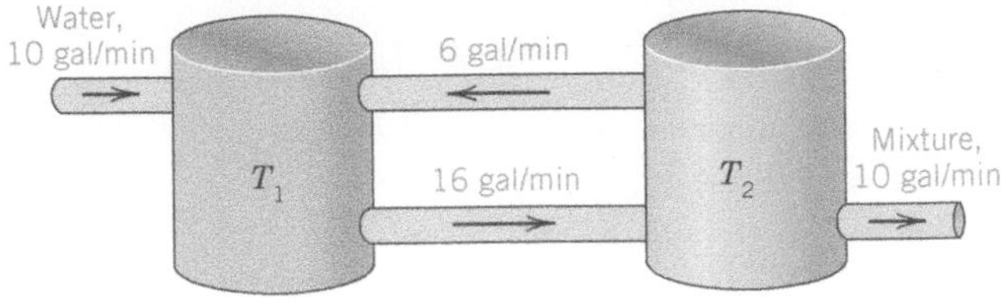

Fig. 101. Tanks in Problem 24

25. Network. Find the currents in Fig. 102 when $R = 2.5\ \Omega$, $L = 1$ H, $C = 0.04$ F, $E(t) = 169 \sin t$ V, $I_1(0) = 0$, $I_2(0) = 0$.

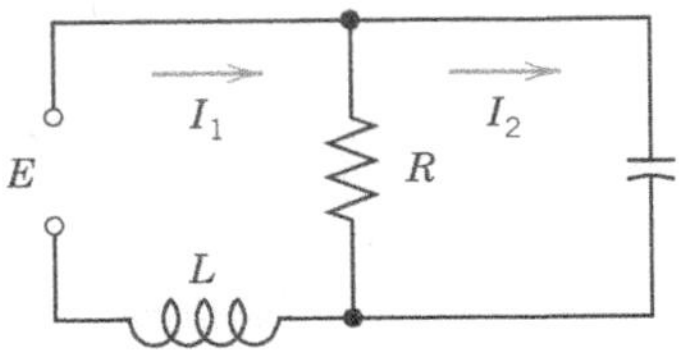

Fig. 102. Network in Problem 25

26. Network. Find the currents in Fig. 103 when $R = 1\ \Omega$, $L = 1.25$ H, $C = 0.2$ F, $I_1(0) = 1$ A, $I_2(0) = 1$ A.

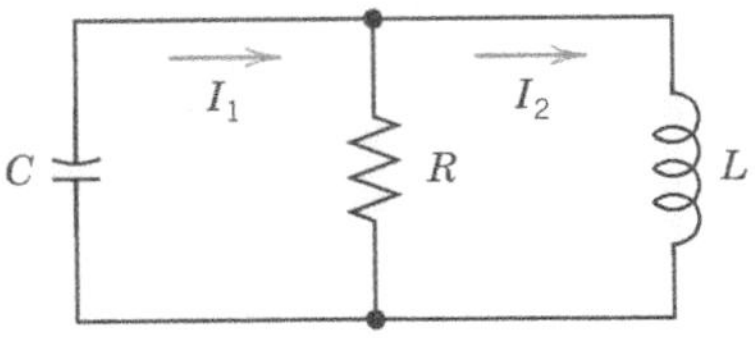

Fig. 103. Network in Problem 26

27–30 LINEARIZATION

Find the location and kind of all critical points of the given nonlinear system by linearization.

27. $y_1' = y_2$, $y_2' = y_1 - y_1^3$

28. $y_1' = \cos y_2$, $y_2' = 3y_1$

29. $y_1' = -4y_2$, $y_2' = \sin y_1$

30. $y_1' = 2y_2 + 2y_2^2$, $y_2' = -8y_1$

SUMMARY OF CHAPTER 4

Systems of ODEs. Phase Plane. Qualitative Methods

Whereas single electric circuits or single mass–spring systems are modeled by single ODEs (Chap. 2), networks of several circuits, systems of several masses and springs, and other engineering problems lead to **systems of ODEs**, involving several unknown functions $y_1(t), \cdots, y_n(t)$. Of central interest are **first-order systems** (Sec. 4.2):

$$\mathbf{y}' = \mathbf{f}(t, \mathbf{y}), \qquad \text{in components,} \qquad \begin{aligned} y_1' &= f_1(t, y_1, \cdots, y_n) \\ &\vdots \\ y_n' &= f_n(t, y_1, \cdots, y_n), \end{aligned}$$

to which higher order ODEs and systems of ODEs can be reduced (Sec. 4.1). In this summary we let $n = 2$, so that

$$\text{(1)} \qquad \mathbf{y}' = \mathbf{f}(t, \mathbf{y}), \qquad \text{in components,} \qquad \begin{aligned} y_1' &= f_1(t, y_1, y_2) \\ y_2' &= f_2(t, y_1, y_2). \end{aligned}$$

Then we can represent solution curves as ***trajectories*** in ***the phase plane*** (the y_1y_2-plane), investigate their totality [the "***phase portrait***" of (1)], and study the kind and ***stability*** of the ***critical points*** (points at which both f_1 and f_2 are zero), and classify them as ***nodes, saddle points, centers,*** or ***spiral points*** (Secs. 4.3, 4.4). These phase plane methods are **qualitative**; with their use we can discover various general properties of solutions without actually solving the system. They are primarily used for **autonomous systems**, that is, systems in which t does not occur explicitly.

A **linear system** is of the form

$$\text{(2)} \qquad \mathbf{y}' = \mathbf{A}\mathbf{y} + \mathbf{g}, \quad \text{where} \quad \mathbf{A} = \begin{bmatrix} a_{11} & a_{12} \\ a_{21} & a_{22} \end{bmatrix}, \quad \mathbf{y} = \begin{bmatrix} y_1 \\ y_2 \end{bmatrix}, \quad \mathbf{g} = \begin{bmatrix} g_1 \\ g_2 \end{bmatrix}.$$

If $\mathbf{g} = \mathbf{0}$, the system is called **homogeneous** and is of the form

$$\text{(3)} \qquad \mathbf{y}' = \mathbf{A}\mathbf{y}.$$

If $a_{11}, \cdots, a_{22}$ are constants, it has solutions $\mathbf{y} = \mathbf{x}e^{\lambda t}$, where λ is a solution of the quadratic equation

$$\begin{vmatrix} a_{11} - \lambda & a_{12} \\ a_{21} & a_{22} - \lambda \end{vmatrix} = (a_{11} - \lambda)(a_{22} - \lambda) - a_{12}a_{21} = 0$$

and $\mathbf{x} \neq \mathbf{0}$ has components x_1, x_2 determined up to a multiplicative constant by

$$(a_{11} - \lambda)x_1 + a_{12}x_2 = 0.$$

(These λ's are called the **eigenvalues** and these vectors $\mathbf{x}$ **eigenvectors** of the matrix $\mathbf{A}$. Further explanation is given in Sec. 4.0.)

A system (2) with $\mathbf{g} \neq \mathbf{0}$ is called **nonhomogeneous.** Its general solution is of the form $\mathbf{y} = \mathbf{y}_h + \mathbf{y}_p$, where $\mathbf{y}_h$ is a general solution of (3) and $\mathbf{y}_p$ a particular solution of (2). Methods of determining the latter are discussed in Sec. 4.6.

The discussion of critical points of linear systems based on eigenvalues is summarized in Tables 4.1 and 4.2 in Sec. 4.4. It also applies to nonlinear systems if the latter are first linearized. The key theorem for this is Theorem 1 in Sec. 4.5, which also includes three famous applications, namely the pendulum and van der Pol equations and the Lotka–Volterra predator–prey population model.

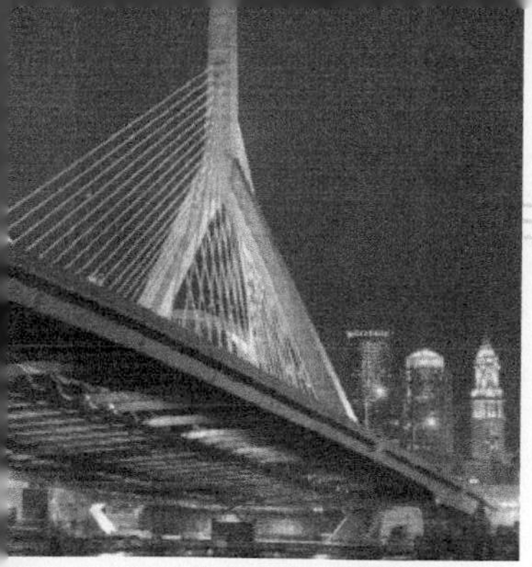

CHAPTER 5

Series Solutions of ODEs. Special Functions

In the previous chapters, we have seen that linear ODEs with *constant coefficients* can be solved by algebraic methods, and that their solutions are elementary functions known from calculus. For ODEs with *variable coefficients* the situation is more complicated, and their solutions may be nonelementary functions. *Legendre's, Bessel's*, and the *hypergeometric equations* are important ODEs of this kind. Since these ODEs and their solutions, the *Legendre polynomials, Bessel functions*, and *hypergeometric functions*, play an important role in engineering modeling, we shall consider the two standard methods for solving such ODEs.

The first method is called the **power series method** because it gives solutions in the form of a power series $a_0 + a_1x + a_2x^2 + a_3x^3 + \cdots$.

The second method is called the **Frobenius method** and generalizes the first; it gives solutions in power series, multiplied by a logarithmic term $\ln x$ or a fractional power x^r, in cases such as Bessel's equation, in which the first method is not general enough.

All those more advanced solutions and various other functions not appearing in calculus are known as *higher functions* or **special functions**, which has become a technical term. Each of these functions is important enough to give it a name and investigate its properties and relations to other functions in great detail (take a look into Refs. [GenRef1], [GenRef10], or [All] in App. 1). Your CAS knows practically all functions you will ever need in industry or research labs, but it is up to you to find your way through this vast terrain of formulas. The present chapter may give you some help in this task.

COMMENT. You can study this chapter directly after Chap. 2 because it needs no material from Chaps. 3 or 4.

Prerequisite: Chap. 2.
Section that may be omitted in a shorter course: 5.5.
References and Answers to Problems: App. 1 Part A, and App. 2.

5.1 Power Series Method

The **power series method** is the standard method for solving linear ODEs with ***variable*** coefficients. It gives solutions in the form of power series. These series can be used for computing values, graphing curves, proving formulas, and exploring properties of solutions, as we shall see. In this section we begin by explaining the idea of the power series method.

From calculus we remember that a **power series** (in powers of $x - x_0$) is an infinite series of the form

$$(1)\qquad \sum_{m=0}^{\infty} a_m(x - x_0)^m = a_0 + a_1(x - x_0) + a_2(x - x_0)^2 + \cdots .$$

Here, x is a variable. $a_0, a_1, a_2, \cdots$ are constants, called the **coefficients** of the series. x_0 is a constant, called the **center** of the series. In particular, if $x_0 = 0$, we obtain a **power series in powers of x**

$$(2)\qquad \sum_{m=0}^{\infty} a_m x^m = a_0 + a_1 x + a_2 x^2 + a_3 x^3 + \cdots .$$

We shall assume that all variables and constants are real.

We note that the term "power series" usually refers to a series of the form (1) [or (2)] but ***does not include*** series of negative or fractional powers of x. We use m as the summation letter, reserving n as a standard notation in the Legendre and Bessel equations for integer values of the parameter.

EXAMPLE 1 **Familiar Power Series** are the Maclaurin series

$$\frac{1}{1-x} = \sum_{m=0}^{\infty} x^m = 1 + x + x^2 + \cdots \qquad (|x| < 1, \textbf{ geometric series})$$

$$e^x = \sum_{m=0}^{\infty} \frac{x^m}{m!} = 1 + x + \frac{x^2}{2!} + \frac{x^3}{3!} + \cdots$$

$$\cos x = \sum_{m=0}^{\infty} \frac{(-1)^m x^{2m}}{(2m)!} = 1 - \frac{x^2}{2!} + \frac{x^4}{4!} - + \cdots$$

$$\sin x = \sum_{m=0}^{\infty} \frac{(-1)^m x^{2m+1}}{(2m+1)!} = x - \frac{x^3}{3!} + \frac{x^5}{5!} - + \cdots .$$

■

Idea and Technique of the Power Series Method

The idea of the power series method for solving linear ODEs seems natural, once we know that the most important ODEs in applied mathematics have solutions of this form. We explain the idea by an ODE that can readily be solved otherwise.

EXAMPLE 2 **Power Series Solution.** Solve $y' - y = 0$.

Solution. In the first step we insert

$$(2)\qquad y = a_0 + a_1 x + a_2 x^2 + a_3 x^3 + \cdots = \sum_{m=0}^{\infty} a_m x^m$$

and the series obtained by termwise differentiation

(3)
$$y' = a_1 + 2a_2x + 3a_3x^2 + \cdots = \sum_{m=1}^{\infty} ma_m x^{m-1}$$

into the ODE:

$$(a_1 + 2a_2x + 3a_3x^2 + \cdots) - (a_0 + a_1x + a_2x^2 + \cdots) = 0.$$

Then we collect like powers of x, finding

$$(a_1 - a_0) + (2a_2 - a_1)x + (3a_3 - a_2)x^2 + \cdots = 0.$$

Equating the coefficient of each power of x to zero, we have

$$a_1 - a_0 = 0, \qquad 2a_2 - a_1 = 0, \qquad 3a_3 - a_2 = 0, \cdots.$$

Solving these equations, we may express $a_1, a_2, \cdots$ in terms of a_0, which remains arbitrary:

$$a_1 = a_0, \qquad a_2 = \frac{a_1}{2} = \frac{a_0}{2!}, \qquad a_3 = \frac{a_2}{3} = \frac{a_0}{3!}, \cdots.$$

With these values of the coefficients, the series solution becomes the familiar general solution

$$y = a_0 + a_0x + \frac{a_0}{2!}x^2 + \frac{a_0}{3!}x^3 + \cdots = a_0\left(1 + x + \frac{x^2}{2!} + \frac{x^3}{3!}\right) = a_0e^x.$$

Test your comprehension by solving $y'' + y = 0$ by power series. You should get the result $y = a_0 \cos x + a_1 \sin x$. ■

We now describe the method in general and justify it after the next example. For a given ODE

(4)
$$y'' + p(x)y' + q(x)y = 0$$

we first represent $p(x)$ and $q(x)$ by power series in powers of x (or of $x - x_0$ if solutions in powers of $x - x_0$ are wanted). Often $p(x)$ and $q(x)$ are polynomials, and then nothing needs to be done in this first step. Next we assume a solution in the form of a power series (2) with unknown coefficients and insert it as well as (3) and

(5)
$$y'' = 2a_2 + 3 \cdot 2a_3x + 4 \cdot 3a_4x^2 + \cdots = \sum_{m=2}^{\infty} m(m-1)a_m x^{m-2}$$

into the ODE. Then we collect like powers of x and equate the sum of the coefficients of each occurring power of x to zero, starting with the constant terms, then taking the terms containing x, then the terms in x^2, and so on. This gives equations from which we can determine the unknown coefficients of (3) successively.

EXAMPLE 3   **A Special Legendre Equation.** The ODE

$$(1 - x^2)y'' - 2xy' + 2y = 0$$

occurs in models exhibiting spherical symmetry. Solve it.

Solution. Substitute (2), (3), and (5) into the ODE. $(1 - x^2)y''$ gives two series, one for y'' and one for $-x^2y''$. In the term $-2xy'$ use (3) and in $2y$ use (2). Write like powers of x vertically aligned. This gives

$$\begin{aligned}
y'' &= 2a_2 + 6a_3x + 12a_4x^2 + 20a_5x^3 + 30a_6x^4 + \cdots \\
-x^2y'' &= \qquad\qquad\qquad - 2a_2x^2 - 6a_3x^3 - 12a_4x^4 - \cdots \\
-2xy' &= \qquad\quad - 2a_1x - 4a_2x^2 - 6a_3x^3 - 8a_4x^4 - \cdots \\
2y &= 2a_0 + 2a_1x + 2a_2x^2 + 2a_3x^3 + 2a_4x^4 + \cdots.
\end{aligned}$$

Add terms of like powers of x. For each power $x^0, x, x^2, \cdots$ equate the sum obtained to zero. Denote these sums by [0] (constant terms), [1] (first power of x), and so on:

Sum	Power	Equations	
[0]	$[x^0]$	$a_2 = -a_0$	
[1]	$[x]$	$a_3 = 0$	
[2]	$[x^2]$	$12a_4 = 4a_2,$	$a_4 = \frac{4}{12}a_2 = -\frac{1}{3}a_0$
[3]	$[x^3]$	$a_5 = 0$	since $a_3 = 0$
[4]	$[x^4]$	$30a_6 = 18a_4,$	$a_6 = \frac{18}{30}a_4 = \frac{18}{30}(-\frac{1}{3})a_0 = -\frac{1}{5}a_0.$

This gives the solution

$$y = a_1x + a_0(1 - x^2 - \tfrac{1}{3}x^4 - \tfrac{1}{5}x^6 - \cdots).$$

a_0 and a_1 remain arbitrary. Hence, this is a general solution that consists of two solutions: x and $1 - x^2 - \frac{1}{3}x^4 - \frac{1}{5}x^6 - \cdots$. These two solutions are members of families of functions called *Legendre polynomials* $P_n(x)$ and *Legendre functions* $Q_n(x)$; here we have $x = P_1(x)$ and $1 - x^2 - \frac{1}{3}x^4 - \frac{1}{5}x^6 - \cdots = -Q_1(x)$. The minus is by convention. The index 1 is called the *order* of these two functions and here the order is 1. More on Legendre polynomials in the next section. ■

Theory of the Power Series Method

The ***n*th partial sum** of (1) is

$$s_n(x) = a_0 + a_1(x - x_0) + a_2(x - x_0)^2 + \cdots + a_n(x - x_0)^n \tag{6}$$

where $n = 0, 1, \cdots$. If we omit the terms of s_n from (1), the remaining expression is

$$R_n(x) = a_{n+1}(x - x_0)^{n+1} + a_{n+2}(x - x_0)^{n+2} + \cdots. \tag{7}$$

This expression is called the **remainder** *of* (1) *after the term* $a_n(x - x_0)^n$.

For example, in the case of the geometric series

$$1 + x + x^2 + \cdots + x^n + \cdots$$

we have

$$\begin{aligned}
s_0 &= 1, & R_0 &= x + x^2 + x^3 + \cdots, \\
s_1 &= 1 + x, & R_1 &= x^2 + x^3 + x^4 + \cdots, \\
s_2 &= 1 + x + x^2, & R_2 &= x^3 + x^4 + x^5 + \cdots, \qquad \text{etc.}
\end{aligned}$$

In this way we have now associated with (1) the sequence of the partial sums $s_0(x), s_1(x), s_2(x), \cdots$. If for some $x = x_1$ this sequence converges, say,

$$\lim_{n\to\infty} s_n(x_1) = s(x_1),$$

then the series (1) is called **convergent** *at* $x = x_1$, the number $s(x_1)$ is called the **value** or *sum* of (1) at x_1, and we write

$$s(x_1) = \sum_{m=0}^{\infty} a_m(x_1 - x_0)^m.$$

Then we have for every n,

$$s(x_1) = s_n(x_1) + R_n(x_1). \tag{8}$$

If that sequence diverges at $x = x_1$, the series (1) is called **divergent** at $x = x_1$.

In the case of convergence, for any positive ϵ there is an N (depending on ϵ) such that, by (8)

$$|R_n(x_1)| = |s(x_1) - s_n(x_1)| < \epsilon \qquad \text{for all } n > N. \tag{9}$$

Geometrically, this means that all $s_n(x_1)$ with $n > N$ lie between $s(x_1) - \epsilon$ and $s(x_1) + \epsilon$ (Fig. 104). Practically, this means that in the case of convergence we can approximate the sum $s(x_1)$ of (1) at x_1 by $s_n(x_1)$ as accurately as we please, by taking n large enough.

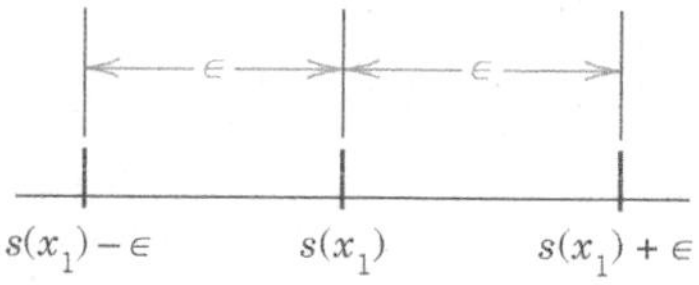

Fig. 104. Inequality (9)

Where does a power series converge? Now if we choose $x = x_0$ in (1), the series reduces to the single term a_0 because the other terms are zero. Hence the series converges at x_0. In some cases this may be the only value of x for which (1) converges. If there are other values of x for which the series converges, these values form an interval, the **convergence interval**. This interval may be finite, as in Fig. 105, with midpoint x_0. Then the series (1) converges for all x in the interior of the interval, that is, for all x for which

$$|x - x_0| < R \tag{10}$$

and diverges for $|x - x_0| > R$. The interval may also be infinite, that is, the series may converge for all x.

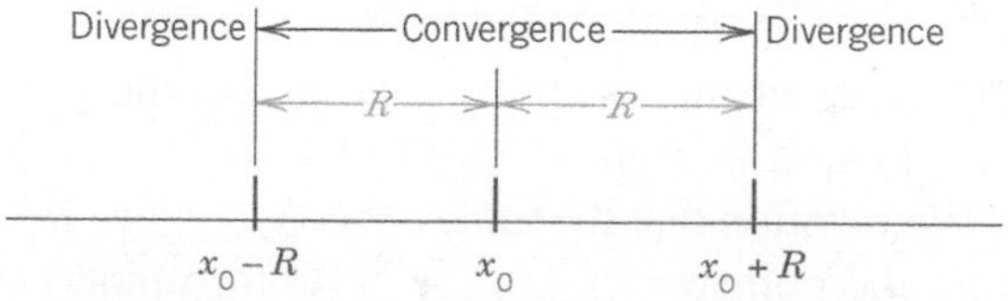

Fig. 105. Convergence interval (10) of a power series with center x_0

The quantity R in Fig. 105 is called the **radius of convergence** (because for a *complex* power series it is the radius of *disk* of convergence). If the series converges for all x, we set $R = \infty$ (and $1/R = 0$).

The radius of convergence can be determined from the coefficients of the series by means of each of the formulas

$$\text{(11)} \qquad \text{(a)}\quad R = 1\Big/\lim_{m\to\infty} \sqrt[m]{|a_m|} \qquad \text{(b)}\quad R = 1\Big/\lim_{m\to\infty}\left|\frac{a_{m+1}}{a_m}\right|$$

provided these limits exist and are not zero. [If these limits are infinite, then (1) converges only at the center x_0.]

EXAMPLE 4 Convergence Radius $R = \infty, 1, 0$

For all three series let $m \to \infty$

$$e^x = \sum_{m=0}^{\infty} \frac{x^m}{m!} = 1 + x + \frac{x^2}{2!} + \cdots, \qquad \left|\frac{a_{m+1}}{a_m}\right| = \frac{1/(m+1)!}{1/m!} = \frac{1}{m+1} \to 0, \qquad R = \infty$$

$$\frac{1}{1-x} = \sum_{m=0}^{\infty} x^m = 1 + x + x^2 + \cdots, \qquad \left|\frac{a_{m+1}}{a_m}\right| = \frac{1}{1} = 1, \qquad R = 1$$

$$\sum_{m=0}^{\infty} m!x^m = 1 + x + 2x^2 + \cdots, \qquad \left|\frac{a_{m+1}}{a_m}\right| = \frac{(m+1)!}{m!} = m + 1 \to \infty, \qquad R = 0.$$

Convergence for all x ($R = \infty$) is the best possible case, convergence in some finite interval the usual, and convergence only at the center ($R = 0$) is useless. ■

When do power series solutions exist? *Answer:* if p, q, r in the ODEs

$$\text{(12)} \qquad y'' + p(x)y' + q(x)y = r(x)$$

have power series representations (Taylor series). More precisely, a function $f(x)$ is called **analytic** at a point $x = x_0$ if it can be represented by a power series in powers of $x - x_0$ with positive radius of convergence. Using this concept, we can state the following basic theorem, in which the ODE (12) is **in standard form**, that is, it begins with the y''. If your ODE begins with, say, $h(x)y''$, divide it first by $h(x)$ and then apply the theorem to the resulting new ODE.

THEOREM 1 Existence of Power Series Solutions

If p, q, and r in (12) *are analytic at $x = x_0$, then every solution of* (12) *is analytic at $x = x_0$ and can thus be represented by a power series in powers of $x - x_0$ with radius of convergence $R > 0$.*

The proof of this theorem requires advanced complex analysis and can be found in Ref. [A11] listed in App. 1.

We mention that the radius of convergence R in Theorem 1 is at least equal to the distance from the point $x = x_0$ to the point (or points) closest to x_0 at which one of the functions p, q, r, as functions of a *complex variable*, is not analytic. (Note that that point may not lie on the x-axis but somewhere in the complex plane.)

Further Theory: Operations on Power Series

In the power series method we differentiate, add, and multiply power series, and we obtain coefficient recursions (as, for instance, in Example 3) by equating the sum of the coefficients of each occurring power of x to zero. These four operations are permissible in the sense explained in what follows. Proofs can be found in Sec. 15.3.

1. Termwise Differentiation. *A power series may be differentiated term by term.* More precisely: if

$$y(x) = \sum_{m=0}^{\infty} a_m(x - x_0)^m$$

converges for $|x - x_0| < R$, where $R > 0$, then the series obtained by differentiating term by term also converges for those x and represents the derivative y' of y for those x:

$$y'(x) = \sum_{m=1}^{\infty} m a_m(x - x_0)^{m-1} \qquad (|x - x_0| < R).$$

Similarly for the second and further derivatives.

2. Termwise Addition. *Two power series may be added term by term.* More precisely: if the series

$$\sum_{m=0}^{\infty} a_m(x - x_0)^m \qquad \text{and} \qquad \sum_{m=0}^{\infty} b_m(x - x_0)^m \tag{13}$$

have positive radii of convergence and their sums are $f(x)$ and $g(x)$, then the series

$$\sum_{m=0}^{\infty} (a_m + b_m)(x - x_0)^m$$

converges and represents $f(x) + g(x)$ for each x that lies in the interior of the convergence interval common to each of the two given series.

3. Termwise Multiplication. *Two power series may be multiplied term by term.* More precisely: Suppose that the series (13) have positive radii of convergence and let $f(x)$ and $g(x)$ be their sums. Then the series obtained by multiplying each term of the first series by each term of the second series and collecting like powers of $x - x_0$, that is,

$$a_0b_0 + (a_0b_1 + a_1b_0)(x - x_0) + (a_0b_2 + a_1b_1 + a_2b_0)(x - x_0)^2 + \cdots$$

$$= \sum_{m=0}^{\infty} (a_0b_m + a_1b_{m-1} + \cdots + a_mb_0)(x - x_0)^m$$

converges and represents $f(x)g(x)$ for each x in the interior of the convergence interval of each of the two given series.

4. Vanishing of All Coefficients ("*Identity Theorem for Power Series.*") *If a power series has a positive radius of convergent convergence and a sum that is identically zero throughout its interval of convergence, then each coefficient of the series must be zero.*

PROBLEM SET 5.1

1. **WRITING AND LITERATURE PROJECT. Power Series in Calculus. (a)** Write a review (2–3 pages) on power series in calculus. Use your own formulations and examples—do not just copy from textbooks. No proofs. **(b)** Collect and arrange Maclaurin series in a systematic list that you can use for your work.

2–5 REVIEW: RADIUS OF CONVERGENCE

Determine the radius of convergence. Show the details of your work.

2. $\sum_{m=0}^{\infty} (m+1)mx^m$

3. $\sum_{m=0}^{\infty} \frac{(-1)^m}{k^m} x^{2m}$

4. $\sum_{m=0}^{\infty} \frac{x^{2m+1}}{(2m+1)!}$

5. $\sum_{m=0}^{\infty} \left(\frac{2}{3}\right)^m x^{2m}$

6–9 SERIES SOLUTIONS BY HAND

Apply the power series method. Do this by hand, not by a CAS, to get a feel for the method, e.g., why a series may terminate, or has even powers only, etc. Show the details.

6. $(1+x)y' = y$

7. $y' = -2xy$

8. $xy' - 3y = k\ (= \text{const})$

9. $y'' + y = 0$

10–14 SERIES SOLUTIONS

Find a power series solution in powers of x. Show the details.

10. $y'' - y' + xy = 0$

11. $y'' - y' + x^2y = 0$

12. $(1-x^2)y'' - 2xy' + 2y = 0$

13. $y'' + (1+x^2)y = 0$

14. $y'' - 4xy' + (4x^2 - 2)y = 0$

15. **Shifting summation indices** is often convenient or necessary in the power series method. Shift the index so that the power under the summation sign is x^m. Check by writing the first few terms explicity.

$$\sum_{s=2}^{\infty} \frac{s(s+1)}{s^2+1} x^{s-1}, \qquad \sum_{p=1}^{\infty} \frac{p^2}{(p+1)!} x^{p+4}$$

16–19 CAS PROBLEMS. IVPs

Solve the initial value problem by a power series. Graph the partial sums of the powers up to and including x^5. Find the value of the sum s (5 digits) at x_1.

16. $y' + 4y = 1, \quad y(0) = 1.25, \quad x_1 = 0.2$

17. $y'' + 3xy' + 2y = 0, \quad y(0) = 1, \quad y'(0) = 1,$ $x = 0.5$

18. $(1-x^2)y'' - 2xy' + 30y = 0, \quad y(0) = 0,$ $y'(0) = 1.875, \quad x_1 = 0.5$

19. $(x-2)y' = xy, \quad y(0) = 4, \quad x_1 = 2$

20. **CAS Experiment. Information from Graphs of Partial Sums.** In numerics we use partial sums of power series. To get a feel for the accuracy for various x, experiment with $\sin x$. Graph partial sums of the Maclaurin series of an increasing number of terms, describing qualitatively the "breakaway points" of these graphs from the graph of $\sin x$. Consider other Maclaurin series of your choice.

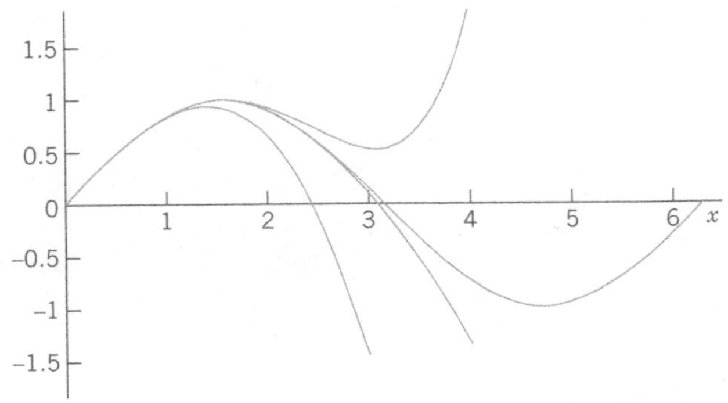

Fig. 106. CAS Experiment 20. $\sin x$ and partial sums s_3, s_5, s_7

5.2 Legendre's Equation. Legendre Polynomials $P_n(x)$

Legendre's differential equation[1]

$$(1) \qquad (1 - x^2)y'' - 2xy' + n(n + 1)y = 0 \qquad (n \text{ constant})$$

is one of the most important ODEs in physics. It arises in numerous problems, particularly in boundary value problems for spheres (take a quick look at Example 1 in Sec. 12.10).

The equation involves a **parameter** n, whose value depends on the physical or engineering problem. So (1) is actually a whole family of ODEs. For $n = 1$ we solved it in Example 3 of Sec. 5.1 (look back at it). Any solution of (1) is called a **Legendre function**. The study of these and other "higher" functions not occurring in calculus is called the **theory of special functions**. Further special functions will occur in the next sections.

Dividing (1) by $1 - x^2$, we obtain the standard form needed in Theorem 1 of Sec. 5.1 and we see that the coefficients $-2x/(1 - x^2)$ and $n(n + 1)/(1 - x^2)$ of the new equation are analytic at $x = 0$, so that we may apply the power series method. Substituting

$$(2) \qquad y = \sum_{m=0}^{\infty} a_m x^m$$

and its derivatives into (1), and denoting the constant $n(n + 1)$ simply by k, we obtain

$$(1 - x^2) \sum_{m=2}^{\infty} m(m - 1)a_m x^{m-2} - 2x \sum_{m=1}^{\infty} m a_m x^{m-1} + k \sum_{m=0}^{\infty} a_m x^m = 0.$$

By writing the first expression as two separate series we have the equation

$$\sum_{m=2}^{\infty} m(m - 1)a_m x^{m-2} - \sum_{m=2}^{\infty} m(m - 1)a_m x^m - \sum_{m=1}^{\infty} 2m a_m x^m + \sum_{m=0}^{\infty} k a_m x^m = 0.$$

It may help you to write out the first few terms of each series explicitly, as in Example 3 of Sec. 5.1; or you may continue as follows. To obtain the same general power x^s in all four series, set $m - 2 = s$ (thus $m = s + 2$) in the first series and simply write s instead of m in the other three series. This gives

$$\sum_{s=0}^{\infty} (s + 2)(s + 1)a_{s+2} x^s - \sum_{s=2}^{\infty} s(s - 1)a_s x^s - \sum_{s=1}^{\infty} 2s a_s x^s + \sum_{s=0}^{\infty} k a_s x^s = 0.$$

---

[1] ADRIEN-MARIE LEGENDRE (1752–1833), French mathematician, who became a professor in Paris in 1775 and made important contributions to special functions, elliptic integrals, number theory, and the calculus of variations. His book *Éléments de géométrie* (1794) became very famous and had 12 editions in less than 30 years.

Formulas on Legendre functions may be found in Refs. [GenRef1] and [GenRef10].

(Note that in the first series the summation begins with $s = 0$.) Since this equation with the right side 0 must be an identity in x if (2) is to be a solution of (1), the sum of the coefficients of each power of x on the left must be zero. Now x^0 occurs in the first and fourth series only, and gives [remember that $k = n(n + 1)$]

$$2 \cdot 1a_2 + n(n+1)a_0 = 0. \tag{3a}$$

x^1 occurs in the first, third, and fourth series and gives

$$3 \cdot 2a_3 + [-2 + n(n+1)]a_1 = 0. \tag{3b}$$

The higher powers $x^2, x^3, \cdots$ occur in all four series and give

$$(s+2)(s+1)a_{s+2} + [-s(s-1) - 2s + n(n+1)]a_s = 0. \tag{3c}$$

The expression in the brackets $[\cdots]$ can be written $(n - s)(n + s + 1)$, as you may readily verify. Solving (3a) for a_2 and (3b) for a_3 as well as (3c) for a_{s+2}, we obtain the general formula

$$a_{s+2} = -\frac{(n-s)(n+s+1)}{(s+2)(s+1)} a_s \qquad (s = 0, 1, \cdots). \tag{4}$$

This is called a **recurrence relation** or **recursion formula**. (Its derivation you may verify with your CAS.) It gives each coefficient in terms of the second one preceding it, except for a_0 and a_1, which are left as arbitrary constants. We find successively

$$a_2 = -\frac{n(n+1)}{2!} a_0 \qquad\qquad a_3 = -\frac{(n-1)(n+2)}{3!} a_1$$

$$a_4 = -\frac{(n-2)(n+3)}{4 \cdot 3} a_2 \qquad\qquad a_5 = -\frac{(n-3)(n+4)}{5 \cdot 4} a_3$$

$$= \frac{(n-2)n(n+1)(n+3)}{4!} a_0 \qquad\qquad = \frac{(n-3)(n-1)(n+2)(n+4)}{5!} a_1$$

and so on. By inserting these expressions for the coefficients into (2) we obtain

$$y(x) = a_0 y_1(x) + a_1 y_2(x) \tag{5}$$

where

$$y_1(x) = 1 - \frac{n(n+1)}{2!} x^2 + \frac{(n-2)n(n+1)(n+3)}{4!} x^4 - + \cdots \tag{6}$$

$$y_2(x) = x - \frac{(n-1)(n+2)}{3!} x^3 + \frac{(n-3)(n-1)(n+2)(n+4)}{5!} x^5 - + \cdots. \tag{7}$$

These series converge for $|x| < 1$ (see Prob. 4; or they may terminate, see below). Since (6) contains even powers of x only, while (7) contains odd powers of x only, the ratio y_1/y_2 is not a constant, so that y_1 and y_2 are not proportional and are thus linearly independent solutions. Hence (5) is a general solution of (1) on the interval $-1 < x < 1$.

Note that $x = \pm 1$ are the points at which $1 - x^2 = 0$, so that the coefficients of the standardized ODE are no longer analytic. So it should not surprise you that we do not get a longer convergence interval of (6) and (7), unless these series terminate after finitely many powers. In that case, the series become polynomials.

Polynomial Solutions. Legendre Polynomials $P_n(x)$

The reduction of power series to polynomials is a great advantage because then we have solutions for all x, without convergence restrictions. For special functions arising as solutions of ODEs this happens quite frequently, leading to various important families of polynomials; see Refs. [GenRef1], [GenRef10] in App. 1. For Legendre's equation this happens when the parameter n is a nonnegative integer because then the right side of (4) is zero for $s = n$, so that $a_{n+2} = 0$, $a_{n+4} = 0$, $a_{n+6} = 0, \cdots$. Hence if n is even, $y_1(x)$ reduces to a polynomial of degree n. If n is odd, the same is true for $y_2(x)$. These polynomials, multiplied by some constants, are called **Legendre polynomials** and are denoted by $P_n(x)$. The standard choice of such constants is done as follows. We choose the coefficient a_n of the highest power x^n as

$$a_n = \frac{(2n)!}{2^n (n!)^2} = \frac{1 \cdot 3 \cdot 5 \cdots (2n-1)}{n!} \qquad (n \text{ a positive integer}) \tag{8}$$

(and $a_n = 1$ if $n = 0$). Then we calculate the other coefficients from (4), solved for a_s in terms of a_{s+2}, that is,

$$a_s = -\frac{(s+2)(s+1)}{(n-s)(n+s+1)} a_{s+2} \qquad (s \leqq n-2). \tag{9}$$

The choice (8) makes $p_n(1) = 1$ for every n (see Fig. 107); this motivates (8). From (9) with $s = n - 2$ and (8) we obtain

$$a_{n-2} = -\frac{n(n-1)}{2(2n-1)} a_n = -\frac{n(n-1)}{2(2n-1)} \cdot \frac{(2n)!}{2^n (n!)^2}$$

Using $(2n)! = 2n(2n-1)(2n-2)!$ in the numerator and $n! = n(n-1)!$ and $n! = n(n-1)(n-2)!$ in the denominator, we obtain

$$a_{n-2} = -\frac{n(n-1)2n(2n-1)(2n-2)!}{2(2n-1)2^n n(n-1)!\, n(n-1)(n-2)!}.$$

$n(n-1)2n(2n-1)$ cancels, so that we get

$$a_{n-2} = -\frac{(2n-2)!}{2^n (n-1)!\,(n-2)!}.$$

Similarly,

$$a_{n-4} = -\frac{(n-2)(n-3)}{4(2n-3)} a_{n-2} = \frac{(2n-4)!}{2^n 2!\,(n-2)!\,(n-4)!}$$

and so on, and in general, when $n - 2m \geqq 0$,

$$a_{n-2m} = (-1)^m \frac{(2n-2m)!}{2^n m!\,(n-m)!\,(n-2m)!}. \tag{10}$$

The resulting solution of Legendre's differential equation (1) is called the **Legendre polynomial** *of degree n* and is denoted by $P_n(x)$.

From (10) we obtain

$$\begin{aligned} P_n(x) &= \sum_{m=0}^{M} (-1)^m \frac{(2n-2m)!}{2^n m!\,(n-m)!\,(n-2m)!} x^{n-2m} \\ &= \frac{(2n)!}{2^n (n!)^2} x^n - \frac{(2n-2)!}{2^n 1!\,(n-1)!\,(n-2)!} x^{n-2} + - \cdots \end{aligned} \tag{11}$$

where $M = n/2$ or $(n-1)/2$, whichever is an integer. The first few of these functions are (Fig. 107)

$$\begin{aligned} &P_0(x) = 1, && P_1(x) = x \\ &P_2(x) = \tfrac{1}{2}(3x^2 - 1), && P_3(x) = \tfrac{1}{2}(5x^3 - 3x) \\ &P_4(x) = \tfrac{1}{8}(35x^4 - 30x^2 + 3), && P_5(x) = \tfrac{1}{8}(63x^5 - 70x^3 + 15x) \end{aligned} \tag{11$'$}$$

and so on. You may now program (11) on your CAS and calculate $P_n(x)$ as needed.

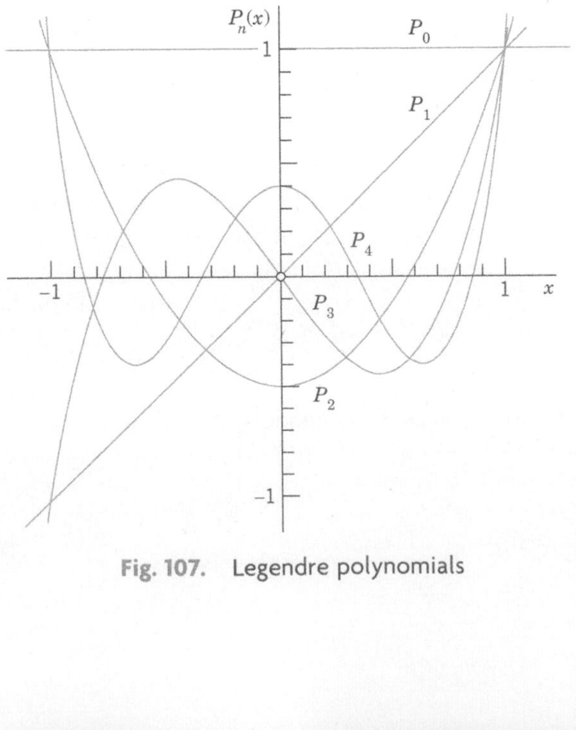

Fig. 107. Legendre polynomials

The Legendre polynomials $P_n(x)$ are **orthogonal** on the interval $-1 \leqq x \leqq 1$, a basic property to be defined and used in making up "Fourier–Legendre series" in the chapter on Fourier series (see Secs. 11.5–11.6).

PROBLEM SET 5.2

1–5 LEGENDRE POLYNOMIALS AND FUNCTIONS

1. Legendre functions for $n = 0$. Show that (6) with $n = 0$ gives $P_0(x) = 1$ and (7) gives (use $\ln(1 + x) = x - \frac{1}{2}x^2 + \frac{1}{3}x^3 + \cdots$)

$$y_2(x) = x + \frac{1}{3}x^3 + \frac{1}{5}x^5 + \cdots = \frac{1}{2}\ln\frac{1+x}{1-x}.$$

Verify this by solving (1) with $n = 0$, setting $z = y'$ and separating variables.

2. Legendre functions for $n = 1$. Show that (7) with $n = 1$ gives $y_2(x) = P_1(x) = x$ and (6) gives

$$y_1 = 1 - x^2 - \frac{1}{3}x^4 - \frac{1}{5}x^6 - \cdots$$
$$= 1 - \frac{1}{2}x \ln\frac{1+x}{1-x}.$$

3. Special n. Derive (11′) from (11).

4. Legendre's ODE. Verify that the polynomials in (11′) satisfy (1).

5. Obtain P_6 and P_7.

6–9 CAS PROBLEMS

6. Graph $P_2(x), \cdots, P_{10}(x)$ on common axes. For what x (approximately) and $n = 2, \cdots, 10$ is $|P_n(x)| < \frac{1}{2}$?

7. From what n on will your CAS no longer produce faithful graphs of $P_n(x)$? Why?

8. Graph $Q_0(x)$, $Q_1(x)$, and some further Legendre functions.

9. Substitute $a_s x^s + a_{s+1}x^{s+1} + a_{s+2}x^{s+2}$ into Legendre's equation and obtain the coefficient recursion (4).

10. TEAM PROJECT. Generating Functions. Generating functions play a significant role in modern applied mathematics (see [GenRef5]). The idea is simple. If we want to study a certain sequence $(f_n(x))$ and can find a function

$$G(u, x) = \sum_{n=0}^{\infty} f_n(x)u^n,$$

we may obtain properties of $(f_n(x))$ from those of G, which "generates" this sequence and is called a **generating function** of the sequence.

(a) Legendre polynomials. Show that

$$G(u, x) = \frac{1}{\sqrt{1 - 2xu + u^2}} = \sum_{n=0}^{\infty} P_n(x)u^n \qquad (12)$$

is a generating function of the Legendre polynomials. *Hint:* Start from the binomial expansion of $1/\sqrt{1 - v}$, then set $v = 2xu - u^2$, multiply the powers of $2xu - u^2$ out, collect all the terms involving u^n, and verify that the sum of these terms is $P_n(x)u^n$.

(b) Potential theory. Let A_1 and A_2 be two points in space (Fig. 108, $r_2 > 0$). Using (12), show that

$$\frac{1}{r} = \frac{1}{\sqrt{r_1^2 + r_2^2 - 2r_1r_2\cos\theta}}$$
$$= \frac{1}{r_2}\sum_{m=0}^{\infty} P_m(\cos\theta)\left(\frac{r_1}{r_2}\right)^m.$$

This formula has applications in potential theory. (Q/r is the electrostatic potential at A_2 due to a charge Q located at A_1. And the series expresses $1/r$ in terms of the distances of A_1 and A_2 from any origin O and the angle θ between the segments OA_1 and OA_2.)

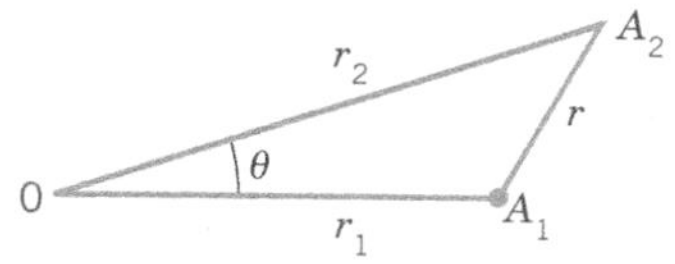

Fig. 108. Team Project 10

(c) Further applications of (12). Show that $P_n(1) = 1$, $P_n(-1) = (-1)^n$, $P_{2n+1}(0) = 0$, and $P_{2n}(0) = (-1)^n \cdot 1 \cdot 3 \cdots (2n - 1)/[2 \cdot 4 \cdots (2n)]$.

11–15 FURTHER FORMULAS

11. ODE. Find a solution of $(a^2 - x^2)y'' - 2xy' + n(n + 1)y = 0$, $a \neq 0$, by reduction to the Legendre equation.

12. Rodrigues's formula (13)[2] Applying the binomial theorem to $(x^2 - 1)^n$, differentiating it n times term by term, and comparing the result with (11), show that

$$P_n(x) = \frac{1}{2^n n!}\frac{d^n}{dx^n}[(x^2 - 1)^n]. \qquad (13)$$

[2] OLINDE RODRIGUES (1794–1851), French mathematician and economist.

13. Rodrigues's formula. Obtain (11′) from (13).

14. Bonnet's recursion.[3] Differentiating (13) with respect to u, using (13) in the resulting formula, and comparing coefficients of u^n, obtain the *Bonnet recursion.*

(14) $(n + 1)P_{n+1}(x) = (2n + 1)xP_n(x) - np_{n-1}(x)$,

where $n = 1, 2, \cdots$. This formula is useful for computations, the loss of significant digits being small (except near zeros). Try (14) out for a few computations of your own choice.

15. Associated Legendre functions $P_n^k(x)$ are needed, e.g., in quantum physics. They are defined by

$$P_n^k(x) = (1 - x^2)^{k/2} \frac{d^k p_n(x)}{dx^k} \tag{15}$$

and are solutions of the ODE

$$(1 - x^2)y'' - 2xy' + q(x)y = 0 \tag{16}$$

where $q(x) = n(n + 1) - k^2/(1 - x^2)$. Find $P_1^1(x)$, $P_2^1(x)$, $P_2^2(x)$, and $P_4^2(x)$ and verify that they satisfy (16).

5.3 Extended Power Series Method: Frobenius Method

Several second-order ODEs of considerable practical importance—the famous Bessel equation among them—have coefficients that are not analytic (definition in Sec. 5.1), but are "not too bad," so that these ODEs can still be solved by series (power series times a logarithm or times a fractional power of x, etc.). Indeed, the following theorem permits an extension of the power series method. The new method is called the **Frobenius method.**[4] Both methods, that is, the power series method and the Frobenius method, have gained in significance due to the use of software in actual calculations.

THEOREM 1

Frobenius Method

Let $b(x)$ and $c(x)$ be any functions that are analytic at $x = 0$. Then the ODE

$$y'' + \frac{b(x)}{x} y' + \frac{c(x)}{x^2} y = 0 \tag{1}$$

has at least one solution that can be represented in the form

$$y(x) = x^r \sum_{m=0}^{\infty} a_m x^m = x^r(a_0 + a_1x + a_2x^2 + \cdots) \qquad (a_0 \neq 0) \tag{2}$$

where the exponent r may be any (real or complex) number (and r is chosen so that $a_0 \neq 0$).

The ODE (1) also has a second solution (such that these two solutions are linearly independent) that may be similar to (2) (with a different r and different coefficients) or may contain a logarithmic term. (Details in Theorem 2 below.)

---

[3] OSSIAN BONNET (1819–1892), French mathematician, whose main work was in differential geometry.

[4] GEORG FROBENIUS (1849–1917), German mathematician, professor at ETH Zurich and University of Berlin, student of Karl Weierstrass (see footnote, Sect. 15.5). He is also known for his work on matrices and in group theory.

In this theorem we may replace x by $x - x_0$ with any number x_0. The condition $a_0 \neq 0$ is no restriction; it simply means that we factor out the highest possible power of x.

The singular point of (1) at $x = 0$ is often called a **regular singular point**, a term confusing to the student, which we shall not use.

For example, Bessel's equation (to be discussed in the next section)

$$y'' + \frac{1}{x} y' + \left(\frac{x^2 - \nu^2}{x^2}\right) y = 0 \qquad (\nu \text{ a parameter})$$

is of the form (1) with $b(x) = 1$ and $c(x) = x^2 - \nu^2$ analytic at $x = 0$, so that the theorem applies. This ODE could not be handled in full generality by the power series method.

Similarly, the so-called hypergeometric differential equation (see Problem Set 5.3) also requires the Frobenius method.

The point is that in (2) we have a power series times a single power of x whose exponent r is not restricted to be a nonnegative integer. (The latter restriction would make the whole expression a power series, by definition; see Sec. 5.1.)

The proof of the theorem requires advanced methods of complex analysis and can be found in Ref. [A11] listed in App. 1.

Regular and Singular Points. The following terms are practical and commonly used. A **regular point** of the ODE

$$y'' + p(x)y' + q(x)y = 0$$

is a point x_0 at which the coefficients p and q are analytic. Similarly, a **regular point** of the ODE

$$\widetilde{h}(x)y'' + \widetilde{p}(x)y'(x) + \widetilde{q}(x)y = 0$$

is an x_0 at which $\widetilde{h}, \widetilde{p}, \widetilde{q}$ are analytic and $\widetilde{h}(x_0) \neq 0$ (so what we can divide by $\widetilde{h}$ and get the previous standard form). Then the power series method can be applied. If x_0 is not a regular point, it is called a **singular point**.

Indicial Equation, Indicating the Form of Solutions

We shall now explain the Frobenius method for solving (1). Multiplication of (1) by x^2 gives the more convenient form

$$x^2 y'' + xb(x)y' + c(x)y = 0. \tag{1'}$$

We first expand $b(x)$ and $c(x)$ in power series,

$$b(x) = b_0 + b_1 x + b_2 x^2 + \cdots, \qquad c(x) = c_0 + c_1 x + c_2 x^2 + \cdots$$

or we do nothing if $b(x)$ and $c(x)$ are polynomials. Then we differentiate (2) term by term, finding

$$y'(x) = \sum_{m=0}^{\infty} (m + r)a_m x^{m+r-1} = x^{r-1}[ra_0 + (r + 1)a_1 x + \cdots]$$

$$\begin{aligned} y''(x) &= \sum_{m=0}^{\infty} (m + r)(m + r - 1)a_m x^{m+r-2} \\ &= x^{r-2}[r(r - 1)a_0 + (r + 1)ra_1 x + \cdots]. \end{aligned} \tag{2*}$$

By inserting all these series into (1′) we obtain

$$x^r[r(r-1)a_0 + \cdots] + (b_0 + b_1x + \cdots)x^r(ra_0 + \cdots)$$
$$+ (c_0 + c_1x + \cdots)x^r(a_0 + a_1x + \cdots) = 0. \qquad (3)$$

We now equate the sum of the coefficients of each power $x^r, x^{r+1}, x^{r+2}, \cdots$ to zero. This yields a system of equations involving the unknown coefficients a_m. The smallest power is x^r and the corresponding equation is

$$[r(r-1) + b_0r + c_0]a_0 = 0.$$

Since by assumption $a_0 \neq 0$, the expression in the brackets $[\cdots]$ must be zero. This gives

$$r(r-1) + b_0r + c_0 = 0. \qquad (4)$$

This important quadratic equation is called the **indicial equation** of the ODE (1). Its role is as follows.

The Frobenius method yields a basis of solutions. One of the two solutions will always be of the form (2), where r is a root of (4). The other solution will be of a form indicated by the indicial equation. There are three cases:

Case 1. Distinct roots not differing by an integer $1, 2, 3, \cdots$.

Case 2. A double root.

Case 3. Roots differing by an integer $1, 2, 3, \cdots$.

Cases 1 and 2 are not unexpected because of the Euler–Cauchy equation (Sec. 2.5), the simplest ODE of the form (1). Case 1 includes complex conjugate roots r_1 and $r_2 = \bar{r}_1$ because $r_1 - r_2 = r_1 - \bar{r}_1 = 2i \operatorname{Im} r_1$ is imaginary, so it cannot be a *real* integer. The form of a basis will be given in Theorem 2 (which is proved in App. 4), without a general theory of convergence, but convergence of the occurring series can be tested in each individual case as usual. Note that in Case 2 *we **must*** have a logarithm, whereas in Case 3 we *may* or *may not.*

THEOREM 2

Frobenius Method. Basis of Solutions. Three Cases

Suppose that the ODE (1) *satisfies the assumptions in Theorem* 1. *Let* r_1 *and* r_2 *be the roots of the indicial equation* (4). *Then we have the following three cases.*

Case* 1. *Distinct Roots Not Differing by an Integer. *A basis is*

$$y_1(x) = x^{r_1}(a_0 + a_1x + a_2x^2 + \cdots) \qquad (5)$$

and

$$y_2(x) = x^{r_2}(A_0 + A_1x + A_2x^2 + \cdots) \qquad (6)$$

with coefficients obtained successively from (3) *with* $r = r_1$ *and* $r = r_2$, *respectively.*

Case 2. Double Root $r_1 = r_2 = r$. *A basis is*

$$y_1(x) = x^r(a_0 + a_1x + a_2x^2 + \cdots) \qquad [r = \tfrac{1}{2}(1 - b_0)] \tag{7}$$

(of the same general form as before) and

$$y_2(x) = y_1(x) \ln x + x^r(A_1x + A_2x^2 + \cdots) \qquad (x > 0). \tag{8}$$

Case 3. Roots Differing by an Integer. *A basis is*

$$y_1(x) = x^{r_1}(a_0 + a_1x + a_2x^2 + \cdots) \tag{9}$$

(of the same general form as before) and

$$y_2(x) = ky_1(x) \ln x + x^{r_2}(A_0 + A_1x + A_2x^2 + \cdots), \tag{10}$$

where the roots are so denoted that $r_1 - r_2 > 0$ *and* k *may turn out to be zero.*

Typical Applications

Technically, the Frobenius method is similar to the power series method, once the roots of the indicial equation have been determined. However, (5)–(10) merely indicate the general form of a basis, and a second solution can often be obtained more rapidly by reduction of order (Sec. 2.1).

EXAMPLE 1 Euler–Cauchy Equation, Illustrating Cases 1 and 2 and Case 3 without a Logarithm

For the Euler–Cauchy equation (Sec. 2.5)

$$x^2y'' + b_0xy' + c_0y = 0 \qquad (b_0,\ c_0 \text{ constant})$$

substitution of $y = x^r$ gives the auxiliary equation

$$r(r-1) + b_0r + c_0 = 0,$$

which is the indicial equation [and $y = x^r$ is a very special form of (2)!]. For different roots r_1, r_2 we get a basis $y_1 = x^{r_1}$, $y_2 = x^{r_2}$, and for a double root r we get a basis x^r, $x^r \ln x$. Accordingly, for this simple ODE, Case 3 plays no extra role. ■

EXAMPLE 2 Illustration of Case 2 (Double Root)

Solve the ODE

$$x(x-1)y'' + (3x-1)y' + y = 0. \tag{11}$$

(This is a special hypergeometric equation, as we shall see in the problem set.)

Solution. Writing (11) in the standard form (1), we see that it satisfies the assumptions in Theorem 1. [What are $b(x)$ and $c(x)$ in (11)?] By inserting (2) and its derivatives (2*) into (11) we obtain

$$\begin{aligned} &\sum_{m=0}^{\infty} (m+r)(m+r-1)a_m x^{m+r} - \sum_{m=0}^{\infty} (m+r)(m+r-1)a_m x^{m+r-1} \\ &+ 3\sum_{m=0}^{\infty} (m+r)a_m x^{m+r} - \sum_{m=0}^{\infty} (m+r)a_m x^{m+r-1} + \sum_{m=0}^{\infty} a_m x^{m+r} = 0. \end{aligned} \tag{12}$$

The smallest power is x^{r-1}, occurring in the second and the fourth series; by equating the sum of its coefficients to zero we have

$$[-r(r-1) - r]a_0 = 0, \qquad \text{thus} \qquad r^2 = 0.$$

Hence this indicial equation has the double root $r = 0$.

First Solution. We insert this value $r = 0$ into (12) and equate the sum of the coefficients of the power x^s to zero, obtaining

$$s(s-1)a_s - (s+1)sa_{s+1} + 3sa_s - (s+1)a_{s+1} + a_s = 0$$

thus $a_{s+1} = a_s$. Hence $a_0 = a_1 = a_2 = \cdots$, and by choosing $a_0 = 1$ we obtain the solution

$$y_1(x) = \sum_{m=0}^{\infty} x^m = \frac{1}{1-x} \qquad (|x| < 1).$$

Second Solution. We get a second independent solution y_2 by the method of reduction of order (Sec. 2.1), substituting $y_2 = uy_1$ and its derivatives into the equation. This leads to (9), Sec. 2.1, which we shall use in this example, instead of starting reduction of order from scratch (as we shall do in the next example). In (9) of Sec. 2.1 we have $p = (3x-1)/(x^2-x)$, the coefficient of y' in (11) ***in standard form***. By partial fractions,

$$-\int p\,dx = -\int \frac{3x-1}{x(x-1)}\,dx = -\int\left(\frac{2}{x-1} + \frac{1}{x}\right)dx = -2\ln(x-1) - \ln x.$$

Hence (9), Sec. 2.1, becomes

$$u' = U = y_1^{-2}e^{-\int p\,dx} = \frac{(x-1)^2}{(x-1)^2 x} = \frac{1}{x}, \qquad u = \ln x, \qquad y_2 = uy_1 = \frac{\ln x}{1-x}.$$

y_1 and y_2 are shown in Fig. 109. These functions are linearly independent and thus form a basis on the interval $0 < x < 1$ (as well as on $1 < x < \infty$). ■

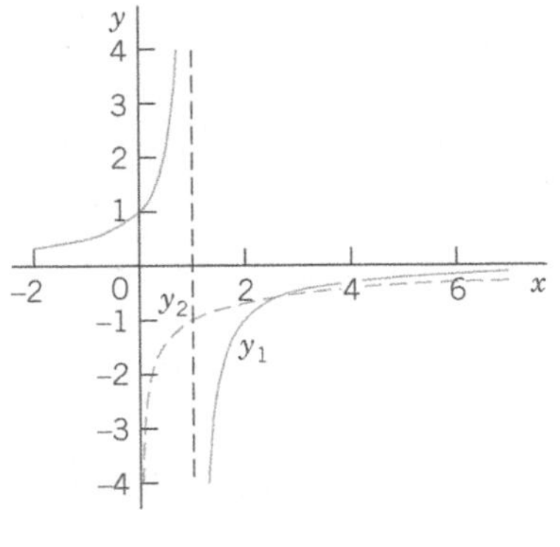

Fig. 109. Solutions in Example 2

EXAMPLE 3 Case 3, Second Solution with Logarithmic Term

Solve the ODE

$$(x^2 - x)y'' - xy' + y = 0. \tag{13}$$

Solution. Substituting (2) and (2*) into (13), we have

$$(x^2-x)\sum_{m=0}^{\infty}(m+r)(m+r-1)a_m x^{m+r-2} - x\sum_{m=0}^{\infty}(m+r)a_m x^{m+r-1} + \sum_{m=0}^{\infty} a_m x^{m+r} = 0.$$

We now take x^2, x, and x inside the summations and collect all terms with power x^{m+r} and simplify algebraically,

$$\sum_{m=0}^{\infty} (m + r - 1)^2 a_m x^{m+r} - \sum_{m=0}^{\infty} (m + r)(m + r - 1)a_m x^{m+r-1} = 0.$$

In the first series we set $m = s$ and in the second $m = s + 1$, thus $s = m - 1$. Then

(14)
$$\sum_{s=0}^{\infty}(s + r - 1)^2 a_s x^{s+r} - \sum_{s=-1}^{\infty} (s + r + 1)(s + r)a_{s+1}x^{s+r} = 0.$$

The lowest power is x^{r-1} (take $s = -1$ in the second series) and gives the indicial equation

$$r(r - 1) = 0.$$

The roots are $r_1 = 1$ and $r_2 = 0$. They differ by an integer. This is Case 3.

First Solution.   From (14) with $r = r_1 = 1$ we have

$$\sum_{s=0}^{\infty}[s^2 a_s - (s + 2)(s + 1)a_{s+1}]x^{s+1} = 0.$$

This gives the recurrence relation

$$a_{s+1} = \frac{s^2}{(s + 2)(s + 1)} a_s \qquad\qquad (s = 0, 1, \cdots).$$

Hence $a_1 = 0, a_2 = 0, \cdots$ successively. Taking $a_0 = 1$, we get as a first solution $y_1 = x^{r_1}a_0 = x$.

Second Solution. Applying reduction of order (Sec. 2.1), we substitute $y_2 = y_1 u = xu$, $y_2' = xu' + u$ and $y_2'' = xu'' + 2u'$ into the ODE, obtaining

$$(x^2 - x)(xu'' + 2u') - x(xu' + u) + xu = 0.$$

xu drops out. Division by x and simplification give

$$(x^2 - x)u'' + (x - 2)u' = 0.$$

From this, using partial fractions and integrating (taking the integration constant zero), we get

$$\frac{u''}{u'} = -\frac{x - 2}{x^2 - x} = -\frac{2}{x} + \frac{1}{1 - x}, \qquad \ln u' = \ln \left| \frac{x - 1}{x^2} \right|.$$

Taking exponents and integrating (again taking the integration constant zero), we obtain

$$u' = \frac{x - 1}{x^2} = \frac{1}{x} - \frac{1}{x^2}, \qquad u = \ln x + \frac{1}{x}, \qquad y_2 = xu = x \ln x + 1.$$

y_1 and y_2 are linearly independent, and y_2 has a logarithmic term. Hence y_1 and y_2 constitute a basis of solutions for positive x.   ■

The Frobenius method solves the **hypergeometric equation**, whose solutions include many known functions as special cases (see the problem set). In the next section we use the method for solving Bessel's equation.

PROBLEM SET 5.3

1. **WRITING PROJECT. Power Series Method and Frobenius Method.** Write a report of 2–3 pages explaining the difference between the two methods. No proofs. Give simple examples of your own.

2–13 FROBENIUS METHOD

Find a basis of solutions by the Frobenius method. Try to identify the series as expansions of known functions. Show the details of your work.

2. $(x+2)^2y'' + (x+2)y' - y = 0$
3. $xy'' + 2y' + xy = 0$
4. $xy'' + y = 0$
5. $xy'' + (2x+1)y' + (x+1)y = 0$
6. $xy'' + 2x^3y' + (x^2-2)y = 0$
7. $y'' + (x-1)y = 0$
8. $xy'' + y' - xy = 0$
9. $2x(x-1)y'' - (x+1)y' + y = 0$
10. $xy'' + 2y' + 4xy = 0$
11. $xy'' + (2-2x)y' + (x-2)y = 0$
12. $x^2y'' + 6xy' + (4x^2+6)y = 0$
13. $xy'' + (1-2x)y' + (x-1)y = 0$
14. **TEAM PROJECT. Hypergeometric Equation, Series, and Function. Gauss's hypergeometric ODE**[5] is

$$x(1-x)y'' + [c - (a+b+1)x]y' - aby = 0. \tag{15}$$

Here, a, b, c are constants. This ODE is of the form $p_2y'' + p_1y' + p_0y = 0$, where p_2, p_1, p_0 are polynomials of degree 2, 1, 0, respectively. These polynomials are written so that the series solution takes a most practical form, namely,

$$y_1(x) = 1 + \frac{ab}{1!\,c}x + \frac{a(a+1)b(b+1)}{2!\,c(c+1)}x^2 + \frac{a(a+1)(a+2)b(b+1)(b+2)}{3!\,c(c+1)(c+2)}x^3 + \cdots. \tag{16}$$

This series is called the **hypergeometric series**. Its sum $y_1(x)$ is called the **hypergeometric function** and is denoted by $F(a, b, c; x)$. Here, $c \neq 0, -1, -2, \cdots$. By choosing specific values of a, b, c we can obtain an incredibly large number of special functions as solutions of (15) [see the small sample of elementary functions in part (c)]. This accounts for the importance of (15).

(a) Hypergeometric series and function. Show that the indicial equation of (15) has the roots $r_1 = 0$ and $r_2 = 1 - c$. Show that for $r_1 = 0$ the Frobenius method gives (16). Motivate the name for (16) by showing that

$$F(1, 1, 1; x) = F(1, b, b; x) = F(a, 1, a; x) = \frac{1}{1-x}.$$

(b) Convergence. For what a or b will (16) reduce to a polynomial? Show that for any other a, b, c ($c \neq 0, -1, -2, \cdots$) the series (16) converges when $|x| < 1$.

(c) Special cases. Show that

$$(1+x)^n = F(-n, b, b; -x),$$
$$(1-x)^n = 1 - nxF(1-n, 1, 2; x),$$
$$\arctan x = xF(\tfrac{1}{2}, 1, \tfrac{3}{2}; -x^2)$$
$$\arcsin x = xF(\tfrac{1}{2}, \tfrac{1}{2}, \tfrac{3}{2}; x^2),$$
$$\ln(1+x) = xF(1, 1, 2; -x),$$
$$\ln\frac{1+x}{1-x} = 2xF(\tfrac{1}{2}, 1, \tfrac{3}{2}; x^2).$$

Find more such relations from the literature on special functions, for instance, from [GenRef1] in App. 1.

(d) Second solution. Show that for $r_2 = 1 - c$ the Frobenius method yields the following solution (where $c \neq 2, 3, 4, \cdots$):

$$y_2(x) = x^{1-c}\left(1 + \frac{(a-c+1)(b-c+1)}{1!\,(-c+2)}x + \frac{(a-c+1)(a-c+2)(b-c+1)(b-c+2)}{2!\,(-c+2)(-c+3)}x^2 + \cdots\right). \tag{17}$$

Show that

$$y_2(x) = x^{1-c}F(a-c+1, b-c+1, 2-c; x).$$

(e) On the generality of the hypergeometric equation. Show that

$$(t^2 + At + B)\ddot{y} + (Ct + D)\dot{y} + Ky = 0 \tag{18}$$

[5] CARL FRIEDRICH GAUSS (1777–1855), great German mathematician. He already made the first of his great discoveries as a student at Helmstedt and Göttingen. In 1807 he became a professor and director of the Observatory at Göttingen. His work was of basic importance in algebra, number theory, differential equations, differential geometry, non-Euclidean geometry, complex analysis, numeric analysis, astronomy, geodesy, electromagnetism, and theoretical mechanics. He also paved the way for a general and systematic use of complex numbers.

with $\dot{y} = dy/dt$, etc., constant A, B, C, D, K, and $t^2 + At + B = (t - t_1)(t - t_2)$, $t_1 \neq t_2$, can be reduced to the hypergeometric equation with independent variable

$$x = \frac{t - t_1}{t_2 - t_1}$$

and parameters related by $Ct_1 + D = -c(t_2 - t_1)$, $C = a + b + 1$, $K = ab$. From this you see that (15) is a "normalized form" of the more general (18) and that various cases of (18) can thus be solved in terms of hypergeometric functions.

15–20 HYPERGEOMETRIC ODE

Find a general solution in terms of hypergeometric functions.

15. $2x(1 - x)y'' - (1 + 6x)y' - 2y = 0$

16. $x(1 - x)y'' + (\frac{1}{2} + 2x)y' - 2y = 0$

17. $4x(1 - x)y'' + y' + 8y = 0$

18. $4(t^2 - 3t + 2)\ddot{y} - 2\dot{y} + y = 0$

19. $2(t^2 - 5t + 6)\ddot{y} + (2t - 3)\dot{y} - 8y = 0$

20. $3t(1 + t)\ddot{y} + t\dot{y} - y = 0$

5.4 Bessel's Equation. Bessel Functions $J_\nu(x)$

One of the most important ODEs in applied mathematics in **Bessel's equation,**[6]

$$x^2y'' + xy' + (x^2 - \nu^2)y = 0 \tag{1}$$

where the parameter ν (nu) is a given real number which is positive or zero. Bessel's equation often appears if a problem shows cylindrical symmetry, for example, as the membranes in Sec.12.9. The equation satisfies the assumptions of Theorem 1. To see this, divide (1) by x^2 to get the standard form $y'' + y'/x + (1 - \nu^2/x^2)y = 0$. Hence, according to the Frobenius theory, it has a solution of the form

$$y(x) = \sum_{m=0}^{\infty} a_m x^{m+r} \qquad (a_0 \neq 0). \tag{2}$$

Substituting (2) and its first and second derivatives into Bessel's equation, we obtain

$$\sum_{m=0}^{\infty} (m + r)(m + r - 1)a_m x^{m+r} + \sum_{m=0}^{\infty} (m + r)a_m x^{m+r}$$
$$+ \sum_{m=0}^{\infty} a_m x^{m+r+2} - \nu^2 \sum_{m=0}^{\infty} a_m x^{m+r} = 0.$$

We equate the sum of the coefficients of x^{s+r} to zero. Note that this power x^{s+r} corresponds to $m = s$ in the first, second, and fourth series, and to $m = s - 2$ in the third series. Hence for $s = 0$ and $s = 1$, the third series does not contribute since $m \geqq 0$.

---

[6]FRIEDRICH WILHELM BESSEL (1784–1846), German astronomer and mathematician, studied astronomy on his own in his spare time as an apprentice of a trade company and finally became director of the new Königsberg Observatory.

Formulas on Bessel functions are contained in Ref. [GenRef10] and the standard treatise [A13].

For $s = 2, 3, \cdots$ all four series contribute, so that we get a general formula for all these s. We find

$$
\begin{aligned}
&\text{(a)} && r(r-1)a_0 + ra_0 - \nu^2 a_0 = 0 && (s=0)\\
\text{(3)}\quad &\text{(b)} && (r+1)ra_1 + (r+1)a_1 - \nu^2 a_1 = 0 && (s=1)\\
&\text{(c)} && (s+r)(s+r-1)a_s + (s+r)a_s + a_{s-2} - \nu^2 a_s = 0 && (s = 2, 3, \cdots).
\end{aligned}
$$

From (3a) we obtain the **indicial equation** by dropping a_0,

$$(r+\nu)(r-\nu) = 0. \tag{4}$$

The roots are $r_1 = \nu\ (\geqq 0)$ and $r_2 = -\nu$.

Coefficient Recursion for $r = r_1 = \nu$. For $r = \nu$, Eq. (3b) reduces to $(2\nu + 1)a_1 = 0$. Hence $a_1 = 0$ since $\nu \geqq 0$. Substituting $r = \nu$ in (3c) and combining the three terms containing a_s gives simply

$$(s + 2\nu)sa_s + a_{s-2} = 0. \tag{5}$$

Since $a_1 = 0$ and $\nu \geqq 0$, it follows from (5) that $a_3 = 0, a_5 = 0, \cdots$. Hence we have to deal only with *even-numbered* coefficients a_s with $s = 2m$. For $s = 2m$, Eq. (5) becomes

$$(2m + 2\nu)2ma_{2m} + a_{2m-2} = 0.$$

Solving for a_{2m} gives the recursion formula

$$a_{2m} = -\frac{1}{2^2 m(\nu + m)} a_{2m-2}, \qquad m = 1, 2, \cdots. \tag{6}$$

From (6) we can now determine $a_2, a_4, \cdots$ successively. This gives

$$a_2 = -\frac{a_0}{2^2(\nu+1)}$$

$$a_4 = -\frac{a_2}{2^2 2(\nu+2)} = \frac{a_0}{2^4 2!\,(\nu+1)(\nu+2)}$$

and so on, and in general

$$a_{2m} = \frac{(-1)^m a_0}{2^{2m} m!\,(\nu+1)(\nu+2)\cdots(\nu+m)}, \qquad m = 1, 2, \cdots. \tag{7}$$

Bessel Functions $J_n(x)$ for Integer $\nu = n$

Integer values of ν are denoted by n. This is standard. For $\nu = n$ the relation (7) becomes

$$a_{2m} = \frac{(-1)^m a_0}{2^{2m} m!\,(n+1)(n+2)\cdots(n+m)}, \qquad m = 1, 2, \cdots. \tag{8}$$

a_0 is still arbitrary, so that the series (2) with these coefficients would contain this arbitrary factor a_0. This would be a highly impractical situation for developing formulas or computing values of this new function. Accordingly, we have to make a choice. The choice $a_0 = 1$ would be possible. A simpler series (2) could be obtained if we could absorb the growing product $(n + 1)(n + 2) \cdots (n + m)$ into a factorial function $(n + m)!$ What should be our choice? Our choice should be

$$a_0 = \frac{1}{2^n n!} \tag{9}$$

because then $n!\,(n + 1) \cdots (n + m) = (n + m)!$ in (8), so that (8) simply becomes

$$a_{2m} = \frac{(-1)^m}{2^{2m+n} m!\,(n + m)!}, \qquad m = 1, 2, \cdots . \tag{10}$$

By inserting these coefficients into (2) and remembering that $c_1 = 0$, $c_3 = 0, \cdots$ we obtain a particular solution of Bessel's equation that is denoted by $J_n(x)$:

$$J_n(x) = x^n \sum_{m=0}^{\infty} \frac{(-1)^m x^{2m}}{2^{2m+n} m!\,(n + m)!} \qquad (n \geqq 0). \tag{11}$$

$J_n(x)$ is called the **Bessel function of the first kind** *of order n*. The series (11) converges for all x, as the ratio test shows. Hence $J_n(x)$ is defined for all x. The series converges very rapidly because of the factorials in the denominator.

EXAMPLE 1 Bessel Functions $J_0(x)$ and $J_1(x)$

For $n = 0$ we obtain from (11) the **Bessel function of order 0**

$$J_0(x) = \sum_{m=0}^{\infty} \frac{(-1)^m x^{2m}}{2^{2m}(m!)^2} = 1 - \frac{x^2}{2^2(1!)^2} + \frac{x^4}{2^4(2!)^2} - \frac{x^6}{2^6(3!)^2} + - \cdots \tag{12}$$

which looks similar to a cosine (Fig. 110). For $n = 1$ we obtain the **Bessel function of order 1**

$$J_1(x) = \sum_{m=0}^{\infty} \frac{(-1)^m x^{2m+1}}{2^{2m+1} m!\,(m + 1)!} = \frac{x}{2} - \frac{x^3}{2^3 1!\,2!} + \frac{x^5}{2^5 2!\,3!} - \frac{x^7}{2^7 3!\,4!} + - \cdots, \tag{13}$$

which looks similar to a sine (Fig. 110). But the zeros of these functions are not completely regularly spaced (see also Table A1 in App. 5) and the height of the "waves" decreases with increasing x. Heuristically, n^2/x^2 in (1) in standard form [(1) divided by x^2] is zero (if $n = 0$) or small in absolute value for large x, and so is y'/x, so that then Bessel's equation comes close to $y'' + y = 0$, the equation of $\cos x$ and $\sin x$; also y'/x acts as a "damping term," in part responsible for the decrease in height. One can show that for large x,

$$J_n(x) \sim \sqrt{\frac{2}{\pi x}} \cos\left(x - \frac{n\pi}{2} - \frac{\pi}{4}\right) \tag{14}$$

where $\sim$ is read "**asymptotically equal**" and means that *for fixed n* the quotient of the two sides approaches 1 as $x \to \infty$.

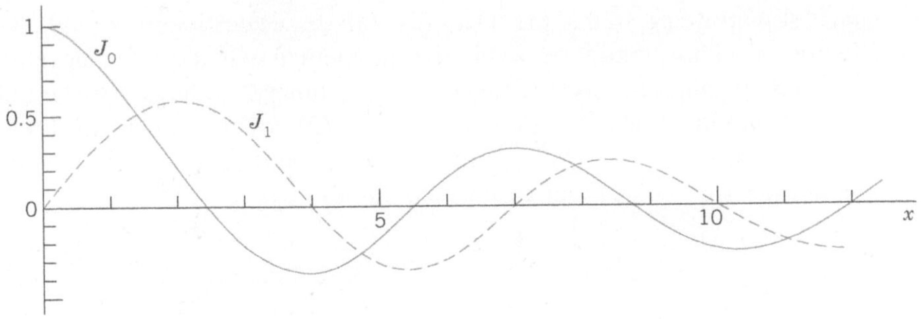

Fig. 110. Bessel functions of the first kind J_0 and J_1

Formula (14) is surprisingly accurate even for smaller x (>0). For instance, it will give you good starting values in a computer program for the basic task of computing zeros. For example, for the first three zeros of J_0 you obtain the values 2.356 (2.405 exact to 3 decimals, error 0.049), 5.498 (5.520, error 0.022), 8.639 (8.654, error 0.015), etc. ■

Bessel Functions $J_\nu(x)$ for any $\nu \geqq 0$. Gamma Function

We now proceed from integer $\nu = n$ to any $\nu \geqq 0$. We had $a_0 = 1/(2^n n!)$ in (9). So we have to extend the factorial function $n!$ to any $\nu \geqq 0$. For this we choose

$$a_0 = \frac{1}{2^\nu \Gamma(\nu + 1)} \tag{15}$$

with the **gamma function** $\Gamma(\nu + 1)$ defined by

$$\Gamma(\nu + 1) = \int_0^\infty e^{-t} t^\nu \, dt \qquad (\nu > -1). \tag{16}$$

(**CAUTION!** Note the convention $\nu + 1$ on the left but ν in the integral.) Integration by parts gives

$$\Gamma(\nu + 1) = -e^{-t} t^\nu \Big|_0^\infty + \nu \int_0^\infty e^{-t} t^{\nu - 1} \, dt = 0 + \nu\Gamma(\nu).$$

This is the basic functional relation of the gamma function

$$\Gamma(\nu + 1) = \nu\Gamma(\nu). \tag{17}$$

Now from (16) with $\nu = 0$ and then by (17) we obtain

$$\Gamma(1) = \int_0^\infty e^{-t} \, dt = -e^{-t} \Big|_0^\infty = 0 - (-1) = 1$$

and then $\Gamma(2) = 1 \cdot \Gamma(1) = 1!$, $\Gamma(3) = 2\Gamma(1) = 2!$ and in general

$$\Gamma(n + 1) = n! \qquad (n = 0, 1, \cdots). \tag{18}$$

Hence ***the gamma function generalizes the factorial function*** **to arbitrary positive** ν. Thus (15) with $\nu = n$ agrees with (9).

Furthermore, from (7) with a_0 given by (15) we first have

$$a_{2m} = \frac{(-1)^m}{2^{2m} m!\,(\nu+1)(\nu+2)\cdots(\nu+m)2^\nu \Gamma(\nu+1)}.$$

Now (17) gives $(\nu+1)\Gamma(\nu+1) = \Gamma(\nu+2)$, $(\nu+2)\Gamma(\nu+2) = \Gamma(\nu+3)$ and so on, so that

$$(\nu+1)(\nu+2)\cdots(\nu+m)\Gamma(\nu+1) = \Gamma(\nu+m+1).$$

Hence because of our (standard!) choice (15) of a_0 the coefficients (7) are simply

$$a_{2m} = \frac{(-1)^m}{2^{2m+\nu} m!\,\Gamma(\nu+m+1)}. \tag{19}$$

With these coefficients and $r = r_1 = \nu$ we get from (2) a particular solution of (1), denoted by $J_\nu(x)$ and given by

$$J_\nu(x) = x^\nu \sum_{m=0}^{\infty} \frac{(-1)^m x^{2m}}{2^{2m+\nu} m!\,\Gamma(\nu+m+1)}. \tag{20}$$

$J_\nu(x)$ is called the **Bessel function of the first kind of order ν**. The series (20) converges for all x, as one can verify by the ratio test.

Discovery of Properties from Series

Bessel functions are a model case for showing how to discover properties and relations of functions from series by which they are *defined.* Bessel functions satisfy an incredibly large number of relationships—look at Ref. [A13] in App. 1; also, find out what your CAS knows. In Theorem 3 we shall discuss four formulas that are backbones in applications and theory.

THEOREM 1

Derivatives, Recursions

The derivative of $J_\nu(x)$ with respect to x can be expressed by $J_{\nu-1}(x)$ or $J_{\nu+1}(x)$ by the formulas

(21)

$$\textbf{(a)}\quad [x^\nu J_\nu(x)]' = x^\nu J_{\nu-1}(x)$$

$$\textbf{(b)}\quad [x^{-\nu} J_\nu(x)]' = -x^{-\nu} J_{\nu+1}(x).$$

Furthermore, $J_\nu(x)$ and its derivative satisfy the recurrence relations

(21)

$$\textbf{(c)}\quad J_{\nu-1}(x) + J_{\nu+1}(x) = \frac{2\nu}{x} J_\nu(x)$$

$$\textbf{(d)}\quad J_{\nu-1}(x) - J_{\nu+1}(x) = 2J'_\nu(x).$$

PROOF **(a)** We multiply (20) by x^ν and take $x^{2\nu}$ under the summation sign. Then we have

$$x^\nu J_\nu(x) = \sum_{m=0}^{\infty} \frac{(-1)^m x^{2m+2\nu}}{2^{2m+\nu} m!\, \Gamma(\nu + m + 1)}.$$

We now differentiate this, cancel a factor 2, pull $x^{2\nu-1}$ out, and use the functional relationship $\Gamma(\nu + m + 1) = (\nu + m)\Gamma(\nu + m)$ [see (17)]. Then (20) with $\nu - 1$ instead of ν shows that we obtain the right side of (21a). Indeed,

$$(x^\nu J_\nu)' = \sum_{m=0}^{\infty} \frac{(-1)^m 2(m + \nu) x^{2m+2\nu-1}}{2^{2m+\nu} m!\, \Gamma(\nu + m + 1)} = x^\nu x^{\nu-1} \sum_{m=0}^{\infty} \frac{(-1)^m x^{2m}}{2^{2m+\nu-1} m!\, \Gamma(\nu + m)}.$$

(b) Similarly, we multiply (20) by $x^{-\nu}$, so that x^ν in (20) cancels. Then we differentiate, cancel $2m$, and use $m! = m(m - 1)!$. This gives, with $m = s + 1$,

$$(x^{-\nu} J_\nu)' = \sum_{m=1}^{\infty} \frac{(-1)^m x^{2m-1}}{2^{2m+\nu-1} (m - 1)!\, \Gamma(\nu + m + 1)} = \sum_{s=0}^{\infty} \frac{(-1)^{s+1} x^{2s+1}}{2^{2s+\nu+1} s!\, \Gamma(\nu + s + 2)}.$$

Equation (20) with $\nu + 1$ instead of ν and s instead of m shows that the expression on the right is $-x^{-\nu} J_{\nu+1}(x)$. This proves (21b).

(c), (d) We perform the differentiation in (21a). Then we do the same in (21b) and multiply the result on both sides by $x^{2\nu}$. This gives

$$\text{(a*)} \qquad \nu x^{\nu-1} J_\nu + x^\nu J_\nu' = x^\nu J_{\nu-1}$$

$$\text{(b*)} \qquad -\nu x^{\nu-1} J_\nu + x^\nu J_\nu' = -x^\nu J_{\nu+1}.$$

Substracting (b*) from (a*) and dividing the result by x^ν gives (21c). Adding (a*) and (b*) and dividing the result by x^ν gives (21d). ■

EXAMPLE 2 **Application of Theorem 1 in Evaluation and Integration**

Formula (21c) can be used recursively in the form

$$J_{\nu+1}(x) = \frac{2\nu}{x} J_\nu(x) - J_{\nu-1}(x)$$

for calculating Bessel functions of higher order from those of lower order. For instance, $J_2(x) = 2J_1(x)/x - J_0(x)$, so that J_2 can be obtained from tables of J_0 and J_1 (in App. 5 or, more accurately, in Ref. [GenRef1] in App. 1).

To illustrate how Theorem 1 helps in integration, we use (21b) with $\nu = 3$ integrated on both sides. This evaluates, for instance, the integral

$$I = \int_1^2 x^{-3} J_4(x)\, dx = -x^{-3} J_3(x) \Big|_1^2 = -\frac{1}{8} J_3(2) + J_3(1).$$

A table of J_3 (on p. 398 of Ref. [GenRef1]) or your CAS will give you

$$-\tfrac{1}{8} \cdot 0.128943 + 0.019563 = 0.003445.$$

Your CAS (or a human computer in precomputer times) obtains J_3 from (21), first using (21c) with $\nu = 2$, that is, $J_3 = 4x^{-1} J_2 - J_1$, then (21c) with $\nu = 1$, that is, $J_2 = 2x^{-1} J_1 - J_0$. Together,

$$I = x^{-3}(4x^{-1}(2x^{-1}J_1 - J_0) - J_1)\Big|_1^2$$
$$= -\tfrac{1}{8}[2J_1(2) - 2J_0(2) - J_1(2)] + [8J_1(1) - 4J_0(1) - J_1(1)]$$
$$= -\tfrac{1}{8}J_1(2) + \tfrac{1}{4}J_0(2) + 7J_1(1) - 4J_0(1).$$

This is what you get, for instance, with Maple if you type int($\cdots$). And if you type evalf(int($\cdots$)), you obtain 0.003445448, in agreement with the result near the beginning of the example. ■

Bessel Functions J_ν with Half-Integer ν Are Elementary

We discover this remarkable fact as another property obtained from the series (20) and confirm it in the problem set by using Bessel's ODE.

EXAMPLE 3 Elementary Bessel Functions J_ν with $\nu = \pm\frac{1}{2}, \pm\frac{3}{2}, \pm\frac{5}{2}, \cdots$. The Value $\Gamma(\frac{1}{2})$

We first prove (Fig. 111)

$$\text{(22)} \qquad \text{(a)}\quad J_{1/2}(x) = \sqrt{\frac{2}{\pi x}}\sin x, \qquad \text{(b)}\quad J_{-1/2}(x) = \sqrt{\frac{2}{\pi x}}\cos x.$$

The series (20) with $\nu = \frac{1}{2}$ is

$$J_{1/2}(x) = \sqrt{x}\sum_{m=0}^{\infty}\frac{(-1)^m x^{2m}}{2^{2m+1/2}m!\,\Gamma(m+\frac{3}{2})} = \sqrt{\frac{2}{x}}\sum_{m=0}^{\infty}\frac{(-1)^m x^{2m+1}}{2^{2m+1}m!\,\Gamma(m+\frac{3}{2})}.$$

The denominator can be written as a product AB, where (use (16) in B)

$$A = 2^m m! = 2m(2m-2)(2m-4)\cdots 4\cdot 2,$$
$$B = 2^{m+1}\Gamma(m+\tfrac{3}{2}) = 2^{m+1}(m+\tfrac{1}{2})(m-\tfrac{1}{2})\cdots\tfrac{3}{2}\cdot\tfrac{1}{2}\Gamma(\tfrac{1}{2})$$
$$= (2m+1)(2m-1)\cdots 3\cdot 1\cdot\sqrt{\pi};$$

here we used (proof below)

$$\text{(23)} \qquad \Gamma(\tfrac{1}{2}) = \sqrt{\pi}.$$

The product of the right sides of A and B can be written

$$AB = (2m+1)2m(2m-1)\cdots 3\cdot 2\cdot 1\sqrt{\pi} = (2m+1)!\sqrt{\pi}.$$

Hence

$$J_{1/2}(x) = \sqrt{\frac{2}{\pi x}}\sum_{m=0}^{\infty}\frac{(-1)^m x^{2m+1}}{(2m+1)!} = \sqrt{\frac{2}{\pi x}}\sin x.$$

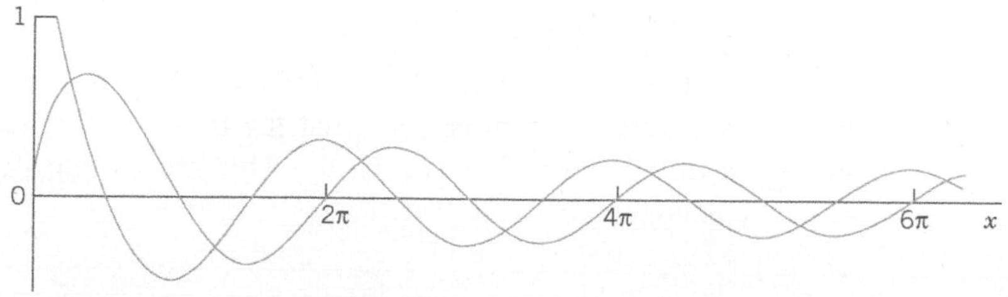

Fig. 111. Bessel functions $J_{1/2}$ and $J_{-1/2}$

This proves (22a). Differentiation and the use of (21a) with $\nu = \frac{1}{2}$ now gives

$$[\sqrt{x}J_{1/2}(x)]' = \sqrt{\frac{2}{\pi}}\cos x = x^{1/2}J_{-1/2}(x).$$

This proves (22b). From (22) follow further formulas successively by (21c), used as in Example 2.

We finally prove $\Gamma(\frac{1}{2}) = \sqrt{\pi}$ by a standard trick worth remembering. In (15) we set $t = u^2$. Then $dt = 2u\,du$ and

$$\Gamma\left(\frac{1}{2}\right) = \int_0^\infty e^{-t}t^{-1/2}\,dt = 2\int_0^\infty e^{-u^2}\,du.$$

We square on both sides, write v instead of u in the second integral, and then write the product of the integrals as a double integral:

$$\Gamma\left(\frac{1}{2}\right)^2 = 4\int_0^\infty e^{-u^2}\,du\int_0^\infty e^{-v^2}\,dv = 4\int_0^\infty\int_0^\infty e^{-(u^2+v^2)}\,du\,dv.$$

We now use polar coordinates r, θ by setting $u = r\cos\theta$, $v = r\sin\theta$. Then the element of area is $du\,dv = r\,dr\,d\theta$ and we have to integrate over r from 0 to ∞ and over θ from 0 to $\pi/2$ (that is, over the first quadrant of the uv-plane):

$$\Gamma\left(\frac{1}{2}\right)^2 = 4\int_0^{\pi/2}\int_0^\infty e^{-r^2}r\,dr\,d\theta = 4\cdot\frac{\pi}{2}\int_0^\infty e^{-r^2}r\,dr = 2\left(-\frac{1}{2}\right)e^{-r^2}\Big|_0^\infty = \pi.$$

By taking the square root on both sides we obtain (23). ■

General Solution. Linear Dependence

For a general solution of Bessel's equation (1) in addition to J_ν we need a second linearly independent solution. For ν not an integer this is easy. Replacing ν by $-\nu$ in (20), we have

$$J_{-\nu}(x) = x^{-\nu}\sum_{m=0}^{\infty}\frac{(-1)^m x^{2m}}{2^{2m-\nu}m!\,\Gamma(m-\nu+1)}. \tag{24}$$

Since Bessel's equation involves ν^2, the functions J_ν and $J_{-\nu}$ are solutions of the equation for the same ν. If ν is not an integer, they are linearly independent, because the first terms in (20) and in (24) are finite nonzero multiples of x^ν and $x^{-\nu}$. Thus, if ν is not an integer, a general solution of Bessel's equation for all $x \neq 0$ is

$$y(x) = c_1J_\nu(x) + c_2J_{-\nu}(x)$$

This cannot be the general solution for an integer $\nu = n$ because, in that case, we have linear dependence. It can be seen that the first terms in (20) and (24) are finite nonzero multiples of x^ν and $x^{-\nu}$, respectively. This means that, for any integer $\nu = n$, we have linear dependence because

$$J_{-n}(x) = (-1)^n J_n(x) \qquad (n = 1, 2, \cdots). \tag{25}$$

PROOF To prove (25), we use (24) and let ν approach a positive integer n. Then the gamma function in the coefficients of the first n terms becomes infinite (see Fig. 553 in App. A3.1), the coefficients become zero, and the summation starts with $m = n$. Since in this case $\Gamma(m - n + 1) = (m - n)!$ by (18), we obtain

$$(26) \qquad J_{-n}(x) = \sum_{m=n}^{\infty} \frac{(-1)^m x^{2m-n}}{2^{2m-n} m!\,(m-n)!} = \sum_{s=0}^{\infty} \frac{(-1)^{n+s} x^{2s+n}}{2^{2s+n}(n+s)!\,s!} \qquad (m = n + s).$$

The last series represents $(-1)^n J_n(x)$, as you can see from (11) with m replaced by s. This completes the proof. ■

The difficulty caused by (25) will be overcome in the next section by introducing further Bessel functions, called *of the second kind* and denoted by Y_ν.

PROBLEM SET 5.4

1. Convergence. Show that the series (11) converges for all x. Why is the convergence very rapid?

2–10 ODEs REDUCIBLE TO BESSEL'S ODE

This is just a sample of such ODEs; some more follow in the next problem set. Find a general solution in terms of J_ν and $J_{-\nu}$ or indicate when this is not possible. Use the indicated substitutions. Show the details of your work.

2. $x^2y'' + xy' + (x^2 - \frac{4}{49})y = 0$

3. $xy'' + y' + \frac{1}{4}y = 0 \quad (\sqrt{x} = z)$

4. $y'' + (e^{-2x} - \frac{1}{9})y = 0 \quad (e^{-x} = z)$

5. Two-parameter ODE
$x^2y'' + xy' + (\lambda^2x^2 - \nu^2)y = 0 \quad (\lambda x = z)$

6. $x^2y'' + \frac{1}{4}(x + \frac{3}{4})\,y = 0 \quad (y = u\sqrt{x},\ \sqrt{x} = z)$

7. $x^2y'' + xy' + \frac{1}{4}(x^2 - 1)y = 0 \quad (x = 2z)$

8. $(2x + 1)^2y'' + 2(2x + 1)y' + 16x(x + 1)y = 0$
$(2x + 1 = z)$

9. $xy'' + (2\nu + 1)y' + xy = 0 \quad (y = x^{-\nu}u)$

10. $x^2y'' + (1 - 2\nu)xy' + \nu^2(x^{2\nu} + 1 - \nu^2)y = 0$
$(y = x^\nu u,\ x^\nu = z)$

11. CAS EXPERIMENT. Change of Coefficient. Find and graph (on common axes) the solutions of

$$y'' + kx^{-1}y' + y = 0,\ y(0) = 1,\ y'(0) = 0,$$

for $k = 0, 1, 2, \cdots, 10$ (or as far as you get useful graphs). For what k do you get elementary functions? Why? Try for noninteger k, particularly between 0 and 2, to see the continuous change of the curve. Describe the change of the location of the zeros and of the extrema as k increases from 0. Can you interpret the ODE as a model in mechanics, thereby explaining your observations?

12. CAS EXPERIMENT. Bessel Functions for Large x.

(a) Graph $J_n(x)$ for $n = 0, \cdots, 5$ on common axes.

(b) Experiment with (14) for integer n. Using graphs, find out from which $x = x_n$ on the curves of (11) and (14) practically coincide. How does x_n change with n?

(c) What happens in (b) if $n = \pm\frac{1}{2}$? (Our usual notation in this case would be ν.)

(d) How does the error of (14) behave as a function of x for fixed n? [Error = exact value minus approximation (14).]

(e) Show from the graphs that $J_0(x)$ has extrema where $J_1(x) = 0$. Which formula proves this? Find further relations between zeros and extrema.

13–15 ZEROS of Bessel functions play a key role in modeling (e.g. of vibrations; see Sec. 12.9).

13. Interlacing of zeros. Using (21) and Rolle's theorem, show that between any two consecutive positive zeros of $J_n(x)$ there is precisely one zero of $J_{n+1}(x)$.

14. Zeros. Compute the first four positive zeros of $J_0(x)$ and $J_1(x)$ from (14). Determine the error and comment.

15. Interlacing of zeros. Using (21) and Rolle's theorem, show that between any two consecutive zeros of $J_0(x)$ there is precisely one zero of $J_1(x)$.

16–18 HALF-INTEGER PARAMETER: APPROACH BY THE ODE

16. Elimination of first derivative. Show that $y = uv$ with $v(x) = \exp\left(-\frac{1}{2}\int p(x)\,dx\right)$ gives from the ODE $y'' + p(x)y' + q(x)y = 0$ the ODE

$$u'' + \left[q(x) - \tfrac{1}{4}p(x)^2 - \tfrac{1}{2}p'(x)\right]u = 0,$$

not containing the first derivative of u.

17. Bessel's equation. Show that for (1) the substitution in Prob. 16 is $y = ux^{-1/2}$ and gives

$$x^2u'' + (x^2 + \tfrac{1}{4} - \nu^2)u = 0. \tag{27}$$

18. Elementary Bessel functions. Derive (22) in Example 3 from (27).

19–25 APPLICATION OF (21): DERIVATIVES, INTEGRALS

Use the powerful formulas (21) to do Probs. 19–25. Show the details of your work.

19. Derivatives. Show that $J_0'(x) = -J_1(x)$, $J_1'(x) = J_0(x) - J_1(x)/x$, $J_2'(x) = \frac{1}{2}[J_1(x) - J_3(x)]$.

20. Bessel's equation. Derive (1) from (21).

21. Basic integral formula. Show that

$$\int x^\nu J_{\nu-1}(x)\,dx = x^\nu J_\nu(x) + c.$$

22. Basic integral formulas. Show that

$$\int x^{-\nu} J_{\nu+1}(x)\,dx = -x^{-\nu} J_\nu(x) + c,$$

$$\int J_{\nu+1}(x)\,dx = \int J_{\nu-1}(x)\,dx - 2J_\nu(x).$$

23. Integration. Show that $\int x^2 J_0(x)\,dx = x^2J_1(x) + xJ_0(x) - \int J_0(x)\,dx$. (The last integral is nonelementary; tables exist, e.g., in Ref. [A13] in App. 1.)

24. Integration. Evaluate $\int x^{-1}J_4(x)\,dx$.

25. Integration. Evaluate $\int J_5(x)\,dx$.

5.5 Bessel Functions $Y_\nu(x)$. General Solution

To obtain a general solution of Bessel's equation (1), Sec. 5.4, for any ν, we now introduce **Bessel functions of the second kind** $Y_\nu(x)$, beginning with the case $\nu = n = 0$.

When $n = 0$, Bessel's equation can be written (divide by x)

$$xy'' + y' + xy = 0. \tag{1}$$

Then the indicial equation (4) in Sec. 5.4 has a double root $r = 0$. This is Case 2 in Sec. 5.3. In this case we first have only one solution, $J_0(x)$. From (8) in Sec. 5.3 we see that the desired second solution must be of the form

$$y_2(x) = J_0(x)\ln x + \sum_{m=1}^{\infty} A_m x^m. \tag{2}$$

We substitute y_2 and its derivatives

$$y_2' = J_0' \ln x + \frac{J_0}{x} + \sum_{m=1}^{\infty} mA_m x^{m-1}$$

$$y_2'' = J_0'' \ln x + \frac{2J_0'}{x} - \frac{J_0}{x^2} + \sum_{m=1}^{\infty} m(m-1)A_m x^{m-2}$$

into (1). Then the sum of the three logarithmic terms $xJ_0'' \ln x$, $J_0' \ln x$, and $xJ_0 \ln x$ is zero because J_0 is a solution of (1). The terms $-J_0/x$ and J_0/x (from xy'' and y') cancel. Hence we are left with

$$2J_0' + \sum_{m=1}^{\infty} m(m-1)A_m x^{m-1} + \sum_{m=1}^{\infty} mA_m x^{m-1} + \sum_{m=1}^{\infty} A_m x^{m+1} = 0.$$

Addition of the first and second series gives $\Sigma m^2 A_m x^{m-1}$. The power series of $J_0'(x)$ is obtained from (12) in Sec. 5.4 and the use of $m!/m = (m-1)!$ in the form

$$J_0'(x) = \sum_{m=1}^{\infty} \frac{(-1)^m 2m x^{2m-1}}{2^{2m}(m!)^2} = \sum_{m=1}^{\infty} \frac{(-1)^m x^{2m-1}}{2^{2m-1} m!\,(m-1)!}.$$

Together with $\Sigma m^2 A_m x^{m-1}$ and $\Sigma A_m x^{m+1}$ this gives

$$(3^*) \qquad \sum_{m=1}^{\infty} \frac{(-1)^m x^{2m-1}}{2^{2m-2} m!\,(m-1)!} + \sum_{m=1}^{\infty} m^2 A_m x^{m-1} + \sum_{m=1}^{\infty} A_m x^{m+1} = 0.$$

First, we show that the A_m with odd subscripts are all zero. The power x^0 occurs only in the second series, with coefficient A_1. Hence $A_1 = 0$. Next, we consider the even powers x^{2s}. The first series contains none. In the second series, $m - 1 = 2s$ gives the term $(2s+1)^2 A_{2s+1} x^{2s}$. In the third series, $m + 1 = 2s$. Hence by equating the sum of the coefficients of x^{2s} to zero we have

$$(2s+1)^2 A_{2s+1} + A_{2s-1} = 0, \qquad s = 1, 2, \cdots.$$

Since $A_1 = 0$, we thus obtain $A_3 = 0, A_5 = 0, \cdots$, successively.

We now equate the sum of the coefficients of x^{2s+1} to zero. For $s = 0$ this gives

$$-1 + 4A_2 = 0, \qquad \text{thus} \qquad A_2 = \tfrac{1}{4}.$$

For the other values of s we have in the first series in (3*) $2m - 1 = 2s + 1$, hence $m = s + 1$, in the second $m - 1 = 2s + 1$, and in the third $m + 1 = 2s + 1$. We thus obtain

$$\frac{(-1)^{s+1}}{2^{2s}(s+1)!\,s!} + (2s+2)^2 A_{2s+2} + A_{2s} = 0.$$

For $s = 1$ this yields

$$\tfrac{1}{8} + 16A_4 + A_2 = 0, \qquad \text{thus} \qquad A_4 = -\tfrac{3}{128}$$

and in general

$$(3) \qquad A_{2m} = \frac{(-1)^{m-1}}{2^{2m}(m!)^2}\left(1 + \frac{1}{2} + \frac{1}{3} + \cdots + \frac{1}{m}\right), \qquad m = 1, 2, \cdots.$$

Using the short notations

$$(4) \qquad h_1 = 1 \qquad h_m = 1 + \frac{1}{2} + \cdots + \frac{1}{m} \qquad m = 2, 3, \cdots$$

and inserting (4) and $A_1 = A_3 = \cdots = 0$ into (2), we obtain the result

$$y_2(x) = J_0(x)\ln x + \sum_{m=1}^{\infty} \frac{(-1)^{m-1} h_m}{2^{2m}(m!)^2} x^{2m}$$

$$(5) \qquad = J_0(x)\ln x + \frac{1}{4}x^2 - \frac{3}{128}x^4 + \frac{11}{13{,}824}x^6 - + \cdots.$$

Since J_0 and y_2 are linearly independent functions, they form a basis of (1) for $x > 0$. Of course, another basis is obtained if we replace y_2 by an independent particular solution of the form $a(y_2 + bJ_0)$, where $a\ (\neq 0)$ and b are constants. It is customary to choose $a = 2/\pi$ and $b = \gamma - \ln 2$, where the number $\gamma = 0.57721566490 \cdots$ is the so-called **Euler constant**, which is defined as the limit of

$$1 + \frac{1}{2} + \cdots + \frac{1}{s} - \ln s$$

as s approaches infinity. The standard particular solution thus obtained is called the **Bessel function of the second kind** *of order zero* (Fig. 112) or **Neumann's function** *of order zero* and is denoted by $Y_0(x)$. Thus [see (4)]

$$Y_0(x) = \frac{2}{\pi}\left[J_0(x)\left(\ln\frac{x}{2} + \gamma\right) + \sum_{m=1}^{\infty} \frac{(-1)^{m-1}h_m}{2^{2m}(m!)^2} x^{2m}\right]. \tag{6}$$

For small $x > 0$ the function $Y_0(x)$ behaves about like $\ln x$ (see Fig. 112, why?), and $Y_0(x) \to -\infty$ as $x \to 0$.

Bessel Functions of the Second Kind $Y_n(x)$

For $\nu = n = 1, 2, \cdots$ a second solution can be obtained by manipulations similar to those for $n = 0$, starting from (10), Sec. 5.4. It turns out that in these cases the solution also contains a logarithmic term.

The situation is not yet completely satisfactory, because the second solution is defined differently, depending on whether the order ν is an integer or not. To provide uniformity of formalism, it is desirable to adopt a form of the second solution that is valid for all values of the order. For this reason we introduce a standard second solution $Y_\nu(x)$ defined for all ν by the formula

(7)

$$\text{(a)} \quad Y_\nu(x) = \frac{1}{\sin \nu\pi}\left[J_\nu(x) \cos \nu\pi - J_{-\nu}(x)\right]$$

$$\text{(b)} \quad Y_n(x) = \lim_{\nu \to n} Y_\nu(x).$$

This function is called the **Bessel function of the second kind** *of order* ν or **Neumann's function**[7] *of order* ν. Figure 112 shows $Y_0(x)$ and $Y_1(x)$.

Let us show that J_ν and Y_ν are indeed linearly independent for all ν (and $x > 0$).

For noninteger order ν, the function $Y_\nu(x)$ is evidently a solution of Bessel's equation because $J_\nu(x)$ and $J_{-\nu}(x)$ are solutions of that equation. Since for those ν the solutions J_ν and $J_{-\nu}$ are linearly independent and Y_ν involves $J_{-\nu}$, the functions J_ν and Y_ν are

[7] CARL NEUMANN (1832–1925), German mathematician and physicist. His work on potential theory using integer equation methods inspired VITO VOLTERRA (1800–1940) of Rome, ERIK IVAR FREDHOLM (1866–1927) of Stockholm, and DAVID HILBERT (1962–1943) of Göttingen (see the footnote in Sec. 7.9) to develop the field of integral equations. For details see Birkhoff, G. and E. Kreyszig, The Establishment of Functional Analysis, *Historia Mathematica* 11 (1984), pp. 258–321.

The solutions $Y_\nu(x)$ are sometimes denoted by $N_\nu(x)$; in Ref. [A13] they are called **Weber's functions**; Euler's constant in (6) is often denoted by C or $\ln \gamma$.

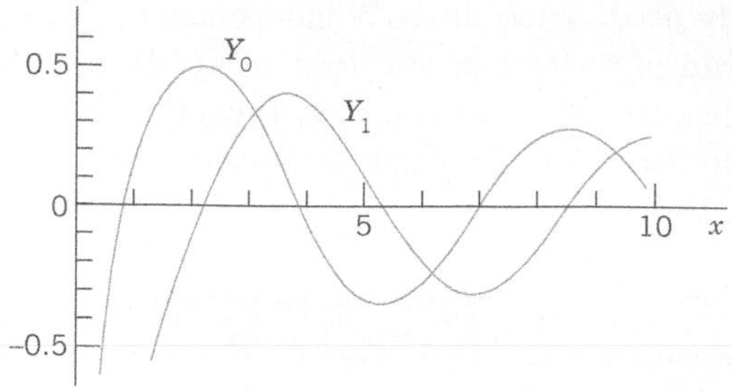

Fig. 112. Bessel functions of the second kind Y_0 and Y_1.
(For a small table, see App. 5.)

linearly independent. Furthermore, it can be shown that the limit in (7b) exists and Y_n is a solution of Bessel's equation for integer order; see Ref. [A13] in App. 1. We shall see that the series development of $Y_n(x)$ contains a logarithmic term. Hence $J_n(x)$ and $Y_n(x)$ are linearly independent solutions of Bessel's equation. The series development of $Y_n(x)$ can be obtained if we insert the series (20) in Sec. 5.4 and (2) in this section for $J_\nu(x)$ and $J_{-\nu}(x)$ into (7a) and then let ν approach n; for details see Ref. [A13]. The result is

$$\begin{aligned} Y_n(x) = {} & \frac{2}{\pi} J_n(x)\left(\ln\frac{x}{2} + \gamma\right) + \frac{x^n}{\pi}\sum_{m=0}^{\infty}\frac{(-1)^{m-1}(h_m + h_{m+n})}{2^{2m+n}m!\,(m+n)!}x^{2m} \\ & - \frac{x^{-n}}{\pi}\sum_{m=0}^{n-1}\frac{(n-m-1)!}{2^{2m-n}m!}x^{2m} \end{aligned} \tag{8}$$

where $x > 0$, $n = 0, 1, \cdots$, and [as in (4)] $h_0 = 0$, $h_1 = 1$,

$$h_m = 1 + \frac{1}{2} + \cdots + \frac{1}{m}, \qquad h_{m+n} = 1 + \frac{1}{2} + \cdots + \frac{1}{m+n}.$$

For $n = 0$ the last sum in (8) is to be replaced by 0 [giving agreement with (6)].

Furthermore, it can be shown that

$$Y_{-n}(x) = (-1)^n Y_n(x).$$

Our main result may now be formulated as follows.

THEOREM 1

General Solution of Bessel's Equation

A general solution of Bessel's equation for all values of ν (and $x > 0$) is

$$y(x) = C_1 J_\nu(x) + C_2 Y_\nu(x). \tag{9}$$

We finally mention that there is a practical need for solutions of Bessel's equation that are complex for real values of x. For this purpose the solutions

$$\begin{aligned} H_\nu^{(1)}(x) &= J_\nu(x) + iY_\nu(x) \\ H_\nu^{(2)}(x) &= J_\nu(x) - iY_\nu(x) \end{aligned} \tag{10}$$

are frequently used. These linearly independent functions are called **Bessel functions of the third kind** *of order* ν or *first and second* **Hankel functions**[8] *of order* ν.

This finishes our discussion on Bessel functions, except for their "orthogonality," which we explain in Sec. 11.6. Applications to vibrations follow in Sec. 12.10.

PROBLEM SET 5.5

1–9 FURTHER ODE's REDUCIBLE TO BESSEL'S ODE

Find a general solution in terms of J_ν and Y_ν. Indicate whether you could also use $J_{-\nu}$ instead of Y_ν. Use the indicated substitution. Show the details of your work.

1. $x^2y'' + xy' + (x^2 - 16)y = 0$
2. $xy'' + 5y' + xy = 0 \quad (y = u/x^2)$
3. $9x^2y'' + 9xy' + (36x^4 - 16)y = 0 \quad (x^2 = z)$
4. $y'' + xy = 0 \quad (y = u\sqrt{x}, \frac{2}{3}x^{3/2} = z)$
5. $4xy'' + 4y' + y = 0 \quad (\sqrt{x} = z)$
6. $xy'' + y' + 36y = 0 \quad (12\sqrt{x} = z)$
7. $y'' + k^2x^2y = 0 \quad (y = u\sqrt{x}, \frac{1}{2}kx^2 = z)$
8. $y'' + k^2x^4y = 0 \quad (y = u\sqrt{x}, \frac{1}{3}kx^3 = z)$
9. $xy'' - 5y' + xy = 0 \quad (y = x^3u)$
10. **CAS EXPERIMENT. Bessel Functions for Large x.** It can be shown that for large x,

$$Y_n(x) \sim \sqrt{2/(\pi x)} \sin\left(x - \tfrac{1}{2}n\pi - \tfrac{1}{4}\pi\right) \tag{11}$$

with $\sim$ defined as in (14) of Sec. 5.4.

(a) Graph $Y_n(x)$ for $n = 0, \cdots, 5$ on common axes. Are there relations between zeros of one function and extrema of another? For what functions?

(b) Find out from graphs from which $x = x_n$ on the curves of (8) and (11) (both obtained from your CAS) practically coincide. How does x_n change with n?

(c) Calculate the first ten zeros x_m, $m = 1, \cdots, 10$, of $Y_0(x)$ from your CAS and from (11). How does the error behave as m increases?

(d) Do (c) for $Y_1(x)$ and $Y_2(x)$. How do the errors compare to those in (c)?

11–15 HANKEL AND MODIFIED BESSEL FUNCTIONS

11. **Hankel functions.** Show that the Hankel functions (10) form a basis of solutions of Bessel's equation for any ν.
12. **Modified Bessel functions of the first kind of order** ν are defined by $I_\nu(x) = i^{-\nu}J_\nu(ix)$, $i = \sqrt{-1}$. Show that I_ν satisfies the ODE

$$x^2y'' + xy' - (x^2 + \nu^2)y = 0. \tag{12}$$

13. **Modified Bessel functions.** Show that $I_\nu(x)$ has the representation

$$I_\nu(x) = \sum_{m=0}^{\infty} \frac{x^{2m+\nu}}{2^{2m+\nu} m!\, \Gamma(m + \nu + 1)}. \tag{13}$$

14. **Reality of I_ν.** Show that $I_\nu(x)$ is real for all real x (and real ν), $I_\nu(x) \neq 0$ for all real $x \neq 0$, and $I_{-n}(x) = I_n(x)$, where n is any integer.
15. **Modified Bessel functions of the third kind** (sometimes called *of the second kind*) are defined by the formula (14) below. Show that they satisfy the ODE (12).

$$K_\nu(x) = \frac{\pi}{2 \sin \nu\pi}\left[I_{-\nu}(x) - I_\nu(x)\right]. \tag{14}$$

CHAPTER 5 REVIEW QUESTIONS AND PROBLEMS

1. Why are we looking for power series solutions of ODEs?
2. What is the difference between the two methods in this chapter? Why do we need two methods?
3. What is the indicial equation? Why is it needed?
4. List the three cases of the Frobenius method, and give examples of your own.
5. Write down the most important ODEs in this chapter from memory.
6. Can a power series solution reduce to a polynomial? When? Why is this important?
7. What is the hypergeometric equation? Where does the name come from?
8. List some properties of the Legendre polynomials.
9. Why did we introduce two kinds of Bessel functions?
10. Can a Bessel function reduce to an elementary function? When?

[8]HERMANN HANKEL (1839–1873), German mathematician.

11–20 **POWER SERIES METHOD OR FROBENIUS METHOD**

Find a basis of solutions. Try to identify the series as expansions of known functions. Show the details of your work.

11. $y'' + 4y = 0$

12. $xy'' + (1 - 2x)y' + (x - 1)y = 0$

13. $(x - 1)^2 y'' - (x - 1)y' - 35y = 0$

14. $16(x + 1)^2 y'' + 3y = 0$

15. $x^2 y'' + xy' + (x^2 - 5)y = 0$

16. $x^2 y'' + 2x^3 y' + (x^2 - 2)y = 0$

17. $xy'' - (x + 1)y' + y = 0$

18. $xy'' + 3y' + 4x^3 y = 0$

19. $y'' + \frac{1}{4x} y = 0$

20. $xy'' + y' - xy = 0$

SUMMARY OF CHAPTER 5

Series Solution of ODEs. Special Functions

The **power series method** gives solutions of linear ODEs

$$y'' + p(x)y' + q(x)y = 0 \tag{1}$$

with ***variable coefficients*** p and q in the form of a power series (with any center x_0, e.g., $x_0 = 0$)

$$y(x) = \sum_{m=0}^{\infty} a_m(x - x_0)^m = a_0 + a_1(x - x_0) + a_2(x - x_0)^2 + \cdots. \tag{2}$$

Such a solution is obtained by substituting (2) and its derivatives into (1). This gives a ***recurrence formula*** for the coefficients. You may program this formula (or even obtain and graph the whole solution) on your CAS.

If p and q are **analytic** at x_0 (that is, representable by a power series in powers of $x - x_0$ with positive radius of convergence; Sec. 5.1), then (1) has solutions of this form (2). The same holds if $\tilde{h}$, $\tilde{p}$, $\tilde{q}$ in

$$\tilde{h}(x)y'' + \tilde{p}(x)y' + \tilde{q}(x)y = 0$$

are analytic at x_0 and $\tilde{h}(x_0) \neq 0$, so that we can divide by $\tilde{h}$ and obtain the standard form (1). ***Legendre's equation*** is solved by the power series method in Sec. 5.2.

The ***Frobenius method*** (Sec. 5.3) extends the power series method to ODEs

$$y'' + \frac{a(x)}{x - x_0} y' + \frac{b(x)}{(x - x_0)^2} y = 0 \tag{3}$$

whose coefficients are **singular** (i.e., not analytic) at x_0, but are "not too bad," namely, such that a and b are analytic at x_0. Then (3) has at least one solution of the form

$$y(x) = (x - x_0)^r \sum_{m=0}^{\infty} a_m(x - x_0)^m = a_0(x - x_0)^r + a_1(x - x_0)^{r+1} + \cdots \tag{4}$$

where r can be any real (or even complex) number and is determined by substituting (4) into (3) from the ***indicial equation*** (Sec. 5.3), along with the coefficients of (4). A second linearly independent solution of (3) may be of a similar form (with different r and a_m's) or may involve a logarithmic term. ***Bessel's equation*** is solved by the Frobenius method in Secs. 5.4 and 5.5.

"**Special functions**" is a common name for higher functions, as opposed to the usual functions of calculus. Most of them arise either as nonelementary integrals [see (24)–(44) in App. 3.1] or as solutions of (1) or (3). They get a name and notation and are included in the usual CASs if they are important in application or in theory. Of this kind, and particularly useful to the engineer and physicist, are ***Legendre's equation and polynomials*** $P_0, P_1, \cdots$ (Sec. 5.2), ***Gauss's hypergeometric equation and functions*** $F(a, b, c; x)$ (Sec. 5.3), and ***Bessel's equation and functions*** J_ν ***and*** Y_ν (Secs. 5.4, 5.5).

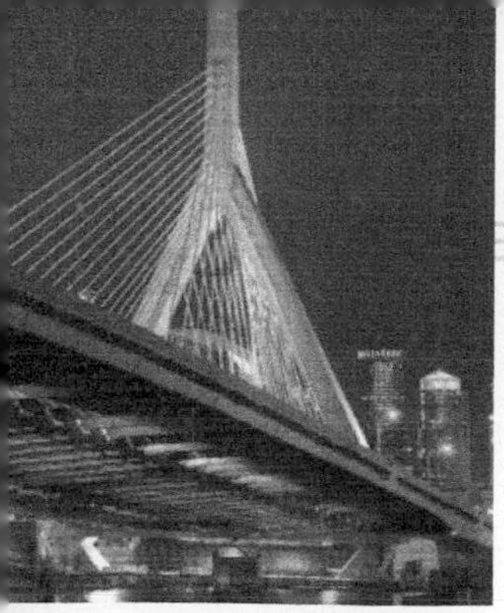

CHAPTER 6

Laplace Transforms

Laplace transforms are invaluable for any engineer's mathematical toolbox as they make solving linear ODEs and related initial value problems, as well as systems of linear ODEs, much easier. Applications abound: electrical networks, springs, mixing problems, signal processing, and other areas of engineering and physics.

The process of solving an ODE using the Laplace transform method consists of three steps, shown schematically in Fig. 113:

Step 1. The given ODE is transformed into an algebraic equation, called the **subsidiary equation**.

Step 2. The subsidiary equation is solved by purely algebraic manipulations.

Step 3. The solution in Step 2 is transformed back, resulting in the solution of the given problem.

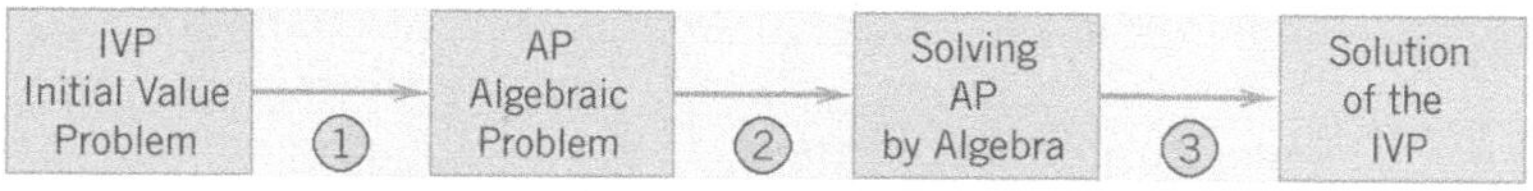

Fig. 113. Solving an IVP by Laplace transforms

The key motivation for learning about Laplace transforms is that the process of solving an ODE is simplified to an algebraic problem (and transformations). This type of mathematics that converts problems of calculus to algebraic problems is known as *operational calculus*. The Laplace transform method has two main advantages over the methods discussed in Chaps. 1–4:

I. Problems are solved more directly: Initial value problems are solved without first determining a general solution. Nonhomogenous ODEs are solved without first solving the corresponding homogeneous ODE.

II. More importantly, the use of the ***unit step function*** (***Heaviside function*** in Sec. 6.3) and ***Dirac's delta*** (in Sec. 6.4) make the method particularly powerful for problems with inputs (driving forces) that have discontinuities or represent short impulses or complicated periodic functions.

The following chart shows where to find information on the Laplace transform in this book.

Topic	Where to find it
ODEs, engineering applications and Laplace transforms	Chapter 6
PDEs, engineering applications and Laplace transforms	Section 12.11
List of general formulas of Laplace transforms	Section 6.8
List of Laplace transforms and inverses	Section 6.9
Note: Your CAS can handle most Laplace transforms.	

Prerequisite: Chap. 2
Sections that may be omitted in a shorter course: 6.5, 6.7
References and Answers to Problems: App. 1 Part A, App. 2.

6.1 Laplace Transform. Linearity. First Shifting Theorem (*s*-Shifting)

In this section, we learn about Laplace transforms and some of their properties. Because Laplace transforms are of basic importance to the engineer, the student should pay close attention to the material. Applications to ODEs follow in the next section.

Roughly speaking, the Laplace transform, when applied to a function, changes that function into a new function by using a process that involves integration. Details are as follows.

If $f(t)$ is a function defined for all $t \geqq 0$, its **Laplace transform**[1] is the integral of $f(t)$ times e^{-st} from $t = 0$ to ∞. It is a function of s, say, $F(s)$, and is denoted by $\mathcal{L}(f)$; thus

$$F(s) = \mathcal{L}(f) = \int_0^\infty e^{-st} f(t)\, dt. \tag{1}$$

Here we must assume that $f(t)$ is such that the integral exists (that is, has some finite value). This assumption is usually satisfied in applications—we shall discuss this near the end of the section.

[1] PIERRE SIMON MARQUIS DE LAPLACE (1749–1827), great French mathematician, was a professor in Paris. He developed the foundation of potential theory and made important contributions to celestial mechanics, astronomy in general, special functions, and probability theory. Napoléon Bonaparte was his student for a year. For Laplace's interesting political involvements, see Ref. [GenRef2], listed in App. 1.

The powerful practical Laplace transform techniques were developed over a century later by the English electrical engineer OLIVER HEAVISIDE (1850–1925) and were often called "Heaviside calculus."

We shall drop variables when this simplifies formulas without causing confusion. For instance, in (1) we wrote $\mathcal{L}(f)$ instead of $\mathcal{L}(f)(s)$ and in (1*) $\mathcal{L}^{-1}(F)$ instead of $\mathcal{L}^{-1}(F)(t)$.

Not only is the result $F(s)$ called the Laplace transform, but the operation just described, which yields $F(s)$ from a given $f(t)$, is also called the **Laplace transform**. It is an "**integral transform**"

$$F(s) = \int_0^\infty k(s, t) f(t)\, dt$$

with "**kernel**" $k(s, t) = e^{-st}$.

Note that the Laplace transform is called an integral transform because it transforms (changes) a function in one space to a function in another space by a *process of integration* that involves a kernel. The kernel or kernel function is a function of the variables in the two spaces and defines the integral transform.

Furthermore, the given function $f(t)$ in (1) is called the **inverse transform** of $F(s)$ and is denoted by $\mathcal{L}^{-1}(F)$; that is, we shall write

$$f(t) = \mathcal{L}^{-1}(F). \tag{1*}$$

Note that (1) and (1*) together imply $\mathcal{L}^{-1}(\mathcal{L}(f)) = f$ and $\mathcal{L}(\mathcal{L}^{-1}(F)) = F$.

Notation

Original functions depend on t and their transforms on s—keep this in mind! Original functions are denoted by ***lowercase letters*** and their transforms by the same ***letters in capital***, so that $F(s)$ denotes the transform of $f(t)$, and $Y(s)$ denotes the transform of $y(t)$, and so on.

EXAMPLE 1 Laplace Transform

Let $f(t) = 1$ when $t \geqq 0$. Find $F(s)$.

Solution. From (1) we obtain by integration

$$\mathcal{L}(f) = \mathcal{L}(1) = \int_0^\infty e^{-st}\, dt = -\frac{1}{s} e^{-st} \bigg|_0^\infty = \frac{1}{s} \qquad (s > 0).$$

Such an integral is called an **improper integral** and, by definition, is evaluated according to the rule

$$\int_0^\infty e^{-st} f(t)\, dt = \lim_{T\to\infty} \int_0^T e^{-st} f(t)\, dt.$$

Hence our convenient notation means

$$\int_0^\infty e^{-st}\, dt = \lim_{T\to\infty} \left[-\frac{1}{s} e^{-st}\right]_0^T = \lim_{T\to\infty}\left[-\frac{1}{s}e^{-sT} + \frac{1}{s}e^0\right] = \frac{1}{s} \qquad (s > 0).$$

We shall use this notation throughout this chapter. ■

EXAMPLE 2 Laplace Transform $\mathcal{L}(e^{at})$ of the Exponential Function e^{at}

Let $f(t) = e^{at}$ when $t \geqq 0$, where a is a constant. Find $\mathcal{L}(f)$.

Solution. Again by (1),

$$\mathcal{L}(e^{at}) = \int_0^\infty e^{-st} e^{at}\, dt = \frac{1}{a-s} e^{-(s-a)t} \bigg|_0^\infty ;$$

hence, when $s - a > 0$,

$$\mathcal{L}(e^{at}) = \frac{1}{s-a}.$$ ■

Must we go on in this fashion and obtain the transform of one function after another directly from the definition? No! We can obtain new transforms from known ones by the use of the many general properties of the Laplace transform. Above all, the Laplace transform is a "linear operation," just as are differentiation and integration. By this we mean the following.

THEOREM 1

Linearity of the Laplace Transform

The Laplace transform is a linear operation; that is, for any functions $f(t)$ and $g(t)$ whose transforms exist and any constants a and b the transform of $af(t) + bg(t)$ exists, and

$$\mathcal{L}\{af(t) + bg(t)\} = a\mathcal{L}\{f(t)\} + b\mathcal{L}\{g(t)\}.$$

PROOF This is true because integration is a linear operation so that (1) gives

$$\mathcal{L}\{af(t) + bg(t)\} = \int_0^\infty e^{-st}[af(t) + bg(t)]\, dt$$

$$= a\int_0^\infty e^{-st}f(t)\, dt + b\int_0^\infty e^{-st}g(t)\, dt = a\mathcal{L}\{f(t)\} + b\mathcal{L}\{g(t)\}. \blacksquare$$

EXAMPLE 3 **Application of Theorem 1: Hyperbolic Functions**

Find the transforms of $\cosh at$ and $\sinh at$.

Solution. Since $\cosh at = \frac{1}{2}(e^{at} + e^{-at})$ and $\sinh at = \frac{1}{2}(e^{at} - e^{-at})$, we obtain from Example 2 and Theorem 1

$$\mathcal{L}(\cosh at) = \frac{1}{2}(\mathcal{L}(e^{at}) + \mathcal{L}(e^{-at})) = \frac{1}{2}\left(\frac{1}{s-a} + \frac{1}{s+a}\right) = \frac{s}{s^2 - a^2}$$

$$\mathcal{L}(\sinh at) = \frac{1}{2}(\mathcal{L}(e^{at}) - \mathcal{L}(e^{-at})) = \frac{1}{2}\left(\frac{1}{s-a} - \frac{1}{s+a}\right) = \frac{a}{s^2 - a^2}. \blacksquare$$

EXAMPLE 4 **Cosine and Sine**

Derive the formulas

$$\mathcal{L}(\cos \omega t) = \frac{s}{s^2 + \omega^2}, \qquad \mathcal{L}(\sin \omega t) = \frac{\omega}{s^2 + \omega^2}.$$

Solution. We write $L_c = \mathcal{L}(\cos \omega t)$ and $L_s = \mathcal{L}(\sin \omega t)$. Integrating by parts and noting that the integral-free parts give no contribution from the upper limit ∞, we obtain

$$L_c = \int_0^\infty e^{-st} \cos \omega t\, dt = \frac{e^{-st}}{-s} \cos \omega t \Big|_0^\infty - \frac{\omega}{s}\int_0^\infty e^{-st} \sin \omega t\, dt = \frac{1}{s} - \frac{\omega}{s} L_s,$$

$$L_s = \int_0^\infty e^{-st} \sin \omega t\, dt = \frac{e^{-st}}{-s} \sin \omega t \Big|_0^\infty + \frac{\omega}{s}\int_0^\infty e^{-st} \cos \omega t\, dt = \frac{\omega}{s} L_c.$$

By substituting L_s into the formula for L_c on the right and then by substituting L_c into the formula for L_s on the right, we obtain

$$L_c = \frac{1}{s} - \frac{\omega}{s}\left(\frac{\omega}{s}L_c\right), \qquad L_c\left(1 + \frac{\omega^2}{s^2}\right) = \frac{1}{s}, \qquad L_c = \frac{s}{s^2 + \omega^2},$$

$$L_s = \frac{\omega}{s}\left(\frac{1}{s} - \frac{\omega}{s}L_s\right), \qquad L_s\left(1 + \frac{\omega^2}{s^2}\right) = \frac{\omega}{s^2}, \qquad L_s = \frac{\omega}{s^2 + \omega^2}. \qquad \blacksquare$$

Basic transforms are listed in Table 6.1. We shall see that from these almost all the others can be obtained by the use of the general properties of the Laplace transform. Formulas 1–3 are special cases of formula 4, which is proved by induction. Indeed, it is true for $n = 0$ because of Example 1 and $0! = 1$. We make the induction hypothesis that it holds for any integer $n \geqq 0$ and then get it for $n + 1$ directly from (1). Indeed, integration by parts first gives

$$\mathscr{L}(t^{n+1}) = \int_0^\infty e^{-st}t^{n+1}\,dt = -\frac{1}{s}e^{-st}t^{n+1}\Big|_0^\infty + \frac{n+1}{s}\int_0^\infty e^{-st}t^n\,dt.$$

Now the integral-free part is zero and the last part is $(n + 1)/s$ times $\mathscr{L}(t^n)$. From this and the induction hypothesis,

$$\mathscr{L}(t^{n+1}) = \frac{n+1}{s}\mathscr{L}(t^n) = \frac{n+1}{s} \cdot \frac{n!}{s^{n+1}} = \frac{(n+1)!}{s^{n+2}}.$$

This proves formula 4.

Table 6.1 Some Functions $f(t)$ and Their Laplace Transforms $\mathscr{L}(f)$

	$f(t)$	$\mathscr{L}(f)$		$f(t)$	$\mathscr{L}(f)$
1	1	$1/s$	7	$\cos \omega t$	$\dfrac{s}{s^2 + \omega^2}$
2	t	$1/s^2$	8	$\sin \omega t$	$\dfrac{\omega}{s^2 + \omega^2}$
3	t^2	$2!/s^3$	9	$\cosh at$	$\dfrac{s}{s^2 - a^2}$
4	t^n ($n = 0, 1, \cdots$)	$\dfrac{n!}{s^{n+1}}$	10	$\sinh at$	$\dfrac{a}{s^2 - a^2}$
5	t^a (a positive)	$\dfrac{\Gamma(a + 1)}{s^{a+1}}$	11	$e^{at}\cos \omega t$	$\dfrac{s - a}{(s - a)^2 + \omega^2}$
6	e^{at}	$\dfrac{1}{s - a}$	12	$e^{at}\sin \omega t$	$\dfrac{\omega}{(s - a)^2 + \omega^2}$

$\Gamma(a + 1)$ in formula 5 is the so-called *gamma function* [(15) in Sec. 5.5 or (24) in App. A3.1]. We get formula 5 from (1), setting $st = x$:

$$\mathcal{L}(t^a) = \int_0^\infty e^{-st} t^a \, dt = \int_0^\infty e^{-x} \left(\frac{x}{s}\right)^a \frac{dx}{s} = \frac{1}{s^{a+1}} \int_0^\infty e^{-x} x^a \, dx$$

where $s > 0$. The last integral is precisely that defining $\Gamma(a + 1)$, so we have $\Gamma(a + 1)/s^{a+1}$, as claimed. (**CAUTION!** $\Gamma(a + 1)$ has x^a in the integral, not x^{a+1}.)

Note the formula 4 also follows from 5 because $\Gamma(n + 1) = n!$ for integer $n \geqq 0$.

Formulas 6–10 were proved in Examples 2–4. Formulas 11 and 12 will follow from 7 and 8 by "shifting," to which we turn next.

s-Shifting: Replacing *s* by *s* − *a* in the Transform

The Laplace transform has the very useful property that, if we know the transform of $f(t)$, we can immediately get that of $e^{at}f(t)$, as follows.

THEOREM 2

First Shifting Theorem, *s*-Shifting

If $f(t)$ has the transform $F(s)$ (where $s > k$ for some k), then $e^{at}f(t)$ has the transform $F(s - a)$ (where $s - a > k$). In formulas,

$$\mathcal{L}\{e^{at}f(t)\} = F(s - a)$$

or, if we take the inverse on both sides,

$$e^{at}f(t) = \mathcal{L}^{-1}\{F(s - a)\}.$$

PROOF We obtain $F(s - a)$ by replacing s with $s - a$ in the integral in (1), so that

$$F(s - a) = \int_0^\infty e^{-(s-a)t} f(t) \, dt = \int_0^\infty e^{-st} [e^{at} f(t)] \, dt = \mathcal{L}\{e^{at}f(t)\}.$$

If $F(s)$ exists (i.e., is finite) for s greater than some k, then our first integral exists for $s - a > k$. Now take the inverse on both sides of this formula to obtain the second formula in the theorem. (**CAUTION!** $-a$ in $F(s - a)$ but $+a$ in $e^{at}f(t)$.) ■

EXAMPLE 5 **s-Shifting: Damped Vibrations. Completing the Square**

From Example 4 and the first shifting theorem we immediately obtain formulas 11 and 12 in Table 6.1,

$$\mathcal{L}\{e^{at}\cos \omega t\} = \frac{s - a}{(s - a)^2 + \omega^2}, \qquad \mathcal{L}\{e^{at}\sin \omega t\} = \frac{\omega}{(s - a)^2 + \omega^2}.$$

For instance, use these formulas to find the inverse of the transform

$$\mathcal{L}(f) = \frac{3s - 137}{s^2 + 2s + 401}.$$

Solution. Applying the inverse transform, using its linearity (Prob. 24), and completing the square, we obtain

$$f = \mathcal{L}^{-1}\left\{\frac{3(s+1)-140}{(s+1)^2+400}\right\} = 3\mathcal{L}^{-1}\left\{\frac{s+1}{(s+1)^2+20^2}\right\} - 7\mathcal{L}^{-1}\left\{\frac{20}{(s+1)^2+20^2}\right\}.$$

We now see that the inverse of the right side is the damped vibration (Fig. 114)

$$f(t) = e^{-t}(3\cos 20t - 7\sin 20t).$$ ■

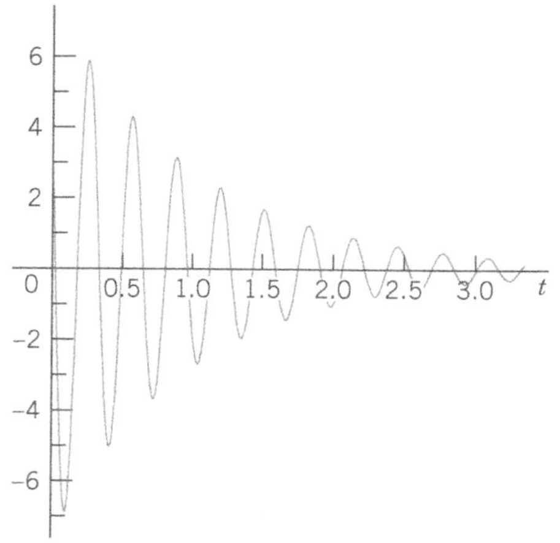

Fig. 114. Vibrations in Example 5

Existence and Uniqueness of Laplace Transforms

This is not a big *practical* problem because in most cases we can check the solution of an ODE without too much trouble. Nevertheless we should be aware of some basic facts.

A function $f(t)$ has a Laplace transform if it does not grow too fast, say, if for all $t \geqq 0$ and some constants M and k it satisfies the "**growth restriction**"

$$|f(t)| \leqq Me^{kt}. \tag{2}$$

(The growth restriction (2) is sometimes called "growth of exponential order," which may be misleading since it hides that the exponent must be kt, not kt^2 or similar.)

$f(t)$ need not be continuous, but it should not be too bad. The technical term (generally used in mathematics) is *piecewise continuity*. $f(t)$ is **piecewise continuous** on a finite interval $a \leqq t \leqq b$ where f is defined, if this interval can be divided into *finitely many* subintervals in each of which f is continuous and has finite limits as t approaches either endpoint of such a subinterval from the interior. This then gives **finite jumps** as in Fig. 115 as the only possible discontinuities, but this suffices in most applications, and so does the following theorem.

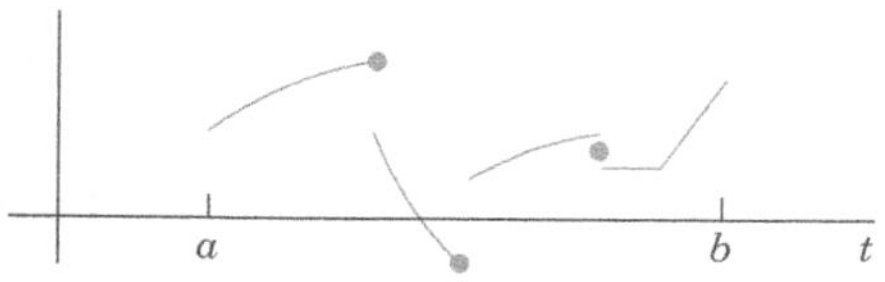

Fig. 115. Example of a piecewise continuous function $f(t)$. (The dots mark the function values at the jumps.)

PROOF We prove (1) first under the *additional assumption* that f' is continuous. Then, by the definition and integration by parts,

$$\mathcal{L}(f') = \int_0^\infty e^{-st} f'(t)\, dt = \left[e^{-st} f(t)\right]\Big|_0^\infty + s\int_0^\infty e^{-st} f(t)\, dt.$$

Since f satisfies (2) in Sec. 6.1, the integrated part on the right is zero at the upper limit when $s > k$, and at the lower limit it contributes $-f(0)$. The last integral is $\mathcal{L}(f)$. It exists for $s > k$ because of Theorem 3 in Sec. 6.1. Hence $\mathcal{L}(f')$ exists when $s > k$ and (1) holds.

If f' is merely piecewise continuous, the proof is similar. In this case the interval of integration of f' must be broken up into parts such that f' is continuous in each such part.

The proof of (2) now follows by applying (1) to f'' and then substituting (1), that is

$$\mathcal{L}(f'') = s\mathcal{L}(f') - f'(0) = s[s\mathcal{L}(f) - f(0)] = s^2\mathcal{L}(f) - sf(0) - f'(0). \qquad ■$$

Continuing by substitution as in the proof of (2) and using induction, we obtain the following extension of Theorem 1.

THEOREM 2

Laplace Transform of the Derivative $f^{(n)}$ of Any Order

Let $f, f', \cdots, f^{(n-1)}$ be continuous for all $t \geqq 0$ and satisfy the growth restriction (2) in Sec. 6.1. Furthermore, let $f^{(n)}$ be piecewise continuous on every finite interval on the semi-axis $t \geqq 0$. Then the transform of $f^{(n)}$ satisfies

$$\text{(3)} \qquad \mathcal{L}(f^{(n)}) = s^n\mathcal{L}(f) - s^{n-1}f(0) - s^{n-2}f'(0) - \cdots - f^{(n-1)}(0).$$

EXAMPLE 1 **Transform of a Resonance Term (Sec. 2.8)**

Let $f(t) = t \sin \omega t$. Then $f(0) = 0$, $f'(t) = \sin \omega t + \omega t \cos \omega t$, $f'(0) = 0$, $f'' = 2\omega \cos \omega t - \omega^2 t \sin \omega t$. Hence by (2),

$$\mathcal{L}(f'') = 2\omega \frac{s}{s^2 + \omega^2} - \omega^2\mathcal{L}(f) = s^2\mathcal{L}(f), \qquad \text{thus} \qquad \mathcal{L}(f) = \mathcal{L}(t \sin \omega t) = \frac{2\omega s}{(s^2 + \omega^2)^2}. \qquad ■$$

EXAMPLE 2 **Formulas 7 and 8 in Table 6.1, Sec. 6.1**

This is a third derivation of $\mathcal{L}(\cos \omega t)$ and $\mathcal{L}(\sin \omega t)$; cf. Example 4 in Sec. 6.1. Let $f(t) = \cos \omega t$. Then $f(0) = 1$, $f'(0) = 0$, $f''(t) = -\omega^2 \cos \omega t$. From this and (2) we obtain

$$\mathcal{L}(f'') = s^2\mathcal{L}(f) - s = -\omega^2\mathcal{L}(f). \qquad \text{By algebra,} \qquad \mathcal{L}(\cos \omega t) = \frac{s}{s^2 + \omega^2}.$$

Similarly, let $g = \sin \omega t$. Then $g(0) = 0$, $g' = \omega \cos \omega t$. From this and (1) we obtain

$$\mathcal{L}(g') = s\mathcal{L}(g) = \omega\mathcal{L}(\cos \omega t). \qquad \text{Hence,} \qquad \mathcal{L}(\sin \omega t) = \frac{\omega}{s}\mathcal{L}(\cos \omega t) = \frac{\omega}{s^2 + \omega^2}. \qquad ■$$

Laplace Transform of the Integral of a Function

Differentiation and integration are inverse operations, and so are multiplication and division. Since differentiation of a function $f(t)$ (roughly) corresponds to multiplication of its transform $\mathcal{L}(f)$ by s, we expect integration of $f(t)$ to correspond to division of $\mathcal{L}(f)$ by s:

THEOREM 3

Laplace Transform of Integral

Let $F(s)$ denote the transform of a function $f(t)$ which is piecewise continuous for $t \geqq 0$ and satisfies a growth restriction (2), Sec. 6.1. Then, for $s > 0$, $s > k$, and $t > 0$,

$$\mathcal{L}\left\{\int_0^t f(\tau)\,d\tau\right\} = \frac{1}{s}F(s), \qquad \text{thus} \qquad \int_0^t f(\tau)\,d\tau = \mathcal{L}^{-1}\left\{\frac{1}{s}F(s)\right\}. \tag{4}$$

PROOF Denote the integral in (4) by $g(t)$. Since $f(t)$ is piecewise continuous, $g(t)$ is continuous, and (2), Sec. 6.1, gives

$$|g(t)| = \left|\int_0^t f(\tau)\,d\tau\right| \leqq \int_0^t |f(\tau)|\,d\tau \leqq M\int_0^t e^{k\tau}\,d\tau = \frac{M}{k}(e^{kt}-1) \leqq \frac{M}{k}e^{kt} \qquad (k > 0).$$

This shows that $g(t)$ also satisfies a growth restriction. Also, $g'(t) = f(t)$, except at points at which $f(t)$ is discontinuous. Hence $g'(t)$ is piecewise continuous on each finite interval and, by Theorem 1, since $g(0) = 0$ (the integral from 0 to 0 is zero)

$$\mathcal{L}\{f(t)\} = \mathcal{L}\{g'(t)\} = s\mathcal{L}\{g(t)\} - g(0) = s\mathcal{L}\{g(t)\}.$$

Division by s and interchange of the left and right sides gives the first formula in (4), from which the second follows by taking the inverse transform on both sides. ■

EXAMPLE 3 **Application of Theorem 3: Formulas 19 and 20 in the Table of Sec. 6.9**

Using Theorem 3, find the inverse of $\dfrac{1}{s(s^2+\omega^2)}$ and $\dfrac{1}{s^2(s^2+\omega^2)}$.

Solution. From Table 6.1 in Sec. 6.1 and the integration in (4) (second formula with the sides interchanged) we obtain

$$\mathcal{L}^{-1}\left\{\frac{1}{s^2+\omega^2}\right\} = \frac{\sin\omega t}{\omega}, \qquad \mathcal{L}^{-1}\left\{\frac{1}{s(s^2+\omega^2)}\right\} = \int_0^t \frac{\sin\omega\tau}{\omega}\,d\tau = \frac{1}{\omega^2}(1-\cos\omega t).$$

This is formula 19 in Sec. 6.9. Integrating this result again and using (4) as before, we obtain formula 20 in Sec. 6.9:

$$\mathcal{L}^{-1}\left\{\frac{1}{s^2(s^2+\omega^2)}\right\} = \frac{1}{\omega^2}\int_0^t (1-\cos\omega\tau)\,d\tau = \left[\frac{\tau}{\omega^2} - \frac{\sin\omega\tau}{\omega^3}\right]_0^t = \frac{t}{\omega^2} - \frac{\sin\omega t}{\omega^3}.$$

It is typical that results such as these can be found in several ways. In this example, try partial fraction reduction. ■

Differential Equations, Initial Value Problems

Let us now discuss how the Laplace transform method solves ODEs and initial value problems. We consider an initial value problem

$$y'' + ay' + by = r(t), \qquad y(0) = K_0, \qquad y'(0) = K_1 \tag{5}$$

where a and b are constant. Here $r(t)$ is the given **input** (*driving force*) applied to the mechanical or electrical system and $y(t)$ is the **output** (*response to the input*) to be obtained. In Laplace's method we do three steps:

Step 1. Setting up the subsidiary equation. This is an algebraic equation for the transform $Y = \mathcal{L}(y)$ obtained by transforming (5) by means of (1) and (2), namely,

$$[s^2Y - sy(0) - y'(0)] + a[sY - y(0)] + bY = R(s)$$

where $R(s) = \mathcal{L}(r)$. Collecting the Y-terms, we have the subsidiary equation

$$(s^2 + as + b)Y = (s + a)y(0) + y'(0) + R(s).$$

Step 2. Solution of the subsidiary equation by algebra. We divide by $s^2 + as + b$ and use the so-called **transfer function**

$$Q(s) = \frac{1}{s^2 + as + b} = \frac{1}{(s + \frac{1}{2}a)^2 + b - \frac{1}{4}a^2}. \tag{6}$$

(Q is often denoted by H, but we need H much more frequently for other purposes.) This gives the solution

$$Y(s) = [(s + a)y(0) + y'(0)]Q(s) + R(s)Q(s). \tag{7}$$

If $y(0) = y'(0) = 0$, this is simply $Y = RQ$; hence

$$Q = \frac{Y}{R} = \frac{\mathcal{L}(\text{output})}{\mathcal{L}(\text{input})}$$

and this explains the name of Q. Note that ***Q depends neither on r(t) nor on the initial conditions*** (but only on a and b).

Step 3. Inversion of Y to obtain $y = \mathcal{L}^{-1}(Y)$. We reduce (7) (usually by ***partial fractions*** as in calculus) to a sum of terms whose inverses can be found from the tables (e.g., in Sec. 6.1 or Sec. 6.9) or by a CAS, so that we obtain the solution $y(t) = \mathcal{L}^{-1}(Y)$ *of* (5).

EXAMPLE 4 **Initial Value Problem: The Basic Laplace Steps**

Solve

$$y'' - y = t, \qquad y(0) = 1, \qquad y'(0) = 1.$$

Solution. *Step 1.* From (2) and Table 6.1 we get the subsidiary equation [with $Y = \mathcal{L}(y)$]

$$s^2Y - sy(0) - y'(0) - Y = 1/s^2, \qquad \text{thus} \qquad (s^2 - 1)Y = s + 1 + 1/s^2.$$

Step 2. The transfer function is $Q = 1/(s^2 - 1)$, and (7) becomes

$$Y = (s + 1)Q + \frac{1}{s^2}Q = \frac{s + 1}{s^2 - 1} + \frac{1}{s^2(s^2 - 1)}.$$

Simplification of the first fraction and an expansion of the last fraction gives

$$Y = \frac{1}{s - 1} + \left(\frac{1}{s^2 - 1} - \frac{1}{s^2}\right).$$

Step 3. From this expression for Y and Table 6.1 we obtain the solution

$$y(t) = \mathcal{L}^{-1}(Y) = \mathcal{L}^{-1}\left\{\frac{1}{s-1}\right\} + \mathcal{L}^{-1}\left\{\frac{1}{s^2-1}\right\} - \mathcal{L}^{-1}\left\{\frac{1}{s^2}\right\} = e^t + \sinh t - t.$$

The diagram in Fig. 116 summarizes our approach. ■

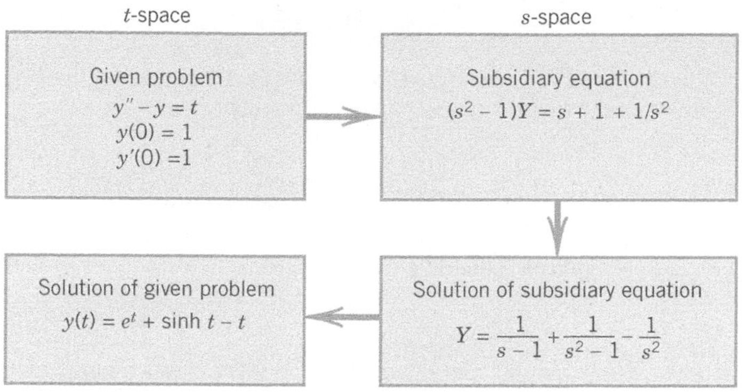

Fig. 116. Steps of the Laplace transform method

EXAMPLE 5 Comparison with the Usual Method

Solve the initial value problem

$$y'' + y' + 9y = 0. \qquad y(0) = 0.16, \qquad y'(0) = 0.$$

Solution. From (1) and (2) we see that the subsidiary equation is

$$s^2Y - 0.16s + sY - 0.16 + 9Y = 0, \qquad \text{thus} \qquad (s^2 + s + 9)Y = 0.16(s + 1).$$

The solution is

$$Y = \frac{0.16(s+1)}{s^2 + s + 9} = \frac{0.16(s + \frac{1}{2}) + 0.08}{(s + \frac{1}{2})^2 + \frac{35}{4}}.$$

Hence by the first shifting theorem and the formulas for cos and sin in Table 6.1 we obtain

$$\begin{aligned} y(t) = \mathcal{L}^{-1}(Y) &= e^{-t/2}\left(0.16 \cos \sqrt{\frac{35}{4}}\, t + \frac{0.08}{\frac{1}{2}\sqrt{35}} \sin \sqrt{\frac{35}{4}}\, t\right) \\ &= e^{-0.5t}(0.16 \cos 2.96t + 0.027 \sin 2.96t). \end{aligned}$$

This agrees with Example 2, Case (III) in Sec. 2.4. The work was less. ■

Advantages of the Laplace Method

1. *Solving a nonhomogeneous ODE does not require first solving the homogeneous ODE.* See Example 4.
2. *Initial values are automatically taken care of.* See Examples 4 and 5.
3. *Complicated inputs* $r(t)$ (right sides of linear ODEs) *can be handled very efficiently,* as we show in the next sections.

EXAMPLE 6 **Shifted Data Problems**

This means initial value problems with initial conditions given at some $t = t_0 > 0$ instead of $t = 0$. For such a problem set $t = \tilde{t} + t_0$, so that $t = t_0$ gives $\tilde{t} = 0$ and the Laplace transform can be applied. For instance, solve

$$y'' + y = 2t, \qquad y(\tfrac{1}{4}\pi) = \tfrac{1}{2}\pi, \qquad y'(\tfrac{1}{4}\pi) = 2 - \sqrt{2}.$$

Solution. We have $t_0 = \tfrac{1}{4}\pi$ and we set $t = \tilde{t} + \tfrac{1}{4}\pi$. Then the problem is

$$\tilde{y}'' + \tilde{y} = 2(\tilde{t} + \tfrac{1}{4}\pi), \qquad \tilde{y}(0) = \tfrac{1}{2}\pi, \qquad \tilde{y}'(0) = 2 - \sqrt{2}$$

where $\tilde{y}(\tilde{t}) = y(t)$. Using (2) and Table 6.1 and denoting the transform of $\tilde{y}$ by $\tilde{Y}$, we see that the subsidiary equation of the "shifted" initial value problem is

$$s^2\tilde{Y} - s \cdot \tfrac{1}{2}\pi - (2 - \sqrt{2}) + \tilde{Y} = \frac{2}{s^2} + \frac{\tfrac{1}{2}\pi}{s}, \qquad \text{thus} \qquad (s^2 + 1)\tilde{Y} = \frac{2}{s^2} + \frac{\tfrac{1}{2}\pi}{s} + \tfrac{1}{2}\pi s + 2 - \sqrt{2}.$$

Solving this algebraically for $\tilde{Y}$, we obtain

$$\tilde{Y} = \frac{2}{(s^2 + 1)s^2} + \frac{\tfrac{1}{2}\pi}{(s^2 + 1)s} + \frac{\tfrac{1}{2}\pi s}{s^2 + 1} + \frac{2 - \sqrt{2}}{s^2 + 1}.$$

The inverse of the first two terms can be seen from Example 3 (with $\omega = 1$), and the last two terms give cos and sin,

$$\tilde{y} = \mathcal{L}^{-1}(\tilde{Y}) = 2(\tilde{t} - \sin \tilde{t}) + \tfrac{1}{2}\pi(1 - \cos \tilde{t}) + \tfrac{1}{2}\pi \cos \tilde{t} + (2 - \sqrt{2}) \sin \tilde{t}$$

$$= 2\tilde{t} + \tfrac{1}{2}\pi - \sqrt{2} \sin \tilde{t}.$$

Now $\tilde{t} = t - \tfrac{1}{4}\pi$, $\sin \tilde{t} = \dfrac{1}{\sqrt{2}} (\sin t - \cos t)$, so that the answer (the solution) is

$$y = 2t - \sin t + \cos t. \qquad \blacksquare$$

PROBLEM SET 6.2

1–11   INITIAL VALUE PROBLEMS (IVPS)

Solve the IVPs by the Laplace transform. If necessary, use partial fraction expansion as in Example 4 of the text. Show all details.

1. $y' + 5.2y = 19.4 \sin 2t, \quad y(0) = 0$
2. $y' + 2y = 0, \quad y(0) = 1.5$
3. $y'' - y' - 6y = 0, \quad y(0) = 11, \quad y'(0) = 28$
4. $y'' + 9y = 10e^{-t}, \quad y(0) = 0, \quad y'(0) = 0$
5. $y'' - \tfrac{1}{4}y = 0, \quad y(0) = 12, \quad y'(0) = 0$
6. $y'' - 6y' + 5y = 29 \cos 2t, \quad y(0) = 3.2,$
   $y'(0) = 6.2$
7. $y'' + 7y' + 12y = 21e^{3t}, \quad y(0) = 3.5,$
   $y'(0) = -10$
8. $y'' - 4y' + 4y = 0, \quad y(0) = 8.1, \quad y'(0) = 3.9$
9. $y'' - 4y' + 3y = 6t - 8, \quad y(0) = 0, \quad y'(0) = 0$
10. $y'' + 0.04y = 0.02t^2, \quad y(0) = -25, \quad y'(0) = 0$
11. $y'' + 3y' + 2.25y = 9t^3 + 64, \quad y(0) = 1,$
    $y'(0) = 31.5$

12–15   SHIFTED DATA PROBLEMS

Solve the shifted data IVPs by the Laplace transform. Show the details.

12. $y'' - 2y' - 3y = 0, \quad y(4) = -3,$
    $y'(4) = -17$
13. $y' - 6y = 0, \quad y(-1) = 4$
14. $y'' + 2y' + 5y = 50t - 100, \quad y(2) = -4,$
    $y'(2) = 14$
15. $y'' + 3y' - 4y = 6e^{2t-3}, \quad y(1.5) = 4,$
    $y'(1.5) = 5$

16–21   OBTAINING TRANSFORMS BY DIFFERENTIATION

Using (1) or (2), find $\mathcal{L}(f)$ if $f(t)$ equals:

16. $t \cos 4t$
17. te^{-at}
18. $\cos^2 2t$
19. $\sin^2 \omega t$
20. $\sin^4 t$. Use Prob. 19.
21. $\cosh^2 t$

22. PROJECT. Further Results by Differentiation. Proceeding as in Example 1, obtain

$$\textbf{(a)}\quad \mathcal{L}(t\cos\omega t) = \frac{s^2-\omega^2}{(s^2+\omega^2)^2}$$

and from this and Example 1: **(b)** formula 21, **(c)** 22, **(d)** 23 in Sec. 6.9,

$$\textbf{(e)}\quad \mathcal{L}(t\cosh at) = \frac{s^2+a^2}{(s^2-a^2)^2},$$

$$\textbf{(f)}\quad \mathcal{L}(t\sinh at) = \frac{2as}{(s^2-a^2)^2}.$$

23–29 INVERSE TRANSFORMS BY INTEGRATION

Using Theorem 3, find $f(t)$ if $\mathcal{L}(F)$ equals:

23. $\dfrac{3}{s^2+s/4}$

24. $\dfrac{20}{s^3-2\pi s^2}$

25. $\dfrac{1}{s(s^2+\omega^2)}$

26. $\dfrac{1}{s^4-s^2}$

27. $\dfrac{s+1}{s^4+9s^2}$

28. $\dfrac{3s+4}{s^4+k^2s^2}$

29. $\dfrac{1}{s^3+as^2}$

30. PROJECT. Comments on Sec. 6.2. **(a)** Give reasons why Theorems 1 and 2 are more important than Theorem 3.

(b) Extend Theorem 1 by showing that if $f(t)$ is continuous, except for an ordinary discontinuity (finite jump) at some $t = a$ (>0), the other conditions remaining as in Theorem 1, then (see Fig. 117)

$$\textbf{(1*)}\quad \mathcal{L}(f') = s\mathcal{L}(f) - f(0) - [f(a+0) - f(a-0)]e^{-as}.$$

(c) Verify (1*) for $f(t) = e^{-t}$ if $0 < t < 1$ and 0 if $t > 1$.

(d) Compare the Laplace transform of solving ODEs with the method in Chap. 2. Give examples of your own to illustrate the advantages of the present method (to the extent we have seen them so far).

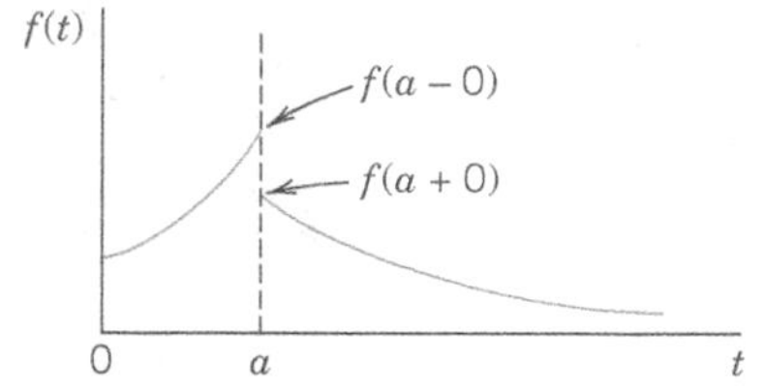

Fig. 117. Formula (1*)

6.3 Unit Step Function (Heaviside Function). Second Shifting Theorem (t-Shifting)

This section and the next one are extremely important because we shall now reach the point where the Laplace transform method shows its real power in applications and its superiority over the classical approach of Chap. 2. The reason is that we shall introduce two auxiliary functions, the *unit step function* or *Heaviside function* $u(t-a)$ (below) and *Dirac's delta* $\delta(t-a)$ (in Sec. 6.4). These functions are suitable for solving ODEs with complicated right sides of considerable engineering interest, such as single waves, inputs (driving forces) that are discontinuous or act for some time only, periodic inputs more general than just cosine and sine, or impulsive forces acting for an instant (hammerblows, for example).

Unit Step Function (Heaviside Function) $u(t-a)$

The **unit step function** or **Heaviside function** $u(t-a)$ is 0 for $t < a$, has a jump of size 1 at $t = a$ (where we can leave it undefined), and is 1 for $t > a$, in a formula:

$$\textbf{(1)}\qquad u(t-a) = \begin{cases} 0 & \text{if } t < a \\ 1 & \text{if } t > a \end{cases} \qquad (a \geqq 0).$$

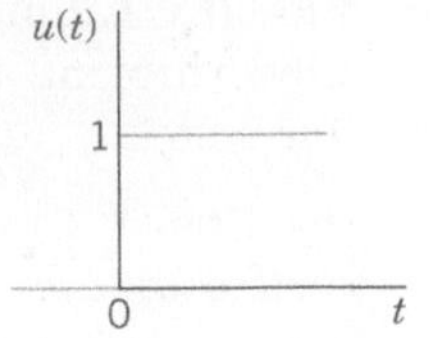

Fig. 118.   Unit step function $u(t)$ **Fig. 119.**   Unit step function $u(t - a)$

Figure 118 shows the special case $u(t)$, which has its jump at zero, and Fig. 119 the general case $u(t - a)$ for an arbitrary positive a. (For Heaviside, see Sec. 6.1.)

The transform of $u(t - a)$ follows directly from the defining integral in Sec. 6.1,

$$\mathcal{L}\{u(t - a)\} = \int_0^\infty e^{-st} u(t - a)\, dt = \int_0^\infty e^{-st} \cdot 1 \, dt = -\frac{e^{-st}}{s}\Big|_{t=a}^{\infty} ;$$

here the integration begins at $t = a\, (\geqq 0)$ because $u(t - a)$ is 0 for $t < a$. Hence

(2)
$$\mathcal{L}\{u(t - a)\} = \frac{e^{-as}}{s} \qquad\qquad (s > 0).$$

The unit step function is a typical "engineering function" made to measure for engineering applications, which often involve functions (mechanical or electrical driving forces) that are either "off" or "on." Multiplying functions $f(t)$ with $u(t - a)$, we can produce all sorts of effects. The simple basic idea is illustrated in Figs. 120 and 121. In Fig. 120 the given function is shown in (A). In (B) it is switched off between $t = 0$ and $t = 2$ (because $u(t - 2) = 0$ when $t < 2$) and is switched on beginning at $t = 2$. In (C) it is shifted to the right by 2 units, say, for instance, by 2 sec, so that it begins 2 sec later in the same fashion as before. More generally we have the following.

*Let $f(t) = 0$ for all negative t. Then $f(t - a)u(t - a)$ with $a > 0$ is $f(t)$ **shifted** (translated) to the right by the amount a.*

Figure 121 shows the effect of many unit step functions, three of them in (A) and infinitely many in (B) when continued periodically to the right; this is the effect of a rectifier that clips off the negative half-waves of a sinuosidal voltage. **CAUTION!** Make sure that you fully understand these figures, in particular the difference between parts (B) and (C) of Fig. 120. Figure 120(C) will be applied next.

(A) $f(t) = 5 \sin t$ (B) $f(t)u(t - 2)$ (C) $f(t - 2)u(t - 2)$

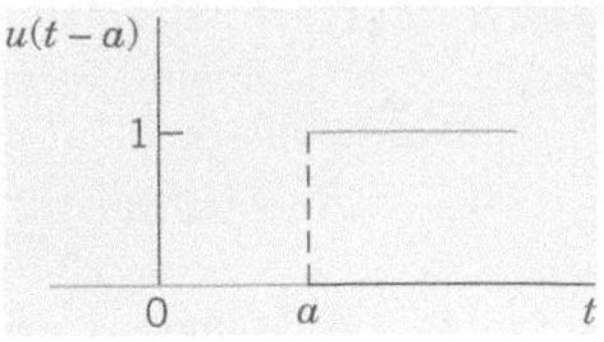

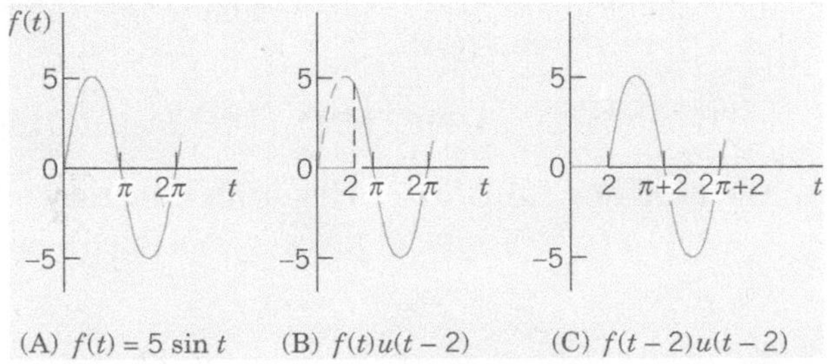

Fig. 120.   Effects of the unit step function: (A) Given function.
(B) Switching off and on. (C) Shift.

(A) $k[u(t-1) - 2u(t-4) + u(t-6)]$ (B) $4 \sin (\frac{1}{2}\pi t)[u(t) - u(t-2) + u(t-4) - + \cdots]$

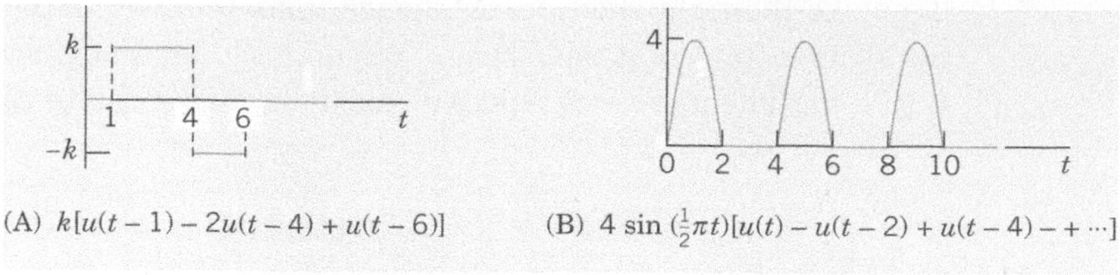

Fig. 121. Use of many unit step functions.

Time Shifting (t-Shifting): Replacing t by $t - a$ in $f(t)$

The first shifting theorem ("s-shifting") in Sec. 6.1 concerned transforms $F(s) = \mathcal{L}\{f(t)\}$ and $F(s - a) = \mathcal{L}\{e^{at}f(t)\}$. The second shifting theorem will concern functions $f(t)$ and $f(t - a)$. Unit step functions are just tools, and the theorem will be needed to apply them in connection with any other functions.

THEOREM 1

> **Second Shifting Theorem; Time Shifting**
>
> *If $f(t)$ has the transform $F(s)$, then the "**shifted function**"*
>
> (3) $$\tilde{f}(t) = f(t - a)u(t - a) = \begin{cases} 0 & \text{if } t < a \\ f(t - a) & \text{if } t > a \end{cases}$$
>
> *has the transform $e^{-as}F(s)$. That is, if $\mathcal{L}\{f(t)\} = F(s)$, then*
>
> (4) $$\mathcal{L}\{f(t - a)u(t - a)\} = e^{-as}F(s).$$
>
> *Or, if we take the inverse on both sides, we can write*
>
> (4*) $$f(t - a)u(t - a) = \mathcal{L}^{-1}\{e^{-as}F(s)\}.$$

Practically speaking, if we know $F(s)$, we can obtain the transform of (3) by multiplying $F(s)$ by e^{-as}. In Fig. 120, the transform of $5 \sin t$ is $F(s) = 5/(s^2 + 1)$, hence the shifted function $5 \sin (t - 2)u(t - 2)$ shown in Fig. 120(C) has the transform

$$e^{-2s}F(s) = 5e^{-2s}/(s^2 + 1).$$

PROOF   We prove Theorem 1. In (4), on the right, we use the definition of the Laplace transform, writing τ for t (to have t available later). Then, taking e^{-as} inside the integral, we have

$$e^{-as}F(s) = e^{-as}\int_0^\infty e^{-s\tau}f(\tau)\,d\tau = \int_0^\infty e^{-s(\tau + a)}f(\tau)\,d\tau.$$

Substituting $\tau + a = t$, thus $\tau = t - a$, $d\tau = dt$ in the integral (**CAUTION**, the lower limit changes!), we obtain

$$e^{-as}F(s) = \int_a^\infty e^{-st}f(t - a)\,dt.$$

To make the right side into a Laplace transform, we must have an integral from 0 to ∞, not from a to ∞. But this is easy. We multiply the integrand by $u(t-a)$. Then for t from 0 to a the integrand is 0, and we can write, with $\tilde{f}$ as in (3),

$$e^{-as}F(s) = \int_0^\infty e^{-st} f(t-a)u(t-a)\,dt = \int_0^\infty e^{-st}\tilde{f}(t)\,dt.$$

(Do you now see why $u(t-a)$ appears?) This integral is the left side of (4), the Laplace transform of $\tilde{f}(t)$ in (3). This completes the proof. ■

EXAMPLE 1 **Application of Theorem 1. Use of Unit Step Functions**

Write the following function using unit step functions and find its transform.

$$f(t) = \begin{cases} 2 & \text{if } 0 < t < 1 \\ \frac{1}{2}t^2 & \text{if } 1 < t < \frac{1}{2}\pi \\ \cos t & \text{if } \quad t > \frac{1}{2}\pi. \end{cases} \qquad \text{(Fig. 122)}$$

Solution. ***Step 1.*** In terms of unit step functions,

$$f(t) = 2(1 - u(t-1)) + \tfrac{1}{2}t^2(u(t-1) - u(t - \tfrac{1}{2}\pi)) + (\cos t)u(t - \tfrac{1}{2}\pi).$$

Indeed, $2(1 - u(t-1))$ gives $f(t)$ for $0 < t < 1$, and so on.

Step 2. To apply Theorem 1, we must write each term in $f(t)$ in the form $f(t-a)u(t-a)$. Thus, $2(1 - u(t-1))$ remains as it is and gives the transform $2(1 - e^{-s})/s$. Then

$$\mathcal{L}\left\{\frac{1}{2}t^2 u(t-1)\right\} = \mathcal{L}\left(\frac{1}{2}(t-1)^2 + (t-1) + \frac{1}{2}\right)u(t-1)\right\} = \left(\frac{1}{s^3} + \frac{1}{s^2} + \frac{1}{2s}\right)e^{-s}$$

$$\mathcal{L}\left\{\frac{1}{2}t^2 u\left(t - \frac{1}{2}\pi\right)\right\} = \mathcal{L}\left\{\frac{1}{2}\left(t - \frac{1}{2}\pi\right)^2 + \frac{\pi}{2}\left(t - \frac{1}{2}\pi\right) + \frac{\pi^2}{8}\right)u\left(t - \frac{1}{2}\pi\right)\right\}$$

$$= \left(\frac{1}{s^3} + \frac{\pi}{2s^2} + \frac{\pi^2}{8s}\right)e^{-\pi s/2}$$

$$\mathcal{L}\left\{(\cos t)\,u\left(t - \frac{1}{2}\pi\right)\right\} = \mathcal{L}\left\{-\left(\sin\left(t - \frac{1}{2}\pi\right)\right)u\left(t - \frac{1}{2}\pi\right)\right\} = -\frac{1}{s^2+1}e^{-\pi s/2}.$$

Together,

$$\mathcal{L}(f) = \frac{2}{s} - \frac{2}{s}e^{-s} + \left(\frac{1}{s^3} + \frac{1}{s^2} + \frac{1}{2s}\right)e^{-s} - \left(\frac{1}{s^3} + \frac{\pi}{2s^2} + \frac{\pi^2}{8s}\right)e^{-\pi s/2} - \frac{1}{s^2+1}e^{-\pi s/2}.$$

If the conversion of $f(t)$ to $f(t-a)$ is inconvenient, replace it by

$$\text{(4**)} \qquad \mathcal{L}\{f(t)u(t-a)\} = e^{-as}\mathcal{L}\{f(t+a)\}.$$

(4**) follows from (4) by writing $f(t-a) = g(t)$, hence $f(t) = g(t+a)$ and then again writing f for g. Thus,

$$\mathcal{L}\left\{\frac{1}{2}t^2 u(t-1)\right\} = e^{-s}\mathcal{L}\left\{\frac{1}{2}(t+1)^2\right\} = e^{-s}\mathcal{L}\left\{\frac{1}{2}t^2 + t + \frac{1}{2}\right\} = e^{-s}\left(\frac{1}{s^3} + \frac{1}{s^2} + \frac{1}{2s}\right)$$

as before. Similarly for $\mathcal{L}\{\frac{1}{2}t^2 u(t - \frac{1}{2}\pi)\}$. Finally, by (4**),

$$\mathcal{L}\left\{\cos t\, u\left(t - \frac{1}{2}\pi\right)\right\} = e^{-\pi s/2}\mathcal{L}\left\{\cos\left(t + \frac{1}{2}\pi\right)\right\} = e^{-\pi s/2}\mathcal{L}\{-\sin t\} = -e^{-\pi s/2}\frac{1}{s^2+1}. \quad ■$$

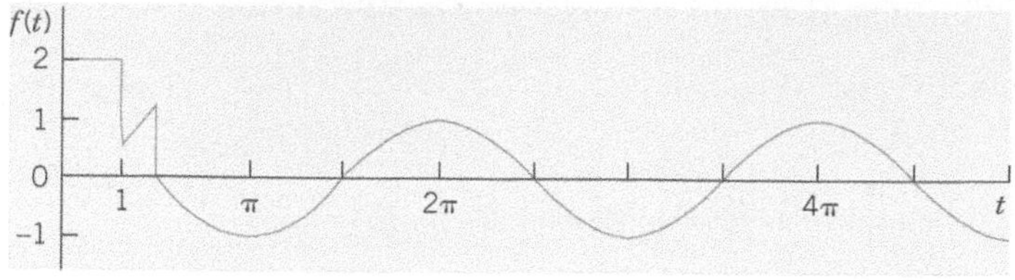

Fig. 122. $f(t)$ in Example 1

EXAMPLE 2 Application of Both Shifting Theorems. Inverse Transform

Find the inverse transform $f(t)$ of

$$F(s) = \frac{e^{-s}}{s^2 + \pi^2} + \frac{e^{-2s}}{s^2 + \pi^2} + \frac{e^{-3s}}{(s + 2)^2}.$$

Solution. Without the exponential functions in the numerator the three terms of $F(s)$ would have the inverses $(\sin \pi t)/\pi$, $(\sin \pi t)/\pi$, and te^{-2t} because $1/s^2$ has the inverse t, so that $1/(s + 2)^2$ has the inverse te^{-2t} by the first shifting theorem in Sec. 6.1. Hence by the second shifting theorem (t-shifting),

$$f(t) = \frac{1}{\pi} \sin (\pi(t - 1))\, u(t - 1) + \frac{1}{\pi} \sin (\pi(t - 2))\, u(t - 2) + (t - 3)e^{-2(t-3)}\, u(t - 3).$$

Now $\sin (\pi t - \pi) = -\sin \pi t$ and $\sin (\pi t - 2\pi) = \sin \pi t$, so that the first and second terms cancel each other when $t > 2$. Hence we obtain $f(t) = 0$ if $0 < t < 1$, $-(\sin \pi t)/\pi$ if $1 < t < 2$, 0 if $2 < t < 3$, and $(t - 3)e^{-2(t-3)}$ if $t > 3$. See Fig. 123. ■

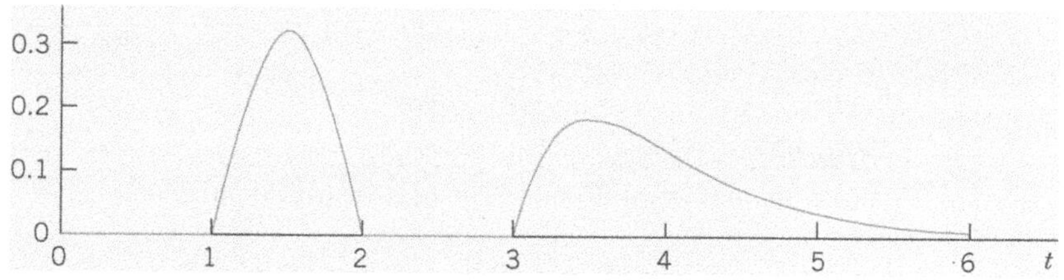

Fig. 123. $f(t)$ in Example 2

EXAMPLE 3 Response of an *RC*-Circuit to a Single Rectangular Wave

Find the current $i(t)$ in the RC-circuit in Fig. 124 if a single rectangular wave with voltage V_0 is applied. The circuit is assumed to be quiescent before the wave is applied.

Solution. The input is $V_0[u(t - a) - u(t - b)]$. Hence the circuit is modeled by the integro-differential equation (see Sec. 2.9 and Fig. 124)

$$Ri(t) + \frac{q(t)}{C} = Ri(t) + \frac{1}{C}\int_0^t i(\tau)\, d\tau = v(t) = V_0[u(t - a) - u(t - b)].$$

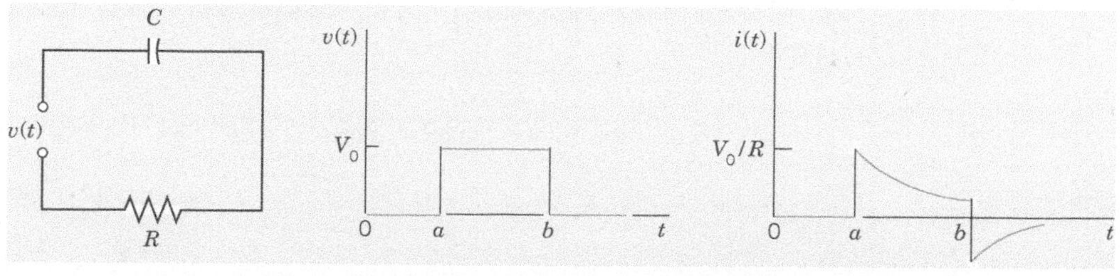

Fig. 124. RC-circuit, electromotive force $v(t)$, and current in Example 3

Using Theorem 3 in Sec. 6.2 and formula (1) in this section, we obtain the subsidiary equation

$$RI(s) + \frac{I(s)}{sC} = \frac{V_0}{s}[e^{-as} - e^{-bs}].$$

Solving this equation algebraically for $I(s)$, we get

$$I(s) = F(s)(e^{-as} - e^{-bs}) \quad \text{where} \quad F(s) = \frac{V_0 I R}{s + 1/(RC)} \quad \text{and} \quad \mathcal{L}^{-1}(F) = \frac{V_0}{R} e^{-t/(RC)},$$

the last expression being obtained from Table 6.1 in Sec. 6.1. Hence Theorem 1 yields the solution (Fig. 124)

$$i(t) = \mathcal{L}^{-1}(I) = \mathcal{L}^{-1}\{e^{-as}F(s)\} - \mathcal{L}^{-1}\{e^{-bs}F(s)\} = \frac{V_0}{R}[e^{-(t-a)/(RC)}u(t-a) - e^{-(t-b)/(RC)}u(t-b)];$$

that is, $i(t) = 0$ if $t < a$, and

$$i(t) = \begin{cases} K_1 e^{-t/(RC)} & \text{if } a < t < b \\ (K_1 - K_2)e^{-t/(RC)} & \text{if } a > b \end{cases}$$

where $K_1 = V_0 e^{a/(RC)}/R$ and $K_2 = V_0 e^{b/(RC)}/R$. ■

EXAMPLE 4 Response of an *RLC*-Circuit to a Sinusoidal Input Acting Over a Time Interval

Find the response (the current) of the *RLC*-circuit in Fig. 125, where $E(t)$ is sinusoidal, acting for a short time interval only, say,

$$E(t) = 100 \sin 400t \quad \text{if } 0 < t < 2\pi \qquad \text{and} \qquad E(t) = 0 \text{ if } t > 2\pi$$

and current and charge are initially zero.

Solution. The electromotive force $E(t)$ can be represented by $(100 \sin 400t)(1 - u(t - 2\pi))$. Hence the model for the current $i(t)$ in the circuit is the integro-differential equation (see Sec. 2.9)

$$0.1i' + 11i + 100\int_0^t i(\tau)\,d\tau = (100 \sin 400t)(1 - u(t - 2\pi)). \qquad i(0) = 0, \quad i'(0) = 0.$$

From Theorems 2 and 3 in Sec. 6.2 we obtain the subsidiary equation for $I(s) = \mathcal{L}(i)$

$$0.1sI + 11I + 100\frac{I}{s} = \frac{100 \cdot 400s}{s^2 + 400^2}\left(\frac{1}{s} - \frac{e^{-2\pi s}}{s}\right).$$

Solving it algebraically and noting that $s^2 + 110s + 1000 = (s + 10)(s + 100)$, we obtain

$$I(s) = \frac{1000 \cdot 400}{(s + 10)(s + 100)}\left(\frac{s}{s^2 + 400^2} - \frac{se^{-2\pi s}}{s^2 + 400^2}\right).$$

For the first term in the parentheses $(\cdots)$ times the factor in front of them we use the partial fraction expansion

$$\frac{400{,}000s}{(s + 10)(s + 100)(s^2 + 400^2)} = \frac{A}{s + 10} + \frac{B}{s + 100} + \frac{Ds + K}{s^2 + 400^2}.$$

Now determine A, B, D, K by your favorite method or by a CAS or as follows. Multiplication by the common denominator gives

$$400{,}000s = A(s + 100)(s^2 + 400^2) + B(s + 10)(s^2 + 400^2) + (Ds + K)(s + 10)(s + 100).$$

We set $s = -10$ and -100 and then equate the sums of the s^3 and s^2 terms to zero, obtaining (all values rounded)

$$\begin{array}{llr} (s = -10) & -4{,}000{,}000 = 90(10^2 + 400^2)A, & A = -0.27760 \\ (s = -100) & -40{,}000{,}000 = -90(100^2 + 400^2)B, & B = 2.6144 \\ (s^3\text{-terms}) & 0 = A + B + D, & D = -2.3368 \\ (s^2\text{-terms}) & 0 = 100A + 10B + 110D + K, & K = 258.66. \end{array}$$

Since $K = 258.66 = 0.6467 \cdot 400$, we thus obtain for the first term I_1 in $I = I_1 - I_2$

$$I_1 = -\frac{0.2776}{s + 10} + \frac{2.6144}{s + 100} - \frac{2.3368s}{s^2 + 400^2} + \frac{0.6467 \cdot 400}{s^2 + 400^2}.$$

From Table 6.1 in Sec. 6.1 we see that its inverse is

$$i_1(t) = -0.2776e^{-10t} + 2.6144e^{-100t} - 2.3368 \cos 400t + 0.6467 \sin 400t.$$

This is the current $i(t)$ when $0 < t < 2\pi$. It agrees for $0 < t < 2\pi$ with that in Example 1 of Sec. 2.9 (except for notation), which concerned the same *RLC*-circuit. Its graph in Fig. 63 in Sec. 2.9 shows that the exponential terms decrease very rapidly. Note that the present amount of work was substantially less.

The second term I_1 of I differs from the first term by the factor $e^{-2\pi s}$. Since $\cos 400(t - 2\pi) = \cos 400t$ and $\sin 400(t - 2\pi) = \sin 400t$, the second shifting theorem (Theorem 1) gives the inverse $i_2(t) = 0$ if $0 < t < 2\pi$, and for $> 2\pi$ it gives

$$i_2(t) = -0.2776e^{-10(t-2\pi)} + 2.6144e^{-100(t-2\pi)} - 2.3368 \cos 400t + 0.6467 \sin 400t.$$

Hence in $i(t)$ the cosine and sine terms cancel, and the current for $t > 2\pi$ is

$$i(t) = -0.2776(e^{-10t} - e^{-10(t-2\pi)}) + 2.6144(e^{-100t} - e^{-100(t-2\pi)}).$$

It goes to zero very rapidly, practically within 0.5 sec. ■

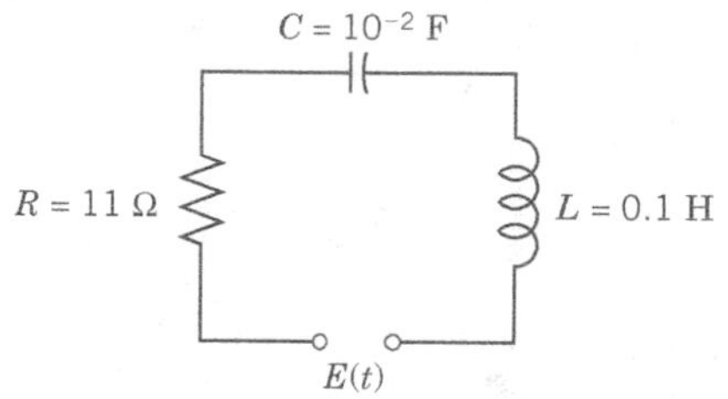

Fig. 125. *RLC*-circuit in Example 4

PROBLEM SET 6.3

1. **Report on Shifting Theorems.** Explain and compare the different roles of the two shifting theorems, using your own formulations and simple examples. Give no proofs.

2–11 SECOND SHIFTING THEOREM, UNIT STEP FUNCTION

Sketch or graph the given function, which is assumed to be zero outside the given interval. Represent it, using unit step functions. Find its transform. Show the details of your work.

2. $t\ (0 < t < 2)$
3. $t - 2\ (t > 2)$
4. $\cos 4t\ (0 < t < \pi)$
5. $e^t\ (0 < t < \pi/2)$
6. $\sin \pi t\ (2 < t < 4)$
7. $e^{-\pi t}\ (2 < t < 4)$
8. $t^2\ (1 < t < 2)$
9. $t^2\ (t > \frac{3}{2})$
10. $\sinh t\ (0 < t < 2)$
11. $\sin t\ (\pi/2 < t < \pi)$

12–17 INVERSE TRANSFORMS BY THE 2ND SHIFTING THEOREM

Find and sketch or graph $f(t)$ if $\mathcal{L}(f)$ equals

12. $e^{-3s}/(s - 1)^3$
13. $6(1 - e^{-\pi s})/(s^2 + 9)$
14. $4(e^{-2s} - 2e^{-5s})/s$
15. e^{-3s}/s^4
16. $2(e^{-s} - e^{-3s})/(s^2 - 4)$
17. $(1 + e^{-2\pi(s+1)})(s + 1)/((s + 1)^2 + 1)$

18–27 IVPs, SOME WITH DISCONTINUOUS INPUT

Using the Laplace transform and showing the details, solve

18. $9y'' - 6y' + y = 0, \quad y(0) = 3, y'(0) = 1$

19. $y'' + 6y' + 8y = e^{-3t} - e^{-5t}, \quad y(0) = 0, y'(0) = 0$

20. $y'' + 10y' + 24y = 144t^2, \quad y(0) = 19/12, y'(0) = -5$

21. $y'' + 9y = 8 \sin t$ if $0 < t < \pi$ and 0 if $t > \pi$; $y(0) = 0, y'(0) = 4$

22. $y'' + 3y' + 2y = 4t$ if $0 < t < 1$ and 8 if $t > 1$; $y(0) = 0, y'(0) = 0$

23. $y'' + y' - 2y = 3 \sin t - \cos t$ if $0 < t < 2\pi$ and $3 \sin 2t - \cos 2t$ if $t > 2\pi$; $y(0) = 1, y'(0) = 0$

24. $y'' + 3y' + 2y = 1$ if $0 < t < 1$ and 0 if $t > 1$; $y(0) = 0, y'(0) = 0$

25. $y'' + y = t$ if $0 < t < 1$ and 0 if $t > 1$; $y(0) = 0, y'(0) = 0$

26. Shifted data. $y'' + 2y' + 5y = 10 \sin t$ if $0 < t < 2\pi$ and 0 if $t > 2\pi$; $y(\pi) = 1, y'(\pi) = 2e^{-\pi} - 2$

27. Shifted data. $y'' + 4y = 8t^2$ if $0 < t < 5$ and 0 if $t > 5$; $y(1) = 1 + \cos 2, y'(1) = 4 - 2 \sin 2$

28–40 MODELS OF ELECTRIC CIRCUITS

28–30 *RL*-CIRCUIT

Using the Laplace transform and showing the details, find the current $i(t)$ in the circuit in Fig. 126, assuming $i(0) = 0$ and:

28. $R = 1$ kΩ ($=1000$ Ω), $L = 1$ H, $v = 0$ if $0 < t < \pi$, and $40 \sin t$ V if $t > \pi$

29. $R = 25$ Ω, $L = 0.1$ H, $v = 490\, e^{-5t}$ V if $0 < t < 1$ and 0 if $t > 1$

30. $R = 10$ Ω, $L = 0.5$ H, $v = 200t$ V if $0 < t < 2$ and 0 if $t > 2$

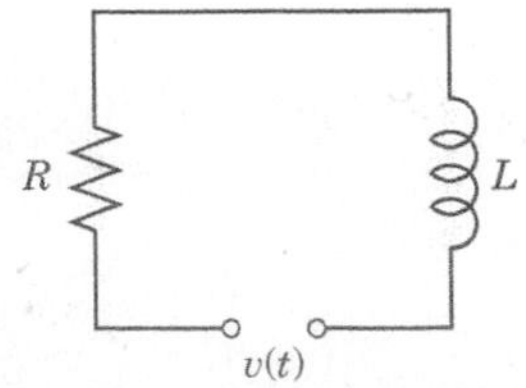

Fig. 126. Problems 28–30

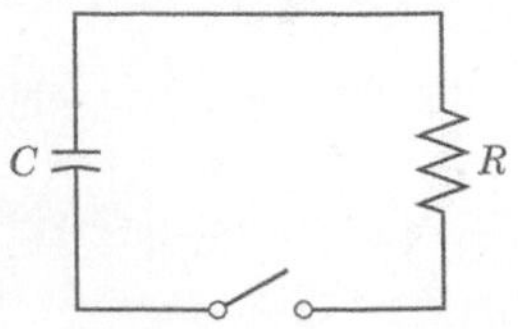

Fig. 127. Problem 31

31. Discharge in *RC*-circuit. Using the Laplace transform, find the charge $q(t)$ on the capacitor of capacitance C in Fig. 127 if the capacitor is charged so that its potential is V_0 and the switch is closed at $t = 0$.

32–34 *RC*-CIRCUIT

Using the Laplace transform and showing the details, find the current $i(t)$ in the circuit in Fig. 128 with $R = 10$ Ω and $C = 10^{-2}$ F, where the current at $t = 0$ is assumed to be zero, and:

32. $v = 0$ if $t < 4$ and $14 \cdot 10^6 e^{-3t}$ V if $t > 4$

33. $v = 0$ if $t < 2$ and $100(t - 2)$ V if $t > 2$

34. $v(t) = 100$ V if $0.5 < t < 0.6$ and 0 otherwise. Why does $i(t)$ have jumps?

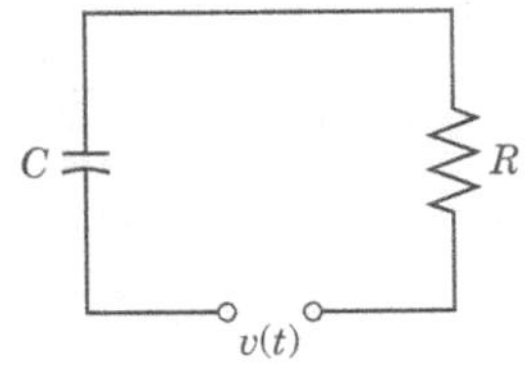

Fig. 128. Problems 32–34

35–37 *LC*-CIRCUIT

Using the Laplace transform and showing the details, find the current $i(t)$ in the circuit in Fig. 129, assuming zero initial current and charge on the capacitor and:

35. $L = 1$ H, $C = 10^{-2}$ F, $v = -9900 \cos t$ V if $\pi < t < 3\pi$ and 0 otherwise

36. $L = 1$ H, $C = 0.25$ F, $v = 200\,(t - \frac{1}{3}t^3)$ V if $0 < t < 1$ and 0 if $t > 1$

37. $L = 0.5$ H, $C = 0.05$ F, $v = 78 \sin t$ V if $0 < t < \pi$ and 0 if $t > \pi$

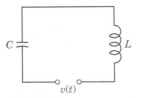

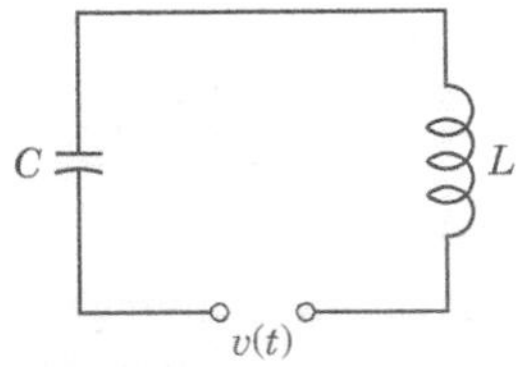

Fig. 129. Problems 35–37

38–40 *RLC*-CIRCUIT

Using the Laplace transform and showing the details, find the current $i(t)$ in the circuit in Fig. 130, assuming zero initial current and charge and:

38. $R = 4$ Ω, $L = 1$ H, $C = 0.05$ F, $v = 34e^{-t}$ V if $0 < t < 4$ and 0 if $t > 4$

39. $R = 2\ \Omega$, $L = 1$ H, $C = 0.5$ F, $v(t) = 1$ kV if $0 < t < 2$ and 0 if $t > 2$

40. $R = 2\ \Omega$, $L = 1$ H, $C = 0.1$ F, $v = 255 \sin t$ V if $0 < t < 2\pi$ and 0 if $t > 2\pi$

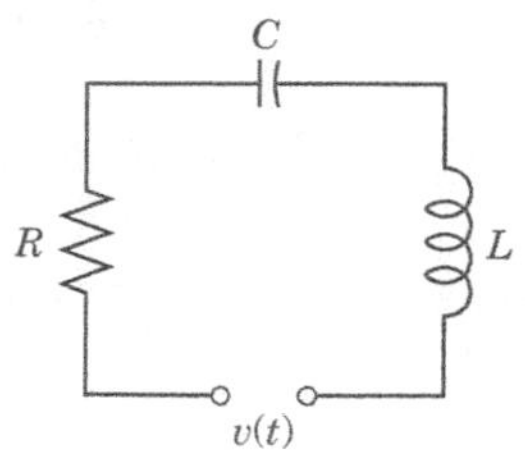

Fig. 130. Problems 38–40

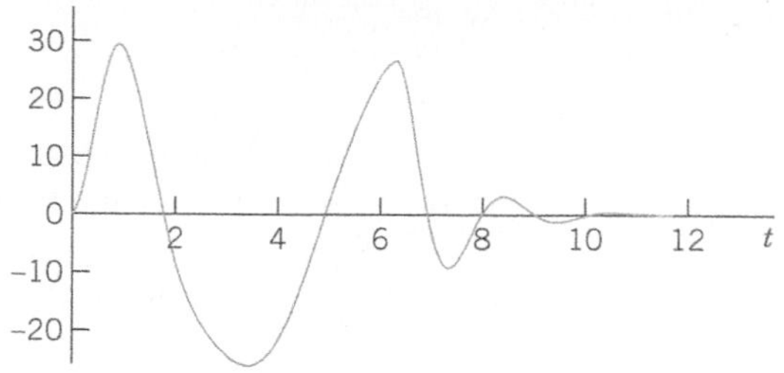

Fig. 131. Current in Problem 40

6.4 Short Impulses. Dirac's Delta Function. Partial Fractions

An airplane making a "hard" landing, a mechanical system being hit by a hammerblow, a ship being hit by a single high wave, a tennis ball being hit by a racket, and many other similar examples appear in everyday life. They are phenomena of an impulsive nature where actions of forces—mechanical, electrical, etc.—are applied over short intervals of time.

We can model such phenomena and problems by "Dirac's delta function," and solve them very effecively by the Laplace transform.

To model situations of that type, we consider the function

$$f_k(t-a) = \begin{cases} 1/k & \text{if } a \leqq t \leqq a+k \\ 0 & \text{otherwise} \end{cases} \tag{1}$$

(Fig. 132)

(and later its limit as $k \to 0$). This function represents, for instance, a force of magnitude $1/k$ acting from $t = a$ to $t = a + k$, where k is positive and small. In mechanics, the integral of a force acting over a time interval $a \leqq t \leqq a + k$ is called the **impulse** of the force; similarly for electromotive forces $E(t)$ acting on circuits. Since the blue rectangle in Fig. 132 has area 1, the impulse of f_k in (1) is

$$I_k = \int_0^\infty f_k(t-a)\, dt = \int_a^{a+k} \frac{1}{k}\, dt = 1. \tag{2}$$

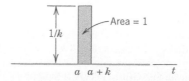

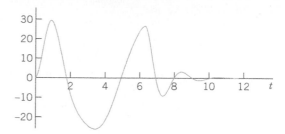

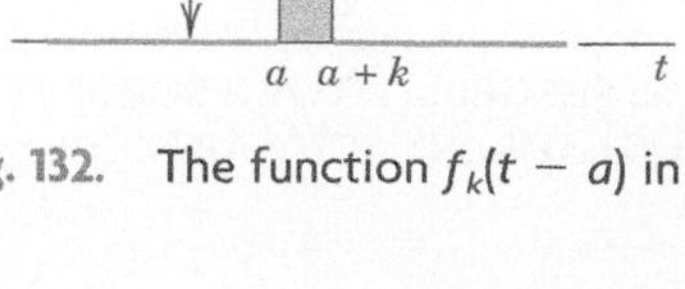

Fig. 132. The function $f_k(t - a)$ in (1)

To find out what will happen if k becomes smaller and smaller, we take the limit of f_k as $k \to 0$ $(k > 0)$. This limit is denoted by $\delta(t - a)$, that is,

$$\delta(t-a) = \lim_{k\to 0} f_k(t-a).$$

$\delta(t - a)$ is called the **Dirac delta function**[2] or the **unit impulse function**.

$\delta(t - a)$ is not a function in the ordinary sense as used in calculus, but a so-called *generalized function*.[2] To see this, we note that the impulse I_k of f_k is 1, so that from (1) and (2) by taking the limit as $k \to 0$ we obtain

$$\text{(3)} \qquad \delta(t-a) = \begin{cases} \infty & \text{if } t = a \\ 0 & \text{otherwise} \end{cases} \qquad \text{and} \qquad \int_0^\infty \delta(t-a)\,dt = 1,$$

but from calculus we know that a function which is everywhere 0 except at a single point must have the integral equal to 0. Nevertheless, in impulse problems, it is convenient to operate on $\delta(t - a)$ as though it were an ordinary function. In particular, for a *continuous* function $g(t)$ one uses the property [often called the **sifting property** of $\delta(t - a)$, not to be confused with *shifting*]

$$\textbf{(4)} \qquad \int_0^\infty g(t)\delta(t-a)\,dt = g(a)$$

which is plausible by (2).

To obtain the Laplace transform of $\delta(t - a)$, we write

$$f_k(t-a) = \frac{1}{k}\left[u(t-a) - u(t-(a+k))\right]$$

and take the transform [see (2)]

$$\mathcal{L}\{f_k(t-a)\} = \frac{1}{ks}\left[e^{-as} - e^{-(a+k)s}\right] = e^{-as}\,\frac{1-e^{-ks}}{ks}.$$

We now take the limit as $k \to 0$. By l'Hôpital's rule the quotient on the right has the limit 1 (differentiate the numerator and the denominator separately with respect to k, obtaining se^{-ks} and s, respectively, and use $se^{-ks}/s \to 1$ as $k \to 0$). Hence the right side has the limit e^{-as}. This suggests defining the transform of $\delta(t - a)$ by this limit, that is,

$$\textbf{(5)} \qquad \mathcal{L}\{\delta(t-a)\} = e^{-as}.$$

The unit step and unit impulse functions can now be used on the right side of ODEs modeling mechanical or electrical systems, as we illustrate next.

---

[2]PAUL DIRAC (1902–1984), English physicist, was awarded the Nobel Prize [jointly with the Austrian ERWIN SCHRÖDINGER (1887–1961)] in 1933 for his work in quantum mechanics.

Generalized functions are also called **distributions**. Their theory was created in 1936 by the Russian mathematician SERGEI L'VOVICH SOBOLEV (1908–1989), and in 1945, under wider aspects, by the French mathematician LAURENT SCHWARTZ (1915–2002).

EXAMPLE 1 **Mass–Spring System Under a Square Wave**

Determine the response of the damped mass–spring system (see Sec. 2.8) under a square wave, modeled by (see Fig. 133)

$$y'' + 3y' + 2y = r(t) = u(t-1) - u(t-2), \qquad y(0) = 0, \qquad y'(0) = 0.$$

Solution. From (1) and (2) in Sec. 6.2 and (2) and (4) in this section we obtain the subsidiary equation

$$s^2Y + 3sY + 2Y = \frac{1}{s}(e^{-s} - e^{-2s}). \qquad \text{Solution} \qquad Y(s) = \frac{1}{s(s^2+3s+2)}(e^{-s} - e^{-2s}).$$

Using the notation $F(s)$ and partial fractions, we obtain

$$F(s) = \frac{1}{s(s^2+3s+2)} = \frac{1}{s(s+1)(s+2)} = \frac{\frac{1}{2}}{s} - \frac{1}{s+1} + \frac{\frac{1}{2}}{s+2}.$$

From Table 6.1 in Sec. 6.1, we see that the inverse is

$$f(t) = \mathcal{L}^{-1}(F) = \tfrac{1}{2} - e^{-t} + \tfrac{1}{2}e^{-2t}.$$

Therefore, by Theorem 1 in Sec. 6.3 (t-shifting) we obtain the square-wave response shown in Fig. 133,

$$\begin{aligned} y &= \mathcal{L}^{-1}(F(s)e^{-s} - F(s)e^{-2s}) \\ &= f(t-1)u(t-1) - f(t-2)u(t-2) \\ &= \begin{cases} 0 & (0 < t < 1) \\ \frac{1}{2} - e^{-(t-1)} + \frac{1}{2}e^{-2(t-1)} & (1 < t < 2) \\ -e^{-(t-1)} + e^{-(t-2)} + \frac{1}{2}e^{-2(t-1)} - \frac{1}{2}e^{-2(t-2)} & (t > 2). \end{cases} \end{aligned}$$

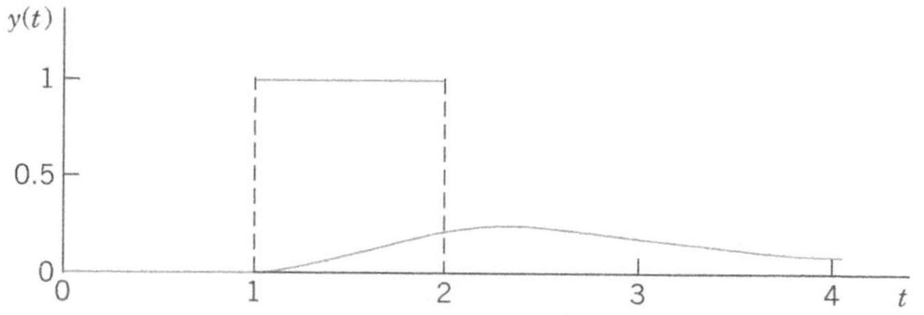

Fig. 133. Square wave and response in Example 1

EXAMPLE 2 **Hammerblow Response of a Mass–Spring System**

Find the response of the system in Example 1 with the square wave replaced by a unit impulse at time $t = 1$.

Solution. We now have the ODE and the subsidiary equation

$$y'' + 3y' + 2y = \delta(t-1), \qquad \text{and} \qquad (s^2 + 3s + 2)Y = e^{-s}.$$

Solving algebraically gives

$$Y(s) = \frac{e^{-s}}{(s+1)(s+2)} = \left(\frac{1}{s+1} - \frac{1}{s+2}\right)e^{-s}.$$

By Theorem 1 the inverse is

$$y(t) = \mathcal{L}^{-1}(Y) = \begin{cases} 0 & \text{if } 0 < t < 1 \\ e^{-(t-1)} - e^{-2(t-1)} & \text{if } \quad t > 1. \end{cases}$$

$y(t)$ is shown in Fig. 134. Can you imagine how Fig. 133 approaches Fig. 134 as the wave becomes shorter and shorter, the area of the rectangle remaining 1? ■

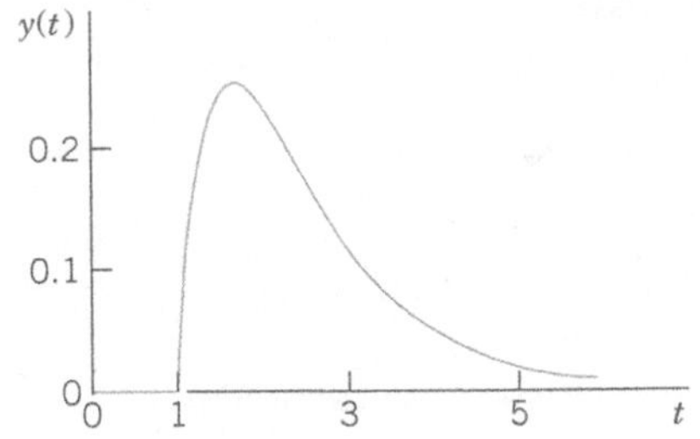

Fig. 134. Response to a hammerblow in Example 2

EXAMPLE 3 Four-Terminal *RLC*-Network

Find the output voltage response in Fig. 135 if $R = 20\ \Omega$, $L = 1$ H, $C = 10^{-4}$ F, the input is $\delta(t)$ (a unit impulse at time $t = 0$), and current and charge are zero at time $t = 0$.

Solution. To understand what is going on, note that the network is an *RLC*-circuit to which two wires at A and B are attached for recording the voltage $v(t)$ on the capacitor. Recalling from Sec. 2.9 that current $i(t)$ and charge $q(t)$ are related by $i = q' = dq/dt$, we obtain the model

$$Li' + Ri + \frac{q}{C} = Lq'' + Rq' + \frac{q}{C} = q'' + 20q' + 10{,}000q = \delta(t).$$

From (1) and (2) in Sec. 6.2 and (5) in this section we obtain the subsidiary equation for $Q(s) = \mathcal{L}(q)$

$$(s^2 + 20s + 10{,}000)Q = 1. \qquad \text{Solution} \qquad Q = \frac{1}{(s + 10)^2 + 9900}.$$

By the first shifting theorem in Sec. 6.1 we obtain from Q damped oscillations for q and v; rounding $9900 \approx 99.50^2$, we get (Fig. 135)

$$q = \mathcal{L}^{-1}(Q) = \frac{1}{99.50}e^{-10t}\sin 99.50t \qquad \text{and} \qquad v = \frac{q}{C} = 100.5e^{-10t}\sin 99.50t.$$ ■

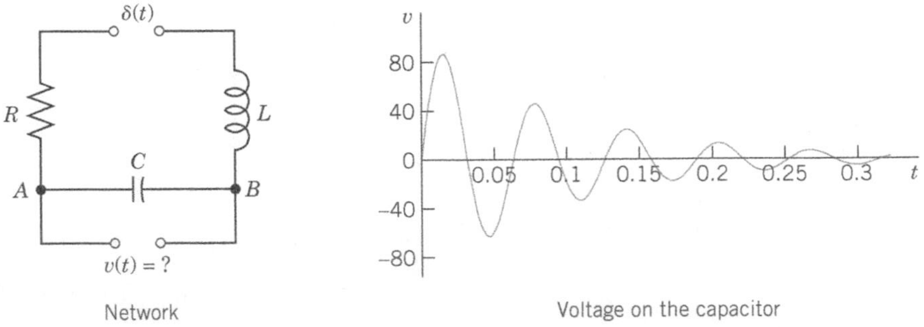

Fig. 135. Network and output voltage in Example 3

More on Partial Fractions

We have seen that the solution Y of a subsidiary equation usually appears as a quotient of polynomials $Y(s) = F(s)/G(s)$, so that a partial fraction representation leads to a sum of expressions whose inverses we can obtain from a table, aided by the first shifting theorem (Sec. 6.1). These representations are sometimes called **Heaviside expansions.**

An *unrepeated factor* $s - a$ in $G(s)$ requires a single partial fraction $A/(s - a)$. See Examples 1 and 2. *Repeated real factors* $(s - a)^2$, $(s - a)^3$, etc., require partial fractions

$$\frac{A_2}{(s-a)^2} + \frac{A_1}{s-a}, \qquad \frac{A_3}{(s-a)^3} + \frac{A_2}{(s-a)^2} + \frac{A_1}{s-a}, \quad \text{etc.,}$$

The inverses are $(A_2t + A_1)e^{at}$, $(\frac{1}{2}A_3t^2 + A_2t + A_1)e^{at}$, etc.

Unrepeated complex factors $(s - a)(s - \bar{a})$, $a = \alpha + i\beta$, $\bar{a} = \alpha - i\beta$, require a partial fraction $(As + B)/[(s - \alpha)^2 + \beta^2]$. For an application, see Example 4 in Sec. 6.3. A further one is the following.

EXAMPLE 4 Unrepeated Complex Factors. Damped Forced Vibrations

Solve the initial value problem for a damped mass–spring system acted upon by a sinusoidal force for some time interval (Fig. 136),

$$y'' + 2y' + 2y = r(t), \quad r(t) = 10 \sin 2t \text{ if } 0 < t < \pi \text{ and } 0 \text{ if } t > \pi; \qquad y(0) = 1, \quad y'(0) = -5.$$

Solution. From Table 6.1, (1), (2) in Sec. 6.2, and the second shifting theorem in Sec. 6.3, we obtain the subsidiary equation

$$(s^2Y - s + 5) + 2(sY - 1) + 2Y = 10\frac{2}{s^2+4}(1 - e^{-\pi s}).$$

We collect the Y-terms, $(s^2 + 2s + 2)Y$, take $-s + 5 - 2 = -s + 3$ to the right, and solve,

$$Y = \frac{20}{(s^2+4)(s^2+2s+2)} - \frac{20e^{-\pi s}}{(s^2+4)(s^2+2s+2)} + \frac{s-3}{s^2+2s+2}. \tag{6}$$

For the last fraction we get from Table 6.1 and the first shifting theorem

$$\mathcal{L}^{-1}\left\{\frac{s+1-4}{(s+1)^2+1}\right\} = e^{-t}(\cos t - 4 \sin t). \tag{7}$$

In the first fraction in (6) we have unrepeated complex roots, hence a partial fraction representation

$$\frac{20}{(s^2+4)(s^2+2s+2)} = \frac{As+B}{s^2+4} + \frac{Ms+N}{s^2+2s+2}.$$

Multiplication by the common denominator gives

$$20 = (As + B)(s^2 + 2s + 2) + (Ms + N)(s^2 + 4).$$

We determine A, B, M, N. Equating the coefficients of each power of s on both sides gives the four equations

$$\text{(a)}\ [s^3]:\ 0 = A + M \qquad \text{(b)}\ [s^2]:\ 0 = 2A + B + N$$

$$\text{(c)}\ [s]:\ 0 = 2A + 2B + 4M \qquad \text{(d)}\ [s^0]:\ 20 = 2B + 4N.$$

We can solve this, for instance, obtaining $M = -A$ from (a), then $A = B$ from (c), then $N = -3A$ from (b), and finally $A = -2$ from (d). Hence $A = -2$, $B = -2$, $M = 2$, $N = 6$, and the first fraction in (6) has the representation

$$\frac{-2s-2}{s^2+4} + \frac{2(s+1)+6-2}{(s+1)^2+1}. \quad \text{Inverse transform:} \quad -2\cos 2t - \sin 2t + e^{-t}(2\cos t + 4 \sin t). \tag{8}$$

The sum of this inverse and (7) is the solution of the problem for $0 < t < \pi$, namely (the sines cancel),

$$y(t) = 3e^{-t}\cos t - 2\cos 2t - \sin 2t \qquad \text{if } 0 < t < \pi. \tag{9}$$

In the second fraction in (6), taken with the minus sign, we have the factor $e^{-\pi s}$, so that from (8) and the second shifting theorem (Sec. 6.3) we get the inverse transform of this fraction for $t > 0$ in the form

$$\begin{aligned} &+2\cos(2t - 2\pi) + \sin(2t - 2\pi) - e^{-(t-\pi)}[2\cos(t-\pi) + 4\sin(t-\pi)] \\ &= 2\cos 2t + \sin 2t + e^{-(t-\pi)}(2\cos t + 4\sin t). \end{aligned}$$

The sum of this and (9) is the solution for $t > \pi$,

$$y(t) = e^{-t}[(3 + 2e^{\pi})\cos t + 4e^{\pi}\sin t] \qquad \text{if } t > \pi. \tag{10}$$

Figure 136 shows (9) (for $0 < t < \pi$) and (10) (for $t > \pi$), a beginning vibration, which goes to zero rapidly because of the damping and the absence of a driving force after $t = \pi$. ■

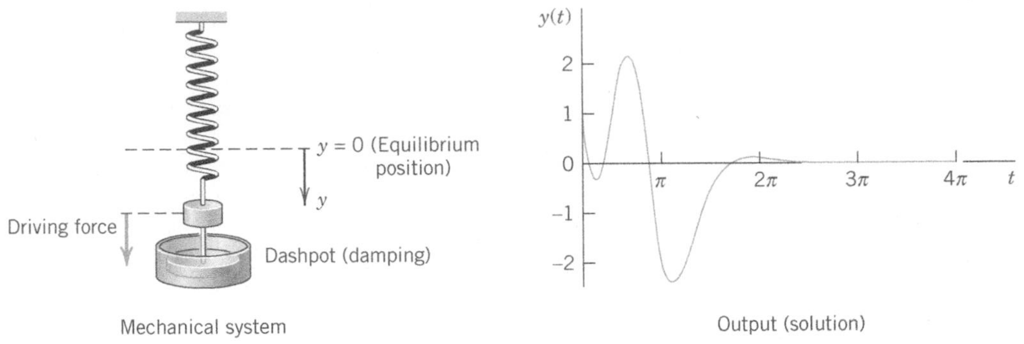

Fig. 136. Example 4

The case of repeated complex factors $[(s - a)(s - \bar{a})]^2$, which is important in connection with resonance, will be handled by "convolution" in the next section.

PROBLEM SET 6.4

1. **CAS PROJECT. Effect of Damping.** Consider a vibrating system of your choice modeled by

$$y'' + cy' + ky = \delta(t).$$

(a) Using graphs of the solution, describe the effect of continuously decreasing the damping to 0, keeping k constant.

(b) What happens if c is kept constant and k is continuously increased, starting from 0?

(c) Extend your results to a system with two δ-functions on the right, acting at different times.

2. **CAS EXPERIMENT. Limit of a Rectangular Wave. Effects of Impulse.**

(a) In Example 1 in the text, take a rectangular wave of area 1 from 1 to $1 + k$. Graph the responses for a sequence of values of k approaching zero, illustrating that for smaller and smaller k those curves approach the curve shown in Fig. 134. *Hint:* If your CAS gives no solution for the differential equation, involving k, take specific k's from the beginning.

(b) Experiment on the response of the ODE in Example 1 (or of another ODE of your choice) to an impulse $\delta(t - a)$ for various systematically chosen a (> 0); choose initial conditions $y(0) \neq 0$, $y'(0) = 0$. Also consider the solution if no impulse is applied. Is there a dependence of the response on a? On b if you choose $b\delta(t - a)$? Would $-\delta(t - \tilde{a})$ with $\tilde{a} > a$ annihilate the effect of $\delta(t - a)$? Can you think of other questions that one could consider experimentally by inspecting graphs?

3–12 EFFECT OF DELTA (IMPULSE) ON VIBRATING SYSTEMS

Find and graph or sketch the solution of the IVP. Show the details.

3. $y'' + 4y = \delta(t - \pi)$, $y(0) = 8$, $y'(0) = 0$

4. $y'' + 16y = 4\delta(t - 3\pi)$, $y(0) = 2, y'(0) = 0$

5. $y'' + y = \delta(t - \pi) - \delta(t - 2\pi)$,
$y(0) = 0, y'(0) = 1$

6. $y'' + 4y' + 5y = \delta(t - 1)$, $y(0) = 0, y'(0) = 3$

7. $4y'' + 24y' + 37y = 17e^{-t} + \delta(t - \frac{1}{2})$,
$y(0) = 1, y'(0) = 1$

8. $y'' + 3y' + 2y = 10(\sin t + \delta(t - 1))$, $y(0) = 1$,
$y'(0) = -1$

9. $y'' + 4y' + 5y = [1 - u(t - 10)]e^t - e^{10}\delta(t - 10)$,
$y(0) = 0, y'(0) = 1$

10. $y'' + 5y' + 6y = \delta(t - \frac{1}{2}\pi) + u(t - \pi)\cos t$,
$y(0) = 0, y'(0) = 0$

11. $y'' + 5y' + 6y = u(t - 1) + \delta(t - 2)$,
$y(0) = 0, y'(0) = 1$

12. $y'' + 2y' + 5y = 25t - 100\delta(t - \pi)$, $y(0) = -2$,
$y'(0) = 5$

13. PROJECT. Heaviside Formulas. (a) Show that for a simple root a and fraction $A/(s - a)$ in $F(s)/G(s)$ we have the *Heaviside formula*

$$A = \lim_{s \to a} \frac{(s - a)F(s)}{G(s)}.$$

(b) Similarly, show that for a root a of order m and fractions in

$$\frac{F(s)}{G(s)} = \frac{A_m}{(s - a)^m} + \frac{A_{m-1}}{(s - a)^{m-1}} + \cdots$$
$$+ \frac{A_1}{s - a} + \text{further fractions}$$

we have the *Heaviside formulas* for the first coefficient

$$A_m = \lim_{s \to a} \frac{(s - a)^m F(s)}{G(s)}$$

and for the other coefficients

$$A_k = \frac{1}{(m - k)!} \lim_{s \to a} \frac{d^{m-k}}{ds^{m-k}} \left[\frac{(s - a)^m F(s)}{G(s)}\right],$$
$$k = 1, \cdots, m - 1.$$

14. TEAM PROJECT. Laplace Transform of Periodic Functions

(a) Theorem. *The Laplace transform of a piecewise continuous function $f(t)$ with period p is*

$$\mathcal{L}(f) = \frac{1}{1 - e^{-ps}} \int_0^p e^{-st} f(t)\, dt \quad (s > 0). \tag{11}$$

Prove this theorem. *Hint:* Write $\int_0^\infty = \int_0^p + \int_p^{2p} + \cdots$. Set $t = (n - 1)p$ in the nth integral. Take out $e^{-(n-1)p}$ from under the integral sign. Use the sum formula for the geometric series.

(b) Half-wave rectifier. Using (11), show that the half-wave rectification of $\sin \omega t$ in Fig. 137 has the Laplace transform

$$\mathcal{L}(f) = \frac{\omega(1 + e^{-\pi s/\omega})}{(s^2 + \omega^2)(1 - e^{-2\pi s/\omega})}$$
$$= \frac{\omega}{(s^2 + \omega^2)(1 - e^{-\pi s/\omega})}.$$

(A *half-wave rectifier* clips the negative portions of the curve. A *full-wave rectifier* converts them to positive; see Fig. 138.)

(c) Full-wave rectifier. Show that the Laplace transform of the full-wave rectification of $\sin \omega t$ is

$$\frac{\omega}{s^2 + \omega^2} \coth \frac{\pi s}{2\omega}.$$

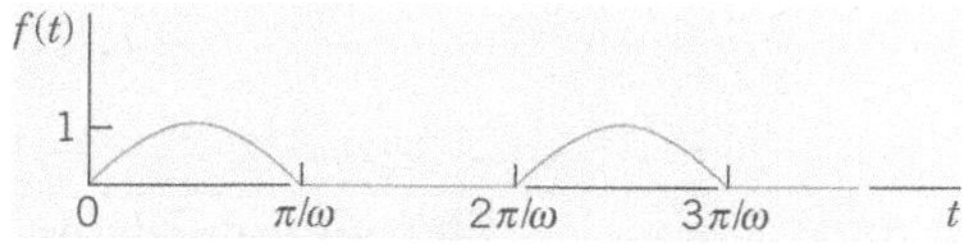

Fig. 137. Half-wave rectification

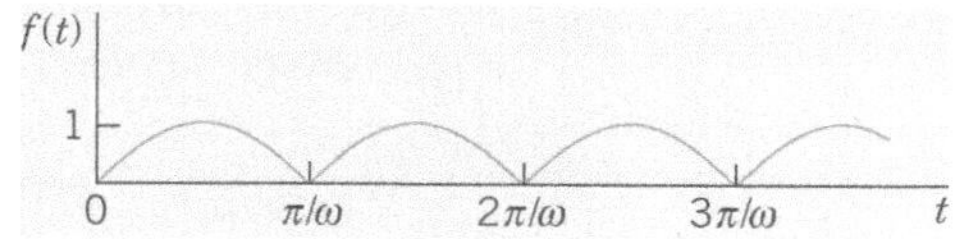

Fig. 138. Full-wave rectification

(d) Saw-tooth wave. Find the Laplace transform of the saw-tooth wave in Fig. 139.

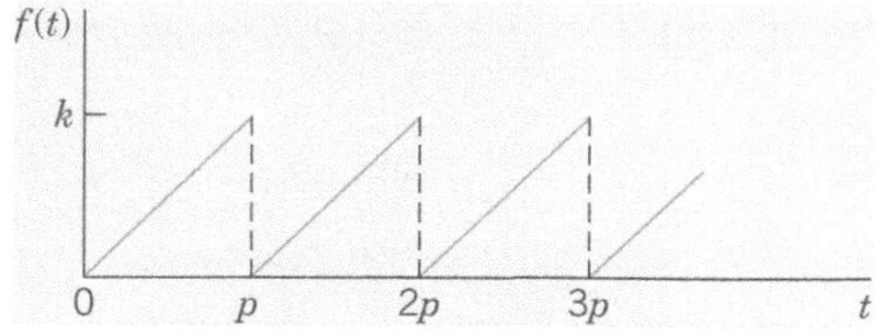

Fig. 139. Saw-tooth wave

15. Staircase function. Find the Laplace transform of the staircase function in Fig. 140 by noting that it is the difference of kt/p and the function in 14(d).

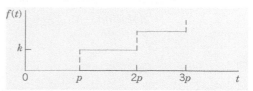

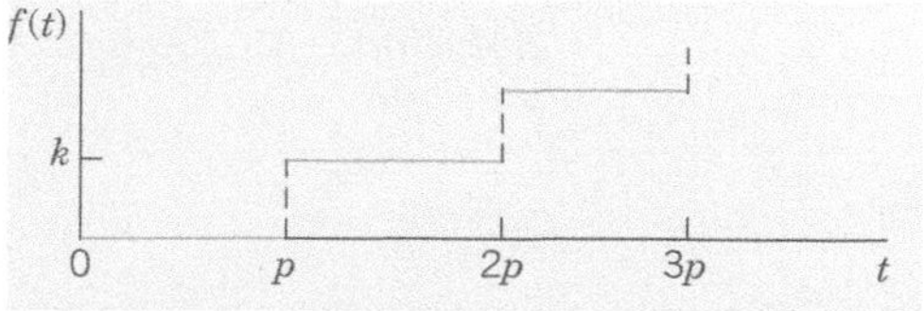

Fig. 140. Staircase function

6.5 Convolution. Integral Equations

Convolution has to do with the multiplication of transforms. The situation is as follows. *Addition* of transforms provides no problem; we know that $\mathcal{L}(f+g) = \mathcal{L}(f) + \mathcal{L}(g)$. Now **multiplication of transforms** occurs frequently in connection with ODEs, integral equations, and elsewhere. Then we usually know $\mathcal{L}(f)$ and $\mathcal{L}(g)$ and would like to know the function whose transform is the product $\mathcal{L}(f)\mathcal{L}(g)$. We might perhaps guess that it is fg, but this is false. *The transform of a product is generally different from the product of the transforms of the factors,*

$$\mathcal{L}(fg) \neq \mathcal{L}(f)\mathcal{L}(g) \qquad \text{in general.}$$

To see this take $f = e^t$ and $g = 1$. Then $fg = e^t$, $\mathcal{L}(fg) = 1/(s-1)$, but $\mathcal{L}(f) = 1/(s-1)$ and $\mathcal{L}(1) = 1/s$ give $\mathcal{L}(f)\mathcal{L}(g) = 1/(s^2 - s)$.

According to the next theorem, the correct answer is that $\mathcal{L}(f)\mathcal{L}(g)$ is the transform of the **convolution** of f and g, denoted by the standard notation $f * g$ and defined by the integral

$$h(t) = (f * g)(t) = \int_0^t f(\tau)g(t-\tau)\,d\tau. \tag{1}$$

THEOREM 1 **Convolution Theorem**

If two functions f and g satisfy the assumption in the existence theorem in Sec. 6.1, *so that their transforms F and G exist, the product $H = FG$ is the transform of h given by* (1). (Proof after Example 2.)

EXAMPLE 1 **Convolution**

Let $H(s) = 1/[(s-a)s]$. Find $h(t)$.

Solution. $1/(s-a)$ has the inverse $f(t) = e^{at}$, and $1/s$ has the inverse $g(t) = 1$. With $f(\tau) = e^{a\tau}$ and $g(t-\tau) \equiv 1$ we thus obtain from (1) the answer

$$h(t) = e^{at} * 1 = \int_0^t e^{a\tau} \cdot 1\,d\tau = \frac{1}{a}(e^{at} - 1).$$

To check, calculate

$$H(s) = \mathcal{L}(h)(s) = \frac{1}{a}\left(\frac{1}{s-a} - \frac{1}{s}\right) = \frac{1}{a} \cdot \frac{a}{s^2 - as} = \frac{1}{s-a} \cdot \frac{1}{s} = \mathcal{L}(e^{at})\mathcal{L}(1). \qquad \blacksquare$$

EXAMPLE 2 **Convolution**

Let $H(s) = 1/(s^2 + \omega^2)^2$. Find $h(t)$.

Solution. The inverse of $1/(s^2 + \omega^2)$ is $(\sin \omega t)/\omega$. Hence from (1) and the first formula in (11) in App. 3.1 we obtain

$$\begin{aligned} h(t) = \frac{\sin \omega t}{\omega} * \frac{\sin \omega t}{\omega} &= \frac{1}{\omega^2}\int_0^t \sin \omega\tau \sin \omega(t-\tau)\,d\tau \\ &= \frac{1}{2\omega^2}\int_0^t [-\cos \omega t + \cos(2\omega\tau - \omega t)]\,d\tau \end{aligned}$$

$$= \frac{1}{2\omega^2}\left[-\tau\cos\omega t + \frac{\sin\omega\tau}{\omega}\right]_{\tau=0}^{t}$$

$$= \frac{1}{2\omega^2}\left[-t\cos\omega t + \frac{\sin\omega t}{\omega}\right]$$

in agreement with formula 21 in the table in Sec. 6.9. ■

PROOF We prove the Convolution Theorem 1. CAUTION! Note which ones are the variables of integration! We can denote them as we want, for instance, by τ and p, and write

$$F(s) = \int_0^\infty e^{-s\tau} f(\tau)\, d\tau \qquad \text{and} \qquad G(s) = \int_0^\infty e^{-sp} g(p)\, dp.$$

We now set $t = p + \tau$, where τ is at first constant. Then $p = t - \tau$, and t varies from τ to ∞. Thus

$$G(s) = \int_\tau^\infty e^{-s(t-\tau)} g(t-\tau)\, dt = e^{s\tau}\int_\tau^\infty e^{-st} g(t-\tau)\, dt.$$

τ in F and t in G vary independently. Hence we can insert the G-integral into the F-integral. Cancellation of $e^{-s\tau}$ and $e^{s\tau}$ then gives

$$F(s)G(s) = \int_0^\infty e^{-s\tau} f(\tau) e^{s\tau} \int_\tau^\infty e^{-st} g(t-\tau)\, dt\, d\tau = \int_0^\infty f(\tau) \int_\tau^\infty e^{-st} g(t-\tau)\, dt\, d\tau.$$

Here we integrate for fixed τ over t from τ to ∞ and then over τ from 0 to ∞. This is the blue region in Fig. 141. Under the assumption on f and g the order of integration can be reversed (see Ref. [A5] for a proof using uniform convergence). We then integrate first over τ from 0 to t and then over t from 0 to ∞, that is,

$$F(s)G(s) = \int_0^\infty e^{-st} \int_0^t f(\tau) g(t-\tau)\, d\tau\, dt = \int_0^\infty e^{-st} h(t)\, dt = \mathcal{L}(h) = H(s).$$

This completes the proof. ■

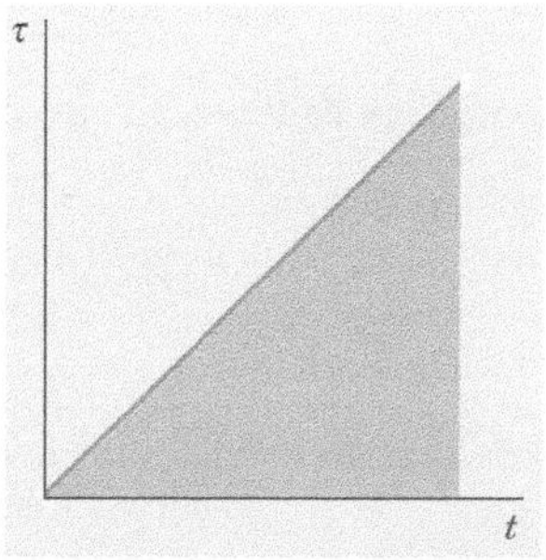

Fig. 141. Region of integration in the $t\tau$-plane in the proof of Theorem 1

From the definition it follows almost immediately that convolution has the properties

$$f * g = g * f \qquad \text{(commutative law)}$$

$$f * (g_1 + g_2) = f * g_1 + f * g_2 \qquad \text{(distributive law)}$$

$$(f * g) * v = f * (g * v) \qquad \text{(associative law)}$$

$$f * 0 = 0 * f = 0$$

similar to those of the multiplication of numbers. However, there are differences of which you should be aware.

EXAMPLE 3 Unusual Properties of Convolution

$f * 1 \neq f$ in general. For instance,

$$t * 1 = \int_0^t \tau \cdot 1 \, d\tau = \frac{1}{2} t^2 \neq t.$$

$(f * f)(t) \geqq 0$ may not hold. For instance, Example 2 with $\omega = 1$ gives

$$\sin t * \sin t = -\tfrac{1}{2} t \cos t + \tfrac{1}{2} \sin t \qquad \text{(Fig. 142).} \quad ■$$

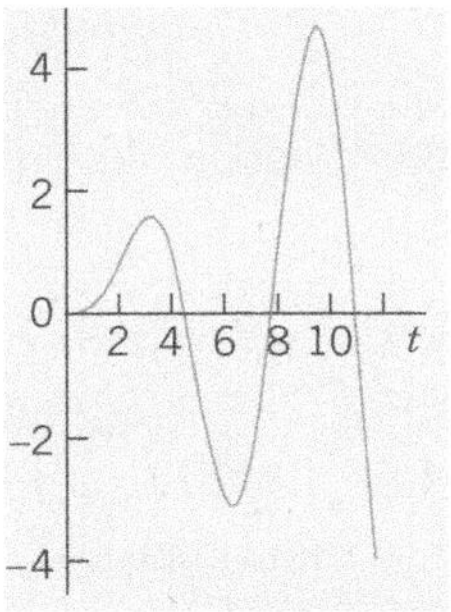

Fig. 142. Example 3

We shall now take up the case of a complex double root (left aside in the last section in connection with partial fractions) and find the solution (the inverse transform) directly by convolution.

EXAMPLE 4 Repeated Complex Factors. Resonance

In an undamped mass–spring system, resonance occurs if the frequency of the driving force equals the natural frequency of the system. Then the model is (see Sec. 2.8)

$$y'' + \omega_0^2 y = K \sin \omega_0 t$$

where $\omega_0^2 = k/m$, k is the spring constant, and m is the mass of the body attached to the spring. We assume $y(0) = 0$ and $y'(0) = 0$, for simplicity. Then the subsidiary equation is

$$s^2 Y + \omega_0^2 Y = \frac{K\omega_0}{s^2 + \omega_0^2}. \qquad \text{Its solution is} \qquad Y = \frac{K\omega_0}{(s^2 + \omega_0^2)^2}.$$

This is a transform as in Example 2 with $\omega = \omega_0$ and multiplied by $K\omega_0$. Hence from Example 2 we can see directly that the solution of our problem is

$$y(t) = \frac{K\omega_0}{2\omega_0^2}\left(-t\cos\omega_0 t + \frac{\sin\omega_0 t}{\omega_0}\right) = \frac{K}{2\omega_0^2}(-\omega_0 t\cos\omega_0 t + \sin\omega_0 t).$$

We see that the first term grows without bound. Clearly, in the case of resonance such a term must occur. (See also a similar kind of solution in Fig. 55 in Sec. 2.8.) ■

Application to Nonhomogeneous Linear ODEs

Nonhomogeneous linear ODEs can now be solved by a general method based on convolution by which the solution is obtained in the form of an integral. To see this, recall from Sec. 6.2 that the subsidiary equation of the ODE

$$y'' + ay' + by = r(t) \qquad (a, b \text{ constant}) \tag{2}$$

has the solution [(7) in Sec. 6.2]

$$Y(s) = [(s + a)y(0) + y'(0)]Q(s) + R(s)Q(s)$$

with $R(s) = \mathcal{L}(r)$ and $Q(s) = 1/(s^2 + as + b)$ the transfer function. Inversion of the first term $[\cdots]$ provides no difficulty; depending on whether $\frac{1}{4}a^2 - b$ is positive, zero, or negative, its inverse will be a linear combination of two exponential functions, or of the form $(c_1 + c_2t)e^{-at/2}$, or a damped oscillation, respectively. The interesting term is $R(s)Q(s)$ because $r(t)$ can have various forms of practical importance, as we shall see. If $y(0) = 0$ and $y'(0) = 0$, then $Y = RQ$, and the convolution theorem gives the solution

$$y(t) = \int_0^t q(t - \tau)r(\tau)\,d\tau. \tag{3}$$

EXAMPLE 5 **Response of a Damped Vibrating System to a Single Square Wave**

Using convolution, determine the response of the damped mass–spring system modeled by

$$y'' + 3y' + 2y = r(t), \qquad r(t) = 1 \text{ if } 1 < t < 2 \text{ and } 0 \text{ otherwise}, \qquad y(0) = y'(0) = 0.$$

This system with an **input** (a driving force) ***that acts for some time only*** (Fig. 143) has been solved by partial fraction reduction in Sec. 6.4 (Example 1).

Solution by Convolution. The transfer function and its inverse are

$$Q(s) = \frac{1}{s^2 + 3s + 2} = \frac{1}{(s + 1)(s + 2)} = \frac{1}{s + 1} - \frac{1}{s + 2}, \qquad \text{hence} \qquad q(t) = e^{-t} - e^{-2t}.$$

Hence the convolution integral (3) is (except for the limits of integration)

$$y(t) = \int q(t - \tau)\cdot 1\,d\tau = \int \left[e^{-(t-\tau)} - e^{-2(t-\tau)}\right] d\tau = e^{-(t-\tau)} - \frac{1}{2}e^{-2(t-\tau)}.$$

Now comes an important point in handling convolution. $r(\tau) = 1$ if $1 < \tau < 2$ only. Hence if $t < 1$, the integral is zero. If $1 < t < 2$, we have to integrate from $\tau = 1$ (not 0) to t. This gives (with the first two terms from the upper limit)

$$y(t) = e^{-0} - \tfrac{1}{2}e^{-0} - (e^{-(t-1)} - \tfrac{1}{2}e^{-2(t-1)}) = \tfrac{1}{2} - e^{-(t-1)} + \tfrac{1}{2}e^{-2(t-1)}.$$

If $t > 2$, we have to integrate from $\tau = 1$ to 2 (not to t). This gives

$$y(t) = e^{-(t-2)} - \tfrac{1}{2}e^{-2(t-2)} - (e^{-(t-1)} - \tfrac{1}{2}e^{-2(t-1)}).$$

Figure 143 shows the input (the square wave) and the interesting output, which is zero from 0 to 1, then increases, reaches a maximum (near 2.6) after the input has become zero (why?), and finally decreases to zero in a monotone fashion. ■

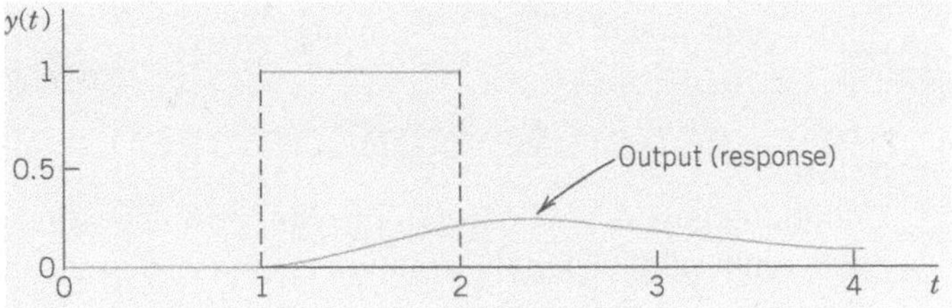

Fig. 143. Square wave and response in Example 5

Integral Equations

Convolution also helps in solving certain **integral equations**, that is, equations in which the unknown function $y(t)$ appears in an integral (and perhaps also outside of it). This concerns equations with an integral of the form of a convolution. Hence these are special and it suffices to explain the idea in terms of two examples and add a few problems in the problem set.

EXAMPLE 6 A Volterra Integral Equation of the Second Kind

Solve the Volterra integral equation of the second kind[3]

$$y(t) - \int_0^t y(\tau) \sin(t-\tau)\, d\tau = t.$$

Solution. From (1) we see that the given equation can be written as a convolution, $y - y * \sin t = t$. Writing $Y = \mathcal{L}(y)$ and applying the convolution theorem, we obtain

$$Y(s) - Y(s)\frac{1}{s^2+1} = Y(s)\frac{s^2}{s^2+1} = \frac{1}{s^2}.$$

The solution is

$$Y(s) = \frac{s^2+1}{s^4} = \frac{1}{s^2} + \frac{1}{s^4} \quad \text{and gives the answer} \quad y(t) = t + \frac{t^3}{6}.$$

Check the result by a CAS or by substitution and repeated integration by parts (which will need patience). ■

EXAMPLE 7 Another Volterra Integral Equation of the Second Kind

Solve the Volterra integral equation

$$y(t) - \int_0^t (1+\tau)y(t-\tau)\, d\tau = 1 - \sinh t.$$

---

[3]If the upper limit of integration is *variable*, the equation is named after the Italian mathematician VITO VOLTERRA (1860–1940), and if that limit is *constant*, the equation is named after the Swedish mathematician ERIK IVAR FREDHOLM (1866–1927). "Of the second kind (first kind)" indicates that y occurs (does not occur) outside of the integral.

Solution. By (1) we can write $y - (1 + t) * y = 1 - \sinh t$. Writing $Y = \mathcal{L}(y)$, we obtain by using the convolution theorem and then taking common denominators

$$Y(s)\left[1 - \left(\frac{1}{s} + \frac{1}{s^2}\right)\right] = \frac{1}{s} - \frac{1}{s^2 - 1}, \qquad \text{hence} \qquad Y(s) \cdot \frac{s^2 - s - 1}{s^2} = \frac{s^2 - 1 - s}{s(s^2 - 1)}.$$

$(s^2 - s - 1)/s$ cancels on both sides, so that solving for Y simply gives

$$Y(s) = \frac{s}{s^2 - 1} \qquad \text{and the solution is} \qquad y(t) = \cosh t. \qquad \blacksquare$$

PROBLEM SET 6.5

1–7 CONVOLUTIONS BY INTEGRATION

Find:

1. $1 * 1$
2. $1 * \sin \omega t$
3. $e^t * e^{-t}$
4. $(\cos \omega t) * (\cos \omega t)$
5. $(\sin \omega t) * (\cos \omega t)$
6. $e^{at} * e^{bt} \; (a \neq b)$
7. $t * e^t$

8–14 INTEGRAL EQUATIONS

Solve by the Laplace transform, showing the details:

8. $y(t) + 4\int_0^t y(\tau)(t - \tau)\, d\tau = 2t$
9. $y(t) - \int_0^t y(\tau)\, d\tau = 1$
10. $y(t) - \int_0^t y(\tau) \sin 2(t - \tau)\, d\tau = \sin 2t$
11. $y(t) + \int_0^t (t - \tau)y(\tau)\, d\tau = 1$
12. $y(t) + \int_0^t y(\tau) \cosh (t - \tau)\, d\tau = t + e^t$
13. $y(t) + 2e^t \int_0^t y(\tau)e^{-\tau}\, d\tau = te^t$
14. $y(t) - \int_0^t y(\tau)(t - \tau)\, d\tau = 2 - \frac{1}{2}t^2$

15. **CAS EXPERIMENT. Variation of a Parameter.**
(a) Replace 2 in Prob. 13 by a parameter k and investigate graphically how the solution curve changes if you vary k, in particular near $k = -2$.
(b) Make similar experiments with an integral equation of your choice whose solution is oscillating.

16. **TEAM PROJECT. Properties of Convolution.** Prove:
(a) Commutativity, $f * g = g * f$
(b) Associativity, $(f * g) * v = f * (g * v)$
(c) Distributivity, $f * (g_1 + g_2) = f * g_1 + f * g_2$
(d) Dirac's delta. Derive the sifting formula (4) in Sec. 6.4 by using f_k with $a = 0$ [(1), Sec. 6.4] and applying the mean value theorem for integrals.
(e) Unspecified driving force. Show that forced vibrations governed by

$$y'' + \omega^2 y = r(t), \quad y(0) = K_1, \quad y'(0) = K_2$$

with $\omega \neq 0$ and an unspecified driving force $r(t)$ can be written in convolution form,

$$y = \frac{1}{\omega} \sin \omega t * r(t) + K_1 \cos \omega t + \frac{K_2}{\omega} \sin \omega t.$$

17–26 INVERSE TRANSFORMS BY CONVOLUTION

Showing details, find $f(t)$ if $\mathcal{L}(f)$ equals:

17. $\dfrac{5.5}{(s + 1.5)(s - 4)}$
18. $\dfrac{1}{(s - a)^2}$
19. $\dfrac{2\pi s}{(s^2 + \pi^2)^2}$
20. $\dfrac{9}{s(s + 3)}$
21. $\dfrac{\omega}{s^2(s^2 + \omega^2)}$
22. $\dfrac{e^{-as}}{s(s - 2)}$
23. $\dfrac{40.5}{s(s^2 - 9)}$
24. $\dfrac{240}{(s^2 + 1)(s^2 + 25)}$
25. $\dfrac{18s}{(s^2 + 36)^2}$
26. **Partial Fractions.** Solve Probs. 17, 21, and 23 by partial fraction reduction.

6.6 Differentiation and Integration of Transforms. ODEs with Variable Coefficients

The variety of methods for obtaining transforms and inverse transforms and their application in solving ODEs is surprisingly large. We have seen that they include direct integration, the use of linearity (Sec. 6.1), shifting (Secs. 6.1, 6.3), convolution (Sec. 6.5), and differentiation and integration of functions $f(t)$ (Sec. 6.2). In this section, we shall consider operations of somewhat lesser importance. They are the differentiation and integration of transforms $F(s)$ and corresponding operations for functions $f(t)$. We show how they are applied to ODEs with variable coefficients.

Differentiation of Transforms

It can be shown that, if a function $f(t)$ satisfies the conditions of the existence theorem in Sec. 6.1, then the derivative $F'(s) = dF/ds$ of the transform $F(s) = \mathcal{L}(f)$ can be obtained by differentiating $F(s)$ under the integral sign with respect to s (proof in Ref. [GenRef4] listed in App. 1). Thus, if

$$F(s) = \int_0^\infty e^{-st} f(t)\, dt, \qquad \text{then} \qquad F'(s) = -\int_0^\infty e^{-st} t f(t)\, dt.$$

Consequently, if $\mathcal{L}(f) = F(s)$, then

$$\text{(1)} \qquad \mathcal{L}\{tf(t)\} = -F'(s), \qquad \text{hence} \qquad \mathcal{L}^{-1}\{F'(s)\} = -tf(t)$$

where the second formula is obtained by applying $\mathcal{L}^{-1}$ on both sides of the first formula. In this way, *differentiation of the transform of a function corresponds to the multiplication of the function by* $-t$.

EXAMPLE 1 Differentiation of Transforms. Formulas 21–23 in Sec. 6.9

We shall derive the following three formulas.

	$\mathcal{L}(f)$	$f(t)$
(2)	$\dfrac{1}{(s^2+\beta^2)^2}$	$\dfrac{1}{2\beta^3}(\sin \beta t - \beta t \cos \beta t)$
(3)	$\dfrac{s}{(s^2+\beta^2)^2}$	$\dfrac{1}{2\beta}\sin \beta t$
(4)	$\dfrac{s^2}{(s^2+\beta^2)^2}$	$\dfrac{1}{2\beta}(\sin \beta t + \beta t \cos \beta t)$

Solution. From (1) and formula 8 (with $\omega = \beta$) in Table 6.1 of Sec. 6.1 we obtain by differentiation (CAUTION! Chain rule!)

$$\mathcal{L}(t \sin \beta t) = \frac{2\beta s}{(s^2+\beta^2)^2}.$$

Dividing by 2β and using the linearity of $\mathcal{L}$, we obtain (3).

Formulas (2) and (4) are obtained as follows. From (1) and formula 7 (with $\omega = \beta$) in Table 6.1 we find

$$\mathcal{L}(t\cos\beta t) = -\frac{(s^2+\beta^2)-2s^2}{(s^2+\beta^2)^2} = \frac{s^2-\beta^2}{(s^2+\beta^2)^2}. \tag{5}$$

From this and formula 8 (with $\omega = \beta$) in Table 6.1 we have

$$\mathcal{L}\left(t\cos\beta t \pm \frac{1}{\beta}\sin\beta t\right) = \frac{s^2-\beta^2}{(s^2+\beta^2)^2} \pm \frac{1}{s^2+\beta^2}.$$

On the right we now take the common denominator. Then we see that for the plus sign the numerator becomes $s^2 - \beta^2 + s^2 + \beta^2 = 2s^2$, so that (4) follows by division by 2. Similarly, for the minus sign the numerator takes the form $s^2 - \beta^2 - s^2 - \beta^2 = -2\beta^2$, and we obtain (2). This agrees with Example 2 in Sec. 6.5. ■

Integration of Transforms

Similarly, if $f(t)$ satisfies the conditions of the existence theorem in Sec. 6.1 and the limit of $f(t)/t$, as t approaches 0 from the right, exists, then for $s > k$,

$$\mathcal{L}\left\{\frac{f(t)}{t}\right\} = \int_s^\infty F(\tilde{s})\,d\tilde{s} \qquad \text{hence} \qquad \mathcal{L}^{-1}\left\{\int_s^\infty F(\tilde{s})\,d\tilde{s}\right\} = \frac{f(t)}{t}. \tag{6}$$

In this way, *integration of the transform of a function $f(t)$ corresponds to the division of $f(t)$ by t.*

We indicate how (6) is obtained. From the definition it follows that

$$\int_s^\infty F(\tilde{s})\,d\tilde{s} = \int_s^\infty \left[\int_0^\infty e^{-\tilde{s}t} f(t)\,dt\right] d\tilde{s},$$

and it can be shown (see Ref. [GenRef4] in App. 1) that under the above assumptions we may reverse the order of integration, that is,

$$\int_s^\infty F(\tilde{s})\,d\tilde{s} = \int_0^\infty \left[\int_s^\infty e^{-\tilde{s}t} f(t)\,d\tilde{s}\right] dt = \int_0^\infty f(t)\left[\int_s^\infty e^{-\tilde{s}t}\,d\tilde{s}\right] dt.$$

Integration of $e^{-\tilde{s}t}$ with respect to $\tilde{s}$ gives $e^{-\tilde{s}t}/(-t)$. Here the integral over $\tilde{s}$ on the right equals e^{-st}/t. Therefore,

$$\int_s^\infty F(\tilde{s})\,d\tilde{s} = \int_0^\infty e^{-st}\frac{f(t)}{t}\,dt = \mathcal{L}\left\{\frac{f(t)}{t}\right\} \qquad (s > k). \quad ■$$

EXAMPLE 2 Differentiation and Integration of Transforms

Find the inverse transform of $\ln\left(1 + \dfrac{\omega^2}{s^2}\right) = \ln\dfrac{s^2+\omega^2}{s^2}$.

Solution. Denote the given transform by $F(s)$. Its derivative is

$$F'(s) = \frac{d}{ds}(\ln(s^2+\omega^2) - \ln s^2) = \frac{2s}{s^2+\omega^2} - \frac{2s}{s^2}.$$

Taking the inverse transform and using (1), we obtain

$$\mathcal{L}^{-}\{F'(s)\} = \mathcal{L}^{-1}\left\{\frac{2s}{s^2+\omega^2} - \frac{2}{s}\right\} = 2\cos\omega t - 2 = -tf(t).$$

Hence the inverse $f(t)$ of $F(s)$ is $f(t) = 2(1 - \cos\omega t)/t$. This agrees with formula 42 in Sec. 6.9.

Alternatively, if we let

$$G(s) = \frac{2s}{s^2+\omega^2} - \frac{2}{s}, \qquad \text{then} \qquad g(t) = \mathcal{L}^{-1}(G) - 2(\cos\omega t - 1).$$

From this and (6) we get, in agreement with the answer just obtained,

$$\mathcal{L}^{-1}\left\{\ln\frac{s^2+\omega^2}{s^2}\right\} = \mathcal{L}^{-1}\left\{\int_s^{\infty} G(s)\,ds\right\} = -\frac{g(t)}{t} = \frac{2}{t}(1 - \cos\omega t),$$

the minus occurring since s is the lower limit of integration.

In a similar way we obtain formula 43 in Sec. 6.9,

$$\mathcal{L}^{-1}\left\{\ln\left(1 - \frac{a^2}{s^2}\right)\right\} = \frac{2}{t}(1 - \cosh at).$$ ■

Special Linear ODEs with Variable Coefficients

Formula (1) can be used to solve certain ODEs with variable coefficients. The idea is this. Let $\mathcal{L}(y) = Y$. Then $\mathcal{L}(y') = sY - y(0)$ (see Sec. 6.2). Hence by (1),

$$\mathcal{L}(ty') = -\frac{d}{ds}[sY - y(0)] = -Y - s\frac{dY}{ds}. \tag{7}$$

Similarly, $\mathcal{L}(y'') = s^2Y - sy(0) - y'(0)$ and by (1)

$$\mathcal{L}(ty'') = -\frac{d}{ds}[s^2Y - sy(0) - y'(0)] = -2sY - s^2\frac{dY}{ds} + y(0). \tag{8}$$

Hence if an ODE has coefficients such as $at + b$, the subsidiary equation is a first-order ODE for Y, which is sometimes simpler than the given second-order ODE. But if the latter has coefficients $at^2 + bt + c$, then two applications of (1) would give a second-order ODE for Y, and this shows that the present method works well only for rather special ODEs with variable coefficients. An important ODE for which the method is advantageous is the following.

EXAMPLE 3 Laguerre's Equation. Laguerre Polynomials

Laguerre's ODE is

$$ty'' + (1 - t)y' + ny = 0. \tag{9}$$

We determine a solution of (9) with $n = 0, 1, 2, \cdots$. From (7)–(9) we get the subsidiary equation

$$\left[-2sY - s^2\frac{dY}{ds} + y(0)\right] + sY - y(0) - \left(-Y - s\frac{dY}{ds}\right) + nY = 0.$$

Simplification gives

$$(s - s^2)\frac{dY}{ds} + (n + 1 - s)Y = 0.$$

Separating variables, using partial fractions, integrating (with the constant of integration taken to be zero), and taking exponentials, we get

$$(10^*) \qquad \frac{dY}{Y} = -\frac{n + 1 - s}{s - s^2}ds = \left(\frac{n}{s - 1} - \frac{n + 1}{s}\right)ds \qquad \text{and} \qquad Y = \frac{(s - 1)^n}{s^{n+1}}.$$

We write $l_n = \mathcal{L}^{-1}(Y)$ and prove **Rodrigues's formula**

$$(10) \qquad l_0 = 1, \qquad l_n(t) = \frac{e^t}{n!}\frac{d^n}{dt^n}(t^n e^{-t}), \qquad n = 1, 2, \cdots.$$

These are polynomials because the exponential terms cancel if we perform the indicated differentiations. They are called **Laguerre polynomials** and are usually denoted by L_n (see Problem Set 5.7, but we continue to reserve capital letters for transforms). We prove (10). By Table 6.1 and the first shifting theorem (s-shifting),

$$\mathcal{L}(t^n e^{-t}) = \frac{n!}{(s + 1)^{n+1}}, \qquad \text{hence by (3) in Sec. 6.2} \qquad \mathcal{L}\left\{\frac{d^n}{dt^n}(t^n e^{-t})\right\} = \frac{n!s^n}{(s + 1)^{n+1}}$$

because the derivatives up to the order $n - 1$ are zero at 0. Now make another shift and divide by $n!$ to get [see (10) and then (10*)]

$$\mathcal{L}(l_n) = \frac{(s - 1)^n}{s^{n+1}} = Y. \qquad ■$$

PROBLEM SET 6.6

1. **REVIEW REPORT. Differentiation and Integration of Functions and Transforms.** Make a draft of these four operations from memory. Then compare your draft with the text and write a 2- to 3-page report on these operations and their significance in applications.

2–11 TRANSFORMS BY DIFFERENTIATION

Showing the details of your work, find $\mathcal{L}(f)$ if $f(t)$ equals:

2. $3t \sinh 4t$
3. $\frac{1}{2}te^{-3t}$
4. $te^{-t} \cos t$
5. $t \cos \omega t$
6. $t^2 \sin 3t$
7. $t^2 \cosh 2t$
8. $te^{-kt} \sin t$
9. $\frac{1}{2}t^2 \sin \pi t$
10. $t^n e^{kt}$
11. $4t \cos \frac{1}{2}\pi t$
12. **CAS PROJECT. Laguerre Polynomials. (a)** Write a CAS program for finding $l_n(t)$ in explicit form from (10). Apply it to calculate $l_0, \cdots, l_{10}$. Verify that $l_0, \cdots, l_{10}$ satisfy Laguerre's differential equation (9).

(b) Show that

$$l_n(t) = \sum_{m=0}^{n} \frac{(-1)^m}{m!}\binom{n}{m}t^m$$

and calculate $l_0, \cdots, l_{10}$ from this formula.

(c) Calculate $l_0, \cdots, l_{10}$ recursively from $l_0 = 1$, $l_1 = 1 - t$ by

$$(n + 1)l_{n+1} = (2n + 1 - t)l_n - nl_{n-1}.$$

(d) A **generating function** (definition in Problem Set 5.2) for the Laguerre polynomials is

$$\sum_{n=0}^{\infty} l_n(t)x^n = (1 - x)^{-1}e^{tx/(x-1)}.$$

Obtain $l_0, \cdots, l_{10}$ from the corresponding partial sum of this power series in x and compare the l_n with those in (a), (b), or (c).

13. **CAS EXPERIMENT. Laguerre Polynomials.** Experiment with the graphs of $l_0, \cdots, l_{10}$, finding out empirically how the first maximum, first minimum, $\cdots$ is moving with respect to its location as a function of n. Write a short report on this.

14–20 INVERSE TRANSFORMS

Using differentiation, integration, s-shifting, or convolution, and showing the details, find $f(t)$ if $\mathcal{L}(f)$ equals:

14. $\dfrac{s}{(s^2+16)^2}$

15. $\dfrac{s}{(s^2-9)^2}$

16. $\dfrac{2s+6}{(s^2+6s+10)^2}$

17. $\ln \dfrac{s}{s-1}$

18. $\operatorname{arccot} \dfrac{s}{\pi}$

19. $\ln \dfrac{s^2+1}{(s-1)^2}$

20. $\ln \dfrac{s+a}{s+b}$

6.7 Systems of ODEs

The Laplace transform method may also be used for solving systems of ODEs, as we shall explain in terms of typical applications. We consider a first-order linear system with constant coefficients (as discussed in Sec. 4.1)

$$\begin{aligned} y_1' &= a_{11}y_1 + a_{12}y_2 + g_1(t) \\ y_2' &= a_{21}y_1 + a_{22}y_2 + g_2(t). \end{aligned} \tag{1}$$

Writing $Y_1 = \mathcal{L}(y_1)$, $Y_2 = \mathcal{L}(y_2)$, $G_1 = \mathcal{L}(g_1)$, $G_2 = \mathcal{L}(g_2)$, we obtain from (1) in Sec. 6.2 the subsidiary system

$$\begin{aligned} sY_1 - y_1(0) &= a_{11}Y_1 + a_{12}Y_2 + G_1(s) \\ sY_2 - y_2(0) &= a_{21}Y_1 + a_{22}Y_2 + G_2(s). \end{aligned}$$

By collecting the Y_1- and Y_2-terms we have

$$\begin{aligned} (a_{11}-s)Y_1 + \quad a_{12}Y_2 \quad &= -y_1(0) - G_1(s) \\ a_{21}Y_1 \quad + (a_{22}-s)Y_2 &= -y_2(0) - G_2(s). \end{aligned} \tag{2}$$

By solving this system algebraically for $Y_1(s), Y_2(s)$ and taking the inverse transform we obtain the solution $y_1 = \mathcal{L}^{-1}(Y_1)$, $y_2 = \mathcal{L}^{-1}(Y_2)$ of the given system (1).

Note that (1) and (2) may be written in vector form (and similarly for the systems in the examples); thus, setting $\mathbf{y} = [y_1 \quad y_2]^\mathsf{T}$, $\mathbf{A} = [a_{jk}]$, $\mathbf{g} = [g_1 \quad g_2]^\mathsf{T}$, $\mathbf{Y} = [Y_1 \quad Y_2]^\mathsf{T}$, $\mathbf{G} = [G_1 \quad G_2]^\mathsf{T}$ we have

$$\mathbf{y}' = \mathbf{A}\mathbf{y} + \mathbf{g} \qquad \text{and} \qquad (\mathbf{A} - s\mathbf{I})\mathbf{Y} = -\mathbf{y}(0) - \mathbf{G}.$$

EXAMPLE 1 Mixing Problem Involving Two Tanks

Tank T_1 in Fig. 144 initially contains 100 gal of pure water. Tank T_2 initially contains 100 gal of water in which 150 lb of salt are dissolved. The inflow into T_1 is 2 gal/min from T_2 and 6 gal/min containing 6 lb of salt from the outside. The inflow into T_2 is 8 gal/min from T_1. The outflow from T_2 is $2 + 6 = 8$ gal/min, as shown in the figure. The mixtures are kept uniform by stirring. Find and plot the salt contents $y_1(t)$ and $y_2(t)$ in T_1 and T_2, respectively.

Solution. The model is obtained in the form of two equations

$$\text{Time rate of change} = \text{Inflow/min} - \text{Outflow/min}$$

for the two tanks (see Sec. 4.1). Thus,

$$y_1' = -\tfrac{8}{100}y_1 + \tfrac{2}{100}y_2 + 6. \qquad y_2' = \tfrac{8}{100}y_1 - \tfrac{8}{100}y_2.$$

The initial conditions are $y_1(0) = 0$, $y_2(0) = 150$. From this we see that the subsidiary system (2) is

$$\begin{aligned}(-0.08 - s)Y_1 + \quad 0.02Y_2 \quad &= -\frac{6}{s}\\ 0.08Y_1 \quad + (-0.08 - s)Y_2 &= -150.\end{aligned}$$

We solve this algebraically for Y_1 and Y_2 by elimination (or by Cramer's rule in Sec. 7.7), and we write the solutions in terms of partial fractions,

$$\begin{aligned}Y_1 &= \frac{9s + 0.48}{s(s + 0.12)(s + 0.04)} = \frac{100}{s} - \frac{62.5}{s + 0.12} - \frac{37.5}{s + 0.04}\\ Y_2 &= \frac{150s^2 + 12s + 0.48}{s(s + 0.12)(s + 0.04)} = \frac{100}{s} + \frac{125}{s + 0.12} - \frac{75}{s + 0.04}.\end{aligned}$$

By taking the inverse transform we arrive at the solution

$$\begin{aligned}y_1 &= 100 - 62.5e^{-0.12t} - 37.5e^{-0.04t}\\ y_2 &= 100 + 125e^{-0.12t} - 75e^{-0.04t}.\end{aligned}$$

Figure 144 shows the interesting plot of these functions. Can you give physical explanations for their main features? Why do they have the limit 100? Why is y_2 not monotone, whereas y_1 is? Why is y_1 from some time on suddenly larger than y_2? Etc. ■

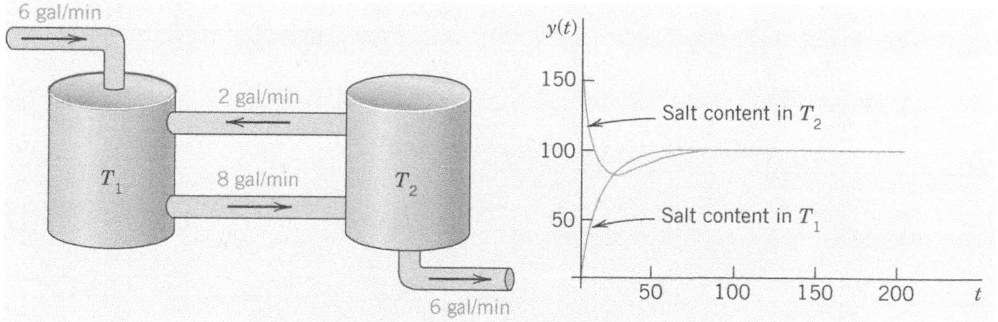

Fig. 144. Mixing problem in Example 1

Other systems of ODEs of practical importance can be solved by the Laplace transform method in a similar way, and eigenvalues and eigenvectors, as we had to determine them in Chap. 4, will come out automatically, as we have seen in Example 1.

EXAMPLE 2 Electrical Network

Find the currents $i_1(t)$ and $i_2(t)$ in the network in Fig. 145 with L and R measured in terms of the usual units (see Sec. 2.9), $v(t) = 100$ volts if $0 \leqq t \leqq 0.5$ sec and 0 thereafter, and $i(0) = 0$, $i'(0) = 0$.

Solution. The model of the network is obtained from Kirchhoff's Voltage Law as in Sec. 2.9. For the lower circuit we obtain

$$0.8i_1' + 1(i_1 - i_2) + 1.4i_1 = 100[1 - u(t - \tfrac{1}{2})]$$

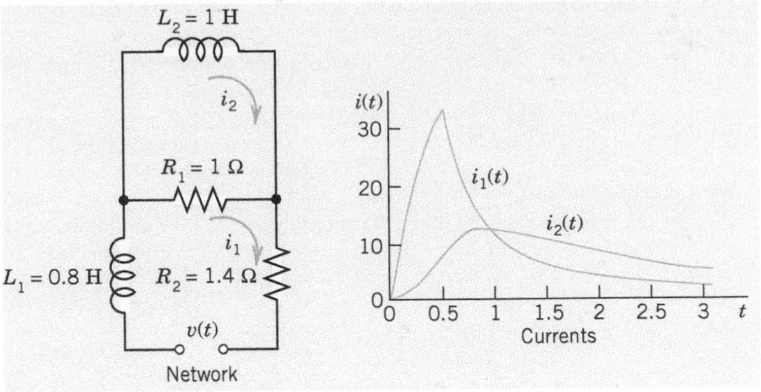

Fig. 145. Electrical network in Example 2

and for the upper

$$1 \cdot i_2' + 1(i_2 - i_1) \qquad = 0.$$

Division by 0.8 and ordering gives for the lower circuit

$$i_1' + 3i_1 - 1.25i_2 = 125[1 - u(t - \tfrac{1}{2})]$$

and for the upper

$$i_2' - i_1 + \qquad i_2 = 0.$$

With $i_1(0) = 0$, $i_2(0) = 0$ we obtain from (1) in Sec. 6.2 and the second shifting theorem the subsidiary system

$$(s + 3)I_1 - 1.25I_2 = 125\left(\frac{1}{s} - \frac{e^{-s/2}}{s}\right)$$
$$-I_1 + (s + 1)I_2 = 0.$$

Solving algebraically for I_1 and I_2 gives

$$I_1 = \frac{125(s + 1)}{s(s + \frac{1}{2})(s + \frac{7}{2})}(1 - e^{-s/2}),$$

$$I_2 = \frac{125}{s(s + \frac{1}{2})(s + \frac{7}{2})}(1 - e^{-s/2}).$$

The right sides, without the factor $1 - e^{-s/2}$, have the partial fraction expansions

$$\frac{500}{7s} - \frac{125}{3(s + \frac{1}{2})} - \frac{625}{21(s + \frac{7}{2})}$$

and

$$\frac{500}{7s} - \frac{250}{3(s + \frac{1}{2})} + \frac{250}{21(s + \frac{7}{2})},$$

respectively. The inverse transform of this gives the solution for $0 \leqq t \leqq \frac{1}{2}$,

$$i_1(t) = -\tfrac{125}{3}e^{-t/2} - \tfrac{625}{21}e^{-7t/2} + \tfrac{500}{7}$$
$$i_2(t) = -\tfrac{250}{3}e^{-t/2} + \tfrac{250}{21}e^{-7t/2} + \tfrac{500}{7} \qquad (0 \leqq t \leqq \tfrac{1}{2}).$$

According to the second shifting theorem the solution for $t > \frac{1}{2}$ is $i_1(t) - i_1(t - \frac{1}{2})$ and $i_2(t) - i_2(t - \frac{1}{2})$, that is,

$$\begin{aligned} i_1(t) &= -\tfrac{125}{3}(1 - e^{1/4})e^{-t/2} - \tfrac{625}{21}(1 - e^{7/4})e^{-7t/2} \\ i_2(t) &= -\tfrac{250}{3}(1 - e^{1/4})e^{-t/2} + \tfrac{250}{21}(1 - e^{7/4})e^{-7t/2} \end{aligned} \qquad (t > \tfrac{1}{2}).$$

Can you explain physically why both currents eventually go to zero, and why $i_1(t)$ has a sharp cusp whereas $i_2(t)$ has a continuous tangent direction at $t = \frac{1}{2}$? ■

Systems of ODEs of higher order can be solved by the Laplace transform method in a similar fashion. As an important application, typical of many similar mechanical systems, we consider coupled vibrating masses on springs.

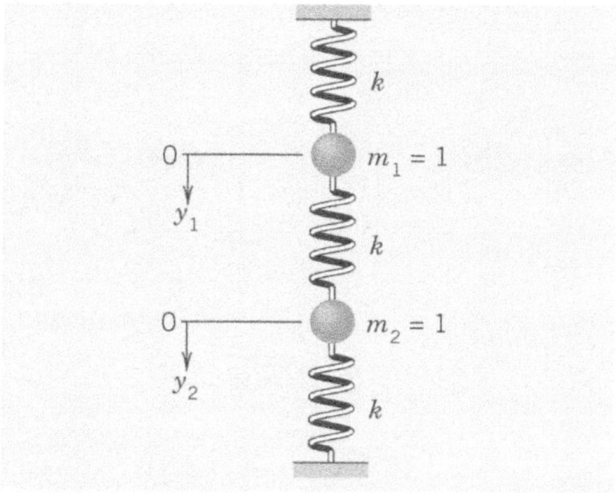

Fig. 146. Example 3

EXAMPLE 3 Model of Two Masses on Springs (Fig. 146)

The mechanical system in Fig. 146 consists of two bodies of mass 1 on three springs of the same spring constant k and of negligibly small masses of the springs. Also damping is assumed to be practically zero. Then the model of the physical system is the system of ODEs

$$\begin{aligned} y_1'' &= -ky_1 + k(y_2 - y_1) \\ y_2'' &= -k(y_2 - y_1) - ky_2. \end{aligned} \tag{3}$$

Here y_1 and y_2 are the displacements of the bodies from their positions of static equilibrium. These ODEs follow from **Newton's second law**, *Mass* × *Acceleration* = *Force*, as in Sec. 2.4 for a single body. We again regard downward forces as positive and upward as negative. On the upper body, $-ky_1$ is the force of the upper spring and $k(y_2 - y_1)$ that of the middle spring, $y_2 - y_1$ being the net change in spring length—think this over before going on. On the lower body, $-k(y_2 - y_1)$ is the force of the middle spring and $-ky_2$ that of the lower spring.

We shall determine the solution corresponding to the initial conditions $y_1(0) = 1$, $y_2(0) = 1$, $y_1'(0) = \sqrt{3k}$, $y_2'(0) = -\sqrt{3k}$. Let $Y_1 = \mathcal{L}(y_1)$ and $Y_2 = \mathcal{L}(y_2)$. Then from (2) in Sec. 6.2 and the initial conditions we obtain the subsidiary system

$$\begin{aligned} s^2Y_1 - s - \sqrt{3k} &= -kY_1 + k(Y_2 - Y_1) \\ s^2Y_2 - s + \sqrt{3k} &= -k(Y_2 - Y_1) - kY_2. \end{aligned}$$

This system of linear algebraic equations in the unknowns Y_1 and Y_2 may be written

$$\begin{aligned} (s^2 + 2k)Y_1 - \quad kY_2 \qquad &= s + \sqrt{3k} \\ -ky_1 \quad + (s^2 + 2k)Y_2 &= s - \sqrt{3k}. \end{aligned}$$

Elimination (or Cramer's rule in Sec. 7.7) yields the solution, which we can expand in terms of partial fractions,

$$Y_1 = \frac{(s + \sqrt{3k})(s^2 + 2k) + k(s - \sqrt{3k})}{(s^2 + 2k)^2 - k^2} = \frac{s}{s^2 + k} + \frac{\sqrt{3k}}{s^2 + 3k}$$

$$Y_2 = \frac{(s^2 + 2k)(s - \sqrt{3k}) + k(s + \sqrt{3k})}{(s^2 + 2k)^2 - k^2} = \frac{s}{s^2 + k} - \frac{\sqrt{3k}}{s^2 + 3k}.$$

Hence the solution of our initial value problem is (Fig. 147)

$$y_1(t) = \mathcal{L}^{-1}(Y_1) = \cos \sqrt{k}t + \sin \sqrt{3k}t$$

$$y_2(t) = \mathcal{L}^{-1}(Y_2) = \cos \sqrt{k}t - \sin \sqrt{3k}t.$$

We see that the motion of each mass is harmonic (the system is undamped!), being the superposition of a "slow" oscillation and a "rapid" oscillation. ■

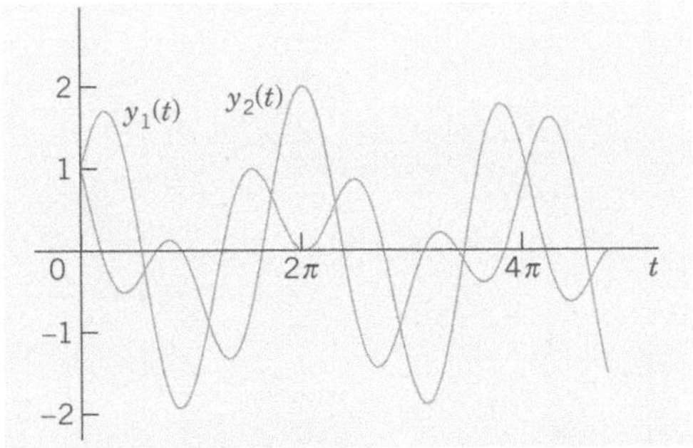

Fig. 147. Solutions in Example 3

PROBLEM SET 6.7

1. TEAM PROJECT. Comparison of Methods for Linear Systems of ODEs

(a) Models. Solve the models in Examples 1 and 2 of Sec. 4.1 by Laplace transforms and compare the amount of work with that in Sec. 4.1. Show the details of your work.

(b) Homogeneous Systems. Solve the systems (8), (11)–(13) in Sec. 4.3 by Laplace transforms. Show the details.

(c) Nonhomogeneous System. Solve the system (3) in Sec. 4.6 by Laplace transforms. Show the details.

2–15 SYSTEMS OF ODES

Using the Laplace transform and showing the details of your work, solve the IVP:

2. $y_1' + y_2 = 0, \quad y_1 + y_2' = 2 \cos t,$
$y_1(0) = 1, \quad y_2(0) = 0$

3. $y_1' = -y_1 + 4y_2, \quad y_2' = 3y_1 - 2y_2,$
$y_1(0) = 3, \quad y_2(0) = 4$

4. $y_1' = 4y_2 - 8 \cos 4t, \quad y_2' = -3y_1 - 9 \sin 4t,$
$y_1(0) = 0, \quad y_2(0) = 3$

5. $y_1' = y_2 + 1 - u(t - 1), \quad y_2' = -y_1 + 1 - u(t - 1),$
$y_1(0) = 0, \quad y_2(0) = 0$

6. $y_1' = 5y_1 + y_2, \quad y_2' = y_1 + 5y_2,$
$y_1(0) = 1, \quad y_2(0) = -3$

7. $y_1' = 2y_1 - 4y_2 + u(t - 1)e^t,$
$y_2' = y_1 - 3y_2 + u(t - 1)e^t, \quad y_1(0) = 3, \quad y_2(0) = 0$

8. $y_1' = -2y_1 + 3y_2, \quad y_2' = 4y_1 - y_2,$
$y_1(0) = 4, \quad y_2(0) = 3$

9. $y_1' = 4y_1 + y_2, \quad y_2' = -y_1 + 2y_2, \quad y_1(0) = 3,$
$y_2(0) = 1$

10. $y_1' = -y_2, \quad y_2' = -y_1 + 2[1 - u(t - 2\pi)] \cos t,$
$y_1(0) = 1, \quad y_2(0) = 0$

11. $y_1'' = y_1 + 3y_2, \quad y_2'' = 4y_1 - 4e^t,$
$y_1(0) = 2, \quad y_1'(0) = 3, \quad y_2(0) = 1, \quad y_2'(0) = 2$

12. $y_1'' = -2y_1 + 2y_2, \quad y_2'' = 2y_1 - 5y_2,$
$y_1(0) = 1, \quad y_1'(0) = 0, \quad y_2(0) = 3, \quad y_2'(0) = 0$

13. $y_1'' + y_2 = -101 \sin 10t, \quad y_2'' + y_1 = 101 \sin 10t,$
$y_1(0) = 0, \quad y_1'(0) = 6, \quad y_2(0) = 8, \quad y_2'(0) = -6$

14. $4y_1' + y_2^{r} - 2y_3' = 0, \quad -2y_1' + y_3' = 1,$
$2y_2' - 4y_3' = -16t$
$y_1(0) = 2, \quad y_2(0) = 0, \quad y_3(0) = 0$

15. $y_1' + y_2' = 2 \sinh t, \quad y_2' + y_3' = e^t,$
$y_3' + y_1' = 2e^t + e^{-t}, \quad y_1(0) = 1, \quad y_2(0) = 1,$
$y_3(0) = 0$

FURTHER APPLICATIONS

16. Forced vibrations of two masses. Solve the model in Example 3 with $k = 4$ and initial conditions $y_1(0) = 1$, $y_1'(0) = 1$, $y_2(0) = 1$, $y_2' = -1$ under the assumption that the force $11 \sin t$ is acting on the first body and the force $-11 \sin t$ on the second. Graph the two curves on common axes and explain the motion physically.

17. CAS Experiment. Effect of Initial Conditions. In Prob. 16, vary the initial conditions systematically, describe and explain the graphs physically. The great variety of curves will surprise you. Are they always periodic? Can you find empirical laws for the changes in terms of continuous changes of those conditions?

18. Mixing problem. What will happen in Example 1 if you double all flows (in particular, an increase to 12 gal/min containing 12 lb of salt from the outside), leaving the size of the tanks and the initial conditions as before? First guess, then calculate. Can you relate the new solution to the old one?

19. Electrical network. Using Laplace transforms, find the currents $i_1(t)$ and $i_2(t)$ in Fig. 148, where $v(t) = 390 \cos t$ and $i_1(0) = 0$, $i_2(0) = 0$. How soon will the currents practically reach their steady state?

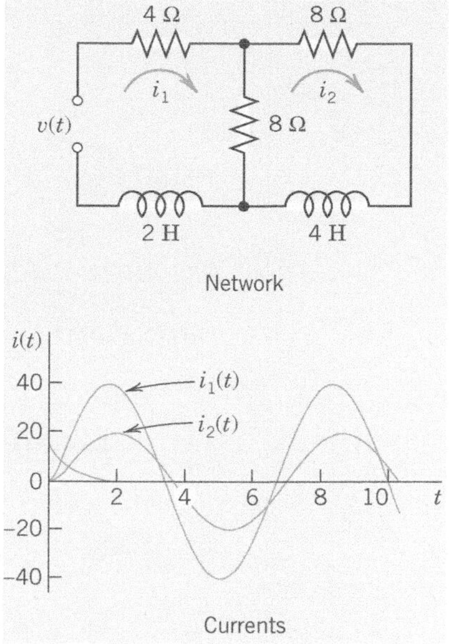

Fig. 148. Electrical network and currents in Problem 19

20. Single cosine wave. Solve Prob. 19 when the EMF (electromotive force) is acting from 0 to 2π only. Can you do this just by looking at Prob. 19, practically without calculation?

6.8 Laplace Transform: General Formulas

Formula	Name, Comments	Sec.
$F(s) = \mathcal{L}\{f(t)\} = \int_0^\infty e^{-st} f(t)\, dt$ $f(t) = \mathcal{L}^{-1}\{F(s)\}$	Definition of Transform Inverse Transform	6.1
$\mathcal{L}\{af(t) + bg(t)\} = a\mathcal{L}\{f(t)\} + b\mathcal{L}\{g(t)\}$	Linearity	6.1
$\mathcal{L}\{e^{at}f(t)\} = F(s-a)$ $\mathcal{L}^{-1}\{F(s-a)\} = e^{at}f(t)$	*s*-Shifting (First Shifting Theorem)	6.1
$\mathcal{L}(f') = s\mathcal{L}(f) - f(0)$ $\mathcal{L}(f'') = s^2\mathcal{L}(f) - sf(0) - f'(0)$ $\mathcal{L}(f^{(n)}) = s^n\mathcal{L}(f) - s^{(n-1)}f(0) - \cdots$ $\cdots - f^{(n-1)}(0)$ $\mathcal{L}\left\{\int_0^t f(\tau)\, d\tau\right\} = \frac{1}{s}\mathcal{L}(f)$	Differentiation of Function Integration of Function	6.2
$(f * g)(t) = \int_0^t f(\tau)g(t-\tau)\, d\tau$ $= \int_0^t f(t-\tau)g(\tau)\, d\tau$ $\mathcal{L}(f * g) = \mathcal{L}(f)\mathcal{L}(g)$	Convolution	6.5
$\mathcal{L}\{f(t-a)u(t-a)\} = e^{-as}F(s)$ $\mathcal{L}^{-1}\{e^{-as}F(s)\} = f(t-a)u(t-a)$	*t*-Shifting (Second Shifting Theorem)	6.3
$\mathcal{L}\{tf(t)\} = -F'(s)$ $\mathcal{L}\left\{\frac{f(t)}{t}\right\} = \int_s^\infty F(\tilde{s})\, d\tilde{s}$	Differentiation of Transform Integration of Transform	6.6
$\mathcal{L}(f) = \frac{1}{1 - e^{-ps}} \int_0^p e^{-st} f(t)\, dt$	*f* Periodic with Period *p*	6.4 Project 16

6.9 Table of Laplace Transforms

For more extensive tables, see Ref.* [A9] *in Appendix 1.

	$F(s) = \mathcal{L}\{f(t)\}$	$f(t)$	Sec.
1	$1/s$	1	6.1
2	$1/s^2$	t	
3	$1/s^n \quad (n = 1, 2, \cdots)$	$t^{n-1}/(n-1)!$	
4	$1/\sqrt{s}$	$1/\sqrt{\pi t}$	
5	$1/s^{3/2}$	$2\sqrt{t/\pi}$	
6	$1/s^a \quad (a > 0)$	$t^{a-1}/\Gamma(a)$	
7	$\dfrac{1}{s-a}$	e^{at}	6.1
8	$\dfrac{1}{(s-a)^2}$	te^{at}	
9	$\dfrac{1}{(s-a)^n} \quad (n = 1, 2, \cdots)$	$\dfrac{1}{(n-1)!}t^{n-1}e^{at}$	
10	$\dfrac{1}{(s-a)^k} \quad (k > 0)$	$\dfrac{1}{\Gamma(k)}t^{k-1}e^{at}$	
11	$\dfrac{1}{(s-a)(s-b)} \quad (a \neq b)$	$\dfrac{1}{a-b}(e^{at} - e^{bt})$	
12	$\dfrac{s}{(s-a)(s-b)} \quad (a \neq b)$	$\dfrac{1}{a-b}(ae^{at} - be^{bt})$	
13	$\dfrac{1}{s^2+\omega^2}$	$\dfrac{1}{\omega}\sin \omega t$	6.1
14	$\dfrac{s}{s^2+\omega^2}$	$\cos \omega t$	
15	$\dfrac{1}{s^2-a^2}$	$\dfrac{1}{a}\sinh at$	
16	$\dfrac{s}{s^2-a^2}$	$\cosh at$	
17	$\dfrac{1}{(s-a)^2+\omega^2}$	$\dfrac{1}{\omega}e^{at}\sinh \omega t$	
18	$\dfrac{s-a}{(s-a)^2+\omega^2}$	$e^{at}\cos \omega t$	
19	$\dfrac{1}{s(s^2+\omega^2)}$	$\dfrac{1}{\omega^2}(1 - \cos \omega t)$	6.2
20	$\dfrac{1}{s^2(s^2+\omega^2)}$	$\dfrac{1}{\omega^3}(\omega t - \sin \omega t)$	

(continued)

Table of Laplace Transforms (*continued*)

	$F(s) = \mathcal{L}\{f(t)\}$	$f(t)$	Sec.
21	$\dfrac{1}{(s^2+\omega^2)^2}$	$\dfrac{1}{2\omega^3}(\sin \omega t - \omega t \cos \omega t)$	
22	$\dfrac{s}{(s^2+\omega^2)^2}$	$\dfrac{t}{2\omega} \sin \omega t$	6.6 (21–23)
23	$\dfrac{s^2}{(s^2+\omega^2)^2}$	$\dfrac{1}{2\omega}(\sin \omega t + \omega t \cos \omega t)$	
24	$\dfrac{s}{(s^2+a^2)(s^2+b^2)} \quad (a^2 \neq b^2)$	$\dfrac{1}{b^2-a^2}(\cos at - \cos bt)$	
25	$\dfrac{1}{s^4+4k^4}$	$\dfrac{1}{4k^3}(\sin kt \cos kt - \cos kt \sinh kt)$	
26	$\dfrac{s}{s^4+4k^4}$	$\dfrac{1}{2k^2} \sin kt \sinh kt$	
27	$\dfrac{1}{s^4-k^4}$	$\dfrac{1}{2k^3}(\sinh kt - \sin kt)$	
28	$\dfrac{s}{s^4-k^4}$	$\dfrac{1}{2k^2}(\cosh kt - \cos kt)$	
29	$\sqrt{s-a} - \sqrt{s-b}$	$\dfrac{1}{2\sqrt{\pi t^3}}(e^{bt} - e^{at})$	
30	$\dfrac{1}{\sqrt{s+a}\,\sqrt{s+b}}$	$e^{-(a+b)t/2} I_0\left(\dfrac{a-b}{2}t\right)$	*I* 5.5
31	$\dfrac{1}{\sqrt{s^2+a^2}}$	$J_0(at)$	*J* 5.4
32	$\dfrac{s}{(s-a)^{3/2}}$	$\dfrac{1}{\sqrt{\pi t}} e^{at}(1+2at)$	
33	$\dfrac{1}{(s^2-a^2)^k} \quad (k > 0)$	$\dfrac{\sqrt{\pi}}{\Gamma(k)}\left(\dfrac{t}{2a}\right)^{k-1/2} I_{k-1/2}(at)$	*I* 5.5
34	e^{-as}/s	$u(t-a)$	6.3
35	e^{-as}	$\delta(t-a)$	6.4
36	$\dfrac{1}{s} e^{-k/s}$	$J_0(2\sqrt{kt})$	*J* 5.4
37	$\dfrac{1}{\sqrt{s}} e^{-k/s}$	$\dfrac{1}{\sqrt{\pi t}} \cos 2\sqrt{kt}$	
38	$\dfrac{1}{s^{3/2}} e^{k/s}$	$\dfrac{1}{\sqrt{\pi k}} \sinh 2\sqrt{kt}$	
39	$e^{-k\sqrt{s}} \quad (k > 0)$	$\dfrac{k}{2\sqrt{\pi t^3}} e^{-k^2/4t}$	

(*continued*)

Table of Laplace Transforms (*continued*)

	$F(s) = \mathcal{L}\{f(t)\}$	$f(t)$	Sec.
40	$\frac{1}{s} \ln s$	$-\ln t - \gamma \quad (\gamma \approx 0.5772)$	γ 5.5
41	$\ln \frac{s-a}{s-b}$	$\frac{1}{t}(e^{bt} - e^{at})$	
42	$\ln \frac{s^2 + \omega^2}{s^2}$	$\frac{2}{t}(1 - \cos \omega t)$	6.6
43	$\ln \frac{s^2 - a^2}{s^2}$	$\frac{2}{t}(1 - \cosh at)$	
44	$\arctan \frac{\omega}{s}$	$\frac{1}{t} \sin \omega t$	
45	$\frac{1}{s} \operatorname{arccot} s$	$\operatorname{Si}(t)$	App. A3.1

CHAPTER 6 REVIEW QUESTIONS AND PROBLEMS

1. State the Laplace transforms of a few simple functions from memory.
2. What are the steps of solving an ODE by the Laplace transform?
3. In what cases of solving ODEs is the present method preferable to that in Chap. 2?
4. What property of the Laplace transform is crucial in solving ODEs?
5. Is $\mathcal{L}\{f(t) + g(t)\} = \mathcal{L}\{f(t)\} + \mathcal{L}\{g(t)\}$? $\mathcal{L}\{f(t)g(t)\} = \mathcal{L}\{f(t)\}\mathcal{L}\{g(t)\}$? Explain.
6. When and how do you use the unit step function and Dirac's delta?
7. If you know $f(t) = \mathcal{L}^{-1}\{F(s)\}$, how would you find $\mathcal{L}^{-1}\{F(s)/s^2\}$?
8. Explain the use of the two shifting theorems from memory.
9. Can a discontinuous function have a Laplace transform? Give reason.
10. If two different continuous functions have transforms, the latter are different. Why is this practically important?

11–19 LAPLACE TRANSFORMS

Find the transform, indicating the method used and showing the details.

11. $5 \cosh 2t - 3 \sinh t$
12. $e^{-t}(\cos 4t - 2 \sin 4t)$
13. $\sin^2 (\frac{1}{2}\pi t)$
14. $16t^2 u(t - \frac{1}{4})$
15. $e^{t/2} u(t - 3)$
16. $u(t - 2\pi) \sin t$
17. $t \cos t + \sin t$
18. $(\sin \omega t) * (\cos \omega t)$
19. $12t * e^{-3t}$

20–28 INVERSE LAPLACE TRANSFORM

Find the inverse transform, indicating the method used and showing the details:

20. $\dfrac{7.5}{s^2 - 2s - 8}$
21. $\dfrac{s+1}{s^2} e^{-s}$
22. $\dfrac{\frac{1}{16}}{s^2 + s + \frac{1}{2}}$
23. $\dfrac{\omega \cos \theta + s \sin \theta}{s^2 + \omega^2}$
24. $\dfrac{s^2 - 6.25}{(s^2 + 6.25)^2}$
25. $\dfrac{6(s+1)}{s^4}$
26. $\dfrac{2s - 10}{s^3} e^{-5s}$
27. $\dfrac{3s + 4}{s^2 + 4s + 5}$
28. $\dfrac{3s}{s^2 - 2s + 2}$

29–37 ODEs AND SYSTEMS

Solve by the Laplace transform, showing the details and graphing the solution:

29. $y'' + 4y' + 5y = 50t, \quad y(0) = 5, \quad y'(0) = -5$
30. $y'' + 16y = 4\delta(t - \pi), \quad y(0) = -1, \quad y'(0) = 0$

31. $y'' - y' - 2y = 12u(t - \pi) \sin t, \quad y(0) = 1, \quad y'(0) = -1$

32. $y'' + 4y = \delta(t - \pi) - \delta(t - 2\pi), \quad y(0) = 1, \quad y'(0) = 0$

33. $y'' + 3y' + 2y = 2u(t - 2), \quad y(0) = 0, \quad y'(0) = 0$

34. $y_1' = y_2, \quad y_2' = -4y_1 + \delta(t - \pi), \quad y_1(0) = 0, \quad y_2(0) = 0$

35. $y_1' = 2y_1 - 4y_2, \quad y_2' = y_1 - 3y_2, \quad y_1(0) = 3, \quad y_2(0) = 0$

36. $y_1' = 2y_1 + 4y_2, \quad y_2' = y_1 + 2y_2, \quad y_1(0) = -4, \quad y_2(0) = -4$

37. $y_1' = y_2 + u(t - \pi), \quad y_2' = -y_1 + u(t - 2\pi), \quad y_1(0) = 1, \quad y_2(0) = 0$

38–45 MASS–SPRING SYSTEMS, CIRCUITS, NETWORKS

Model and solve by the Laplace transform:

38. Show that the model of the mechanical system in Fig. 149 (no friction, no damping) is

$$m_1 y_1'' = -k_1 y_1 + k_2(y_2 - y_1)$$
$$m_2 y_2'' = -k_2(y_2 - y_1) - k_3 y_2.$$

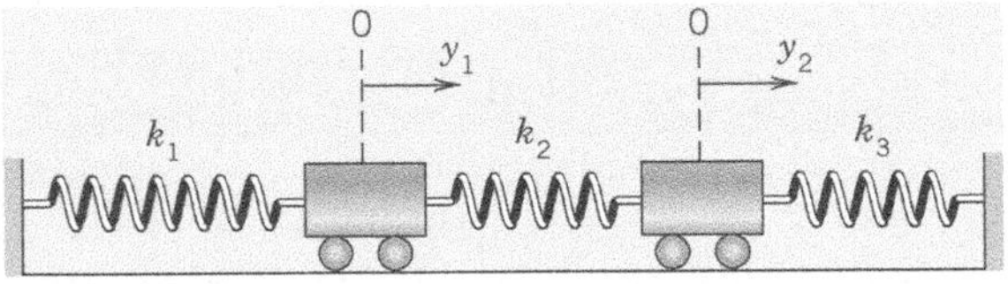

Fig. 149. System in Problems 38 and 39

39. In Prob. 38, let $m_1 = m_2 = 10$ kg, $k_1 = k_3 = 20$ kg/sec^2, $k_2 = 40$ kg/sec^2. Find the solution satisfying the initial conditions $y_1(0) = y_2(0) = 0$, $y_1'(0) = 1$ meter/sec, $y_2'(0) = -1$ meter/sec.

40. Find the model (the system of ODEs) in Prob. 38 extended by adding another mass m_3 and another spring of modulus k_4 in series.

41. Find the current $i(t)$ in the RC-circuit in Fig. 150, where $R = 10\ \Omega$, $C = 0.1$ F, $v(t) = 10t$ V if $0 < t < 4$, $v(t) = 40$ V if $t > 4$, and the initial charge on the capacitor is 0.

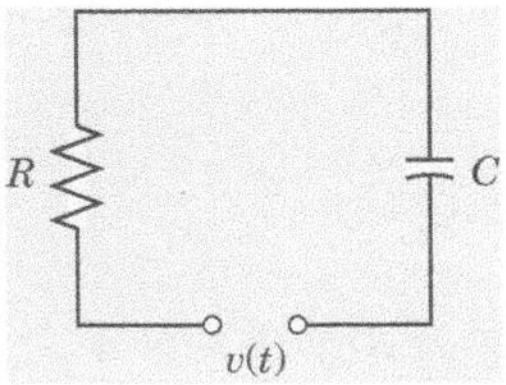

Fig. 150. RC-circuit

42. Find and graph the charge $q(t)$ and the current $i(t)$ in the LC-circuit in Fig. 151, assuming $L = 1$ H, $C = 1$ F, $v(t) = 1 - e^{-t}$ if $0 < t < \pi$, $v(t) = 0$ if $t > \pi$, and zero initial current and charge.

43. Find the current $i(t)$ in the RLC-circuit in Fig. 152, where $R = 160\ \Omega$, $L = 20$ H, $C = 0.002$ F, $v(t) = 37 \sin 10t$ V, and current and charge at $t = 0$ are zero.

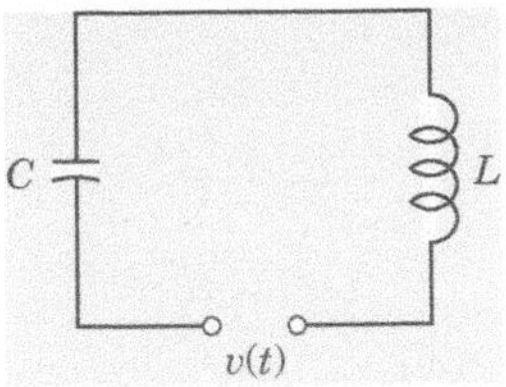

Fig. 151. LC-circuit

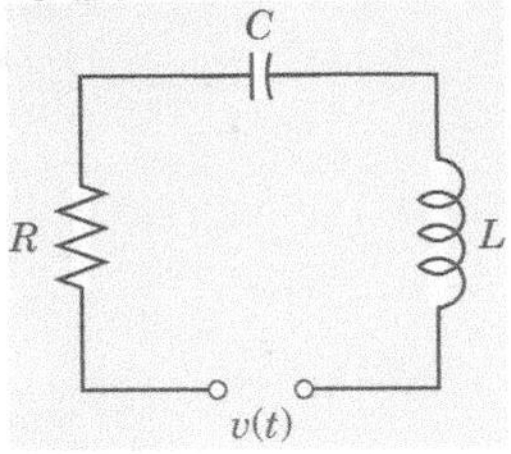

Fig. 152. RLC-circuit

44. Show that, by Kirchhoff's Voltage Law (Sec. 2.9), the currents in the network in Fig. 153 are obtained from the system

$$Li_1' + R(i_1 - i_2) = v(t)$$
$$R(i_2' - i_1') + \frac{1}{C} i_2 = 0.$$

Solve this system, assuming that $R = 10\ \Omega$, $L = 20$ H, $C = 0.05$ F, $v = 20$ V, $i_1(0) = 0$, $i_2(0) = 2$ A.

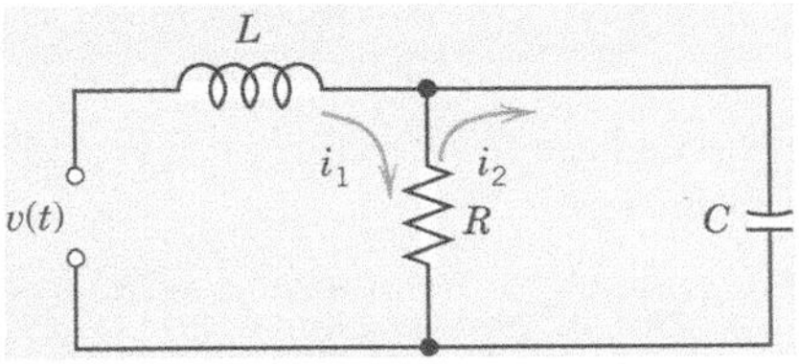

Fig. 153. Network in Problem 44

45. Set up the model of the network in Fig. 154 and find the solution, assuming that all charges and currents are 0 when the switch is closed at $t = 0$. Find the limits of $i_1(t)$ and $i_2(t)$ as $t \to \infty$, (i) from the solution, (ii) directly from the given network.

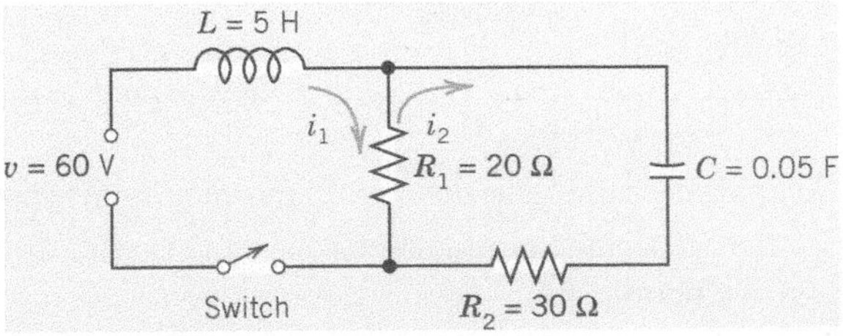

Fig. 154. Network in Problem 45

SUMMARY OF CHAPTER 6

Laplace Transforms

The main purpose of Laplace transforms is the solution of differential equations and systems of such equations, as well as corresponding initial value problems. The **Laplace transform** $F(s) = \mathcal{L}(f)$ of a function $f(t)$ is defined by

$$F(s) = \mathcal{L}(f) = \int_0^\infty e^{-st} f(t)\, dt \qquad \text{(Sec. 6.1).} \tag{1}$$

This definition is motivated by the property that the differentiation of f with respect to t corresponds to the multiplication of the transform F by s; more precisely,

$$\begin{aligned} \mathcal{L}(f') &= s\mathcal{L}(f) - f(0) \\ \mathcal{L}(f'') &= s^2\mathcal{L}(f) - sf(0) - f'(0) \end{aligned} \qquad \text{(Sec. 6.2)} \tag{2}$$

etc. Hence by taking the transform of a given differential equation

$$y'' + ay' + by = r(t) \qquad (a, b \text{ constant}) \tag{3}$$

and writing $\mathcal{L}(y) = Y(s)$, we obtain the **subsidiary equation**

$$(s^2 + as + b)Y = \mathcal{L}(r) + sf(0) + f'(0) + af(0). \tag{4}$$

Here, in obtaining the transform $\mathcal{L}(r)$ we can get help from the small table in Sec. 6.1 or the larger table in Sec. 6.9. This is the first step. In the second step we solve the subsidiary equation *algebraically* for $Y(s)$. In the third step we determine the **inverse transform** $y(t) = \mathcal{L}^{-1}(Y)$, that is, the solution of the problem. This is generally the hardest step, and in it we may again use one of those two tables. $Y(s)$ will often be a rational function, so that we can obtain the inverse $\mathcal{L}^{-1}(Y)$ by partial fraction reduction (Sec. 6.4) if we see no simpler way.

The Laplace method avoids the determination of a general solution of the homogeneous ODE, and we also need not determine values of arbitrary constants in a general solution from initial conditions; instead, we can insert the latter directly into (4). Two further facts account for the practical importance of the Laplace transform. First, it has some basic properties and resulting techniques that simplify the determination of transforms and inverses. The most important of these properties are listed in Sec. 6.8, together with references to the corresponding sections. More on the use of unit step functions and Dirac's delta can be found in Secs. 6.3 and 6.4, and more on convolution in Sec. 6.5. Second, due to these properties, the present method is particularly suitable for handling right sides $r(t)$ given by different expressions over different intervals of time, for instance, when $r(t)$ is a square wave or an impulse or of a form such as $r(t) = \cos t$ if $0 \leqq t \leqq 4\pi$ and 0 elsewhere.

The application of the Laplace transform to systems of ODEs is shown in Sec. 6.7. (The application to PDEs follows in Sec. 12.12.)

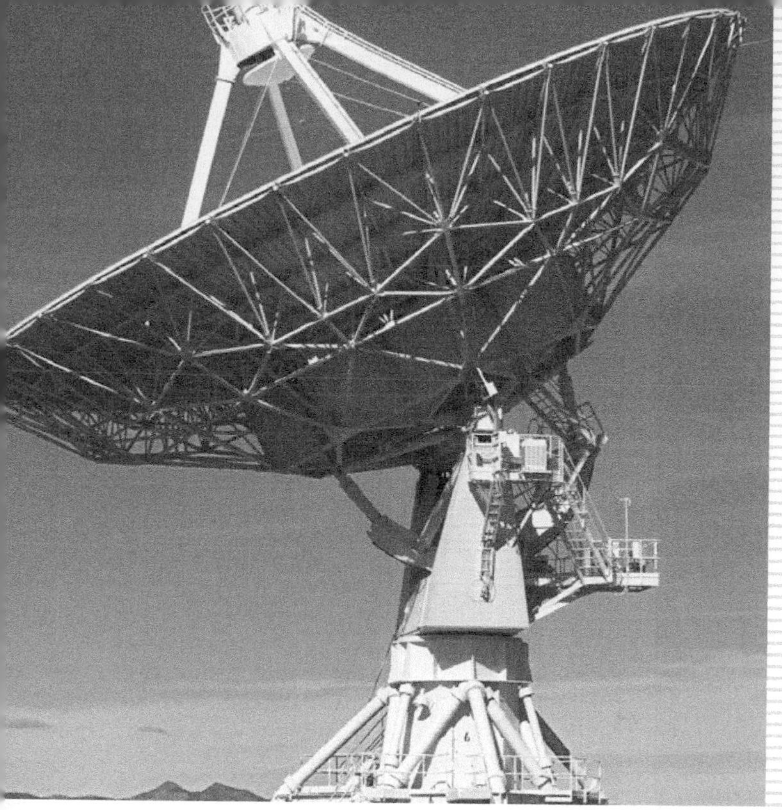

PART B

Linear Algebra. Vector Calculus

Matrices and *vectors*, which underlie **linear algebra** (Chaps. 7 and 8), allow us to represent numbers or functions in an ordered and compact form. Matrices can hold enormous amounts of data—think of a network of millions of computer connections or cell phone connections—in a form that can be rapidly processed by computers. The main topic of Chap. 7 is *how to solve systems of linear equations using matrices*. Concepts of rank, basis, linear transformations, and vector spaces are closely related. Chapter 8 deals with eigenvalue problems. Linear algebra is an active field that has many applications in engineering physics, numerics (see Chaps. 20–22), economics, and others.

Chapters 9 and 10 extend calculus to ***vector calculus***. We start with vectors from linear algebra and develop vector differential calculus. We differentiate functions of several variables and discuss vector differential operations such as grad, div, and curl. Chapter 10 extends regular integration to integration over curves, surfaces, and solids, thereby obtaining new types of integrals. Ingenious theorems by Gauss, Green, and Stokes allow us to transform these integrals into one another.

Software suitable for linear algebra (Lapack, Maple, Mathematica, Matlab) can be found in the list at the opening of Part E of the book if needed.

Numeric linear algebra (Chap. 20) ***can be studied directly after Chap. 7 or 8*** because Chap. 20 is independent of the other chapters in Part E on numerics.

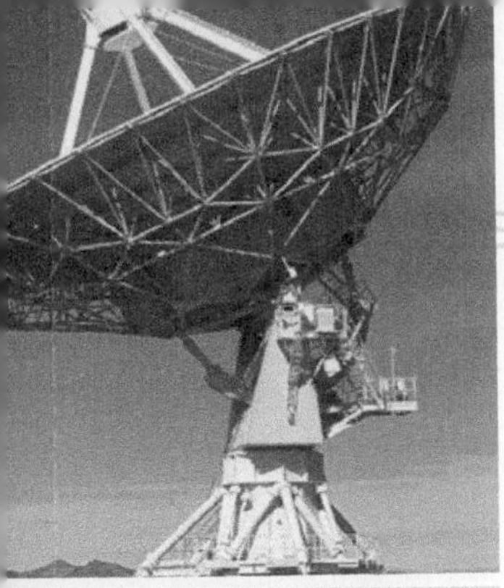

CHAPTER 7

Linear Algebra: Matrices, Vectors, Determinants. Linear Systems

Linear algebra is a fairly extensive subject that covers vectors and matrices, determinants, systems of linear equations, vector spaces and linear transformations, eigenvalue problems, and other topics. As an area of study it has a broad appeal in that it has many applications in engineering, physics, geometry, computer science, economics, and other areas. It also contributes to a deeper understanding of mathematics itself.

Matrices, which are rectangular arrays of numbers or functions, and **vectors** are the main tools of linear algebra. Matrices are important because they let us express large amounts of data and functions in an organized and concise form. Furthermore, since matrices are single objects, we denote them by single letters and calculate with them directly. All these features have made matrices and vectors very popular for expressing scientific and mathematical ideas.

The chapter keeps a good mix between applications (electric networks, Markov processes, traffic flow, etc.) and theory. Chapter 7 is structured as follows: Sections 7.1 and 7.2 provide an intuitive introduction to matrices and vectors and their operations, including matrix multiplication. The next block of sections, that is, Secs. 7.3–7.5 provide the most important method for ***solving systems of linear equations*** by the Gauss elimination method. This method is a cornerstone of linear algebra, and the method itself and variants of it appear in different areas of mathematics and in many applications. It leads to a consideration of the behavior of solutions and concepts such as rank of a matrix, linear independence, and bases. We shift to determinants, a topic that has declined in importance, in Secs. 7.6 and 7.7. Section 7.8 covers inverses of matrices. The chapter ends with vector spaces, inner product spaces, linear transformations, and composition of linear transformations. Eigenvalue problems follow in Chap. 8.

COMMENT. ***Numeric linear algebra (Secs. 20.1–20.5) can be studied immediately after this chapter.***

Prerequisite: None.
Sections that may be omitted in a short course: 7.5, 7.9.
References and Answers to Problems: App. 1 Part B, and App. 2.

7.1 Matrices, Vectors: Addition and Scalar Multiplication

The basic concepts and rules of matrix and vector algebra are introduced in Secs. 7.1 and 7.2 and are followed by **linear systems** (systems of linear equations), a main application, in Sec. 7.3.

Let us first take a leisurely look at matrices before we formalize our discussion. A **matrix** is a rectangular array of numbers or functions which we will enclose in brackets. For example,

$$(1)\qquad \begin{bmatrix} 0.3 & 1 & -5 \\ 0 & -0.2 & 16 \end{bmatrix}, \quad \begin{bmatrix} a_{11} & a_{12} & a_{13} \\ a_{21} & a_{22} & a_{23} \\ a_{31} & a_{32} & a_{33} \end{bmatrix},$$
$$\begin{bmatrix} e^{-x} & 2x^2 \\ e^{6x} & 4x \end{bmatrix}, \quad [a_1 \quad a_2 \quad a_3], \quad \begin{bmatrix} 4 \\ \frac{1}{2} \end{bmatrix}$$

are matrices. The numbers (or functions) are called **entries** or, less commonly, *elements* of the matrix. The first matrix in (1) has two **rows**, which are the horizontal lines of entries. Furthermore, it has three **columns**, which are the vertical lines of entries. The second and third matrices are **square matrices**, which means that each has as many rows as columns—3 and 2, respectively. The entries of the second matrix have two indices, signifying their location within the matrix. The first index is the number of the row and the second is the number of the column, so that together the entry's position is uniquely identified. For example, a_{23} (read *a two three*) is in Row 2 and Column 3, etc. The notation is standard and applies to all matrices, including those that are not square.

Matrices having just a single row or column are called **vectors**. Thus, the fourth matrix in (1) has just one row and is called a **row vector**. The last matrix in (1) has just one column and is called a **column vector**. Because the goal of the indexing of entries was to uniquely identify the position of an element within a matrix, one index suffices for vectors, whether they are row or column vectors. Thus, the third entry of the row vector in (1) is denoted by a_3.

Matrices are handy for storing and processing data in applications. Consider the following two common examples.

EXAMPLE 1 Linear Systems, a Major Application of Matrices

We are given a system of linear equations, briefly a **linear system**, such as

$$\begin{aligned} 4x_1 + 6x_2 + 9x_3 &= 6 \\ 6x_1 \qquad\quad - 2x_3 &= 20 \\ 5x_1 - 8x_2 + \ \ x_3 &= 10 \end{aligned}$$

where x_1, x_2, x_3 are the **unknowns**. We form the **coefficient matrix**, call it **A**, by listing the coefficients of the unknowns in the position in which they appear in the linear equations. In the second equation, there is no unknown x_2, which means that the coefficient of x_2 is 0 and hence in matrix **A**, $a_{22} = 0$, Thus,

$$\mathbf{A} = \begin{bmatrix} 4 & 6 & 9 \\ 6 & 0 & -2 \\ 5 & -8 & 1 \end{bmatrix}. \quad \text{We form another matrix} \quad \tilde{\mathbf{A}} = \begin{bmatrix} 4 & 6 & 9 & 6 \\ 6 & 0 & -2 & 20 \\ 5 & -8 & 1 & 10 \end{bmatrix}$$

by augmenting **A** with the right sides of the linear system and call it the augmented matrix of the system.

Since we can go back and recapture the system of linear equations directly from the augmented matrix $\tilde{\mathbf{A}}$, $\tilde{\mathbf{A}}$ contains all the information of the system and can thus be used to solve the linear system. This means that we can just use the augmented matrix to do the calculations needed to solve the system. We shall explain this in detail in Sec. 7.3. Meanwhile you may verify by substitution that the solution is $x_1 = 3, x_2 = \frac{1}{2}, x_3 = -1$.

The notation x_1, x_2, x_3 for the unknowns is practical but not essential; we could choose x, y, z or some other letters. ■

EXAMPLE 2 **Sales Figures in Matrix Form**

Sales figures for three products I, II, III in a store on Monday (Mon), Tuesday (Tues), $\cdots$ may for each week be arranged in a matrix

$$\mathbf{A} = \begin{array}{c} \\ \end{array}\begin{array}{ccccccc} \text{Mon} & \text{Tues} & \text{Wed} & \text{Thur} & \text{Fri} & \text{Sat} & \text{Sun} \end{array}$$

$$\mathbf{A} = \begin{bmatrix} 40 & 33 & 81 & 0 & 21 & 47 & 33 \\ 0 & 12 & 78 & 50 & 50 & 96 & 90 \\ 10 & 0 & 0 & 27 & 43 & 78 & 56 \end{bmatrix} \cdot \begin{array}{c} \text{I} \\ \text{II} \\ \text{III} \end{array}$$

If the company has 10 stores, we can set up 10 such matrices, one for each store. Then, by adding corresponding entries of these matrices, we can get a matrix showing the total sales of each product on each day. Can you think of other data which can be stored in matrix form? For instance, in transportation or storage problems? Or in listing distances in a network of roads? ■

General Concepts and Notations

Let us formalize what we just have discussed. We shall denote matrices by capital boldface letters **A**, **B**, **C**, $\cdots$, or by writing the general entry in brackets; thus $\mathbf{A} = [a_{jk}]$, and so on. By an $m \times n$ **matrix** (read *m by n matrix*) we mean a matrix with m rows and n columns—rows always come first! $m \times n$ is called the **size** of the matrix. Thus an $m \times n$ matrix is of the form

$$\mathbf{A} = [a_{jk}] = \begin{bmatrix} a_{11} & a_{12} & \cdots & a_{1n} \\ a_{21} & a_{22} & \cdots & a_{2n} \\ \cdot & \cdot & \cdots & \cdot \\ a_{m1} & a_{m2} & \cdots & a_{mn} \end{bmatrix}. \tag{2}$$

The matrices in (1) are of sizes 2×3, 3×3, 2×2, 1×3, and 2×1, respectively.

Each entry in (2) has two subscripts. The first is the *row number* and the second is the *column number.* Thus a_{21} is the entry in Row 2 and Column 1.

If $m = n$, we call **A** an $n \times n$ **square matrix**. Then its diagonal containing the entries $a_{11}, a_{22}, \cdots, a_{nn}$ is called the **main diagonal** of **A**. Thus the main diagonals of the two square matrices in (1) are a_{11}, a_{22}, a_{33} and e^{-x}, $4x$, respectively.

Square matrices are particularly important, as we shall see. A matrix of any size $m \times n$ is called a **rectangular matrix**; this includes square matrices as a special case.

Vectors

A **vector** is a matrix with only one row or column. Its entries are called the **components** of the vector. We shall denote vectors by ***lowercase*** boldface letters **a**, **b**, $\cdots$ or by its general component in brackets, $\mathbf{a} = [a_j]$, and so on. Our special vectors in (1) suggest that a (general) **row vector** is of the form

$$\mathbf{a} = [a_1 \quad a_2 \quad \cdots \quad a_n]. \qquad \text{For instance,} \qquad \mathbf{a} = [-2 \quad 5 \quad 0.8 \quad 0 \quad 1].$$

A **column vector** is of the form

$$\mathbf{b} = \begin{bmatrix} b_1 \\ b_2 \\ \vdots \\ b_m \end{bmatrix}. \qquad \text{For instance,} \qquad \mathbf{b} = \begin{bmatrix} 4 \\ 0 \\ -7 \end{bmatrix}.$$

Addition and Scalar Multiplication of Matrices and Vectors

What makes matrices and vectors really useful and particularly suitable for computers is the fact that we can calculate with them almost as easily as with numbers. Indeed, we now introduce rules for addition and for scalar multiplication (multiplication by numbers) that were suggested by practical applications. (Multiplication of matrices by matrices follows in the next section.) We first need the concept of equality.

DEFINITION

Equality of Matrices

Two matrices $\mathbf{A} = [a_{jk}]$ and $\mathbf{B} = [b_{jk}]$ are **equal**, written $\mathbf{A} = \mathbf{B}$, if and only if they have the same size and the corresponding entries are equal, that is, $a_{11} = b_{11}$, $a_{12} = b_{12}$, and so on. Matrices that are not equal are called **different**. Thus, matrices of different sizes are always different.

EXAMPLE 3 **Equality of Matrices**

Let

$$\mathbf{A} = \begin{bmatrix} a_{11} & a_{12} \\ a_{21} & a_{22} \end{bmatrix} \qquad \text{and} \qquad \mathbf{B} = \begin{bmatrix} 4 & 0 \\ 3 & -1 \end{bmatrix}.$$

Then

$$\mathbf{A} = \mathbf{B} \qquad \text{if and only if} \qquad \begin{matrix} a_{11} = 4, & a_{12} = 0, \\ a_{21} = 3, & a_{22} = -1. \end{matrix}$$

The following matrices are all different. Explain!

$$\begin{bmatrix} 1 & 3 \\ 4 & 2 \end{bmatrix} \quad \begin{bmatrix} 4 & 2 \\ 1 & 3 \end{bmatrix} \quad \begin{bmatrix} 4 & 1 \\ 2 & 3 \end{bmatrix} \quad \begin{bmatrix} 1 & 3 & 0 \\ 4 & 2 & 0 \end{bmatrix} \quad \begin{bmatrix} 0 & 1 & 3 \\ 0 & 4 & 2 \end{bmatrix}$$ ■

DEFINITION

Addition of Matrices

The **sum** of two matrices $\mathbf{A} = [a_{jk}]$ and $\mathbf{B} = [b_{jk}]$ ***of the same size*** is written $\mathbf{A} + \mathbf{B}$ and has the entries $a_{jk} + b_{jk}$ obtained by adding the corresponding entries of $\mathbf{A}$ and $\mathbf{B}$. Matrices of different sizes cannot be added.

As a special case, the **sum** $\mathbf{a} + \mathbf{b}$ of two row vectors or two column vectors, which must have the same number of components, is obtained by adding the corresponding components.

EXAMPLE 4 **Addition of Matrices and Vectors**

$$\text{If}\quad \mathbf{A} = \begin{bmatrix} -4 & 6 & 3 \\ 0 & 1 & 2 \end{bmatrix} \quad\text{and}\quad \mathbf{B} = \begin{bmatrix} 5 & -1 & 0 \\ 3 & 1 & 0 \end{bmatrix}, \quad\text{then}\quad \mathbf{A} + \mathbf{B} = \begin{bmatrix} 1 & 5 & 3 \\ 3 & 2 & 2 \end{bmatrix}.$$

$\mathbf{A}$ in Example 3 and our present $\mathbf{A}$ cannot be added. If $\mathbf{a} = [5 \quad 7 \quad 2]$ and $\mathbf{b} = [-6 \quad 2 \quad 0]$, then $\mathbf{a} + \mathbf{b} = [-1 \quad 9 \quad 2]$.

An application of matrix addition was suggested in Example 2. Many others will follow. ■

DEFINITION

Scalar Multiplication (Multiplication by a Number)

The **product** of any $m \times n$ matrix $\mathbf{A} = [a_{jk}]$ and any **scalar** c (number c) is written $c\mathbf{A}$ and is the $m \times n$ matrix $c\mathbf{A} = [ca_{jk}]$ obtained by multiplying each entry of $\mathbf{A}$ by c.

Here $(-1)\mathbf{A}$ is simply written $-\mathbf{A}$ and is called the **negative** of $\mathbf{A}$. Similarly, $(-k)\mathbf{A}$ is written $-k\mathbf{A}$. Also, $\mathbf{A} + (-\mathbf{B})$ is written $\mathbf{A} - \mathbf{B}$ and is called the **difference** of $\mathbf{A}$ and $\mathbf{B}$ (which must have the same size!).

EXAMPLE 5 **Scalar Multiplication**

$$\text{If}\quad \mathbf{A} = \begin{bmatrix} 2.7 & -1.8 \\ 0 & 0.9 \\ 9.0 & -4.5 \end{bmatrix}, \quad\text{then}\quad -\mathbf{A} = \begin{bmatrix} -2.7 & 1.8 \\ 0 & -0.9 \\ -9.0 & 4.5 \end{bmatrix}, \quad \frac{10}{9}\mathbf{A} = \begin{bmatrix} 3 & -2 \\ 0 & 1 \\ 10 & -5 \end{bmatrix}, \quad 0\mathbf{A} = \begin{bmatrix} 0 & 0 \\ 0 & 0 \\ 0 & 0 \end{bmatrix}.$$

If a matrix $\mathbf{B}$ shows the distances between some cities in miles, $1.609\mathbf{B}$ gives these distances in kilometers. ■

Rules for Matrix Addition and Scalar Multiplication. From the familiar laws for the addition of numbers we obtain similar laws for the addition of matrices of the same size $m \times n$, namely,

$$\text{(3)}\qquad \begin{aligned} &\text{(a)} && \mathbf{A} + \mathbf{B} = \mathbf{B} + \mathbf{A} \\ &\text{(b)} && (\mathbf{A} + \mathbf{B}) + \mathbf{C} = \mathbf{A} + (\mathbf{B} + \mathbf{C}) \qquad (\text{written } \mathbf{A} + \mathbf{B} + \mathbf{C}) \\ &\text{(c)} && \mathbf{A} + \mathbf{0} = \mathbf{A} \\ &\text{(d)} && \mathbf{A} + (-\mathbf{A}) = \mathbf{0}. \end{aligned}$$

Here $\mathbf{0}$ denotes the **zero matrix** (of size $m \times n$), that is, the $m \times n$ matrix with all entries zero. If $m = 1$ or $n = 1$, this is a vector, called a **zero vector**.

Hence matrix addition is *commutative* and *associative* [by (3a) and (3b)]. Similarly, for scalar multiplication we obtain the rules

$$
\textbf{(4)}\qquad
\begin{aligned}
&\text{(a)} \quad c(\mathbf{A}+\mathbf{B}) = c\mathbf{A}+c\mathbf{B}\\
&\text{(b)} \quad (c+k)\mathbf{A} = c\mathbf{A}+k\mathbf{A}\\
&\text{(c)} \quad c(k\mathbf{A}) = (ck)\mathbf{A} \qquad \text{(written } ck\mathbf{A}\text{)}\\
&\text{(d)} \quad 1\mathbf{A} = \mathbf{A}.
\end{aligned}
$$

PROBLEM SET 7.1

1–7 GENERAL QUESTIONS

1. **Equality.** Give reasons why the five matrices in Example 3 are all different.
2. **Double subscript notation.** If you write the matrix in Example 2 in the form $\mathbf{A} = [a_{jk}]$, what is a_{31}? a_{13}? a_{26}? a_{33}?
3. **Sizes.** What sizes do the matrices in Examples 1, 2, 3, and 5 have?
4. **Main diagonal.** What is the main diagonal of **A** in Example 1? Of **A** and **B** in Example 3?
5. **Scalar multiplication.** If **A** in Example 2 shows the number of items sold, what is the matrix **B** of units sold if a unit consists of **(a)** 5 items and **(b)** 10 items?
6. If a 12×12 matrix **A** shows the distances between 12 cities in kilometers, how can you obtain from **A** the matrix **B** showing these distances in miles?
7. **Addition of vectors.** Can you add: A row and a column vector with different numbers of components? With the same number of components? Two row vectors with the same number of components but different numbers of zeros? A vector and a scalar? A vector with four components and a 2×2 matrix?

8–16 ADDITION AND SCALAR MULTIPLICATION OF MATRICES AND VECTORS

Let

$$
\mathbf{A} = \begin{bmatrix} 0 & 2 & 4 \\ 6 & 5 & 5 \\ 1 & 0 & -3 \end{bmatrix}, \quad
\mathbf{B} = \begin{bmatrix} 0 & 5 & 2 \\ 5 & 3 & 4 \\ -2 & 4 & -2 \end{bmatrix}
$$

$$
\mathbf{C} = \begin{bmatrix} 5 & 2 \\ -2 & 4 \\ 1 & 0 \end{bmatrix}, \quad
\mathbf{D} = \begin{bmatrix} -4 & 1 \\ 5 & 0 \\ 2 & -1 \end{bmatrix},
$$

$$
\mathbf{E} = \begin{bmatrix} 0 & 2 \\ 3 & 4 \\ 3 & -1 \end{bmatrix}
$$

$$
\mathbf{u} = \begin{bmatrix} 1.5 \\ 0 \\ -3.0 \end{bmatrix}, \quad
\mathbf{v} = \begin{bmatrix} -1 \\ 3 \\ 2 \end{bmatrix}, \quad
\mathbf{w} = \begin{bmatrix} -5 \\ -30 \\ 10 \end{bmatrix}.
$$

Find the following expressions, indicating which of the rules in (3) or (4) they illustrate, or give reasons why they are not defined.

8. $2\mathbf{A} + 4\mathbf{B}$, $4\mathbf{B} + 2\mathbf{A}$, $0\mathbf{A} + \mathbf{B}$, $0.4\mathbf{B} - 4.2\mathbf{A}$
9. $3\mathbf{A}$, $0.5\mathbf{B}$, $3\mathbf{A} + 0.5\mathbf{B}$, $3\mathbf{A} + 0.5\mathbf{B} + \mathbf{C}$
10. $(4 \cdot 3)\mathbf{A}$, $4(3\mathbf{A})$, $14\mathbf{B} - 3\mathbf{B}$, $11\mathbf{B}$
11. $8\mathbf{C} + 10\mathbf{D}$, $2(5\mathbf{D} + 4\mathbf{C})$, $0.6\mathbf{C} - 0.6\mathbf{D}$, $0.6(\mathbf{C} - \mathbf{D})$
12. $(\mathbf{C} + \mathbf{D}) + \mathbf{E}$, $(\mathbf{D} + \mathbf{E}) + \mathbf{C}$, $0(\mathbf{C} - \mathbf{E}) + 4\mathbf{D}$, $\mathbf{A} - 0\mathbf{C}$
13. $(2 \cdot 7)\mathbf{C}$, $2(7\mathbf{C})$, $-\mathbf{D} + 0\mathbf{E}$, $\mathbf{E} - \mathbf{D} + \mathbf{C} + \mathbf{u}$
14. $(5\mathbf{u} + 5\mathbf{v}) - \frac{1}{2}\mathbf{w}$, $-20(\mathbf{u} + \mathbf{v}) + 2\mathbf{w}$, $\mathbf{E} - (\mathbf{u} + \mathbf{v})$, $10(\mathbf{u} + \mathbf{v}) + \mathbf{w}$
15. $(\mathbf{u} + \mathbf{v}) - \mathbf{w}$, $\mathbf{u} + (\mathbf{v} - \mathbf{w})$, $\mathbf{C} + 0\mathbf{w}$, $0\mathbf{E} + \mathbf{u} - \mathbf{v}$
16. $15\mathbf{v} - 3\mathbf{w} - 0\mathbf{u}$, $-3\mathbf{w} + 15\mathbf{v}$, $\mathbf{D} - \mathbf{u} + 3\mathbf{C}$, $8.5\mathbf{w} - 11.1\mathbf{u} + 0.4\mathbf{v}$
17. **Resultant of forces.** If the above vectors **u**, **v**, **w** represent forces in space, their sum is called their *resultant*. Calculate it.
18. **Equilibrium.** By definition, forces are *in equilibrium* if their resultant is the zero vector. Find a force **p** such that the above **u**, **v**, **w**, and **p** are in equilibrium.
19. **General rules.** Prove (3) and (4) for general 2×3 matrices and scalars c and k.

20. TEAM PROJECT. Matrices for Networks. Matrices have various engineering applications, as we shall see. For instance, they can be used to characterize connections in electrical networks, in nets of roads, in production processes, etc., as follows.

(a) Nodal Incidence Matrix. The network in Fig. 155 consists of six *branches* (connections) and four *nodes* (points where two or more branches come together). One node is the *reference node* (grounded node, whose voltage is zero). We number the other nodes and number and direct the branches. This we do arbitrarily. The network can now be described by a matrix $\mathbf{A} = [a_{jk}]$, where

$$a_{jk} = \begin{cases} +1 \text{ if branch } k \text{ leaves node } (j) \\ -1 \text{ if branch } k \text{ enters node } (j) \\ \;\;0 \text{ if branch } k \text{ does not touch node } (j). \end{cases}$$

A is called the *nodal incidence matrix* of the network. Show that for the network in Fig. 155 the matrix **A** has the given form.

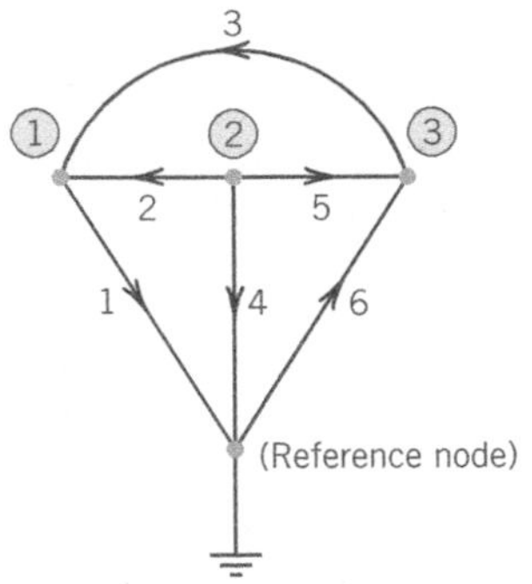

Branch	1	2	3	4	5	6
Node (1)	1	–1	–1	0	0	0
Node (2)	0	1	0	1	1	0
Node (3)	0	0	1	0	–1	–1

Fig. 155. Network and nodal incidence matrix in Team Project 20(a)

(b) Find the nodal incidence matrices of the networks in Fig. 156.

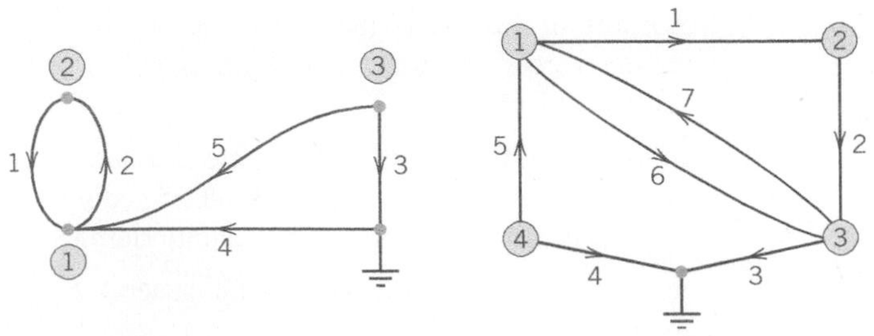

Fig. 156. Electrical networks in Team Project 20(b)

(c) Sketch the three networks corresponding to the nodal incidence matrices

$$\begin{bmatrix} 1 & 0 & 0 & 1 \\ -1 & 1 & 0 & 0 \\ 0 & -1 & 1 & 0 \end{bmatrix}, \quad \begin{bmatrix} 1 & -1 & 0 & 0 & 1 \\ -1 & 1 & -1 & 1 & 0 \\ 0 & 0 & 1 & -1 & 0 \end{bmatrix},$$

$$\begin{bmatrix} 1 & 0 & 1 & 0 & 0 \\ -1 & 1 & 0 & 1 & 0 \\ 0 & -1 & -1 & 0 & 1 \end{bmatrix}.$$

(d) Mesh Incidence Matrix. A network can also be characterized by the *mesh incidence matrix* $\mathbf{M} = [m_{jk}]$, where

$$m_{jk} = \begin{cases} +1 \text{ if branch } k \text{ is in mesh } \boxed{j} \\ \quad \text{and has the same orientation} \\ -1 \text{ if branch } k \text{ is in mesh } \boxed{j} \\ \quad \text{and has the opposite orientation} \\ \;\;0 \text{ if branch } k \text{ is not in mesh } \boxed{j} \end{cases}$$

and a mesh is a loop with no branch in its interior (or in its exterior). Here, the meshes are numbered and directed (oriented) in an arbitrary fashion. Show that for the network in Fig. 157, the matrix **M** has the given form, where Row 1 corresponds to mesh 1, etc.

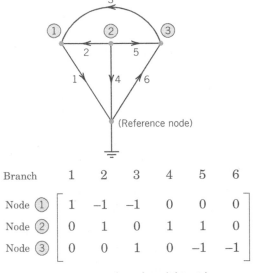

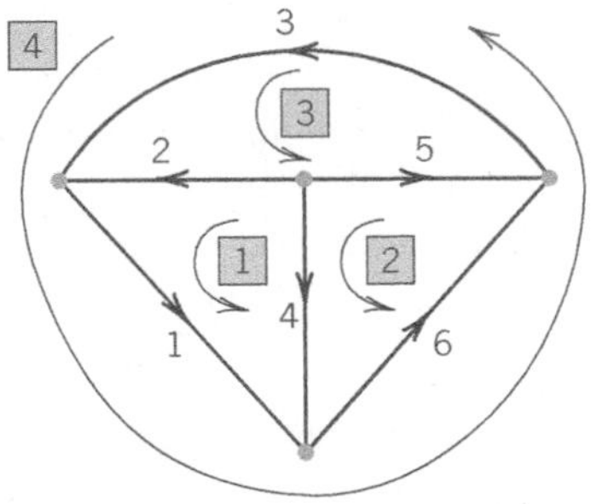

$$\mathbf{M} = \begin{bmatrix} 1 & 1 & 0 & -1 & 0 & 0 \\ 0 & 0 & 0 & 1 & -1 & 1 \\ 0 & -1 & 1 & 0 & 1 & 0 \\ 1 & 0 & 1 & 0 & 0 & 1 \end{bmatrix}$$

Fig. 157. Network and matrix **M** in Team Project 20(d)

7.2 Matrix Multiplication

Matrix multiplication means that one multiplies matrices by matrices. Its definition is standard but it looks artificial. *Thus you have to study matrix multiplication carefully*, multiply a few matrices together for practice until you can understand how to do it. Here then is the definition. (Motivation follows later.)

DEFINITION

Multiplication of a Matrix by a Matrix

The **product** $\mathbf{C} = \mathbf{AB}$ (in this order) of an $m \times n$ matrix $\mathbf{A} = [a_{jk}]$ times an $r \times p$ matrix $\mathbf{B} = [b_{jk}]$ is defined if and only if $r = n$ and is then the $m \times p$ matrix $\mathbf{C} = [c_{jk}]$ with entries

$$\text{(1)} \qquad c_{jk} = \sum_{l=1}^{n} a_{jl}b_{lk} = a_{j1}b_{1k} + a_{j2}b_{2k} + \cdots + a_{jn}b_{nk} \qquad \begin{array}{l} j = 1, \cdots, m \\ k = 1, \cdots, p. \end{array}$$

The condition $r = n$ means that the second factor, **B**, must have as many rows as the first factor has columns, namely n. A diagram of sizes that shows when matrix multiplication is possible is as follows:

$$\begin{array}{ccc} \mathbf{A} & \mathbf{B} & = \quad \mathbf{C} \\ [m \times n] & [n \times p] & = [m \times p]. \end{array}$$

The entry c_{jk} in (1) is obtained by multiplying each entry in the jth row of **A** by the corresponding entry in the kth column of **B** and then adding these n products. For instance, $c_{21} = a_{21}b_{11} + a_{22}b_{21} + \cdots + a_{2n}b_{n1}$, and so on. One calls this briefly a ***multiplication of rows into columns***. For $n = 3$, this is illustrated by

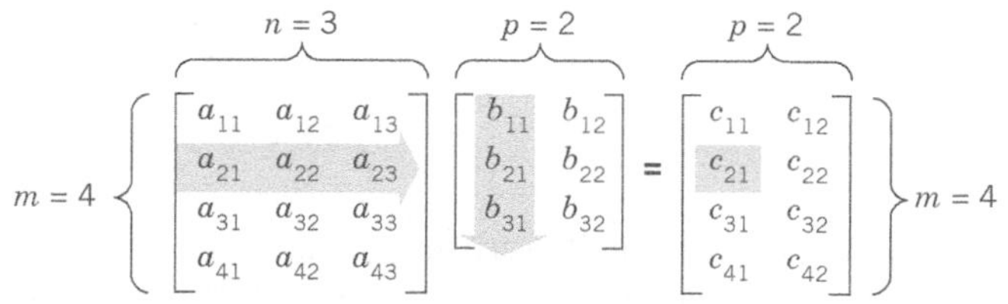

Notations in a product **AB** = **C**

where we shaded the entries that contribute to the calculation of entry c_{21} just discussed.

Matrix multiplication will be motivated by its use in *linear transformations* in this section and more fully in Sec. 7.9.

Let us illustrate the main points of matrix multiplication by some examples. Note that matrix multiplication also includes multiplying a matrix by a vector, since, after all, a vector is a special matrix.

EXAMPLE 1 **Matrix Multiplication**

$$\mathbf{AB} = \begin{bmatrix} 3 & 5 & -1 \\ 4 & 0 & 2 \\ -6 & -3 & 2 \end{bmatrix} \begin{bmatrix} 2 & -2 & 3 & 1 \\ 5 & 0 & 7 & 8 \\ 9 & -4 & 1 & 1 \end{bmatrix} = \begin{bmatrix} 22 & -2 & 43 & 42 \\ 26 & -16 & 14 & 6 \\ -9 & 4 & -37 & -28 \end{bmatrix}$$

Here $c_{11} = 3 \cdot 2 + 5 \cdot 5 + (-1) \cdot 9 = 22$, and so on. The entry in the box is $c_{23} = 4 \cdot 3 + 0 \cdot 7 + 2 \cdot 1 = 14$. The product **BA** is not defined. ■

EXAMPLE 2 Multiplication of a Matrix and a Vector

$$\begin{bmatrix} 4 & 2 \\ 1 & 8 \end{bmatrix}\begin{bmatrix} 3 \\ 5 \end{bmatrix} = \begin{bmatrix} 4 \cdot 3 + 2 \cdot 5 \\ 1 \cdot 3 + 8 \cdot 5 \end{bmatrix} = \begin{bmatrix} 22 \\ 43 \end{bmatrix} \quad \text{whereas} \quad \begin{bmatrix} 3 \\ 5 \end{bmatrix}\begin{bmatrix} 4 & 2 \\ 1 & 8 \end{bmatrix} \quad \text{is undefined.}$$ ■

EXAMPLE 3 Products of Row and Column Vectors

$$[3 \quad 6 \quad 1]\begin{bmatrix} 1 \\ 2 \\ 4 \end{bmatrix} = [19], \qquad \begin{bmatrix} 1 \\ 2 \\ 4 \end{bmatrix}[3 \quad 6 \quad 1] = \begin{bmatrix} 3 & 6 & 1 \\ 6 & 12 & 2 \\ 12 & 24 & 4 \end{bmatrix}.$$ ■

EXAMPLE 4 CAUTION! Matrix Multiplication Is Not Commutative, AB ≠ BA in General

This is illustrated by Examples 1 and 2, where one of the two products is not even defined, and by Example 3, where the two products have different sizes. But it also holds for square matrices. For instance,

$$\begin{bmatrix} 1 & 1 \\ 100 & 100 \end{bmatrix}\begin{bmatrix} -1 & 1 \\ 1 & -1 \end{bmatrix} = \begin{bmatrix} 0 & 0 \\ 0 & 0 \end{bmatrix} \quad \text{but} \quad \begin{bmatrix} -1 & 1 \\ 1 & -1 \end{bmatrix}\begin{bmatrix} 1 & 1 \\ 100 & 100 \end{bmatrix} = \begin{bmatrix} 99 & 99 \\ -99 & -99 \end{bmatrix}.$$

It is interesting that this also shows that $\mathbf{AB} = \mathbf{0}$ does ***not*** necessarily imply $\mathbf{BA} = \mathbf{0}$ or $\mathbf{A} = \mathbf{0}$ or $\mathbf{B} = \mathbf{0}$. We shall discuss this further in Sec. 7.8, along with reasons when this happens. ■

Our examples show that in matrix products ***the order of factors must always be observed very carefully***. Otherwise matrix multiplication satisfies rules similar to those for numbers, namely.

$$(2) \qquad \begin{aligned} &\text{(a)} \quad (k\mathbf{A})\mathbf{B} = k(\mathbf{AB}) = \mathbf{A}(k\mathbf{B}) \quad \textit{written } k\mathbf{AB} \textit{ or } \mathbf{A}k\mathbf{B} \\ &\text{(b)} \quad \mathbf{A}(\mathbf{BC}) = (\mathbf{AB})\mathbf{C} \quad \textit{written } \mathbf{ABC} \\ &\text{(c)} \quad (\mathbf{A} + \mathbf{B})\mathbf{C} = \mathbf{AC} + \mathbf{BC} \\ &\text{(d)} \quad \mathbf{C}(\mathbf{A} + \mathbf{B}) = \mathbf{CA} + \mathbf{CB} \end{aligned}$$

provided **A**, **B**, and **C** are such that the expressions on the left are defined; here, k is any scalar. (2b) is called the **associative law**. (2c) and (2d) are called the **distributive laws**.

Since matrix multiplication is a multiplication of rows into columns, we can write the defining formula (1) more compactly as

$$(3) \qquad c_{jk} = \mathbf{a}_j\mathbf{b}_k, \qquad j = 1, \cdots, m; \quad k = 1, \cdots, p,$$

where $\mathbf{a}_j$ is the jth row vector of **A** and $\mathbf{b}_k$ is the kth column vector of **B**, so that in agreement with (1),

$$\mathbf{a}_j\mathbf{b}_k = [a_{j1} \quad a_{j2} \quad \cdots \quad a_{jn}]\begin{bmatrix} b_{1k} \\ \vdots \\ b_{nk} \end{bmatrix} = a_{j1}b_{1k} + a_{j2}b_{2k} + \cdots + a_{jn}b_{nk}.$$

EXAMPLE 5 **Product in Terms of Row and Column Vectors**

If $\mathbf{A} = [a_{jk}]$ is of size 3×3 and $\mathbf{B} = [b_{jk}]$ is of size 3×4, then

(4)
$$\mathbf{AB} = \begin{bmatrix} \mathbf{a_1b_1} & \mathbf{a_1b_2} & \mathbf{a_1b_3} & \mathbf{a_1b_4} \\ \mathbf{a_2b_1} & \mathbf{a_2b_2} & \mathbf{a_2b_3} & \mathbf{a_2b_4} \\ \mathbf{a_3b_1} & \mathbf{a_3b_2} & \mathbf{a_3b_3} & \mathbf{a_3b_4} \end{bmatrix}.$$

Taking $\mathbf{a_1} = [3 \quad 5 \quad -1]$, $\mathbf{a_2} = [4 \quad 0 \quad 2]$, etc., verify (4) for the product in Example 1.

Parallel processing of products on the computer is facilitated by a variant of (3) for computing $\mathbf{C} = \mathbf{AB}$, which is used by standard algorithms (such as in Lapack). In this method, $\mathbf{A}$ is used as given, $\mathbf{B}$ is taken in terms of its column vectors, and the product is computed columnwise; thus,

(5)
$$\mathbf{AB} = \mathbf{A}[\mathbf{b_1} \quad \mathbf{b_2} \quad \cdots \quad \mathbf{b_p}] = [\mathbf{Ab_1} \quad \mathbf{Ab_2} \quad \cdots \quad \mathbf{Ab_p}].$$

Columns of $\mathbf{B}$ are then assigned to different processors (individually or several to each processor), which simultaneously compute the columns of the product matrix $\mathbf{Ab_1}$, $\mathbf{Ab_2}$, etc.

EXAMPLE 6 **Computing Products Columnwise by (5)**

To obtain

$$\mathbf{AB} = \begin{bmatrix} 4 & 1 \\ -5 & 2 \end{bmatrix} \begin{bmatrix} 3 & 0 & 7 \\ -1 & 4 & 6 \end{bmatrix} = \begin{bmatrix} 11 & 4 & 34 \\ -17 & 8 & -23 \end{bmatrix}$$

from (5), calculate the columns

$$\begin{bmatrix} 4 & 1 \\ -5 & 2 \end{bmatrix}\begin{bmatrix} 3 \\ -1 \end{bmatrix} = \begin{bmatrix} 11 \\ -17 \end{bmatrix}, \quad \begin{bmatrix} 4 & 1 \\ -5 & 2 \end{bmatrix}\begin{bmatrix} 0 \\ 4 \end{bmatrix} = \begin{bmatrix} 4 \\ 8 \end{bmatrix}, \quad \begin{bmatrix} 4 & 1 \\ -5 & 2 \end{bmatrix}\begin{bmatrix} 7 \\ 6 \end{bmatrix} = \begin{bmatrix} 34 \\ -23 \end{bmatrix}$$

of $\mathbf{AB}$ and then write them as a single matrix, as shown in the first formula on the right.

Motivation of Multiplication by Linear Transformations

Let us now motivate the "unnatural" matrix multiplication by its use in **linear transformations**. For $n = 2$ variables these transformations are of the form

(6*)
$$y_1 = a_{11}x_1 + a_{12}x_2$$
$$y_2 = a_{21}x_1 + a_{22}x_2$$

and suffice to explain the idea. (For general n they will be discussed in Sec. 7.9.) For instance, (6*) may relate an x_1x_2-coordinate system to a y_1y_2-coordinate system in the plane. In vectorial form we can write (6*) as

(6)
$$\mathbf{y} = \begin{bmatrix} y_1 \\ y_2 \end{bmatrix} = \mathbf{Ax} = \begin{bmatrix} a_{11} & a_{12} \\ a_{21} & a_{22} \end{bmatrix}\begin{bmatrix} x_1 \\ x_2 \end{bmatrix} = \begin{bmatrix} a_{11}x_1 + a_{12}x_2 \\ a_{21}x_1 + a_{22}x_2 \end{bmatrix}.$$

Now suppose further that the x_1x_2-system is related to a w_1w_2-system by another linear transformation, say,

$$\mathbf{x} = \begin{bmatrix} x_1 \\ x_2 \end{bmatrix} = \mathbf{Bw} = \begin{bmatrix} b_{11} & b_{12} \\ b_{21} & b_{22} \end{bmatrix} \begin{bmatrix} w_1 \\ w_2 \end{bmatrix} = \begin{bmatrix} b_{11}w_1 + b_{12}w_2 \\ b_{21}w_1 + b_{22}w_2 \end{bmatrix}. \tag{7}$$

Then the y_1y_2-system is related to the w_1w_2-system indirectly via the x_1x_2-system, and we wish to express this relation directly. Substitution will show that this direct relation is a linear transformation, too, say,

$$\mathbf{y} = \mathbf{Cw} = \begin{bmatrix} c_{11} & c_{12} \\ c_{21} & c_{22} \end{bmatrix} \begin{bmatrix} w_1 \\ w_2 \end{bmatrix} = \begin{bmatrix} c_{11}w_1 + c_{12}w_2 \\ c_{21}w_1 + c_{22}w_2 \end{bmatrix}. \tag{8}$$

Indeed, substituting (7) into (6), we obtain

$$\begin{aligned} y_1 &= a_{11}(b_{11}w_1 + b_{12}w_2) + a_{12}(b_{21}w_1 + b_{22}w_2) \\ &= (a_{11}b_{11} + a_{12}b_{21})w_1 + (a_{11}b_{12} + a_{12}b_{22})w_2 \\ y_2 &= a_{21}(b_{11}w_1 + b_{12}w_2) + a_{22}(b_{21}w_1 + b_{22}w_2) \\ &= (a_{21}b_{11} + a_{22}b_{21})w_1 + (a_{21}b_{12} + a_{22}b_{22})w_2. \end{aligned}$$

Comparing this with (8), we see that

$$\begin{aligned} c_{11} &= a_{11}b_{11} + a_{12}b_{21} \qquad & c_{12} &= a_{11}b_{12} + a_{12}b_{22} \\ c_{21} &= a_{21}b_{11} + a_{22}b_{21} \qquad & c_{22} &= a_{21}b_{12} + a_{22}b_{22}. \end{aligned}$$

This proves that $\mathbf{C} = \mathbf{AB}$ with the product defined as in (1). For larger matrix sizes the idea and result are exactly the same. Only the number of variables changes. We then have m variables y and n variables x and p variables w. The matrices $\mathbf{A}$, $\mathbf{B}$, and $\mathbf{C} = \mathbf{AB}$ then have sizes $m \times n$, $n \times p$, and $m \times p$, respectively. And the requirement that $\mathbf{C}$ be the product $\mathbf{AB}$ leads to formula (1) in its general form. *This motivates matrix multiplication.*

Transposition

We obtain the transpose of a matrix by writing its rows as columns (or equivalently its columns as rows). This also applies to the transpose of vectors. Thus, a row vector becomes a column vector and vice versa. In addition, for square matrices, we can also "reflect" the elements along the main diagonal, that is, interchange entries that are symmetrically positioned with respect to the main diagonal to obtain the transpose. Hence a_{12} becomes a_{21}, a_{31} becomes a_{13}, and so forth. Example 7 illustrates these ideas. Also note that, if $\mathbf{A}$ is the given matrix, then we denote its transpose by $\mathbf{A}^\mathsf{T}$.

EXAMPLE 7 Transposition of Matrices and Vectors

If $\qquad \mathbf{A} = \begin{bmatrix} 5 & -8 & 1 \\ 4 & 0 & 0 \end{bmatrix}, \qquad$ then $\qquad \mathbf{A}^\mathsf{T} = \begin{bmatrix} 5 & 4 \\ -8 & 0 \\ 1 & 0 \end{bmatrix}.$

A little more compactly, we can write

$$\begin{bmatrix} 5 & -8 & 1 \\ 4 & 0 & 0 \end{bmatrix}^{\mathsf{T}} = \begin{bmatrix} 5 & 4 \\ -8 & 0 \\ 1 & 0 \end{bmatrix}, \qquad \begin{bmatrix} 3 & 0 & 7 \\ 8 & -1 & 5 \\ 1 & -9 & 4 \end{bmatrix}^{\mathsf{T}} = \begin{bmatrix} 3 & 8 & 1 \\ 0 & -1 & -9 \\ 7 & 5 & 4 \end{bmatrix},$$

Furthermore, the transpose $[6 \quad 2 \quad 3]^{\mathsf{T}}$ of the row vector $[6 \quad 2 \quad 3]$ is the column vector

$$[6 \quad 2 \quad 3]^{\mathsf{T}} = \begin{bmatrix} 6 \\ 2 \\ 3 \end{bmatrix} \cdot \qquad \text{Conversely,} \qquad \begin{bmatrix} 6 \\ 2 \\ 3 \end{bmatrix}^{\mathsf{T}} = [6 \quad 2 \quad 3]. \qquad \blacksquare$$

DEFINITION

Transposition of Matrices and Vectors

The transpose of an $m \times n$ matrix $\mathbf{A} = [a_{jk}]$ is the $n \times m$ matrix $\mathbf{A}^{\mathsf{T}}$ (read *A transpose*) that has the first *row* of $\mathbf{A}$ as its first *column*, the second *row* of $\mathbf{A}$ as its second *column*, and so on. Thus the transpose of $\mathbf{A}$ in (2) is $\mathbf{A}^{\mathsf{T}} = [a_{kj}]$, written out

$$\text{(9)} \qquad \mathbf{A}^{\mathsf{T}} = [a_{kj}] = \begin{bmatrix} a_{11} & a_{21} & \cdots & a_{m1} \\ a_{12} & a_{22} & \cdots & a_{m2} \\ \cdot & \cdot & \cdots & \cdot \\ a_{1n} & a_{2n} & \cdots & a_{mn} \end{bmatrix}.$$

As a special case, transposition converts row vectors to column vectors and conversely.

Transposition gives us a choice in that we can work either with the matrix or its transpose, whichever is more convenient.

Rules for transposition are

$$\textbf{(10)} \qquad \begin{aligned} &\text{(a)} \quad (\mathbf{A}^{\mathsf{T}})^{\mathsf{T}} = \mathbf{A} \\ &\text{(b)} \quad (\mathbf{A} + \mathbf{B})^{\mathsf{T}} = \mathbf{A}^{\mathsf{T}} + \mathbf{B}^{\mathsf{T}} \\ &\text{(c)} \quad (c\mathbf{A})^{\mathsf{T}} = c\mathbf{A}^{\mathsf{T}} \\ &\text{(d)} \quad (\mathbf{A}\mathbf{B})^{\mathsf{T}} = \mathbf{B}^{\mathsf{T}}\mathbf{A}^{\mathsf{T}}. \end{aligned}$$

CAUTION! Note that in (10d) the transposed matrices are ***in reversed order.*** We leave the proofs as an exercise in Probs. 9 and 10.

Special Matrices

Certain kinds of matrices will occur quite frequently in our work, and we now list the most important ones of them.

Symmetric and Skew-Symmetric Matrices. Transposition gives rise to two useful classes of matrices. **Symmetric** matrices are square matrices whose transpose equals the matrix itself. **Skew-symmetric** matrices are square matrices whose transpose equals ***minus*** the matrix. Both cases are defined in (11) and illustrated by Example 8.

$$\mathbf{A}^{\mathsf{T}} = \mathbf{A} \quad (\text{thus } a_{kj} = a_{jk}), \qquad \mathbf{A}^{\mathsf{T}} = -\mathbf{A} \quad (\text{thus } a_{kj} = -a_{jk}, \text{ hence } a_{jj} = 0). \tag{11}$$

Symmetric Matrix — Skew-Symmetric Matrix

EXAMPLE 8 Symmetric and Skew-Symmetric Matrices

$$\mathbf{A} = \begin{bmatrix} 20 & 120 & 200 \\ 120 & 10 & 150 \\ 200 & 150 & 30 \end{bmatrix} \text{ is symmetric, and } \quad \mathbf{B} = \begin{bmatrix} 0 & 1 & -3 \\ -1 & 0 & -2 \\ 3 & 2 & 0 \end{bmatrix} \text{ is skew-symmetric.}$$

For instance, if a company has three building supply centers C_1, C_2, C_3, then $\mathbf{A}$ could show costs, say, a_{jj} for handling 1000 bags of cement at center C_j, and a_{jk} ($j \neq k$) the cost of shipping 1000 bags from C_j to C_k. Clearly, $a_{jk} = a_{kj}$ if we assume shipping in the opposite direction will cost the same.

Symmetric matrices have several general properties which make them important. This will be seen as we proceed. ■

Triangular Matrices. Upper triangular matrices are square matrices that can have nonzero entries only on and ***above*** the main diagonal, whereas any entry below the diagonal must be zero. Similarly, **lower triangular matrices** can have nonzero entries only on and ***below*** the main diagonal. Any entry on the main diagonal of a triangular matrix may be zero or not.

EXAMPLE 9 Upper and Lower Triangular Matrices

$$\begin{bmatrix} 1 & 3 \\ 0 & 2 \end{bmatrix}, \quad \begin{bmatrix} 1 & 4 & 2 \\ 0 & 3 & 2 \\ 0 & 0 & 6 \end{bmatrix}, \qquad \begin{bmatrix} 2 & 0 & 0 \\ 8 & -1 & 0 \\ 7 & 6 & 8 \end{bmatrix}, \quad \begin{bmatrix} 3 & 0 & 0 & 0 \\ 9 & -3 & 0 & 0 \\ 1 & 0 & 2 & 0 \\ 1 & 9 & 3 & 6 \end{bmatrix}.$$

Upper triangular — Lower triangular ■

Diagonal Matrices. These are square matrices that can have nonzero entries only on the main diagonal. Any entry above or below the main diagonal must be zero.

If all the diagonal entries of a diagonal matrix $\mathbf{S}$ are equal, say, c, we call $\mathbf{S}$ a **scalar matrix** because multiplication of any square matrix $\mathbf{A}$ of the same size by $\mathbf{S}$ has the same effect as the multiplication by a scalar, that is,

$$\mathbf{AS} = \mathbf{SA} = c\mathbf{A}. \tag{12}$$

In particular, a scalar matrix, whose entries on the main diagonal are all 1, is called a **unit matrix** (or **identity matrix**) and is denoted by $\mathbf{I}_n$ or simply by $\mathbf{I}$. For $\mathbf{I}$, formula (12) becomes

$$\mathbf{AI} = \mathbf{IA} = \mathbf{A}. \tag{13}$$

EXAMPLE 10 Diagonal Matrix D. Scalar Matrix S. Unit Matrix I

$$\mathbf{D} = \begin{bmatrix} 2 & 0 & 0 \\ 0 & -3 & 0 \\ 0 & 0 & 0 \end{bmatrix}, \quad \mathbf{S} = \begin{bmatrix} c & 0 & 0 \\ 0 & c & 0 \\ 0 & 0 & c \end{bmatrix}, \quad \mathbf{I} = \begin{bmatrix} 1 & 0 & 0 \\ 0 & 1 & 0 \\ 0 & 0 & 1 \end{bmatrix}$$ ■

Some Applications of Matrix Multiplication

EXAMPLE 11 **Computer Production. Matrix Times Matrix**

Supercomp Ltd produces two computer models PC1086 and PC1186. The matrix **A** shows the cost per computer (in thousands of dollars) and **B** the production figures for the year 2010 (in multiples of 10,000 units.) Find a matrix **C** that shows the shareholders the cost per quarter (in millions of dollars) for raw material, labor, and miscellaneous.

$$
\mathbf{A} = \begin{bmatrix} 1.2 & 1.6 \\ 0.3 & 0.4 \\ 0.5 & 0.6 \end{bmatrix} \begin{matrix} \text{Raw Components} \\ \text{Labor} \\ \text{Miscellaneous} \end{matrix}
\qquad
\mathbf{B} = \begin{bmatrix} 3 & 8 & 6 & 9 \\ 6 & 2 & 4 & 3 \end{bmatrix} \begin{matrix} \text{PC1086} \\ \text{PC1186} \end{matrix}
$$

with column headings PC1086 PC1186 over **A** and Quarter 1 2 3 4 over **B**.

Solution.

$$
\mathbf{C} = \mathbf{AB} = \begin{bmatrix} 13.2 & 12.8 & 13.6 & 15.6 \\ 3.3 & 3.2 & 3.4 & 3.9 \\ 5.1 & 5.2 & 5.4 & 6.3 \end{bmatrix} \begin{matrix} \text{Raw Components} \\ \text{Labor} \\ \text{Miscellaneous} \end{matrix}
$$

with column headings Quarter 1 2 3 4.

Since cost is given in multiples of $1000 and production in multiples of 10,000 units, the entries of **C** are multiples of $10 millions; thus $c_{11} = 13.2$ means $132 million, etc.

EXAMPLE 12 **Weight Watching. Matrix Times Vector**

Suppose that in a weight-watching program, a person of 185 lb burns 350 cal/hr in walking (3 mph), 500 in bicycling (13 mph), and 950 in jogging (5.5 mph). Bill, weighing 185 lb, plans to exercise according to the matrix shown. Verify the calculations (W = Walking, B = Bicycling, J = Jogging).

$$
\begin{matrix} \text{MON} \\ \text{WED} \\ \text{FRI} \\ \text{SAT} \end{matrix}
\begin{bmatrix} 1.0 & 0 & 0.5 \\ 1.0 & 1.0 & 0.5 \\ 1.5 & 0 & 0.5 \\ 2.0 & 1.5 & 1.0 \end{bmatrix}
\begin{bmatrix} 350 \\ 500 \\ 950 \end{bmatrix}
=
\begin{bmatrix} 825 \\ 1325 \\ 1000 \\ 2400 \end{bmatrix}
\begin{matrix} \text{MON} \\ \text{WED} \\ \text{FRI} \\ \text{SAT} \end{matrix}
$$

with column headings W B J over the first matrix.

EXAMPLE 13 **Markov Process. Powers of a Matrix. Stochastic Matrix**

Suppose that the 2004 state of land use in a city of 60 mi^2 of built-up area is

C: Commercially Used 25%   I: Industrially Used 20%   R: Residentially Used 55%.

Find the states in 2009, 2014, and 2019, assuming that the transition probabilities for 5-year intervals are given by the matrix **A** and remain practically the same over the time considered.

$$
\mathbf{A} = \begin{bmatrix} 0.7 & 0.1 & 0 \\ 0.2 & 0.9 & 0.2 \\ 0.1 & 0 & 0.8 \end{bmatrix} \begin{matrix} \text{To C} \\ \text{To I} \\ \text{To R} \end{matrix}
$$

with column headings From C From I From R.

A is a **stochastic matrix**, that is, a square matrix with all entries nonnegative and all column sums equal to 1. Our example concerns a **Markov process**,[1] that is, a process for which the probability of entering a certain state depends only on the last state occupied (and the matrix **A**), not on any earlier state.

Solution. From the matrix **A** and the 2004 state we can compute the 2009 state,

$$\begin{matrix} \text{C} \\ \text{I} \\ \text{R} \end{matrix} \quad \begin{bmatrix} 0.7 \cdot 25 + 0.1 \cdot 20 + 0 \cdot 55 \\ 0.2 \cdot 25 + 0.9 \cdot 20 + 0.2 \cdot 55 \\ 0.1 \cdot 25 + 0 \cdot 20 + 0.8 \cdot 55 \end{bmatrix} = \begin{bmatrix} 0.7 & 0.1 & 0 \\ 0.2 & 0.9 & 0.2 \\ 0.1 & 0 & 0.8 \end{bmatrix} \begin{bmatrix} 25 \\ 20 \\ 55 \end{bmatrix} = \begin{bmatrix} 19.5 \\ 34.0 \\ 46.5 \end{bmatrix}.$$

To explain: The 2009 figure for C equals 25% times the probability 0.7 that C goes into C, plus 20% times the probability 0.1 that I goes into C, plus 55% times the probability 0 that R goes into C. Together,

$$25 \cdot 0.7 + 20 \cdot 0.1 + 55 \cdot 0 = 19.5\,[\%]. \qquad \text{Also} \qquad 25 \cdot 0.2 + 20 \cdot 0.9 + 55 \cdot 0.2 = 34\,[\%].$$

Similarly, the new R is 46.5%. We see that the 2009 state vector is the column vector

$$\mathbf{y} = [19.5 \quad 34.0 \quad 46.5]^{\mathsf{T}} = \mathbf{Ax} = \mathbf{A}\,[25 \quad 20 \quad 55]^{\mathsf{T}}$$

where the column vector $\mathbf{x} = [25 \quad 20 \quad 55]^{\mathsf{T}}$ is the given 2004 state vector. Note that the sum of the entries of $\mathbf{y}$ is 100 [%]. Similarly, you may verify that for 2014 and 2019 we get the state vectors

$$\mathbf{z} = \mathbf{Ay} = \mathbf{A}(\mathbf{Ax}) = \mathbf{A}^2\mathbf{x} = [17.05 \quad 43.80 \quad 39.15]^{\mathsf{T}}$$

$$\mathbf{u} = \mathbf{Az} = \mathbf{A}^2\mathbf{y} = \mathbf{A}^3\mathbf{x} = [16.315 \quad 50.660 \quad 33.025]^{\mathsf{T}}.$$

Answer. In 2009 the commercial area will be 19.5% (11.7 mi^2), the industrial 34% (20.4 mi^2), and the residential 46.5% (27.9 mi^2). For 2014 the corresponding figures are 17.05%, 43.80%, and 39.15%. For 2019 they are 16.315%, 50.660%, and 33.025%. (In Sec. 8.2 we shall see what happens in the limit, assuming that those probabilities remain the same. In the meantime, can you experiment or guess?) ■

PROBLEM SET 7.2

1–10 GENERAL QUESTIONS

1. **Multiplication.** Why is multiplication of matrices restricted by conditions on the factors?
2. **Square matrix.** What form does a 3 × 3 matrix have if it is symmetric as well as skew-symmetric?
3. **Product of vectors.** Can every 3 × 3 matrix be represented by two vectors as in Example 3?
4. **Skew-symmetric matrix.** How many different entries can a 4 × 4 skew-symmetric matrix have? An $n \times n$ skew-symmetric matrix?
5. Same questions as in Prob. 4 for symmetric matrices.
6. **Triangular matrix.** If $\mathbf{U}_1$, $\mathbf{U}_2$ are upper triangular and $\mathbf{L}_1$, $\mathbf{L}_2$ are lower triangular, which of the following are triangular?

   $\mathbf{U}_1 + \mathbf{U}_2$, $\mathbf{U}_1\mathbf{U}_2$, $\mathbf{U}_1^2$, $\mathbf{U}_1 + \mathbf{L}_1$, $\mathbf{U}_1\mathbf{L}_1$, $\mathbf{L}_1 + \mathbf{L}_2$

7. **Idempotent matrix**, defined by $\mathbf{A}^2 = \mathbf{A}$. Can you find four 2 × 2 idempotent matrices?
8. **Nilpotent matrix**, defined by $\mathbf{B}^m = \mathbf{0}$ for some m. Can you find three 2 × 2 nilpotent matrices?
9. **Transposition.** Can you prove (10a)–(10c) for 3 × 3 matrices? For $m \times n$ matrices?
10. **Transposition. (a)** Illustrate (10d) by simple examples. **(b)** Prove (10d).

11–20 MULTIPLICATION, ADDITION, AND TRANSPOSITION OF MATRICES AND VECTORS

Let

$$\mathbf{A} = \begin{bmatrix} 4 & -2 & 3 \\ -2 & 1 & 6 \\ 1 & 2 & 2 \end{bmatrix}, \qquad \mathbf{B} = \begin{bmatrix} 1 & -3 & 0 \\ -3 & 1 & 0 \\ 0 & 0 & -2 \end{bmatrix}$$

$$\mathbf{C} = \begin{bmatrix} 0 & 1 \\ 3 & 2 \\ -2 & 0 \end{bmatrix}, \qquad \mathbf{a} = [1 \quad -2 \quad 0], \quad \mathbf{b} = \begin{bmatrix} 3 \\ 1 \\ -1 \end{bmatrix}.$$

[1] ANDREI ANDREJEVITCH MARKOV (1856–1922), Russian mathematician, known for his work in probability theory.

Showing all intermediate results, calculate the following expressions or give reasons why they are undefined:

11. $\mathbf{AB}$, $\mathbf{AB}^\mathsf{T}$, $\mathbf{BA}$, $\mathbf{B}^\mathsf{T}\mathbf{A}$

12. $\mathbf{AA}^\mathsf{T}$, $\mathbf{A}^2$, $\mathbf{BB}^\mathsf{T}$, $\mathbf{B}^2$

13. $\mathbf{CC}^\mathsf{T}$, $\mathbf{BC}$, $\mathbf{CB}$, $\mathbf{C}^\mathsf{T}\mathbf{B}$

14. $3\mathbf{A} - 2\mathbf{B}$, $(3\mathbf{A} - 2\mathbf{B})^\mathsf{T}$, $3\mathbf{A}^\mathsf{T} - 2\mathbf{B}^\mathsf{T}$, $(3\mathbf{A} - 2\mathbf{B})^\mathsf{T}\mathbf{a}^\mathsf{T}$

15. $\mathbf{Aa}$, $\mathbf{Aa}^\mathsf{T}$, $(\mathbf{Ab})^\mathsf{T}$, $\mathbf{b}^\mathsf{T}\mathbf{A}^\mathsf{T}$

16. $\mathbf{BC}$, $\mathbf{BC}^\mathsf{T}$, $\mathbf{Bb}$, $\mathbf{b}^\mathsf{T}\mathbf{B}$

17. $\mathbf{ABC}$, $\mathbf{ABa}$, $\mathbf{ABb}$, $\mathbf{Ca}^\mathsf{T}$

18. $\mathbf{ab}$, $\mathbf{ba}$, $\mathbf{aA}$, $\mathbf{Bb}$

19. $1.5\mathbf{a} + 3.0\mathbf{b}$, $1.5\mathbf{a}^\mathsf{T} + 3.0\mathbf{b}$, $(\mathbf{A} - \mathbf{B})\mathbf{b}$, $\mathbf{Ab} - \mathbf{Bb}$

20. $\mathbf{b}^\mathsf{T}\mathbf{Ab}$, $\mathbf{aBa}^\mathsf{T}$, $\mathbf{aCC}^\mathsf{T}$, $\mathbf{C}^\mathsf{T}\mathbf{ba}$

21. General rules. Prove (2) for 2×2 matrices $\mathbf{A} = [a_{jk}]$, $\mathbf{B} = [b_{jk}]$, $\mathbf{C} = [c_{jk}]$, and a general scalar.

22. Product. Write $\mathbf{AB}$ in Prob. 11 in terms of row and column vectors.

23. Product. Calculate $\mathbf{AB}$ in Prob. 11 columnwise. See Example 1.

24. Commutativity. Find all 2×2 matrices $\mathbf{A} = [a_{jk}]$ that commute with $\mathbf{B} = [b_{jk}]$, where $b_{jk} = j + k$.

25. TEAM PROJECT. Symmetric and Skew-Symmetric Matrices. These matrices occur quite frequently in applications, so it is worthwhile to study some of their most important properties.

(a) Verify the claims in (11) that $a_{kj} = a_{jk}$ for a symmetric matrix, and $a_{kj} = -a_{jk}$ for a skew-symmetric matrix. Give examples.

(b) Show that for every square matrix $\mathbf{C}$ the matrix $\mathbf{C} + \mathbf{C}^\mathsf{T}$ is symmetric and $\mathbf{C} - \mathbf{C}^\mathsf{T}$ is skew-symmetric. Write $\mathbf{C}$ in the form $\mathbf{C} = \mathbf{S} + \mathbf{T}$, where $\mathbf{S}$ is symmetric and $\mathbf{T}$ is skew-symmetric and find $\mathbf{S}$ and $\mathbf{T}$ in terms of $\mathbf{C}$. Represent $\mathbf{A}$ and $\mathbf{B}$ in Probs. 11–20 in this form.

(c) A **linear combination** of matrices $\mathbf{A}, \mathbf{B}, \mathbf{C}, \cdots, \mathbf{M}$ of the same size is an expression of the form

$$(14) \qquad a\mathbf{A} + b\mathbf{B} + c\mathbf{C} + \cdots + m\mathbf{M},$$

where $a, \cdots, m$ are any scalars. Show that if these matrices are square and symmetric, so is (14); similarly, if they are skew-symmetric, so is (14).

(d) Show that $\mathbf{AB}$ with symmetric $\mathbf{A}$ and $\mathbf{B}$ is symmetric if and only if $\mathbf{A}$ and $\mathbf{B}$ **commute**, that is, $\mathbf{AB} = \mathbf{BA}$.

(e) Under what condition is the product of skew-symmetric matrices skew-symmetric?

26–30 FURTHER APPLICATIONS

26. Production. In a production process, let N mean "no trouble" and T "trouble." Let the transition probabilities from one day to the next be 0.8 for $N \to N$, hence 0.2 for $N \to T$, and 0.5 for $T \to N$, hence 0.5 for $T \to T$.

If today there is no trouble, what is the probability of N two days after today? Three days after today?

27. CAS Experiment. Markov Process. Write a program for a Markov process. Use it to calculate further steps in Example 13 of the text. Experiment with other stochastic 3×3 matrices, also using different starting values.

28. Concert subscription. In a community of 100,000 adults, subscribers to a concert series tend to renew their subscription with probability 90% and persons presently not subscribing will subscribe for the next season with probability 0.2%. If the present number of subscribers is 1200, can one predict an increase, decrease, or no change over each of the next three seasons?

29. Profit vector. Two factory outlets F_1 and F_2 in New York and Los Angeles sell sofas (S), chairs (C), and tables (T) with a profit of \$35, \$62, and \$30, respectively. Let the sales in a certain week be given by the matrix

$$\mathbf{A} = \begin{array}{c} \begin{matrix} \text{S} & \text{C} & \text{T} \end{matrix} \\ \begin{bmatrix} 400 & 60 & 240 \\ 100 & 120 & 500 \end{bmatrix} \end{array} \begin{matrix} F_1 \\ F_2 \end{matrix}$$

Introduce a "profit vector" $\mathbf{p}$ such that the components of $\mathbf{v} = \mathbf{Ap}$ give the total profits of F_1 and F_2.

30. TEAM PROJECT. Special Linear Transformations. Rotations have various applications. We show in this project how they can be handled by matrices.

(a) Rotation in the plane. Show that the linear transformation $\mathbf{y} = \mathbf{Ax}$ with

$$\mathbf{A} = \begin{bmatrix} \cos\theta & -\sin\theta \\ \sin\theta & \cos\theta \end{bmatrix}, \quad \mathbf{x} = \begin{bmatrix} x_1 \\ x_2 \end{bmatrix}, \quad \mathbf{y} = \begin{bmatrix} y_1 \\ y_2 \end{bmatrix}$$

is a counterclockwise rotation of the Cartesian x_1x_2-coordinate system in the plane about the origin, where θ is the angle of rotation.

(b) Rotation through $n\theta$. Show that in (a)

$$\mathbf{A}^n = \begin{bmatrix} \cos n\theta & -\sin n\theta \\ \sin n\theta & \cos n\theta \end{bmatrix}.$$

Is this plausible? Explain this in words.

(c) Addition formulas for cosine and sine. By geometry we should have

$$\begin{bmatrix} \cos\alpha & -\sin\alpha \\ \sin\alpha & \cos\alpha \end{bmatrix} \begin{bmatrix} \cos\beta & -\sin\beta \\ \sin\beta & \cos\beta \end{bmatrix} = \begin{bmatrix} \cos(\alpha+\beta) & -\sin(\alpha+\beta) \\ \sin(\alpha+\beta) & \cos(\alpha+\beta) \end{bmatrix}.$$

Derive from this the addition formulas (6) in App. A3.1.

(d) Computer graphics. To visualize a three-dimensional object with plane faces (e.g., a cube), we may store the position vectors of the vertices with respect to a suitable $x_1x_2x_3$-coordinate system (and a list of the connecting edges) and then obtain a two-dimensional image on a video screen by projecting the object onto a coordinate plane, for instance, onto the x_1x_2-plane by setting $x_3 = 0$. To change the appearance of the image, we can impose a linear transformation on the position vectors stored. Show that a diagonal matrix **D** with main diagonal entries 3, 1, $\frac{1}{2}$ gives from an $\mathbf{x} = [x_j]$ the new position vector $\mathbf{y} = \mathbf{D}\mathbf{x}$, where $y_1 = 3x_1$ (stretch in the x_1-direction by a factor 3), $y_2 = x_2$ (unchanged), $y_3 = \frac{1}{2}x_3$ (contraction in the x_3-direction). What effect would a scalar matrix have?

(e) Rotations in space. Explain $\mathbf{y} = \mathbf{A}\mathbf{x}$ geometrically when **A** is one of the three matrices

$$\begin{bmatrix} 1 & 0 & 0 \\ 0 & \cos\theta & -\sin\theta \\ 0 & \sin\theta & \cos\theta \end{bmatrix},$$

$$\begin{bmatrix} \cos\varphi & 0 & -\sin\varphi \\ 0 & 1 & 0 \\ \sin\varphi & 0 & \cos\varphi \end{bmatrix}, \quad \begin{bmatrix} \cos\psi & -\sin\psi & 0 \\ \sin\psi & \cos\psi & 0 \\ 0 & 0 & 1 \end{bmatrix}.$$

What effect would these transformations have in situations such as that described in (d)?

7.3 Linear Systems of Equations. Gauss Elimination

We now come to one of the most important use of matrices, that is, using matrices to solve systems of linear equations. We showed informally, in Example 1 of Sec. 7.1, how to represent the information contained in a system of linear equations by a matrix, called the augmented matrix. This matrix will then be used in solving the linear system of equations. Our approach to solving linear systems is called the Gauss elimination method. Since this method is so fundamental to linear algebra, the student should be alert.

A shorter term for systems of linear equations is just **linear systems**. Linear systems model many applications in engineering, economics, statistics, and many other areas. Electrical networks, traffic flow, and commodity markets may serve as specific examples of applications.

Linear System, Coefficient Matrix, Augmented Matrix

A **linear system of m equations in n unknowns** $x_1, \cdots, x_n$ is a set of equations of the form

$$\begin{aligned} a_{11}x_1 + \cdots + a_{1n}x_n &= b_1 \\ a_{21}x_1 + \cdots + a_{2n}x_n &= b_2 \\ \cdots\cdots\cdots\cdots\cdots\cdots\cdots\cdots \\ a_{m1}x_1 + \cdots + a_{mn}x_n &= b_m. \end{aligned} \tag{1}$$

The system is called *linear* because each variable x_j appears in the first power only, just as in the equation of a straight line. $a_{11}, \cdots, a_{mn}$ are given numbers, called the **coefficients** of the system. $b_1, \cdots, b_m$ on the right are also given numbers. If all the b_j are zero, then (1) is called a **homogeneous system**. If at least one b_j is not zero, then (1) is called a **nonhomogeneous system**.

A **solution** of (1) is a set of numbers $x_1, \cdots, x_n$ that satisfies all the m equations. A **solution vector** of (1) is a vector $\mathbf{x}$ whose components form a solution of (1). If the system (1) is homogeneous, it always has at least the **trivial solution** $x_1 = 0, \cdots, x_n = 0$.

Matrix Form of the Linear System (1). From the definition of matrix multiplication we see that the m equations of (1) may be written as a single vector equation

$$\mathbf{Ax} = \mathbf{b} \tag{2}$$

where the **coefficient matrix** $\mathbf{A} = [a_{jk}]$ is the $m \times n$ matrix

$$\mathbf{A} = \begin{bmatrix} a_{11} & a_{12} & \cdots & a_{1n} \\ a_{21} & a_{22} & \cdots & a_{2n} \\ \cdot & \cdot & \cdots & \cdot \\ a_{m1} & a_{m2} & \cdots & a_{mn} \end{bmatrix}, \quad \text{and} \quad \mathbf{x} = \begin{bmatrix} x_1 \\ \cdot \\ \cdot \\ \cdot \\ x_n \end{bmatrix} \quad \text{and} \quad \mathbf{b} = \begin{bmatrix} b_1 \\ \vdots \\ b_m \end{bmatrix}$$

are column vectors. We assume that the coefficients a_{jk} are not all zero, so that $\mathbf{A}$ is not a zero matrix. Note that $\mathbf{x}$ has n components, whereas $\mathbf{b}$ has m components. The matrix

$$\tilde{\mathbf{A}} = \left[\begin{array}{ccc|c} a_{11} & \cdots & a_{1n} & b_1 \\ \cdot & \cdots & \cdot & \cdot \\ \cdot & \cdots & \cdot & \cdot \\ a_{m1} & \cdots & a_{mn} & b_m \end{array}\right]$$

is called the **augmented matrix** of the system (1). The dashed vertical line could be omitted, as we shall do later. It is merely a reminder that the last column of $\tilde{\mathbf{A}}$ did not come from matrix $\mathbf{A}$ but came from vector $\mathbf{b}$. Thus, we *augmented* the matrix $\mathbf{A}$.

*Note that **the augmented matrix*** $\tilde{\mathbf{A}}$ ***determines the system* (1) *completely*** because it contains all the given numbers appearing in (1).

EXAMPLE 1 Geometric Interpretation. Existence and Uniqueness of Solutions

If $m = n = 2$, we have two equations in two unknowns x_1, x_2

$$a_{11}x_1 + a_{12}x_2 = b_1$$
$$a_{21}x_1 + a_{22}x_2 = b_2.$$

If we interpret x_1, x_2 as coordinates in the x_1x_2-plane, then each of the two equations represents a straight line, and (x_1, x_2) is a solution if and only if the point P with coordinates x_1, x_2 lies on both lines. Hence there are three possible cases (see Fig. 158 on next page):

(*a*) Precisely one solution if the lines intersect

(*b*) Infinitely many solutions if the lines coincide

(*c*) No solution if the lines are parallel

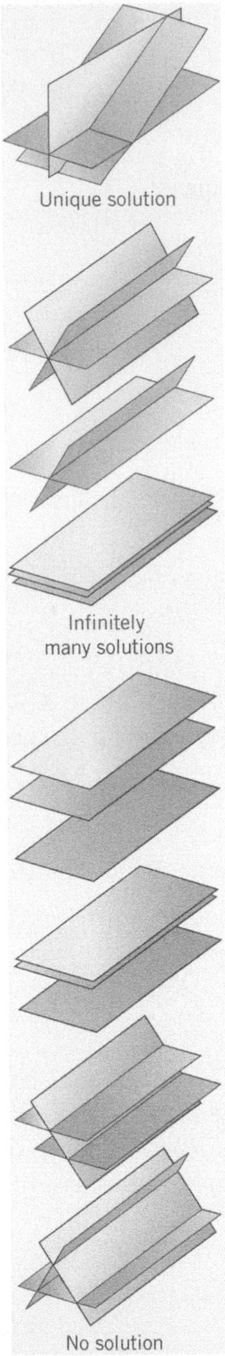

Fig. 158. Three equations in three unknowns interpreted as planes in space

For instance,

Case (a)	Case (b)	Case (c)
$x_1 + x_2 = 1$	$x_1 + x_2 = 1$	$x_1 + x_2 = 1$
$2x_1 - x_2 = 0$	$2x_1 + 2x_2 = 2$	$x_1 + x_2 = 0$

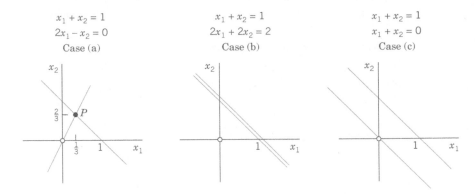

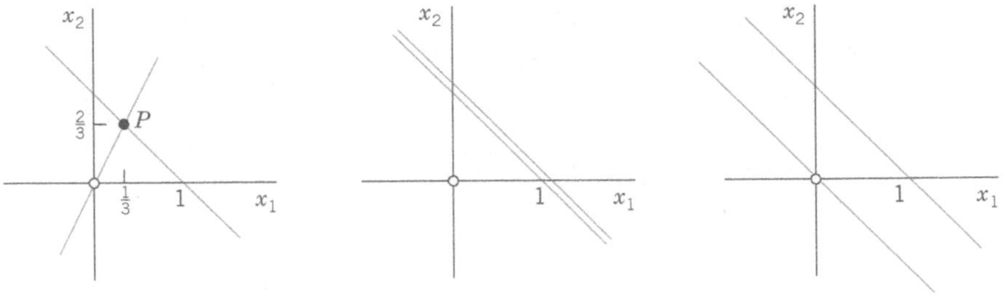

If the system is homogenous, Case (c) cannot happen, because then those two straight lines pass through the origin, whose coordinates (0, 0) constitute the trivial solution. Similarly, our present discussion can be extended from two equations in two unknowns to three equations in three unknowns. We give the geometric interpretation of three possible cases concerning solutions in Fig. 158. Instead of straight lines we have planes and the solution depends on the positioning of these planes in space relative to each other. The student may wish to come up with some specific examples. ■

Our simple example illustrated that a system (1) may have no solution. This leads to such questions as: Does a given system (1) have a solution? Under what conditions does it have precisely one solution? If it has more than one solution, how can we characterize the set of all solutions? We shall consider such questions in Sec. 7.5.

First, however, let us discuss an important systematic method for solving linear systems.

Gauss Elimination and Back Substitution

The Gauss elimination method can be motivated as follows. Consider a linear system that is in ***triangular form*** (in full, *upper* triangular form) such as

$$\begin{aligned} 2x_1 + 5x_2 &= 2 \\ 13x_2 &= -26 \end{aligned}$$

(*Triangular* means that all the nonzero entries of the corresponding coefficient matrix lie above the diagonal and form an upside-down 90° triangle.) Then we can solve the system by **back substitution**, that is, we solve the last equation for the variable, $x_2 = -26/13 = -2$, and then work backward, substituting $x_2 = -2$ into the first equation and solving it for x_1, obtaining $x_1 = \frac{1}{2}(2 - 5x_2) = \frac{1}{2}(2 - 5 \cdot (-2)) = 6$. This gives us the idea of first reducing a general system to triangular form. For instance, let the given system be

$$\begin{aligned} 2x_1 + 5x_2 &= 2 \\ -4x_1 + 3x_2 &= -30. \end{aligned} \qquad \text{Its augmented matrix is} \qquad \begin{bmatrix} 2 & 5 & 2 \\ -4 & 3 & -30 \end{bmatrix}.$$

We leave the first equation as it is. We eliminate x_1 from the second equation, to get a triangular system. For this we add twice the first equation to the second, and we do the same operation on the ***rows*** of the augmented matrix. This gives $-4x_1 + 4x_1 + 3x_2 + 10x_2 = -30 + 2 \cdot 2$, that is,

$$\begin{aligned} 2x_1 + 5x_2 &= 2 \\ 13x_2 &= -26 \end{aligned} \qquad \begin{matrix} \\ \text{Row 2 + 2 Row 1} \end{matrix} \qquad \begin{bmatrix} 2 & 5 & 2 \\ 0 & 13 & -26 \end{bmatrix}$$

where Row 2 + 2 Row 1 means "Add twice Row 1 to Row 2" in the original matrix. This is the **Gauss elimination** (for 2 equations in 2 unknowns) giving the triangular form, from which back substitution now yields $x_2 = -2$ and $x_1 = 6$, as before.

Since a linear system is completely determined by its augmented matrix, ***Gauss elimination can be done by merely considering the matrices***, as we have just indicated. We do this again in the next example, emphasizing the matrices by writing them first and the equations behind them, just as a help in order not to lose track.

EXAMPLE 2 Gauss Elimination. Electrical Network

Solve the linear system

$$\begin{aligned} x_1 - x_2 + x_3 &= 0 \\ -x_1 + x_2 - x_3 &= 0 \\ 10x_2 + 25x_3 &= 90 \\ 20x_1 + 10x_2 \qquad &= 80. \end{aligned}$$

Derivation from the circuit in Fig. 159 (Optional). This is the system for the unknown currents $x_1 = i_1, x_2 = i_2, x_3 = i_3$ in the electrical network in Fig. 159. To obtain it, we label the currents as shown, choosing directions arbitrarily; if a current will come out negative, this will simply mean that the current flows against the direction of our arrow. The current entering each battery will be the same as the current leaving it. The equations for the currents result from Kirchhoff's laws:

Kirchhoff's Current Law (KCL). *At any point of a circuit, the sum of the inflowing currents equals the sum of the outflowing currents.*

Kirchhoff's Voltage Law (KVL). *In any closed loop, the sum of all voltage drops equals the impressed electromotive force.*

Node P gives the first equation, node Q the second, the right loop the third, and the left loop the fourth, as indicated in the figure.

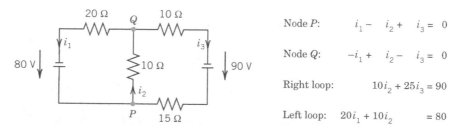

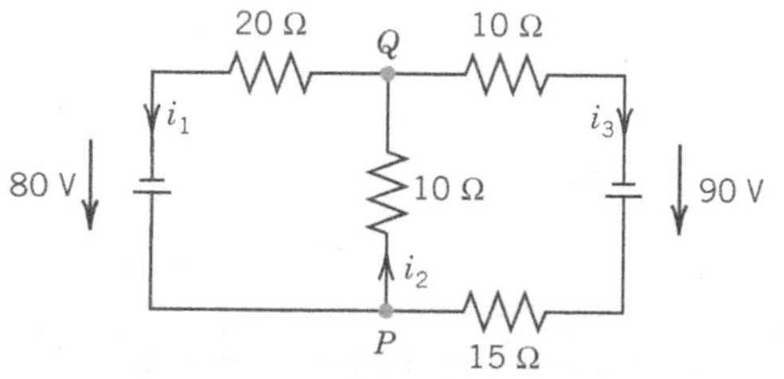

Node P:	$i_1 - i_2 + i_3 = 0$
Node Q:	$-i_1 + i_2 - i_3 = 0$
Right loop:	$10i_2 + 25i_3 = 90$
Left loop:	$20i_1 + 10i_2 = 80$

Fig. 159. Network in Example 2 and equations relating the currents

Solution by Gauss Elimination. This system could be solved rather quickly by noticing its particular form. But this is not the point. The point is that the Gauss elimination is systematic and will work in general, also for large systems. We apply it to our system and then do back substitution. As indicated, let us write the augmented matrix of the system first and then the system itself:

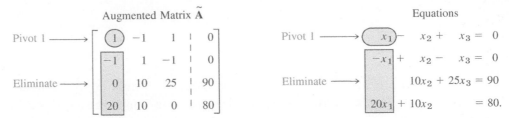

Step 1. Elimination of x_1

Call the first row of **A** the **pivot row** and the first equation the **pivot equation**. Call the coefficient 1 of its x_1-term the **pivot** in this step. Use this equation to eliminate x_1 (get rid of x_1) in the other equations. For this, do:

   Add 1 times the pivot equation to the second equation.

   Add -20 times the pivot equation to the fourth equation.

This corresponds to **row operations** on the augmented matrix as indicated in BLUE behind the ***new matrix*** in (3). So the operations are performed on the ***preceding matrix***. The result is

$$(3) \quad
\begin{bmatrix}
1 & -1 & 1 & | & 0 \\
0 & 0 & 0 & | & 0 \\
0 & 10 & 25 & | & 90 \\
0 & 30 & -20 & | & 80
\end{bmatrix}
\begin{matrix}
\\ \text{Row 2 + Row 1} \\ \\ \text{Row 4 } - \text{ 20 Row 1}
\end{matrix}
\qquad
\begin{matrix}
x_1 - x_2 + x_3 = 0 \\
0 = 0 \\
10x_2 + 25x_3 = 90 \\
30x_2 - 20x_3 = 80.
\end{matrix}$$

Step 2. Elimination of x_2

The first equation remains as it is. We want the new second equation to serve as the next pivot equation. But since it has no x_2-term (in fact, it is $0 = 0$), we must first change the order of the equations and the corresponding rows of the new matrix. We put $0 = 0$ at the end and move the third equation and the fourth equation one place up. This is called **partial pivoting** (as opposed to the rarely used *total pivoting*, in which the order of the unknowns is also changed). It gives

$$
\begin{matrix} \text{Pivot 10} \longrightarrow \\ \text{Eliminate 30} \longrightarrow \end{matrix}
\begin{bmatrix}
1 & -1 & 1 & | & 0 \\
0 & 10 & 25 & | & 90 \\
0 & 30 & -20 & | & 80 \\
0 & 0 & 0 & | & 0
\end{bmatrix}
\qquad
\begin{matrix} \text{Pivot 10} \longrightarrow \\ \text{Eliminate } 30x_2 \longrightarrow \end{matrix}
\begin{matrix}
x_1 - x_2 + x_3 = 0 \\
10x_2 + 25x_3 = 90 \\
30x_2 - 20x_3 = 80 \\
0 = 0.
\end{matrix}$$

To eliminate x_2, do:

   Add -3 times the pivot equation to the third equation.

The result is

$$(4) \quad
\begin{bmatrix}
1 & -1 & 1 & | & 0 \\
0 & 10 & 25 & | & 90 \\
0 & 0 & -95 & | & -190 \\
0 & 0 & 0 & | & 0
\end{bmatrix}
\begin{matrix}
\\ \\ \text{Row 3 } - \text{ 3 Row 2} \\ \\
\end{matrix}
\qquad
\begin{matrix}
x_1 - x_2 + x_3 = 0 \\
10x_2 + 25x_3 = 90 \\
-95x_3 = -190 \\
0 = 0.
\end{matrix}$$

Back Substitution.   *Determination of x_3, x_2, x_1 (in this order)*

Working backward from the last to the first equation of this "triangular" system (4), we can now readily find x_3, then x_2, and then x_1:

$$
\begin{aligned}
-95x_3 &= -190 & x_3 &= i_3 = 2\,[\text{A}] \\
10x_2 + 25x_3 &= 90 & x_2 &= \tfrac{1}{10}(90 - 25x_3) = i_2 = 4\,[\text{A}] \\
x_1 - x_2 + x_3 &= 0 & x_1 &= x_2 - x_3 = i_1 = 2\,[\text{A}]
\end{aligned}$$

where A stands for "amperes." This is the answer to our problem. The solution is unique. ∎

Elementary Row Operations. Row-Equivalent Systems

Example 2 illustrates the operations of the Gauss elimination. These are the first two of three operations, which are called

Elementary Row Operations for Matrices:

Interchange of two rows

Addition of a constant multiple of one row to another row

Multiplication of a row by a ***nonzero*** *constant c*

CAUTION! These operations are for rows, ***not for columns***! They correspond to the following

Elementary Operations for Equations:

Interchange of two equations

Addition of a constant multiple of one equation to another equation

Multiplication of an equation by a ***nonzero*** *constant c*

Clearly, the interchange of two equations does not alter the solution set. Neither does their addition because we can undo it by a corresponding subtraction. Similarly for their multiplication, which we can undo by multiplying the new equation by $1/c$ (since $c \neq 0$), producing the original equation.

We now call a linear system S_1 **row-equivalent** to a linear system S_2 if S_1 can be obtained from S_2 by (finitely many!) row operations. This justifies Gauss elimination and establishes the following result.

THEOREM 1

Row-Equivalent Systems

Row-equivalent linear systems have the same set of solutions.

Because of this theorem, systems having the same solution sets are often called *equivalent systems*. But note well that we are dealing with ***row operations***. No column operations on the augmented matrix are permitted in this context because they would generally alter the solution set.

A linear system (1) is called **overdetermined** if it has more equations than unknowns, as in Example 2, **determined** if $m = n$, as in Example 1, and **underdetermined** if it has fewer equations than unknowns.

Furthermore, a system (1) is called **consistent** if it has at least one solution (thus, one solution or infinitely many solutions), but **inconsistent** if it has no solutions at all, as $x_1 + x_2 = 1, x_1 + x_2 = 0$ in Example 1, Case (c).

Gauss Elimination: The Three Possible Cases of Systems

We have seen, in Example 2, that Gauss elimination can solve linear systems that have a unique solution. This leaves us to apply Gauss elimination to a system with infinitely many solutions (in Example 3) and one with no solution (in Example 4).

EXAMPLE 3 Gauss Elimination if Infinitely Many Solutions Exist

Solve the following linear system of three equations in four unknowns whose augmented matrix is

$$(5)\quad \begin{bmatrix} 3.0 & 2.0 & 2.0 & -5.0 & | & 8.0 \\ 0.6 & 1.5 & 1.5 & -5.4 & | & 2.7 \\ 1.2 & -0.3 & -0.3 & 2.4 & | & 2.1 \end{bmatrix}. \qquad \text{Thus,} \qquad \begin{aligned} \boxed{3.0x_1} + 2.0x_2 + 2.0x_3 - 5.0x_4 &= 8.0 \\ \boxed{0.6x_1} + 1.5x_2 + 1.5x_3 - 5.4x_4 &= 2.7 \\ \boxed{1.2x_1} - 0.3x_2 - 0.3x_3 + 2.4x_4 &= 2.1. \end{aligned}$$

Solution. As in the previous example, we circle pivots and box terms of equations and corresponding entries to be eliminated. We indicate the operations in terms of equations and operate on both equations and matrices.

Step 1. Elimination of x_1 from the second and third equations by adding

$-0.6/3.0 = -0.2$ times the first equation to the second equation,

$-1.2/3.0 = -0.4$ times the first equation to the third equation.

This gives the following, in which the pivot of the next step is circled.

$$(6)\quad \begin{bmatrix} 3.0 & 2.0 & 2.0 & -5.0 & | & 8.0 \\ 0 & 1.1 & 1.1 & -4.4 & | & 1.1 \\ 0 & -1.1 & -1.1 & 4.4 & | & -1.1 \end{bmatrix} \begin{matrix} \\ \text{Row 2} - 0.2 \text{ Row 1} \\ \text{Row 3} - 0.4 \text{ Row 1} \end{matrix} \qquad \begin{aligned} 3.0x_1 + 2.0x_2 + 2.0x_3 - 5.0x_4 &= 8.0 \\ \boxed{1.1x_2} + 1.1x_3 - 4.4x_4 &= 1.1 \\ \boxed{-1.1x_2} - 1.1x_3 + 4.4x_4 &= -1.1. \end{aligned}$$

Step 2. Elimination of x_2 from the third equation of (6) by adding

$1.1/1.1 = 1$ times the second equation to the third equation.

This gives

$$(7)\quad \begin{bmatrix} 3.0 & 2.0 & 2.0 & -5.0 & | & 8.0 \\ 0 & 1.1 & 1.1 & -4.4 & | & 1.1 \\ 0 & 0 & 0 & 0 & | & 0 \end{bmatrix} \begin{matrix} \\ \\ \text{Row 3} + \text{Row 2} \end{matrix} \qquad \begin{aligned} 3.0x_1 + 2.0x_2 + 2.0x_3 - 5.0x_4 &= 8.0 \\ 1.1x_2 + 1.1x_3 - 4.4x_4 &= 1.1 \\ 0 &= 0. \end{aligned}$$

Back Substitution. From the second equation, $x_2 = 1 - x_3 + 4x_4$. From this and the first equation, $x_1 = 2 - x_4$. Since x_3 and x_4 remain arbitrary, we have infinitely many solutions. If we choose a value of x_3 and a value of x_4, then the corresponding values of x_1 and x_2 are uniquely determined.

On Notation. If unknowns remain arbitrary, it is also customary to denote them by other letters $t_1, t_2, \cdots$. In this example we may thus write $x_1 = 2 - x_4 = 2 - t_2$, $x_2 = 1 - x_3 + 4x_4 = 1 - t_1 + 4t_2$, $x_3 = t_1$ (first arbitrary unknown), $x_4 = t_2$ (second arbitrary unknown). ■

EXAMPLE 4 Gauss Elimination if no Solution Exists

What will happen if we apply the Gauss elimination to a linear system that has no solution? The answer is that in this case the method will show this fact by producing a contradiction. For instance, consider

$$\begin{bmatrix} 3 & 2 & 1 & | & 3 \\ 2 & 1 & 1 & | & 0 \\ 6 & 2 & 4 & | & 6 \end{bmatrix} \qquad \begin{aligned} \boxed{3x_1} + 2x_2 + x_3 &= 3 \\ \boxed{2x_1} + x_2 + x_3 &= 0 \\ \boxed{6x_1} + 2x_2 + 4x_3 &= 6. \end{aligned}$$

Step 1. Elimination of x_1 from the second and third equations by adding

$-\frac{2}{3}$ times the first equation to the second equation,

$-\frac{6}{3} = -2$ times the first equation to the third equation.

This gives

$$\begin{bmatrix} 3 & 2 & 1 & | & 3 \\ 0 & -\frac{1}{3} & \frac{1}{3} & | & -2 \\ 0 & -2 & 2 & | & 0 \end{bmatrix} \quad \begin{matrix} \\ \text{Row 2} - \frac{2}{3}\text{ Row 1} \\ \text{Row 3} - 2\text{ Row 1} \end{matrix} \qquad \begin{aligned} 3x_1 + 2x_2 + x_3 &= 3 \\ -\tfrac{1}{3}x_2 + \tfrac{1}{3}x_3 &= -2 \\ -2x_2 + 2x_3 &= 0. \end{aligned}$$

Step 2. Elimination of x_2 from the third equation gives

$$\begin{bmatrix} 3 & 2 & 1 & | & 3 \\ 0 & -\frac{1}{3} & \frac{1}{3} & | & -2 \\ 0 & 0 & 0 & | & 12 \end{bmatrix} \quad \begin{matrix} \\ \\ \text{Row 3} - 6\text{ Row 2} \end{matrix} \qquad \begin{aligned} 3x_1 + 2x_2 + x_3 &= 3 \\ -\tfrac{1}{3}x_2 + \tfrac{1}{3}x_3 &= -2 \\ 0 &= 12. \end{aligned}$$

The false statement $0 = 12$ shows that the system has no solution. ■

Row Echelon Form and Information From It

At the end of the Gauss elimination the form of the coefficient matrix, the augmented matrix, and the system itself are called the **row echelon form**. In it, rows of zeros, if present, are the last rows, and, in each nonzero row, the leftmost nonzero entry is farther to the right than in the previous row. For instance, in Example 4 the coefficient matrix and its augmented in row echelon form are

$$\begin{bmatrix} 3 & 2 & 1 \\ 0 & -\frac{1}{3} & \frac{1}{3} \\ 0 & 0 & 0 \end{bmatrix} \quad \text{and} \quad \begin{bmatrix} 3 & 2 & 1 & | & 3 \\ 0 & -\frac{1}{3} & \frac{1}{3} & | & -2 \\ 0 & 0 & 0 & | & 12 \end{bmatrix}. \tag{8}$$

Note that we do not require that the leftmost nonzero entries be 1 since this would have no theoretic or numeric advantage. (The so-called *reduced echelon form*, in which those entries *are* 1, will be discussed in Sec. 7.8.)

The original system of m equations in n unknowns has augmented matrix $[\mathbf{A}\,|\,\mathbf{b}]$. This is to be row reduced to matrix $[\mathbf{R}\,|\,\mathbf{f}]$. The two systems $\mathbf{Ax} = \mathbf{b}$ and $\mathbf{Rx} = \mathbf{f}$ are equivalent: if either one has a solution, so does the other, and the solutions are identical.

At the end of the Gauss elimination (before the back substitution), the row echelon form of the augmented matrix will be

(9)

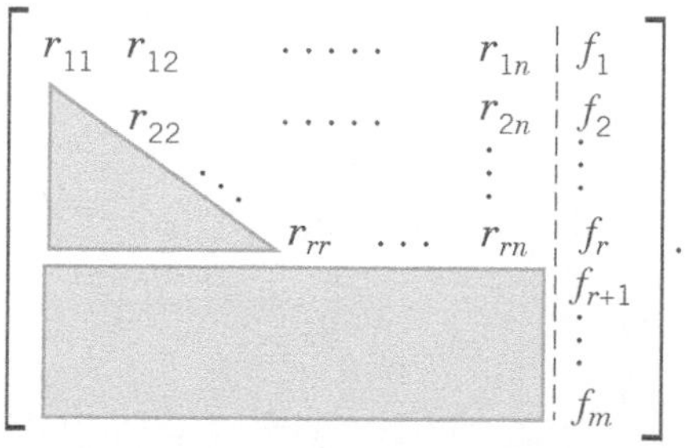

Here, $r \leqq m$, $r_{11} \neq 0$, and all entries in the blue triangle and blue rectangle are zero.

The number of nonzero rows, r, in the row-reduced coefficient matrix $\mathbf{R}$ is called the **rank of R** and also the **rank of A**. Here is the method for determining whether $\mathbf{Ax} = \mathbf{b}$ has solutions and what they are:

(a) No solution. If r is less than m (meaning that $\mathbf{R}$ actually has at least one row of all 0s) *and* at least one of the numbers $f_{r+1}, f_{r+2}, \cdots, f_m$ is not zero, then the system

$\mathbf{Rx} = \mathbf{f}$ is inconsistent: No solution is possible. Therefore the system $\mathbf{Ax} = \mathbf{b}$ is inconsistent as well. See Example 4, where $r = 2 < m = 3$ and $f_{r+1} = f_3 = 12$.

If the system is consistent (either $r = m$, or $r < m$ *and* all the numbers $f_{r+1}, f_{r+2}, \cdots, f_m$ are zero), then there are solutions.

(b) Unique solution. If the system is consistent and $r = n$, there is exactly one solution, which can be found by back substitution. See Example 2, where $r = n = 3$ and $m = 4$.

(c) Infinitely many solutions. To obtain any of these solutions, choose values of $x_{r+1}, \cdots, x_n$ arbitrarily. Then solve the rth equation for x_r (in terms of those arbitrary values), then the $(r - 1)$st equation for x_{r-1}, and so on up the line. See Example 3.

Orientation.   Gauss elimination is reasonable in computing time and storage demand. We shall consider those aspects in Sec. 20.1 in the chapter on numeric linear algebra. Section 7.4 develops fundamental concepts of linear algebra such as linear independence and rank of a matrix. These in turn will be used in Sec. 7.5 to fully characterize the behavior of linear systems in terms of existence and uniqueness of solutions.

PROBLEM SET 7.3

1–14   GAUSS ELIMINATION

Solve the linear system given explicitly or by its augmented matrix. Show details.

1. $4x - 6y = -11$
   $-3x + 8y = 10$

2. $\begin{bmatrix} 3.0 & -0.5 & 0.6 \\ 1.5 & 4.5 & 6.0 \end{bmatrix}$

3. $x + y - z = 9$
   $\phantom{x + {}} 8y + 6z = -6$
   $-2x + 4y - 6z = 40$

4. $\begin{bmatrix} 4 & 1 & 0 & 4 \\ 5 & -3 & 1 & 2 \\ -9 & 2 & -1 & 5 \end{bmatrix}$

5. $\begin{bmatrix} 13 & 12 & -6 \\ -4 & 7 & -73 \\ 11 & -13 & 157 \end{bmatrix}$

6. $\begin{bmatrix} 4 & -8 & 3 & 16 \\ -1 & 2 & -5 & -21 \\ 3 & -6 & 1 & 7 \end{bmatrix}$

7. $\begin{bmatrix} 2 & 4 & 1 & 0 \\ -1 & 1 & -2 & 0 \\ 4 & 0 & 6 & 0 \end{bmatrix}$

8. $ 4y + 3z = 8$
   $2x - z = 2$
   $3x + 2y = 5$

9. $-2y - 2z = -8$
   $3x + 4y - 5z = 13$

10. $\begin{bmatrix} 5 & -7 & 3 & 17 \\ -15 & 21 & -9 & 50 \end{bmatrix}$

11. $\begin{bmatrix} 0 & 5 & 5 & -10 & 0 \\ 2 & -3 & -3 & 6 & 2 \\ 4 & 1 & 1 & -2 & 4 \end{bmatrix}$

12. $\begin{bmatrix} 2 & -2 & 4 & 0 & 0 \\ -3 & 3 & -6 & 5 & 15 \\ 1 & -1 & 2 & 0 & 0 \end{bmatrix}$

13. $\phantom{-3w - {}} 10x + 4y - 2z = -4$
   $-3w - 17x + y + 2z = 2$
   $w + x + y = 6$
   $8w - 34x + 16y - 10z = 4$

14. $\begin{bmatrix} 2 & 3 & 1 & -11 & 1 \\ 5 & -2 & 5 & -4 & 5 \\ 1 & -1 & 3 & -3 & 3 \\ 3 & 4 & -7 & 2 & -7 \end{bmatrix}$

15. Equivalence relation. By definition, an *equivalence relation* on a set is a relation satisfying three conditions: (named as indicated)

  (i) Each element A of the set is equivalent to itself (*Reflexivity*).

  (ii) If A is equivalent to B, then B is equivalent to A (*Symmetry*).

  (iii) If A is equivalent to B and B is equivalent to C, then A is equivalent to C (*Transitivity*).

Show that row equivalence of matrices satisfies these three conditions. *Hint.* Show that for each of the three elementary row operations these conditions hold.

16. CAS PROJECT. Gauss Elimination and Back Substitution. Write a program for Gauss elimination and back substitution **(a)** that does not include pivoting and **(b)** that does include pivoting. Apply the programs to Probs. 11–14 and to some larger systems of your choice.

17–21 MODELS OF NETWORKS

In Probs. 17–19, using Kirchhoff's laws (see Example 2) and showing the details, find the currents:

17.

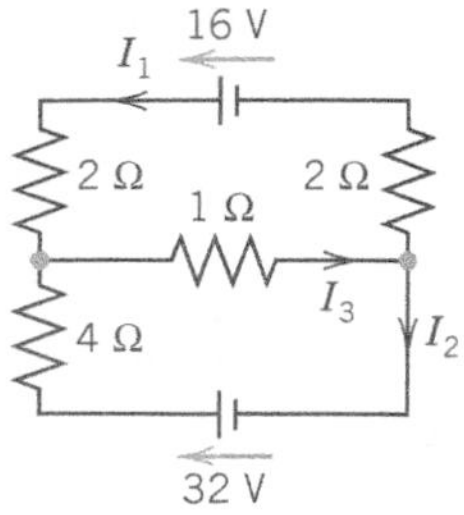

18.

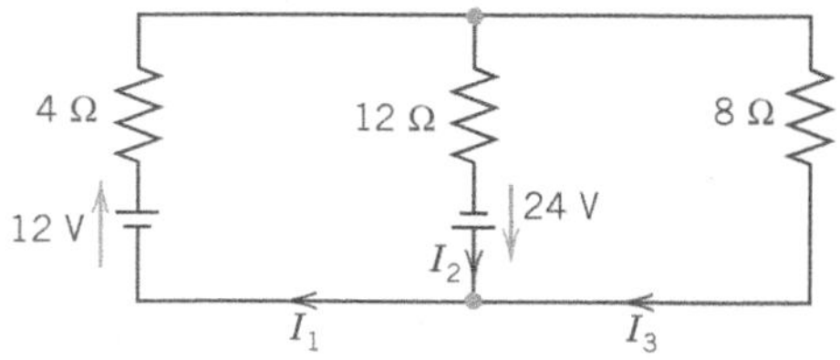

19.

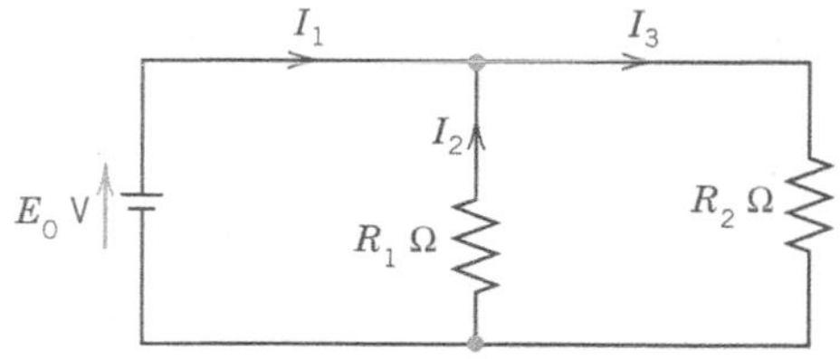

20. Wheatstone bridge. Show that if $R_x/R_3 = R_1/R_2$ in the figure, then $I = 0$. (R_0 is the resistance of the instrument by which I is measured.) This bridge is a method for determining R_x. R_1, R_2, R_3 are known. R_3 is variable. To get R_x, make $I = 0$ by varying R_3. Then calculate $R_x = R_3R_1/R_2$.

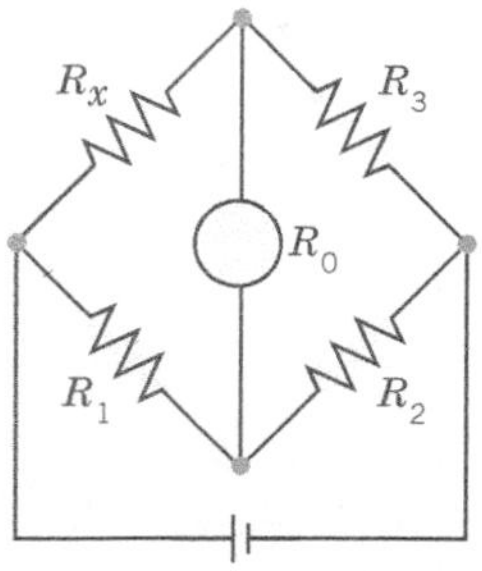

Wheatstone bridge

Problem 20

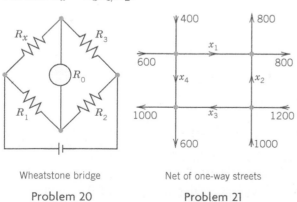

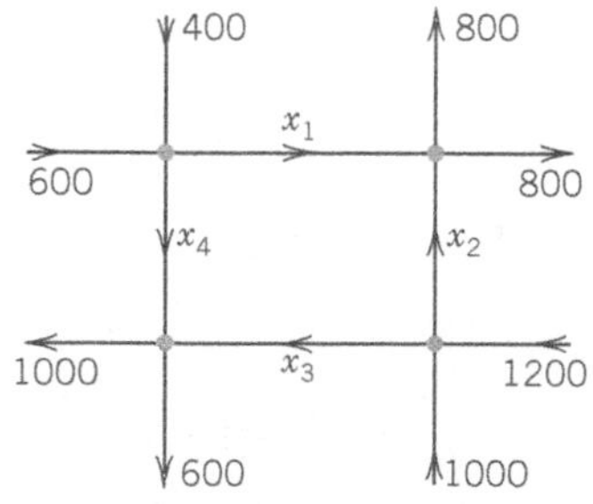

Net of one-way streets

Problem 21

21. Traffic flow. Methods of electrical circuit analysis have applications to other fields. For instance, applying the analog of Kirchhoff's Current Law, find the traffic flow (cars per hour) in the net of one-way streets (in the directions indicated by the arrows) shown in the figure. Is the solution unique?

22. Models of markets. Determine the equilibrium solution ($D_1 = S_1, D_2 = S_2$) of the two-commodity market with linear model (D, S, P = demand, supply, price; index 1 = first commodity, index 2 = second commodity)

$$D_1 = 40 - 2P_1 - P_2, \qquad S_1 = 4P_1 - P_2 + 4,$$

$$D_2 = 5P_1 - 2P_2 + 16, \qquad S_2 = 3P_2 - 4.$$

23. Balancing a chemical equation $x_1C_3H_8 + x_2O_2 \rightarrow x_3CO_2 + x_4H_2O$ means finding integer x_1, x_2, x_3, x_4 such that the numbers of atoms of carbon (C), hydrogen (H), and oxygen (O) are the same on both sides of this reaction, in which propane C_3H_8 and O_2 give carbon dioxide and water. Find the smallest positive integers $x_1, \cdots, x_4$.

24. PROJECT. Elementary Matrices. The idea is that elementary operations can be accomplished by matrix multiplication. If **A** is an $m \times n$ matrix on which we want to do an elementary operation, then there is a matrix **E** such that **EA** is the new matrix after the operation. Such an **E** is called an **elementary matrix**. This idea can be helpful, for instance, in the design of algorithms. (*Computationally*, it is generally preferable to do row operations *directly*, rather than by multiplication by **E**.)

(a) Show that the following are elementary matrices, for interchanging Rows 2 and 3, for adding −5 times the first row to the third, and for multiplying the fourth row by 8.

$$\mathbf{E}_1 = \begin{bmatrix} 1 & 0 & 0 & 0 \\ 0 & 0 & 1 & 0 \\ 0 & 1 & 0 & 0 \\ 0 & 0 & 0 & 1 \end{bmatrix},$$

$$\mathbf{E}_2 = \begin{bmatrix} 1 & 0 & 0 & 0 \\ 0 & 1 & 0 & 0 \\ -5 & 0 & 1 & 0 \\ 0 & 0 & 0 & 1 \end{bmatrix},$$

$$\mathbf{E}_3 = \begin{bmatrix} 1 & 0 & 0 & 0 \\ 0 & 1 & 0 & 0 \\ 0 & 0 & 1 & 0 \\ 0 & 0 & 0 & 8 \end{bmatrix}.$$

Apply $\mathbf{E}_1, \mathbf{E}_2, \mathbf{E}_3$ to a vector and to a 4×3 matrix of your choice. Find $\mathbf{B} = \mathbf{E}_3\mathbf{E}_2\mathbf{E}_1\mathbf{A}$, where $\mathbf{A} = [a_{jk}]$ is the general 4×2 matrix. Is $\mathbf{B}$ equal to $\mathbf{C} = \mathbf{E}_1\mathbf{E}_2\mathbf{E}_3\mathbf{A}$?

(b) Conclude that $\mathbf{E}_1, \mathbf{E}_2, \mathbf{E}_3$ are obtained by doing the corresponding elementary operations on the 4×4 unit matrix. Prove that *if* $\mathbf{M}$ *is obtained from* $\mathbf{A}$ *by an elementary row operation, then*

$$\mathbf{M} = \mathbf{EA},$$

where $\mathbf{E}$ *is obtained from the* $n \times n$ *unit matrix* $\mathbf{I}_n$ *by the same row operation.*

7.4 Linear Independence. Rank of a Matrix. Vector Space

Since our next goal is to fully characterize the behavior of linear systems in terms of existence and uniqueness of solutions (Sec. 7.5), we have to introduce new fundamental linear algebraic concepts that will aid us in doing so. Foremost among these are ***linear independence*** and the ***rank of a matrix***. Keep in mind that these concepts are intimately linked with the important Gauss elimination method and how it works.

Linear Independence and Dependence of Vectors

Given any set of m vectors $\mathbf{a}_{(1)}, \cdots, \mathbf{a}_{(m)}$ (with the same number of components), a **linear combination** of these vectors is an expression of the form

$$c_1\mathbf{a}_{(1)} + c_2\mathbf{a}_{(2)} + \cdots + c_m\mathbf{a}_{(m)}$$

where $c_1, c_2, \cdots, c_m$ are any scalars. Now consider the equation

$$c_1\mathbf{a}_{(1)} + c_2\mathbf{a}_{(2)} + \cdots + c_m\mathbf{a}_{(m)} = \mathbf{0}. \tag{1}$$

Clearly, this vector equation (1) holds if we choose all c_j's zero, because then it becomes $\mathbf{0} = \mathbf{0}$. If this is the only m-tuple of scalars for which (1) holds, then our vectors $\mathbf{a}_{(1)}, \cdots, \mathbf{a}_{(m)}$ are said to form a *linearly independent set* or, more briefly, we call them **linearly independent**. Otherwise, if (1) also holds with scalars not all zero, we call these vectors **linearly dependent**. This means that we can express at least one of the vectors as a linear combination of the other vectors. For instance, if (1) holds with, say, $c_1 \neq 0$, we can solve (1) for $\mathbf{a}_{(1)}$:

$$\mathbf{a}_{(1)} = k_2\mathbf{a}_{(2)} + \cdots + k_m\mathbf{a}_{(m)} \qquad \text{where } k_j = -c_j/c_1.$$

(Some k_j's may be zero. Or even all of them, namely, if $\mathbf{a}_{(1)} = \mathbf{0}$.)

Why is linear independence important? Well, if a set of vectors is linearly dependent, then we can get rid of at least one or perhaps more of the vectors until we get a linearly independent set. This set is then the smallest "truly essential" set with which we can work. Thus, we cannot express any of the vectors, of this set, linearly in terms of the others.

EXAMPLE 1 **Linear Independence and Dependence**

The three vectors

$$\mathbf{a}_{(1)} = [\ 3 \quad 0 \quad 2 \quad 2]$$
$$\mathbf{a}_{(2)} = [-6 \quad 42 \quad 24 \quad 54]$$
$$\mathbf{a}_{(3)} = [\ 21 \quad -21 \quad 0 \quad -15]$$

are linearly dependent because

$$6\mathbf{a}_{(1)} - \tfrac{1}{2}\mathbf{a}_{(2)} - \mathbf{a}_{(3)} = \mathbf{0}.$$

Although this is easily checked by vector arithmetic (do it!), it is not so easy to discover. However, a systematic method for finding out about linear independence and dependence follows below.

The first two of the three vectors are linearly independent because $c_1\mathbf{a}_{(1)} + c_2\mathbf{a}_{(2)} = \mathbf{0}$ implies $c_2 = 0$ (from the second components) and then $c_1 = 0$ (from any other component of $\mathbf{a}_{(1)}$. ■

Rank of a Matrix

DEFINITION

The **rank** of a matrix **A** is the maximum number of linearly independent row vectors of **A**. It is denoted by rank **A**.

Our further discussion will show that the rank of a matrix is an important key concept for understanding general properties of matrices and linear systems of equations.

EXAMPLE 2 **Rank**

The matrix

(2)
$$\mathbf{A} = \begin{bmatrix} 3 & 0 & 2 & 2 \\ -6 & 42 & 24 & 54 \\ 21 & -21 & 0 & -15 \end{bmatrix}$$

has rank 2, because Example 1 shows that the first two row vectors are linearly independent, whereas all three row vectors are linearly dependent.

Note further that rank $\mathbf{A} = 0$ if and only if $\mathbf{A} = \mathbf{0}$. This follows directly from the definition. ■

We call a matrix $\mathbf{A}_1$ **row-equivalent** to a matrix $\mathbf{A}_2$ if $\mathbf{A}_1$ can be obtained from $\mathbf{A}_2$ by (finitely many!) elementary row operations.

Now the maximum number of linearly independent row vectors of a matrix does not change if we change the order of rows or multiply a row by a nonzero c or take a linear combination by adding a multiple of a row to another row. This shows that rank is **invariant** under elementary row operations:

THEOREM 1

Row-Equivalent Matrices

Row-equivalent matrices have the same rank.

Hence we can determine the rank of a matrix by reducing the matrix to row-echelon form, as was done in Sec. 7.3. Once the matrix is in row-echelon form, we count the number of nonzero rows, which is precisely the rank of the matrix.

EXAMPLE 3 Determination of Rank

For the matrix in Example 2 we obtain successively

$$\mathbf{A} = \begin{bmatrix} 3 & 0 & 2 & 2 \\ -6 & 42 & 24 & 54 \\ 21 & -21 & 0 & -15 \end{bmatrix} \quad \text{(given)}$$

$$\begin{bmatrix} 3 & 0 & 2 & 2 \\ 0 & 42 & 28 & 58 \\ 0 & -21 & -14 & -29 \end{bmatrix} \quad \begin{matrix} \\ \text{Row 2} + 2 \text{ Row 1} \\ \text{Row 3} - 7 \text{ Row 1} \end{matrix}$$

$$\begin{bmatrix} 3 & 0 & 2 & 2 \\ 0 & 42 & 28 & 58 \\ 0 & 0 & 0 & 0 \end{bmatrix} \quad \begin{matrix} \\ \\ \text{Row 3} + \frac{1}{2} \text{ Row 2.} \end{matrix}$$

The last matrix is in row-echelon form and has two nonzero rows. Hence rank $\mathbf{A} = 2$, as before. ■

Examples 1–3 illustrate the following useful theorem (with $p = 3$, $n = 3$, and the rank of the matrix $= 2$).

THEOREM 2 Linear Independence and Dependence of Vectors

Consider p vectors that each have n components. Then these vectors are linearly independent if the matrix formed, with these vectors as row vectors, has rank p. However, these vectors are linearly dependent if that matrix has rank less than p.

Further important properties will result from the basic

THEOREM 3 Rank in Terms of Column Vectors

The rank r of a matrix **A** *equals the maximum number of linearly independent* ***column*** *vectors of* **A**.

Hence **A** *and its transpose* $\mathbf{A}^\mathsf{T}$ *have the same rank.*

PROOF In this proof we write simply "rows" and "columns" for row and column vectors. Let **A** be an $m \times n$ matrix of rank $\mathbf{A} = r$. Then by definition of rank, **A** has r linearly independent rows which we denote by $\mathbf{v}_{(1)}, \cdots, \mathbf{v}_{(r)}$ (regardless of their position in **A**), and all the rows $\mathbf{a}_{(1)}, \cdots, \mathbf{a}_{(m)}$ of **A** are linear combinations of those, say,

$$\begin{aligned} \mathbf{a}_{(1)} &= c_{11}\mathbf{v}_{(1)} + c_{12}\mathbf{v}_{(2)} + \cdots + c_{1r}\mathbf{v}_{(r)} \\ \mathbf{a}_{(2)} &= c_{21}\mathbf{v}_{(1)} + c_{22}\mathbf{v}_{(2)} + \cdots + c_{2r}\mathbf{v}_{(r)} \\ &\vdots \\ \mathbf{a}_{(m)} &= c_{m1}\mathbf{v}_{(1)} + c_{m2}\mathbf{v}_{(2)} + \cdots + c_{mr}\mathbf{v}_{(r)}. \end{aligned} \tag{3}$$

These are vector equations for rows. To switch to columns, we write (3) in terms of components as n such systems, with $k = 1, \cdots, n$,

$$
(4) \qquad
\begin{aligned}
a_{1k} &= c_{11}v_{1k} + c_{12}v_{2k} + \cdots + c_{1r}v_{rk} \\
a_{2k} &= c_{21}v_{1k} + c_{22}v_{2k} + \cdots + c_{2r}v_{rk} \\
&\vdots \\
a_{mk} &= c_{m1}v_{1k} + c_{m2}v_{2k} + \cdots + c_{mr}v_{rk}
\end{aligned}
$$

and collect components in columns. Indeed, we can write (4) as

$$
(5) \qquad
\begin{bmatrix} a_{1k} \\ a_{2k} \\ \vdots \\ a_{mk} \end{bmatrix}
= v_{1k} \begin{bmatrix} c_{11} \\ c_{21} \\ \vdots \\ c_{m1} \end{bmatrix}
+ v_{2k} \begin{bmatrix} c_{12} \\ c_{22} \\ \vdots \\ c_{m2} \end{bmatrix}
+ \cdots + v_{rk} \begin{bmatrix} c_{1r} \\ c_{2r} \\ \vdots \\ c_{mr} \end{bmatrix}
$$

where $k = 1, \cdots, n$. Now the vector on the left is the kth column vector of $\mathbf{A}$. We see that each of these n columns is a linear combination of the same r columns on the right. Hence $\mathbf{A}$ cannot have more linearly independent columns than rows, whose number is rank $\mathbf{A} = r$. Now rows of $\mathbf{A}$ are columns of the transpose $\mathbf{A}^\mathsf{T}$. For $\mathbf{A}^\mathsf{T}$ our conclusion is that $\mathbf{A}^\mathsf{T}$ cannot have more linearly independent columns than rows, so that $\mathbf{A}$ cannot have more linearly independent rows than columns. Together, the number of linearly independent columns of $\mathbf{A}$ must be r, the rank of $\mathbf{A}$. This completes the proof. ■

EXAMPLE 4 Illustration of Theorem 3

The matrix in (2) has rank 2. From Example 3 we see that the first two row vectors are linearly independent and by "working backward" we can verify that Row 3 = 6 Row 1 − $\frac{1}{2}$ Row 2. Similarly, the first two columns are linearly independent, and by reducing the last matrix in Example 3 by columns we find that

Column 3 = $\frac{2}{3}$ Column 1 + $\frac{2}{3}$ Column 2 and Column 4 = $\frac{2}{3}$ Column 1 + $\frac{29}{21}$ Column 2. ■

Combining Theorems 2 and 3 we obtain

THEOREM 4 Linear Dependence of Vectors

Consider p vectors each having n components. If $n < p$, then these vectors are linearly dependent.

PROOF The matrix $\mathbf{A}$ with those p vectors as row vectors has p rows and $n < p$ columns; hence by Theorem 3 it has rank $\mathbf{A} \leqq n < p$, which implies linear dependence by Theorem 2. ■

Vector Space

The following related concepts are of general interest in linear algebra. In the present context they provide a clarification of essential properties of matrices and their role in connection with linear systems.

Consider a nonempty set V of vectors where each vector has the same number of components. If, for any two vectors $\mathbf{a}$ and $\mathbf{b}$ in V, we have that all their linear combinations $\alpha\mathbf{a} + \beta\mathbf{b}$ (α, β any real numbers) are also elements of V, and if, furthermore, $\mathbf{a}$ and $\mathbf{b}$ satisfy the laws (3a), (3c), (3d), and (4) in Sec. 7.1, as well as any vectors $\mathbf{a}$, $\mathbf{b}$, $\mathbf{c}$ in V satisfy (3b) then V is a vector space. Note that here we wrote laws (3) and (4) of Sec. 7.1 in lowercase letters $\mathbf{a}$, $\mathbf{b}$, $\mathbf{c}$, which is our notation for vectors. More on vector spaces in Sec. 7.9.

The maximum number of linearly independent vectors in V is called the **dimension** of V and is denoted by dim V. Here we assume the dimension to be finite; infinite dimension will be defined in Sec. 7.9.

A linearly independent set in V consisting of a maximum possible number of vectors in V is called a **basis** for V. In other words, any largest possible set of independent vectors in V forms basis for V. That means, if we add one or more vector to that set, the set will be linearly dependent. (See also the beginning of Sec. 7.4 on linear independence and dependence of vectors.) Thus, the number of vectors of a basis for V equals dim V.

The set of all linear combinations of given vectors $\mathbf{a}_{(1)}, \cdots, \mathbf{a}_{(p)}$ with the same number of components is called the **span** of these vectors. Obviously, a span is a vector space. If in addition, the given vectors $\mathbf{a}_{(1)}, \cdots, \mathbf{a}_{(p)}$ are linearly independent, then they form a basis for that vector space.

This then leads to another equivalent definition of basis. A set of vectors is a **basis** for a vector space V if (1) the vectors in the set are linearly independent, and if (2) any vector in V can be expressed as a linear combination of the vectors in the set. If (2) holds, we also say that the set of vectors **spans** the vector space V.

By a **subspace** of a vector space V we mean a nonempty subset of V (including V itself) that forms a vector space with respect to the two algebraic operations (addition and scalar multiplication) defined for the vectors of V.

EXAMPLE 5 **Vector Space, Dimension, Basis**

The span of the three vectors in Example 1 is a vector space of dimension 2. A basis of this vector space consists of any two of those three vectors, for instance, $\mathbf{a}_{(1)}$, $\mathbf{a}_{(2)}$, or $\mathbf{a}_{(1)}$, $\mathbf{a}_{(3)}$, etc. ■

We further note the simple

THEOREM 5 **Vector Space R^n**

The vector space R^n consisting of all vectors with n components (n real numbers) has dimension n.

PROOF A basis of n vectors is $\mathbf{a}_{(1)} = [1 \quad 0 \quad \cdots \quad 0]$, $\mathbf{a}_{(2)} = [0 \quad 1 \quad 0 \quad \cdots \quad 0]$, $\cdots$, $\mathbf{a}_{(n)} = [0 \quad \cdots \quad 0 \quad 1]$. ■

For a matrix $\mathbf{A}$, we call the span of the row vectors the **row space** of $\mathbf{A}$. Similarly, the span of the column vectors of $\mathbf{A}$ is called the **column space** of $\mathbf{A}$.

Now, Theorem 3 shows that a matrix $\mathbf{A}$ has as many linearly independent rows as columns. By the definition of dimension, their number is the dimension of the row space or the column space of $\mathbf{A}$. This proves

THEOREM 6 **Row Space and Column Space**

The row space and the column space of a matrix $\mathbf{A}$ have the same dimension, equal to rank $\mathbf{A}$.

Finally, for a given matrix **A** the solution set of the homogeneous system $\mathbf{Ax} = \mathbf{0}$ is a vector space, called the **null space** of **A**, and its dimension is called the **nullity** of **A**. In the next section we motivate and prove the basic relation

$$\text{rank } \mathbf{A} + \text{nullity } \mathbf{A} = \text{Number of columns of } \mathbf{A}. \tag{6}$$

PROBLEM SET 7.4

1–10 RANK, ROW SPACE, COLUMN SPACE

Find the rank. Find a basis for the row space. Find a basis for the column space. *Hint.* Row-reduce the matrix and its transpose. (You may omit obvious factors from the vectors of these bases.)

1. $\begin{bmatrix} 4 & -2 & 6 \\ -2 & 1 & -3 \end{bmatrix}$ **2.** $\begin{bmatrix} a & b \\ b & a \end{bmatrix}$

3. $\begin{bmatrix} 0 & 3 & 5 \\ 3 & 5 & 0 \\ 5 & 0 & 10 \end{bmatrix}$ **4.** $\begin{bmatrix} 6 & -4 & 0 \\ -4 & 0 & 2 \\ 0 & 2 & 6 \end{bmatrix}$

5. $\begin{bmatrix} 0.2 & -0.1 & 0.4 \\ 0 & 1.1 & -0.3 \\ 0.1 & 0 & -2.1 \end{bmatrix}$ **6.** $\begin{bmatrix} 0 & 1 & 0 \\ -1 & 0 & -4 \\ 0 & 4 & 0 \end{bmatrix}$

7. $\begin{bmatrix} 8 & 0 & 4 & 0 \\ 0 & 2 & 0 & 4 \\ 4 & 0 & 2 & 0 \end{bmatrix}$ **8.** $\begin{bmatrix} 2 & 4 & 8 & 16 \\ 16 & 8 & 4 & 2 \\ 4 & 8 & 16 & 2 \\ 2 & 16 & 8 & 4 \end{bmatrix}$

9. $\begin{bmatrix} 9 & 0 & 1 & 0 \\ 0 & 0 & 1 & 0 \\ 1 & 1 & 1 & 1 \\ 0 & 0 & 1 & 0 \end{bmatrix}$ **10.** $\begin{bmatrix} 5 & -2 & 1 & 0 \\ -2 & 0 & -4 & 1 \\ 1 & -4 & -11 & 2 \\ 0 & 1 & 2 & 0 \end{bmatrix}$

11. CAS Experiment. Rank. **(a)** Show experimentally that the $n \times n$ matrix $\mathbf{A} = [a_{jk}]$ with $a_{jk} = j + k - 1$ has rank 2 for any n. (Problem 20 shows $n = 4$.) Try to prove it.

(b) Do the same when $a_{jk} = j + k + c$, where c is any positive integer.

(c) What is rank **A** if $a_{jk} = 2^{j+k-2}$? Try to find other large matrices of low rank independent of n.

12–16 GENERAL PROPERTIES OF RANK

Show the following:

12. $\text{rank } \mathbf{B}^\mathsf{T}\mathbf{A}^\mathsf{T} = \text{rank } \mathbf{AB}$. (Note the order!)

13. $\text{rank } \mathbf{A} = \text{rank } \mathbf{B}$ does *not* imply $\text{rank } \mathbf{A}^2 = \text{rank } \mathbf{B}^2$. (Give a counterexample.)

14. If **A** is not square, either the row vectors or the column vectors of **A** are linearly dependent.

15. If the row vectors of a square matrix are linearly independent, so are the column vectors, and conversely.

16. Give examples showing that the rank of a product of matrices cannot exceed the rank of either factor.

17–25 LINEAR INDEPENDENCE

Are the following sets of vectors linearly independent? Show the details of your work.

17. $[3\ \ 4\ \ 0\ \ 2]$, $[2\ \ -1\ \ 3\ \ 7]$, $[1\ \ 16\ \ -12\ \ -22]$

18. $[1\ \ \tfrac{1}{2}\ \ \tfrac{1}{3}\ \ \tfrac{1}{4}]$, $[\tfrac{1}{2}\ \ \tfrac{1}{3}\ \ \tfrac{1}{4}\ \ \tfrac{1}{5}]$, $[\tfrac{1}{3}\ \ \tfrac{1}{4}\ \ \tfrac{1}{5}\ \ \tfrac{1}{6}]$, $[\tfrac{1}{4}\ \ \tfrac{1}{5}\ \ \tfrac{1}{6}\ \ \tfrac{1}{7}]$

19. $[0\ \ 1\ \ 1]$, $[1\ \ 1\ \ 1]$, $[0\ \ 0\ \ 1]$

20. $[1\ \ 2\ \ 3\ \ 4]$, $[2\ \ 3\ \ 4\ \ 5]$, $[3\ \ 4\ \ 5\ \ 6]$, $[4\ \ 5\ \ 6\ \ 7]$

21. $[2\ \ 0\ \ 0\ \ 7]$, $[2\ \ 0\ \ 0\ \ 8]$, $[2\ \ 0\ \ 0\ \ 9]$, $[2\ \ 0\ \ 1\ \ 0]$

22. $[0.4\ \ -0.2\ \ 0.2]$, $[0\ \ 0\ \ 0]$, $[3.0\ \ -0.6\ \ 1.5]$

23. $[9\ \ 8\ \ 7\ \ 6\ \ 5]$, $[9\ \ 7\ \ 5\ \ 3\ \ 1]$

24. $[4\ \ -1\ \ 3]$, $[0\ \ 8\ \ 1]$, $[1\ \ 3\ \ -5]$, $[2\ \ 6\ \ 1]$

25. $[6\ \ 0\ \ -1\ \ 3]$, $[2\ \ 2\ \ 5\ \ 0]$, $[-4\ \ -4\ \ -4\ \ -4]$

26. Linearly independent subset. Beginning with the last of the vectors $[3\ \ 0\ \ 1\ \ 2]$, $[6\ \ 1\ \ 0\ \ 0]$, $[12\ \ 1\ \ 2\ \ 4]$, $[6\ \ 0\ \ 2\ \ 4]$, and $[9\ \ 0\ \ 1\ \ 2]$, omit one after another until you get a linearly independent set.

27–35 VECTOR SPACE

Is the given set of vectors a vector space? Give reasons. If your answer is yes, determine the dimension and find a basis. ($v_1, v_2, \cdots$ denote components.)

27. All vectors in R^3 with $v_1 - v_2 + 2v_3 = 0$

28. All vectors in R^3 with $3v_2 + v_3 = k$

29. All vectors in R^2 with $v_1 \geqq v_2$

30. All vectors in R^n with the first $n - 2$ components zero

31. All vectors in R^5 with positive components

32. All vectors in R^3 with $3v_1 - 2v_2 + v_3 = 0$, $4v_1 + 5v_2 = 0$

33. All vectors in R^3 with $3v_1 - v_3 = 0$, $2v_1 + 3v_2 - 4v_3 = 0$

34. All vectors in R^n with $|v_j| = 1$ for $j = 1, \cdots, n$

35. All vectors in R^4 with $v_1 = 2v_2 = 3v_3 = 4v_4$

7.5 Solutions of Linear Systems: Existence, Uniqueness

Rank, as just defined, gives complete information about existence, uniqueness, and general structure of the solution set of linear systems as follows.

A linear system of equations in n unknowns has a unique solution if the coefficient matrix and the augmented matrix have the same rank n, and infinitely many solutions if that common rank is less than n. The system has no solution if those two matrices have different rank.

To state this precisely and prove it, we shall use the generally important concept of a **submatrix** of **A**. By this we mean any matrix obtained from **A** by omitting some rows or columns (or both). By definition this includes **A** itself (as the matrix obtained by omitting no rows or columns); this is practical.

THEOREM 1

Fundamental Theorem for Linear Systems

(a) Existence. *A linear system of m equations in n unknowns* $x_1, \cdots, x_n$

$$
\begin{aligned}
a_{11}x_1 + a_{12}x_2 + \cdots + a_{1n}x_n &= b_1 \\
a_{21}x_1 + a_{22}x_2 + \cdots + a_{2n}x_n &= b_2 \\
\cdots\cdots\cdots\cdots\cdots\cdots\cdots\cdots & \\
a_{m1}x_1 + a_{m2}x_2 + \cdots + a_{mn}x_n &= b_m
\end{aligned}
\tag{1}
$$

is **consistent**, *that is, has solutions, if and only if the coefficient matrix* **A** *and the augmented matrix* $\tilde{\mathbf{A}}$ *have the same rank. Here,*

$$
\mathbf{A} = \begin{bmatrix} a_{11} & \cdots & a_{1n} \\ \cdot & \cdots & \cdot \\ \cdot & \cdots & \cdot \\ a_{m1} & \cdots & a_{mn} \end{bmatrix} \quad \textit{and} \quad \tilde{\mathbf{A}} = \left[\begin{array}{ccc|c} a_{11} & \cdots & a_{1n} & b_1 \\ \cdot & \cdots & \cdot & \cdot \\ \cdot & \cdots & \cdot & \cdot \\ a_{m1} & \cdots & a_{mn} & b_m \end{array}\right]
$$

(b) Uniqueness. *The system* (1) *has precisely one solution if and only if this common rank r of* **A** *and* $\tilde{\mathbf{A}}$ *equals n.*

(c) Infinitely many solutions. *If this common rank r is less than n, the system (1) has infinitely many solutions. All of these solutions are obtained by determining r suitable unknowns (whose submatrix of coefficients must have rank r) in terms of the remaining $n - r$ unknowns, to which arbitrary values can be assigned.* (See Example 3 in Sec. 7.3.)

(d) Gauss elimination (Sec. 7.3). *If solutions exist, they can all be obtained by the Gauss elimination.* (This method will automatically reveal whether or not solutions exist; see Sec. 7.3.)

PROOF **(a)** We can write the system (1) in vector form $\mathbf{Ax} = \mathbf{b}$ or in terms of column vectors $\mathbf{c}_{(1)}, \cdots, \mathbf{c}_{(n)}$ of $\mathbf{A}$:

$$\mathbf{c}_{(1)}x_1 + \mathbf{c}_{(2)}x_2 + \cdots + \mathbf{c}_{(n)}x_n = \mathbf{b}. \tag{2}$$

$\tilde{\mathbf{A}}$ is obtained by augmenting $\mathbf{A}$ by a single column $\mathbf{b}$. Hence, by Theorem 3 in Sec. 7.4, rank $\tilde{\mathbf{A}}$ equals rank $\mathbf{A}$ or rank $\mathbf{A} + 1$. Now if (1) has a solution $\mathbf{x}$, then (2) shows that $\mathbf{b}$ must be a linear combination of those column vectors, so that $\tilde{\mathbf{A}}$ and $\mathbf{A}$ have the same maximum number of linearly independent column vectors and thus the same rank.

Conversely, if rank $\tilde{\mathbf{A}} =$ rank $\mathbf{A}$, then $\mathbf{b}$ must be a linear combination of the column vectors of $\mathbf{A}$, say,

$$\mathbf{b} = \alpha_1\mathbf{c}_{(1)} + \cdots + \alpha_n\mathbf{c}_{(n)} \tag{2*}$$

since otherwise rank $\tilde{\mathbf{A}} =$ rank $\mathbf{A} + 1$. But (2*) means that (1) has a solution, namely, $x_1 = \alpha_1, \cdots, x_n = \alpha_n$, as can be seen by comparing (2*) and (2).

(b) If rank $\mathbf{A} = n$, the n column vectors in (2) are linearly independent by Theorem 3 in Sec. 7.4. We claim that then the representation (2) of $\mathbf{b}$ is unique because otherwise

$$\mathbf{c}_{(1)}x_1 + \cdots + \mathbf{c}_{(n)}x_n = \mathbf{c}_{(1)}\tilde{x}_1 + \cdots + \mathbf{c}_{(n)}\tilde{x}_n.$$

This would imply (take all terms to the left, with a minus sign)

$$(x_1 - \tilde{x}_1)\mathbf{c}_{(1)} + \cdots + (x_n - \tilde{x}_n)\mathbf{c}_{(n)} = \mathbf{0}$$

and $x_1 - \tilde{x}_1 = 0, \cdots, x_n - \tilde{x}_n = 0$ by linear independence. But this means that the scalars $x_1, \cdots, x_n$ in (2) are uniquely determined, that is, the solution of (1) is unique.

(c) If rank $\mathbf{A} =$ rank $\tilde{\mathbf{A}} = r < n$, then by Theorem 3 in Sec. 7.4 there is a linearly independent set K of r column vectors of $\mathbf{A}$ such that the other $n - r$ column vectors of $\mathbf{A}$ are linear combinations of those vectors. We renumber the columns and unknowns, denoting the renumbered quantities by $\hat{}$, so that $\{\hat{\mathbf{c}}_{(1)}, \cdots, \hat{\mathbf{c}}_{(r)}\}$ is that linearly independent set K. Then (2) becomes

$$\hat{\mathbf{c}}_{(1)}\hat{x}_1 + \cdots + \hat{\mathbf{c}}_{(r)}\hat{x}_r + \hat{\mathbf{c}}_{(r+1)}\hat{x}_{r+1} + \cdots + \hat{\mathbf{c}}_{(n)}\hat{x}_n = \mathbf{b},$$

$\hat{\mathbf{c}}_{(r+1)}, \cdots, \hat{\mathbf{c}}_{(n)}$ are linear combinations of the vectors of K, and so are the vectors $\hat{x}_{r+1}\hat{\mathbf{c}}_{(r+1)}, \cdots, \hat{x}_n\hat{\mathbf{c}}_{(n)}$. Expressing these vectors in terms of the vectors of K and collecting terms, we can thus write the system in the form

$$\hat{\mathbf{c}}_{(1)}y_1 + \cdots + \hat{\mathbf{c}}_{(r)}y_r = \mathbf{b} \tag{3}$$

with $y_j = \hat{x}_j + \beta_j$, where β_j results from the $n - r$ terms $\hat{\mathbf{c}}_{(r+1)}\hat{x}_{r+1}, \cdots, \hat{\mathbf{c}}_{(n)}\hat{x}_n$; here, $j = 1, \cdots, r$. Since the system has a solution, there are $y_1, \cdots, y_r$ satisfying (3). These scalars are unique since K is linearly independent. Choosing $\hat{x}_{r+1}, \cdots, \hat{x}_n$ fixes the β_j and corresponding $\hat{x}_j = y_j - \beta_j$, where $j = 1, \cdots, r$.

(d) This was discussed in Sec. 7.3 and is restated here as a reminder. ■

The theorem is illustrated in Sec. 7.3. In Example 2 there is a unique solution since rank $\tilde{\mathbf{A}}$ = rank $\mathbf{A} = n = 3$ (as can be seen from the last matrix in the example). In Example 3 we have rank $\tilde{\mathbf{A}}$ = rank $\mathbf{A} = 2 < n = 4$ and can choose x_3 and x_4 arbitrarily. In Example 4 there is no solution because rank $\mathbf{A} = 2 <$ rank $\tilde{\mathbf{A}} = 3$.

Homogeneous Linear System

Recall from Sec. 7.3 that a linear system (1) is called **homogeneous** if all the b_j's are zero, and **nonhomogeneous** if one or several b_j's are not zero. For the homogeneous system we obtain from the Fundamental Theorem the following results.

THEOREM 2

Homogeneous Linear System

A homogeneous linear system

$$
\begin{aligned}
a_{11}x_1 + a_{12}x_2 + \cdots + a_{1n}x_n &= 0 \\
a_{21}x_1 + a_{22}x_2 + \cdots + a_{2n}x_n &= 0 \\
\cdots\cdots\cdots\cdots\cdots\cdots &\cdots \\
a_{m1}x_1 + a_{m2}x_2 + \cdots + a_{mn}x_n &= 0
\end{aligned}
\tag{4}
$$

always has the **trivial solution** $x_1 = 0, \cdots, x_n = 0$. *Nontrivial solutions exist if and only if rank* $\mathbf{A} < n$. *If* rank $\mathbf{A} = r < n$, *these solutions, together with* $\mathbf{x} = \mathbf{0}$, *form a vector space* (see Sec. 7.4) *of dimension* $n - r$ called the **solution space** of (4).

In particular, if $\mathbf{x}_{(1)}$ *and* $\mathbf{x}_{(2)}$ *are solution vectors of* (4), *then* $\mathbf{x} = c_1\mathbf{x}_{(1)} + c_2\mathbf{x}_{(2)}$ *with any scalars* c_1 *and* c_2 *is a solution vector of* (4). (This ***does not hold*** for nonhomogeneous systems. Also, the term *solution space is used for homogeneous* systems only.)

PROOF The first proposition can be seen directly from the system. It agrees with the fact that $\mathbf{b} = \mathbf{0}$ implies that rank $\tilde{\mathbf{A}}$ = rank $\mathbf{A}$, so that a homogeneous system is always ***consistent***. If rank $\mathbf{A} = n$, the trivial solution is the unique solution according to (b) in Theorem 1. If rank $\mathbf{A} < n$, there are nontrivial solutions according to (c) in Theorem 1. The solutions form a vector space because if $\mathbf{x}_{(1)}$ and $\mathbf{x}_{(2)}$ are any of them, then $\mathbf{A}\mathbf{x}_{(1)} = \mathbf{0}$, $\mathbf{A}\mathbf{x}_{(2)} = \mathbf{0}$, and this implies $\mathbf{A}(\mathbf{x}_{(1)} + \mathbf{x}_{(2)}) = \mathbf{A}\mathbf{x}_{(1)} + \mathbf{A}\mathbf{x}_{(2)} = \mathbf{0}$ as well as $\mathbf{A}(c\mathbf{x}_{(1)}) = c\mathbf{A}\mathbf{x}_{(1)} = \mathbf{0}$, where c is arbitrary. If rank $\mathbf{A} = r < n$, Theorem 1 (c) implies that we can choose $n - r$ suitable unknowns, call them $x_{r+1}, \cdots, x_n$, in an arbitrary fashion, and every solution is obtained in this way. Hence a basis for the solution space, briefly called a **basis of solutions** of (4), is $\mathbf{y}_{(1)}, \cdots, \mathbf{y}_{(n-r)}$, where the basis vector $\mathbf{y}_{(j)}$ is obtained by choosing $x_{r+j} = 1$ and the other $x_{r+1}, \cdots, x_n$ zero; the corresponding first r components of this solution vector are then determined. Thus the solution space of (4) has dimension $n - r$. This proves Theorem 2. ■

The solution space of (4) is also called the **null space** of **A** because $\mathbf{Ax} = \mathbf{0}$ for every $\mathbf{x}$ in the solution space of (4). Its dimension is called the **nullity** of **A**. Hence Theorem 2 states that

$$\text{rank } \mathbf{A} + \text{nullity } \mathbf{A} = n \tag{5}$$

where n is the number of unknowns (number of columns of **A**).

Furthermore, by the definition of rank we have rank $\mathbf{A} \leqq m$ in (4). Hence if $m < n$, then rank $\mathbf{A} < n$. By Theorem 2 this gives the practically important

THEOREM 3

Homogeneous Linear System with Fewer Equations Than Unknowns

A homogeneous linear system with fewer equations than unknowns always has nontrivial solutions.

Nonhomogeneous Linear Systems

The characterization of all solutions of the linear system (1) is now quite simple, as follows.

THEOREM 4

Nonhomogeneous Linear System

If a nonhomogeneous linear system (1) *is consistent, then all of its solutions are obtained as*

$$\mathbf{x} = \mathbf{x}_0 + \mathbf{x}_h \tag{6}$$

where $\mathbf{x}_0$ *is any* (fixed) *solution of* (1) *and* $\mathbf{x}_h$ *runs through all the solutions of the corresponding homogeneous system* (4).

PROOF The difference $\mathbf{x}_h = \mathbf{x} - \mathbf{x}_0$ of any two solutions of (1) is a solution of (4) because $\mathbf{Ax}_h = \mathbf{A}(\mathbf{x} - \mathbf{x}_0) = \mathbf{Ax} - \mathbf{Ax}_0 = \mathbf{b} - \mathbf{b} = \mathbf{0}$. Since $\mathbf{x}$ is any solution of (1), we get all the solutions of (1) if in (6) we take any solution $\mathbf{x}_0$ of (1) and let $\mathbf{x}_h$ vary throughout the solution space of (4). ■

This covers a main part of our discussion of characterizing the solutions of systems of linear equations. Our next main topic is determinants and their role in linear equations.

7.6 For Reference: Second- and Third-Order Determinants

We created this section as a quick general reference section on second- and third-order determinants. It is completely independent of the theory in Sec. 7.7 and suffices as a reference for many of our examples and problems. Since this section is for reference, ***go on to the next section, consulting this material only when needed.***

A **determinant of second order** is denoted and defined by

$$D = \det \mathbf{A} = \begin{vmatrix} a_{11} & a_{12} \\ a_{21} & a_{22} \end{vmatrix} = a_{11}a_{22} - a_{12}a_{21}. \tag{1}$$

So here we have ***bars*** (whereas a matrix has ***brackets***).

Cramer's rule for solving linear systems of two equations in two unknowns

$$\text{(2)} \qquad \begin{aligned} &\text{(a)} \quad a_{11}x_1 + a_{12}x_2 = b_1 \\ &\text{(b)} \quad a_{21}x_1 + a_{22}x_2 = b_2 \end{aligned}$$

is

$$\text{(3)} \qquad \begin{aligned} x_1 &= \frac{\begin{vmatrix} b_1 & a_{12} \\ b_2 & a_{22} \end{vmatrix}}{D} = \frac{b_1a_{22} - a_{12}b_2}{D}, \\ x_2 &= \frac{\begin{vmatrix} a_{11} & b_1 \\ a_{21} & b_2 \end{vmatrix}}{D} = \frac{a_{11}b_2 - b_1a_{21}}{D} \end{aligned}$$

with D as in (1), provided

$$D \neq 0.$$

The value $D = 0$ appears for homogeneous systems with nontrivial solutions.

PROOF We prove (3). To eliminate x_2 multiply (2a) by a_{22} and (2b) by $-a_{12}$ and add,

$$(a_{11}a_{22} - a_{12}a_{21})x_1 = b_1a_{22} - a_{12}b_2.$$

Similarly, to eliminate x_1 multiply (2a) by $-a_{21}$ and (2b) by a_{11} and add,

$$(a_{11}a_{22} - a_{12}a_{21})x_2 = a_{11}b_2 - b_1a_{21}.$$

Assuming that $D = a_{11}a_{22} - a_{12}a_{21} \neq 0$, dividing, and writing the right sides of these two equations as determinants, we obtain (3). ■

EXAMPLE 1 Cramer's Rule for Two Equations

If $\begin{aligned} 4x_1 + 3x_2 &= 12 \\ 2x_1 + 5x_2 &= -8 \end{aligned}$ then $x_1 = \dfrac{\begin{vmatrix} 12 & 3 \\ -8 & 5 \end{vmatrix}}{\begin{vmatrix} 4 & 3 \\ 2 & 5 \end{vmatrix}} = \dfrac{84}{14} = 6, \quad x_2 = \dfrac{\begin{vmatrix} 4 & 12 \\ 2 & -8 \end{vmatrix}}{\begin{vmatrix} 4 & 3 \\ 2 & 5 \end{vmatrix}} = \dfrac{-56}{14} = -4.$ ■

Third-Order Determinants

A **determinant of third order** can be defined by

$$\text{(4)} \quad D = \begin{vmatrix} a_{11} & a_{12} & a_{13} \\ a_{21} & a_{22} & a_{23} \\ a_{31} & a_{32} & a_{33} \end{vmatrix} = a_{11}\begin{vmatrix} a_{22} & a_{23} \\ a_{32} & a_{33} \end{vmatrix} - a_{21}\begin{vmatrix} a_{12} & a_{13} \\ a_{32} & a_{33} \end{vmatrix} + a_{31}\begin{vmatrix} a_{12} & a_{13} \\ a_{22} & a_{23} \end{vmatrix}.$$

Note the following. The signs on the right are $+ - +$. Each of the three terms on the right is an entry in the first column of D times its **minor**, that is, the second-order determinant obtained from D by deleting the row and column of that entry; thus, for a_{11} delete the first row and first column, and so on.

If we write out the minors in (4), we obtain

$$(4^*) \quad D = a_{11}a_{22}a_{33} - a_{11}a_{23}a_{32} + a_{21}a_{13}a_{32} - a_{21}a_{12}a_{33} + a_{31}a_{12}a_{23} - a_{31}a_{13}a_{22}.$$

Cramer's Rule for Linear Systems of Three Equations

$$(5) \qquad \begin{aligned} a_{11}x_1 + a_{12}x_2 + a_{13}x_3 &= b_1 \\ a_{21}x_1 + a_{22}x_2 + a_{23}x_3 &= b_2 \\ a_{31}x_1 + a_{32}x_2 + a_{33}x_3 &= b_3 \end{aligned}$$

is

$$(6) \qquad x_1 = \frac{D_1}{D}, \quad x_2 = \frac{D_2}{D}, \quad x_3 = \frac{D_3}{D} \qquad (D \neq 0)$$

with the *determinant D of the system* given by (4) and

$$D_1 = \begin{vmatrix} b_1 & a_{12} & a_{13} \\ b_2 & a_{22} & a_{23} \\ b_3 & a_{32} & a_{33} \end{vmatrix}, \quad D_2 = \begin{vmatrix} a_{11} & b_1 & a_{13} \\ a_{21} & b_2 & a_{23} \\ a_{31} & b_3 & a_{33} \end{vmatrix}, \quad D_3 = \begin{vmatrix} a_{11} & a_{12} & b_1 \\ a_{21} & a_{22} & b_2 \\ a_{31} & a_{32} & b_3 \end{vmatrix}.$$

Note that D_1, D_2, D_3 are obtained by replacing Columns 1, 2, 3, respectively, by the column of the right sides of (5).

Cramer's rule (6) can be derived by eliminations similar to those for (3), but it also follows from the general case (Theorem 4) in the next section.

7.7 Determinants. Cramer's Rule

Determinants were originally introduced for solving linear systems. Although ***impractical in computations***, they have important engineering applications in eigenvalue problems (Sec. 8.1), differential equations, vector algebra (Sec. 9.3), and in other areas. They can be introduced in several equivalent ways. Our definition is particularly for dealing with linear systems.

A **determinant of order** n is a scalar associated with an $n \times n$ (hence ***square!***) matrix $\mathbf{A} = [a_{jk}]$, and is denoted by

$$(1) \qquad D = \det \mathbf{A} = \begin{vmatrix} a_{11} & a_{12} & \cdots & a_{1n} \\ a_{21} & a_{22} & \cdots & a_{2n} \\ \cdot & \cdot & \cdots & \cdot \\ \cdot & \cdot & \cdots & \cdot \\ a_{n1} & a_{n2} & \cdots & a_{nn} \end{vmatrix}.$$

For $n = 1$, this determinant is defined by

$$D = a_{11}. \tag{2}$$

For $n \geqq 2$ by

$$D = a_{j1}C_{j1} + a_{j2}C_{j2} + \cdots + a_{jn}C_{jn} \qquad (j = 1, 2, \cdots, \text{ or } n) \tag{3a}$$

or

$$D = a_{1k}C_{1k} + a_{2k}C_{2k} + \cdots + a_{nk}C_{nk} \quad (k = 1, 2, \cdots, \text{ or } n). \tag{3b}$$

Here,

$$C_{jk} = (-1)^{j+k}M_{jk}$$

and M_{jk} is a determinant of order $n - 1$, namely, the determinant of the submatrix of **A** obtained from **A** by omitting the row and column of the entry a_{jk}, that is, the jth row and the kth column.

In this way, D is defined in terms of n determinants of order $n - 1$, each of which is, in turn, defined in terms of $n - 1$ determinants of order $n - 2$, and so on—until we finally arrive at second-order determinants, in which those submatrices consist of single entries whose determinant is defined to be the entry itself.

From the definition it follows that *we may* ***expand*** *D by any row or column,* that is, choose in (3) the entries in any row or column, similarly when expanding the C_{jk}'s in (3), and so on.

This definition is unambiguous, that is, it yields the same value for D no matter which columns or rows we choose in expanding. A proof is given in App. 4.

Terms used in connection with determinants are taken from matrices. In D we have n^2 **entries** a_{jk}, also n **rows** and n **columns**, and a **main diagonal** on which $a_{11}, a_{22}, \cdots, a_{nn}$ stand. Two terms are new:

M_{jk} is called the **minor** *of a_{jk} in D,* and C_{jk} the **cofactor** *of a_{jk} in D.*

For later use we note that (3) may also be written in terms of minors

$$D = \sum_{k=1}^{n}(-1)^{j+k}a_{jk}M_{jk} \qquad (j = 1, 2, \cdots, \text{ or } n) \tag{4a}$$

$$D = \sum_{j=1}^{n}(-1)^{j+k}a_{jk}M_{jk} \qquad (k = 1, 2, \cdots, \text{ or } n). \tag{4b}$$

EXAMPLE 1 Minors and Cofactors of a Third-Order Determinant

In (4) of the previous section the minors and cofactors of the entries in the first column can be seen directly. For the entries in the second row the minors are

$$M_{21} = \begin{vmatrix} a_{12} & a_{13} \\ a_{32} & a_{33} \end{vmatrix}, \qquad M_{22} = \begin{vmatrix} a_{11} & a_{13} \\ a_{31} & a_{33} \end{vmatrix}, \qquad M_{23} = \begin{vmatrix} a_{11} & a_{12} \\ a_{31} & a_{32} \end{vmatrix}$$

and the cofactors are $C_{21} = -M_{21}$, $C_{22} = +M_{22}$, and $C_{23} = -M_{23}$. Similarly for the third row—write these down yourself. And verify that the signs in C_{jk} form a **checkerboard pattern**

$$\begin{matrix} + & - & + \\ - & + & - \\ + & - & + \end{matrix}$$

■

EXAMPLE 2 **Expansions of a Third-Order Determinant**

$$D = \begin{vmatrix} 1 & 3 & 0 \\ 2 & 6 & 4 \\ -1 & 0 & 2 \end{vmatrix} = 1\begin{vmatrix} 6 & 4 \\ 0 & 2 \end{vmatrix} - 3\begin{vmatrix} 2 & 4 \\ -1 & 2 \end{vmatrix} + 0\begin{vmatrix} 2 & 6 \\ -1 & 0 \end{vmatrix}$$

$$= 1(12 - 0) - 3(4 + 4) + 0(0 + 6) = -12.$$

This is the expansion by the first row. The expansion by the third column is

$$D = 0\begin{vmatrix} 2 & 6 \\ -1 & 0 \end{vmatrix} - 4\begin{vmatrix} 1 & 3 \\ -1 & 0 \end{vmatrix} + 2\begin{vmatrix} 1 & 3 \\ 2 & 6 \end{vmatrix} = 0 - 12 + 0 = -12.$$

Verify that the other four expansions also give the value -12. ■

EXAMPLE 3 **Determinant of a Triangular Matrix**

$$\begin{vmatrix} -3 & 0 & 0 \\ 6 & 4 & 0 \\ -1 & 2 & 5 \end{vmatrix} = -3\begin{vmatrix} 4 & 0 \\ 2 & 5 \end{vmatrix} = -3 \cdot 4 \cdot 5 = -60.$$

Inspired by this, can you formulate a little theorem on determinants of triangular matrices? Of diagonal matrices? ■

General Properties of Determinants

There is an attractive way of finding determinants (1) that consists of applying elementary row operations to (1). By doing so we obtain an "upper triangular" determinant (see Sec. 7.1, for definition with "matrix" replaced by "determinant") whose value is then very easy to compute, being just the product of its diagonal entries. This approach is ***similar*** (***but not the same!***) to what we did to matrices in Sec. 7.3. *In particular, be aware that interchanging two rows in a determinant introduces a multiplicative factor of* -1 *to the value of the determinant*! Details are as follows.

THEOREM 1 **Behavior of an *n*th-Order Determinant under Elementary Row Operations**

(a) *Interchange of two rows multiplies the value of the determinant by* -1.

(b) *Addition of a multiple of a row to another row does not alter the value of the determinant.*

(c) *Multiplication of a row by a nonzero constant c multiplies the value of the determinant by c.* (This holds also when $c = 0$, but no longer gives an elementary row operation.)

PROOF **(a)** By induction. The statement holds for $n = 2$ because

$$\begin{vmatrix} a & b \\ c & d \end{vmatrix} = ad - bc, \quad \text{but} \quad \begin{vmatrix} c & d \\ a & b \end{vmatrix} = bc - ad.$$

We now make the induction hypothesis that (a) holds for determinants of order $n - 1 \geqq 2$ and show that it then holds for determinants of order n. Let D be of order n. Let E be obtained from D by the interchange of two rows. Expand D and E by a row that is ***not*** one of those interchanged, call it the jth row. Then by (4a),

$$D = \sum_{k=1}^{n} (-1)^{j+k} a_{jk} M_{jk}, \qquad E = \sum_{k=1}^{n} (-1)^{j+k} a_{jk} N_{jk} \tag{5}$$

where N_{jk} is obtained from the minor M_{jk} of a_{jk} in D by the interchange of those two rows which have been interchanged in D (and which N_{jk} must both contain because we expand by another row!). Now these minors are of order $n - 1$. Hence the induction hypothesis applies and gives $N_{jk} = -M_{jk}$. Thus $E = -D$ by (5).

(b) Add c times Row i to Row j. Let $\tilde{D}$ be the new determinant. Its entries in Row j are $a_{jk} + ca_{ik}$. If we expand $\tilde{D}$ by this Row j, we see that we can write it as $\tilde{D} = D_1 + cD_2$, where $D_1 = D$ has in Row j the a_{jk}, whereas D_2 has in that Row j the a_{jk} from the addition. Hence D_2 has a_{jk} in both Row i and Row j. Interchanging these two rows gives D_2 back, but on the other hand it gives $-D_2$ by (a). Together $D_2 = -D_2 = 0$, so that $\tilde{D} = D_1 = D$.

(c) Expand the determinant by the row that has been multiplied.

CAUTION! $\det(c\mathbf{A}) = c^n \det \mathbf{A}$ (not $c \det \mathbf{A}$). Explain why. ■

EXAMPLE 4 Evaluation of Determinants by Reduction to Triangular Form

Because of Theorem 1 we may evaluate determinants by reduction to triangular form, as in the Gauss elimination for a matrix. For instance (with the blue explanations always referring to the ***preceding determinant***)

$$D = \begin{vmatrix} 2 & 0 & -4 & 6 \\ 4 & 5 & 1 & 0 \\ 0 & 2 & 6 & -1 \\ -3 & 8 & 9 & 1 \end{vmatrix}$$

$$= \begin{vmatrix} 2 & 0 & -4 & 6 \\ 0 & 5 & 9 & -12 \\ 0 & 2 & 6 & -1 \\ 0 & 8 & 3 & 10 \end{vmatrix} \begin{matrix} \\ \text{Row 2} - 2\text{ Row 1} \\ \\ \text{Row 4} + 1.5\text{ Row 1} \end{matrix}$$

$$= \begin{vmatrix} 2 & 0 & -4 & 6 \\ 0 & 5 & 9 & -12 \\ 0 & 0 & 2.4 & 3.8 \\ 0 & 0 & -11.4 & 29.2 \end{vmatrix} \begin{matrix} \\ \\ \text{Row 3} - 0.4\text{ Row 2} \\ \text{Row 4} - 1.6\text{ Row 2} \end{matrix}$$

$$= \begin{vmatrix} 2 & 0 & -4 & 6 \\ 0 & 5 & 9 & -12 \\ 0 & 0 & 2.4 & 3.8 \\ 0 & 0 & -0 & 47.25 \end{vmatrix} \begin{matrix} \\ \\ \\ \text{Row 4} + 4.75\text{ Row 3} \end{matrix}$$

$$= 2 \cdot 5 \cdot 2.4 \cdot 47.25 = 1134.$$ ■

THEOREM 2

Further Properties of *n*th-Order Determinants

(a)–(c) *in Theorem* 1 *hold also for columns.*

(d) ***Transposition*** *leaves the value of a determinant unaltered.*

(e) ***A zero row or column*** *renders the value of a determinant zero.*

(f) ***Proportional rows or columns*** *render the value of a determinant zero. In particular, a determinant with two identical rows or columns has the value zero.*

PROOF **(a)–(e)** follow directly from the fact that a determinant can be expanded by any row column. In **(d)**, transposition is defined as for matrices, that is, the jth row becomes the jth column of the transpose.

(f) If Row $j = c$ times Row i, then $D = cD_1$, where D_1 has Row j = Row i. Hence an interchange of these rows reproduces D_1, but it also gives $-D_1$ by Theorem 1(a). Hence $D_1 = 0$ and $D = cD_1 = 0$. Similarly for columns. ■

It is quite remarkable that the important concept of the rank of a matrix **A**, which is the maximum number of linearly independent row or column vectors of **A** (see Sec. 7.4), can be related to determinants. Here we may assume that rank $\mathbf{A} > 0$ because the only matrices with rank 0 are the zero matrices (see Sec. 7.4).

THEOREM 3

Rank in Terms of Determinants

Consider an $m \times n$ *matrix* $\mathbf{A} = [a_{jk}]$:

(1) **A** *has rank* $r \geqq 1$ *if and only if* **A** *has an* $r \times r$ *submatrix with a nonzero determinant.*

(2) *The determinant of any square submatrix with more than* r *rows, contained in* **A** *(if such a matrix exists!) has a value equal to zero.*

Furthermore, if $m = n$, *we have:*

(3) *An* $n \times n$ *square matrix* **A** *has rank* n *if and only if*

$$\det \mathbf{A} \neq 0.$$

PROOF The key idea is that elementary row operations (Sec. 7.3) alter neither rank (by Theorem 1 in Sec. 7.4) nor the property of a determinant being nonzero (by Theorem 1 in this section). The echelon form $\hat{\mathbf{A}}$ of **A** (see Sec. 7.3) has r nonzero row vectors (which are the first r row vectors) if and only if rank $\mathbf{A} = r$. Without loss of generality, we can assume that $r \geqq 1$. Let $\hat{\mathbf{R}}$ be the $r \times r$ submatrix in the left upper corner of $\hat{\mathbf{A}}$ (so that the entries of $\hat{\mathbf{R}}$ are in both the first r rows and r columns of $\hat{\mathbf{A}}$). Now $\hat{\mathbf{R}}$ is triangular, with all diagonal entries r_{jj} nonzero. Thus, $\det \hat{\mathbf{R}} = r_{11} \cdots r_{rr} \neq 0$. Also $\det \mathbf{R} \neq 0$ for the corresponding $r \times r$ submatrix **R** of **A** because $\hat{\mathbf{R}}$ results from **R** by elementary row operations. This proves part (1).

Similarly, $\det \mathbf{S} = 0$ for any square submatrix **S** of $r + 1$ or more rows perhaps contained in **A** because the corresponding submatrix $\hat{\mathbf{S}}$ of $\hat{\mathbf{A}}$ must contain a row of zeros (otherwise we would have rank $\mathbf{A} \geqq r + 1$), so that $\det \hat{\mathbf{S}} = 0$ by Theorem 2. This proves part (2). Furthermore, we have proven the theorem for an $m \times n$ matrix.

For an $n \times n$ square matrix $\mathbf{A}$ we proceed as follows. To prove (3), we apply part (1) (already proven!). This gives us that rank $\mathbf{A} = n \geqq 1$ if and only if $\mathbf{A}$ contains an $n \times n$ submatrix with nonzero determinant. But the only such submatrix contained in our square matrix $\mathbf{A}$, is $\mathbf{A}$ itself, hence det $\mathbf{A} \neq 0$. This proves part (3). ■

Cramer's Rule

Theorem 3 opens the way to the classical solution formula for linear systems known as **Cramer's rule**,[2] which gives solutions as quotients of determinants. ***Cramer's rule is not practical in computations*** for which the methods in Secs. 7.3 and 20.1–20.3 are suitable. However, Cramer's rule is of ***theoretical interest*** in differential equations (Secs. 2.10 and 3.3) and in other theoretical work that has engineering applications.

THEOREM 4

Cramer's Theorem (Solution of Linear Systems by Determinants)

(a) *If a linear system of n equations in the same number of unknowns* $x_1, \cdots, x_n$

$$\begin{aligned} a_{11}x_1 + a_{12}x_2 + \cdots + a_{1n}x_n &= b_1 \\ a_{21}x_1 + a_{22}x_2 + \cdots + a_{2n}x_n &= b_2 \\ \cdots\cdots\cdots\cdots\cdots\cdots\cdots\cdots \\ a_{n1}x_1 + a_{n2}x_2 + \cdots + a_{nn}x_n &= b_n \end{aligned} \tag{6}$$

has a nonzero coefficient determinant $D = \det \mathbf{A}$, *the system has precisely one solution. This solution is given by the formulas*

$$x_1 = \frac{D_1}{D}, \quad x_2 = \frac{D_2}{D}, \cdots, \quad x_n = \frac{D_n}{D} \qquad \textbf{(Cramer's rule)} \tag{7}$$

where D_k *is the determinant obtained from D by replacing in D the kth column by the column with the entries* $b_1, \cdots, b_n$.

(b) *Hence if the system* (6) *is* **homogeneous** *and* $D \neq 0$, *it has only the trivial solution* $x_1 = 0, x_2 = 0, \cdots, x_n = 0$. *If* $D = 0$, *the homogeneous system also has nontrivial solutions.*

PROOF The augmented matrix $\widetilde{\mathbf{A}}$ of the system (6) is of size $n \times (n + 1)$. Hence its rank can be at most n. Now if

$$D = \det \mathbf{A} = \begin{vmatrix} a_{11} & \cdots & a_{1n} \\ \cdot & \cdots & \cdot \\ \cdot & \cdots & \cdot \\ a_{n1} & \cdots & a_{nn} \end{vmatrix} \neq 0, \tag{8}$$

[2]GABRIEL CRAMER (1704–1752), Swiss mathematician.

then rank $\mathbf{A} = n$ by Theorem 3. Thus rank $\tilde{\mathbf{A}}$ = rank $\mathbf{A}$. Hence, by the Fundamental Theorem in Sec. 7.5, the system (6) has a unique solution.

Let us now prove (7). Expanding D by its kth column, we obtain

$$D = a_{1k}C_{1k} + a_{2k}C_{2k} + \cdots + a_{nk}C_{nk}, \tag{9}$$

where C_{ik} is the cofactor of entry a_{ik} in D. If we replace the entries in the kth column of D by any other numbers, we obtain a new determinant, say, $\hat{D}$. Clearly, its expansion by the kth column will be of the form (9), with $a_{1k}, \cdots, a_{nk}$ replaced by those new numbers and the cofactors C_{ik} as before. In particular, if we choose as new numbers the entries $a_{1l}, \cdots, a_{nl}$ of the lth column of D (where $l \neq k$), we have a new determinant $\hat{D}$ which has the column $[a_{1l} \quad \cdots \quad a_{nl}]^\mathsf{T}$ twice, once as its lth column, and once as its kth because of the replacement. Hence $\hat{D} = 0$ by Theorem 2(f). If we now expand $\hat{D}$ by the column that has been replaced (the kth column), we thus obtain

$$a_{1l}C_{1k} + a_{2l}C_{2k} + \cdots + a_{nl}C_{nk} = 0 \qquad (l \neq k). \tag{10}$$

We now multiply the first equation in (6) by C_{1k} on both sides, the second by $C_{2k}, \cdots,$ the last by C_{nk}, and add the resulting equations. This gives

$$\begin{aligned} C_{1k}(a_{11}x_1 + \cdots + a_{1n}x_n) + \cdots + C_{nk}(a_{n1}x_1 + \cdots + a_{nn}x_n) \\ = b_1C_{1k} + \cdots + b_nC_{nk}. \end{aligned} \tag{11}$$

Collecting terms with the same x_j, we can write the left side as

$$x_1(a_{11}C_{1k} + a_{21}C_{2k} + \cdots + a_{n1}C_{nk}) + \cdots + x_n(a_{1n}C_{1k} + a_{2n}C_{2k} + \cdots + a_{nn}C_{nk}).$$

From this we see that x_k is multiplied by

$$a_{1k}C_{1k} + a_{2k}C_{2k} + \cdots + a_{nk}C_{nk}.$$

Equation (9) shows that this equals D. Similarly, x_1 is multiplied by

$$a_{1l}C_{1k} + a_{2l}C_{2k} + \cdots + a_{nl}C_{nk}.$$

Equation (10) shows that this is zero when $l \neq k$. Accordingly, the left side of (11) equals simply x_kD, so that (11) becomes

$$x_kD = b_1C_{1k} + b_2C_{2k} + \cdots + b_nC_{nk}.$$

Now the right side of this is D_k as defined in the theorem, expanded by its kth column, so that division by D gives (7). This proves Cramer's rule.

If (6) is homogeneous and $D \neq 0$, then each D_k has a column of zeros, so that $D_k = 0$ by Theorem 2(e), and (7) gives the trivial solution.

Finally, if (6) is homogeneous and $D = 0$, then rank $\mathbf{A} < n$ by Theorem 3, so that nontrivial solutions exist by Theorem 2 in Sec. 7.5. ■

EXAMPLE 5 **Illustration of Cramer's Rule (Theorem 4)**

For $n = 2$, see Example 1 of Sec. 7.6. Also, at the end of that section, we give Cramer's rule for a general linear system of three equations. ■

Finally, an important application for Cramer's rule dealing with inverse matrices will be given in the next section.

PROBLEM SET 7.7

1–6 GENERAL PROBLEMS

1. **General Properties of Determinants.** Illustrate each statement in Theorems 1 and 2 with an example of your choice.

2. **Second-Order Determinant.** Expand a general second-order determinant in four possible ways and show that the results agree.

3. **Third-Order Determinant.** Do the task indicated in Theorem 2. Also evaluate D by reduction to triangular form.

4. **Expansion Numerically Impractical.** Show that the computation of an nth-order determinant by expansion involves $n!$ multiplications, which if a multiplication takes 10^{-9} sec would take these times:

n	10	15	20	25
Time	0.004 sec	22 min	77 years	$0.5 \cdot 10^9$ years

5. **Multiplication by Scalar.** Show that $\det(k\mathbf{A}) = k^n \det \mathbf{A}$ (not $k \det \mathbf{A}$). Give an example.

6. **Minors, cofactors.** Complete the list in Example 1.

7–15 EVALUATION OF DETERMINANTS

Showing the details, evaluate:

7. $\begin{vmatrix} \cos\alpha & \sin\alpha \\ \sin\beta & \cos\beta \end{vmatrix}$

8. $\begin{vmatrix} 0.4 & 4.9 \\ 1.5 & -1.3 \end{vmatrix}$

9. $\begin{vmatrix} \cos n\theta & \sin n\theta \\ -\sin n\theta & \cos n\theta \end{vmatrix}$

10. $\begin{vmatrix} \cosh t & \sinh t \\ \sinh t & \cosh t \end{vmatrix}$

11. $\begin{vmatrix} 4 & -1 & 8 \\ 0 & 2 & 3 \\ 0 & 0 & 5 \end{vmatrix}$

12. $\begin{vmatrix} a & b & c \\ c & a & b \\ b & c & a \end{vmatrix}$

13. $\begin{vmatrix} 0 & 4 & -1 & 5 \\ -4 & 0 & 3 & -2 \\ 1 & -3 & 0 & 1 \\ -5 & 2 & -1 & 0 \end{vmatrix}$

14. $\begin{vmatrix} 4 & 7 & 0 & 0 \\ 2 & 8 & 0 & 0 \\ 0 & 0 & 1 & 5 \\ 0 & 0 & -2 & 2 \end{vmatrix}$

15. $\begin{vmatrix} 1 & 2 & 0 & 0 \\ 2 & 4 & 2 & 0 \\ 0 & 2 & 9 & 2 \\ 0 & 0 & 2 & 16 \end{vmatrix}$

16. **CAS EXPERIMENT. Determinant of Zeros and Ones.** Find the value of the determinant of the $n \times n$ matrix A_n with main diagonal entries all 0 and all others 1. Try to find a formula for this. Try to prove it by induction. Interpret A_3 and A_4 as *incidence matrices* (as in Problem Set 7.1 but without the minuses) of a triangle and a tetrahedron, respectively; similarly for an *n-simplex*, having n vertices and $n(n-1)/2$ edges (and spanning R^{n-1}, $n = 5, 6, \cdots$).

17–19 RANK BY DETERMINANTS

Find the rank by Theorem 3 (which is not very practical) and check by row reduction. Show details.

17. $\begin{bmatrix} 4 & 9 \\ -8 & -6 \\ 16 & 12 \end{bmatrix}$

18. $\begin{bmatrix} 0 & 4 & -6 \\ 4 & 0 & 10 \\ -6 & 10 & 0 \end{bmatrix}$

19. $\begin{bmatrix} 1 & 5 & 2 & 2 \\ 1 & 3 & 2 & 6 \\ 4 & 0 & 8 & 48 \end{bmatrix}$

20. **TEAM PROJECT. Geometric Applications: Curves and Surfaces Through Given Points.** The idea is to get an equation from the vanishing of the determinant of a homogeneous linear system as the condition for a nontrivial solution in Cramer's theorem. We explain the trick for obtaining such a system for the case of a line L through two given points P_1: (x_1, y_1) and P_2: (x_2, y_2). The unknown line is $ax + by = -c$, say. We write it as $ax + by + c \cdot 1 = 0$. To get a nontrivial solution a, b, c, the determinant of the "coefficients" $x, y, 1$ must be zero. The system is

$$\begin{aligned} ax + by + c \cdot 1 &= 0 \quad \text{(Line } L\text{)} \\ ax_1 + by_1 + c \cdot 1 &= 0 \quad (P_1 \text{ on } L) \\ ax_2 + by_2 + c \cdot 1 &= 0 \quad (P_2 \text{ on } L). \end{aligned} \tag{12}$$

(a) Line through two points. Derive from $D = 0$ in (12) the familiar formula

$$\frac{x - x_1}{x_1 - x_2} = \frac{y - y_1}{y_1 - y_2}.$$

(b) Plane. Find the analog of (12) for a plane through three given points. Apply it when the points are (1, 1, 1), (3, 2, 6), (5, 0, 5).

(c) Circle. Find a similar formula for a circle in the plane through three given points. Find and sketch the circle through (2, 6), (6, 4), (7, 1).

(d) Sphere. Find the analog of the formula in (c) for a sphere through four given points. Find the sphere through (0, 0, 5), (4, 0, 1), (0, 4, 1), (0, 0, −3) by this formula or by inspection.

(e) General conic section. Find a formula for a general conic section (the vanishing of a determinant of 6th order). Try it out for a quadratic parabola and for a more general conic section of your own choice.

21–25 CRAMER'S RULE

Solve by Cramer's rule. Check by Gauss elimination and back substitution. Show details.

21. $3x - 5y = 15.5$, $6x + 16y = 5.0$

22. $2x - 4y = -24$, $5x + 2y = 0$

23. $3y - 4z = 16$, $2x - 5y + 7z = -27$, $-x - 9z = 9$

24. $3x - 2y + z = 13$, $-2x + y + 4z = 11$, $x + 4y - 5z = -31$

25. $-4w + x + y = -10$, $w - 4x + z = 1$, $w - 4y + z = -7$, $x + y - 4z = 10$

7.8 Inverse of a Matrix. Gauss–Jordan Elimination

In this section we consider* square *matrices exclusively.

The **inverse** of an $n \times n$ matrix $\mathbf{A} = [a_{jk}]$ is denoted by $\mathbf{A}^{-1}$ and is an $n \times n$ matrix such that

$$\text{(1)} \qquad \mathbf{A}\mathbf{A}^{-1} = \mathbf{A}^{-1}\mathbf{A} = \mathbf{I}$$

where **I** is the $n \times n$ unit matrix (see Sec. 7.2).

If **A** has an inverse, then **A** is called a **nonsingular matrix**. If **A** has no inverse, then **A** is called a **singular matrix**.

If **A** *has an inverse, the inverse is unique.*

Indeed, if both **B** and **C** are inverses of **A**, then $\mathbf{AB} = \mathbf{I}$ and $\mathbf{CA} = \mathbf{I}$, so that we obtain the uniqueness from

$$\mathbf{B} = \mathbf{IB} = (\mathbf{CA})\mathbf{B} = \mathbf{C}(\mathbf{AB}) = \mathbf{CI} = \mathbf{C}.$$

We prove next that **A** has an inverse (is nonsingular) if and only if it has maximum possible rank n. The proof will also show that $\mathbf{Ax} = \mathbf{b}$ implies $\mathbf{x} = \mathbf{A}^{-1}\mathbf{b}$ provided $\mathbf{A}^{-1}$ exists, and will thus give a motivation for the inverse as well as a relation to linear systems. (But this will ***not*** give a good method of solving $\mathbf{Ax} = \mathbf{b}$ ***numerically*** because the Gauss elimination in Sec. 7.3 requires fewer computations.)

THEOREM 1

Existence of the Inverse

The inverse $\mathbf{A}^{-1}$ *of an* $n \times n$ *matrix* **A** *exists if and only if* rank $\mathbf{A} = n$, *thus* (by Theorem 3, Sec. 7.7) *if and only if* det $\mathbf{A} \neq 0$. *Hence* **A** *is nonsingular if* rank $\mathbf{A} = n$, *and is singular if* rank $\mathbf{A} < n$.

PROOF Let $\mathbf{A}$ be a given $n \times n$ matrix and consider the linear system

$$\mathbf{Ax} = \mathbf{b}. \tag{2}$$

If the inverse $\mathbf{A}^{-1}$ exists, then multiplication from the left on both sides and use of (1) gives

$$\mathbf{A}^{-1}\mathbf{Ax} = \mathbf{x} = \mathbf{A}^{-1}\mathbf{b}.$$

This shows that (2) has a solution $\mathbf{x}$, which is unique because, for another solution $\mathbf{u}$, we have $\mathbf{Au} = \mathbf{b}$, so that $\mathbf{u} = \mathbf{A}^{-1}\mathbf{b} = \mathbf{x}$. Hence $\mathbf{A}$ must have rank n by the Fundamental Theorem in Sec. 7.5.

Conversely, let rank $\mathbf{A} = n$. Then by the same theorem, the system (2) has a unique solution $\mathbf{x}$ for any $\mathbf{b}$. Now the back substitution following the Gauss elimination (Sec. 7.3) shows that the components x_j of $\mathbf{x}$ are linear combinations of those of $\mathbf{b}$. Hence we can write

$$\mathbf{x} = \mathbf{Bb} \tag{3}$$

with $\mathbf{B}$ to be determined. Substitution into (2) gives

$$\mathbf{Ax} = \mathbf{A}(\mathbf{Bb}) = (\mathbf{AB})\mathbf{b} = \mathbf{Cb} = \mathbf{b} \qquad (\mathbf{C} = \mathbf{AB})$$

for any $\mathbf{b}$. Hence $\mathbf{C} = \mathbf{AB} = \mathbf{I}$, the unit matrix. Similarly, if we substitute (2) into (3) we get

$$\mathbf{x} = \mathbf{Bb} = \mathbf{B}(\mathbf{Ax}) = (\mathbf{BA})\mathbf{x}$$

for any $\mathbf{x}$ (and $\mathbf{b} = \mathbf{Ax}$). Hence $\mathbf{BA} = \mathbf{I}$. Together, $\mathbf{B} = \mathbf{A}^{-1}$ exists. ■

Determination of the Inverse by the Gauss–Jordan Method

To actually determine the inverse $\mathbf{A}^{-1}$ of a nonsingular $n \times n$ matrix $\mathbf{A}$, we can use a variant of the Gauss elimination (Sec. 7.3), called the **Gauss–Jordan elimination**.[3] The idea of the method is as follows.

Using $\mathbf{A}$, we form n linear systems

$$\mathbf{Ax}_{(1)} = \mathbf{e}_{(1)}, \quad \cdots, \quad \mathbf{Ax}_{(n)} = \mathbf{e}_{(n)}$$

where the vectors $\mathbf{e}_{(1)}, \cdots, \mathbf{e}_{(n)}$ are the columns of the $n \times n$ unit matrix $\mathbf{I}$; thus, $\mathbf{e}_{(1)} = [1 \quad 0 \quad \cdots \quad 0]^{\mathsf{T}}$, $\mathbf{e}_{(2)} = [0 \quad 1 \quad 0 \quad \cdots \quad 0]^{\mathsf{T}}$, etc. These are n vector equations in the unknown vectors $\mathbf{x}_{(1)}, \cdots, \mathbf{x}_{(n)}$. We combine them into a single matrix equation

[3] WILHELM JORDAN (1842–1899), German geodesist and mathematician. He did important geodesic work in Africa, where he surveyed oases. [See Althoen, S.C. and R. McLaughlin, Gauss–Jordan reduction: A brief history. *American Mathematical Monthly*, Vol. **94**, No. 2 (1987), pp. 130–142.]

We do ***not recommend*** it as a method for solving systems of linear equations, since the number of operations in addition to those of the Gauss elimination is larger than that for back substitution, which the Gauss–Jordan elimination avoids. See also Sec. 20.1.

$\mathbf{AX} = \mathbf{I}$, with the unknown matrix $\mathbf{X}$ having the columns $\mathbf{x}_{(1)}, \cdots, \mathbf{x}_{(n)}$. Correspondingly, we combine the n augmented matrices $[\mathbf{A} \quad \mathbf{e}_{(1)}], \cdots, [\mathbf{A} \quad \mathbf{e}_{(n)}]$ into one wide $n \times 2n$ "augmented matrix" $\widetilde{\mathbf{A}} = [\mathbf{A} \quad \mathbf{I}]$. Now multiplication of $\mathbf{AX} = \mathbf{I}$ by $\mathbf{A}^{-1}$ from the left gives $\mathbf{X} = \mathbf{A}^{-1}\mathbf{I} = \mathbf{A}^{-1}$. Hence, to solve $\mathbf{AX} = \mathbf{I}$ for $\mathbf{X}$, we can apply the Gauss elimination to $\widetilde{\mathbf{A}} = [\mathbf{A} \quad \mathbf{I}]$. This gives a matrix of the form $[\mathbf{U} \quad \mathbf{H}]$ with upper triangular $\mathbf{U}$ because the Gauss elimination triangularizes systems. The Gauss–Jordan method reduces $\mathbf{U}$ by further elementary row operations to diagonal form, in fact to the unit matrix $\mathbf{I}$. This is done by eliminating the entries of $\mathbf{U}$ above the main diagonal and making the diagonal entries all 1 by multiplication (see Example 1). Of course, the method operates on the entire matrix $[\mathbf{U} \quad \mathbf{H}]$, transforming $\mathbf{H}$ into some matrix $\mathbf{K}$, hence the entire $[\mathbf{U} \quad \mathbf{H}]$ to $[\mathbf{I} \quad \mathbf{K}]$. This is the "augmented matrix" of $\mathbf{IX} = \mathbf{K}$. Now $\mathbf{IX} = \mathbf{X} = \mathbf{A}^{-1}$, as shown before. By comparison, $\mathbf{K} = \mathbf{A}^{-1}$, so that we can read $\mathbf{A}^{-1}$ directly from $[\mathbf{I} \quad \mathbf{K}]$.

The following example illustrates the practical details of the method.

EXAMPLE 1 Finding the Inverse of a Matrix by Gauss–Jordan Elimination

Determine the inverse $\mathbf{A}^{-1}$ of

$$\mathbf{A} = \begin{bmatrix} -1 & 1 & 2 \\ 3 & -1 & 1 \\ -1 & 3 & 4 \end{bmatrix}.$$

Solution. We apply the Gauss elimination (Sec. 7.3) to the following $n \times 2n = 3 \times 6$ matrix, where BLUE always refers to the previous matrix.

$$[\mathbf{A} \quad \mathbf{I}] = \left[\begin{array}{rrr|rrr} -1 & 1 & 2 & 1 & 0 & 0 \\ 3 & -1 & 1 & 0 & 1 & 0 \\ -1 & 3 & 4 & 0 & 0 & 1 \end{array}\right]$$

$$\left[\begin{array}{rrr|rrr} -1 & 1 & 2 & 1 & 0 & 0 \\ 0 & 2 & 7 & 3 & 1 & 0 \\ 0 & 2 & 2 & -1 & 0 & 1 \end{array}\right] \begin{array}{l} \\ \text{Row 2 + 3 Row 1} \\ \text{Row 3 − Row 1} \end{array}$$

$$\left[\begin{array}{rrr|rrr} -1 & 1 & 2 & 1 & 0 & 0 \\ 0 & 2 & 7 & 3 & 1 & 0 \\ 0 & 0 & -5 & -4 & -1 & 1 \end{array}\right] \begin{array}{l} \\ \\ \text{Row 3 − Row 2} \end{array}$$

This is $[\mathbf{U} \quad \mathbf{H}]$ as produced by the Gauss elimination. Now follow the additional Gauss–Jordan steps, reducing $\mathbf{U}$ to $\mathbf{I}$, that is, to diagonal form with entries 1 on the main diagonal.

$$\left[\begin{array}{rrr|rrr} 1 & -1 & -2 & -1 & 0 & 0 \\ 0 & 1 & 3.5 & 1.5 & 0.5 & 0 \\ 0 & 0 & 1 & 0.8 & 0.2 & -0.2 \end{array}\right] \begin{array}{l} \text{−Row 1} \\ \text{0.5 Row 2} \\ \text{−0.2 Row 3} \end{array}$$

$$\left[\begin{array}{rrr|rrr} 1 & -1 & 0 & 0.6 & 0.4 & -0.4 \\ 0 & 1 & 0 & -1.3 & -0.2 & 0.7 \\ 0 & 0 & 1 & 0.8 & 0.2 & -0.2 \end{array}\right] \begin{array}{l} \text{Row 1 + 2 Row 3} \\ \text{Row 2 − 3.5 Row 3} \\ \\ \end{array}$$

$$\left[\begin{array}{rrr|rrr} 1 & 0 & 0 & -0.7 & 0.2 & 0.3 \\ 0 & 1 & 0 & -1.3 & -0.2 & 0.7 \\ 0 & 0 & 1 & 0.8 & 0.2 & -0.2 \end{array}\right] \begin{array}{l} \text{Row 1 + Row 2} \\ \\ \\ \end{array}$$

The last three columns constitute $\mathbf{A}^{-1}$. Check:

$$\begin{bmatrix} -1 & 1 & 2 \\ 3 & -1 & 1 \\ -1 & 3 & 4 \end{bmatrix} \begin{bmatrix} -0.7 & 0.2 & 0.3 \\ -1.3 & -0.2 & 0.7 \\ 0.8 & 0.2 & -0.2 \end{bmatrix} = \begin{bmatrix} 1 & 0 & 0 \\ 0 & 1 & 0 \\ 0 & 0 & 1 \end{bmatrix}.$$

Hence $\mathbf{A}\mathbf{A}^{-1} = \mathbf{I}$. Similarly, $\mathbf{A}^{-1}\mathbf{A} = \mathbf{I}$. ■

Formulas for Inverses

Since finding the inverse of a matrix is really a problem of solving a system of linear equations, it is not surprising that Cramer's rule (Theorem 4, Sec. 7.7) might come into play. And similarly, as Cramer's rule was useful for theoretical study but not for computation, so too is the explicit formula (4) in the following theorem useful for theoretical considerations but not recommended for actually determining inverse matrices, except for the frequently occurring 2×2 case as given in (4*).

THEOREM 2

Inverse of a Matrix by Determinants

The inverse of a nonsingular $n \times n$ matrix $\mathbf{A} = [a_{jk}]$ is given by

$$\text{(4)} \qquad \mathbf{A}^{-1} = \frac{1}{\det \mathbf{A}}[C_{jk}]^{\mathsf{T}} = \frac{1}{\det \mathbf{A}} \begin{bmatrix} C_{11} & C_{21} & \cdots & C_{n1} \\ C_{12} & C_{22} & \cdots & C_{n2} \\ \cdot & \cdot & \cdots & \cdot \\ C_{1n} & C_{2n} & \cdots & C_{nn} \end{bmatrix},$$

where C_{jk} is the cofactor of a_{jk} in $\det \mathbf{A}$ (see Sec. 7.7). (CAUTION! Note well that in $\mathbf{A}^{-1}$, the cofactor C_{jk} occupies the same place as a_{kj} (not a_{jk}) does in $\mathbf{A}$.)

In particular, the inverse of

$$\text{(4*)} \qquad \mathbf{A} = \begin{bmatrix} a_{11} & a_{12} \\ a_{21} & a_{22} \end{bmatrix} \quad \textit{is} \quad \mathbf{A}^{-1} = \frac{1}{\det \mathbf{A}} \begin{bmatrix} a_{22} & -a_{12} \\ -a_{21} & a_{11} \end{bmatrix}.$$

PROOF We denote the right side of (4) by $\mathbf{B}$ and show that $\mathbf{BA} = \mathbf{I}$. We first write

$$\text{(5)} \qquad \mathbf{BA} = \mathbf{G} = [g_{kl}]$$

and then show that $\mathbf{G} = \mathbf{I}$. Now by the definition of matrix multiplication and because of the form of $\mathbf{B}$ in (4), we obtain (CAUTION! C_{sk}, not C_{ks})

$$\text{(6)} \qquad g_{kl} = \sum_{s=1}^{n} \frac{C_{sk}}{\det \mathbf{A}} a_{sl} = \frac{1}{\det \mathbf{A}} (a_{1l}C_{1k} + \cdots + a_{nl}C_{nk}).$$

Now (9) and (10) in Sec. 7.7 show that the sum $(\cdots)$ on the right is $D = \det \mathbf{A}$ when $l = k$, and is zero when $l \neq k$. Hence

$$g_{kk} = \frac{1}{\det \mathbf{A}} \det \mathbf{A} = 1,$$

$$g_{kl} = 0 \quad (l \neq k).$$

In particular, for $n = 2$ we have in (4), in the first row, $C_{11} = a_{22}$, $C_{21} = -a_{12}$ and, in the second row, $C_{12} = -a_{21}$, $C_{22} = a_{11}$. This gives (4*). ■

The special case $n = 2$ occurs quite frequently in geometric and other applications. You may perhaps want to memorize formula (4*). Example 2 gives an illustration of (4*).

EXAMPLE 2 **Inverse of a 2 × 2 Matrix by Determinants**

$$\mathbf{A} = \begin{bmatrix} 3 & 1 \\ 2 & 4 \end{bmatrix}, \quad \mathbf{A}^{-1} = \frac{1}{10}\begin{bmatrix} 4 & -1 \\ -2 & 3 \end{bmatrix} = \begin{bmatrix} 0.4 & -0.1 \\ -0.2 & 0.3 \end{bmatrix}$$ ■

EXAMPLE 3 **Further Illustration of Theorem 2**

Using (4), find the inverse of

$$\mathbf{A} = \begin{bmatrix} -1 & 1 & 2 \\ 3 & -1 & 1 \\ -1 & 3 & 4 \end{bmatrix}.$$

Solution. We obtain $\det \mathbf{A} = -1(-7) - 1 \cdot 13 + 2 \cdot 8 = 10$, and in (4),

$$C_{11} = \begin{vmatrix} -1 & 1 \\ 3 & 4 \end{vmatrix} = -7, \quad C_{21} = -\begin{vmatrix} 1 & 2 \\ 3 & 4 \end{vmatrix} = 2, \quad C_{31} = \begin{vmatrix} 1 & 2 \\ -1 & 1 \end{vmatrix} = 3,$$

$$C_{12} = -\begin{vmatrix} 3 & 1 \\ -1 & 4 \end{vmatrix} = -13, \quad C_{22} = \begin{vmatrix} -1 & 2 \\ -1 & 4 \end{vmatrix} = -2, \quad C_{32} = -\begin{vmatrix} -1 & 2 \\ 3 & 1 \end{vmatrix} = 7,$$

$$C_{13} = \begin{vmatrix} 3 & -1 \\ -1 & 3 \end{vmatrix} = 8, \quad C_{23} = -\begin{vmatrix} -1 & 1 \\ -1 & 3 \end{vmatrix} = 2, \quad C_{33} = \begin{vmatrix} -1 & 1 \\ 3 & -1 \end{vmatrix} = -2,$$

so that by (4), in agreement with Example 1,

$$\mathbf{A}^{-1} = \begin{bmatrix} -0.7 & 0.2 & 0.3 \\ -1.3 & -0.2 & 0.7 \\ 0.8 & 0.2 & -0.2 \end{bmatrix}.$$ ■

Diagonal matrices $\mathbf{A} = [a_{jk}]$, $a_{jk} = 0$ when $j \neq k$, have an inverse if and only if all $a_{jj} \neq 0$. Then $\mathbf{A}^{-1}$ is diagonal, too, with entries $1/a_{11}, \cdots, 1/a_{nn}$.

PROOF For a diagonal matrix we have in (4)

$$\frac{C_{11}}{D} = \frac{a_{22} \cdots a_{nn}}{a_{11}a_{22} \cdots a_{nn}} = \frac{1}{a_{11}}, \quad \text{etc.}$$ ■

EXAMPLE 4 Inverse of a Diagonal Matrix

Let

$$\mathbf{A} = \begin{bmatrix} -0.5 & 0 & 0 \\ 0 & 4 & 0 \\ 0 & 0 & 1 \end{bmatrix}.$$

Then we obtain the inverse $\mathbf{A}^{-1}$ by inverting each individual diagonal element of $\mathbf{A}$, that is, by taking $1/(-0.5)$, $\frac{1}{4}$, and $\frac{1}{1}$ as the diagonal entries of $\mathbf{A}^{-1}$, that is,

$$\mathbf{A}^{-1} = \begin{bmatrix} -2 & 0 & 0 \\ 0 & 0.25 & 0 \\ 0 & 0 & 1 \end{bmatrix}.$$ ■

Products can be inverted by taking the inverse of each factor and multiplying these inverses ***in reverse order***,

$$(\mathbf{AC})^{-1} = \mathbf{C}^{-1}\mathbf{A}^{-1}. \tag{7}$$

Hence for more than two factors,

$$(\mathbf{AC} \cdots \mathbf{PQ})^{-1} = \mathbf{Q}^{-1}\mathbf{P}^{-1} \cdots \mathbf{C}^{-1}\mathbf{A}^{-1}. \tag{8}$$

PROOF The idea is to start from (1) for $\mathbf{AC}$ instead of $\mathbf{A}$, that is, $\mathbf{AC}(\mathbf{AC})^{-1} = \mathbf{I}$, and multiply it on both sides from the left, first by $\mathbf{A}^{-1}$, which because of $\mathbf{A}^{-1}\mathbf{A} = \mathbf{I}$ gives

$$\begin{aligned} \mathbf{A}^{-1}\mathbf{AC}(\mathbf{AC})^{-1} &= \mathbf{C}(\mathbf{AC})^{-1} \\ &= \mathbf{A}^{-1}\mathbf{I} = \mathbf{A}^{-1}, \end{aligned}$$

and then multiplying this on both sides from the left, this time by $\mathbf{C}^{-1}$ and by using $\mathbf{C}^{-1}\mathbf{C} = \mathbf{I}$,

$$\mathbf{C}^{-1}\mathbf{C}(\mathbf{AC})^{-1} = (\mathbf{AC})^{-1} = \mathbf{C}^{-1}\mathbf{A}^{-1}.$$

This proves (7), and from it, (8) follows by induction. ■

We also note that *the inverse of the inverse is the given matrix*, as you may prove,

$$(\mathbf{A}^{-1})^{-1} = \mathbf{A}. \tag{9}$$

Unusual Properties of Matrix Multiplication. Cancellation Laws

Section 7.2 contains warnings that some properties of matrix multiplication deviate from those for numbers, and we are now able to explain the restricted validity of the so-called **cancellation laws** [2] and [3] below, using rank and inverse, concepts that were not yet available in Sec. 7.2. The deviations from the usual are of great practical importance and must be carefully observed. They are as follows.

[1] Matrix multiplication is not commutative, that is, in general we have

$$\mathbf{AB} \neq \mathbf{BA}.$$

[2] $\mathbf{AB} = \mathbf{0}$ does not generally imply $\mathbf{A} = \mathbf{0}$ or $\mathbf{B} = \mathbf{0}$ (or $\mathbf{BA} = \mathbf{0}$); for example,

$$\begin{bmatrix} 1 & 1 \\ 2 & 2 \end{bmatrix} \begin{bmatrix} -1 & 1 \\ 1 & -1 \end{bmatrix} = \begin{bmatrix} 0 & 0 \\ 0 & 0 \end{bmatrix}.$$

[3] $\mathbf{AC} = \mathbf{AD}$ does not generally imply $\mathbf{C} = \mathbf{D}$ (even when $\mathbf{A} \neq \mathbf{0}$).

Complete answers to [2] and [3] are contained in the following theorem.

THEOREM 3

Cancellation Laws

Let $\mathbf{A}, \mathbf{B}, \mathbf{C}$ *be* $n \times n$ *matrices. Then:*

(a) *If* rank $\mathbf{A} = n$ *and* $\mathbf{AB} = \mathbf{AC}$, *then* $\mathbf{B} = \mathbf{C}$.

(b) *If* rank $\mathbf{A} = n$, *then* $\mathbf{AB} = \mathbf{0}$ *implies* $\mathbf{B} = \mathbf{0}$. *Hence if* $\mathbf{AB} = \mathbf{0}$, *but* $\mathbf{A} \neq \mathbf{0}$ *as well as* $\mathbf{B} \neq \mathbf{0}$, *then* rank $\mathbf{A} < n$ *and* rank $\mathbf{B} < n$.

(c) *If* $\mathbf{A}$ *is singular, so are* $\mathbf{BA}$ *and* $\mathbf{AB}$.

PROOF **(a)** The inverse of $\mathbf{A}$ exists by Theorem 1. Multiplication by $\mathbf{A}^{-1}$ from the left gives $\mathbf{A}^{-1}\mathbf{AB} = \mathbf{A}^{-1}\mathbf{AC}$, hence $\mathbf{B} = \mathbf{C}$.

(b) Let rank $\mathbf{A} = n$. Then $\mathbf{A}^{-1}$ exists, and $\mathbf{AB} = \mathbf{0}$ implies $\mathbf{A}^{-1}\mathbf{AB} = \mathbf{B} = \mathbf{0}$. Similarly when rank $\mathbf{B} = n$. This implies the second statement in (b).

($\mathbf{c}_1$) Rank $\mathbf{A} < n$ by Theorem 1. Hence $\mathbf{Ax} = \mathbf{0}$ has nontrivial solutions by Theorem 2 in Sec. 7.5. Multiplication by $\mathbf{B}$ shows that these solutions are also solutions of $\mathbf{BAx} = \mathbf{0}$, so that rank $(\mathbf{BA}) < n$ by Theorem 2 in Sec. 7.5 and $\mathbf{BA}$ is singular by Theorem 1.

($\mathbf{c}_2$) $\mathbf{A}^\mathsf{T}$ is singular by Theorem 2(d) in Sec. 7.7. Hence $\mathbf{B}^\mathsf{T}\mathbf{A}^\mathsf{T}$ is singular by part (c_1), and is equal to $(\mathbf{AB})^\mathsf{T}$ by (10d) in Sec. 7.2. Hence $\mathbf{AB}$ is singular by Theorem 2(d) in Sec. 7.7. ■

Determinants of Matrix Products

The determinant of a matrix product $\mathbf{AB}$ or $\mathbf{BA}$ can be written as the product of the determinants of the factors, and it is interesting that det $\mathbf{AB}$ = det $\mathbf{BA}$, although $\mathbf{AB} \neq \mathbf{BA}$ in general. The corresponding formula (10) is needed occasionally and can be obtained by Gauss–Jordan elimination (see Example 1) and from the theorem just proved.

THEOREM 4

Determinant of a Product of Matrices

For any $n \times n$ *matrices* $\mathbf{A}$ *and* $\mathbf{B}$,

(10) $$\det(\mathbf{AB}) = \det(\mathbf{BA}) = \det \mathbf{A} \det \mathbf{B}.$$

PROOF If **A** or **B** is singular, so are **AB** and **BA** by Theorem 3(c), and (10) reduces to 0 = 0 by Theorem 3 in Sec. 7.7.

Now let **A** and **B** be nonsingular. Then we can reduce **A** to a diagonal matrix $\hat{\mathbf{A}} = [a_{jk}]$ by Gauss–Jordan steps. Under these operations, det **A** retains its value, by Theorem 1 in Sec. 7.7, (a) and (b) [not (c)] except perhaps for a sign reversal in row interchanging when pivoting. But the same operations reduce **AB** to $\hat{\mathbf{A}}\mathbf{B}$ with the same effect on det (**AB**). Hence it remains to prove (10) for $\hat{\mathbf{A}}\mathbf{B}$; written out,

$$\hat{\mathbf{A}}\mathbf{B} = \begin{bmatrix} \hat{a}_{11} & 0 & \cdots & 0 \\ 0 & \hat{a}_{22} & \cdots & 0 \\ & & \ddots & \\ 0 & 0 & \cdots & \hat{a}_{nn} \end{bmatrix} \begin{bmatrix} b_{11} & b_{12} & \cdots & b_{1n} \\ b_{21} & b_{22} & \cdots & b_{2n} \\ & & \vdots & \\ b_{n1} & b_{n2} & \cdots & b_{nn} \end{bmatrix}$$

$$= \begin{bmatrix} \hat{a}_{11}b_{11} & \hat{a}_{11}b_{12} & \cdots & \hat{a}_{11}b_{1n} \\ \hat{a}_{22}b_{21} & \hat{a}_{22}b_{22} & \cdots & \hat{a}_{22}b_{2n} \\ & & \vdots & \\ \hat{a}_{nn}b_{n1} & \hat{a}_{nn}b_{n2} & \cdots & \hat{a}_{nn}b_{nn} \end{bmatrix}.$$

We now take the determinant det $(\hat{\mathbf{A}}\mathbf{B})$. On the right we can take out a factor $\hat{a}_{11}$ from the first row, $\hat{a}_{22}$ from the second, $\cdots$, $\hat{a}_{nn}$ from the nth. But this product $\hat{a}_{11}\hat{a}_{22}\cdots\hat{a}_{nn}$ equals det $\hat{\mathbf{A}}$ because $\hat{\mathbf{A}}$ is diagonal. The remaining determinant is det **B**. This proves (10) for det (**AB**), and the proof for det (**BA**) follows by the same idea. ■

This completes our discussion of linear systems (Secs. 7.3–7.8). Section 7.9 on vector spaces and linear transformations is optional. ***Numeric methods*** are discussed in Secs. 20.1–20.4, which are independent of other sections on numerics.

PROBLEM SET 7.8

1–10 INVERSE

Find the inverse by Gauss–Jordan (or by (4*) if $n = 2$). Check by using (1).

1. $\begin{bmatrix} 1.80 & -2.32 \\ -0.25 & 0.60 \end{bmatrix}$

2. $\begin{bmatrix} \cos 2\theta & \sin 2\theta \\ -\sin 2\theta & \cos 2\theta \end{bmatrix}$

3. $\begin{bmatrix} 0.3 & -0.1 & 0.5 \\ 2 & 6 & 4 \\ 5 & 0 & 9 \end{bmatrix}$

4. $\begin{bmatrix} 0 & 0 & 0.1 \\ 0 & -0.4 & 0 \\ 2.5 & 0 & 0 \end{bmatrix}$

5. $\begin{bmatrix} 1 & 0 & 0 \\ 2 & 1 & 0 \\ 5 & 4 & 1 \end{bmatrix}$

6. $\begin{bmatrix} -4 & 0 & 0 \\ 0 & 8 & 13 \\ 0 & 3 & 5 \end{bmatrix}$

7. $\begin{bmatrix} 0 & 1 & 0 \\ 1 & 0 & 0 \\ 0 & 0 & 1 \end{bmatrix}$

8. $\begin{bmatrix} 1 & 2 & 3 \\ 4 & 5 & 6 \\ 7 & 8 & 9 \end{bmatrix}$

9. $\begin{bmatrix} 0 & 8 & 0 \\ 0 & 0 & 4 \\ 2 & 0 & 0 \end{bmatrix}$

10. $\begin{bmatrix} \frac{2}{3} & \frac{1}{3} & \frac{2}{3} \\ -\frac{2}{3} & \frac{2}{3} & \frac{1}{3} \\ \frac{1}{3} & \frac{2}{3} & -\frac{2}{3} \end{bmatrix}$

11–18 SOME GENERAL FORMULAS

11. Inverse of the square. Verify $(\mathbf{A}^2)^{-1} = (\mathbf{A}^{-1})^2$ for **A** in Prob. 1.

12. Prove the formula in Prob. 11.

13. Inverse of the transpose. Verify $(\mathbf{A}^T)^{-1} = (\mathbf{A}^{-1})^T$ for $\mathbf{A}$ in Prob. 1.

14. Prove the formula in Prob. 13.

15. Inverse of the inverse. Prove that $(\mathbf{A}^{-1})^{-1} = \mathbf{A}$.

16. Rotation. Give an application of the matrix in Prob. 2 that makes the form of the inverse obvious.

17. Triangular matrix. Is the inverse of a triangular matrix always triangular (as in Prob. 5)? Give reason.

18. Row interchange. Same task as in Prob. 16 for the matrix in Prob. 7.

19–20 FORMULA (4)

Formula (4) is occasionally needed in theory. To understand it, apply it and check the result by Gauss–Jordan:

19. In Prob. 3

20. In Prob. 6

7.9 Vector Spaces, Inner Product Spaces, Linear Transformations *Optional*

We have captured the essence of vector spaces in Sec. 7.4. There we dealt with ***special vector spaces*** that arose quite naturally in the context of matrices and linear systems. The elements of these vector spaces, called *vectors*, satisfied rules (3) and (4) of Sec. 7.1 (which were similar to those for numbers). These special vector spaces were generated by *spans*, that is, linear combination of finitely many vectors. Furthermore, each such vector had n real numbers as *components*. Review this material before going on.

We can generalize this idea by taking *all* vectors with n real numbers as components and obtain the very important ***real n-dimensional vector space*** R^n. The vectors are known as "real vectors." Thus, each vector in R^n is an ordered n-tuple of real numbers.

Now we can consider special values for n. For $n = 2$, we obtain R^2, the vector space of all ordered pairs, which correspond to the **vectors in the plane**. For $n = 3$, we obtain R^3, the vector space of all ordered triples, which are the **vectors in 3-space**. These vectors have wide applications in mechanics, geometry, and calculus and are basic to the engineer and physicist.

Similarly, if we take all ordered n-tuples of *complex numbers* as vectors and complex numbers as scalars, we obtain the **complex vector space** C^n, which we shall consider in Sec. 8.5.

Furthermore, there are other sets of practical interest consisting of matrices, functions, transformations, or others for which addition and scalar multiplication can be defined in an almost natural way so that they too form vector spaces.

It is perhaps not too great an intellectual jump to create, from the ***concrete model*** R^n, the ***abstract concept*** of a ***real vector space*** V by taking the basic properties (3) and (4) in Sec. 7.1 as axioms. In this way, the definition of a real vector space arises.

DEFINITION

Real Vector Space

A nonempty set V of elements $\mathbf{a}, \mathbf{b}, \cdots$ is called a **real vector space** (or *real linear space*), and these elements are called **vectors** (regardless of their nature, which will come out from the context or will be left arbitrary) if, in V, there are defined two algebraic operations (called *vector addition* and *scalar multiplication*) as follows.

I. Vector addition associates with every pair of vectors $\mathbf{a}$ and $\mathbf{b}$ of V a unique vector of V, called the *sum* of $\mathbf{a}$ and $\mathbf{b}$ and denoted by $\mathbf{a} + \mathbf{b}$, such that the following axioms are satisfied.

I.1 *Commutativity.* For any two vectors $\mathbf{a}$ and $\mathbf{b}$ of V,

$$\mathbf{a} + \mathbf{b} = \mathbf{b} + \mathbf{a}.$$

I.2 *Associativity.* For any three vectors $\mathbf{a}$, $\mathbf{b}$, $\mathbf{c}$ of V,

$$(\mathbf{a} + \mathbf{b}) + \mathbf{c} = \mathbf{a} + (\mathbf{b} + \mathbf{c}) \qquad \text{(written } \mathbf{a} + \mathbf{b} + \mathbf{c}\text{).}$$

I.3 There is a unique vector in V, called the *zero vector* and denoted by $\mathbf{0}$, such that for every $\mathbf{a}$ in V,

$$\mathbf{a} + \mathbf{0} = \mathbf{a}.$$

I.4 For every $\mathbf{a}$ in V there is a unique vector in V that is denoted by $-\mathbf{a}$ and is such that

$$\mathbf{a} + (-\mathbf{a}) = \mathbf{0}.$$

II. Scalar multiplication. The real numbers are called **scalars**. Scalar multiplication associates with every $\mathbf{a}$ in V and every scalar c a unique vector of V, called the *product* of c and $\mathbf{a}$ and denoted by $c\mathbf{a}$ (or $\mathbf{a}c$) such that the following axioms are satisfied.

II.1 *Distributivity.* For every scalar c and vectors $\mathbf{a}$ and $\mathbf{b}$ in V,

$$c(\mathbf{a} + \mathbf{b}) = c\mathbf{a} + c\mathbf{b}.$$

II.2 *Distributivity.* For all scalars c and k and every $\mathbf{a}$ in V,

$$(c + k)\mathbf{a} = c\mathbf{a} + k\mathbf{a}.$$

II.3 *Associativity.* For all scalars c and k and every $\mathbf{a}$ in V,

$$c(k\mathbf{a}) = (ck)\mathbf{a} \qquad \text{(written } ck\mathbf{a}\text{).}$$

II.4 For every $\mathbf{a}$ in V,

$$1\mathbf{a} = \mathbf{a}.$$

If, in the above definition, we take complex numbers as scalars instead of real numbers, we obtain the axiomatic definition of a **complex vector space**.

Take a look at the axioms in the above definition. Each axiom stands on its own: It is concise, useful, and it expresses a simple property of V. There are as few axioms as possible and together they express ***all*** the desired properties of V. Selecting good axioms is a process of trial and error that often extends over a long period of time. But once agreed upon, axioms become *standard* such as the ones in the definition of a real vector space.

The following concepts related to a vector space are exactly defined as those given in Sec. 7.4. Indeed, a **linear combination** of vectors $a_{(1)}, \cdots, a_{(m)}$ in a vector space V is an expression

$$c_1\mathbf{a}_{(1)} + \cdots + c_m\mathbf{a}_m \qquad (c_1, \cdots, c_m \text{ any scalars}).$$

These vectors form a **linearly independent set** (briefly, they are called **linearly independent**) if

$$(1) \qquad c_1\mathbf{a}_{(1)} + \cdots + c_m\mathbf{a}_{(m)} = \mathbf{0}$$

implies that $c_1 = 0, \cdots, c_m = 0$. Otherwise, if (1) also holds with scalars not all zero, the vectors are called **linearly dependent**.

Note that (1) with $m = 1$ is $c\mathbf{a} = \mathbf{0}$ and shows that a single vector $\mathbf{a}$ is linearly independent if and only if $\mathbf{a} \neq \mathbf{0}$.

V has **dimension *n***, or is ***n*-dimensional**, if it contains a linearly independent set of n vectors, whereas any set of more than n vectors in V is linearly dependent. That set of n linearly independent vectors is called a **basis** for V. Then every vector in V can be written as a linear combination of the basis vectors. Furthermore, for a given basis, this representation is unique (see Prob. 2).

EXAMPLE 1 **Vector Space of Matrices**

The real 2×2 matrices form a four-dimensional real vector space. A basis is

$$\mathbf{B}_{11} = \begin{bmatrix} 1 & 0 \\ 0 & 0 \end{bmatrix}, \quad \mathbf{B}_{12} = \begin{bmatrix} 0 & 1 \\ 0 & 0 \end{bmatrix}, \quad \mathbf{B}_{21} = \begin{bmatrix} 0 & 0 \\ 1 & 0 \end{bmatrix}, \quad \mathbf{B}_{22} = \begin{bmatrix} 0 & 0 \\ 0 & 1 \end{bmatrix}$$

because any 2×2 matrix $\mathbf{A} = [a_{jk}]$ has a unique representation $\mathbf{A} = a_{11}\mathbf{B}_{11} + a_{12}\mathbf{B}_{12} + a_{21}\mathbf{B}_{21} + a_{22}\mathbf{B}_{22}$. Similarly, the real $m \times n$ matrices with fixed m and n form an mn-dimensional vector space. What is the dimension of the vector space of all 3×3 skew-symmetric matrices? Can you find a basis? ■

EXAMPLE 2 **Vector Space of Polynomials**

The set of all constant, linear, and quadratic polynomials in x together is a vector space of dimension 3 with basis $\{1, x, x^2\}$ under the usual addition and multiplication by real numbers because these two operations give polynomials not exceeding degree 2. What is the dimension of the vector space of all polynomials of degree not exceeding a given fixed n? Can you find a basis? ■

If a vector space V contains a linearly independent set of n vectors for every n, no matter how large, then V is called **infinite dimensional**, as opposed to a *finite dimensional* (n-dimensional) vector space just defined. An example of an infinite dimensional vector space is the space of all continuous functions on some interval $[a, b]$ of the x-axis, as we mention without proof.

Inner Product Spaces

If $\mathbf{a}$ and $\mathbf{b}$ are vectors in R^n, regarded as column vectors, we can form the product $\mathbf{a}^\mathsf{T}\mathbf{b}$. This is a 1×1 matrix, which we can identify with its single entry, that is, with a number.

This product is called the **inner product** or **dot product** of **a** and **b**. Other notations for it are (**a**, **b**) and **a** • **b**. Thus

$$\mathbf{a}^\mathsf{T}\mathbf{b} = (\mathbf{a}, \mathbf{b}) = \mathbf{a} \cdot \mathbf{b} = [a_1 \cdots a_n] \begin{bmatrix} b_1 \\ \vdots \\ b_n \end{bmatrix} = \sum_{i=1}^{n} a_l b_l = a_1 b_1 + \cdots + a_n b_n.$$

We now extend this concept to general real vector spaces by taking basic properties of (**a**, **b**) as axioms for an "abstract inner product" (**a**, **b**) as follows.

DEFINITION

Real Inner Product Space

A real vector space V is called a **real inner product space** (or *real pre-Hilbert*[4] *space*) if it has the following property. With every pair of vectors **a** and **b** in V there is associated a real number, which is denoted by (**a**, **b**) and is called the **inner product** of **a** and **b**, such that the following axioms are satisfied.

I. For all scalars q_1 and q_2 and all vectors **a**, **b**, **c** in V,

$$(q_1\mathbf{a} + q_2\mathbf{b}, \mathbf{c}) = q_1(\mathbf{a}, \mathbf{c}) + q_2(\mathbf{b}, \mathbf{c}) \qquad \textit{(Linearity)}.$$

II. For all vectors **a** and **b** in V,

$$(\mathbf{a}, \mathbf{b}) = (\mathbf{b}, \mathbf{a}) \qquad \textit{(Symmetry)}.$$

III. For every **a** in V,

$$\left.\begin{aligned} (\mathbf{a}, \mathbf{a}) &\geqq 0, \\ (\mathbf{a}, \mathbf{a}) = 0 \quad &\text{if and only if} \quad \mathbf{a} = \mathbf{0} \end{aligned}\right\} \qquad \textit{(Positive-definiteness)}.$$

Vectors whose inner product is zero are called **orthogonal**.

The *length* or **norm** of a vector in V is defined by

$$\|\mathbf{a}\| = \sqrt{(\mathbf{a}, \mathbf{a})} \quad (\geqq 0). \tag{2}$$

A vector of norm 1 is called a **unit vector**.

[4]DAVID HILBERT (1862–1943), great German mathematician, taught at Königsberg and Göttingen and was the creator of the famous Göttingen mathematical school. He is known for his basic work in algebra, the calculus of variations, integral equations, functional analysis, and mathematical logic. His "Foundations of Geometry" helped the axiomatic method to gain general recognition. His famous 23 problems (presented in 1900 at the International Congress of Mathematicians in Paris) considerably influenced the development of modern mathematics.

If V is finite dimensional, it is actually a so-called *Hilbert space*; see [GenRef7], p. 128, listed in App. 1.

From these axioms and from (2) one can derive the basic inequality

(3) $$|(\mathbf{a}, \mathbf{b})| \leqq \|\mathbf{a}\| \, \|\mathbf{b}\|$$ (*Cauchy–Schwarz*[5] *inequality*).

From this follows

(4) $$\|\mathbf{a} + \mathbf{b}\| \leqq \|\mathbf{a}\| + \|\mathbf{b}\|$$ (*Triangle inequality*).

A simple direct calculation gives

(5) $$\|\mathbf{a} + \mathbf{b}\|^2 + \|\mathbf{a} - \mathbf{b}\|^2 = 2(\|\mathbf{a}\|^2 + \|\mathbf{b}\|^2)$$ (*Parallelogram equality*).

EXAMPLE 3 *n*-Dimensional Euclidean Space

R^n with the inner product

(6) $$(\mathbf{a}, \mathbf{b}) = \mathbf{a}^\mathsf{T}\mathbf{b} = a_1b_1 + \cdots + a_nb_n$$

(where both **a** and **b** are *column* vectors) is called the ***n*-dimensional Euclidean space** and is denoted by E^n or again simply by R^n. Axioms I–III hold, as direct calculation shows. Equation (2) gives the "**Euclidean norm**"

(7) $$\|\mathbf{a}\| = \sqrt{(\mathbf{a}, \mathbf{a})} = \sqrt{\mathbf{a}^\mathsf{T}\mathbf{a}} = \sqrt{a_1^2 + \cdots + a_n^2}.$$ ■

EXAMPLE 4 An Inner Product for Functions. Function Space

The set of all real-valued continuous functions $f(x), g(x), \cdots$ on a given interval $\alpha \leqq x \leqq \beta$ is a real vector space under the usual addition of functions and multiplication by scalars (real numbers). On this "**function space**" we can define an inner product by the integral

(8) $$(f, g) = \int_\alpha^\beta f(x)\, g(x)\, dx.$$

Axioms I–III can be verified by direct calculation. Equation (2) gives the norm

(9) $$\|f\| = \sqrt{(f, f)} = \sqrt{\int_\alpha^\beta f(x)^2\, dx}.$$ ■

Our examples give a first impression of the great generality of the abstract concepts of vector spaces and inner product spaces. Further details belong to more advanced courses (on functional analysis, meaning abstract modern analysis; see [GenRef7] listed in App. 1) and cannot be discussed here. Instead we now take up a related topic where matrices play a central role.

Linear Transformations

Let X and Y be any vector spaces. To each vector **x** in X we assign a unique vector **y** in Y. Then we say that a **mapping** (or **transformation** or **operator**) of X into Y is given. Such a mapping is denoted by a capital letter, say F. The vector **y** in Y assigned to a vector **x** in X is called the **image** of **x** under F and is denoted by $F(\mathbf{x})$ [or $F\mathbf{x}$, without parentheses].

[5]HERMANN AMANDUS SCHWARZ (1843–1921). German mathematician, known by his work in complex analysis (conformal mapping) and differential geometry. For Cauchy see Sec. 2.5.

F is called a **linear mapping** or **linear transformation** if, for all vectors $\mathbf{v}$ and $\mathbf{x}$ in X and scalars c,

$$
\begin{aligned}
F(\mathbf{v}+\mathbf{x}) &= F(\mathbf{v}) + F(\mathbf{x}) \\
F(c\mathbf{x}) &= cF(\mathbf{x}).
\end{aligned} \tag{10}
$$

Linear Transformation of Space R^n into Space R^m

From now on we let $X = R^n$ and $Y = R^m$. Then any real $m \times n$ matrix $\mathbf{A} = [a_{jk}]$ gives a transformation of R^n into R^m,

$$\mathbf{y} = \mathbf{A}\mathbf{x}. \tag{11}$$

Since $\mathbf{A}(\mathbf{u}+\mathbf{x}) = \mathbf{A}\mathbf{u} + \mathbf{A}\mathbf{x}$ and $\mathbf{A}(c\mathbf{x}) = c\mathbf{A}\mathbf{x}$, this transformation is linear.

We show that, conversely, every linear transformation F of R^n into R^m can be given in terms of an $m \times n$ matrix $\mathbf{A}$, after a basis for R^n and a basis for R^m have been chosen. This can be proved as follows.

Let $\mathbf{e}_{(1)}, \cdots, \mathbf{e}_{(n)}$ be any basis for R^n. Then every $\mathbf{x}$ in R^n has a unique representation

$$\mathbf{x} = x_1\mathbf{e}_{(1)} + \cdots + x_n\mathbf{e}_{(n)}.$$

Since F is linear, this representation implies for the image $F(\mathbf{x})$:

$$F(\mathbf{x}) = F(x_1\mathbf{e}_{(1)} + \cdots + x_n\mathbf{e}_{(n)}) = x_1F(\mathbf{e}_{(1)}) + \cdots + x_nF(\mathbf{e}_{(n)}).$$

Hence F is uniquely determined by the images of the vectors of a basis for R^n. We now choose for R^n the "**standard basis**"

$$
\mathbf{e}_{(1)} = \begin{bmatrix} 1 \\ 0 \\ 0 \\ \vdots \\ 0 \end{bmatrix}, \quad
\mathbf{e}_{(2)} = \begin{bmatrix} 0 \\ 1 \\ 0 \\ \vdots \\ 0 \end{bmatrix}, \quad \cdots, \quad
\mathbf{e}_{(n)} = \begin{bmatrix} 0 \\ 0 \\ 0 \\ \vdots \\ 1 \end{bmatrix} \tag{12}
$$

where $\mathbf{e}_{(j)}$ has its jth component equal to 1 and all others 0. We show that we can now determine an $m \times n$ matrix $\mathbf{A} = [a_{jk}]$ such that for every $\mathbf{x}$ in R^n and image $\mathbf{y} = F(\mathbf{x})$ in R^m,

$$\mathbf{y} = F(\mathbf{x}) = \mathbf{A}\mathbf{x}.$$

Indeed, from the image $\mathbf{y}^{(1)} = F(\mathbf{e}_{(1)})$ of $\mathbf{e}_{(1)}$ we get the condition

$$
\mathbf{y}^{(1)} = \begin{bmatrix} y_1^{(1)} \\ y_2^{(1)} \\ \vdots \\ y_m^{(1)} \end{bmatrix} = \begin{bmatrix} a_{11} & \cdots & a_{1n} \\ a_{21} & \cdots & a_{2n} \\ \vdots & & \vdots \\ a_{m1} & \cdots & a_{mm} \end{bmatrix} \begin{bmatrix} 1 \\ 0 \\ \vdots \\ 0 \end{bmatrix}
$$

from which we can determine the first column of **A**, namely $a_{11} = y_1^{(1)}$, $a_{21} = y_2^{(1)}$, $\cdots$, $a_{m1} = y_m^{(1)}$. Similarly, from the image of $\mathbf{e}_{(2)}$ we get the second column of **A**, and so on. This completes the proof. ■

We say that **A represents** F, or *is a representation of* F, with respect to the bases for R^n and R^m. Quite generally, the purpose of a "**representation**" is the replacement of one object of study by another object whose properties are more readily apparent.

In three-dimensional Euclidean space E^3 the standard basis is usually written $\mathbf{e}_{(1)} = \mathbf{i}$, $\mathbf{e}_{(2)} = \mathbf{j}$, $\mathbf{e}_{(3)} = \mathbf{k}$. Thus,

$$\text{(13)} \qquad \mathbf{i} = \begin{bmatrix} 1 \\ 0 \\ 0 \end{bmatrix}, \quad \mathbf{j} = \begin{bmatrix} 0 \\ 1 \\ 0 \end{bmatrix}, \quad \mathbf{k} = \begin{bmatrix} 0 \\ 0 \\ 1 \end{bmatrix}.$$

These are the three unit vectors in the positive directions of the axes of the **Cartesian coordinate system in space**, that is, the usual coordinate system with the same scale of measurement on the three mutually perpendicular coordinate axes.

EXAMPLE 5 Linear Transformations

Interpreted as transformations of Cartesian coordinates in the plane, the matrices

$$\begin{bmatrix} 0 & 1 \\ 1 & 0 \end{bmatrix}, \quad \begin{bmatrix} 1 & 0 \\ 0 & -1 \end{bmatrix}, \quad \begin{bmatrix} -1 & 0 \\ 0 & -1 \end{bmatrix}, \quad \begin{bmatrix} a & 0 \\ 0 & 1 \end{bmatrix}$$

represent a reflection in the line $x_2 = x_1$, a reflection in the x_1-axis, a reflection in the origin, and a stretch (when $a > 1$, or a contraction when $0 < a < 1$) in the x_1-direction, respectively. ■

EXAMPLE 6 Linear Transformations

Our discussion preceding Example 5 is simpler than it may look at first sight. To see this, find **A** representing the linear transformation that maps (x_1, x_2) onto $(2x_1 - 5x_2, 3x_1 + 4x_2)$.

Solution. Obviously, the transformation is

$$\begin{aligned} y_1 &= 2x_1 - 5x_2 \\ y_2 &= 3x_1 + 4x_2. \end{aligned}$$

From this we can directly see that the matrix is

$$\mathbf{A} = \begin{bmatrix} 2 & -5 \\ 3 & 4 \end{bmatrix}. \qquad \text{Check:} \quad \begin{bmatrix} y_1 \\ y_2 \end{bmatrix} = \begin{bmatrix} 2 & -5 \\ 3 & 4 \end{bmatrix} \begin{bmatrix} x_1 \\ x_2 \end{bmatrix} = \begin{bmatrix} 2x_1 - 5x_2 \\ 3x_1 + 4x_2 \end{bmatrix}.$$ ■

If **A** in (11) is square, $n \times n$, then (11) maps R^n into R^n. If this **A** is nonsingular, so that $\mathbf{A}^{-1}$ exists (see Sec. 7.8), then multiplication of (11) by $\mathbf{A}^{-1}$ from the left and use of $\mathbf{A}^{-1}\mathbf{A} = \mathbf{I}$ gives the **inverse transformation**

$$\text{(14)} \qquad \mathbf{x} = \mathbf{A}^{-1}\mathbf{y}.$$

It maps every $\mathbf{y} = \mathbf{y}_0$ onto that $\mathbf{x}$, which by (11) is mapped onto $\mathbf{y}_0$. *The inverse of a linear transformation is itself linear,* because it is given by a matrix, as (14) shows.

Composition of Linear Transformations

We want to give you a flavor of how linear transformations in general vector spaces work. You will notice, if you read carefully, that definitions and verifications (Example 7) strictly follow the given rules and you can think your way through the material by going in a slow systematic fashion.

The last operation we want to discuss is composition of linear transformations. Let X, Y, W be general vector spaces. As before, let F be a linear transformation from X to Y. Let G be a linear transformation from W to X. Then we denote, by H, the **composition** of F and G, that is,

$$H = F \circ G = FG = F(G),$$

which means *we take transformation G and then apply transformation F to it* (***in that order!***, *i.e. you go from left to right*).

Now, to give this a more concrete meaning, if we let $\mathbf{w}$ be a vector in W, then $G(\mathbf{w})$ is a vector in X and $F(G(\mathbf{w}))$ is a vector in Y. Thus, H maps W to Y, and we can write

$$H(\mathbf{w}) = (F \circ G)(\mathbf{w}) = (FG)(\mathbf{w}) = F(G(\mathbf{w})), \tag{15}$$

which completes the definition of composition in a general vector space setting. But is composition really linear? To check this we have to verify that H, as defined in (15), obeys the two equations of (10).

EXAMPLE 7 The Composition of Linear Transformations Is Linear

To show that H is indeed linear we must show that (10) holds. We have, for two vectors $\mathbf{w}_1$, $\mathbf{w}_2$ in W,

$$\begin{aligned} H(\mathbf{w}_1 + \mathbf{w}_2) &= (F \circ G)(\mathbf{w}_1 + \mathbf{w}_2) \\ &= F(G(\mathbf{w}_1 + \mathbf{w}_2)) \\ &= F(G(\mathbf{w}_1) + G(\mathbf{w}_2)) && \text{(by linearity of } G) \\ &= F(G(\mathbf{w}_1)) + F(G(\mathbf{w}_2)) && \text{(by linearity of } F) \\ &= (F \circ G)(\mathbf{w}_1) + (F \circ G)(\mathbf{w}_2) && \text{(by (15))} \\ &= H(\mathbf{w}_1) + H(\mathbf{w}_2) && \text{(by definition of } H). \end{aligned}$$

Similarly, $H(c\mathbf{w}_2) = (F \circ G)(c\mathbf{w}_2) = F(G(c\mathbf{w}_2)) = F(c(G(\mathbf{w}_2))$

$$= cF(G(\mathbf{w}_2)) = c(F \circ G)(\mathbf{w}_2) = cH(\mathbf{w}_2).$$ ■

We defined composition as a linear transformation in a general vector space setting and showed that the composition of linear transformations is indeed linear.

Next we want to relate composition of linear transformations to matrix multiplication.

To do so we let $X = R^n$, $Y = R^m$, and $W = R^p$. This choice of particular vector spaces allows us to represent the linear transformations as matrices and form matrix equations, as was done in (11). Thus F can be represented by a general real $m \times n$ matrix $\mathbf{A} = [a_{jk}]$ and G by an $n \times p$ matrix $\mathbf{B} = [b_{jk}]$. Then we can write for F, with column vectors $\mathbf{x}$ with n entries, and resulting vector $\mathbf{y}$, with m entries

$$\mathbf{y} = \mathbf{A}\mathbf{x} \tag{16}$$

and similarly for G, with column vector $\mathbf{w}$ with p entries,

$$\mathbf{x} = \mathbf{B}\mathbf{w}. \tag{17}$$

Substituting (17) into (16) gives

$$\mathbf{y} = \mathbf{A}\mathbf{x} = \mathbf{A}(\mathbf{B}\mathbf{w}) = (\mathbf{A}\mathbf{B})\mathbf{w} = \mathbf{A}\mathbf{B}\mathbf{w} = \mathbf{C}\mathbf{w} \qquad \text{where } \mathbf{C} = \mathbf{A}\mathbf{B}. \tag{18}$$

This is (15) in a matrix setting, this is, ***we can define the composition of linear transformations in the Euclidean spaces as multiplication by matrices***. Hence, the real $m \times p$ matrix $\mathbf{C}$ represents a linear transformation H which maps R^p to R^n with vector $\mathbf{w}$, a column vector with p entries.

Remarks. Our discussion is similar to the one in Sec. 7.2, where we motivated the "unnatural" matrix multiplication of matrices. Look back and see that our current, more general, discussion is written out there for the case of dimension $m = 2, n = 2$, and $p = 2$. (You may want to write out our development by picking small *distinct* dimensions, such as $m = 2, n = 3$, and $p = 4$, and writing down the matrices and vectors. This is a trick of the trade of mathematicians in that we like to develop and test theories on smaller examples to see that they work.)

EXAMPLE 8 Linear Transformations. Composition

In Example 5 of Sec. 7.9, let $\mathbf{A}$ be the first matrix and $\mathbf{B}$ be the fourth matrix with $a > 1$. Then, applying $\mathbf{B}$ to a vector $\mathbf{w} = [w_1 \;\; w_2]^\mathsf{T}$, stretches the element w_1 by a in the x_1 direction. Next, when we apply $\mathbf{A}$ to the "stretched" vector, we reflect the vector along the line $x_1 = x_2$, resulting in a vector $\mathbf{y} = [w_2 \;\; aw_1]^\mathsf{T}$. But this represents, precisely, a geometric description for the composition H of two linear transformations F and G represented by matrices $\mathbf{A}$ and $\mathbf{B}$. We now show that, for this example, our result can be obtained by straightforward matrix multiplication, that is,

$$\mathbf{AB} = \begin{bmatrix} 0 & 1 \\ 1 & 0 \end{bmatrix} \begin{bmatrix} a & 0 \\ 0 & 1 \end{bmatrix} = \begin{bmatrix} 0 & 1 \\ a & 0 \end{bmatrix}$$

and as in (18) calculate

$$\mathbf{ABw} = \begin{bmatrix} 0 & 1 \\ a & 0 \end{bmatrix} \begin{bmatrix} w_1 \\ w_2 \end{bmatrix} = \begin{bmatrix} w_2 \\ aw_1 \end{bmatrix},$$

which is the same as before. This shows that indeed $\mathbf{AB} = \mathbf{C}$, and we see the composition of linear transformations can be represented by a linear transformation. It also shows that the order of matrix multiplication is important (!). You may want to try applying $\mathbf{A}$ first and then $\mathbf{B}$, resulting in $\mathbf{BA}$. What do you see? Does it make geometric sense? Is it the same result as $\mathbf{AB}$? ■

We have learned several abstract concepts such as vector space, inner product space, and linear transformation. *The introduction of such concepts allows engineers and scientists to communicate in a concise and common language.* For example, the concept of a vector space encapsulated a lot of ideas in a very concise manner. For the student, learning such concepts provides a foundation for more advanced studies in engineering.

This concludes Chapter 7. The central theme was *the Gaussian elimination of* Sec. 7.3 from which most of the other concepts and theory flowed. The next chapter again has a central theme, that is, *eigenvalue problems,* an area very rich in applications such as in engineering, modern physics, and other areas.

PROBLEM SET 7.9

1. **Basis.** Find three bases of R^2.
2. **Uniqueness.** Show that the representation $\mathbf{v} = c_1\mathbf{a}_{(1)} + \cdots + c_n\mathbf{a}_{(n)}$ of any given vector in an n-dimensional vector space V in terms of a given basis $\mathbf{a}_{(1)}, \cdots, \mathbf{a}_{(n)}$ for V is unique. ***Hint.*** Take two representations and consider the difference.

3–10 VECTOR SPACE

(More problems in Problem Set 9.4.) Is the given set, taken with the usual addition and scalar multiplication, a vector space? Give reason. If your answer is yes, find the dimension and a basis.

3. All vectors in R^3 satisfying $-v_1 + 2v_2 + 3v_3 = 0$, $-4v_1 + v_2 + v_3 = 0$.
4. All skew-symmetric 3×3 matrices.
5. All polynomials in x of degree 4 or less with nonnegative coefficients.
6. All functions $y(x) = a \cos 2x + b \sin 2x$ with arbitrary constants a and b.
7. All functions $y(x) = (ax + b)e^{-x}$ with any constant a and b.
8. All $n \times n$ matrices $\mathbf{A}$ with fixed n and $\det \mathbf{A} = 0$.
9. All 2×2 matrices $[a_{jk}]$ with $a_{11} + a_{22} = 0$.
10. All 3×2 matrices $[a_{jk}]$ with first column any multiple of $[3 \quad 0 \quad -5]^\mathsf{T}$.

11–14 LINEAR TRANSFORMATIONS

Find the inverse transformation. Show the details.

11. $y_1 = 0.5x_1 - 0.5x_2$, $y_2 = 1.5x_1 - 2.5x_2$
12. $y_1 = 3x_1 + 2x_2$, $y_2 = 4x_1 + x_2$
13. $y_1 = 5x_1 + 3x_2 - 3x_3$, $y_2 = 3x_1 + 2x_2 - 2x_3$, $y_3 = 2x_1 - x_2 + 2x_3$
14. $y_1 = 0.2x_1 - 0.1x_2$, $y_2 = -0.2x_2 + 0.1x_3$, $y_3 = 0.1x_1 + 0.1x_3$

15–20 EUCLIDEAN NORM

Find the Euclidean norm of the vectors:

15. $[3 \quad 1 \quad -4]^\mathsf{T}$
16. $[\tfrac{1}{2} \quad \tfrac{1}{3} \quad -\tfrac{1}{2} \quad -\tfrac{1}{3}]^\mathsf{T}$
17. $[1 \quad 0 \quad 0 \quad 1 \quad -1 \quad 0 \quad -1 \quad 1]^\mathsf{T}$
18. $[-4 \quad 8 \quad -1]^\mathsf{T}$
19. $[\tfrac{2}{3} \quad \tfrac{2}{3} \quad \tfrac{1}{3} \quad 0]^\mathsf{T}$
20. $[\tfrac{1}{2} \quad -\tfrac{1}{2} \quad -\tfrac{1}{2} \quad \tfrac{1}{2}]^\mathsf{T}$

21–25 INNER PRODUCT. ORTHOGONALITY

21. **Orthogonality.** For what value(s) of k are the vectors $[2 \quad \tfrac{1}{2} \quad -4 \quad 0]^\mathsf{T}$ and $[5 \quad k \quad 0 \quad \tfrac{1}{4}]^\mathsf{T}$ orthogonal?
22. **Orthogonality.** Find all vectors in R^3 orthogonal to $[2 \quad 0 \quad 1]$. Do they form a vector space?
23. **Triangle inequality.** Verify (4) for the vectors in Probs. 15 and 18.
24. **Cauchy–Schwarz inequality.** Verify (3) for the vectors in Probs. 16 and 19.
25. **Parallelogram equality.** Verify (5) for the first two column vectors of the coefficient matrix in Prob. 13.

CHAPTER 7 REVIEW QUESTIONS AND PROBLEMS

1. What properties of matrix multiplication differ from those of the multiplication of numbers?
2. Let $\mathbf{A}$ be a 100×100 matrix and $\mathbf{B}$ a 100×50 matrix. Are the following expressions defined or not? $\mathbf{A} + \mathbf{B}$, $\mathbf{A}^2$, $\mathbf{B}^2$, $\mathbf{AB}$, $\mathbf{BA}$, $\mathbf{AA}^\mathsf{T}$, $\mathbf{B}^\mathsf{T}\mathbf{A}$, $\mathbf{B}^\mathsf{T}\mathbf{B}$, $\mathbf{BB}^\mathsf{T}$, $\mathbf{B}^\mathsf{T}\mathbf{AB}$. Give reasons.
3. Are there any linear systems without solutions? With one solution? With more than one solution? Give simple examples.
4. Let $\mathbf{C}$ be 10×10 matrix and $\mathbf{a}$ a column vector with 10 components. Are the following expressions defined or not? $\mathbf{Ca}$, $\mathbf{C}^\mathsf{T}\mathbf{a}$, $\mathbf{Ca}^\mathsf{T}$, $\mathbf{aC}$, $\mathbf{a}^\mathsf{T}\mathbf{C}$, $(\mathbf{Ca}^\mathsf{T})^\mathsf{T}$.
5. Motivate the definition of matrix multiplication.
6. Explain the use of matrices in linear transformations.
7. How can you give the rank of a matrix in terms of row vectors? Of column vectors? Of determinants?
8. What is the role of rank in connection with solving linear systems?
9. What is the idea of Gauss elimination and back substitution?
10. What is the inverse of a matrix? When does it exist? How would you determine it?

11–20 MATRIX AND VECTOR CALCULATIONS

Showing the details, calculate the following expressions or give reason why they are not defined, when

$$\mathbf{A} = \begin{bmatrix} 3 & 1 & -3 \\ 1 & 4 & 2 \\ -3 & 2 & 5 \end{bmatrix}, \quad \mathbf{B} = \begin{bmatrix} 0 & 4 & 1 \\ -4 & 0 & -2 \\ -1 & 2 & 0 \end{bmatrix},$$

$$\mathbf{u} = \begin{bmatrix} 2 \\ 0 \\ -5 \end{bmatrix}, \quad \mathbf{v} = \begin{bmatrix} 7 \\ -3 \\ 3 \end{bmatrix}$$

11. $\mathbf{AB}$, $\mathbf{BA}$

12. $\mathbf{A}^{\mathsf{T}}$, $\mathbf{B}^{\mathsf{T}}$

13. $\mathbf{Au}$, $\mathbf{u}^{\mathsf{T}}\mathbf{A}$

14. $\mathbf{u}^{\mathsf{T}}\mathbf{v}$, $\mathbf{uv}^{\mathsf{T}}$

15. $\mathbf{u}^{\mathsf{T}}\mathbf{Au}$, $\mathbf{v}^{\mathsf{T}}\mathbf{Bv}$

16. $\mathbf{A}^{-1}$, $\mathbf{B}^{-1}$

17. $\det \mathbf{A}$, $\det \mathbf{A}^2$, $(\det \mathbf{A})^2$, $\det \mathbf{B}$

18. $(\mathbf{A}^2)^{-1}$, $(\mathbf{A}^{-1})^2$

19. $\mathbf{AB} - \mathbf{BA}$

20. $(\mathbf{A} + \mathbf{A}^{\mathsf{T}})(\mathbf{B} - \mathbf{B}^{\mathsf{T}})$

21–28 LINEAR SYSTEMS

Showing the details, find all solutions or indicate that no solution exists.

21.
$$\begin{aligned} 4y + z &= 0 \\ 12x - 5y - 3z &= 34 \\ -6x + 4z &= 8 \end{aligned}$$

22.
$$\begin{aligned} 5x - 3y + z &= 7 \\ 2x + 3y - z &= 0 \\ 8x + 9y - 3z &= 2 \end{aligned}$$

23.
$$\begin{aligned} 9x + 3y - 6z &= 60 \\ 2x - 4y + 8z &= 4 \end{aligned}$$

24.
$$\begin{aligned} -6x + 39y - 9z &= -12 \\ 2x - 13y + 3z &= 4 \end{aligned}$$

25.
$$\begin{aligned} 0.3x - 0.7y + 1.3z &= 3.24 \\ 0.9y - 0.8z &= -2.53 \\ 0.7z &= 1.19 \end{aligned}$$

26.
$$\begin{aligned} 2x + 3y - 7z &= 3 \\ -4x - 6y + 14z &= 7 \end{aligned}$$

27.
$$\begin{aligned} x + 2y &= 6 \\ 3x + 5y &= 20 \\ -4x + y &= -42 \end{aligned}$$

28.
$$\begin{aligned} -8x + 2z &= 1 \\ 6y + 4z &= 3 \\ 12x + 2y &= 2 \end{aligned}$$

29–32 RANK

Determine the ranks of the coefficient matrix and the augmented matrix and state how many solutions the linear system will have.

29. In Prob. 23

30. In Prob. 24

31. In Prob. 27

32. In Prob. 26

33–35 NETWORKS

Find the currents.

33.

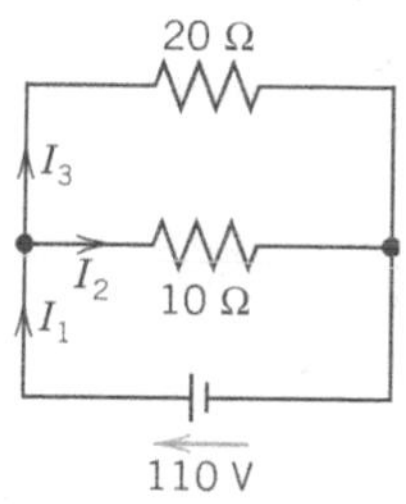

34.

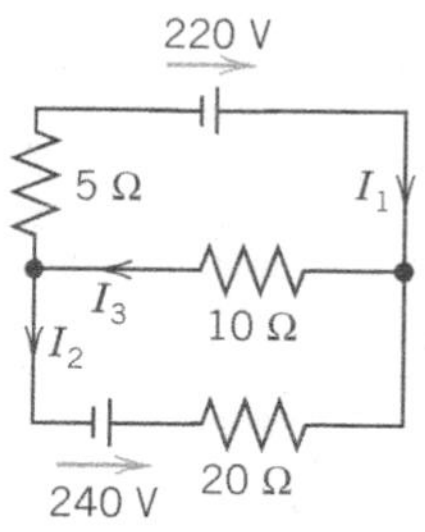

35.

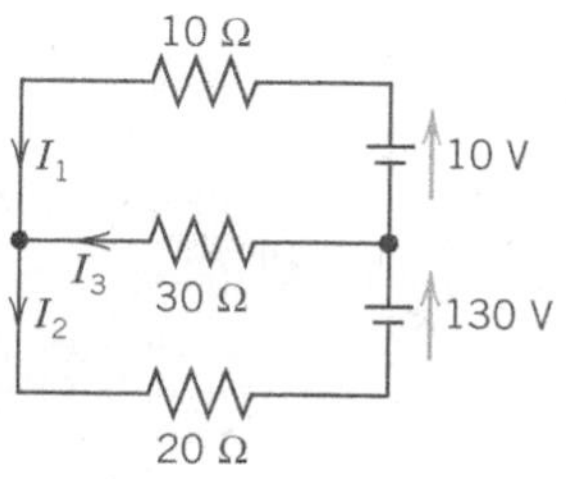

SUMMARY OF CHAPTER 7

Linear Algebra: Matrices, Vectors, Determinants. Linear Systems

An $m \times n$ **matrix** $\mathbf{A} = [a_{jk}]$ is a rectangular array of numbers or functions ("entries," "elements") arranged in m horizontal **rows** and n vertical **columns**. If $m = n$, the matrix is called **square**. A $1 \times n$ matrix is called a **row vector** and an $m \times 1$ matrix a **column vector** (Sec. 7.1).

The **sum** $\mathbf{A} + \mathbf{B}$ of matrices of the same **size** (i.e., both $m \times n$) is obtained by adding corresponding entries. The **product** of $\mathbf{A}$ by a scalar c is obtained by multiplying each a_{jk} by c (Sec. 7.1).

The **product** $\mathbf{C} = \mathbf{AB}$ of an $m \times n$ matrix $\mathbf{A}$ by an $r \times p$ matrix $\mathbf{B} = [b_{jk}]$ is defined only when $r = n$, and is the $m \times p$ matrix $\mathbf{C} = [c_{jk}]$ with entries

$$c_{jk} = a_{j1}b_{1k} + a_{j2}b_{2k} + \cdots + a_{jn}b_{nk} \qquad \text{(row } j \text{ of } \mathbf{A} \text{ times column } k \text{ of } \mathbf{B}\text{)}. \tag{1}$$

This multiplication is motivated by the composition of **linear transformations** (Secs. 7.2, 7.9). It is associative, but is ***not commutative:*** if $\mathbf{AB}$ is defined, $\mathbf{BA}$ may not be defined, but even if $\mathbf{BA}$ is defined, $\mathbf{AB} \neq \mathbf{BA}$ in general. Also $\mathbf{AB} = \mathbf{0}$ may not imply $\mathbf{A} = \mathbf{0}$ or $\mathbf{B} = \mathbf{0}$ or $\mathbf{BA} = \mathbf{0}$ (Secs. 7.2, 7.8). Illustrations:

$$\begin{bmatrix} 1 & 1 \\ 2 & 2 \end{bmatrix}\begin{bmatrix} -1 & 1 \\ 1 & -1 \end{bmatrix} = \begin{bmatrix} 0 & 0 \\ 0 & 0 \end{bmatrix}$$

$$\begin{bmatrix} -1 & 1 \\ 1 & -1 \end{bmatrix}\begin{bmatrix} 1 & 1 \\ 2 & 2 \end{bmatrix} = \begin{bmatrix} 1 & 1 \\ -1 & -1 \end{bmatrix}$$

$$[1 \quad 2]\begin{bmatrix} 3 \\ 4 \end{bmatrix} = [11], \qquad \begin{bmatrix} 3 \\ 4 \end{bmatrix}[1 \quad 2] = \begin{bmatrix} 3 & 6 \\ 4 & 8 \end{bmatrix}.$$

The **transpose** $\mathbf{A}^\mathsf{T}$ of a matrix $\mathbf{A} = [a_{jk}]$ is $\mathbf{A}^\mathsf{T} = [a_{kj}]$; rows become columns and conversely (Sec. 7.2). Here, $\mathbf{A}$ need not be square. If it is and $\mathbf{A} = \mathbf{A}^\mathsf{T}$, then $\mathbf{A}$ is called **symmetric;** if $\mathbf{A} = -\mathbf{A}^\mathsf{T}$, it is called **skew-symmetric**. For a product, $(\mathbf{AB})^\mathsf{T} = \mathbf{B}^\mathsf{T}\mathbf{A}^\mathsf{T}$ (Sec. 7.2).

A main application of matrices concerns **linear systems of equations**

$$\mathbf{Ax} = \mathbf{b} \qquad \text{(Sec. 7.3)} \tag{2}$$

(m equations in n unknowns $x_1, \cdots, x_n$; $\mathbf{A}$ and $\mathbf{b}$ given). The most important method of solution is the **Gauss elimination** (Sec. 7.3), which reduces the system to "triangular" form by *elementary row operations,* which leave the set of solutions unchanged. (Numeric aspects and variants, such as *Doolittle's* and *Cholesky's methods,* are discussed in Secs. 20.1 and 20.2.)

Cramer's rule (Secs. 7.6, 7.7) represents the unknowns in a system (2) of n equations in n unknowns as quotients of determinants; for numeric work it is impractical. **Determinants** (Sec. 7.7) have decreased in importance, but will retain their place in eigenvalue problems, elementary geometry, etc.

The **inverse** $\mathbf{A}^{-1}$ of a square matrix satisfies $\mathbf{AA}^{-1} = \mathbf{A}^{-1}\mathbf{A} = \mathbf{I}$. It exists if and only if $\det \mathbf{A} \neq 0$. It can be computed by the *Gauss–Jordan elimination* (Sec. 7.8).

The **rank** r of a matrix $\mathbf{A}$ is the maximum number of linearly independent rows or columns of $\mathbf{A}$ or, equivalently, the number of rows of the largest square submatrix of $\mathbf{A}$ with nonzero determinant (Secs. 7.4, 7.7).

The system (2) has solutions if and only if $\operatorname{rank} \mathbf{A} = \operatorname{rank} [\mathbf{A} \quad \mathbf{b}]$, where $[\mathbf{A} \quad \mathbf{b}]$ is the **augmented matrix** (Fundamental Theorem, Sec. 7.5).

The **homogeneous system**

$$\mathbf{Ax} = \mathbf{0} \tag{3}$$

has solutions $\mathbf{x} \neq \mathbf{0}$ ("nontrivial solutions") if and only if $\operatorname{rank} \mathbf{A} < n$, in the case $m = n$ equivalently if and only if $\det \mathbf{A} = \mathbf{0}$ (Secs. 7.6, 7.7).

Vector spaces, inner product spaces, and linear transformations are discussed in Sec. 7.9. See also Sec. 7.4.

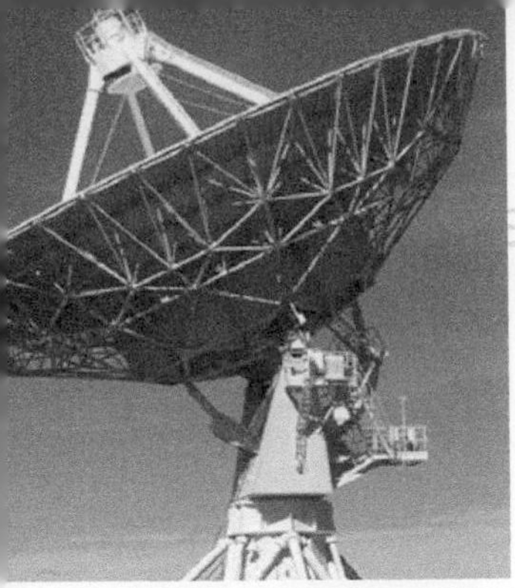

CHAPTER 8

Linear Algebra: Matrix Eigenvalue Problems

A matrix eigenvalue problem considers the vector equation

$$\mathbf{A}\mathbf{x} = \lambda \mathbf{x}. \qquad (1)$$

Here **A** is a given square matrix, λ an unknown scalar, and **x** an unknown vector. In a matrix eigenvalue problem, the task is to determine λ's and **x**'s that satisfy (1). Since $\mathbf{x} = \mathbf{0}$ is always a solution for any λ and thus not interesting, we only admit solutions with $\mathbf{x} \neq \mathbf{0}$.

The solutions to (1) are given the following names: The λ's that satisfy (1) are called **eigenvalues of A** and the corresponding nonzero **x**'s that also satisfy (1) are called **eigenvectors of A**.

From this rather innocent looking vector equation flows an amazing amount of relevant theory and an incredible richness of applications. Indeed, eigenvalue problems come up all the time in engineering, physics, geometry, numerics, theoretical mathematics, biology, environmental science, urban planning, economics, psychology, and other areas. Thus, in your career you are likely to encounter eigenvalue problems.

We start with a basic and thorough introduction to eigenvalue problems in Sec. 8.1 and explain (1) with several simple matrices. This is followed by a section devoted entirely to applications ranging from mass–spring systems of physics to population control models of environmental science. We show you these diverse examples to train your skills in modeling and solving eigenvalue problems. Eigenvalue problems for real symmetric, skew-symmetric, and orthogonal matrices are discussed in Sec. 8.3 and their complex counterparts (which are important in modern physics) in Sec. 8.5. In Sec. 8.4 we show how by diagonalizing a matrix, we obtain its eigenvalues.

COMMENT. ***Numerics for eigenvalues* (Secs. 20.6–20.9) *can be studied immediately after this chapter*.**

Prerequisite: Chap. 7.
Sections that may be omitted in a shorter course: 8.4, 8.5.
References and Answers to Problems: App. 1 Part B, App. 2.

The following chart identifies where different types of eigenvalue problems appear in the book.

Topic	Where to find it
Matrix Eigenvalue Problem (algebraic eigenvalue problem)	Chap. 8
Eigenvalue Problems in Numerics	Secs. 20.6–20.9
Eigenvalue Problem for ODEs (Sturm–Liouville problems)	Secs. 11.5, 11.6
Eigenvalue Problems for Systems of ODEs	Chap. 4
Eigenvalue Problems for PDEs	Secs. 12.3–12.11

8.1 The Matrix Eigenvalue Problem. Determining Eigenvalues and Eigenvectors

Consider multiplying nonzero vectors by a given square matrix, such as

$$\begin{bmatrix} 6 & 3 \\ 4 & 7 \end{bmatrix}\begin{bmatrix} 5 \\ 1 \end{bmatrix} = \begin{bmatrix} 33 \\ 27 \end{bmatrix}, \qquad \begin{bmatrix} 6 & 3 \\ 4 & 7 \end{bmatrix}\begin{bmatrix} 3 \\ 4 \end{bmatrix} = \begin{bmatrix} 30 \\ 40 \end{bmatrix}.$$

We want to see what influence the multiplication of the given matrix has on the vectors. In the first case, we get a totally new vector with a different direction and different length when compared to the original vector. This is what usually happens and is of no interest here. In the second case something interesting happens. The multiplication produces a vector $[30 \quad 40]^T = 10\,[3 \quad 4]^T$, which means the new vector has the same direction as the original vector. The scale constant, which we denote by λ is 10. *The problem of systematically finding such* λ*'s and nonzero vectors for a given square matrix will be the theme of this chapter.* It is called the *matrix eigenvalue* problem or, more commonly, the *eigenvalue* problem.

We formalize our observation. Let $\mathbf{A} = [a_{jk}]$ be a given nonzero square matrix of dimension $n \times n$. Consider the following vector equation:

$$\mathbf{Ax} = \lambda\mathbf{x}. \tag{1}$$

The problem of finding nonzero $\mathbf{x}$'s and λ's that satisfy equation (1) is called an eigenvalue problem.

Remark. So $\mathbf{A}$ is a given square (!) matrix, $\mathbf{x}$ is an unknown vector, and λ is an unknown scalar. Our task is to find λ's and nonzero $\mathbf{x}$'s that satisfy (1). Geometrically, we are looking for vectors, $\mathbf{x}$, for which the multiplication by $\mathbf{A}$ has the same effect as the multiplication by a scalar λ; in other words, $\mathbf{Ax}$ should be proportional to $\mathbf{x}$. Thus, the multiplication has the effect of producing, from the original vector $\mathbf{x}$, a new vector $\lambda\mathbf{x}$ that has the same or opposite (minus sign) direction as the original vector. (This was all demonstrated in our intuitive opening example. Can you see that the second equation in that example satisfies (1) with $\lambda = 10$ and $\mathbf{x} = [3 \quad 4]^T$, and $\mathbf{A}$ the given 2×2 matrix? Write it out.) Now why do we require $\mathbf{x}$ to be nonzero? The reason is that $\mathbf{x} = \mathbf{0}$ is always a solution of (1) for any value of λ, because $\mathbf{A0} = \mathbf{0}$. This is of no interest.

We introduce more terminology. A value of λ, for which (1) has a solution $\mathbf{x} \neq \mathbf{0}$, is called an **eigenvalue** or *characteristic value* of the matrix **A**. Another term for λ is a *latent* root. ("Eigen" is German and means "proper" or "characteristic."). The corresponding solutions $\mathbf{x} \neq \mathbf{0}$ of (1) are called the **eigenvectors** or *characteristic vectors* of **A** corresponding to that eigenvalue λ. The set of all the eigenvalues of **A** is called the **spectrum** of **A**. We shall see that the spectrum consists of at least one eigenvalue and at most of n numerically different eigenvalues. The largest of the absolute values of the eigenvalues of **A** is called the *spectral radius* of **A**, a name to be motivated later.

How to Find Eigenvalues and Eigenvectors

Now, with the new terminology for (1), we can just say that the problem of determining the eigenvalues and eigenvectors of a matrix is called an eigenvalue problem. (However, more precisely, we are considering an algebraic eigenvalue problem, as opposed to an eigenvalue problem involving an ODE or PDE, as considered in Secs. 11.5 and 12.3, or an integral equation.)

Eigenvalues have a very large number of applications in diverse fields such as in engineering, geometry, physics, mathematics, biology, environmental science, economics, psychology, and other areas. You will encounter applications for elastic membranes, Markov processes, population models, and others in this chapter.

Since, from the viewpoint of engineering applications, eigenvalue problems are the most important problems in connection with matrices, the student should carefully follow our discussion.

Example 1 demonstrates how to systematically solve a simple eigenvalue problem.

EXAMPLE 1 Determination of Eigenvalues and Eigenvectors

We illustrate all the steps in terms of the matrix

$$\mathbf{A} = \begin{bmatrix} -5 & 2 \\ 2 & -2 \end{bmatrix}.$$

Solution. **(a)** ***Eigenvalues.*** These must be determined ***first***. Equation (1) is

$$\mathbf{Ax} = \begin{bmatrix} -5 & 2 \\ 2 & -2 \end{bmatrix}\begin{bmatrix} x_1 \\ x_2 \end{bmatrix} = \lambda \begin{bmatrix} x_1 \\ x_2 \end{bmatrix}; \qquad \text{in components,} \qquad \begin{aligned} -5x_1 + 2x_2 &= \lambda x_1 \\ 2x_1 - 2x_2 &= \lambda x_2. \end{aligned}$$

Transferring the terms on the right to the left, we get

$$\begin{aligned} (-5 - \lambda)x_1 + \quad 2x_2 \quad &= 0 \\ 2x_1 \quad + (-2 - \lambda)x_2 &= 0. \end{aligned} \tag{2*}$$

This can be written in matrix notation

$$(\mathbf{A} - \lambda\mathbf{I})\mathbf{x} = \mathbf{0} \tag{3*}$$

because (1) is $\mathbf{Ax} - \lambda\mathbf{x} = \mathbf{Ax} - \lambda\mathbf{Ix} = (\mathbf{A} - \lambda\mathbf{I})\mathbf{x} = \mathbf{0}$, which gives (3*). We see that this is a ***homogeneous*** linear system. By Cramer's theorem in Sec. 7.7 it has a nontrivial solution $\mathbf{x} \neq \mathbf{0}$ (an eigenvector of **A** we are looking for) if and only if its coefficient determinant is zero, that is,

$$D(\lambda) = \det(\mathbf{A} - \lambda\mathbf{I}) = \begin{vmatrix} -5 - \lambda & 2 \\ 2 & -2 - \lambda \end{vmatrix} = (-5 - \lambda)(-2 - \lambda) - 4 = \lambda^2 + 7\lambda + 6 = 0. \tag{4*}$$

We call $D(\lambda)$ the **characteristic determinant** or, if expanded, the **characteristic polynomial**, and $D(\lambda) = 0$ the **characteristic equation** of **A**. The solutions of this quadratic equation are $\lambda_1 = -1$ and $\lambda_2 = -6$. These are the eigenvalues of **A**.

(b₁) ***Eigenvector of* A *corresponding to*** λ_1. This vector is obtained from (2*) with $\lambda = \lambda_1 = -1$, that is,

$$-4x_1 + 2x_2 = 0$$
$$2x_1 - x_2 = 0.$$

A solution is $x_2 = 2x_1$, as we see from either of the two equations, so that we need only one of them. This determines an eigenvector corresponding to $\lambda_1 = -1$ up to a scalar multiple. If we choose $x_1 = 1$, we obtain the eigenvector

$$\mathbf{x}_1 = \begin{bmatrix} 1 \\ 2 \end{bmatrix}, \qquad \text{Check:} \qquad \mathbf{A}\mathbf{x}_1 = \begin{bmatrix} -5 & 2 \\ 2 & -2 \end{bmatrix}\begin{bmatrix} 1 \\ 2 \end{bmatrix} = \begin{bmatrix} -1 \\ -2 \end{bmatrix} = (-1)\mathbf{x}_1 = \lambda_1\mathbf{x}_1.$$

(b₂) ***Eigenvector of* A *corresponding to*** λ_2. For $\lambda = \lambda_2 = -6$, equation (2*) becomes

$$x_1 + 2x_2 = 0$$
$$2x_1 + 4x_2 = 0.$$

A solution is $x_2 = -x_1/2$ with arbitrary x_1. If we choose $x_1 = 2$, we get $x_2 = -1$. Thus an eigenvector of **A** corresponding to $\lambda_2 = -6$ is

$$\mathbf{x}_2 = \begin{bmatrix} 2 \\ -1 \end{bmatrix}, \qquad \text{Check:} \qquad \mathbf{A}\mathbf{x}_2 = \begin{bmatrix} -5 & 2 \\ 2 & -2 \end{bmatrix}\begin{bmatrix} 2 \\ -1 \end{bmatrix} = \begin{bmatrix} -12 \\ 6 \end{bmatrix} = (-6)\mathbf{x}_2 = \lambda_2\mathbf{x}_2.$$

For the matrix in the intuitive opening example at the start of Sec. 8.1, the characteristic equation is $\lambda^2 - 13\lambda + 30 = (\lambda - 10)(\lambda - 3) = 0$. The eigenvalues are $\{10, \ 3\}$. Corresponding eigenvectors are $[3 \quad 4]^T$ and $[-1 \quad 1]^T$, respectively. The reader may want to verify this. ■

This example illustrates the general case as follows. Equation (1) written in components is

$$\begin{aligned} a_{11}x_1 + \cdots + a_{1n}x_n &= \lambda x_1 \\ a_{21}x_1 + \cdots + a_{2n}x_n &= \lambda x_2 \\ \cdots\cdots\cdots\cdots\cdots\cdots\cdots\cdots \\ a_{n1}x_1 + \cdots + a_{nn}x_n &= \lambda x_n. \end{aligned}$$

Transferring the terms on the right side to the left side, we have

$$\begin{aligned} (a_{11} - \lambda)x_1 + \quad a_{12}x_2 \quad + \cdots + \quad a_{1n}x_n &= 0 \\ a_{21}x_1 \quad + (a_{22} - \lambda)x_2 + \cdots + \quad a_{2n}x_n &= 0 \\ \cdots\cdots\cdots\cdots\cdots\cdots\cdots\cdots\cdots\cdots\cdots\cdots \\ a_{n1}x_1 \quad + \quad a_{n2}x_2 \quad + \cdots + (a_{nn} - \lambda)x_n &= 0. \end{aligned} \tag{2}$$

In matrix notation,

$$(\mathbf{A} - \lambda\mathbf{I})\mathbf{x} = \mathbf{0}. \tag{3}$$

By Cramer's theorem in Sec. 7.7, this homogeneous linear system of equations has a nontrivial solution if and only if the corresponding determinant of the coefficients is zero:

$$(4)\qquad D(\lambda) = \det(\mathbf{A} - \lambda\mathbf{I}) = \begin{vmatrix} a_{11} - \lambda & a_{12} & \cdots & a_{1n} \\ a_{21} & a_{22} - \lambda & \cdots & a_{2n} \\ \cdot & \cdot & \cdots & \cdot \\ a_{n1} & a_{n2} & \cdots & a_{nn} - \lambda \end{vmatrix} = 0.$$

$\mathbf{A} - \lambda\mathbf{I}$ is called the **characteristic matrix** and $D(\lambda)$ the **characteristic determinant** of $\mathbf{A}$. Equation (4) is called the **characteristic equation** of $\mathbf{A}$. By developing $D(\lambda)$ we obtain a polynomial of nth degree in λ. This is called the **characteristic polynomial** of $\mathbf{A}$.

This proves the following important theorem.

THEOREM 1

Eigenvalues

The eigenvalues of a square matrix $\mathbf{A}$ *are the roots of the characteristic equation* (4) *of* $\mathbf{A}$.

Hence an $n \times n$ *matrix has at least one eigenvalue and at most* n *numerically different eigenvalues.*

For larger n, the actual computation of eigenvalues will, in general, require the use of Newton's method (Sec. 19.2) or another numeric approximation method in Secs. 20.7–20.9.

The eigen**values** ***must be determined first.*** Once these are known, corresponding eigen*vectors* are obtained from the system (2), for instance, by the Gauss elimination, where λ is the eigenvalue for which an eigenvector is wanted. This is what we did in Example 1 and shall do again in the examples below. (To prevent misunderstandings: *numeric approximation methods*, such as in Sec. 20.8, may determine eigen*vectors first*.)

Eigenvectors have the following properties.

THEOREM 2

Eigenvectors, Eigenspace

If $\mathbf{w}$ *and* $\mathbf{x}$ *are eigenvectors of a matrix* $\mathbf{A}$ *corresponding to* **the same** *eigenvalue* λ, *so are* $\mathbf{w} + \mathbf{x}$ (provided $\mathbf{x} \neq -\mathbf{w}$) *and* $k\mathbf{x}$ *for any* $k \neq 0$.

Hence the eigenvectors corresponding to one and the same eigenvalue λ *of* $\mathbf{A}$, *together with* $\mathbf{0}$, *form a vector space* (cf. Sec. 7.4), *called the* **eigenspace** *of* $\mathbf{A}$ *corresponding to that* λ.

PROOF $\mathbf{A}\mathbf{w} = \lambda\mathbf{w}$ and $\mathbf{A}\mathbf{x} = \lambda\mathbf{x}$ *imply* $\mathbf{A}(\mathbf{w} + \mathbf{x}) = \mathbf{A}\mathbf{w} + \mathbf{A}\mathbf{x} = \lambda\mathbf{w} + \lambda\mathbf{x} = \lambda(\mathbf{w} + \mathbf{x})$ and $\mathbf{A}(k\mathbf{w}) = k(\mathbf{A}\mathbf{w}) = k(\lambda\mathbf{w}) = \lambda(k\mathbf{w})$; hence $\mathbf{A}(k\mathbf{w} + \ell\mathbf{x}) = \lambda(k\mathbf{w} + \ell\mathbf{x})$. ∎

In particular, *an eigenvector* $\mathbf{x}$ *is determined only up to a constant factor.* Hence we can **normalize** $\mathbf{x}$, that is, multiply it by a scalar to get a unit vector (see Sec. 7.9). For instance, $\mathbf{x}_1 = [1 \quad 2]^T$ in Example 1 has the length $\|\mathbf{x}_1\| = \sqrt{1^2 + 2^2} = \sqrt{5}$; hence $[1/\sqrt{5} \quad 2/\sqrt{5}]^T$ is a normalized eigenvector (a unit eigenvector).

Examples 2 and 3 will illustrate that an $n \times n$ matrix may have n linearly independent eigenvectors, or it may have fewer than n. In Example 4 we shall see that a *real* matrix may have *complex* eigenvalues and eigenvectors.

EXAMPLE 2 Multiple Eigenvalues

Find the eigenvalues and eigenvectors of

$$\mathbf{A} = \begin{bmatrix} -2 & 2 & -3 \\ 2 & 1 & -6 \\ -1 & -2 & 0 \end{bmatrix}.$$

Solution. For our matrix, the characteristic determinant gives the characteristic equation

$$-\lambda^3 - \lambda^2 + 21\lambda + 45 = 0.$$

The roots (eigenvalues of $\mathbf{A}$) are $\lambda_1 = 5$, $\lambda_2 = \lambda_3 = -3$. (If you have trouble finding roots, you may want to use a root finding algorithm such as Newton's method (Sec. 19.2). Your CAS or scientific calculator can find roots. However, to really learn and remember this material, you have to do some exercises with paper and pencil.) To find eigenvectors, we apply the Gauss elimination (Sec. 7.3) to the system $(\mathbf{A} - \lambda\mathbf{I})\mathbf{x} = \mathbf{0}$, first with $\lambda = 5$ and then with $\lambda = -3$. For $\lambda = 5$ the characteristic matrix is

$$\mathbf{A} - \lambda\mathbf{I} = \mathbf{A} - 5\mathbf{I} = \begin{bmatrix} -7 & 2 & -3 \\ 2 & -4 & -6 \\ -1 & -2 & -5 \end{bmatrix}. \quad \text{It row-reduces to} \quad \begin{bmatrix} -7 & 2 & -3 \\ 0 & -\frac{24}{7} & -\frac{48}{7} \\ 0 & 0 & 0 \end{bmatrix}.$$

Hence it has rank 2. Choosing $x_3 = -1$ we have $x_2 = 2$ from $-\frac{24}{7}x_2 - \frac{48}{7}x_3 = 0$ and then $x_1 = 1$ from $-7x_1 + 2x_2 - 3x_3 = 0$. Hence an eigenvector of $\mathbf{A}$ corresponding to $\lambda = 5$ is $\mathbf{x}_1 = [1 \quad 2 \quad -1]^\mathsf{T}$.

For $\lambda = -3$ the characteristic matrix

$$\mathbf{A} - \lambda\mathbf{I} = \mathbf{A} + 3\mathbf{I} = \begin{bmatrix} 1 & 2 & -3 \\ 2 & 4 & -6 \\ -1 & -2 & 3 \end{bmatrix} \quad \text{row-reduces to} \quad \begin{bmatrix} 1 & 2 & -3 \\ 0 & 0 & 0 \\ 0 & 0 & 0 \end{bmatrix}.$$

Hence it has rank 1. From $x_1 + 2x_2 - 3x_3 = 0$ we have $x_1 = -2x_2 + 3x_3$. Choosing $x_2 = 1, x_3 = 0$ and $x_2 = 0, x_3 = 1$, we obtain two linearly independent eigenvectors of $\mathbf{A}$ corresponding to $\lambda = -3$ [as they must exist by (5), Sec. 7.5, with rank $= 1$ and $n = 3$],

$$\mathbf{x}_2 = \begin{bmatrix} -2 \\ 1 \\ 0 \end{bmatrix}$$

and

$$\mathbf{x}_3 = \begin{bmatrix} 3 \\ 0 \\ 1 \end{bmatrix}.$$ ■

The order M_λ of an eigenvalue λ as a root of the characteristic polynomial is called the **algebraic multiplicity** of λ. The number m_λ of linearly independent eigenvectors corresponding to λ is called the **geometric multiplicity** of λ. Thus m_λ is the dimension of the eigenspace corresponding to this λ.

Since the characteristic polynomial has degree n, the sum of all the algebraic multiplicities must equal n. In Example 2 for $\lambda = -3$ we have $m_\lambda = M_\lambda = 2$. In general, $m_\lambda \leqq M_\lambda$, as can be shown. The difference $\Delta_\lambda = M_\lambda - m_\lambda$ is called the **defect** of λ. Thus $\Delta_{-3} = 0$ in Example 2, but positive defects Δ_λ can easily occur:

EXAMPLE 3 Algebraic Multiplicity, Geometric Multiplicity. Positive Defect

The characteristic equation of the matrix

$$\mathbf{A} = \begin{bmatrix} 0 & 1 \\ 0 & 0 \end{bmatrix} \quad \text{is} \quad \det(\mathbf{A} - \lambda\mathbf{I}) = \begin{vmatrix} -\lambda & 1 \\ 0 & -\lambda \end{vmatrix} = \lambda^2 = 0.$$

Hence $\lambda = 0$ is an eigenvalue of algebraic multiplicity $M_0 = 2$. But its geometric multiplicity is only $m_0 = 1$, since eigenvectors result from $-0x_1 + x_2 = 0$, hence $x_2 = 0$, in the form $[x_1 \quad 0]^\mathsf{T}$. Hence for $\lambda = 0$ the defect is $\Delta_0 = 1$.

Similarly, the characteristic equation of the matrix

$$\mathbf{A} = \begin{bmatrix} 3 & 2 \\ 0 & 3 \end{bmatrix} \quad \text{is} \quad \det(\mathbf{A} - \lambda\mathbf{I}) = \begin{vmatrix} 3-\lambda & 2 \\ 0 & 3-\lambda \end{vmatrix} = (3-\lambda)^2 = 0.$$

Hence $\lambda = 3$ is an eigenvalue of algebraic multiplicity $M_3 = 2$, but its geometric multiplicity is only $m_3 = 1$, since eigenvectors result from $0x_1 + 2x_2 = 0$ in the form $[x_1 \quad 0]^\mathsf{T}$. ■

EXAMPLE 4 Real Matrices with Complex Eigenvalues and Eigenvectors

Since real polynomials may have complex roots (which then occur in conjugate pairs), a real matrix may have complex eigenvalues and eigenvectors. For instance, the characteristic equation of the skew-symmetric matrix

$$\mathbf{A} = \begin{bmatrix} 0 & 1 \\ -1 & 0 \end{bmatrix} \quad \text{is} \quad \det(\mathbf{A} - \lambda\mathbf{I}) = \begin{vmatrix} -\lambda & 1 \\ -1 & -\lambda \end{vmatrix} = \lambda^2 + 1 = 0.$$

It gives the eigenvalues $\lambda_1 = i\,(= \sqrt{-1})$, $\lambda_2 = -i$. Eigenvectors are obtained from $-ix_1 + x_2 = 0$ and $ix_1 + x_2 = 0$, respectively, and we can choose $x_1 = 1$ to get

$$\begin{bmatrix} 1 \\ i \end{bmatrix} \quad \text{and} \quad \begin{bmatrix} 1 \\ -i \end{bmatrix}. \quad ■$$

In the next section we shall need the following simple theorem.

THEOREM 3 Eigenvalues of the Transpose

The transpose $\mathbf{A}^\mathsf{T}$ of a square matrix $\mathbf{A}$ has the same eigenvalues as $\mathbf{A}$.

PROOF Transposition does not change the value of the characteristic determinant, as follows from Theorem 2d in Sec. 7.7. ■

Having gained a first impression of matrix eigenvalue problems, we shall illustrate their importance with some typical applications in Sec. 8.2.

PROBLEM SET 8.1

1–16 EIGENVALUES, EIGENVECTORS

Find the eigenvalues. Find the corresponding eigenvectors. Use the given λ or factor in Probs. 11 and 15.

1. $\begin{bmatrix} 3.0 & 0 \\ 0 & -0.6 \end{bmatrix}$

2. $\begin{bmatrix} 0 & 0 \\ 0 & 0 \end{bmatrix}$

3. $\begin{bmatrix} 5 & -2 \\ 9 & -6 \end{bmatrix}$

4. $\begin{bmatrix} 1 & 2 \\ 2 & 4 \end{bmatrix}$

5. $\begin{bmatrix} 0 & 3 \\ -3 & 0 \end{bmatrix}$

6. $\begin{bmatrix} 1 & 2 \\ 0 & 3 \end{bmatrix}$

7. $\begin{bmatrix} 0 & 1 \\ 0 & 0 \end{bmatrix}$

8. $\begin{bmatrix} a & b \\ -b & a \end{bmatrix}$

9. $\begin{bmatrix} 0.8 & -0.6 \\ 0.6 & 0.8 \end{bmatrix}$

10. $\begin{bmatrix} \cos\theta & -\sin\theta \\ \sin\theta & \cos\theta \end{bmatrix}$

11. $\begin{bmatrix} 6 & 2 & -2 \\ 2 & 5 & 0 \\ -2 & 0 & 7 \end{bmatrix}, \quad \lambda = 3$

12. $\begin{bmatrix} 3 & 5 & 3 \\ 0 & 4 & 6 \\ 0 & 0 & 1 \end{bmatrix}$

13. $\begin{bmatrix} 13 & 5 & 2 \\ 2 & 7 & -8 \\ 5 & 4 & 7 \end{bmatrix}$

14. $\begin{bmatrix} 2 & 0 & -1 \\ 0 & \frac{1}{2} & 0 \\ 1 & 0 & 4 \end{bmatrix}$

15. $\begin{bmatrix} -1 & 0 & 12 & 0 \\ 0 & -1 & 0 & 12 \\ 0 & 0 & -1 & -4 \\ 0 & 0 & -4 & -1 \end{bmatrix}, \quad (\lambda + 1)^2$

16. $\begin{bmatrix} -3 & 0 & 4 & 2 \\ 0 & 1 & -2 & 4 \\ 2 & 4 & -1 & -2 \\ 0 & 2 & -2 & 3 \end{bmatrix}$

17–20 LINEAR TRANSFORMATIONS AND EIGENVALUES

Find the matrix $\mathbf{A}$ in the linear transformation $\mathbf{y} = \mathbf{A}\mathbf{x}$, where $\mathbf{x} = [x_1 \ \ x_2]^\mathsf{T}$ ($\mathbf{x} = [x_1 \ \ x_2 \ \ x_3]^\mathsf{T}$) are Cartesian coordinates. Find the eigenvalues and eigenvectors and explain their geometric meaning.

17. Counterclockwise rotation through the angle $\pi/2$ about the origin in R^2.

18. Reflection about the x_1-axis in R^2.

19. Orthogonal projection (perpendicular projection) of R^2 onto the x_2-axis.

20. Orthogonal projection of R^3 onto the plane $x_2 = x_1$.

21–25 GENERAL PROBLEMS

21. Nonzero defect. Find further 2×2 and 3×3 matrices with positive defect. See Example 3.

22. Multiple eigenvalues. Find further 2×2 and 3×3 matrices with multiple eigenvalues. See Example 2.

23. Complex eigenvalues. Show that the eigenvalues of a real matrix are real or complex conjugate in pairs.

24. Inverse matrix. Show that $\mathbf{A}^{-1}$ exists if and only if the eigenvalues $\lambda_1, \cdots, \lambda_n$ are all nonzero, and then $\mathbf{A}^{-1}$ has the eigenvalues $1/\lambda_1, \cdots, 1/\lambda_n$.

25. Transpose. Illustrate Theorem 3 with examples of your own.

8.2 Some Applications of Eigenvalue Problems

We have selected some typical examples from the wide range of applications of matrix eigenvalue problems. The last example, that is, Example 4, shows an application involving vibrating springs and ODEs. It falls into the domain of Chapter 4, which covers matrix eigenvalue problems related to ODE's modeling mechanical systems and electrical networks. Example 4 is included to keep our discussion independent of Chapter 4. (However, the reader not interested in ODEs may want to skip Example 4 without loss of continuity.)

EXAMPLE 1 Stretching of an Elastic Membrane

An elastic membrane in the x_1x_2-plane with boundary circle $x_1^2 + x_2^2 = 1$ (Fig. 160) is stretched so that a point P: (x_1, x_2) goes over into the point Q: (y_1, y_2) given by

$$(1) \qquad \mathbf{y} = \begin{bmatrix} y_1 \\ y_2 \end{bmatrix} = \mathbf{A}\mathbf{x} = \begin{bmatrix} 5 & 3 \\ 3 & 5 \end{bmatrix} \begin{bmatrix} x_1 \\ x_2 \end{bmatrix}; \qquad \text{in components,} \qquad \begin{aligned} y_1 &= 5x_1 + 3x_2 \\ y_2 &= 3x_1 + 5x_2. \end{aligned}$$

Find the **principal directions**, that is, the directions of the position vector $\mathbf{x}$ of P for which the direction of the position vector $\mathbf{y}$ of Q is the same or exactly opposite. What shape does the boundary circle take under this deformation?

Solution. We are looking for vectors $\mathbf{x}$ such that $\mathbf{y} = \lambda\mathbf{x}$. Since $\mathbf{y} = \mathbf{A}\mathbf{x}$, this gives $\mathbf{A}\mathbf{x} = \lambda\mathbf{x}$, the equation of an eigenvalue problem. In components, $\mathbf{A}\mathbf{x} = \lambda\mathbf{x}$ is

$$(2) \qquad \begin{aligned} 5x_1 + 3x_2 &= \lambda x_1 \\ 3x_1 + 5x_2 &= \lambda x_2 \end{aligned} \qquad \text{or} \qquad \begin{aligned} (5-\lambda)x_1 + 3x_2 &= 0 \\ 3x_1 + (5-\lambda)x_2 &= 0. \end{aligned}$$

The characteristic equation is

$$(3) \qquad \begin{vmatrix} 5-\lambda & 3 \\ 3 & 5-\lambda \end{vmatrix} = (5-\lambda)^2 - 9 = 0.$$

Its solutions are $\lambda_1 = 8$ and $\lambda_2 = 2$. These are the eigenvalues of our problem. For $\lambda = \lambda_1 = 8$, our system (2) becomes

$$\begin{aligned} -3x_1 + 3x_2 &= 0, \\ 3x_1 - 3x_2 &= 0. \end{aligned} \qquad \begin{aligned} &\text{Solution } x_2 = x_1,\ x_1 \text{ arbitrary,} \\ &\text{for instance, } x_1 = x_2 = 1. \end{aligned}$$

For $\lambda_2 = 2$, our system (2) becomes

$$\begin{aligned} 3x_1 + 3x_2 &= 0, \\ 3x_1 + 3x_2 &= 0. \end{aligned} \qquad \begin{aligned} &\text{Solution } x_2 = -x_1,\ x_1 \text{ arbitrary,} \\ &\text{for instance, } x_1 = 1, x_2 = -1. \end{aligned}$$

We thus obtain as eigenvectors of $\mathbf{A}$, for instance, $[1 \quad 1]^\mathsf{T}$ corresponding to λ_1 and $[1 \quad -1]^\mathsf{T}$ corresponding to λ_2 (or a nonzero scalar multiple of these). These vectors make 45° and 135° angles with the positive x_1-direction. They give the principal directions, the answer to our problem. The eigenvalues show that in the principal directions the membrane is stretched by factors 8 and 2, respectively; see Fig. 160.

Accordingly, if we choose the principal directions as directions of a new Cartesian u_1u_2-coordinate system, say, with the positive u_1-semi-axis in the first quadrant and the positive u_2-semi-axis in the second quadrant of the x_1x_2-system, and if we set $u_1 = r\cos\phi$, $u_2 = r\sin\phi$, then a boundary point of the unstretched circular membrane has coordinates $\cos\phi$, $\sin\phi$. Hence, after the stretch we have

$$z_1 = 8\cos\phi, \qquad z_2 = 2\sin\phi.$$

Since $\cos^2\phi + \sin^2\phi = 1$, this shows that the deformed boundary is an ellipse (Fig. 160)

$$(4) \qquad \frac{z_1^2}{8^2} + \frac{z_2^2}{2^2} = 1. \qquad ■$$

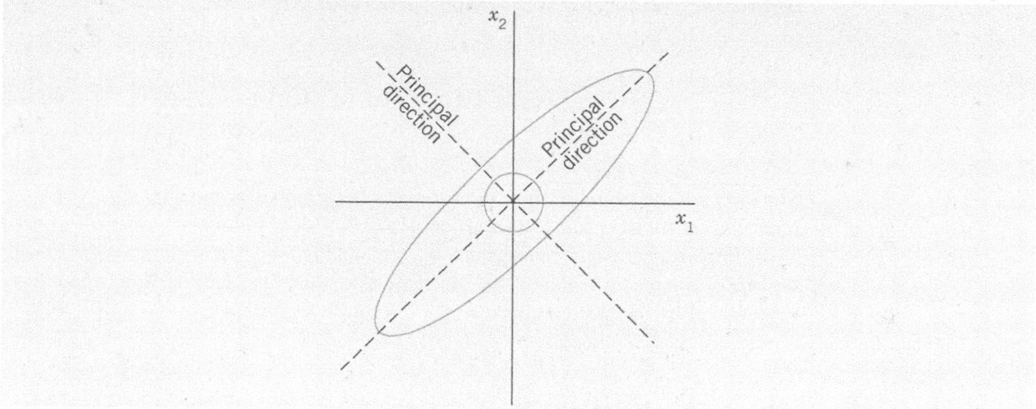

Fig. 160. Undeformed and deformed membrane in Example 1

EXAMPLE 2 Eigenvalue Problems Arising from Markov Processes

Markov processes as considered in Example 13 of Sec. 7.2 lead to eigenvalue problems if we ask for the limit state of the process in which the state vector **x** is reproduced under the multiplication by the stochastic matrix **A** governing the process, that is, $\mathbf{Ax} = \mathbf{x}$. Hence **A** should have the eigenvalue 1, and **x** should be a corresponding eigenvector. This is of practical interest because it shows the long-term tendency of the development modeled by the process.

In that example,

$$\mathbf{A} = \begin{bmatrix} 0.7 & 0.1 & 0 \\ 0.2 & 0.9 & 0.2 \\ 0.1 & 0 & 0.8 \end{bmatrix}. \qquad \text{For the transpose,} \qquad \begin{bmatrix} 0.7 & 0.2 & 0.1 \\ 0.1 & 0.9 & 0 \\ 0 & 0.2 & 0.8 \end{bmatrix} \begin{bmatrix} 1 \\ 1 \\ 1 \end{bmatrix} = \begin{bmatrix} 1 \\ 1 \\ 1 \end{bmatrix}.$$

Hence $\mathbf{A}^\mathsf{T}$ has the eigenvalue 1, and the same is true for **A** by Theorem 3 in Sec. 8.1. An eigenvector **x** of **A** for $\lambda = 1$ is obtained from

$$\mathbf{A} - \mathbf{I} = \begin{bmatrix} -0.3 & 0.1 & 0 \\ 0.2 & -0.1 & 0.2 \\ 0.1 & 0 & -0.2 \end{bmatrix}, \qquad \text{row-reduced to} \qquad \begin{bmatrix} -\frac{3}{10} & \frac{1}{10} & 0 \\ 0 & -\frac{1}{30} & \frac{1}{5} \\ 0 & 0 & 0 \end{bmatrix}.$$

Taking $x_3 = 1$, we get $x_2 = 6$ from $-x_2/30 + x_3/5 = 0$ and then $x_1 = 2$ from $-3x_1/10 + x_2/10 = 0$. This gives $\mathbf{x} = [2 \quad 6 \quad 1]^\mathsf{T}$. It means that in the long run, the ratio Commercial:Industrial:Residential will approach 2:6:1, provided that the probabilities given by **A** remain (about) the same. (We switched to ordinary fractions to avoid rounding errors.) ■

EXAMPLE 3 Eigenvalue Problems Arising from Population Models. Leslie Model

The Leslie model describes age-specified population growth, as follows. Let the oldest age attained by the females in some animal population be 9 years. Divide the population into three age classes of 3 years each. Let the "*Leslie matrix*" be

$$\mathbf{L} = [l_{jk}] = \begin{bmatrix} 0 & 2.3 & 0.4 \\ 0.6 & 0 & 0 \\ 0 & 0.3 & 0 \end{bmatrix} \tag{5}$$

where l_{1k} is the average number of daughters born to a single female during the time she is in age class k, and $l_{j,j-1} (j = 2, 3)$ is the fraction of females in age class $j - 1$ that will survive and pass into class j. (a) What is the number of females in each class after 3, 6, 9 years if each class initially consists of 400 females? (b) For what initial distribution will the number of females in each class change by the same proportion? What is this rate of change?

Solution. (a) Initially, $\mathbf{x}_{(0)}^{\mathsf{T}} = [400 \quad 400 \quad 400]$. After 3 years,

$$\mathbf{x}_{(3)} = \mathbf{L}\mathbf{x}_{(0)} = \begin{bmatrix} 0 & 2.3 & 0.4 \\ 0.6 & 0 & 0 \\ 0 & 0.3 & 0 \end{bmatrix} \begin{bmatrix} 400 \\ 400 \\ 400 \end{bmatrix} = \begin{bmatrix} 1080 \\ 240 \\ 120 \end{bmatrix}.$$

Similarly, after 6 years the number of females in each class is given by $\mathbf{x}_{(6)}^{\mathsf{T}} = (\mathbf{L}\mathbf{x}_{(3)})^{\mathsf{T}} = [600 \quad 648 \quad 72]$, and after 9 years we have $\mathbf{x}_{(9)}^{\mathsf{T}} = (\mathbf{L}\mathbf{x}_{(6)})^{\mathsf{T}} = [1519.2 \quad 360 \quad 194.4]$.

(b) Proportional change means that we are looking for a distribution vector $\mathbf{x}$ such that $\mathbf{L}\mathbf{x} = \lambda\mathbf{x}$, where λ is the rate of change (growth if $\lambda > 1$, decrease if $\lambda < 1$). The characteristic equation is (develop the characteristic determinant by the first column)

$$\det(\mathbf{L} - \lambda\mathbf{I}) = -\lambda^3 - 0.6(-2.3\lambda - 0.3 \cdot 0.4) = -\lambda^3 + 1.38\lambda + 0.072 = 0.$$

A positive root is found to be (for instance, by Newton's method, Sec. 19.2) $\lambda = 1.2$. A corresponding eigenvector $\mathbf{x}$ can be determined from the characteristic matrix

$$\mathbf{A} - 1.2\mathbf{I} = \begin{bmatrix} -1.2 & 2.3 & 0.4 \\ 0.6 & -1.2 & 0 \\ 0 & 0.3 & -1.2 \end{bmatrix}, \qquad \text{say,} \qquad \mathbf{x} = \begin{bmatrix} 1 \\ 0.5 \\ 0.125 \end{bmatrix}$$

where $x_3 = 0.125$ is chosen, $x_2 = 0.5$ then follows from $0.3x_2 - 1.2x_3 = 0$, and $x_1 = 1$ from $-1.2x_1 + 2.3x_2 + 0.4x_3 = 0$. To get an initial population of 1200 as before, we multiply $\mathbf{x}$ by $1200/(1 + 0.5 + 0.125) = 738$. *Answer:* Proportional growth of the numbers of females in the three classes will occur if the initial values are 738, 369, 92 in classes 1, 2, 3, respectively. The growth rate will be 1.2 per 3 years. ■

EXAMPLE 4 **Vibrating System of Two Masses on Two Springs (Fig. 161)**

Mass–spring systems involving several masses and springs can be treated as eigenvalue problems. For instance, the mechanical system in Fig. 161 is governed by the system of ODEs

$$(6) \qquad \begin{aligned} y_1'' &= -3y_1 - 2(y_1 - y_2) = -5y_1 + 2y_2 \\ y_2'' &= \qquad\quad\; -2(y_2 - y_1) = 2y_1 - 2y_2 \end{aligned}$$

where y_1 and y_2 are the displacements of the masses from rest, as shown in the figure, and primes denote derivatives with respect to time t. In vector form, this becomes

$$(7) \qquad \mathbf{y}'' = \begin{bmatrix} y_1'' \\ y_2'' \end{bmatrix} = \mathbf{A}\mathbf{y} = \begin{bmatrix} -5 & 2 \\ 2 & -2 \end{bmatrix} \begin{bmatrix} y_1 \\ y_2 \end{bmatrix}.$$

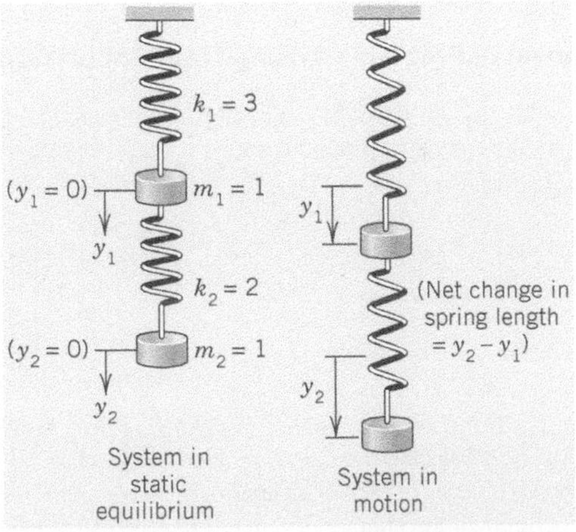

Fig. 161. Masses on springs in Example 4

We try a vector solution of the form

$$\mathbf{y} = \mathbf{x}e^{\omega t}. \tag{8}$$

This is suggested by a mechanical system of a single mass on a spring (Sec. 2.4), whose motion is given by exponential functions (and sines and cosines). Substitution into (7) gives

$$\omega^2 \mathbf{x} e^{\omega t} = \mathbf{A}\mathbf{x}e^{\omega t}.$$

Dividing by $e^{\omega t}$ and writing $\omega^2 = \lambda$, we see that our mechanical system leads to the eigenvalue problem

$$\mathbf{A}\mathbf{x} = \lambda \mathbf{x} \qquad \text{where } \lambda = \omega^2. \tag{9}$$

From Example 1 in Sec. 8.1 we see that $\mathbf{A}$ has the eigenvalues $\lambda_1 = -1$ and $\lambda_2 = -6$. Consequently, $\omega = \pm\sqrt{-1} = \pm i$ and $\sqrt{-6} = \pm i\sqrt{6}$, respectively. Corresponding eigenvectors are

$$\mathbf{x}_1 = \begin{bmatrix} 1 \\ 2 \end{bmatrix} \quad \text{and} \quad \mathbf{x}_2 = \begin{bmatrix} 2 \\ -1 \end{bmatrix}. \tag{10}$$

From (8) we thus obtain the four complex solutions [see (10), Sec. 2.2]

$$\mathbf{x}_1 e^{\pm it} = \mathbf{x}_1(\cos t \pm i \sin t),$$
$$\mathbf{x}_2 e^{\pm i\sqrt{6}t} = \mathbf{x}_2(\cos \sqrt{6}\, t \pm i \sin \sqrt{6}\, t).$$

By addition and subtraction (see Sec. 2.2) we get the four real solutions

$$\mathbf{x}_1 \cos t, \qquad \mathbf{x}_1 \sin t, \qquad \mathbf{x}_2 \cos \sqrt{6}\, t, \qquad \mathbf{x}_2 \sin \sqrt{6}\, t.$$

A general solution is obtained by taking a linear combination of these,

$$\mathbf{y} = \mathbf{x}_1(a_1 \cos t + b_1 \sin t) + \mathbf{x}_2(a_2 \cos \sqrt{6}\, t + b_2 \sin \sqrt{6}\, t)$$

with arbitrary constants a_1, b_1, a_2, b_2 (to which values can be assigned by prescribing initial displacement and initial velocity of each of the two masses). By (10), the components of $\mathbf{y}$ are

$$y_1 = a_1 \cos t + b_1 \sin t + 2a_2 \cos \sqrt{6}\, t + 2b_2 \sin \sqrt{6}\, t$$
$$y_2 = 2a_1 \cos t + 2b_1 \sin t - a_2 \cos \sqrt{6}\, t - b_2 \sin \sqrt{6}\, t.$$

These functions describe harmonic oscillations of the two masses. Physically, this had to be expected because we have neglected damping. ■

PROBLEM SET 8.2

1–6 ELASTIC DEFORMATIONS

Given $\mathbf{A}$ in a deformation $\mathbf{y} = \mathbf{A}\mathbf{x}$, find the principal directions and corresponding factors of extension or contraction. Show the details.

1. $\begin{bmatrix} 3.0 & 1.5 \\ 1.5 & 3.0 \end{bmatrix}$

2. $\begin{bmatrix} 2.0 & 0.4 \\ 0.4 & 2.0 \end{bmatrix}$

3. $\begin{bmatrix} 7 & \sqrt{6} \\ \sqrt{6} & 2 \end{bmatrix}$

4. $\begin{bmatrix} 5 & 2 \\ 2 & 13 \end{bmatrix}$

5. $\begin{bmatrix} 1 & \frac{1}{2} \\ \frac{1}{2} & 1 \end{bmatrix}$

6. $\begin{bmatrix} 1.25 & 0.75 \\ 0.75 & 1.25 \end{bmatrix}$

7–9 MARKOV PROCESSES

Find the limit state of the Markov process modeled by the given matrix. Show the details.

7. $\begin{bmatrix} 0.2 & 0.5 \\ 0.8 & 0.5 \end{bmatrix}$

8. $\begin{bmatrix} 0.4 & 0.3 & 0.3 \\ 0.3 & 0.6 & 0.1 \\ 0.3 & 0.1 & 0.6 \end{bmatrix}$

9. $\begin{bmatrix} 0.6 & 0.1 & 0.2 \\ 0.4 & 0.1 & 0.4 \\ 0 & 0.8 & 0.4 \end{bmatrix}$

10–12 AGE-SPECIFIC POPULATION

Find the growth rate in the Leslie model (see Example 3) with the matrix as given. Show the details.

10. $\begin{bmatrix} 0 & 9.0 & 5.0 \\ 0.4 & 0 & 0 \\ 0 & 0.4 & 0 \end{bmatrix}$ 11. $\begin{bmatrix} 0 & 3.45 & 0.60 \\ 0.90 & 0 & 0 \\ 0 & 0.45 & 0 \end{bmatrix}$

12. $\begin{bmatrix} 0 & 3.0 & 2.0 & 2.0 \\ 0.5 & 0 & 0 & 0 \\ 0 & 0.5 & 0 & 0 \\ 0 & 0 & 0.1 & 0 \end{bmatrix}$

13–15 LEONTIEF MODELS[1]

13. **Leontief input–output model.** Suppose that three industries are interrelated so that their outputs are used as inputs by themselves, according to the 3×3 **consumption matrix**

$$\mathbf{A} = [a_{jk}] = \begin{bmatrix} 0.1 & 0.5 & 0 \\ 0.8 & 0 & 0.4 \\ 0.1 & 0.5 & 0.6 \end{bmatrix}$$

where a_{jk} is the fraction of the output of industry k consumed (purchased) by industry j. Let p_j be the price charged by industry j for its total output. A problem is to find prices so that for each industry, total expenditures equal total income. Show that this leads to $\mathbf{Ap} = \mathbf{p}$, where $\mathbf{p} = [p_1 \quad p_2 \quad p_3]^\mathsf{T}$, and find a solution $\mathbf{p}$ with nonnegative p_1, p_2, p_3.

14. Show that a consumption matrix as considered in Prob. 13 must have column sums 1 and always has the eigenvalue 1.

15. **Open Leontief input–output model.** If not the whole output but only a portion of it is consumed by the industries themselves, then instead of $\mathbf{Ax} = \mathbf{x}$ (as in Prob. 13), we have $\mathbf{x} - \mathbf{Ax} = \mathbf{y}$, where $\mathbf{x} = [x_1 \quad x_2 \quad x_3]^\mathsf{T}$ is produced, $\mathbf{Ax}$ is consumed by the industries, and, thus, $\mathbf{y}$ is the net production available for other consumers. Find for what production $\mathbf{x}$ a given **demand vector** $\mathbf{y} = [0.1 \quad 0.3 \quad 0.1]^\mathsf{T}$ can be achieved if the consumption matrix is

$$\mathbf{A} = \begin{bmatrix} 0.1 & 0.4 & 0.2 \\ 0.5 & 0 & 0.1 \\ 0.1 & 0.4 & 0.4 \end{bmatrix}.$$

16–20 GENERAL PROPERTIES OF EIGENVALUE PROBLEMS

Let $\mathbf{A} = [a_{jk}]$ be an $n \times n$ matrix with (not necessarily distinct) eigenvalues $\lambda_1, \cdots, \lambda_n$. Show.

16. **Trace.** The sum of the main diagonal entries, called the *trace* of $\mathbf{A}$, equals the sum of the eigenvalues of $\mathbf{A}$.

17. **"Spectral shift."** $\mathbf{A} - k\mathbf{I}$ has the eigenvalues $\lambda_1 - k, \cdots, \lambda_n - k$ and the same eigenvectors as $\mathbf{A}$.

18. **Scalar multiples, powers.** $k\mathbf{A}$ has the eigenvalues $k\lambda_1, \cdots, k\lambda_n$. $\mathbf{A}^m (m = 1, 2, \cdots)$ has the eigenvalues $\lambda_1^m, \cdots, \lambda_n^m$. The eigenvectors are those of $\mathbf{A}$.

19. **Spectral mapping theorem.** The **"polynomial matrix"**

$$p(\mathbf{A}) = k_m \mathbf{A}^m + k_{m-1}\mathbf{A}^{m-1} + \cdots + k_1\mathbf{A} + k_0\mathbf{I}$$

has the eigenvalues

$$p(\lambda_j) = k_m \lambda_j^m + k_{m-1}\lambda_j^{m-1} + \cdots + k_1\lambda_j + k_0$$

where $j = 1, \cdots, n$, and the same eigenvectors as $\mathbf{A}$.

20. **Perron's theorem.** A Leslie matrix $\mathbf{L}$ with positive $l_{12}, l_{13}, l_{21}, l_{32}$ has a positive eigenvalue. (This is a special case of the Perron–Frobenius theorem in Sec. 20.7, which is difficult to prove in its general form.)

8.3 Symmetric, Skew-Symmetric, and Orthogonal Matrices

We consider three classes of real square matrices that, because of their remarkable properties, occur quite frequently in applications. The first two matrices have already been mentioned in Sec. 7.2. The goal of Sec. 8.3 is to show their remarkable properties.

[1]WASSILY LEONTIEF (1906–1999). American economist at New York University. For his input–output analysis he was awarded the Nobel Prize in 1973.

DEFINITIONS

Symmetric, Skew-Symmetric, and Orthogonal Matrices

A ***real*** square matrix $\mathbf{A} = [a_{jk}]$ is called

symmetric if transposition leaves it unchanged,

$$\text{(1)} \qquad \mathbf{A}^\mathsf{T} = \mathbf{A}, \qquad \text{thus} \qquad a_{kj} = a_{jk},$$

skew-symmetric if transposition gives the negative of $\mathbf{A}$,

$$\text{(2)} \qquad \mathbf{A}^\mathsf{T} = -\mathbf{A}, \qquad \text{thus} \qquad a_{kj} = -a_{jk},$$

orthogonal if transposition gives the inverse of $\mathbf{A}$,

$$\text{(3)} \qquad \mathbf{A}^\mathsf{T} = \mathbf{A}^{-1}.$$

EXAMPLE 1 **Symmetric, Skew-Symmetric, and Orthogonal Matrices**

The matrices

$$\begin{bmatrix} -3 & 1 & 5 \\ 1 & 0 & -2 \\ 5 & -2 & 4 \end{bmatrix}, \quad \begin{bmatrix} 0 & 9 & -12 \\ -9 & 0 & 20 \\ 12 & -20 & 0 \end{bmatrix}, \quad \begin{bmatrix} \frac{2}{3} & \frac{1}{3} & \frac{2}{3} \\ -\frac{2}{3} & \frac{2}{3} & \frac{1}{3} \\ \frac{1}{3} & \frac{2}{3} & -\frac{2}{3} \end{bmatrix}$$

are symmetric, skew-symmetric, and orthogonal, respectively, as you should verify. Every skew-symmetric matrix has all main diagonal entries zero. (Can you prove this?) ■

Any real square matrix $\mathbf{A}$ may be written as the sum of a symmetric matrix $\mathbf{R}$ and a skew-symmetric matrix $\mathbf{S}$, where

$$\text{(4)} \qquad \mathbf{R} = \tfrac{1}{2}(\mathbf{A} + \mathbf{A}^\mathsf{T}) \qquad \text{and} \qquad \mathbf{S} = \tfrac{1}{2}(\mathbf{A} - \mathbf{A}^\mathsf{T}).$$

EXAMPLE 2 **Illustration of Formula (4)**

$$\mathbf{A} = \begin{bmatrix} 9 & 5 & 2 \\ 2 & 3 & -8 \\ 5 & 4 & 3 \end{bmatrix} = \mathbf{R} + \mathbf{S} = \begin{bmatrix} 9.0 & 3.5 & 3.5 \\ 3.5 & 3.0 & -2.0 \\ 3.5 & -2.0 & 3.0 \end{bmatrix} + \begin{bmatrix} 0 & 1.5 & -1.5 \\ -1.5 & 0 & -6.0 \\ 1.5 & 6.0 & 0 \end{bmatrix}$$ ■

THEOREM 1

Eigenvalues of Symmetric and Skew-Symmetric Matrices

(a) *The eigenvalues of a symmetric matrix are real.*

(b) *The eigenvalues of a skew-symmetric matrix are pure imaginary or zero.*

This basic theorem (and an extension of it) will be proved in Sec. 8.5.

EXAMPLE 3 **Eigenvalues of Symmetric and Skew-Symmetric Matrices**

The matrices in (1) and (7) of Sec. 8.2 are symmetric and have real eigenvalues. The skew-symmetric matrix in Example 1 has the eigenvalues 0, $-25i$, and $25i$. (Verify this.) The following matrix has the real eigenvalues 1 and 5 but is not symmetric. Does this contradict Theorem 1?

$$\begin{bmatrix} 3 & 4 \\ 1 & 3 \end{bmatrix}$$ ■

Orthogonal Transformations and Orthogonal Matrices

Orthogonal transformations are transformations

$$\text{(5)} \qquad \mathbf{y} = \mathbf{A}\mathbf{x} \qquad \text{where } \mathbf{A} \text{ is an orthogonal matrix.}$$

With each vector $\mathbf{x}$ in R^n such a transformation assigns a vector $\mathbf{y}$ in R^n. For instance, the plane rotation through an angle θ

$$\text{(6)} \qquad \mathbf{y} = \begin{bmatrix} y_1 \\ y_2 \end{bmatrix} = \begin{bmatrix} \cos\theta & -\sin\theta \\ \sin\theta & \cos\theta \end{bmatrix} \begin{bmatrix} x_1 \\ x_2 \end{bmatrix}$$

is an orthogonal transformation. It can be shown that any orthogonal transformation in the plane or in three-dimensional space is a **rotation** (possibly combined with a reflection in a straight line or a plane, respectively).

The main reason for the importance of orthogonal matrices is as follows.

THEOREM 2 **Invariance of Inner Product**

An orthogonal transformation preserves the value of the **inner product** *of vectors* $\mathbf{a}$ *and* $\mathbf{b}$ *in* R^n, *defined by*

$$\text{(7)} \qquad \mathbf{a} \bullet \mathbf{b} = \mathbf{a}^{\mathsf{T}}\mathbf{b} = [a_1 \quad \cdots \quad a_n] \begin{bmatrix} b_1 \\ \vdots \\ b_n \end{bmatrix}.$$

That is, for any $\mathbf{a}$ *and* $\mathbf{b}$ *in* R^n, *orthogonal* $n \times n$ *matrix* $\mathbf{A}$, *and* $\mathbf{u} = \mathbf{A}\mathbf{a}, \mathbf{v} = \mathbf{A}\mathbf{b}$ *we have* $\mathbf{u} \bullet \mathbf{v} = \mathbf{a} \bullet \mathbf{b}$.

Hence the transformation also preserves the **length** *or* **norm** *of any vector* $\mathbf{a}$ *in* R^n *given by*

$$\text{(8)} \qquad \|\mathbf{a}\| = \sqrt{\mathbf{a} \bullet \mathbf{a}} = \sqrt{\mathbf{a}^{\mathsf{T}}\mathbf{a}}.$$

PROOF Let $\mathbf{A}$ be orthogonal. Let $\mathbf{u} = \mathbf{A}\mathbf{a}$ and $\mathbf{v} = \mathbf{A}\mathbf{b}$. We must show that $\mathbf{u} \bullet \mathbf{v} = \mathbf{a} \bullet \mathbf{b}$. Now $(\mathbf{A}\mathbf{a})^{\mathsf{T}} = \mathbf{a}^{\mathsf{T}}\mathbf{A}^{\mathsf{T}}$ by (10d) in Sec. 7.2 and $\mathbf{A}^{\mathsf{T}}\mathbf{A} = \mathbf{A}^{-1}\mathbf{A} = \mathbf{I}$ by (3). Hence

$$\text{(9)} \qquad \mathbf{u} \bullet \mathbf{v} = \mathbf{u}^{\mathsf{T}}\mathbf{v} = (\mathbf{A}\mathbf{a})^{\mathsf{T}}\mathbf{A}\mathbf{b} = \mathbf{a}^{\mathsf{T}}\mathbf{A}^{\mathsf{T}}\mathbf{A}\mathbf{b} = \mathbf{a}^{\mathsf{T}}\mathbf{I}\mathbf{b} = \mathbf{a}^{\mathsf{T}}\mathbf{b} = \mathbf{a} \bullet \mathbf{b}.$$

From this the invariance of $\|\mathbf{a}\|$ follows if we set $\mathbf{b} = \mathbf{a}$. ■

Orthogonal matrices have further interesting properties as follows.

THEOREM 3

Orthonormality of Column and Row Vectors

A real square matrix is orthogonal if and only if its column vectors $\mathbf{a}_1, \cdots, \mathbf{a}_n$ *(and also its row vectors) form an* **orthonormal system**, *that is,*

$$\mathbf{a}_j \bullet \mathbf{a}_k = \mathbf{a}_j^{\mathsf{T}}\mathbf{a}_k = \begin{cases} 0 & \text{if } j \neq k \\ 1 & \text{if } j = k. \end{cases} \tag{10}$$

PROOF **(a)** Let $\mathbf{A}$ be orthogonal. Then $\mathbf{A}^{-1}\mathbf{A} = \mathbf{A}^{\mathsf{T}}\mathbf{A} = \mathbf{I}$. In terms of column vectors $\mathbf{a}_1, \cdots, \mathbf{a}_n$,

$$\mathbf{I} = \mathbf{A}^{-1}\mathbf{A} = \mathbf{A}^{\mathsf{T}}\mathbf{A} = \begin{bmatrix} \mathbf{a}_1^{\mathsf{T}} \\ \vdots \\ \mathbf{a}_n^{\mathsf{T}} \end{bmatrix} [\mathbf{a}_1 \cdots \mathbf{a}_n] = \begin{bmatrix} \mathbf{a}_1^{\mathsf{T}}\mathbf{a}_1 & \mathbf{a}_1^{\mathsf{T}}\mathbf{a}_2 & \cdots & \mathbf{a}_1^{\mathsf{T}}\mathbf{a}_n \\ \cdot & \cdot & \cdots & \cdot \\ \mathbf{a}_n^{\mathsf{T}}\mathbf{a}_1 & \mathbf{a}_n^{\mathsf{T}}\mathbf{a}_2 & \cdots & \mathbf{a}_n^{\mathsf{T}}\mathbf{a}_n \end{bmatrix}. \tag{11}$$

The last equality implies (10), by the definition of the $n \times n$ unit matrix $\mathbf{I}$. From (3) it follows that the inverse of an orthogonal matrix is orthogonal (see CAS Experiment 12). Now the column vectors of $\mathbf{A}^{-1}(=\mathbf{A}^{\mathsf{T}})$ are the row vectors of $\mathbf{A}$. Hence the row vectors of $\mathbf{A}$ also form an orthonormal system.

(b) Conversely, if the column vectors of $\mathbf{A}$ satisfy (10), the off-diagonal entries in (11) must be 0 and the diagonal entries 1. Hence $\mathbf{A}^{\mathsf{T}}\mathbf{A} = \mathbf{I}$, as (11) shows. Similarly, $\mathbf{A}\mathbf{A}^{\mathsf{T}} = \mathbf{I}$. This implies $\mathbf{A}^{\mathsf{T}} = \mathbf{A}^{-1}$ because also $\mathbf{A}^{-1}\mathbf{A} = \mathbf{A}\mathbf{A}^{-1} = \mathbf{I}$ and the inverse is unique. Hence $\mathbf{A}$ is orthogonal. Similarly when the row vectors of $\mathbf{A}$ form an orthonormal system, by what has been said at the end of part **(a)**. ■

THEOREM 4

Determinant of an Orthogonal Matrix

The determinant of an orthogonal matrix has the value $+1$ *or* -1.

PROOF From $\det \mathbf{AB} = \det \mathbf{A} \det \mathbf{B}$ (Sec. 7.8, Theorem 4) and $\det \mathbf{A}^{\mathsf{T}} = \det \mathbf{A}$ (Sec. 7.7, Theorem 2d), we get for an orthogonal matrix

$$1 = \det \mathbf{I} = \det(\mathbf{A}\mathbf{A}^{-1}) = \det(\mathbf{A}\mathbf{A}^{\mathsf{T}}) = \det \mathbf{A} \det \mathbf{A}^{\mathsf{T}} = (\det \mathbf{A})^2.$$ ■

EXAMPLE 4 **Illustration of Theorems 3 and 4**

The last matrix in Example 1 and the matrix in (6) illustrate Theorems 3 and 4 because their determinants are -1 and $+1$, as you should verify. ■

THEOREM 5

Eigenvalues of an Orthogonal Matrix

The eigenvalues of an orthogonal matrix $\mathbf{A}$ *are real or complex conjugates in pairs and have absolute value* 1.

PROOF The first part of the statement holds for any real matrix **A** because its characteristic polynomial has real coefficients, so that its zeros (the eigenvalues of **A**) must be as indicated. The claim that $|\lambda| = 1$ will be proved in Sec. 8.5. ■

EXAMPLE 5 Eigenvalues of an Orthogonal Matrix

The orthogonal matrix in Example 1 has the characteristic equation

$$-\lambda^3 + \tfrac{2}{3}\lambda^2 + \tfrac{2}{3}\lambda - 1 = 0.$$

Now one of the eigenvalues must be real (why?), hence $+1$ or -1. Trying, we find -1. Division by $\lambda + 1$ gives $-(\lambda^2 - 5\lambda/3 + 1) = 0$ and the two eigenvalues $(5 + i\sqrt{11})/6$ and $(5 - i\sqrt{11})/6$, which have absolute value 1. Verify all of this. ■

Looking back at this section, you will find that the numerous basic results it contains have relatively short, straightforward proofs. This is typical of large portions of matrix eigenvalue theory.

PROBLEM SET 8.3

1–10 SPECTRUM

Are the following matrices symmetric, skew-symmetric, or orthogonal? Find the spectrum of each, thereby illustrating Theorems 1 and 5. Show your work in detail.

1. $\begin{bmatrix} 0.8 & 0.6 \\ -0.6 & 0.8 \end{bmatrix}$

2. $\begin{bmatrix} a & b \\ -b & a \end{bmatrix}$

3. $\begin{bmatrix} 2 & 8 \\ -8 & 2 \end{bmatrix}$

4. $\begin{bmatrix} \cos\theta & -\sin\theta \\ \sin\theta & \cos\theta \end{bmatrix}$

5. $\begin{bmatrix} 6 & 0 & 0 \\ 0 & 2 & -2 \\ 0 & -2 & 5 \end{bmatrix}$

6. $\begin{bmatrix} a & k & k \\ k & a & k \\ k & k & a \end{bmatrix}$

7. $\begin{bmatrix} 0 & 9 & -12 \\ -9 & 0 & 20 \\ 12 & -20 & 0 \end{bmatrix}$

8. $\begin{bmatrix} 1 & 0 & 0 \\ 0 & \cos\theta & -\sin\theta \\ 0 & \sin\theta & \cos\theta \end{bmatrix}$

9. $\begin{bmatrix} 0 & 0 & 1 \\ 0 & 1 & 0 \\ -1 & 0 & 0 \end{bmatrix}$

10. $\begin{bmatrix} \frac{4}{9} & \frac{8}{9} & \frac{1}{9} \\ -\frac{7}{9} & \frac{4}{9} & -\frac{4}{9} \\ -\frac{4}{9} & \frac{1}{9} & \frac{8}{9} \end{bmatrix}$

11. **WRITING PROJECT. Section Summary.** Summarize the main concepts and facts in this section, giving illustrative examples of your own.

12. **CAS EXPERIMENT. Orthogonal Matrices.**

    (a) Products. Inverse. Prove that the product of two orthogonal matrices is orthogonal, and so is the inverse of an orthogonal matrix. What does this mean in terms of rotations?

    (b) Rotation. Show that (6) is an orthogonal transformation. Verify that it satisfies Theorem 3. Find the inverse transformation.

    (c) Powers. Write a program for computing powers $\mathbf{A}^m$ $(m = 1, 2, \cdots)$ of a 2×2 matrix **A** and their spectra. Apply it to the matrix in Prob. 1 (call it **A**). To what rotation does **A** correspond? Do the eigenvalues of $\mathbf{A}^m$ have a limit as $m \to \infty$?

    (d) Compute the eigenvalues of $(0.9\mathbf{A})^m$, where **A** is the matrix in Prob. 1. Plot them as points. What is their limit? Along what kind of curve do these points approach the limit?

    (e) Find **A** such that $\mathbf{y} = \mathbf{Ax}$ is a counterclockwise rotation through 30° in the plane.

13–20 GENERAL PROPERTIES

13. **Verification.** Verify the statements in Example 1.
14. Verify the statements in Examples 3 and 4.
15. **Sum.** Are the eigenvalues of $\mathbf{A} + \mathbf{B}$ sums of the eigenvalues of **A** and of **B**?
16. **Orthogonality.** Prove that eigenvectors of a symmetric matrix corresponding to different eigenvalues are orthogonal. Give examples.
17. **Skew-symmetric matrix.** Show that the inverse of a skew-symmetric matrix is skew-symmetric.
18. Do there exist nonsingular skew-symmetric $n \times n$ matrices with odd n?
19. **Orthogonal matrix.** Do there exist skew-symmetric orthogonal 3×3 matrices?
20. **Symmetric matrix.** Do there exist nondiagonal symmetric 3×3 matrices that are orthogonal?

8.4 Eigenbases. Diagonalization. Quadratic Forms

So far we have emphasized properties of eigen*values*. We now turn to general properties of eigen*vectors*. Eigenvectors of an $n \times n$ matrix $\mathbf{A}$ may (or may not!) form a basis for R^n. If we are interested in a transformation $\mathbf{y} = \mathbf{Ax}$, such an "**eigenbasis**" (basis of eigenvectors)—if it exists—is of great advantage because then we can represent any $\mathbf{x}$ in R^n uniquely as a linear combination of the eigenvectors $\mathbf{x}_1, \cdots, \mathbf{x}_n$, say,

$$\mathbf{x} = c_1\mathbf{x}_1 + c_2\mathbf{x}_2 + \cdots + c_n\mathbf{x}_n.$$

And, denoting the corresponding (not necessarily distinct) eigenvalues of the matrix $\mathbf{A}$ by $\lambda_1, \cdots, \lambda_n$, we have $\mathbf{Ax}_j = \lambda_j\mathbf{x}_j$, so that we simply obtain

$$(1) \qquad \begin{aligned} \mathbf{y} = \mathbf{Ax} &= \mathbf{A}(c_1\mathbf{x}_1 + \cdots + c_n\mathbf{x}_n) \\ &= c_1\mathbf{Ax}_1 + \cdots + c_n\mathbf{Ax}_n \\ &= c_1\lambda_1\mathbf{x}_1 + \cdots + c_n\lambda_n\mathbf{x}_n. \end{aligned}$$

This shows that we have decomposed the complicated action of $\mathbf{A}$ on an arbitrary vector $\mathbf{x}$ into a sum of simple actions (multiplication by scalars) on the eigenvectors of $\mathbf{A}$. This is the point of an eigenbasis.

Now if the n eigenvalues are all different, we do obtain a basis:

THEOREM 1

Basis of Eigenvectors

*If an $n \times n$ matrix $\mathbf{A}$ has n **distinct** eigenvalues, then $\mathbf{A}$ has a basis of eigenvectors $\mathbf{x}_1, \cdots, \mathbf{x}_n$ for R^n.*

PROOF All we have to show is that $\mathbf{x}_1, \cdots, \mathbf{x}_n$ are linearly independent. Suppose they are not. Let r be the largest integer such that $\{\mathbf{x}_1, \cdots, \mathbf{x}_r\}$ is a linearly independent set. Then $r < n$ and the set $\{\mathbf{x}_1, \cdots, \mathbf{x}_r, \mathbf{x}_{r+1}\}$ is linearly dependent. Thus there are scalars $c_1, \cdots, c_{r+1}$, not all zero, such that

$$(2) \qquad c_1\mathbf{x}_1 + \cdots + c_{r+1}\mathbf{x}_{r+1} = \mathbf{0}$$

(see Sec. 7.4). Multiplying both sides by $\mathbf{A}$ and using $\mathbf{Ax}_j = \lambda_j\mathbf{x}_j$, we obtain

$$(3) \qquad \mathbf{A}(c_1\mathbf{x}_1 + \cdots + c_{r+1}\mathbf{x}_{r+1}) = c_1\lambda_1\mathbf{x}_1 + \cdots + c_{r+1}\lambda_{r+1}\mathbf{x}_{r+1} = \mathbf{A0} = \mathbf{0}.$$

To get rid of the last term, we subtract λ_{r+1} times (2) from this, obtaining

$$c_1(\lambda_1 - \lambda_{r+1})\mathbf{x}_1 + \cdots + c_r(\lambda_r - \lambda_{r+1})\mathbf{x}_r = \mathbf{0}.$$

Here $c_1(\lambda_1 - \lambda_{r+1}) = 0, \cdots, c_r(\lambda_r - \lambda_{r+1}) = 0$ since $\{x_1, \cdots, x_r\}$ is linearly independent. Hence $c_1 = \cdots = c_r = 0$, since all the eigenvalues are distinct. But with this, (2) reduces to $c_{r+1}\mathbf{x}_{r+1} = \mathbf{0}$, hence $c_{r+1} = 0$, since $\mathbf{x}_{r+1} \neq \mathbf{0}$ (an eigenvector!). This contradicts the fact that not all scalars in (2) are zero. Hence the conclusion of the theorem must hold. ■

EXAMPLE 1 Eigenbasis. Nondistinct Eigenvalues. Nonexistence

The matrix $\mathbf{A} = \begin{bmatrix} 5 & 3 \\ 3 & 5 \end{bmatrix}$ has a basis of eigenvectors $\begin{bmatrix} 1 \\ 1 \end{bmatrix}, \begin{bmatrix} 1 \\ -1 \end{bmatrix}$ corresponding to the eigenvalues $\lambda_1 = 8$, $\lambda_2 = 2$. (See Example 1 in Sec. 8.2.)

Even if not all n eigenvalues are different, a matrix $\mathbf{A}$ may still provide an eigenbasis for R^n. See Example 2 in Sec. 8.1, where $n = 3$.

On the other hand, $\mathbf{A}$ ***may not have enough linearly independent eigenvectors to make up a basis***. For instance, $\mathbf{A}$ in Example 3 of Sec. 8.1 is

$$\mathbf{A} = \begin{bmatrix} 0 & 1 \\ 0 & 0 \end{bmatrix} \qquad \text{and has only one eigenvector} \qquad \begin{bmatrix} k \\ 0 \end{bmatrix} \qquad (k \neq 0, \text{ arbitrary}).$$ ■

Actually, eigenbases exist under much more general conditions than those in Theorem 1. An important case is the following.

THEOREM 2 Symmetric Matrices

A symmetric matrix has an orthonormal basis of eigenvectors for R^n.

For a proof (which is involved) see Ref. [B3], vol. 1, pp. 270–272.

EXAMPLE 2 Orthonormal Basis of Eigenvectors

The first matrix in Example 1 is symmetric, and an orthonormal basis of eigenvectors is $[1/\sqrt{2} \quad 1/\sqrt{2}]^{\mathsf{T}}$, $[1/\sqrt{2} \quad -1/\sqrt{2}]^{\mathsf{T}}$. ■

Similarity of Matrices. Diagonalization

Eigenbases also play a role in reducing a matrix $\mathbf{A}$ to a diagonal matrix whose entries are the eigenvalues of $\mathbf{A}$. This is done by a "similarity transformation," which is defined as follows (and will have various applications in numerics in Chap. 20).

DEFINITION Similar Matrices. Similarity Transformation

An $n \times n$ matrix $\hat{\mathbf{A}}$ is called **similar** to an $n \times n$ matrix $\mathbf{A}$ if

$$\hat{\mathbf{A}} = \mathbf{P}^{-1}\mathbf{A}\mathbf{P} \tag{4}$$

for some (nonsingular!) $n \times n$ matrix $\mathbf{P}$. This transformation, which gives $\hat{\mathbf{A}}$ from $\mathbf{A}$, is called a **similarity transformation**.

The key property of this transformation is that it preserves the eigenvalues of $\mathbf{A}$:

THEOREM 3 Eigenvalues and Eigenvectors of Similar Matrices

If $\hat{\mathbf{A}}$ is similar to $\mathbf{A}$, then $\hat{\mathbf{A}}$ has the same eigenvalues as $\mathbf{A}$.

Furthermore, if $\mathbf{x}$ is an eigenvector of $\mathbf{A}$, then $\mathbf{y} = \mathbf{P}^{-1}\mathbf{x}$ is an eigenvector of $\hat{\mathbf{A}}$ corresponding to the same eigenvalue.

PROOF From $\mathbf{Ax} = \lambda\mathbf{x}$ (λ an eigenvalue, $\mathbf{x} \neq \mathbf{0}$) we get $\mathbf{P}^{-1}\mathbf{Ax} = \lambda\mathbf{P}^{-1}\mathbf{x}$. Now $\mathbf{I} = \mathbf{PP}^{-1}$. By this ***identity trick*** the equation $\mathbf{P}^{-1}\mathbf{Ax} = \lambda\mathbf{P}^{-1}\mathbf{x}$ gives

$$\mathbf{P}^{-1}\mathbf{Ax} = \mathbf{P}^{-1}\mathbf{AIx} = \mathbf{P}^{-1}\mathbf{APP}^{-1}\mathbf{x} = (\mathbf{P}^{-1}\mathbf{AP})\mathbf{P}^{-1}\mathbf{x} = \hat{\mathbf{A}}(\mathbf{P}^{-1}\mathbf{x}) = \lambda\mathbf{P}^{-1}\mathbf{x}.$$

Hence λ is an eigenvalue of $\hat{\mathbf{A}}$ and $\mathbf{P}^{-1}\mathbf{x}$ a corresponding eigenvector. Indeed, $\mathbf{P}^{-1}\mathbf{x} \neq \mathbf{0}$ because $\mathbf{P}^{-1}\mathbf{x} = \mathbf{0}$ would give $\mathbf{x} = \mathbf{Ix} = \mathbf{PP}^{-1}\mathbf{x} = \mathbf{P0} = \mathbf{0}$, contradicting $\mathbf{x} \neq \mathbf{0}$. ■

EXAMPLE 3 **Eigenvalues and Vectors of Similar Matrices**

Let,

$$\mathbf{A} = \begin{bmatrix} 6 & -3 \\ 4 & -1 \end{bmatrix} \quad \text{and} \quad \mathbf{P} = \begin{bmatrix} 1 & 3 \\ 1 & 4 \end{bmatrix}.$$

Then

$$\hat{\mathbf{A}} = \begin{bmatrix} 4 & -3 \\ -1 & 1 \end{bmatrix}\begin{bmatrix} 6 & -3 \\ 4 & -1 \end{bmatrix}\begin{bmatrix} 1 & 3 \\ 1 & 4 \end{bmatrix} = \begin{bmatrix} 3 & 0 \\ 0 & 2 \end{bmatrix}.$$

Here $\mathbf{P}^{-1}$ was obtained from (4*) in Sec. 7.8 with $\det \mathbf{P} = 1$. We see that $\hat{\mathbf{A}}$ has the eigenvalues $\lambda_1 = 3$, $\lambda_2 = 2$. The characteristic equation of $\mathbf{A}$ is $(6 - \lambda)(-1 - \lambda) + 12 = \lambda^2 - 5\lambda + 6 = 0$. It has the roots (the eigenvalues of $\mathbf{A}$) $\lambda_1 = 3$, $\lambda_2 = 2$, confirming the first part of Theorem 3.

We confirm the second part. From the first component of $(\mathbf{A} - \lambda\mathbf{I})\mathbf{x} = \mathbf{0}$ we have $(6 - \lambda)x_1 - 3x_2 = 0$. For $\lambda = 3$ this gives $3x_1 - 3x_2 = 0$, say, $\mathbf{x}_1 = [1 \quad 1]^\mathsf{T}$. For $\lambda = 2$ it gives $4x_1 - 3x_2 = 0$, say, $\mathbf{x}_2 = [3 \quad 4]^\mathsf{T}$. In Theorem 3 we thus have

$$\mathbf{y}_1 = \mathbf{P}^{-1}\mathbf{x}_1 = \begin{bmatrix} 4 & -3 \\ -1 & 1 \end{bmatrix}\begin{bmatrix} 1 \\ 1 \end{bmatrix} = \begin{bmatrix} 1 \\ 0 \end{bmatrix}, \qquad \mathbf{y}_2 = \mathbf{P}^{-1}\mathbf{x}_2 = \begin{bmatrix} 4 & -3 \\ -1 & 1 \end{bmatrix}\begin{bmatrix} 3 \\ 4 \end{bmatrix} = \begin{bmatrix} 0 \\ 1 \end{bmatrix}.$$

Indeed, these are eigenvectors of the diagonal matrix $\hat{\mathbf{A}}$.

Perhaps we see that $\mathbf{x}_1$ and $\mathbf{x}_2$ are the columns of $\mathbf{P}$. This suggests the general method of transforming a matrix $\mathbf{A}$ to diagonal form $\mathbf{D}$ by using $\mathbf{P} = \mathbf{X}$, the matrix with eigenvectors as columns. ■

By a suitable similarity transformation we can now transform a matrix $\mathbf{A}$ to a diagonal matrix $\mathbf{D}$ whose diagonal entries are the eigenvalues of $\mathbf{A}$:

THEOREM 4 **Diagonalization of a Matrix**

If an $n \times n$ matrix $\mathbf{A}$ *has a basis of eigenvectors, then*

$$\mathbf{D} = \mathbf{X}^{-1}\mathbf{AX} \tag{5}$$

is diagonal, with the eigenvalues of $\mathbf{A}$ *as the entries on the main diagonal. Here* $\mathbf{X}$ *is the matrix with these eigenvectors as column vectors. Also,*

$$\mathbf{D}^m = \mathbf{X}^{-1}\mathbf{A}^m\mathbf{X} \qquad (m = 2, 3, \cdots). \tag{5*}$$

PROOF Let $\mathbf{x}_1, \cdots, \mathbf{x}_n$ be a basis of eigenvectors of $\mathbf{A}$ for R^n. Let the corresponding eigenvalues of $\mathbf{A}$ be $\lambda_1, \cdots, \lambda_n$, respectively, so that $\mathbf{A}\mathbf{x}_1 = \lambda_1\mathbf{x}_1, \cdots, \mathbf{A}\mathbf{x}_n = \lambda_n\mathbf{x}_n$. Then $\mathbf{X} = [\mathbf{x}_1 \cdots \mathbf{x}_n]$ has rank n, by Theorem 3 in Sec. 7.4. Hence $\mathbf{X}^{-1}$ exists by Theorem 1 in Sec. 7.8. We claim that

$$\mathbf{A}\mathbf{X} = \mathbf{A}[\mathbf{x}_1 \quad \cdots \quad \mathbf{x}_n] = [\mathbf{A}\mathbf{x}_1 \quad \cdots \quad \mathbf{A}\mathbf{x}_n] = [\lambda_1\mathbf{x}_1 \quad \cdots \quad \lambda_n\mathbf{x}_n] = \mathbf{X}\mathbf{D} \tag{6}$$

where $\mathbf{D}$ is the diagonal matrix as in (5). The fourth equality in (6) follows by direct calculation. (Try it for $n = 2$ and then for general n.) The third equality uses $\mathbf{A}\mathbf{x}_k = \lambda_k\mathbf{x}_k$. The second equality results if we note that the first column of $\mathbf{AX}$ is $\mathbf{A}$ times the first column of $\mathbf{X}$, which is $\mathbf{x}_1$, and so on. For instance, when $n = 2$ and we write $\mathbf{x}_1 = [x_{11} \quad x_{21}]$, $\mathbf{x}_2 = [x_{12} \quad x_{22}]$, we have

$$\mathbf{AX} = \mathbf{A}[\mathbf{x}_1 \quad \mathbf{x}_2] = \begin{bmatrix} a_{11} & a_{12} \\ a_{21} & a_{22} \end{bmatrix} \begin{bmatrix} x_{11} & x_{12} \\ x_{21} & x_{22} \end{bmatrix}$$

$$= \begin{bmatrix} a_{11}x_{11} + a_{12}x_{21} & a_{11}x_{12} + a_{12}x_{22} \\ a_{21}x_{11} + a_{22}x_{21} & a_{21}x_{12} + a_{22}x_{22} \end{bmatrix} = [\mathbf{A}\mathbf{x}_1 \quad \mathbf{A}\mathbf{x}_2].$$

Column 1 Column 2

If we multiply (6) by $\mathbf{X}^{-1}$ from the left, we obtain (5). Since (5) is a similarity transformation, Theorem 3 implies that $\mathbf{D}$ has the same eigenvalues as $\mathbf{A}$. Equation (5*) follows if we note that

$$\mathbf{D}^2 = \mathbf{DD} = (\mathbf{X}^{-1}\mathbf{AX})(\mathbf{X}^{-1}\mathbf{AX}) = \mathbf{X}^{-1}\mathbf{A}(\mathbf{XX}^{-1})\mathbf{AX} = \mathbf{X}^{-1}\mathbf{AAX} = \mathbf{X}^{-1}\mathbf{A}^2\mathbf{X}, \quad \text{etc.} \blacksquare$$

EXAMPLE 4 **Diagonalization**

Diagonalize

$$\mathbf{A} = \begin{bmatrix} 7.3 & 0.2 & -3.7 \\ -11.5 & 1.0 & 5.5 \\ 17.7 & 1.8 & -9.3 \end{bmatrix}.$$

Solution. The characteristic determinant gives the characteristic equation $-\lambda^3 - \lambda^2 + 12\lambda = 0$. The roots (eigenvalues of $\mathbf{A}$) are $\lambda_1 = 3$, $\lambda_2 = -4$, $\lambda_3 = 0$. By the Gauss elimination applied to $(\mathbf{A} - \lambda\mathbf{I})\mathbf{x} = \mathbf{0}$ with $\lambda = \lambda_1, \lambda_2, \lambda_3$ we find eigenvectors and then $\mathbf{X}^{-1}$ by the Gauss–Jordan elimination (Sec. 7.8, Example 1). The results are

$$\begin{bmatrix} -1 \\ 3 \\ -1 \end{bmatrix}, \quad \begin{bmatrix} 1 \\ -1 \\ 3 \end{bmatrix}, \quad \begin{bmatrix} 2 \\ 1 \\ 4 \end{bmatrix}, \qquad \mathbf{X} = \begin{bmatrix} -1 & 1 & 2 \\ 3 & -1 & 1 \\ -1 & 3 & 4 \end{bmatrix}, \qquad \mathbf{X}^{-1} = \begin{bmatrix} -0.7 & 0.2 & 0.3 \\ -1.3 & -0.2 & 0.7 \\ 0.8 & 0.2 & -0.2 \end{bmatrix}.$$

Calculating $\mathbf{AX}$ and multiplying by $\mathbf{X}^{-1}$ from the left, we thus obtain

$$\mathbf{D} = \mathbf{X}^{-1}\mathbf{AX} = \begin{bmatrix} -0.7 & 0.2 & 0.3 \\ -1.3 & -0.2 & 0.7 \\ 0.8 & 0.2 & -0.2 \end{bmatrix} \begin{bmatrix} -3 & -4 & 0 \\ 9 & 4 & 0 \\ -3 & -12 & 0 \end{bmatrix} = \begin{bmatrix} 3 & 0 & 0 \\ 0 & -4 & 0 \\ 0 & 0 & 0 \end{bmatrix}. \qquad \blacksquare$$

Quadratic Forms. Transformation to Principal Axes

By definition, a **quadratic form** Q in the components $x_1, \cdots, x_n$ of a vector $\mathbf{x}$ is a sum of n^2 terms, namely,

$$
\begin{aligned}
Q = \mathbf{x}^\mathsf{T}\mathbf{A}\mathbf{x} &= \sum_{j=1}^{n}\sum_{k=1}^{n} a_{jk}x_j x_k \\
&= \quad a_{11}x_1^2 + a_{12}x_1x_2 + \cdots + a_{1n}x_1x_n \\
&\quad + a_{21}x_2x_1 + a_{22}x_2^2 + \cdots + a_{2n}x_2x_n \\
&\quad + \cdots\cdots\cdots\cdots\cdots\cdots\cdots\cdots \\
&\quad + a_{n1}x_nx_1 + a_{n2}x_nx_2 + \cdots + a_{nn}x_n^2.
\end{aligned} \tag{7}
$$

$\mathbf{A} = [a_{jk}]$ is called the **coefficient matrix** of the form. We may assume that $\mathbf{A}$ is ***symmetric***, because we can take off-diagonal terms together in pairs and write the result as a sum of two equal terms; see the following example.

EXAMPLE 5 Quadratic Form. Symmetric Coefficient Matrix

Let

$$\mathbf{x}^\mathsf{T}\mathbf{A}\mathbf{x} = [x_1 \quad x_2]\begin{bmatrix} 3 & 4 \\ 6 & 2 \end{bmatrix}\begin{bmatrix} x_1 \\ x_2 \end{bmatrix} = 3x_1^2 + 4x_1x_2 + 6x_2x_1 + 2x_2^2 = 3x_1^2 + 10x_1x_2 + 2x_2^2.$$

Here $4 + 6 = 10 = 5 + 5$. From the corresponding ***symmetric*** matrix $\mathbf{C} = [c_{jk}]$, where $c_{jk} = \frac{1}{2}(a_{jk} + a_{kj})$, thus $c_{11} = 3$, $c_{12} = c_{21} = 5$, $c_{22} = 2$, we get the same result; indeed,

$$\mathbf{x}^\mathsf{T}\mathbf{C}\mathbf{x} = [x_1 \quad x_2]\begin{bmatrix} 3 & 5 \\ 5 & 2 \end{bmatrix}\begin{bmatrix} x_1 \\ x_2 \end{bmatrix} = 3x_1^2 + 5x_1x_2 + 5x_2x_1 + 2x_2^2 = 3x_1^2 + 10x_1x_2 + 2x_2^2. \qquad \blacksquare$$

Quadratic forms occur in physics and geometry, for instance, in connection with conic sections (ellipses $x_1^2/a^2 + x_2^2/b^2 = 1$, etc.) and quadratic surfaces (cones, etc.). Their transformation to principal axes is an important practical task related to the diagonalization of matrices, as follows.

By Theorem 2, the *symmetric* coefficient matrix $\mathbf{A}$ of (7) has an orthonormal basis of eigenvectors. Hence if we take these as column vectors, we obtain a matrix $\mathbf{X}$ that is orthogonal, so that $\mathbf{X}^{-1} = \mathbf{X}^\mathsf{T}$. From (5) we thus have $\mathbf{A} = \mathbf{X}\mathbf{D}\mathbf{X}^{-1} = \mathbf{X}\mathbf{D}\mathbf{X}^\mathsf{T}$. Substitution into (7) gives

$$Q = \mathbf{x}^\mathsf{T}\mathbf{X}\mathbf{D}\mathbf{X}^\mathsf{T}\mathbf{x}. \tag{8}$$

If we set $\mathbf{X}^\mathsf{T}\mathbf{x} = \mathbf{y}$, then, since $\mathbf{X}^\mathsf{T} = \mathbf{X}^{-1}$, we have $\mathbf{X}^{-1}\mathbf{x} = \mathbf{y}$ and thus obtain

$$\mathbf{x} = \mathbf{X}\mathbf{y}. \tag{9}$$

Furthermore, in (8) we have $\mathbf{x}^\mathsf{T}\mathbf{X} = (\mathbf{X}^\mathsf{T}\mathbf{x})^\mathsf{T} = \mathbf{y}^\mathsf{T}$ and $\mathbf{X}^\mathsf{T}\mathbf{x} = \mathbf{y}$, so that Q becomes simply

$$Q = \mathbf{y}^\mathsf{T}\mathbf{D}\mathbf{y} = \lambda_1 y_1^2 + \lambda_2 y_2^2 + \cdots + \lambda_n y_n^2. \tag{10}$$

This proves the following basic theorem.

THEOREM 5

Principal Axes Theorem

The substitution (9) *transforms a quadratic form*

$$Q = \mathbf{x}^{\mathsf{T}}\mathbf{A}\mathbf{x} = \sum_{j=1}^{n}\sum_{k=1}^{n} a_{jk}x_jx_k \qquad (a_{kj} = a_{jk})$$

to the principal axes form or **canonical form** (10), *where* $\lambda_1, \cdots, \lambda_n$ *are the* (*not necessarily distinct*) *eigenvalues of the* (*symmetric!*) *matrix* **A**, *and* **X** *is an orthogonal matrix with corresponding eigenvectors* $\mathbf{x}_1, \cdots, \mathbf{x}_n$, *respectively, as column vectors.*

EXAMPLE 6 **Transformation to Principal Axes. Conic Sections**

Find out what type of conic section the following quadratic form represents and transform it to principal axes:

$$Q = 17x_1^2 - 30x_1x_2 + 17x_2^2 = 128.$$

Solution. We have $Q = \mathbf{x}^{\mathsf{T}}\mathbf{A}\mathbf{x}$, where

$$\mathbf{A} = \begin{bmatrix} 17 & -15 \\ -15 & 17 \end{bmatrix}, \qquad \mathbf{x} = \begin{bmatrix} x_1 \\ x_2 \end{bmatrix}.$$

This gives the characteristic equation $(17 - \lambda)^2 - 15^2 = 0$. It has the roots $\lambda_1 = 2$, $\lambda_2 = 32$. Hence (10) becomes

$$Q = 2y_1^2 + 32y_2^2.$$

We see that $Q = 128$ represents the ellipse $2y_1^2 + 32y_2^2 = 128$, that is,

$$\frac{y_1^2}{8^2} + \frac{y_2^2}{2^2} = 1.$$

If we want to know the direction of the principal axes in the x_1x_2-coordinates, we have to determine normalized eigenvectors from $(\mathbf{A} - \lambda\mathbf{I})\mathbf{x} = \mathbf{0}$ with $\lambda = \lambda_1 = 2$ and $\lambda = \lambda_2 = 32$ and then use (9). We get

$$\begin{bmatrix} 1/\sqrt{2} \\ 1/\sqrt{2} \end{bmatrix} \quad \text{and} \quad \begin{bmatrix} -1/\sqrt{2} \\ 1/\sqrt{2} \end{bmatrix},$$

hence

$$\mathbf{x} = \mathbf{X}\mathbf{y} = \begin{bmatrix} 1/\sqrt{2} & -1/\sqrt{2} \\ 1/\sqrt{2} & 1/\sqrt{2} \end{bmatrix} \begin{bmatrix} y_1 \\ y_2 \end{bmatrix}, \qquad \begin{aligned} x_1 &= y_1/\sqrt{2} - y_2/\sqrt{2} \\ x_2 &= y_1/\sqrt{2} + y_2/\sqrt{2}. \end{aligned}$$

This is a 45° rotation. Our results agree with those in Sec. 8.2, Example 1, except for the notations. See also Fig. 160 in that example. ■

PROBLEM SET 8.4

1–5 SIMILAR MATRICES HAVE EQUAL EIGENVALUES

Verify this for $\mathbf{A}$ and $\hat{\mathbf{A}} = \mathbf{P}^{-1}\mathbf{AP}$. If $\mathbf{y}$ is an eigenvector of $\mathbf{P}$, show that $\mathbf{x} = \mathbf{Py}$ are eigenvectors of $\mathbf{A}$. Show the details of your work.

1. $\mathbf{A} = \begin{bmatrix} 3 & 4 \\ 4 & -3 \end{bmatrix}, \quad \mathbf{P} = \begin{bmatrix} -4 & 2 \\ 3 & -1 \end{bmatrix}$

2. $\mathbf{A} = \begin{bmatrix} 1 & 0 \\ 2 & -1 \end{bmatrix}, \quad \mathbf{P} = \begin{bmatrix} 7 & -5 \\ 10 & -7 \end{bmatrix}$

3. $\mathbf{A} = \begin{bmatrix} 8 & -4 \\ 2 & 2 \end{bmatrix}, \quad \mathbf{P} = \begin{bmatrix} 0.28 & 0.96 \\ -0.96 & 0.28 \end{bmatrix}$

4. $\mathbf{A} = \begin{bmatrix} 0 & 0 & 2 \\ 0 & 3 & 2 \\ 1 & 0 & 1 \end{bmatrix}, \quad \mathbf{P} = \begin{bmatrix} 2 & 0 & 3 \\ 0 & 1 & 0 \\ 3 & 0 & 5 \end{bmatrix},$
$\lambda_1 = 3$

5. $\mathbf{A} = \begin{bmatrix} -5 & 0 & 15 \\ 3 & 4 & -9 \\ -5 & 0 & 15 \end{bmatrix}, \quad \mathbf{P} = \begin{bmatrix} 0 & 1 & 0 \\ 1 & 0 & 0 \\ 0 & 0 & 1 \end{bmatrix}$

6. PROJECT. Similarity of Matrices. Similarity is basic, for instance, in designing numeric methods.

(a) Trace. By definition, the **trace** of an $n \times n$ matrix $\mathbf{A} = [a_{jk}]$ is the sum of the diagonal entries,

$$\text{trace } \mathbf{A} = a_{11} + a_{22} + \cdots + a_{nn}.$$

Show that the trace equals the sum of the eigenvalues, each counted as often as its algebraic multiplicity indicates. Illustrate this with the matrices $\mathbf{A}$ in Probs. 1, 3, and 5.

(b) Trace of product. Let $\mathbf{B} = [b_{jk}]$ be $n \times n$. Show that similar matrices have equal traces, by first proving

$$\text{trace } \mathbf{AB} = \sum_{i=1}^{n} \sum_{l=1}^{n} a_{il}b_{li} = \text{trace } \mathbf{BA}.$$

(c) Find a relationship between $\hat{\mathbf{A}}$ in (4) and $\hat{\mathbf{A}} = \mathbf{PAP}^{-1}$.

(d) Diagonalization. What can you do in (5) if you want to change the order of the eigenvalues in $\mathbf{D}$, for instance, interchange $d_{11} = \lambda_1$ and $d_{22} = \lambda_2$?

7. No basis. Find further 2×2 and 3×3 matrices without eigenbasis.

8. Orthonormal basis. Illustrate Theorem 2 with further examples.

9–16 DIAGONALIZATION OF MATRICES

Find an eigenbasis (a basis of eigenvectors) and diagonalize. Show the details.

9. $\begin{bmatrix} 1 & 2 \\ 2 & 4 \end{bmatrix}$

10. $\begin{bmatrix} 1 & 0 \\ 2 & -1 \end{bmatrix}$

11. $\begin{bmatrix} -19 & 7 \\ -42 & 16 \end{bmatrix}$

12. $\begin{bmatrix} -4.3 & 7.7 \\ 1.3 & 9.3 \end{bmatrix}$

13. $\begin{bmatrix} 4 & 0 & 0 \\ 12 & -2 & 0 \\ 21 & -6 & 1 \end{bmatrix}$

14. $\begin{bmatrix} -5 & -6 & 6 \\ -9 & -8 & 12 \\ -12 & -12 & 16 \end{bmatrix}, \quad \lambda_1 = -2$

15. $\begin{bmatrix} 4 & 3 & 3 \\ 3 & 6 & 1 \\ 3 & 1 & 6 \end{bmatrix}, \quad \lambda_1 = 10$

16. $\begin{bmatrix} 1 & 1 & 0 \\ 1 & 1 & 0 \\ 0 & 0 & -4 \end{bmatrix}$

17–23 PRINCIPAL AXES. CONIC SECTIONS

What kind of conic section (or pair of straight lines) is given by the quadratic form? Transform it to principal axes. Express $\mathbf{x}^\mathsf{T} = [x_1 \;\; x_2]$ in terms of the new coordinate vector $\mathbf{y}^\mathsf{T} = [y_1 \;\; y_2]$, as in Example 6.

17. $7x_1^2 + 6x_1x_2 + 7x_2^2 = 200$

18. $3x_1^2 + 8x_1x_2 - 3x_2^2 = 10$

19. $3x_1^2 + 22x_1x_2 + 3x_2^2 = 0$

20. $9x_1^2 + 6x_1x_2 + x_2^2 = 10$

21. $x_1^2 - 12x_1x_2 + x_2^2 = 70$

22. $4x_1^2 + 12x_1x_2 + 13x_2^2 = 16$

23. $-11x_1^2 + 84x_1x_2 + 24x_2^2 = 156$

24. Definiteness. A quadratic form $Q(\mathbf{x}) = \mathbf{x}^\mathsf{T}\mathbf{A}\mathbf{x}$ and its (symmetric!) matrix $\mathbf{A}$ are called **(a) positive definite** if $Q(\mathbf{x}) > 0$ for all $\mathbf{x} \neq \mathbf{0}$, **(b) negative definite** if $Q(\mathbf{x}) < 0$ for all $\mathbf{x} \neq \mathbf{0}$, **(c) indefinite** if $Q(\mathbf{x})$ takes both positive and negative values. (See Fig. 162.) [$Q(\mathbf{x})$ and $\mathbf{A}$ are called *positive semidefinite* (*negative semidefinite*) if $Q(\mathbf{x}) \geqq 0$ ($Q(\mathbf{x}) \leqq 0$) for all $\mathbf{x}$.] Show that a necessary and sufficient condition for (a), (b), and (c) is that the eigenvalues of $\mathbf{A}$ are (a) all positive, (b) all negative, and (c) both positive and negative.

Hint. Use Theorem 5.

25. Definiteness. A necessary and sufficient condition for positive definiteness of a quadratic form $Q(\mathbf{x}) = \mathbf{x}^\mathsf{T}\mathbf{A}\mathbf{x}$ with *symmetric* matrix $\mathbf{A}$ is that all the **principal minors** are positive (see Ref. [B3], vol. 1, p. 306), that is,

$$a_{11} > 0, \qquad \begin{vmatrix} a_{11} & a_{12} \\ a_{12} & a_{22} \end{vmatrix} > 0,$$

$$\begin{vmatrix} a_{11} & a_{12} & a_{13} \\ a_{12} & a_{22} & a_{23} \\ a_{13} & a_{23} & a_{33} \end{vmatrix} > 0, \qquad \cdots, \qquad \det \mathbf{A} > 0.$$

Show that the form in Prob. 22 is positive definite, whereas that in Prob. 23 is indefinite.

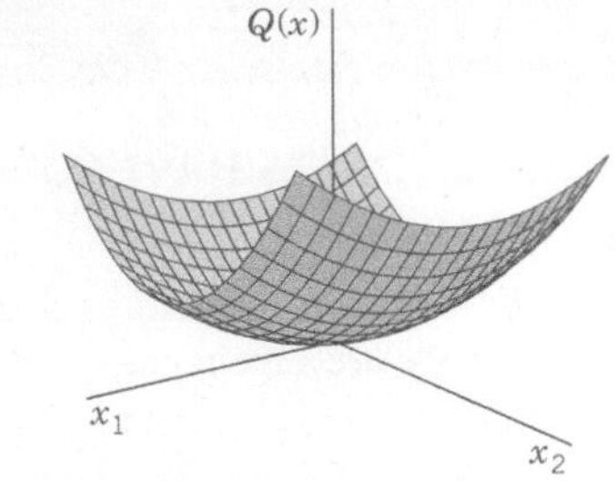

(a) Positive definite form

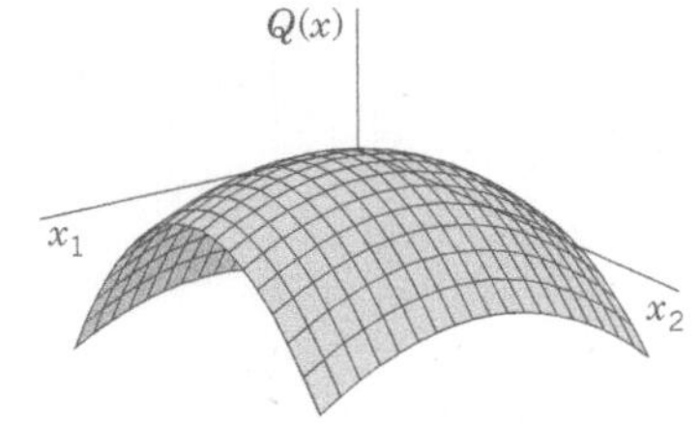

(b) Negative definite form

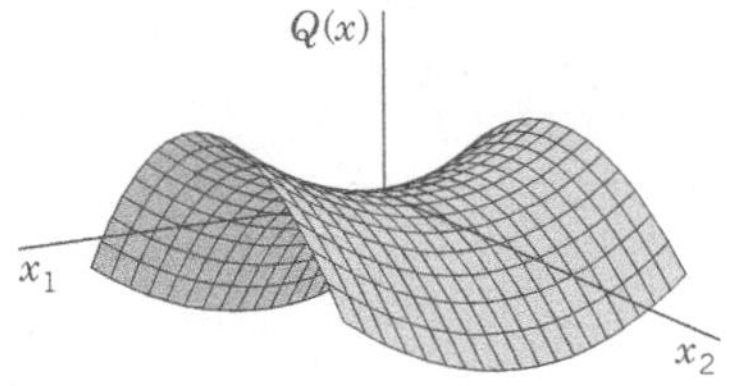

(c) Indefinite form

Fig. 162. Quadratic forms in two variables (Problem 24)

8.5 Complex Matrices and Forms. *Optional*

The three classes of matrices in Sec. 8.3 have complex counterparts which are of practical interest in certain applications, for instance, in quantum mechanics. This is mainly because of their spectra as shown in Theorem 1 in this section. The second topic is about extending quadratic forms of Sec. 8.4 to complex numbers. (The reader who wants to brush up on complex numbers may want to consult Sec. 13.1.)

> **Notations**
>
> $\overline{\mathbf{A}} = [\overline{a}_{jk}]$ is obtained from $\mathbf{A} = [a_{jk}]$ by replacing each entry $a_{jk} = \alpha + i\beta$ (α, β real) with its complex conjugate $\overline{a}_{jk} = \alpha - i\beta$. Also, $\overline{\mathbf{A}}^\mathsf{T} = [\overline{a}_{kj}]$ is the transpose of $\overline{\mathbf{A}}$, hence the conjugate transpose of $\mathbf{A}$.

EXAMPLE 1 Notations

If $\mathbf{A} = \begin{bmatrix} 3+4i & 1-i \\ 6 & 2-5i \end{bmatrix}$, then $\overline{\mathbf{A}} = \begin{bmatrix} 3-4i & 1+i \\ 6 & 2+5i \end{bmatrix}$ and $\overline{\mathbf{A}}^\mathsf{T} = \begin{bmatrix} 3-4i & 6 \\ 1+i & 2+5i \end{bmatrix}$. ■

Hermitian, Skew-Hermitian, and Unitary Matrices

A square matrix $\mathbf{A} = [a_{kj}]$ is called

Hermitian	if	$\overline{\mathbf{A}}^{\mathsf{T}} = \mathbf{A}$,	that is,	$\overline{a}_{kj} = a_{jk}$
skew-Hermitian	if	$\overline{\mathbf{A}}^{\mathsf{T}} = -\mathbf{A}$,	that is,	$\overline{a}_{kj} = -a_{jk}$
unitary	if	$\overline{\mathbf{A}}^{\mathsf{T}} = \mathbf{A}^{-1}$.		

The first two classes are named after Hermite (see footnote 13 in Problem Set 5.8).

From the definitions we see the following. If $\mathbf{A}$ is Hermitian, the entries on the main diagonal must satisfy $\overline{a}_{jj} = a_{jj}$; that is, they are real. Similarly, if $\mathbf{A}$ is skew-Hermitian, then $\overline{a}_{jj} = -a_{jj}$. If we set $a_{jj} = \alpha + i\beta$, this becomes $\alpha - i\beta = -(\alpha + i\beta)$. Hence $\alpha = 0$, so that a_{jj} must be pure imaginary or 0.

EXAMPLE 2   **Hermitian, Skew-Hermitian, and Unitary Matrices**

$$\mathbf{A} = \begin{bmatrix} 4 & 1 - 3i \\ 1 + 3i & 7 \end{bmatrix} \qquad \mathbf{B} = \begin{bmatrix} 3i & 2 + i \\ -2 + i & -i \end{bmatrix} \qquad \mathbf{C} = \begin{bmatrix} \frac{1}{2}i & \frac{1}{2}\sqrt{3} \\ \frac{1}{2}\sqrt{3} & \frac{1}{2}i \end{bmatrix}$$

are Hermitian, skew-Hermitian, and unitary matrices, respectively, as you may verify by using the definitions. ■

If a Hermitian matrix is real, then $\overline{\mathbf{A}}^{\mathsf{T}} = \mathbf{A}^{\mathsf{T}} = \mathbf{A}$. Hence a real Hermitian matrix is a symmetric matrix (Sec. 8.3).

Similarly, if a skew-Hermitian matrix is real, then $\overline{\mathbf{A}}^{\mathsf{T}} = \mathbf{A}^{\mathsf{T}} = -\mathbf{A}$. Hence a real skew-Hermitian matrix is a skew-symmetric matrix.

Finally, if a unitary matrix is real, then $\overline{\mathbf{A}}^{\mathsf{T}} = \mathbf{A}^{\mathsf{T}} = \mathbf{A}^{-1}$. Hence a real unitary matrix is an orthogonal matrix.

This shows that *Hermitian, skew-Hermitian, and unitary matrices generalize symmetric, skew-symmetric, and orthogonal matrices, respectively.*

Eigenvalues

It is quite remarkable that the matrices under consideration have spectra (sets of eigenvalues; see Sec. 8.1) that can be characterized in a general way as follows (see Fig. 163).

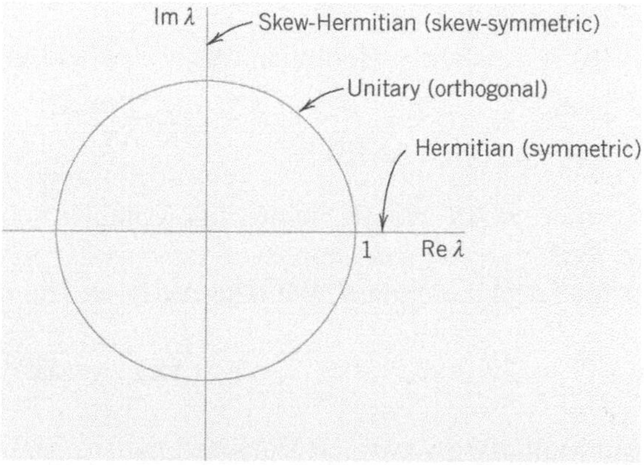

Fig. 163.   Location of the eigenvalues of Hermitian, skew-Hermitian, and unitary matrices in the complex λ-plane

THEOREM 1

Eigenvalues

(a) *The eigenvalues of a Hermitian matrix (and thus of a symmetric matrix) are real.*

(b) *The eigenvalues of a skew-Hermitian matrix (and thus of a skew-symmetric matrix) are pure imaginary or zero.*

(c) *The eigenvalues of a unitary matrix (and thus of an orthogonal matrix) have absolute value* 1.

EXAMPLE 3 **Illustration of Theorem 1**

For the matrices in Example 2 we find by direct calculation

Matrix		Characteristic Equation	Eigenvalues
A	Hermitian	$\lambda^2 - 11\lambda + 18 = 0$	$9, \quad 2$
B	Skew-Hermitian	$\lambda^2 - 2i\lambda + 8 = 0$	$4i, \quad -2i$
C	Unitary	$\lambda^2 - i\lambda - 1 = 0$	$\frac{1}{2}\sqrt{3} + \frac{1}{2}i, \quad -\frac{1}{2}\sqrt{3} + \frac{1}{2}i$

and $|\pm\frac{1}{2}\sqrt{3} + \frac{1}{2}i|^2 = \frac{3}{4} + \frac{1}{4} = 1.$ ■

PROOF We prove Theorem 1. Let λ be an eigenvalue and $\mathbf{x}$ an eigenvector of $\mathbf{A}$. Multiply $\mathbf{Ax} = \lambda\mathbf{x}$ from the left by $\overline{\mathbf{x}}^{\mathsf{T}}$, thus $\overline{\mathbf{x}}^{\mathsf{T}}\mathbf{Ax} = \lambda\overline{\mathbf{x}}^{\mathsf{T}}\mathbf{x}$, and divide by $\overline{\mathbf{x}}^{\mathsf{T}}\mathbf{x} = \overline{x}_1x_1 + \cdots + \overline{x}_nx_n = |x_1|^2 + \cdots + |x_n|^2$, which is real and not 0 because $\mathbf{x} \neq \mathbf{0}$. This gives

$$\lambda = \frac{\overline{\mathbf{x}}^{\mathsf{T}}\mathbf{Ax}}{\overline{\mathbf{x}}^{\mathsf{T}}\mathbf{x}}. \tag{1}$$

(a) If $\mathbf{A}$ is Hermitian, $\overline{\mathbf{A}}^{\mathsf{T}} = \mathbf{A}$ or $\mathbf{A}^{\mathsf{T}} = \overline{\mathbf{A}}$ and we show that then the numerator in (1) is real, which makes λ real. $\overline{\mathbf{x}}^{\mathsf{T}}\mathbf{Ax}$ is a scalar; hence taking the transpose has no effect. Thus

$$\overline{\mathbf{x}}^{\mathsf{T}}\mathbf{Ax} = (\overline{\mathbf{x}}^{\mathsf{T}}\mathbf{Ax})^{\mathsf{T}} = \mathbf{x}^{\mathsf{T}}\mathbf{A}^{\mathsf{T}}\overline{\mathbf{x}} = \mathbf{x}^{\mathsf{T}}\overline{\mathbf{A}}\overline{\mathbf{x}} = (\overline{\overline{\mathbf{x}}^{\mathsf{T}}\mathbf{Ax}}). \tag{2}$$

Hence, $\overline{\mathbf{x}}^{\mathsf{T}}\mathbf{Ax}$ equals its complex conjugate, so that it must be real. ($a + ib = a - ib$ implies $b = 0$.)

(b) If $\mathbf{A}$ is skew-Hermitian, $\mathbf{A}^{\mathsf{T}} = -\overline{\mathbf{A}}$ and instead of (2) we obtain

$$\overline{\mathbf{x}}^{\mathsf{T}}\mathbf{Ax} = -(\overline{\overline{\mathbf{x}}^{\mathsf{T}}\mathbf{Ax}}) \tag{3}$$

so that $\overline{\mathbf{x}}^{\mathsf{T}}\mathbf{Ax}$ equals minus its complex conjugate and is pure imaginary or 0. ($a + ib = -(a - ib)$ implies $a = 0$.)

(c) Let $\mathbf{A}$ be unitary. We take $\mathbf{Ax} = \lambda\mathbf{x}$ and its conjugate transpose

$$(\overline{\mathbf{Ax}})^{\mathsf{T}} = (\overline{\lambda\mathbf{x}})^{\mathsf{T}} = \overline{\lambda}\overline{\mathbf{x}}^{\mathsf{T}}$$

and multiply the two left sides and the two right sides,

$$(\overline{\mathbf{Ax}})^{\mathsf{T}}\mathbf{Ax} = \overline{\lambda}\lambda\overline{\mathbf{x}}^{\mathsf{T}}\mathbf{x} = |\lambda|^2\overline{\mathbf{x}}^{\mathsf{T}}\mathbf{x}.$$

But $\mathbf{A}$ is unitary, $\overline{\mathbf{A}}^{\mathsf{T}} = \mathbf{A}^{-1}$, so that on the left we obtain

$$(\overline{\mathbf{A}\mathbf{x}})^{\mathsf{T}}\mathbf{A}\mathbf{x} = \overline{\mathbf{x}}^{\mathsf{T}}\overline{\mathbf{A}}^{\mathsf{T}}\mathbf{A}\mathbf{x} = \overline{\mathbf{x}}^{\mathsf{T}}\mathbf{A}^{-1}\mathbf{A}\mathbf{x} = \overline{\mathbf{x}}^{\mathsf{T}}\mathbf{I}\mathbf{x} = \overline{\mathbf{x}}^{\mathsf{T}}\mathbf{x}.$$

Together, $\overline{\mathbf{x}}^{\mathsf{T}}\mathbf{x} = |\lambda|^2\overline{\mathbf{x}}^{\mathsf{T}}\mathbf{x}$. We now divide by $\overline{\mathbf{x}}^{\mathsf{T}}\mathbf{x}\ (\neq 0)$ to get $|\lambda|^2 = 1$. Hence $|\lambda| = 1$.
This proves Theorem 1 as well as Theorems 1 and 5 in Sec. 8.3. ■

Key properties of orthogonal matrices (invariance of the inner product, orthonormality of rows and columns; see Sec. 8.3) generalize to unitary matrices in a remarkable way.

To see this, instead of R^n we now use the **complex vector space** C^n of all complex vectors with n complex numbers as components, and complex numbers as scalars. For such complex vectors the **inner product** is defined by (note the overbar for the complex conjugate)

$$\mathbf{a} \bullet \mathbf{b} = \overline{\mathbf{a}}^{\mathsf{T}}\mathbf{b}. \tag{4}$$

The **length** or **norm** of such a complex vector is a *real* number defined by

$$\|\mathbf{a}\| = \sqrt{\mathbf{a} \bullet \mathbf{a}} = \sqrt{\overline{\mathbf{a}}_j^{\mathsf{T}}\mathbf{a}} = \sqrt{\overline{a}_1 a_1 + \cdots + \overline{a}_n a_n} = \sqrt{|a_1|^2 + \cdots + |a_n|^2}. \tag{5}$$

THEOREM 2

Invariance of Inner Product

A **unitary transformation**, *that is,* $\mathbf{y} = \mathbf{A}\mathbf{x}$ *with a unitary matrix* $\mathbf{A}$, *preserves the value of the inner product* (4), *hence also the norm* (5).

PROOF The proof is the same as that of Theorem 2 in Sec. 8.3, which the theorem generalizes. In the analog of (9), Sec. 8.3, we now have bars,

$$\mathbf{u} \bullet \mathbf{v} = \overline{\mathbf{u}}^{\mathsf{T}}\mathbf{v} = (\overline{\mathbf{A}\mathbf{a}})^{\mathsf{T}}\mathbf{A}\mathbf{b} = \overline{\mathbf{a}}^{\mathsf{T}}\overline{\mathbf{A}}^{\mathsf{T}}\mathbf{A}\mathbf{b} = \overline{\mathbf{a}}^{\mathsf{T}}\mathbf{I}\mathbf{b} = \overline{\mathbf{a}}^{\mathsf{T}}\mathbf{b} = \mathbf{a} \bullet \mathbf{b}.$$

The complex analog of an orthonormal system of real vectors (see Sec. 8.3) is defined as follows.

DEFINITION

Unitary System

A *unitary system* is a set of complex vectors satisfying the relationships

$$\mathbf{a}_j \bullet \mathbf{a}_k = \overline{\mathbf{a}}_j^{\mathsf{T}}\mathbf{a}_k = \begin{cases} 0 & \text{if} \quad j \neq k \\ 1 & \text{if} \quad j = k. \end{cases} \tag{6}$$

Theorem 3 in Sec. 8.3 extends to complex as follows.

THEOREM 3

Unitary Systems of Column and Row Vectors

A complex square matrix is unitary if and only if its column vectors (and also its row vectors) form a unitary system.

PROOF The proof is the same as that of Theorem 3 in Sec. 8.3, except for the bars required in $\overline{\mathbf{A}}^{\mathsf{T}} = \mathbf{A}^{-1}$ and in (4) and (6) of the present section.

THEOREM 4

Determinant of a Unitary Matrix

Let $\mathbf{A}$ *be a unitary matrix. Then its determinant has absolute value one, that is,* $|\det \mathbf{A}| = 1$.

PROOF Similarly, as in Sec. 8.3, we obtain

$$1 = \det(\mathbf{A}\mathbf{A}^{-1}) = \det(\mathbf{A}\overline{\mathbf{A}}^{\mathsf{T}}) = \det \mathbf{A} \det \overline{\mathbf{A}}^{\mathsf{T}} = \det \mathbf{A} \det \overline{\mathbf{A}}$$
$$= \det \mathbf{A}\, \overline{\det \mathbf{A}} = |\det \mathbf{A}|^2.$$

Hence $|\det \mathbf{A}| = 1$ (where $\det \mathbf{A}$ may now be complex). ■

EXAMPLE 4

Unitary Matrix Illustrating Theorems 1c and 2–4

For the vectors $\mathbf{a}^{\mathsf{T}} = [2 \quad -i]$ and $\mathbf{b}^{\mathsf{T}} = [1 + i \quad 4i]$ we get $\overline{\mathbf{a}}^{\mathsf{T}} = [2 \quad i]^{\mathsf{T}}$ and $\overline{\mathbf{a}}^{\mathsf{T}}\mathbf{b} = 2(1 + i) - 4 = -2 + 2i$ and with

$$\mathbf{A} = \begin{bmatrix} 0.8i & 0.6 \\ 0.6 & 0.8i \end{bmatrix} \quad \text{also} \quad \mathbf{Aa} = \begin{bmatrix} i \\ 2 \end{bmatrix} \quad \text{and} \quad \mathbf{Ab} = \begin{bmatrix} -0.8 + 3.2i \\ -2.6 + 0.6i \end{bmatrix},$$

as one can readily verify. This gives $(\overline{\mathbf{Aa}})^{\mathsf{T}}\mathbf{Ab} = -2 + 2i$, illustrating Theorem 2. The matrix is unitary. Its columns form a unitary system,

$$\overline{\mathbf{a}}_1^{\mathsf{T}}\mathbf{a}_1 = -0.8i \cdot 0.8i + 0.6^2 = 1, \qquad \overline{\mathbf{a}}_1^{\mathsf{T}}\mathbf{a}_2 = -0.8i \cdot 0.6 + 0.6 \cdot 0.8i = 0,$$
$$\overline{\mathbf{a}}_2^{\mathsf{T}}\mathbf{a}_2 = 0.6^2 + (-0.8i)0.8i = 1$$

and so do its rows. Also, $\det \mathbf{A} = -1$. The eigenvalues are $0.6 + 0.8i$ and $-0.6 + 0.8i$, with eigenvectors $[1 \quad 1]^{\mathsf{T}}$ and $[1 \quad -1]^{\mathsf{T}}$, respectively. ■

Theorem 2 in Sec. 8.4 on the existence of an eigenbasis extends to complex matrices as follows.

THEOREM 5

Basis of Eigenvectors

A Hermitian, skew-Hermitian, or unitary matrix has a basis of eigenvectors for C^n *that is a unitary system.*

For a proof see Ref. [B3], vol. 1, pp. 270–272 and p. 244 (Definition 2).

EXAMPLE 5

Unitary Eigenbases

The matrices **A**, **B**, **C** in Example 2 have the following unitary systems of eigenvectors, as you should verify.

$$\mathbf{A}: \quad \frac{1}{\sqrt{35}}[1 - 3i \quad 5]^{\mathsf{T}} \quad (\lambda = 9), \qquad \frac{1}{\sqrt{14}}[1 - 3i \quad -2]^{\mathsf{T}} \quad (\lambda = 2)$$

$$\mathbf{B}: \quad \frac{1}{\sqrt{30}}[1 - 2i \quad -5]^{\mathsf{T}} \quad (\lambda = -2i), \qquad \frac{1}{\sqrt{30}}[5 \quad 1 + 2i]^{\mathsf{T}} \quad (\lambda = 4i)$$

$$\mathbf{C}: \quad \frac{1}{\sqrt{2}}[1 \quad 1]^{\mathsf{T}} \quad (\lambda = \tfrac{1}{2}(i + \sqrt{3})), \qquad \frac{1}{\sqrt{2}}[1 \quad -1]^{\mathsf{T}} \quad (\lambda = \tfrac{1}{2}(i - \sqrt{3})).$$ ■

Hermitian and Skew-Hermitian Forms

The concept of a quadratic form (Sec. 8.4) can be extended to complex. We call the numerator $\bar{\mathbf{x}}^{\mathsf{T}}\mathbf{A}\mathbf{x}$ in (1) a **form** in the components $x_1, \cdots, x_n$ of $\mathbf{x}$, which may now be complex. This form is again a sum of n^2 terms

$$\begin{aligned}\bar{\mathbf{x}}^{\mathsf{T}}\mathbf{A}\mathbf{x} &= \sum_{j=1}^{n}\sum_{k=1}^{n} a_{jk}\bar{x}_j x_k \\ &= \quad a_{11}\bar{x}_1x_1 + \cdots + a_{1n}\bar{x}_1x_n \\ &\quad + a_{21}\bar{x}_2x_1 + \cdots + a_{2n}\bar{x}_2x_n \\ &\quad + \cdots\cdots\cdots\cdots\cdots\cdots \\ &\quad + a_{n1}\bar{x}_nx_1 + \cdots + a_{nn}\bar{x}_nx_n.\end{aligned} \tag{7}$$

A is called its **coefficient matrix**. The form is called a **Hermitian** or **skew-Hermitian form** if **A** is Hermitian or skew-Hermitian, respectively. *The value of a Hermitian form is real, and that of a skew-Hermitian form is pure imaginary or zero.* This can be seen directly from (2) and (3) and accounts for the importance of these forms in physics. Note that (2) and (3) are valid for any vectors because, in the proof of (2) and (3), we did not use that $\mathbf{x}$ is an eigenvector but only that $\bar{\mathbf{x}}^{\mathsf{T}}\mathbf{x}$ is real and not 0.

EXAMPLE 6 **Hermitian Form**

For **A** in Example 2 and, say, $\mathbf{x} = [1+i \quad 5i]^{\mathsf{T}}$ we get

$$\bar{\mathbf{x}}^{\mathsf{T}}\mathbf{A}\mathbf{x} = [1-i \quad -5i]\begin{bmatrix} 4 & 1-3i \\ 1+3i & 7\end{bmatrix}\begin{bmatrix} 1+i \\ 5i\end{bmatrix} = [1-i \quad -5i]\begin{bmatrix} 4(1+i)+(1-3i)\cdot 5i \\ (1+3i)(1+i)+7\cdot 5i\end{bmatrix} = 223. \quad ■$$

Clearly, if **A** and **x** in (4) are real, then (7) reduces to a quadratic form, as discussed in the last section.

PROBLEM SET 8.5

1–6 EIGENVALUES AND VECTORS

Is the given matrix Hermitian? Skew-Hermitian? Unitary? Find its eigenvalues and eigenvectors.

1. $\begin{bmatrix} 6 & i \\ -i & 6\end{bmatrix}$

2. $\begin{bmatrix} i & 1+i \\ -1+i & 0\end{bmatrix}$

3. $\begin{bmatrix} \frac{1}{2} & i\sqrt{\frac{3}{4}} \\ i\sqrt{\frac{3}{4}} & \frac{1}{2}\end{bmatrix}$

4. $\begin{bmatrix} 0 & i \\ i & 0\end{bmatrix}$

5. $\begin{bmatrix} i & 0 & 0 \\ 0 & 0 & i \\ 0 & i & 0\end{bmatrix}$

6. $\begin{bmatrix} 0 & 2+2i & 0 \\ 2-2i & 0 & 2+2i \\ 0 & 2-2i & 0\end{bmatrix}$

7. **Pauli spin matrices.** Find the eigenvalues and eigenvectors of the so-called *Pauli spin matrices* and show that $\mathbf{S}_x\mathbf{S}_y = i\mathbf{S}_z$, $\mathbf{S}_y\mathbf{S}_x = -i\mathbf{S}_z$, $\mathbf{S}_x^2 = \mathbf{S}_y^2 = \mathbf{S}_z^2 = \mathbf{I}$, where

$$\mathbf{S}_x = \begin{bmatrix} 0 & 1 \\ 1 & 0\end{bmatrix}, \quad \mathbf{S}_y = \begin{bmatrix} 0 & -i \\ i & 0\end{bmatrix},$$

$$\mathbf{S}_z = \begin{bmatrix} 1 & 0 \\ 0 & -1\end{bmatrix}.$$

8. **Eigenvectors.** Find eigenvectors of **A**, **B**, **C** in Examples 2 and 3.

9–12 COMPLEX FORMS

Is the matrix **A** Hermitian or skew-Hermitian? Find $\bar{\mathbf{x}}^\mathsf{T}\mathbf{A}\mathbf{x}$. Show the details.

9. $\mathbf{A} = \begin{bmatrix} 4 & 3-2i \\ 3+2i & -4 \end{bmatrix}, \quad \mathbf{x} = \begin{bmatrix} -4i \\ 2+2i \end{bmatrix}$

10. $\mathbf{A} = \begin{bmatrix} i & -2+3i \\ 2+3i & 0 \end{bmatrix}, \quad \mathbf{x} = \begin{bmatrix} 2i \\ 8 \end{bmatrix}$

11. $\mathbf{A} = \begin{bmatrix} i & 1 & 2+i \\ -1 & 0 & 3i \\ -2+i & 3i & i \end{bmatrix}, \quad \mathbf{x} = \begin{bmatrix} 1 \\ i \\ -i \end{bmatrix}$

12. $\mathbf{A} = \begin{bmatrix} 1 & i & 4 \\ -i & 3 & 0 \\ 4 & 0 & 2 \end{bmatrix}, \quad \mathbf{x} = \begin{bmatrix} 1 \\ i \\ -i \end{bmatrix}$

13–20 GENERAL PROBLEMS

13. Product. Show that $\overline{(\mathbf{ABC})}^\mathsf{T} = -\mathbf{C}^{-1}\mathbf{BA}$ for any $n \times n$ Hermitian **A**, skew-Hermitian **B**, and unitary **C**.

14. Product. Show $\overline{(\mathbf{BA})}^\mathsf{T} = -\mathbf{AB}$ for **A** and **B** in Example 2. For any $n \times n$ Hermitian **A** and skew-Hermitian **B**.

15. Decomposition. Show that any square matrix may be written as the sum of a Hermitian and a skew-Hermitian matrix. Give examples.

16. Unitary matrices. Prove that the product of two unitary $n \times n$ matrices and the inverse of a unitary matrix are unitary. Give examples.

17. Powers of unitary matrices in applications may sometimes be very simple. Show that $\mathbf{C}^{12} = \mathbf{I}$ in Example 2. Find further examples.

18. Normal matrix. This important concept denotes a matrix that commutes with its conjugate transpose, $\mathbf{A}\overline{\mathbf{A}}^\mathsf{T} = \overline{\mathbf{A}}^\mathsf{T}\mathbf{A}$. Prove that Hermitian, skew-Hermitian, and unitary matrices are normal. Give corresponding examples of your own.

19. Normality criterion. Prove that **A** is normal if and only if the Hermitian and skew-Hermitian matrices in Prob. 18 commute.

20. Find a simple matrix that is not normal. Find a normal matrix that is not Hermitian, skew-Hermitian, or unitary.

CHAPTER 8 REVIEW QUESTIONS AND PROBLEMS

1. In solving an eigenvalue problem, what is given and what is sought?
2. Give a few typical applications of eigenvalue problems.
3. Do there exist square matrices without eigenvalues?
4. Can a real matrix have complex eigenvalues? Can a complex matrix have real eigenvalues?
5. Does a 5×5 matrix always have a real eigenvalue?
6. What is algebraic multiplicity of an eigenvalue? Defect?
7. What is an eigenbasis? When does it exist? Why is it important?
8. When can we expect orthogonal eigenvectors?
9. State the definitions and main properties of the three classes of real matrices and of complex matrices that we have discussed.
10. What is diagonalization? Transformation to principal axes?

11–15 SPECTRUM

Find the eigenvalues. Find the eigenvectors.

11. $\begin{bmatrix} 2.5 & 0.5 \\ 0.5 & 2.5 \end{bmatrix}$ **12.** $\begin{bmatrix} -7 & 4 \\ -12 & 7 \end{bmatrix}$

13. $\begin{bmatrix} 8 & -1 \\ 5 & 2 \end{bmatrix}$

14. $\begin{bmatrix} 7 & 2 & -1 \\ 2 & 7 & 1 \\ -1 & 1 & 8.5 \end{bmatrix}$

15. $\begin{bmatrix} 0 & -3 & -6 \\ 3 & 0 & -6 \\ 6 & 6 & 0 \end{bmatrix}$

16–17 SIMILARITY

Verify that **A** and $\hat{\mathbf{A}} = \mathbf{p}^{-1}\mathbf{AP}$ have the same spectrum.

16. $\mathbf{A} = \begin{bmatrix} 19 & 12 \\ 12 & 1 \end{bmatrix}, \quad \mathbf{P} = \begin{bmatrix} 2 & 4 \\ 4 & 2 \end{bmatrix}$

17. $\mathbf{A} = \begin{bmatrix} 7 & -4 \\ 12 & -7 \end{bmatrix}, \quad \mathbf{P} = \begin{bmatrix} 5 & 3 \\ 3 & 5 \end{bmatrix}$

18. $\mathbf{A} = \begin{bmatrix} -4 & 6 & 6 \\ 0 & 2 & 0 \\ -1 & 1 & 1 \end{bmatrix}, \quad \mathbf{P} = \begin{bmatrix} 1 & 8 & -7 \\ 0 & 1 & 3 \\ 0 & 0 & 1 \end{bmatrix}$

19–21 DIAGONALIZATION

Find an eigenbasis and diagonalize.

9. $\begin{bmatrix} -1.4 & 1.0 \\ -1.0 & 1.1 \end{bmatrix}$

20. $\begin{bmatrix} 72 & -56 \\ -56 & 513 \end{bmatrix}$

21. $\begin{bmatrix} -12 & 22 & 6 \\ 8 & 2 & 6 \\ -8 & 20 & 16 \end{bmatrix}$

22–25 CONIC SECTIONS. PRINCIPAL AXES

Transform to canonical form (to principal axes). Express $[x_1 \ \ x_2]^\mathsf{T}$ in terms of the new variables $[y_1 \ \ y_2]^\mathsf{T}$.

22. $9x_1^2 - 6x_1x_2 + 17x_2^2 = 36$

23. $4x_1^2 + 24x_1x_2 - 14x_2^2 = 20$

24. $5x_1^2 + 24x_1x_2 - 5x_2^2 = 0$

25. $3.7x_1^2 + 3.2x_1x_2 + 1.3x_2^2 = 4.5$

SUMMARY OF CHAPTER 8

Linear Algebra: Matrix Eigenvalue Problems

The practical importance of matrix eigenvalue problems can hardly be overrated. The problems are defined by the vector equation

$$\text{(1)} \qquad \mathbf{Ax} = \lambda \mathbf{x}.$$

A is a given square matrix. All matrices in this chapter are ***square***. λ is a scalar. To *solve* the problem (1) means to determine values of λ, called **eigenvalues** (or **characteristic values**) of **A**, such that (1) has a nontrivial solution **x** (that is, $\mathbf{x} \neq \mathbf{0}$), called an **eigenvector** of **A** corresponding to that λ. An $n \times n$ matrix has at least one and at most n numerically different eigenvalues. These are the solutions of the **characteristic equation** (Sec. 8.1)

$$\text{(2)} \qquad D(\lambda) = \det(\mathbf{A} - \lambda\mathbf{I}) = \begin{vmatrix} a_{11} - \lambda & a_{12} & \cdots & a_{1n} \\ a_{21} & a_{22} - \lambda & \cdots & a_{2n} \\ \cdot & \cdot & \cdots & \cdot \\ a_{n1} & a_{n2} & \cdots & a_{nn} - \lambda \end{vmatrix} = 0.$$

$D(\lambda)$ is called the **characteristic determinant** of **A**. By expanding it we get the **characteristic polynomial** of **A**, which is of degree n in λ. Some typical applications are shown in Sec. 8.2.

Section 8.3 is devoted to eigenvalue problems for **symmetric** ($\mathbf{A}^\mathsf{T} = \mathbf{A}$), **skew-symmetric** ($\mathbf{A}^\mathsf{T} = -\mathbf{A}$), and **orthogonal matrices** ($\mathbf{A}^\mathsf{T} = \mathbf{A}^{-1}$). Section 8.4 concerns the diagonalization of matrices and the transformation of quadratic forms to principal axes and its relation to eigenvalues.

Section 8.5 extends Sec. 8.3 to the complex analogs of those real matrices, called **Hermitian** ($\mathbf{A}^\mathsf{T} = \mathbf{A}$), **skew-Hermitian** ($\mathbf{A}^\mathsf{T} = -\mathbf{A}$), and **unitary matrices** ($\bar{\mathbf{A}}^\mathsf{T} = \mathbf{A}^{-1}$). All the eigenvalues of a Hermitian matrix (and a symmetric one) are real. For a skew-Hermitian (and a skew-symmetric) matrix they are pure imaginary or zero. For a unitary (and an orthogonal) matrix they have absolute value 1.

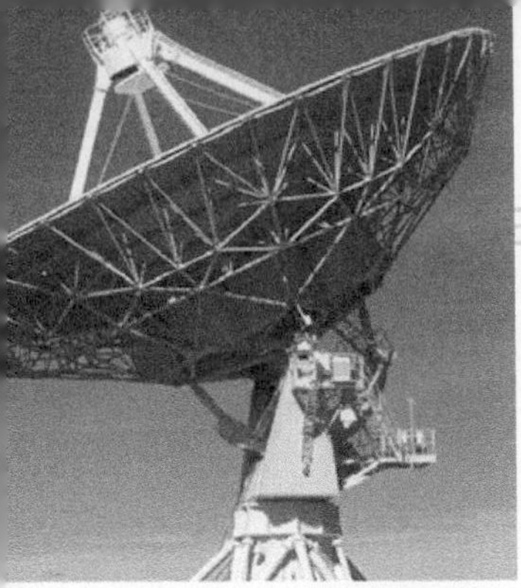

CHAPTER 9

Vector Differential Calculus. Grad, Div, Curl

Engineering, physics, and computer sciences, in general, but particularly solid mechanics, aerodynamics, aeronautics, fluid flow, heat flow, electrostatics, quantum physics, laser technology, robotics as well as other areas have applications that require an understanding of **vector calculus**. This field encompasses vector differential calculus and vector integral calculus. Indeed, the engineer, physicist, and mathematician need a good grounding in these areas as provided by the carefully chosen material of Chaps. 9 and 10.

Forces, velocities, and various other quantities may be thought of as vectors. Vectors appear frequently in the applications above and also in the biological and social sciences, so it is natural that problems are modeled in **3-space**. This is the space of three dimensions with the usual measurement of distance, as given by the Pythagorean theorem. Within that realm, **2-space** (the plane) is a special case. Working in 3-space requires that we extend the common differential calculus to vector differential calculus, that is, the calculus that deals with vector functions and vector fields and is explained in this chapter.

Chapter 9 is arranged in three groups of sections. Sections 9.1–9.3 extend the basic algebraic operations of vectors into 3-space. These operations include the inner product and the cross product. Sections 9.4 and 9.5 form the heart of vector differential calculus. Finally, Secs. 9.7–9.9 discuss three physically important concepts related to scalar and vector fields: gradient (Sec. 9.7), divergence (Sec. 9.8), and curl (Sec. 9.9). They are expressed in Cartesian coordinates in this chapter and, if desired, expressed in ***curvilinear coordinates*** in a short section in App. A3.4.

We shall keep this chapter ***independent of Chaps. 7 and 8***. Our present approach is in harmony with Chap. 7, with the restriction to two and three dimensions providing for a richer theory with basic physical, engineering, and geometric applications.

Prerequisite: Elementary use of second- and third-order determinants in Sec. 9.3.
Sections that may be omitted in a shorter course: 9.5, 9.6.
References and Answers to Problems: App. 1 Part B, App. 2.

9.1 Vectors in 2-Space and 3-Space

In engineering, physics, mathematics, and other areas we encounter two kinds of quantities. They are scalars and vectors.

A **scalar** is a quantity that is determined by its magnitude. It takes on a numerical value, i.e., a number. Examples of scalars are time, temperature, length, distance, speed, density, energy, and voltage.

In contrast, a **vector** is a quantity that has both magnitude and direction. We can say that a vector is an ***arrow*** or a ***directed line segment***. For example, a velocity vector has length or magnitude, which is speed, and direction, which indicates the direction of motion. Typical examples of vectors are displacement, velocity, and force, see Fig. 164 as an illustration.

More formally, we have the following. We denote vectors by lowercase boldface letters **a**, **b**, **v**, etc. In handwriting you may use arrows, for instance, $\vec{a}$ (in place of **a**), $\vec{b}$, etc.

A vector (arrow) has a tail, called its **initial point**, and a tip, called its **terminal point**. This is motivated in the **translation** (displacement without rotation) of the triangle in Fig. 165, where the initial point P of the vector **a** is the original position of a point, and the terminal point Q is the terminal position of that point, its position *after* the translation. The length of the arrow equals the distance between P and Q. This is called the **length** (or *magnitude*) of the vector **a** and is denoted by $|\mathbf{a}|$. Another name for *length* is **norm** (or *Euclidean norm*).

A vector of length 1 is called a **unit vector**.

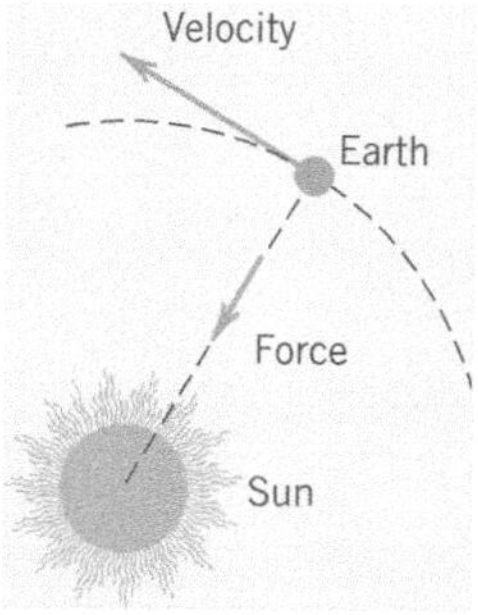

Fig. 164. Force and velocity

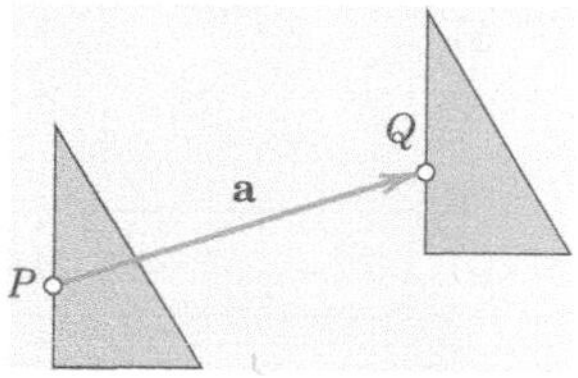

Fig. 165. Translation

Of course, we would like to calculate with vectors. For instance, we want to find the resultant of forces or compare parallel forces of different magnitude. This motivates our next ideas: to define *components* of a vector, and then the two basic algebraic operations of *vector addition* and *scalar multiplication.*

For this we must first define *equality of vectors* in a way that is practical in connection with forces and other applications.

DEFINITION

Equality of Vectors

Two vectors **a** and **b** are equal, written $\mathbf{a} = \mathbf{b}$, if they have the same length and the same direction [as explained in Fig. 166; in particular, note (B)]. Hence a vector can be arbitrarily translated; that is, its initial point can be chosen arbitrarily.

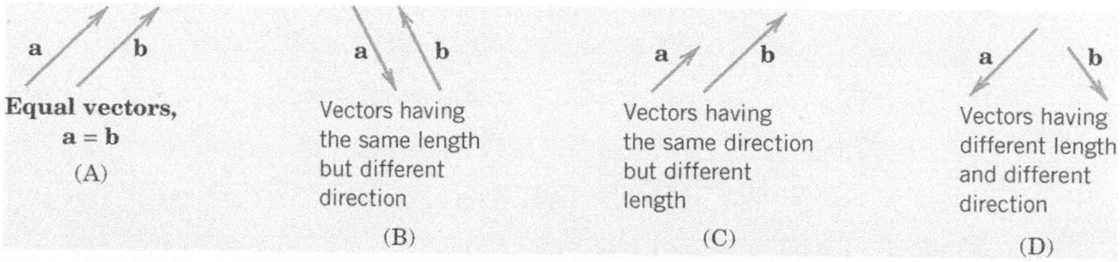

Fig. 166. (A) Equal vectors. (B)–(D) Different vectors

Components of a Vector

We choose an xyz **Cartesian coordinate system**[1] in space (Fig. 167), that is, a usual rectangular coordinate system with the same scale of measurement on the three mutually perpendicular coordinate axes. Let **a** be a given vector with initial point P: (x_1, y_1, z_1) and terminal point Q: (x_2, y_2, z_2). Then the three coordinate differences

$$a_1 = x_2 - x_1, \qquad a_2 = y_2 - y_1, \qquad a_3 = z_2 - z_1 \tag{1}$$

are called the **components** of the vector **a** with respect to that coordinate system, and we write simply $\mathbf{a} = [a_1, a_2, a_3]$. See Fig. 168.

The **length** $|\mathbf{a}|$ of **a** can now readily be expressed in terms of components because from (1) and the Pythagorean theorem we have

$$|\mathbf{a}| = \sqrt{a_1^2 + a_2^2 + a_3^2}. \tag{2}$$

EXAMPLE 1 **Components and Length of a Vector**

The vector **a** with initial point P: (4, 0, 2) and terminal point Q: (6, −1, 2) has the components

$$a_1 = 6 - 4 = 2, \qquad a_2 = -1 - 0 = -1, \qquad a_3 = 2 - 2 = 0.$$

Hence $\mathbf{a} = [2, -1, 0]$. (Can you sketch **a**, as in Fig. 168?) Equation (2) gives the length

$$|\mathbf{a}| = \sqrt{2^2 + (-1)^2 + 0^2} = \sqrt{5}.$$

If we choose (−1, 5, 8) as the initial point of **a**, the corresponding terminal point is (1, 4, 8).

If we choose the origin (0, 0, 0) as the initial point of **a**, the corresponding terminal point is (2, −1, 0); its coordinates equal the components of **a**. This suggests that we can determine each point in space by a vector, called the *position vector* of the point, as follows. ■

A Cartesian coordinate system being given, the **position vector r** of a point A: (x, y, z) is the vector with the origin (0, 0, 0) as the initial point and A as the terminal point (see Fig. 169). Thus in components, $\mathbf{r} = [x, y, z]$. This can be seen directly from (1) with $x_1 = y_1 = z_1 = 0$.

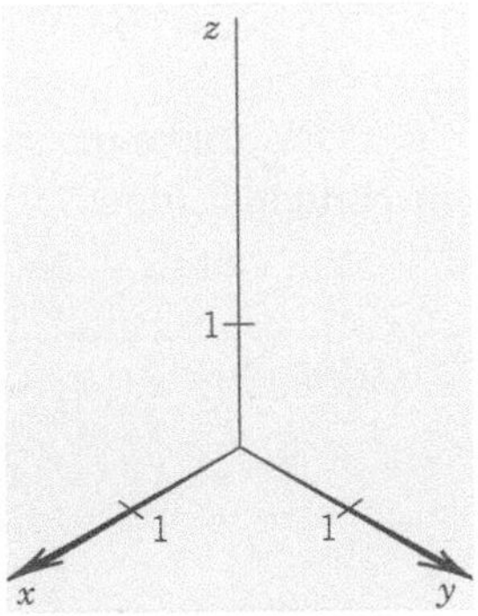

Fig. 167. Cartesian coordinate system

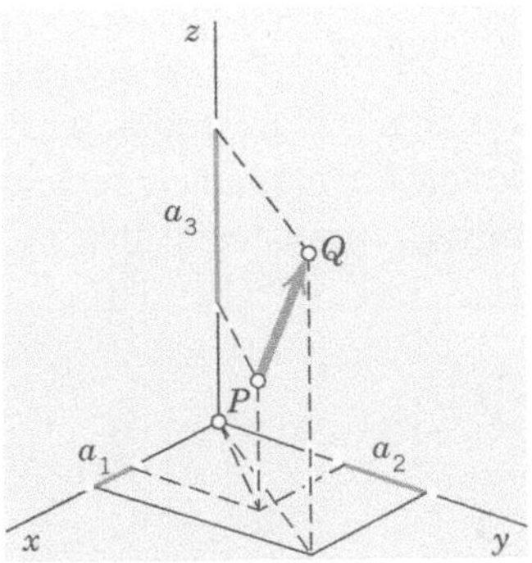

Fig. 168. Components of a vector

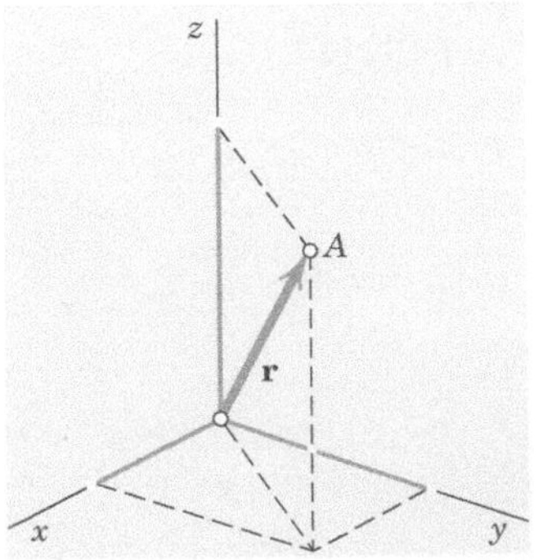

Fig. 169. Position vector **r** of a point A: (x, y, z)

[1]Named after the French philosopher and mathematician RENATUS CARTESIUS, latinized for RENÉ DESCARTES (1596–1650), who invented analytic geometry. His basic work *Géométrie* appeared in 1637, as an appendix to his *Discours de la méthode*.

Furthermore, if we translate a vector **a**, with initial point P and terminal point Q, then corresponding coordinates of P and Q change by the same amount, so that the differences in (1) remain unchanged. This proves

THEOREM 1

Vectors as Ordered Triples of Real Numbers

A fixed Cartesian coordinate system being given, each vector is uniquely determined by its ordered triple of corresponding components. Conversely, to each ordered triple of real numbers (a_1, a_2, a_3) *there corresponds precisely one vector* $\mathbf{a} = [a_1, a_2, a_3]$, *with* $(0, 0, 0)$ *corresponding to the* **zero vector 0**, *which has length* 0 *and no direction.*

Hence a vector equation $\mathbf{a} = \mathbf{b}$ *is equivalent to the three equations* $a_1 = b_1$, $a_2 = b_2$, $a_3 = b_3$ *for the components.*

We now see that from our "geometric" definition of a vector as an arrow we have arrived at an "algebraic" characterization of a vector by Theorem 1. We could have started from the latter and reversed our process. This shows that the two approaches are equivalent.

Vector Addition, Scalar Multiplication

Calculations with vectors are very useful and are almost as simple as the arithmetic for real numbers. Vector arithmetic follows almost naturally from applications. We first define how to add vectors and later on how to multiply a vector by a number.

DEFINITION

Addition of Vectors

The **sum a + b** of two vectors $\mathbf{a} = [a_1, a_2, a_3]$ and $\mathbf{b} = [b_1, b_2, b_3]$ is obtained by adding the corresponding components,

$$\mathbf{a} + \mathbf{b} = [a_1 + b_1, \quad a_2 + b_2, \quad a_3 + b_3]. \tag{3}$$

Geometrically, place the vectors as in Fig. 170 (the initial point of **b** at the terminal point of **a**); then **a** + **b** is the vector drawn from the initial point of **a** to the terminal point of **b**.

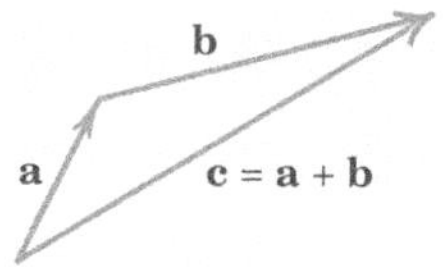

Fig. 170. Vector addition

For forces, this addition is the parallelogram law by which we obtain the **resultant** of two forces in mechanics. See Fig. 171.

Figure 172 shows (for the plane) that the "algebraic" way and the "geometric way" of vector addition give the same vector.

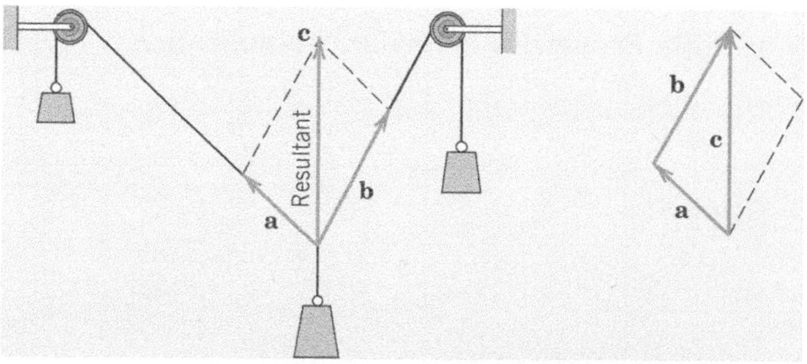

Fig. 171. Resultant of two forces (**parallelogram law**)

Basic Properties of Vector Addition. Familiar laws for real numbers give immediately

$$
\text{(4)}\quad
\begin{aligned}
&\text{(a)} && \mathbf{a} + \mathbf{b} = \mathbf{b} + \mathbf{a} && \textit{(Commutativity)}\\
&\text{(b)} && (\mathbf{u} + \mathbf{v}) + \mathbf{w} = \mathbf{u} + (\mathbf{v} + \mathbf{w}) && \textit{(Associativity)}\\
&\text{(c)} && \mathbf{a} + \mathbf{0} = \mathbf{0} + \mathbf{a} = \mathbf{a} &&\\
&\text{(d)} && \mathbf{a} + (-\mathbf{a}) = \mathbf{0}. &&
\end{aligned}
$$

Properties (a) and (b) are verified geometrically in Figs. 173 and 174. Furthermore, $-\mathbf{a}$ denotes the vector having the length $|\mathbf{a}|$ and the direction opposite to that of $\mathbf{a}$.

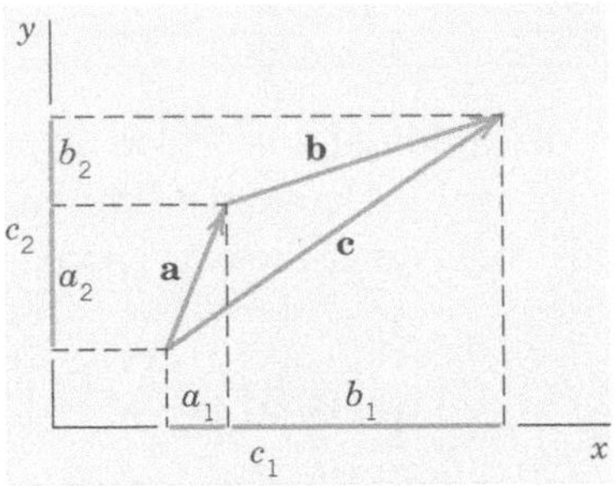

Fig. 172. Vector addition

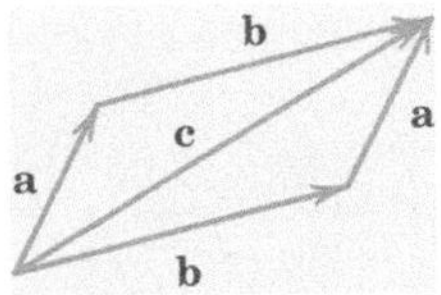

Fig. 173. Cummutativity of vector addition

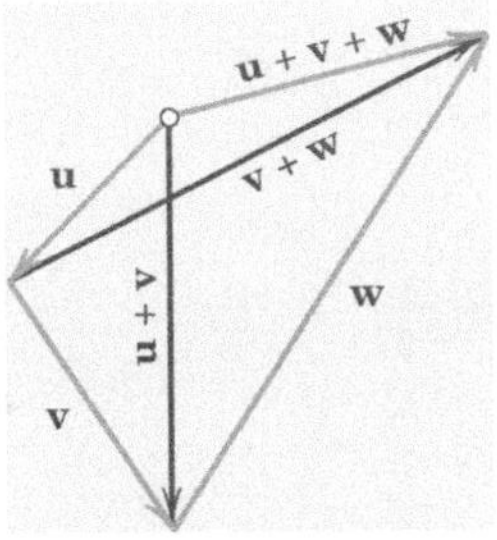

Fig. 174. Associativity of vector addition

In (4b) we may simply write $\mathbf{u} + \mathbf{v} + \mathbf{w}$, and similarly for sums of more than three vectors. Instead of $\mathbf{a} + \mathbf{a}$ we also write $2\mathbf{a}$, and so on. This (and the notation $-\mathbf{a}$ used just before) motivates defining the second algebraic operation for vectors as follows.

DEFINITION

Scalar Multiplication (Multiplication by a Number)

The product $c\mathbf{a}$ of any vector $\mathbf{a} = [a_1, a_2, a_3]$ and any scalar c (real number c) is the vector obtained by multiplying each component of $\mathbf{a}$ by c,

$$
\text{(5)}\qquad c\mathbf{a} = [ca_1, ca_2, ca_3].
$$

Geometrically, if $\mathbf{a} \neq \mathbf{0}$, then $c\mathbf{a}$ with $c > 0$ has the direction of $\mathbf{a}$ and with $c < 0$ the direction opposite to $\mathbf{a}$. In any case, the length of $c\mathbf{a}$ is $|c\mathbf{a}| = |c||\mathbf{a}|$, and $c\mathbf{a} = \mathbf{0}$ if $\mathbf{a} = \mathbf{0}$ or $c = 0$ (or both). (See Fig. 175.)

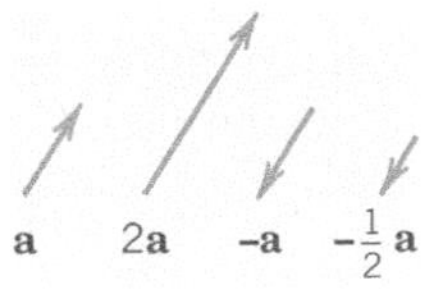

Fig. 175. Scalar multiplication [multiplication of vectors by scalars (numbers)]

Basic Properties of Scalar Multiplication. From the definitions we obtain directly

$$
\text{(6)}\quad
\begin{aligned}
&\text{(a)} && c(\mathbf{a} + \mathbf{b}) = c\mathbf{a} + c\mathbf{b} &&\\
&\text{(b)} && (c + k)\mathbf{a} = c\mathbf{a} + k\mathbf{a} &&\\
&\text{(c)} && c(k\mathbf{a}) = (ck)\mathbf{a} && \text{(written } ck\mathbf{a}\text{)}\\
&\text{(d)} && 1\mathbf{a} = \mathbf{a}. &&
\end{aligned}
$$

You may prove that (4) and (6) imply for any vector **a**

$$\text{(7)} \qquad \begin{aligned} &\text{(a)} \quad 0\mathbf{a} = \mathbf{0} \\ &\text{(b)} \quad (-1)\mathbf{a} = -\mathbf{a}. \end{aligned}$$

Instead of $\mathbf{b} + (-\mathbf{a})$ we simply write $\mathbf{b} - \mathbf{a}$ (Fig. 176).

EXAMPLE 2 Vector Addition. Multiplication by Scalars

With respect to a given coordinate system, let

$$\mathbf{a} = [4, 0, 1] \quad \text{and} \quad \mathbf{b} = [2, -5, \tfrac{1}{3}].$$

Then $-\mathbf{a} = [-4, 0, -1]$, $7\mathbf{a} = [28, 0, 7]$, $\mathbf{a} + \mathbf{b} = [6, -5, \tfrac{4}{3}]$, and

$$2(\mathbf{a} - \mathbf{b}) = 2[2, 5, \tfrac{2}{3}] = [4, 10, \tfrac{4}{3}] = 2\mathbf{a} - 2\mathbf{b}.$$ ■

Unit Vectors i, j, k. Besides $\mathbf{a} = [a_1, a_2, a_3]$ another popular way of writing vectors is

$$\text{(8)} \qquad \mathbf{a} = a_1\mathbf{i} + a_2\mathbf{j} + a_3\mathbf{k}.$$

In this representation, **i**, **j**, **k** are the unit vectors in the positive directions of the axes of a Cartesian coordinate system (Fig. 177). Hence, in components,

$$\text{(9)} \qquad \mathbf{i} = [1, 0, 0], \quad \mathbf{j} = [0, 1, 0], \quad \mathbf{k} = [0, 0, 1]$$

and the right side of (8) is a sum of three vectors parallel to the three axes.

EXAMPLE 3 ijk Notation for Vectors

In Example 2 we have $\mathbf{a} = 4\mathbf{i} + \mathbf{k}$, $\mathbf{b} = 2\mathbf{i} - 5\mathbf{j} + \tfrac{1}{3}\mathbf{k}$, and so on. ■

All the vectors $\mathbf{a} = [a_1, a_2, a_3] = a_1\mathbf{i} + a_2\mathbf{j} + a_3\mathbf{k}$ (with real numbers as components) form the **real vector space** R^3 with the two *algebraic operations* of vector addition and scalar multiplication as just defined. R^3 has **dimension** 3. The triple of vectors **i**, **j**, **k** is called a **standard basis** of R^3. Given a Cartesian coordinate system, the representation (8) of a given vector is unique.

Vector space R^3 is a model of a general vector space, as discussed in Sec. 7.9, but is not needed in this chapter.

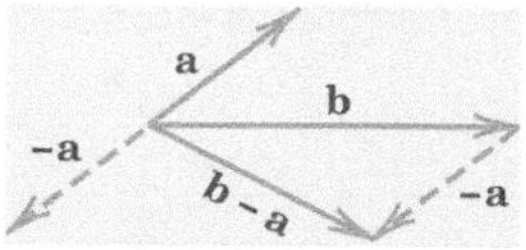

Fig. 176. Difference of vectors

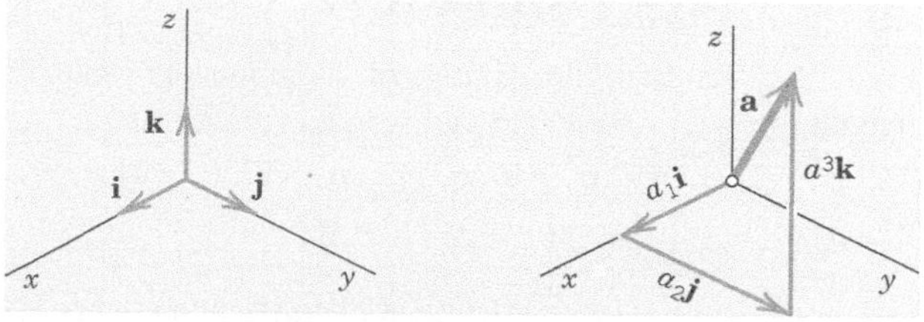

Fig. 177. The unit vectors **i**, **j**, **k** and the representation (8)

PROBLEM SET 9.1

1–5 COMPONENTS AND LENGTH

Find the components of the vector $\mathbf{v}$ with initial point P and terminal point Q. Find $|\mathbf{v}|$. Sketch $|\mathbf{v}|$. Find the unit vector $\mathbf{u}$ in the direction of $\mathbf{v}$.

1. P: $(1, 1, 0)$, Q: $(6, 2, 0)$

2. P: $(1, 1, 1)$, Q: $(2, 2, 0)$

3. P: $(-3.0, 4{,}0, -0.5)$, Q: $(5.5, 0, 1.2)$

4. P: $(1, 4, 2)$, Q: $(-1, -4, -2)$

5. P: $(0, 0, 0)$, Q: $(2, 1, -2)$

6–10

Find the terminal point Q of the vector $\mathbf{v}$ with components as given and initial point P. Find $|\mathbf{v}|$.

6. $4, 0, 0$; P: $(0, 2, 13)$

7. $\frac{1}{2}, 3, -\frac{1}{4}$; P: $(\frac{7}{2}, -3, \frac{3}{4})$

8. $13.1, 0.8, -2.0$; P: $(0, 0, 0)$

9. $6, 1, -4$; P: $(-6, -1, -4)$

10. $0, -3, 3$; P: $(0, 3, -3)$

11–18 ADDITION, SCALAR MULTIPLICATION

Let $\mathbf{a} = [3, 2, 0] = 3\mathbf{i} + 2\mathbf{j}$; $\mathbf{b} = [-4, 6, 0] = 4\mathbf{i} + 6\mathbf{j}$, $\mathbf{c} = [5, -1, 8] = 5\mathbf{i} - \mathbf{j} + 8\mathbf{k}$, $\mathbf{d} = [0, 0, 4] = 4\mathbf{k}$. Find:

11. $2\mathbf{a}$, $\frac{1}{2}\mathbf{a}$, $-\mathbf{a}$

12. $(\mathbf{a} + \mathbf{b}) + \mathbf{c}$, $\mathbf{a} + (\mathbf{b} + \mathbf{c})$

13. $\mathbf{b} + \mathbf{c}$, $\mathbf{c} + \mathbf{b}$

14. $3\mathbf{c} - 6\mathbf{d}$, $3(\mathbf{c} - 2\mathbf{d})$

15. $7(\mathbf{c} - \mathbf{b})$, $7\mathbf{c} - 7\mathbf{b}$

16. $\frac{9}{2}\mathbf{a} - 3\mathbf{c}$, $9(\frac{1}{2}\mathbf{a} - \frac{1}{3}\mathbf{c})$

17. $(7 - 3)\mathbf{a}$, $7\mathbf{a} - 3\mathbf{a}$

18. $4\mathbf{a} + 3\mathbf{b}$, $-4\mathbf{a} - 3\mathbf{b}$

19. What laws do Probs. 12–16 illustrate?

20. Prove Eqs. (4) and (6).

21–25 FORCES, RESULTANT

Find the resultant in terms of components and its magnitude.

21. $\mathbf{p} = [2, 3, 0]$, $\mathbf{q} = [0, 6, 1]$, $\mathbf{u} = [2, 0, -4]$

22. $\mathbf{p} = [1, -2, 3]$, $\mathbf{q} = [3, 21, -16]$, $\mathbf{u} = [-4, -19, 13]$

23. $\mathbf{u} = [8, -1, 0]$, $\mathbf{v} = [\frac{1}{2}, 0, \frac{4}{3}]$, $\mathbf{w} = [-\frac{17}{2}, 1, \frac{11}{3}]$

24. $\mathbf{p} = [-1, 2, -3]$, $\mathbf{q} = [1, 1, 1]$, $\mathbf{u} = [1, -2, 2]$

25. $\mathbf{u} = [3, 1, -6]$, $\mathbf{v} = [0, 2, 5]$, $\mathbf{w} = [3, -1, -13]$

26–37 FORCES, VELOCITIES

26. Equilibrium. Find $\mathbf{v}$ such that $\mathbf{p}$, $\mathbf{q}$, $\mathbf{u}$ in Prob. 21 and $\mathbf{v}$ are in equilibrium.

27. Find $\mathbf{p}$ such that $\mathbf{u}$, $\mathbf{v}$, $\mathbf{w}$ in Prob. 23 and $\mathbf{p}$ are in equilibrium.

28. Unit vector. Find the unit vector in the direction of the resultant in Prob. 24.

29. Restricted resultant. Find all $\mathbf{v}$ such that the resultant of $\mathbf{v}$, $\mathbf{p}$, $\mathbf{q}$, $\mathbf{u}$ with $\mathbf{p}$, $\mathbf{q}$, $\mathbf{u}$ as in Prob. 21 is parallel to the xy-plane.

30. Find $\mathbf{v}$ such that the resultant of $\mathbf{p}$, $\mathbf{q}$, $\mathbf{u}$, $\mathbf{v}$ with $\mathbf{p}$, $\mathbf{q}$, $\mathbf{u}$ as in Prob. 24 has no components in x- and y-directions.

31. For what k is the resultant of $[2, 0, -7]$, $[1, 2, -3]$, and $[0, 3, k]$ parallel to the xy-plane?

32. If $|\mathbf{p}| = 6$ and $|\mathbf{q}| = 4$, what can you say about the magnitude and direction of the resultant? Can you think of an application to robotics?

33. Same question as in Prob. 32 if $|\mathbf{p}| = 9$, $|\mathbf{q}| = 6$, $|\mathbf{u}| = 3$.

34. Relative velocity. If airplanes A and B are moving southwest with speed $|\mathbf{v}_A| = 550$ mph, and northwest with speed $|\mathbf{v}_B| = 450$ mph, respectively, what is the relative velocity $\mathbf{v} = \mathbf{v}_B - \mathbf{v}_A$ of B with respect to A?

35. Same question as in Prob. 34 for two ships moving northeast with speed $|\mathbf{v}_A| = 22$ knots and west with speed $|\mathbf{v}_B| = 19$ knots.

36. Reflection. If a ray of light is reflected once in each of two mutually perpendicular mirrors, what can you say about the reflected ray?

37. Force polygon. Truss. Find the forces in the system of two rods (*truss*) in the figure, where $|\mathbf{p}| = 1000$ nt. *Hint.* Forces in equilibrium form a polygon, the *force polygon*.

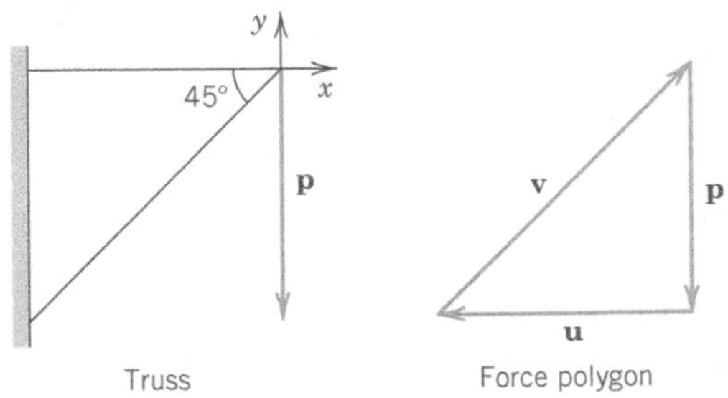

Problem 37

38. TEAM PROJECT. Geometric Applications. To increase your skill in dealing with vectors, use vectors to prove the following (see the figures).

(a) The diagonals of a parallelogram bisect each other.

(b) The line through the midpoints of adjacent sides of a parallelogram bisects one of the diagonals in the ratio 1 : 3.

(c) Obtain (b) from (a).

(d) The three medians of a triangle (the segments from a vertex to the midpoint of the opposite side) meet at a single point, which divides the medians in the ratio 2 : 1.

(e) The quadrilateral whose vertices are the midpoints of the sides of an arbitrary quadrilateral is a parallelogram.

(f) The four space diagonals of a parallelepiped meet and bisect each other.

(g) The sum of the vectors drawn from the center of a regular polygon to its vertices is the zero vector.

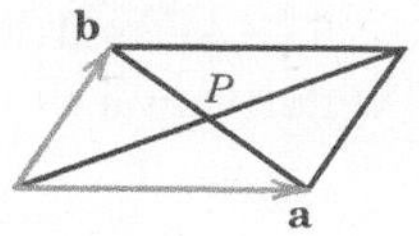

Team Project 38(a)

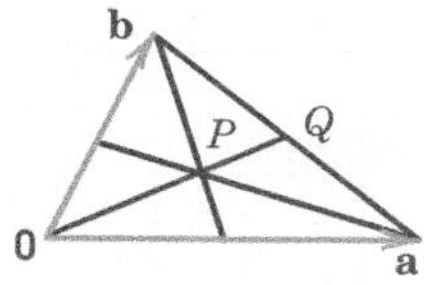

Team Project 38(d)

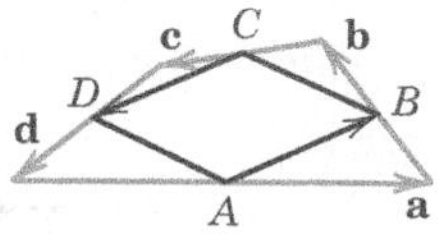

Team Project 38(e)

9.2 Inner Product (Dot Product)

Orthogonality

The inner product or dot product can be motivated by calculating work done by a constant force, determining components of forces, or other applications. It involves the length of vectors and the angle between them. The inner product is a kind of multiplication of two vectors, defined in such a way that the outcome is a scalar. Indeed, another term for inner product is scalar product, a term we shall not use here. The definition of the inner product is as follows.

DEFINITION

Inner Product (Dot Product) of Vectors

The **inner product** or **dot product** $\mathbf{a} \cdot \mathbf{b}$ (read "**a** dot **b**") of two vectors **a** and **b** is the product of their lengths times the cosine of their angle (see Fig. 178),

$$\text{(1)} \qquad \begin{aligned} \mathbf{a} \cdot \mathbf{b} &= |\mathbf{a}||\mathbf{b}| \cos \gamma \qquad &\text{if} \quad \mathbf{a} \neq \mathbf{0}, \mathbf{b} \neq \mathbf{0} \\ \mathbf{a} \cdot \mathbf{b} &= 0 &\text{if} \quad \mathbf{a} = \mathbf{0} \text{ or } \mathbf{b} = \mathbf{0}. \end{aligned}$$

The angle γ, $0 \leqq \gamma \leqq \pi$, between **a** and **b** is measured when the initial points of the vectors coincide, as in Fig. 178. In components, $\mathbf{a} = [a_1, a_2, a_3]$, $\mathbf{b} = [b_1, b_2, b_3]$, and

$$\text{(2)} \qquad \mathbf{a} \cdot \mathbf{b} = a_1 b_1 + a_2 b_2 + a_3 b_3.$$

The second line in (1) is needed because γ is undefined when $\mathbf{a} = \mathbf{0}$ or $\mathbf{b} = \mathbf{0}$. The derivation of (2) from (1) is shown below.

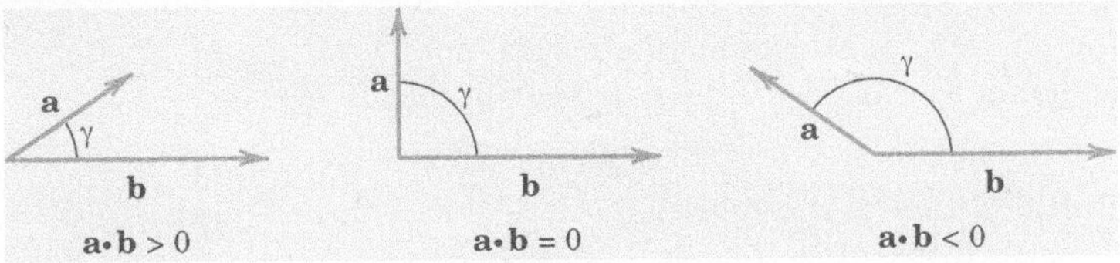

Fig. 178. Angle between vectors and value of inner product

Orthogonality. Since the cosine in (1) may be positive, 0, or negative, so may be the inner product (Fig. 178). The case that the inner product is zero is of particular practical interest and suggests the following concept.

A vector **a** is called **orthogonal** to a vector **b** if $\mathbf{a} \cdot \mathbf{b} = 0$. Then **b** is also orthogonal to **a**, and we call **a** and **b** **orthogonal vectors**. Clearly, this happens for nonzero vectors if and only if $\cos \gamma = 0$; thus $\gamma = \pi/2$ (90°). This proves the important

THEOREM 1

Orthogonality Criterion

The inner product of two nonzero vectors is 0 *if and only if these vectors are perpendicular.*

Length and Angle. Equation (1) with $\mathbf{b} = \mathbf{a}$ gives $\mathbf{a} \cdot \mathbf{a} = |\mathbf{a}|^2$. Hence

$$|\mathbf{a}| = \sqrt{\mathbf{a} \cdot \mathbf{a}}. \tag{3}$$

From (3) and (1) we obtain for the angle γ between two nonzero vectors

$$\cos \gamma = \frac{\mathbf{a} \cdot \mathbf{b}}{|\mathbf{a}||\mathbf{b}|} = \frac{\mathbf{a} \cdot \mathbf{b}}{\sqrt{\mathbf{a} \cdot \mathbf{a}}\sqrt{\mathbf{b} \cdot \mathbf{b}}}. \tag{4}$$

EXAMPLE 1

Inner Product. Angle Between Vectors

Find the inner product and the lengths of $\mathbf{a} = [1, 2, 0]$ and $\mathbf{b} = [3, -2, 1]$ as well as the angle between these vectors.

Solution. $\mathbf{a} \cdot \mathbf{b} = 1 \cdot 3 + 2 \cdot (-2) + 0 \cdot 1 = -1$, $|\mathbf{a}| = \sqrt{\mathbf{a} \cdot \mathbf{a}} = \sqrt{5}$, $|\mathbf{b}| = \sqrt{\mathbf{b} \cdot \mathbf{b}} = \sqrt{14}$, and (4) gives the angle

$$\gamma = \arccos \frac{\mathbf{a} \cdot \mathbf{b}}{|\mathbf{a}||\mathbf{b}|} = \arccos(-0.11952) = 1.69061 = 96.865°. \qquad \blacksquare$$

From the definition we see that the inner product has the following properties. For any vectors **a**, **b**, **c** and scalars q_1, q_2,

$$(5)\quad \begin{aligned} &\text{(a)} && (q_1\mathbf{a} + q_2\mathbf{b}) \bullet \mathbf{c} = q_1\mathbf{a} \bullet \mathbf{c} + q_1\mathbf{b} \bullet \mathbf{c} && \textit{(Linearity)} \\ &\text{(b)} && \mathbf{a} \bullet \mathbf{b} = \mathbf{b} \bullet \mathbf{a} && \textit{(Symmetry)} \\ &\text{(c)} && \left.\begin{aligned} &\mathbf{a} \bullet \mathbf{a} \geqq 0 \\ &\mathbf{a} \bullet \mathbf{a} = 0 \quad \text{if and only if} \quad \mathbf{a} = \mathbf{0} \end{aligned}\right\} && \textit{(Positive-definiteness).} \end{aligned}$$

Hence *dot multiplication is commutative* as shown by (5b). Furthermore, it is *distributive with respect to vector addition.* This follows from (5a) with $q_1 = 1$ and $q_2 = 1$:

$$\text{(5a*)} \qquad (\mathbf{a} + \mathbf{b}) \bullet \mathbf{c} = \mathbf{a} \bullet \mathbf{c} + \mathbf{b} \bullet \mathbf{c} \qquad \textit{(Distributivity).}$$

Furthermore, from (1) and $|\cos \gamma| \leqq 1$ we see that

$$(6) \qquad |\mathbf{a} \bullet \mathbf{b}| \leqq |\mathbf{a}||\mathbf{b}| \qquad \textit{(Cauchy–Schwarz inequality).}$$

Using this and (3), you may prove (see Prob. 16)

$$(7) \qquad |\mathbf{a} + \mathbf{b}| \leqq |\mathbf{a}| + |\mathbf{b}| \qquad \textbf{(Triangle inequality).}$$

Geometrically, (7) with $<$ says that one side of a triangle must be shorter than the other two sides together; this motivates the name of (7).

A simple direct calculation with inner products shows that

$$(8) \qquad |\mathbf{a} + \mathbf{b}|^2 + |\mathbf{a} - \mathbf{b}|^2 = 2(|\mathbf{a}|^2 + |\mathbf{b}|^2) \quad \textit{(Parallelogram equality).}$$

Equations (6)–(8) play a basic role in so-called *Hilbert spaces*, which are abstract inner product spaces. Hilbert spaces form the basis of quantum mechanics, for details see [GenRef7] listed in App. 1.

Derivation of (2) from (1). We write $\mathbf{a} = a_1\mathbf{i} + a_2\mathbf{j} + a_3\mathbf{k}$ and $\mathbf{b} = b_1\mathbf{i} + b_2\mathbf{j} + b_3\mathbf{k}$, as in (8) of Sec. 9.1. If we substitute this into $\mathbf{a} \bullet \mathbf{b}$ and use (5a*), we first have a sum of $3 \times 3 = 9$ products

$$\mathbf{a} \bullet \mathbf{b} = a_1b_1\mathbf{i} \bullet \mathbf{i} + a_1b_2\mathbf{i} \bullet \mathbf{j} + \cdots + a_3b_3\mathbf{k} \bullet \mathbf{k}.$$

Now **i**, **j**, **k** are unit vectors, so that $\mathbf{i} \bullet \mathbf{i} = \mathbf{j} \bullet \mathbf{j} = \mathbf{k} \bullet \mathbf{k} = 1$ by (3). Since the coordinate axes are perpendicular, so are **i**, **j**, **k**, and Theorem 1 implies that the other six of those nine products are 0, namely, $\mathbf{i} \bullet \mathbf{j} = \mathbf{j} \bullet \mathbf{i} = \mathbf{j} \bullet \mathbf{k} = \mathbf{k} \bullet \mathbf{j} = \mathbf{k} \bullet \mathbf{i} = \mathbf{i} \bullet \mathbf{k} = 0$. But this reduces our sum for $\mathbf{a} \bullet \mathbf{b}$ to (2). ■

Applications of Inner Products

Typical applications of inner products are shown in the following examples and in Problem Set 9.2.

EXAMPLE 2 Work Done by a Force Expressed as an Inner Product

This is a major application. It concerns a body on which a *constant* force $\mathbf{p}$ acts. (For a *variable* force, see Sec. 10.1.) Let the body be given a displacement $\mathbf{d}$. Then the work done by $\mathbf{p}$ in the displacement is defined as

$$W = |\mathbf{p}||\mathbf{d}| \cos \alpha = \mathbf{p} \cdot \mathbf{d}, \tag{9}$$

that is, magnitude $|\mathbf{p}|$ of the force times length $|\mathbf{d}|$ of the displacement times the cosine of the angle α between $\mathbf{p}$ and $\mathbf{d}$ (Fig. 179). If $\alpha < 90°$, as in Fig. 179, then $W > 0$. If $\mathbf{p}$ and $\mathbf{d}$ are orthogonal, then the work is zero (why?). If $\alpha > 90°$, then $W < 0$, which means that in the displacement one has to do work against the force. For example, think of swimming across a river at some angle α against the current.

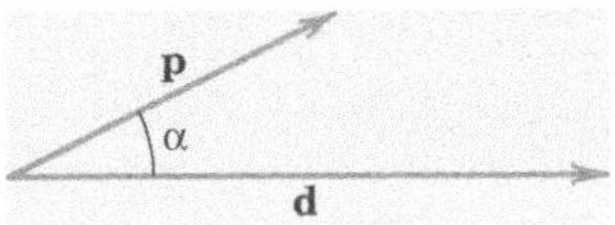

Fig. 179. Work done by a force

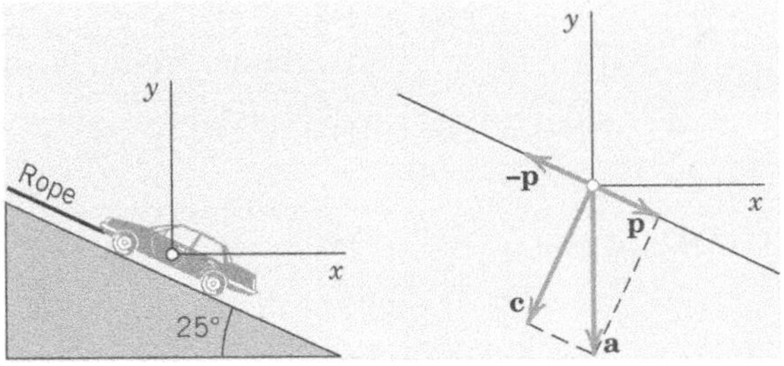

Fig. 180. Example 3

EXAMPLE 3 Component of a Force in a Given Direction

What force in the rope in Fig. 180 will hold a car of 5000 lb in equilibrium if the ramp makes an angle of 25° with the horizontal?

Solution. Introducing coordinates as shown, the weight is $\mathbf{a} = [0, -5000]$ because this force points downward, in the negative y-direction. We have to represent $\mathbf{a}$ as a sum (resultant) of two forces, $\mathbf{a} = \mathbf{c} + \mathbf{p}$, where $\mathbf{c}$ is the force the car exerts on the ramp, which is of no interest to us, and $\mathbf{p}$ is parallel to the rope. A vector in the direction of the rope is (see Fig. 180)

$$\mathbf{b} = [-1, \tan 25°] = [-1, 0.46631], \qquad \text{thus} \qquad |\mathbf{b}| = 1.10338,$$

The direction of the unit vector $\mathbf{u}$ is opposite to the direction of the rope so that

$$\mathbf{u} = -\frac{1}{|\mathbf{b}|}\mathbf{b} = [0.90631, -0.42262].$$

Since $|\mathbf{u}| = 1$ and $\cos \gamma > 0$, we see that we can write our result as

$$|\mathbf{p}| = (|\mathbf{a}| \cos \gamma)|\mathbf{u}| = \mathbf{a} \cdot \mathbf{u} = -\frac{\mathbf{a} \cdot \mathbf{b}}{|\mathbf{b}|} = \frac{5000 \cdot 0.46631}{1.10338} = 2113 \text{ [lb]}.$$

We can also note that $\gamma = 90° - 25° = 65°$ is the angle between $\mathbf{a}$ and $\mathbf{p}$ so that

$$|\mathbf{p}| = |\mathbf{a}| \cos \gamma = 5000 \cos 65° = 2113 \text{ [lb]}.$$

Answer: About 2100 lb. ■

Example 3 is typical of applications that deal with the **component** or **projection** *of a vector* **a** *in the direction of a vector* **b** ($\neq \mathbf{0}$). If we denote by p the length of the orthogonal projection of **a** on a straight line l parallel to **b** as shown in Fig. 181, then

$$p = |\mathbf{a}| \cos \gamma. \tag{10}$$

Here p is taken with the plus sign if $p\mathbf{b}$ has the direction of **b** and with the minus sign if $p\mathbf{b}$ has the direction opposite to **b**.

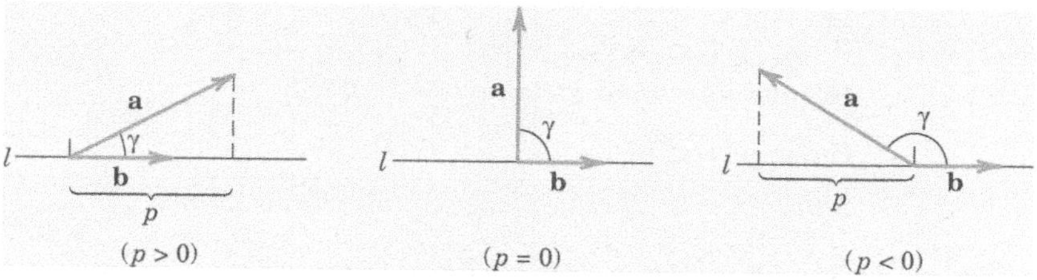

Fig. 181. Component of a vector **a** in the direction of a vector **b**

Multiplying (10) by $|\mathbf{b}|/|\mathbf{b}| = 1$, we have $\mathbf{a} \cdot \mathbf{b}$ in the numerator and thus

$$p = \frac{\mathbf{a} \cdot \mathbf{b}}{|\mathbf{b}|} \qquad (\mathbf{b} \neq \mathbf{0}). \tag{11}$$

If **b** is a unit vector, as it is often used for fixing a direction, then (11) simply gives

$$p = \mathbf{a} \cdot \mathbf{b} \qquad (|\mathbf{b}| = 1). \tag{12}$$

Figure 182 shows the projection p of **a** in the direction of **b** (as in Fig. 181) and the projection $q = |\mathbf{b}| \cos \gamma$ of **b** in the direction of **a**.

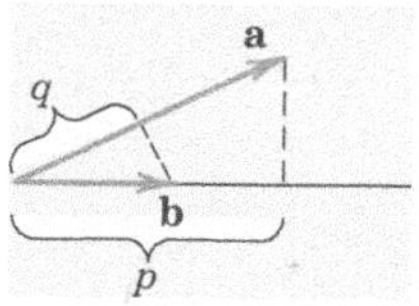

Fig. 182. Projections p of **a** on **b** and q of **b** on **a**

EXAMPLE 4 Orthonormal Basis

By definition, an *orthonormal basis* for 3-space is a basis $\{\mathbf{a}, \mathbf{b}, \mathbf{c}\}$ consisting of orthogonal unit vectors. It has the great advantage that the determination of the coefficients in representations $\mathbf{v} = l_1\mathbf{a} + l_2\mathbf{b} + l_3\mathbf{c}$ of a given vector **v** is very simple. We claim that $l_1 = \mathbf{a} \cdot \mathbf{v}$, $l_2 = \mathbf{b} \cdot \mathbf{v}$, $l_3 = \mathbf{c} \cdot \mathbf{v}$. Indeed, this follows simply by taking the inner products of the representation with **a**, **b**, **c**, respectively, and using the orthonormality of the basis, $\mathbf{a} \cdot \mathbf{v} = l_1\mathbf{a} \cdot \mathbf{a} + l_2\mathbf{a} \cdot \mathbf{b} + l_3\mathbf{a} \cdot \mathbf{c} = l_1$, etc.

For example, the unit vectors **i**, **j**, **k** in (8), Sec. 9.1, associated with a Cartesian coordinate system form an orthonormal basis, called the **standard basis** with respect to the given coordinate system. ■

EXAMPLE 5 Orthogonal Straight Lines in the Plane

Find the straight line L_1 through the point P: (1, 3) in the xy-plane and perpendicular to the straight line $L_2: x - 2y + 2 = 0$; see Fig. 183.

Solution. The idea is to write a general straight line $L_1: a_1x + a_2y = c$ as $\mathbf{a} \cdot \mathbf{r} = c$ with $\mathbf{a} = [a_1, a_2] \neq \mathbf{0}$ and $\mathbf{r} = [x, y]$, according to (2). Now the line L_1^* through the origin and parallel to L_1 is $\mathbf{a} \cdot \mathbf{r} = 0$. Hence, by Theorem 1, the vector $\mathbf{a}$ is perpendicular to $\mathbf{r}$. Hence it is perpendicular to L_1^* and also to L_1 because L_1 and L_1^* are parallel. $\mathbf{a}$ is called a **normal vector** of L_1 (and of L_1^*).

Now a normal vector of the given line $x - 2y + 2 = 0$ is $\mathbf{b} = [1, -2]$. Thus L_1 is perpendicular to L_2 if $\mathbf{b} \cdot \mathbf{a} = a_1 - 2a_2 = 0$, for instance, if $\mathbf{a} = [2, 1]$. Hence L_1 is given by $2x + y = c$. It passes through P: (1, 3) when $2 \cdot 1 + 3 = c = 5$. *Answer:* $y = -2x + 5$. Show that the point of intersection is $(x, y) = (1.6, 1.8)$. ■

EXAMPLE 6 Normal Vector to a Plane

Find a unit vector perpendicular to the plane $4x + 2y + 4z = -7$.

Solution. Using (2), we may write any plane in space as

$$\mathbf{a} \cdot \mathbf{r} = a_1x + a_2y + a_3z = c \tag{13}$$

where $\mathbf{a} = [a_1, a_2, a_3] \neq \mathbf{0}$ and $\mathbf{r} = [x, y, z]$. The unit vector in the direction of $\mathbf{a}$ is (Fig. 184)

$$\mathbf{n} = \frac{1}{|\mathbf{a}|}\mathbf{a}.$$

Dividing by $|\mathbf{a}|$, we obtain from (13)

$$\mathbf{n} \cdot \mathbf{r} = p \qquad \text{where} \qquad p = \frac{c}{|\mathbf{a}|}. \tag{14}$$

From (12) we see that p is the projection of $\mathbf{r}$ in the direction of $\mathbf{n}$. This projection has the same constant value $c/|\mathbf{a}|$ for the position vector $\mathbf{r}$ of any point in the plane. Clearly this holds if and only if $\mathbf{n}$ is perpendicular to the plane. $\mathbf{n}$ is called a **unit normal vector** of the plane (the other being $-\mathbf{n}$).

Furthermore, from this and the definition of projection, it follows that $|p|$ is the distance of the plane from the origin. Representation (14) is called **Hesse's[2] normal form** of a plane. In our case, $\mathbf{a} = [4, 2, 4]$, $c = -7$, $|\mathbf{a}| = 6$, $\mathbf{n} = \frac{1}{6}\mathbf{a} = [\frac{2}{3}, \frac{1}{3}, \frac{2}{3}]$, and the plane has the distance $\frac{7}{6}$ from the origin. ■

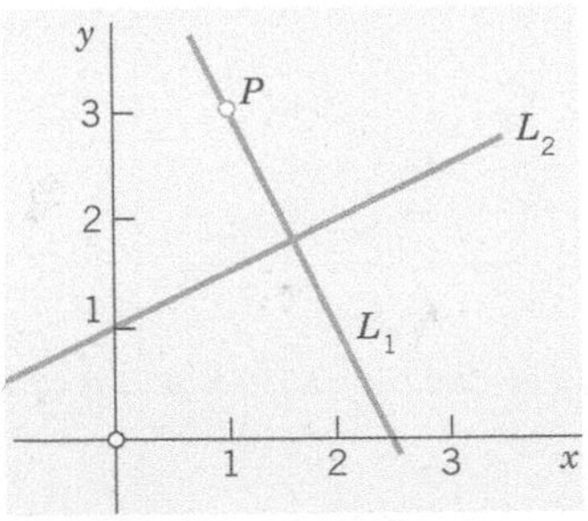

Fig. 183. Example 5

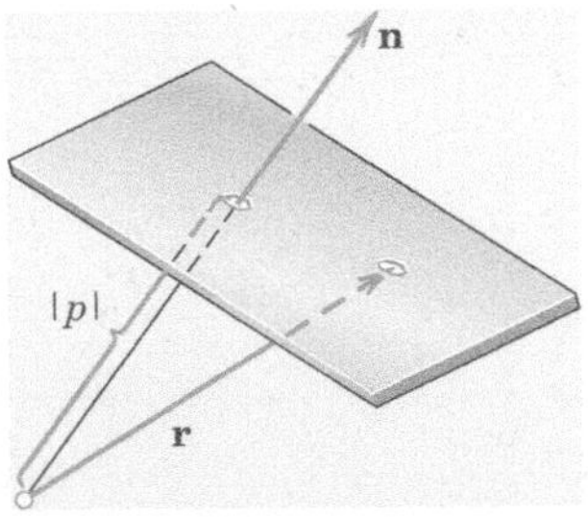

Fig. 184. Normal vector to a plane

[2]LUDWIG OTTO HESSE (1811–1874), German mathematician who contributed to the theory of curves and surfaces.

PROBLEM SET 9.2

1–10 INNER PRODUCT

Let $\mathbf{a} = [1, -3, 5]$, $\mathbf{b} = [4, 0, 8]$, $\mathbf{c} = [-2, 9, 1]$. Find:

1. $\mathbf{a} \cdot \mathbf{b}$, $\mathbf{b} \cdot \mathbf{a}$, $\mathbf{b} \cdot \mathbf{c}$
2. $(-3\mathbf{a} + 5\mathbf{c}) \cdot \mathbf{b}$, $15(\mathbf{a} - \mathbf{c}) \cdot \mathbf{b}$
3. $|\mathbf{a}|$, $|2\mathbf{b}|$, $|-\mathbf{c}|$
4. $|\mathbf{a} + \mathbf{b}|$, $|\mathbf{a}| + |\mathbf{b}|$
5. $|\mathbf{b} + \mathbf{c}|$, $|\mathbf{b}| + |\mathbf{c}|$
6. $|\mathbf{a} + \mathbf{c}|^2 + |\mathbf{a} - \mathbf{c}|^2 - 2(|\mathbf{a}|^2 + |\mathbf{c}|^2)$
7. $|\mathbf{a} \cdot \mathbf{c}|$, $|\mathbf{a}||\mathbf{c}|$
8. $5\mathbf{a} \cdot 13\mathbf{b}$, $65\mathbf{a} \cdot \mathbf{b}$
9. $15\mathbf{a} \cdot \mathbf{b} + 15\mathbf{a} \cdot \mathbf{c}$, $15\mathbf{a} \cdot (\mathbf{b} + \mathbf{c})$
10. $\mathbf{a} \cdot (\mathbf{b} - \mathbf{c})$, $(\mathbf{a} - \mathbf{b}) \cdot \mathbf{c}$

11–16 GENERAL PROBLEMS

11. What laws do Probs. 1 and 4–7 illustrate?
12. What does $\mathbf{u} \cdot \mathbf{v} = \mathbf{u} \cdot \mathbf{w}$ imply if $\mathbf{u} = \mathbf{0}$? If $\mathbf{u} \neq \mathbf{0}$?
13. Prove the Cauchy–Schwarz inequality.
14. Verify the Cauchy–Schwarz and triangle inequalities for the above **a** and **b**.
15. Prove the parallelogram equality. Explain its name.
16. **Triangle inequality.** Prove Eq. (7). *Hint.* Use Eq. (3) for $|\mathbf{a} + \mathbf{b}|$ and Eq. (6) to prove the square of Eq. (7), then take roots.

17–20 WORK

Find the work done by a force **p** acting on a body if the body is displaced along the straight segment $\overline{AB}$ from A to B. Sketch $\overline{AB}$ and **p**. Show the details.

17. $\mathbf{p} = [2, 5, 0]$, A: (1, 3, 3), B: (3, 5, 5)
18. $\mathbf{p} = [-1, -2, 4]$, A: (0, 0, 0), B: (6, 7, 5)
19. $\mathbf{p} = [0, 4, 3]$, A: (4, 5, −1), B: (1, 3, 0)
20. $\mathbf{p} = [6, -3, -3]$, A: (1, 5, 2), B: (3, 4, 1)
21. **Resultant.** Is the work done by the resultant of two forces in a displacement the sum of the work done by each of the forces separately? Give proof or counterexample.

22–30 ANGLE BETWEEN VECTORS

Let $\mathbf{a} = [1, 1, 0]$, $\mathbf{b} = [3, 2, 1]$, and $\mathbf{c} = [1, 0, 2]$. Find the angle between:

22. **a**, **b**
23. **b**, **c**
24. $\mathbf{a} + \mathbf{c}$, $\mathbf{b} + \mathbf{c}$
25. What will happen to the angle in Prob. 24 if we replace **c** by $n\mathbf{c}$ with larger and larger n?
26. **Cosine law.** Deduce the law of cosines by using vectors **a**, **b**, and $\mathbf{a} - \mathbf{b}$.
27. **Addition law.** $\cos(\alpha - \beta) = \cos\alpha\cos\beta + \sin\alpha \sin\beta$. Obtain this by using $\mathbf{a} = [\cos\alpha, \sin\alpha]$, $\mathbf{b} = [\cos\beta, \sin\beta]$ where $0 \leqq \alpha \leqq \beta \leqq 2\pi$.
28. **Triangle.** Find the angles of the triangle with vertices A: (0, 0, 2), B: (3, 0, 2), and C: (1, 1, 1). Sketch the triangle.
29. **Parallelogram.** Find the angles if the vertices are (0, 0), (6, 0), (8, 3), and (2, 3).
30. **Distance.** Find the distance of the point A: (1, 0, 2) from the plane P: $3x + y + z = 9$. Make a sketch.

31–35 ORTHOGONALITY is particularly important, mainly because of orthogonal coordinates, such as ***Cartesian coordinates***, whose ***natural basis*** [Eq. (9), Sec. 9.1], consists of three orthogonal unit vectors.

31. For what values of a_1 are $[a_1, 4, 3]$ and $[3, -2, 12]$ orthogonal?
32. **Planes.** For what c are $3x + z = 5$ and $8x - y + cz = 9$ orthogonal?
33. **Unit vectors.** Find all unit vectors $\mathbf{a} = [a_1, a_2]$ in the plane orthogonal to [4, 3].
34. **Corner reflector.** Find the angle between a light ray and its reflection in three orthogonal plane mirrors, known as *corner reflector*.
35. **Parallelogram.** When will the diagonals be orthogonal? Give a proof.

36–40 COMPONENT IN THE DIRECTION OF A VECTOR

Find the component of **a** in the direction of **b**. Make a sketch.

36. $\mathbf{a} = [1, 1, 1]$, $\mathbf{b} = [2, 1, 3]$
37. $\mathbf{a} = [3, 4, 0]$, $\mathbf{b} = [4, -3, 2]$
38. $\mathbf{a} = [8, 2, 0]$, $\mathbf{b} = [-4, -1, 0]$
39. When will the component (the projection) of **a** in the direction of **b** be equal to the component (the projection) of **b** in the direction of **a**? First guess.
40. What happens to the component of **a** in the direction of **b** if you change the length of **b**?

9.3 Vector Product (Cross Product)

We shall define another form of multiplication of vectors, inspired by applications, whose result will be a *vector*. This is in contrast to the dot product of Sec. 9.2 where multiplication resulted in a *scalar*. We can construct a vector $\mathbf{v}$ that is perpendicular to two vectors $\mathbf{a}$ and $\mathbf{b}$, which are two sides of a parallelogram on a plane in space as indicated in Fig. 185, such that the length $|\mathbf{v}|$ is numerically equal to the area of that parallelogram. Here then is the new concept.

DEFINITION

Vector Product (Cross Product, Outer Product) of Vectors

The **vector product** or **cross product** $\mathbf{a} \times \mathbf{b}$ (read "$\mathbf{a}$ cross $\mathbf{b}$") of two vectors $\mathbf{a}$ and $\mathbf{b}$ is the vector $\mathbf{v}$ denoted by

$$\mathbf{v} = \mathbf{a} \times \mathbf{b}$$

I. If $\mathbf{a} = \mathbf{0}$ or $\mathbf{b} = \mathbf{0}$, then we define $\mathbf{v} = \mathbf{a} \times \mathbf{b} = \mathbf{0}$.

II. If both vectors are nonzero vectors, then vector $\mathbf{v}$ has the length

$$|\mathbf{v}| = |\mathbf{a} \times \mathbf{b}| = |\mathbf{a}||\mathbf{b}| \sin \gamma, \tag{1}$$

where γ is the angle between $\mathbf{a}$ and $\mathbf{b}$ as in Sec. 9.2.

Furthermore, by design, $\mathbf{a}$ and $\mathbf{b}$ form the sides of a parallelogram on a plane in space. The parallelogram is shaded in blue in Fig. 185. The area of this blue parallelogram is precisely given by Eq. (1), so that the length $|\mathbf{v}|$ of the vector $\mathbf{v}$ is equal to the area of that parallelogram.

III. If $\mathbf{a}$ and $\mathbf{b}$ lie in the same straight line, i.e., $\mathbf{a}$ and $\mathbf{b}$ have the same or opposite directions, then γ is $0°$ or $180°$ so that $\sin \gamma = 0$. In that case $|\mathbf{v}| = 0$ so that $\mathbf{v} = \mathbf{a} \times \mathbf{b} = \mathbf{0}$.

IV. If cases I and III do not occur, then $\mathbf{v}$ is a nonzero vector. The direction of $\mathbf{v} = \mathbf{a} \times \mathbf{b}$ is perpendicular to both $\mathbf{a}$ and $\mathbf{b}$ such that $\mathbf{a}$, $\mathbf{b}$, $\mathbf{v}$—precisely in this order (!)—form a right-handed triple as shown in Figs. 185–187 and explained below.

Another term for vector product is outer product.

Remark. Note that I and III completely characterize the exceptional case when the cross product is equal to the zero vector, and II and IV the regular case where the cross product is perpendicular to two vectors.

Just as we did with the dot product, we would also like to express the cross product in components. Let $\mathbf{a} = [a_1, a_2, a_3]$ and $\mathbf{b} = [b_1, b_2, b_3]$. Then $\mathbf{v} = [v_1, v_2, v_3] = \mathbf{a} \times \mathbf{b}$ has the components

$$v_1 = a_2b_3 - a_3b_2, \qquad v_2 = a_3b_1 - a_1b_3, \qquad v_3 = a_1b_2 - a_2b_1. \tag{2}$$

Here the Cartesian coordinate system is *right-handed*, as explained below (see also Fig. 188). (For a left-handed system, each component of $\mathbf{v}$ must be multiplied by -1. Derivation of (2) in App. 4.)

Right-Handed Triple. A triple of vectors **a**, **b**, **v** is *right-handed* if the vectors in the given order assume the same sort of orientation as the thumb, index finger, and middle finger of the right hand when these are held as in Fig. 186. We may also say that if **a** is rotated into the direction of **b** through the angle $\gamma\ (<\pi)$, then **v** advances in the same direction as a right-handed screw would if turned in the same way (Fig. 187).

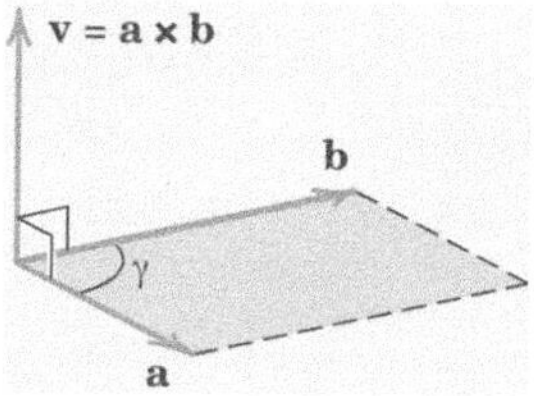

Fig. 185. Vector product

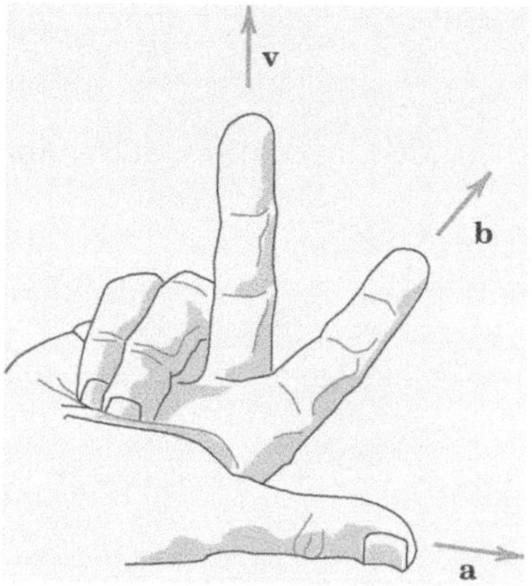

Fig. 186. Right-handed triple of vectors **a, b, v**

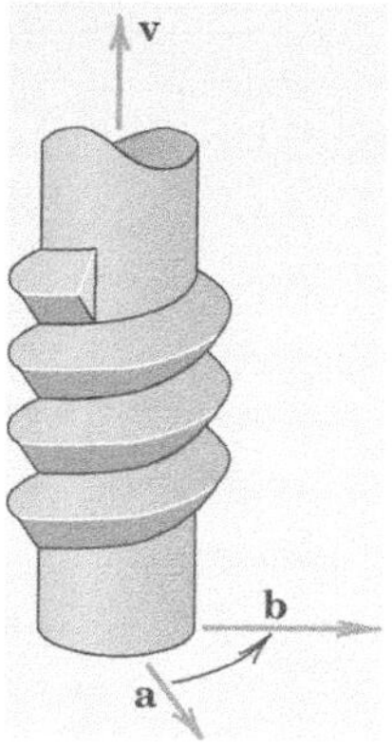

Fig. 187. Right-handed screw

Right-Handed Cartesian Coordinate System. The system is called **right-handed** if the corresponding unit vectors **i**, **j**, **k** in the positive directions of the axes (see Sec. 9.1) form a right-handed triple as in Fig. 188a. The system is called **left-handed** if the sense of **k** is reversed, as in Fig. 188b. In applications, we prefer right-handed systems.

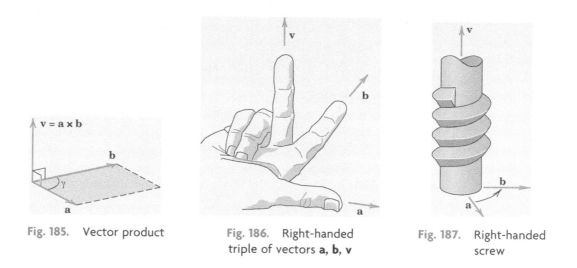

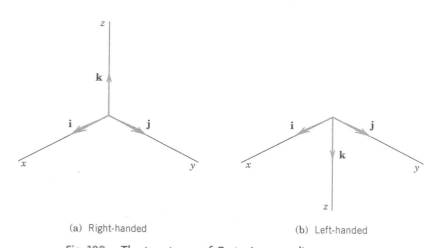

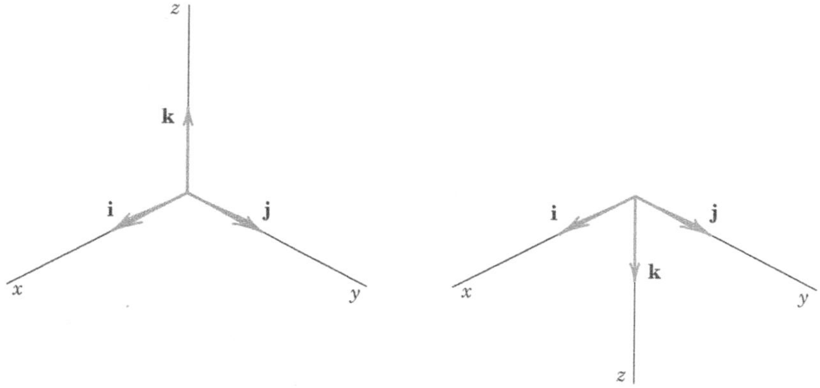

(a) Right-handed (b) Left-handed

Fig. 188. The two types of Cartesian coordinate systems

How to Memorize (2). If you know second- and third-order determinants, you see that (2) can be written

$$(2^*)\qquad v_1 = \begin{vmatrix} a_2 & a_3 \\ b_2 & b_3 \end{vmatrix}, \qquad v_2 = -\begin{vmatrix} a_1 & a_3 \\ b_1 & b_3 \end{vmatrix} = +\begin{vmatrix} a_3 & a_1 \\ b_3 & b_1 \end{vmatrix}, \qquad v_3 = \begin{vmatrix} a_1 & a_2 \\ b_1 & b_2 \end{vmatrix}$$

and $\mathbf{v} = [v_1, v_2, v_3] = v_1\mathbf{i} + v_2\mathbf{j} + v_3\mathbf{k}$ is the expansion of the following symbolic determinant by its first row. (We call the determinant "symbolic" because the first row consists of vectors rather than of numbers.)

$$(2^{**}) \qquad \mathbf{v} = \mathbf{a} \times \mathbf{b} = \begin{vmatrix} \mathbf{i} & \mathbf{j} & \mathbf{k} \\ a_1 & a_2 & a_3 \\ b_1 & b_2 & b_3 \end{vmatrix} = \begin{vmatrix} a_2 & a_3 \\ b_2 & b_3 \end{vmatrix} \mathbf{i} - \begin{vmatrix} a_1 & a_3 \\ b_1 & b_3 \end{vmatrix} \mathbf{j} + \begin{vmatrix} a_1 & a_2 \\ b_1 & b_2 \end{vmatrix} \mathbf{k}.$$

For a left-handed system the determinant has a minus sign in front.

EXAMPLE 1 Vector Product

For the vector product $\mathbf{v} = \mathbf{a} \times \mathbf{b}$ of $\mathbf{a} = [1, 1, 0]$ and $\mathbf{b} = [3, 0, 0]$ in right-handed coordinates we obtain from (2)

$$v_1 = 0, \qquad v_2 = 0, \qquad v_3 = 1 \cdot 0 - 1 \cdot 3 = -3.$$

We confirm this by (2**):

$$\mathbf{v} = \mathbf{a} \times \mathbf{b} = \begin{vmatrix} \mathbf{i} & \mathbf{j} & \mathbf{k} \\ 1 & 1 & 0 \\ 3 & 0 & 0 \end{vmatrix} = \begin{vmatrix} 1 & 0 \\ 0 & 0 \end{vmatrix} \mathbf{i} - \begin{vmatrix} 1 & 0 \\ 3 & 0 \end{vmatrix} \mathbf{j} + \begin{vmatrix} 1 & 1 \\ 3 & 0 \end{vmatrix} \mathbf{k} = -3\mathbf{k} = [0, 0, -3].$$

To check the result in this simple case, sketch **a**, **b**, and **v**. Can you see that two vectors in the xy-plane must always have their vector product parallel to the z-axis (or equal to the zero vector)? ■

EXAMPLE 2 Vector Products of the Standard Basis Vectors

$$(3) \qquad \begin{aligned} \mathbf{i} \times \mathbf{j} &= \mathbf{k}, & \mathbf{j} \times \mathbf{k} &= \mathbf{i}, & \mathbf{k} \times \mathbf{i} &= \mathbf{j} \\ \mathbf{j} \times \mathbf{i} &= -\mathbf{k}, & \mathbf{k} \times \mathbf{j} &= -\mathbf{i}, & \mathbf{i} \times \mathbf{k} &= -\mathbf{j}. \end{aligned}$$

We shall use this in the next proof. ■

THEOREM 1 General Properties of Vector Products

(a) *For every scalar* l,

$$(4) \qquad (l\mathbf{a}) \times \mathbf{b} = l(\mathbf{a} \times \mathbf{b}) = \mathbf{a} \times (l\mathbf{b}).$$

(b) *Cross multiplication is distributive with respect to vector addition; that is,*

$$(5) \qquad \begin{aligned} (\alpha) \quad & \mathbf{a} \times (\mathbf{b} + \mathbf{c}) = (\mathbf{a} \times \mathbf{b}) + (\mathbf{a} \times \mathbf{c}), \\ (\beta) \quad & (\mathbf{a} + \mathbf{b}) \times \mathbf{c} = (\mathbf{a} \times \mathbf{c}) + (\mathbf{b} \times \mathbf{c}). \end{aligned}$$

(c) *Cross multiplication is* ***not commutative*** *but* ***anticommutative;*** *that is,*

$$(6) \qquad \mathbf{b} \times \mathbf{a} = -(\mathbf{a} \times \mathbf{b}) \qquad \text{(Fig. 189).}$$

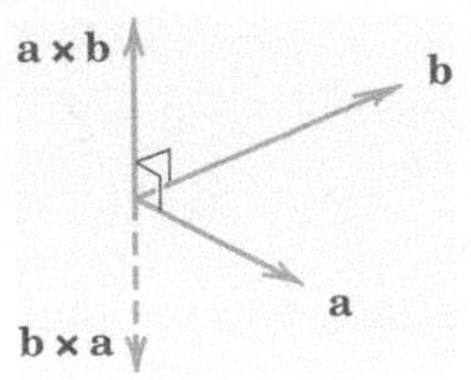

Fig. 189. Anticommutativity of cross multiplication

(d) *Cross multiplication is* **not associative**; *that is, in general,*

(7) $$\mathbf{a} \times (\mathbf{b} \times \mathbf{c}) \neq (\mathbf{a} \times \mathbf{b}) \times \mathbf{c}$$

so that the parentheses cannot be omitted.

PROOF Equation (4) follows directly from the definition. In (5α), formula (2*) gives for the first component on the left

$$\begin{vmatrix} a_2 & a_3 \\ b_2 + c_2 & b_3 + c_3 \end{vmatrix} = a_2(b_3 + c_3) - a_3(b_2 + c_2)$$

$$= (a_2b_3 - a_3b_2) + (a_2c_3 - a_3c_2)$$

$$= \begin{vmatrix} a_2 & a_3 \\ b_2 & b_3 \end{vmatrix} + \begin{vmatrix} a_2 & a_3 \\ c_2 & c_3 \end{vmatrix}.$$

By (2*) the sum of the two determinants is the first component of $(\mathbf{a} \times \mathbf{b}) + (\mathbf{a} \times \mathbf{c})$, the right side of (5$\alpha$). For the other components in (5α) and in 5(β), equality follows by the same idea.

Anticommutativity (6) follows from (2**) by noting that the interchange of Rows 2 and 3 multiplies the determinant by -1. We can confirm this geometrically if we set $\mathbf{a} \times \mathbf{b} = \mathbf{v}$ and $\mathbf{b} \times \mathbf{a} = \mathbf{w}$; then $|\mathbf{v}| = |\mathbf{w}|$ by (1), and for $\mathbf{b}$, $\mathbf{a}$, $\mathbf{w}$ to form a *right-handed* triple, we must have $\mathbf{w} = -\mathbf{v}$.

Finally, $\mathbf{i} \times (\mathbf{i} \times \mathbf{j}) = \mathbf{i} \times \mathbf{k} = -\mathbf{j}$, whereas $(\mathbf{i} \times \mathbf{i}) \times \mathbf{j} = \mathbf{0} \times \mathbf{j} = \mathbf{0}$ (see Example 2). This proves (7). ■

Typical Applications of Vector Products

EXAMPLE 3 **Moment of a Force**

In mechanics the moment m of a force $\mathbf{p}$ about a point Q is defined as the product $m = |\mathbf{p}|d$, where d is the (perpendicular) distance between Q and the line of action L of $\mathbf{p}$ (Fig. 190). If $\mathbf{r}$ is the vector from Q to any point A on L, then $d = |\mathbf{r}| \sin \gamma$, as shown in Fig. 190, and

$$m = |\mathbf{r}||\mathbf{p}| \sin \gamma.$$

Since γ is the angle between $\mathbf{r}$ and $\mathbf{p}$, we see from (1) that $m = |\mathbf{r} \times \mathbf{p}|$. The vector

(8) $$\mathbf{m} = \mathbf{r} \times \mathbf{p}$$

is called the **moment vector** or **vector moment** of $\mathbf{p}$ about Q. Its magnitude is m. If $\mathbf{m} \neq \mathbf{0}$, its direction is that of the axis of the rotation about Q that $\mathbf{p}$ has the tendency to produce. This axis is perpendicular to both $\mathbf{r}$ and $\mathbf{p}$. ■

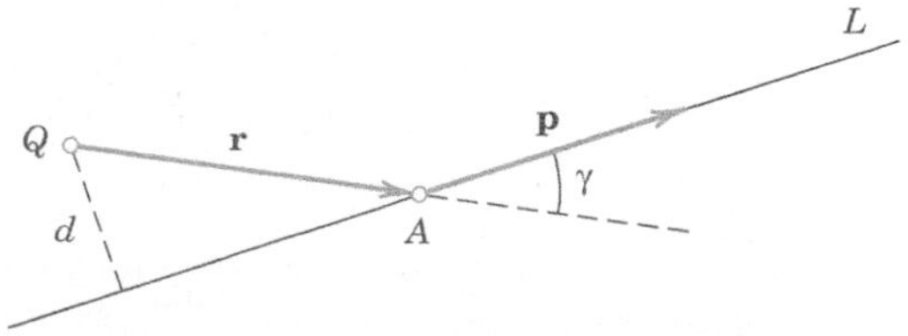

Fig. 190. Moment of a force **p**

EXAMPLE 4 Moment of a Force

Find the moment of the force **p** about the center Q of a wheel, as given in Fig. 191.

Solution. Introducing coordinates as shown in Fig. 191, we have

$$\mathbf{p} = [1000 \cos 30°, \quad 1000 \sin 30°, \quad 0] = [866, \quad 500, \quad 0], \qquad \mathbf{r} = [0, \quad 1.5, \quad 0].$$

(Note that the center of the wheel is at $y = -1.5$ on the y-axis.) Hence (8) and (2**) give

$$\mathbf{m} = \mathbf{r} \times \mathbf{p} = \begin{vmatrix} \mathbf{i} & \mathbf{j} & \mathbf{k} \\ 0 & 1.5 & 0 \\ 866 & 500 & 0 \end{vmatrix} = 0\mathbf{i} - 0\mathbf{j} + \begin{vmatrix} 0 & 1.5 \\ 866 & 500 \end{vmatrix} \mathbf{k} = [0, 0, -1299].$$

This moment vector **m** is normal, i.e., perpendicular to the plane of the wheel. Hence it has the direction of the axis of rotation about the center Q of the wheel that the force **p** has the tendency to produce. The moment **m** points in the negative z-direction, This is, the direction in which a right-handed screw would advance if turned in that way. ■

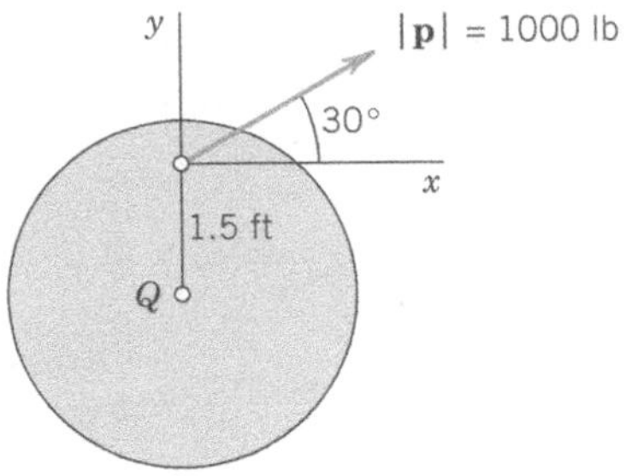

Fig. 191. Moment of a force **p**

EXAMPLE 5 Velocity of a Rotating Body

A rotation of a rigid body B in space can be simply and uniquely described by a vector **w** as follows. The direction of **w** is that of the axis of rotation and such that the rotation appears clockwise if one looks from the initial point of **w** to its terminal point. The length of **w** is equal to the **angular speed** $\omega(>0)$ of the rotation, that is, the linear (or tangential) speed of a point of B divided by its distance from the axis of rotation.

Let P be any point of B and d its distance from the axis. Then P has the speed ωd. Let **r** be the position vector of P referred to a coordinate system with origin 0 on the axis of rotation. Then $d = |\mathbf{r}| \sin \gamma$, where γ is the angle between **w** and **r**. Therefore,

$$\omega d = |\mathbf{w}||\mathbf{r}| \sin \gamma = |\mathbf{w} \times \mathbf{r}|.$$

From this and the definition of vector product we see that the velocity vector **v** of P can be represented in the form (Fig. 192)

$$\mathbf{v} = \mathbf{w} \times \mathbf{r}. \tag{9}$$

This simple formula is useful for determining **v** at any point of B. ■

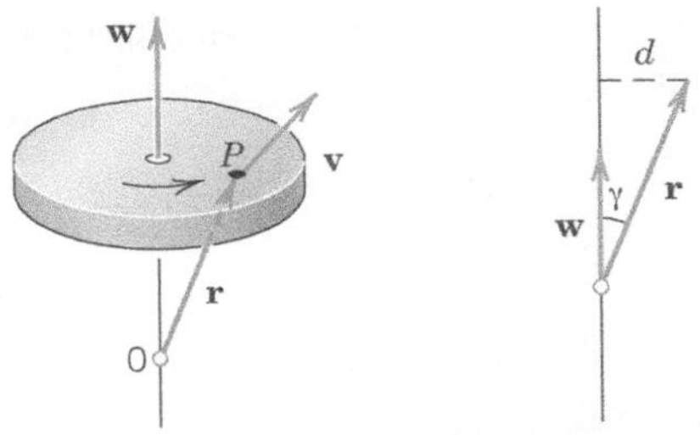

Fig. 192. Rotation of a rigid body

Scalar Triple Product

Certain products of vectors, having three or more factors, occur in applications. The most important of these products is the scalar triple product or mixed product of three vectors **a**, **b**, **c**.

$$(\mathbf{a}\ \ \mathbf{b}\ \ \mathbf{c}) = \mathbf{a} \bullet (\mathbf{b} \times \mathbf{c}). \tag{10*}$$

The scalar triple product is indeed a scalar since (10*) involves a dot product, which in turn is a scalar. We want to express the scalar triple product in components and as a third-order determinant. To this end, let $\mathbf{a} = [a_1, a_2, a_3]$, $\mathbf{b} = [b_1, b_2, b_3]$, and $\mathbf{c} = [c_1, c_2, c_3]$. Also set $\mathbf{b} \times \mathbf{c} = \mathbf{v} = [v_1, v_2, v_3]$. Then from the dot product in components [formula (2) in Sec. 9.2] and from (2*) with **b** and **c** instead of **a** and **b** we first obtain

$$\begin{aligned}\mathbf{a} \bullet (\mathbf{b} \times \mathbf{c}) &= \mathbf{a} \bullet \mathbf{v} = a_1 v_1 + a_2 v_2 + a_3 v_3 \\ &= a_1 \begin{vmatrix} b_2 & b_3 \\ c_2 & c_3 \end{vmatrix} + a_2 \begin{vmatrix} b_3 & b_1 \\ c_3 & c_1 \end{vmatrix} + a_3 \begin{vmatrix} b_1 & b_2 \\ c_1 & c_2 \end{vmatrix}.\end{aligned}$$

The sum on the right is the expansion of a third-order determinant by its first row. Thus we obtain the desired formula for the scalar triple product, that is,

$$(\mathbf{a}\ \ \mathbf{b}\ \ \mathbf{c}) = \mathbf{a} \bullet (\mathbf{b} \times \mathbf{c}) = \begin{vmatrix} a_1 & a_2 & a_3 \\ b_1 & b_2 & b_3 \\ c_1 & c_2 & c_3 \end{vmatrix}. \tag{10}$$

The most important properties of the scalar triple product are as follows.

THEOREM 2

Properties and Applications of Scalar Triple Products

(a) *In* (10) *the dot and cross can be interchanged:*

$$(\mathbf{a}\ \ \mathbf{b}\ \ \mathbf{c}) = \mathbf{a} \bullet (\mathbf{b} \times \mathbf{c}) = (\mathbf{a} \times \mathbf{b}) \bullet \mathbf{c}. \tag{11}$$

(b) Geometric interpretation. *The absolute value* $|(\mathbf{a}\ \ \mathbf{b}\ \ \mathbf{c})|$ *of* (10) *is the volume of the parallelepiped* (oblique box) *with* **a**, **b**, **c** *as edge vectors* (Fig. 193).

(c) Linear independence. *Three vectors in* R^3 *are linearly independent if and only if their scalar triple product is not zero.*

PROOF **(a)** Dot multiplication is commutative, so that by (10)

$$(\mathbf{a} \times \mathbf{b}) \bullet \mathbf{c} = \mathbf{c} \bullet (\mathbf{a} \times \mathbf{b}) = \begin{vmatrix} c_1 & c_2 & c_3 \\ a_1 & a_2 & a_3 \\ b_1 & b_2 & b_3 \end{vmatrix}.$$

From this we obtain the determinant in (10) by interchanging Rows 1 and 2 and in the result Rows 2 and 3. But this does not change the value of the determinant because each interchange produces a factor -1, and $(-1)(-1) = 1$. This proves (11).

(b) The volume of that box equals the height $h = |\mathbf{a}||\cos\gamma|$ (Fig. 193) times the area of the base, which is the area $|\mathbf{b} \times \mathbf{c}|$ of the parallelogram with sides **b** and **c**. Hence the volume is

$$|\mathbf{a}||\mathbf{b} \times \mathbf{c}|\,|\cos\gamma| = |\mathbf{a} \cdot (\mathbf{b} \times \mathbf{c})| \qquad \text{(Fig. 193)}$$

as given by the absolute value of (11).

(c) Three nonzero vectors, whose initial points coincide, are linearly independent if and only if the vectors do not lie in the same plane nor lie on the same straight line.

This happens if and only if the triple product in (b) is not zero, so that the independence criterion follows. (The case of one of the vectors being the zero vector is trivial.)

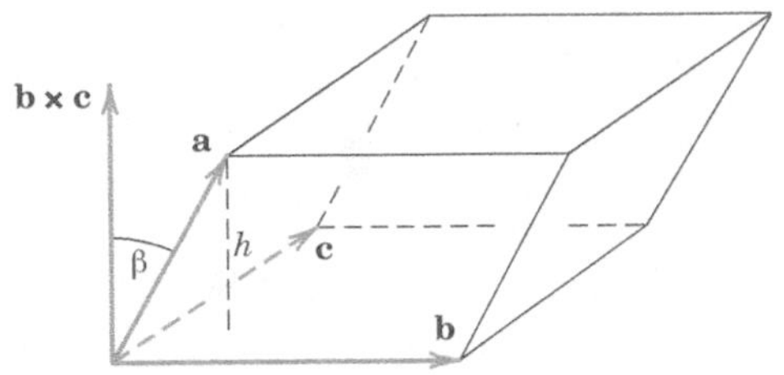

Fig. 193. Geometric interpretation of a scalar triple product

EXAMPLE 6 Tetrahedron

A tetrahedron is determined by three edge vectors **a**, **b**, **c**, as indicated in Fig. 194. Find the volume of the tetrahedron in Fig. 194, when $\mathbf{a} = [2, 0, 3]$, $\mathbf{b} = [0, 4, 1]$, $c = [5, 6, 0]$.

Solution. The volume V of the parallelepiped with these vectors as edge vectors is the absolute value of the scalar triple product

$$(\mathbf{a}\ \ \mathbf{b}\ \ \mathbf{c}) = \begin{vmatrix} 2 & 0 & 3 \\ 0 & 4 & 1 \\ 5 & 6 & 0 \end{vmatrix} = 2\begin{vmatrix} 4 & 1 \\ 6 & 0 \end{vmatrix} + 3\begin{vmatrix} 0 & 4 \\ 5 & 6 \end{vmatrix} = -12 - 60 = -72.$$

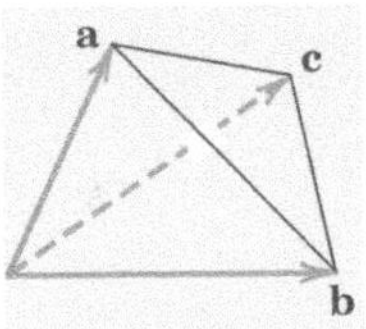

Fig. 194. Tetrahedron

Hence $V = 72$. The minus sign indicates that if the coordinates are right-handed, the triple **a**, **b**, **c** is left-handed. The volume of a tetrahedron is $\frac{1}{6}$ of that of the parallelepiped (can you prove it?), hence 12.

Can you sketch the tetrahedron, choosing the origin as the common initial point of the vectors? What are the coordinates of the four vertices? ■

This is the end of vector *algebra* (in space R^3 and in the plane). Vector *calculus* (differentiation) begins in the next section.

PROBLEM SET 9.3

1–10 GENERAL PROBLEMS

1. Give the details of the proofs of Eqs. (4) and (5).
2. What does $\mathbf{a} \times \mathbf{b} = \mathbf{a} \times \mathbf{c}$ with $\mathbf{a} \neq \mathbf{0}$ imply?
3. Give the details of the proofs of Eqs. (6) and (11).
4. **Lagrange's identity for $|\mathbf{a} \times \mathbf{b}|$.** Verify it for $\mathbf{a} = [3, 4, 2]$ and $\mathbf{b} = [1, 0, 2]$. Prove it, using $\sin^2\gamma = 1 - \cos^2\gamma$. The identity is

$$\text{(12)} \qquad |\mathbf{a} \times \mathbf{b}| = \sqrt{(\mathbf{a} \cdot \mathbf{a})(\mathbf{b} \cdot \mathbf{b}) - (\mathbf{a} \cdot \mathbf{b})^2}.$$

5. What happens in Example 3 of the text if you replace $\mathbf{p}$ by $-\mathbf{p}$?

6. What happens in Example 5 if you choose a P at distance $2d$ from the axis of rotation?

7. Rotation. A wheel is rotating about the y-axis with angular speed $\omega = 20 \text{ sec}^{-1}$. The rotation appears clockwise if one looks from the origin in the positive y-direction. Find the velocity and speed at the point $[8, 6, 0]$. Make a sketch.

8. Rotation. What are the velocity and speed in Prob. 7 at the point $(4, 2, -2)$ if the wheel rotates about the line $y = x, z = 0$ with $\omega = 10 \text{ sec}^{-1}$?

9. Scalar triple product. What does $(\mathbf{a}\ \mathbf{b}\ \mathbf{c}) = 0$ imply with respect to these vectors?

10. WRITING REPORT. Summarize the most important applications discussed in this section. Give examples. No proofs.

11–23 VECTOR AND SCALAR TRIPLE PRODUCTS

With respect to right-handed Cartesian coordinates, let $\mathbf{a} = [2, 1, 0]$, $\mathbf{b} = [-3, 2, 0]$, $\mathbf{c} = [1, 4, -2]$, and $\mathbf{d} = [5, -1, 3]$. Showing details, find:

11. $\mathbf{a} \times \mathbf{b}$, $\mathbf{b} \times \mathbf{a}$, $\mathbf{a} \bullet \mathbf{b}$

12. $3\mathbf{c} \times 5\mathbf{d}$, $15\mathbf{d} \times \mathbf{c}$, $15\mathbf{d} \bullet \mathbf{c}$, $15\mathbf{c} \bullet \mathbf{d}$

13. $\mathbf{c} \times (\mathbf{a} + \mathbf{b})$, $\mathbf{a} \times \mathbf{c} + \mathbf{b} \times \mathbf{c}$

14. $4\mathbf{b} \times 3\mathbf{c} + 12\mathbf{c} \times \mathbf{b}$

15. $(\mathbf{a} + \mathbf{d}) \times (\mathbf{d} + \mathbf{a})$

16. $(\mathbf{b} \times \mathbf{c}) \bullet \mathbf{d}$, $\mathbf{b} \bullet (\mathbf{c} \times \mathbf{d})$

17. $(\mathbf{b} \times \mathbf{c}) \times \mathbf{d}$, $\mathbf{b} \times (\mathbf{c} \times \mathbf{d})$

18. $(\mathbf{a} \times \mathbf{b}) \times \mathbf{a}$, $\mathbf{a} \times (\mathbf{b} \times \mathbf{a})$

19. $(\mathbf{i}\ \mathbf{j}\ \mathbf{k})$, $(\mathbf{i}\ \mathbf{k}\ \mathbf{j})$

20. $(\mathbf{a} \times \mathbf{b}) \times (\mathbf{c} \times \mathbf{d})$, $(\mathbf{a}\ \mathbf{b}\ \mathbf{d})\mathbf{c} - (\mathbf{a}\ \mathbf{b}\ \mathbf{c})\mathbf{d}$

21. $4\mathbf{b} \times 3\mathbf{c}$, $12|\mathbf{b} \times \mathbf{c}|$, $12|\mathbf{c} \times \mathbf{b}|$

22. $(\mathbf{a} - \mathbf{b}\ \ \mathbf{c} - \mathbf{b}\ \ \mathbf{d} - \mathbf{b})$, $(\mathbf{a}\ \mathbf{c}\ \mathbf{d})$

23. $\mathbf{b} \times \mathbf{b}$, $(\mathbf{b} - \mathbf{c}) \times (\mathbf{c} - \mathbf{b})$, $\mathbf{b} \bullet \mathbf{b}$

24. TEAM PROJECT. Useful Formulas for Three and Four Vectors. Prove (13)–(16), which are often useful in practical work, and illustrate each formula with two examples. *Hint.* For (13) choose Cartesian coordinates such that $\mathbf{d} = [d_1, 0, 0]$ and $\mathbf{c} = [c_1, c_2, 0]$. Show that each side of (13) then equals $[-b_2c_2d_1, b_1c_2d_1, 0]$, and give reasons why the two sides are then equal in any Cartesian coordinate system. For (14) and (15) use (13).

(13) $\mathbf{b} \times (\mathbf{c} \times \mathbf{d}) = (\mathbf{b} \bullet \mathbf{d})\mathbf{c} - (\mathbf{b} \bullet \mathbf{c})\mathbf{d}$

(14) $(\mathbf{a} \times \mathbf{b}) \times (\mathbf{c} \times \mathbf{d}) = (\mathbf{a}\ \mathbf{b}\ \mathbf{d})\mathbf{c} - (\mathbf{a}\ \mathbf{b}\ \mathbf{c})\mathbf{d}$

(15) $(\mathbf{a} \times \mathbf{b}) \bullet (\mathbf{c} \times \mathbf{d}) = (\mathbf{a} \bullet \mathbf{c})(\mathbf{b} \bullet \mathbf{d}) - (\mathbf{a} \bullet \mathbf{d})(\mathbf{b} \bullet \mathbf{c})$

(16) $(\mathbf{a}\ \mathbf{b}\ \mathbf{c}) = (\mathbf{b}\ \mathbf{c}\ \mathbf{a}) = (\mathbf{c}\ \mathbf{a}\ \mathbf{b})$
$= -(\mathbf{c}\ \mathbf{b}\ \mathbf{a}) = -(\mathbf{a}\ \mathbf{c}\ \mathbf{b})$

25–35 APPLICATIONS

25. Moment m of a force p. Find the moment vector $\mathbf{m}$ and m of $\mathbf{p} = [2, 3, 0]$ about Q: $(2, 1, 0)$ acting on a line through A: $(0, 3, 0)$. Make a sketch.

26. Moment. Solve Prob. 25 if $\mathbf{p} = [1, 0, 3]$, Q: $(2, 0, 3)$, and A: $(4, 3, 5)$.

27. Parallelogram. Find the area if the vertices are $(4, 2, 0)$, $(10, 4, 0)$, $(5, 4, 0)$, and $(11, 6, 0)$. Make a sketch.

28. A remarkable parallelogram. Find the area of the quadrangle Q whose vertices are the midpoints of the sides of the quadrangle P with vertices A: $(2, 1, 0)$, B: $(5, -1. 0)$, C: $(8, 2, 0)$, and D: $(4, 3, 0)$. Verify that Q is a parallelogram.

29. Triangle. Find the area if the vertices are $(0, 0, 1)$, $(2, 0, 5)$, and $(2, 3, 4)$.

30. Plane. Find the plane through the points A: $(1, 2, \frac{1}{4})$, B: $(4, 2, -2)$, and C: $(0, 8, 4)$.

31. Plane. Find the plane through $(1, 3, 4)$, $(1, -2, 6)$, and $(4, 0, 7)$.

32. Parallelepiped. Find the volume if the edge vectors are $\mathbf{i} + \mathbf{j}$, $-2\mathbf{i} + 2\mathbf{k}$, and $-2\mathbf{i} - 3\mathbf{k}$. Make a sketch.

33. Tetrahedron. Find the volume if the vertices are $(1, 1, 1)$, $(5, -7, 3)$, $(7, 4, 8)$, and $(10, 7, 4)$.

34. Tetrahedron. Find the volume if the vertices are $(1, 3, 6)$, $(3, 7, 12)$, $(8, 8, 9)$, and $(2, 2, 8)$.

35. WRITING PROJECT. Applications of Cross Products. Summarize the most important applications we have discussed in this section and give a few simple examples. No proofs.

9.4 Vector and Scalar Functions and Their Fields. Vector Calculus: Derivatives

Our discussion of vector calculus begins with identifying the two types of functions on which it operates. Let P be any point in a domain of definition. Typical domains in applications are three-dimensional, or a surface or a curve in space. Then we define a **vector function** $\mathbf{v}$, whose values are vectors, that is,

$$\mathbf{v} = \mathbf{v}(P) = [v_1(P), v_2(P), v_3(P)]$$

that depends on points P in space. We say that a vector function defines a **vector field** in a domain of definition. Typical domains were just mentioned. Examples of vector fields are the field of tangent vectors of a curve (shown in Fig. 195), normal vectors of a surface (Fig. 196), and velocity field of a rotating body (Fig. 197). Note that vector functions may also depend on time t or on some other parameters.

Similarly, we define a **scalar function** f, whose values are scalars, that is,

$$f = f(P)$$

that depends on P. We say that a scalar function defines a scalar field in that three-dimensional domain or surface or curve in space. Two representative examples of scalar fields are the temperature field of a body and the pressure field of the air in Earth's atmosphere. Note that scalar functions may also depend on some parameter such as time t.

Notation. If we introduce Cartesian coordinates x, y, z, then, instead of writing $\mathbf{v}(P)$ for the vector function, we can write

$$\mathbf{v}(x, y, z) = [v_1(x, y, z), \quad v_2(x, y, z), \quad v_3(x, y, z)].$$

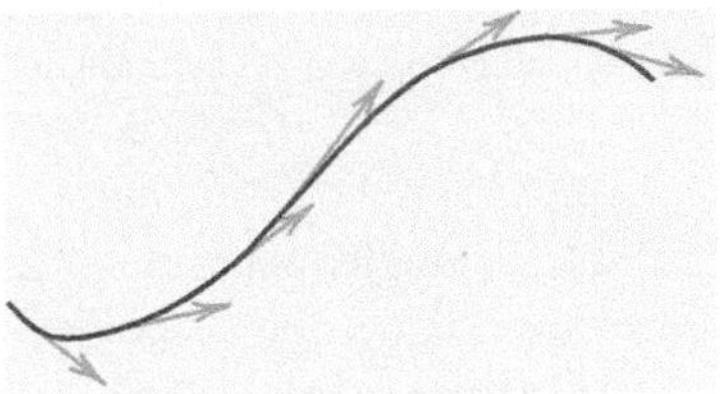

Fig. 195. Field of tangent vectors of a curve

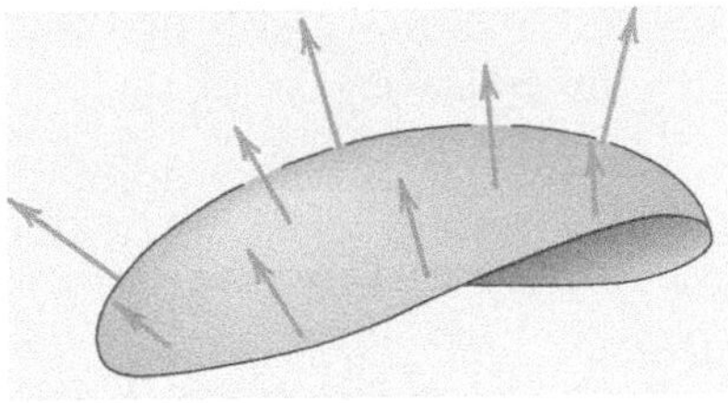

Fig. 196. Field of normal vectors of a surface

We have to keep in mind that the components depend on our choice of coordinate system, whereas a vector field that has a physical or geometric meaning should have magnitude and direction depending only on P, not on the choice of coordinate system.

Similarly, for a scalar function, we write

$$f(P) = f(x, y, z).$$

We illustrate our discussion of vector functions, scalar functions, vector fields, and scalar fields by the following three examples.

EXAMPLE 1 **Scalar Function (Euclidean Distance in Space)**

The distance $f(P)$ of any point P from a fixed point P_0 in space is a scalar function whose domain of definition is the whole space. $f(P)$ defines a scalar field in space. If we introduce a Cartesian coordinate system and P_0 has the coordinates x_0, y_0, z_0, then f is given by the well-known formula

$$f(P) = f(x, y, z) = \sqrt{(x - x_0)^2 + (y - y_0)^2 + (z - z_0)^2}$$

where x, y, z are the coordinates of P. If we replace the given Cartesian coordinate system with another such system by translating and rotating the given system, then the values of the coordinates of P and P_0 will in general change, but $f(P)$ will have the same value as before. Hence $f(P)$ is a scalar function. The direction cosines of the straight line through P and P_0 are not scalars because their values depend on the choice of the coordinate system. ■

EXAMPLE 2 **Vector Field (Velocity Field)**

At any instant the velocity vectors $\mathbf{v}(P)$ of a rotating body B constitute a vector field, called the **velocity field** of the rotation. If we introduce a Cartesian coordinate system having the origin on the axis of rotation, then (see Example 5 in Sec. 9.3)

$$\mathbf{v}(x, y, z) = \mathbf{w} \times \mathbf{r} = \mathbf{w} \times [x, y, z] = \mathbf{w} \times (x\mathbf{i} + y\mathbf{j} + z\mathbf{k}) \tag{1}$$

where x, y, z are the coordinates of any point P of B at the instant under consideration. If the coordinates are such that the z-axis is the axis of rotation and $\mathbf{w}$ points in the positive z-direction, then $\mathbf{w} = \omega\mathbf{k}$ and

$$\mathbf{v} = \begin{vmatrix} \mathbf{i} & \mathbf{j} & \mathbf{k} \\ 0 & 0 & \omega \\ x & y & z \end{vmatrix} = \omega[-y, x, 0] = \omega(-y\mathbf{i} + x\mathbf{j}).$$

An example of a rotating body and the corresponding velocity field are shown in Fig. 197. ■

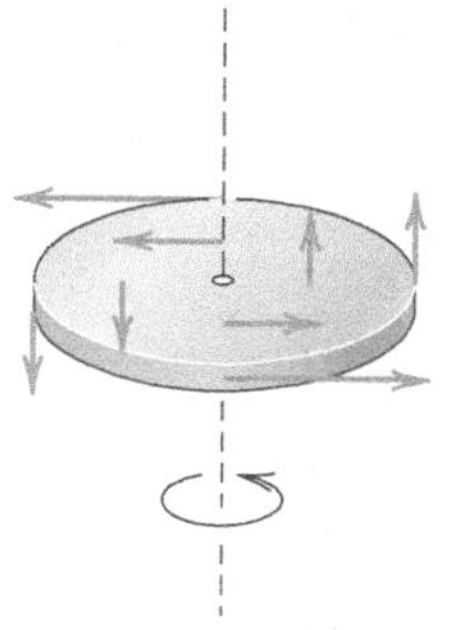

Fig. 197. Velocity field of a rotating body

EXAMPLE 3 **Vector Field (Field of Force, Gravitational Field)**

Let a particle A of mass M be fixed at a point P_0 and let a particle B of mass m be free to take up various positions P in space. Then A attracts B. According to **Newton's law of gravitation** the corresponding gravitational force $\mathbf{p}$ is directed from P to P_0, and its magnitude is proportional to $1/r^2$, where r is the distance between P and P_0, say,

$$|\mathbf{p}| = \frac{c}{r^2}, \qquad c = GMm. \tag{2}$$

Here $G = 6.67 \cdot 10^{-8}\ \mathrm{cm}^3/(\mathrm{g} \cdot \mathrm{sec}^2)$ is the gravitational constant. Hence $\mathbf{p}$ defines a vector field in space. If we introduce Cartesian coordinates such that P_0 has the coordinates x_0, y_0, z_0 and P has the coordinates x, y, z, then by the Pythagorean theorem,

$$r = \sqrt{(x - x_0)^2 + (y - y_0)^2 + (z - z_0)^2} \qquad (\geqq 0).$$

Assuming that $r > 0$ and introducing the vector

$$\mathbf{r} = [x - x_0,\ \ y - y_0,\ \ z - z_0] = (x - x_0)\mathbf{i} + (y - y_0)\mathbf{j} + (z - z_0)\mathbf{k},$$

we have $|\mathbf{r}| = r$, and $(-1/r)\mathbf{r}$ is a unit vector in the direction of $\mathbf{p}$; the minus sign indicates that $\mathbf{p}$ is directed from P to P_0 (Fig. 198). From this and (2) we obtain

$$\begin{aligned} \mathbf{p} = |\mathbf{p}|\left(-\frac{1}{r}\mathbf{r}\right) &= -\frac{c}{r^3}\mathbf{r} = \left[-c\,\frac{x - x_0}{r^3},\ \ -c\,\frac{y - y_0}{r^3},\ \ -c\,\frac{z - z_0}{r^3}\right] \\ &= -c\,\frac{x - x_0}{r^3}\mathbf{i} - c\,\frac{y - y_0}{r^3}\mathbf{j} - c\,\frac{z - z_0}{r^3}\mathbf{k}. \end{aligned} \tag{3}$$

This vector function describes the gravitational force acting on B. ■

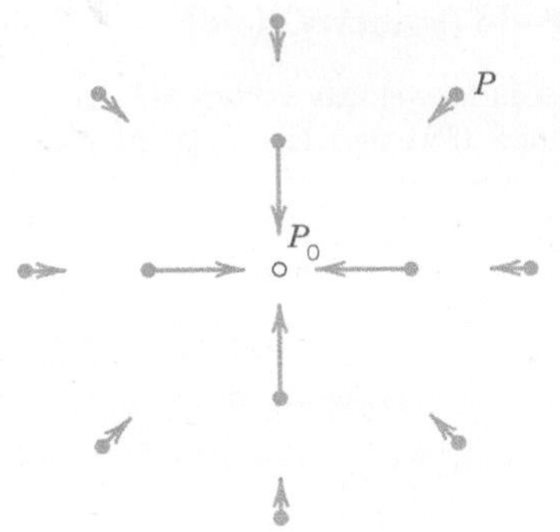

Fig. 198. Gravitational field in Example 3

Vector Calculus

The student may be pleased to learn that many of the concepts covered in (regular) calculus carry over to vector calculus. Indeed, we show how the basic concepts of convergence, continuity, and differentiability from calculus can be defined for vector functions in a simple and natural way. Most important of these is the derivative of a vector function.

Convergence. An infinite sequence of vectors $\mathbf{a}_{(n)}$, $n = 1, 2, \cdots$, is said to **converge** if there is a vector $\mathbf{a}$ such that

$$\lim_{n\to\infty} |\mathbf{a}_{(n)} - \mathbf{a}| = 0. \tag{4}$$

$\mathbf{a}$ is called the **limit vector** of that sequence, and we write

$$\lim_{n\to\infty} \mathbf{a}_{(n)} = \mathbf{a}. \tag{5}$$

If the vectors are given in Cartesian coordinates, then this sequence of vectors converges to $\mathbf{a}$ if and only if the three sequences of components of the vectors converge to the corresponding components of $\mathbf{a}$. We leave the simple proof to the student.

Similarly, a vector function $\mathbf{v}(t)$ of a real variable t is said to have the **limit** $\boldsymbol{l}$ as t approaches t_0, if $\mathbf{v}(t)$ is defined in some neighborhood of t_0 (possibly except at t_0) and

$$\lim_{t\to t_0} |\mathbf{v}(t) - \boldsymbol{l}| = 0. \tag{6}$$

Then we write

$$\lim_{t\to t_0} \mathbf{v}(t) = \boldsymbol{l}. \tag{7}$$

Here, a *neighborhood* of t_0 is an interval (segment) on the t-axis containing t_0 as an interior point (not as an endpoint).

Continuity. A vector function $\mathbf{v}(t)$ is said to be **continuous** at $t = t_0$ if it is defined in some neighborhood of t_0 (including at t_0 itself!) and

$$\lim_{t\to t_0} \mathbf{v}(t) = \mathbf{v}(t_0). \tag{8}$$

If we introduce a Cartesian coordinate system, we may write

$$\mathbf{v}(t) = [v_1(t), v_2(t), v_3(t)] = v_1(t)\mathbf{i} + v_2(t)\mathbf{j} + v_3(t)\mathbf{k}.$$

Then $\mathbf{v}(t)$ is continuous at t_0 if and only if its three components are continuous at t_0.

We now state the most important of these definitions.

DEFINITION

Derivative of a Vector Function

A vector function $\mathbf{v}(t)$ is said to be **differentiable** at a point t if the following limit exists:

(9)
$$\mathbf{v}'(t) = \lim_{\Delta t \to 0} \frac{\mathbf{v}(t + \Delta t) - \mathbf{v}(t)}{\Delta t}.$$

This vector $\mathbf{v}'(t)$ is called the **derivative** of $\mathbf{v}(t)$. See Fig. 199.

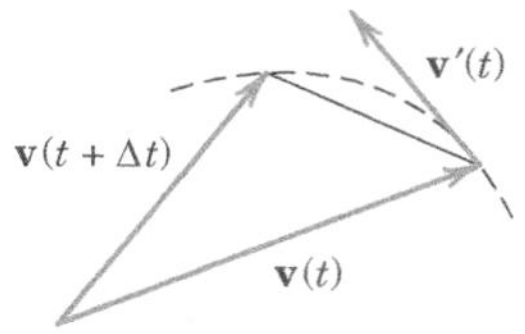

Fig. 199. Derivative of a vector function

In components with respect to a given Cartesian coordinate system,

(10)
$$\mathbf{v}'(t) = [v_1'(t), \quad v_2'(t), \quad v_3'(t)].$$

Hence the derivative $\mathbf{v}'(t)$ ***is obtained by differentiating each component separately.*** For instance, if $\mathbf{v} = [t, t^2, 0]$, then $\mathbf{v}' = [1, 2t, 0]$.

Equation (10) follows from (9) and conversely because (9) is a "vector form" of the usual formula of calculus by which the derivative of a function of a single variable is defined. [The curve in Fig. 199 is the locus of the terminal points representing $\mathbf{v}(t)$ for values of the independent variable in some interval containing t and $t + \Delta t$ in (9)]. It follows that the familiar differentiation rules continue to hold for differentiating vector functions, for instance,

$$(c\mathbf{v})' = c\mathbf{v}' \qquad (c \text{ constant}),$$

$$(\mathbf{u} + \mathbf{v})' = \mathbf{u}' + \mathbf{v}'$$

and in particular

(11)
$$(\mathbf{u} \bullet \mathbf{v})' = \mathbf{u}' \bullet \mathbf{v} + \mathbf{u} \bullet \mathbf{v}'$$

(12)
$$(\mathbf{u} \times \mathbf{v})' = \mathbf{u}' \times \mathbf{v} + \mathbf{u} \times \mathbf{v}'$$

(13)
$$(\mathbf{u} \;\; \mathbf{v} \;\; \mathbf{w})' = (\mathbf{u}' \;\; \mathbf{v} \;\; \mathbf{w}) + (\mathbf{u} \;\; \mathbf{v}' \;\; \mathbf{w}) + (\mathbf{u} \;\; \mathbf{v} \;\; \mathbf{w}').$$

The simple proofs are left to the student. In (12), note the order of the vectors carefully because cross multiplication is not commutative.

EXAMPLE 4 **Derivative of a Vector Function of Constant Length**

Let $\mathbf{v}(t)$ be a vector function whose length is constant, say, $|\mathbf{v}(t)| = c$. Then $|\mathbf{v}|^2 = \mathbf{v} \cdot \mathbf{v} = c^2$, and $(\mathbf{v} \cdot \mathbf{v})' = 2\mathbf{v} \cdot \mathbf{v}' = 0$, by differentiation [see (11)]. This yields the following result. *The derivative of a vector function* $\mathbf{v}(t)$ *of constant length is either the zero vector or is perpendicular to* $\mathbf{v}(t)$. ■

Partial Derivatives of a Vector Function

Our present discussion shows that partial differentiation of vector functions of two or more variables can be introduced as follows. Suppose that the components of a vector function

$$\mathbf{v} = [v_1, \quad v_2, \quad v_3] = v_1\mathbf{i} + v_2\mathbf{j} + v_3\mathbf{k}$$

are differentiable functions of n variables $t_1, \cdots, t_n$. Then the **partial derivative** of $\mathbf{v}$ with respect to t_m is denoted by $\partial\mathbf{v}/\partial t_m$ and is defined as the vector function

$$\frac{\partial \mathbf{v}}{\partial t_m} = \frac{\partial v_1}{\partial t_m}\mathbf{i} + \frac{\partial v_2}{\partial t_m}\mathbf{j} + \frac{\partial v_3}{\partial t_m}\mathbf{k}.$$

Similarly, second partial derivatives are

$$\frac{\partial^2 \mathbf{v}}{\partial t_l \partial t_m} = \frac{\partial^2 v_1}{\partial t_l \partial t_m}\mathbf{i} + \frac{\partial^2 v_2}{\partial t_l \partial t_m}\mathbf{j} + \frac{\partial^2 v_3}{\partial t_l \partial t_m}\mathbf{k},$$

and so on.

EXAMPLE 5 **Partial Derivatives**

Let $\mathbf{r}(t_1, t_2) = a \cos t_1\,\mathbf{i} + a \sin t_1\,\mathbf{j} + t_2\,\mathbf{k}$. Then $\dfrac{\partial \mathbf{r}}{\partial t_1} = -a \sin t_1\,\mathbf{i} + a \cos t_1\,\mathbf{j}$ and $\dfrac{\partial \mathbf{r}}{\partial t_2} = \mathbf{k}$. ■

Various physical and geometric applications of derivatives of vector functions will be discussed in the next sections as well as in Chap. 10.

PROBLEM SET 9.4

1–8 **SCALAR FIELDS IN THE PLANE**

Let the temperature T in a body be independent of z so that it is given by a scalar function $T = T(x, t)$. Identify the isotherms $T(x, y) =$ const. Sketch some of them.

1. $T = x^2 - y^2$
2. $T = xy$
3. $T = 3x - 4y$
4. $T = \arctan(y/x)$
5. $T = y/(x^2 + y^2)$
6. $T = x/(x^2 + y^2)$
7. $T = 9x^2 + 4y^2$
8. CAS PROJECT. Scalar Fields in the Plane. Sketch or graph isotherms of the following fields and describe what they look like.

(a) $x^2 - 4x - y^2$
(b) $x^2y - y^3/3$
(c) $\cos x \sinh y$
(d) $\sin x \sinh y$
(e) $e^x \sin y$
(f) $e^{2x} \cos 2y$
(g) $x^4 - 6x^2y^2 + y^4$
(h) $x^2 - 2x - y^2$

9–14 **SCALAR FIELDS IN SPACE**

What kind of surfaces are the **level surfaces** $f(x, y, z) =$ const?

9. $f = 4x - 3y + 2z$
10. $f = 9(x^2 + y^2) + z^2$
11. $f = 5x^2 + 2y^2$
12. $f = z - \sqrt{x^2 + y^2}$
13. $f = z - (x^2 + y^2)$
14. $f = x - y^2$

15–20 VECTOR FIELDS

Sketch figures similar to Fig. 198. Try to interpet the field of **v** as a velocity field.

15. $\mathbf{v} = \mathbf{i} + \mathbf{j}$

16. $\mathbf{v} = -y\mathbf{i} + x\mathbf{j}$

17. $\mathbf{v} = x\mathbf{j}$

18. $\mathbf{v} = x\mathbf{i} + y\mathbf{j}$

19. $\mathbf{v} = x\mathbf{i} - y\mathbf{j}$

20. $\mathbf{v} = y\mathbf{i} - x\mathbf{j}$

21. CAS PROJECT. Vector Fields. Plot by arrows:

(a) $\mathbf{v} = [x, x^2]$

(b) $\mathbf{v} = [1/y, 1/x]$

(c) $\mathbf{v} = [\cos x, \sin x]$

(d) $\mathbf{v} = e^{-(x^2+y^2)}[x, -y]$

22–25 DIFFERENTIATION

22. Find the first and second derivatives of $\mathbf{r} = [3 \cos 2t, 3 \sin 2t, 4t]$.

23. Prove (11)–(13). Give two typical examples for each formula.

24. Find the first partial derivatives of $\mathbf{v}_1 = [e^x \cos y, e^x \sin y]$ and $\mathbf{v}_2 = [\cos x \cosh y, -\sin x \sinh y]$.

25. WRITING PROJECT. Differentiation of Vector Functions. Summarize the essential ideas and facts and give examples of your own.

9.5 Curves. Arc Length. Curvature. Torsion

Vector calculus has important applications to curves (Sec. 9.5) and surfaces (to be covered in Sec. 10.5) in physics and geometry. The application of vector calculus to geometry is a field known as **differential geometry**. Differential geometric methods are applied to problems in mechanics, computer-aided as well as traditional engineering design, geodesy, geography, space travel, and relativity theory. For details, see [GenRef8] and [GenRef9] in App. 1.

Bodies that move in space form paths that may be represented by curves C. This and other applications show the need for **parametric representations** of C with **parameter** t, which may denote time or something else (see Fig. 200). A typical parametric representation is given by

$$\mathbf{r}(t) = [x(t), \quad y(t), \quad z(t)] = x(t)\mathbf{i} + y(t)\mathbf{j} + z(t)\mathbf{k}. \tag{1}$$

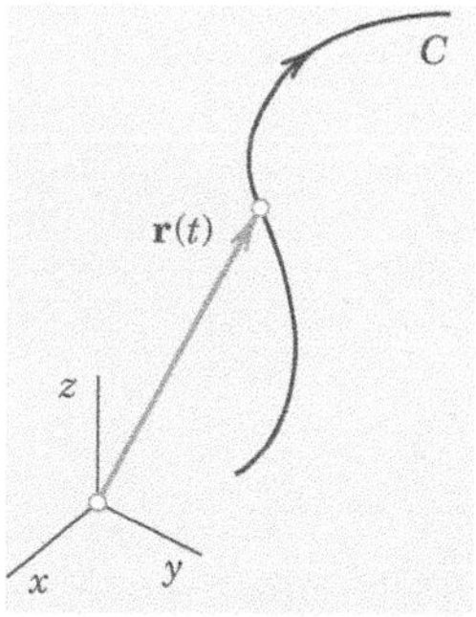

Fig. 200. Parametric representation of a curve

Here t is the parameter and x, y, z are Cartesian coordinates, that is, the usual rectangular coordinates as shown in Sec. 9.1. To each value $t = t_0$, there corresponds a point of C with position vector $\mathbf{r}(t_0)$ whose coordinates are $x(t_0), y(t_0), z(t_0)$. This is illustrated in Figs. 201 and 202.

The use of parametric representations has key advantages over other representations that involve projections into the xy-plane and xz-plane or involve a pair of equations with y or with z as independent variable. The projections look like this:

$$y = f(x), \qquad z = g(x). \tag{2}$$

The advantages of using (1) instead of (2) are that, in (1), the coordinates x, y, z all play an equal role, that is, all three coordinates are dependent variables. Moreover, the parametric representation (1) induces an orientation on C. This means that as we increase t, we travel along the curve C in a certain direction. The sense of increasing t is called the positive sense on C. The sense of decreasing t is then called the negative sense on C, given by (1).

Examples 1–4 give parametric representations of several important curves.

EXAMPLE 1 Circle. Parametric Representation. Positive Sense

The circle $x^2 + y^2 = 4$, $z = 0$ in the xy-plane with center 0 and radius 2 can be represented parametrically by

$$\mathbf{r}(t) = [2\cos t, 2\sin t, 0] \qquad \text{or simply by} \qquad \mathbf{r}(t) = [2\cos t, 2\sin t] \qquad \text{(Fig. 201)}$$

where $0 \leqq t \leqq 2\pi$. Indeed, $x^2 + y^2 = (2\cos t)^2 + (2\sin t)^2 = 4(\cos^2 t + \sin^2 t) = 4$, For $t = 0$ we have $\mathbf{r}(0) = [2, 0]$, for $t = \frac{1}{2}\pi$ we get $\mathbf{r}(\frac{1}{2}\pi) = [0, 2]$, and so on. The positive sense induced by this representation is the counterclockwise sense.

If we replace t with $t^* = -t$, we have $t = -t^*$ and get

$$\mathbf{r}^*(t^*) = [2\cos(-t^*), 2\sin(-t^*)] = [2\cos t^*, -2\sin t^*].$$

This has reversed the orientation, and the circle is now oriented clockwise. ■

EXAMPLE 2 Ellipse

The vector function

$$\mathbf{r}(t) = [a\cos t, \quad b\sin t, \quad 0] = a\cos t\,\mathbf{i} + b\sin t\,\mathbf{j} \qquad (3) \qquad \text{(Fig. 202)}$$

represents an ellipse in the xy-plane with center at the origin and principal axes in the direction of the x- and y-axes. In fact, since $\cos^2 t + \sin^2 t = 1$, we obtain from (3)

$$\frac{x^2}{a^2} + \frac{y^2}{b^2} = 1, \qquad z = 0.$$

If $b = a$, then (3) represents a *circle* of radius a. ■

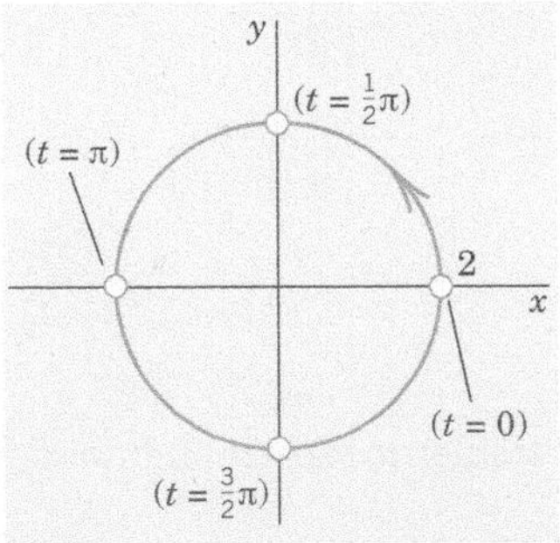

Fig. 201. Circle in Example 1

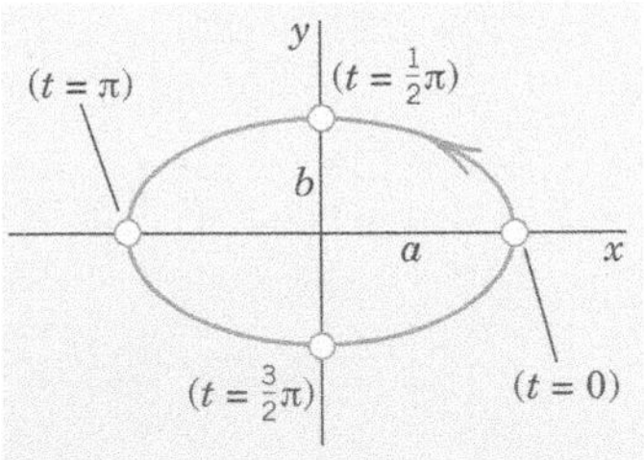

Fig. 202. Ellipse in Example 2

EXAMPLE 3 Straight Line

A straight line L through a point A with position vector $\mathbf{a}$ in the direction of a constant vector $\mathbf{b}$ (see Fig. 203) can be represented parametrically in the form

$$\mathbf{r}(t) = \mathbf{a} + t\mathbf{b} = [a_1 + tb_1, \quad a_2 + tb_2, \quad a_3 + tb_3]. \qquad (4)$$

If **b** is a unit vector, its components are the **direction cosines** of L. In this case, $|t|$ measures the distance of the points of L from A. For instance, the straight line in the xy-plane through A: (3, 2) having slope 1 is (sketch it)

$$\mathbf{r}(t) = [3,\quad 2,\quad 0] + t[1,\quad 1,\quad 0] = [3 + t,\quad 2 + t,\quad 0].$$ ■

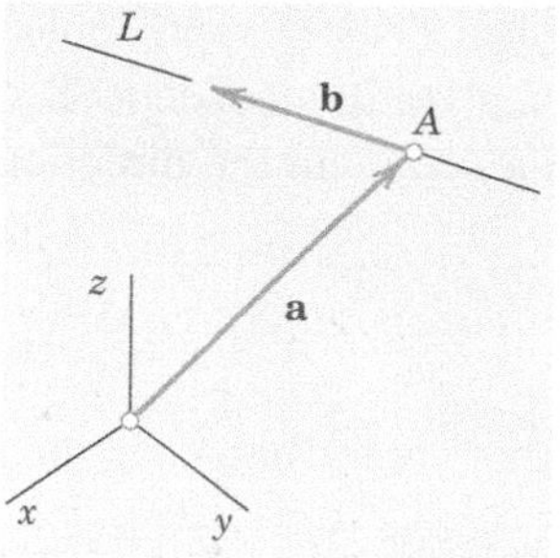

Fig. 203. Parametric representation of a straight line

A **plane curve** is a curve that lies in a plane in space. A curve that is not plane is called a **twisted curve**. A standard example of a twisted curve is the following.

EXAMPLE 4 Circular Helix

The twisted curve C represented by the vector function

$$\mathbf{r}(t) = [a \cos t,\quad a \sin t,\quad ct] = a \cos t\,\mathbf{i} + a \sin t\,\mathbf{j} + ct\,\mathbf{k} \qquad (c \neq 0) \tag{5}$$

is called a *circular helix*. It lies on the cylinder $x^2 + y^2 = a^2$. If $c > 0$, the helix is shaped like a right-handed screw (Fig. 204). If $c < 0$, it looks like a left-handed screw (Fig. 205). If $c = 0$, then (5) is a circle. ■

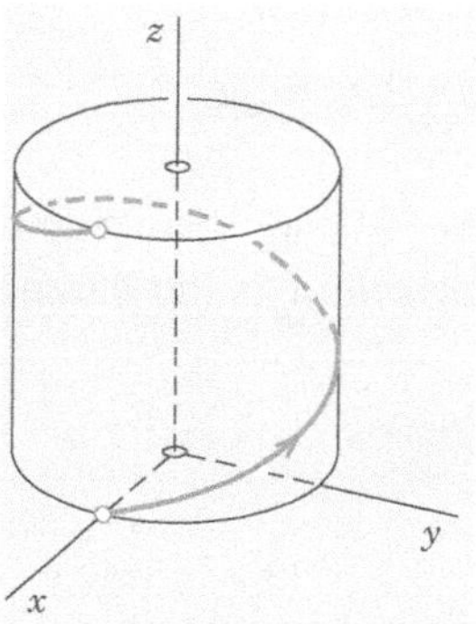

Fig. 204. Right-handed circular helix

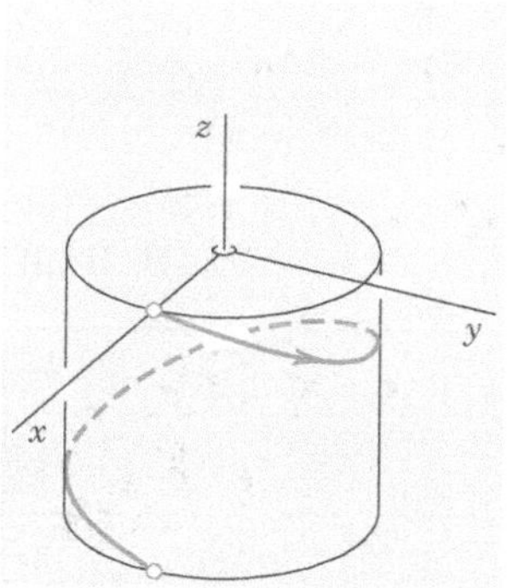

Fig. 205. Left-handed circular helix

A **simple curve** is a curve without **multiple points**, that is, without points at which the curve intersects or touches itself. Circle and helix are simple curves. Figure 206 shows curves that are not simple. An example is $[\sin 2t,\quad \cos t,\quad 0]$. Can you sketch it?

An **arc** of a curve is the portion between any two points of the curve. For simplicity, we say "curve" for curves as well as for arcs.

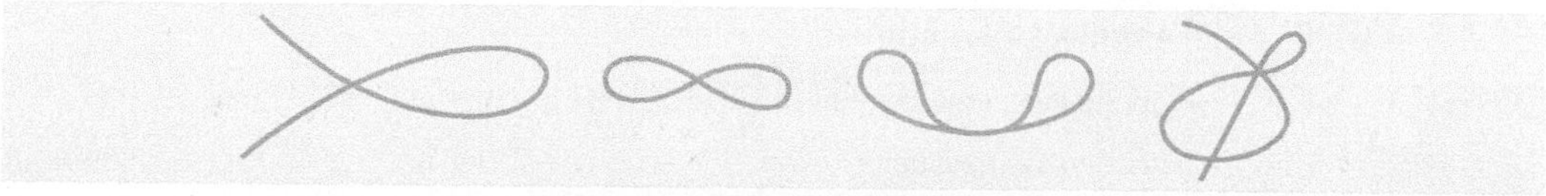

Fig. 206. Curves with multiple points

Tangent to a Curve

The next idea is the approximation of a curve by straight lines, leading to tangents and to a definition of length. Tangents are straight lines touching a curve. The **tangent** to a simple curve C at a point P of C is the limiting position of a straight line L through P and a point Q of C as Q approaches P along C. See Fig. 207.

Let us formalize this concept. If C is given by $\mathbf{r}(t)$, and P and Q correspond to t and $t + \Delta t$, then a vector in the direction of L is

$$\frac{1}{\Delta t}[\mathbf{r}(t + \Delta t) - \mathbf{r}(t)]. \tag{6}$$

In the limit this vector becomes the derivative

$$\mathbf{r}'(t) = \lim_{\Delta t \to 0} \frac{1}{\Delta t}[\mathbf{r}(t + \Delta t) - \mathbf{r}(t)], \tag{7}$$

provided $\mathbf{r}(t)$ is differentiable, as we shall assume from now on. If $\mathbf{r}'(t) \neq \mathbf{0}$, we call $\mathbf{r}'(t)$ a **tangent vector** of C at P because it has the direction of the tangent. The corresponding unit vector is the **unit tangent vector** (see Fig. 207)

$$\mathbf{u} = \frac{1}{|\mathbf{r}'|}\mathbf{r}'. \tag{8}$$

Note that both $\mathbf{r}'$ and $\mathbf{u}$ point in the direction of increasing t. Hence their sense depends on the orientation of C. It is reversed if we reverse the orientation.

It is now easy to see that the **tangent** to C at P is given by

$$\mathbf{q}(w) = \mathbf{r} + w\mathbf{r}' \qquad \text{(Fig. 208)}. \tag{9}$$

This is the sum of the position vector $\mathbf{r}$ of P and a multiple of the tangent vector $\mathbf{r}'$ of C at P. Both vectors depend on P. The variable w is the parameter in (9).

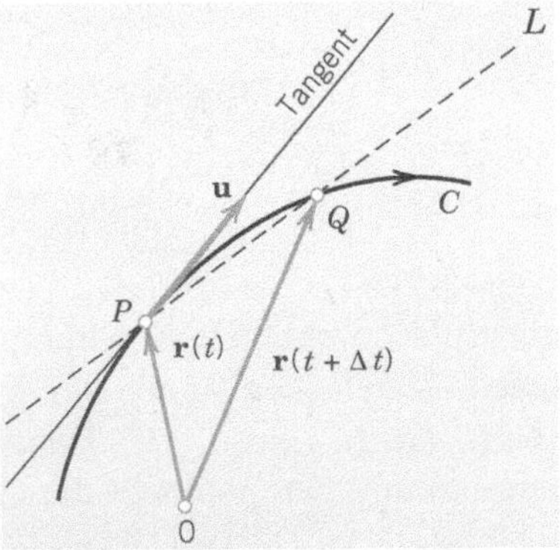

Fig. 207. Tangent to a curve

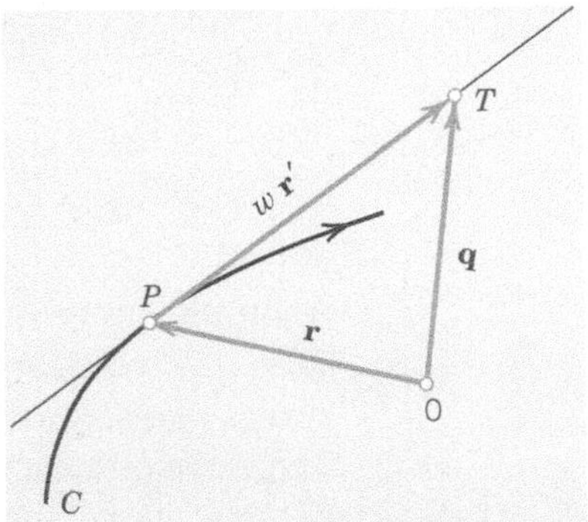

Fig. 208. Formula (9) for the tangent to a curve

EXAMPLE 5 Tangent to an Ellipse

Find the tangent to the ellipse $\frac{1}{4}x^2 + y^2 = 1$ at P: $(\sqrt{2}, 1/\sqrt{2})$.

Solution. Equation (3) with semi-axes $a = 2$ and $b = 1$ gives $\mathbf{r}(t) = [2 \cos t, \sin t]$. The derivative is $\mathbf{r}'(t) = [-2 \sin t, \cos t]$. Now P corresponds to $t = \pi/4$ because

$$\mathbf{r}(\pi/4) = [2 \cos (\pi/4), \sin (\pi/4)] = [\sqrt{2}, \ 1/\sqrt{2}].$$

Hence $\mathbf{r}'(\pi/4) = [-\sqrt{2},\ \ 1/\sqrt{2}]$. From (9) we thus get the *answer*

$$\mathbf{q}(w) = [\sqrt{2},\ \ 1/\sqrt{2}] + w[-\sqrt{2},\ \ 1/\sqrt{2}] = [\sqrt{2}(1-w),\ \ (1/\sqrt{2})(1+w)].$$

To check the result, sketch or graph the ellipse and the tangent. ■

Length of a Curve

We are now ready to define the length l of a curve. l will be the limit of the lengths of broken lines of n chords (see Fig. 209, where $n = 5$) with larger and larger n. For this, let $\mathbf{r}(t)$, $a \leqq t \leqq b$, represent C. For each $n = 1, 2, \cdots$, we subdivide ("partition") the interval $a \leqq t \leqq b$ by points

$$t_0 (= a),\quad t_1, \cdots, t_{n-1},\quad t_n (= b),\qquad \text{where}\qquad t_0 < t_1 < \cdots < t_n.$$

This gives a broken line of chords with endpoints $\mathbf{r}(t_0), \cdots, \mathbf{r}(t_n)$. We do this arbitrarily but so that the greatest $|\Delta t_m| = |t_m - t_{m-1}|$ approaches 0 as $n \to \infty$. The lengths $l_1, l_2, \cdots$ of these chords can be obtained from the Pythagorean theorem. If $\mathbf{r}(t)$ has a continuous derivative $\mathbf{r}'(t)$, it can be shown that the sequence $l_1, l_2, \cdots$ has a limit, which is independent of the particular choice of the representation of C and of the choice of subdivisions. This limit is given by the integral

$$l = \int_a^b \sqrt{\mathbf{r}' \bullet \mathbf{r}'}\, dt \qquad \left(\mathbf{r}' = \frac{d\mathbf{r}}{dt}\right). \tag{10}$$

l is called the **length** of C, and C is called **rectifiable**. Formula (10) is made plausible in calculus for plane curves and is proved for curves in space in [GenRef8] listed in App. 1. The actual evaluation of the integral (10) will, in general, be difficult. However, some simple cases are given in the problem set.

Arc Length s of a Curve

The length (10) of a curve C is a constant, a positive number. But if we replace the fixed b in (10) with a variable t, the integral becomes a function of t, denoted by $s(t)$ and called the *arc length function* or simply the **arc length** of C. Thus

$$s(t) = \int_a^t \sqrt{\mathbf{r}' \bullet \mathbf{r}'}\, d\tilde{t} \qquad \left(\mathbf{r}' = \frac{d\mathbf{r}}{d\tilde{t}}\right). \tag{11}$$

Here the variable of integration is denoted by $\tilde{t}$ because t is now used in the upper limit.

Geometrically, $s(t_0)$ with some $t_0 > a$ is the length of the arc of C between the points with parametric values a and t_0. The choice of a (the point $s = 0$) is arbitrary; changing a means changing s by a constant.

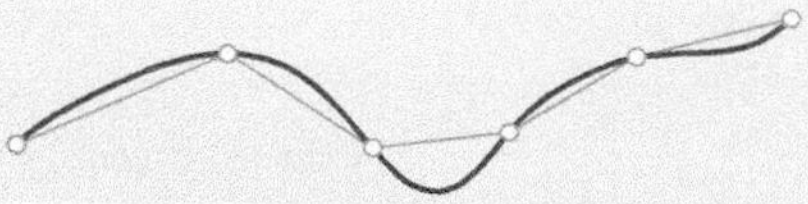

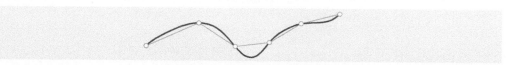

Fig. 209. Length of a curve

Linear Element ds.   If we differentiate (11) and square, we have

(12)
$$\left(\frac{ds}{dt}\right)^2 = \frac{d\mathbf{r}}{dt} \bullet \frac{d\mathbf{r}}{dt} = |\mathbf{r}'(t)|^2 = \left(\frac{dx}{dt}\right)^2 + \left(\frac{dy}{dt}\right)^2 + \left(\frac{dz}{dt}\right)^2.$$

It is customary to write

(13*)
$$d\mathbf{r} = [dx, dy, dz] = dx\,\mathbf{i} + dy\,\mathbf{j} + dz\,\mathbf{k}$$

and

(13)
$$ds^2 = d\mathbf{r} \bullet d\mathbf{r} = dx^2 + dy^2 + dz^2.$$

ds is called the **linear element** of C.

Arc Length as Parameter.   The use of s in (1) instead of an arbitrary t simplifies various formulas. For the unit tangent vector (8) we simply obtain

(14)
$$\mathbf{u}(s) = \mathbf{r}'(s).$$

Indeed, $|\mathbf{r}'(s)| = (ds/ds) = 1$ in (12) shows that $\mathbf{r}'(s)$ is a unit vector. Even greater simplifications due to the use of s will occur in curvature and torsion (below).

EXAMPLE 6   **Circular Helix. Circle. Arc Length as Parameter**

The **helix** $\mathbf{r}(t) = [a \cos t, a \sin t, ct]$ in (5) has the derivative $\mathbf{r}'(t) = [-a \sin t, a \cos t, c]$. Hence $\mathbf{r}' \bullet \mathbf{r}' = a^2 + c^2$, a constant, which we denote by K^2. Hence the integrand in (11) is constant, equal to K, and the integral is $s = Kt$. Thus $t = s/K$, so that a representation of the helix with the arc length s as parameter is

(15)
$$\mathbf{r}^*(s) = \mathbf{r}\left(\frac{s}{K}\right) = \left[a \cos \frac{s}{K},\ a \sin \frac{s}{K},\ \frac{cs}{K}\right], \qquad K = \sqrt{a^2 + c^2}.$$

A **circle** is obtained if we set $c = 0$. Then $K = a$, $t = s/a$, and a representation with arc length s as parameter is

$$\mathbf{r}^*(s) = \mathbf{r}\left(\frac{s}{a}\right) = \left[a \cos \frac{s}{a},\ a \sin \frac{s}{a}\right]. \qquad \blacksquare$$

Curves in Mechanics. Velocity. Acceleration

Curves play a basic role in mechanics, where they may serve as paths of moving bodies. Then such a curve C should be represented by a parametric representation $\mathbf{r}(t)$ with **time** t as parameter. The tangent vector (7) of C is then called the **velocity vector** $\mathbf{v}$ because, being tangent, it points in the instantaneous direction of motion and its length gives the **speed** $|\mathbf{v}| = |\mathbf{r}'| = \sqrt{\mathbf{r}' \bullet \mathbf{r}'} = ds/dt$; see (12). The second derivative of $\mathbf{r}(t)$ is called the **acceleration vector** and is denoted by $\mathbf{a}$. Its length $|\mathbf{a}|$ is called the **acceleration** of the motion. Thus

(16)
$$\mathbf{v}(t) = \mathbf{r}'(t), \qquad \mathbf{a}(t) = \mathbf{v}'(t) = \mathbf{r}''(t).$$

Tangential and Normal Acceleration. Whereas the velocity vector is always tangent to the path of motion, the acceleration vector will generally have another direction. We can split the acceleration vector into two directional components, that is,

$$\mathbf{a} = \mathbf{a}_{\text{tan}} + \mathbf{a}_{\text{norm}}, \tag{17}$$

where the **tangential acceleration vector** $\mathbf{a}_{\text{tan}}$ is tangent to the path (or, sometimes, $\mathbf{0}$) and the **normal acceleration vector** $\mathbf{a}_{\text{norm}}$ is normal (perpendicular) to the path (or, sometimes, $\mathbf{0}$).

Expressions for the vectors in (17) are obtained from (16) by the chain rule. We first have

$$\mathbf{v}(t) = \frac{d\mathbf{r}}{dt} = \frac{d\mathbf{r}}{ds}\frac{ds}{dt} = \mathbf{u}(s)\frac{ds}{dt}$$

where $\mathbf{u}(s)$ is the unit tangent vector (14). Another differentiation gives

$$\mathbf{a}(t) = \frac{d\mathbf{v}}{dt} = \frac{d}{dt}\left(\mathbf{u}(s)\frac{ds}{dt}\right) = \frac{d\mathbf{u}}{ds}\left(\frac{ds}{dt}\right)^2 + \mathbf{u}(s)\frac{d^2s}{dt^2}. \tag{18}$$

Since the tangent vector $\mathbf{u}(s)$ has constant length (length one), its derivative $d\mathbf{u}/ds$ is perpendicular to $\mathbf{u}(s)$, from the result in Example 4 in Sec. 9.4. Hence the first term on the right of (18) is the normal acceleration vector, and the second term on the right is the tangential acceleration vector, so that (18) is of the form (17).

Now the length $|\mathbf{a}_{\text{tan}}|$ is the absolute value of the projection of $\mathbf{a}$ in the direction of $\mathbf{v}$, given by (11) in Sec. 9.2 with $\mathbf{b} = \mathbf{v}$; that is, $|\mathbf{a}_{\text{tan}}| = |\mathbf{a} \cdot \mathbf{v}|/|\mathbf{v}|$. Hence $\mathbf{a}_{\text{tan}}$ is this expression times the unit vector $(1/|\mathbf{v}|)\mathbf{v}$ in the direction of $\mathbf{v}$, that is,

$$\mathbf{a}_{\text{tan}} = \frac{\mathbf{a} \cdot \mathbf{v}}{\mathbf{v} \cdot \mathbf{v}}\mathbf{v}. \qquad \text{Also,} \qquad \mathbf{a}_{\text{norm}} = \mathbf{a} - \mathbf{a}_{\text{tan}}. \tag{18*}$$

We now turn to two examples that are relevant to applications in space travel. They deal with the *centripetal* and *centrifugal* accelerations, as well as the *Coriolis acceleration*.

EXAMPLE 7 Centripetal Acceleration. Centrifugal Force

The vector function

$$\mathbf{r}(t) = [R\cos\omega t, \quad R\sin\omega t] = R\cos\omega t\,\mathbf{i} + R\sin\omega t\,\mathbf{j} \qquad \text{(Fig. 210)}$$

(with fixed $\mathbf{i}$ and $\mathbf{j}$) represents a circle C of radius R with center at the origin of the xy-plane and describes the motion of a small body B counterclockwise around the circle. Differentiation gives the velocity vector

$$\mathbf{v} = \mathbf{r}' = [-R\omega\sin\omega t, \quad R\omega\cos\omega t] = -R\omega\sin\omega t\,\mathbf{i} + R\omega\cos\omega t\,\mathbf{j} \qquad \text{(Fig. 210)}$$

$\mathbf{v}$ is tangent to C. Its magnitude, the speed, is

$$|\mathbf{v}| = |\mathbf{r}'| = \sqrt{\mathbf{r}' \cdot \mathbf{r}'} = R\omega.$$

Hence it is constant. The speed divided by the distance R from the center is called the **angular speed**. It equals ω, so that it is constant, too. Differentiating the velocity vector, we obtain the acceleration vector

$$\mathbf{a} = \mathbf{v}' = [-R\omega^2\cos\omega t, \quad -R\omega^2\sin\omega t] = -R\omega^2\cos\omega t\,\mathbf{i} - R\omega^2\sin\omega t\,\mathbf{j}. \tag{19}$$

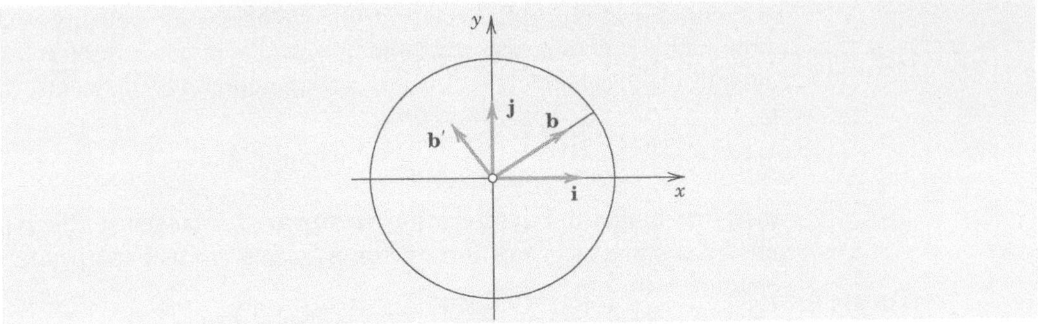

Fig. 210. Centripetal acceleration **a**

This shows that $\mathbf{a} = -\omega^2\mathbf{r}$ (Fig. 210), so that there is an acceleration toward the center, called the **centripetal acceleration** of the motion. It occurs because the velocity vector is changing direction at a constant rate. Its magnitude is constant, $|\mathbf{a}| = \omega^2|\mathbf{r}| = \omega^2 R$. Multiplying **a** by the mass m of B, we get the **centripetal force** $m\mathbf{a}$. The opposite vector $-m\mathbf{a}$ is called the **centrifugal force**. At each instant these two forces are in equilibrium.

We see that in this motion the acceleration vector is normal (perpendicular) to C; hence there is no tangential acceleration. ■

EXAMPLE 8 **Superposition of Rotations. Coriolis Acceleration**

A projectile is moving with constant speed along a meridian of the rotating earth in Fig. 211. Find its acceleration.

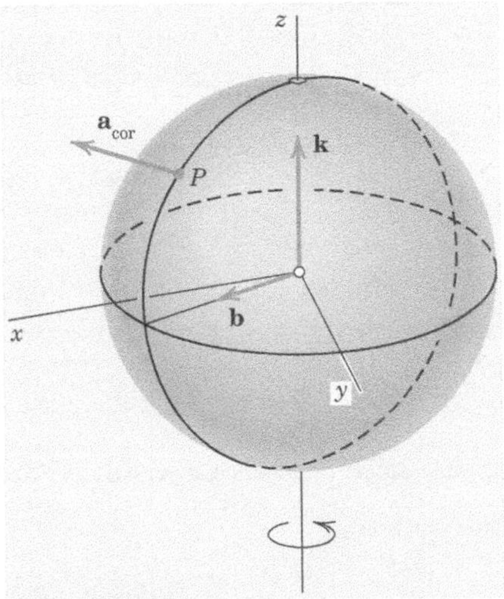

Fig. 211. Example 8. Superposition of two rotations

Solution. Let x, y, z be a fixed Cartesian coordinate system in space, with unit vectors **i**, **j**, **k** in the directions of the axes. Let the Earth, together with a unit vector **b**, be rotating about the z-axis with angular speed $\omega > 0$ (see Example 7). Since **b** is rotating together with the Earth, it is of the form

$$\mathbf{b}(t) = \cos \omega t\, \mathbf{i} + \sin \omega t\, \mathbf{j}.$$

Let the projectile be moving on the meridian whose plane is spanned by **b** and **k** (Fig. 211) with constant angular speed $\omega > 0$. Then its position vector in terms of **b** and **k** is

$$\mathbf{r}(t) = R \cos \gamma t\, \mathbf{b}(t) + R \sin \gamma t\, \mathbf{k} \qquad (R = \text{Radius of the Earth}).$$

We have finished setting up the model. Next, we apply vector calculus to obtain the desired acceleration of the projectile. Our result will be unexpected—and highly relevant for air and space travel. The first and second derivatives of **b** with respect to t are

$$\begin{aligned}\mathbf{b}'(t) &= -\omega \sin \omega t\, \mathbf{i} + \omega \cos \omega t\, \mathbf{j} \\ \mathbf{b}''(t) &= -\omega^2 \cos \omega t\, \mathbf{i} - \omega^2 \sin \omega t\, \mathbf{j} = -\omega^2 \mathbf{b}(t).\end{aligned} \tag{20}$$

The first and second derivatives of $\mathbf{r}(t)$ with respect to t are

$$\begin{aligned}\mathbf{v} = \mathbf{r}'(t) &= R \cos \gamma t\, \mathbf{b}' - \gamma R \sin \gamma t\, \mathbf{b} + \gamma R \cos \gamma t\, \mathbf{k} \\ \mathbf{a} = \mathbf{v}' &= R \cos \gamma t\, \mathbf{b}'' - 2\gamma R \sin \gamma t\, \mathbf{b}' - \gamma^2 R \cos \gamma t\, \mathbf{b} - \gamma^2 R \sin \gamma t\, \mathbf{k} \\ &= R \cos \gamma t\, \mathbf{b}'' - 2\gamma R \sin \gamma t\, \mathbf{b}' - \gamma^2 \mathbf{r}.\end{aligned} \tag{21}$$

By analogy with Example 7 and because of $\mathbf{b}'' = -\omega^2 \mathbf{b}$ in (20) we conclude that the first term in **a** (involving ω in $\mathbf{b}''$!) is the centripetal acceleration due to the rotation of the Earth. Similarly, the third term in the last line (involving γ!) is the centripetal acceleration due to the motion of the projectile on the meridian M of the rotating Earth.

The second, unexpected term $-2\gamma R \sin \gamma t\, \mathbf{b}'$ in **a** is called the **Coriolis acceleration**[3] (Fig. 211) and is due to the interaction of the two rotations. On the Northern Hemisphere, $\sin \gamma t > 0$ (for $t > 0$; also $\gamma > 0$ by assumption), so that $\mathbf{a}_{\text{cor}}$ has the direction of $-\mathbf{b}'$, that is, opposite to the rotation of the Earth. $|\mathbf{a}_{\text{cor}}|$ is maximum at the North Pole and zero at the equator. The projectile B of mass m_0 experiences a force $-m_0 \mathbf{a}_{\text{cor}}$ opposite to $m_0 \mathbf{a}_{\text{cor}}$, which tends to let B deviate from M to the right (and in the Southern Hemisphere, where $\sin \gamma t < 0$, to the left). This deviation has been observed for missiles, rockets, shells, and atmospheric airflow. ■

Curvature and Torsion. *Optional*

This last topic of Sec. 9.5 is optional but completes our discussion of curves relevant to vector calculus.

The **curvature** $\kappa(s)$ of a curve C: $\mathbf{r}(s)$ (s the arc length) at a point P of C measures the rate of change $|\mathbf{u}'(s)|$ of the unit tangent vector $\mathbf{u}(s)$ at P. Hence $\kappa(s)$ measures the deviation of C at P from a straight line (its tangent at P). Since $\mathbf{u}(s) = \mathbf{r}'(s)$, the definition is

$$\kappa(s) = |\mathbf{u}'(s)| = |\mathbf{r}''(s)| \qquad ({}' = d/ds). \tag{22}$$

The **torsion** $\tau(s)$ of C at P measures the rate of change of the **osculating plane** O of curve C at point P. Note that this plane is spanned by **u** and $\mathbf{u}'$ and shown in Fig. 212. Hence $\tau(s)$ measures the deviation of C at P from a plane (from O at P). Now the rate of change is also measured by the derivative $\mathbf{b}'$ of a normal vector **b** at O. By the definition of vector product, a unit normal vector of O is $\mathbf{b} = \mathbf{u} \times (1/\kappa)\mathbf{u}' = \mathbf{u} \times \mathbf{p}$. Here $\mathbf{p} = (1/\kappa)\mathbf{u}'$ is called the **unit principal normal vector** and **b** is called the **unit binormal vector** of C at P. The vectors are labeled in Fig. 212. Here we must assume that $\kappa \neq 0$; hence $\kappa > 0$. The absolute value of the torsion is now defined by

$$|\tau(s)| = |\mathbf{b}'(s)|. \tag{23*}$$

Whereas $\kappa(s)$ is nonnegative, it is practical to give the torsion a sign, motivated by "right-handed" and "left-handed" (see Figs. 204 and 205). This needs a little further calculation. Since **b** is a unit vector, it has constant length. Hence $\mathbf{b}'$ is perpendicular

[3]GUSTAVE GASPARD CORIOLIS (1792–1843), French engineer who did research in mechanics.

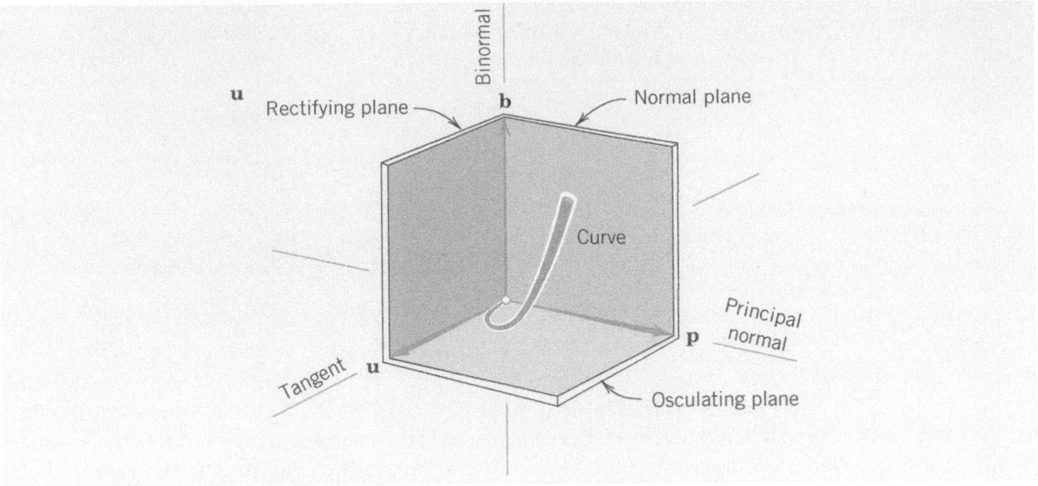

Fig. 212. Trihedron. Unit vectors **u**, **p**, **b** and planes

to **b** (see Example 4 in Sec. 9.4). Now $\mathbf{b}'$ is also perpendicular to **u** because, by the definition of vector product, we have $\mathbf{b} \cdot \mathbf{u} = 0$, $\mathbf{b} \cdot \mathbf{u}' = 0$. This implies

$$(\mathbf{b} \cdot \mathbf{u})' = 0; \qquad \text{that is,} \qquad \mathbf{b}' \cdot \mathbf{u} + \mathbf{b} \cdot \mathbf{u}' = \mathbf{b}' \cdot \mathbf{u} + 0 = 0.$$

Hence if $\mathbf{b}' \neq \mathbf{0}$ at P, it must have the direction of **p** or $-\mathbf{p}$, so that it must be of the form $\mathbf{b}' = -\tau\mathbf{p}$. Taking the dot product of this by **p** and using $\mathbf{p} \cdot \mathbf{p} = 1$ gives

$$\tau(s) = -\mathbf{p}(s) \cdot \mathbf{b}'(s). \tag{23}$$

The minus sign is chosen to make the torsion of a *right-handed* helix *positive* and that of a *left-handed* helix *negative* (Figs. 204 and 205). The orthonormal vector triple **u**, **p**, **b** is called the **trihedron** of C. Figure 212 also shows the names of the three straight lines in the directions of **u**, **p**, **b**, which are the intersections of the **osculating plane**, the **normal plane**, and the **rectifying plane**.

PROBLEM SET 9.5

1–10 PARAMETRIC REPRESENTATIONS

What curves are represented by the following? Sketch them.

1. $[3 + 2\cos t, 2\sin t, 0]$

2. $[a + t, b + 3t, c - 5t]$

3. $[0, t, t^3]$

4. $[-2, 2 + 5\cos t, -1 + 5\sin t]$

5. $[2 + 4\cos t, 1 + \sin t, 0]$

6. $[a + 3\cos \pi t, b - 2\sin \pi t, 0]$

7. $[4\cos t, 4\sin t, 3t]$

8. $[\cosh t, \sinh t, 2]$

9. $[\cos t, \sin 2t, 0]$

10. $[t, 2, 1/t]$

11–20 FIND A PARAMETRIC REPRESENTATION

11. Circle in the plane $z = 1$ with center $(3, 2)$ and passing through the origin.

12. Circle in the yz-plane with center $(4, 0)$ and passing through $(0, 3)$. Sketch it.

13. Straight line through $(2, 1, 3)$ in the direction of $\mathbf{i} + 2\mathbf{j}$.

14. Straight line through $(1, 1, 1)$ and $(4, 0, 2)$. Sketch it.

15. Straight line $y = 4x - 1$, $z = 5x$.

16. The intersection of the circular cylinder of radius 1 about the z-axis and the plane $z = y$.

17. Circle $\frac{1}{2}x^2 + y^2 = 1$, $z = y$.

18. Helix $x^2 + y^2 = 25$, $z = 2\arctan(y/x)$.

19. Hyperbola $4x^2 - 3y^2 = 4$, $z = -2$.

20. Intersection of $2x - y + 3z = 2$ and $x + 2y - z = 3$.

21. Orientation. Explain why setting $t = -t^*$ reverses the orientation of $[a \cos t, a \sin t, 0]$.

22. CAS PROJECT. Curves. Graph the following more complicated curves:

(a) $\mathbf{r}(t) = [2 \cos t + \cos 2t, 2 \sin t - \sin 2t]$ (*Steiner's hypocycloid*).

(b) $\mathbf{r}(t) = [\cos t + k \cos 2t, \sin t - k \sin 2t]$ with $k = 10, 2, 1, \frac{1}{2}, 0, -\frac{1}{2}, -1$.

(c) $\mathbf{r}(t) = [\cos t, \ \sin 5t]$ (a *Lissajous curve*).

(d) $\mathbf{r}(t) = [\cos t, \ \sin kt]$. For what k's will it be closed?

(e) $\mathbf{r}(t) = [R \sin \omega t + \omega R t, \ \ R \cos \omega t + R]$ (*cycloid*).

23. CAS PROJECT. Famous Curves in Polar Form. Use your CAS to graph the following curves[4] given in polar form $\rho = \rho(\theta)$, $\rho^2 = x^2 + y^2$, $\tan \theta = y/x$, and investigate their form depending on parameters a and b.

$\rho = a\theta$ *Spiral of Archimedes*

$\rho = ae^{b\theta}$ *Logarithmic spiral*

$\rho = \dfrac{2a \sin^2 \theta}{\cos \theta}$ *Cissoid of Diocles*

$\rho = \dfrac{a}{\cos \theta} + b$ *Conchoid of Nicomedes*

$\rho = a/\theta$ *Hyperbolic spiral*

$\rho = \dfrac{3a \sin 2\theta}{\cos^3 \theta + \sin^3 \theta}$ *Folium of Descartes*

$\rho = 2a \dfrac{\sin 3\theta}{\sin 2\theta}$ *Maclaurin's trisectrix*

$\rho = 2a \cos \theta + b$ *Pascal's snail*

24–28 TANGENT

Given a curve C: $\mathbf{r}(t)$, find a tangent vector $\mathbf{r}'(t)$, a unit tangent vector $\mathbf{u}'(t)$, and the tangent of C at P. Sketch curve and tangent.

24. $\mathbf{r}(t) = [t, \frac{1}{2}t^2, 1]$, P: (2, 2, 1)

25. $\mathbf{r}(t) = [10 \cos t, 1, 10 \sin t]$, P: (6, 1, 8)

26. $\mathbf{r}(t) = [\cos t, \sin t, 9t]$, P: $(1, 0, 18\pi)$

27. $\mathbf{r}(t) = [t, 1/t, 0]$, P: $(2, \frac{1}{2}, 0)$

28. $\mathbf{r}(t) = [t, t^2, t^3]$, P: (1, 1, 1)

29–32 LENGTH

Find the length and sketch the curve.

29. Catenary $\mathbf{r}(t) = [t, \cosh t]$ from $t = 0$ to $t = 1$.

30. Circular helix $\mathbf{r}(t) = [4 \cos t, 4 \sin t, 5t]$ from (4, 0, 0) to $(4, 0, 10\pi)$.

31. Circle $\mathbf{r}(t) = [a \cos t, a \sin t]$ from $(a, 0)$ to $(0, a)$.

32. Hypocycloid $\mathbf{r}(t) = [a \cos^3 t, a \sin^3 t]$, total length.

33. Plane curve. Show that Eq. (10) implies $\ell = \int_a^b \sqrt{1 + y'^2}\, dx$ for the length of a plane curve C: $y = f(x)$, $z = 0$, and $a = x = b$.

34. Polar coordinates $\rho = \sqrt{x^2 + y^2}$, $\theta = \arctan(y/x)$ give

$$\ell = \int_\alpha^\beta \sqrt{\rho^2 + \rho'^2}\, d\theta,$$

where $\rho' = d\rho/d\theta$. Derive this. Use it to find the total length of the **cardioid** $\rho = a(1 - \cos \theta)$. Sketch this curve. *Hint.* Use (10) in App. 3.1.

35–46 CURVES IN MECHANICS

Forces acting on moving objects (cars, airplanes, ships, etc.) require the engineer to know corresponding ***tangential*** and ***normal accelerations***. In Probs. 35–38 find them, along with the ***velocity*** and ***speed***. Sketch the path.

35. Parabola $\mathbf{r}(t) = [t, t^2, 0]$. Find $\mathbf{v}$ and $\mathbf{a}$.

36. Straight line $\mathbf{r}(t) = [8t, 6t, 0]$. Find $\mathbf{v}$ and $\mathbf{a}$.

37. Cycloid $\mathbf{r}(t) = (R \sin \omega t + Rt)\mathbf{i} + (R \cos \omega t + R)\mathbf{j}$. This is the path of a point on the rim of a wheel of radius R that rolls without slipping along the x-axis. Find $\mathbf{v}$ and $\mathbf{a}$ at the maximum y-values of the curve.

38. Ellipse $\mathbf{r} = [\cos t, 2 \sin t, 0]$.

39–42 THE USE OF A CAS may greatly facilitate the investigation of more complicated paths, as they occur in gear transmissions and other constructions. To grasp the idea, using a CAS, graph the path and find velocity, speed, and tangential and normal acceleration.

39. $\mathbf{r}(t) = [\cos t + \cos 2t, \sin t - \sin 2t]$

40. $\mathbf{r}(t) = [2 \cos t + \cos 2t, 2 \sin t - \sin 2t]$

41. $\mathbf{r}(t) = [\cos t, \sin 2t, \cos 2t]$

42. $\mathbf{r}(t) = [ct \cos t, ct \sin t, ct]$ $(c \neq 0)$

43. Sun and Earth. Find the acceleration of the Earth toward the sun from (19) and the fact that Earth revolves about the sun in a nearly circular orbit with an almost constant speed of 30 km/s.

44. Earth and moon. Find the centripetal acceleration of the moon toward Earth, assuming that the orbit of the moon is a circle of radius 239,000 miles = $3.85 \cdot 10^8$ m, and the time for one complete revolution is 27.3 days = $2.36 \cdot 10^6$ s.

---

[4]Named after ARCHIMEDES (c. 287–212 B.C.), DESCARTES (Sec. 9.1), DIOCLES (200 B.C.), MACLAURIN (Sec. 15.4), NICOMEDES (250? B.C.) ÉTIENNE PASCAL (1588–1651), father of BLAISE PASCAL (1623–1662).

45. Satellite. Find the speed of an artificial Earth satellite traveling at an altitude of 80 miles above Earth's surface, where $g = 31 \text{ ft/sec}^2$. (The radius of the Earth is 3960 miles.)

46. Satellite. A satellite moves in a circular orbit 450 miles above Earth's surface and completes 1 revolution in 100 min. Find the acceleration of gravity at the orbit from these data and from the radius of Earth (3960 miles).

47–55 CURVATURE AND TORSION

47. Circle. Show that a circle of radius a has curvature $1/a$.

48. Curvature. Using (22), show that if C is represented by $\mathbf{r}(t)$ with arbitrary t, then

$$(22^*) \quad \kappa(t) = \frac{\sqrt{(\mathbf{r}' \bullet \mathbf{r}')(\mathbf{r}'' \bullet \mathbf{r}'') - (\mathbf{r}' \bullet \mathbf{r}'')^2}}{(\mathbf{r}' \bullet \mathbf{r}')^{3/2}}.$$

49. Plane curve. Using (22*), show that for a curve $y = f(x)$,

$$(22^{**}) \quad \kappa(x) = \frac{|y''|}{(1 + y'^2)^{3/2}} \qquad \left(y' = \frac{dy}{dx}, \text{ etc.}\right).$$

50. Torsion. Using $\mathbf{b} = \mathbf{u} \times \mathbf{p}$ and (23), show that (when $\kappa > 0$)

$$(23^{**}) \quad \tau(s) = (\mathbf{u} \quad \mathbf{p} \quad \mathbf{p}') = (\mathbf{r}' \quad \mathbf{r}'' \quad \mathbf{r}''')/\kappa^2.$$

51. Torsion. Show that if C is represented by $\mathbf{r}(t)$ with arbitrary parameter t, then, assuming $\kappa > 0$ as before,

$$(23^{***}) \quad \tau(t) = \frac{(\mathbf{r}' \quad \mathbf{r}'' \quad \mathbf{r}''')}{(\mathbf{r}' \bullet \mathbf{r}')(\mathbf{r}'' \bullet \mathbf{r}'') - (\mathbf{r}' \bullet \mathbf{r}'')^2}.$$

52. Helix. Show that the helix $[a \cos t, \quad a \sin t, \quad ct]$ can be represented by $[a \cos(s/K), \quad a \sin(s/K), \quad cs/K]$, where $K = \sqrt{a^2 + c^2}$ and s is the arc length. Show that it has constant curvature $\kappa = a/K^2$ and torsion $\tau = c/K^2$.

53. Find the torsion of C: $\mathbf{r}(t) = [t, t^2, t^3]$, which looks similar to the curve in Fig. 212.

54. Frenet[5] formulas. Show that

$$\mathbf{u}' = \kappa \mathbf{p}, \quad \mathbf{p}' = -\kappa \mathbf{u} + \tau \mathbf{b}, \quad \mathbf{b}' = -\tau \mathbf{p}.$$

55. Obtain κ and τ in Prob. 52 from (22*) and (23***) and the original representation in Prob. 54 with parameter t.

9.6 Calculus Review: Functions of Several Variables. *Optional*

The parametric representations of curves C required vector functions that depended on a *single* variable x, s, or t. We now want to systematically cover vector functions of *several variables*. This optional section is inserted into the book for your convenience and to make the book reasonably self-contained. **Go onto Sec. 9.7 and consult Sec. 9.6 only when needed**. *For partial derivatives*, see App. A3.2.

Chain Rules

Figure 213 shows the notations in the following basic theorem.

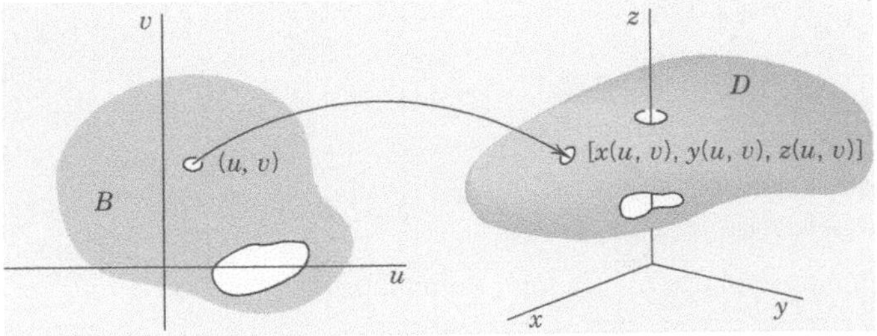

Fig. 213. Notations in Theorem 1

[5]JEAN-FRÉDÉRIC FRENET (1816–1900), French mathematician.

THEOREM 1

Chain Rule

Let $w = f(x, y, z)$ be continuous and have continuous first partial derivatives in a domain D in xyz-space. Let $x = x(u, v)$, $y = y(u, v)$, $z = z(u, v)$ be functions that are continuous and have first partial derivatives in a domain B in the uv-plane, where B is such that for every point (u, v) in B, the corresponding point $[x(u, v), y(u, v), z(u, v)]$ lies in D. See Fig. 213. *Then the function*

$$w = f(x(u, v), y(u, v), z(u, v))$$

is defined in B, has first partial derivatives with respect to u and v in B, and

$$\text{(1)} \qquad \begin{aligned} \frac{\partial w}{\partial u} &= \frac{\partial w}{\partial x}\frac{\partial x}{\partial u} + \frac{\partial w}{\partial y}\frac{\partial y}{\partial u} + \frac{\partial w}{\partial z}\frac{\partial z}{\partial u} \\ \frac{\partial w}{\partial v} &= \frac{\partial w}{\partial x}\frac{\partial x}{\partial v} + \frac{\partial w}{\partial y}\frac{\partial y}{\partial v} + \frac{\partial w}{\partial z}\frac{\partial z}{\partial v}. \end{aligned}$$

In this theorem, a **domain** D is an open connected point set in xyz-space, where "connected" means that any two points of D can be joined by a broken line of finitely many linear segments all of whose points belong to D. "Open" means that every point P of D has a neighborhood (a little ball with center P) all of whose points belong to D. For example, the interior of a cube or of an ellipsoid (the solid without the boundary surface) is a domain.

In calculus, x, y, z are often called the **intermediate variables**, in contrast with the **independent variables** u, v and the **dependent variable** w.

Special Cases of Practical Interest

If $w = f(x, y)$ and $x = x(u, v)$, $y = y(u, v)$ as before, then (1) becomes

$$\text{(2)} \qquad \begin{aligned} \frac{\partial w}{\partial u} &= \frac{\partial w}{\partial x}\frac{\partial x}{\partial u} + \frac{\partial w}{\partial y}\frac{\partial y}{\partial u} \\ \frac{\partial w}{\partial v} &= \frac{\partial w}{\partial x}\frac{\partial x}{\partial v} + \frac{\partial w}{\partial y}\frac{\partial y}{\partial v}. \end{aligned}$$

If $w = f(x, y, z)$ and $x = x(t)$, $y = y(t)$, $z = z(t)$, then (1) gives

$$\text{(3)} \qquad \frac{dw}{dt} = \frac{\partial w}{\partial x}\frac{dx}{dt} + \frac{\partial w}{\partial y}\frac{dy}{dt} + \frac{\partial w}{\partial z}\frac{dz}{dt}.$$

If $w = f(x, y)$ and $x = x(t)$, $y = y(t)$, then (3) reduces to

$$\frac{dw}{dt} = \frac{\partial w}{\partial x}\frac{dx}{dt} + \frac{\partial w}{\partial y}\frac{dy}{dt}. \tag{4}$$

Finally, the simplest case $w = f(x)$, $x = x(t)$ gives

$$\frac{dw}{dt} = \frac{dw}{dx}\frac{dx}{dt}. \tag{5}$$

EXAMPLE 1 Chain Rule

If $w = x^2 - y^2$ and we define polar coordinates r, θ by $x = r\cos\theta$, $y = r\sin\theta$, then (2) gives

$$\frac{\partial w}{\partial r} = 2x\cos\theta - 2y\sin\theta = 2r\cos^2\theta - 2r\sin^2\theta = 2r\cos 2\theta$$

$$\frac{\partial w}{\partial \theta} = 2x(-r\sin\theta) - 2y(r\cos\theta) = -2r^2\cos\theta\sin\theta - 2r^2\sin\theta\cos\theta = -2r^2\sin 2\theta.$$ ■

Partial Derivatives on a Surface $z = g(x, y)$

Let $w = f(x, y, z)$ and let $z = g(x, y)$ represent a surface S in space. Then on S the function becomes

$$\tilde{w}(x, y) = f(x, y, g(x, y)).$$

Hence, by (1), the partial derivatives are

$$\frac{\partial \tilde{w}}{\partial x} = \frac{\partial f}{\partial x} + \frac{\partial f}{\partial z}\frac{\partial g}{\partial x}, \qquad \frac{\partial \tilde{w}}{\partial y} = \frac{\partial f}{\partial y} + \frac{\partial f}{\partial z}\frac{\partial g}{\partial y} \qquad [z = g(x, y)]. \tag{6}$$

We shall need this formula in Sec. 10.9.

EXAMPLE 2 Partial Derivatives on Surface

Let $w = f = x^3 + y^3 + z^3$ and let $z = g = x^2 + y^2$. Then (6) gives

$$\frac{\partial \tilde{w}}{\partial x} = 3x^2 + 3z^2 \cdot 2x = 3x^2 + 3(x^2 + y^2)^2 \cdot 2x,$$

$$\frac{\partial \tilde{w}}{\partial y} = 3y^2 + 3z^2 \cdot 2y = 3y^2 + 3(x^2 + y^2)^2 \cdot 2y.$$

We confirm this by substitution, using $w(x, y) = x^3 + y^3 + (x^2 + y^2)^3$, that is,

$$\frac{\partial \tilde{w}}{\partial x} = 3x^2 + 3(x^2 + y^2)^2 \cdot 2x, \qquad \frac{\partial \tilde{w}}{\partial y} = 3y^2 + 3(x^2 + y^2)^2 \cdot 2y.$$ ■

Mean Value Theorems

THEOREM 2

Mean Value Theorem

Let $f(x, y, z)$ be continuous and have continuous first partial derivatives in a domain D in xyz-space. Let P_0: (x_0, y_0, z_0) and P: $(x_0 + h, y_0 + k, z_0 + l)$ be points in D such that the straight line segment P_0P joining these points lies entirely in D. Then

$$f(x_0 + h, y_0 + k, z_0 + l) - f(x_0, y_0, z_0) = h\frac{\partial f}{\partial x} + k\frac{\partial f}{\partial y} + l\frac{\partial f}{\partial z}, \tag{7}$$

the partial derivatives being evaluated at a suitable point of that segment.

Special Cases

For a function $f(x, y)$ of two variables (satisfying assumptions as in the theorem), formula (7) reduces to (Fig. 214)

$$f(x_0 + h, y_0 + k) - f(x_0, y_0) = h\frac{\partial f}{\partial x} + k\frac{\partial f}{\partial y}, \tag{8}$$

and, for a function $f(x)$ of a single variable, (7) becomes

$$f(x_0 + h) - f(x_0) = h\frac{\partial f}{\partial x}, \tag{9}$$

where in (9), the domain D is a segment of the x-axis and the derivative is taken at a suitable point between x_0 and $x_0 + h$.

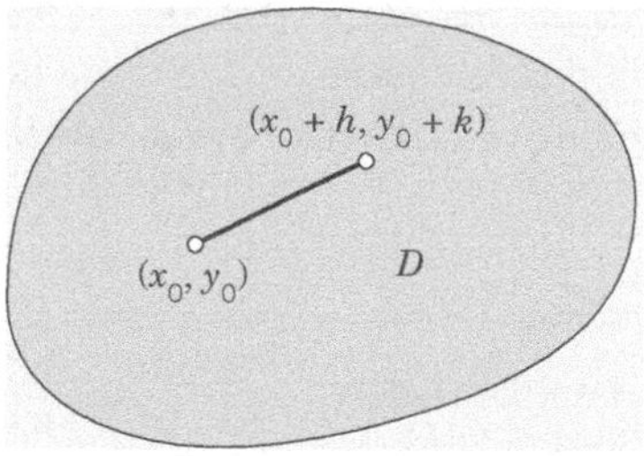

Fig. 214. Mean value theorem for a function of two variables [Formula (8)]

9.7 Gradient of a Scalar Field. Directional Derivative

We shall see that *some* of the vector fields that occur in applications—not all of them!—can be obtained from scalar fields. Using scalar fields instead of vector fields is of a considerable advantage because scalar fields are easier to use than vector fields. It is the "gradient" that allows us to obtain vector fields from scalar fields, and thus the gradient is of great practical importance to the engineer.

DEFINITION 1

Gradient

The setting is that we are given a scalar function $f(x, y, z)$ that is defined and differentiable in a domain in 3-space with Cartesian coordinates x, y, z. We denote the **gradient** of that function by grad f or ∇f (read **nabla** f). Then the qradient of $f(x, y, z)$ is defined as the vector function

$$\text{(1)} \qquad \operatorname{grad} f = \nabla f = \left[\frac{\partial f}{\partial x}, \frac{\partial f}{\partial y}, \frac{\partial f}{\partial z}\right] = \frac{\partial f}{\partial x}\mathbf{i} + \frac{\partial f}{\partial y}\mathbf{j} + \frac{\partial f}{\partial z}\mathbf{k}.$$

Remarks. For a definition of the gradient in curvilinear coordinates, see App. 3.4. As a quick example, if $f(x, y, z) = 2y^3 + 4xz + 3x$, then grad $f = [4z + 3, 6y^2, 4x]$. Furthermore, we will show later in this section that (1) actually does define a vector.

The notation ∇f is suggested by the *differential operator* ∇ (read *nabla*) defined by

$$\text{(1*)} \qquad \nabla = \frac{\partial}{\partial x}\mathbf{i} + \frac{\partial}{\partial y}\mathbf{j} + \frac{\partial}{\partial z}\mathbf{k}.$$

Gradients are useful in several ways, notably in giving the rate of change of $f(x, y, z)$ in any direction in space, in obtaining surface normal vectors, and in deriving vector fields from scalar fields, as we are going to show in this section.

Directional Derivative

From calculus we know that the partial derivatives in (1) give the rates of change of $f(x, y, z)$ in the directions of the three coordinate axes. It seems natural to extend this and ask for the rate of change of *f in an arbitrary direction* in space. This leads to the following concept.

DEFINITION 2

Directional Derivative

The directional derivative $D_{\mathbf{b}}f$ or df/ds of a function $f(x, y, z)$ at a point P in the direction of a vector **b** is defined by (see Fig. 215)

$$\text{(2)} \qquad D_{\mathbf{b}}f = \frac{df}{ds} = \lim_{s \to 0} \frac{f(Q) - f(P)}{s}.$$

Here Q is a variable point on the straight line L in the direction of **b**, and $|s|$ is the distance between P and Q. Also, $s > 0$ if Q lies in the direction of **b** (as in Fig. 215), $s < 0$ if Q lies in the direction of $-\mathbf{b}$, and $s = 0$ if $Q = P$.

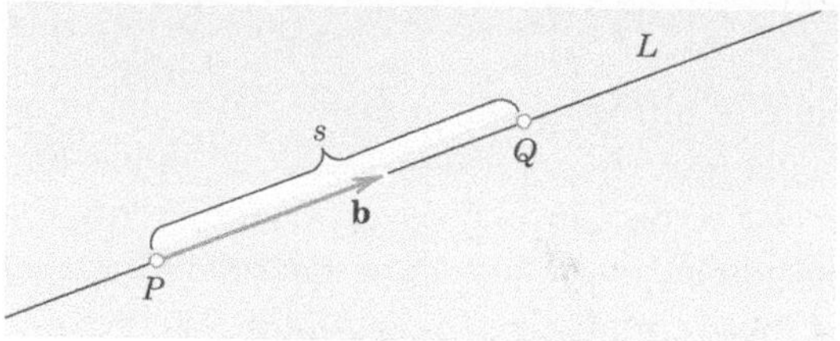

Fig. 215. Directional derivative

The next idea is to use Cartesian xyz-coordinates and for **b** a unit vector. Then the line L is given by

(3) $$\mathbf{r}(s) = x(s)\mathbf{i} + y(s)\mathbf{j} + z(s)\mathbf{k} = \mathbf{p}_0 + s\mathbf{b} \qquad (|\mathbf{b}| = 1)$$

where $\mathbf{p}_0$ the position vector of P. Equation (2) now shows that $D_{\mathbf{b}}f = df/ds$ is the derivative of the function $f(x(s), y(s), z(s))$ with respect to the arc length s of L. Hence, assuming that f has continuous partial derivatives and applying the chain rule [formula (3) in the previous section], we obtain

(4) $$D_{\mathbf{b}}f = \frac{df}{ds} = \frac{\partial f}{\partial x}x' + \frac{\partial f}{\partial y}y' + \frac{\partial f}{\partial z}z'$$

where primes denote derivatives with respect to s (which are taken at $s = 0$). But here, differentiating (3) gives $\mathbf{r}' = x'\mathbf{i} + y'\mathbf{j} + z'\mathbf{k} = \mathbf{b}$. Hence (4) is simply the inner product of grad f and **b** [see (2), Sec. 9.2]; that is,

(5) $$D_{\mathbf{b}}f = \frac{df}{ds} = \mathbf{b} \cdot \operatorname{grad} f \qquad (|\mathbf{b}| = 1).$$

ATTENTION! If the direction is given by a vector **a** of any length ($\neq 0$), then

(5*) $$D_{\mathbf{a}}f = \frac{df}{ds} = \frac{1}{|\mathbf{a}|}\mathbf{a} \cdot \operatorname{grad} f.$$

EXAMPLE 1 **Gradient. Directional Derivative**

Find the directional derivative of $f(x, y, z) = 2x^2 + 3y^2 + z^2$ at P: $(2, 1, 3)$ in the direction of $\mathbf{a} = [1, 0, -2]$.

Solution. grad $f = [4x, 6y, 2z]$ gives at P the vector grad $f(P) = [8, 6, 6]$. From this and (5*) we obtain, since $|\mathbf{a}| = \sqrt{5}$,

$$D_{\mathbf{a}}f(P) = \frac{1}{\sqrt{5}}[1, 0, -2] \cdot [8, 6, 6] = \frac{1}{\sqrt{5}}(8 + 0 - 12) = -\frac{4}{\sqrt{5}} = -1.789.$$

The minus sign indicates that at P the function f is decreasing in the direction of **a**. ■

Gradient Is a Vector. Maximum Increase

Here is a finer point of mathematics that concerns the consistency of our theory: grad f in (1) *looks* like a vector—after all, it has three components! But to prove that it *actually is* a vector, since it is defined in terms of components depending on the Cartesian coordinates, we must show that grad f has a length and direction independent of the choice of those coordinates. See proof of Theorem 1. In contrast, $[\partial f/\partial x, 2\partial f/\partial y, \partial f/\partial z]$ also looks like a vector but does not have a length and direction independent of the choice of Cartesian coordinates.

Incidentally, the direction makes the gradient eminently useful: grad f *points in the direction of maximum increase of* f.

THEOREM 1

Use of Gradient: Direction of Maximum Increase

Let $f(P) = f(x, y, z)$ be a scalar function having continuous first partial derivatives in some domain B in space. Then grad f *exists in B and is a vector, that is, its length and direction are independent of the particular choice of Cartesian coordinates. If* grad $f(P) \neq \mathbf{0}$ *at some point P, it has the direction of maximum increase of f at P.*

PROOF From (5) and the definition of inner product [(1) in Sec. 9.2] we have

$$D_{\mathbf{b}}f = |\mathbf{b}||\text{grad}\, f| \cos\gamma = |\text{grad}\, f| \cos\gamma \tag{6}$$

where γ is the angle between $\mathbf{b}$ and grad f. Now f is a scalar function. Hence its value at a point P depends on P but not on the particular choice of coordinates. The same holds for the arc length s of the line L in Fig. 215, hence also for $D_{\mathbf{b}}f$. Now (6) shows that $D_{\mathbf{b}}f$ is maximum when $\cos\gamma = 1$, $\gamma = 0$, and then $D_{\mathbf{b}}f = |\text{grad}\, f|$. It follows that the length and direction of grad f are independent of the choice of coordinates. Since $\gamma = 0$ if and only if $\mathbf{b}$ has the direction of grad f, the latter is the direction of maximum increase of f at P, provided grad $f \neq \mathbf{0}$ at P. Make sure that you understood the proof to get a good feel for mathematics.

Gradient as Surface Normal Vector

Gradients have an important application in connection with surfaces, namely, as surface normal vectors, as follows. Let S be a surface represented by $f(x, y, z) = c = \text{const}$, where f is differentiable. Such a surface is called a **level surface** of f, and for different c we get different level surfaces. Now let C be a curve on S through a point P of S. As a curve in space, C has a representation $\mathbf{r}(t) = [x(t), y(t), z(t)]$. For C to lie on the surface S, the components of $\mathbf{r}(t)$ must satisfy $f(x, y, z) = c$, that is,

$$f(x(t), y(t), z(t)) = c. \tag{7}$$

Now a tangent vector of C is $\mathbf{r}'(t) = [x'(t), y'(t), z'(t)]$. And the tangent vectors of all curves on S passing through P will generally form a plane, called the **tangent plane** of S at P. (Exceptions occur at edges or cusps of S, for instance, at the apex of the cone in Fig. 217.) The normal of this plane (the straight line through P perpendicular to the tangent plane) is called the **surface normal** to S at P. A vector in the direction of the surface normal is called a **surface normal vector** of S at P. We can obtain such a vector quite simply by differentiating (7) with respect to t. By the chain rule,

$$\frac{\partial f}{\partial x} x' + \frac{\partial f}{\partial y} y' + \frac{\partial f}{\partial z} z' = (\text{grad}\, f) \cdot \mathbf{r}' = 0.$$

Hence grad f is orthogonal to all the vectors $\mathbf{r}'$ in the tangent plane, so that it is a normal vector of S at P. Our result is as follows (see Fig. 216).

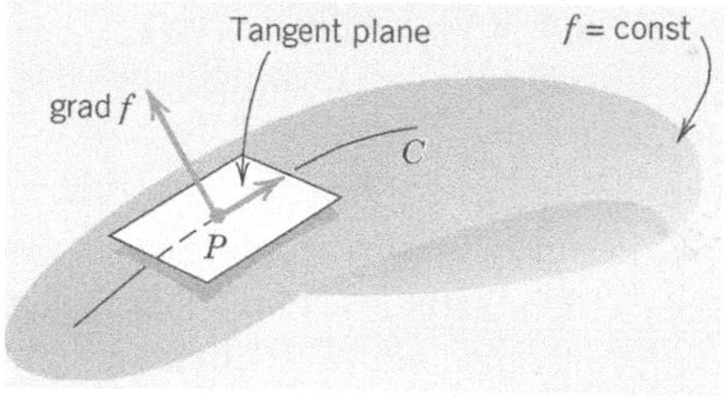

Fig. 216. Gradient as surface normal vector

THEOREM 2

Gradient as Surface Normal Vector

Let f be a differentiable scalar function in space. Let $f(x, y, z) = c = const$ represent a surface S. Then if the gradient of f at a point P of S is not the zero vector, it is a normal vector of S at P.

EXAMPLE 2 **Gradient as Surface Normal Vector. Cone**

Find a unit normal vector $\mathbf{n}$ of the cone of revolution $z^2 = 4(x^2 + y^2)$ at the point P: (1, 0, 2).

Solution. The cone is the level surface $f = 0$ of $f(x, y, z) = 4(x^2 + y^2) - z^2$. Thus (Fig. 217)

$$\text{grad}\, f = [8x, \quad 8y, \quad -2z], \qquad \text{grad}\, f(P) = [8, \quad 0, \quad -4]$$

$$\mathbf{n} = \frac{1}{|\text{grad}\, f(P)|} \text{grad}\, f(P) = \left[\frac{2}{\sqrt{5}}, \quad 0, \quad -\frac{1}{\sqrt{5}}\right].$$

$\mathbf{n}$ points downward since it has a negative z-component. The other unit normal vector of the cone at P is $-\mathbf{n}$. ■

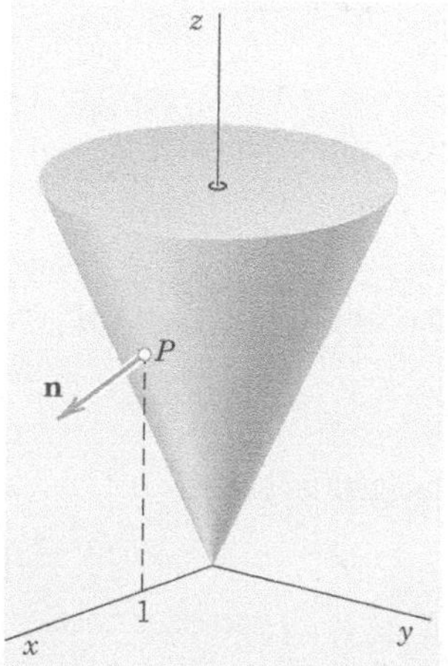

Fig. 217. Cone and unit normal vector $\mathbf{n}$

Vector Fields That Are Gradients of Scalar Fields ("Potentials")

At the beginning of this section we mentioned that some vector fields have the advantage that they can be obtained from scalar fields, which can be worked with more easily. Such a vector field is given by a vector function $\mathbf{v}(P)$, which is obtained as the gradient of a scalar function, say, $\mathbf{v}(P) = \text{grad} f(P)$. The function $f(P)$ is called a *potential function* or a **potential** of $\mathbf{v}(P)$. Such a $\mathbf{v}(P)$ and the corresponding vector field are called **conservative** because in such a vector field, energy is conserved; that is, no energy is lost (or gained) in displacing a body (or a charge in the case of an electrical field) from a point P to another point in the field and back to P. We show this in Sec. 10.2.

Conservative fields play a central role in physics and engineering. A basic application concerns the gravitational force (see Example 3 in Sec. 9.4) and we show that it has a potential which satisfies Laplace's equation, the most important partial differential equation in physics and its applications.

THEOREM 3

Gravitational Field. Laplace's Equation

The force of attraction

$$\mathbf{p} = -\frac{c}{r^3}\mathbf{r} = -c\left[\frac{x-x_0}{r^3}, \frac{y-y_0}{r^3}, \frac{z-z_0}{r^3}\right] \tag{8}$$

between two particles at points P_0: (x_0, y_0, z_0) and P: (x, y, z) (as given by Newton's law of gravitation) *has the potential* $f(x, y, z) = c/r$, *where* $r\ (> 0)$ *is the distance between* P_0 *and* P.

Thus $\mathbf{p} = \text{grad} f = \text{grad}\,(c/r)$. *This potential* f *is a solution of* **Laplace's equation**

$$\nabla^2 f = \frac{\partial^2 f}{\partial x^2} + \frac{\partial^2 f}{\partial y^2} + \frac{\partial^2 f}{\partial z^2} = 0. \tag{9}$$

[$\nabla^2 f$ (read *nabla squared* f) is called the **Laplacian** of f.]

PROOF That distance is $r = ((x - x_0)^2 + (y - y_0)^2 + (z - z_2)^2)^{1/2}$. The key observation now is that for the components of $\mathbf{p} = [p_1, p_2, p_3]$ we obtain by partial differentiation

$$\frac{\partial}{\partial x}\left(\frac{1}{r}\right) = \frac{-2(x-x_0)}{2[(x-x_0)^2 + (y-y_0)^2 + (z-z_0)^2]^{3/2}} = -\frac{x-x_0}{r^3} \tag{10a}$$

and similarly

$$\frac{\partial}{\partial y}\left(\frac{1}{r}\right) = -\frac{y-y_0}{r^3}, \qquad \frac{\partial}{\partial z}\left(\frac{1}{r}\right) = -\frac{z-z_0}{r^3}. \tag{10b}$$

From this we see that, indeed, **p** is the gradient of the scalar function $f = c/r$. The second statement of the theorem follows by partially differentiating (10), that is,

$$\frac{\partial^2}{\partial x^2}\left(\frac{1}{r}\right) = -\frac{1}{r^3} + \frac{3(x - x_0)^2}{r^5},$$

$$\frac{\partial^2}{\partial y^2}\left(\frac{1}{r}\right) = -\frac{1}{r^3} + \frac{3(y - y_0)^2}{r^5},$$

$$\frac{\partial^2}{\partial z^2}\left(\frac{1}{r}\right) = -\frac{1}{r^3} + \frac{3(z - z_0)^2}{r^5},$$

and then adding these three expressions. Their common denominator is r^5. Hence the three terms $-1/r^3$ contribute $-3r^2$ to the numerator, and the three other terms give the sum

$$3(x - x_0)^2 + 3(y - y_0)^2 + 3(z - z_0)^2 = 3r^2,$$

so that the numerator is 0, and we obtain (9). ■

$\nabla^2 f$ is also denoted by Δf. The differential operator

$$\text{(11)} \qquad \nabla^2 = \Delta = \frac{\partial^2}{\partial x^2} + \frac{\partial^2}{\partial y^2} + \frac{\partial^2}{\partial z^2}$$

(read "nabla squared" or "delta") is called the **Laplace operator**. It can be shown that the field of force produced by any distribution of masses is given by a vector function that is the gradient of a scalar function f, and f satisfies (9) in any region that is free of matter.

The great importance of the Laplace equation also results from the fact that there are other laws in physics that are of the same form as Newton's law of gravitation. For instance, in electrostatics the force of attraction (or repulsion) between two particles of opposite (or like) charge Q_1 and Q_2 is

$$\text{(12)} \qquad \mathbf{p} = \frac{k}{r^3}\mathbf{r} \qquad \textbf{(Coulomb's law[6])}.$$

Laplace's equation will be discussed in detail in Chaps. 12 and 18.

A method for finding out whether a given vector field has a potential will be explained in Sec. 9.9.

---

[6]CHARLES AUGUSTIN DE COULOMB (1736–1806), French physicist and engineer. Coulomb's law was derived by him from his own very precise measurements.

PROBLEM SET 9.7

1–6 CALCULATION OF GRADIENTS

Find grad f. Graph some level curves f = const. Indicate ∇f by arrows at some points of these curves.

1. $f = (x+1)(2y-1)$
2. $f = 9x^2 + 4y^2$
3. $f = y/x$
4. $(y+6)^2 + (x-4)^2$
5. $f = x^4 + y^4$
6. $f = (x^2 - y^2)/(x^2 + y^2)$

7–10 USEFUL FORMULAS FOR GRADIENT AND LAPLACIAN

Prove and illustrate by an example.

7. $\nabla(f^n) = nf^{n-1}\nabla f$
8. $\nabla(fg) = f\nabla g + g\nabla f$
9. $\nabla(f/g) = (1/g^2)(g\nabla f - f\nabla g)$
10. $\nabla^2(fg) = g\nabla^2 f + 2\nabla f \cdot \nabla g + f\nabla^2 g$

11–15 USE OF GRADIENTS. ELECTRIC FORCE

The force in an electrostatic field given by $f(x, y, z)$ has the direction of the gradient. Find ∇f and its value at P.

11. $f = xy$, P: $(-4, 5)$
12. $f = x/(x^2 + y^2)$, P: $(1, 1)$
13. $f = \ln(x^2 + y^2)$, P: $(8, 6)$
14. $f = (x^2 + y^2 + z^2)^{-1/2}$ P: $(12, 0, 16)$
15. $f = 4x^2 + 9y^2 + z^2$, P: $(5, -1, -11)$
16. For what points P: (x, y, z) does ∇f with $f = 25x^2 + 9y^2 + 16z^2$ have the direction from P to the origin?
17. Same question as in Prob. 16 when $f = 25x^2 + 4y^2$.

18–23 VELOCITY FIELDS

Given the velocity potential f of a flow, find the velocity $\mathbf{v} = \nabla f$ of the field and its value $\mathbf{v}(P)$ at P. Sketch $\mathbf{v}(P)$ and the curve f = const passing through P.

18. $f = x^2 - 6x - y^2$, P: $(-1, 5)$
19. $f = \cos x \cosh y$, P: $(\frac{1}{2}\pi, \ln 2)$
20. $f = x(1 + (x^2 + y^2)^{-1})$, P: $(1, 1)$
21. $f = e^x \cos y$, P: $(1, \frac{1}{2}\pi)$
22. At what points is the flow in Prob. 21 directed vertically upward?
23. At what points is the flow in Prob. 21 horizontal?

24–27 HEAT FLOW

Experiments show that in a temperature field, heat flows in the direction of maximum decrease of temperature T. Find this direction in general and at the given point P. Sketch that direction at P as an arrow.

24. $T = 3x^2 - 2y^2$, P: $(2.5, 1.8)$
25. $T = z/(x^2 + y^2)$, P: $(0, 1, 2)$
26. $T = x^2 + y^2 + 4z^2$, P: $(2, -1, 2)$
27. **CAS PROJECT. Isotherms.** Graph some curves of constant temperature ("isotherms") and indicate directions of heat flow by arrows when the temperature equals **(a)** $x^3 - 3xy^2$, **(b)** $\sin x \sinh y$, and **(c)** $e^x \cos y$.
28. **Steepest ascent.** If $z(x, y) = 3000 - x^2 - 9y^2$ [meters] gives the elevation of a mountain at sea level, what is the direction of steepest ascent at P: $(4, 1)$?
29. **Gradient.** What does it mean if $|\nabla f(P)| > |\nabla f(Q)|$ at two points P and Q in a scalar field?

9.8 Divergence of a Vector Field

Vector calculus owes much of its importance in engineering and physics to the gradient, divergence, and curl. From a scalar field we can obtain a vector field by the gradient (Sec. 9.7). Conversely, from a vector field we can obtain a scalar field by the divergence or another vector field by the curl (to be discussed in Sec. 9.9). These concepts were suggested by basic physical applications. This will be evident from our examples.

To begin, let $\mathbf{v}(x, y, z)$ be a differentiable vector function, where x, y, z are Cartesian coordinates, and let v_1, v_2, v_3 be the components of $\mathbf{v}$. Then the function

$$\text{div}\,\mathbf{v} = \frac{\partial v_1}{\partial x} + \frac{\partial v_2}{\partial y} + \frac{\partial v_3}{\partial z} \tag{1}$$

is called the **divergence** of $\mathbf{v}$ or the *divergence of the vector field defined by* $\mathbf{v}$. For example, if

$$\mathbf{v} = [3xz, 2xy, -yz^2] = 3xz\mathbf{i} + 2xy\mathbf{j} - yz^2\mathbf{k}, \qquad \text{then} \qquad \operatorname{div} \mathbf{v} = 3z + 2x - 2yz.$$

Another common notation for the divergence is

$$\begin{aligned}\operatorname{div} \mathbf{v} = \nabla \cdot \mathbf{v} &= \left[\frac{\partial}{\partial x}, \frac{\partial}{\partial \mathbf{y}}, \frac{\partial}{\partial \mathbf{z}}\right] \cdot [v_1, v_2, v_3] \\ &= \left(\frac{\partial}{\partial x}\mathbf{i} + \frac{\partial}{\partial y}\mathbf{j} + \frac{\partial}{\partial z}\mathbf{k}\right) \cdot (v_1\mathbf{i} + v_2\mathbf{j} + v_3\mathbf{k}) \\ &= \frac{\partial v_1}{\partial x} + \frac{\partial v_2}{\partial y} + \frac{\partial v_3}{\partial z},\end{aligned}$$

with the understanding that the "product" $(\partial/\partial x)v_1$ in the dot product means the partial derivative $\partial v_1/\partial x$, etc. This is a convenient notation, but nothing more. Note that $\nabla \cdot \mathbf{v}$ means the scalar div $\mathbf{v}$, whereas ∇f means the vector grad f defined in Sec. 9.7.

In Example 2 we shall see that the divergence has an important physical meaning. Clearly, the values of a function that characterizes a physical or geometric property must be independent of the particular choice of coordinates. In other words, these values must be invariant with respect to coordinate transformations. Accordingly, the following theorem should hold.

THEOREM 1

Invariance of the Divergence

The divergence div $\mathbf{v}$ *is a scalar function, that is, its values depend only on the points in space (and, of course, on* $\mathbf{v}$*) but not on the choice of the coordinates in* (1), *so that with respect to other Cartesian coordinates* x^*, y^*, z^* *and corresponding components* $v_1{}^*$, $v_2{}^*$, $v_3{}^*$ *of* $\mathbf{v}$,

(2) $$\operatorname{div} \mathbf{v} = \frac{\partial v_1^*}{\partial x^*} + \frac{\partial v_2^*}{\partial y^*} + \frac{\partial v_3^*}{\partial z^*}.$$

We shall prove this theorem in Sec. 10.7, using integrals.

Presently, let us turn to the more immediate practical task of gaining a feel for the significance of the divergence. Let $f(x, y, z)$ be a twice differentiable scalar function. Then its gradient exists,

$$\mathbf{v} = \operatorname{grad} f = \left[\frac{\partial f}{\partial x}, \frac{\partial f}{\partial y}, \frac{\partial f}{\partial z}\right] = \frac{\partial f}{\partial x}\mathbf{i} + \frac{\partial f}{\partial y}\mathbf{j} + \frac{\partial f}{\partial z}\mathbf{k}$$

and we can differentiate once more, the first component with respect to x, the second with respect to y, the third with respect to z, and then form the divergence,

$$\operatorname{div} \mathbf{v} = \operatorname{div} (\operatorname{grad} f) = \frac{\partial^2 f}{\partial x^2} + \frac{\partial^2 f}{\partial y^2} + \frac{\partial^2 f}{\partial z^2}.$$

Hence we have the basic result that *the divergence of the gradient is the Laplacian* (Sec. 9.7),

$$\text{div}\,(\text{grad}\, f) = \nabla^2 f. \tag{3}$$

EXAMPLE 1 Gravitational Force. Laplace's Equation

The gravitational force $\mathbf{p}$ in Theorem 3 of the last section is the gradient of the scalar function $f(x, y, z) = c/r$, which satisfies Laplaces equation $\nabla^2 f = 0$. According to (3) this implies that div $\mathbf{p} = 0$ $(r > 0)$. ■

The following example from hydrodynamics shows the physical significance of the divergence of a vector field. We shall get back to this topic in Sec. 10.8 and add further physical details.

EXAMPLE 2 Flow of a Compressible Fluid. Physical Meaning of the Divergence

We consider the motion of a fluid in a region R having no **sources** or **sinks** in R, that is, no points at which fluid is produced or disappears. The concept of **fluid state** is meant to cover also gases and vapors. Fluids in the restricted sense, or liquids, such as water or oil, have very small compressibility, which can be neglected in many problems. In contrast, gases and vapors have high compressibility. Their density ρ (= mass per unit volume) depends on the coordinates x, y, z in space and may also depend on time t. We assume that our fluid is compressible. We consider the flow through a rectangular box B of small edges Δx, Δy, Δz parallel to the coordinate axes as shown in Fig. 218. (Here Δ is a standard notation for small quantities and, of course, has nothing to do with the notation for the Laplacian in (11) of Sec. 9.7.) The box B has the volume $\Delta V = \Delta x \,\Delta y\, \Delta z$. Let $\mathbf{v} = [v_1, v_2, v_3] = v_1\mathbf{i} + v_2\mathbf{j} + v_3\mathbf{k}$ be the velocity vector of the motion. We set

$$\mathbf{u} = \rho\mathbf{v} = [u_1, u_2, u_3] = u_1\mathbf{i} + u_2\mathbf{j} + u_3\mathbf{k} \tag{4}$$

and assume that $\mathbf{u}$ and $\mathbf{v}$ are continuously differentiable vector functions of x, y, z, and t, that is, they have first partial derivatives which are continuous. Let us calculate the change in the mass included in B by considering the **flux** across the boundary, that is, the total loss of mass leaving B per unit time. Consider the flow through the left of the three faces of B that are visible in Fig. 218, whose area is $\Delta x \Delta z$. Since the vectors $v_1\mathbf{i}$ and $v_3\,\mathbf{k}$ are parallel to that face, the components v_1 and v_3 of $\mathbf{v}$ contribute nothing to this flow. Hence the mass of fluid entering through that face during a short time interval Δt is given approximately by

$$(\rho v_2)_y\, \Delta x\, \Delta z\, \Delta t = (u_2)_y\, \Delta x\, \Delta z\, \Delta t,$$

where the subscript y indicates that this expression refers to the left face. The mass of fluid leaving the box B through the opposite face during the same time interval is approximately $(u_2)_{y+\Delta y}\, \Delta x\, \Delta z\, \Delta t$, where the subscript $y + \Delta y$ indicates that this expression refers to the right face (which is not visible in Fig. 218). The difference

$$\Delta u_2\, \Delta x\, \Delta z\, \Delta t = \frac{\Delta u_2}{\Delta y}\Delta V\, \Delta t \qquad [\Delta u_2 = (u_2)_{y+\Delta y} - (u_2)_y]$$

is the approximate loss of mass. Two similar expressions are obtained by considering the other two pairs of parallel faces of B. If we add these three expressions, we find that the total loss of mass in B during the time interval Δt is approximately

$$\left(\frac{\Delta u_1}{\Delta x} + \frac{\Delta u_2}{\Delta y} + \frac{\Delta u_3}{\Delta z}\right)\Delta V\, \Delta t,$$

where

$$\Delta u_1 = (u_1)_{x+\Delta x} - (u_1)_x \qquad \text{and} \qquad \Delta u_3 = (u_3)_{z+\Delta z} - (u_3)_z.$$

This loss of mass in B is caused by the time rate of change of the density and is thus equal to

$$-\frac{\partial\rho}{\partial t}\Delta V\, \Delta t.$$

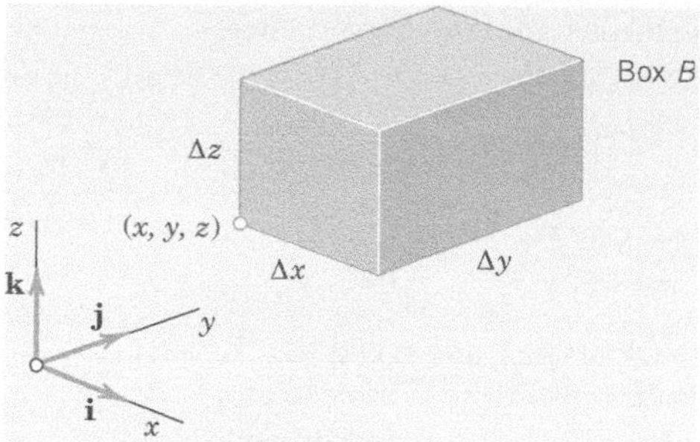

Fig. 218. Physical interpretation of the divergence

If we equate both expressions, divide the resulting equation by $\Delta V \Delta t$, and let Δx, Δy, Δz, and Δt approach zero, then we obtain

$$\operatorname{div} \mathbf{u} = \operatorname{div} (\rho \mathbf{v}) = -\frac{\partial \rho}{\partial t}$$

or

$$\frac{\partial \rho}{\partial t} + \operatorname{div} (\rho \mathbf{v}) = 0. \tag{5}$$

This important relation is called the *condition for the conservation of mass* or the **continuity equation** *of a compressible fluid flow.*

If the flow is **steady**, that is, independent of time, then $\partial \rho / \partial t = 0$ and the continuity equation is

$$\operatorname{div} (\rho \mathbf{v}) = 0. \tag{6}$$

If the density ρ is constant, so that the fluid is incompressible, then equation (6) becomes

$$\operatorname{div} \mathbf{v} = 0. \tag{7}$$

This relation is known as the **condition of incompressibility**. It expresses the fact that the balance of outflow and inflow for a given volume element is zero at any time. Clearly, the assumption that the flow has no sources or sinks in R is essential to our argument. $\mathbf{v}$ is also referred to as **solenoidal**.

From this discussion you should conclude and remember that, roughly speaking, ***the divergence measures outflow minus inflow***. ■

Comment. The **divergence theorem** of Gauss, an integral theorem involving the divergence, follows in the next chapter (Sec. 10.7).

PROBLEM SET 9.8

1–6 CALCULATION OF THE DIVERGENCE

Find div **v** and its value at P.

1. $\mathbf{v} = [x^2, 4y^2, 9z^2]$, P: $(-1, 0, \frac{1}{2})$
2. $\mathbf{v} = [0, \cos xyz, \sin xyz]$, P: $(2, \frac{1}{2}\pi, 0)$
3. $\mathbf{v} = (x^2 + y^2)^{-1}[x, y]$
4. $\mathbf{v} = [v_1(y, z), v_2(z, x), v_3(x, y)]$, P: $(3, 1, -1)$]
5. $\mathbf{v} = x^2y^2z^2[x, y, z]$, P: $(3, -1, 4)$
6. $\mathbf{v} = (x^2 + y^2 + z^2)^{-3/2}[x, y, z]$
7. For what v_3 is $\mathbf{v} = [e^x \cos y, e^x \sin y, v_3]$ solenoidal?
8. Let $\mathbf{v} = [x, y, v_3]$. Find a v_3 such that **(a)** div $\mathbf{v} > 0$ everywhere, **(b)** div $\mathbf{v} > 0$ if $|z| < 1$ and div $\mathbf{v} < 0$ if $|z| > 1$.

9. PROJECT. Useful Formulas for the Divergence. Prove

(a) $\text{div}\,(k\mathbf{v}) = k\,\text{div}\,\mathbf{v}$ (k constant)

(b) $\text{div}\,(f\mathbf{v}) = f\,\text{div}\,\mathbf{v} + \mathbf{v} \bullet \nabla f$

(c) $\text{div}\,(f\nabla g) = f\nabla^2 g + \nabla f \bullet \nabla g$

(d) $\text{div}\,(f\nabla g) - \text{div}\,(g\nabla f) = f\nabla^2 g - g\nabla^2 f$

Verify (b) for $f = e^{xyz}$ and $\mathbf{v} = ax\mathbf{i} + by\mathbf{j} + cz\mathbf{k}$. Obtain the answer to Prob. 6 from (b). Verify (c) for $f = x^2 - y^2$ and $g = e^{x+y}$. Give examples of your own for which (a)–(d) are advantageous.

10. CAS EXPERIMENT. Visualizing the Divergence. Graph the given velocity field **v** of a fluid flow in a square centered at the origin with sides parallel to the coordinate axes. Recall that the divergence measures outflow minus inflow. By looking at the flow near the sides of the square, can you see whether div **v** must be positive or negative or may perhaps be zero? Then calculate div **v**. First do the given flows and then do some of your own. Enjoy it.

(a) $\mathbf{v} = \mathbf{i}$

(b) $\mathbf{v} = x\mathbf{i}$

(c) $\mathbf{v} = x\mathbf{i} - y\mathbf{j}$

(d) $\mathbf{v} = x\mathbf{i} + y\mathbf{j}$

(e) $\mathbf{v} = -x\mathbf{i} - y\mathbf{j}$

(f) $\mathbf{v} = (x^2 + y^2)^{-1}(-y\mathbf{i} + x\mathbf{j})$

11. Incompressible flow. Show that the flow with velocity vector $\mathbf{v} = y\mathbf{i}$ is incompressible. Show that the particles that at time $t = 0$ are in the cube whose faces are portions of the planes $x = 0, x = 1, y = 0, y = 1, z = 0, z = 1$ occupy at $t = 1$ the volume 1.

12. Compressible flow. Consider the flow with velocity vector $\mathbf{v} = x\mathbf{i}$. Show that the individual particles have the position vectors $\mathbf{r}(t) = c_1e^t\mathbf{i} + c_2\mathbf{j} + c_3\mathbf{k}$ with constant c_1, c_2, c_3. Show that the particles that at $t = 0$ are in the cube of Prob. 11 at $t = 1$ occupy the volume e.

13. Rotational flow. The velocity vector $\mathbf{v}(x, y, z)$ of an incompressible fluid rotating in a cylindrical vessel is of the form $\mathbf{v} = \mathbf{w} \times \mathbf{r}$, where **w** is the (constant) rotation vector; see Example 5 in Sec. 9.3. Show that $\text{div}\,\mathbf{v} = 0$. Is this plausible because of our present Example 2?

14. Does $\text{div}\,\mathbf{u} = \text{div}\,\mathbf{v}$ imply $\mathbf{u} = \mathbf{v}$ or $\mathbf{u} = \mathbf{v} + \mathbf{k}$ (**k** constant)? Give reason.

15–20 LAPLACIAN

Calculate $\nabla^2 f$ by Eq. (3). Check by direct differentiation. Indicate when (3) is simpler. Show the details of your work.

15. $f = \cos^2 x + \sin^2 y$

16. $f = e^{xyz}$

17. $f = \ln (x^2 + y^2)$

18. $f = z - \sqrt{x^2 + y^2}$

19. $f = 1/(x^2 + y^2 + z^2)$

20. $f = e^{2x} \cosh 2y$

9.9 Curl of a Vector Field

The concepts of gradient (Sec. 9.7), divergence (Sec. 9.8), and curl are of fundamental importance in vector calculus and frequently applied in vector fields. In this section we define and discuss the concept of the curl and apply it to several engineering problems.

Let $\mathbf{v}(x, y, z) = [v_1, v_2, v_3] = v_1\mathbf{i} + v_2\mathbf{j} + v_3\mathbf{k}$ be a differentiable vector function of the Cartesian coordinates x, y, z. Then the **curl** *of the vector function* **v** or *of the vector field given by* **v** is defined by the "symbolic" determinant

$$\text{(1)} \qquad \text{curl}\,\mathbf{v} = \nabla \times \mathbf{v} = \begin{vmatrix} \mathbf{i} & \mathbf{j} & \mathbf{k} \\ \dfrac{\partial}{\partial x} & \dfrac{\partial}{\partial y} & \dfrac{\partial}{\partial z} \\ v_1 & v_2 & v_3 \end{vmatrix}$$

$$= \left(\frac{\partial v_3}{\partial y} - \frac{\partial v_2}{\partial z}\right)\mathbf{i} + \left(\frac{\partial v_1}{\partial z} - \frac{\partial v_3}{\partial x}\right)\mathbf{j} + \left(\frac{\partial v_2}{\partial x} - \frac{\partial v_1}{\partial y}\right)\mathbf{k}.$$

This is the formula when x, y, z are *right-handed.* If they are *left-handed*, the determinant has a minus sign in front (just as in (2**) in Sec. 9.3).

Instead of curl $\mathbf{v}$ one also uses the notation rot $\mathbf{v}$. This is suggested by "rotation," an application explored in Example 2. Note that curl $\mathbf{v}$ is a vector, as shown in Theorem 3.

EXAMPLE 1 Curl of a Vector Function

Let $\mathbf{v} = [yz, \quad 3zx, \quad z] = yz\mathbf{i} + 3zx\mathbf{j} + z\mathbf{k}$ with right-handed x, y, z. Then (1) gives

$$\operatorname{curl} \mathbf{v} = \begin{vmatrix} \mathbf{i} & \mathbf{j} & \mathbf{k} \\ \dfrac{\partial}{\partial x} & \dfrac{\partial}{\partial y} & \dfrac{\partial}{\partial z} \\ yz & 3zx & z \end{vmatrix} = -3x\mathbf{i} + y\mathbf{j} + (3z - z)\mathbf{k} = -3x\mathbf{i} + y\mathbf{j} + 2z\mathbf{k}.$$ ■

The curl has many applications. A typical example follows. More about the nature and significance of the curl will be considered in Sec. 10.9.

EXAMPLE 2 Rotation of a Rigid Body. Relation to the Curl

We have seen in Example 5, Sec. 9.3, that a rotation of a rigid body B about a fixed axis in space can be described by a vector $\mathbf{w}$ of magnitude ω in the direction of the axis of rotation, where ω (>0) is the angular speed of the rotation, and $\mathbf{w}$ is directed so that the rotation appears clockwise if we look in the direction of $\mathbf{w}$. According to (9), Sec. 9.3, the velocity field of the rotation can be represented in the form

$$\mathbf{v} = \mathbf{w} \times \mathbf{r}$$

where $\mathbf{r}$ is the position vector of a moving point with respect to a Cartesian coordinate system *having the origin on the axis of rotation.* Let us choose right-handed Cartesian coordinates such that the axis of rotation is the z-axis. Then (see Example 2 in Sec. 9.4)

$$\mathbf{w} = [0, \quad 0, \quad \omega] = \omega\mathbf{k}, \qquad \mathbf{v} = \mathbf{w} \times \mathbf{r} = [-\omega y, \quad \omega x, \quad 0] = -\omega y\mathbf{i} + \omega x\mathbf{j}.$$

Hence

$$\operatorname{curl} \mathbf{v} = \begin{vmatrix} \mathbf{i} & \mathbf{j} & \mathbf{k} \\ \dfrac{\partial}{\partial x} & \dfrac{\partial}{\partial y} & \dfrac{\partial}{\partial z} \\ -\omega y & \omega x & 0 \end{vmatrix} = [0, \quad 0, \quad 2\omega] = 2\omega\mathbf{k} = 2\mathbf{w}.$$

This proves the following theorem. ■

THEOREM 1 Rotating Body and Curl

The curl of the velocity field of a rotating rigid body has the direction of the axis of the rotation, and its magnitude equals twice the angular speed of the rotation.

Next we show how the grad, div, and curl are interrelated, thereby shedding further light on the nature of the curl.

THEOREM 2

Grad, Div, Curl

Gradient fields are **irrotational**. *That is, if a continuously differentiable vector function is the gradient of a scalar function f, then its curl is the zero vector,*

$$\text{curl}\,(\text{grad}\, f) = \mathbf{0}. \quad (2)$$

Furthermore, the divergence of the curl of a twice continuously differentiable vector function **v** *is zero,*

$$\text{div}\,(\text{curl}\, \mathbf{v}) = 0. \quad (3)$$

PROOF Both (2) and (3) follow directly from the definitions by straightforward calculation. In the proof of (3) the six terms cancel in pairs. ■

EXAMPLE 3 **Rotational and Irrotational Fields**

The field in Example 2 is not irrotational. A similar velocity field is obtained by stirring tea or coffee in a cup. The gravitational field in Theorem 3 of Sec. 9.7 has curl $\mathbf{p} = \mathbf{0}$. It is an irrotational gradient field. ■

The term "irrotational" for curl $\mathbf{v} = \mathbf{0}$ is suggested by the use of the curl for characterizing the rotation in a field. If a gradient field occurs elsewhere, not as a velocity field, it is usually called **conservative** (see Sec. 9.7). Relation (3) is plausible because of the interpretation of the curl as a rotation and of the divergence as a flux (see Example 2 in Sec. 9.8).

Finally, since the curl is defined in terms of coordinates, we should do what we did for the gradient in Sec. 9.7, namely, to find out whether the curl is a vector. This is true, as follows.

THEOREM 3

Invariance of the Curl

curl **v** is a vector. It has a length and a direction that are independent of the particular choice of a Cartesian coordinate system in space.

PROOF The proof is quite involved and shown in App. 4.

We have completed our discussion of vector differential calculus. The companion Chap. 10 on vector integral calculus follows and makes use of many concepts covered in this chapter, including dot and cross products, parametric representation of curves C, along with grad, div, and curl.

PROBLEM SET 9.9

1. **WRITING REPORT. Grad, div, curl.** List the definitions and most important facts and formulas for grad, div, curl, and ∇^2. Use your list to write a corresponding report of 3–4 pages, with examples of your own. No proofs.
2. **(a)** What direction does curl **v** have if **v** is parallel to the yz-plane? **(b)** If, moreover, **v** is independent of x?
3. Prove Theorem 2. Give two examples for (2) and (3) each.

4–8 CALCULUTION OF CURL

Find curl **v** for **v** given with respect to right-handed Cartesian coordinates. Show the details of your work.

4. $\mathbf{v} = [2y^2, 5x, 0]$

5. $\mathbf{v} = xyz\,[x, y, z]$

6. $\mathbf{v} = (x^2 + y^2 + z^2)^{-3/2}\,[x, y, z]$

7. $\mathbf{v} = [0, 0, e^{-x}\sin y]$

8. $\mathbf{v} = [e^{-z^2}, e^{-x^2}, e^{-y^2}]$

9–13 FLUID FLOW

Let **v** be the velocity vector of a steady fluid flow. Is the flow irrotational? Incompressible? Find the streamlines (the paths of the particles). *Hint.* See the answers to Probs. 9 and 11 for a determination of a path.

9. $\mathbf{v} = [0, 3z^2, 0]$

10. $\mathbf{v} = [\sec x, \csc x, 0]$

11. $\mathbf{v} = [y, -2x, 0]$

12. $\mathbf{v} = [-y, x, \pi]$

13. $\mathbf{v} = [x, y, -z]$

14. PROJECT. Useful Formulas for the Curl. Assuming sufficient differentiability, show that

(a) $\operatorname{curl}(\mathbf{u} + \mathbf{v}) = \operatorname{curl}\mathbf{u} + \operatorname{curl}\mathbf{v}$

(b) $\operatorname{div}(\operatorname{curl}\mathbf{v}) = 0$

(c) $\operatorname{curl}(f\mathbf{v}) = (\operatorname{grad} f) \times \mathbf{v} + f\operatorname{curl}\mathbf{v}$

(d) $\operatorname{curl}(\operatorname{grad} f) = \mathbf{0}$

(e) $\operatorname{div}(\mathbf{u} \times \mathbf{v}) = \mathbf{v} \cdot \operatorname{curl}\mathbf{u} - \mathbf{u} \cdot \operatorname{curl}\mathbf{v}$

15–20 DIV AND CURL

With respect to right-handed coordinates, let $\mathbf{u} = [y, z, x]$, $\mathbf{v} = [yz, zx, xy]$, $f = xyz$, and $g = x + y + z$. Find the given expressions. Check your result by a formula in Proj. 14 if applicable.

15. $\operatorname{curl}(\mathbf{u} + \mathbf{v})$, $\operatorname{curl}\mathbf{v}$

16. $\operatorname{curl}(g\mathbf{v})$

17. $\mathbf{v} \cdot \operatorname{curl}\mathbf{u}$, $\mathbf{u} \cdot \operatorname{curl}\mathbf{v}$, $\mathbf{u} \cdot \operatorname{curl}\mathbf{u}$

18. $\operatorname{div}(\mathbf{u} \times \mathbf{v})$

19. $\operatorname{curl}(g\mathbf{u} + \mathbf{v})$, $\operatorname{curl}(g\mathbf{u})$

20. $\operatorname{div}(\operatorname{grad}(fg))$

CHAPTER 9 REVIEW QUESTIONS AND PROBLEMS

1. What is a vector? A vector function? A vector field? A scalar? A scalar function? A scalar field? Give examples.

2. What is an inner product, a vector product, a scalar triple product? What applications motivate these products?

3. What are right-handed and left-handed coordinates? When is this distinction important?

4. When is a vector product the zero vector? What is orthogonality?

5. How is the derivative of a vector function defined? What is its significance in geometry and mechanics?

6. If $\mathbf{r}(t)$ represents a motion, what are $\mathbf{r}'(t)$, $|\mathbf{r}'(t)|$, $\mathbf{r}''(t)$, and $|\mathbf{r}''(t)|$?

7. Can a moving body have constant speed but variable velocity? Nonzero acceleration?

8. What do you know about directional derivatives? Their relation to the gradient?

9. Write down the definitions and explain the significance of grad, div, and curl.

10. Granted sufficient differentiability, which of the following expressions make sense? $f \operatorname{curl}\mathbf{v}$, $\mathbf{v}\operatorname{curl} f$, $\mathbf{u} \times \mathbf{v}$, $\mathbf{u} \times \mathbf{v} \times \mathbf{w}$, $f \cdot \mathbf{v}$, $f \cdot (\mathbf{v} \times \mathbf{w})$, $\mathbf{u} \cdot (\mathbf{v} \times \mathbf{w})$, $\mathbf{v} \times \operatorname{curl}\mathbf{v}$, $\operatorname{div}(f\mathbf{v})$, $\operatorname{curl}(f\mathbf{v})$, and $\operatorname{curl}(f \cdot \mathbf{v})$.

11–19 ALGEBRAIC OPERATIONS FOR VECTORS

Let $\mathbf{a} = [4, 7, 0]$, $\mathbf{b} = [3, -1, 5]$, $\mathbf{c} = [-6, 2, 0]$, and $\mathbf{d} = [1, -2, 8]$. Calculate the following expressions. Try to make a sketch.

11. $\mathbf{a} \cdot \mathbf{c}$, $3\mathbf{b} \cdot 8\mathbf{d}$, $24\mathbf{d} \cdot \mathbf{b}$, $\mathbf{a} \cdot \mathbf{a}$

12. $\mathbf{a} \times \mathbf{c}$, $\mathbf{b} \times \mathbf{d}$, $\mathbf{d} \times \mathbf{b}$, $\mathbf{a} \times \mathbf{a}$

13. $\mathbf{b} \times \mathbf{c}$, $\mathbf{c} \times \mathbf{b}$, $\mathbf{c} \times \mathbf{c}$, $\mathbf{c} \cdot \mathbf{c}$

14. $5(\mathbf{a} \times \mathbf{b}) \cdot \mathbf{c}$, $\mathbf{a} \cdot (5\mathbf{b} \times \mathbf{c})$, $(5\mathbf{a}\ \mathbf{b}\ \mathbf{c})$, $5(\mathbf{a} \cdot \mathbf{b}) \times \mathbf{c}$

15. $6(\mathbf{a} \times \mathbf{b}) \times \mathbf{d}$, $\mathbf{a} \times 6(\mathbf{b} \times \mathbf{d})$, $2\mathbf{a} \times 3\mathbf{b} \times \mathbf{d}$

16. $(1/|\mathbf{a}|)\mathbf{a}$, $(1/|\mathbf{b}|)\mathbf{b}$, $\mathbf{a} \cdot \mathbf{b}/|\mathbf{b}|$, $\mathbf{a} \cdot \mathbf{b}/|\mathbf{a}|$

17. $(\mathbf{a}\ \mathbf{b}\ \mathbf{d})$, $(\mathbf{b}\ \mathbf{a}\ \mathbf{d})$, $(\mathbf{b}\ \mathbf{d}\ \mathbf{a})$

18. $|\mathbf{a} + \mathbf{b}|$, $|\mathbf{a}| + |\mathbf{b}|$

19. $\mathbf{a} \times \mathbf{b} - \mathbf{b} \times \mathbf{a}$, $(\mathbf{a} \times \mathbf{c}) \cdot \mathbf{c}$, $|\mathbf{a} \times \mathbf{b}|$

20. Commutativity. When is $\mathbf{u} \times \mathbf{v} = \mathbf{v} \times \mathbf{u}$? When is $\mathbf{u} \cdot \mathbf{v} = \mathbf{v} \cdot \mathbf{u}$?

21. Resultant, equilibrium. Find **u** such that **u** and **a**, **b**, **c**, **d** above and **u** are in equilibrium.

22. Resultant. Find the most general **v** such that the resultant of **v**, **a**, **b**, **c** (see above) is parallel to the *yz*-plane.

23. Angle. Find the angle between **a** and **c**. Between **b** and **d**. Sketch **a** and **c**.

24. Planes. Find the angle between the two planes $P_1\colon 4x - y + 3z = 12$ and $P_2\colon x + 2y + 4z = 4$. Make a sketch.

25. Work. Find the work done by $\mathbf{q} = [5, 2, 0]$ in the displacement from $(1, 1, 0)$ to $(4, 3, 0)$.

26. Component. When is the component of a vector **v** in the direction of a vector **w** equal to the component of **w** in the direction of **v**?

27. Component. Find the component of $\mathbf{v} = [4, 7, 0]$ in the direction of $\mathbf{w} = [2, 2, 0]$. Sketch it.

28. Moment. When is the moment of a force equal to zero?

29. Moment. A force $\mathbf{p} = [4, 2, 0]$ is acting in a line through (2, 3, 0). Find its moment vector about the center (5, 1, 0) of a wheel.

30. Velocity, acceleration. Find the velocity, speed, and acceleration of the motion given by $\mathbf{r}(t) = [3 \cos t, 3 \sin t, 4t]$ (t = time) at the point $P: (3/\sqrt{2}, 3/\sqrt{2}, \pi)$.

31. Tetrahedron. Find the volume if the vertices are (0, 0, 0), (3, 1, 2), (2, 4, 0), (5, 4, 0).

32–40 GRAD, DIV, CURL, ∇^2, $D_v f$

Let $f = xy - yz$, $\mathbf{v} = [2y, 2z, 4x + z]$, and $\mathbf{w} = [3z^2, x^2 - y^2, y^2]$. Find:

32. grad f and f grad f at P: (2, 7, 0)

33. div $\mathbf{v}$, div $\mathbf{w}$

34. curl $\mathbf{v}$, curl $\mathbf{w}$

35. div (grad f), $\nabla^2 f$, $\nabla^2(xyf)$

36. (curl $\mathbf{w}$) • $\mathbf{v}$ at (4, 0, 2)

37. grad (div $\mathbf{w}$)

38. $D_v f$ at P: (1, 1, 2)

39. $D_w f$ at P: (3, 0, 2)

40. $\mathbf{v} \cdot ((\text{curl } \mathbf{w}) \times \mathbf{v})$

SUMMARY OF CHAPTER 9

Vector Differential Calculus. Grad, Div, Curl

All vectors of the form $\mathbf{a} = [a_1, a_2, a_3] = a_1\mathbf{i} + a_2\mathbf{j} + a_3\mathbf{k}$ constitute the **real vector space** R^3 with componentwise vector addition

$$(1) \qquad [a_1, a_2, a_3] + [b_1, b_2, b_3] = [a_1 + b_1, a_2 + b_2, a_3 + b_3]$$

and componentwise scalar multiplication (c a scalar, a real number)

$$(2) \qquad c[a_1, a_2, a_3] = [ca_1, ca_2, ca_3] \qquad \text{(Sec. 9.1).}$$

For instance, the *resultant* of forces **a** and **b** is the sum **a** + **b**.

The **inner product** or **dot product** of two vectors is defined by

$$(3) \qquad \mathbf{a} \cdot \mathbf{b} = |\mathbf{a}||\mathbf{b}| \cos \gamma = a_1b_1 + a_2b_2 + a_3b_3 \qquad \text{(Sec. 9.2)}$$

where γ is the angle between **a** and **b**. This gives for the **norm** or **length** $|\mathbf{a}|$ of **a**

$$(4) \qquad |\mathbf{a}| = \sqrt{\mathbf{a} \cdot \mathbf{a}} = \sqrt{a_1^2 + a_2^2 + a_3^2}$$

as well as a formula for γ. If $\mathbf{a} \cdot \mathbf{b} = 0$, we call **a** and **b orthogonal**. The dot product is suggested by the *work* $W = \mathbf{p} \cdot \mathbf{d}$ done by a force **p** in a displacement **d**.

The **vector product** or **cross product** $\mathbf{v} = \mathbf{a} \times \mathbf{b}$ is a vector of length

$$(5) \qquad |\mathbf{a} \times \mathbf{b}| = |\mathbf{a}||\mathbf{b}| \sin \gamma \qquad \text{(Sec. 9.3)}$$

and perpendicular to both **a** and **b** such that **a**, **b**, **v** form a *right-handed* triple. In terms of components with respect to right-handed coordinates,

$$(6) \qquad \mathbf{a} \times \mathbf{b} = \begin{vmatrix} \mathbf{i} & \mathbf{j} & \mathbf{k} \\ a_1 & a_2 & a_3 \\ b_1 & b_2 & b_3 \end{vmatrix} \qquad \text{(Sec. 9.3).}$$

The vector product is suggested, for instance, by moments of forces or by rotations. CAUTION! This multiplication is *anti*commutative, $\mathbf{a} \times \mathbf{b} = -\mathbf{b} \times \mathbf{a}$, and is *not* associative.

An (oblique) box with edges **a**, **b**, **c** has volume equal to the absolute value of the **scalar triple product**

$$(\mathbf{a} \quad \mathbf{b} \quad \mathbf{c}) = \mathbf{a} \cdot (\mathbf{b} \times \mathbf{c}) = (\mathbf{a} \times \mathbf{b}) \cdot \mathbf{c}. \tag{7}$$

Sections 9.4–9.9 extend differential calculus to vector functions

$$\mathbf{v}(t) = [v_1(t), v_2(t), v_3(t)] = v_1(t)\mathbf{i} + v_2(t)\mathbf{j} + v_3(t)\mathbf{k}$$

and to vector functions of more than one variable (see below). The derivative of $\mathbf{v}(t)$ is

$$\mathbf{v}' = \frac{d\mathbf{v}}{dt} = \lim_{\Delta t \to 0} \frac{\mathbf{v}(t + \Delta t) - \mathbf{v}(t)}{\Delta t} = [v_1', v_2', v_3'] = v_1'\mathbf{i} + v_2'\mathbf{j} + v_3'\mathbf{k}. \tag{8}$$

Differentiation rules are as in calculus. They imply (Sec. 9.4)

$$(\mathbf{u} \cdot \mathbf{v})' = \mathbf{u}' \cdot \mathbf{v} + \mathbf{u} \cdot \mathbf{v}', \qquad (\mathbf{u} \times \mathbf{v})' = \mathbf{u}' \times \mathbf{v} + \mathbf{u} \times \mathbf{v}'.$$

Curves C in space represented by the position vector $\mathbf{r}(t)$ have $\mathbf{r}'(t)$ as a **tangent vector** (the **velocity** in mechanics when t is time), $\mathbf{r}'(s)$ (s arc length, Sec. 9.5) as the *unit tangent vector,* and $|\mathbf{r}''(s)| = \kappa$ as the *curvature* (the *acceleration* in mechanics).

Vector functions $\mathbf{v}(x, y, z) = [v_1(x, y, z), v_2(x, y, z), v_3(x, y, z)]$ represent vector fields in space. Partial derivatives with respect to the Cartesian coordinates x, y, z are obtained componentwise, for instance,

$$\frac{\partial \mathbf{v}}{\partial x} = \left[\frac{\partial v_1}{\partial x}, \frac{\partial v_2}{\partial x}, \frac{\partial v_3}{\partial x}\right] = \frac{\partial v_1}{\partial x}\mathbf{i} + \frac{\partial v_2}{\partial x}\mathbf{j} + \frac{\partial v_3}{\partial x}\mathbf{k} \qquad \text{(Sec. 9.6).}$$

The **gradient** of a scalar function f is

$$\operatorname{grad} f = \nabla f = \left[\frac{\partial f}{\partial x}, \frac{\partial f}{\partial y}, \frac{\partial f}{\partial z}\right] \qquad \text{(Sec. 9.7).} \tag{9}$$

The **directional derivative** of f in the direction of a vector **a** is

$$D_{\mathbf{a}}f = \frac{df}{ds} = \frac{1}{|\mathbf{a}|}\mathbf{a} \cdot \nabla f \qquad \text{(Sec. 9.7).} \tag{10}$$

The **divergence** of a vector function **v** is

$$\operatorname{div} \mathbf{v} = \nabla \cdot \mathbf{v} = \frac{\partial v_1}{\partial x} + \frac{\partial v_2}{\partial y} + \frac{\partial v_3}{\partial z}. \qquad \text{(Sec. 9.8).} \tag{11}$$

The **curl** of **v** is

$$\text{curl}\,\mathbf{v} = \nabla \times \mathbf{v} = \begin{vmatrix} \mathbf{i} & \mathbf{j} & \mathbf{k} \\ \dfrac{\partial}{\partial x} & \dfrac{\partial}{\partial y} & \dfrac{\partial}{\partial z} \\ v_1 & v_2 & v_3 \end{vmatrix} \tag{12}$$ (Sec. 9.9)

or minus the determinant if the coordinates are left-handed.

Some basic formulas for grad, div, curl are (Secs. 9.7–9.9)

$$\begin{aligned} \nabla(fg) &= f\nabla g + g\nabla f \\ \nabla(f/g) &= (1/g^2)(g\nabla f - f\nabla g) \end{aligned} \tag{13}$$

$$\begin{aligned} \text{div}\,(f\mathbf{v}) &= f\,\text{div}\,\mathbf{v} + \mathbf{v}\cdot\nabla f \\ \text{div}\,(f\nabla g) &= f\nabla^2 g + \nabla f\cdot\nabla g \end{aligned} \tag{14}$$

$$\begin{aligned} \nabla^2 f &= \text{div}\,(\nabla f) \\ \nabla^2(fg) &= g\nabla^2 f + 2\nabla f\cdot\nabla g + f\nabla^2 g \end{aligned} \tag{15}$$

$$\begin{aligned} \text{curl}\,(f\mathbf{v}) &= \nabla f \times \mathbf{v} + f\,\text{curl}\,\mathbf{v} \\ \text{div}\,(\mathbf{u}\times\mathbf{v}) &= \mathbf{v}\cdot\text{curl}\,\mathbf{u} - \mathbf{u}\cdot\text{curl}\,\mathbf{v} \end{aligned} \tag{16}$$

$$\begin{aligned} \text{curl}\,(\nabla f) &= \mathbf{0} \\ \text{div}\,(\text{curl}\,\mathbf{v}) &= 0. \end{aligned} \tag{17}$$

For grad, div, curl, and ∇^2 in **curvilinear coordinates** see App. A3.4.

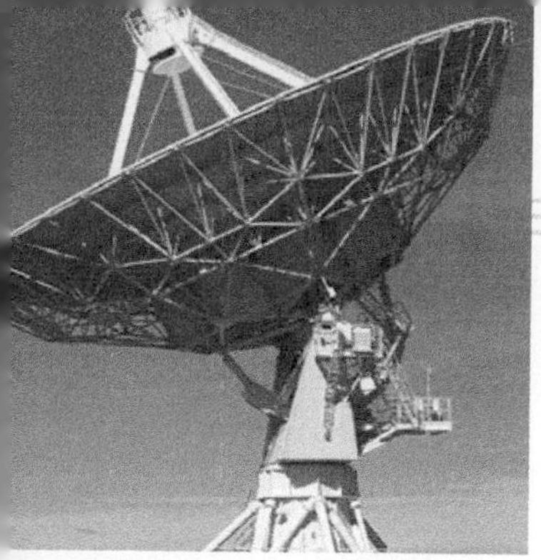

CHAPTER 10

Vector Integral Calculus. Integral Theorems

Vector integral calculus can be seen as a generalization of regular integral calculus. You may wish to review integration. (To refresh your memory, there is an optional review section on double integrals; see Sec. 10.3.)

Indeed, vector integral calculus extends integrals as known from regular calculus to integrals over curves, called *line integrals* (Secs. 10.1, 10.2), surfaces, called *surface integrals* (Sec. 10.6), and solids, called *triple integrals* (Sec. 10.7). The beauty of vector integral calculus is that we can transform these different integrals into one another. You do this to simplify evaluations, that is, one type of integral might be easier to solve than another, such as in potential theory (Sec. 10.8). More specifically, Green's theorem in the plane allows you to transform line integrals into double integrals, or conversely, double integrals into line integrals, as shown in Sec. 10.4. Gauss's convergence theorem (Sec. 10.7) converts surface integrals into triple integrals, and vice-versa, and Stokes's theorem deals with converting line integrals into surface integrals, and vice-versa.

This chapter is a companion to Chapter 9 on vector differential calculus. From Chapter 9, you will need to know inner product, curl, and divergence and how to parameterize curves. The root of the transformation of the integrals was largely physical intuition. Since the corresponding formulas involve the divergence and the curl, the study of this material will lead to a deeper physical understanding of these two operations.

Vector integral calculus is very important to the engineer and physicist and has many applications in solid mechanics, in fluid flow, in heat problems, and others.

Prerequisite: Elementary integral calculus, Secs. 9.7–9.9
Sections that may be omitted in a shorter course: 10.3, 10.5, 10.8
References and Answers to Problems: App. 1 Part B, App. 2

10.1 Line Integrals

The concept of a line integral is a simple and natural generalization of a definite integral

$$\int_a^b f(x)\,dx. \tag{1}$$

Recall that, in (1), we integrate the function $f(x)$, also known as the integrand, from $x = a$ along the x-axis to $x = b$. Now, in a line integral, we shall integrate a given function, also

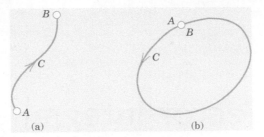

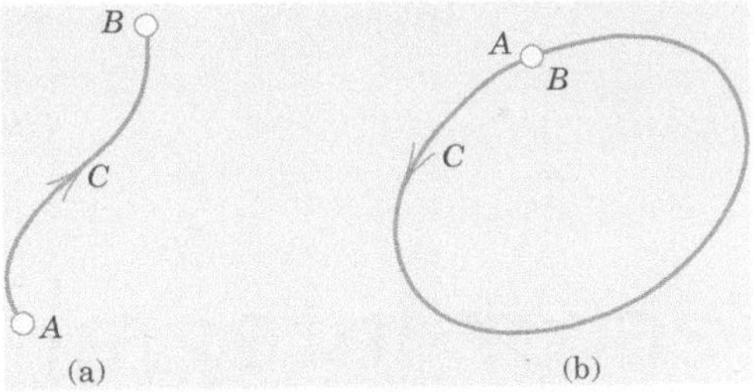

Fig. 219. Oriented curve

called the **integrand**, along a curve C in space or in the plane. (Hence curve integral would be a better name but line integral is standard).

This requires that we represent the curve C by a parametric representation (as in Sec. 9.5)

$$\mathbf{r}(t) = [x(t), y(t), z(t)] = x(t)\mathbf{i} + y(t)\mathbf{j} + z(t)\mathbf{k} \qquad (a \leqq t \leqq b). \tag{2}$$

The curve C is called the **path of integration**. Look at Fig. 219a. The path of integration goes from A to B. Thus A: $\mathbf{r}(a)$ is its initial point and B: $\mathbf{r}(b)$ is its terminal point. C is now *oriented.* The direction from A to B, in which t increases is called the positive direction on C. We mark it by an arrow. The points A and B may coincide, as it happens in Fig. 219b. Then C is called a **closed path.**

C is called a **smooth curve** if it has at each point a unique tangent whose direction varies continuously as we move along C. We note that $\mathbf{r}(t)$ in (2) is differentiable. Its derivative $\mathbf{r}'(t) = d\mathbf{r}/dt$ is continuous and different from the zero vector at every point of C.

General Assumption

In this book, every path of integration of a line integral is assumed to be **piecewise smooth**, *that is, it consists of* ***finitely many*** *smooth curves.*

For example, the boundary curve of a square is piecewise smooth. It consists of four smooth curves or, in this case, line segments which are the four sides of the square.

Definition and Evaluation of Line Integrals

A **line integral** of a vector function $\mathbf{F}(\mathbf{r})$ over a curve C: $\mathbf{r}(t)$ is defined by

$$\int_C \mathbf{F}(\mathbf{r}) \cdot d\mathbf{r} = \int_a^b \mathbf{F}(\mathbf{r}(t)) \cdot \mathbf{r}'(t)\, dt \qquad \mathbf{r}' = \frac{d\mathbf{r}}{dt} \tag{3}$$

where $\mathbf{r}(t)$ is the parametric representation of C as given in (2). (The dot product was defined in Sec. 9.2.) Writing (3) in terms of components, with $d\mathbf{r} = [dx, \quad dy, \quad dz]$ as in Sec. 9.5 and $' = d/dt$, we get

$$\begin{aligned} \int_C \mathbf{F}(\mathbf{r}) \cdot d\mathbf{r} &= \int_C (F_1\, dx + F_2\, dy + F_3\, dz) \\ &= \int_a^b (F_1 x' + F_2 y' + F_3 z')\, dt. \end{aligned} \tag{3'}$$

If the path of integration C in (3) is a *closed* curve, then instead of

$$\int_C \quad \text{we also write} \quad \oint_C .$$

Note that the integrand in (3) is a scalar, not a vector, because we take the dot product. Indeed, $\mathbf{F} \cdot \mathbf{r}'/|\mathbf{r}'|$ is the tangential component of $\mathbf{F}$. (For "component" see (11) in Sec. 9.2.)

We see that the integral in (3) on the right is a definite integral of a function of t taken over the interval $a \leqq t \leqq b$ on the t-axis in the ***positive*** direction: The direction of increasing t. This definite integral exists for continuous $\mathbf{F}$ and piecewise smooth C, because this makes $\mathbf{F} \cdot \mathbf{r}'$ piecewise continuous.

Line integrals (3) arise naturally in mechanics, where they give the work done by a force $\mathbf{F}$ in a displacement along C. This will be explained in detail below. We may thus call the line integral (3) the **work integral**. Other forms of the line integral will be discussed later in this section.

EXAMPLE 1 Evaluation of a Line Integral in the Plane

Find the value of the line integral (3) when $\mathbf{F}(\mathbf{r}) = [-y, -xy] = -y\mathbf{i} - xy\mathbf{j}$ and C is the circular arc in Fig. 220 from A to B.

Solution. We may represent C by $\mathbf{r}(t) = [\cos t, \sin t] = \cos t\,\mathbf{i} + \sin t\,\mathbf{j}$, where $0 \leqq t \leqq \pi/2$. Then $x(t) = \cos t$, $y(t) = \sin t$, and

$$\mathbf{F}(\mathbf{r}(t)) = -y(t)\mathbf{i} - x(t)y(t)\mathbf{j} = [-\sin t, \quad -\cos t \sin t] = -\sin t\,\mathbf{i} - \cos t \sin t\,\mathbf{j}.$$

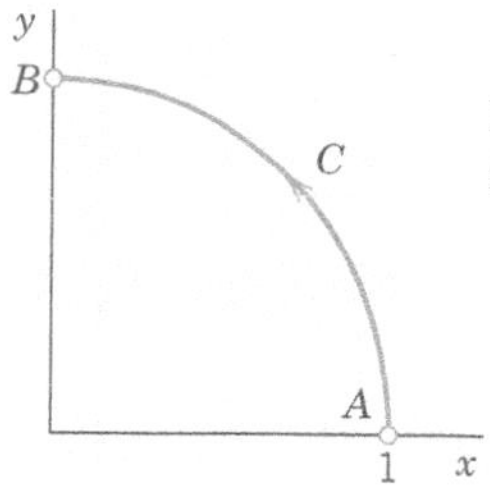

Fig. 220. Example 1

By differentiation, $\mathbf{r}'(t) = [-\sin t, \quad \cos t] = -\sin t\,\mathbf{i} + \cos t\,\mathbf{j}$, so that by (3) [use (10) in App. 3.1; set $\cos t = u$ in the second term]

$$\int_C \mathbf{F}(\mathbf{r}) \cdot d\mathbf{r} = \int_0^{\pi/2} [-\sin t, -\cos t \sin t] \cdot [-\sin t, \cos t]\, dt = \int_0^{\pi/2} (\sin^2 t - \cos^2 t \sin t)\, dt$$

$$= \int_0^{\pi/2} \frac{1}{2}(1 - \cos 2t)\, dt - \int_1^0 u^2(-du) = \frac{\pi}{4} - 0 - \frac{1}{3} \approx 0.4521. \qquad \blacksquare$$

EXAMPLE 2 Line Integral in Space

The evaluation of line integrals in space is practically the same as it is in the plane. To see this, find the value of (3) when $\mathbf{F}(\mathbf{r}) = [z, x, y] = z\mathbf{i} + x\mathbf{j} + y\mathbf{k}$ and C is the helix (Fig. 221)

$$\mathbf{r}(t) = [\cos t, \sin t, 3t] = \cos t\,\mathbf{i} + \sin t\,\mathbf{j} + 3t\mathbf{k} \qquad (0 \leqq t \leqq 2\pi). \tag{4}$$

Solution. From (4) we have $x(t) = \cos t$, $y(t) = \sin t$, $z(t) = 3t$. Thus

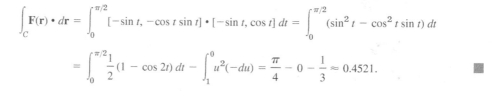

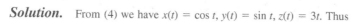

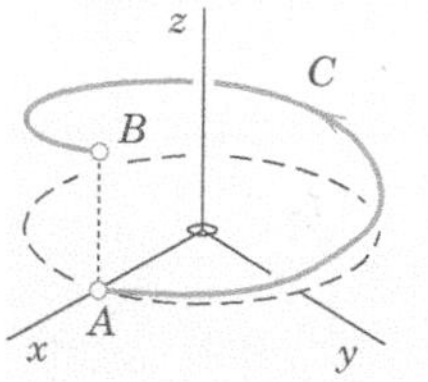

Fig. 221. Example 2

$$\mathbf{F}(\mathbf{r}(t)) \cdot \mathbf{r}'(t) = (3t\,\mathbf{i} + \cos t\,\mathbf{j} + \sin t\,\mathbf{k}) \cdot (-\sin t\,\mathbf{i} + \cos t\,\mathbf{j} + 3\mathbf{k}).$$

The dot product is $3t(-\sin t) + \cos^2 t + 3 \sin t$. Hence (3) gives

$$\int_C \mathbf{F}(\mathbf{r}) \cdot d\mathbf{r} = \int_0^{2\pi} (-3t \sin t + \cos^2 t + 3 \sin t)\, dt = 6\pi + \pi + 0 = 7\pi \approx 21.99. \qquad \blacksquare$$

Simple general properties of the line integral (3) follow directly from corresponding properties of the definite integral in calculus, namely,

$$\int_C k\mathbf{F} \cdot d\mathbf{r} = k \int_C \mathbf{F} \cdot d\mathbf{r} \qquad (k \text{ constant}) \tag{5a}$$

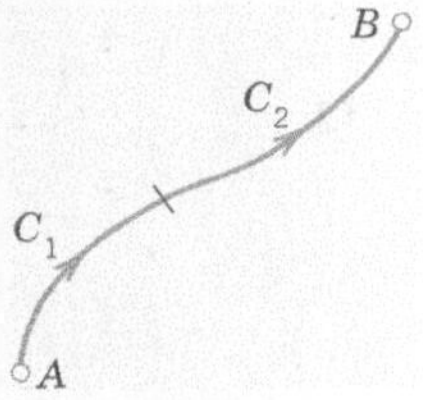

Fig. 222. Formula (5c)

$$\int_C (\mathbf{F} + \mathbf{G}) \bullet d\mathbf{r} = \int_C \mathbf{F} \bullet d\mathbf{r} + \int_C \mathbf{G} \bullet d\mathbf{r} \tag{5b}$$

$$\int_C \mathbf{F} \bullet d\mathbf{r} = \int_{C_1} \mathbf{F} \bullet d\mathbf{r} + \int_{C_2} \mathbf{F} \bullet d\mathbf{r} \qquad \text{(Fig. 222)} \tag{5c}$$

where in (5c) the path C is subdivided into two arcs C_1 and C_2 that have the same orientation as C (Fig. 222). In (5b) the orientation of C is the same in all three integrals. If the sense of integration along C is reversed, the value of the integral is multiplied by -1. However, we note the following independence if the sense is preserved.

THEOREM 1

Direction-Preserving Parametric Transformations

Any representations of C that give the same positive direction on C also yield the same value of the line integral (3).

PROOF The proof follows by the chain rule. Let $\mathbf{r}(t)$ be the given representation with $a \leqq t \leqq b$ as in (3). Consider the transformation $t = \phi(t^*)$ which transforms the t interval to $a^* \leqq t^* \leqq b^*$ and has a positive derivative dt/dt^*. We write $\mathbf{r}(t) = \mathbf{r}(\phi(t^*)) = \mathbf{r}^*(t^*)$. Then $dt = (dt/dt^*)\, dt^*$ and

$$\int_C \mathbf{F}(\mathbf{r}^*) \bullet d\mathbf{r}^* = \int_{a^*}^{b^*} \mathbf{F}(\mathbf{r}(\phi(t^*))) \bullet \frac{d\mathbf{r}}{dt}\frac{dt}{dt^*}\, dt^*$$

$$= \int_a^b \mathbf{F}(\mathbf{r}(t)) \bullet \frac{d\mathbf{r}}{dt}\, dt = \int_C \mathbf{F}(\mathbf{r}) \bullet d\mathbf{r}. \qquad \blacksquare$$

Motivation of the Line Integral (3): Work Done by a Force

The work W done by a *constant* force $\mathbf{F}$ in the displacement along a *straight* segment $\mathbf{d}$ is $W = \mathbf{F} \bullet \mathbf{d}$; see Example 2 in Sec. 9.2. This suggests that we define the work W done by a *variable* force $\mathbf{F}$ in the displacement along a curve C: $\mathbf{r}(t)$ as the limit of sums of works done in displacements along small chords of C. We show that this definition amounts to defining W by the line integral (3).

For this we choose points $t_0\,(=a) < t_1 < \cdots < t_n\,(=b)$. Then the work ΔW_m done by $\mathbf{F}(\mathbf{r}(t_m))$ in the straight displacement from $\mathbf{r}(t_m)$ to $\mathbf{r}(t_{m+1})$ is

$$\Delta W_m = \mathbf{F}(\mathbf{r}(t_m)) \bullet [\mathbf{r}(t_{m+1}) - \mathbf{r}(t_m)] \approx \mathbf{F}(\mathbf{r}(t_m)) \bullet \mathbf{r}'(t_m)\Delta t_m \qquad (\Delta t_m = \Delta t_{m+1} - t_m).$$

The sum of these n works is $W_n = \Delta W_0 + \cdots + \Delta W_{n-1}$. If we choose points and consider W_n for every n arbitrarily but so that the greatest Δt_m approaches zero as $n \to \infty$, then the limit of W_n as $n \to \infty$ is the line integral (3). This integral exists because of our general assumption that $\mathbf{F}$ is continuous and C is piecewise smooth; this makes $\mathbf{r}'(t)$ continuous, except at finitely many points where C may have corners or cusps. $\blacksquare$

EXAMPLE 3 **Work Done by a Variable Force**

If $\mathbf{F}$ in Example 1 is a force, the work done by $\mathbf{F}$ in the displacement along the quarter-circle is 0.4521, measured in suitable units, say, newton-meters (nt · m, also called joules, abbreviation J; see also inside front cover). Similarly in Example 2. ■

EXAMPLE 4 **Work Done Equals the Gain in Kinetic Energy**

Let $\mathbf{F}$ be a force, so that (3) is work. Let t be time, so that $d\mathbf{r}/dt = \mathbf{v}$, velocity. Then we can write (3) as

$$W = \int_C \mathbf{F} \bullet d\mathbf{r} = \int_a^b \mathbf{F}(\mathbf{r}(t)) \bullet \mathbf{v}(t)\, dt. \tag{6}$$

Now by Newton's second law, that is, force = mass × acceleration, we get

$$\mathbf{F} = m\mathbf{r}''(t) = m\mathbf{v}'(t),$$

where m is the mass of the body displaced. Substitution into (5) gives [see (11), Sec. 9.4]

$$W = \int_a^b m\mathbf{v}' \bullet \mathbf{v}\, dt = \int_a^b m\left(\frac{\mathbf{v} \bullet \mathbf{v}}{2}\right)' dt = \frac{m}{2}|\mathbf{v}|^2 \Big|_{t=a}^{t=b}.$$

On the right, $m|\mathbf{v}|^2/2$ is the kinetic energy. Hence *the work done equals the gain in kinetic energy*. This is a basic law in mechanics. ■

Other Forms of Line Integrals

The line integrals

$$\int_C F_1\, dx, \qquad \int_C F_2\, dy, \qquad \int_C F_3\, dz \tag{7}$$

are special cases of (3) when $\mathbf{F} = F_1\mathbf{i}$ or $F_2\mathbf{j}$ or $F_3\mathbf{k}$, respectively.

Furthermore, without taking a dot product as in (3) we can obtain a line integral whose value is a vector rather than a scalar, namely,

$$\int_C \mathbf{F}(\mathbf{r})\, dt = \int_a^b \mathbf{F}(\mathbf{r}(t))\, dt = \int_a^b [F_1(\mathbf{r}(t)),\quad F_2(\mathbf{r}(t)),\quad F_3(\mathbf{r}(t))]\, dt. \tag{8}$$

Obviously, a special case of (7) is obtained by taking $F_1 = f$, $F_2 = F_3 = 0$. Then

$$\int_C f(\mathbf{r})\, dt = \int_a^b f(\mathbf{r}(t))\, dt \tag{8*}$$

with C as in (2). The evaluation is similar to that before.

EXAMPLE 5 **A Line Integral of the Form (8)**

Integrate $\mathbf{F}(\mathbf{r}) = [xy, yz, z]$ along the helix in Example 2.

Solution. $\mathbf{F}(\mathbf{r}(t)) = [\cos t \sin t,\ 3t \sin t,\ 3t]$ integrated with respect to t from 0 to 2π gives

$$\int_0^{2\pi} \mathbf{F}(\mathbf{r}(t))\, dt = \left[-\frac{1}{2}\cos^2 t,\quad 3\sin t - 3t\cos t,\quad \frac{3}{2}t^2\right]\Bigg|_0^{2\pi} = [0,\quad -6\pi,\quad 6\pi^2]. \qquad ■$$

Path Dependence

Path dependence of line integrals is practically and theoretically so important that we formulate it as a theorem. And a whole section (Sec. 10.2) will be devoted to conditions under which path dependence does not occur.

THEOREM 2

Path Dependence

The line integral (3) *generally depends not only on* **F** *and on the endpoints A and B of the path, but also on the path itself along which the integral is taken.*

PROOF Almost any example will show this. Take, for instance, the straight segment C_1: $\mathbf{r}_1(t) = [t, t, 0]$ and the parabola C_2: $\mathbf{r}_2(t) = [t, t^2, 0]$ with $0 \leqq t \leqq 1$ (Fig. 223) and integrate $\mathbf{F} = [0, xy, 0]$. Then $\mathbf{F}(\mathbf{r}_1(t)) \bullet \mathbf{r}_1'(t) = t^2$, $\mathbf{F}(\mathbf{r}_2(t)) \bullet \mathbf{r}_2'(t) = 2t^4$, so that integration gives 1/3 and 2/5, respectively. ■

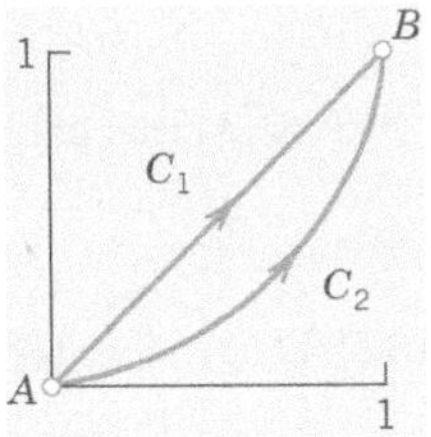

Fig. 223. Proof of Theorem 2

PROBLEM SET 10.1

1. WRITING PROJECT. From Definite Integrals to Line Integrals. Write a short report (1–2 pages) with examples on line integrals as generalizations of definite integrals. The latter give the area under a curve. Explain the corresponding geometric interpretation of a line integral.

2–11 LINE INTEGRAL. WORK

Calculate $\int_C \mathbf{F}(\mathbf{r}) \bullet d\mathbf{r}$ for the given data. If **F** is a force, this gives the work done by the force in the displacement along C. Show the details.

2. $\mathbf{F} = [y^2, -x^2]$, C: $y = 4x^2$ from (0, 0) to (1, 4)

3. **F** as in Prob. 2, C from (0, 0) straight to (1, 4). Compare.

4. $\mathbf{F} = [xy, x^2y^2]$, C from (2, 0) straight to (0, 2)

5. **F** as in Prob. 4, C the quarter-circle from (2, 0) to (0, 2) with center (0, 0)

6. $\mathbf{F} = [x - y, y - z, z - x]$, C: $\mathbf{r} = [2 \cos t, t, 2 \sin t]$ from (2, 0, 0) to $(2, 2\pi, 0)$

7. $\mathbf{F} = [x^2, y^2, z^2]$, C: $\mathbf{r} = [\cos t, \sin t, e^t]$ from (1, 0, 1) to $(1, 0, e^{2\pi})$. Sketch C.

8. $\mathbf{F} = [e^x, \cosh y, \sinh z]$, C: $\mathbf{r} = [t, t^2, t^3]$ from (0, 0, 0) to $(\frac{1}{2}, \frac{1}{4}, \frac{1}{8})$. Sketch C.

9. $\mathbf{F} = [x + y, y + z, z + x]$, C: $\mathbf{r} = [2t, 5t, t]$ from $t = 0$ to 1. Also from $t = -1$ to 1.

10. $\mathbf{F} = [x, -z, 2y]$ from (0, 0, 0) straight to (1, 1, 0), then to (1, 1, 1), back to (0, 0, 0)

11. $\mathbf{F} = [e^{-x}, e^{-y}, e^{-z}]$, C: $\mathbf{r} = [t, t^2, t]$ from (0, 0, 0) to (2, 4, 2). Sketch C.

12. PROJECT. Change of Parameter. Path Dependence. Consider the integral $\int_C \mathbf{F}(\mathbf{r}) \bullet d\mathbf{r}$, where $\mathbf{F} = [xy, -y^2]$.

(a) One path, several representations. Find the value of the integral when $\mathbf{r} = [\cos t, \sin t]$, $0 \leqq t \leqq \pi/2$. Show that the value remains the same if you set $t = -p$ or $t = p^2$ or apply two other parametric transformations of your own choice.

(b) Several paths. Evaluate the integral when C: $y = x^n$, thus $\mathbf{r} = [t, t^n]$, $0 \leqq t \leqq 1$, where $n = 1, 2, 3, \cdots$. Note that these infinitely many paths have the same endpoints.

(c) Limit. What is the limit in (b) as $n \to \infty$? Can you confirm your result by direct integration without referring to (b)?

(d) Show path dependence with a simple example of your choice involving two paths.

13. ***ML*-Inequality, Estimation of Line Integrals.** Let $\mathbf{F}$ be a vector function defined on a curve C. Let $|\mathbf{F}|$ be bounded, say, $|\mathbf{F}| \leqq M$ on C, where M is some positive number. Show that

$$\left| \int_C \mathbf{F} \cdot d\mathbf{r} \right| \leqq ML \qquad (L = \text{Length of } C). \tag{9}$$

14. Using (9), find a bound for the absolute value of the work W done by the force $\mathbf{F} = [x^2, y]$ in the displacement from (0, 0) straight to (3, 4). Integrate exactly and compare.

15–20 INTEGRALS (8) AND (8*)

Evaluate them with $\mathbf{F}$ or f and C as follows.

15. $\mathbf{F} = [y^2, z^2, x^2]$, C: $\mathbf{r} = [3 \cos t, 3 \sin t, 2t]$, $0 \leqq t \leqq 4\pi$

16. $f = 3x + y + 5z$, C: $\mathbf{r} = [t, \cosh t, \sinh t]$, $0 \leqq t \leqq 1$. Sketch C.

17. $\mathbf{F} = [x + y, y + z, z + x]$, C: $\mathbf{r} = [4 \cos t, \sin t, 0]$, $0 \leqq t \leqq \pi$

18. $\mathbf{F} = [y^{1/3}, x^{1/3}, 0]$, C the hypocycloid $\mathbf{r} = [\cos^3 t, \sin^3 t, 0]$, $0 \leqq t \leqq \pi/4$

19. $f = xyz$, C: $\mathbf{r} = [4t, 3t^2, 12t]$, $-2 \leqq t \leqq 2$. Sketch C.

20. $\mathbf{F} = [xz, yz, x^2y^2]$, C: $\mathbf{r} = [t, t, e^t]$, $0 \leqq t \leqq 5$. Sketch C.

10.2 Path Independence of Line Integrals

We want to find out under what conditions, in some domain, a line integral takes on the same value no matter what path of integration is taken (in that domain). As before we consider line integrals

$$\int_C \mathbf{F}(\mathbf{r}) \cdot d\mathbf{r} = \int_C (F_1\,dx + F_2\,dy + F_3\,dz) \qquad (d\mathbf{r} = [dx, \; dy, \; dz]) \tag{1}$$

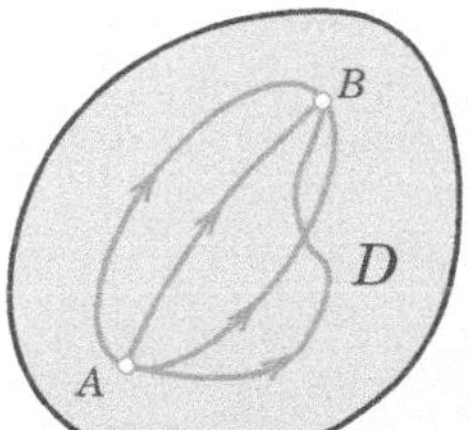

Fig. 224. Path independence

The line integral (1) is said to be **path independent in a domain *D* in space** if for every pair of endpoints A, B in domain D, (1) has the same value for all paths in D that begin at A and end at B. This is illustrated in Fig. 224. (See Sec. 9.6 for "domain.")

Path independence is important. For instance, in mechanics it may mean that we have to do the same amount of work regardless of the path to the mountaintop, be it short and steep or long and gentle. Or it may mean that in releasing an elastic spring we get back the work done in expanding it. Not all forces are of this type—think of swimming in a big round pool in which the water is rotating as in a whirlpool.

We shall follow up with three ideas about path independence. We shall see that path independence of (1) in a domain D holds if and only if:

(*Theorem* 1) $\mathbf{F} = \text{grad} f$, where grad f is the gradient of f as explained in Sec. 9.7.

(*Theorem* 2) Integration around closed curves C in D always gives 0.

(*Theorem* 3) curl $\mathbf{F} = \mathbf{0}$, provided D is simply connected, as defined below.

Do you see that these theorems can help in understanding the examples and counterexample just mentioned?

Let us begin our discussion with the following very practical criterion for path independence.

THEOREM 1

Path Independence

A line integral (1) with continuous F_1, F_2, F_3 in a domain D in space is path independent in D if and only if $\mathbf{F} = [F_1,\ \ F_2,\ \ F_3]$ is the gradient of some function f in D,

$$\text{(2)} \qquad \mathbf{F} = \operatorname{grad} f, \qquad \text{thus,} \qquad F_1 = \frac{\partial f}{\partial x}, \qquad F_2 = \frac{\partial f}{\partial y}, \qquad F_3 = \frac{\partial f}{\partial z}.$$

PROOF **(a)** We assume that (2) holds for some function f in D and show that this implies path independence. Let C be any path in D from any point A to any point B in D, given by $\mathbf{r}(t) = [x(t),\ \ y(t),\ \ z(t)]$, where $a \leqq t \leqq b$. Then from (2), the chain rule in Sec. 9.6, and (3′) in the last section we obtain

$$\begin{aligned} \int_C (F_1\,dx + F_2\,dy + F_3\,dz) &= \int_C \left(\frac{\partial f}{\partial x}dx + \frac{\partial f}{\partial y}dy + \frac{\partial f}{\partial z}dz\right) \\ &= \int_a^b \left(\frac{\partial f}{\partial x}\frac{dx}{dt} + \frac{\partial f}{\partial y}\frac{dy}{dt} + \frac{\partial f}{\partial z}\frac{dz}{dt}\right) dt \\ &= \int_a^b \frac{df}{dt}\,dt = f[x(t), y(t), z(t)]\Big|_{t=a}^{t=b} \\ &= f(x(b), y(b), z(b)) - f(x(a), y(a), z(a)) \\ &= f(B) - f(A). \end{aligned}$$

(b) The more complicated proof of the converse, that path independence implies (2) for some f, is given in App. 4. ■

The last formula in part (a) of the proof,

$$\text{(3)} \qquad \int_A^B (F_1\,dx + F_2\,dy + F_3\,dz) = f(B) - f(A) \qquad [\mathbf{F} = \operatorname{grad} f]$$

is the analog of the usual formula for definite integrals in calculus,

$$\int_a^b g(x)\,dx = G(x)\Big|_a^b = G(b) - G(a) \qquad [G'(x) = g(x)].$$

Formula (3) should be applied whenever a line integral is independent of path.

Potential theory relates to our present discussion if we remember from Sec. 9.7 that when $\mathbf{F} = \operatorname{grad} f$, then f is called a potential of $\mathbf{F}$. Thus the integral (1) is independent of path in D if and only if $\mathbf{F}$ is the gradient of a potential in D.

EXAMPLE 1 Path Independence

Show that the integral $\int_C \mathbf{F} \cdot d\mathbf{r} = \int_C (2x\,dx + 2y\,dy + 4z\,dz)$ is path independent in any domain in space and find its value in the integration from A: (0, 0, 0) to B: (2, 2, 2).

Solution. $\mathbf{F} = [2x, 2y, 4z] = \text{grad}\, f$, where $f = x^2 + y^2 + 2z^2$ because $\partial f/\partial x = 2x = F_1$, $\partial f/\partial y = 2y = F_2$, $\partial f/\partial z = 4z = F_3$. Hence the integral is independent of path according to Theorem 1, and (3) gives $f(B) - f(A) = f(2, 2, 2) - f(0, 0, 0) = 4 + 4 + 8 = 16$.

If you want to check this, use the most convenient path C: $\mathbf{r}(t) = [t, \ t, \ t], 0 \leqq t \leqq 2$, on which $\mathbf{F}(\mathbf{r}(t)) = [2t, \ 2t, \ 4t]$, so that $\mathbf{F}(\mathbf{r}(t)) \cdot \mathbf{r}'(t) = 2t + 2t + 4t = 8t$, and integration from 0 to 2 gives $8 \cdot 2^2/2 = 16$.

If you did not see the potential by inspection, use the method in the next example. ■

EXAMPLE 2 Path Independence. Determination of a Potential

Evaluate the integral $I = \int_C (3x^2\,dx + 2yz\,dy + y^2\,dz)$ from A: (0, 1, 2) to B: (1, −1, 7) by showing that $\mathbf{F}$ has a potential and applying (3).

Solution. If $\mathbf{F}$ has a potential f, we should have

$$f_x = F_1 = 3x^2, \qquad f_y = F_2 = 2yz, \qquad f_z = F_3 = y^2.$$

We show that we can satisfy these conditions. By integration of f_x and differentiation,

$$\begin{aligned} &f = x^3 + g(y, z), \qquad f_y = g_y = 2yz, \qquad g = y^2 z + h(z), \qquad f = x^3 + y^2 z + h(z) \\ &f_z = y^2 + h' = y^2, \qquad h' = 0 \qquad\qquad h = 0, \ \text{say}. \end{aligned}$$

This gives $f(x,y,z) = x^3 + y^3 z$ and by (3),

$$I = f(1, -1, 7) - f(0, 1, 2) = 1 + 7 - (0 + 2) = 6.$$ ■

Path Independence and Integration Around Closed Curves

The simple idea is that two paths with common endpoints (Fig. 225) make up a single closed curve. This gives almost immediately

THEOREM 2

Path Independence

The integral (1) is path independent in a domain D if and only if its value around every closed path in D is zero.

PROOF If we have path independence, then integration from A to B along C_1 and along C_2 in Fig. 225 gives the same value. Now C_1 and C_2 together make up a closed curve C, and if we integrate from A along C_1 to B as before, but then in the opposite sense along C_2 back to A (so that this second integral is multiplied by -1), the sum of the two integrals is zero, but this is the integral around the closed curve C.

Conversely, assume that the integral around any closed path C in D is zero. Given any points A and B and any two curves C_1 and C_2 from A to B in D, we see that C_1 with the orientation reversed and C_2 together form a closed path C. By assumption, the integral over C is zero. Hence the integrals over C_1 and C_2, both taken from A to B, must be equal. This proves the theorem. ■

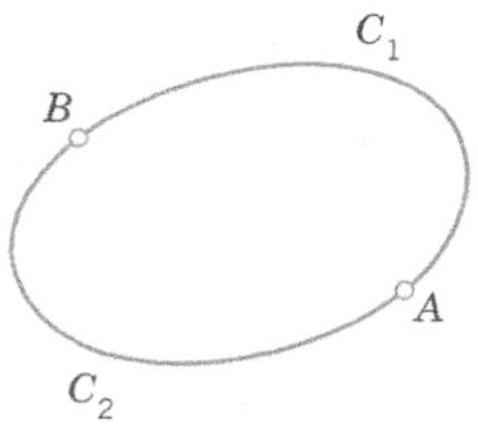

Fig. 225. Proof of Theorem 2

Work. Conservative and Nonconservative (Dissipative) Physical Systems

Recall from the last section that in mechanics, the integral (1) gives the work done by a force **F** in the displacement of a body along the curve C. Then Theorem 2 states that work is path independent in D if and only if its value is zero for displacement around every closed path in D. Furthermore, Theorem 1 tells us that this happens if and only if **F** is the gradient of a potential in D. In this case, **F** and the vector field defined by **F** are called **conservative** in D because in this case mechanical energy is conserved; that is, no work is done in the displacement from a point A and back to A. Similarly for the displacement of an electrical charge (an electron, for instance) in a conservative electrostatic field.

Physically, the kinetic energy of a body can be interpreted as the ability of the body to do work by virtue of its motion, and if the body moves in a conservative field of force, after the completion of a round trip the body will return to its initial position with the same kinetic energy it had originally. For instance, the gravitational force is conservative; if we throw a ball vertically up, it will (if we assume air resistance to be negligible) return to our hand with the same kinetic energy it had when it left our hand.

Friction, air resistance, and water resistance always act against the direction of motion. They tend to diminish the total mechanical energy of a system, usually converting it into heat or mechanical energy of the surrounding medium (possibly both). Furthermore, if during the motion of a body, these forces become so large that they can no longer be neglected, then the resultant force **F** of the forces acting on the body is no longer conservative. This leads to the following terms. A physical system is called **conservative** if all the forces acting in it are conservative. If this does not hold, then the physical system is called **nonconservative** or **dissipative**.

Path Independence and Exactness of Differential Forms

Theorem 1 relates path independence of the line integral (1) to the gradient and Theorem 2 to integration around closed curves. A third idea (leading to Theorems 3* and 3, below) relates path independence to the exactness of the **differential form** or *Pfaffian form*[1]

$$\mathbf{F} \cdot d\mathbf{r} = F_1\,dx + F_2\,dy + F_3\,dz \tag{4}$$

under the integral sign in (1). This form (4) is called **exact** in a domain D in space if it is the differential

$$df = \frac{\partial f}{\partial x}dx + \frac{\partial f}{\partial y}dy + \frac{\partial f}{\partial z}dz = (\operatorname{grad} f) \cdot d\mathbf{r}$$

of a differentiable function $f(x, y, z)$ everywhere in D, that is, if we have

$$\mathbf{F} \cdot d\mathbf{r} = df.$$

Comparing these two formulas, we see that the form (4) is exact if and only if there is a differentiable function $f(x, y, z)$ in D such that everywhere in D,

$$\mathbf{F} = \operatorname{grad} f, \qquad \text{thus,} \qquad F_1 = \frac{\partial f}{\partial x}, \qquad F_2 = \frac{\partial f}{\partial y}, \qquad F_3 = \frac{\partial f}{\partial z}. \tag{5}$$

[1]JOHANN FRIEDRICH PFAFF (1765–1825). German mathematician.

Hence Theorem 1 implies

THEOREM 3*

Path Independence

The integral (1) *is path independent in a domain D in space if and only if the differential form* (4) *has continuous coefficient functions F_1, F_2, F_3 and is exact in D.*

This theorem is of practical importance because it leads to a useful exactness criterion. First we need the following concept, which is of general interest.

A domain D is called **simply connected** if every closed curve in D can be continuously shrunk to any point in D without leaving D.

For example, the interior of a sphere or a cube, the interior of a sphere with finitely many points removed, and the domain between two concentric spheres are simply connected. On the other hand, the interior of a torus, which is a doughnut as shown in Fig. 249 in Sec. 10.6 is not simply connected. Neither is the interior of a cube with one space diagonal removed.

The criterion for exactness (and path independence by Theorem 3*) is now as follows.

THEOREM 3

Criterion for Exactness and Path Independence

Let F_1, F_2, F_3 in the line integral (1),

$$\int_C \mathbf{F}(\mathbf{r}) \cdot d\mathbf{r} = \int_C (F_1\,dx + F_2\,dy + F_3\,dz),$$

be continuous and have continuous first partial derivatives in a domain D in space. Then:

(a) *If the differential form* (4) *is exact in D—and thus* (1) *is path independent by Theorem 3*—, then in D,*

$$\text{curl}\,\mathbf{F} = \mathbf{0}; \tag{6}$$

in components (see Sec. 9.9)

$$\frac{\partial F_3}{\partial y} = \frac{\partial F_2}{\partial z}, \qquad \frac{\partial F_1}{\partial z} = \frac{\partial F_3}{\partial x}, \qquad \frac{\partial F_2}{\partial x} = \frac{\partial F_1}{\partial y}. \tag{6'}$$

(b) *If* (6) *holds in D and D is simply connected, then* (4) *is exact in D—and thus* (1) *is path independent by Theorem 3*.*

PROOF (a) If (4) is exact in D, then $\mathbf{F} = \text{grad}\, f$ in D by Theorem 3*, and, furthermore, $\text{curl}\,\mathbf{F} = \text{curl}\,(\text{grad}\, f) = \mathbf{0}$ by (2) in Sec. 9.9, so that (6) holds.

(b) The proof needs "Stokes's theorem" and will be given in Sec. 10.9. ■

Line Integral in the Plane. For $\int_C \mathbf{F}(\mathbf{r}) \cdot d\mathbf{r} = \int_C (F_1\,dx + F_2\,dy)$ the curl has only one component (the z-component), so that (6′) reduces to the single relation

$$\frac{\partial F_2}{\partial x} = \frac{\partial F_1}{\partial y} \tag{6''}$$

(which also occurs in (5) of Sec. 1.4 on exact ODEs).

EXAMPLE 3 **Exactness and Independence of Path. Determination of a Potential**

Using (6′), show that the differential form under the integral sign of

$$I = \int_C [2xyz^2\,dx + (x^2z^2 + z\cos yz)\,dy + (2x^2yz + y\cos yz)\,dz]$$

is exact, so that we have independence of path in any domain, and find the value of I from A: (0, 0, 1) to B: (1, $\pi/4$, 2).

Solution. Exactness follows from (6′), which gives

$$(F_3)_y = 2x^2z + \cos yz - yz\sin yz = (F_2)_z$$

$$(F_1)_z = 4xyz = (F_3)_x$$

$$(F_2)_x = 2xz^2 = (F_1)_y.$$

To find f, we integrate F_2 (which is "long," so that we save work) and then differentiate to compare with F_1 and F_3,

$$f = \int F_2\,dy = \int (x^2z^2 + z\cos yz)\,dy = x^2z^2y + \sin yz + g(x, z)$$

$$f_x = 2xz^2y + g_x = F_1 = 2xyz^2, \qquad g_x = 0, \qquad g = h(z)$$

$$f_z = 2x^2zy + y\cos yz + h' = F_3 = 2x^2zy + y\cos yz, \qquad h' = 0.$$

$h' = 0$ implies $h = \text{const}$ and we can take $h = 0$, so that $g = 0$ in the first line. This gives, by (3),

$$f(x, y, z) = x^2yz^2 + \sin yz, \qquad f(B) - f(A) = 1 \cdot \frac{\pi}{4} \cdot 4 + \sin\frac{\pi}{2} - 0 = \pi + 1.$$ ■

The assumption in Theorem 3 that D is simply connected is essential and cannot be omitted. Perhaps the simplest example to see this is the following.

EXAMPLE 4 **On the Assumption of Simple Connectedness in Theorem 3**

Let

$$F_1 = -\frac{y}{x^2 + y^2}, \qquad F_2 = \frac{x}{x^2 + y^2}, \qquad F_3 = 0. \tag{7}$$

Differentiation shows that (6′) is satisfied in any domain of the xy-plane not containing the origin, for example, in the domain $D: \frac{1}{2} < \sqrt{x^2 + y^2} < \frac{3}{2}$ shown in Fig. 226. Indeed, F_1 and F_2 do not depend on z, and $F_3 = 0$, so that the first two relations in (6′) are trivially true, and the third is verified by differentiation:

$$\frac{\partial F_2}{\partial x} = \frac{x^2 + y^2 - x \cdot 2x}{(x^2 + y^2)^2} = \frac{y^2 - x^2}{(x^2 + y^2)^2},$$

$$\frac{\partial F_1}{\partial y} = -\frac{x^2 + y^2 - y \cdot 2y}{(x^2 + y^2)^2} = \frac{y^2 - x^2}{(x^2 + y^2)^2}.$$

Clearly, D in Fig. 226 is not simply connected. If the integral

$$I = \int_C (F_1\,dx + F_2\,dy) = \int_C \frac{-y\,dx + x\,dy}{x^2 + y^2}$$

were independent of path in D, then $I = 0$ on any closed curve in D, for example, on the circle $x^2 + y^2 = 1$. But setting $x = r\cos\theta$, $y = r\sin\theta$ and noting that the circle is represented by $r = 1$, we have

$$x = \cos\theta, \qquad dx = -\sin\theta\,d\theta, \qquad y = \sin\theta, \qquad dy = \cos\theta\,d\theta,$$

so that $-y\,dx + x\,dy = \sin^2\theta\,d\theta + \cos^2\theta\,d\theta = d\theta$ and counterclockwise integration gives

$$I = \int_0^{2\pi} \frac{d\theta}{1} = 2\pi.$$

Since D is not simply connected, we cannot apply Theorem 3 and cannot conclude that I is independent of path in D.

Although $\mathbf{F} = \operatorname{grad} f$, where $f = \arctan(y/x)$ (verify!), we cannot apply Theorem 1 either because the polar angle $f = \theta = \arctan(y/x)$ is not single-valued, as it is required for a function in calculus. ■

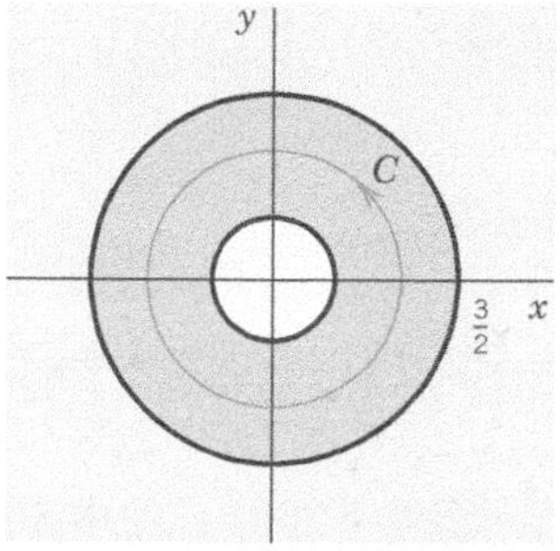

Fig. 226. Example 4

PROBLEM SET 10.2

1. **WRITING PROJECT. Report on Path Independence.** Make a list of the main ideas and facts on path independence and dependence in this section. Then work this list into a report. Explain the definitions and the practical usefulness of the theorems, with illustrative examples of your own. No proofs.

2. **On Example 4.** Does the situation in Example 4 of the text change if you take the domain $0 < \sqrt{x^2 + y^2} < 3/2$?

3–9 PATH INDEPENDENT INTEGRALS

Show that the form under the integral sign is exact in the plane (Probs. 3–4) or in space (Probs. 5–9) and evaluate the integral. Show the details of your work.

3. $\displaystyle\int_{(\pi/2,\,\pi)}^{(\pi,\,0)} (\tfrac{1}{2}\cos\tfrac{1}{2}x\cos 2y\,dx - 2\sin\tfrac{1}{2}x\sin 2y\,dy)$

4. $\displaystyle\int_{(4,\,0)}^{(6,\,1)} e^{4y}(2x\,dx + 4x^2\,dy)$

5. $\displaystyle\int_{(0,\,0,\,\pi)}^{(2,\,1/2,\,\pi/2)} e^{xy}(y\sin z\,dx + x\sin z\,dy + \cos z\,dz)$

6. $\displaystyle\int_{(0,\,0,\,0)}^{(1,\,1,\,0)} e^{x^2+y^2+z^2}(x\,dx + y\,dy + z\,dz)$

7. $\displaystyle\int_{(0,\,2,\,3)}^{(1,\,1,\,1)} (yz\sinh xz\,dx + \cosh xz\,dy + xy\sinh xz\,dz)$

8. $\displaystyle\int_{(5,\,3,\,\pi)}^{(3,\,\pi,\,3)} (\cos yz\,dx - xz\sin yz\,dy - xy\sin yz\,dz)$

9. $\displaystyle\int_{(0,\,1,\,0)}^{(1,\,0,\,1)} (e^x\cosh y\,dx + (e^x\sinh y + e^z\cosh y)\,dy + e^z\sinh y\,dz)$

10. **PROJECT. Path Dependence.** **(a)** Show that $I = \int_C (x^2y\,dx + 2xy^2\,dy)$ is path dependent in the xy-plane.

(b) Integrate from (0, 0) along the straight-line segment to $(1, b)$, $0 \leqq b \leqq 1$, and then vertically up to (1, 1); see the figure. For which b is I maximum? What is its maximum value?

(c) Integrate I from (0, 0) along the straight-line segment to $(c, 1)$, $0 \leqq c \leqq 1$, and then horizontally to (1, 1). For $c = 1$, do you get the same value as for $b = 1$ in (b)? For which c is I maximum? What is its maximum value?

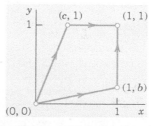

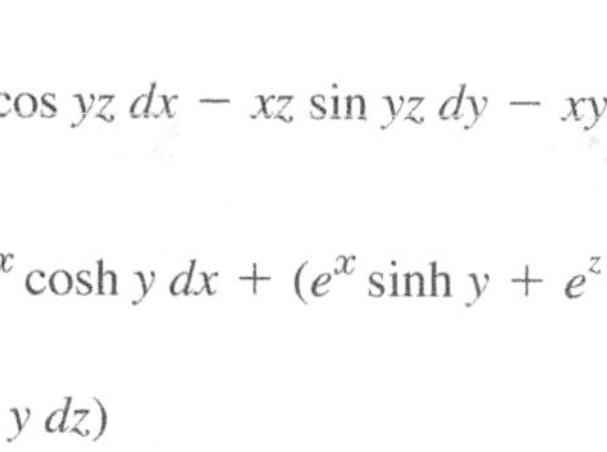

Project 10. Path Dependence

11. On Example 4. Show that in Example 4 of the text, $\mathbf{F} = \text{grad}\,(\arctan\,(y/x))$. Give examples of domains in which the integral is path independent.

12. CAS EXPERIMENT. Extension of Project 10. Integrate $x^2y\,dx + 2xy^2\,dy$ over various circles through the points (0, 0) and (1, 1). Find experimentally the smallest value of the integral and the approximate location of the center of the circle.

13–19 PATH INDEPENDENCE?

Check, and if independent, integrate from (0, 0, 0) to (a, b, c).

13. $2e^{x^2}(x \cos 2y\,dx - \sin 2y\,dy)$

14. $(\sinh xy)\,(z\,dx - x\,dz)$

15. $x^2y\,dx - 4xy^2\,dy + 8z^2x\,dz$

16. $e^y\,dx + (xe^y - e^z)\,dy - ye^z\,dz$

17. $4y\,dx + z\,dy + (y - 2z)\,dz$

18. $(\cos xy)(yz\,dx + xz\,dy) - 2 \sin xy\,dz$

19. $(\cos\,(x^2 + 2y^2 + z^2))\,(2x\,dx + 4y\,dy + 2z\,dz)$

20. Path Dependence. Construct three simple examples in each of which two equations (6′) are satisfied, but the third is not.

10.3 Calculus Review: Double Integrals. *Optional*

This section is optional. Students familiar with double integrals from calculus should skip this review and go on to Sec. 10.4. This section is included in the book to make it reasonably self-contained.

In a definite integral (1), Sec. 10.1, we integrate a function $f(x)$ over an interval (a segment) of the x-axis. In a double integral we integrate a function $f(x, y)$, called the *integrand,* over a closed bounded region[2] R in the xy-plane, whose boundary curve has a unique tangent at almost every point, but may perhaps have finitely many cusps (such as the vertices of a triangle or rectangle).

The definition of the double integral is quite similar to that of the definite integral. We subdivide the region R by drawing parallels to the x- and y-axes (Fig. 227). We number the rectangles that are entirely within R from 1 to n. In each such rectangle we choose a point, say, (x_k, y_k) in the kth rectangle, whose area we denote by ΔA_k. Then we form the sum

$$J_n = \sum_{k=1}^{n} f(x_k, y_k)\,\Delta A_k.$$

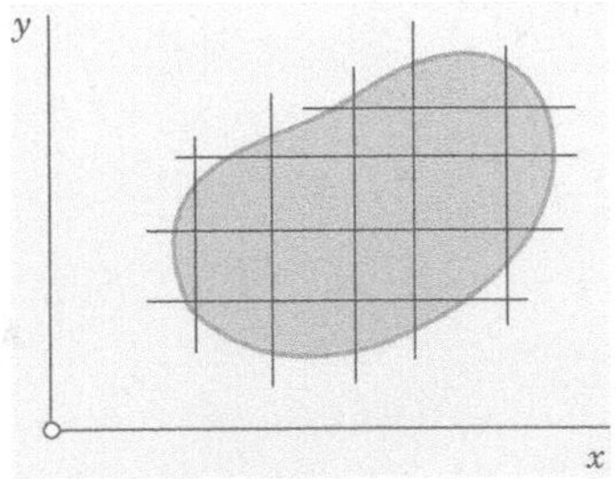

Fig. 227. Subdivision of a region R

[2]A **region** R is a domain (Sec. 9.6) plus, perhaps, some or all of its boundary points. R is **closed** if its **boundary** (all its boundary points) are regarded as belonging to R; and R is **bounded** if it can be enclosed in a circle of sufficiently large radius. A **boundary point** P of R is a point (of R or not) such that every disk with center P contains points of R and also points not of R.

This we do for larger and larger positive integers n in a completely independent manner, but so that the length of the maximum diagonal of the rectangles approaches zero as n approaches infinity. In this fashion we obtain a sequence of real numbers $J_{n_1}, J_{n_2}, \cdots$. Assuming that $f(x, y)$ is continuous in R and R is bounded by finitely many smooth curves (see Sec. 10.1), one can show (see Ref. [GenRef4] in App. 1) that this sequence converges and its limit is independent of the choice of subdivisions and corresponding points (x_k, y_k). This limit is called the **double integral** *of* $f(x, y)$ *over the region* R, and is denoted by

$$\iint_R f(x, y)\, dx\, dy \qquad \text{or} \qquad \iint_R f(x, y)\, dA.$$

Double integrals have properties quite similar to those of definite integrals. Indeed, for any functions f and g of (x, y), defined and continuous in a region R,

$$\iint_R kf\, dx\, dy = k\iint_R f\, dx\, dy \qquad (k \text{ constant})$$

(1)
$$\iint_R (f + g)\, dx\, dy = \iint_R f\, dx\, dy + \iint_R g\, dx\, dy$$

$$\iint_R f\, dx\, dy = \iint_{R_1} f\, dx\, dy + \iint_{R_2} f\, dx\, dy \qquad \text{(Fig. 228)}.$$

Furthermore, if R is simply connected (see Sec. 10.2), then there exists at least one point (x_0, y_0) in R such that we have

(2)
$$\iint_R f(x, y)\, dx\, dy = f(x_0, y_0)A,$$

where A is the area of R. This is called the **mean value theorem** *for double integrals.*

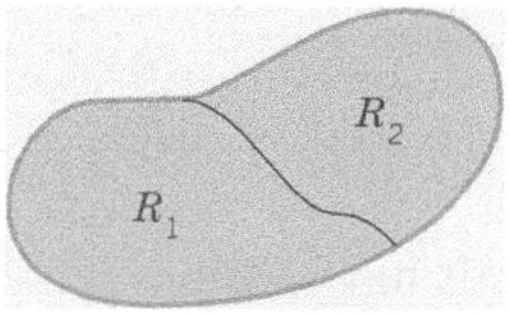

Fig. 228. Formula (1)

Evaluation of Double Integrals by Two Successive Integrations

Double integrals over a region R may be evaluated by two successive integrations. We may integrate first over y and then over x. Then the formula is

(3)
$$\iint_R f(x, y)\, dx\, dy = \int_a^b \left[\int_{g(x)}^{h(x)} f(x, y)\, dy\right] dx \qquad \text{(Fig. 229)}.$$

Here $y = g(x)$ and $y = h(x)$ represent the boundary curve of R (see Fig. 229) and, keeping x constant, we integrate $f(x, y)$ over y from $g(x)$ to $h(x)$. The result is a function of x, and we integrate it from $x = a$ to $x = b$ (Fig. 229).

Similarly, for integrating first over x and then over y the formula is

$$\iint_R f(x, y)\, dx\, dy = \int_c^d \left[\int_{p(y)}^{q(y)} f(x, y)\, dx \right] dy \qquad \text{(Fig. 230).} \tag{4}$$

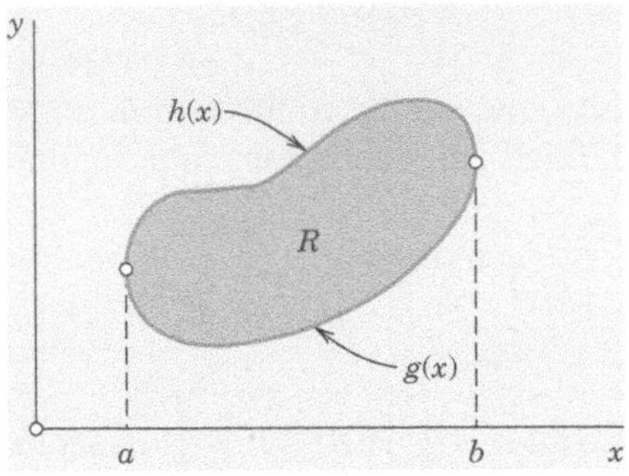

Fig. 229. Evaluation of a double integral

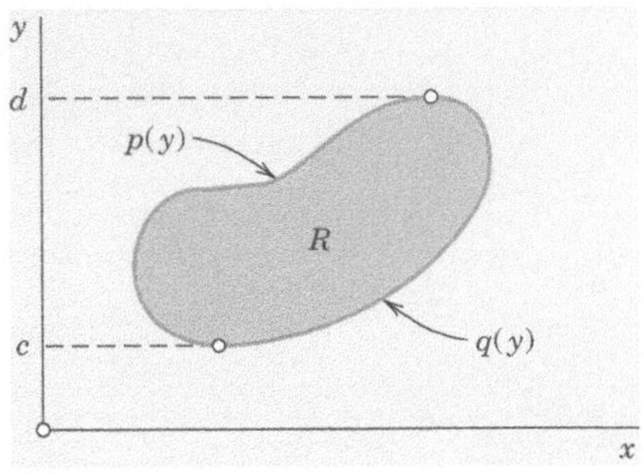

Fig. 230. Evaluation of a double integral

The boundary curve of R is now represented by $x = p(y)$ and $x = q(y)$. Treating y as a constant, we first integrate $f(x, y)$ over x from $p(y)$ to $q(y)$ (see Fig. 230) and then the resulting function of y from $y = c$ to $y = d$.

In (3) we assumed that R can be given by inequalities $a \leqq x \leqq b$ and $g(x) \leqq y \leqq h(x)$. Similarly in (4) by $c \leqq y \leqq d$ and $p(y) \leqq x \leqq q(y)$. If a region R has no such representation, then, in any practical case, it will at least be possible to subdivide R into finitely many portions each of which can be given by those inequalities. Then we integrate $f(x, y)$ over each portion and take the sum of the results. This will give the value of the integral of $f(x, y)$ over the entire region R.

Applications of Double Integrals

Double integrals have various physical and geometric applications. For instance, the **area** A of a region R in the xy-plane is given by the double integral

$$A = \iint_R dx\, dy.$$

The **volume** V beneath the surface $z = f(x, y)\ (>0)$ and above a region R in the xy-plane is (Fig. 231)

$$V = \iint_R f(x, y)\, dx\, dy$$

because the term $f(x_k, y_k)\Delta A_k$ in J_n at the beginning of this section represents the volume of a rectangular box with base of area ΔA_k and altitude $f(x_k, y_k)$.

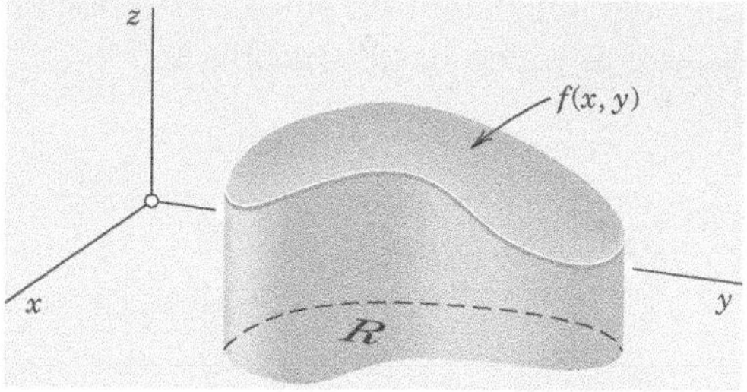

Fig. 231. Double integral as volume

As another application, let $f(x, y)$ be the density (= mass per unit area) of a distribution of mass in the xy-plane. Then the **total mass** M in R is

$$M = \iint_R f(x, y)\, dx\, dy;$$

the **center of gravity** of the mass in R has the coordinates $\bar{x}, \bar{y}$, where

$$\bar{x} = \frac{1}{M}\iint_R xf(x, y)\, dx\, dy \quad \text{and} \quad \bar{y} = \frac{1}{M}\iint_R yf(x, y)\, dx\, dy;$$

the **moments of inertia** I_x and I_y of the mass in R about the x- and y-axes, respectively, are

$$I_x = \iint_R y^2 f(x, y)\, dx\, dy, \qquad I_y = \iint_R x^2 f(x, y)\, dx\, dy;$$

and the **polar moment of inertia** I_0 about the origin of the mass in R is

$$I_0 = I_x + I_y = \iint_R (x^2 + y^2) f(x, y)\, dx\, dy.$$

An example is given below.

Change of Variables in Double Integrals. Jacobian

Practical problems often require a change of the variables of integration in double integrals. Recall from calculus that for a definite integral the formula for the change from x to u is

$$\int_a^b f(x)\, dx = \int_\alpha^\beta f(x(u)) \frac{dx}{du}\, du. \tag{5}$$

Here we assume that $x = x(u)$ is continuous and has a continuous derivative in some interval $\alpha \leqq u \leqq \beta$ such that $x(\alpha) = a$, $x(\beta) = b$ [or $x(\alpha) = b$, $x(\beta) = a$] and $x(u)$ varies between a and b when u varies between α and β.

The formula for a change of variables in double integrals from x, y to u, v is

$$\iint_R f(x, y)\, dx\, dy = \iint_{R^*} f(x(u, v), y(u, v)) \left| \frac{\partial(x, y)}{\partial(u, v)} \right| du\, dv; \tag{6}$$

that is, the integrand is expressed in terms of u and v, and $dx\,dy$ is replaced by $du\,dv$ times the absolute value of the **Jacobian**[3]

$$(7)\qquad J = \frac{\partial(x, y)}{\partial(u, v)} = \begin{vmatrix} \dfrac{\partial x}{\partial u} & \dfrac{\partial x}{\partial v} \\ \dfrac{\partial y}{\partial u} & \dfrac{\partial y}{\partial v} \end{vmatrix} = \frac{\partial x}{\partial u}\frac{\partial y}{\partial v} - \frac{\partial x}{\partial v}\frac{\partial y}{\partial u}.$$

Here we assume the following. The functions

$$x = x(u, v), \qquad y = y(u, v)$$

effecting the change are continuous and have continuous partial derivatives in some region R^* in the uv-plane such that for every (u, v) in R^* the corresponding point (x, y) lies in R and, conversely, to every (x, y) in R there corresponds one and only one (u, v) in R^*; furthermore, the Jacobian J is either positive throughout R^* or negative throughout R^*. For a proof, see Ref. [GenRef4] in App. 1.

EXAMPLE 1 **Change of Variables in a Double Integral**

Evaluate the following double integral over the square R in Fig. 232.

$$\iint_R (x^2 + y^2)\,dx\,dy$$

Solution. The shape of R suggests the transformation $x + y = u$, $x - y = v$. Then $x = \frac{1}{2}(u + v)$, $y = \frac{1}{2}(u - v)$. The Jacobian is

$$J = \frac{\partial(x, y)}{\partial(u, v)} = \begin{vmatrix} \frac{1}{2} & \frac{1}{2} \\ \frac{1}{2} & -\frac{1}{2} \end{vmatrix} = -\frac{1}{2}.$$

R corresponds to the square $0 \leqq u \leqq 2$, $0 \leqq v \leqq 2$. Therefore,

$$\iint_R (x^2 + y^2)\,dx\,dy = \int_0^2\int_0^2 \frac{1}{2}(u^2 + v^2)\frac{1}{2}\,du\,dv = \frac{8}{3}. \qquad \blacksquare$$

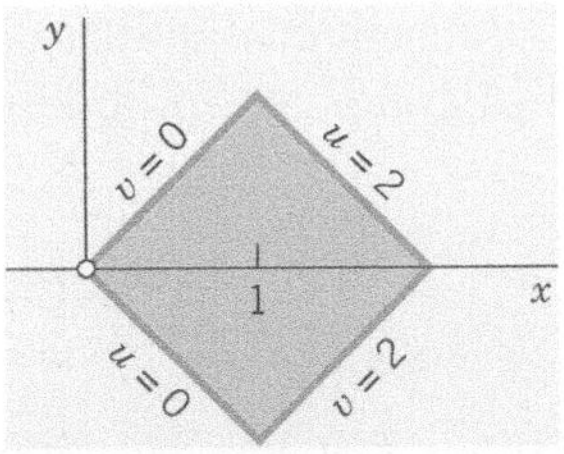

Fig. 232. Region R in Example 1

[3]Named after the German mathematician CARL GUSTAV JACOB JACOBI (1804–1851), known for his contributions to elliptic functions, partial differential equations, and mechanics.

Of particular practical interest are **polar coordinates** r and θ, which can be introduced by setting $x = r\cos\theta$, $y = r\sin\theta$. Then

$$J = \frac{\partial(x, y)}{\partial(r, \theta)} = \begin{vmatrix} \cos\theta & -r\sin\theta \\ \sin\theta & r\cos\theta \end{vmatrix} = r$$

and

$$\textbf{(8)} \qquad \iint_R f(x, y)\,dx\,dy = \iint_{R^*} f(r\cos\theta, r\sin\theta)\,r\,dr\,d\theta$$

where R^* is the region in the $r\theta$-plane corresponding to R in the xy-plane.

EXAMPLE 2 **Double Integrals in Polar Coordinates. Center of Gravity. Moments of Inertia**

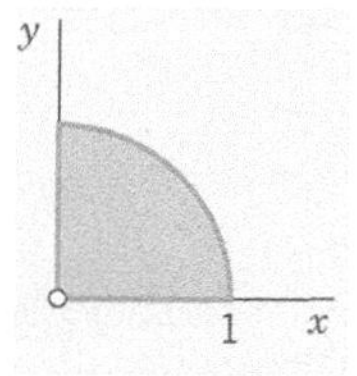

Fig. 233. Example 2

Let $f(x, y) = 1$ be the mass density in the region in Fig. 233. Find the total mass, the center of gravity, and the moments of inertia I_x, I_y, I_0.

Solution. We use the polar coordinates just defined and formula (8). This gives the total mass

$$M = \iint_R dx\,dy = \int_0^{\pi/2}\int_0^1 r\,dr\,d\theta = \int_0^{\pi/2} \frac{1}{2}\,d\theta = \frac{\pi}{4}.$$

The center of gravity has the coordinates

$$\bar{x} = \frac{4}{\pi}\int_0^{\pi/2}\int_0^1 r\cos\theta\, r\,dr\,d\theta = \frac{4}{\pi}\int_0^{\pi/2}\frac{1}{3}\cos\theta\,d\theta = \frac{4}{3\pi} = 0.4244$$

$$\bar{y} = \frac{4}{3\pi} \qquad \text{for reasons of symmetry.}$$

The moments of inertia are

$$I_x = \iint_R y^2\,dx\,dy = \int_0^{\pi/2}\int_0^1 r^2\sin^2\theta\, r\,dr\,d\theta = \int_0^{\pi/2}\frac{1}{4}\sin^2\theta\,d\theta$$

$$= \int_0^{\pi/2}\frac{1}{8}(1 - \cos 2\theta)\,d\theta = \frac{1}{8}\left(\frac{\pi}{2} - 0\right) = \frac{\pi}{16} = 0.1963$$

$$I_y = \frac{\pi}{16} \quad \text{for reasons of symmetry,} \qquad I_0 = I_x + I_y = \frac{\pi}{8} = 0.3927.$$

Why are $\bar{x}$ and $\bar{y}$ less than $\frac{1}{2}$? ■

This is the end of our review on double integrals. These integrals will be needed in this chapter, beginning in the next section.

PROBLEM SET 10.3

1. **Mean value theorem.** Illustrate (2) with an example.

2–8 DOUBLE INTEGRALS

Describe the region of integration and evaluate.

2. $\int_0^2 \int_x^{2x} (x + y)^2 \, dy \, dx$

3. $\int_0^3 \int_{-y}^{y} (x^2 + y^2) \, dx \, dy$

4. Prob. 3, order reversed.

5. $\int_0^1 \int_{x^2}^{x} (1 - 2xy) \, dy \, dx$

6. $\int_0^2 \int_0^y \sinh (x + y) \, dx \, dy$

7. Prob. 6, order reversed.

8. $\int_0^{\pi/4} \int_0^{\cos y} x^2 \sin y \, dx \, dy$

9–11 VOLUME

Find the volume of the given region in space.

9. The region beneath $z = 4x^2 + 9y^2$ and above the rectangle with vertices (0, 0), (3, 0), (3, 2), (0, 2) in the xy-plane.

10. The first octant region bounded by the coordinate planes and the surfaces $y = 1 - x^2$, $z = 1 - x^2$. Sketch it.

11. The region above the xy-plane and below the paraboloid $z = 1 - (x^2 + y^2)$.

12–16 CENTER OF GRAVITY

Find the center of gravity $(\bar{x}, \bar{y})$ of a mass of density $f(x, y) = 1$ in the given region R.

12.
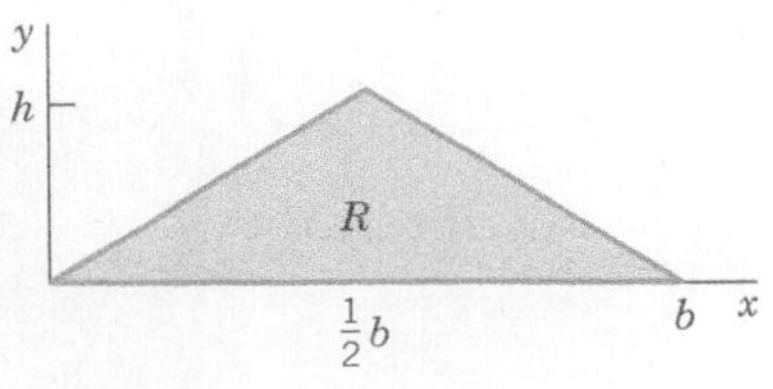

13.
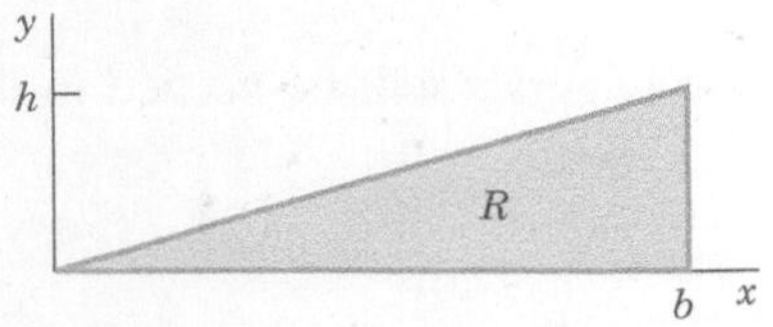

14.
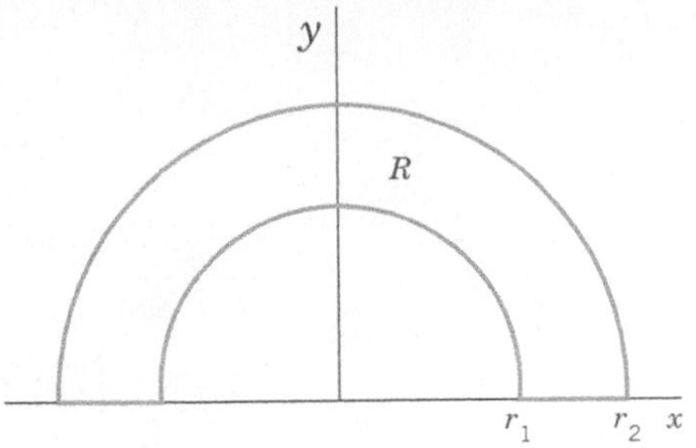

15.
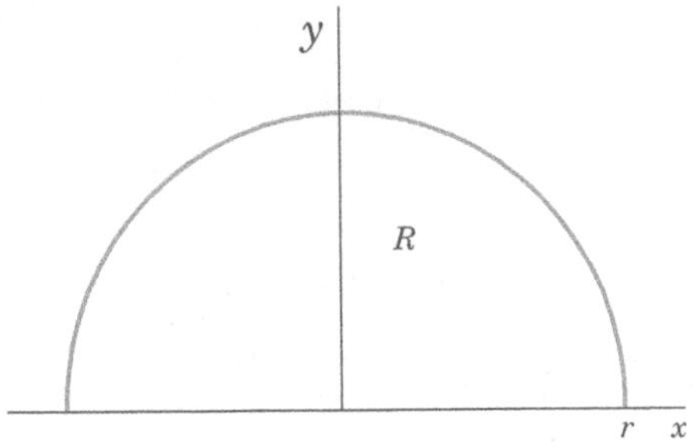

16.
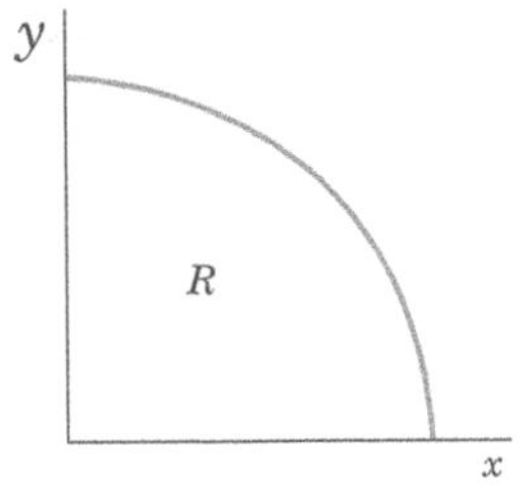

17–20 MOMENTS OF INERTIA

Find I_x, I_y, I_0 of a mass of density $f(x, y) = 1$ in the region R in the figures, which the engineer is likely to need, along with other profiles listed in engineering handbooks.

17. R as in Prob. 13.

18. R as in Prob. 12.

19.
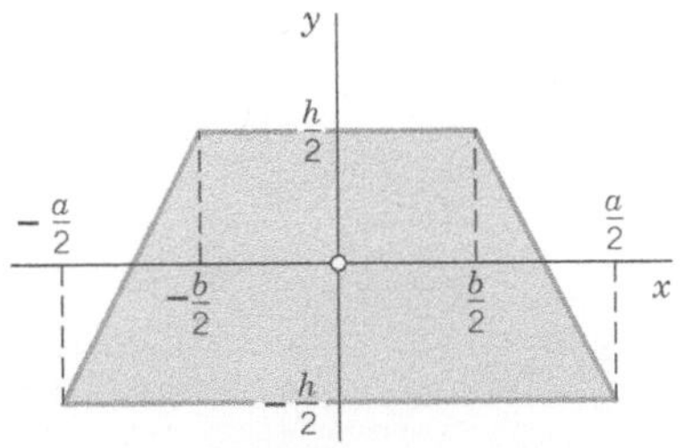

20.
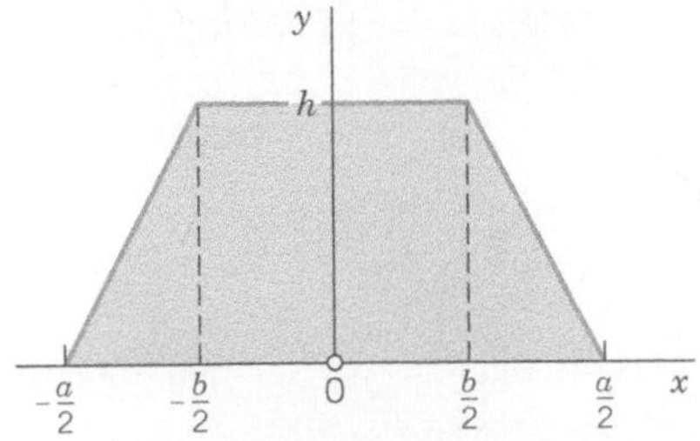

10.4 Green's Theorem in the Plane

Double integrals over a plane region may be transformed into line integrals over the boundary of the region and conversely. This is of practical interest because it may simplify the evaluation of an integral. It also helps in theoretical work when we want to switch from one kind of integral to the other. The transformation can be done by the following theorem.

THEOREM 1

Green's Theorem in the Plane[4]
(Transformation between Double Integrals and Line Integrals)

Let R be a closed bounded region (see Sec. 10.3) *in the xy-plane whose boundary C consists of finitely many smooth curves* (see Sec. 10.1). *Let $F_1(x, y)$ and $F_2(x, y)$ be functions that are continuous and have continuous partial derivatives $\partial F_1/\partial y$ and $\partial F_2/\partial x$ everywhere in some domain containing R. Then*

$$\iint_R \left(\frac{\partial F_2}{\partial x} - \frac{\partial F_1}{\partial y}\right) dx\, dy = \oint_C (F_1\, dx + F_2\, dy). \tag{1}$$

Here we integrate along the entire boundary C of R in such a sense that R is on the left as we advance in the direction of integration (see Fig. 234).

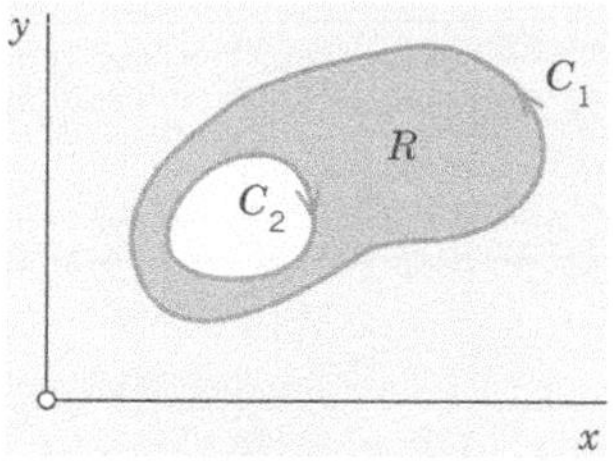

Fig. 234. Region R whose boundary C consists of two parts: C_1 is traversed counterclockwise, while C_2 is traversed clockwise in such a way that R is on the left for both curves

Setting $\mathbf{F} = [F_1, \ F_2] = F_1\mathbf{i} + F_2\mathbf{j}$ *and using* (1) *in* Sec. 9.9, we obtain (1) in vectorial form,

$$\iint_R (\text{curl } \mathbf{F}) \cdot \mathbf{k}\, dx\, dy = \oint_C \mathbf{F} \cdot d\mathbf{r}. \tag{1'}$$

The proof follows after the first example. For $\oint$ see Sec. 10.1.

[4]GEORGE GREEN (1793–1841), English mathematician who was self-educated, started out as a baker, and at his death was fellow of Caius College, Cambridge. His work concerned potential theory in connection with electricity and magnetism, vibrations, waves, and elasticity theory. It remained almost unknown, even in England, until after his death.

A "domain containing R" in the theorem guarantees that the assumptions about F_1 and F_2 at boundary points of R are the same as at other points of R.

EXAMPLE 1   **Verification of Green's Theorem in the Plane**

Green's theorem in the plane will be quite important in our further work. Before proving it, let us get used to it by verifying it for $F_1 = y^2 - 7y$, $F_2 = 2xy + 2x$ and C the circle $x^2 + y^2 = 1$.

Solution.   In (1) on the left we get

$$\int\int_R \left(\frac{\partial F_2}{\partial x} - \frac{\partial F_1}{\partial y}\right) dx\, dy = \int\int_R [(2y + 2) - (2y - 7)] dx\, dy = 9 \int\int_R dx\, dy = 9\pi$$

since the circular disk R has area π.

We now show that the line integral in (1) on the right gives the same value, 9π. We must orient C counterclockwise, say, $\mathbf{r}(t) = [\cos t, \sin t]$. Then $\mathbf{r}'(t) = [-\sin t, \cos t]$, and on C,

$$F_1 = y^2 - 7y = \sin^2 t - 7 \sin t, \qquad F_2 = 2xy + 2x = 2 \cos t \sin t + 2 \cos t.$$

Hence the line integral in (1) becomes, verifying Green's theorem,

$$\oint_C (F_1 x' + F_2 y')\, dt = \int_0^{2\pi} [(\sin^2 t - 7 \sin t)(-\sin t) + 2(\cos t \sin t + \cos t)(\cos t)]\, dt$$

$$= \int_0^{2\pi} (-\sin^3 t + 7 \sin^2 t + 2 \cos^2 t \sin t + 2 \cos^2 t)\, dt$$

$$= 0 + 7\pi - 0 + 2\pi = 9\pi. \qquad \blacksquare$$

PROOF   We prove Green's theorem in the plane, first for a *special region R* that can be represented in both forms

$$a \leqq x \leqq b, \qquad u(x) \leqq y \leqq v(x) \qquad\qquad \text{(Fig. 235)}$$

and

$$c \leqq y \leqq d, \qquad p(y) \leqq x \leqq q(y) \qquad\qquad \text{(Fig. 236)}$$

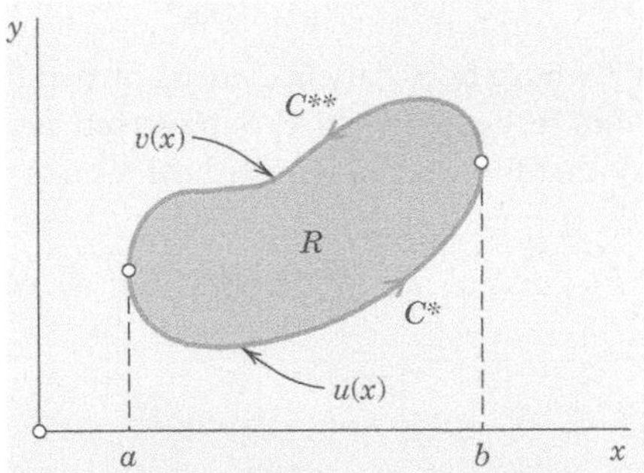

Fig. 235.   Example of a special region

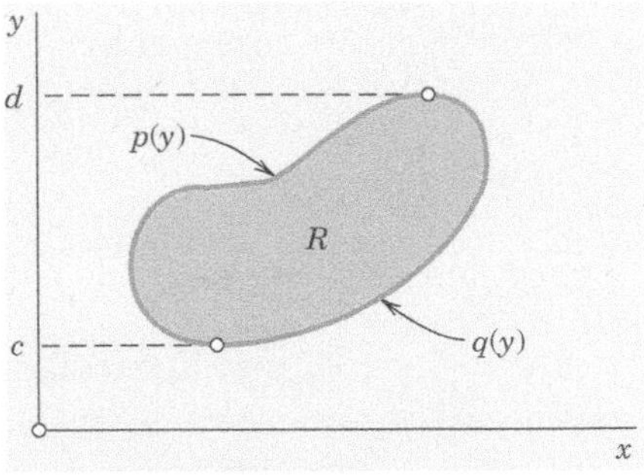

Fig. 236.   Example of a special region

Using (3) in the last section, we obtain for the second term on the left side of (1) taken without the minus sign

$$(2) \qquad\qquad \int\int_R \frac{\partial F_1}{\partial y} dx\, dy = \int_a^b \left[\int_{u(x)}^{v(x)} \frac{\partial F_1}{\partial y} dy \right] dx \qquad\qquad \text{(see Fig. 235).}$$

(The first term will be considered later.) We integrate the inner integral:

$$\int_{u(x)}^{v(x)} \frac{\partial F_1}{\partial y}\, dy = F_1(x, y)\Big|_{y=u(x)}^{y=v(x)} = F_1[x, v(x)] - F_1[x, u(x)].$$

By inserting this into (2) we find (changing a direction of integration)

$$\begin{aligned}\iint_R \frac{\partial F_1}{\partial y}\, dx\, dy &= \int_a^b F_1[x, v(x)]\, dx - \int_a^b F_1[x, u(x)]\, dx \\ &= -\int_b^a F_1[x, v(x)]\, dx - \int_a^b F_1[x, u(x)]\, dx.\end{aligned}$$

Since $y = v(x)$ represents the curve C^{**} (Fig. 235) and $y = u(x)$ represents C^*, the last two integrals may be written as line integrals over C^{**} and C^* (oriented as in Fig. 235); therefore,

$$\begin{aligned}\iint_R \frac{\partial F_1}{\partial y}\, dx\, dy &= -\int_{C^{**}} F_1(x, y)\, dx - \int_{C^*} F_1(x, y)\, dx \\ &= -\oint_C F_1(x, y)\, dx.\end{aligned} \tag{3}$$

This proves (1) in Green's theorem if $F_2 = 0$.

The result remains valid if C has portions parallel to the y-axis (such as $\tilde{C}$ and $\tilde{\tilde{C}}$ in Fig. 237). Indeed, the integrals over these portions are zero because in (3) on the right we integrate with respect to x. Hence we may add these integrals to the integrals over C^* and C^{**} to obtain the integral over the whole boundary C in (3).

We now treat the first term in (1) on the left in the same way. Instead of (3) in the last section we use (4), and the second representation of the special region (see Fig. 236). Then (again changing a direction of integration)

$$\begin{aligned}\iint_R \frac{\partial F_2}{\partial x}\, dx\, dy &= \int_c^d \left[\int_{p(y)}^{q(y)} \frac{\partial F_2}{\partial x}\, dx\right] dy \\ &= \int_c^d F_2(q(y), y)\, dy + \int_d^c F_2(p(y), y)\, dy \\ &= \oint_C F_2(x, y)\, dy.\end{aligned}$$

Together with (3) this gives (1) and proves Green's theorem for special regions.

We now prove the theorem for a region R that itself is not a special region but can be subdivided into finitely many special regions as shown in Fig. 238. In this case we apply the theorem to each subregion and then add the results; the left-hand members add up to the integral over R while the right-hand members add up to the line integral over C plus

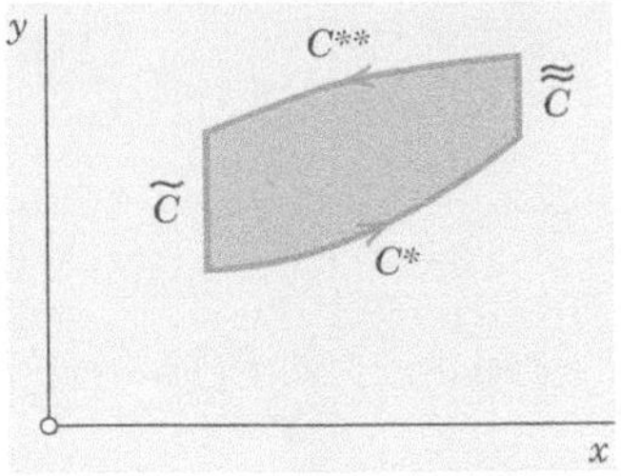

Fig. 237. Proof of Green's theorem

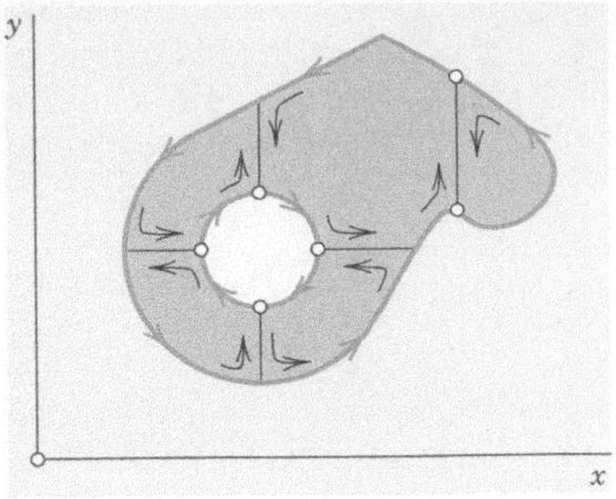

Fig. 238. Proof of Green's theorem

integrals over the curves introduced for subdividing R. The simple ***key observation*** now is that each of the latter integrals occurs twice, taken once in each direction. Hence they cancel each other, leaving us with the line integral over C.

The proof thus far covers all regions that are of interest in practical problems. To prove the theorem for a most general region R satisfying the conditions in the theorem, we must approximate R by a region of the type just considered and then use a limiting process. For details of this see Ref. [GenRef4] in App. 1.

Some Applications of Green's Theorem

EXAMPLE 2 Area of a Plane Region as a Line Integral Over the Boundary

In (1) we first choose $F_1 = 0$, $F_2 = x$ and then $F_1 = -y$, $F_2 = 0$. This gives

$$\iint_R dx\,dy = \oint_C x\,dy \qquad \text{and} \qquad \iint_R dx\,dy = -\oint_C y\,dx$$

respectively. The double integral is the area A of R. By addition we have

$$A = \frac{1}{2}\oint_C (x\,dy - y\,dx) \tag{4}$$

where we integrate as indicated in Green's theorem. This interesting formula expresses the area of R in terms of a line integral over the boundary. It is used, for instance, in the theory of certain **planimeters** (mechanical instruments for measuring area). See also Prob. 11.

For an ***ellipse*** $x^2/a^2 + y^2/b^2 = 1$ or $x = a\cos t$, $y = b\sin t$ we get $x' = -a\sin t$, $y' = b\cos t$; thus from (4) we obtain the familiar formula for the area of the region bounded by an ellipse,

$$A = \frac{1}{2}\int_0^{2\pi} (xy' - yx')\,dt = \frac{1}{2}\int_0^{2\pi} [ab\cos^2 t - (-ab\sin^2 t)]\,dt = \pi ab. \qquad \blacksquare$$

EXAMPLE 3 Area of a Plane Region in Polar Coordinates

Let r and θ be polar coordinates defined by $x = r\cos\theta$, $y = r\sin\theta$. Then

$$dx = \cos\theta\,dr - r\sin\theta\,d\theta, \qquad dy = \sin\theta\,dr + r\cos\theta\,d\theta,$$

and (4) becomes a formula that is well known from calculus, namely,

$$A = \frac{1}{2}\oint_C r^2\, d\theta. \tag{5}$$

As an application of (5), we consider the **cardioid** $r = a(1 - \cos\theta)$, where $0 \leqq \theta \leqq 2\pi$ (Fig. 239). We find

$$A = \frac{a^2}{2}\int_0^{2\pi} (1 - \cos\theta)^2\, d\theta = \frac{3\pi}{2}a^2.$$

■

EXAMPLE 4 **Transformation of a Double Integral of the Laplacian of a Function into a Line Integral of Its Normal Derivative**

The Laplacian plays an important role in physics and engineering. A first impression of this was obtained in Sec. 9.7, and we shall discuss this further in Chap. 12. At present, let us use Green's theorem for deriving a basic integral formula involving the Laplacian.

We take a function $w(x, y)$ that is continuous and has continuous first and second partial derivatives in a domain of the xy-plane containing a region R of the type indicated in Green's theorem. We set $F_1 = -\partial w/\partial y$ and $F_2 = \partial w/\partial x$. Then $\partial F_1/\partial y$ and $\partial F_2/\partial x$ are continuous in R, and in (1) on the left we obtain

$$\frac{\partial F_2}{\partial x} - \frac{\partial F_1}{\partial y} = \frac{\partial^2 w}{\partial x^2} + \frac{\partial^2 w}{\partial y^2} = \nabla^2 w, \tag{6}$$

the Laplacian of w (see Sec. 9.7). Furthermore, using those expressions for F_1 and F_2, we get in (1) on the right

$$\oint_C (F_1\, dx + F_2\, dy) = \oint_C \left(F_1 \frac{dx}{ds} + F_2 \frac{dy}{ds}\right) ds = \oint_C \left(-\frac{\partial w}{\partial y}\frac{dx}{ds} + \frac{\partial w}{\partial x}\frac{dy}{ds}\right) ds \tag{7}$$

where s is the arc length of C, and C is oriented as shown in Fig. 240. The integrand of the last integral may be written as the dot product

$$(\text{grad } w) \bullet \mathbf{n} = \left[\frac{\partial w}{\partial x}, \frac{\partial w}{\partial y}\right] \bullet \left[\frac{dy}{ds}, -\frac{dx}{ds}\right] = \frac{\partial w}{\partial x}\frac{dy}{ds} - \frac{\partial w}{\partial y}\frac{dx}{ds}. \tag{8}$$

The vector $\mathbf{n}$ is a unit normal vector to C, because the vector $\mathbf{r}'(s) = d\mathbf{r}/ds = [dx/ds,\ dy/ds]$ is the unit tangent vector of C, and $\mathbf{r}' \bullet \mathbf{n} = 0$, so that $\mathbf{n}$ is perpendicular to $\mathbf{r}'$. Also, $\mathbf{n}$ is directed to the *exterior* of C because in Fig. 240 the positive x-component dx/ds of $\mathbf{r}'$ is the negative y-component of $\mathbf{n}$, and similarly at other points. From this and (4) in Sec. 9.7 we see that the left side of (8) is the derivative of w in the direction of the outward normal of C. This derivative is called the **normal derivative** of w and is denoted by $\partial w/\partial n$; that is, $\partial w/\partial n = (\text{grad } w) \bullet \mathbf{n}$. Because of (6), (7), and (8), Green's theorem gives the desired formula relating the Laplacian to the normal derivative,

$$\iint_R \nabla^2 w\, dx\, dy = \oint_C \frac{\partial w}{\partial n}\, ds. \tag{9}$$

For instance, $w = x^2 - y^2$ satisfies Laplace's equation $\nabla^2 w = 0$. Hence its normal derivative integrated over a closed curve must give 0. Can you verify this directly by integration, say, for the square $0 \leqq x \leqq 1, 0 \leqq y \leqq 1$? ■

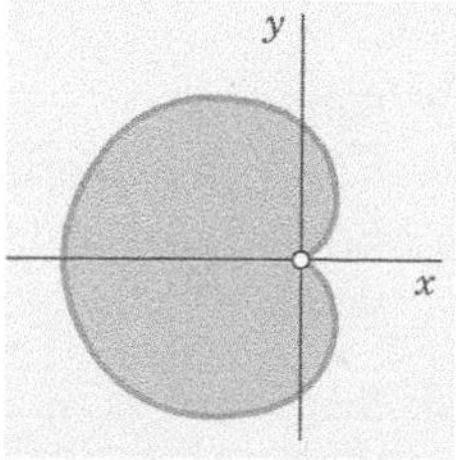

Fig. 239. Cardioid

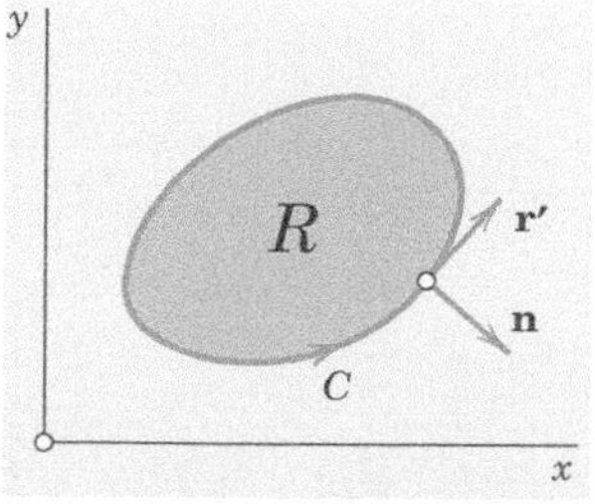

Fig. 240. Example 4

Green's theorem in the plane can be used in both directions, and thus may aid in the evaluation of a given integral by transforming the given integral into another integral that is easier to solve. This is illustrated further in the problem set. Moreover, and perhaps more fundamentally, Green's theorem will be the essential tool in the proof of a very important integral theorem, namely, Stokes's theorem in Sec. 10.9.

PROBLEM SET 10.4

1–10 LINE INTEGRALS: EVALUATION BY GREEN'S THEOREM

Evaluate $\int_C \mathbf{F}(\mathbf{r}) \bullet d\mathbf{r}$ counterclockwise around the boundary C of the region R by Green's theorem, where

1. $\mathbf{F} = [y, -x]$, C the circle $x^2 + y^2 = 1/4$
2. $\mathbf{F} = [6y^2, 2x - 2y^4]$, R the square with vertices $\pm(2, 2)$, $\pm(2, -2)$
3. $\mathbf{F} = [x^2e^y, y^2e^x]$, R the rectangle with vertices $(0, 0)$, $(2, 0)$, $(2, 3)$, $(0, 3)$
4. $\mathbf{F} = [x \cosh 2y, 2x^2 \sinh 2y]$, $R: x^2 \leqq y \leqq x$
5. $\mathbf{F} = [x^2 + y^2, x^2 - y^2]$, $R: 1 \leqq y \leqq 2 - x^2$
6. $\mathbf{F} = [\cosh y, -\sinh x]$, $R: 1 \leqq x \leqq 3, x \leqq y \leqq 3x$
7. $\mathbf{F} = \operatorname{grad}(x^3 \cos^2(xy))$, R as in Prob. 5
8. $\mathbf{F} = [-e^{-x} \cos y, -e^{-x} \sin y]$, R the semidisk $x^2 + y^2 \leqq 16, x \geqq 0$
9. $\mathbf{F} = [e^{y/x}, e^y \ln x + 2x]$, $R: 1 + x^4 \leqq y \leqq 2$
10. $\mathbf{F} = [x^2y^2, -x/y^2]$, $R: 1 \leqq x^2 + y^2 \leqq 4, x \geqq 0$, $y \geqq x$. Sketch R.
11. **CAS EXPERIMENT.** Apply (4) to figures of your choice whose area can also be obtained by another method and compare the results.
12. **PROJECT. Other Forms of Green's Theorem in the Plane.** Let R and C be as in Green's theorem, $\mathbf{r}'$ a unit tangent vector, and $\mathbf{n}$ the outer unit normal vector of C (Fig. 240 in Example 4). Show that (1) may be written

$$\iint_R \operatorname{div} \mathbf{F} \, dx \, dy = \oint_C \mathbf{F} \bullet \mathbf{n} \, ds \tag{10}$$

or

$$\iint_R (\operatorname{curl} \mathbf{F}) \bullet \mathbf{k} \, dx \, dy = \oint_C \mathbf{F} \bullet \mathbf{r}' \, ds \tag{11}$$

where $\mathbf{k}$ is a unit vector perpendicular to the xy-plane. **Verify** (10) and (11) for $\mathbf{F} = [7x, -3y]$ and C the circle $x^2 + y^2 = 4$ as well as for an example of your own choice.

13–17 INTEGRAL OF THE NORMAL DERIVATIVE

Using (9), find the value of $\int_C \frac{\partial w}{\partial n} \, ds$ taken counterclockwise over the boundary C of the region R.

13. $w = \cosh x$, R the triangle with vertices $(0, 0)$, $(4, 2)$, $(0, 2)$.
14. $w = x^2y + xy^2$, $R: x^2 + y^2 \leqq 1, x \geqq 0, y \geqq 0$
15. $w = e^x \cos y + xy^3$, $R: 1 \leqq y \leqq 10 - x^2, x \geqq 0$
16. $W = x^2 + y^2$, $C: x^2 + y^2 = 4$. Confirm the answer by direct integration.
17. $w = x^3 - y^3$, $0 \leqq y \leqq x^2$, $|x| \leqq 2$
18. **Laplace's equation.** Show that for a solution $w(x, y)$ of Laplace's equation $\nabla^2 w = 0$ in a region R with boundary curve C and outer unit normal vector $\mathbf{n}$,

$$\iint_R \left[\left(\frac{\partial w}{\partial x}\right)^2 + \left(\frac{\partial w}{\partial y}\right)^2\right] dx \, dy = \oint_C w \frac{\partial w}{\partial n} \, ds. \tag{12}$$

19. Show that $w = e^x \sin y$ satisfies Laplace's equation $\nabla^2 w = 0$ and, using (12), integrate $w(\partial w/\partial n)$ counterclockwise around the boundary curve C of the rectangle $0 \leqq x \leqq 2, 0 \leqq y \leqq 5$.
20. Same task as in Prob. 19 when $w = x^2 + y^2$ and C the boundary curve of the triangle with vertices $(0, 0)$, $(1, 0)$, $(0, 1)$.

10.5 Surfaces for Surface Integrals

Whereas, with line integrals, we integrate over *curves* in space (Secs. 10.1, 10.2), with surface integrals we integrate over *surfaces* in space. Each curve in space is represented by a parametric equation (Secs. 9.5, 10.1). This suggests that we should also find parametric representations for the surfaces in space. This is indeed one of the goals of this section. The surfaces considered are cylinders, spheres, cones, and others. The second goal is to learn about surface normals. Both goals prepare us for Sec. 10.6 on surface integrals. Note that for simplicity, we shall say "surface" also for a portion of a surface.

Representation of Surfaces

Representations of a surface S in xyz-space are

$$z = f(x, y) \qquad \text{or} \qquad g(x, y, z) = 0. \tag{1}$$

For example, $z = +\sqrt{a^2 - x^2 - y^2}$ or $x^2 + y^2 + z^2 - a^2 = 0$ $(z \geqq 0)$ represents a hemisphere of radius a and center 0.

Now for *curves* C in line integrals, it was more practical and gave greater flexibility to use a *parametric* representation $\mathbf{r} = \mathbf{r}(t)$, where $a \leqq t \leqq b$. This is a mapping of the interval $a \leqq t \leqq b$, located on the t-axis, onto the curve C (actually a portion of it) in xyz-space. It maps every t in that interval onto the point of C with position vector $\mathbf{r}(t)$. See Fig. 241A.

Similarly, for surfaces S in surface integrals, it will often be more practical to use a *parametric* representation. Surfaces are *two*-dimensional. Hence we need *two* parameters,

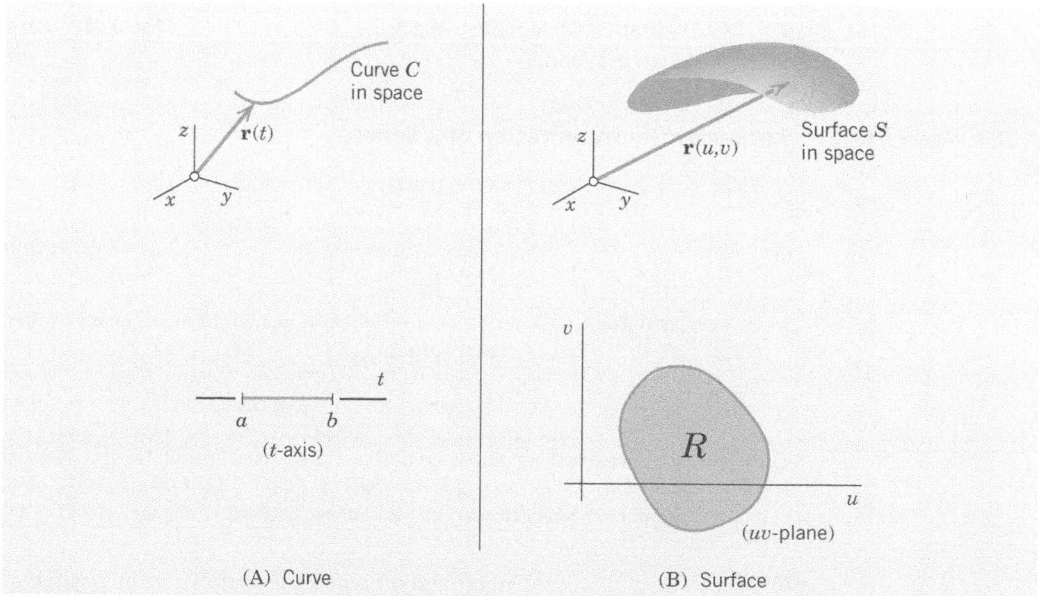

Fig. 241. Parametric representations of a curve and a surface

which we call u and v. Thus a **parametric representation** of a surface S in space is of the form

(2) $$\mathbf{r}(u, v) = [x(u, v), y(u, v), z(u, v)] = x(u, v)\mathbf{i} + y(u, v)\mathbf{j} + z(u, v)\mathbf{k}$$

where (u, v) varies in some region R of the uv-plane. This mapping (2) maps every point (u, v) in R onto the point of S with position vector $\mathbf{r}(u, v)$. See Fig. 241B.

EXAMPLE 1 **Parametric Representation of a Cylinder**

The circular cylinder $x^2 + y^2 = a^2$, $-1 \leqq z \leqq 1$, has radius a, height 2, and the z-axis as axis. A parametric representation is

$$\mathbf{r}(u, v) = [a \cos u, a \sin u, v] = a \cos u\,\mathbf{i} + a \sin u\,\mathbf{j} + v\mathbf{k} \qquad \text{(Fig. 242).}$$

The components of $\mathbf{r}$ are $x = a \cos u$, $y = a \sin u$, $z = v$. The parameters u, v vary in the rectangle R: $0 \leqq u \leqq 2\pi$, $-1 \leqq v \leqq 1$in the uv-plane. The curves u = const are vertical straight lines. The curves v = const are parallel circles. The point P in Fig. 242 corresponds to $u = \pi/3 = 60°$, $v = 0.7$. ■

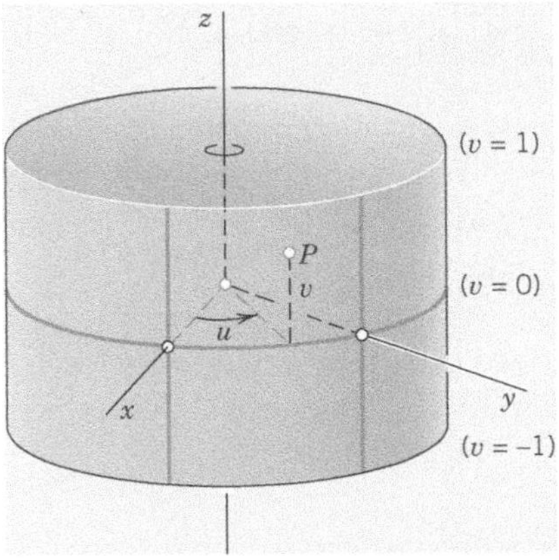

Fig. 242. Parametric representation of a cylinder

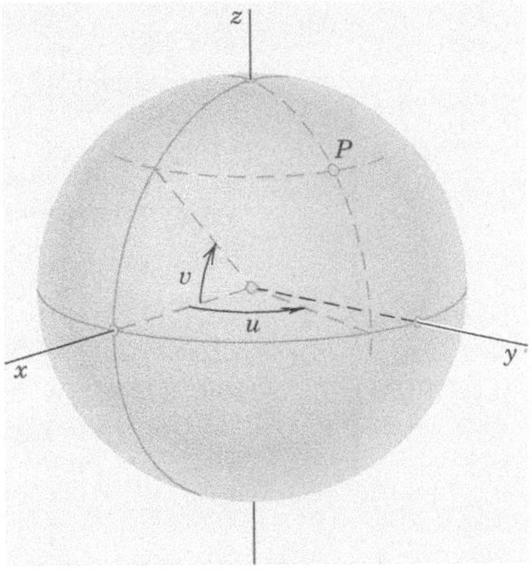

Fig. 243. Parametric representation of a sphere

EXAMPLE 2 **Parametric Representation of a Sphere**

A sphere $x^2 + y^2 + z^2 = a^2$ can be represented in the form

(3) $$\mathbf{r}(u, v) = a \cos v \cos u\,\mathbf{i} + a \cos v \sin u\,\mathbf{j} + a \sin v\mathbf{k}$$

where the parameters u, v vary in the rectangle R in the uv-plane given by the inequalities $0 \leqq u \leqq 2\pi$, $-\pi/2 \leqq v \leqq \pi/2$. The components of $\mathbf{r}$ are

$$x = a \cos v \cos u, \qquad y = a \cos v \sin u, \qquad z = a \sin v.$$

The curves u = const and v = const are the "meridians" and "parallels" on S (see Fig. 243). *This representation is used in* ***geography*** *for measuring the latitude and longitude of points on the globe.*

Another parametric representation of the sphere also used in mathematics is

(3*) $$\mathbf{r}(u, v) = a \cos u \sin v\,\mathbf{i} + a \sin u \sin v\,\mathbf{j} + a \cos v\,\mathbf{k}$$

where $0 \leqq u \leqq 2\pi$, $0 \leqq v \leqq \pi$. ■

EXAMPLE 3 Parametric Representation of a Cone

A circular cone $z = \sqrt{x^2 + y^2}$, $0 \leqq t \leqq H$ can be represented by

$$\mathbf{r}(u, v) = [u \cos v, u \sin v, u] = u \cos v\,\mathbf{i} + u \sin v\,\mathbf{j} + u\mathbf{k},$$

in components $x = u \cos v$, $y = u \sin v$, $z = u$. The parameters vary in the rectangle R: $0 \leqq u \leqq H$, $0 \leqq v \leqq 2\pi$. Check that $x^2 + y^2 = z^2$, as it should be. What are the curves $u = \text{const}$ and $v = \text{const}$? ■

Tangent Plane and Surface Normal

Recall from Sec. 9.7 that the tangent vectors of all the curves on a surface S through a point P of S form a plane, called the **tangent plane** of S at P (Fig. 244). Exceptions are points where S has an edge or a cusp (like a cone), so that S cannot have a tangent plane at such a point. Furthermore, a vector perpendicular to the tangent plane is called a **normal vector** of S at P.

Now since S can be given by $\mathbf{r} = \mathbf{r}(u, v)$ in (2), the new idea is that we get a curve C on S by taking a pair of differentiable functions

$$u = u(t), \qquad v = v(t)$$

whose derivatives $u' = du/dt$ and $v' = dv/dt$ are continuous. Then C has the position vector $\tilde{\mathbf{r}}(t) = \mathbf{r}(u(t), v(t))$. By differentiation and the use of the chain rule (Sec. 9.6) we obtain a tangent vector of C on S

$$\tilde{\mathbf{r}}'(t) = \frac{d\tilde{\mathbf{r}}}{dt} = \frac{\partial \mathbf{r}}{\partial u} u' + \frac{\partial \mathbf{r}}{\partial v} v'.$$

Hence the partial derivatives $\mathbf{r}_u$ and $\mathbf{r}_v$ *at P are tangential to S at P.* We assume that they are linearly independent, which geometrically means that the curves $u = \text{const}$ and $v = \text{const}$ on S intersect at P at a nonzero angle. Then $\mathbf{r}_u$ and $\mathbf{r}_v$ span the tangent plane of S at P. Hence their cross product gives a **normal vector N** of S at P.

$$\mathbf{N} = \mathbf{r}_u \times \mathbf{r}_v \neq \mathbf{0}. \tag{4}$$

The corresponding **unit normal vector n** of S at P is (Fig. 244)

$$\mathbf{n} = \frac{1}{|\mathbf{N}|}\mathbf{N} = \frac{1}{|\mathbf{r}_u \times \mathbf{r}_v|}\mathbf{r}_u \times \mathbf{r}_v. \tag{5}$$

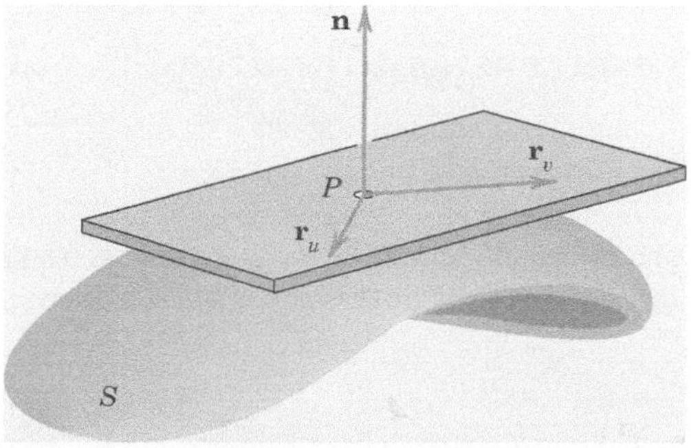

Fig. 244. Tangent plane and normal vector

Also, if S is represented by $g(x, y, z) = 0$, then, by Theorem 2 in Sec. 9.7,

$$\mathbf{n} = \frac{1}{|\text{grad } g|} \text{grad } g. \tag{5*}$$

A surface S is called a **smooth surface** if its surface normal depends continuously on the points of S.

S is called **piecewise smooth** if it consists of finitely many smooth portions.

For instance, a sphere is smooth, and the surface of a cube is piecewise smooth (explain!). We can now summarize our discussion as follows.

THEOREM 1

Tangent Plane and Surface Normal

If a surface S is given by (2) with continuous $\mathbf{r}_u = \partial\mathbf{r}/\partial u$ and $\mathbf{r}_v = \partial\mathbf{r}/\partial v$ satisfying (4) at every point of S, then S has, at every point P, a unique tangent plane passing through P and spanned by $\mathbf{r}_u$ and $\mathbf{r}_v$, and a unique normal whose direction depends continuously on the points of S. A normal vector is given by (4) and the corresponding unit normal vector by (5). (See Fig. 244.)

EXAMPLE 4 Unit Normal Vector of a Sphere

From (5*) we find that the sphere $g(x, y, z) = x^2 + y^2 + z^2 - a^2 = 0$ has the unit normal vector

$$\mathbf{n}(x, y, z) = \left[\frac{x}{a}, \frac{y}{a}, \frac{z}{a}\right] = \frac{x}{a}\mathbf{i} + \frac{y}{a}\mathbf{j} + \frac{z}{a}\mathbf{k}.$$

We see that $\mathbf{n}$ has the direction of the position vector $[x, y, z]$ of the corresponding point. Is it obvious that this must be the case? ■

EXAMPLE 5 Unit Normal Vector of a Cone

At the apex of the cone $g(x, y, z) = -z + \sqrt{x^2 + y^2} = 0$ in Example 3, the unit normal vector $\mathbf{n}$ becomes undetermined because from (5*) we get

$$\mathbf{n} = \left[\frac{x}{\sqrt{2(x^2 + y^2)}}, \frac{y}{\sqrt{2(x^2 + y^2)}}, \frac{-1}{\sqrt{2}}\right] = \frac{1}{\sqrt{2}}\left(\frac{x}{\sqrt{x^2 + y^2}}\mathbf{i} + \frac{y}{\sqrt{x^2 + y^2}}\mathbf{j} - \mathbf{k}\right). \quad ■$$

We are now ready to discuss surface integrals and their applications, beginning in the next section.

PROBLEM SET 10.5

1–8 PARAMETRIC SURFACE REPRESENTATION

Familiarize yourself with parametric representations of important surfaces by deriving a representation (1), by finding the **parameter curves** (curves u = const and v = const) of the surface and a normal vector $\mathbf{N} = \mathbf{r}_u \times \mathbf{r}_v$ of the surface. Show the details of your work.

1. xy-plane $\mathbf{r}(u, v) = (u, \ v)$ (thus $u\mathbf{i} + v\mathbf{j}$; similarly in Probs. 2–8).
2. xy-plane in polar coordinates $\mathbf{r}(u, v) = [u \cos v, \ u \sin v]$ (thus $u = r, v = \theta$)
3. Cone $\mathbf{r}(u, v) = [u \cos v, \ u \sin v, \ cu]$
4. Elliptic cylinder $\mathbf{r}(u, v) = [a \cos v, \ b \sin v, \ u]$
5. Paraboloid of revolution $\mathbf{r}(u, v) = [u \cos v, \ u \sin v, \ u^2]$
6. Helicoid $\mathbf{r}(u, v) = [u \cos v, \ u \sin v, \ v]$. Explain the name.
7. Ellipsoid $\mathbf{r}(u, v) = [a \cos v \cos u, \ b \cos v \sin u, \ c \sin v]$
8. Hyperbolic paraboloid $\mathbf{r}(u, v) = [au \cosh v, \ bu \sinh v, \ u^2]$

9. CAS EXPERIMENT. Graphing Surfaces, Dependence on a, b, c. Graph the surfaces in Probs. 3–8. In Prob. 6 generalize the surface by introducing parameters a, b. Then find out in Probs. 4 and 6–8 how the shape of the surfaces depends on a, b, c.

10. Orthogonal parameter curves $u = \text{const}$ and $v = \text{const}$ on $\mathbf{r}(u, v)$ occur if and only if $\mathbf{r}_u \bullet \mathbf{r}_v = 0$. Give examples. Prove it.

11. Satisfying (4). Represent the paraboloid in Prob. 5 so that $\tilde{\mathbf{N}}(0, 0) \neq \mathbf{0}$ and show $\tilde{\mathbf{N}}$.

12. Condition (4). Find the points in Probs. 1–8 at which (4) $\mathbf{N} \neq \mathbf{0}$ does not hold. Indicate whether this results from the shape of the surface or from the choice of the representation.

13. Representation $z = f(x, y)$. Show that $z = f(x, y)$ or $g = z - f(x, y) = 0$ can be written ($f_u = \partial f/\partial u$, etc.)

$$\text{(6)} \qquad \mathbf{r}(u, v) = [u, \quad v, \quad f(u, v)] \quad \text{and} \quad \mathbf{N} = \text{grad } g = [-f_u, \quad -f_v, \quad 1].$$

14–19 DERIVE A PARAMETRIC REPRESENTATION

Find a normal vector. The answer gives *one* representation; there are many. Sketch the surface and parameter curves.

14. Plane $4x + 3y + 2z = 12$

15. Cylinder of revolution $(x - 2)^2 + (y + 1)^2 = 25$

16. Ellipsoid $x^2 + y^2 + \frac{1}{9}z^2 = 1$

17. Sphere $x^2 + (y + 2.8)^2 + (z - 3.2)^2 = 2.25$

18. Elliptic cone $z = \sqrt{x^2 + 4y^2}$

19. Hyperbolic cylinder $x^2 - y^2 = 1$

20. PROJECT. Tangent Planes $T(P)$ will be less important in our work, but you should know how to represent them.

(a) If S: $\mathbf{r}(u, v)$, then $T(P)$: $(\mathbf{r}^* - \mathbf{r} \quad \mathbf{r}_u \quad \mathbf{r}_v) = 0$ (a scalar triple product) or

$$\mathbf{r}^*(p, q) = \mathbf{r}(P) + p\mathbf{r}_u(P) + q\mathbf{r}_v(P).$$

(b) If S: $g(x, y, z) = 0$, then

$$T(P): (\mathbf{r}^* - \mathbf{r}(P)) \cdot \nabla g = 0.$$

(c) If S: $z = f(x, y)$, then

$$T(P): z^* - z = (x^* - x)f_x(P) + (y^* - y)f_y(P).$$

Interpret (a)–(c) geometrically. Give two examples for (a), two for (b), and two for (c).

10.6 Surface Integrals

To define a surface integral, we take a surface S, given by a parametric representation as just discussed,

$$\text{(1)} \qquad \mathbf{r}(u, v) = [x(u, v), y(u, v), z(u, v)] = x(u, v)\mathbf{i} + y(u, v)\mathbf{j} + z(u, v)\mathbf{k}$$

where (u, v) varies over a region R in the uv-plane. We assume S to be piecewise smooth (Sec. 10.5), so that S has a normal vector

$$\text{(2)} \qquad \mathbf{N} = \mathbf{r}_u \times \mathbf{r}_v \quad \text{and unit normal vector} \quad \mathbf{n} = \frac{1}{|\mathbf{N}|}\mathbf{N}$$

at every point (except perhaps for some edges or cusps, as for a cube or cone). For a given vector function $\mathbf{F}$ we can now define the **surface integral** over S by

$$\text{(3)} \qquad \iint_S \mathbf{F} \bullet \mathbf{n}\, dA = \iint_R \mathbf{F}(\mathbf{r}(u, v)) \bullet \mathbf{N}(u, v)\, du\, dv.$$

Here $\mathbf{N} = |\mathbf{N}|\mathbf{n}$ by (2), and $|\mathbf{N}| = |\mathbf{r}_u \times \mathbf{r}_v|$ is the area of the parallelogram with sides $\mathbf{r}_u$ and $\mathbf{r}_v$, by the definition of cross product. Hence

$$\text{(3*)} \qquad \mathbf{n}\, dA = \mathbf{n}|\mathbf{N}|\, du\, dv = \mathbf{N}\, du\, dv.$$

And we see that $dA = |\mathbf{N}| du\, dv$ is the element of area of S.

Also $\mathbf{F} \cdot \mathbf{n}$ is the normal component of $\mathbf{F}$. This integral arises naturally in flow problems, where it gives the **flux** across S when $\mathbf{F} = \rho \mathbf{v}$. Recall, from Sec. 9.8, that the flux across S is the mass of fluid crossing S per unit time. Furthermore, ρ is the density of the fluid and $\mathbf{v}$ the velocity vector of the flow, as illustrated by Example 1 below. We may thus call the surface integral (3) the **flux integral**.

We can write (3) in components, using $\mathbf{F} = [F_1, \ F_2, \ F_3]$, $\mathbf{N} = [N_1, \ N_2, \ N_3]$, and $\mathbf{n} = [\cos \alpha, \cos \beta, \cos \gamma]$. Here, α, β, γ are the angles between $\mathbf{n}$ and the coordinate axes; indeed, for the angle between $\mathbf{n}$ and $\mathbf{i}$, formula (4) in Sec. 9.2 gives $\cos \alpha = \mathbf{n} \cdot \mathbf{i}/|\mathbf{n}||\mathbf{i}| = \mathbf{n} \cdot \mathbf{i}$, and so on. We thus obtain from (3)

$$\text{(4)} \qquad \begin{aligned} \iint_S \mathbf{F} \cdot \mathbf{n} \, dA &= \iint_S (F_1 \cos \alpha + F_2 \cos \beta + F_3 \cos \gamma) \, dA \\ &= \iint_R (F_1 N_1 + F_2 N_2 + F_3 N_3) \, du \, dv. \end{aligned}$$

In (4) we can write $\cos \alpha \, dA = dy \, dz$, $\cos \beta \, dA = dz \, dx$, $\cos \gamma \, dA = dx \, dy$. Then (4) becomes the following integral for the flux:

$$\text{(5)} \qquad \iint_S \mathbf{F} \cdot \mathbf{n} \, dA = \iint_S (F_1 \, dy \, dz + F_2 \, dz \, dx + F_3 \, dx \, dy).$$

We can use this formula to evaluate surface integrals by converting them to double integrals over regions in the coordinate planes of the xyz-coordinate system. But we must carefully take into account the orientation of S (the choice of $\mathbf{n}$). We explain this for the integrals of the F_3-terms,

$$\text{(5')} \qquad \iint_S F_3 \cos \gamma \, dA = \iint_S F_3 \, dx \, dy.$$

If the surface S is given by $z = h(x, y)$ with (x, y) varying in a region $\bar{R}$ in the xy-plane, and if S is oriented so that $\cos \gamma > 0$, then (5′) gives

$$\text{(5'')} \qquad \iint_S F_3 \cos \gamma \, dA = + \iint_{\bar{R}} F_3(x, y, h(x, y)) \, dx \, dy.$$

But if $\cos \gamma < 0$, the integral on the right of (5″) gets a minus sign in front. This follows if we note that the element of area $dx \, dy$ in the xy-plane is the projection $|\cos \gamma| \, dA$ of the element of area dA of S; and we have $\cos \gamma = +|\cos \gamma|$ when $\cos \gamma > 0$, but $\cos \gamma = -|\cos \gamma|$ when $\cos \gamma < 0$. Similarly for the other two terms in (5). At the same time, this justifies the notations in (5).

Other forms of surface integrals will be discussed later in this section.

EXAMPLE 1 Flux Through a Surface

Compute the flux of water through the parabolic cylinder S: $y = x^2$, $0 \leqq x \leqq 2$, $0 \leqq z \leqq 3$ (Fig. 245) if the velocity vector is $\mathbf{v} = \mathbf{F} = [3z^2, \ 6, \ 6xz]$, speed being measured in meters/sec. (Generally, $\mathbf{F} = \rho \mathbf{v}$, but water has the density $\rho = 1 \text{ g/cm}^3 = 1 \text{ ton/m}^3$.)

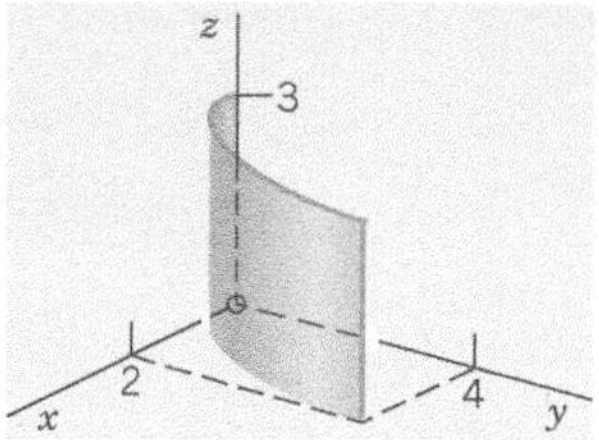

Fig. 245. Surface S in Example 1

Solution. Writing $x = u$ and $z = v$, we have $y = x^2 = u^2$. Hence a representation of S is

$$S: \quad \mathbf{r} = [u, u^2, v] \qquad (0 \leqq u \leqq 2, 0 \leqq v \leqq 3).$$

By differentiation and by the definition of the cross product,

$$\mathbf{N} = \mathbf{r}_u \times \mathbf{r}_v = [1, \quad 2u, \quad 0] \times [0, \quad 0, \quad 1] = [2u, \quad -1, \quad 0].$$

On S, writing simply $\mathbf{F}(S)$ for $\mathbf{F}[\mathbf{r}(u, v)]$, we have $\mathbf{F}(S) = [3v^2, 6, 6uv]$. Hence $\mathbf{F}(S) \cdot \mathbf{N} = 6uv^2 - 6$. By integration we thus get from (3) the flux

$$\iint_S \mathbf{F} \cdot \mathbf{n}\, dA = \int_0^3 \int_0^2 (6uv^2 - 6)\, du\, dv = \int_0^3 (3u^2v^2 - 6u)\Big|_{u=0}^{2} dv$$

$$= \int_0^3 (12v^2 - 12)\, dv = (4v^3 - 12v)\Big|_{v=0}^{3} = 108 - 36 = 72 \text{ [m}^3\text{/sec]}$$

or 72,000 liters/sec. Note that the y-component of $\mathbf{F}$ is positive (equal to 6), so that in Fig. 245 the flow goes from left to right.

Let us confirm this result by (5). Since

$$\mathbf{N} = |\mathbf{N}|\mathbf{n} = |\mathbf{N}|[\cos\alpha, \quad \cos\beta, \quad \cos\gamma] = [2u, \quad -1, \quad 0] = [2x, \quad -1, \quad 0]$$

we see that $\cos\alpha > 0$, $\cos\beta < 0$, and $\cos\gamma = 0$. Hence the second term of (5) on the right gets a minus sign, and the last term is absent. This gives, in agreement with the previous result,

$$\iint_S \mathbf{F} \cdot \mathbf{n}\, dA = \int_0^3 \int_0^4 3z^2\, dy\, dz - \int_0^2 \int_0^3 6\, dz\, dx = \int_0^3 4(3z^2)\, dz - \int_0^2 6 \cdot 3\, dx = 4 \cdot 3^3 - 6 \cdot 3 \cdot 2 = 72. \quad \blacksquare$$

EXAMPLE 2 Surface Integral

Evaluate (3) when $\mathbf{F} = [x^2, 0, 3y^2]$ and S is the portion of the plane $x + y + z = 1$ in the first octant (Fig. 246).

Solution. Writing $x = u$ and $y = v$, we have $z = 1 - x - y = 1 - u - v$. Hence we can represent the plane $x + y + z = 1$ in the form $\mathbf{r}(u, v) = [u, v, 1 - u - v]$. We obtain the first-octant portion S of this plane by restricting $x = u$ and $y = v$ to the projection R of S in the xy-plane. R is the triangle bounded by the two coordinate axes and the straight line $x + y = 1$, obtained from $x + y + z = 1$ by setting $z = 0$. Thus $0 \leqq x \leqq 1 - y, 0 \leqq y \leqq 1$.

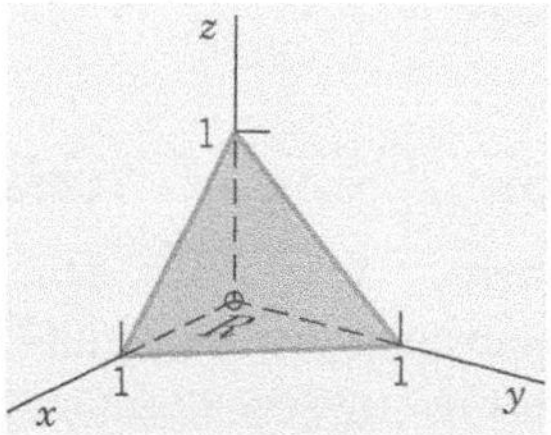

Fig. 246. Portion of a plane in Example 2

By inspection or by differentiation,

$$\mathbf{N} = \mathbf{r}_u \times \mathbf{r}_v = [1, 0, -1] \times [0, 1, -1] = [1, 1, 1].$$

Hence $\mathbf{F}(S) \cdot \mathbf{N} = [u^2, 0, 3v^2] \cdot [1, 1, 1] = u^2 + 3v^2$. By (3),

$$\iint_S \mathbf{F} \cdot \mathbf{n}\, dA = \iint_R (u^2 + 3v^2)\, du\, dv = \int_0^1 \int_0^{1-v} (u^2 + 3v^2)\, du\, dv$$

$$= \int_0^1 \left[\frac{1}{3}(1 - v)^3 + 3v^2(1 - v)\right] dv = \frac{1}{3}.$$

Orientation of Surfaces

From (3) or (4) we see that the value of the integral depends on the choice of the unit normal vector $\mathbf{n}$. (Instead of $\mathbf{n}$ we could choose $-\mathbf{n}$.) We express this by saying that such an integral is an *integral over an* **oriented surface** S, that is, over a surface S on which we have chosen one of the two possible unit normal vectors in a continuous fashion. (For a piecewise smooth surface, this needs some further discussion, which we give below.) If we change the orientation of S, this means that we replace $\mathbf{n}$ with $-\mathbf{n}$. Then each component of $\mathbf{n}$ in (4) is multiplied by -1, so that we have

THEOREM 1

Change of Orientation in a Surface Integral

The replacement of $\mathbf{n}$ *by* $-\mathbf{n}$ *(hence of* $\mathbf{N}$ *by* $-\mathbf{N}$*) corresponds to the multiplication of the integral in* (3) *or* (4) *by* -1.

In practice, how do we make such a change of $\mathbf{N}$ happen, if S is given in the form (1)? The easiest way is to interchange u and v, because then $\mathbf{r}_u$ becomes $\mathbf{r}_v$ and conversely, so that $\mathbf{N} = \mathbf{r}_u \times \mathbf{r}_v$ becomes $\mathbf{r}_v \times \mathbf{r}_u = -\mathbf{r}_u \times \mathbf{r}_v = -\mathbf{N}$, as wanted. Let us illustrate this.

EXAMPLE 3 **Change of Orientation in a Surface Integral**

In Example 1 we now represent S by $\tilde{\mathbf{r}} = [v, v^2, u]$, $0 \leqq v \leqq 2$, $0 \leqq u \leqq 3$. Then

$$\tilde{\mathbf{N}} = \tilde{\mathbf{r}}_u \times \tilde{\mathbf{r}}_v = [0, 0, 1] \times [1, 2v, 0] = [-2v, 1, 0].$$

For $\mathbf{F} = [3z^2, 6, 6xz]$ we now get $\tilde{\mathbf{F}}(S) = [3u^2, 6, 6uv]$. Hence $\tilde{\mathbf{F}}(S) \cdot \tilde{\mathbf{N}} = -6u^2v + 6$ and integration gives the old result times -1,

$$\iint_R \tilde{\mathbf{F}}(S) \cdot \tilde{\mathbf{N}}\, dv\, du = \int_0^3 \int_0^2 (-6u^2v + 6)\, dv\, du = \int_0^3 (-12u^2 + 12)\, du = -72.$$

Orientation of Smooth Surfaces

A smooth surface S (see Sec. 10.5) is called **orientable** if the positive normal direction, when given at an arbitrary point P_0 of S, can be continued in a unique and continuous way to the entire surface. In many practical applications, the surfaces are smooth and thus orientable.

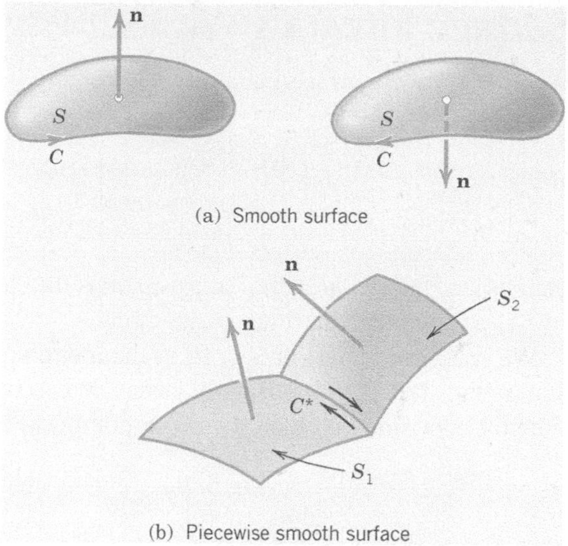

Fig. 247. Orientation of a surface

Orientation of Piecewise Smooth Surfaces

Here the following idea will do it. For a *smooth* orientable surface S with boundary curve C we may associate with each of the two possible orientations of S an orientation of C, as shown in Fig. 247a. Then a *piecewise smooth* surface is called **orientable** if we can orient each smooth piece of S so that along each curve C^* which is a common boundary of two pieces S_1 and S_2 the positive direction of C^* relative to S_1 is opposite to the direction of C^* relative to S_2. See Fig. 247b for two adjacent pieces; note the arrows along C^*.

Theory: Nonorientable Surfaces

A sufficiently small piece of a smooth surface is always orientable. This may not hold for entire surfaces. A well-known example is the **Möbius strip**,[5] shown in Fig. 248. To make a model, take the rectangular paper in Fig. 248, make a half-twist, and join the short sides together so that A goes onto A, and B onto B. At P_0 take a normal vector pointing, say, to the *left*. Displace it along C to the right (in the lower part of the figure) around the strip until you return to P_0 and see that you get a normal vector pointing to the *right*, opposite to the given one. See also Prob. 17.

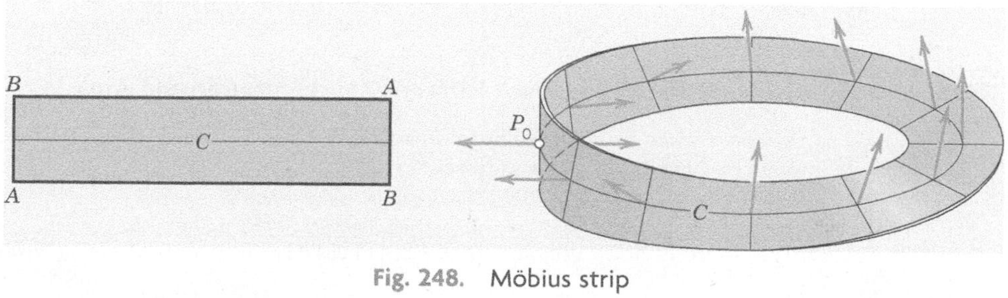

Fig. 248. Möbius strip

[5]AUGUST FERDINAND MÖBIUS (1790–1868), German mathematician, student of Gauss, known for his work in surface theory, geometry, and complex analysis (see Sec. 17.2).

Surface Integrals Without Regard to Orientation

Another type of surface integral is

$$\textbf{(6)} \qquad \iint_S G(\mathbf{r})\,dA = \iint_R G(\mathbf{r}(u, v))|\mathbf{N}(u, v)|\,du\,dv.$$

Here $dA = |\mathbf{N}|\,du\,dv = |\mathbf{r}_u \times \mathbf{r}_v|\,du\,dv$ is the element of area of the surface S represented by (1) and we disregard the orientation.

We shall need later (in Sec. 10.9) the **mean value theorem** for surface integrals, which states that if R in (6) is simply connected (see Sec. 10.2) and $G(\mathbf{r})$ is continuous in a domain containing R, then there is a point (u_0, v_0) in R such that

$$(7) \qquad \iint_S G(\mathbf{r})\,dA = G(\mathbf{r}(u_0, v_0))A \qquad (A = \text{Area of } S).$$

As for applications, if $G(\mathbf{r})$ is the mass density of S, then (6) is the total mass of S. If $G = 1$, then (6) gives the **area** $A(S)$ of S,

$$\textbf{(8)} \qquad A(S) = \iint_S dA = \iint_R |\mathbf{r}_u \times \mathbf{r}_v|\,du\,dv.$$

Examples 4 and 5 show how to apply (8) to a sphere and a torus. The final example, Example 6, explains how to calculate moments of inertia for a surface.

EXAMPLE 4 Area of a Sphere

For a sphere $\mathbf{r}(u, v) = [a \cos v \cos u, \quad a \cos v \sin u, \quad a \sin v]$, $\quad 0 \leqq u \leqq 2\pi, \quad -\pi/2 \leqq v \leqq \pi/2$ [see (3) in Sec. 10.5], we obtain by direct calculation (verify!)

$$\mathbf{r}_u \times \mathbf{r}_v = [a^2 \cos^2 v \cos u, \qquad a^2 \cos^2 v \sin u, \qquad a^2 \cos v \sin v].$$

Using $\cos^2 u + \sin^2 u = 1$ and then $\cos^2 v + \sin^2 v = 1$, we obtain

$$|\mathbf{r}_u \times \mathbf{r}_v| = a^2(\cos^4 v \cos^2 u + \cos^4 v \sin^2 u + \cos^2 v \sin^2 v)^{1/2} = a^2|\cos v|.$$

With this, (8) gives the familiar formula (note that $|\cos v| = \cos v$ when $-\pi/2 \leqq v \leqq \pi/2$)

$$A(S) = a^2 \int_{-\pi/2}^{\pi/2} \int_0^{2\pi} |\cos v|\,du\,dv = 2\pi a^2 \int_{-\pi/2}^{\pi/2} \cos v\,dv = 4\pi a^2. \qquad \blacksquare$$

EXAMPLE 5 Torus Surface (Doughnut Surface): Representation and Area

A *torus surface* S is obtained by rotating a circle C about a straight line L in space so that C does not intersect or touch L but its plane always passes through L. If L is the z-axis and C has radius b and its center has distance $a\ (> b)$ from L, as in Fig. 249, then S can be represented by

$$\mathbf{r}(u, v) = (a + b \cos v) \cos u\,\mathbf{i} + (a + b \cos v) \sin u\,\mathbf{j} + b \sin v\,\mathbf{k}$$

where $0 \leqq u \leqq 2\pi, 0 \leqq v \leqq 2\pi$. Thus

$$\mathbf{r}_u = -(a + b \cos v) \sin u\,\mathbf{i} + (a + b \cos v) \cos u\,\mathbf{j}$$

$$\mathbf{r}_v = -b \sin v \cos u\,\mathbf{i} - b \sin v \sin u\,\mathbf{j} + b \cos v\,\mathbf{k}$$

$$\mathbf{r}_u \times \mathbf{r}_v = b(a + b \cos v)(\cos u \cos v\,\mathbf{i} + \sin u \cos v\,\mathbf{j} + \sin v\,\mathbf{k}).$$

Hence $|\mathbf{r}_u \times \mathbf{r}_v| = b(a + b\cos v)$, and (8) gives the total area of the torus,

$$A(S) = \int_0^{2\pi}\int_0^{2\pi} b(a + b\cos v)\,du\,dv = 4\pi^2 ab. \tag{9}$$

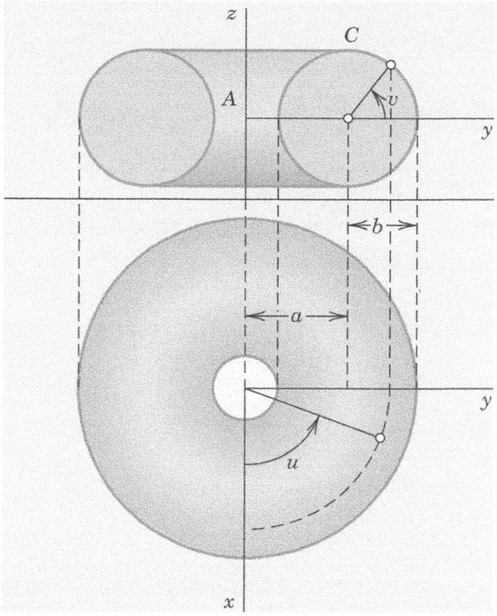

Fig. 249. Torus in Example 5

EXAMPLE 6 Moment of Inertia of a Surface

Find the moment of inertia I of a spherical lamina $S{:} = x^2 + y^2 + z^2 = a^2$ of constant mass density and total mass M about the z-axis.

Solution. If a mass is distributed over a surface S and $\mu(x, y, z)$ is the density of the mass (= mass per unit area), then the moment of inertia I of the mass with respect to a given axis L is defined by the surface integral

$$I = \iint_S \mu D^2\,dA \tag{10}$$

where $D(x, y, z)$ is the distance of the point (x, y, z) from L. Since, in the present example, μ is constant and S has the area $A = 4\pi a^2$, we have $\mu = M/A = M/(4\pi a^2)$.

For S we use the same representation as in Example 4. Then $D^2 = x^2 + y^2 = a^2\cos^2 v$. Also, as in that example, $dA = a^2\cos v\,du\,dv$. This gives the following result. [In the integration, use $\cos^3 v = \cos v\,(1 - \sin^2 v)$.]

$$I = \iint_S \mu D^2\,dA = \frac{M}{4\pi a^2}\int_{-\pi/2}^{\pi/2}\int_0^{2\pi} a^4\cos^3 v\,du\,dv = \frac{Ma^2}{2}\int_{-\pi/2}^{\pi/2}\cos^3 v\,dv = \frac{2Ma^2}{3}.$$

Representations $z = f(x, y)$. If a surface S is given by $z = f(x, y)$, then setting $u = x$, $v = y$, $\mathbf{r} = [u, v, f]$ gives

$$|\mathbf{N}| = |\mathbf{r}_u \times \mathbf{r}_v| = |[1, 0, f_u] \times [0, 1, f_v]| = |[-f_u, -f_v, 1]| = \sqrt{1 + f_u^2 + f_v^2}$$

and, since $f_u = f_x, f_v = f_y$, formula (6) becomes

$$\iint_S G(\mathbf{r})\,dA = \iint_{R^*} G(x, y, f(x, y))\sqrt{1 + \left(\frac{\partial f}{\partial x}\right)^2 + \left(\frac{\partial f}{\partial y}\right)^2}\,dx\,dy. \tag{11}$$

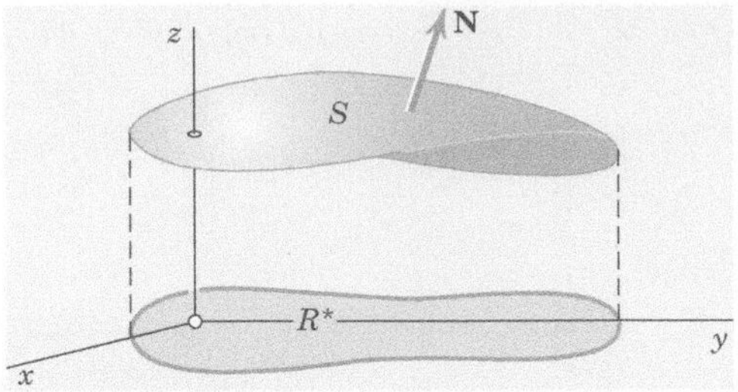

Fig. 250. Formula (11)

Here R^* is the projection of S into the xy-plane (Fig. 250) and the normal vector $\mathbf{N}$ on S points *up*. If it points *down*, the integral on the right is preceded by a minus sign.

From (11) with $G = 1$ we obtain for the **area** $A(S)$ of S: $z = f(x, y)$ the formula

$$A(S) = \iint_{R^*} \sqrt{1 + \left(\frac{\partial f}{\partial x}\right)^2 + \left(\frac{\partial f}{\partial y}\right)^2}\, dx\, dy \tag{12}$$

where R^* is the projection of S into the xy-plane, as before.

PROBLEM SET 10.6

1–10 FLUX INTEGRALS (3) $\int_S \mathbf{F} \cdot \mathbf{n}\, dA$

Evaluate the integral for the given data. Describe the kind of surface. Show the details of your work.

1. $\mathbf{F} = [-x^2, y^2, 0]$, S: $\mathbf{r} = [u, v, 3u - 2v]$, $0 \leqq u \leqq 1.5$, $-2 \leqq v \leqq 2$
2. $\mathbf{F} = [e^y, e^x, 1]$, S: $x + y + z = 1$, $x \geqq 0$, $y \geqq 0$, $z \geqq 0$
3. $\mathbf{F} = [0, x, 0]$, S: $x^2 + y^2 + z^2 = 1$, $x \geqq 0$, $y \geqq 0$, $z \geqq 0$
4. $\mathbf{F} = [e^y, -e^z, e^x]$, S: $x^2 + y^2 = 25$, $x \geqq 0$, $y \geqq 0$, $0 \leqq z \leqq 2$
5. $\mathbf{F} = [x, y, z]$, S: $\mathbf{r} = [u \cos v, u \sin v, u^2]$, $0 \leqq u \leqq 4$, $-\pi \leqq v \leqq \pi$
6. $\mathbf{F} = [\cosh y, 0, \sinh x]$, S: $z = x + y^2$, $0 \leqq y \leqq x$, $0 \leqq x \leqq 1$
7. $\mathbf{F} = [0, \sin y, \cos z]$, S the cylinder $x = y^2$, where $0 \leqq y \leqq \pi/4$ and $0 \leqq z \leqq y$
8. $\mathbf{F} = [\tan xy, x, y]$, S: $y^2 + z^2 = 1$, $2 \leqq x \leqq 5$, $y \geqq 0$, $z \geqq 0$
9. $\mathbf{F} = [0, \sinh z, \cosh x]$, S: $x^2 + z^2 = 4$, $0 \leqq x \leqq 1/\sqrt{2}$, $0 \leqq y \leqq 5$, $z \geqq 0$
10. $\mathbf{F} = [y^2, x^2, z^4]$, S: $z = 4\sqrt{x^2 + y^2}$, $0 \leqq z \leqq 8$, $y \geqq 0$
11. **CAS EXPERIMENT. Flux Integral.** Write a program for evaluating surface integrals (3) that prints intermediate results ($\mathbf{F}$, $\mathbf{F} \cdot \mathbf{N}$, the integral over one of the two variables). Can you obtain experimentally some rules on functions and surfaces giving integrals that can be evaluated by the usual methods of calculus? Make a list of positive and negative results.

12–16 SURFACE INTEGRALS (6) $\iint_S G(\mathbf{r})\, dA$

Evaluate these integrals for the following data. Indicate the kind of surface. Show the details.

12. $G = \cos x + \sin x$, S the portion of $x + y + z = 1$ in the first octant
13. $G = x + y + z$, $z = x + 2y$, $0 \leqq x \leqq \pi$, $0 \leqq y \leqq x$
14. $G = ax + by + cz$, S: $x^2 + y^2 + z^2 = 1$, $y = 0$, $z = 0$
15. $G = (1 + 9xz)^{3/2}$, S: $\mathbf{r} = [u, v, u^3]$, $0 \leqq u \leqq 1$, $-2 \leqq v \leqq 2$
16. $G = \arctan(y/x)$, S: $z = x^2 + y^2$, $1 \leqq z \leqq 9$, $x \geqq 0$, $y \geqq 0$
17. **Fun with Möbius.** Make Möbius strips from long slim rectangles R of grid paper (graph paper) by pasting the short sides together after giving the paper a half-twist. In each case count the number of parts obtained by cutting along lines parallel to the edge. **(a)** Make R three squares wide and cut until you reach the beginning. **(b)** Make R four squares wide. Begin cutting one square away from the edge until you reach the beginning. Then cut the portion that is still two squares wide. **(c)** Make

R five squares wide and cut similarly. **(d)** Make R six squares wide and cut. Formulate a conjecture about the number of parts obtained.

18. Gauss "Double Ring" (See Möbius, *Works* **2**, 518–559). Make a paper cross (Fig. 251) into a "double ring" by joining opposite arms along their outer edges (without twist), one ring below the plane of the cross and the other above. Show experimentally that one can choose any four boundary points A, B, C, D and join A and C as well as B and D by two nonintersecting curves. What happens if you cut along the two curves? If you make a half-twist in each of the two rings and then cut? (Cf. E. Kreyszig, Proc. CSHPM 13 (2000), 23–43.)

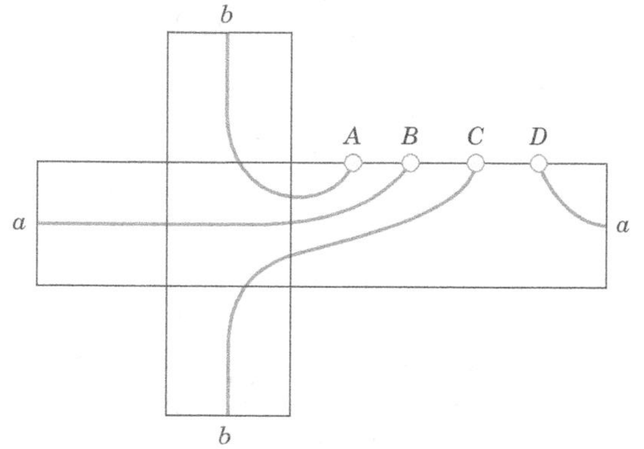

Fig. 251. **Problem 18.** Gauss "Double Ring"

APPLICATIONS

19. Center of gravity. Justify the following formulas for the mass M and the center of gravity $(\bar{x}, \bar{y}, \bar{z})$ of a lamina S of density (mass per unit area) $\sigma(x, y, z)$ in space:

$$M = \iint_S \sigma\, dA, \quad \bar{x} = \frac{1}{M}\iint_S x\sigma\, dA,$$

$$\bar{y} = \frac{1}{M}\iint_S y\sigma\, dA, \quad \bar{z} = \frac{1}{M}\iint_S z\sigma\, dA.$$

20. Moments of inertia. Justify the following formulas for the moments of inertia of the lamina in Prob. 19 about the x-, y-, and z-axes, respectively:

$$I_x = \iint_S (y^2 + z^2)\sigma\, dA, \quad I_y = \iint_S (x^2 + z^2)\sigma\, dA,$$

$$I_z = \iint_S (x^2 + y^2)\sigma\, dA.$$

21. Find a formula for the moment of inertia of the lamina in Prob. 20 about the line $y = x, z = 0$.

22–23 Find the moment of inertia of a lamina S of density 1 about an axis B, where

22. $S: x^2 + y^2 = 1, 0 \leqq z \leqq h$, B: the line $z = h/2$ in the xz-plane

23. $S: x^2 + y^2 = z^2, \quad 0 \leqq z \leqq h$, B: the z-axis

24. Steiner's theorem.[6] If I_B is the moment of inertia of a mass distribution of total mass M with respect to a line B through the center of gravity, show that its moment of inertia I_K with respect to a line K, which is parallel to B and has the distance k from it is

$$I_K = I_B + k^2 M.$$

25. Using Steiner's theorem, find the moment of inertia of a mass of density 1 on the sphere $S: x^2 + y^2 + z^2 = 1$ about the line $K: x = 1, y = 0$ from the moment of inertia of the mass about a suitable line B, which you must first calculate.

26. TEAM PROJECT. First Fundamental Form of S. Given a surface S: $\mathbf{r}(u, v)$, the differential form

$$ds^2 = E\, du^2 + 2F\, du\, dv + G\, dv^2 \tag{13}$$

with coefficients (in standard notation, unrelated to F, G elsewhere in this chapter)

$$E = \mathbf{r}_u \cdot \mathbf{r}_u, \quad F = \mathbf{r}_u \cdot \mathbf{r}_v, \quad G = \mathbf{r}_v \cdot \mathbf{r}_v \tag{14}$$

is called the **first fundamental form** of S. This form is basic because it permits us to calculate lengths, angles, and areas on S. To show this prove (a)–(c):

(a) For a curve $C: u = u(t), v = v(t), a \leqq t \leqq b$, on S, formulas (10), Sec. 9.5, and (14) give the length

$$\begin{aligned} l &= \int_a^b \sqrt{\mathbf{r}'(t) \cdot \mathbf{r}'(t)}\, dt \\ &= \int_a^b \sqrt{Eu'^2 + 2Fu'v' + Gv'^2}\, dt. \end{aligned} \tag{15}$$

(b) The angle γ between two intersecting curves $C_1: u = g(t), v = h(t)$ and $C_2: u = p(t), v = q(t)$ on $S: \mathbf{r}(u, v)$ is obtained from

$$\cos \gamma = \frac{\mathbf{a} \cdot \mathbf{b}}{|\mathbf{a}||\mathbf{b}|} \tag{16}$$

where $\mathbf{a} = \mathbf{r}_u g' + \mathbf{r}_v h'$ and $\mathbf{b} = \mathbf{r}_u p' + \mathbf{r}_v q'$ are tangent vectors of C_1 and C_2.

[6] JACOB STEINER (1796–1863), Swiss geometer, born in a small village, learned to write only at age 14, became a pupil of Pestalozzi at 18, later studied at Heidelberg and Berlin and, finally, because of his outstanding research, was appointed professor at Berlin University.

(c) The square of the length of the normal vector **N** can be written

$$(17) \qquad |\mathbf{N}|^2 = |\mathbf{r}_u \times \mathbf{r}_v|^2 = EG - F^2,$$

so that formula (8) for the area $A(S)$ of S becomes

$$(18) \qquad \begin{aligned} A(S) &= \iint_S dA = \iint_R |\mathbf{N}|\, du\, dv \\ &= \iint_R \sqrt{EG - F^2}\, du\, dv. \end{aligned}$$

(d) For polar coordinates $u\ (= r)$ and $v\ (= \theta)$ defined by $x = u \cos v$, $y = u \sin v$ we have $E = 1$, $F = 0$, $G = u^2$, so that

$$ds^2 = du^2 + u^2\, dv^2 = dr^2 + r^2\, d\theta^2.$$

Calculate from this and (18) the area of a disk of radius a.

(e) Find the first fundamental form of the torus in Example 5. Use it to calculate the area A of the torus. Show that A can also be obtained by the **theorem of Pappus**,[7] which states that the area of a surface of revolution equals the product of the length of a meridian C and the length of the path of the center of gravity of C when C is rotated through the angle 2π.

(f) Calculate the first fundamental form for the usual representations of important surfaces of your own choice (cylinder, cone, etc.) and apply them to the calculation of lengths and areas on these surfaces.

10.7 Triple Integrals. Divergence Theorem of Gauss

In this section we discuss another "big" integral theorem, the divergence theorem, which transforms surface integrals into triple integrals. So let us begin with a review of the latter.

A **triple integral** is an integral of a function $f(x, y, z)$ taken over a closed bounded, three-dimensional region T in space. (Note that "closed" and "bounded" are defined in the same way as in footnote 2 of Sec. 10.3, with "sphere" substituted for "circle"). We subdivide T by planes parallel to the coordinate planes. Then we consider those boxes of the subdivision that lie entirely inside T, and number them from 1 to n. Here each box consists of a rectangular parallelepiped. In each such box we choose an arbitrary point, say, (x_k, y_k, z_k) in box k. The volume of box k we denote by ΔV_k. We now form the sum

$$J_n = \sum_{k=1}^{n} f(x_k, y_k, z_k)\, \Delta V_k.$$

This we do for larger and larger positive integers n arbitrarily but so that the maximum length of all the edges of those n boxes approaches zero as n approaches infinity. This gives a sequence of real numbers $J_{n_1}, J_{n_2}, \cdots$. We assume that $f(x, y, z)$ is continuous in a domain containing T, and T is bounded by finitely many *smooth surfaces* (see Sec. 10.5). Then it can be shown (see Ref. [GenRef4] in App. 1) that the sequence converges to a limit that is independent of the choice of subdivisions and corresponding points

[7] PAPPUS OF ALEXANDRIA (about A.D. 300), Greek mathematician. The theorem is also called Guldin's theorem. HABAKUK GULDIN (1577–1643) was born in St. Gallen, Switzerland, and later became professor in Graz and Vienna.

(x_k, y_k, z_k). This limit is called the **triple integral** *of* $f(x, y, z)$ *over the region* T and is denoted by

$$\iiint_T f(x, y, z)\, dx\, dy\, dz \qquad \text{or by} \qquad \iiint_T f(x, y, z)\, dV.$$

Triple integrals can be evaluated by three successive integrations. This is similar to the evaluation of double integrals by two successive integrations, as discussed in Sec. 10.3. Example 1 below explains this.

Divergence Theorem of Gauss

Triple integrals can be transformed into surface integrals over the boundary surface of a region in space and conversely. Such a transformation is of practical interest because one of the two kinds of integral is often simpler than the other. It also helps in establishing fundamental equations in fluid flow, heat conduction, etc., as we shall see. The transformation is done by the *divergence theorem,* which involves the **divergence** of a vector function $\mathbf{F} = [F_1, F_2, F_3] = F_1\mathbf{i} + F_2\mathbf{j} + F_3\mathbf{k}$, namely,

$$\text{(1)} \qquad \operatorname{div} \mathbf{F} = \frac{\partial F_1}{\partial x} + \frac{\partial F_2}{\partial y} + \frac{\partial F_3}{\partial z} \qquad \text{(Sec. 9.8).}$$

THEOREM 1

Divergence Theorem of Gauss
(Transformation Between Triple and Surface Integrals)

Let T be a closed bounded region in space whose boundary is a piecewise smooth orientable surface S. Let $\mathbf{F}(x, y, z)$ be a vector function that is continuous and has continuous first partial derivatives in some domain containing T. Then

$$\text{(2)} \qquad \iiint_T \operatorname{div} \mathbf{F}\, dV = \iint_S \mathbf{F} \cdot \mathbf{n}\, dA.$$

In components *of* $\mathbf{F} = [F_1, \ F_2, \ F_3]$ *and of the outer unit normal vector* $\mathbf{n} = [\cos\alpha, \ \cos\beta, \ \cos\gamma]$ *of* S (as in Fig. 253), *formula* (2) *becomes*

$$\text{(2*)} \qquad \begin{aligned} &\iiint_T \left(\frac{\partial F_1}{\partial x} + \frac{\partial F_2}{\partial y} + \frac{\partial F_3}{\partial z}\right) dx\, dy\, dz \\ &\qquad = \iint_S (F_1 \cos\alpha + F_2 \cos\beta + F_3 \cos\gamma)\, dA \\ &\qquad = \iint_S (F_1\, dy\, dz + F_2\, dz\, dx + F_3\, dx\, dy). \end{aligned}$$

"Closed bounded region" is explained above, "piecewise smooth orientable" in Sec. 10.5, and "domain containing T" in footnote 4, Sec. 10.4, for the two-dimensional case.

Before we prove the theorem, let us show a standard application.

EXAMPLE 1 Evaluation of a Surface Integral by the Divergence Theorem

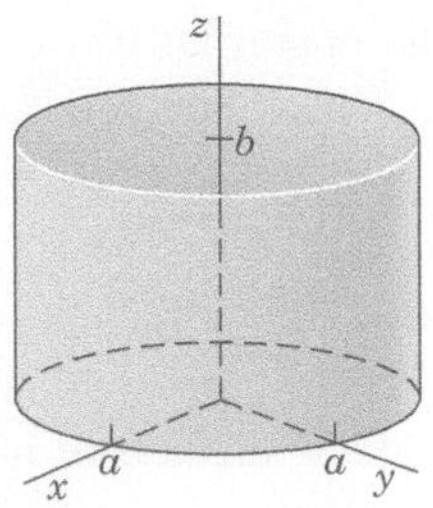

Fig. 252. Surface S in Example 1

Before we prove the theorem, let us show a typical application. Evaluate

$$I = \iint_S (x^3\,dy\,dz + x^2y\,dz\,dx + x^2z\,dx\,dy)$$

where S is the closed surface in Fig. 252 consisting of the cylinder $x^2 + y^2 = a^2$ $(0 \leqq z \leqq b)$ and the circular disks $z = 0$ and $z = b$ $(x^2 + y^2 \leqq a^2)$.

Solution. $F_1 = x^3$, $F_2 = x^2y$, $F_3 = x^2z$. Hence $\operatorname{div} \mathbf{F} = 3x^2 + x^2 + x^2 = 5x^2$. The form of the surface suggests that we introduce polar coordinates r, θ defined by $x = r\cos\theta$, $y = r\sin\theta$ (thus cylindrical coordinates r, θ, z). Then the volume element is $dx\,dy\,dz = r\,dr\,d\theta\,dz$, and we obtain

$$I = \iiint_T 5x^2\,dx\,dy\,dz = \int_{z=0}^{b}\int_{\theta=0}^{2\pi}\int_{r=0}^{a} (5r^2\cos^2\theta)\,r\,dr\,d\theta\,dz$$

$$= 5\int_{z=0}^{b}\int_{\theta=0}^{2\pi} \frac{a^4}{4}\cos^2\theta\,d\theta\,dz = 5\int_{z=0}^{b} \frac{a^4\pi}{4}\,dz = \frac{5\pi}{4}a^4b.$$ ■

PROOF We prove the divergence theorem, beginning with the first equation in (2*). This equation is true if and only if the integrals of each component on both sides are equal; that is,

$$\iiint_T \frac{\partial F_1}{\partial x}\,dx\,dy\,dz = \iint_S F_1\cos\alpha\,dA, \tag{3}$$

$$\iiint_T \frac{\partial F_2}{\partial y}\,dx\,dy\,dz = \iint_S F_2\cos\beta\,dA, \tag{4}$$

$$\iiint_T \frac{\partial F_3}{\partial z}\,dx\,dy\,dz = \iint_S F_3\cos\gamma\,dA. \tag{5}$$

We first prove (5) for a ***special region*** T that is bounded by a piecewise smooth orientable surface S and has the property that any straight line parallel to any one of the coordinate axes and intersecting T has at most *one* segment (or a single point) in common with T. This implies that T can be represented in the form

$$g(x, y) \leqq z \leqq h(x, y) \tag{6}$$

where (x, y) varies in the orthogonal projection $\overline{R}$ of T in the xy-plane. Clearly, $z = g(x, y)$ represents the "bottom" S_2 of S (Fig. 253), whereas $z = h(x, y)$ represents the "top" S_1 of S, and there may be a remaining vertical portion S_3 of S. (The portion S_3 may degenerate into a curve, as for a sphere.)

To prove (5), we use (6). Since **F** is continuously differentiable in some domain containing T, we have

$$\iiint_T \frac{\partial F_3}{\partial z}\, dx\, dy\, dz = \iint_{\bar{R}} \left[\int_{g(x,y)}^{h(x,y)} \frac{\partial F_3}{\partial z}\, dz \right] dx\, dy. \tag{7}$$

Integration of the inner integral $[\cdots]$ gives $F_3[x, y, h(x, y)] - F_3[x, y, g(x, y)]$. Hence the triple integral in (7) equals

$$\iint_{\bar{R}} F_3[x, y, h(x, y)]\, dx\, dy - \iint_{\bar{R}} F_3[x, y, g(x, y)]\, dx\, dy. \tag{8}$$

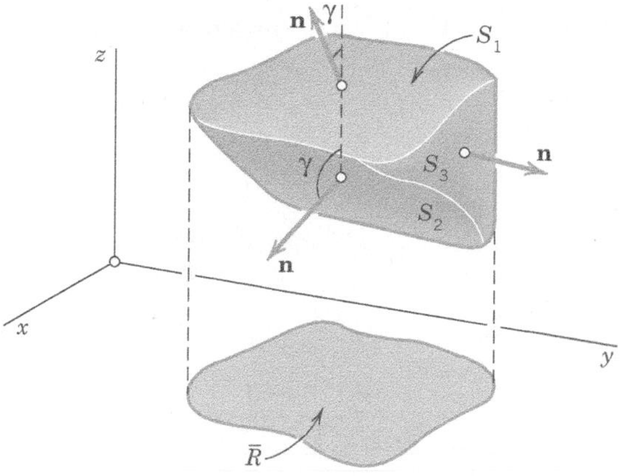

Fig. 253. Example of a special region

But the same result is also obtained by evaluating the right side of (5); that is [see also the last line of (2*)],

$$\begin{aligned} \iint_S F_3 \cos\gamma\, dA &= \iint_S F_3\, dx\, dy \\ &= +\iint_{\bar{R}} F_3[x, y, h(x, y)]\, dx\, dy - \iint_{\bar{R}} F_3[x, y, g(x, y)]\, dx\, dy, \end{aligned}$$

where the first integral over $\bar{R}$ gets a plus sign because $\cos\gamma > 0$ on S_1 in Fig. 253 [as in (5″), Sec. 10.6], and the second integral gets a minus sign because $\cos\gamma < 0$ on S_2. This proves (5).

The relations (3) and (4) now follow by merely relabeling the variables and using the fact that, by assumption, T has representations similar to (6), namely,

$$\tilde{g}(y, z) \leqq x \leqq \tilde{h}(y, z) \qquad \text{and} \qquad \tilde{\tilde{g}}(z, x) \leqq y \leqq \tilde{\tilde{h}}(z, x).$$

This proves the first equation in (2*) for special regions. It implies (2) because the left side of (2*) is just the definition of the divergence, and the right sides of (2) and of the first equation in (2*) are equal, as was shown in the first line of (4) in the last section. Finally, equality of the right sides of (2) and (2*), last line, is seen from (5) in the last section.

This establishes the divergence theorem for special regions.

For any region T that can be subdivided into *finitely many* special regions by means of auxiliary surfaces, the theorem follows by adding the result for each part separately. This procedure is analogous to that in the proof of Green's theorem in Sec. 10.4. The surface integrals over the auxiliary surfaces cancel in pairs, and the sum of the remaining surface integrals is the surface integral over the whole boundary surface S of T; the triple integrals over the parts of T add up to the triple integral over T.

The divergence theorem is now proved for any bounded region that is of interest in practical problems. The extension to a most general region T of the type indicated in the theorem would require a certain limit process; this is similar to the situation in the case of Green's theorem in Sec. 10.4. ■

EXAMPLE 2 Verification of the Divergence Theorem

Evaluate $\iint_S (7x\mathbf{i} - z\mathbf{k}) \cdot \mathbf{n}\, dA$ over the sphere $S: x^2 + y^2 + z^2 = 4$ (a) by (2), (b) directly.

Solution. **(a)** $\operatorname{div} \mathbf{F} = \operatorname{div} [7x, 0, -z] = \operatorname{div} [7x\mathbf{i} - z\mathbf{k}] = 7 - 1 = 6$. *Answer:* $6 \cdot (\frac{4}{3})\pi \cdot 2^3 = 64\pi$.

(b) We can represent S by (3), Sec. 10.5 (with $a = 2$), and we shall use $\mathbf{n}\, dA = \mathbf{N}\, du\, dv$ [see (3*), Sec. 10.6]. Accordingly,

$$S: \quad \mathbf{r} = [2 \cos v \cos u, \quad 2 \cos v \sin u, \quad 2 \sin u] \quad .$$

Then

$$\mathbf{r}_u = [-2 \cos v \sin u, \quad 2 \cos v \cos u, \quad 0]$$

$$\mathbf{r}_v = [-2 \sin v \cos u, \quad -2 \sin v \sin u, \quad 2 \cos v]$$

$$\mathbf{N} = \mathbf{r}_u \times \mathbf{r}_v = [4 \cos^2 v \cos u, \quad 4 \cos^2 v \sin u, \quad 4 \cos v \sin v].$$

Now on S we have $x = 2 \cos v \cos u$, $z = 2 \sin v$, so that $\mathbf{F} = [7x, 0, -z]$ becomes on S

$$\mathbf{F}(S) = [14 \cos v \cos u, \quad 0, \quad -2 \sin v]$$

and

$$\mathbf{F}(S) \cdot \mathbf{N} = (14 \cos v \cos u) \cdot 4 \cos^2 v \cos u + (-2 \sin v) \cdot 4 \cos v \sin v$$

$$= 56 \cos^3 v \cos^2 u - 8 \cos v \sin^2 v.$$

On S we have to integrate over u from 0 to 2π. This gives

$$\pi \cdot 56 \cos^3 v - 2\pi \cdot 8 \cos v \sin^2 v.$$

The integral of $\cos v \sin^2 v$ equals $(\sin^3 v)/3$, and that of $\cos^3 v = \cos v\,(1 - \sin^2 v)$ equals $\sin v - (\sin^3 v)/3$. On S we have $-\pi/2 \leqq v \leqq \pi/2$, so that by substituting these limits we get

$$56\pi(2 - \tfrac{2}{3}) - 16\pi \cdot \tfrac{2}{3} = 64\pi$$

as hoped for. To see the point of Gauss's theorem, compare the amounts of work. ■

Coordinate Invariance of the Divergence. The divergence (1) is defined in terms of coordinates, but we can use the divergence theorem to show that div $\mathbf{F}$ has a meaning independent of coordinates.

For this purpose we first note that triple integrals have properties quite similar to those of double integrals in Sec. 10.3. In particular, the **mean value theorem for triple integrals** asserts that for any continuous function $f(x, y, z)$ in a bounded and simply connected region T there is a point $Q: (x_0, y_0, z_0)$ in T such that

$$\iiint_T f(x, y, z)\, dV = f(x_0, y_0, z_0)\, V(T) \qquad (V(T) = \text{volume of } T). \tag{9}$$

In this formula we interchange the two sides, divide by $V(T)$, and set $f = \text{div } \mathbf{F}$. Then by the divergence theorem we obtain for the divergence an integral over the boundary surface $S(T)$ of T,

$$\text{div } \mathbf{F}(x_0, y_0, z_0) = \frac{1}{V(T)} \iiint_T \text{div } \mathbf{F} \, dV = \frac{1}{V(T)} \iint_{S(T)} \mathbf{F} \cdot \mathbf{n} \, dA. \tag{10}$$

We now choose a point $P: (x_1, y_1, z_1)$ in T and let T shrink down onto P so that the maximum distance $d(T)$ of the points of T from P goes to zero. Then $Q: (x_0, y_0, z_0)$ must approach P. Hence (10) becomes

$$\text{div } \mathbf{F}(P) = \lim_{d(T) \to 0} \frac{1}{V(T)} \iint_{S(T)} \mathbf{F} \cdot \mathbf{n} \, dA. \tag{11}$$

This proves

THEOREM 2

Invariance of the Divergence

The divergence of a vector function $\mathbf{F}$ *with continuous first partial derivatives in a region* T *is independent of the particular choice of Cartesian coordinates. For any* P *in* T *it is given by* (11).

Equation (11) is sometimes used as a *definition* of the divergence. Then the representation (1) in Cartesian coordinates can be derived from (11).

Further applications of the divergence theorem follow in the problem set and in the next section. The examples in the next section will also shed further light on the nature of the divergence.

PROBLEM SET 10.7

1–8 APPLICATION: MASS DISTRIBUTION

Find the total mass of a mass distribution of density σ in a region T in space.

1. $\sigma = x^2 + y^2 + z^2$, T the box $|x| \leqq 4$, $|y| \leqq 1$, $0 \leqq z \leqq 2$

2. $\sigma = xyz$, T the box $0 \leqq x \leqq a$, $0 \leqq y \leqq b$, $0 \leqq z \leqq c$

3. $\sigma = e^{-x-y-z}$, $T: 0 \leqq x \leqq 1 - y$, $0 \leqq y \leqq 1$, $0 \leqq z \leqq 2$

4. σ as in Prob. 3, T the tetrahedron with vertices (0, 0, 0), (3, 0, 0), (0, 3, 0), (0, 0, 3)

5. $\sigma = \sin 2x \cos 2y$, $T: 0 \leqq x \leqq \frac{1}{4}\pi$, $\frac{1}{4}\pi - x \leqq y \leqq \frac{1}{4}\pi$, $0 \leqq z \leqq 6$

6. $\sigma = x^2y^2z^2$, T the cylindrical region $x^2 + z^2 \leqq 16$, $|y| \leqq 4$

7. $\sigma = \arctan(y/x)$, $T: x^2 + y^2 + z^2 \leqq a^2$, $z \geqq 0$

8. $\sigma = x^2 + y^2$, T as in Prob. 7

9–18 APPLICATION OF THE DIVERGENCE THEOREM

Evaluate the surface integral $\iint_S \mathbf{F} \cdot \mathbf{n} \, dA$ by the divergence theorem. Show the details.

9. $\mathbf{F} = [x^2, 0, z^2]$, S the surface of the box $|x| \leqq 1$, $|y| \leqq 3$, $0 \leqq z \leqq 2$

10. Solve Prob. 9 by direct integration.

11. $\mathbf{F} = [e^x, e^y, e^z]$, S the surface of the cube $|x| \leqq 1$, $|y| \leqq 1$, $|z| \leqq 1$

12. $\mathbf{F} = [x^3 - y^3, y^3 - z^3, z^3 - x^3]$, S the surface of $x^2 + y^2 + z^2 \leqq 25$, $z \geqq 0$

13. $\mathbf{F} = [\sin y, \cos x, \cos z]$, S, the surface of $x^2 + y^2 \leqq 4$, $|z| \leqq 2$ (a cylinder and two disks!)

14. $\mathbf{F}$ as in Prob. 13, S the surface of $x^2 + y^2 \leqq 9$, $0 \leqq z \leqq 2$

15. $\mathbf{F} = [2x^2, \frac{1}{2}y^2, \sin \pi z]$, S the surface of the tetrahedron with vertices (0, 0, 0), (1, 0, 0), (0, 1, 0), (0, 0, 1)

16. $\mathbf{F} = [\cosh x, z, y]$, S as in Prob. 15

17. $\mathbf{F} = [x^2, y^2, z^2]$, S the surface of the cone $x^2 + y^2 \leqq z^2$, $0 \leqq z \leqq h$

18. $\mathbf{F} = [xy, yz, zx]$, S the surface of the cone $x^2 + y^2 \leqq 4z^2$, $0 \leqq z \leqq 2$

19–23 APPLICATION: MOMENT OF INERTIA

Given a mass of density 1 in a region T of space, find the moment of intertia about the x-axis

$$I_x = \iiint_T (y^2 + z^2)\, dx\, dy\, dz.$$

19. The box $-a \leqq x \leqq a, \quad -b \leqq y \leqq b, \quad -c \leqq z \leqq c$

20. The ball $x^2 + y^2 + z^2 \leqq a^2$

21. The cylinder $y^2 + z^2 \leqq a^2, \quad 0 \leqq x \leqq h$

22. The paraboloid $y^2 + z^2 \leqq x, \quad 0 \leqq x \leqq h$

23. The cone $y^2 + z^2 \leqq x^2, \quad 0 \leqq x \leqq h$

24. Why is I_x in Prob. 23 for large h larger than I_x in Prob. 22 (and the same h)? Why is it smaller for $h = 1$? Give physical reason.

25. Show that for a solid of revolution, $I_x = \dfrac{\pi}{2}\displaystyle\int_0^h r^4(x)\, dx.$

Solve Probs. 20–23 by this formula.

10.8 Further Applications of the Divergence Theorem

The divergence theorem has many important applications: In *fluid flow*, it helps characterize *sources* and *sinks* of fluids. In *heat flow*, it leads to the *heat equation.* In *potential theory*, it gives properties of the solutions of *Laplace's equation.* In this section, we assume that the region T and its boundary surface S are such that the divergence theorem applies.

EXAMPLE 1 Fluid Flow. Physical Interpretation of the Divergence

From the divergence theorem we may obtain an intuitive interpretation of the divergence of a vector. For this purpose we consider the flow of an incompressible fluid (see Sec. 9.8) of constant density $\rho = 1$ which is **steady**, that is, does not vary with time. Such a flow is determined by the field of its velocity vector $\mathbf{v}(P)$ at any point P.

Let S be the boundary surface of a region T in space, and let $\mathbf{n}$ be the outer unit normal vector of S. Then $\mathbf{v} \cdot \mathbf{n}$ is the normal component of $\mathbf{v}$ in the direction of $\mathbf{n}$, and $|\mathbf{v} \cdot \mathbf{n}\, dA|$ is the mass of fluid *leaving* T (if $\mathbf{v} \cdot \mathbf{n} > 0$ at some P) or *entering* T (if $\mathbf{v} \cdot \mathbf{n} < 0$ at P) per unit time at some point P of S through a small portion ΔS of S of area ΔA. Hence the total mass of fluid that flows across S from T to the outside per unit time is given by the surface integral

$$\iint_S \mathbf{v} \cdot \mathbf{n}\, dA.$$

Division by the volume V of T gives the **average flow** out of T:

$$\frac{1}{V}\iint_S \mathbf{v} \cdot \mathbf{n}\, dA. \tag{1}$$

Since the flow is steady and the fluid is incompressible, the amount of fluid flowing outward must be continuously supplied. Hence, if the value of the integral (1) is different from zero, there must be **sources** (*positive sources and negative sources, called* **sinks**) in T, that is, points where fluid is produced or disappears.

If we let T shrink down to a fixed point P in T, we obtain from (1) the **source intensity** at P given by the right side of (11) in the last section with $\mathbf{F} \cdot \mathbf{n}$ replaced by $\mathbf{v} \cdot \mathbf{n}$, that is,

$$\operatorname{div} \mathbf{v}(P) = \lim_{d(T) \to 0} \frac{1}{V(T)} \iint_{S(T)} \mathbf{v} \cdot \mathbf{n}\, dA. \tag{2}$$

Hence *the divergence of the velocity vector* **v** *of a steady incompressible flow is the source intensity of the flow at the corresponding point.*

There are no sources in T if and only if div **v** is zero everywhere in T. Then for any closed surface S in T we have

$$\iint_S \mathbf{v} \cdot \mathbf{n}\, dA = 0. \qquad \blacksquare$$

EXAMPLE 2 **Modeling of Heat Flow. Heat or Diffusion Equation**

Physical experiments show that in a body, heat flows in the direction of decreasing temperature, and the rate of flow is proportional to the gradient of the temperature. This means that the velocity **v** of the heat flow in a body is of the form

$$\mathbf{v} = -K \operatorname{grad} U \tag{3}$$

where $U(x, y, z, t)$ is temperature, t is time, and K is called the *thermal conductivity* of the body; in ordinary physical circumstances K is a constant. Using this information, set up the mathematical model of heat flow, the so-called **heat equation** or **diffusion equation**.

Solution. Let T be a region in the body bounded by a surface S with outer unit normal vector **n** such that the divergence theorem applies. Then $\mathbf{v} \cdot \mathbf{n}$ is the component of **v** in the direction of **n**, and the amount of heat leaving T per unit time is

$$\iint_S \mathbf{v} \cdot \mathbf{n}\, dA.$$

This expression is obtained similarly to the corresponding surface integral in the last example. Using

$$\operatorname{div}(\operatorname{grad} U) = \nabla^2 U = U_{xx} + U_{yy} + U_{zz}$$

(the Laplacian; see (3) in Sec. 9.8), we have by the divergence theorem and (3)

$$\begin{aligned} \iint_S \mathbf{v} \cdot \mathbf{n}\, dA &= -K \iiint_T \operatorname{div}(\operatorname{grad} U)\, dx\, dy\, dz \\ &= -K \iiint_T \nabla^2 U\, dx\, dy\, dz. \end{aligned} \tag{4}$$

On the other hand, the total amount of heat H in T is

$$H = \iiint_T \sigma \rho U\, dx\, dy\, dz$$

where the constant σ is the specific heat of the material of the body and ρ is the density (= mass per unit volume) of the material. Hence the time rate of decrease of H is

$$-\frac{\partial H}{\partial t} = -\iiint_T \sigma \rho \frac{\partial U}{\partial t}\, dx\, dy\, dz$$

and this must be equal to the above amount of heat leaving T. From (4) we thus have

$$-\iiint_T \sigma \rho \frac{\partial U}{\partial t}\, dx\, dy\, dz = -K \iiint_T \nabla^2 U\, dx\, dy\, dz$$

or

$$\iiint_T \left(\sigma \rho \frac{\partial U}{\partial t} - K \nabla^2 U\right) dx\, dy\, dz = 0.$$

Since this holds for any region T in the body, the integrand (if continuous) must be zero everywhere; that is,

$$\frac{\partial U}{\partial t} = c^2 \nabla^2 U \qquad c^2 = \frac{K}{\sigma \rho} \tag{5}$$

where c^2 is called the *thermal diffusivity* of the material. This partial differential equation is called the **heat equation**. It is the fundamental equation for heat conduction. And our derivation is another impressive demonstration of the great importance of the divergence theorem. Methods for solving heat problems will be shown in Chap. 12.

The heat equation is also called the **diffusion equation** because it also models diffusion processes of motions of molecules tending to level off differences in density or pressure in gases or liquids.

If heat flow does not depend on time, it is called **steady-state heat flow**. Then $\partial U/\partial t = 0$, so that (5) reduces to *Laplace's equation* $\nabla^2 U = 0$. We met this equation in Secs. 9.7 and 9.8, and we shall now see that the divergence theorem adds basic insights into the nature of solutions of this equation. ■

Potential Theory. Harmonic Functions

The theory of solutions of **Laplace's equation**

$$\nabla^2 f = \frac{\partial^2 f}{\partial x^2} + \frac{\partial^2 f}{\partial y^2} + \frac{\partial^2 f}{\partial z^2} = 0 \tag{6}$$

is called **potential theory**. A solution of (6) with ***continuous*** second-order partial derivatives is called a **harmonic function**. That continuity is needed for application of the divergence theorem in potential theory, where the theorem plays a key role that we want to explore. Further details of potential theory follow in Chaps. 12 and 18.

EXAMPLE 3 **A Basic Property of Solutions of Laplace's Equation**

The integrands in the divergence theorem are div $\mathbf{F}$ and $\mathbf{F} \cdot \mathbf{n}$ (Sec. 10.7). If $\mathbf{F}$ is the gradient of a scalar function, say, $\mathbf{F} = \operatorname{grad} f$, then div $\mathbf{F} = \operatorname{div}(\operatorname{grad} f) = \nabla^2 f$; see (3), Sec. 9.8. Also, $\mathbf{F} \cdot \mathbf{n} = \mathbf{n} \cdot \mathbf{F} = \mathbf{n} \cdot \operatorname{grad} f$. This is the directional derivative of f in the outer normal direction of S, the boundary surface of the region T in the theorem. This derivative is called the (outer) **normal derivative** of f and is denoted by $\partial f/\partial n$. Thus the formula in the divergence theorem becomes

$$\iiint_T \nabla^2 f \, dV = \iint_S \frac{\partial f}{\partial n} \, dA. \tag{7}$$

This is the three-dimensional analog of (9) in Sec. 10.4. Because of the assumptions in the divergence theorem this gives the following result. ■

THEOREM 1 **A Basic Property of Harmonic Functions**

Let $f(x, y, z)$ be a harmonic function in some domain D is space. Let S be any piecewise smooth closed orientable surface in D whose entire region it encloses belongs to D. Then the integral of the normal derivative of f taken over S is zero. (For "piecewise smooth" see Sec. 10.5.)

EXAMPLE 4 **Green's Theorems**

Let f and g be scalar functions such that $F = f \operatorname{grad} g$ satisfies the assumptions of the divergence theorem in some region T. Then

$$\begin{aligned}
\operatorname{div} \mathbf{F} &= \operatorname{div} (f \operatorname{grad} g) \\
&= \operatorname{div}\left(\left[f\frac{\partial g}{\partial x}, f\frac{\partial g}{\partial y}, f\frac{\partial g}{\partial z}\right]\right) \\
&= \left(\frac{\partial f}{\partial x}\frac{\partial g}{\partial x} + f\frac{\partial^2 g}{\partial x^2}\right) + \left(\frac{\partial f}{\partial y}\frac{\partial g}{\partial y} + f\frac{\partial^2 g}{\partial y^2}\right) + \left(\frac{\partial f}{\partial z}\frac{\partial g}{\partial z} + f\frac{\partial^2 g}{\partial z^2}\right) \\
&= f\nabla^2 g + \operatorname{grad} f \cdot \operatorname{grad} g.
\end{aligned}$$

Also, since f is a scalar function,

$$\begin{aligned}
\mathbf{F} \cdot \mathbf{n} &= \mathbf{n} \cdot \mathbf{F} \\
&= \mathbf{n} \cdot (f \operatorname{grad} g) \\
&= (\mathbf{n} \cdot \operatorname{grad} g) f.
\end{aligned}$$

Now $\mathbf{n} \cdot \operatorname{grad} g$ is the directional derivative $\partial g/\partial n$ of g in the outer normal direction of S. Hence the formula in the divergence theorem becomes "**Green's first formula**"

$$\iiint_T (f\nabla^2 g + \operatorname{grad} f \cdot \operatorname{grad} g)\, dV = \iint_S f\frac{\partial g}{\partial n}\, dA. \tag{8}$$

Formula (8) together with the assumptions is known as the *first form of Green's theorem.*

Interchanging f and g we obtain a similar formula. Subtracting this formula from (8) we find

$$\iiint_T (f\nabla^2 g - g\nabla^2 f)\, dV = \iint_S \left(f\frac{\partial g}{\partial n} - g\frac{\partial f}{\partial n}\right) dA. \tag{9}$$

This formula is called **Green's second formula** or (together with the assumptions) the *second form of Green's theorem.* ■

EXAMPLE 5 **Uniqueness of Solutions of Laplace's Equation**

Let f be harmonic in a domain D and let f be zero everywhere on a piecewise smooth closed orientable surface S in D whose entire region T it encloses belongs to D. Then $\nabla^2 g$ is zero in T, and the surface integral in (8) is zero, so that (8) with $g = f$ gives

$$\iiint_T \operatorname{grad} f \cdot \operatorname{grad} f\, dV = \iiint_T |\operatorname{grad} f|^2\, dV = 0.$$

Since f is harmonic, $\operatorname{grad} f$ and thus $|\operatorname{grad} f|$ are continuous in T and on S, and since $|\operatorname{grad} f|$ is nonnegative, to make the integral over T zero, $\operatorname{grad} f$ must be the zero vector everywhere in T. Hence $f_x = f_y = f_z = 0$, and f is constant in T and, because of continuity, it is equal to its value 0 on S. This proves the following theorem.

THEOREM 2

Harmonic Functions

Let $f(x, y, z)$ be harmonic in some domain D and zero at every point of a piecewise smooth closed orientable surface S in D whose entire region T it encloses belongs to D. Then f is identically zero in T.

This theorem has an important consequence. Let f_1 and f_2 be functions that satisfy the assumptions of Theorem 1 and take on the same values on S. Then their difference $f_1 - f_2$ satisfies those assumptions and has the value 0 everywhere on S. Hence, Theorem 2 implies that

$$f_1 - f_2 = 0 \qquad \text{throughout} \qquad T,$$

and we have the following fundamental result.

THEOREM 3

Uniqueness Theorem for Laplace's Equation

Let T be a region that satisfies the assumptions of the divergence theorem, and let $f(x, y, z)$ be a harmonic function in a domain D that contains T and its boundary surface S. Then f is uniquely determined in T by its values on S.

The problem of determining a solution u of a partial differential equation in a region T such that u assumes given values on the boundary surface S of T is called the **Dirichlet problem.**[8] We may thus reformulate Theorem 3 as follows.

THEOREM 3*

Uniqueness Theorem for the Dirichlet Problem

If the assumptions in Theorem 3 are satisfied and the Dirichlet problem for the Laplace equation has a solution in T, then this solution is unique.

These theorems demonstrate the extreme importance of the divergence theorem in potential theory. ■

PROBLEM SET 10.8

1–6 VERIFICATIONS

1. **Harmonic functions.** Verify Theorem 1 for $f = 2z^2 - x^2 - y^2$ and S the surface of the box $0 \leqq x \leqq a$, $0 \leqq y \leqq b$, $0 \leqq z \leqq c$.
2. **Harmonic functions.** Verify Theorem 1 for $f = x^2 - y^2$ and the surface of the cylinder $x^2 + y^2 = 4$, $0 \leqq z \leqq h$.
3. **Green's first identity.** Verify (8) for $f = 4y^2$, $g = x^2$, S the surface of the "unit cube" $0 \leqq x \leqq 1$, $0 \leqq y \leqq 1$, $0 \leqq z \leqq 1$. What are the assumptions on f and g in (8)? Must f and g be harmonic?
4. **Green's first identity.** Verify (8) for $f = x$, $g = y^2 + z^2$, S the surface of the box $0 \leqq x \leqq 1$, $0 \leqq y \leqq 2$, $0 \leqq z \leqq 3$.

[8]PETER GUSTAV LEJEUNE DIRICHLET (1805–1859), German mathematician, studied in Paris under Cauchy and others and succeeded Gauss at Göttingen in 1855. He became known by his important research on Fourier series (he knew Fourier personally) and in number theory.

5. Green's second identity. Verify (9) for $f = 6y^2$, $g = 2x^2$, S the unit cube in Prob. 3.

6. Green's second identity. Verify (9) for $f = x^2$, $g = y^4$, S the unit cube in Prob. 3.

7–11 VOLUME

Use the divergence theorem, assuming that the assumptions on T and S are satisfied.

7. Show that a region T with boundary surface S has the volume

$$V = \iint_S x\, dy\, dz = \iint_S y\, dz\, dx = \iint_S z\, dx\, dy$$
$$= \frac{1}{3}\iint_S (x\, dy\, dz + y\, dz\, dx + z\, dx\, dy).$$

8. Cone. Using the third expression for v in Prob. 7, verify $V = \pi a^2 h/3$ for the volume of a circular cone of height h and radius of base a.

9. Ball. Find the volume under a hemisphere of radius a from in Prob. 7.

10. Volume. Show that a region T with boundary surface S has the volume

$$V = \frac{1}{3}\iint_S r \cos\phi\, dA$$

where r is the distance of a variable point $P: (x, y, z)$ on S from the origin O and ϕ is the angle between the directed line OP and the outer normal of S at P. Make a sketch. *Hint.* Use (2) in Sec. 10.7 with $\mathbf{F} = [x, y, z]$.

11. Ball. Find the volume of a ball of radius a from Prob. 10.

12. TEAM PROJECT. Divergence Theorem and Potential Theory. The importance of the divergence theorem in potential theory is obvious from (7)–(9) and Theorems 1–3. To emphasize it further, consider functions f and g that are harmonic in some domain D containing a region T with boundary surface S such that T satisfies the assumptions in the divergence theorem. Prove, and illustrate by examples, that then:

(a) $\displaystyle\iint_S g\frac{\partial g}{\partial n}\, dA = \iiint_T |\text{grad}\, g|^2\, dV.$

(b) If $\partial g/\partial n = 0$ on S, then g is constant in T.

(c) $\displaystyle\iint_S \left(f\frac{\partial g}{\partial n} - g\frac{\partial f}{\partial n}\right) dA = 0.$

(d) If $\partial f/\partial n = \partial g/\partial n$ on S, then $f = g + c$ in T, where c is a constant.

(e) The **Laplacian** can be represented ***independently of coordinate systems*** in the form

$$\nabla^2 f = \lim_{d(T)\to 0} \frac{1}{V(T)} \iint_{S(T)} \frac{\partial f}{\partial n}\, dA$$

where $d(T)$ is the maximum distance of the points of a region T bounded by $S(T)$ from the point at which the Laplacian is evaluated and $V(T)$ is the volume of T.

10.9 Stokes's Theorem

Let us review some of the material covered so far. Double integrals over a region in the plane can be transformed into line integrals over the boundary curve of that region and conversely, line integrals into double integrals. This important result is known as *Green's theorem in the plane* and was explained in Sec. 10.4. We also learned that we can transform triple integrals into surface integrals and vice versa, that is, surface integrals into triple integrals. This "big" theorem is called *Gauss's divergence theorem* and was shown in Sec. 10.7.

To complete our discussion on transforming integrals, we now introduce another "big" theorem that allows us to transform surface integrals into line integrals and conversely, line integrals into surface integrals. It is called **Stokes's Theorem**, and it generalizes Green's theorem in the plane (see Example 2 below for this immediate observation). Recall from Sec. 9.9 that

$$\text{(1)} \qquad \text{curl}\,\mathbf{F} = \begin{vmatrix} \mathbf{i} & \mathbf{j} & \mathbf{k} \\ \partial/\partial x & \partial/\partial y & \partial/\partial z \\ F_1 & F_2 & F_3 \end{vmatrix}$$

which we will need immediately.

THEOREM 1

Stokes's Theorem[9]
(Transformation Between Surface and Line Integrals)

Let S be a piecewise smooth[9] *oriented surface in space and let the boundary of S be a piecewise smooth simple closed curve C. Let* $\mathbf{F}(x, y, z)$ *be a continuous vector function that has continuous first partial derivatives in a domain in space containing S. Then*

$$\iint_S (\text{curl } \mathbf{F}) \cdot \mathbf{n} \, dA = \oint_C \mathbf{F} \cdot \mathbf{r}'(s) \, ds. \tag{2}$$

Here $\mathbf{n}$ *is a unit normal vector of S and, depending on* $\mathbf{n}$, *the integration around C is taken in the sense shown in Fig. 254. Furthermore,* $\mathbf{r}' = d\mathbf{r}/ds$ *is the unit tangent vector and s the arc length of C.*

In components, formula (2) becomes

$$\begin{aligned} &\iint_R \left[\left(\frac{\partial F_3}{\partial y} - \frac{\partial F_2}{\partial z}\right) N_1 + \left(\frac{\partial F_1}{\partial z} - \frac{\partial F_3}{\partial x}\right) N_2 + \left(\frac{\partial F_2}{\partial x} - \frac{\partial F_1}{\partial y}\right) N_3\right] du \, dv \\ &\qquad = \oint_{\bar{C}} (F_1 \, dx + F_2 \, dy + F_3 \, dz). \end{aligned} \tag{2*}$$

Here, $\mathbf{F} = [F_1, \ F_2, \ F_3]$, $\mathbf{N} = [N_1, \ N_2, \ N_3]$, $\mathbf{n} \, dA = \mathbf{N} \, du \, dv$, $\mathbf{r}' \, ds = [dx, \ dy, \ dz]$, *and R is the region with boundary curve* $\bar{C}$ *in the uv-plane corresponding to S represented by* $\mathbf{r}(u, v)$.

The proof follows after Example 1.

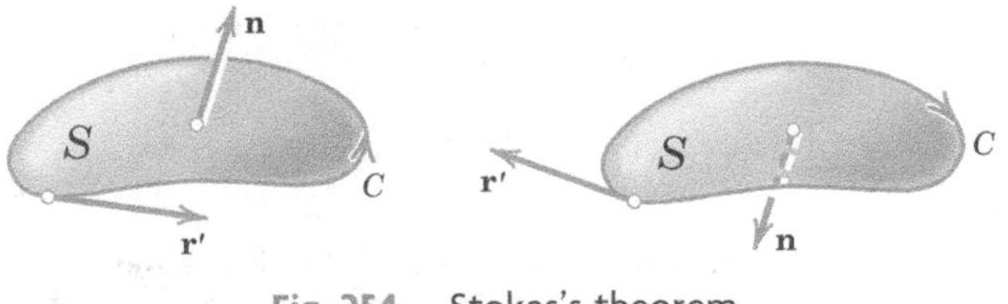

Fig. 254. Stokes's theorem

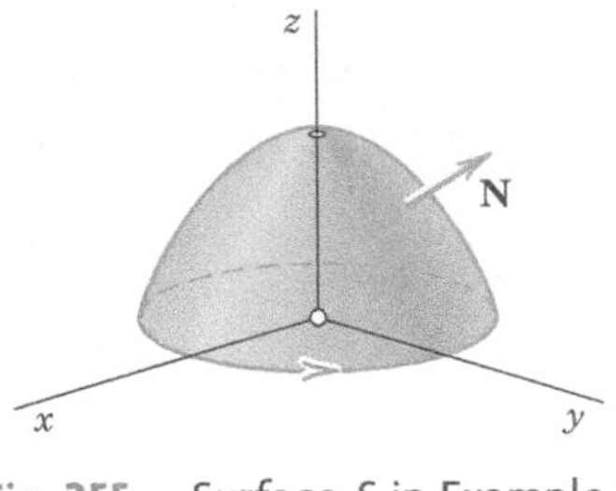

Fig. 255. Surface *S* in Example 1

EXAMPLE 1

Verification of Stokes's Theorem

Before we prove Stokes's theorem, let us first get used to it by verifying it for $\mathbf{F} = [y, z, x]$ and S the paraboloid (Fig. 255)

$$z = f(x, y) = 1 - (x^2 + y^2), \qquad z \geqq 0.$$

[9]Sir GEORGE GABRIEL STOKES (1819–1903), Irish mathematician and physicist who became a professor in Cambridge in 1849. He is also known for his important contribution to the theory of infinite series and to viscous flow (Navier–Stokes equations), geodesy, and optics.

"Piecewise smooth" curves and surfaces are defined in Secs. 10.1 and 10.5.

Solution. The curve C, oriented as in Fig. 255, is the circle $\mathbf{r}(s) = [\cos s, \sin s, 0]$. Its unit tangent vector is $\mathbf{r}'(s) = [-\sin s, \cos s, 0]$. The function $\mathbf{F} = [y, z, x]$ on C is $\mathbf{F}(\mathbf{r}(s)) = [\sin s, 0, \cos s]$. Hence

$$\oint_C \mathbf{F} \cdot d\mathbf{r} = \int_0^{2\pi} \mathbf{F}(\mathbf{r}(s)) \cdot \mathbf{r}'(s)\, ds = \int_0^{2\pi} [(\sin s)(-\sin s) + 0 + 0]\, ds = -\pi.$$

We now consider the surface integral. We have $F_1 = y$, $F_2 = z$, $F_3 = x$, so that in (2*) we obtain

$$\operatorname{curl} \mathbf{F} = \operatorname{curl} [F_1, \quad F_2, \quad F_3] = \operatorname{curl} [y, \quad z, \quad x] = [-1, \quad -1, \quad -1].$$

A normal vector of S is $\mathbf{N} = \operatorname{grad}(z - f(x, y)) = [2x, 2y, 1]$. Hence $(\operatorname{curl} \mathbf{F}) \cdot \mathbf{N} = -2x - 2y - 1$. Now $\mathbf{n}\, dA = \mathbf{N}\, dx\, dy$ (see (3*) in Sec. 10.6 with x, y instead of u, v). Using polar coordinates r, θ defined by $x = r\cos\theta$, $y = r\sin\theta$ and denoting the projection of S into the xy-plane by R, we thus obtain

$$\iint_S (\operatorname{curl} \mathbf{F}) \cdot \mathbf{n}\, dA = \iint_R (\operatorname{curl} \mathbf{F}) \cdot \mathbf{N}\, dx\, dy = \iint_R (-2x - 2y - 1)\, dx\, dy$$

$$= \int_{\theta=0}^{2\pi} \int_{r=0}^{1} (-2r(\cos\theta + \sin\theta) - 1) r\, dr\, d\theta$$

$$= \int_{\theta=0}^{2\pi} \left(-\frac{2}{3}(\cos\theta + \sin\theta) - \frac{1}{2}\right) d\theta = 0 + 0 - \frac{1}{2}(2\pi) = -\pi. \quad \blacksquare$$

PROOF We prove Stokes's theorem. Obviously, (2) holds if the integrals of each component on both sides of (2*) are equal; that is,

$$\iint_R \left(\frac{\partial F_1}{\partial z} N_2 - \frac{\partial F_1}{\partial y} N_3\right) du\, dv = \oint_{\bar{C}} F_1\, dx \tag{3}$$

$$\iint_R \left(-\frac{\partial F_2}{\partial z} N_1 + \frac{\partial F_2}{\partial x} N_3\right) du\, dv = \oint_{\bar{C}} F_2\, dy \tag{4}$$

$$\iint_R \left(\frac{\partial F_3}{\partial y} N_1 - \frac{\partial F_3}{\partial x} N_2\right) du\, dv = \oint_{\bar{C}} F_3\, dz. \tag{5}$$

We prove this first for a surface S that can be represented simultaneously in the forms

$$\text{(a)} \quad z = f(x, y), \qquad \text{(b)} \quad y = g(x, z), \qquad \text{(c)} \quad x = h(y, z). \tag{6}$$

We prove (3), using (6a). Setting $u = x$, $v = y$, we have from (6a)

$$\mathbf{r}(u, v) = \mathbf{r}(x, y) = [x, y, f(x, y)] = x\mathbf{i} + y\mathbf{j} + f\mathbf{k}$$

and in (2), Sec. 10.6, by direct calculation

$$\mathbf{N} = \mathbf{r}_u \times \mathbf{r}_v = \mathbf{r}_x \times \mathbf{r}_y = [-f_x, \quad -f_y, \quad 1] = -f_x\mathbf{i} - f_y\mathbf{j} + \mathbf{k}.$$

Note that $\mathbf{N}$ is an *upper* normal vector of S, since it has a *positive* z-component. Also, $R = S^*$, the projection of S into the xy-plane, with boundary curve $\overline{C} = C^*$ (Fig. 256). Hence the left side of (3) is

$$\iint_{S^*} \left[\frac{\partial F_1}{\partial z}(-f_y) - \frac{\partial F_1}{\partial y} \right] dx\, dy. \tag{7}$$

We now consider the right side of (3). We transform this line integral over $\overline{C} = C^*$ into a double integral over S^* by applying Green's theorem [formula (1) in Sec. 10.4 with $F_2 = 0$]. This gives

$$\oint_{C^*} F_1\, dx = \iint_{S^*} -\frac{\partial F_1}{\partial y}\, dx\, dy.$$

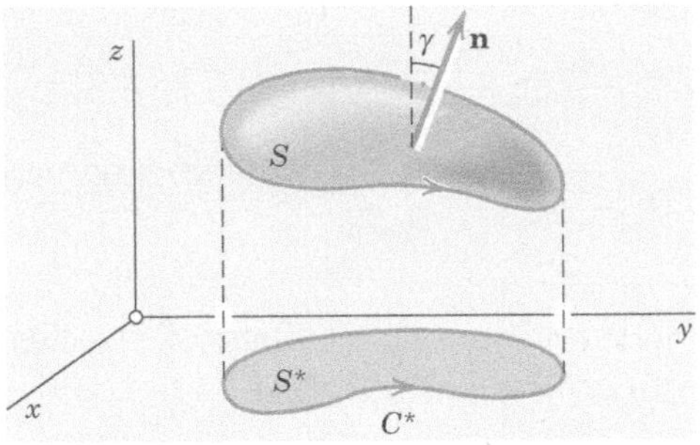

Fig. 256. Proof of Stokes's theorem

Here, $F_1 = F_1(x, y, f(x, y))$. Hence by the chain rule (see also Prob. 10 in Problem Set 9.6),

$$-\frac{\partial F_1(x, y, f(x, y))}{\partial y} = -\frac{\partial F_1(x, y, z)}{\partial y} - \frac{\partial F_1(x, y, z)}{\partial z}\frac{\partial f}{\partial y} \qquad [z = f(x, y)].$$

We see that the right side of this equals the integrand in (7). This proves (3). Relations (4) and (5) follow in the same way if we use (6b) and (6c), respectively. By addition we obtain (2*). This proves Stokes's theorem for a surface S that can be represented simultaneously in the forms (6a), (6b), (6c).

As in the proof of the divergence theorem, our result may be immediately extended to a surface S that can be decomposed into finitely many pieces, each of which is of the kind just considered. This covers most of the cases of practical interest. The proof in the case of a most general surface S satisfying the assumptions of the theorem would require a limit process; this is similar to the situation in the case of Green's theorem in Sec. 10.4. ■

EXAMPLE 2 Green's Theorem in the Plane as a Special Case of Stokes's Theorem

Let $\mathbf{F} = [F_1, F_2] = F_1\mathbf{i} + F_2\mathbf{j}$ be a vector function that is continuously differentiable in a domain in the xy-plane containing a simply connected bounded closed region S whose boundary C is a piecewise smooth simple closed curve. Then, according to (1),

$$(\text{curl } \mathbf{F}) \bullet \mathbf{n} = (\text{curl } \mathbf{F}) \bullet \mathbf{k} = \frac{\partial F_2}{\partial x} - \frac{\partial F_1}{\partial y}.$$

Hence the formula in Stokes's theorem now takes the form

$$\iint_S \left(\frac{\partial F_2}{\partial x} - \frac{\partial F_1}{\partial y}\right) dA = \oint_C (F_1\, dx + F_2\, dy).$$

This shows that Green's theorem in the plane (Sec. 10.4) is a special case of Stokes's theorem (which we needed in the proof of the latter!). ■

EXAMPLE 3 **Evaluation of a Line Integral by Stokes's Theorem**

Evaluate $\int_C \mathbf{F} \cdot \mathbf{r}'\, ds$, where C is the circle $x^2 + y^2 = 4$, $z = -3$, oriented counterclockwise as seen by a person standing at the origin, and, with respect to right-handed Cartesian coordinates,

$$\mathbf{F} = [y,\ \ xz^3,\ \ -zy^3] = y\mathbf{i} + xz^3\mathbf{j} - zy^3\mathbf{k}.$$

Solution. As a surface S bounded by C we can take the plane circular disk $x^2 + y^2 \leqq 4$ in the plane $z = -3$. Then $\mathbf{n}$ in Stokes's theorem points in the positive z-direction; thus $\mathbf{n} = \mathbf{k}$. Hence $(\text{curl } \mathbf{F}) \cdot \mathbf{n}$ is simply the component of curl $\mathbf{F}$ in the positive z-direction. Since $\mathbf{F}$ with $z = -3$ has the components $F_1 = y$, $F_2 = -27x$, $F_3 = 3y^3$, we thus obtain

$$(\text{curl } \mathbf{F}) \cdot \mathbf{n} = \frac{\partial F_2}{\partial x} - \frac{\partial F_1}{\partial y} = -27 - 1 = -28.$$

Hence the integral over S in Stokes's theorem equals -28 times the area 4π of the disk S. This yields the answer $-28 \cdot 4\pi = -112\pi \approx -352$. Confirm this by direct calculation, which involves somewhat more work. ■

EXAMPLE 4 **Physical Meaning of the Curl in Fluid Motion. Circulation**

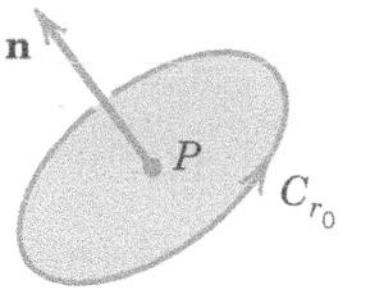

Fig. 257. Example 4

Let S_{r_0} be a circular disk of radius r_0 and center P bounded by the circle C_{r_0} (Fig. 257), and let $\mathbf{F}(Q) \equiv \mathbf{F}(x, y, z)$ be a continuously differentiable vector function in a domain containing S_{r_0}. Then by Stokes's theorem and the mean value theorem for surface integrals (see Sec. 10.6),

$$\oint_{C_{r_0}} \mathbf{F} \cdot \mathbf{r}'\, ds = \iint_{S_{r_0}} (\text{curl } \mathbf{F}) \cdot \mathbf{n}\, dA = (\text{curl } \mathbf{F}) \cdot \mathbf{n}(P^*) A_{r_0}$$

where A_{r_0} is the area of S_{r_0} and P^* is a suitable point of S_{r_0}. This may be written in the form

$$(\text{curl } \mathbf{F}) \cdot \mathbf{n}(P^*) = \frac{1}{A_{r_0}} \oint_{C_{r_0}} \mathbf{F} \cdot \mathbf{r}'\, ds.$$

In the case of a fluid motion with velocity vector $\mathbf{F} = \mathbf{v}$, the integral

$$\oint_{C_{r_0}} \mathbf{v} \cdot \mathbf{r}'\, ds$$

is called the **circulation** of the flow around C_{r_0}. It measures the extent to which the corresponding fluid motion is a rotation around the circle C_{r_0}. If we now let r_0 approach zero, we find

$$(8) \qquad (\text{curl } \mathbf{v}) \cdot \mathbf{n}(P) = \lim_{r_0 \to 0} \frac{1}{A_{r_0}} \oint_{C_{r_0}} \mathbf{v} \cdot \mathbf{r}'\, ds;$$

that is, the component of the curl in the positive normal direction can be regarded as the **specific circulation** (circulation per unit area) of the flow in the surface at the corresponding point. ■

EXAMPLE 5 **Work Done in the Displacement around a Closed Curve**

Find the work done by the force $\mathbf{F} = 2xy^3 \sin z\, \mathbf{i} + 3x^2y^2 \sin z\, \mathbf{j} + x^2y^3 \cos z\, \mathbf{k}$ in the displacement around the curve of intersection of the paraboloid $z = x^2 + y^2$ and the cylinder $(x - 1)^2 + y^2 = 1$.

Solution. This work is given by the line integral in Stokes's theorem. Now $\mathbf{F} = \text{grad} f$, where $f = x^2y^3 \sin z$ and $\text{curl}(\text{grad} f) = \mathbf{0}$ (see (2) in Sec. 9.9), so that $(\text{curl } \mathbf{F}) \cdot \mathbf{n} = 0$ and the work is 0 by Stokes's theorem. This agrees with the fact that the present field is conservative (definition in Sec. 9.7). ■

Stokes's Theorem Applied to Path Independence

We emphasized in Sec. 10.2 that the value of a line integral generally depends not only on the function to be integrated and on the two endpoints A and B of the path of integration C, but also on the particular choice of a path from A to B. In Theorem 3 of Sec. 10.2 we proved that if a line integral

$$\int_C \mathbf{F}(\mathbf{r}) \cdot d\mathbf{r} = \int_C (F_1\,dx + F_2\,dy + F_3\,dz) \tag{9}$$

(involving continuous F_1, F_2, F_3 that have continuous first partial derivatives) is path independent in a domain D, then curl $\mathbf{F} = \mathbf{0}$ in D. And we claimed in Sec. 10.2 that, conversely, curl $\mathbf{F} = \mathbf{0}$ everywhere in D implies path independence of (9) in D provided D is *simply connected.* A proof of this needs Stokes's theorem and can now be given as follows.

Let C be any closed path in D. Since D is simply connected, we can find a surface S in D bounded by C. Stokes's theorem applies and gives

$$\oint_C (F_1\,dx + F_2\,dy + F_3\,dz) = \oint_C \mathbf{F} \cdot \mathbf{r}'\,ds = \iint_S (\text{curl } \mathbf{F}) \cdot \mathbf{n}\,dA$$

for proper direction on C and normal vector $\mathbf{n}$ on S. Since curl $\mathbf{F} = \mathbf{0}$ in D, the surface integral and hence the line integral are zero. This and Theorem 2 of Sec. 10.2 imply that the integral (9) is path independent in D. This completes the proof. ■

PROBLEM SET 10.9

1–10 DIRECT INTEGRATION OF SURFACE INTEGRALS

Evaluate the surface integral $\iint_S (\text{curl } \mathbf{F}) \cdot \mathbf{n}\,dA$ directly for the given $\mathbf{F}$ and S.

1. $\mathbf{F} = [z^2, -x^2, 0]$, S the rectangle with vertices $(0, 0, 0)$, $(1, 0, 0)$, $(0, 4, 4)$, $(1, 4, 4)$

2. $\mathbf{F} = [-13 \sin y, 3 \sinh z, x]$, S the rectangle with vertices $(0, 0, 2)$, $(4, 0, 2)$, $(4, \pi/2, 2)$, $(0, \pi/2, 2)$

3. $\mathbf{F} = [e^{-z}, e^{-z} \cos y, e^{-z} \sin y]$, S: $z = y^2/2$, $-1 \leqq x \leqq 1$, $0 \leqq y \leqq 1$

4. $\mathbf{F}$ as in Prob. 1, $z = xy$ $(0 \leqq x \leqq 1,\ 0 \leqq y \leqq 4)$. Compare with Prob. 1.

5. $\mathbf{F} = [z^2, \frac{3}{2}x, 0]$, S: $0 \leqq x \leqq a$, $0 \leqq y \leqq a$, $z = 1$

6. $\mathbf{F} = [y^3, -x^3, 0]$, S: $x^2 + y^2 \leqq 1$, $z = 0$

7. $\mathbf{F} = [e^y, e^z, e^x]$, S: $z = x^2$ $(0 \leqq x \leqq 2,\ 0 \leqq y \leqq 1)$

8. $\mathbf{F} = [z^2, x^2, y^2]$, S: $z = \sqrt{x^2 + y^2}$, $y \geqq 0$, $0 \leqq z \leqq h$

9. Verify Stokes's theorem for $\mathbf{F}$ and S in Prob. 5.

10. Verify Stokes's theorem for $\mathbf{F}$ and S in Prob. 6.

11. Stokes's theorem not applicable. Evaluate $\oint_C \mathbf{F} \cdot \mathbf{r}'\,ds$, $\mathbf{F} = (x^2 + y^2)^{-1}[-y, x]$, C: $x^2 + y^2 = 1$, $z = 0$, oriented clockwise. Why can Stokes's theorem not be applied? What (false) result would it give?

12. WRITING PROJECT. Grad, Div, Curl in Connection with Integrals. Make a list of ideas and results on this topic in this chapter. See whether you can rearrange or combine parts of your material. Then subdivide the material into 3–5 portions and work out the details of each portion. Include no proofs but simple typical examples of your own that lead to a better understanding of the material.

13–20 EVALUATION OF $\oint_C \mathbf{F} \cdot \mathbf{r}'\,ds$

Calculate this line integral by Stokes's theorem for the given $\mathbf{F}$ and C. Assume the Cartesian coordinates to be right-handed and the z-component of the surface normal to be nonnegative.

13. $\mathbf{F} = [-5y, 4x, z]$, C the circle $x^2 + y^2 = 16$, $z = 4$

14. $\mathbf{F} = [z^3, x^3, y^3]$, C the circle $x = 2,\ \ y^2 + z^2 = 9$

15. $\mathbf{F} = [y^2, x^2, z + x]$ around the triangle with vertices $(0, 0, 0)$, $(1, 0, 0)$, $(1, 1, 0)$

16. $\mathbf{F} = [e^y, 0, e^x]$, C as in Prob. 15

17. $\mathbf{F} = [0, z^3, 0]$, C the boundary curve of the cylinder $x^2 + y^2 = 1, x \geqq 0,\ \ y \geqq 0,\ \ 0 \leqq z \leqq 1$

18. $\mathbf{F} = [-y, 2z, 0]$, C the boundary curve of $y^2 + z^2 = 4$, $z \geqq 0,\ \ 0 \leqq x \leqq h$

19. $\mathbf{F} = [z, e^z, 0]$, C the boundary curve of the portion of the cone $z = \sqrt{x^2 + y^2},\ \ x \geqq 0,\ \ y \geqq 0,\ \ 0 \leqq z \leqq 1$

20. $\mathbf{F} = [0,\ \cos x, 0]$, C the boundary curve of $y^2 + z^2 = 4$, $y \geqq 0,\ \ z \geqq 0,\ \ 0 \leqq x \leqq \pi$

CHAPTER 10 REVIEW QUESTIONS AND PROBLEMS

1. State from memory how to evaluate a line integral. A surface integral.

2. What is path independence of a line integral? What is its physical meaning and importance?

3. What line integrals can be converted to surface integrals and conversely? How?

4. What surface integrals can be converted to volume integrals? How?

5. What role did the gradient play in this chapter? The curl? State the definitions from memory.

6. What are typical applications of line integrals? Of surface integrals?

7. Where did orientation of a surface play a role? Explain.

8. State the divergence theorem from memory. Give applications.

9. In some line and surface integrals we started from vector functions, but integrands were scalar functions. How and why did we proceed in this way?

10. State Laplace's equation. Explain its physical importance. Summarize our discussion on harmonic functions.

11–20 LINE INTEGRALS (WORK INTEGRALS)

Evaluate $\int_C \mathbf{F}(\mathbf{r}) \cdot d\mathbf{r}$ for given $\mathbf{F}$ and C by the method that seems most suitable. Remember that if $\mathbf{F}$ is a force, the integral gives the work done in the displacement along C. Show details.

11. $\mathbf{F} = [2x^2, -4y^2]$, C the straight-line segment from $(4, 2)$ to $(-6, 10)$

12. $\mathbf{F} = [y \cos xy, x \cos xy, e^z]$, C the straight-line segment from $(\pi, 1, 0)$ to $(\frac{1}{2}, \pi, 1)$

13. $\mathbf{F} = [y^2, 2xy + 5 \sin x, 0]$, C the boundary of $0 \leqq x \leqq \pi/2,\ \ 0 \leqq y \leqq 2,\ \ z = 0$

14. $\mathbf{F} = [-y^3, x^3 + e^{-y}, 0]$, C the circle $x^2 + y^2 = 25$, $z = 2$

15. $\mathbf{F} = [x^3, e^{2y}, e^{-4z}]$, C: $x^2 + 9y^2 = 9,\ \ z = x^2$

16. $\mathbf{F} = [x^2, y^2, y^2x]$, C the helix $\mathbf{r} = [2 \cos t, 2 \sin t, 3t]$ from $(2, 0, 0)$ to $(-2, 0, 3\pi)$

17. $\mathbf{F} = [9z, 5x, 3y]$, C the ellipse $x^2 + y^2 = 9$, $z = x + 2$

18. $\mathbf{F} = [\sin \pi y, \cos \pi x, \sin \pi x]$, C the boundary curve of $0 \leqq x \leqq 1,\ \ 0 \leqq y \leqq 2,\ \ z = x$

19. $\mathbf{F} = [z, 2y, x]$, C the helix $r = [\cos t, \sin t, t]$ from $(1, 0, 0)$ to $(1, 0, 2\pi)$

20. $\mathbf{F} = [ze^{xz}, 2 \sinh 2y, xe^{xz}]$, C the parabola $y = x$, $z = x^2,\ -1 \leqq x \leqq 1$

21–25 DOUBLE INTEGRALS, CENTER OF GRAVITY

Find the coordinates $\bar{x}, \bar{y}$ of the center of gravity of a mass of density $f(x, y)$ in the region R. Sketch R, show details.

21. $f = xy$, R the triangle with vertices $(0, 0)$, $(2, 0)$, $(2, 2)$

22. $f = x^2 + y^2$, R: $x^2 + y^2 \leqq a^2,\ \ y \geqq 0$

23. $f = x^2$, R: $-1 \leqq x \leqq 2,\ \ x^2 \leqq y \leqq x + 2$. Why is $\bar{x} > 0$?

24. $f = 1$, R: $0 \leqq y \leqq 1 - x^4$

25. $f = ky$, $k > 0$, arbitrary, $0 \leqq y \leqq 1 - x^2$, $0 \leqq x \leqq 1$

26. Why are $\bar{x}$ and $\bar{y}$ in Prob. 25 independent of k?

27–35 SURFACE INTEGRALS $\iint_S \mathbf{F} \cdot \mathbf{n}\, dA$. DIVERGENCE THEOREM

Evaluate the integral diectly or, if possible, by the divergence theorem. Show details.

27. $\mathbf{F} = [ax, by, cz]$, S the sphere $x^2 + y^2 + z^2 = 36$

28. $\mathbf{F} = [x + y^2, y + z^2, z + x^2]$, S the ellipsoid with semi-axes of lengths a, b, c

29. $\mathbf{F} = [y + z, 20y, 2z^3]$, S the surface of $0 \leqq x \leqq 2$, $0 \leqq y \leqq 1,\ \ 0 \leqq z \leqq y$

30. $\mathbf{F} = [1, 1, 1]$, S: $x^2 + y^2 + 4z^2 = 4,\ \ z \geqq 0$

31. $\mathbf{F} = [e^x, e^y, e^z]$, S the surface of the box $|x| \leqq 1$, $|y| \leqq 1,\ \ |z| \leqq 1$

32. $\mathbf{F} = [y^2, x^2, z^2]$, S the portion of the paraboloid $z = x^2 + y^2$, $z \leqq 9$

33. $\mathbf{F} = [y^2, x^2, z^2]$, S: $r = [u, u^2, v]$, $0 \leqq u \leqq 2$, $-2 \leqq v \leqq 2$

34. $\mathbf{F} = [x, xy, z]$, S the boundary of $x^2 + y^2 \leqq 1$, $0 \leqq z \leqq 5$

35. $\mathbf{F} = [x + z, y + z, x + y]$, S the sphere of radius 3 with center 0

SUMMARY OF CHAPTER 10

Vector Integral Calculus. Integral Theorems

Chapter 9 extended *differential* calculus to vectors, that is, to vector functions $\mathbf{v}(x, y, z)$ or $\mathbf{v}(t)$. Similarly, Chapter 10 extends *integral* calculus to vector functions. This involves *line integrals* (Sec. 10.1), *double integrals* (Sec. 10.3), *surface integrals* (Sec. 10.6), and *triple integrals* (Sec. 10.7) and the three "big" theorems for transforming these integrals into one another, the theorems of Green (Sec. 10.4), Gauss (Sec. 10.7), and Stokes (Sec. 10.9).

The analog of the definite integral of calculus is the **line integral** (Sec. 10.1)

$$\text{(1)} \qquad \int_C \mathbf{F}(\mathbf{r}) \cdot d\mathbf{r} = \int_C (F_1\,dx + F_2\,dy + F_3\,dz) = \int_a^b \mathbf{F}(\mathbf{r}(t)) \cdot \frac{d\mathbf{r}}{dt}\,dt$$

where C: $\mathbf{r}(t) = [x(t), y(t), z(t)] = x(t)\mathbf{i} + y(t)\mathbf{j} + z(t)\mathbf{k}$ $(a \leqq t \leqq b)$ is a curve in space (or in the plane). Physically, (1) may represent the work done by a (variable) force in a displacement. Other kinds of line integrals and their applications are also discussed in Sec. 10.1.

Independence of path of a line integral in a domain D means that the integral of a given function over any path C with endpoints P and Q has the same value for all paths from P to Q that lie in D; here P and Q are fixed. An integral (1) is independent of path in D if and only if the differential form $F_1\,dx + F_2\,dy + F_3\,dz$ with continuous F_1, F_2, F_3 is **exact** in D (Sec. 10.2). Also, if curl $\mathbf{F} = \mathbf{0}$, where $\mathbf{F} = [F_1, F_2, F_3]$, has continuous first partial derivatives in a *simply connected* domain D, then the integral (1) is independent of path in D (Sec. 10.2).

Integral Theorems. The formula of **Green's theorem in the plane** (Sec. 10.4)

$$\text{(2)} \qquad \iint_R \left(\frac{\partial F_2}{\partial x} - \frac{\partial F_1}{\partial y}\right) dx\,dy = \oint_C (F_1\,dx + F_2\,dy)$$

transforms **double integrals** over a region R in the xy-plane into line integrals over the boundary curve C of R and conversely. For other forms of (2) see Sec. 10.4.

Similarly, the formula of the **divergence theorem of Gauss** (Sec. 10.7)

$$\text{(3)} \qquad \iiint_T \operatorname{div} \mathbf{F}\,dV = \iint_S \mathbf{F} \cdot \mathbf{n}\,dA$$

transforms **triple integrals** over a region T in space into surface integrals over the boundary surface S of T, and conversely. Formula (3) implies **Green's formulas**

$$\iiint_T (f\nabla^2 g + \nabla f \cdot \nabla g)\, dV = \iint_S f \frac{\partial g}{\partial n}\, dA, \tag{4}$$

$$\iiint_T (f\nabla^2 g - g\nabla^2 f)\, dV = \iint_S \left(f \frac{\partial g}{\partial n} - g \frac{\partial f}{\partial n}\right) dA. \tag{5}$$

Finally, the formula of **Stokes's theorem** (Sec. 10.9)

$$\iint_S (\text{curl}\ \mathbf{F}) \cdot \mathbf{n}\, dA = \oint_C \mathbf{F} \cdot \mathbf{r}'(s)\, ds \tag{6}$$

transforms **surface integrals** over a surface S into line integrals over the boundary curve C of S and conversely.

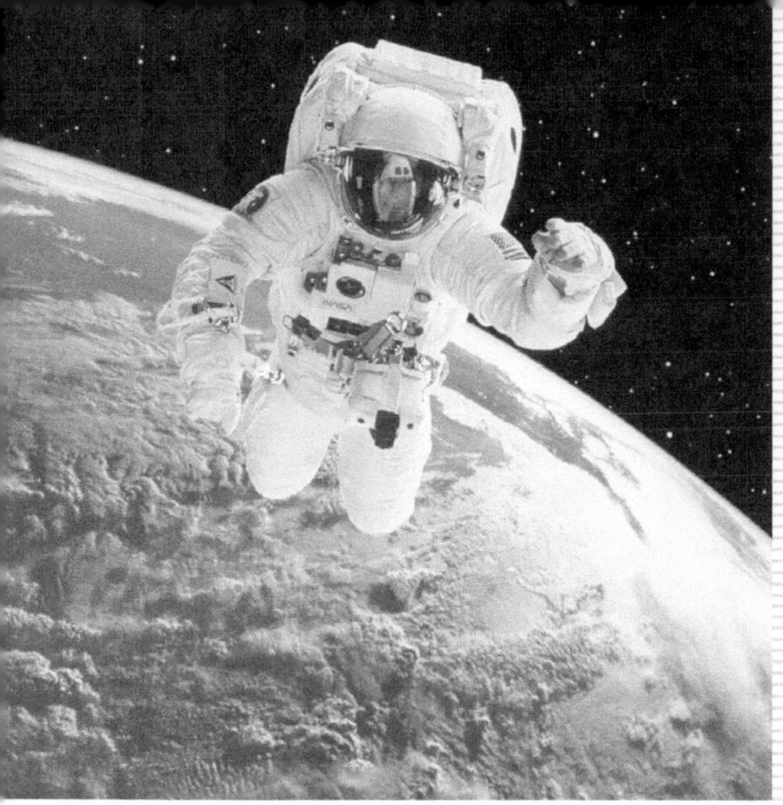

PART C

Fourier Analysis. Partial Differential Equations (PDEs)

CHAPTER 11 Fourier Analysis

CHAPTER 12 Partial Differential Equations (PDEs)

Chapter 11 and Chapter 12 are directly related to each other in that **Fourier analysis** has its most important applications in modeling and solving partial differential equations (PDEs) related to boundary and initial value problems of mechanics, heat flow, electrostatics, and other fields. However, the study of PDEs is a study in its own right. Indeed, PDEs are the subject of much ongoing research.

Fourier analysis allows us to model periodic phenomena which appear frequently in engineering and elsewhere—think of rotating parts of machines, alternating electric currents or the motion of planets. Related period functions may be complicated. Now, the ingeneous idea of Fourier analysis is to represent complicated functions in terms of simple periodic functions, namely cosines and sines. The representations will be infinite series called **Fourier series.**[1] This idea can be generalized to more general series (see Sec. 11.5) and to integral representations (see Sec. 11.7).

The discovery of Fourier series had a huge impetus on applied mathematics as well as on mathematics as a whole. Indeed, its influence on the concept of a function, on integration theory, on convergence theory, and other theories of mathematics has been substantial (see [GenRef7] in App. 1).

Chapter 12 deals with the most important partial differential equations (PDEs) of physics and engineering, such as the wave equation, the heat equation, and the Laplace equation. These equations can model a vibrating string/membrane, temperatures on a bar, and electrostatic potentials, respectively. PDEs are very important in many areas of physics and engineering and have many more applications than ODEs.

---

[1]JEAN-BAPTISTE JOSEPH FOURIER (1768–1830), French physicist and mathematician, lived and taught in Paris, accompanied Napoléon in the Egyptian War, and was later made prefect of Grenoble. The beginnings on Fourier series can be found in works by Euler and by Daniel Bernoulli, but it was Fourier who employed them in a systematic and general manner in his main work, *Théorie analytique de la chaleur* (*Analytic Theory of Heat,* Paris, 1822), in which he developed the theory of heat conduction (heat equation; see Sec. 12.5), making these series a most important tool in applied mathematics.

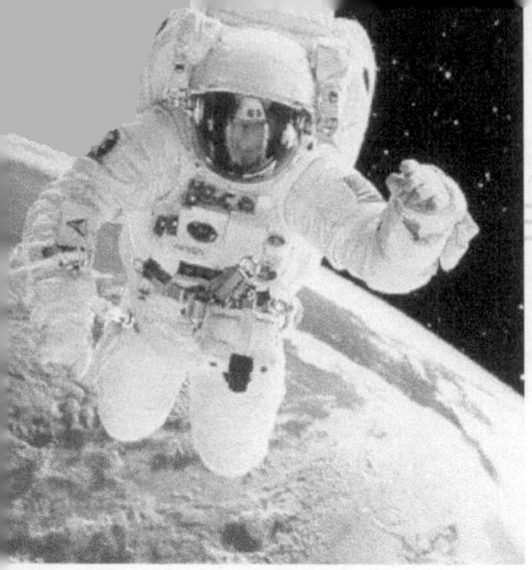

CHAPTER 11

Fourier Analysis

This chapter on Fourier analysis covers three broad areas: Fourier series in Secs. 11.1–11.4, more general orthonormal series called Sturm–Liouville expansions in Secs. 11.5 and 11.6 and Fourier integrals and transforms in Secs. 11.7–11.9.

The central starting point of Fourier analysis is **Fourier series**. They are infinite series designed to represent general periodic functions in terms of simple ones, namely, cosines and sines. This trigonometric system is ***orthogonal***, allowing the computation of the coefficients of the Fourier series by use of the well-known Euler formulas, as shown in Sec. 11.1. Fourier series are very important to the engineer and physicist because they allow the solution of ODEs in connection with forced oscillations (Sec. 11.3) and the approximation of periodic functions (Sec. 11.4). Moreover, applications of Fourier analysis to PDEs are given in Chap. 12. Fourier series are, in a certain sense, more universal than the familiar Taylor series in calculus because many *discontinuous* periodic functions that come up in applications can be developed in Fourier series but do not have Taylor series expansions.

The underlying idea of the Fourier series can be extended in two important ways. We can replace the trigonometric system by other families of orthogonal functions, e.g., Bessel functions and obtain the **Sturm–Liouville expansions**. Note that related Secs. 11.5 and 11.6 used to be part of Chap. 5 but, for greater readability and logical coherence, are now part of Chap. 11. The second expansion is applying Fourier series to nonperiodic phenomena and obtaining Fourier integrals and Fourier transforms. Both extensions have important applications to solving PDEs as will be shown in Chap. 12.

In a digital age, the ***discrete Fourier transform*** plays an important role. Signals, such as voice or music, are sampled and analyzed for frequencies. An important algorithm, in this context, is the ***fast Fourier transform***. This is discussed in Sec. 11.9.

Note that the two extensions of Fourier series are independent of each other and may be studied in the order suggested in this chapter or by studying Fourier integrals and transforms first and then Sturm–Liouville expansions.

Prerequisite: Elementary integral calculus (needed for Fourier coefficients).
Sections that may be omitted in a shorter course: 11.4–11.9.
References and Answers to Problems: App. 1 Part C, App. 2.

11.1 Fourier Series

Fourier series are infinite series that represent periodic functions in terms of cosines and sines. As such, Fourier series are of greatest importance to the engineer and applied mathematician. To define Fourier series, we first need some background material. A function $f(x)$ is called a **periodic function** if $f(x)$ is defined for all real x, except

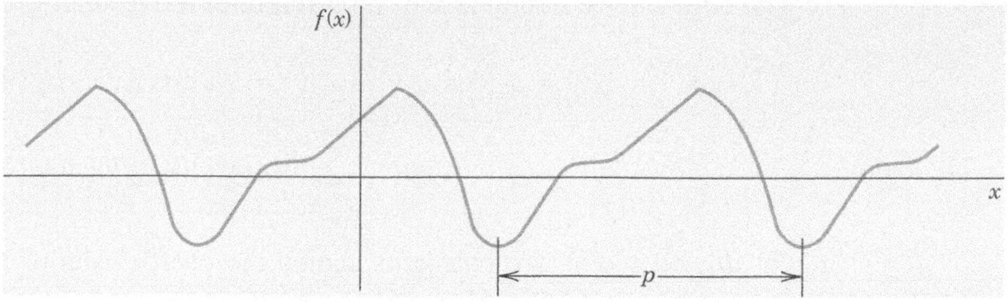

Fig. 258. Periodic function of period p

possibly at some points, and if there is some positive number p, called a **period** of $f(x)$, such that

$$f(x + p) = f(x) \qquad \text{for all } x. \tag{1}$$

(The function $f(x) = \tan x$ is a periodic function that is not defined for all real x but undefined for some points (more precisely, countably many points), that is $x = \pm\pi/2, \pm 3\pi/2, \cdots$.)

The graph of a periodic function has the characteristic that it can be obtained by periodic repetition of its graph in any interval of length p (Fig. 258).

The smallest positive period is often called the *fundamental period.* (See Probs. 2–4.)

Familiar periodic functions are the cosine, sine, tangent, and cotangent. Examples of functions that are not periodic are $x, x^2, x^3, e^x, \cosh x$, and $\ln x$, to mention just a few.

If $f(x)$ has period p, it also has the period $2p$ because (1) implies $f(x + 2p) = f([x + p] + p) = f(x + p) = f(x)$, etc.; thus for any integer $n = 1, 2, 3, \cdots,$

$$f(x + np) = f(x) \qquad \text{for all } x. \tag{2}$$

Furthermore if $f(x)$ and $g(x)$ have period p, then $af(x) + bg(x)$ with any constants a and b also has the period p.

Our problem in the first few sections of this chapter will be the representation of various ***functions f(x) of period 2π*** in terms of the simple functions

$$1, \quad \cos x, \quad \sin x, \quad \cos 2x, \quad \sin 2x, \cdots, \quad \cos nx, \quad \sin nx, \cdots. \tag{3}$$

All these functions have the period 2π. They form the so-called **trigonometric system.** Figure 259 shows the first few of them (except for the constant 1, which is periodic with any period).

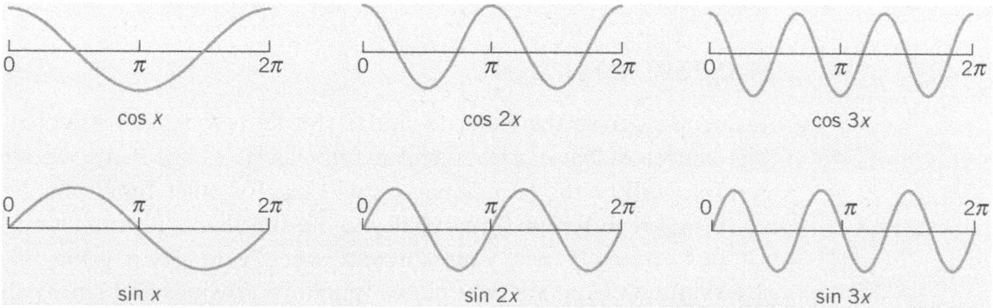

Fig. 259. Cosine and sine functions having the period 2π (the first few members of the trigonometric system (3), except for the constant 1)

The series to be obtained will be a **trigonometric series**, that is, a series of the form

$$a_0 + a_1 \cos x + b_1 \sin x + a_2 \cos 2x + b_2 \sin 2x + \cdots$$

$$= a_0 + \sum_{n=1}^{\infty} (a_n \cos nx + b_n \sin nx). \tag{4}$$

$a_0, a_1, b_1, a_2, b_2, \cdots$ are constants, called the **coefficients** of the series. We see that each term has the period 2π. Hence *if the coefficients are such that the series converges, its sum will be a function of period* 2π.

Expressions such as (4) will occur frequently in Fourier analysis. To compare the expression on the right with that on the left, simply write the terms in the summation. Convergence of one side implies convergence of the other and the sums will be the same.

Now suppose that $f(x)$ is a given function of period 2π and is such that it can be **represented** by a series (4), that is, (4) converges and, moreover, has the sum $f(x)$. Then, using the equality sign, we write

$$f(x) = a_0 + \sum_{n=1}^{\infty} (a_n \cos nx + b_n \sin nx) \tag{5}$$

and call (5) the **Fourier series** of $f(x)$. We shall prove that in this case the coefficients of (5) are the so-called **Fourier coefficients** of $f(x)$, given by the **Euler formulas**

(6)

$$\text{(0)} \quad a_0 = \frac{1}{2\pi} \int_{-\pi}^{\pi} f(x)\, dx$$

$$\text{(a)} \quad a_n = \frac{1}{\pi} \int_{-\pi}^{\pi} f(x) \cos nx\, dx \qquad n = 1, 2, \cdots$$

$$\text{(b)} \quad b_n = \frac{1}{\pi} \int_{-\pi}^{\pi} f(x) \sin nx\, dx \qquad n = 1, 2, \cdots.$$

The name "Fourier series" is sometimes also used in the exceptional case that (5) with coefficients (6) does not converge or does not have the sum $f(x)$—this may happen but is merely of theoretical interest. (For Euler see footnote 4 in Sec. 2.5.)

A Basic Example

Before we derive the Euler formulas (6), let us consider how (5) and (6) are applied in this important basic example. Be fully alert, as the way we approach and solve this example will be the technique you will use for other functions. Note that the integration is a little bit different from what you are familiar with in calculus because of the n. Do not just routinely use your software but try to get a good understanding and make observations: How are continuous functions (cosines and sines) able to represent a given discontinuous function? How does the quality of the approximation increase if you take more and more terms of the series? Why are the approximating functions, called the

partial sums of the series, in this example always zero at 0 and π? Why is the factor $1/n$ (obtained in the integration) important?

EXAMPLE 1 Periodic Rectangular Wave (Fig. 260)

Find the Fourier coefficients of the periodic function $f(x)$ in Fig. 260. The formula is

$$f(x) = \begin{cases} -k & \text{if} \quad -\pi < x < 0 \\ k & \text{if} \quad 0 < x < \pi \end{cases} \qquad \text{and} \qquad f(x + 2\pi) = f(x). \tag{7}$$

Functions of this kind occur as external forces acting on mechanical systems, electromotive forces in electric circuits, etc. (The value of $f(x)$ at a single point does not affect the integral; hence we can leave $f(x)$ undefined at $x = 0$ and $x = \pm\pi$.)

Solution. From (6.0) we obtain $a_0 = 0$. This can also be seen without integration, since the area under the curve of $f(x)$ between $-\pi$ and π (taken with a minus sign where $f(x)$ is negative) is zero. From (6a) we obtain the coefficients $a_1, a_2, \cdots$ of the cosine terms. Since $f(x)$ is given by two expressions, the integrals from $-\pi$ to π split into two integrals:

$$a_n = \frac{1}{\pi}\int_{-\pi}^{\pi} f(x) \cos nx\, dx = \frac{1}{\pi}\left[\int_{-\pi}^{0} (-k) \cos nx\, dx + \int_{0}^{\pi} k \cos nx\, dx\right]$$

$$= \frac{1}{\pi}\left[-k \frac{\sin nx}{n}\Big|_{-\pi}^{0} + k \frac{\sin nx}{n}\Big|_{0}^{\pi}\right] = 0$$

because $\sin nx = 0$ at $-\pi$, 0, and π for all $n = 1, 2, \cdots$. We see that all these cosine coefficients are zero. That is, the Fourier series of (7) has no cosine terms, just sine terms, it is a **Fourier sine series** with coefficients $b_1, b_2, \cdots$ obtained from (6b);

$$b_n = \frac{1}{\pi}\int_{-\pi}^{\pi} f(x) \sin nx\, dx = \frac{1}{\pi}\left[\int_{-\pi}^{0} (-k) \sin nx\, dx + \int_{0}^{\pi} k \sin nx\, dx\right]$$

$$= \frac{1}{\pi}\left[k \frac{\cos nx}{n}\Big|_{-\pi}^{0} - k \frac{\cos nx}{n}\Big|_{0}^{\pi}\right].$$

Since $\cos(-\alpha) = \cos \alpha$ and $\cos 0 = 1$, this yields

$$b_n = \frac{k}{n\pi}[\cos 0 - \cos(-n\pi) - \cos n\pi + \cos 0] = \frac{2k}{n\pi}(1 - \cos n\pi).$$

Now, $\cos \pi = -1$, $\cos 2\pi = 1$, $\cos 3\pi = -1$, etc.; in general,

$$\cos n\pi = \begin{cases} -1 & \text{for odd } n, \\ 1 & \text{for even } n, \end{cases} \qquad \text{and thus} \qquad 1 - \cos n\pi = \begin{cases} 2 & \text{for odd } n, \\ 0 & \text{for even } n. \end{cases}$$

Hence the Fourier coefficients b_n of our function are

$$b_1 = \frac{4k}{\pi}, \qquad b_2 = 0, \qquad b_3 = \frac{4k}{3\pi}, \qquad b_4 = 0, \qquad b_5 = \frac{4k}{5\pi}, \cdots.$$

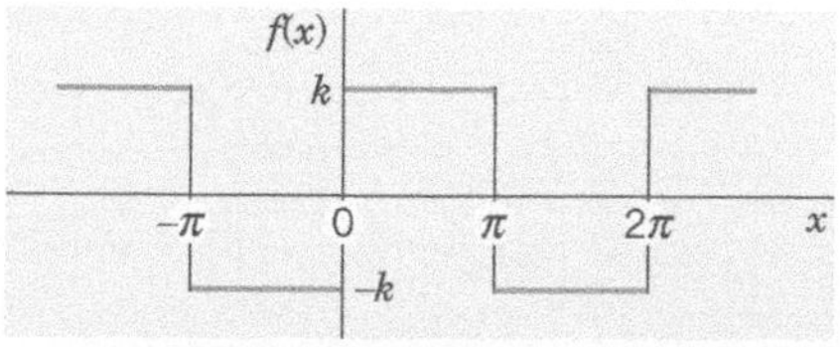

Fig. 260. Given function $f(x)$ (Periodic reactangular wave)

Since the a_n are zero, the Fourier series of $f(x)$ is

$$\frac{4k}{\pi}\left(\sin x + \tfrac{1}{3}\sin 3x + \tfrac{1}{5}\sin 5x + \cdots\right). \tag{8}$$

The partial sums are

$$S_1 = \frac{4k}{\pi}\sin x, \qquad S_2 = \frac{4k}{\pi}\left(\sin x + \frac{1}{3}\sin 3x\right). \qquad \text{etc.}$$

Their graphs in Fig. 261 seem to indicate that the series is convergent and has the sum $f(x)$, the given function. We notice that at $x = 0$ and $x = \pi$, the points of discontinuity of $f(x)$, all partial sums have the value zero, the arithmetic mean of the limits $-k$ and k of our function, at these points. This is typical.

Furthermore, assuming that $f(x)$ is the sum of the series and setting $x = \pi/2$, we have

$$f\left(\frac{\pi}{2}\right) = k = \frac{4k}{\pi}\left(1 - \frac{1}{3} + \frac{1}{5} - + \cdots\right).$$

Thus

$$1 - \frac{1}{3} + \frac{1}{5} - \frac{1}{7} + - \cdots = \frac{\pi}{4}.$$

This is a famous result obtained by Leibniz in 1673 from geometric considerations. It illustrates that the values of various series with constant terms can be obtained by evaluating Fourier series at specific points. ■

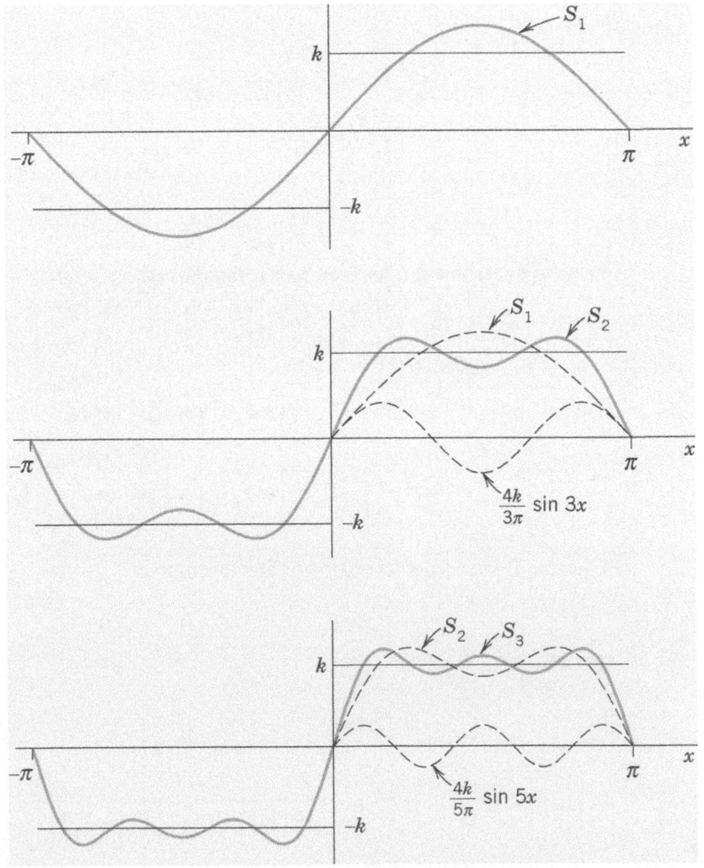

Fig. 261. First three partial sums of the corresponding Fourier series

Derivation of the Euler Formulas (6)

The key to the Euler formulas (6) is the **orthogonality** of (3), a concept of basic importance, as follows. Here we generalize the concept of inner product (Sec. 9.3) to functions.

THEOREM 1

Orthogonality of the Trigonometric System (3)

The trigonometric system (3) *is orthogonal on the interval* $-\pi \leqq x \leqq \pi$ (hence also on $0 \leqq x \leqq 2\pi$ or any other interval of length 2π because of periodicity); *that is, the integral of the product of any two functions in* (3) *over that interval is* 0, *so that for any integers n and m,*

$$\text{(9)}\qquad \begin{aligned} &\textbf{(a)}\quad \int_{-\pi}^{\pi} \cos nx \cos mx \, dx = 0 \qquad (n \neq m)\\ &\textbf{(b)}\quad \int_{-\pi}^{\pi} \sin nx \sin mx \, dx = 0 \qquad (n \neq m)\\ &\textbf{(c)}\quad \int_{-\pi}^{\pi} \sin nx \cos mx \, dx = 0 \qquad (n \neq m \text{ or } n = m). \end{aligned}$$

PROOF This follows simply by transforming the integrands trigonometrically from products into sums. In (9a) and (9b), by (11) in App. A3.1,

$$\int_{-\pi}^{\pi} \cos nx \cos mx \, dx = \frac{1}{2}\int_{-\pi}^{\pi} \cos (n + m)x \, dx + \frac{1}{2}\int_{-\pi}^{\pi} \cos (n - m)x \, dx$$

$$\int_{-\pi}^{\pi} \sin nx \sin mx \, dx = \frac{1}{2}\int_{-\pi}^{\pi} \cos (n - m)x \, dx - \frac{1}{2}\int_{-\pi}^{\pi} \cos (n + m)x \, dx.$$

Since $m \neq n$ (integer!), the integrals on the right are all 0. Similarly, in (9c), for all integer m and n (without exception; do you see why?)

$$\int_{-\pi}^{\pi} \sin nx \cos mx \, dx = \frac{1}{2}\int_{-\pi}^{\pi} \sin (n + m)x \, dx + \frac{1}{2}\int_{-\pi}^{\pi} \sin (n - m)x \, dx = 0 + 0. \qquad \blacksquare$$

Application of Theorem 1 to the Fourier Series (5)

We prove (6.0). Integrating on both sides of (5) from $-\pi$ to π, we get

$$\int_{-\pi}^{\pi} f(x) \, dx = \int_{-\pi}^{\pi} \left[a_0 + \sum_{n=1}^{\infty} (a_n \cos nx + b_n \sin nx) \right] dx.$$

We now assume that termwise integration is allowed. (We shall say in the proof of Theorem 2 when this is true.) Then we obtain

$$\int_{-\pi}^{\pi} f(x) \, dx = a_0 \int_{-\pi}^{\pi} dx + \sum_{n=1}^{\infty} \left(a_n \int_{-\pi}^{\pi} \cos nx \, dx + b_n \int_{-\pi}^{\pi} \sin nx \, dx \right).$$

The first term on the right equals $2\pi a_0$. Integration shows that all the other integrals are 0. Hence division by 2π gives (6.0).

We prove (6a). Multiplying (5) on both sides by $\cos mx$ with any ***fixed*** positive integer m and integrating from $-\pi$ to π, we have

$$(10) \qquad \int_{-\pi}^{\pi} f(x)\cos mx\,dx = \int_{-\pi}^{\pi}\left[a_0 + \sum_{n=1}^{\infty}(a_n \cos nx + b_n \sin nx)\right]\cos mx\,dx.$$

We now integrate term by term. Then on the right we obtain an integral of $a_0 \cos mx$, which is 0; an integral of $a_n \cos nx \cos mx$, which is $a_m\pi$ for $n = m$ and 0 for $n \neq m$ by (9a); and an integral of $b_n \sin nx \cos mx$, which is 0 for all n and m by (9c). Hence the right side of (10) equals $a_m\pi$. Division by π gives (6a) (with m instead of n).

We finally prove (6b). Multiplying (5) on both sides by $\sin mx$ with any ***fixed*** positive integer m and integrating from $-\pi$ to π, we get

$$(11) \qquad \int_{-\pi}^{\pi} f(x)\sin mx\,dx = \int_{-\pi}^{\pi}\left[a_0 + \sum_{n=1}^{\infty}(a_n \cos nx + b_n \sin nx)\right]\sin mx\,dx.$$

Integrating term by term, we obtain on the right an integral of $a_0 \sin mx$, which is 0; an integral of $a_n \cos nx \sin mx$, which is 0 by (9c); and an integral of $b_n \sin nx \sin mx$, which is $b_m\pi$ if $n = m$ and 0 if $n \neq m$, by (9b). This implies (6b) (with n denoted by m). This completes the proof of the Euler formulas (6) for the Fourier coefficients. ■

Convergence and Sum of a Fourier Series

The class of functions that can be represented by Fourier series is surprisingly large and general. Sufficient conditions valid in most applications are as follows.

THEOREM 2

Representation by a Fourier Series

Let $f(x)$ be periodic with period 2π and piecewise continuous (see Sec. 6.1) *in the interval $-\pi \leqq x \leqq \pi$. Furthermore, let $f(x)$ have a left-hand derivative and a right-hand derivative at each point of that interval. Then the Fourier series* (5) *of $f(x)$* [with coefficients (6)] *converges. Its sum is $f(x)$, except at points x_0 where $f(x)$ is discontinuous. There the sum of the series is the average of the left- and right-hand limits*[2] *of $f(x)$ at x_0.*

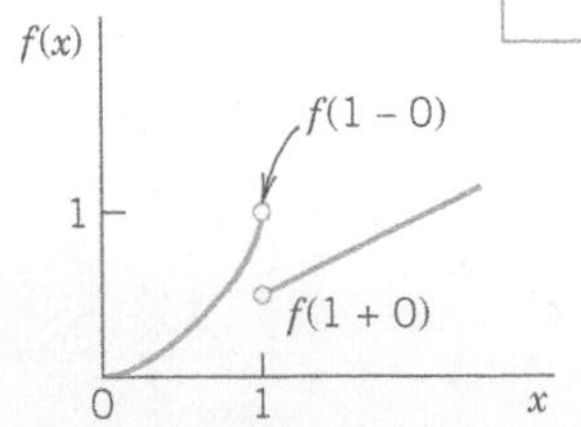

Fig. 262. Left- and right-hand limits

$f(1 - 0) = 1,$

$f(1 + 0) = \frac{1}{2}$

of the function

$$f(x) = \begin{cases} x^2 & \text{if } x < 1 \\ x/2 & \text{if } x \geqq 1 \end{cases}$$

[2] The **left-hand limit** of $f(x)$ at x_0 is defined as the limit of $f(x)$ as x approaches x_0 from the left and is commonly denoted by $f(x_0 - 0)$. Thus

$$f(x_0 - 0) = \lim_{h \to 0} f(x_0 - h) \text{ as } h \to 0 \text{ through positive values.}$$

The **right-hand limit** is denoted by $f(x_0 + 0)$ and

$$f(x_0 + 0) = \lim_{h \to 0} f(x_0 + h) \text{ as } h \to 0 \text{ through positive values.}$$

The **left-** and **right-hand derivatives** of $f(x)$ at x_0 are defined as the limits of

$$\frac{f(x_0 - h) - f(x_0 - 0)}{-h} \quad \text{and} \quad \frac{f(x_0 + h) - f(x_0 + 0)}{-h},$$

respectively, as $h \to 0$ through positive values. Of course if $f(x)$ is continuous at x_0, the last term in both numerators is simply $f(x_0)$.

PROOF We prove convergence, but only for a continuous function $f(x)$ having continuous first and second derivatives. And we do not prove that the sum of the series is $f(x)$ because these proofs are much more advanced; see, for instance, Ref. [C12] listed in App. 1. Integrating (6a) by parts, we obtain

$$a_n = \frac{1}{\pi}\int_{-\pi}^{\pi} f(x)\cos nx\,dx = \frac{f(x)\sin nx}{n\pi}\bigg|_{-\pi}^{\pi} - \frac{1}{n\pi}\int_{-\pi}^{\pi} f'(x)\sin nx\,dx.$$

The first term on the right is zero. Another integration by parts gives

$$a_n = \frac{f'(x)\cos nx}{n^2\pi}\bigg|_{-\pi}^{\pi} - \frac{1}{n^2\pi}\int_{-\pi}^{\pi} f''(x)\cos nx\,dx.$$

The first term on the right is zero because of the periodicity and continuity of $f'(x)$. Since f'' is continuous in the interval of integration, we have

$$|f''(x)| < M$$

for an appropriate constant M. Furthermore, $|\cos nx| \leqq 1$. It follows that

$$|a_n| = \frac{1}{n^2\pi}\left|\int_{-\pi}^{\pi} f''(x)\cos nx\,dx\right| < \frac{1}{n^2\pi}\int_{-\pi}^{\pi} M\,dx = \frac{2M}{n^2}.$$

Similarly, $|b_n| < 2\,M/n^2$ for all n. Hence the absolute value of each term of the Fourier series of $f(x)$ is at most equal to the corresponding term of the series

$$|a_0| + 2M\left(1 + 1 + \frac{1}{2^2} + \frac{1}{2^2} + \frac{1}{3^2} + \frac{1}{3^2} + \cdots\right)$$

which is convergent. Hence that Fourier series converges and the proof is complete. (Readers already familiar with uniform convergence will see that, by the Weierstrass test in Sec. 15.5, under our present assumptions the Fourier series converges uniformly, and our derivation of (6) by integrating term by term is then justified by Theorem 3 of Sec. 15.5.) ■

EXAMPLE 2 **Convergence at a Jump as Indicated in Theorem 2**

The rectangular wave in Example 1 has a jump at $x = 0$. Its left-hand limit there is $-k$ and its right-hand limit is k (Fig. 261). Hence the average of these limits is 0. The Fourier series (8) of the wave does indeed converge to this value when $x = 0$ because then all its terms are 0. Similarly for the other jumps. This is in agreement with Theorem 2. ■

Summary. A Fourier series of a given function $f(x)$ of period 2π is a series of the form (5) with coefficients given by the Euler formulas (6). Theorem 2 gives conditions that are sufficient for this series to converge and at each x to have the value $f(x)$, except at discontinuities of $f(x)$, where the series equals the arithmetic mean of the left-hand and right-hand limits of $f(x)$ at that point.

PROBLEM SET 11.1

1–5 PERIOD, FUNDAMENTAL PERIOD

The *fundamental period* is the smallest positive period. Find it for

1. $\cos x$, $\sin x$, $\cos 2x$, $\sin 2x$, $\cos \pi x$, $\sin \pi x$, $\cos 2\pi x$, $\sin 2\pi x$
2. $\cos nx$, $\sin nx$, $\cos \dfrac{2\pi x}{k}$, $\sin \dfrac{2\pi x}{k}$, $\cos \dfrac{2\pi n x}{k}$, $\sin \dfrac{2\pi n x}{k}$
3. If $f(x)$ and $g(x)$ have period p, show that $h(x) = af(x) + bg(x)$ (a, b, constant) has the period p. Thus all functions of period p form a **vector space**.
4. **Change of scale.** If $f(x)$ has period p, show that $f(ax)$, $a \neq 0$, and $f(x/b)$, $b \neq 0$, are periodic functions of x of periods p/a and bp, respectively. Give examples.
5. Show that $f =$ const is periodic with any period but has no fundamental period.

6–10 GRAPHS OF 2π–PERIODIC FUNCTIONS

Sketch or graph $f(x)$ which for $-\pi < x < \pi$ is given as follows.

6. $f(x) = |x|$
7. $f(x) = |\sin x|$, $f(x) = \sin |x|$
8. $f(x) = e^{-|x|}$, $f(x) = |e^{-x}|$
9. $f(x) = \begin{cases} x & \text{if } -\pi < x < 0 \\ \pi - x & \text{if } 0 < x < \pi \end{cases}$
10. $f(x) = \begin{cases} -\cos^2 x & \text{if } -\pi < x < 0 \\ \cos^2 x & \text{if } 0 < x < \pi \end{cases}$
11. **Calculus review.** Review integration techniques for integrals as they are likely to arise from the Euler formulas, for instance, definite integrals of $x \cos nx$, $x^2 \sin nx$, $e^{-2x} \cos nx$, etc.

12–21 FOURIER SERIES

Find the Fourier series of the given function $f(x)$, which is assumed to have the period 2π. Show the details of your work. Sketch or graph the partial sums up to that including $\cos 5x$ and $\sin 5x$.

12. $f(x)$ in Prob. 6
13. $f(x)$ in Prob. 9
14. $f(x) = x^2$ $(-\pi < x < \pi)$
15. $f(x) = x^2$ $(0 < x < 2\pi)$
16.

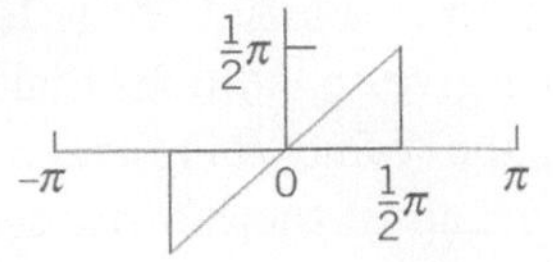

17.

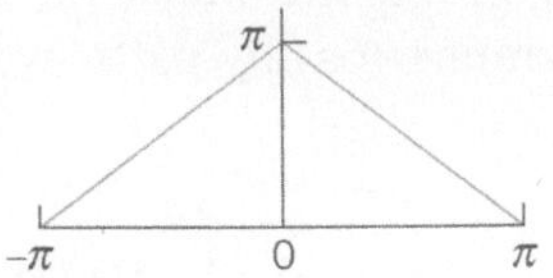

18.

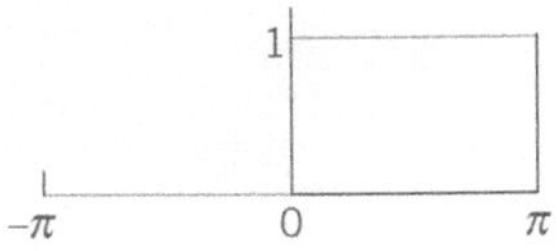

19.

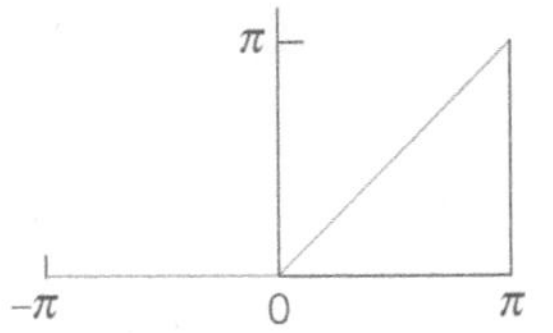

20.

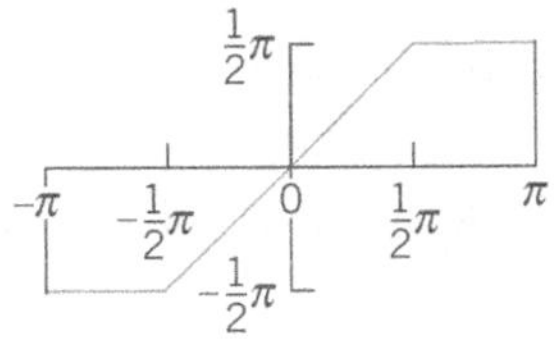

21.

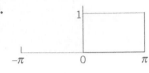

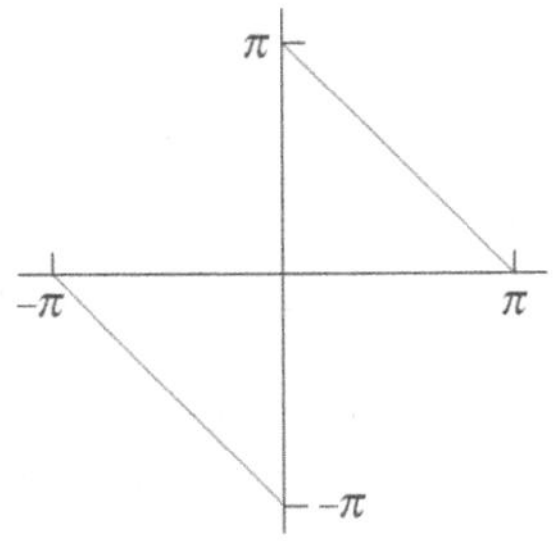

22. **CAS EXPERIMENT. Graphing.** Write a program for graphing partial sums of the following series. Guess from the graph what $f(x)$ the series may represent. Confirm or disprove your guess by using the Euler formulas.

    (a) $2(\sin x + \frac{1}{3}\sin 3x + \frac{1}{5}\sin 5x + \cdots)$
    $\quad - 2(\frac{1}{2}\sin 2x + \frac{1}{4}\sin 4x + \frac{1}{6}\sin 6x \cdots)$

    (b) $\dfrac{1}{2} + \dfrac{4}{\pi^2}\left(\cos x + \dfrac{1}{9}\cos 3x + \dfrac{1}{25}\cos 5x + \cdots\right)$

    (c) $\frac{2}{3}\pi^2 + 4(\cos x - \frac{1}{4}\cos 2x + \frac{1}{9}\cos 3x - \frac{1}{16}\cos 4x + - \cdots)$

23. **Discontinuities.** Verify the last statement in Theorem 2 for the discontinuities of $f(x)$ in Prob. 21.
24. **CAS EXPERIMENT. Orthogonality.** Integrate and graph the integral of the product $\cos mx \cos nx$ (with various integer m and n of your choice) from $-a$ to a as a function of a and conclude orthogonality of $\cos mx$

and cos nx ($m \neq n$) for $a = \pi$ from the graph. For what m and n will you get orthogonality for $a = \pi/2, \pi/3, \pi/4$? Other a? Extend the experiment to cos mx sin nx and sin mx sin nx.

25. CAS EXPERIMENT. Order of Fourier Coefficients. The order seems to be $1/n$ if f is discontinous, and $1/n^2$ if f is continuous but $f' = df/dx$ is discontinuous, $1/n^3$ if f and f' are continuous but f'' is discontinuous, etc. Try to verify this for examples. Try to prove it by integrating the Euler formulas by parts. What is the practical significance of this?

11.2 Arbitrary Period. Even and Odd Functions. Half-Range Expansions

We now expand our initial basic discussion of Fourier series.

Orientation. This section concerns three topics:

1. Transition from period 2π to any period $2L$, for the function f, simply by a transformation of scale on the x-axis.
2. Simplifications. Only cosine terms if f is even ("Fourier cosine series"). Only sine terms if f is odd ("Fourier sine series").
3. Expansion of f given for $0 \leqq x \leqq L$ in two Fourier series, one having only cosine terms and the other only sine terms ("half-range expansions").

1. From Period 2π to Any Period $p = 2L$

Clearly, periodic functions in applications may have any period, not just 2π as in the last section (chosen to have simple formulas). The notation $p = 2L$ for the period is practical because L will be a length of a violin string in Sec. 12.2, of a rod in heat conduction in Sec. 12.5, and so on.

The transition from period 2π to be period $p = 2L$ is effected by a suitable change of scale, as follows. Let $f(x)$ have period $p = 2L$. Then we can introduce a new variable v such that $f(x)$, as a function of v, has period 2π. If we set

$$\text{(1)} \qquad \text{(a)} \quad x = \frac{p}{2\pi}v, \qquad \text{so that} \qquad \text{(b)} \quad v = \frac{2\pi}{p}x = \frac{\pi}{L}x$$

then $v = \pm\pi$ corresponds to $x = \pm L$. This means that f, as a function of v, has period 2π and, therefore, a Fourier series of the form

$$\text{(2)} \qquad f(x) = f\left(\frac{L}{\pi}v\right) = a_0 + \sum_{n=1}^{\infty} (a_n \cos nv + b_n \sin nv)$$

with coefficients obtained from (6) in the last section

$$\text{(3)} \qquad \begin{aligned} a_0 &= \frac{1}{2\pi}\int_{-\pi}^{\pi} f\left(\frac{L}{\pi}v\right) dv, \qquad a_n = \frac{1}{\pi}\int_{-\pi}^{\pi} f\left(\frac{L}{\pi}v\right) \cos nv \, dv, \\ b_n &= \frac{1}{\pi}\int_{-\pi}^{\pi} f\left(\frac{L}{\pi}v\right) \sin nv \, dv. \end{aligned}$$

We could use these formulas directly, but the change to x simplifies calculations. Since

$$(4) \qquad v = \frac{\pi}{L}x, \qquad \text{we have} \qquad dv = \frac{\pi}{L}\,dx$$

and we integrate over x from $-L$ to L. Consequently, we obtain for a function $f(x)$ of period $2L$ the Fourier series

$$\textbf{(5)} \qquad f(x) = a_0 + \sum_{n=1}^{\infty}\left(a_n \cos\frac{n\pi}{L}x + b_n \sin\frac{n\pi}{L}x\right)$$

with the **Fourier coefficients** of $f(x)$ given by the **Euler formulas** (π/L in dx cancels $1/\pi$ in (3))

$$\textbf{(6)} \qquad \begin{aligned} &\textbf{(0)} \quad a_0 = \frac{1}{2L}\int_{-L}^{L} f(x)\,dx \\ &\textbf{(a)} \quad a_n = \frac{1}{L}\int_{-L}^{L} f(x)\cos\frac{n\pi x}{L}\,dx \qquad n = 1, 2, \cdots \\ &\textbf{(b)} \quad b_n = \frac{1}{L}\int_{-L}^{L} f(x)\sin\frac{n\pi x}{L}\,dx \qquad n = 1, 2, \cdots . \end{aligned}$$

Just as in Sec. 11.1, we continue to call (5) with any coefficients a **trigonometric series**. And we can integrate from 0 to $2L$ or over any other interval of length $p = 2L$.

EXAMPLE 1 Periodic Rectangular Wave

Find the Fourier series of the function (Fig. 263)

$$f(x) = \begin{cases} 0 & \text{if} \quad -2 < x < -1 \\ k & \text{if} \quad -1 < x < \ \ 1 \\ 0 & \text{if} \quad \ \ 1 < x < \ \ 2 \end{cases} \qquad p = 2L = 4, \quad L = 2.$$

Solution. From (6.0) we obtain $a_0 = k/2$ (verify!). From (6a) we obtain

$$a_n = \frac{1}{2}\int_{-2}^{2} f(x)\cos\frac{n\pi x}{2}\,dx = \frac{1}{2}\int_{-1}^{1} k\cos\frac{n\pi x}{2}\,dx = \frac{2k}{n\pi}\sin\frac{n\pi}{2}.$$

Thus $a_n = 0$ if n is even and

$$a_n = 2k/n\pi \quad \text{if} \quad n = 1, 5, 9, \cdots, \qquad a_n = -2k/n\pi \quad \text{if} \quad n = 3, 7, 11, \cdots .$$

From (6b) we find that $b_n = 0$ for $n = 1, 2, \cdots$. Hence the Fourier series is a **Fourier cosine series** (that is, it has no sine terms)

$$f(x) = \frac{k}{2} + \frac{2k}{\pi}\left(\cos\frac{\pi}{2}x - \frac{1}{3}\cos\frac{3\pi}{2}x + \frac{1}{5}\cos\frac{5\pi}{2}x - + \cdots\right). \qquad ■$$

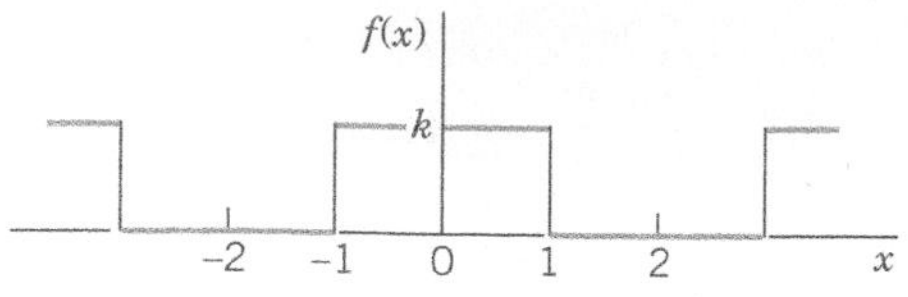

Fig. 263. Example 1

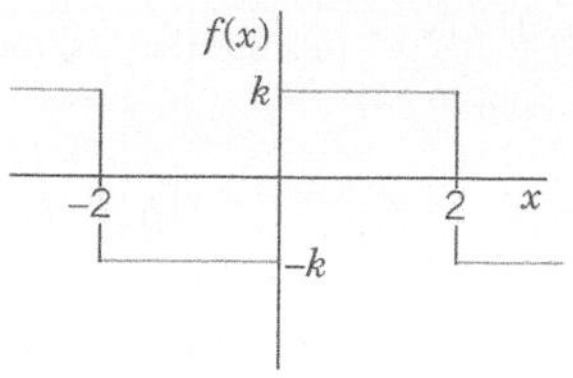

Fig. 264. Example 2

EXAMPLE 2 Periodic Rectangular Wave. Change of Scale

Find the Fourier series of the function (Fig. 264)

$$f(x) = \begin{cases} -k & \text{if} \quad -2 < x < 0 \\ k & \text{if} \quad 0 < x < 2 \end{cases} \qquad p = 2L = 4, \qquad L = 2.$$

Solution. Since $L = 2$, we have in (3) $v = \pi x/2$ and obtain from (8) in Sec. 11.1 with v instead of x, that is,

$$g(v) = \frac{4k}{\pi}\left(\sin v + \frac{1}{3}\sin 3v + \frac{1}{5}\sin 5v + \cdots\right)$$

the present Fourier series

$$f(x) = \frac{4k}{\pi}\left(\sin\frac{\pi}{2}x + \frac{1}{3}\sin\frac{3\pi}{2}x + \frac{1}{5}\sin\frac{5\pi}{2}x + \cdots\right).$$

Confirm this by using (6) and integrating. ■

EXAMPLE 3 Half-Wave Rectifier

A sinusoidal voltage $E \sin \omega t$, where t is time, is passed through a half-wave rectifier that clips the negative portion of the wave (Fig. 265). Find the Fourier series of the resulting periodic function

$$u(t) = \begin{cases} 0 & \text{if} \quad -L < t < 0, \\ E\sin\omega t & \text{if} \quad 0 < t < L \end{cases} \qquad p = 2L = \frac{2\pi}{\omega}, \qquad L = \frac{\pi}{\omega}.$$

Solution. Since $u = 0$ when $-L < t < 0$, we obtain from (6.0), with t instead of x,

$$a_0 = \frac{\omega}{2\pi}\int_0^{\pi/\omega} E\sin\omega t\, dt = \frac{E}{\pi}$$

and from (6a), by using formula (11) in App. A3.1 with $x = \omega t$ and $y = n\omega t$,

$$a_n = \frac{\omega}{\pi}\int_0^{\pi/\omega} E\sin\omega t\cos n\omega t\, dt = \frac{\omega E}{2\pi}\int_0^{\pi/\omega} [\sin(1+n)\,\omega t + \sin(1-n)\,\omega t]\, dt.$$

If $n = 1$, the integral on the right is zero, and if $n = 2, 3, \cdots$, we readily obtain

$$a_n = \frac{\omega E}{2\pi}\left[-\frac{\cos(1+n)\,\omega t}{(1+n)\,\omega} - \frac{\cos(1-n)\,\omega t}{(1-n)\,\omega}\right]_0^{\pi/\omega}$$

$$= \frac{E}{2\pi}\left(\frac{-\cos(1+n)\pi + 1}{1+n} + \frac{-\cos(1-n)\pi + 1}{1-n}\right).$$

If n is odd, this is equal to zero, and for even n we have

$$a_n = \frac{E}{2\pi}\left(\frac{2}{1+n} + \frac{2}{1-n}\right) = -\frac{2E}{(n-1)(n+1)\pi} \qquad (n = 2, 4, \cdots).$$

In a similar fashion we find from (6b) that $b_1 = E/2$ and $b_n = 0$ for $n = 2, 3, \cdots$. Consequently,

$$u(t) = \frac{E}{\pi} + \frac{E}{2}\sin \omega t - \frac{2E}{\pi}\left(\frac{1}{1 \cdot 3}\cos 2\omega t + \frac{1}{3 \cdot 5}\cos 4\omega t + \cdots\right).$$

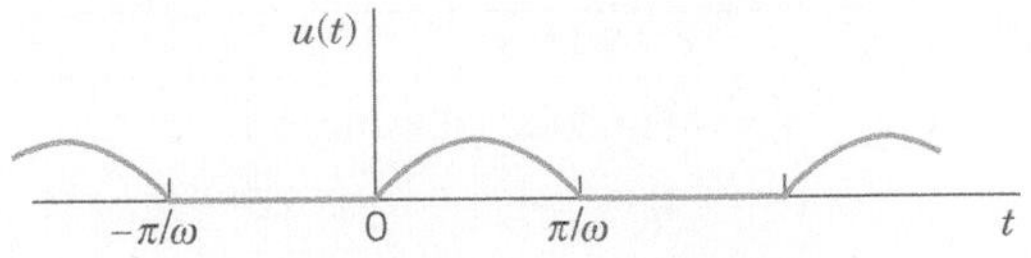

Fig. 265. Half-wave rectifier

2. Simplifications: Even and Odd Functions

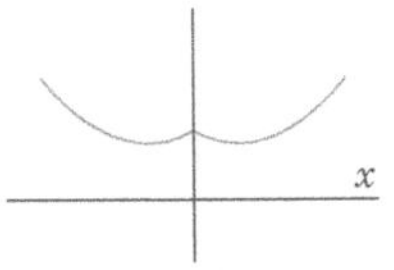

Fig. 266. Even function

If $f(x)$ is an **even function**, that is, $f(-x) = f(x)$ (see Fig. 266), its Fourier series (5) reduces to a **Fourier cosine series**

(5*)
$$f(x) = a_0 + \sum_{n=1}^{\infty} a_n \cos \frac{n\pi}{L} x \qquad (f \text{ even})$$

with coefficients (note: integration from 0 to L only!)

(6*)
$$a_0 = \frac{1}{L}\int_0^L f(x)\, dx, \qquad a_n = \frac{2}{L}\int_0^L f(x) \cos \frac{n\pi x}{L}\, dx, \qquad n = 1, 2, \cdots.$$

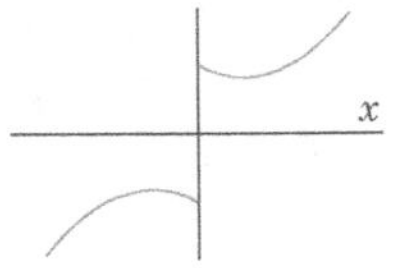

Fig. 267. Odd function

If $f(x)$ is an **odd function**, that is, $f(-x) = -f(x)$ (see Fig. 267), its Fourier series (5) reduces to a **Fourier sine series**

(5)**
$$f(x) = \sum_{n=1}^{\infty} b_n \sin \frac{n\pi}{L} x \qquad (f \text{ odd})$$

with coefficients

(6**)
$$b_n = \frac{2}{L}\int_0^L f(x) \sin \frac{n\pi x}{L}\, dx.$$

These formulas follow from (5) and (6) by remembering from calculus that the definite integral gives the net area (= area above the axis minus area below the axis) under the curve of a function between the limits of integration. This implies

(7)
$$\text{(a)} \quad \int_{-L}^{L} g(x)\, dx = 2\int_0^L g(x)\, dx \qquad \text{for even } g$$

$$\text{(b)} \quad \int_{-L}^{L} h(x)\, dx = 0 \qquad \text{for odd } h$$

Formula (7b) implies the reduction to the cosine series (even f makes $f(x) \sin (n\pi x/L)$ odd since sin is odd) and to the sine series (odd f makes $f(x) \cos (n\pi x/L)$ odd since cos is even). Similarly, (7a) reduces the integrals in (6*) and (6**) to integrals from 0 to L. These reductions are obvious from the graphs of an even and an odd function. (Give a formal proof.)

Summary

Even Function of Period 2π. If f is even and $L = \pi$, then

$$f(x) = a_0 + \sum_{n=1}^{\infty} a_n \cos nx$$

with coefficients

$$a_0 = \frac{1}{\pi}\int_0^{\pi} f(x)\,dx, \qquad a_n = \frac{2}{\pi}\int_0^{\pi} f(x)\cos nx\,dx, \quad n = 1, 2, \cdots$$

Odd Function of Period 2π. If f is odd and $L = \pi$, then

$$f(x) = \sum_{n=1}^{\infty} b_n \sin nx$$

with coefficients

$$b_n = \frac{2}{\pi}\int_0^{\pi} f(x)\sin nx\,dx, \qquad n = 1, 2, \cdots.$$

EXAMPLE 4 Fourier Cosine and Sine Series

The rectangular wave in Example 1 is even. Hence it follows without calculation that its Fourier series is a Fourier cosine series, the b_n are all zero. Similarly, it follows that the Fourier series of the odd function in Example 2 is a Fourier sine series.

In Example 3 you can see that the Fourier cosine series represents $u(t) - E/\pi - \frac{1}{2}E \sin \omega t$. Can you prove that this is an even function? ■

Further simplifications result from the following property, whose very simple proof is left to the student.

THEOREM 1 Sum and Scalar Multiple

The Fourier coefficients of a sum $f_1 + f_2$ are the sums of the corresponding Fourier coefficients of f_1 and f_2.

The Fourier coefficients of cf are c times the corresponding Fourier coefficients of f.

EXAMPLE 5 Sawtooth Wave

Find the Fourier series of the function (Fig. 268)

$$f(x) = x + \pi \quad \text{if} \quad -\pi < x < \pi \qquad \text{and} \qquad f(x + 2\pi) = f(x).$$

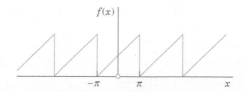

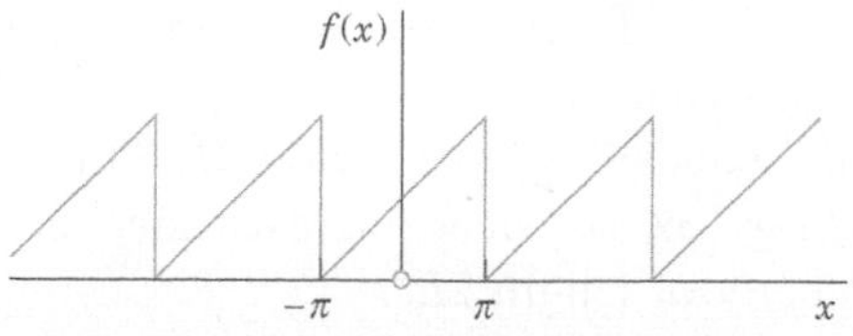

Fig. 268. The function $f(x)$. Sawtooth wave

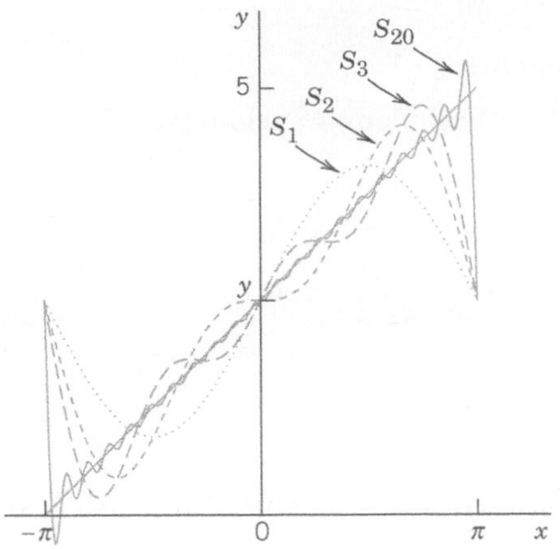

Fig. 269. Partial sums S_1, S_2, S_3, S_{20} in Example 5

Solution. We have $f = f_1 + f_2$, where $f_1 = x$ and $f_2 = \pi$. The Fourier coefficients of f_2 are zero, except for the first one (the constant term), which is π. Hence, by Theorem 1, the Fourier coefficients a_n, b_n are those of f_1, except for a_0, which is π. Since f_1 is odd, $a_n = 0$ for $n = 1, 2, \cdots$, and

$$b_n = \frac{2}{\pi}\int_0^{\pi} f_1(x) \sin nx\, dx = \frac{2}{\pi}\int_0^{\pi} x \sin nx\, dx.$$

Integrating by parts, we obtain

$$b_n = \frac{2}{\pi}\left[\frac{-x\cos nx}{n}\bigg|_0^{\pi} + \frac{1}{n}\int_0^{\pi} \cos nx\, dx\right] = -\frac{2}{n}\cos n\pi.$$

Hence $b_1 = 2$, $b_2 = -\frac{2}{2}$, $b_3 = \frac{2}{3}$, $b_4 = -\frac{2}{4}, \cdots$, and the Fourier series of $f(x)$ is

$$f(x) = \pi + 2\left(\sin x - \frac{1}{2}\sin 2x + \frac{1}{3}\sin 3x - + \cdots\right). \qquad \text{(Fig. 269)}$$ ■

3. Half-Range Expansions

Half-range expansions are Fourier series. The idea is simple and useful. Figure 270 explains it. We want to represent $f(x)$ in Fig. 270.0 by a Fourier series, where $f(x)$ may be the shape of a distorted violin string or the temperature in a metal bar of length L, for example. (Corresponding problems will be discussed in Chap. 12.) Now comes the idea.

We could extend $f(x)$ as a function of period L and develop the extended function into a Fourier series. But this series would, in general, contain *both* cosine *and* sine terms. We can do better and get simpler series. Indeed, for our given f we can calculate Fourier coefficients from (6*) or from (6**). And we have a choice and can take what seems more practical. If we use (6*), we get (5*). This is the **even periodic extension** f_1 of f in Fig. 270a. If we choose (6**) instead, we get (5**), the **odd periodic extension** f_2 of f in Fig. 270b.

Both extensions have period $2L$. This motivates the name **half-range expansions**: f is given (and of physical interest) only on half the range, that is, on half the interval of periodicity of length $2L$.

Let us illustrate these ideas with an example that we shall also need in Chap. 12.

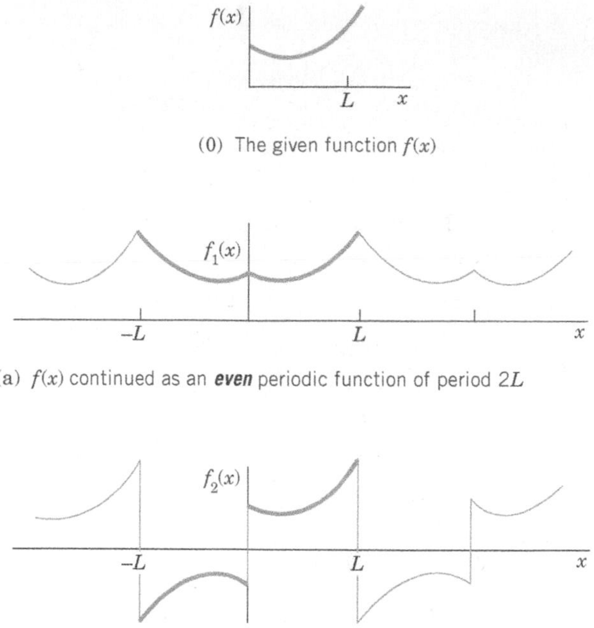

(0) The given function $f(x)$

(a) $f(x)$ continued as an ***even*** periodic function of period $2L$

(b) $f(x)$ continued as an ***odd*** periodic function of period $2L$

Fig. 270. Even and odd extensions of period $2L$

EXAMPLE 6 "Triangle" and Its Half-Range Expansions

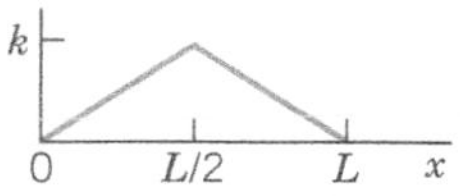

Fig. 271. The given function in Example 6

Find the two half-range expansions of the function (Fig. 271)

$$f(x) = \begin{cases} \dfrac{2k}{L}x & \text{if} \quad 0 < x < \dfrac{L}{2} \\[2ex] \dfrac{2k}{L}(L - x) & \text{if} \quad \dfrac{L}{2} < x < L. \end{cases}$$

Solution. (a) ***Even periodic extension.*** From (6*) we obtain

$$a_0 = \frac{1}{L}\left[\frac{2k}{L}\int_0^{L/2} x\,dx + \frac{2k}{L}\int_{L/2}^{L}(L - x)\,dx\right] = \frac{k}{2},$$

$$a_n = \frac{2}{L}\left[\frac{2k}{L}\int_L^{L/2} x\cos\frac{n\pi}{L}x\,dx + \frac{2k}{L}\int_{L/2}^{L}(L - x)\cos\frac{n\pi}{L}x\,dx\right].$$

We consider a_n. For the first integral we obtain by integration by parts

$$\int_0^{L/2} x\cos\frac{n\pi}{L}x\,dx = \frac{Lx}{n\pi}\sin\frac{n\pi}{L}x\Bigg|_0^{L/2} - \frac{L}{n\pi}\int_0^{L/2}\sin\frac{n\pi}{L}x\,dx$$

$$= \frac{L^2}{2n\pi}\sin\frac{n\pi}{2} + \frac{L^2}{n^2\pi^2}\left(\cos\frac{n\pi}{2} - 1\right).$$

Similarly, for the second integral we obtain

$$\int_{L/2}^{L}(L - x)\cos\frac{n\pi}{L}x\,dx = \frac{L}{n\pi}(L - x)\sin\frac{n\pi}{L}x\Bigg|_{L/2}^{L} + \frac{L}{n\pi}\int_{L/2}^{L}\sin\frac{n\pi}{L}x\,dx$$

$$= \left(0 - \frac{L}{n\pi}\left(L - \frac{L}{2}\right)\sin\frac{n\pi}{2}\right) - \frac{L^2}{n^2\pi^2}\left(\cos n\pi - \cos\frac{n\pi}{2}\right).$$

We insert these two results into the formula for a_n. The sine terms cancel and so does a factor L^2. This gives

$$a_n = \frac{4k}{n^2\pi^2}\left(2\cos\frac{n\pi}{2} - \cos n\pi - 1\right).$$

Thus,

$$a_2 = -16k/(2^2\pi^2), \qquad a_6 = -16k/(6^2\pi^2), \qquad a_{10} = -16k/(10^2\pi^2), \cdots$$

and $a_n = 0$ if $n \neq 2, 6, 10, 14, \cdots$. Hence the first half-range expansion of $f(x)$ is (Fig. 272a)

$$f(x) = \frac{k}{2} - \frac{16k}{\pi^2}\left(\frac{1}{2^2}\cos\frac{2\pi}{L}x + \frac{1}{6^2}\cos\frac{6\pi}{L}x + \cdots\right).$$

This Fourier cosine series represents the even periodic extension of the given function $f(x)$, of period $2L$.

(b) ***Odd periodic extension.*** Similarly, from (6**) we obtain

$$b_n = \frac{8k}{n^2\pi^2}\sin\frac{n\pi}{2}. \tag{5}$$

Hence the other half-range expansion of $f(x)$ is (Fig. 272b)

$$f(x) = \frac{8k}{\pi^2}\left(\frac{1}{1^2}\sin\frac{\pi}{L}x - \frac{1}{3^2}\sin\frac{3\pi}{L}x + \frac{1}{5^2}\sin\frac{5\pi}{L}x - + \cdots\right).$$

The series represents the odd periodic extension of $f(x)$, of period $2L$.

Basic applications of these results will be shown in Secs. 12.3 and 12.5. ■

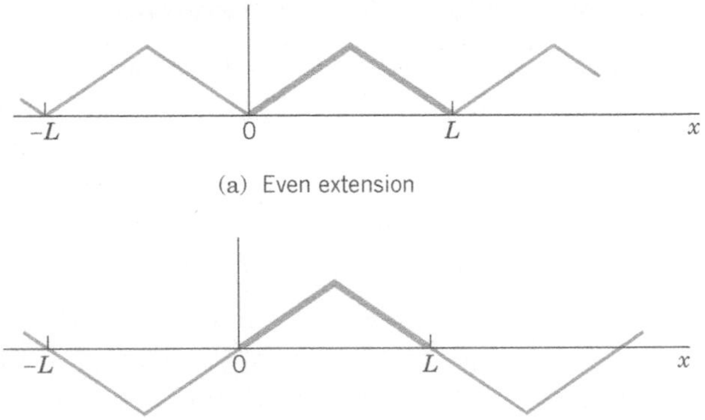

(a) Even extension

(b) Odd extension

Fig. 272. Periodic extensions of $f(x)$ in Example 6

PROBLEM SET 11.2

1–7 EVEN AND ODD FUNCTIONS

Are the following functions even or odd or neither even nor odd?

1. e^x, $e^{-|x|}$, $x^3 \cos nx$, $x^2 \tan \pi x$, $\sinh x - \cosh x$

2. $\sin^2 x$, $\sin(x^2)$, $\ln x$, $x/(x^2+1)$, $x \cot x$

3. Sums and products of even functions

4. Sums and products of odd functions

5. Absolute values of odd functions

6. Product of an odd times an even function

7. Find all functions that are both even and odd.

8–17 FOURIER SERIES FOR PERIOD $p = 2L$

Is the given function even or odd or neither even nor odd? Find its Fourier series. Show details of your work.

8.

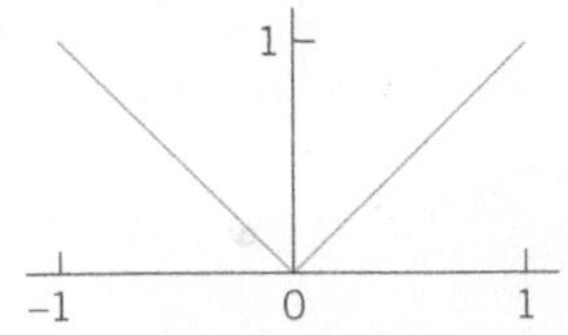

9.

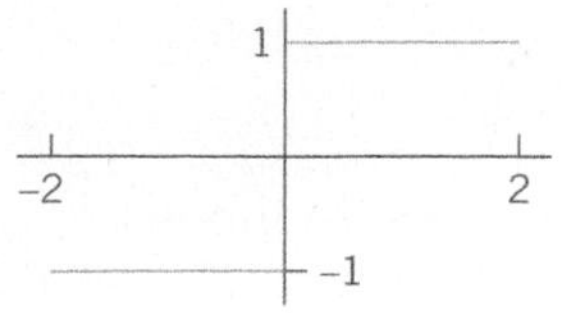

10.

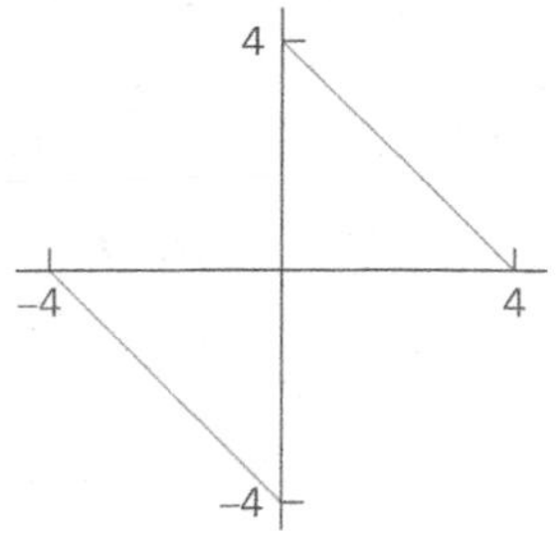

11. $f(x) = x^2 \quad (-1 < x < 1), \quad p = 2$

12. $f(x) = 1 - x^2/4 \quad (-2 < x < 2), \quad p = 4$

13.

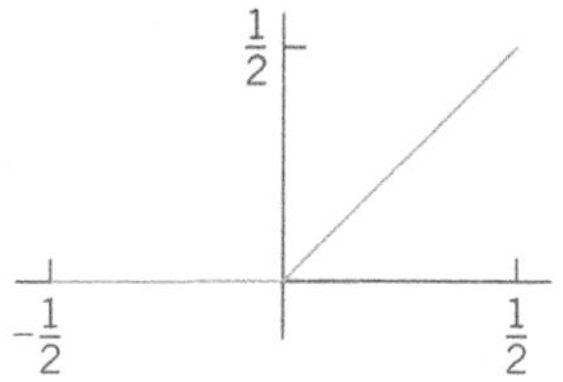

14. $f(x) = \cos \pi x \quad (-\frac{1}{2} < x < \frac{1}{2}), \quad p = 1$

15.

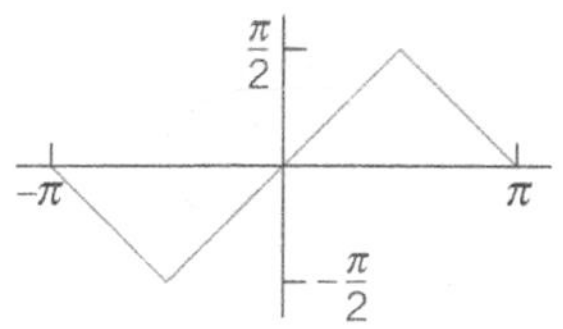

16. $f(x) = x|x| \quad (-1 < x < 1), \quad p = 2$

17.

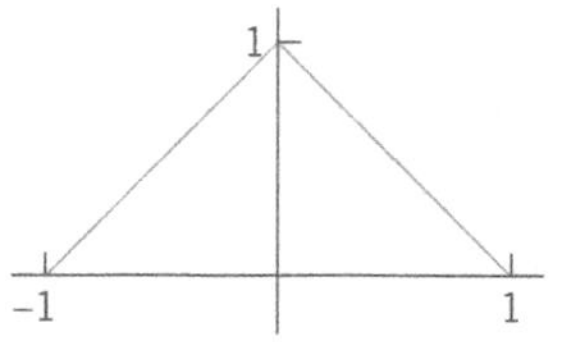

18. Rectifier. Find the Fourier series of the function obtained by passing the voltage $v(t) = V_0 \cos 100\pi t$ through a half-wave rectifier that clips the negative half-waves.

19. Trigonometric Identities. Show that the familiar identities $\cos^3 x = \frac{3}{4}\cos x + \frac{1}{4}\cos 3x$ and $\sin^3 x = \frac{3}{4}\sin x - \frac{1}{4}\sin 3x$ can be interpreted as Fourier series expansions. Develop $\cos^4 x$.

20. Numeric Values. Using Prob. 11, show that $1 + \frac{1}{4} + \frac{1}{9} + \frac{1}{16} + \cdots = \frac{1}{6}\pi^2$.

21. CAS PROJECT. Fourier Series of 2*L*-Periodic Functions. (a) Write a program for obtaining partial sums of a Fourier series (5).

(b) Apply the program to Probs. 8–11, graphing the first few partial sums of each of the four series on common axes. Choose the first five or more partial sums until they approximate the given function reasonably well. Compare and comment.

22. Obtain the Fourier series in Prob. 8 from that in Prob. 17.

23–29 HALF-RANGE EXPANSIONS

Find **(a)** the Fourier cosine series, **(b)** the Fourier sine series. Sketch $f(x)$ and its two periodic extensions. Show the details.

23.

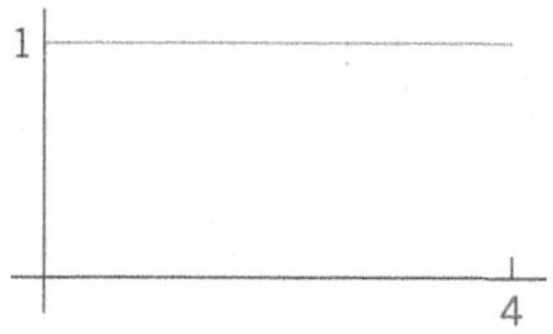

24.

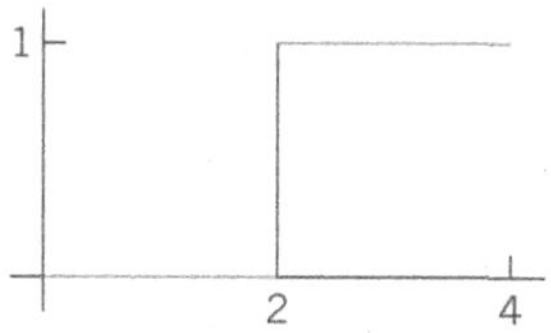

25.

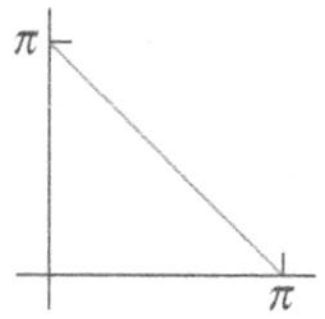

26.

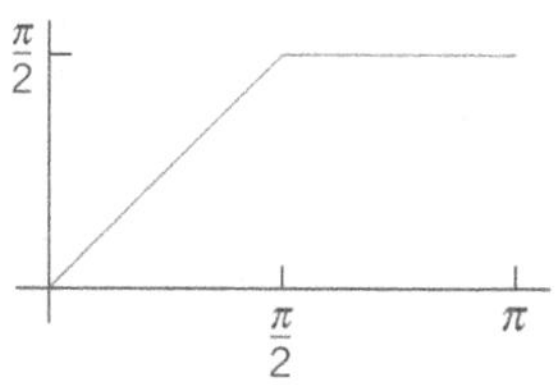

27.

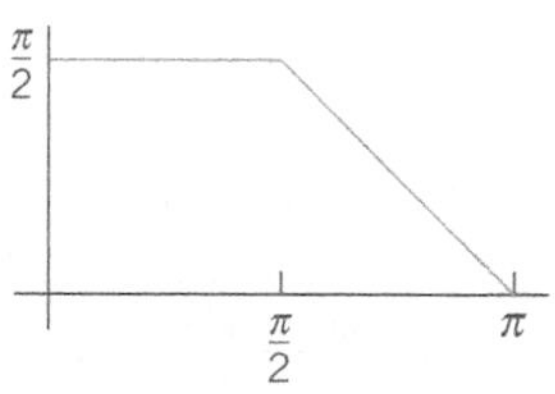

28.

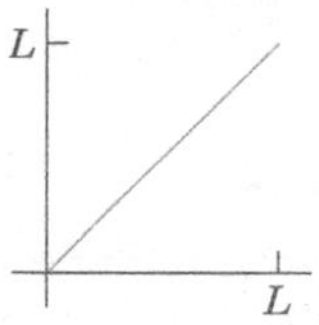

29. $f(x) = \sin x \; (0 < x < \pi)$

30. Obtain the solution to Prob. 26 from that of Prob. 27.

11.3 Forced Oscillations

Fourier series have important applications for both ODEs and PDEs. In this section we shall focus on ODEs and cover similar applications for PDEs in Chap. 12. All these applications will show our indebtedness to Euler's and Fourier's ingenious idea of splitting up periodic functions into the simplest ones possible.

From Sec. 2.8 we know that forced oscillations of a body of mass m on a spring of modulus k are governed by the ODE

$$my'' + cy' + ky = r(t) \tag{1}$$

where $y = y(t)$ is the displacement from rest, c the damping constant, k the spring constant (spring modulus), and $r(t)$ the external force depending on time t. Figure 274 shows the model and Fig. 275 its electrical analog, an RLC-circuit governed by

$$LI'' + RI' + \frac{1}{C}I = E'(t) \qquad \text{(Sec. 2.9).} \tag{1*}$$

We consider (1). If $r(t)$ is a sine or cosine function and if there is damping $(c > 0)$, then the steady-state solution is a harmonic oscillation with frequency equal to that of $r(t)$. However, if $r(t)$ is not a pure sine or cosine function but is any other periodic function, then the steady-state solution will be a superposition of harmonic oscillations with frequencies equal to that of $r(t)$ and integer multiples of these frequencies. And if one of these frequencies is close to the (practical) resonant frequency of the vibrating system (see Sec. 2.8), then the corresponding oscillation may be the dominant part of the response of the system to the external force. This is what the use of Fourier series will show us. Of course, this is quite surprising to an observer unfamiliar with Fourier series, which are highly important in the study of vibrating systems and resonance. Let us discuss the entire situation in terms of a typical example.

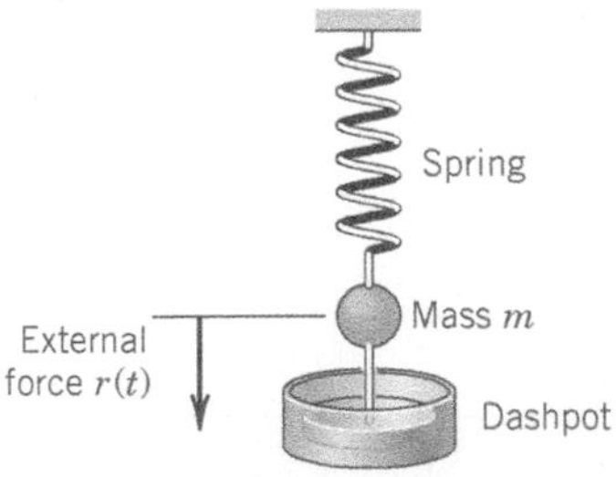

Fig. 274. Vibrating system under consideration

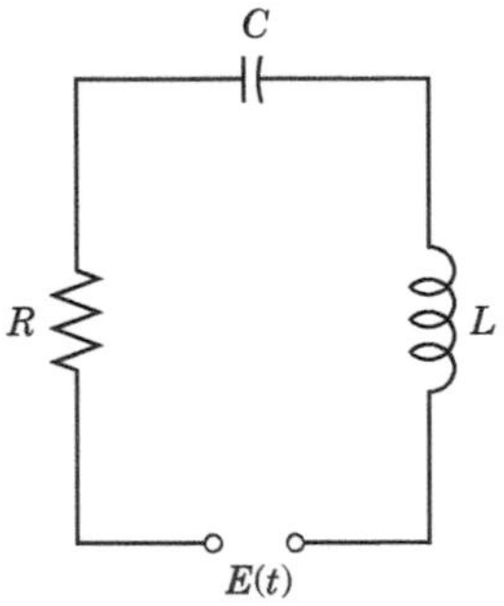

Fig. 275. Electrical analog of the system in Fig. 274 (*RLC-circuit*)

EXAMPLE 1 Forced Oscillations under a Nonsinusoidal Periodic Driving Force

In (1), let $m = 1$ (g), $c = 0.05$ (g/sec), and $k = 25$ (g/sec^2), so that (1) becomes

$$y'' + 0.05y' + 25y = r(t) \tag{2}$$

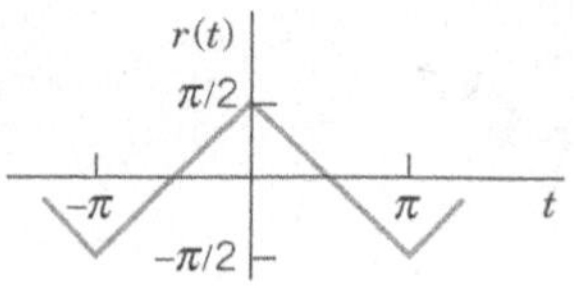

Fig. 276. Force in Example 1

where $r(t)$ is measured in $g \cdot cm/sec^2$. Let (Fig. 276)

$$r(t) = \begin{cases} t + \dfrac{\pi}{2} & \text{if } -\pi < t < 0, \\ -t + \dfrac{\pi}{2} & \text{if } \quad 0 < t < \pi, \end{cases} \qquad r(t + 2\pi) = r(t).$$

Find the steady-state solution $y(t)$.

Solution. We represent $r(t)$ by a Fourier series, finding

$$r(t) = \frac{4}{\pi}\left(\cos t + \frac{1}{3^2}\cos 3t + \frac{1}{5^2}\cos 5t + \cdots\right). \tag{3}$$

Then we consider the ODE

$$y'' + 0.05y' + 25y = \frac{4}{n^2\pi}\cos nt \qquad (n = 1, 3, \cdots) \tag{4}$$

whose right side is a single term of the series (3). From Sec. 2.8 we know that the steady-state solution $y_n(t)$ of (4) is of the form

$$y_n = A_n \cos nt + B_n \sin nt. \tag{5}$$

By substituting this into (4) we find that

$$A_n = \frac{4(25 - n^2)}{n^2\pi D_n}, \qquad B_n = \frac{0.2}{n\pi D_n}, \qquad \text{where} \qquad D_n = (25 - n^2)^2 + (0.05n)^2. \tag{6}$$

Since the ODE (2) is linear, we may expect the steady-state solution to be

$$y = y_1 + y_3 + y_5 + \cdots \tag{7}$$

where y_n is given by (5) and (6). In fact, this follows readily by substituting (7) into (2) and using the Fourier series of $r(t)$, provided that termwise differentiation of (7) is permissible. (Readers already familiar with the notion of uniform convergence [Sec. 15.5] may prove that (7) may be differentiated term by term.)

From (6) we find that the amplitude of (5) is (a factor $\sqrt{D_n}$ cancels out)

$$C_n = \sqrt{A_n^2 + B_n^2} = \frac{4}{n^2\pi\sqrt{D_n}}.$$

Values of the first few amplitudes are

$$C_1 = 0.0531 \quad C_3 = 0.0088 \quad C_5 = 0.2037 \quad C_7 = 0.0011 \quad C_9 = 0.0003.$$

Figure 277 shows the input (multiplied by 0.1) and the output. For $n = 5$ the quantity D_n is very small, the denominator of C_5 is small, and C_5 is so large that y_5 is the dominating term in (7). Hence the output is almost a harmonic oscillation of five times the frequency of the driving force, a little distorted due to the term y_1, whose amplitude is about 25% of that of y_5. You could make the situation still more extreme by decreasing the damping constant c. Try it. ■

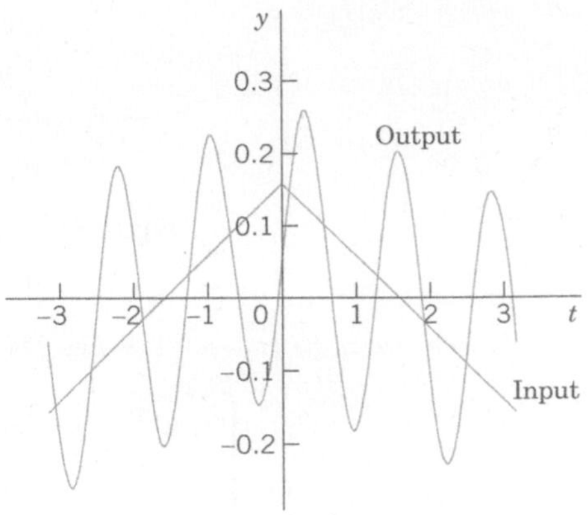

Fig. 277. Input and steady-state output in Example 1

PROBLEM SET 11.3

1. **Coefficients C_n.** Derive the formula for C_n from A_n and B_n.
2. **Change of spring and damping.** In Example 1, what happens to the amplitudes C_n if we take a stiffer spring, say, of $k = 49$? If we increase the damping?
3. **Phase shift.** Explain the role of the B_n's. What happens if we let $c \to 0$?
4. **Differentiation of input.** In Example 1, what happens if we replace $r(t)$ with its derivative, the rectangular wave? What is the ratio of the new C_n to the old ones?
5. **Sign of coefficients.** Some of the A_n in Example 1 are positive, some negative. All B_n are positive. Is this physically understandable?

6–11 GENERAL SOLUTION

Find a general solution of the ODE $y'' + \omega^2 y = r(t)$ with $r(t)$ as given. Show the details of your work.

6. $r(t) = \sin \alpha t + \sin \beta t$, $\omega^2 \neq \alpha^2, \beta^2$
7. $r(t) = \sin t$, $\omega = 0.5, 0.9, 1.1, 1.5, 10$
8. **Rectifier.** $r(t) = \pi/4\ |\cos t|$ if $-\pi < t < \pi$ and $r(t + 2\pi) = r(t)$, $|\omega| \neq 0, 2, 4, \cdots$
9. What kind of solution is excluded in Prob. 8 by $|\omega| \neq 0, 2, 4, \cdots$?
10. **Rectifier.** $r(t) = \pi/4\ |\sin t|$ if $0 < t < 2\pi$ and $r(t + 2\pi) = r(t)$, $|\omega| \neq 0, 2, 4, \cdots$
11. $r(t) = \begin{cases} -1 & \text{if} \quad -\pi < t < 0 \\ 1 & \text{if} \quad 0 < t < \pi, \end{cases} \qquad |\omega| \neq 1, 3, 5, \cdots$
12. **CAS Program.** Write a program for solving the ODE just considered and for jointly graphing input and output of an initial value problem involving that ODE. Apply the program to Probs. 7 and 11 with initial values of your choice.

13–16 STEADY-STATE DAMPED OSCILLATIONS

Find the steady-state oscillations of $y'' + cy' + y = r(t)$ with $c > 0$ and $r(t)$ as given. Note that the spring constant is $k = 1$. Show the details. In Probs. 14–16 sketch $r(t)$.

13. $r(t) = \sum_{n=1}^{N} (a_n \cos nt + b_n \sin nt)$
14. $r(t) = \begin{cases} -1 & \text{if} \quad -\pi < t < 0 \\ 1 & \text{if} \quad 0 < t < \pi \end{cases}$ and $r(t + 2\pi) = r(t)$
15. $r(t) = t(\pi^2 - t^2)$ if $-\pi < t < \pi$ and $r(t + 2\pi) = r(t)$
16. $r(t) = \begin{cases} t & \text{if} \quad -\pi/2 < t < \pi/2 \\ \pi - t & \text{if} \quad \pi/2 < t < 3\pi/2 \end{cases}$ and $r(t + 2\pi) = r(t)$

17–19 *RLC*-CIRCUIT

Find the steady-state current $I(t)$ in the *RLC*-circuit in Fig. 275, where $R = 10\ \Omega$, $L = 1$ H, $C = 10^{-1}$ F and with $E(t)$ V as follows and periodic with period 2π. Graph or sketch the first four partial sums. Note that the coefficients of the solution decrease rapidly. *Hint.* Remember that the ODE contains $E'(t)$, not $E(t)$, cf. Sec. 2.9.

17. $E(t) = \begin{cases} -50t^2 & \text{if} \quad -\pi < t < 0 \\ 50t^2 & \text{if} \quad 0 < t < \pi \end{cases}$

18. $E(t) = \begin{cases} 100\,(t - t^2) & \text{if} \quad -\pi < t < 0 \\ 100\,(t + t^2) & \text{if} \quad 0 < t < \pi \end{cases}$

19. $E(t) = 200t(\pi^2 - t^2) \quad (-\pi < t < \pi)$

20. CAS EXPERIMENT. Maximum Output Term. Graph and discuss outputs of $y'' + cy' + ky = r(t)$ with $r(t)$ as in Example 1 for various c and k with emphasis on the maximum C_n and its ratio to the second largest $|C_n|$.

11.4 Approximation by Trigonometric Polynomials

Fourier series play a prominent role not only in differential equations but also in **approximation theory**, an area that is concerned with approximating functions by other functions—usually simpler functions. Here is how Fourier series come into the picture.

Let $f(x)$ be a function on the interval $-\pi \leqq x \leqq \pi$ that can be represented on this interval by a Fourier series. Then the **Nth partial sum** of the Fourier series

$$f(x) \approx a_0 + \sum_{n=1}^{N} (a_n \cos nx + b_n \sin nx) \tag{1}$$

is an approximation of the given $f(x)$. In (1) we choose an arbitrary N and keep it fixed. Then we ask whether (1) is the "best" approximation of f by a **trigonometric polynomial of the same degree N**, that is, by a function of the form

$$F(x) = A_0 + \sum_{n=1}^{N} (A_n \cos nx + B_n \sin nx) \qquad (N \text{ fixed}). \tag{2}$$

Here, "best" means that the "error" of the approximation is as small as possible.

Of course we must first define what we mean by the **error** of such an approximation. We could choose the maximum of $|f(x) - F(x)|$. But in connection with Fourier series it is better to choose a definition of error that measures the goodness of agreement between f and F *on the whole interval* $-\pi \leqq x \leqq \pi$. This is preferable since the sum f of a Fourier series may have jumps: F in Fig. 278 is a good overall approximation of f, but the maximum of $|f(x) - F(x)|$ (more precisely, the *supremum*) is large. We choose

$$E = \int_{-\pi}^{\pi} (f - F)^2 \, dx. \tag{3}$$

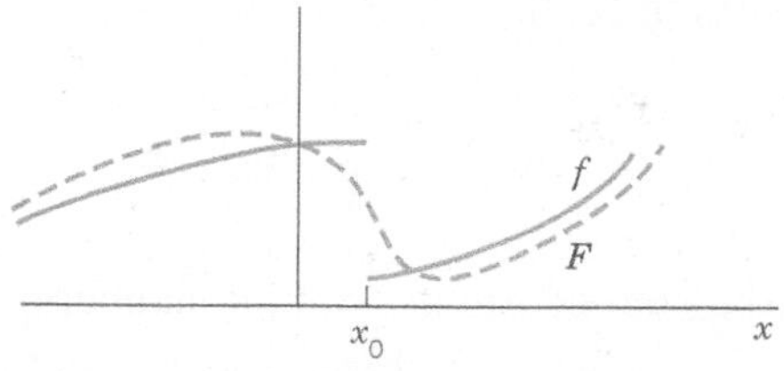

Fig. 278. Error of approximation

This is called the **square error** of F relative to the function f on the interval $-\pi \leqq x \leqq \pi$. Clearly, $E \geqq 0$.

N being fixed, we want to determine the coefficients in (2) such that E is minimum. Since $(f - F)^2 = f^2 - 2fF + F^2$, we have

$$E = \int_{-\pi}^{\pi} f^2\,dx - 2\int_{-\pi}^{\pi} fF\,dx + \int_{-\pi}^{\pi} F^2\,dx. \tag{4}$$

We square (2), insert it into the last integral in (4), and evaluate the occurring integrals. This gives integrals of $\cos^2 nx$ and $\sin^2 nx$ $(n \geqq 1)$, which equal π, and integrals of $\cos nx$, $\sin nx$, and $(\cos nx)(\sin mx)$, which are zero (just as in Sec. 11.1). Thus

$$\begin{aligned}\int_{-\pi}^{\pi} F^2\,dx &= \int_{-\pi}^{\pi} \left[A_0 + \sum_{n=1}^{N} (A_n \cos nx + B_n \sin nx)\right]^2 dx \\ &= \pi(2A_0^2 + A_1^2 + \cdots + A_N^2 + B_1^2 + \cdots + B_N^2).\end{aligned}$$

We now insert (2) into the integral of fF in (4). This gives integrals of $f \cos nx$ as well as $f \sin nx$, just as in Euler's formulas, Sec. 11.1, for a_n and b_n (each multiplied by A_n or B_n). Hence

$$\int_{-\pi}^{\pi} fF\,dx = \pi(2A_0a_0 + A_1a_1 + \cdots + A_Na_N + B_1b_1 + \cdots + B_Nb_N).$$

With these expressions, (4) becomes

$$\begin{aligned}E = \int_{-\pi}^{\pi} f^2\,dx &- 2\pi\left[2A_0a_0 + \sum_{n=1}^{N} (A_na_n + B_nb_n)\right] \\ &+ \pi\left[2A_0^2 + \sum_{n=1}^{N} (A_n^2 + B_n^2)\right].\end{aligned} \tag{5}$$

We now take $A_n = a_n$ and $B_n = b_n$ in (2). Then in (5) the second line cancels half of the integral-free expression in the first line. Hence for this choice of the coefficients of F the square error, call it E^*, is

$$E^* = \int_{-\pi}^{\pi} f^2\,dx - \pi\left[2a_0^2 + \sum_{n=1}^{N} (a_n^2 + b_n^2)\right]. \tag{6}$$

We finally subtract (6) from (5). Then the integrals drop out and we get terms $A_n^2 - 2A_na_n + a_n^2 = (A_n - an)^2$ and similar terms $(B_n - b_n)^2$:

$$E - E^* = \pi\left\{2(A_0 - a_0)^2 + \sum_{n=1}^{N}[(A_n - a_n)^2 + (B_n - b_n)^2]\right\}.$$

Since the sum of squares of real numbers on the right cannot be negative,

$$E - E^* \geqq 0, \qquad \text{thus} \qquad E \geqq E^*,$$

and $E = E^*$ if and only if $A_0 = a_0, \cdots, B_N = b_N$. This proves the following fundamental minimum property of the partial sums of Fourier series.

THEOREM 1

Minimum Square Error

The square error of F in (2) (with fixed N) relative to f on the interval $-\pi \leqq x \leqq \pi$ is minimum if and only if the coefficients of F in (2) are the Fourier coefficients of f. This minimum value E^* is given by (6).

From (6) we see that E^* cannot increase as N increases, but may decrease. Hence *with increasing N the partial sums of the Fourier series of f yield better and better approximations to f,* considered from the viewpoint of the square error.

Since $E^* \geqq 0$ and (6) holds for every N, we obtain from (6) the important **Bessel's inequality**

$$2a_0^2 + \sum_{n=1}^{\infty} (a_n^2 + b_n^2) \leqq \frac{1}{\pi}\int_{-\pi}^{\pi} f(x)^2\,dx \tag{7}$$

for the Fourier coefficients of any function f for which integral on the right exists. (For F. W. Bessel see Sec. 5.5.)

It can be shown (see [C12] in App. 1) that for such a function f, **Parseval's theorem** holds; that is, formula (7) holds with the equality sign, so that it becomes **Parseval's identity**[3]

$$2a_0^2 + \sum_{n=1}^{\infty} (a_n^2 + b_n^2) = \frac{1}{\pi}\int_{-\pi}^{\pi} f(x)^2\,dx. \tag{8}$$

EXAMPLE 1

Minimum Square Error for the Sawtooth Wave

Compute the minimum square error E^* of $F(x)$ with $N = 1, 2, \cdots, 10, 20, \cdots, 100$ and 1000 relative to

$$f(x) = x + \pi \qquad (-\pi < x < \pi)$$

on the interval $-\pi \leqq x \leqq \pi$.

Solution. $F(x) = \pi + 2\left(\sin x - \frac{1}{2}\sin 2x + \frac{1}{3}\sin 3x - + \cdots + \frac{(-1)^{N+1}}{N}\sin Nx\right)$ by Example 3 in Sec. 11.3. From this and (6),

$$E^* = \int_{-\pi}^{\pi} (x + \pi)^2\,dx - \pi\left(2\pi^2 + 4\sum_{n=1}^{N}\frac{1}{n^2}\right).$$

Numeric values are:

N	E^*	N	E^*	N	E^*	N	E^*
1	8.1045	6	1.9295	20	0.6129	70	0.1782
2	4.9629	7	1.6730	30	0.4120	80	0.1561
3	3.5666	8	1.4767	40	0.3103	90	0.1389
4	2.7812	9	1.3216	50	0.2488	100	0.1250
5	2.2786	10	1.1959	60	0.2077	1000	0.0126

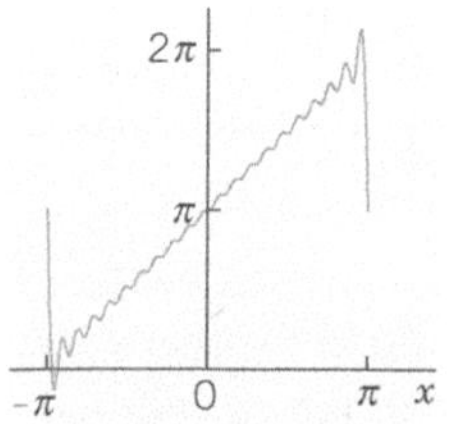

Fig. 279. F with $N = 20$ in Example 1

[3]MARC ANTOINE PARSEVAL (1755–1836), French mathematician. A physical interpretation of the identity follows in the next section.

$F = S_1, S_2, S_3$ are shown in Fig. 269 in Sec. 11.2, and $F = S_{20}$ is shown in Fig. 279. Although $|f(x) - F(x)|$ is large at $\pm\pi$ (how large?), where f is discontinuous, F approximates f quite well on the whole interval, except near $\pm\pi$, where "waves" remain owing to the "Gibbs phenomenon," which we shall discuss in the next section.

Can you think of functions f for which E^* decreases more quickly with increasing N? ■

PROBLEM SET 11.4

1. **CAS Problem.** Do the numeric and graphic work in Example 1 in the text.

2–5 MINIMUM SQUARE ERROR

Find the trigonometric polynomial $F(x)$ of the form (2) for which the square error with respect to the given $f(x)$ on the interval $-\pi < x < \pi$ is minimum. Compute the minimum value for $N = 1, 2, \cdots, 5$ (or also for larger values if you have a CAS).

2. $f(x) = x \quad (-\pi < x < \pi)$
3. $f(x) = |x| \quad (-\pi < x < \pi)$
4. $f(x) = x^2 \quad (-\pi < x < \pi)$
5. $f(x) = \begin{cases} -1 & \text{if} \quad -\pi < x < 0 \\ 1 & \text{if} \quad 0 < x < \pi \end{cases}$
6. Why are the square errors in Prob. 5 substantially larger than in Prob. 3?
7. $f(x) = x^3 \quad (-\pi < x < \pi)$
8. $f(x) = |\sin x| \quad (-\pi < x < \pi)$, full-wave rectifier
9. **Monotonicity.** Show that the minimum square error (6) is a monotone decreasing function of N. How can you use this in practice?
10. **CAS EXPERIMENT. Size and Decrease of E^*.** Compare the size of the minimum square error E^* for functions of your choice. Find experimentally the factors on which the decrease of E^* with N depends. For each function considered find the smallest N such that $E^* < 0.1$.

11–15 PARSEVALS'S IDENTITY

Using (8), prove that the series has the indicated sum. Compute the first few partial sums to see that the convergence is rapid.

11. $1 + \dfrac{1}{3^2} + \dfrac{1}{5^2} + \cdots = \dfrac{\pi^2}{8} = 1.233700550$

    Use Example 1 in Sec. 11.1.
12. $1 + \dfrac{1}{2^4} + \dfrac{1}{3^4} + \cdots = \dfrac{\pi^4}{90} = 1.082323234$

    Use Prob. 14 in Sec. 11.1.
13. $1 + \dfrac{1}{3^4} + \dfrac{1}{5^4} + \dfrac{1}{7^4} + \cdots = \dfrac{\pi^4}{96} = 1.014678032$

    Use Prob. 17 in Sec. 11.1.
14. $\displaystyle\int_{-\pi}^{\pi} \cos^4 x\, dx = \frac{3\pi}{4}$
15. $\displaystyle\int_{-\pi}^{\pi} \cos^6 x\, dx = \frac{5\pi}{8}$

11.5 Sturm–Liouville Problems. Orthogonal Functions

The idea of the Fourier series was to represent general periodic functions in terms of cosines and sines. The latter formed a *trigonometric system.* This trigonometric system has the desirable property of orthogonality which allows us to compute the coefficient of the Fourier series by the Euler formulas.

The question then arises, can this approach be generalized? That is, can we replace the trigonometric system of Sec. 11.1 by other *orthogonal systems* (*sets of other orthogonal functions*)? The answer is "yes" and will lead to generalized Fourier series, including the Fourier–Legendre series and the Fourier–Bessel series in Sec. 11.6.

To prepare for this generalization, we first have to introduce the concept of a Sturm–Liouville Problem. (The motivation for this approach will become clear as you read on.) Consider a second-order ODE of the form

$$(1) \qquad [p(x)y']' + [q(x) + \lambda r(x)]y = 0$$

on some interval $a \leqq x \leqq b$, satisfying conditions of the form

$$(2) \qquad \begin{aligned} &\textbf{(a)} \quad k_1 y + k_2 y' = 0 \quad \text{at } x = a \\ &\textbf{(b)} \quad l_1 y + l_2 y' = 0 \quad \text{at } x = b. \end{aligned}$$

Here λ is a parameter, and k_1, k_2, l_1, l_2 are given real constants. Furthermore, at least one of each constant in each condition (2) must be different from zero. (We will see in Example 1 that, if $p(x) = r(x) = 1$ and $q(x) = 0$, then $\sin \sqrt{\lambda}x$ and $\cos \sqrt{\lambda}x$ satisfy (1) and constants can be found to satisfy (2).) Equation (1) is known as a **Sturm–Liouville equation.**[4] Together with conditions 2(a), 2(b) it is know as the **Sturm–Liouville problem**. It is an example of a boundary value problem.

A **boundary value problem** consists of an ODE and given boundary conditions referring to the two boundary points (endpoints) $x = a$ and $x = b$ of a given interval $a \leqq x \leqq b$.

The goal is to solve these type of problems. To do so, we have to consider

Eigenvalues, Eigenfunctions

Clearly, $y \equiv 0$ is a solution—the "**trivial solution**"—of the problem (1), (2) for any λ because (1) is homogeneous and (2) has zeros on the right. This is of no interest. We want to find **eigenfunctions** $y(x)$, that is, solutions of (1) satisfying (2) without being identically zero. We call a number λ for which an eigenfunction exists an **eigenvalue** of the Sturm–Liouville problem (1), (2).

Many important ODEs in engineering can be written as Sturm–Liouville equations. The following example serves as a case in point.

EXAMPLE 1 Trigonometric Functions as Eigenfunctions. Vibrating String

Find the eigenvalues and eigenfunctions of the Sturm–Liouville problem

$$(3) \qquad y'' + \lambda y = 0, \qquad y(0) = 0, \quad y(\pi) = 0.$$

This problem arises, for instance, if an elastic string (a violin string, for example) is stretched a little and fixed at its ends $x = 0$ and $x = \pi$ and then allowed to vibrate. Then $y(x)$ is the "space function" of the deflection $u(x, t)$ of the string, assumed in the form $u(x, t) = y(x)w(t)$, where t is time. (This model will be discussed in great detail in Secs, 12.2–12.4.)

Solution. From (1) nad (2) we see that $p = 1, q = 0, r = 1$ in (1), and $a = 0, b = \pi, k_1 = l_1 = 1$, $k_2 = l_2 = 0$ in (2). For negative $\lambda = -\nu^2$ a general solution of the ODE in (3) is $y(x) = c_1 e^{\nu x} + c_2 e^{-\nu x}$. From the boundary conditions we obtain $c_1 = c_2 = 0$, so that $y \equiv 0$, which is not an eigenfunction. For $\lambda = 0$ the situation is similar. For positive $\lambda = \nu^2$ a general solution is

$$y(x) = A \cos \nu x + B \sin \nu x.$$

[4]JACQUES CHARLES FRANÇOIS STURM (1803–1855) was born and studied in Switzerland and then moved to Paris, where he later became the successor of Poisson in the chair of mechanics at the Sorbonne (the University of Paris).

JOSEPH LIOUVILLE (1809–1882), French mathematician and professor in Paris, contributed to various fields in mathematics and is particularly known by his important work in complex analysis (Liouville's theorem; Sec. 14.4), special functions, differential geometry, and number theory.

From the first boundary condition we obtain $y(0) = A = 0$. The second boundary condition then yields

$$y(\pi) = B \sin \nu\pi = 0, \qquad \text{thus} \qquad \nu = 0, \pm 1, \pm 2, \cdots.$$

For $\nu = 0$ we have $y \equiv 0$. For $\lambda = \nu^2 = 1, 4, 9, 16, \cdots$, taking $B = 1$, we obtain

$$y(x) = \sin \nu x \qquad (\nu = \sqrt{\lambda} = 1, 2, \cdots).$$

Hence the eigenvalues of the problem are $\lambda = \nu^2$, where $\nu = 1, 2, \cdots$, and corresponding eigenfunctions are $y(x) = \sin \nu x$, where$\nu = 1, 2 \cdots$. ■

Note that the solution to this problem is precisely the trigonometric system of the Fourier series considered earlier. It can be shown that, under rather general conditions on the functions p, q, r in (1), the Sturm–Liouville problem (1), (2) has infinitely many eigenvalues. The corresponding rather complicated theory can be found in Ref. [All] listed in App. 1.

Furthermore, if p, q, r, and p' in (1) are real-valued and continuous on the interval $a \leqq x \leqq b$ and r is positive throughout that interval (or negative throughout that interval), then all the eigenvalues of the Sturm–Liouville problem (1), (2) are real. (Proof in App. 4.) This is what the engineer would expect since eigenvalues are often related to frequencies, energies, or other physical quantities that must be real.

The most remarkable and important property of eigenfunctions of Sturm–Liouville problems is their ***orthogonality***, which will be crucial in series developments in terms of eigenfunctions, as we shall see in the next section. This suggests that we should next consider orthogonal functions.

Orthogonal Functions

Functions $y_1(x), y_2(x), \cdots$ defined on some interval $a \leqq x \leqq b$ are called **orthogonal** on this interval with respect to the **weight function** $r(x) > 0$ if for all m and all n different from m,

$$\textbf{(4)} \qquad (y_m, y_n) = \int_a^b r(x) y_m(x) y_n(x)\, dx = 0 \qquad (m \neq n).$$

(y_m, y_n) is a ***standard notation*** for this integral. **The norm** $\|y_m\|$ of y_m is defined by

$$\textbf{(5)} \qquad \|y_m\| = \sqrt{(y_m, y_m)} = \sqrt{\int_a^b r(x) y_m^2(x)\, dx}.$$

Note that this is the square root of the integral in (4) with $n = m$.

The functions $y_1, y_2, \cdots$ are called **orthonormal** on $a \leqq x \leqq b$ if they are orthogonal on this interval and all have norm 1. Then we can write (4), (5) jointly by using the **Kronecker symbol**[5] δ_{mn}, namely,

$$(y_m, y_n) = \int_a^b r(x) y_m(x) y_n(x)\, dx = \delta_{mn} = \begin{cases} 0 & \text{if} \quad m \neq n \\ 1 & \text{if} \quad m = n. \end{cases}$$

[5]LEOPOLD KRONECKER (1823–1891). German mathematician at Berlin University, who made important contributions to algebra, group theory, and number theory.

If $r(x) = 1$, we more briefly call the functions *orthogonal* instead of orthogonal with respect to $r(x) = 1$; similarly for orthognormality. Then

$$(y_m, y_n) = \int_a^b y_m(x)y_n(x)\,dx = 0 \quad (m \neq n), \qquad \|y_m\| = \sqrt{(y_m, y_n)} = \sqrt{\int_a^b y_m^2(x)\,dx}.$$

The next example serves as an illustration of the material on orthogonal functions just discussed.

EXAMPLE 2 Orthogonal Functions. Orthonormal Functions. Notation

The functions $y_m(x) = \sin mx$, $m = 1, 2, \cdots$ form an orthogonal set on the interval $-\pi \leqq x \leqq \pi$, because for $m \neq n$ we obtain by integration [see (11) in App. A3.1]

$$(y_m, y_n) = \int_{-\pi}^{\pi} \sin mx \sin nx\,dx = \frac{1}{2}\int_{-\pi}^{\pi} \cos(m-n)x\,dx - \frac{1}{2}\int_{-\pi}^{\pi} \cos(m+n)x\,dx = 0, \quad (m \neq n).$$

The norm $\|y_m\| = \sqrt{(y_m, y_m)}$ equals $\sqrt{\pi}$ because

$$\|y_m\|^2 = (y_m, y_m) = \int_{-\pi}^{\pi} \sin^2 mx\,dx = \pi \qquad (m = 1, 2, \cdots)$$

Hence the corresponding orthonormal set, obtained by division by the norm, is

$$\frac{\sin x}{\sqrt{\pi}}, \quad \frac{\sin 2x}{\sqrt{\pi}}, \quad \frac{\sin 3x}{\sqrt{\pi}}, \quad \cdots.$$ ■

Theorem 1 shows that for any Sturm–Liouville problem, the eigenfunctions associated with these problems are orthogonal. This means, in practice, if we can formulate a problem as a Sturm–Liouville problem, then by this theorem we are guaranteed orthogonality.

THEOREM 1 Orthogonality of Eigenfunctions of Sturm–Liouville Problems

Suppose that the functions p, q, r, and p' in the Sturm–Liouville equation (1) are real-valued and continuous and $r(x) > 0$ on the interval $a \leqq x \leqq b$. Let $y_m(x)$ and $y_n(x)$ be eigenfunctions of the Sturm–Liouville problem (1), (2) that correspond to different eigenvalues λ_m and λ_n, respectively. Then y_m, y_n are orthogonal on that interval with respect to the weight function r, that is,

$$(6) \qquad (y_m, y_n) = \int_a^b r(x)y_m(x)y_n(x)\,dx = 0 \qquad (m \neq n).$$

If $p(a) = 0$, then (2a) *can be dropped from the problem. If $p(b) = 0$, then* (2b) can be dropped. [It is then required that y and y' remain bounded at such a point, and the problem is called **singular**, as opposed to a **regular problem** in which (2) is used.]

If $p(a) = p(b)$, then (2) *can be replaced by the* "**periodic boundary conditions**"

$$(7) \qquad y(a) = y(b), \qquad y'(a) = y'(b).$$

The boundary value problem consisting of the Sturm–Liouville equation (1) and the periodic boundary conditions (7) is called a **periodic Sturm–Liouville problem**.

PROOF By assumption, y_m and y_n satisfy the Sturm–Liouville equations

$$(py_m')' + (q + \lambda_m r)y_m = 0$$
$$(py_n')' + (q + \lambda_n r)y_n = 0$$

respectively. We multiply the first equation by y_n, the second by $-y_m$, and add,

$$(\lambda_m - \lambda_n)ry_m y_n = y_m(py_n')' - y_n(py_m')' = [(py_n')y_m - [(py_m')y_n]'$$

where the last equality can be readily verified by performing the indicated differentiation of the last expression in brackets. This expression is continuous on $a \leqq x \leqq b$ since p and p' are continuous by assumption and y_m, y_n are solutions of (1). Integrating over x from a to b, we thus obtain

$$(8) \qquad (\lambda_m - \lambda_n)\int_a^b ry_m y_n\, dx = [p(y'_n y_m - y'_m y_n)]_a^b \qquad (a < b).$$

The expression on the right equals the sum of the subsequent Lines 1 and 2,

$$(9) \qquad \begin{aligned} &p(b)[y_n'(b)y_m(b) - y_m'(b)y_n(b)] && \text{(Line 1)} \\ -&p(a)[y_n'(a)y_m(a) - y_m'(a)y_n(a)] && \text{(Line 2).} \end{aligned}$$

Hence if (9) is zero, (8) with $\lambda_m - \lambda_n \neq 0$ implies the orthogonality (6). Accordingly, we have to show that (9) is zero, using the boundary conditions (2) as needed.

Case 1. $\boldsymbol{p(a) = p(b) = 0}$**.** Clearly, (9) is zero, and (2) is not needed.
Case 2. $\boldsymbol{p(a) \neq 0, p(b) = 0}$**.** Line 1 of (9) is zero. Consider Line 2. From (2a) we have

$$k_1 y_n(a) + k_2 y_n'(a) = 0,$$
$$k_1 y_m(a) + k_2 y_m'(a) = 0.$$

Let $k_2 \neq 0$. We multiply the first equation by $y_m(a)$, the last by $-y_n(a)$ and add,

$$k_2[y_n'(a)y_m(a) - y_m'(a)y_n(a)] = 0.$$

This is k_2 times Line 2 of (9), which thus is zero since $k_2 \neq 0$. If $k_2 = 0$, then $k_1 \neq 0$ by assumption, and the argument of proof is similar.

Case 3. $\boldsymbol{p(a) = 0, p(b) \neq 0}$**.** Line 2 of (9) is zero. From (2b) it follows that Line 1 of (9) is zero; this is similar to Case 2.
Case 4. $\boldsymbol{p(a) \neq 0, p(b) \neq 0}$**.** We use both (2a) and (2b) and proceed as in Cases 2 and 3.
Case 5. $\boldsymbol{p(a) = p(b)}$**.** Then (9) becomes

$$p(b)[y_n'(b)y_m(b) - y_m'(b)y_n(b) - y_n'(a)y_m(a) + y_m'(a)y_n(a)].$$

The expression in brackets $[\cdots]$ is zero, either by (2) used as before, or more directly by (7). Hence in this case, (7) can be used instead of (2), as claimed. This completes the proof of Theorem 1. ■

EXAMPLE 3 **Application of Theorem 1. Vibrating String**

The ODE in Example 1 is a Sturm–Liouville equation with $p = 1, q = 0$, and $r = 1$. From Theorem 1 it follows that the eigenfunctions $y_m = \sin mx$ $(m = 1, 2, \cdots)$ are orthogonal on the interval $0 \leqq x \leqq \pi$. ■

Example 3 confirms, from this new perspective, that the trigonometric system underlying the Fourier series is orthogonal, as we knew from Sec. 11.1.

EXAMPLE 4 Application of Theorem 1. Orthogonlity of the Legendre Polynomials

Legendre's equation $(1 - x^2)y'' - 2xy' + n(n + 1)y = 0$ may be written

$$[(1 - x^2)y']' + \lambda y = 0 \qquad \lambda = n(n + 1).$$

Hence, this is a Sturm–Liouville equation (1) with $p = 1 - x^2$, $q = 0$, and $r = 1$. Since $p(-1) = p(1) = 0$, we need no boundary conditions, but have a ***"singular" Sturm–Liouville problem*** on the interval $-1 \leqq x \leqq 1$. We know that for $n = 0, 1, \cdots$, hence $\lambda = 0, 1 \cdot 2, 2 \cdot 3, \cdots$, the Legendre polynomials $P_n(x)$ are solutions of the problem. Hence these are the eigenfunctions. From Theorem 1 it follows that they are orthogonal on that interval, that is,

$$(10) \qquad \int_{-1}^{1} P_m(x)P_n(x)\,dx = 0 \qquad (m \neq n).$$ ■

What we have seen is that the trigonometric system, underlying the Fourier series, is a solution to a Sturm–Liouville problem, as shown in Example 1, and that this trigonometric system is orthogonal, which we knew from Sec. 11.1 and confirmed in Example 3.

PROBLEM SET 11.5

1. **Proof of Theorem 1.** Carry out the details in Cases 3 and 4.

2–6 ORTHOGONALITY

2. **Normalization of eigenfunctions** y_m of (1), (2) means that we multiply y_m by a nonzero constant c_m such that $c_m y_m$ has norm 1. Show that $z_m = cy_m$ with *any* $c \neq 0$ is an eigenfunction for the eigenvalue corresponding to y_m.
3. **Change of x.** Show that if the functions $y_0(x), y_1(x), \cdots$ form an orthogonal set on an interval $a \leqq x \leqq b$ (with $r(x) = 1$), then the functions $y_0(ct + k), y_1(ct + k), \cdots, c > 0$, form an orthogonal set on the interval $(a - k)/c \leqq t \leqq (b - k)/c$.
4. **Change of x.** Using Prob. 3, derive the orthogonality of 1, $\cos \pi x$, $\sin \pi x$, $\cos 2\pi x$, $\sin 2\pi x$, $\cdots$ on $-1 \leqq x \leqq 1$ $(r(x) = 1)$ from that of 1, $\cos x$, $\sin x$, $\cos 2x$, $\sin 2x$, $\cdots$ on $-\pi \leqq x \leqq \pi$.
5. **Legendre polynomials.** Show that the functions $P_n(\cos \theta)$, $n = 0, 1, \cdots$, from an orthogonal set on the interval $0 \leqq \theta \leqq \pi$ with respect to the weight function $\sin \theta$.
6. **Tranformation to Sturm–Liouville form.** Show that $y'' + fy' + (g + \lambda h)\,y = 0$ takes the form (1) if you set $p = \exp(\int f\,dx)$, $q = pg$, $r = hp$. Why would you do such a transformation?

7–15 STURM–LIOUVILLE PROBLEMS

Find the eigenvalues and eigenfunctions. Verify orthogonality. Start by writing the ODE in the form (1), using Prob. 6. Show details of your work.

7. $y'' + \lambda y = 0, \quad y(0) = 0, \quad y(10) = 0$
8. $y'' + \lambda y = 0, \quad y(0) = 0, \quad y(L) = 0$
9. $y'' + \lambda y = 0, \quad y(0) = 0, \quad y'(L) = 0$
10. $y'' + \lambda y = 0, \quad y(0) = y(1), \quad y'(0) = y'(1)$
11. $(y'/x)' + (\lambda + 1)y/x^3 = 0, \quad y(1) = 0, \quad y(e^\pi) = 0.$ (Set $x = e^t$.)
12. $y'' - 2y' + (\lambda + 1)y = 0, \quad y(0) = 0, \quad y(1) = 0$
13. $y'' + 8y' + (\lambda + 16)y = 0, \quad y(0) = 0, \quad y(\pi) = 0$
14. **TEAM PROJECT. Special Functions. Orthogonal polynomials** play a great role in applications. For this reason, Legendre polynomials and various other orthogonal polynomials have been studied extensively; see Refs. [GenRef1], [GenRef10] in App. 1. Consider some of the most important ones as follows.

(a) Chebyshev polynomials[6] of the first and second kind are defined by

$$T_n(x) = \cos(n \arccos x)$$
$$U_n(x) = \frac{\sin[(n+1)\arccos x]}{\sqrt{1-x^2}}$$

respectively, where $n = 0, 1, \cdots$. Show that

$$T_0 = 1, \qquad T_1(x) = x, \qquad T_2(x) = 2x^2 - 1.$$
$$T_3(x) = 4x^3 - 3x,$$
$$U_0 = 1, \qquad U_1(x) = 2x, \qquad U_2(x) = 4x^2 - 1,$$
$$U_3(x) = 8x^3 - 4x.$$

Show that the Chebyshev polynomials $T_n(x)$ are orthogonal on the interval $-1 \leqq x \leqq 1$ with respect to the weight function $r(x) = 1/\sqrt{1-x^2}$. (*Hint.* To evaluate the integral, set $\arccos x = \theta$.) Verify that $T_n(x)$, $n = 0, 1, 2, 3$, satisfy the **Chebyshev equation**

$$(1 - x^2)y'' - xy' + n^2 y = 0.$$

(b) Orthogonality on an infinite interval: Laguerre polynomials[7] are defined by $L_0 = 1$, and

$$L_n(x) = \frac{e^x}{n!}\frac{d^n(x^n e^{-x})}{dx^n}, \qquad n = 1, 2, \cdots.$$

Show that

$$L_n(x) = 1 - x, \qquad L_2(x) = 1 - 2x + x^2/2,$$
$$L_3(x) = 1 - 3x + 3x^2/2 - x^3/6.$$

Prove that the Laguerre polynomials are orthogonal on the positive axis $0 \leqq x < \infty$ with respect to the weight function $r(x) = e^{-x}$. *Hint.* Since the highest power in L_m is x^m, it suffices to show that $\int e^{-x} x^k L_n \, dx = 0$ for $k < n$. Do this by k integrations by parts.

11.6 Orthogonal Series. Generalized Fourier Series

Fourier series are made up of the trigonometric system (Sec. 11.1), which is orthogonal, and orthogonality was essential in obtaining the Euler formulas for the Fourier coefficients. Orthogonality will also give us coefficient formulas for the desired generalized Fourier series, including the Fourier–Legendre series and the Fourier–Bessel series. This generalization is as follows.

Let $y_0, y_1, y_2, \cdots$ be orthogonal with respect to a weight function $r(x)$ on an interval $a \leqq x \leqq b$, and let $f(x)$ be a function that can be represented by a convergent series

$$f(x) = \sum_{m=0}^{\infty} a_m y_m(x) = a_0 y_0(x) + a_1 y_1(x) + \cdots. \tag{1}$$

This is called an **orthogonal series, orthogonal expansion**, or **generalized Fourier series**. If the y_m are the eigenfunctions of a Sturm–Liouville problem, we call (1) an **eigenfunction expansion**. In (1) we use again m for summation since n will be used as a fixed order of Bessel functions.

Given $f(x)$, we have to determine the coefficients in (1), called the **Fourier constants** *of $f(x)$ with respect to $y_0, y_1, \cdots$*. Because of the orthogonality, this is simple. Similarly to Sec. 11.1, we multiply both sides of (1) by $r(x)y_n(x)$ (*n* ***fixed***) and then integrate on

[6] PAFNUTI CHEBYSHEV (1821–1894), Russian mathematician, is known for his work in approximation theory and the theory of numbers. Another transliteration of the name is TCHEBICHEF.

[7] EDMOND LAGUERRE (1834–1886), French mathematician, who did research work in geometry and in the theory of infinite series.

both sides from a to b. We assume that term-by-term integration is permissible. (This is justified, for instance, in the case of "uniform convergence," as is shown in Sec. 15.5.) Then we obtain

$$(f, y_n) = \int_a^b r f y_n \, dx = \int_a^b r \left(\sum_{m=0}^{\infty} a_m y_m \right) y_n \, dx = \sum_{m=0}^{\infty} a_m \int_a^b r y_m y_n \, dx = \sum_{m=0}^{\infty} a_m (y_m, y_n).$$

Because of the orthogonality all the integrals on the right are zero, except when $m = n$. Hence the whole infinite series reduces to the single term

$$a_n (y_n, y_n) = a_n \|y_n\|^2. \qquad \text{Thus} \qquad (f, y_n) = a_n \|y_n\|^2.$$

Assuming that all the functions y_n have nonzero norm, we can divide by $\|y_n\|^2$; writing again m for n, to be in agreement with (1), we get the desired formula for the Fourier constants

$$\text{(2)} \qquad a_m = \frac{(f, y_m)}{\|y_m\|^2} = \frac{1}{\|y_m\|^2} \int_a^b r(x) f(x) y_m(x) \, dx \qquad (n = 0, 1, \cdots).$$

This formula generalizes the Euler formulas (6) in Sec. 11.1 as well as the principle of their derivation, namely, by orthogonality.

EXAMPLE 1 Fourier–Legendre Series

A **Fourier–Legendre series** is an eigenfunction expansion

$$f(x) = \sum_{m=0}^{\infty} a_m P_m(x) = a_0 P_0 + a_1 P_1(x) + a_2 P_2(x) + \cdots = a_0 + a_1 x + a_2 (\tfrac{3}{2} x^2 - \tfrac{1}{2}) + \cdots$$

in terms of Legendre polynomials (Sec. 5.3). The latter are the eigenfunctions of the Sturm–Liouville problem in Example 4 of Sec. 11.5 on the interval $-1 \leqq x \leqq 1$. We have $r(x) = 1$ for Legendre's equation, and (2) gives

$$\text{(3)} \qquad a_m = \frac{2m + 1}{2} \int_{-1}^{1} f(x) P_m(x) \, dx, \qquad m = 0, 1, \cdots$$

because the norm is

$$\text{(4)} \qquad \|P_m\| = \sqrt{\int_{-1}^{1} P_m(x)^2 \, dx} = \sqrt{\frac{2}{2m + 1}} \qquad (m = 0, 1, \cdots)$$

as we state without proof. The proof of (4) is tricky; it uses Rodrigues's formula in Problem Set 5.2 and a reduction of the resulting integral to a quotient of gamma functions.

For instance, let $f(x) = \sin \pi x$. Then we obtain the coefficients

$$a_m = \frac{2m + 1}{2} \int_{-1}^{1} (\sin \pi x) P_m(x) \, dx, \qquad \text{thus} \qquad a_1 = \frac{3}{2} \int_{-1}^{1} x \sin \pi x \, dx = \frac{3}{\pi} = 0.95493, \qquad \text{etc.}$$

Hence the Fourier–Legendre series of $\sin \pi x$ is

$$\sin \pi x = 0.95493 P_1(x) - 1.15824 P_3(x) + 0.21929 P_5(x) - 0.01664 P_7(x) + 0.00068 P_9(x)$$
$$- 0.00002 P_{11}(x) + \cdots .$$

The coefficient of P_{13} is about $3 \cdot 10^{-7}$. The sum of the first three nonzero terms gives a curve that practically coincides with the sine curve. Can you see why the even-numbered coefficients are zero? Why a_3 is the absolutely biggest coefficient? ■

EXAMPLE 2   Fourier–Bessel Series

These series model vibrating membranes (Sec. 12.9) and other physical systems of circular symmetry. We derive these series in three steps.

Step 1. **Bessel's equation as a Sturm–Liouville equation.** The Bessel function $J_n(x)$ with fixed integer $n \geqq 0$ satisfies Bessel's equation (Sec. 5.5)

$$\tilde{x}^2 \ddot{J}_n(\tilde{x}) + \tilde{x} \dot{J}_n(\tilde{x}) + (\tilde{x}^2 - n^2) J_n(\tilde{x}) = 0$$

where $\dot{J}_n = dJ_n/d\tilde{x}$ and $\ddot{J}_n = d^2 J_n/d\tilde{x}^2$. We set $\tilde{x} = kx$. Then $x = \tilde{x}/k$ and by the chain rule, $\dot{J}_n = dJ_n/d\tilde{x} = (dJ_n/dx)/k$ and $\ddot{J}_n = J_n''/k^2$. In the first two terms of Bessel's equation, k^2 and k drop out and we obtain

$$x^2 J_n''(kx) + x J_n'(kx) + (k^2 x^2 - n^2) J_n(kx) = 0.$$

Dividing by x and using $(xJ_n'(kx))' = xJ_n''(kx) + J_n'(kx)$ gives the Sturm–Liouville equation

$$(5) \qquad\qquad [xJ_n'(kx)]' + \left(-\frac{n^2}{x} + \lambda x \right) J_n(kx) = 0 \qquad\qquad \lambda = k^2$$

with $p(x) = x$, $q(x) = -n^2/x$, $r(x) = x$, and parameter $\lambda = k^2$. Since $p(0) = 0$, Theorem 1 in Sec. 11.5 implies orthogonality on an interval $0 \leqq x \leqq R$ (R given, fixed) of those solutions $J_n(kx)$ that are zero at $x = R$, that is,

$$(6) \qquad\qquad\qquad J_n(kR) = 0 \qquad\qquad\qquad (n \text{ fixed}).$$

Note that $q(x) = -n^2/x$ is discontinuous at 0, but this does not affect the proof of Theorem 1.

Step 2. **Orthogonality.** It can be shown (see Ref. [A13]) that $J_n(\tilde{x})$ has infinitely many zeros, say, $\tilde{x} = a_{n,1} < a_{n,2} < \cdots$ (see Fig. 110 in Sec. 5.4 for $n = 0$ and 1). Hence we must have

$$(7) \qquad\qquad kR = \alpha_{n,m} \qquad \text{thus} \qquad k_{n,m} = \alpha_{n,m}/R \qquad\qquad (m = 1, 2, \cdots).$$

This proves the following orthogonality property.

THEOREM 1

Orthogonality of Bessel Functions

For each fixed nonnegative integer n the sequence of Bessel functions of the first kind $J_n(k_{n,1}x)$, $J_n(k_{n,2}x)$, $\cdots$ with $k_{n,m}$ as in (7) forms an orthogonal set on the interval $0 \leqq x \leqq R$ with respect to the weight function $r(x) = x$, that is,

$$(8) \qquad \int_0^R x J_n(k_{n,m}x) J_n(k_{n,j}x)\, dx = 0 \qquad (j \neq m, n \text{ fixed}).$$

Hence we have obtained *infinitely many orthogonal sets* of Bessel functions, one for each of $J_0, J_1, J_2, \cdots$. Each set is orthogonal on an interval $0 \leqq x \leqq R$ with a fixed positive R of our choice and with respect to the weight x. The orthogonal set for J_n is $J_n(k_{n,1}x), J_n(k_{n,2}x), J_n(k_{n,3}x), \cdots$, where n is *fixed* and $k_{n,m}$ is given by (7).

Step 3. **Fourier–Bessel series.** The Fourier–Bessel series corresponding to J_n (n fixed) is

$$f(x) = \sum_{m=1}^{\infty} a_m J_n(k_{n,m}x) = a_1 J_n(k_{n,1}x) + a_2 J_n(k_{n,2}x) + a_3 J_n(k_{n,3}x) + \cdots \qquad (n \text{ fixed}). \tag{9}$$

The coefficients are (with $\alpha_{n,m} = k_{n,m}R$)

$$a_m = \frac{2}{R^2 J_{n+1}^2(\alpha_{n,m})} \int_0^R x f(x) J_n(k_{n,m}x)\, dx, \qquad m = 1, 2, \cdots \tag{10}$$

because the square of the norm is

$$\|J_n(k_{n,m}x)\|^2 = \int_0^R x J_n^2(k_{n,m}x)\, dx = \frac{R^2}{2} J_{n+1}^2(k_{n,m}R) \tag{11}$$

as we state without proof (which is tricky; see the discussion beginning on p. 576 of [A13]). ■

EXAMPLE 3 **Special Fourier–Bessel Series**

For instance, let us consider $f(x) = 1 - x^2$ and take $R = 1$ and $n = 0$ in the series (9), simply writing λ for $\alpha_{0,m}$. Then $k_{n,m} = \alpha_{0,m} = \lambda = 2.405, 5.520, 8.654, 11.792$, etc. (use a CAS or Table A1 in App. 5). Next we calculate the coefficients a_m by (10)

$$a_m = \frac{2}{J_1^2(\lambda)} \int_0^1 x(1 - x^2) J_0(\lambda x)\, dx.$$

This can be integrated by a CAS or by formulas as follows. First use $[xJ_1(\lambda x)]' = \lambda x J_0(\lambda x)$ from Theorem 1 in Sec. 5.4 and then integration by parts,

$$a_m = \frac{2}{J_1^2(\lambda)} \int_0^1 x(1 - x^2) J_0(\lambda x)\, dx = \frac{2}{J_1^2(\lambda)} \left[\frac{1}{\lambda}(1 - x^2) x J_1(\lambda x) \Big|_0^1 - \frac{1}{\lambda} \int_0^1 x J_1(\lambda x)(-2x)\, dx \right].$$

The integral-free part is zero. The remaining integral can be evaluated by $[x^2 J_2(\lambda x)]' = \lambda x^2 J_1(\lambda x)$ from Theorem 1 in Sec. 5.4. This gives

$$a_m = \frac{4 J_2(\lambda)}{\lambda^2 J_1^2(\lambda)} \qquad (\lambda = \alpha_{0,m}).$$

Numeric values can be obtained from a CAS (or from the table on p. 409 of Ref. [GenRef1] in App. 1, together with the formula $J_2 = 2x^{-1}J_1 - J_0$ in Theorem 1 of Sec. 5.4). This gives the eigenfunction expansion of $1 - x^2$ in terms of Bessel functions J_0, that is,

$$1 - x^2 = 1.1081 J_0(2.405x) - 0.1398 J_0(5.520x) + 0.0455 J_0(8.654x) - 0.0210 J_0(11.792x) + \cdots.$$

A graph would show that the curve of $1 - x^2$ and that of the sum of first three terms practically coincide. ■

Mean Square Convergence. Completeness

Ideas on approximation in the last section generalize from Fourier series to orthogonal series (1) that are made up of an orthonormal set that is "complete," that is, consists of "sufficiently many" functions so that (1) can represent large classes of other functions (definition below).

In this connection, convergence is **convergence in the norm**, also called **mean-square convergence**; that is, a sequence of functions f_k is called **convergent** *with the limit* f if

$$\lim_{k \to \infty} \|f_k - f\| = 0; \tag{12*}$$

written out by (5) in Sec. 11.5 (where we can drop the square root, as this does not affect the limit)

$$\lim_{k\to\infty} \int_a^b r(x)[f_k(x) - f(x)]^2\,dx = 0. \tag{12}$$

Accordingly, the series (1) converges and represents f if

$$\lim_{k\to\infty} \int_a^b r(x)[s_k(x) - f(x)]^2\,dx = 0 \tag{13}$$

where s_k is the kth partial sum of (1).

$$s_k(x) = \sum_{m=0}^{k} a_m y_m(x). \tag{14}$$

Note that the integral in (13) generalizes (3) in Sec. 11.4.

We now define completeness. An **orthonormal** set $y_0, y_1, \cdots$ on an interval $a \leqq x \leqq b$ is **complete** *in a set of functions S* defined on $a \leqq x \leqq b$ if we can approximate every f belonging to S arbitrarily closely in the norm by a linear combination $a_0y_0 + a_1y_1 + \cdots + a_ky_k$, that is, technically, if for every $\epsilon > 0$ we can find constants $a_0, \cdots, a_k$ (with k large enough) such that

$$\|f - (a_0y_0 + \cdots + a_ky_k)\| < \epsilon. \tag{15}$$

Ref. [GenRef7] in App. 1 uses the more modern term **total** for *complete*.

We can now extend the ideas in Sec. 11.4 that guided us from (3) in Sec. 11.4 to Bessel's and Parseval's formulas (7) and (8) in that section. Performing the square in (13) and using (14), we first have (analog of (4) in Sec. 11.4)

$$\int_a^b r(x)[s_k(x) - f(x)]^2\,dx = \int_a^b rs_k^2\,dx - 2\int_a^b rfs_k\,dx + \int_a^b rf^2\,dx$$

$$= \int_a^b r\left[\sum_{m=0}^{k} a_m y_m\right]^2 dx - 2\sum_{m=0}^{k} a_m \int_a^b rfy_m\,dx + \int_a^b rf^2\,dx.$$

The first integral on the right equals Σa_m^2 because $\int ry_my_l\,dx = 0$ for $m \neq l$, and $\int ry_m^2\,dx = 1$. In the second sum on the right, the integral equals a_m, by (2) with $\|y_m\|^2 = 1$. Hence the first term on the right cancels half of the second term, so that the right side reduces to (analog of (6) in Sec. 11.4)

$$-\sum_{m=0}^{k} a_m^2 + \int_a^b rf^2\,dx.$$

This is nonnegative because in the previous formula the integrand on the left is nonnegative (recall that the weight $r(x)$ is positive!) and so is the integral on the left. This proves the important **Bessel's inequality** (analog of (7) in Sec. 11.4)

$$\sum_{m=0}^{k} a_m^2 \leqq \|f\|^2 = \int_a^b r(x)f(x)^2\,dx \qquad (k = 1, 2, \cdots), \tag{16}$$

Here we can let $k \to \infty$, because the left sides form a monotone increasing sequence that is bounded by the right side, so that we have convergence by the familiar Theorem 1 in App. A.3.3 Hence

$$\sum_{m=0}^{\infty} a_m^2 \leqq \|f\|^2. \tag{17}$$

Furthermore, if $y_0, y_1, \cdots$ is complete in a set of functions S, then (13) holds for every f belonging to S. By (13) this implies equality in (16) with $k \to \infty$. Hence in the case of completeness every f in S saisfies the so-called **Parseval equality** (analog of (8) in Sec. 11.4)

$$\sum_{m=0}^{\infty} a_m^2 = \|f\|^2 = \int_a^b r(x) f(x)^2 \, dx. \tag{18}$$

As a consequence of (18) we prove that in the case of *completeness* there is no function orthogonal to *every* function of the orthonormal set, with the trivial exception of a function of zero norm:

THEOREM 2

Completeness

Let $y_0, y_1, \cdots$ be a complete orthonormal set on $a \leqq x \leqq b$ in a set of functions S. Then if a function f belongs to S and is orthogonal to every y_m, it must have norm zero. In particular, if f is continuous, then f must be identically zero.

PROOF Since f is orthogonal to every y_m, the left side of (18) must be zero. If f is continuous, then $\|f\| = 0$ implies $f(x) \equiv 0$, as can be seen directly from (5) in Sec. 11.5 with f instead of y_m because $r(x) > 0$ by assumption. ■

PROBLEM SET 11.6

1–7 FOURIER–LEGENDRE SERIES

Showing the details, develop

1. $63x^5 - 90x^3 + 35x$
2. $(x + 1)^2$
3. $1 - x^4$
4. $1, \ x, \ x^2, \ x^3, \ x^4$
5. Prove that if $f(x)$ is even (is odd, respectively), its Fourier–Legendre series contains only $P_m(x)$ with even m (only $P_m(x)$ with odd m, respectively). Give examples.
6. What can you say about the coefficients of the Fourier–Legendre series of $f(x)$ if the Maclaurin series of $f(x)$ contains only powers x^{4m} $(m = 0, 1, 2, \cdots)$?
7. What happens to the Fourier–Legendre series of a *polynomial* $f(x)$ if you change a coefficient of $f(x)$? Experiment. Try to prove your answer.

8–13 CAS EXPERIMENT

FOURIER–LEGENDRE SERIES. Find and graph (on common axes) the partial sums up to S_{m_0} whose graph practically coincides with that of $f(x)$ within graphical accuracy. State m_0. On what does the size of m_0 seem to depend?

8. $f(x) = \sin \pi x$
9. $f(x) = \sin 2\pi x$
10. $f(x) = e^{-x^2}$
11. $f(x) = (1 + x^2)^{-1}$
12. $f(x) = J_0(\alpha_{0,1} x)$, $\alpha_{0,1}$ = the first positive zero of $J_0(x)$
13. $f(x) = J_0(\alpha_{0,2} x)$, $\alpha_{0,2}$ = the second positive zero of $J_0(x)$

14. TEAM PROJECT. Orthogonality on the Entire Real Axis. Hermite Polynomials.[8] These orthogonal polynomials are defined by $He_0(1) = 1$ and

$$(19)\quad He_n(x) = (-1)^n e^{x^2/2} \frac{d^n}{dx^n}(e^{-x^2/2}), \qquad n = 1, 2, \cdots.$$

REMARK. As is true for many special functions, the literature contains more than one notation, and one sometimes defines as Hermite polynomials the functions

$$H_0^* = 1, \qquad H_n^*(x) = (-1)^n e^{x^2} \frac{d^n e^{-x^2}}{dx^n}.$$

This differs from our definition, which is preferred in applications.

(a) Small Values of *n*. Show that

$$He_1(x) = x, \qquad He_2(x) = x^2 - 1,$$
$$He_3(x) = x^3 - 3x, \qquad He_4(x) = x^4 - 6x^2 + 3.$$

(b) Generating Function. A generating function of the Hermite polynomials is

$$(20)\qquad e^{tx - t^2/2} = \sum_{n=0}^{\infty} a_n(x) t^n$$

because $He_n(x) = n!\, a_n(x)$. Prove this. *Hint*: Use the formula for the coefficients of a Maclaurin series and note that $tx - \frac{1}{2}t^2 = \frac{1}{2}x^2 - \frac{1}{2}(x - t)^2$.

(c) Derivative. Differentiating the generating function with respect to x, show that

$$(21)\qquad He_n'(x) = nHe_{n-1}(x).$$

(d) Orthogonality on the *x*-Axis needs a weight function that goes to zero sufficiently fast as $x \to \pm\infty$, (Why?)

Show that the Hermite polynomials are orthogonal on $-\infty < x < \infty$ with respect to the weight function $r(x) = e^{-x^2/2}$. *Hint*. Use integration by parts and (21).

(e) ODEs. Show that

$$(22)\qquad He_n'(x) = xHe_n(x) - He_{n+1}(x).$$

Using this with $n - 1$ instead of n and (21), show that $y = He_n(x)$ satisfies the ODE

$$(23)\qquad y'' = xy' + ny = 0.$$

Show that $w = e^{-x^2/4}y$ is a solution of **Weber's equation**

$$(24)\quad w'' + (n + \tfrac{1}{2} - \tfrac{1}{4}x^2)w = 0 \qquad (n = 0, 1, \cdots).$$

15. CAS EXPERIMENT. Fourier–Bessel Series. Use Example 2 and $R = 1$, so that you get the series

$$(25)\qquad f(x) = a_1 J_0(\alpha_{0,1}x) + a_2 J_0(\alpha_{0,2}x) + a_3 J_0(\alpha_{0,3}x) + \cdots$$

With the zeros $\alpha_{0,1}\alpha_{0,2}, \cdots$ from your CAS (see also Table A1 in App. 5).

(a) Graph the terms $J_0(\alpha_{0,1}x), \cdots, J_0(\alpha_{0,10}x)$ for $0 \leqq x \leqq 1$ on common axes.

(b) Write a program for calculating partial sums of (25). Find out for what $f(x)$ your CAS can evaluate the integrals. Take two such $f(x)$ and comment empirically on the speed of convergence by observing the decrease of the coefficients.

(c) Take $f(x) = 1$ in (25) and evaluate the integrals for the coefficients analytically by (21a), Sec. 5.4, with $v = 1$. Graph the first few partial sums on common axes.

11.7 Fourier Integral

Fourier series are powerful tools for problems involving functions that are periodic or are of interest on a finite interval only. Sections 11.2 and 11.3 first illustrated this, and various further applications follow in Chap. 12. Since, of course, many problems involve functions that are ***nonperiodic*** *and are of interest on the whole x-axis,* we ask what can be done to extend the method of Fourier series to such functions. This idea will lead to "Fourier integrals."

In Example 1 we start from a special function f_L of period $2L$ and see what happens to its Fourier series if we let $L \to \infty$. Then we do the same for an *arbitrary* function f_L of period $2L$. This will motivate and suggest the main result of this section, which is an integral representation given in Theorem 1 below.

[8] CHARLES HERMITE (1822–1901), French mathematician, is known for his work in algebra and number theory. The great HENRI POINCARÉ (1854–1912) was one of his students.

EXAMPLE 1 Rectangular Wave

Consider the periodic rectangular wave $f_L(x)$ of period $2L > 2$ given by

$$f_L(x) = \begin{cases} 0 & \text{if} \quad -L < x < -1 \\ 1 & \text{if} \quad -1 < x < \ \ 1 \\ 0 & \text{if} \quad \ \ 1 < x < L. \end{cases}$$

The left part of Fig. 280 shows this function for $2L = 4, 8, 16$ as well as the nonperiodic function $f(x)$, which we obtain from f_L if we let $L \to \infty$,

$$f(x) = \lim_{L\to\infty} f_L(x) = \begin{cases} 1 & \text{if } -1 < x < 1 \\ 0 & \text{otherwise.} \end{cases}$$

We now explore what happens to the Fourier coefficients of f_L as L increases. Since f_L is even, $b_n = 0$ for all n. For a_n the Euler formulas (6), Sec. 11.2, give

$$a_0 = \frac{1}{2L}\int_{-1}^{1} dx = \frac{1}{L}, \qquad a_n = \frac{1}{L}\int_{-1}^{1} \cos\frac{n\pi x}{L}\,dx = \frac{2}{L}\int_{0}^{1} \cos\frac{n\pi x}{L}\,dx = \frac{2}{L}\,\frac{\sin(n\pi/L)}{n\pi/L}.$$

This sequence of Fourier coefficients is called the **amplitude spectrum** of f_L because $|a_n|$ is the maximum amplitude of the wave $a_n \cos(n\pi x/L)$. Figure 280 shows this spectrum for the periods $2L = 4, 8, 16$. We see that for increasing L these amplitudes become more and more dense on the positive w_n-axis, where $w_n = n\pi/L$. Indeed, for $2L = 4, 8, 16$ we have 1, 3, 7 amplitudes per "half-wave" of the function $(2 \sin w_n)/(Lw_n)$ (dashed in the figure). Hence for $2L = 2^k$ we have $2^{k-1} - 1$ amplitudes per half-wave, so that these amplitudes will eventually be everywhere dense on the positive w_n-axis (and will decrease to zero).

The outcome of this example gives an intuitive impression of what about to expect if we turn from our special function to an arbitrary one, as we shall do next. ■

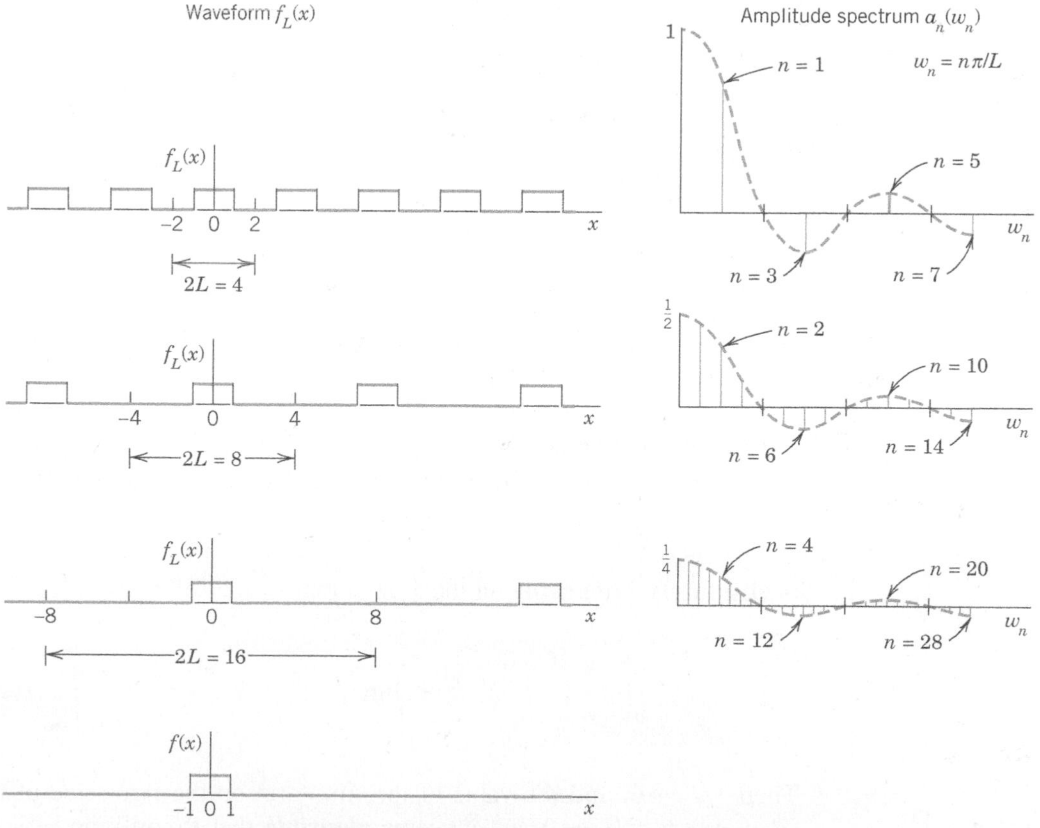

Fig. 280. Waveforms and amplitude spectra in Example 1

From Fourier Series to Fourier Integral

We now consider any periodic function $f_L(x)$ of period $2L$ that can be represented by a Fourier series

$$f_L(x) = a_0 + \sum_{n=1}^{\infty} (a_n \cos w_n x + b_n \sin w_n x), \qquad w_n = \frac{n\pi}{L}$$

and find out what happens if we let $L \to \infty$. Together with Example 1 the present calculation will suggest that we should expect an integral (instead of a series) involving $\cos wx$ and $\sin wx$ with w no longer restricted to integer multiples $w = w_n = n\pi/L$ of π/L but taking *all* values. We shall also see what form such an integral might have.

If we insert a_n and b_n from the Euler formulas (6), Sec. 11.2, and denote the variable of integration by v, the Fourier series of $f_L(x)$ becomes

$$f_L(x) = \frac{1}{2L}\int_{-L}^{L} f_L(v)\,dv + \frac{1}{L}\sum_{n=1}^{\infty}\left[\cos w_n x \int_{-L}^{L} f_L(v)\cos w_n v\,dv + \sin w_n x \int_{-L}^{L} f_L(v)\sin w_n v\,dv\right].$$

We now set

$$\Delta w = w_{n+1} - w_n = \frac{(n+1)\pi}{L} - \frac{n\pi}{L} = \frac{\pi}{L}.$$

Then $1/L = \Delta w/\pi$, and we may write the Fourier series in the form

$$f_L(x) = \frac{1}{2L}\int_{-L}^{L} f_L(v)\,dv + \frac{1}{\pi}\sum_{n=1}^{\infty}\left[(\cos w_n x)\,\Delta w \int_{-L}^{L} f_L(v)\cos w_n v\,dv + (\sin w_n x)\Delta w \int_{-L}^{L} f_L(v)\sin w_n v\,dv\right]. \tag{1}$$

This representation is valid for any fixed L, arbitrarily large, but finite.

We now let $L \to \infty$ and assume that the resulting nonperiodic function

$$f(x) = \lim_{L\to\infty} f_L(x)$$

is **absolutely integrable** on the x-axis; that is, the following (finite!) limits exist:

$$\lim_{a\to-\infty}\int_{a}^{0} |f(x)|\,dx + \lim_{b\to\infty}\int_{0}^{b} |f(x)|\,dx \quad \left(\text{written } \int_{-\infty}^{\infty} |f(x)|\,dx\right). \tag{2}$$

Then $1/L \to 0$, and the value of the first term on the right side of (1) approaches zero. Also $\Delta w = \pi/L \to 0$ and it seems ***plausible*** that the infinite series in (1) becomes an integral from 0 to ∞, which represents $f(x)$, namely,

$$(3) \qquad f(x) = \frac{1}{\pi}\int_0^\infty \left[\cos wx \int_{-\infty}^{\infty} f(v)\cos wv \, dv + \sin wx \int_{-\infty}^{\infty} f(v)\sin wv \, dv\right] dw.$$

If we introduce the notations

$$\textbf{(4)} \qquad A(w) = \frac{1}{\pi}\int_{-\infty}^{\infty} f(v)\cos wv \, dv, \qquad B(w) = \frac{1}{\pi}\int_{-\infty}^{\infty} f(v)\sin wv \, dv$$

we can write this in the form

$$\textbf{(5)} \qquad f(x) = \int_0^\infty [A(w)\cos wx + B(w)\sin wx] \, dw.$$

This is called a representation of $f(x)$ by a **Fourier integral**.

It is clear that our naive approach merely ***suggests*** the representation (5), but by no means establishes it; in fact, the limit of the series in (1) as Δw approaches zero is not the definition of the integral (3). Sufficient conditions for the validity of (5) are as follows.

THEOREM 1

Fourier Integral

If $f(x)$ is piecewise continuous (see Sec. 6.1) *in every finite interval and has a right-hand derivative and a left-hand derivative at every point* (see Sec 11.1) *and if the integral* (2) *exists, then $f(x)$ can be represented by a Fourier integral* (5) *with A and B given by* (4). *At a point where $f(x)$ is discontinuous the value of the Fourier integral equals the average of the left- and right-hand limits of $f(x)$ at that point* (see Sec. 11.1). (Proof in Ref. [C12]; see App. 1.)

Applications of Fourier Integrals

The main application of Fourier integrals is in solving ODEs and PDEs, as we shall see for PDEs in Sec. 12.6. However, we can also use Fourier integrals in integration and in discussing functions defined by integrals, as the next example.

EXAMPLE 2 **Single Pulse, Sine Integral. Dirichlet's Discontinuous Factor. Gibbs Phenomenon**

Find the Fourier integral representation of the function

$$f(x) = \begin{cases} 1 & \text{if } |x| < 1 \\ 0 & \text{if } |x| > 1 \end{cases} \qquad \text{(Fig. 281)}$$

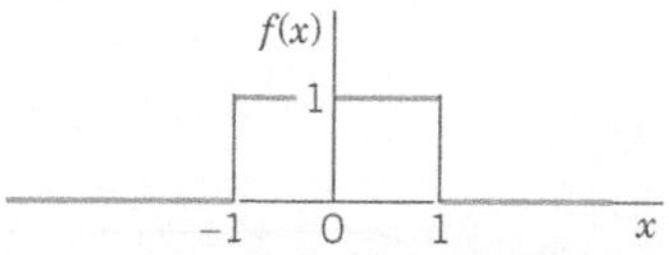

Fig. 281. Example 2

Solution. From (4) we obtain

$$A(w) = \frac{1}{\pi}\int_{-\infty}^{\infty} f(v)\cos wv\,dv = \frac{1}{\pi}\int_{-1}^{1}\cos wv\,dv = \frac{\sin wv}{\pi w}\Big|_{-1}^{1} = \frac{2\sin w}{\pi w}$$

$$B(w) = \frac{1}{\pi}\int_{-1}^{1}\sin wv\,dv = 0$$

and (5) gives the *answer*

$$f(x) = \frac{2}{\pi}\int_0^{\infty}\frac{\cos wx \sin w}{w}\,dw. \tag{6}$$

The average of the left- and right-hand limits of $f(x)$ at $x = 1$ is equal to $(1 + 0)/2$, that is, $\frac{1}{2}$.
Furthermore, from (6) and Theorem 1 we obtain (multiply by $\pi/2$)

$$\int_0^{\infty}\frac{\cos wx \sin w}{w}\,dw = \begin{cases} \pi/2 & \text{if } 0 \leqq x < 1, \\ \pi/4 & \text{if } x = 1, \\ 0 & \text{if } x > 1. \end{cases} \tag{7}$$

We mention that this integral is called **Dirichlet's discontinous factor**. (For P. L. Dirichlet see Sec. 10.8.)
The case $x = 0$ is of particular interest. If $x = 0$, then (7) gives

$$\int_0^{\infty}\frac{\sin w}{w}\,dw = \frac{\pi}{2}. \tag{8*}$$

We see that this integral is the limit of the so-called **sine integral**

$$\text{Si}(u) = \int_0^{u}\frac{\sin w}{w}\,dw \tag{8}$$

as $u \to \infty$. The graphs of Si(u) and of the integrand are shown in Fig. 282.
In the case of a Fourier series the graphs of the partial sums are approximation curves of the curve of the periodic function represented by the series. Similarly, in the case of the Fourier integral (5), approximations are obtained by replacing ∞ by numbers a. Hence the integral

$$\frac{2}{\pi}\int_0^{a}\frac{\cos wx \sin w}{w}\,dw \tag{9}$$

approximates the right side in (6) and therefore $f(x)$.

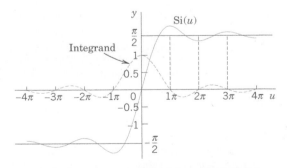

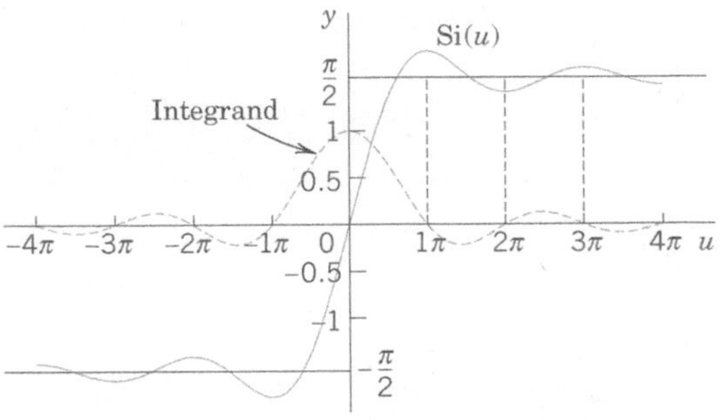

Fig. 282. Sine integral Si(u) and integrand

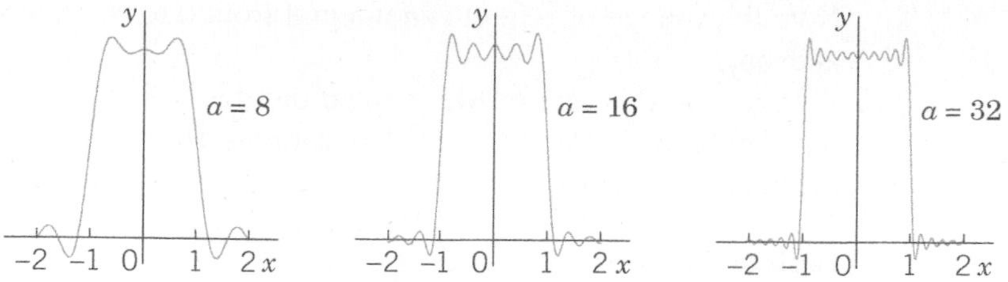

Fig. 283. The integral (9) for $a = 8, 16$, and 32, illustrating the development of the Gibbs phenomenon

Figure 283 shows oscillations near the points of discontinuity of $f(x)$. We might expect that these oscillations disappear as a approaches infinity. But this is not true; with increasing a, they are shifted closer to the points $x = \pm 1$. This unexpected behavior, which also occurs in connection with Fourier series (see Sec. 11.2), is known as the **Gibbs phenomenon**. We can explain it by representing (9) in terms of sine integrals as follows. Using (11) in App. A3.1, we have

$$\frac{2}{\pi}\int_0^a \frac{\cos wx \sin w}{w}\,dw = \frac{1}{\pi}\int_0^a \frac{\sin (w + wx)}{w}\,dw + \frac{1}{\pi}\int_0^a \frac{\sin (w - wx)}{w}\,dw.$$

In the first integral on the right we set $w + wx = t$. Then $dw/w = dt/t$, and $0 \leqq w \leqq a$ corresponds to $0 \leqq t \leqq (x + 1)a$. In the last integral we set $w - wx = -t$. Then $dw/w = dt/t$, and $0 \leqq w \leqq a$ corresponds to $0 \leqq t \leqq (x - 1)a$. Since $\sin(-t) = -\sin t$, we thus obtain

$$\frac{2}{\pi}\int_0^a \frac{\cos wx \sin w}{w}\,dw = \frac{1}{\pi}\int_0^{(x+1)a} \frac{\sin t}{t}\,dt - \frac{1}{\pi}\int_0^{(x-1)a} \frac{\sin t}{t}\,dt.$$

From this and (8) we see that our integral (9) equals

$$\frac{1}{\pi}\operatorname{Si}(a[x + 1]) - \frac{1}{\pi}\operatorname{Si}(a[x - 1])$$

and the oscillations in Fig. 283 result from those in Fig. 282. The increase of a amounts to a transformation of the scale on the axis and causes the shift of the oscillations (the waves) toward the points of discontinuity -1 and 1. ■

Fourier Cosine Integral and Fourier Sine Integral

Just as Fourier *series* simplify if a function is even or odd (see Sec. 11.2), so do Fourier *integrals*, and you can save work. Indeed, if f has a Fourier integral representation and is *even*, then $B(w) = 0$ in (4). This holds because the integrand of $B(w)$ is odd. Then (5) reduces to a **Fourier cosine integral**

(10) $$f(x) = \int_0^\infty A(w) \cos wx \, dw \quad \text{where} \quad A(w) = \frac{2}{\pi}\int_0^\infty f(v) \cos wv \, dv.$$

Note the change in $A(w)$: for even f the integrand is even, hence the integral from $-\infty$ to ∞ equals twice the integral from 0 to ∞, just as in (7a) of Sec. 11.2.

Similarly, if f has a Fourier integral representation and is *odd*, then $A(w) = 0$ in (4). This is true because the integrand of $A(w)$ is odd. Then (5) becomes a **Fourier sine integral**

(11) $$f(x) = \int_0^\infty B(w) \sin wx \, dw \quad \text{where} \quad B(w) = \frac{2}{\pi}\int_0^\infty f(v) \sin wv \, dv.$$

Note the change of $B(w)$ to an integral from 0 to ∞ because $B(w)$ is even (odd times odd is even).

Earlier in this section we pointed out that the main application of the Fourier integral representation is in differential equations. However, these representations also help in evaluating integrals, as the following example shows for integrals from 0 to ∞.

EXAMPLE 3 Laplace Integrals

We shall derive the Fourier cosine and Fourier sine integrals of $f(x) = e^{-kx}$, where $x > 0$ and $k > 0$ (Fig. 284). The result will be used to evaluate the so-called Laplace integrals.

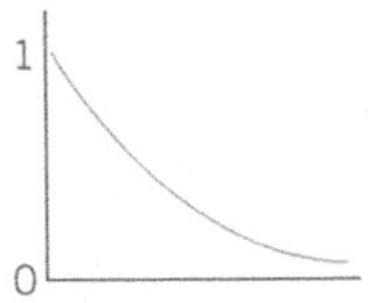

Fig. 284. f(x) in Example 3

Solution. (a) From (10) we have $A(w) = \frac{2}{\pi}\int_0^\infty e^{-kv}\cos wv\, dv$. Now, by integration by parts,

$$\int e^{-kv}\cos wv\, dv = -\frac{k}{k^2+w^2}e^{-kv}\left(-\frac{w}{k}\sin wv + \cos wv\right).$$

If $v = 0$, the expression on the right equals $-k/(k^2 + w^2)$. If v approaches infinity, that expression approaches zero because of the exponential factor. Thus $2/\pi$ times the integral from 0 to ∞ gives

$$A(w) = \frac{2k/\pi}{k^2+w^2}. \tag{12}$$

By substituting this into the first integral in (10) we thus obtain the Fourier cosine integral representation

$$f(x) = e^{-kx} = \frac{2k}{\pi}\int_0^\infty \frac{\cos wx}{k^2+w^2}\, dw \qquad (x > 0, \quad k > 0).$$

From this representation we see that

$$\int_0^\infty \frac{\cos wx}{k^2+w^2}\, dw = \frac{\pi}{2k}e^{-kx} \qquad (x > 0, \quad k > 0). \tag{13}$$

(b) Similarly, from (11) we have $B(w) = \frac{2}{\pi}\int_0^\infty e^{-kv}\sin wv\, dv$. By integration by parts,

$$\int e^{-kv}\sin wv\, dv = -\frac{w}{k^2+w^2}e^{-kv}\left(\frac{k}{w}\sin wv + \cos wv\right).$$

This equals $-w/(k^2 + w^2)$ if $v = 0$, and approaches 0 as $v \to \infty$. Thus

$$B(w) = \frac{2w/\pi}{k^2+w^2}. \tag{14}$$

From (14) we thus obtain the Fourier sine integral representation

$$f(x) = e^{-kx} = \frac{2}{\pi}\int_0^\infty \frac{w\sin wx}{k^2+w^2}\, dw.$$

From this we see that

$$\int_0^\infty \frac{w\sin wx}{k^2+w^2}\, dw = \frac{\pi}{2}e^{-kx} \qquad (x > 0, \quad k > 0). \tag{15}$$

The integrals (13) and (15) are called the **Laplace integrals**. ■

PROBLEM SET 11.7

1–6 EVALUATION OF INTEGRALS

Show that the integral represents the indicated function. *Hint.* Use (5), (10), or (11); the integral tells you which one, and its value tells you what function to consider. Show your work in detail.

1. $$\int_0^\infty \frac{\cos xw + w \sin xw}{1 + w^2}\, dx = \begin{cases} 0 & \text{if } x < 0 \\ \pi/2 & \text{if } x = 0 \\ \pi e^{-x} & \text{if } x > 0 \end{cases}$$

2. $$\int_0^\infty \frac{\sin \pi w \sin xw}{1 - w^2}\, dw = \begin{cases} \frac{\pi}{2} \sin x & \text{if } 0 \leqq x \leqq \pi \\ 0 & \text{if } x > \pi \end{cases}$$

3. $$\int_0^\infty \frac{1 - \cos \pi w}{w} \sin xw\, dw = \begin{cases} \frac{1}{2}\pi & \text{if } 0 < x < \pi \\ 0 & \text{if } x > \pi \end{cases}$$

4. $$\int_0^\infty \frac{\cos \frac{1}{2}\pi w}{1 - w^2} \cos xw\, dw = \begin{cases} \frac{1}{2}\pi \cos x & \text{if } 0 < |x| < \frac{1}{2}\pi \\ 0 & \text{if } |x| \geqq \frac{1}{2}\pi \end{cases}$$

5. $$\int_0^\infty \frac{\sin w - w \cos w}{w^2} \sin xw\, dw = \begin{cases} \frac{1}{2}\pi x & \text{if } 0 < x < 1 \\ \frac{1}{4}\pi & \text{if } x = 1 \\ 0 & \text{if } x > 1 \end{cases}$$

6. $$\int_0^\infty \frac{w^3 \sin xw}{w^4 + 4}\, dw = \tfrac{1}{2}\pi e^{-x} \cos x \quad \text{if } x > 0$$

7–12 FOURIER COSINE INTEGRAL REPRESENTATIONS

Represent $f(x)$ as an integral (10).

7. $f(x) = \begin{cases} 1 & \text{if } 0 < x < 1 \\ 0 & \text{if } x > 1 \end{cases}$

8. $f(x) = \begin{cases} x^2 & \text{if } 0 < x < 1 \\ 0 & \text{if } x > 1 \end{cases}$

9. $f(x) = 1/(1 + x^2)$ [$x > 0$. *Hint.* See (13).]

10. $f(x) = \begin{cases} a^2 - x^2 & \text{if } 0 < x < a \\ 0 & \text{if } x > a \end{cases}$

11. $f(x) = \begin{cases} \sin x & \text{if } 0 < x < \pi \\ 0 & \text{if } x > \pi \end{cases}$

12. $f(x) = \begin{cases} e^{-x} & \text{if } 0 < x < a \\ 0 & \text{if } x > a \end{cases}$

13. CAS EXPERIMENT. Approximate Fourier Cosine Integrals. Graph the integrals in Prob. 7, 9, and 11 as functions of x. Graph approximations obtained by replacing ∞ with finite upper limits of your choice. Compare the quality of the approximations. Write a short report on your empirical results and observations.

14. PROJECT. Properties of Fourier Integrals

(a) Fourier cosine integral. Show that (10) implies

(a1) $$f(ax) = \frac{1}{a}\int_0^\infty A\left(\frac{w}{a}\right) \cos xw\, dw$$
$(a > 0)$ *(Scale change)*

(a2) $$xf(x) = \int_0^\infty B^*(w) \sin xw\, dw,$$
$$B^* = -\frac{dA}{dw}, \quad A \text{ as in (10)}$$

(a3) $$x^2 f(x) = \int_0^\infty A^*(w) \cos xw\, dw,$$
$$A^* = -\frac{d^2A}{dw^2}.$$

(b) Solve Prob. 8 by applying (a3) to the result of Prob. 7.

(c) Verify (a2) for $f(x) = 1$ if $0 < x < a$ and $f(x) = 0$ if $x > a$.

(d) Fourier sine integral. Find formulas for the Fourier sine integral similar to those in (a).

15. CAS EXPERIMENT. Sine Integral. Plot $\text{Si}(u)$ for positive u. Does the sequence of the maximum and minimum values give the impression that it converges and has the limit $\pi/2$? Investigate the Gibbs phenomenon graphically.

16–20 FOURIER SINE INTEGRAL REPRESENTATIONS

Represent $f(x)$ as an integral (11).

16. $f(x) = \begin{cases} x & \text{if } 0 < x < a \\ 0 & \text{if } x > a \end{cases}$

17. $f(x) = \begin{cases} 1 & \text{if } 0 < x < 1 \\ 0 & \text{if } x > 1 \end{cases}$

18. $f(x) = \begin{cases} \cos x & \text{if } 0 < x < \pi \\ 0 & \text{if } x > \pi \end{cases}$

19. $f(x) = \begin{cases} e^{x} & \text{if } 0 < x < 1 \\ 0 & \text{if } x > 1 \end{cases}$

20. $f(x) = \begin{cases} e^{-x} & \text{if } 0 < x < 1 \\ 0 & \text{if } x > 1 \end{cases}$

11.8 Fourier Cosine and Sine Transforms

An **integral transform** is a transformation in the form of an integral that produces from given functions new functions depending on a different variable. One is mainly interested in these transforms because they can be used as tools in solving ODEs, PDEs, and integral equations and can often be of help in handling and applying special functions. The Laplace transform of Chap. 6 serves as an example and is by far the most important integral transform in engineering.

Next in order of importance are Fourier transforms. They can be obtained from the Fourier integral in Sec. 11.7 in a straightforward way. In this section we derive two such transforms that are real, and in Sec. 11.9 a complex one.

Fourier Cosine Transform

The Fourier cosine transform concerns ***even functions*** $f(x)$. We obtain it from the Fourier cosine integral [(10) in Sec. 10.7]

$$f(x) = \int_0^\infty A(w) \cos wx \, dw, \qquad \text{where} \qquad A(w) = \frac{2}{\pi}\int_0^\infty f(v) \cos wv \, dv.$$

Namely, we set $A(w) = \sqrt{2/\pi}\, \hat{f}_c(w)$, where c suggests "cosine." Then, writing $v = x$ in the formula for $A(w)$, we have

$$\textbf{(1a)} \qquad \hat{f}_c(w) = \sqrt{\frac{2}{\pi}} \int_0^\infty f(x) \cos wx \, dx$$

and

$$\textbf{(1b)} \qquad f(x) = \sqrt{\frac{2}{\pi}} \int_0^\infty \hat{f}_c(w) \cos wx \, dw.$$

Formula (1a) gives from $f(x)$ a new function $\hat{f}_c(w)$, called the **Fourier cosine transform** of $f(x)$. Formula (1b) gives us back $f(x)$ from $\hat{f}_c(w)$, and we therefore call $f(x)$ the **inverse Fourier cosine transform** of $\hat{f}_c(w)$.

The process of obtaining the transform $\hat{f}_c$ from a given f is also called the **Fourier cosine transform** or the *Fourier cosine transform method.*

Fourier Sine Transform

Similarly, in (11), Sec. 11.7, we set $B(w) = \sqrt{2/\pi}\, \hat{f}_s(w)$, where s suggests "sine." Then, writing $v = x$, we have from (11), Sec. 11.7, the **Fourier sine transform**, of $f(x)$ given by

$$\textbf{(2a)} \qquad \hat{f}_s(w) = \sqrt{\frac{2}{\pi}} \int_0^\infty f(x) \sin wx \, dx,$$

and the **inverse Fourier sine transform** of $\hat{f}_s(w)$, given by

(2b)
$$f(x) = \sqrt{\frac{2}{\pi}} \int_0^\infty \hat{f}_s(w) \sin wx \, dw.$$

The process of obtaining $f_s(w)$ from $f(x)$ is also called the **Fourier sine transform** or the *Fourier sine transform method.*

Other notations are

$$\mathcal{F}_c(f) = \hat{f}_c, \qquad \mathcal{F}_s(f) = \hat{f}_s$$

and $\mathcal{F}_c^{-1}$ and $\mathcal{F}_s^{-1}$ for the inverses of $\mathcal{F}_c$ and $\mathcal{F}_s$, respectively.

EXAMPLE 1 Fourier Cosine and Fourier Sine Transforms

Find the Fourier cosine and Fourier sine transforms of the function

$$f(x) = \begin{cases} k & \text{if} \quad 0 < x < a \\ 0 & \text{if} \quad x > a \end{cases} \qquad \text{(Fig. 285).}$$

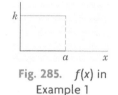

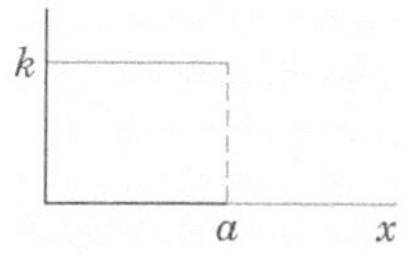

Fig. 285. $f(x)$ in Example 1

Solution. From the definitions (1a) and (2a) we obtain by integration

$$\hat{f}_c(w) = \sqrt{\frac{2}{\pi}}\, k \int_0^a \cos wx \, dx = \sqrt{\frac{2}{\pi}}\, k \left(\frac{\sin aw}{w}\right)$$

$$\hat{f}_s(w) = \sqrt{\frac{2}{\pi}}\, k \int_0^a \sin wx \, dx = \sqrt{\frac{2}{\pi}}\, k \left(\frac{1 - \cos aw}{w}\right).$$

This agrees with formulas 1 in the first two tables in Sec. 11.10 (where $k = 1$).

Note that for $f(x) = k = \text{const}$ $(0 < x < \infty)$, these transforms do not exist. (Why?) ■

EXAMPLE 2 Fourier Cosine Transform of the Exponential Function

Find $\mathcal{F}_c(e^{-x})$.

Solution. By integration by parts and recursion,

$$\mathcal{F}_c(e^{-x}) = \sqrt{\frac{2}{\pi}} \int_0^\infty e^{-x} \cos wx \, dx = \sqrt{\frac{2}{\pi}} \frac{e^{-x}}{1 + w^2} (-\cos wx + w \sin wx) \Big|_0^\infty = \frac{\sqrt{2/\pi}}{1 + w^2}.$$

This agrees with formula 3 in Table I, Sec. 11.10, with $a = 1$. See also the next example. ■

What did we do to introduce the two integral transforms under consideration? Actually not much: We changed the notations A and B to get a "symmetric" distribution of the constant $2/\pi$ in the original formulas (1) and (2). This redistribution is a standard convenience, but it is not essential. One could do without it.

What have we gained? We show next that these transforms have operational properties that permit them to convert differentiations into algebraic operations (just as the Laplace transform does). This is the key to their application in solving differential equations.

Linearity, Transforms of Derivatives

If $f(x)$ is absolutely integrable (see Sec. 11.7) on the positive x-axis and piecewise continuous (see Sec. 6.1) on every finite interval, then the Fourier cosine and sine transforms of f exist.

Furthermore, if f and g have Fourier cosine and sine transforms, so does $af + bg$ for any constants a and b, and by (1a)

$$\mathscr{F}_c(af + bg) = \sqrt{\frac{2}{\pi}} \int_0^\infty [af(x) + bg(x)] \cos wx \, dx$$

$$= a\sqrt{\frac{2}{\pi}} \int_0^\infty f(x) \cos wx \, dx + b\sqrt{\frac{2}{\pi}} \int_0^\infty g(x) \cos wx \, dx.$$

The right side is $a\mathscr{F}_c(f) + b\mathscr{F}_c(g)$. Similarly for $\mathscr{F}_s$, by (2). This shows that the Fourier cosine and sine transforms are **linear operations**,

$$(3) \qquad
\begin{aligned}
&\text{(a)} \quad \mathscr{F}_c(af + bg) = a\mathscr{F}_c(f) + b\mathscr{F}_c(g), \\
&\text{(b)} \quad \mathscr{F}_s(af + bg) = a\mathscr{F}_s(f) + b\mathscr{F}_s(g).
\end{aligned}$$

THEOREM 1

Cosine and Sine Transforms of Derivatives

Let $f(x)$ be continuous and absolutely integrable on the x-axis, let $f'(x)$ be piecewise continuous on every finite interval, and let $f(x) \to 0$ as $x \to \infty$. Then

$$(4) \qquad
\begin{aligned}
&\text{(a)} \quad \mathscr{F}_c\{f'(x)\} = w\mathscr{F}_s\{f(x)\} - \sqrt{\frac{2}{\pi}}\, f(0), \\
&\text{(b)} \quad \mathscr{F}_s\{f'(x)\} = -w\mathscr{F}_c\{f(x)\}.
\end{aligned}$$

PROOF   This follows from the definitions and by using integration by parts, namely,

$$\mathscr{F}_c\{f'(x)\} = \sqrt{\frac{2}{\pi}} \int_0^\infty f'(x) \cos wx \, dx$$

$$= \sqrt{\frac{2}{\pi}} \left[f(x) \cos wx \Big|_0^\infty + w \int_0^\infty f(x) \sin wx \, dx \right]$$

$$= -\sqrt{\frac{2}{\pi}}\, f(0) + w\mathscr{F}_s\{f(x)\};$$

and similarly,

$$\mathscr{F}_s\{f'(x)\} = \sqrt{\frac{2}{\pi}} \int_0^\infty f'(x) \sin wx \, dx$$

$$= \sqrt{\frac{2}{\pi}} \left[f(x) \sin wx \Big|_0^\infty - w \int_0^\infty f(x) \cos wx \, dx \right]$$

$$= 0 - w\mathscr{F}_c\{f(x)\}. \qquad \blacksquare$$

Formula (4a) with f' instead of f gives (when f', f'' satisfy the respective assumptions for f, f' in Theorem 1)

$$\mathcal{F}_c\{f''(x)\} = w\mathcal{F}_s\{f'(x)\} - \sqrt{\frac{2}{\pi}} f'(0);$$

hence by (4b)

(5a)
$$\mathcal{F}_c\{f''(x)\} = -w^2\mathcal{F}_c\{f(x)\} - \sqrt{\frac{2}{\pi}} f'(0).$$

Similarly,

(5b)
$$\mathcal{F}_s\{f''(x)\} = -w^2\mathcal{F}_s\{f(x)\} + \sqrt{\frac{2}{\pi}} wf(0).$$

A basic application of (5) to PDEs will be given in Sec. 12.7. For the time being we show how (5) can be used for deriving transforms.

EXAMPLE 3 An Application of the Operational Formula (5)

Find the Fourier cosine transform $\mathcal{F}_c(e^{-ax})$ of $f(x) = e^{-ax}$, where $a > 0$.

Solution. By differentiation, $(e^{-ax})'' = a^2e^{-ax}$; thus

$$a^2 f(x) = f''(x).$$

From this, (5a), and the linearity (3a),

$$\begin{aligned} a^2\mathcal{F}_c(f) &= \mathcal{F}_c(f'') \\ &= -w^2\mathcal{F}_c(f) - \sqrt{\frac{2}{\pi}} f'(0) \\ &= -w^2\mathcal{F}_c(f) + a\sqrt{\frac{2}{\pi}}. \end{aligned}$$

Hence

$$(a^2 + w^2)\mathcal{F}_c(f) = a\sqrt{2/\pi}.$$

The *answer* is (see Table I, Sec. 11.10)

$$\mathcal{F}_c(e^{-ax}) = \sqrt{\frac{2}{\pi}}\left(\frac{a}{a^2 + w^2}\right) \qquad (a > 0).$$ ■

Tables of Fourier cosine and sine transforms are included in Sec. 11.10.

PROBLEM SET 11.8

1–8 FOURIER COSINE TRANSFORM

1. Find the cosine transform $\hat{f}_c(w)$ of $f(x) = 1$ if $0 < x < 1$, $f(x) = -1$ if $1 < x < 2$, $f(x) = 0$ if $x > 2$.
2. Find f in Prob. 1 from the answer $\hat{f}_c$.
3. Find $\hat{f}_c(w)$ for $f(x) = x$ if $0 < x < 2$, $f(x) = 0$ if $x > 2$.
4. Derive formula 3 in Table I of Sec. 11.10 by integration.
5. Find $\hat{f}_c(w)$ for $f(x) = x^2$ if $0 < x < 1$, $f(x) = 0$ if $x > 1$.
6. **Continuity assumptions.** Find $\hat{g}_c(w)$ for $g(x) = 2$ if $0 < x < 1$, $g(x) = 0$ if $x > 1$. Try to obtain from it $\hat{f}_c(w)$ for $f(x)$ in Prob. 5 by using (5a).
7. **Existence?** Does the Fourier cosine transform of $x^{-1} \sin x$ $(0 < x < \infty)$ exist? Of $x^{-1} \cos x$? Give reasons.
8. **Existence?** Does the Fourier cosine transform of $f(x) = k = \text{const}$ $(0 < x < \infty)$ exist? The Fourier sine transform?

9–15 FOURIER SINE TRANSFORM

9. Find $\mathcal{F}_s(e^{-ax})$, $a > 0$, by integration.
10. Obtain the answer to Prob. 9 from (5b).
11. Find $\hat{f}_s(w)$ for $f(x) = x^2$ if $0 < x < 1$, $f(x) = 0$ if $x > 1$.
12. Find $\mathcal{F}_s(xe^{-x^2/2})$ from (4b) and a suitable formula in Table I of Sec. 11.10.
13. Find $\mathcal{F}_s(e^{-x})$ from (4a) and formula 3 of Table I in Sec. 11.10.
14. **Gamma function.** Using formulas 2 and 4 in Table II of Sec. 11.10, prove $\Gamma(\frac{1}{2}) = \sqrt{\pi}$ [(30) in App. A3.1], a value needed for Bessel functions and other applications.
15. **WRITING PROJECT. Finding Fourier Cosine and Sine Transforms.** Write a short report on ways of obtaining these transforms, with illustrations by examples of your own.

11.9 Fourier Transform. Discrete and Fast Fourier Transforms

In Sec. 11.8 we derived two real transforms. Now we want to derive a complex transform that is called the **Fourier transform**. It will be obtained from the complex Fourier integral, which will be discussed next.

Complex Form of the Fourier Integral

The (real) Fourier integral is [see (4), (5), Sec. 11.7]

$$f(x) = \int_0^{\infty} [A(w) \cos wx + B(w) \sin wx] \, dw$$

where

$$A(w) = \frac{1}{\pi} \int_{-\infty}^{\infty} f(v) \cos wv \, dv, \qquad B(w) = \frac{1}{\pi} \int_{-\infty}^{\infty} f(v) \sin wv \, dv.$$

Substituting A and B into the integral for f, we have

$$f(x) = \frac{1}{\pi} \int_0^{\infty} \int_{-\infty}^{\infty} f(v)[\cos wv \cos wx + \sin wv \sin wx] \, dv \, dw.$$

By the addition formula for the cosine [(6) in App. A3.1] the expression in the brackets $[\cdots]$ equals $\cos(wv - wx)$ or, since the cosine is even, $\cos(wx - wv)$. We thus obtain

$$f(x) = \frac{1}{\pi}\int_0^\infty \left[\int_{-\infty}^\infty f(v)\cos(wx - wv)dv\right]dw. \tag{1*}$$

The integral in brackets is an *even* function of w, call it $F(w)$, because $\cos(wx - wv)$ is an even function of w, the function f does not depend on w, and we integrate with respect to v (not w). Hence the integral of $F(w)$ from $w = 0$ to ∞ is $\frac{1}{2}$ times the integral of $F(w)$ from $-\infty$ to ∞. Thus (note the change of the integration limit!)

$$f(x) = \frac{1}{2\pi}\int_{-\infty}^\infty \left[\int_{-\infty}^\infty f(v)\cos(wx - wv)\,dv\right]dw. \tag{1}$$

We claim that the integral of the form (1) with sin instead of cos is zero:

$$\frac{1}{2\pi}\int_{-\infty}^\infty \left[\int_{-\infty}^\infty f(v)\sin(wx - wv)\,dv\right]dw = 0. \tag{2}$$

This is true since $\sin(wx - wv)$ is an odd function of w, which makes the integral in brackets an odd function of w, call it $G(w)$. Hence the integral of $G(w)$ from $-\infty$ to ∞ is zero, as claimed.

We now take the integrand of (1) plus $i\,(= \sqrt{-1})$ times the integrand of (2) and use the **Euler formula** [(11) in Sec. 2.2]

$$e^{ix} = \cos x + i\sin x. \tag{3}$$

Taking $wx - wv$ instead of x in (3) and multiplying by $f(v)$ gives

$$f(v)\cos(wx - wv) + if(v)\sin(wx - wv) = f(v)e^{i(wx-wv)}.$$

Hence the result of adding (1) plus i times (2), called the **complex Fourier integral**, is

$$f(x) = \frac{1}{2\pi}\int_{-\infty}^\infty \int_{-\infty}^\infty f(v)e^{iw(x-v)}\,dv\,dw \qquad (i = \sqrt{-1}). \tag{4}$$

To obtain the desired Fourier transform will take only a very short step from here.

Fourier Transform and Its Inverse

Writing the exponential function in (4) as a product of exponential functions, we have

$$f(x) = \frac{1}{\sqrt{2\pi}}\int_{-\infty}^\infty \left[\frac{1}{\sqrt{2\pi}}\int_{-\infty}^\infty f(v)e^{-iwv}\,dv\right]e^{iwx}\,dw. \tag{5}$$

The expression in brackets is a function of w, is denoted by $\hat{f}(w)$, and is called the **Fourier transform** of f; writing $v = x$, we have

$$\hat{f}(w) = \frac{1}{\sqrt{2\pi}}\int_{-\infty}^\infty f(x)e^{-iwx}\,dx. \tag{6}$$

With this, (5) becomes

$$f(x) = \frac{1}{\sqrt{2\pi}} \int_{-\infty}^{\infty} \hat{f}(w) e^{iwx} \, dw \tag{7}$$

and is called the **inverse Fourier transform** of $\hat{f}(w)$.

Another notation for the Fourier transform is

$$\hat{f} = \mathcal{F}(f),$$

so that

$$f = \mathcal{F}^{-1}(\hat{f}).$$

The process of obtaining the Fourier transform $\mathcal{F}(f) = \hat{f}$ from a given f is also called the **Fourier transform** or the *Fourier transform method.*

Using concepts defined in Secs. 6.1 and 11.7 we now state (without proof) conditions that are sufficient for the existence of the Fourier transform.

THEOREM 1

Existence of the Fourier Transform

If $f(x)$ is absolutely integrable on the x-axis and piecewise continuous on every finite interval, then the Fourier transform $\hat{f}(w)$ of $f(x)$ given by (6) exists.

EXAMPLE 1 **Fourier Transform**

Find the Fourier transform of $f(x) = 1$ if $|x| < 1$ and $f(x) = 0$ otherwise.

Solution. Using (6) and integrating, we obtain

$$\hat{f}(w) = \frac{1}{\sqrt{2\pi}} \int_{-1}^{1} e^{-iwx} \, dx = \frac{1}{\sqrt{2\pi}} \cdot \frac{e^{-iwx}}{-iw} \bigg|_{-1}^{1} = \frac{1}{-iw\sqrt{2\pi}} (e^{-iw} - e^{iw}).$$

As in (3) we have $e^{iw} = \cos w + i \sin w$, $e^{-iw} = \cos w - i \sin w$, and by subtraction

$$e^{iw} - e^{-iw} = 2i \sin w.$$

Substituting this in the previous formula on the right, we see that i drops out and we obtain the answer

$$\hat{f}(w) = \sqrt{\frac{\pi}{2}} \frac{\sin w}{w}.$$ ■

EXAMPLE 2 **Fourier Transform**

Find the Fourier transform $\mathcal{F}(e^{-ax})$ of $f(x) = e^{-ax}$ if $x > 0$ and $f(x) = 0$ if $x < 0$; here $a > 0$.

Solution. From the definition (6) we obtain by integration

$$\mathcal{F}(e^{-ax}) = \frac{1}{\sqrt{2\pi}} \int_{0}^{\infty} e^{-ax} e^{-iwx} \, dx$$

$$= \frac{1}{\sqrt{2\pi}} \frac{e^{-(a+iw)x}}{-(a + iw)} \bigg|_{x=0}^{\infty} = \frac{1}{\sqrt{2\pi}(a + iw)}.$$

This proves formula 5 of Table III in Sec. 11.10. ■

Physical Interpretation: Spectrum

The nature of the representation (7) of $f(x)$ becomes clear if we think of it as a superposition of sinusoidal oscillations of all possible frequencies, called a **spectral representation**. This name is suggested by optics, where light is such a superposition of colors (frequencies). In (7), the "**spectral density**" $\hat{f}(w)$ measures the intensity of $f(x)$ in the frequency interval between w and $w + \Delta w$ (Δw small, fixed). We claim that, in connection with vibrations, the integral

$$\int_{-\infty}^{\infty} |\hat{f}(w)|^2 \, dw$$

can be interpreted as the **total energy** of the physical system. Hence an integral of $|\hat{f}(w)|^2$ from a to b gives the contribution of the frequencies w between a and b to the total energy.

To make this plausible, we begin with a mechanical system giving a single frequency, namely, the harmonic oscillator (mass on a spring, Sec. 2.4)

$$my'' + ky = 0.$$

Here we denote time t by x. Multiplication by y' gives $my'y'' + ky'y = 0$. By integration,

$$\tfrac{1}{2}mv^2 + \tfrac{1}{2}ky^2 = E_0 = \text{const}$$

where $v = y'$ is the velocity. The first term is the kinetic energy, the second the potential energy, and E_0 the total energy of the system. Now a general solution is (use (3) in Sec. 11.4 with $t = x$)

$$y = a_1 \cos w_0 x + b_1 \sin w_0 x = c_1 e^{iw_0 x} + c_{-1} e^{-iw_0 x}, \qquad w_0^2 = k/m$$

where $c_1 = (a_1 - ib_1)/2$, $c_{-1} = \bar{c}_1 = (a_1 + ib_1)/2$. We write simply $A = c_1 e^{iw_0 x}$, $B = c_{-1} e^{-iw_0 x}$. Then $y = A + B$. By differentiation, $v = y' = A' + B' = iw_0(A - B)$. Substitution of v and y on the left side of the equation for E_0 gives

$$E_0 = \tfrac{1}{2}mv^2 + \tfrac{1}{2}ky^2 = \tfrac{1}{2}m(iw_0)^2(A - B)^2 + \tfrac{1}{2}k(A + B)^2.$$

Here $w_0^2 = k/m$, as just stated; hence $mw_0^2 = k$. Also $i^2 = -1$, so that

$$E_0 = \tfrac{1}{2}k[-(A - B)^2 + (A + B)^2] = 2kAB = 2kc_1 e^{iw_0 x} c_{-1} e^{-iw_0 x} = 2kc_1 c_{-1} = 2k|c_1|^2.$$

Hence *the energy is proportional to the square of the amplitude* $|c_1|$.

As the next step, if a more complicated system leads to a periodic solution $y = f(x)$ that can be represented by a Fourier series, then instead of the single energy term $|c_1|^2$ we get a series of squares $|c_n|^2$ of Fourier coefficients c_n given by (6), Sec. 11.4. In this case we have a "**discrete spectrum**" (or "**point spectrum**") consisting of countably many isolated frequencies (infinitely many, in general), the corresponding $|c_n|^2$ being the contributions to the total energy.

Finally, a system whose solution can be represented by an integral (7) leads to the above integral for the energy, as is plausible from the cases just discussed.

Linearity. Fourier Transform of Derivatives

New transforms can be obtained from given ones by using

THEOREM 2

Linearity of the Fourier Transform

The Fourier transform is a ***linear operation****; that is, for any functions $f(x)$ and $g(x)$ whose Fourier transforms exist and any constants a and b, the Fourier transform of $af + bg$ exists, and*

$$\mathcal{F}(af + bg) = a\mathcal{F}(f) + b\mathcal{F}(g). \tag{8}$$

PROOF This is true because integration is a linear operation, so that (6) gives

$$\begin{aligned}\mathcal{F}\{af(x) + bg(x)\} &= \frac{1}{\sqrt{2\pi}} \int_{-\infty}^{\infty} [af(x) + bg(x)]\, e^{-iwx}\, dx \\ &= a \frac{1}{\sqrt{2\pi}} \int_{-\infty}^{\infty} f(x)e^{-iwx}\, dx + b \frac{1}{\sqrt{2\pi}} \int_{-\infty}^{\infty} g(x)e^{-iwx}\, dx \\ &= a\mathcal{F}\{f(x)\} + b\mathcal{F}\{g(x)\}.\end{aligned}$$

■

In applying the Fourier transform to differential equations, the key property is that differentiation of functions corresponds to multiplication of transforms by iw:

THEOREM 3

Fourier Transform of the Derivative of $f(x)$

Let $f(x)$ be continuous on the x-axis and $f(x) \to 0$ as $|x| \to \infty$. Furthermore, let $f'(x)$ be absolutely integrable on the x-axis. Then

$$\mathcal{F}\{f'(x)\} = iw\mathcal{F}\{f(x)\}. \tag{9}$$

PROOF From the definition of the Fourier transform we have

$$\mathcal{F}\{f'(x)\} = \frac{1}{\sqrt{2\pi}} \int_{-\infty}^{\infty} f'(x)e^{-iwx}\, dx.$$

Integrating by parts, we obtain

$$\mathcal{F}\{f'(x)\} = \frac{1}{\sqrt{2\pi}} \left[f(x)e^{-iwx} \Big|_{-\infty}^{\infty} - (-iw) \int_{-\infty}^{\infty} f(x)e^{-iwx}\, dx \right].$$

Since $f(x) \to 0$ as $|x| \to \infty$, the desired result follows, namely,

$$\mathcal{F}\{f'(x)\} = 0 + iw\mathcal{F}\{f(x)\}.$$

■

Two successive applications of (9) give

$$\mathcal{F}(f'') = iw\mathcal{F}(f') = (iw)^2\mathcal{F}(f).$$

Since $(iw)^2 = -w^2$, we have for the transform of the second derivative of f

(10)
$$\mathcal{F}\{f''(x)\} = -w^2\mathcal{F}\{f(x)\}.$$

Similarly for higher derivatives.

An application of (10) to differential equations will be given in Sec. 12.6. For the time being we show how (9) can be used to derive transforms.

EXAMPLE 3 **Application of the Operational Formula (9)**

Find the Fourier transform of xe^{-x^2} from Table III, Sec 11.10.

Solution.   We use (9). By formula 9 in Table III

$$\mathcal{F}(xe^{-x^2}) = \mathcal{F}\{-\tfrac{1}{2}(e^{-x^2})'\}$$

$$= -\tfrac{1}{2}\mathcal{F}\{(e^{-x^2})'\}$$

$$= -\tfrac{1}{2}iw\mathcal{F}(e^{-x^2})$$

$$= -\frac{1}{2}iw\frac{1}{\sqrt{2}}e^{-w^2/4}$$

$$= -\frac{iw}{2\sqrt{2}}e^{-w^2/4}. \qquad\blacksquare$$

Convolution

The **convolution** $f * g$ of functions f and g is defined by

(11)
$$h(x) = (f * g)(x) = \int_{-\infty}^{\infty} f(p)g(x - p)\,dp = \int_{-\infty}^{\infty} f(x - p)g(p)\,dp.$$

The purpose is the same as in the case of Laplace transforms (Sec. 6.5): taking the convolution of two functions and then taking the transform of the convolution is the same as multiplying the transforms of these functions (and multiplying them by $\sqrt{2\pi}$):

THEOREM 4

Convolution Theorem

Suppose that $f(x)$ and $g(x)$ are piecewise continuous, bounded, and absolutely integrable on the x-axis. Then

(12)
$$\mathcal{F}(f * g) = \sqrt{2\pi}\,\mathcal{F}(f)\,\mathcal{F}(g).$$

PROOF By the definition,

$$\mathcal{F}(f * g) = \frac{1}{\sqrt{2\pi}} \int_{-\infty}^{\infty} \int_{-\infty}^{\infty} f(p) g(x-p)\, dp\, e^{-iwx}\, dx.$$

An interchange of the order of integration gives

$$\mathcal{F}(f * g) = \frac{1}{\sqrt{2\pi}} \int_{-\infty}^{\infty} \int_{-\infty}^{\infty} f(p) g(x-p) e^{-iwx}\, dx\, dp.$$

Instead of x we now take $x - p = q$ as a new variable of integration. Then $x = p + q$ and

$$\mathcal{F}(f * g) = \frac{1}{\sqrt{2\pi}} \int_{-\infty}^{\infty} \int_{-\infty}^{\infty} f(p) g(q) e^{-iw(p+q)}\, dq\, dp.$$

This double integral can be written as a product of two integrals and gives the desired result

$$\mathcal{F}(f * g) = \frac{1}{\sqrt{2\pi}} \int_{-\infty}^{\infty} f(p) e^{-iwp}\, dp \int_{-\infty}^{\infty} g(q) e^{-iwq}\, dq$$

$$= \frac{1}{\sqrt{2\pi}} [\sqrt{2\pi}\, \mathcal{F}(f)][\sqrt{2\pi}\, \mathcal{F}(g)] = \sqrt{2\pi}\, \mathcal{F}(f)\mathcal{F}(g). \qquad \blacksquare$$

By taking the inverse Fourier transform on both sides of (12), writing $\hat{f} = \mathcal{F}(f)$ and $\hat{g} = \mathcal{F}(g)$ as before, and noting that $\sqrt{2\pi}$ and $1/\sqrt{2\pi}$ in (12) and (7) cancel each other, we obtain

(13) $$(f * g)(x) = \int_{-\infty}^{\infty} \hat{f}(w)\hat{g}(w) e^{iwx}\, dw,$$

a formula that will help us in solving partial differential equations (Sec. 12.6).

Discrete Fourier Transform (DFT), Fast Fourier Transform (FFT)

In using Fourier series, Fourier transforms, and trigonometric approximations (Sec. 11.6) we have to assume that a function $f(x)$, to be developed or transformed, is given on some interval, over which we integrate in the Euler formulas, etc. Now very often a function $f(x)$ is given only in terms of values at finitely many points, and one is interested in extending Fourier analysis to this case. The main application of such a "discrete Fourier analysis" concerns large amounts of equally spaced data, as they occur in telecommunication, time series analysis, and various simulation problems. In these situations, dealing with sampled values rather than with functions, we can replace the Fourier transform by the so-called **discrete Fourier transform (DFT)** as follows.

Let $f(x)$ be periodic, for simplicity of period 2π. We assume that N measurements of $f(x)$ are taken over the interval $0 \leqq x \leqq 2\pi$ at regularly spaced points

$$x_k = \frac{2\pi k}{N}, \qquad k = 0, 1, \cdots, N-1. \tag{14}$$

We also say that $f(x)$ is being **sampled** at these points. We now want to determine a **complex trigonometric polynomial**

$$q(x) = \sum_{n=0}^{N-1} c_n e^{inx_k} \tag{15}$$

that **interpolates** $f(x)$ at the nodes (14), that is, $q(x_k) = f(x_k)$, written out, with f_k denoting $f(x_k)$,

$$f_k = f(x_k) = q(x_k) = \sum_{n=0}^{N-1} c_n e^{inx_k}, \qquad k = 0, 1, \cdots, N-1. \tag{16}$$

Hence we must determine the coefficients $c_0, \cdots, c_{N-1}$ such that (16) holds. We do this by an idea similar to that in Sec. 11.1 for deriving the Fourier coefficients by using the orthogonality of the trigonometric system. Instead of integrals we now take sums. Namely, we multiply (16) by e^{-imx_k} (note the minus!) and sum over k from 0 to $N-1$. Then we interchange the order of the two summations and insert x_k from (14). This gives

$$\sum_{k=0}^{N-1} f_k e^{-imx_k} = \sum_{k=0}^{N-1}\sum_{n=0}^{N-1} c_n e^{i(n-m)x_k} = \sum_{n=0}^{N-1} c_n \sum_{k=0}^{N-1} e^{i(n-m)2\pi k/N}. \tag{17}$$

Now

$$e^{i(n-m)2\pi k/N} = [e^{i(n-m)2\pi/N}]^k.$$

We donote $[\cdots]$ by r. For $n = m$ we have $r = e^0 = 1$. The sum of *these* terms over k equals N, the number of these terms. For $n \neq m$ we have $r \neq 1$ and by the formula for a geometric sum [(6) in Sec. 15.1 with $q = r$ and $n = N-1$]

$$\sum_{k=0}^{N-1} r^k = \frac{1 - r^N}{1 - r} = 0$$

because $r^N = 1$; indeed, since k, m, and n are integers,

$$r^N = e^{i(n-m)2\pi k} = \cos 2\pi k(n-m) + i \sin 2\pi k(n-m) = 1 + 0 = 1.$$

This shows that the right side of (17) equals $c_m N$. Writing n for m and dividing by N, we thus obtain the desired coefficient formula

$$c_n = \frac{1}{N}\sum_{k=0}^{N-1} f_k e^{-inx_k} \qquad f_k = f(x_k), \quad n = 0, 1, \cdots, N-1. \tag{18*}$$

Since computation of the c_n (by the fast Fourier transform, below) involves successive halfing of the problem size N, it is practical to drop the factor $1/N$ from c_n and define the

discrete Fourier transform of the given signal $\mathbf{f} = [f_0 \quad \cdots \quad f_{N-1}]^\mathsf{T}$ to be the vector $\hat{\mathbf{f}} = [\hat{f}_0 \quad \cdots \quad \hat{f}_{N-1}]$ with components

$$\textbf{(18)} \qquad \hat{f}_n = Nc_n = \sum_{k=0}^{N-1} f_k e^{-inx_k}, \qquad f_k = f(x_k), \quad n = 0, \cdots, N-1.$$

This is the frequency spectrum of the signal.

In vector notation, $\hat{\mathbf{f}} = \mathbf{F}_N\mathbf{f}$, where the $N \times N$ **Fourier matrix** $\mathbf{F}_N = [e_{nk}]$ has the entries [given in (18)]

$$(19) \qquad e_{nk} = e^{-inx_k} = e^{-2\pi ink/N} = w^{nk}, \qquad w = w_N = e^{-2\pi i/N},$$

where $n, k = 0, \cdots, N-1$.

EXAMPLE 4 Discrete Fourier Transform (DFT). Sample of $N = 4$ Values

Let $N = 4$ measurements (sample values) be given. Then $w = e^{-2\pi i/N} = e^{-\pi i/2} = -i$ and thus $w^{nk} = (-i)^{nk}$. Let the sample values be, say $\mathbf{f} = [0 \quad 1 \quad 4 \quad 9]^\mathsf{T}$. Then by (18) and (19),

$$(20) \qquad \hat{\mathbf{f}} = \mathbf{F}_4\mathbf{f} = \begin{bmatrix} w^0 & w^0 & w^0 & w^0 \\ w^0 & w^1 & w^2 & w^3 \\ w^0 & w^2 & w^4 & w^6 \\ w^0 & w^3 & w^6 & w^9 \end{bmatrix} \mathbf{f} = \begin{bmatrix} 1 & 1 & 1 & 1 \\ 1 & -i & -1 & i \\ 1 & -1 & 1 & -1 \\ 1 & i & -1 & -i \end{bmatrix} \begin{bmatrix} 0 \\ 1 \\ 4 \\ 9 \end{bmatrix} = \begin{bmatrix} 14 \\ -4+8i \\ -6 \\ -4-8i \end{bmatrix}.$$

From the first matrix in (20) it is easy to infer what $\mathbf{F}_N$ looks like for arbitrary N, which in practice may be 1000 or more, for reasons given below. ■

From the DFT (the frequency spectrum) $\hat{\mathbf{f}} = \mathbf{F}_N\mathbf{f}$ we can recreate the given signal $\hat{\mathbf{f}} = \mathbf{F}_N^{-1}\mathbf{f}$, as we shall now prove. Here $\mathbf{F}_N$ and its complex conjugate $\overline{\mathbf{F}}_N = \frac{1}{N}[\overline{w}^{nk}]$ satisfy

$$(21\text{a}) \qquad \overline{\mathbf{F}}_N\mathbf{F}_N = \mathbf{F}_N\overline{\mathbf{F}}_N = N\mathbf{I}$$

where $\mathbf{I}$ is the $N \times N$ unit matrix; hence $\mathbf{F}_N$ has the inverse

$$(21\text{b}) \qquad \mathbf{F}_N^{-1} = \frac{1}{N}\overline{\mathbf{F}}_N.$$

PROOF We prove (21). By the multiplication rule (row times column) the product matrix $\mathbf{G}_N = \overline{\mathbf{F}}_N\mathbf{F}_N = [g_{jk}]$ in (21a) has the entries $g_{jk} =$ *Row j of* $\overline{\mathbf{F}}_N$ *times Column k of* $\mathbf{F}_N$. That is, writing $W = \overline{w}^j w^k$, we prove that

$$\begin{aligned} g_{jk} &= (\overline{w}^j w^k)^0 + (\overline{w}^j w^k)^1 + \cdots + (\overline{w}^j w^k)^{N-1} \\ &= W^0 + W^1 + \cdots + W^{N-1} = \begin{cases} 0 & \text{if } j \neq k \\ N & \text{if } j = k. \end{cases} \end{aligned}$$

Indeed, when $j = k$, then $\overline{w}^k w^k = (\overline{w}w)^k = (e^{2\pi i/N}e^{-2\pi i/N})^k = 1^k = 1$, so that the sum of *these* N terms equals N; these are the diagonal entries of $\mathbf{G}_N$. Also, when $j \neq k$, then $W \neq 1$ and we have a geometric sum (whose value is given by (6) in Sec. 15.1 with $q = W$ and $n = N - 1$)

$$W^0 + W^1 + \cdots + W^{N-1} = \frac{1 - W^N}{1 - W} = 0$$

because $W^N = (\overline{w}^j w^k)^N = (e^{2\pi i})^j (e^{-2\pi i})^k = 1^j \cdot 1^k = 1$. ■

We have seen that $\hat{\mathbf{f}}$ is the frequency spectrum of the signal $f(x)$. Thus the components $\hat{f}_n$ of $\hat{\mathbf{f}}$ give a resolution of the 2π-periodic function $f(x)$ into simple (complex) harmonics. Here one should use only n's that are much smaller than $N/2$, to avoid **aliasing**. By this we mean the effect caused by sampling at too few (equally spaced) points, so that, for instance, in a motion picture, rotating wheels appear as rotating too slowly or even in the wrong sense. Hence in applications, N is usually large. But this poses a problem. Eq. (18) requires $O(N)$ operations for any particular n, hence $O(N^2)$ operations for, say, all $n < N/2$. Thus, already for 1000 sample points the straightforward calculation would involve millions of operations. However, this difficulty can be overcome by the so-called **fast Fourier transform (FFT)**, for which codes are readily available (e.g., in Maple). The FFT is a computational method for the DFT that needs only $O(N) \log_2 N$ operations instead of $O(N^2)$. It makes the DFT a practical tool for large N. Here one chooses $N = 2^p$ (p integer) and uses the special form of the Fourier matrix to break down the given problem into smaller problems. For instance, when $N = 1000$, those operations are reduced by a factor $1000/\log_2 1000 \approx 100$.

The breakdown produces two problems of size $M = N/2$. This breakdown is possible because for $N = 2M$ we have in (19)

$$w_N^2 = w_{2M}^2 = (e^{-2\pi i/N})^2 = e^{-4\pi i/(2M)} = e^{-2\pi i/(M)} = w_M.$$

The given vector $\mathbf{f} = [f_0 \;\cdots\; f_{N-1}]^\mathsf{T}$ is split into two vectors with M components each, namely, $\mathbf{f}_{ev} = [f_0 \; f_2 \;\cdots\; f_{N-2}]^\mathsf{T}$ containing the even components of $\mathbf{f}$, and $\mathbf{f}_{od} = [f_1 \; f_3 \;\cdots\; f_{N-1}]^\mathsf{T}$ containing the odd components of $\mathbf{f}$. For $\mathbf{f}_{ev}$ and $\mathbf{f}_{od}$ we determine the DFTs

$$\hat{\mathbf{f}}_{ev} = [\hat{f}_{ev,0} \;\; \hat{f}_{ev,2} \;\; \cdots \;\; \hat{f}_{ev,N-2}]^\mathsf{T} = \mathbf{F}_M \mathbf{f}_{ev}$$

and

$$\hat{\mathbf{f}}_{od} = [\hat{f}_{od,1} \;\; \hat{f}_{od,3} \;\; \cdots \;\; \hat{f}_{od,N-1}]^\mathsf{T} = \mathbf{F}_M \mathbf{f}_{od}$$

involving the same $M \times M$ matrix $\mathbf{F}_M$. From these vectors we obtain the components of the DFT of the given vector f by the formulas

$$\text{(22)} \qquad \begin{aligned} &\text{(a)} \quad \hat{f}_n = \hat{f}_{ev,n} + w_N^n \hat{f}_{od,n} && n = 0, \cdots, M-1 \\ &\text{(b)} \quad \hat{f}_{n+M} = \hat{f}_{ev,n} - w_N^n \hat{f}_{od,n} && n = 0, \cdots, M-1. \end{aligned}$$

For $N = 2^p$ this breakdown can be repeated $p - 1$ times in order to finally arrive at $N/2$ problems of size 2 each, so that the number of multiplications is reduced as indicated above.

We show the reduction from $N = 4$ to $M = N/2 = 2$ and then prove (22).

EXAMPLE 5 Fast Fourier Transform (FFT). Sample of $N = 4$ Values

When $N = 4$, then $w = w_N = -i$ as in Example 4 and $M = N/2 = 2$, hence $w = w_M = e^{-2\pi i/2} = e^{-\pi i} = -1$. Consequently,

$$\hat{\mathbf{f}}_{\text{ev}} = \begin{bmatrix} \hat{f}_0 \\ \hat{f}_2 \end{bmatrix} = \mathbf{F}_2\mathbf{f}_{\text{ev}} = \begin{bmatrix} 1 & 1 \\ 1 & -1 \end{bmatrix}\begin{bmatrix} f_0 \\ f_2 \end{bmatrix} = \begin{bmatrix} f_0 + f_2 \\ f_0 - f_2 \end{bmatrix}$$

$$\hat{\mathbf{f}}_{\text{od}} = \begin{bmatrix} \hat{f}_1 \\ \hat{f}_3 \end{bmatrix} = \mathbf{F}_2\mathbf{f}_{\text{od}} = \begin{bmatrix} 1 & 1 \\ 1 & -1 \end{bmatrix}\begin{bmatrix} f_1 \\ f_3 \end{bmatrix} = \begin{bmatrix} f_1 + f_3 \\ f_1 - f_3 \end{bmatrix}.$$

From this and (22a) we obtain

$$\hat{f}_0 = \hat{f}_{\text{ev},0} + w_N^0 \hat{f}_{\text{od},0} = (f_0 + f_2) + (f_1 + f_3) = f_0 + f_1 + f_2 + f_3$$

$$\hat{f}_1 = \hat{f}_{\text{ev},1} + w_N^1 \hat{f}_{\text{od},1} = (f_0 - f_2) - i(f_1 + f_3) = f_0 - if_1 - f_2 + if_3.$$

Similarly, by (22b),

$$\hat{f}_2 = \hat{f}_{\text{ev},0} - w_N^0 \hat{f}_{\text{od},0} = (f_0 + f_2) - (f_1 + f_3) = f_0 - f_1 + f_2 - f_3$$

$$\hat{f}_3 = \hat{f}_{\text{ev},1} - w_N^1 \hat{f}_{\text{od},1} = (f_0 - f_2) - (-i)(f_1 - f_3) = f_0 + if_1 - f_2 - if_3.$$

This agrees with Example 4, as can be seen by replacing 0, 1, 4, 9 with f_0, f_1, f_2, f_3. ■

We prove (22). From (18) and (19) we have for the components of the DFT

$$\hat{f}_n = \sum_{k=0}^{N-1} w_N^{kn} f_k.$$

Splitting into two sums of $M = N/2$ terms each gives

$$\hat{f}_n = \sum_{k=0}^{M-1} w_N^{2kn} f_{2k} + \sum_{k=0}^{M-1} w_N^{(2k+1)n} f_{2k+1}.$$

We now use $w_N^2 = w_M$ and pull out w_N^n from under the second sum, obtaining

$$\hat{f}_n = \sum_{k=0}^{M-1} w_M^{kn} f_{\text{ev},k} + w_N^n \sum_{k=0}^{M-1} w_M^{kn} f_{\text{od},k}. \tag{23}$$

The two sums are $f_{\text{ev},n}$ and $f_{\text{od},n}$, the components of the "half-size" transforms $\mathbf{Ff}_{\text{ev}}$ and $\mathbf{Ff}_{\text{od}}$.

Formula (22a) is the same as (23). In (22b) we have $n + M$ instead of n. This causes a sign changes in (23), namely $-w_N^n$ before the second sum because

$$w_N^M = e^{-2\pi iM/N} = e^{-2\pi i/2} = e^{-\pi i} = -1.$$

This gives the minus in (22b) and completes the proof. ■

PROBLEM SET 11.9

1. Review in complex. Show that $1/i = -i$, $e^{-ix} = \cos x - i \sin x$, $e^{ix} + e^{-ix} = 2 \cos x$, $e^{ix} - e^{-ix} = 2i \sin x$, $e^{ikx} = \cos kx + i \sin kx$.

2–11 FOURIER TRANSFORMS BY INTEGRATION

Find the Fourier transform of $f(x)$ (without using Table III in Sec. 11.10). Show details.

2. $f(x) = \begin{cases} e^{2ix} & \text{if } -1 < x < 1 \\ 0 & \text{otherwise} \end{cases}$

3. $f(x) = \begin{cases} 1 & \text{if } a < x < b \\ 0 & \text{otherwise} \end{cases}$

4. $f(x) = \begin{cases} e^{kx} & \text{if } x < 0 \quad (k > 0) \\ 0 & \text{if } x > 0 \end{cases}$

5. $f(x) = \begin{cases} e^{x} & \text{if } -a < x < a \\ 0 & \text{otherwise} \end{cases}$

6. $f(x) = e^{-|x|} \quad (-\infty < x < \infty)$

7. $f(x) = \begin{cases} x & \text{if } 0 < x < a \\ 0 & \text{otherwise} \end{cases}$

8. $f(x) = \begin{cases} xe^{-x} & \text{if } -1 < x < 0 \\ 0 & \text{otherwise} \end{cases}$

9. $f(x) = \begin{cases} |x| & \text{if } -1 < x < 1 \\ 0 & \text{otherwise} \end{cases}$

10. $f(x) = \begin{cases} x & \text{if } -1 < x < 1 \\ 0 & \text{otherwise} \end{cases}$

11. $f(x) = \begin{cases} -1 & \text{if } -1 < x < 0 \\ 1 & \text{if } 0 < x < 1 \\ 0 & \text{otherwise} \end{cases}$

12–17 USE OF TABLE III IN SEC. 11.10. OTHER METHODS

12. Find $\mathcal{F}(f(x))$ for $f(x) = xe^{-x}$ if $x > 0$, $f(x) = 0$ if $x < 0$, by (9) in the text and formula 5 in Table III (with $a = 1$). *Hint.* Consider xe^{-x} and e^{-x}.

13. Obtain $\mathcal{F}(e^{-x^2/2})$ from Table III.

14. In Table III obtain formula 7 from formula 8.

15. In Table III obtain formula 1 from formula 2.

16. TEAM PROJECT. Shifting (a) Show that if $f(x)$ has a Fourier transform, so does $f(x - a)$, and $\mathcal{F}\{f(x - a)\} = e^{-iwa}\mathcal{F}\{f(x)\}$.

(b) Using (a), obtain formula 1 in Table III, Sec. 11.10, from formula 2.

(c) Shifting on the *w*-Axis. Show that if $\hat{f}(w)$ is the Fourier transform of $f(x)$, then $\hat{f}(w - a)$ is the Fourier transform of $e^{iax}f(x)$.

(d) Using (c), obtain formula 7 in Table III from 1 and formula 8 from 2.

17. What could give you the idea to solve Prob. 11 by using the solution of Prob. 9 and formula (9) in the text? Would this work?

18–25 DISCRETE FOURIER TRANSFORM

18. Verify the calculations in Example 4 of the text.

19. Find the transform of a general signal $f = [f_1 \ \ f_2 \ \ f_3 \ \ f_4]^{\mathsf{T}}$ of four values.

20. Find the inverse matrix in Example 4 of the text and use it to recover the given signal.

21. Find the transform (the frequency spectrum) of a general signal of two values $[f_1 \ \ f_2]^{\mathsf{T}}$.

22. Recreate the given signal in Prob. 21 from the frequency spectrum obtained.

23. Show that for a signal of eight sample values, $w = e^{-i/4} = (1 - i)/\sqrt{2}$. Check by squaring.

24. Write the Fourier matrix $\mathbf{F}$ for a sample of eight values explicitly.

25. CAS Problem. Calculate the inverse of the 8×8 Fourier matrix. Transform a general sample of eight values and transform it back to the given data.

11.10 Tables of Transforms

Table I. Fourier Cosine Transforms

See (2) in Sec. 11.8.

	$f(x)$	$\hat{f}_c(w) = \mathcal{F}_c(f)$	
1	$\begin{cases} 1 & \text{if } 0 < x < a \\ 0 & \text{otherwise} \end{cases}$	$\sqrt{\frac{2}{\pi}} \frac{\sin aw}{w}$	
2	$x^{a-1} \quad (0 < a < 1)$	$\sqrt{\frac{2}{\pi}} \frac{\Gamma(a)}{w^a} \cos \frac{a\pi}{2}$	($\Gamma(a)$ see App. A3.1.)
3	$e^{-ax} \quad (a > 0)$	$\sqrt{\frac{2}{\pi}} \left(\frac{a}{a^2 + w^2} \right)$	
4	$e^{-x^2/2}$	$e^{-w^2/2}$	
5	$e^{-ax^2} \quad (a > 0)$	$\frac{1}{\sqrt{2a}} e^{-w^2/(4a)}$	
6	$x^n e^{-ax} \quad (a > 0)$	$\sqrt{\frac{2}{\pi}} \frac{n!}{(a^2 + w^2)^{n+1}} \operatorname{Re}(a + iw)^{n+1}$	Re = Real part
7	$\begin{cases} \cos x & \text{if } 0 < x < a \\ 0 & \text{otherwise} \end{cases}$	$\frac{1}{\sqrt{2\pi}} \left[\frac{\sin a(1 - w)}{1 - w} + \frac{\sin a(1 + w)}{1 + w} \right]$	
8	$\cos(ax^2) \quad (a > 0)$	$\frac{1}{\sqrt{2a}} \cos\left(\frac{w^2}{4a} - \frac{\pi}{4} \right)$	
9	$\sin(ax^2) \quad (a > 0)$	$\frac{1}{\sqrt{2a}} \cos\left(\frac{w^2}{4a} + \frac{\pi}{4} \right)$	
10	$\frac{\sin ax}{x} \quad (a > 0)$	$\sqrt{\frac{\pi}{2}} (1 - u(w - a))$	(See Sec. 6.3.)
11	$\frac{e^{-x} \sin x}{x}$	$\frac{1}{\sqrt{2\pi}} \arctan \frac{2}{w^2}$	
12	$J_0(ax) \quad (a > 0)$	$\sqrt{\frac{2}{\pi}} \frac{1}{\sqrt{a^2 - w^2}} (1 - u(w - a))$	(See Secs. 5.5, 6.3.)

Table II. Fourier Sine Transforms

See (5) in Sec. 11.8.

	$f(x)$	$\hat{f}_s(w) = \mathcal{F}_s(f)$	
1	$\begin{cases} 1 & \text{if } 0 < x < a \\ 0 & \text{otherwise} \end{cases}$	$\sqrt{\frac{2}{\pi}}\left[\frac{1 - \cos aw}{w}\right]$	
2	$1/\sqrt{x}$	$1/\sqrt{w}$	
3	$1/x^{3/2}$	$2\sqrt{w}$	
4	$x^{a-1} \quad (0 < a < 1)$	$\sqrt{\frac{2}{\pi}}\frac{\Gamma(a)}{w^a}\sin\frac{a\pi}{2}$	($\Gamma(a)$ see App. A3.1.)
5	$e^{-ax} \quad (a > 0)$	$\sqrt{\frac{2}{\pi}}\left(\frac{w}{a^2 + w^2}\right)$	
6	$\frac{e^{-ax}}{x} \quad (a > 0)$	$\sqrt{\frac{2}{\pi}}\arctan\frac{w}{a}$	
7	$x^n e^{-ax} \quad (a > 0)$	$\sqrt{\frac{2}{\pi}}\frac{n!}{(a^2 + w^2)^{n+1}}\operatorname{Im}(a + iw)^{n+1}$	Im = Imaginary part
8	$xe^{-x^2/2}$	$we^{-w^2/2}$	
9	$xe^{-ax^2} \quad (a > 0)$	$\frac{w}{(2a)^{3/2}}e^{-w^2/4a}$	
10	$\begin{cases} \sin x & \text{if } 0 < x < a \\ 0 & \text{otherwise} \end{cases}$	$\frac{1}{\sqrt{2\pi}}\left[\frac{\sin a(1 - w)}{1 - w} - \frac{\sin a(1 + w)}{1 + w}\right]$	
11	$\frac{\cos ax}{x} \quad (a > 0)$	$\sqrt{\frac{\pi}{2}}\,u(w - a)$	(See Sec. 6.3.)
12	$\arctan\frac{2a}{x} \quad (a > 0)$	$\sqrt{2\pi}\,\frac{\sin aw}{w}e^{-aw}$	

Table III. Fourier Transforms

See (6) in Sec. 11.9.

	$f(x)$	$\hat{f}(w) = \mathcal{F}(f)$
1	$\begin{cases} 1 & \text{if } -b < x < b \\ 0 & \text{otherwise} \end{cases}$	$\sqrt{\frac{2}{\pi}}\, \frac{\sin bw}{w}$
2	$\begin{cases} 1 & \text{if } b < x < c \\ 0 & \text{otherwise} \end{cases}$	$\frac{e^{-ibw} - e^{-icw}}{iw\sqrt{2\pi}}$
3	$\frac{1}{x^2 + a^2} \quad (a > 0)$	$\sqrt{\frac{\pi}{2}}\, \frac{e^{-a\lvert w\rvert}}{a}$
4	$\begin{cases} x & \text{if } 0 < x < b \\ 2x - b & \text{if } b < x < 2b \\ 0 & \text{otherwise} \end{cases}$	$\frac{-1 + 2e^{ibw} - e^{-2ibw}}{\sqrt{2\pi}w^2}$
5	$\begin{cases} e^{-ax} & \text{if } x > 0 \\ 0 & \text{otherwise} \end{cases} \quad (a > 0)$	$\frac{1}{\sqrt{2\pi}(a + iw)}$
6	$\begin{cases} e^{ax} & \text{if } b < x < c \\ 0 & \text{otherwise} \end{cases}$	$\frac{e^{(a-iw)c} - e^{(a-iw)b}}{\sqrt{2\pi}(a - iw)}$
7	$\begin{cases} e^{iax} & \text{if } -b < x < b \\ 0 & \text{otherwise} \end{cases}$	$\sqrt{\frac{2}{\pi}}\, \frac{\sin b(w - a)}{w - a}$
8	$\begin{cases} e^{iax} & \text{if } b < x < c \\ 0 & \text{otherwise} \end{cases}$	$\frac{i}{\sqrt{2\pi}}\, \frac{e^{ib(a-w)} - e^{ic(a-w)}}{a - w}$
9	$e^{-ax^2} \quad (a > 0)$	$\frac{1}{\sqrt{2a}}\, e^{-w^2/4a}$
10	$\frac{\sin ax}{x} \quad (a > 0)$	$\sqrt{\frac{\pi}{2}}$ if $\lvert w\rvert < a$; 0 if $\lvert w\rvert > a$

CHAPTER 11 REVIEW QUESTIONS AND PROBLEMS

1. What is a Fourier series? A Fourier cosine series? A half-range expansion? Answer from memory.
2. What are the Euler formulas? By what very important idea did we obtain them?
3. How did we proceed from 2π-periodic to general-periodic functions?
4. Can a discontinuous function have a Fourier series? A Taylor series? Why are such functions of interest to the engineer?
5. What do you know about convergence of a Fourier series? About the Gibbs phenomenon?
6. The output of an ODE can oscillate several times as fast as the input. How come?
7. What is approximation by trigonometric polynomials? What is the minimum square error?
8. What is a Fourier integral? A Fourier sine integral? Give simple examples.
9. What is the Fourier transform? The discrete Fourier transform?
10. What are Sturm–Liouville problems? By what idea are they related to Fourier series?

11–20 FOURIER SERIES. In Probs. 11, 13, 16, 20 find the Fourier series of $f(x)$ as given over one period and sketch $f(x)$ and partial sums. In Probs. 12, 14, 15, 17–19 give answers, with reasons. Show your work detail.

11. $f(x) = \begin{cases} 0 & \text{if} \quad -2 < x < 0 \\ 2 & \text{if} \quad 0 < x < 2 \end{cases}$
12. Why does the series in Prob. 11 have no cosine terms?
13. $f(x) = \begin{cases} 0 & \text{if} \quad -1 < x < 0 \\ x & \text{if} \quad 0 < x < 1 \end{cases}$
14. What function does the series of the cosine terms in Prob. 13 represent? The series of the sine terms?
15. What function do the series of the cosine terms and the series of the sine terms in the Fourier series of e^x $(-5 < x < 5)$ represent?
16. $f(x) = |x| \quad (-\pi < x < \pi)$
17. Find a Fourier series from which you can conclude that $1 - 1/3 + 1/5 - 1/7 + - \cdots = \pi/4$.
18. What function and series do you obtain in Prob. 16 by (termwise) differentiation?
19. Find the half-range expansions of $f(x) = x$ $(0 < x < 1)$.
20. $f(x) = 3x^2 \quad (-\pi < x < \pi)$

21–22 GENERAL SOLUTION

Solve, $y'' + \omega^2 y = r(t)$, where $|\omega| \neq 0, 1, 2, \cdots, r(t)$ is 2π-periodic and

21. $r(t) = 3t^2 \ (-\pi < t < \pi)$
22. $r(t) = |t| \ (-\pi < t < \pi)$

23–25 MINIMUM SQUARE ERROR

23. Compute the minimum square error for $f(x) = x/\pi$ $(-\pi < x < \pi)$ and trigonometric polynomials of degree $N = 1, \cdots, 5$.
24. How does the minimum square error change if you multiply $f(x)$ by a constant k?
25. Same task as in Prob. 23, for $f(x) = |x|/\pi$ $(-\pi < x < \pi)$. Why is E^* now much smaller (by a factor 100, approximately!)?

26–30 FOURIER INTEGRALS AND TRANSFORMS

Sketch the given function and represent it as indicated. If you have a CAS, graph approximate curves obtained by replacing ∞ with finite limits; also look for Gibbs phenomena.

26. $f(x) = x + 1$ if $0 < x < 1$ and 0 otherwise; by the Fourier sine transform
27. $f(x) = x$ if $0 < x < 1$ and 0 otherwise; by the Fourier integral
28. $f(x) = kx$ if $a < x < b$ and 0 otherwise; by the Fourier transform
29. $f(x) = x$ if $1 < x < a$ and 0 otherwise; by the Fourier cosine transform
30. $f(x) = e^{-2x}$ if $x > 0$ and 0 otherwise; by the Fourier transform

SUMMARY OF CHAPTER 11

Fourier Analysis. Partial Differential Equations (PDEs)

Fourier series concern **periodic functions** $f(x)$ of period $p = 2L$, that is, by definition $f(x + p) = f(x)$ for all x and some fixed $p > 0$; thus, $f(x + np) = f(x)$ for any integer n. These series are of the form

$$f(x) = a_0 + \sum_{n=1}^{\infty} \left(a_n \cos \frac{n\pi}{L} x + b_n \sin \frac{n\pi}{L} x \right) \tag{1}$$ (Sec. 11.2)

with coefficients, called the **Fourier coefficients** of $f(x)$, given by the Euler formulas (Sec. 11.2)

$$\begin{aligned} a_0 &= \frac{1}{2L} \int_{-L}^{L} f(x)\, dx, \qquad a_n = \frac{1}{L} \int_{-L}^{L} f(x) \cos \frac{n\pi x}{L}\, dx \\ b_n &= \frac{1}{L} \int_{-L}^{L} f(x) \sin \frac{n\pi x}{L}\, dx \end{aligned} \tag{2}$$

where $n = 1, 2, \cdots$. For period 2π we simply have (Sec. 11.1)

$$f(x) = a_0 + \sum_{n=1}^{\infty} (a_n \cos nx + b_n \sin nx) \tag{1*}$$

with the *Fourier coefficients* of $f(x)$ (Sec. 11.1)

$$a_0 = \frac{1}{2\pi} \int_{-\pi}^{\pi} f(x)\, dx, \quad a_n = \frac{1}{\pi} \int_{-\pi}^{\pi} f(x) \cos nx\, dx, \quad b_n = \frac{1}{\pi} \int_{-\pi}^{\pi} f(x) \sin nx\, dx.$$

Fourier series are fundamental in connection with periodic phenomena, particularly in models involving differential equations (Sec. 11.3, Chap, 12). If $f(x)$ is even $[f(-x) = f(x)]$ or odd $[f(-x) = -f(x)]$, they reduce to **Fourier cosine** or **Fourier sine series**, respectively (Sec. 11.2). If $f(x)$ is given for $0 \leqq x \leqq L$ only, it has two **half-range expansions** of period $2L$, namely, a cosine and a sine series (Sec. 11.2).

The set of cosine and sine functions in (1) is called the **trigonometric system**. Its most basic property is its **orthogonality** on an interval of length $2L$; that is, for all integers m and $n \neq m$ we have

$$\int_{-L}^{L} \cos \frac{m\pi x}{L} \cos \frac{n\pi x}{L}\, dx = 0, \qquad \int_{-L}^{L} \sin \frac{m\pi x}{L} \sin \frac{n\pi x}{L}\, dx = 0$$

and for all integers m and n,

$$\int_{-L}^{L} \cos \frac{m\pi x}{L} \sin \frac{n\pi x}{L}\, dx = 0.$$

This orthogonality was crucial in deriving the Euler formulas (2).

Partial sums of Fourier series minimize the **square error** (Sec. 11.4).

Replacing the trigonometric system in (1) by other orthogonal systems first leads to ***Sturm–Liouville problems*** (Sec. 11.5), which are boundary value problems for ODEs. These problems are ***eigenvalue problems*** and as such involve a parameter λ that is often related to frequencies and energies. The solutions to Sturm–Liouville problems are called ***eigenfunctions***. Similar considerations lead to other orthogonal series such as ***Fourier–Legendre series*** and ***Fourier–Bessel series*** classified as ***generalized Fourier series*** (Sec. 11.6).

Ideas and techniques of Fourier series extend to nonperiodic functions $f(x)$ defined on the entire real line; this leads to the **Fourier integral**

$$f(x) = \int_0^\infty [A(w)\cos wx + B(w)\sin wx]\,dw \qquad \text{(Sec. 11.7)} \tag{3}$$

where

$$A(w) = \frac{1}{\pi}\int_{-\infty}^\infty f(v)\cos wv\,dv, \qquad B(w) = \frac{1}{\pi}\int_{-\infty}^\infty f(v)\sin wv\,dv \tag{4}$$

or, in complex form (Sec. 11.9),

$$f(x) = \frac{1}{\sqrt{2\pi}}\int_{-\infty}^\infty \hat{f}(w)e^{iwx}\,dw \qquad (i = \sqrt{-1}) \tag{5}$$

where

$$\hat{f}(w) = \frac{1}{\sqrt{2\pi}}\int_{-\infty}^\infty f(x)e^{-iwx}\,dx. \tag{6}$$

Formula (6) transforms $f(x)$ into its **Fourier transform** $\hat{f}(w)$, and (5) is the inverse transform.

Related to this are the **Fourier cosine transform** (Sec. 11.8)

$$\hat{f}_c(w) = \sqrt{\frac{2}{\pi}}\int_0^\infty f(x)\cos wx\,dx \tag{7}$$

and the **Fourier sine transform** (Sec. 11.8)

$$\hat{f}_s(w) = \sqrt{\frac{2}{\pi}}\int_0^\infty f(x)\sin wx\,dx. \tag{8}$$

The ***discrete Fourier transform (DFT)*** and a practical method of computing it, called the ***fast Fourier transform (FFT)***, are discussed in Sec. 11.9.

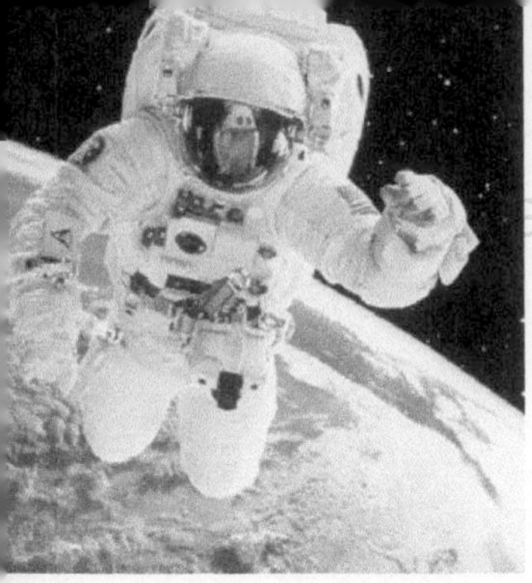

CHAPTER 12

Partial Differential Equations (PDEs)

A PDE is an equation that contains one or more partial derivatives of an unknown function that depends on at least two variables. Usually one of these deals with time t and the remaining with space (spatial variable(s)). The most important PDEs are the wave equations that can model the vibrating string (Secs. 12.2, 12.3, 12.4, 12.12) and the vibrating membrane (Secs. 12.8, 12.9, 12.10), the heat equation for temperature in a bar or wire (Secs. 12.5, 12.6), and the Laplace equation for electrostatic potentials (Secs. 12.6, 12.10, 12.11). PDEs are very important in dynamics, elasticity, heat transfer, electromagnetic theory, and quantum mechanics. They have a much wider range of applications than ODEs, which can model only the simplest physical systems. Thus PDEs are subjects of many ongoing research and development projects.

Realizing that modeling with PDEs is more involved than modeling with ODEs, we take a gradual, well-planned approach to modeling with PDEs. To do this we carefully derive the PDE that models the phenomena, such as the one-dimensional wave equation for a vibrating elastic string (say a violin string) in Sec. 12.2, and then solve the PDE in a separate section, that is, Sec. 12.3. In a similar vein, we derive the heat equation in Sec. 12.5 and then solve and generalize it in Sec. 12.6.

We derive these PDEs from physics and consider methods for solving initial and boundary value problems, that is, methods of obtaining solutions which satisfy the conditions required by the physical situations. In Secs. 12.7 and 12.12 we show how PDEs can also be solved by Fourier and Laplace transform methods.

COMMENT. ***Numerics for PDEs*** is explained in Secs. 21.4–21.7, which, for greater teaching flexibility, is designed to be independent of the other sections on numerics in Part E.

Prerequisites: Linear ODEs (Chap. 2), Fourier series (Chap. 11).
Sections that may be omitted in a shorter course: 12.7, 12.10–12.12.
References and Answers to Problems: App. 1 Part C, App. 2.

12.1 Basic Concepts of PDEs

A **partial differential equation (PDE)** is an equation involving one or more partial derivatives of an (unknown) function, call it u, that depends on two or more variables, often time t and one or several variables in space. The order of the highest derivative is called the **order** of the PDE. Just as was the case for ODEs, second-order PDEs will be the most important ones in applications.

Just as for ordinary differential equations (ODEs) we say that a PDE is **linear** if it is of the first degree in the unknown function u and its partial derivatives. Otherwise we call it **nonlinear**. Thus, all the equations in Example 1 are linear. We call a *linear* PDE **homogeneous** if each of its terms contains either u or one of its partial derivatives. Otherwise we call the equation **nonhomogeneous**. Thus, (4) in Example 1 (with f not identically zero) is nonhomogeneous, whereas the other equations are homogeneous.

EXAMPLE 1 Important Second-Order PDEs

(1) $$\frac{\partial^2 u}{\partial t^2} = c^2 \frac{\partial^2 u}{\partial x^2}$$ ***One-dimensional wave equation***

(2) $$\frac{\partial u}{\partial t} = c^2 \frac{\partial^2 u}{\partial x^2}$$ ***One-dimensional heat equation***

(3) $$\frac{\partial^2 u}{\partial x^2} + \frac{\partial^2 u}{\partial y^2} = 0$$ ***Two-dimensional Laplace equation***

(4) $$\frac{\partial^2 u}{\partial x^2} + \frac{\partial^2 u}{\partial y^2} = f(x, y)$$ ***Two-dimensional Poisson equation***

(5) $$\frac{\partial^2 u}{\partial t^2} = c^2 \left(\frac{\partial^2 u}{\partial x^2} + \frac{\partial^2 u}{\partial y^2}\right)$$ ***Two-dimensional wave equation***

(6) $$\frac{\partial^2 u}{\partial x^2} + \frac{\partial^2 u}{\partial y^2} + \frac{\partial^2 u}{\partial z^2} = 0$$ ***Three-dimensional Laplace equation***

Here c is a positive constant, t is time, x, y, z are Cartesian coordinates, and *dimension* is the number of these coordinates in the equation. ■

A **solution** of a PDE in some region R of the space of the independent variables is a function that has all the partial derivatives appearing in the PDE in some domain D (definition in Sec. 9.6) containing R, and satisfies the PDE everywhere in R.

Often one merely requires that the function is continuous on the boundary of R, has those derivatives in the interior of R, and satisfies the PDE in the interior of R. Letting R lie in D simplifies the situation regarding derivatives on the boundary of R, which is then the same on the boundary as it is in the interior of R.

In general, the totality of solutions of a PDE is very large. For example, the functions

(7) $$u = x^2 - y^2, \qquad u = e^x \cos y, \qquad u = \sin x \cosh y, \qquad u = \ln (x^2 + y^2)$$

which are entirely different from each other, are solutions of (3), as you may verify. We shall see later that the unique solution of a PDE corresponding to a given physical problem will be obtained by the use of ***additional conditions*** arising from the problem. For instance, this may be the condition that the solution u assume given values on the boundary of the region R ("**boundary conditions**"). Or, when time t is one of the variables, u (or $u_t = \partial u/\partial t$ or both) may be prescribed at $t = 0$ ("**initial conditions**").

We know that if an ODE is linear and homogeneous, then from known solutions we can obtain further solutions by superposition. For PDEs the situation is quite similar:

THEOREM 1

Fundamental Theorem on Superposition

*If u_1 and u_2 are solutions of a **homogeneous linear** PDE in some region R, then*

$$u = c_1 u_1 + c_2 u_2$$

with any constants c_1 and c_2 is also a solution of that PDE in the region R.

The simple proof of this important theorem is quite similar to that of Theorem 1 in Sec. 2.1 and is left to the student.

Verification of solutions in Probs. 2–13 proceeds as for ODEs. Problems 16–23 concern PDEs solvable like ODEs. To help the student with them, we consider two typical examples.

EXAMPLE 2 Solving $u_{xx} - u = 0$ Like an ODE

Find solutions u of the PDE $u_{xx} - u = 0$ depending on x and y.

Solution. Since no y-derivatives occur, we can solve this PDE like $u'' - u = 0$. In Sec. 2.2 we would have obtained $u = Ae^x + Be^{-x}$ with constant A and B. Here A and B may be functions of y, so that the answer is

$$u(x, y) = A(y)e^x + B(y)e^{-x}$$

with arbitrary functions A and B. We thus have a great variety of solutions. Check the result by differentiation. ■

EXAMPLE 3 Solving $u_{xy} = -u_x$ Like an ODE

Find solutions $u = u(x, y)$ of this PDE.

Solution. Setting $u_x = p$, we have $p_y = -p$, $p_y/p = -1$, $\ln|p| = -y + \tilde{c}(x)$, $p = c(x)e^{-y}$ and by integration with respect to x,

$$u(x, y) = f(x)e^{-y} + g(y) \qquad \text{where} \qquad f(x) = \int c(x)\,dx,$$

here, $f(x)$ and $g(y)$ are arbitrary. ■

PROBLEM SET 12.1

1. Fundamental theorem. Prove it for second-order PDEs in two and three independent variables. *Hint.* Prove it by substitution.

2–13 VERIFICATION OF SOLUTIONS

Verifiy (by substitution) that the given function is a solution of the PDE. Sketch or graph the solution as a surface in space.

2–5 Wave Equation (1) with suitable c

2. $u = x^2 + t^2$

3. $u = \cos 4t \sin 2x$

4. $u = \sin kct \cos kx$

5. $u = \sin at \sin bx$

6–9 Heat Equation (2) with suitable c

6. $u = e^{-t} \sin x$

7. $u = e^{-\omega^2 c^2 t} \cos \omega x$

8. $u = e^{-9t} \sin \omega x$

9. $u = e^{-\pi^2 t} \cos 25x$

10–13 Laplace Equation (3)

10. $u = e^x \cos y,\ e^x \sin y$

11. $u = \arctan (y/x)$

12. $u = \cos y \sinh x,\ \sin y \cosh x$

13. $u = x/(x^2 + y^2),\ y/(x^2 + y^2)$

14. TEAM PROJECT. Verification of Solutions

(a) Wave equation. Verify that $u(x, t) = v(x + ct) + w(x - ct)$ with any twice differentiable functions v and w satisfies (1).

(b) Poisson equation. Verify that each u satisfies (4) with $f(x, y)$ as indicated.

$u = y/x \qquad f = 2y/x^3$

$u = \sin xy \qquad f = (x^2 + y^2) \sin xy$

$u = e^{x^2 - y^2} \qquad f = 4(x^2 + y^2)e^{x^2 - y^2}$

$u = 1/\sqrt{x^2 + y^2} \qquad f = (x^2 + y^2)^{-3/2}$

(c) Laplace equation. Verify that $u = 1/\sqrt{x^2 + y^2 + z^2}$ satisfies (6) and $u = \ln (x^2 + y^2)$ satisfies (3). Is $u = 1/\sqrt{x^2 + y^2}$ a solution of (3)? Of what Poisson equation?

(d) Verify that u with any (sufficiently often differentiable) v and w satisfies the given PDE.

$u = v(x) + w(y) \qquad u_{xy} = 0$

$u = v(x)w(y) \qquad uu_{xy} = u_x u_y$

$u = v(x + 2t) + w(x - 2t) \qquad u_{tt} = 4u_{xx}$

15. Boundary value problem. Verify that the function $u(x, y) = a \ln (x^2 + y^2) + b$ satisfies Laplace's equation

(3) and determine a and b so that u satisfies the boundary conditions $u = 110$ on the circle $x^2 + y^2 = 1$ and $u = 0$ on the circle $x^2 + y^2 = 100$.

16–23 PDEs SOLVABLE AS ODEs

This happens if a PDE involves derivatives with respect to one variable only (or can be transformed to such a form), so that the other variable(s) can be treated as parameter(s). Solve for $u = u(x, y)$:

16. $u_{yy} = 0$ **17.** $u_{xx} + 16\pi^2 u = 0$

18. $25u_{yy} - 4u = 0$ **19.** $u_y + y^2 u = 0$

20. $2u_{xx} + 9u_x + 4u = -3\cos x - 29\sin x$

21. $u_{yy} + 6u_y + 13u = 4e^{3y}$

22. $u_{xy} = u_x$ **23.** $x^2 u_{xx} + 2xu_x - 2u = 0$

24. Surface of revolution. Show that the solutions $z = z(x, y)$ of $yz_x = xz_y$ represent surfaces of revolution. Give examples. *Hint.* Use polar coordinates r, θ and show that the equation becomes $z_\theta = 0$.

25. System of PDEs. Solve $u_{xx} = 0$, $u_{yy} = 0$

12.2 Modeling: Vibrating String, Wave Equation

In this section we model a vibrating string, which will lead to our first important PDE, that is, equation (3) which will then be solved in Sec. 12.3. *The student should pay very close attention to this delicate modeling process and detailed derivation starting from scratch,* as the skills learned can be applied to modeling other phenomena in general and in particular to modeling a vibrating membrane (Sec. 12.7).

We want to derive the PDE modeling small transverse vibrations of an elastic string, such as a violin string. We place the string along the x-axis, stretch it to length L, and fasten it at the ends $x = 0$ and $x = L$. We then distort the string, and at some instant, call it $t = 0$, we release it and allow it to vibrate. The problem is to determine the vibrations of the string, that is, to find its deflection $u(x, t)$ at any point x and at any time $t > 0$; see Fig. 286.

$u(x, t)$ will be the solution of a PDE that is the model of our physical system to be derived. This PDE should not be too complicated, so that we can solve it. Reasonable simplifying assumptions (just as for ODEs modeling vibrations in Chap. 2) are as follows.

Physical Assumptions

1. The mass of the string per unit length is constant ("homogeneous string"). The string is perfectly elastic and does not offer any resistance to bending.
2. The tension caused by stretching the string before fastening it at the ends is so large that the action of the gravitational force on the string (trying to pull the string down a little) can be neglected.
3. The string performs small transverse motions in a vertical plane; that is, every particle of the string moves strictly vertically and so that the deflection and the slope at every point of the string always remain small in absolute value.

Under these assumptions we may expect solutions $u(x, t)$ that describe the physical reality sufficiently well.

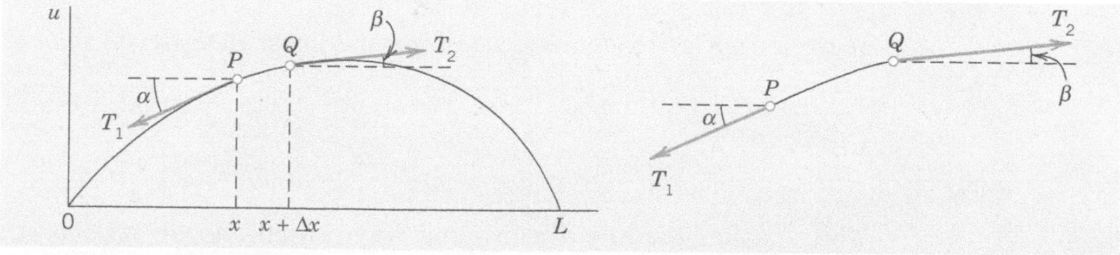

Fig. 286. Deflected string at fixed time t. Explanation on p. 544

Derivation of the PDE of the Model ("Wave Equation") from Forces

The model of the vibrating string will consist of a PDE ("wave equation") and additional conditions. To obtain the PDE, we consider the ***forces acting on a small portion of the string*** (Fig. 286). This method is typical of modeling in mechanics and elsewhere.

Since the string offers no resistance to bending, the tension is tangential to the curve of the string at each point. Let T_1 and T_2 be the tension at the endpoints P and Q of that portion. Since the points of the string move vertically, there is no motion in the horizontal direction. Hence the horizontal components of the tension must be constant. Using the notation shown in Fig. 286, we thus obtain

$$T_1 \cos \alpha = T_2 \cos \beta = T = \text{const.} \tag{1}$$

In the vertical direction we have two forces, namely, the vertical components $-T_1 \sin \alpha$ and $T_2 \sin \beta$ of T_1 and T_2; here the minus sign appears because the component at P is directed downward. By **Newton's second law** (Sec. 2.4) the resultant of these two forces is equal to the mass $\rho \Delta x$ of the portion times the acceleration $\partial^2 u/\partial t^2$, evaluated at some point between x and $x + \Delta x$; here ρ is the mass of the undeflected string per unit length, and Δx is the length of the portion of the undeflected string. (Δ is generally used to denote small quantities; this has nothing to do with the Laplacian ∇^2, which is sometimes also denoted by Δ.) Hence

$$T_2 \sin \beta - T_1 \sin \alpha = \rho \Delta x \frac{\partial^2 u}{\partial t^2}.$$

Using (1), we can divide this by $T_2 \cos \beta = T_1 \cos \alpha = T$, obtaining

$$\frac{T_2 \sin \beta}{T_2 \cos \beta} - \frac{T_1 \sin \alpha}{T_1 \cos \alpha} = \tan \beta - \tan \alpha = \frac{\rho \Delta x}{T} \frac{\partial^2 u}{\partial t^2}. \tag{2}$$

Now $\tan \alpha$ and $\tan \beta$ are the slopes of the string at x and $x + \Delta x$:

$$\tan \alpha = \left(\frac{\partial u}{\partial x}\right)\bigg|_x \qquad \text{and} \qquad \tan \beta = \left(\frac{\partial u}{\partial x}\right)\bigg|_{x+\Delta x}.$$

Here we have to write *partial* derivatives because u also depends on time t. Dividing (2) by Δx, we thus have

$$\frac{1}{\Delta x}\left[\left(\frac{\partial u}{\partial x}\right)\bigg|_{x+\Delta x} - \left(\frac{\partial u}{\partial x}\right)\bigg|_x\right] = \frac{\rho}{T} \frac{\partial^2 u}{\partial t^2}.$$

If we let Δx approach zero, we obtain the linear PDE

$$\textbf{(3)} \qquad \frac{\partial^2 u}{\partial t^2} = c^2 \frac{\partial^2 u}{\partial x^2}, \qquad c^2 = \frac{T}{\rho}.$$

This is called the **one-dimensional wave equation**. We see that it is homogeneous and of the second order. The physical constant T/ρ is denoted by c^2 (instead of c) to indicate that this constant is *positive,* a fact that will be essential to the form of the solutions. "One-dimensional" means that the equation involves only one space variable, x. In the next section we shall complete setting up the model and then show how to solve it by a general method that is probably the most important one for PDEs in engineering mathematics.

12.3 Solution by Separating Variables. Use of Fourier Series

We continue our work from Sec. 12.2, where we modeled a vibrating string and obtained the one-dimensional wave equation. We now have to complete the model by adding additional conditions and then solving the resulting model.

The model of a vibrating elastic string (a violin string, for instance) consists of the **one-dimensional wave equation**

$$\frac{\partial^2 u}{\partial t^2} = c^2 \frac{\partial^2 u}{\partial x^2} \qquad c^2 = \frac{T}{\rho} \tag{1}$$

for the unknown deflection $u(x, t)$ of the string, a PDE that we have just obtained, and some ***additional conditions***, which we shall now derive.

Since the string is fastened at the ends $x = 0$ and $x = L$ (see Sec. 12.2), we have the two **boundary conditions**

$$\text{(a)}\quad u(0, t) = 0, \qquad \text{(b)}\quad u(L, t) = 0, \qquad \text{for all } t \geqq 0. \tag{2}$$

Furthermore, the form of the motion of the string will depend on its *initial deflection* (deflection at time $t = 0$), call it $f(x)$, and on its *initial velocity* (velocity at $t = 0$), call it $g(x)$. We thus have the two **initial conditions**

$$\text{(a)}\quad u(x, 0) = f(x), \qquad \text{(b)}\quad u_t(x, 0) = g(x) \qquad (0 \leqq x \leqq L) \tag{3}$$

where $u_t = \partial u/\partial t$. We now have to find a solution of the PDE (1) satisfying the conditions (2) and (3). This will be the solution of our problem. We shall do this in three steps, as follows.

Step 1. By the "**method of separating variables**" or *product method,* setting $u(x, t) = F(x)G(t)$, we obtain from (1) two ODEs, one for $F(x)$ and the other one for $G(t)$.

Step 2. We determine solutions of these ODEs that satisfy the boundary conditions (2).

Step 3. Finally, using **Fourier series**, we compose the solutions found in Step 2 to obtain a solution of (1) satisfying both (2) and (3), that is, the solution of our model of the vibrating string.

Step 1. Two ODEs from the Wave Equation (1)

In the **method of separating variables**, or *product method*, we determine solutions of the wave equation (1) of the form

$$u(x, t) = F(x)G(t) \tag{4}$$

which are a product of two functions, each depending on only one of the variables x and t. This is a powerful general method that has various applications in engineering mathematics, as we shall see in this chapter. Differentiating (4), we obtain

$$\frac{\partial^2 u}{\partial t^2} = F\ddot{G} \qquad \text{and} \qquad \frac{\partial^2 u}{\partial x^2} = F''G$$

where dots denote derivatives with respect to t and primes derivatives with respect to x. By inserting this into the wave equation (1) we have

$$F\ddot{G} = c^2 F''G.$$

Dividing by c^2FG and simplifying gives

$$\frac{\ddot{G}}{c^2G} = \frac{F''}{F}.$$

The variables are now separated, the left side depending only on t and the right side only on x. Hence both sides must be constant because, if they were variable, then changing t or x would affect only one side, leaving the other unaltered. Thus, say,

$$\frac{\ddot{G}}{c^2G} = \frac{F''}{F} = k.$$

Multiplying by the denominators gives immediately two ***ordinary*** DEs

(5) $$F'' - kF = 0$$

and

(6) $$\ddot{G} - c^2kG = 0.$$

Here, the **separation constant** k is still arbitrary.

Step 2. Satisfying the Boundary Conditions (2)

We now determine solutions F and G of (5) and (6) so that $u = FG$ satisfies the boundary conditions (2), that is,

(7) $$u(0, t) = F(0)G(t) = 0, \qquad u(L, t) = F(L)G(t) = 0 \qquad \text{for all } t.$$

We first solve (5). If $G \equiv 0$, then $u = FG \equiv 0$, which is of no interest. Hence $G \not\equiv 0$ and then by (7),

(8) $$\text{(a)} \quad F(0) = 0, \qquad \text{(b)} \quad F(L) = 0.$$

We show that k must be negative. For $k = 0$ the general solution of (5) is $F = ax + b$, and from (8) we obtain $a = b = 0$, so that $F \equiv 0$ and $u = FG \equiv 0$, which is of no interest. For positive $k = \mu^2$ a general solution of (5) is

$$F = Ae^{\mu x} + Be^{-\mu x}$$

and from (8) we obtain $F \equiv 0$ as before (verify!). Hence we are left with the possibility of choosing k negative, say, $k = -p^2$. Then (5) becomes $F'' + p^2F = 0$ and has as a general solution

$$F(x) = A \cos px + B \sin px.$$

From this and (8) we have

$$F(0) = A = 0 \qquad \text{and then} \qquad F(L) = B \sin pL = 0.$$

We must take $B \neq 0$ since otherwise $F \equiv 0$. Hence $\sin pL = 0$. Thus

$$\text{(9)} \qquad pL = n\pi, \qquad \text{so that} \qquad p = \frac{n\pi}{L} \qquad (n \text{ integer}).$$

Setting $B = 1$, we thus obtain infinitely many solutions $F(x) = F_n(x)$, where

$$\text{(10)} \qquad F_n(x) = \sin \frac{n\pi}{L} x \qquad (n = 1, 2, \cdots).$$

These solutions satisfy (8). [For negative integer n we obtain essentially the same solutions, except for a minus sign, because $\sin(-\alpha) = -\sin \alpha$.]

We now solve (6) with $k = -p^2 = -(n\pi/L)^2$ resulting from (9), that is,

$$\text{(11*)} \qquad \ddot{G} + \lambda_n^2 G = 0 \qquad \text{where} \qquad \lambda_n = cp = \frac{cn\pi}{L}.$$

A general solution is

$$G_n(t) = B_n \cos \lambda_n t + B_n^* \sin \lambda_n t.$$

Hence solutions of (1) satisfying (2) are $u_n(x, t) = F_n(x)G_n(t) = G_n(t)F_n(x)$, written out

$$\text{(11)} \qquad u_n(x, t) = (B_n \cos \lambda_n t + B_n^* \sin \lambda_n t) \sin \frac{n\pi}{L} x \qquad (n = 1, 2, \cdots).$$

These functions are called the **eigenfunctions**, or *characteristic functions*, and the values $\lambda_n = cn\pi/L$ are called the **eigenvalues**, or *characteristic values*, of the vibrating string. The set $\{\lambda_1, \lambda_2, \cdots\}$ is called the **spectrum**.

Discussion of Eigenfunctions. We see that each u_n represents a harmonic motion having the **frequency** $\lambda_n/2\pi = cn/2L$ cycles per unit time. This motion is called the nth **normal mode** of the string. The first normal mode is known as the *fundamental mode* ($n = 1$), and the others are known as *overtones;* musically they give the octave, octave plus fifth, etc. Since in (11)

$$\sin \frac{n\pi x}{L} = 0 \qquad \text{at} \qquad x = \frac{L}{n}, \frac{2L}{n}, \cdots, \frac{n-1}{n} L,$$

the nth normal mode has $n - 1$ **nodes**, that is, points of the string that do not move (in addition to the fixed endpoints); see Fig. 287.

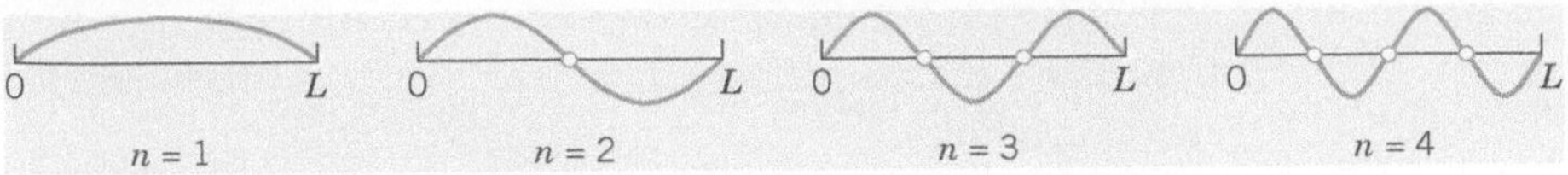

Fig. 287. Normal modes of the vibrating string

Figure 288 shows the second normal mode for various values of t. At any instant the string has the form of a sine wave. When the left part of the string is moving down, the other half is moving up, and conversely. For the other modes the situation is similar.

Tuning is done by changing the tension T. Our formula for the frequency $\lambda_n/2\pi = cn/2L$ of u_n with $c = \sqrt{T/\rho}$ [see (3), Sec. 12.2] confirms that effect because it shows that the frequency is proportional to the tension. T cannot be increased indefinitely, but can you see what to do to get a string with a high fundamental mode? (Think of both L and ρ.) Why is a violin smaller than a double-bass?

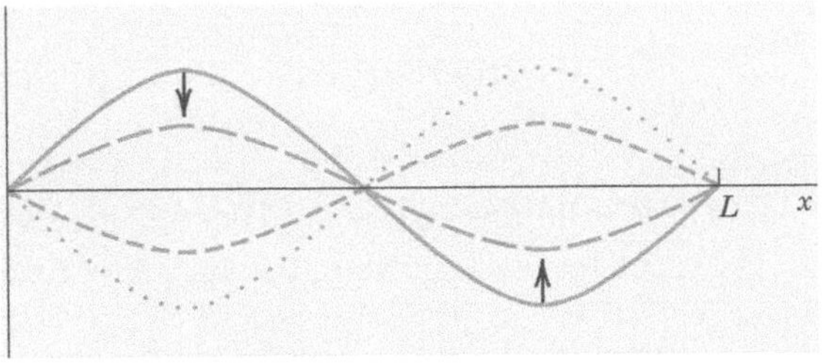

Fig. 288. Second normal mode for various values of t

Step 3. Solution of the Entire Problem. Fourier Series

The eigenfunctions (11) satisfy the wave equation (1) and the boundary conditions (2) (string fixed at the ends). A single u_n will generally not satisfy the initial conditions (3). But since the wave equation (1) is linear and homogeneous, it follows from Fundamental Theorem 1 in Sec. 12.1 that the sum of finitely many solutions u_n is a solution of (1). To obtain a solution that also satisfies the initial conditions (3), we consider the infinite series (with $\lambda_n = cn\pi/L$ as before)

$$u(x, t) = \sum_{n=1}^{\infty} u_n(x, t) = \sum_{n=1}^{\infty} (B_n \cos \lambda_n t + B_n^* \sin \lambda_n t) \sin \frac{n\pi}{L} x. \tag{12}$$

Satisfying Initial Condition (3a) (Given Initial Displacement). From (12) and (3a) we obtain

$$u(x, 0) = \sum_{n=1}^{\infty} B_n \sin \frac{n\pi}{L} x = f(x). \qquad (0 \leqq x \leqq L). \tag{13}$$

Hence we must choose the B_n's so that $u(x, 0)$ becomes the **Fourier sine series** of $f(x)$. Thus, by (4) in Sec. 11.3,

$$B_n = \frac{2}{L} \int_0^L f(x) \sin \frac{n\pi x}{L} \, dx, \qquad n = 1, 2, \cdots. \tag{14}$$

Satisfying Initial Condition (3b) (Given Initial Velocity). Similarly, by differentiating (12) with respect to t and using (3b), we obtain

$$\left.\frac{\partial u}{\partial t}\right|_{t=0} = \left[\sum_{n=1}^{\infty} (-B_n\lambda_n \sin \lambda_n t + B_n^*\lambda_n \cos \lambda_n t) \sin \frac{n\pi x}{L}\right]_{t=0}$$

$$= \sum_{n=1}^{\infty} B_n^* \lambda_n \sin \frac{n\pi x}{L} = g(x).$$

Hence we must choose the B_n^*'s so that for $t = 0$ the derivative $\partial u/\partial t$ becomes the Fourier sine series of $g(x)$. Thus, again by (4) in Sec. 11.3,

$$B_n^* \lambda_n = \frac{2}{L} \int_0^L g(x) \sin \frac{n\pi x}{L}\, dx.$$

Since $\lambda_n = cn\pi/L$, we obtain by division

(15) $$B_n^* = \frac{2}{cn\pi} \int_0^L g(x) \sin \frac{n\pi x}{L}\, dx, \qquad n = 1, 2, \cdots.$$

Result. Our discussion shows that $u(x, t)$ given by (12) with coefficients (14) and (15) is a solution of (1) that satisfies all the conditions in (2) and (3), provided the series (12) converges and so do the series obtained by differentiating (12) twice termwise with respect to x and t and have the sums $\partial^2 u/\partial x^2$ and $\partial^2 u/\partial t^2$, respectively, which are continuous.

Solution (12) Established. According to our derivation, the solution (12) is at first a purely formal expression, but we shall now establish it. For the sake of simplicity we consider only the case when the initial velocity $g(x)$ is identically zero. Then the B_n^* are zero, and (12) reduces to

(16) $$u(x, t) = \sum_{n=1}^{\infty} B_n \cos \lambda_n t \sin \frac{n\pi x}{L}, \qquad \lambda_n = \frac{cn\pi}{L}.$$

It is possible to ***sum this series***, that is, to write the result in a closed or finite form. For this purpose we use the formula [see (11), App. A3.1]

$$\cos \frac{cn\pi}{L} t \sin \frac{n\pi}{L} x = \frac{1}{2}\left[\sin \left\{\frac{n\pi}{L}(x - ct)\right\} + \sin \left\{\frac{n\pi}{L}(x + ct)\right\}\right].$$

Consequently, we may write (16) in the form

$$u(x, t) = \frac{1}{2} \sum_{n=1}^{\infty} B_n \sin \left\{\frac{n\pi}{L}(x - ct)\right\} + \frac{1}{2} \sum_{n=1}^{\infty} B_n \sin \left\{\frac{n\pi}{L}(x + ct)\right\}.$$

These two series are those obtained by substituting $x - ct$ and $x + ct$, respectively, for the variable x in the Fourier sine series (13) for $f(x)$. Thus

(17) $$u(x, t) = \tfrac{1}{2}[f^*(x - ct) + f^*(x + ct)]$$

where f^* is the odd periodic extension of f with the period $2L$ (Fig. 289). Since the initial deflection $f(x)$ is continuous on the interval $0 \leqq x \leqq L$ and zero at the endpoints, it follows from (17) that $u(x, t)$ is a continuous function of both variables x and t for all values of the variables. By differentiating (17) we see that $u(x, t)$ is a solution of (1), provided $f(x)$ is twice differentiable on the interval $0 < x < L$, and has one-sided second derivatives at $x = 0$ and $x = L$, which are zero. Under these conditions $u(x, t)$ is established as a solution of (1), satisfying (2) and (3) with $g(x) \equiv 0$. ■

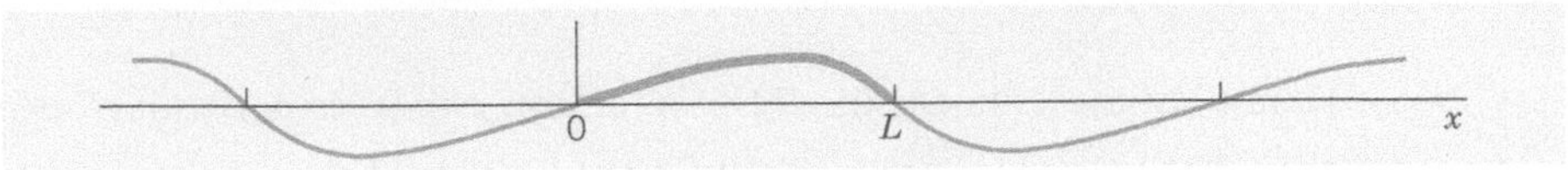

Fig. 289. Odd periodic extension of $f(x)$

Generalized Solution. If $f'(x)$ and $f''(x)$ are merely piecewise continuous (see Sec. 6.1), or if those one-sided derivatives are not zero, then for each t there will be finitely many values of x at which the second derivatives of u appearing in (1) do not exist. Except at these points the wave equation will still be satisfied. We may then regard $u(x, t)$ as a "**generalized solution,**" as it is called, that is, as a solution in a broader sense. For instance, a triangular initial deflection as in Example 1 (below) leads to a generalized solution.

Physical Interpretation of the Solution (17). The graph of $f^*(x - ct)$ is obtained from the graph of $f^*(x)$ by shifting the latter ct units to the right (Fig. 290). This means that $f^*(x - ct)(c > 0)$ represents a wave that is traveling to the right as t increases. Similarly, $f^*(x + ct)$ represents a wave that is traveling to the left, and $u(x, t)$ is the superposition of these two waves.

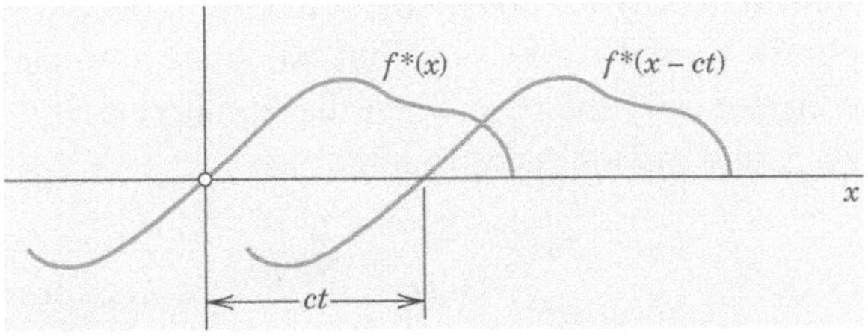

Fig. 290. Interpretation of (17)

EXAMPLE 1 Vibrating String if the Initial Deflection Is Triangular

Find the solution of the wave equation (1) satisfying (2) and corresponding to the triangular initial deflection

$$f(x) = \begin{cases} \dfrac{2k}{L}x & \text{if} \quad 0 < x < \dfrac{L}{2} \\[2ex] \dfrac{2k}{L}(L - x) & \text{if} \quad \dfrac{L}{2} < x < L \end{cases}$$

and initial velocity zero. (Figure 291 shows $f(x) = u(x, 0)$ at the top.)

Solution. Since $g(x) \equiv 0$, we have $B_n^* = 0$ in (12), and from Example 4 in Sec. 11.3 we see that the B_n are given by (5), Sec. 11.3. Thus (12) takes the form

$$u(x, t) = \frac{8k}{\pi^2}\left[\frac{1}{1^2}\sin\frac{\pi}{L}x\cos\frac{\pi c}{L}t - \frac{1}{3^2}\sin\frac{3\pi}{L}x\cos\frac{3\pi c}{L}t + - \cdots\right].$$

For graphing the solution we may use $u(x, 0) = f(x)$ and the above interpretation of the two functions in the representation (17). This leads to the graph shown in Fig. 291. ■

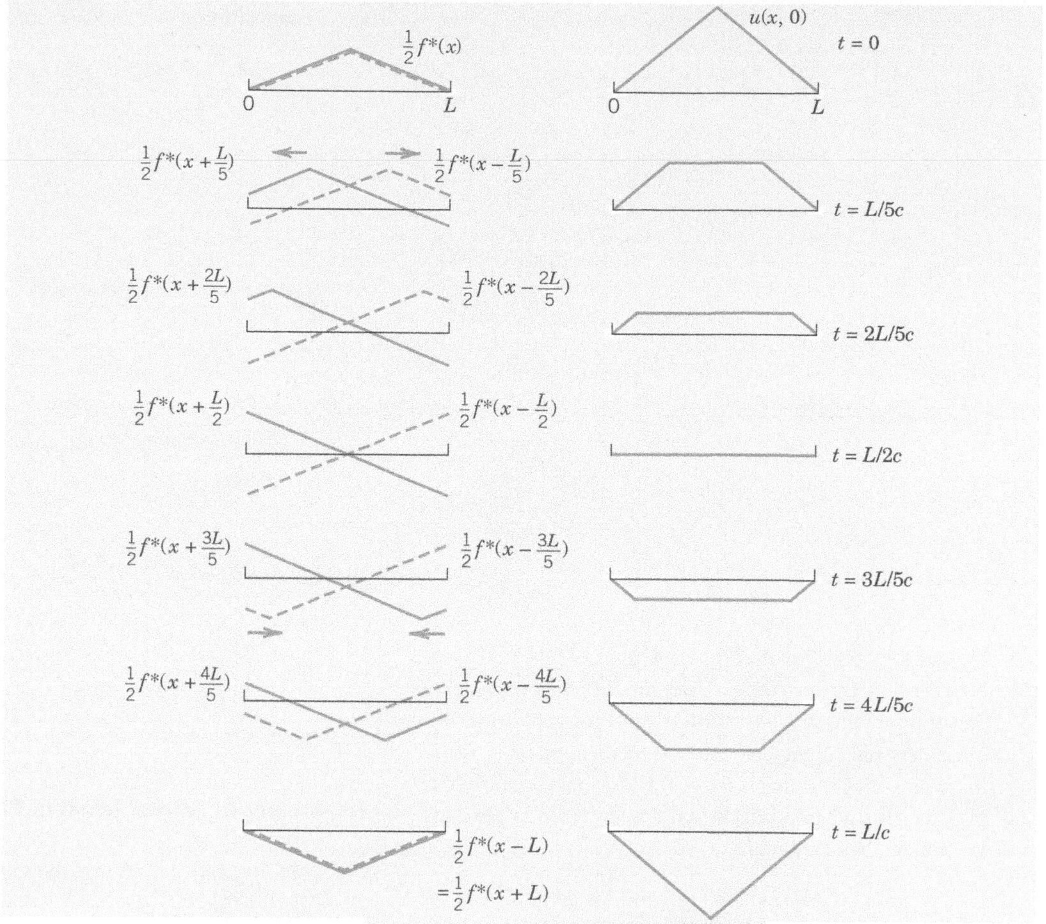

Fig. 291. Solution $u(x, t)$ in Example 1 for various values of t (right part of the figure) obtained as the superposition of a wave traveling to the right (dashed) and a wave traveling to the left (left part of the figure)

PROBLEM SET 12.3

1. **Frequency.** How does the frequency of the fundamental mode of the vibrating string depend on the length of the string? On the mass per unit length? What happens if we double the tension? Why is a contrabass larger than a violin?

2. **Physical Assumptions.** How would the motion of the string change if Assumption 3 were violated? Assumption 2? The second part of Assumption 1? The first part? Do we really need all these assumptions?

3. **String of length π.** Write down the derivation in this section for length $L = \pi$, to see the very substantial simplification of formulas in this case that may show ideas more clearly.

4. **CAS PROJECT. Graphing Normal Modes.** Write a program for graphing u_n with $L = \pi$ and c^2 of your choice similarly as in Fig. 287. Apply the program to u_2, u_3, u_4. Also graph these solutions as surfaces over the xt-plane. Explain the connection between these two kinds of graphs.

5–13 DEFLECTION OF THE STRING

Find $u(x, t)$ for the string of length $L = 1$ and $c^2 = 1$ when the initial velocity is zero and the initial deflection with small k (say, 0.01) is as follows. Sketch or graph $u(x, t)$ as in Fig. 291 in the text.

5. $k \sin 3\pi x$

6. $k(\sin \pi x - \frac{1}{2}\sin 2\pi x)$

7. $kx(1-x)$ **8.** $kx^2(1-x)$

9.

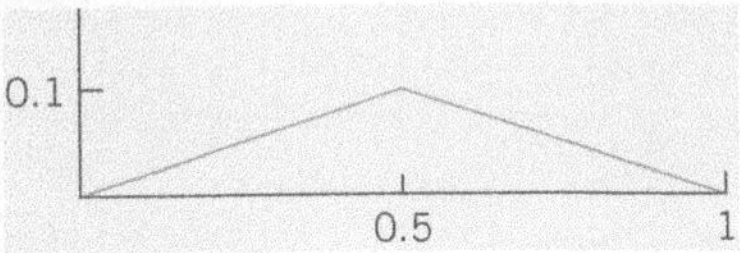

10.

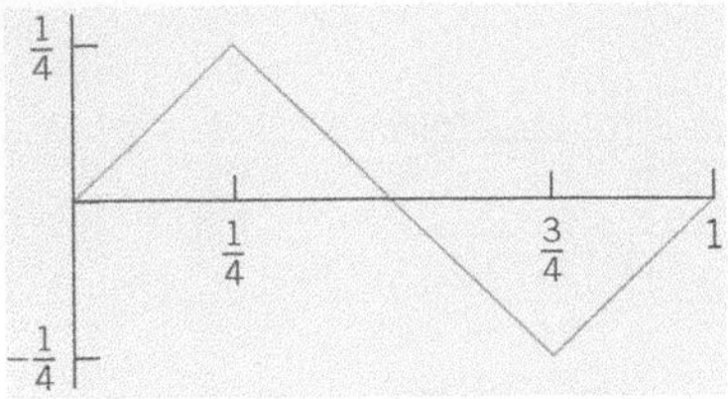

11.

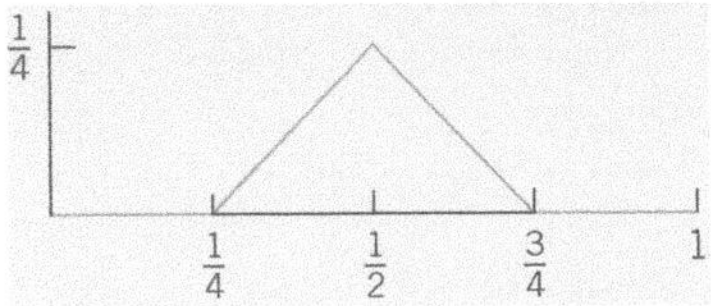

12.

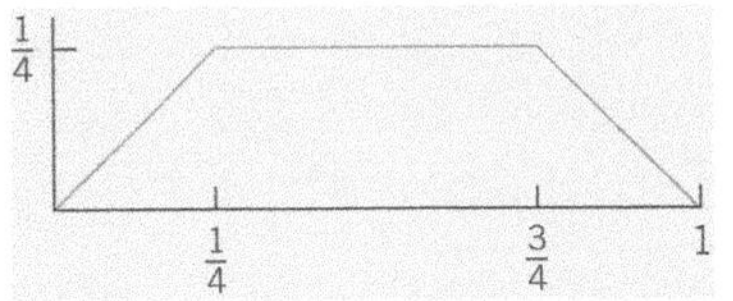

13. $2x - 4x^2$ if $0 < x < \frac{1}{2}$, 0 if $\frac{1}{2} < x < 1$

14. Nonzero initial velocity. Find the deflection $u(x, t)$ of the string of length $L = \pi$ and $c^2 = 1$ for zero initial displacement and "triangular" initial velocity $u_t(x, 0) = 0.01x$ if $0 \leqq x \leqq \frac{1}{2}\pi$, $u_t(x, 0) = 0.01(\pi - x)$ if $\frac{1}{2}\pi \leqq x \leqq \pi$. (Initial conditions with $u_t(x, 0) \neq 0$ are hard to realize experimentally.)

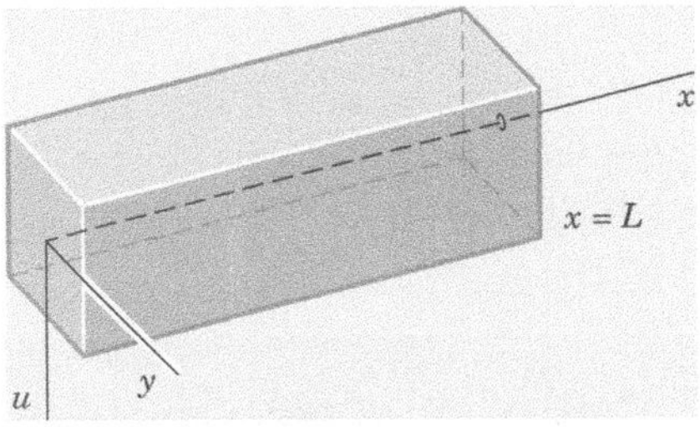

Fig. 292. Elastic beam

15–20 SEPARATION OF A FOURTH-ORDER PDE. VIBRATING BEAM

By the principles used in modeling the string it can be shown that small free vertical vibrations of a uniform elastic beam (Fig. 292) are modeled by the fourth-order PDE

$$\frac{\partial^2 u}{\partial t^2} = -c^2 \frac{\partial^4 u}{\partial x^4} \qquad \text{(21)} \qquad \text{(Ref. [C11])}$$

where $c^2 = EI/\rho A$ (E = Young's modulus of elasticity, I = moment of intertia of the cross section with respect to the y-axis in the figure, ρ = density, A = cross-sectional area). (*Bending* of a beam under a load is discussed in Sec. 3.3.)

15. Substituting $u = F(x)G(t)$ into (21), show that

$$F^{(4)}/F = -\ddot{G}/c^2 G = \beta^4 = \text{const},$$
$$F(x) = A \cos \beta x + B \sin \beta x + C \cosh \beta x + D \sinh \beta x,$$
$$G(t) = a \cos c\beta^2 t + b \sin c\beta^2 t.$$

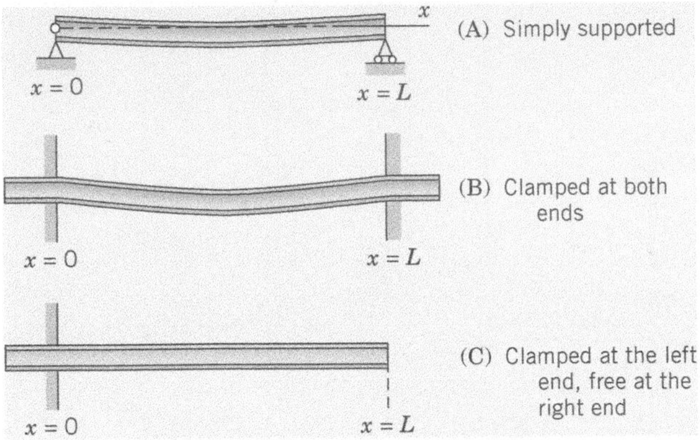

Fig. 293. Supports of a beam

16. Simply supported beam in Fig. 293A. Find solutions $u_n = F_n(x)G_n(t)$ of (21) corresponding to zero initial velocity and satisfying the boundary conditions (see Fig. 293A)

$$u(0, t) = 0, \; u(L, t) = 0$$
(ends simply supported for all times t),
$$u_{xx}(0, t) = 0, \; u_{xx}(L, t) = 0$$
(zero moments, hence zero curvature, at the ends).

17. Find the solution of (21) that satisfies the conditions in Prob. 16 as well as the initial condition

$$u(x, 0) = f(x) = x(L - x).$$

18. Compare the results of Probs. 17 and 7. What is the basic difference between the frequencies of the normal modes of the vibrating string and the vibrating beam?

19. Clamped beam in Fig. 293B. What are the boundary conditions for the clamped beam in Fig. 293B? Show that F in Prob. 15 satisfies these conditions if βL is a solution of the equation

$$\cosh \beta L \cos \beta L = 1. \qquad \text{(22)}$$

Determine approximate solutions of (22), for instance, graphically from the intersections of the curves of $\cos \beta L$ and $1/\cosh \beta L$.

20. Clamped-free beam in Fig. 293C. If the beam is clamped at the left and free at the right (Fig. 293C), the boundary conditions are

$$u(0, t) = 0, \qquad u_x(0, t) = 0,$$
$$u_{xx}(L, t) = 0, \qquad u_{xxx}(L, t) = 0.$$

Show that F in Prob. 15 satisfies these conditions if βL is a solution of the equation

$$\cosh \beta L \cos \beta L = -1. \tag{23}$$

Find approximate solutions of (23).

12.4 D'Alembert's Solution of the Wave Equation. Characteristics

It is interesting that the solution (17), Sec. 12.3, of the wave equation

$$\frac{\partial^2 u}{\partial t^2} = c^2 \frac{\partial^2 u}{\partial x^2}, \qquad c^2 = \frac{T}{\rho}, \tag{1}$$

can be immediately obtained by transforming (1) in a suitable way, namely, by introducing the new independent variables

$$v = x + ct, \qquad w = x - ct. \tag{2}$$

Then u becomes a function of v and w. The derivatives in (1) can now be expressed in terms of derivatives with respect to v and w by the use of the chain rule in Sec. 9.6. Denoting partial derivatives by subscripts, we see from (2) that $v_x = 1$ and $w_x = 1$. For simplicity let us denote $u(x, t)$, as a function of v and w, by the same letter u. Then

$$u_x = u_v v_x + u_w w_x = u_v + u_w.$$

We now apply the chain rule to the right side of this equation. We assume that all the partial derivatives involved are continuous, so that $u_{wv} = u_{vw}$. Since $v_x = 1$ and $w_x = 1$, we obtain

$$u_{xx} = (u_v + u_w)_x = (u_v + u_w)_v v_x + (u_v + u_w)_w w_x = u_{vv} + 2u_{vw} + u_{ww}.$$

Transforming the other derivative in (1) by the same procedure, we find

$$u_{tt} = c^2(u_{vv} - 2u_{vw} + u_{ww}).$$

By inserting these two results in (1) we get (see footnote 2 in App. A3.2)

$$u_{vw} \equiv \frac{\partial^2 u}{\partial w \, \partial v} = 0. \tag{3}$$

The point of the present method is that (3) can be readily solved by two successive integrations, first with respect to w and then with respect to v. This gives

$$\frac{\partial u}{\partial v} = h(v) \qquad \text{and} \qquad u = \int h(v)\, dv + \psi(w).$$

Here $h(v)$ and $\psi(w)$ are arbitrary functions of v and w, respectively. Since the integral is a function of v, say, $\phi(v)$, the solution is of the form $u = \phi(v) + \psi(w)$. In terms of x and t, by (2), we thus have

$$u(x, t) = \phi(x + ct) + \psi(x - ct). \tag{4}$$

This is known as **d'Alembert's solution**[1] of the wave equation (1).

Its derivation was much more elegant than the method in Sec. 12.3, but d'Alembert's method is special, whereas the use of Fourier series applies to various equations, as we shall see.

D'Alembert's Solution Satisfying the Initial Conditions

$$\text{(a)}\quad u(x, 0) = f(x), \qquad \text{(b)}\quad u_t(x, 0) = g(x). \tag{5}$$

These are the same as (3) in Sec. 12.3. By differentiating (4) we have

$$u_t(x, t) = c\phi'(x + ct) - c\psi'(x - ct) \tag{6}$$

where primes denote derivatives with respect to the *entire* arguments $x + ct$ and $x - ct$, respectively, and the minus sign comes from the chain rule. From (4)–(6) we have

$$u(x, 0) = \phi(x) + \psi(x) = f(x), \tag{7}$$

$$u_t(x, 0) = c\phi'(x) + c\psi'(x) = g(x). \tag{8}$$

Dividing (8) by c and integrating with respect to x, we obtain

$$\phi(x) - \psi(x) = k(x_0) + \frac{1}{c}\int_{x_0}^{x} g(s)\, ds, \qquad k(x_0) = \phi(x_0) - \psi(x_0). \tag{9}$$

If we add this to (7), then ψ drops out and division by 2 gives

$$\phi(x) = \frac{1}{2} f(x) + \frac{1}{2c}\int_{x_0}^{x} g(s)\, ds + \frac{1}{2} k(x_0). \tag{10}$$

Similarly, subtraction of (9) from (7) and division by 2 gives

$$\psi(x) = \frac{1}{2} f(x) - \frac{1}{2c}\int_{x_0}^{x} g(s)\, ds - \frac{1}{2} k(x_0). \tag{11}$$

In (10) we replace x by $x + ct$; we then get an integral from x_0 to $x + ct$. In (11) we replace x by $x - ct$ and get minus an integral from x_0 to $x - ct$ or plus an integral from $x - ct$ to x_0. Hence addition of $\phi(x + ct)$ and $\psi(x - ct)$ gives $u(x, t)$ [see (4)] in the form

$$u(x, t) = \frac{1}{2}[f(x + ct) + f(x - ct)] + \frac{1}{2c}\int_{x-ct}^{x+ct} g(s)\, ds. \tag{12}$$

---

[1]JEAN LE ROND D'ALEMBERT (1717–1783), French mathematician, also known for his important work in mechanics.

We mention that the general theory of PDEs provides a systematic way for finding the transformation (2) that simplifies (1). See Ref. [C8] in App. 1.

If the initial velocity is zero, we see that this reduces to

$$u(x, t) = \tfrac{1}{2}[f(x + ct) + f(x - ct)], \tag{13}$$

in agreement with (17) in Sec. 12.3. You may show that because of the boundary conditions (2) in that section the function f must be odd and must have the period $2L$.

Our result shows that the two initial conditions [the functions $f(x)$ and $g(x)$ in (5)] determine the solution uniquely.

The solution of the wave equation by the Laplace transform method will be shown in Sec. 12.11.

Characteristics. Types and Normal Forms of PDEs

The idea of d'Alembert's solution is just a special instance of the **method of characteristics.** This concerns PDEs of the form

$$Au_{xx} + 2Bu_{xy} + Cu_{yy} = F(x, y, u, u_x, u_y) \tag{14}$$

(as well as PDEs in more than two variables). Equation (14) is called **quasilinear** because it is linear in the highest derivatives (but may be arbitrary otherwise). There are three types of PDEs (14), depending on the discriminant $AC - B^2$, as follows.

Type	Defining Condition	Example in Sec. 12.1
Hyperbolic	$AC - B^2 < 0$	Wave equation (1)
Parabolic	$AC - B^2 = 0$	Heat equation (2)
Elliptic	$AC - B^2 > 0$	Laplace equation (3)

Note that (1) and (2) in Sec. 12.1 involve t, but to have y as in (14), we set $y = ct$ in (1), obtaining $u_{tt} - c^2u_{xx} = c^2(u_{yy} - u_{xx}) = 0$. And in (2) we set $y = c^2t$, so that $u_t - c^2u_{xx} = c^2(u_y - u_{xx})$.

A, B, C may be functions of x, y, so that a PDE may be **of mixed type**, that is, of different type in different regions of the xy-plane. An important mixed-type PDE is the **Tricomi equation** (see Prob. 10).

Transformation of (14) to Normal Form. The normal forms of (14) and the corresponding transformations depend on the type of the PDE. They are obtained by solving the **characteristic equation** of (14), which is the ODE

$$Ay'^2 - 2By' + C = 0 \tag{15}$$

where $y' = dy/dx$ (note $-2B$, not $+2B$). The solutions of (15) are called the **characteristics** of (14), and we write them in the form $\Phi(x, y) = \text{const}$ and $\Psi(x, y) = \text{const}$. Then the transformations giving new variables v, w instead of x, y and the normal forms of (14) are as follows.

Type	New Variables		Normal Form
Hyperbolic	$v = \Phi$	$w = \Psi$	$u_{vw} = F_1$
Parabolic	$v = x$	$w = \Phi = \Psi$	$u_{ww} = F_2$
Elliptic	$v = \frac{1}{2}(\Phi + \Psi)$	$w = \frac{1}{2i}(\Phi - \Psi)$	$u_{vv} + u_{ww} = F_3$

Here, $\Phi = \Phi(x, y), \Psi = \Psi(x, y), F_1 = F_1(v, w, u, u_v, u_w)$, etc., and we denote u as function of v, w again by u, for simplicity. We see that the normal form of a hyperbolic PDE is as in d'Alembert's solution. In the parabolic case we get just one family of solutions $\Phi = \Psi$. In the elliptic case, $i = \sqrt{-1}$, and the characteristics are complex and are of minor interest. For derivation, see Ref. [GenRef3] in App. 1.

EXAMPLE 1 D'Alembert's Solution Obtained Systematically

The theory of characteristics gives d'Alembert's solution in a systematic fashion. To see this, we write the wave equation $u_{tt} - c^2 u_{xx} = 0$ in the form (14) by setting $y = ct$. By the chain rule, $u_t = u_y y_t = c u_y$ and $u_{tt} = c^2 u_{yy}$. Division by c^2 gives $u_{xx} - u_{yy} = 0$, as stated before. Hence the characteristic equation is $y'^2 - 1 = (y' + 1)(y' - 1) = 0$. The two families of solutions (characteristics) are $\Phi(x, y) = y + x = \text{const}$ and $\Psi(x, y) = y - x = \text{const}$. This gives the new variables $v = \Phi = y + x = ct + x$ and $w = \Psi = y - x = ct - x$ and d'Alembert's solution $u = f_1(x + ct) + f_2(x - ct)$. ■

PROBLEM SET 12.4

1. Show that c is the speed of each of the two waves given by (4).
2. Show that, because of the boundary conditions (2), Sec. 12.3, the function f in (13) of this section must be odd and of period $2L$.
3. If a steel wire 2 m in length weighs 0.9 nt (about 0.20 lb) and is stretched by a tensile force of 300 nt (about 67.4 lb), what is the corresponding speed of transverse waves?
4. What are the frequencies of the eigenfunctions in Prob. 3?

5–8 GRAPHING SOLUTIONS

Using (13) sketch or graph a figure (similar to Fig. 291 in Sec. 12.3) of the deflection $u(x, t)$ of a vibrating string (length $L = 1$, ends fixed, $c = 1$) starting with initial velocity 0 and initial deflection (k small, say, $k = 0.01$).

5. $f(x) = k \sin \pi x$
6. $f(x) = k(1 - \cos \pi x)$
7. $f(x) = k \sin 2\pi x$
8. $f(x) = kx(1 - x)$

9–18 NORMAL FORMS

Find the type, transform to normal form, and solve. Show your work in detail.

9. $u_{xx} + 4u_{yy} = 0$
10. $u_{xx} - 16u_{yy} = 0$
11. $u_{xx} + 2u_{xy} + u_{yy} = 0$
12. $u_{xx} - 2u_{xy} + u_{yy} = 0$
13. $u_{xx} + 5u_{xy} + 4u_{yy} = 0$
14. $xu_{xy} - yu_{yy} = 0$
15. $xu_{xx} - yu_{xy} = 0$
16. $u_{xx} + 2u_{xy} + 10u_{yy} = 0$
17. $u_{xx} - 4u_{xy} + 5u_{yy} = 0$
18. $u_{xx} - 6u_{xy} + 9u_{yy} = 0$
19. **Longitudinal Vibrations of an Elastic Bar or Rod.** These vibrations in the direction of the x-axis are modeled by the wave equation $u_{tt} = c^2 u_{xx}$, $c^2 = E/\rho$ (see Tolstov [C9], p. 275). If the rod is fastened at one end, $x = 0$, and free at the other, $x = L$, we have $u(0, t) = 0$ and $u_x(L, t) = 0$. Show that the motion corresponding to initial displacement $u(x, 0) = f(x)$ and initial velocity zero is

$$u = \sum_{n=0}^{\infty} A_n \sin p_n x \cos p_n ct,$$

$$A_n = \frac{2}{L}\int_0^L f(x) \sin p_n x \, dx, \qquad p_n = \frac{(2n + 1)\pi}{2L}.$$

20. **Tricomi and Airy equations.**[2] Show that the *Tricomi equation* $yu_{xx} + u_{yy} = 0$ is of mixed type. Obtain the **Airy equation** $G'' - yG = 0$ from the Tricomi equation by separation. (For solutions, see p. 446 of Ref. [GenRef1] listed in App. 1.)

[2]Sir GEORGE BIDELL AIRY (1801–1892), English mathematician, known for his work in elasticity. FRANCESCO TRICOMI (1897–1978), Italian mathematician, who worked in integral equations and functional analysis.

12.5 Modeling: Heat Flow from a Body in Space. Heat Equation

After the wave equation (Sec. 12.2) we now derive and discuss the next "big" PDE, the **heat equation**, which governs the temperature u in a body in space. We obtain this model of temperature distribution under the following.

Physical Assumptions

1. The *specific heat* σ and the *density* ρ of the material of the body are constant. No heat is produced or disappears in the body.
2. Experiments show that, in a body, heat flows in the direction of decreasing temperature, and the rate of flow is proportional to the gradient (cf. Sec. 9.7) of the temperature; that is, the velocity $\mathbf{v}$ of the heat flow in the body is of the form

$$\mathbf{v} = -K \operatorname{grad} u \tag{1}$$

   where $u(x, y, z, t)$ is the temperature at a point (x, y, z) and time t.
3. The *thermal conductivity* K is constant, as is the case for homogeneous material and nonextreme temperatures.

Under these assumptions we can model heat flow as follows.

Let T be a region in the body bounded by a surface S with outer unit normal vector $\mathbf{n}$ such that the divergence theorem (Sec. 10.7) applies. Then

$$\mathbf{v} \cdot \mathbf{n}$$

is the component of $\mathbf{v}$ in the direction of $\mathbf{n}$. Hence $|\mathbf{v} \cdot \mathbf{n}\, \Delta A|$ is the amount of heat *leaving* T (if $\mathbf{v} \cdot \mathbf{n} > 0$ at some point P) or *entering* T (if $\mathbf{v} \cdot \mathbf{n} < 0$ at P) per unit time at some point P of S through a small portion ΔS of S of area ΔA. Hence the total amount of heat that flows across S from T is given by the surface integral

$$\iint_S \mathbf{v} \cdot \mathbf{n}\, dA.$$

Note that, so far, this parallels the derivation on fluid flow in Example 1 of Sec. 10.8.

Using Gauss's theorem (Sec. 10.7), we now convert our surface integral into a volume integral over the region T. Because of (1) this gives [use (3) in Sec. 9.8]

$$\begin{aligned} \iint_S \mathbf{v} \cdot \mathbf{n}\, dA &= -K \iint_S (\operatorname{grad} u) \cdot \mathbf{n}\, dA = -K \iiint_T \operatorname{div} (\operatorname{grad} u)\, dx\, dy\, dz \\ &= -K \iiint_T \nabla^2 u\, dx\, dy\, dz. \end{aligned} \tag{2}$$

Here,

$$\nabla^2 u = \frac{\partial^2 u}{\partial x^2} + \frac{\partial^2 u}{\partial y^2} + \frac{\partial^2 u}{\partial z^2}$$

is the **Laplacian** of u.

On the other hand, the total amount of heat in T is

$$H = \iiint_T \sigma \rho u \, dx \, dy \, dz$$

with σ and ρ as before. Hence the time rate of decrease of H is

$$-\frac{\partial H}{\partial t} = -\iiint_T \sigma \rho \frac{\partial u}{\partial t} \, dx \, dy \, dz.$$

This must be equal to the amount of heat leaving T because no heat is produced or disappears in the body. From (2) we thus obtain

$$-\iiint_T \sigma \rho \frac{\partial u}{\partial t} \, dx \, dy \, dz = -K \iiint_T \nabla^2 u \, dx \, dy \, dz$$

or (divide by $-\sigma\rho$)

$$\iiint_T \left(\frac{\partial u}{\partial t} - c^2 \nabla^2 u \right) dx \, dy \, dz = 0 \qquad c^2 = \frac{K}{\sigma \rho}.$$

Since this holds for any region T in the body, the integrand (if continuous) must be zero everywhere. That is,

$$\textbf{(3)} \qquad \frac{\partial u}{\partial t} = c^2 \nabla^2 u. \qquad c^2 = K/\rho\sigma$$

This is the **heat equation**, the fundamental PDE modeling heat flow. It gives the temperature $u(x, y, z, t)$ in a body of homogeneous material in space. The constant c^2 is the *thermal diffusivity*. K is the *thermal conductivity*, σ the *specific heat*, and ρ the *density* of the material of the body. $\nabla^2 u$ is the Laplacian of u and, with respect to the Cartesian coordinates x, y, z, is

$$\nabla^2 u = \frac{\partial^2 u}{\partial x^2} + \frac{\partial^2 u}{\partial y^2} + \frac{\partial^2 u}{\partial z^2}.$$

The heat equation is also called the **diffusion equation** because it also models chemical diffusion processes of one substance or gas into another.

12.6 Heat Equation: Solution by Fourier Series. Steady Two-Dimensional Heat Problems. Dirichlet Problem

We want to solve the (one-dimensional) heat equation just developed in Sec. 12.5 and give several applications. This is followed much later in this section by an extension of the heat equation to two dimensions.

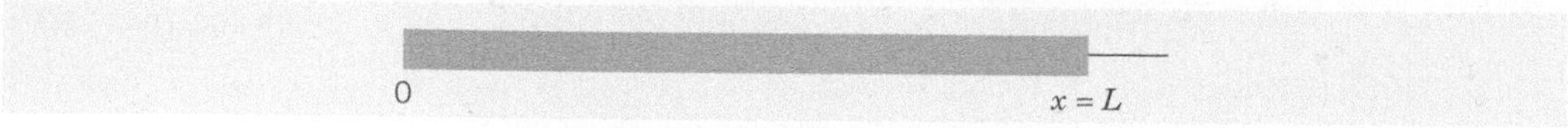

Fig. 294. Bar under consideration

As an important application of the heat equation, let us first consider the temperature in a long thin metal bar or wire of constant cross section and homogeneous material, which is oriented along the x-axis (Fig. 294) and is perfectly insulated laterally, so that heat flows in the x-direction only. Then besides time, u depends only on x, so that the Laplacian reduces to $u_{xx} = \partial^2 u/\partial x^2$, and the heat equation becomes the **one-dimensional heat equation**

$$\frac{\partial u}{\partial t} = c^2 \frac{\partial^2 u}{\partial x^2}. \tag{1}$$

This PDE seems to differ only very little from the wave equation, which has a term u_{tt} instead of u_t, but we shall see that this will make the solutions of (1) behave quite differently from those of the wave equation.

We shall solve (1) for some important types of boundary and initial conditions. We begin with the case in which the ends $x = 0$ and $x = L$ of the bar are kept at temperature zero, so that we have the **boundary conditions**

$$u(0, t) = 0, \qquad u(L, t) = 0 \qquad \text{for all } t \geqq 0. \tag{2}$$

Furthermore, the initial temperature in the bar at time $t = 0$ is given, say, $f(x)$, so that we have the **initial condition**

$$u(x, 0) = f(x) \qquad [f(x) \text{ given}]. \tag{3}$$

Here we must have $f(0) = 0$ and $f(L) = 0$ because of (2).

We shall determine a solution $u(x, t)$ of (1) satisfying (2) and (3)—one initial condition will be enough, as opposed to two initial conditions for the wave equation. Technically, our method will parallel that for the wave equation in Sec. 12.3: a separation of variables, followed by the use of Fourier series. You may find a step-by-step comparison worthwhile.

***Step 1.* Two ODEs from the heat equation (1).** Substitution of a product $u(x, t) = F(x)G(t)$ into (1) gives $F\dot{G} = c^2 F'' G$ with $\dot{G} = dG/dt$ and $F'' = d^2F/dx^2$. To separate the variables, we divide by $c^2 FG$, obtaining

$$\frac{\dot{G}}{c^2 G} = \frac{F''}{F}. \tag{4}$$

The left side depends only on t and the right side only on x, so that both sides must equal a constant k (as in Sec. 12.3). You may show that for $k = 0$ or $k > 0$ the only solution $u = FG$ satisfying (2) is $u \equiv 0$. For negative $k = -p^2$ we have from (4)

$$\frac{\dot{G}}{c^2 G} = \frac{F''}{F} = -p^2.$$

Multiplication by the denominators immediately gives the two ODEs

$$F'' + p^2F = 0 \tag{5}$$

and

$$\dot{G} + c^2p^2G = 0. \tag{6}$$

***Step 2.* Satisfying the boundary conditions (2).** We first solve (5). A general solution is

$$F(x) = A \cos px + B \sin px. \tag{7}$$

From the boundary conditions (2) it follows that

$$u(0, t) = F(0)G(t) = 0 \qquad \text{and} \qquad u(L, t) = F(L)G(t) = 0.$$

Since $G \equiv 0$ would give $u \equiv 0$, we require $F(0) = 0$, $F(L) = 0$ and get $F(0) = A = 0$ by (7) and then $F(L) = B \sin pL = 0$, with $B \neq 0$ (to avoid $F \equiv 0$); thus,

$$\sin pL = 0, \qquad \text{hence} \qquad p = \frac{n\pi}{L}, \qquad n = 1, 2, \cdots.$$

Setting $B = 1$, we thus obtain the following solutions of (5) satisfying (2):

$$F_n(x) = \sin \frac{n\pi x}{L}, \qquad n = 1, 2, \cdots.$$

(As in Sec. 12.3, we need not consider *negative* integer values of n.)

All this was literally the same as in Sec. 12.3. From now on it differs since (6) differs from (6) in Sec. 12.3. We now solve (6). For $p = n\pi/L$, as just obtained, (6) becomes

$$\dot{G} + \lambda_n^2 G = 0 \qquad \text{where} \qquad \lambda_n = \frac{cn\pi}{L}.$$

It has the general solution

$$G_n(t) = B_n e^{-\lambda_n^2 t}, \qquad n = 1, 2, \cdots$$

where B_n is a constant. Hence the functions

$$u_n(x, t) = F_n(x)G_n(t) = B_n \sin \frac{n\pi x}{L} e^{-\lambda_n^2 t} \qquad (n = 1, 2, \cdots) \tag{8}$$

are solutions of the heat equation (1), satisfying (2). These are the **eigenfunctions** of the problem, corresponding to the **eigenvalues** $\lambda_n = cn\pi/L$.

***Step 3.* Solution of the entire problem. Fourier series.** So far we have solutions (8) satisfying the boundary conditions (2). To obtain a solution that also satisfies the initial condition (3), we consider a series of these eigenfunctions,

$$u(x, t) = \sum_{n=1}^{\infty} u_n(x, t) = \sum_{n=1}^{\infty} B_n \sin \frac{n\pi x}{L} e^{-\lambda_n^2 t} \qquad \left(\lambda_n = \frac{cn\pi}{L}\right). \tag{9}$$

From this and (3) we have

$$u(x, 0) = \sum_{n=1}^{\infty} B_n \sin \frac{n\pi x}{L} = f(x).$$

Hence for (9) to satisfy (3), the B_n's must be the coefficients of the **Fourier sine series**, as given by (4) in Sec. 11.3; thus

$$B_n = \frac{2}{L}\int_0^L f(x) \sin \frac{n\pi x}{L}\, dx \qquad (n = 1, 2, \cdots). \tag{10}$$

The solution of our problem can be established, assuming that $f(x)$ is piecewise continuous (see Sec. 6.1) on the interval $0 \leqq x \leqq L$ and has one-sided derivatives (see Sec. 11.1) at all interior points of that interval; that is, under these assumptions the series (9) with coefficients (10) is the solution of our physical problem. A proof requires knowledge of uniform convergence and will be given at a later occasion (Probs. 19, 20 in Problem Set 15.5).

Because of the exponential factor, all the terms in (9) approach zero as t approaches infinity. The rate of decay increases with n.

EXAMPLE 1 Sinusoidal Initial Temperature

Find the temperature $u(x, t)$ in a laterally insulated copper bar 80 cm long if the initial temperature is $100 \sin(\pi x/80)$ °C and the ends are kept at 0°C. How long will it take for the maximum temperature in the bar to drop to 50°C? First guess, then calculate. *Physical data for copper:* density 8.92 g/cm^3, specific heat 0.092 cal/(g °C), thermal conductivity 0.95 cal/(cm sec °C).

Solution. The initial condition gives

$$u(x, 0) = \sum_{n=1}^{\infty} B_n \sin \frac{n\pi x}{80} = f(x) = 100 \sin \frac{\pi x}{80}.$$

Hence, by inspection or from (9), we get $B_1 = 100$, $B_2 = B_3 = \cdots = 0$. In (9) we need $\lambda_1^2 = c^2\pi^2/L^2$, where $c^2 = K/(\sigma\rho) = 0.95/(0.092 \cdot 8.92) = 1.158$ [cm^2/sec]. Hence we obtain

$$\lambda_1^2 = 1.158 \cdot 9.870/80^2 = 0.001785 \text{ [sec}^{-1}\text{]}.$$

The solution (9) is

$$u(x, t) = 100 \sin \frac{\pi x}{80} e^{-0.001785t}.$$

Also, $100e^{-0.001785t} = 50$ when $t = (\ln 0.5)/(-0.001785) = 388$ [sec] ≈ 6.5 [min]. Does your guess, or at least its order of magnitude, agree with this result? ■

EXAMPLE 2 Speed of Decay

Solve the problem in Example 1 when the initial temperature is $100 \sin(3\pi x/80)$ °C and the other data are as before.

Solution. In (9), instead of $n = 1$ we now have $n = 3$, and $\lambda_3^2 = 3^2\lambda_1^2 = 9 \cdot 0.001785 = 0.01607$, so that the solution now is

$$u(x, t) = 100 \sin \frac{3\pi x}{80} e^{-0.01607t}.$$

Hence the maximum temperature drops to 50°C in $t = (\ln 0.5)/(-0.01607) \approx 43$ [sec], which is much faster (9 times as fast as in Example 1; why?).

Had we chosen a bigger n, the decay would have been still faster, and in a sum or series of such terms, each term has its own rate of decay, and terms with large n are practically 0 after a very short time. Our next example is of this type, and the curve in Fig. 295 corresponding to $t = 0.5$ looks almost like a sine curve; that is, it is practically the graph of the first term of the solution. ■

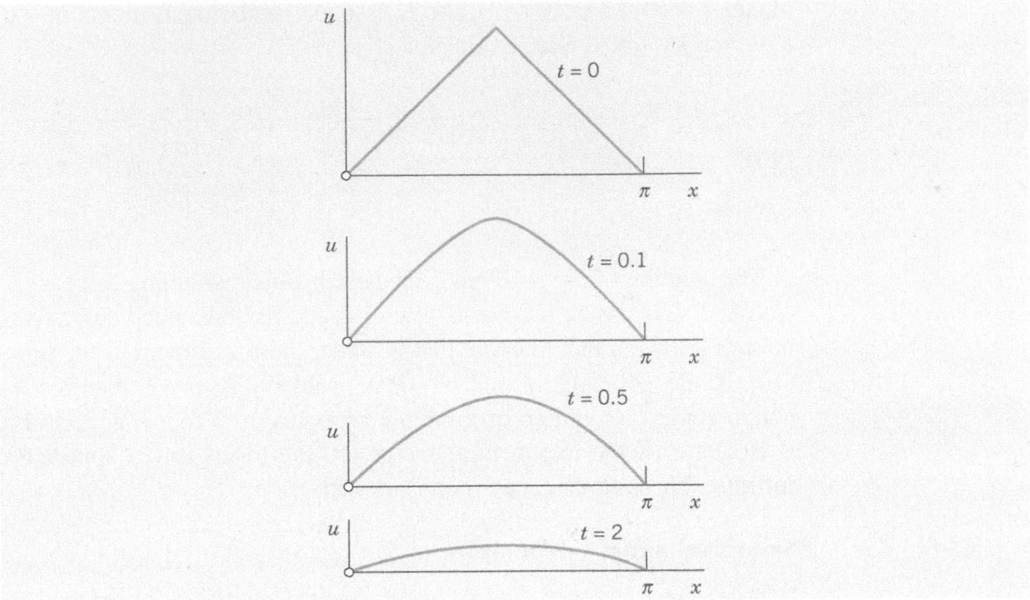

Fig. 295. Example 3. Decrease of temperature with time t for $L = \pi$ and $c = 1$

EXAMPLE 3 "Triangular" Initial Temperature in a Bar

Find the temperature in a laterally insulated bar of length L whose ends are kept at temperature 0, assuming that the initial temperature is

$$f(x) = \begin{cases} x & \text{if} \quad 0 < x < L/2, \\ L - x & \text{if} \quad L/2 < x < L. \end{cases}$$

(The uppermost part of Fig. 295 shows this function for the special $L = \pi$.)

Solution. From (10) we get

$$B_n = \frac{2}{L}\left(\int_0^{L/2} x \sin\frac{n\pi x}{L}\,dx + \int_{L/2}^{L} (L - x)\sin\frac{n\pi x}{L}\,dx\right). \tag{10*}$$

Integration gives $B_n = 0$ if n is even,

$$B_n = \frac{4L}{n^2\pi^2} \quad (n = 1, 5, 9, \cdots) \qquad \text{and} \qquad B_n = -\frac{4L}{n^2\pi^2} \quad (n = 3, 7, 11, \cdots).$$

(see also Example 4 in Sec. 11.3 with $k = L/2$). Hence the solution is

$$u(x, t) = \frac{4L}{\pi^2}\left[\sin\frac{\pi x}{L}\exp\left[-\left(\frac{c\pi}{L}\right)^2 t\right] - \frac{1}{9}\sin\frac{3\pi x}{L}\exp\left[-\left(\frac{3c\pi}{L}\right)^2 t\right] + - \cdots\right].$$

Figure 295 shows that the temperature decreases with increasing t, because of the heat loss due to the cooling of the ends.

Compare Fig. 295 and Fig. 291 in Sec. 12.3 and comment. ■

EXAMPLE 4 **Bar with Insulated Ends. Eigenvalue 0**

Find a solution formula of (1), (3) with (2) replaced by the condition that both ends of the bar are insulated.

Solution. Physical experiments show that the rate of heat flow is proportional to the gradient of the temperature. Hence if the ends $x = 0$ and $x = L$ of the bar are insulated, so that no heat can flow through the ends, we have grad $u = u_x = \partial u/\partial x$ and the boundary conditions

$$(2^*) \qquad u_x(0, t) = 0, \qquad u_x(L, t) = 0 \qquad \text{for all } t.$$

Since $u(x, t) = F(x)G(t)$, this gives $u_x(0, t) = F'(0)G(t) = 0$ and $u_x(L, t) = F'(L)G(t) = 0$. Differentiating (7), we have $F'(x) = -Ap \sin px + Bp \cos px$, so that

$$F'(0) = Bp = 0 \qquad \text{and then} \qquad F'(L) = -Ap \sin pL = 0.$$

The second of these conditions gives $p = p_n = n\pi/L$, $(n = 0, 1, 2, \cdots)$. From this and (7) with $A = 1$ and $B = 0$ we get $F_n(x) = \cos(n\pi x/L)$, $(n = 0, 1, 2, \cdots)$. With G_n as before, this yields the eigenfunctions

$$(11) \qquad u_n(x, t) = F_n(x)G_n(t) = A_n \cos\frac{n\pi x}{L} e^{-\lambda_n^2 t} \qquad (n = 0, 1, \cdots)$$

corresponding to the eigenvalues $\lambda_n = cn\pi/L$. The latter are as before, but we now have the additional eigenvalue $\lambda_0 = 0$ and eigenfunction $u_0 = \text{const}$, which is the solution of the problem if the initial temperature $f(x)$ is constant. This shows the remarkable fact that ***a separation constant can very well be zero, and zero can be an eigenvalue.***

Furthermore, whereas (8) gave a Fourier sine series, we now get from (11) a Fourier cosine series

$$\textbf{(12)} \qquad u(x, t) = \sum_{n=0}^{\infty} u_n(x, t) = \sum_{n=0}^{\infty} A_n \cos\frac{n\pi x}{L} e^{-\lambda_n^2 t} \qquad \left(\lambda_n = \frac{cn\pi}{L}\right).$$

Its coefficients result from the initial condition (3),

$$u(x, 0) = \sum_{n=0}^{\infty} A_n \cos\frac{n\pi x}{L} = f(x),$$

in the form (2), Sec. 11.3, that is,

$$(13) \qquad A_0 = \frac{1}{L}\int_0^L f(x)\,dx, \qquad A_n = \frac{2}{L}\int_0^L f(x)\cos\frac{n\pi x}{L}\,dx, \qquad n = 1, 2, \cdots.$$

■

EXAMPLE 5 **"Triangular" Initial Temperature in a Bar with Insulated Ends**

Find the temperature in the bar in Example 3, assuming that the ends are insulated (instead of being kept at temperature 0).

Solution. For the triangular initial temperature, (13) gives $A_0 = L/4$ and (see also Example 4 in Sec. 11.3 with $k = L/2$)

$$A_n = \frac{2}{L}\left[\int_0^{L/2} x\cos\frac{n\pi x}{L}\,dx + \int_{L/2}^{L} (L - x)\cos\frac{n\pi x}{L}\,dx\right] = \frac{2L}{n^2\pi^2}\left(2\cos\frac{n\pi}{2} - \cos n\pi - 1\right).$$

Hence the solution (12) is

$$u(x, t) = \frac{L}{4} - \frac{8L}{\pi^2}\left\{\frac{1}{2^2}\cos\frac{2\pi x}{L}\exp\left[-\left(\frac{2c\pi}{L}\right)^2 t\right] + \frac{1}{6^2}\cos\frac{6\pi x}{L}\exp\left[-\left(\frac{6c\pi}{L}\right)^2 t\right] + \cdots\right\}.$$

We see that the terms decrease with increasing t, and $u \to L/4$ as $t \to \infty$; this is the mean value of the initial temperature. This is plausible because no heat can escape from this totally insulated bar. In contrast, the cooling of the ends in Example 3 led to heat loss and $u \to 0$, the temperature at which the ends were kept. ■

Steady Two-Dimensional Heat Problems. Laplace's Equation

We shall now extend our discussion from one to two space dimensions and consider the two-dimensional heat equation

$$\frac{\partial u}{\partial t} = c^2\nabla^2 u = c^2\left(\frac{\partial^2 u}{\partial x^2} + \frac{\partial^2 u}{\partial y^2}\right)$$

for **steady** (that is, *time-independent*) problems. Then $\partial u/\partial t = 0$ and the heat equation reduces to **Laplace's equation**

$$\nabla^2 u = \frac{\partial^2 u}{\partial x^2} + \frac{\partial^2 u}{\partial y^2} = 0 \qquad (14)$$

(which has already occurred in Sec. 10.8 and will be considered further in Secs. 12.8–12.11). A heat problem then consists of this PDE to be considered in some region R of the xy-plane and a given boundary condition on the boundary curve C of R. This is a **boundary value problem (BVP)**. One calls it:

First BVP or **Dirichlet Problem** if u is prescribed on C (**"Dirichlet boundary condition"**)

Second BVP or **Neumann Problem** if the normal derivative $u_n = \partial u/\partial n$ is prescribed on C (**"Neumann boundary condition"**)

Third BVP, Mixed BVP, or **Robin Problem** if u is prescribed on a portion of C and u_n on the rest of C (**"Mixed boundary condition"**).

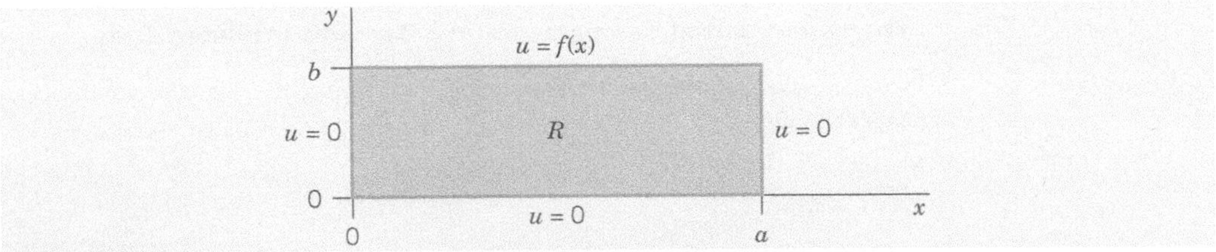

Fig. 296. Rectangle R and given boundary values

Dirichlet Problem in a Rectangle R (Fig. 296). We consider a Dirichlet problem for Laplace's equation (14) in a rectangle R, assuming that the temperature $u(x, y)$ equals a given function $f(x)$ on the upper side and 0 on the other three sides of the rectangle.

We solve this problem by separating variables. Substituting $u(x, y) = F(x)G(y)$ into (14) written as $u_{xx} = -u_{yy}$, dividing by FG, and equating both sides to a negative constant, we obtain

$$\frac{1}{F} \cdot \frac{d^2F}{dx^2} = -\frac{1}{G} \cdot \frac{d^2G}{dy^2} = -k.$$

From this we get

$$\frac{d^2F}{dx^2} + kF = 0,$$

and the left and right boundary conditions imply

$$F(0) = 0, \qquad \text{and} \qquad F(a) = 0.$$

This gives $k = (n\pi/a)^2$ and corresponding nonzero solutions

$$F(x) = F_n(x) = \sin \frac{n\pi}{a} x, \qquad n = 1, 2, \cdots. \tag{15}$$

The ODE for G with $k = (n\pi/a)^2$ then becomes

$$\frac{d^2G}{dy^2} - \left(\frac{n\pi}{a}\right)^2 G = 0.$$

Solutions are

$$G(y) = G_n(y) = A_n e^{n\pi y/a} + B_n e^{-n\pi y/a}.$$

Now the boundary condition $u = 0$ on the lower side of R implies that $G_n(0) = 0$; that is, $G_n(0) = A_n + B_n = 0$ or $B_n = -A_n$. This gives

$$G_n(y) = A_n(e^{n\pi y/a} - e^{-n\pi y/a}) = 2A_n \sinh \frac{n\pi y}{a}.$$

From this and (15), writing $2A_n = A_n^*$, we obtain as the **eigenfunctions** of our problem

$$u_n(x, y) = F_n(x)G_n(y) = A_n^* \sin \frac{n\pi x}{a} \sinh \frac{n\pi y}{a}. \tag{16}$$

These solutions satisfy the boundary condition $u = 0$ on the left, right, and lower sides.

To get a solution also satisfying the boundary condition $u(x, b) = f(x)$ on the upper side, we consider the infinite series

$$u(x, y) = \sum_{n=1}^{\infty} u_n(x, y).$$

From this and (16) with $y = b$ we obtain

$$u(x, b) = f(x) = \sum_{n=1}^{\infty} A_n^* \sin \frac{n\pi x}{a} \sinh \frac{n\pi b}{a}.$$

We can write this in the form

$$u(x, b) = \sum_{n=1}^{\infty} \left(A_n^* \sinh \frac{n\pi b}{a}\right) \sin \frac{n\pi x}{a}.$$

This shows that the expressions in the parentheses must be the Fourier coefficients b_n of $f(x)$; that is, by (4) in Sec. 11.3,

$$b_n = A_n^* \sinh \frac{n\pi b}{a} = \frac{2}{a} \int_0^a f(x) \sin \frac{n\pi x}{a}\, dx.$$

From this and (16) we see that the solution of our problem is

$$u(x, y) = \sum_{n=1}^{\infty} u_n(x, y) = \sum_{n=1}^{\infty} A_n^* \sin \frac{n\pi x}{a} \sinh \frac{n\pi y}{a} \tag{17}$$

where

$$A_n^* = \frac{2}{a \sinh (n\pi b/a)} \int_0^a f(x) \sin \frac{n\pi x}{a}\, dx. \tag{18}$$

We have obtained this solution formally, neither considering convergence nor showing that the series for u, u_{xx}, and u_{yy} have the right sums. This can be proved if one assumes that f and f' are continuous and f'' is piecewise continuous on the interval $0 \leqq x \leqq a$. The proof is somewhat involved and relies on uniform convergence. It can be found in [C4] listed in App. 1.

Unifying Power of Methods. Electrostatics, Elasticity

The Laplace equation (14) also governs the electrostatic potential of electrical charges in any region that is free of these charges. Thus our steady-state heat problem can also be interpreted as an electrostatic potential problem. Then (17), (18) is the potential in the rectangle R when the upper side of R is at potential $f(x)$ and the other three sides are grounded.

Actually, in the steady-state case, the two-dimensional wave equation (to be considered in Secs. 12.8, 12.9) also reduces to (14). Then (17), (18) is the displacement of a rectangular elastic membrane (rubber sheet, drumhead) that is fixed along its boundary, with three sides lying in the xy-plane and the fourth side given the displacement $f(x)$.

This is another impressive demonstration of the ***unifying power*** of mathematics. It illustrates that ***entirely different physical systems may have the same mathematical model*** and can thus be treated by the same mathematical methods.

PROBLEM SET 12.6

1. **Decay.** How does the rate of decay of (8) with fixed n depend on the specific heat, the density, and the thermal conductivity of the material?

2. **Decay.** If the first eigenfunction (8) of the bar decreases to half its value within 20 sec, what is the value of the diffusivity?

3. **Eigenfunctions.** Sketch or graph and compare the first three eigenfunctions (8) with $B_n = 1, c = 1$, and $L = \pi$ for $t = 0, 0.1, 0.2, \cdots, 1.0$.

4. **WRITING PROJECT. Wave and Heat Equations.** Compare these PDEs with respect to general behavior of eigenfunctions and kind of boundary and initial

conditions. State the difference between Fig. 291 in Sec. 12.3 and Fig. 295.

5–7 LATERALLY INSULATED BAR

Find the temperature $u(x, t)$ in a bar of silver of length 10 cm and constant cross section of area 1 cm^2 (density 10.6 g/cm^3, thermal conductivity 1.04 cal/(cm sec °C), specific heat 0.056 cal/(g °C) that is perfectly insulated laterally, with ends kept at temperature 0°C and initial temperature $f(x)$ °C, where

5. $f(x) = \sin 0.1\pi x$

6. $f(x) = 4 - 0.8|x - 5|$

7. $f(x) = x(10 - x)$

8. Arbitrary temperatures at ends. If the ends $x = 0$ and $x = L$ of the bar in the text are kept at constant temperatures U_1 and U_2, respectively, what is the temperature $u_1(x)$ in the bar after a long time (theoretically, as $t \to \infty$)? First guess, then calculate.

9. In Prob. 8 find the temperature at any time.

10. Change of end temperatures. Assume that the ends of the bar in Probs. 5–7 have been kept at 100°C for a long time. Then at some instant, call it $t = 0$, the temperature at $x = L$ is suddenly changed to 0°C and kept at 0°C, whereas the temperature at $x = 0$ is kept at 100°C. Find the temperature in the middle of the bar at $t = 1, 2, 3, 10, 50$ sec. First guess, then calculate.

BAR UNDER ADIABATIC CONDITIONS

"Adiabatic" means no heat exchange with the neighborhood, because the bar is completely insulated, also at the ends. *Physical Information:* The heat flux at the ends is proportional to the value of $\partial u/\partial x$ there.

11. Show that for the completely insulated bar, $u_x(0, t) = 0$, $u_x(L, t) = 0$, $u(x, t) = f(x)$ and separation of variables gives the following solution, with A_n given by (2) in Sec. 11.3.

$$u(x, t) = A_0 + \sum_{n=1}^{\infty} A_n \cos\frac{n\pi x}{L} e^{-(cn\pi/L)^2 t}$$

12–15 Find the temperature in Prob. 11 with $L = \pi$, $c = 1$, and

12. $f(x) = x$

13. $f(x) = 1$

14. $f(x) = \cos 2x$

15. $f(x) = 1 - x/\pi$

16. A bar with heat generation of constant rate $H\ (> 0)$ is modeled by $u_t = c^2 u_{xx} + H$. Solve this problem if $L = \pi$ and the ends of the bar are kept at 0°C. *Hint.* Set $u = v - Hx(x - \pi)/(2c^2)$.

17. Heat flux. The *heat flux* of a solution $u(x, t)$ across $x = 0$ is defined by $\phi(t) = -Ku_x(0, t)$. Find $\phi(t)$ for the solution (9). Explain the name. Is it physically understandable that ϕ goes to 0 as $t \to \infty$?

18–25 TWO-DIMENSIONAL PROBLEMS

18. Laplace equation. Find the potential in the rectangle $0 \leqq x \leqq 20$, $0 \leqq y \leqq 40$ whose upper side is kept at potential 110 V and whose other sides are grounded.

19. Find the potential in the square $0 \leqq x \leqq 2$, $0 \leqq y \leqq 2$ if the upper side is kept at the potential $1000 \sin \frac{1}{2}\pi x$ and the other sides are grounded.

20. CAS PROJECT. Isotherms. Find the steady-state solutions (temperatures) in the square plate in Fig. 297 with $a = 2$ satisfying the following boundary conditions. Graph isotherms.

(a) $u = 80 \sin \pi x$ on the upper side, 0 on the others.

(b) $u = 0$ on the vertical sides, assuming that the other sides are perfectly insulated.

(c) Boundary conditions of your choice (such that the solution is not identically zero).

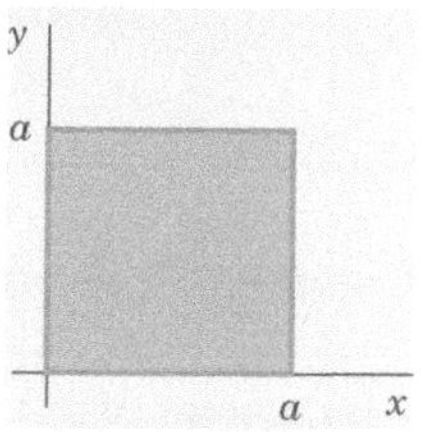

Fig. 297. Square plate

21. Heat flow in a plate. The faces of the thin square plate in Fig. 297 with side $a = 24$ are perfectly insulated. The upper side is kept at 25°C and the other sides are kept at 0°C. Find the steady-state temperature $u(x, y)$ in the plate.

22. Find the steady-state temperature in the plate in Prob. 21 if the lower side is kept at U_0°C, the upper side at U_1°C, and the other sides are kept at 0°C. *Hint:* Split into two problems in which the boundary temperature is 0 on three sides for each problem.

23. Mixed boundary value problem. Find the steady-state temperature in the plate in Prob. 21 with the upper and lower sides perfectly insulated, the left side kept at 0°C, and the right side kept at $f(y)$°C.

24. Radiation. Find steady-state temperatures in the rectangle in Fig. 296 with the upper and left sides perfectly insulated and the right side radiating into a medium at 0°C according to $u_x(a, y) + hu(a, y) = 0$, $h > 0$ constant. (You will get many solutions since no condition on the lower side is given.)

25. Find formulas similar to (17), (18) for the temperature in the rectangle R of the text when the lower side of R is kept at temperature $f(x)$ and the other sides are kept at 0°C.

12.7 Heat Equation: Modeling Very Long Bars. Solution by Fourier Integrals and Transforms

Our discussion of the heat equation

$$\frac{\partial u}{\partial t} = c^2 \frac{\partial^2 u}{\partial x^2} \tag{1}$$

in the last section extends to bars of infinite length, which are good models of very long bars or wires (such as a wire of length, say, 300 ft). Then the role of Fourier series in the solution process will be taken by **Fourier integrals** (Sec. 11.7).

Let us illustrate the method by solving (1) for a bar that extends to infinity on both sides (and is laterally insulated as before). Then we do not have boundary conditions, but only the **initial condition**

$$u(x, 0) = f(x) \qquad (-\infty < x < \infty) \tag{2}$$

where $f(x)$ is the given initial temperature of the bar.

To solve this problem, we start as in the last section, substituting $u(x, t) = F(x)G(t)$ into (1). This gives the two ODEs

$$F'' + p^2 F = 0 \qquad \text{[see (5), Sec. 12.6]} \tag{3}$$

and

$$\dot{G} + c^2 p^2 G = 0 \qquad \text{[see (6), Sec. 12.6].} \tag{4}$$

Solutions are

$$F(x) = A \cos px + B \sin px \qquad \text{and} \qquad G(t) = e^{-c^2 p^2 t},$$

respectively, where A and B are any constants. Hence a solution of (1) is

$$u(x, t; p) = FG = (A \cos px + B \sin px)\, e^{-c^2 p^2 t}. \tag{5}$$

Here we had to choose the separation constant k negative, $k = -p^2$, because positive values of k would lead to an increasing exponential function in (5), which has no physical meaning.

Use of Fourier Integrals

Any series of functions (5), found in the usual manner by taking p as multiples of a fixed number, would lead to a function that is periodic in x when $t = 0$. However, since $f(x)$

in (2) is not assumed to be periodic, it is natural to use **Fourier integrals** instead of Fourier series. Also, A and B in (5) are arbitrary and we may regard them as functions of p, writing $A = A(p)$ and $B = B(P)$. Now, since the heat equation (1) is linear and homogeneous, the function

$$\textbf{(6)} \qquad u(x, t) = \int_0^\infty u(x, t; p)\, dp = \int_0^\infty [A(p) \cos px + B(p) \sin px]\, e^{-c^2p^2t}\, dp$$

is then a solution of (1), provided this integral exists and can be differentiated twice with respect to x and once with respect to t.

Determination of $A(p)$ and $B(p)$ from the Initial Condition. From (6) and (2) we get

$$(7) \qquad u(x, 0) = \int_0^\infty [A(p) \cos px + B(p) \sin px]\, dp = f(x).$$

This gives $A(p)$ and $B(p)$ in terms of $f(x)$; indeed, from (4) in Sec. 11.7 we have

$$(8) \qquad A(p) = \frac{1}{\pi} \int_{-\infty}^{\infty} f(v) \cos pv\, dv, \qquad B(p) = \frac{1}{\pi} \int_{-\infty}^{\infty} f(v) \sin pv\, dv.$$

According to (1*), Sec. 11.9, our Fourier integral (7) with these $A(p)$ and $B(p)$ can be written

$$u(x, 0) = \frac{1}{\pi} \int_0^\infty \left[\int_{-\infty}^{\infty} f(v) \cos(px - pv)\, dv \right] dp.$$

Similarly, (6) in this section becomes

$$u(x, t) = \frac{1}{\pi} \int_0^\infty \left[\int_{-\infty}^{\infty} f(v) \cos(px - pv) e^{-c^2p^2t}\, dv \right] dp.$$

Assuming that we may reverse the order of integration, we obtain

$$\textbf{(9)} \qquad u(x, t) = \frac{1}{\pi} \int_{-\infty}^{\infty} f(v) \left[\int_0^\infty e^{-c^2p^2t} \cos(px - pv)\, dp \right] dv.$$

Then we can evaluate the inner integral by using the formula

$$(10) \qquad \int_0^\infty e^{-s^2} \cos 2bs\, ds = \frac{\sqrt{\pi}}{2} e^{-b^2}.$$

[A derivation of (10) is given in Problem Set 16.4 (Team Project 24).] This takes the form of our inner integral if we choose $p = s/(c\sqrt{t})$ as a new variable of integration and set

$$b = \frac{x - v}{2c\sqrt{t}}.$$

Then $2bs = (x - v)p$ and $ds = c\sqrt{t}\,dp$, so that (10) becomes

$$\int_0^\infty e^{-c^2p^2t}\cos(px - pv)\,dp = \frac{\sqrt{\pi}}{2c\sqrt{t}}\exp\left\{-\frac{(x-v)^2}{4c^2t}\right\}.$$

By inserting this result into (9) we obtain the representation

$$u(x, t) = \frac{1}{2c\sqrt{\pi t}}\int_{-\infty}^{\infty} f(v)\exp\left\{-\frac{(x-v)^2}{4c^2t}\right\}dv. \tag{11}$$

Taking $z = (v - x)/(2c\sqrt{t})$ as a variable of integration, we get the alternative form

$$u(x, t) = \frac{1}{\sqrt{\pi}}\int_{-\infty}^{\infty} f(x + 2cz\sqrt{t})\,e^{-z^2}\,dz. \tag{12}$$

If $f(x)$ is bounded for all values of x and integrable in every finite interval, it can be shown (see Ref. [C10]) that the function (11) or (12) satisfies (1) and (2). Hence this function is the required solution in the present case.

EXAMPLE 1 Temperature in an Infinite Bar

Find the temperature in the infinite bar if the initial temperature is (Fig. 298)

$$f(x) = \begin{cases} U_0 = \text{const} & \text{if } |x| < 1, \\ 0 & \text{if } |x| > 1. \end{cases}$$

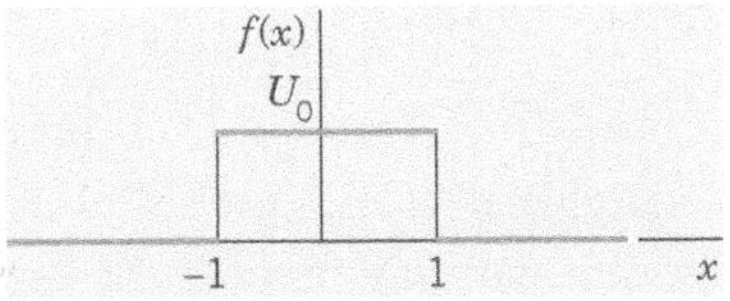

Fig. 298. Initial temperature in Example 1

Solution. From (11) we have

$$u(x, t) = \frac{U_0}{2c\sqrt{\pi t}}\int_{-1}^{1}\exp\left\{-\frac{(x-v)^2}{4c^2t}\right\}dv.$$

If we introduce the above variable of integration z, then the integration over v from -1 to 1 corresponds to the integration over z from $(-1 - x)/(2c\sqrt{t})$ to $(1 - x)/(2c\sqrt{t})$, and

$$u(x, t) = \frac{U_0}{\sqrt{\pi}}\int_{-(1+x)/(2c\sqrt{t})}^{(1-x)/(2c\sqrt{t})} e^{-z^2}\,dz \qquad (t > 0). \tag{13}$$

We mention that this integral is not an elementary function, but can be expressed in terms of the error function, whose values have been tabulated. (Table A4 in App. 5 contains a few values; larger tables are listed in Ref. [GenRef1] in App. 1. See also CAS Project 1, p. 574.) Figure 299 shows $u(x, t)$ for $U_0 = 100°\text{C}$, $c^2 = 1\ \text{cm}^2/\text{sec}$, and several values of t. ■

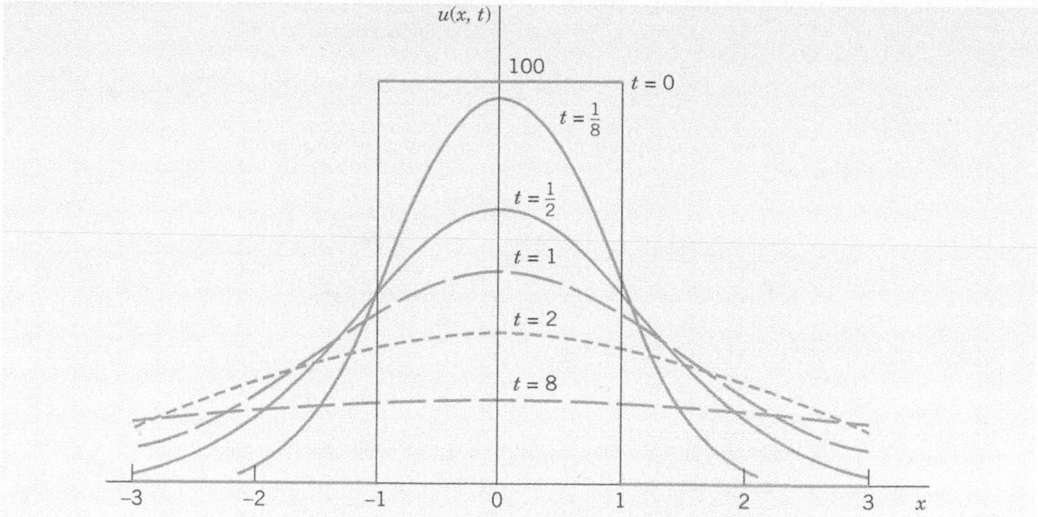

Fig. 299. Solution $u(x, t)$ in Example 1 for $U_0 = 100°C$, $c^2 = 1 \text{ cm}^2/\text{sec}$, and several values of t

Use of Fourier Transforms

The Fourier transform is closely related to the Fourier integral, from which we obtained the transform in Sec. 11.9. And the transition to the Fourier cosine and sine transform in Sec. 11.8 was even simpler. (You may perhaps wish to review this before going on.) Hence it should not surprise you that we can use these transforms for solving our present or similar problems. The Fourier transform applies to problems concerning the entire axis, and the Fourier cosine and sine transforms to problems involving the positive half-axis. Let us explain these transform methods by typical applications that fit our present discussion.

EXAMPLE 2 Temperature in the Infinite Bar in Example 1

Solve Example 1 using the Fourier transform.

Solution. The problem consists of the heat equation (1) and the initial condition (2), which in this example is

$$f(x) = U_0 = \text{const} \quad \text{if } |x| < 1 \qquad \text{and 0 otherwise.}$$

Our strategy is to take the Fourier transform with respect to x and then to solve the resulting ***ordinary*** DE in t. The details are as follows.

Let $\hat{u} = \mathcal{F}(u)$ denote the Fourier transform of u, ***regarded as a function of x***. From (10) in Sec. 11.9 we see that the heat equation (1) gives

$$\mathcal{F}(u_t) = c^2\mathcal{F}(u_{xx}) = c^2(-w^2)\mathcal{F}(u) = -c^2w^2\hat{u}.$$

On the left, assuming that we may interchange the order of differentiation and integration, we have

$$\mathcal{F}(u_t) = \frac{1}{\sqrt{2\pi}}\int_{-\infty}^{\infty} u_t e^{-iwx}\,dx = \frac{1}{\sqrt{2\pi}}\frac{\partial}{\partial t}\int_{-\infty}^{\infty} ue^{-iwx}\,dx = \frac{\partial \hat{u}}{\partial t}.$$

Thus

$$\frac{\partial \hat{u}}{\partial t} = -c^2w^2\hat{u}.$$

Since this equation involves only a derivative with respect to t but none with respect to w, this is a first-order ***ordinary DE***, with t as the independent variable and w as a parameter. By separating variables (Sec. 1.3) we get the general solution

$$\hat{u}(w, t) = C(w)e^{-c^2w^2t}$$

with the arbitrary "constant" $C(w)$ depending on the parameter w. The initial condition (2) yields the relationship $\hat{u}(w, 0) = C(w) = \hat{f}(w) = \mathcal{F}(f)$. Our intermediate result is

$$\hat{u}(w, t) = \hat{f}(w)e^{-c^2w^2t}.$$

The inversion formula (7), Sec. 11.9, now gives the solution

$$u(x, t) = \frac{1}{\sqrt{2\pi}} \int_{-\infty}^{\infty} \hat{f}(w)\, e^{-c^2w^2t}\, e^{iwx}\, dw. \tag{14}$$

In this solution we may insert the Fourier transform

$$\hat{f}(w) = \frac{1}{\sqrt{2\pi}} \int_{-\infty}^{\infty} f(v)e^{ivw}dv.$$

Assuming that we may invert the order of integration, we then obtain

$$u(x, t) = \frac{1}{2\pi} \int_{-\infty}^{\infty} f(v)\left[\int_{-\infty}^{\infty} e^{-c^2w^2t}\, e^{i(wx-wv)}dw\right] dv.$$

By the Euler formula (3). Sec. 11.9, the integrand of the inner integral equals

$$e^{-c^2w^2t} \cos(wx - wv) + ie^{-c^2w^2t} \sin(wx - wv).$$

We see that its imaginary part is an odd function of w, so that its integral is 0. (More precisely, this is the principal part of the integral; see Sec. 16.4.) The real part is an even function of w, so that its integral from $-\infty$ to ∞ equals twice the integral from 0 to ∞:

$$u(x, t) = \frac{1}{\pi} \int_{-\infty}^{\infty} f(v)\left[\int_{0}^{\infty} e^{-c^2w^2t} \cos(wx - wv)\, dw\right] dv.$$

This agrees with (9) (with $p = w$) and leads to the further formulas (11) and (13). ■

EXAMPLE 3 **Solution in Example 1 by the Method of Convolution**

Solve the heat problem in Example 1 by the method of convolution.

Solution. The beginning is as in Example 2 and leads to (14), that is,

$$u(x, t) = \frac{1}{\sqrt{2\pi}} \int_{-\infty}^{\infty} \hat{f}(w)e^{-c^2w^2t}e^{iwx}\, dw. \tag{15}$$

Now comes the crucial idea. We recognize that this is of the form (13) in Sec. 11.9, that is,

$$u(x, t) = (f * g)(x) = \int_{-\infty}^{\infty} \hat{f}(w)\hat{g}(w)e^{iwx}\, dw \tag{16}$$

where

$$\hat{g}(w) = \frac{1}{\sqrt{2\pi}} e^{-c^2w^2t}. \tag{17}$$

Since, by the definition of convolution [(11), Sec. 11.9],

$$(f * g)(x) = \int_{-\infty}^{\infty} f(p)g(x - p)\, dp, \tag{18}$$

as our next and last step we must determine the inverse Fourier transform g of $\hat{g}$. For this we can use formula 9 in Table III of Sec. 11.10,

$$\mathcal{F}(e^{-ax^2}) = \frac{1}{\sqrt{2a}} e^{-w^2/(4a)}$$

with a suitable a. With $c^2 t = 1/(4a)$ or $a = 1/(4c^2 t)$, using (17) we obtain

$$\mathcal{F}(e^{-x^2/(4c^2t)}) = \sqrt{2c^2 t}\, e^{-c^2w^2t} = \sqrt{2c^2 t}\, \sqrt{2\pi}\, \hat{g}(w).$$

Hence $\hat{g}$ has the inverse

$$\frac{1}{\sqrt{2c^2 t}\, \sqrt{2\pi}} e^{-x^2/(4c^2t)}.$$

Replacing x with $x - p$ and substituting this into (18) we finally have

$$u(x, t) = (f * g)(x) = \frac{1}{2c\sqrt{\pi t}} \int_{-\infty}^{\infty} f(p) \exp\left\{-\frac{(x-p)^2}{4c^2 t}\right\} dp. \tag{19}$$

This solution formula of our problem agrees with (11). We wrote $(f * g)(x)$, without indicating the parameter t with respect to which we did not integrate. ■

EXAMPLE 4 Fourier Sine Transform Applied to the Heat Equation

If a laterally insulated bar extends from $x = 0$ to infinity, we can use the Fourier sine transform. We let the initial temperature be $u(x, 0) = f(x)$ and impose the boundary condition $u(0, t) = 0$. Then from the heat equation and (9b) in Sec. 11.8, since $f(0) = u(0, 0) = 0$, we obtain

$$\mathcal{F}_s(u_t) = \frac{\partial \hat{u}_s}{\partial t} = c^2 \mathcal{F}_s(u_{xx}) = -c^2 w^2 \mathcal{F}_s(u) = -c^2 w^2 \hat{u}_s(w, t).$$

This is a first-order ODE $\partial \hat{u}_s/\partial t + c^2 w^2 \hat{u}_s = 0$. Its solution is

$$\hat{u}_s(w, t) = C(w) e^{-c^2w^2t}.$$

From the initial condition $u(x, 0) = f(x)$ we have $\hat{u}_s(w, 0) = \hat{f}_s(w) = C(w)$. Hence

$$\hat{u}_s(w, t) = \hat{f}_s(w) e^{-c^2w^2t}.$$

Taking the inverse Fourier sine transform and substituting

$$\hat{f}_s(w) = \sqrt{\frac{2}{\pi}} \int_0^{\infty} f(p) \sin wp \, dp$$

on the right, we obtain the solution formula

$$u(x, t) = \frac{2}{\pi} \int_0^{\infty} \int_0^{\infty} f(p) \sin wp \, e^{-c^2w^2t} \sin wx \, dp \, dw. \tag{20}$$

Figure 300 shows (20) with $c = 1$ for $f(x) = 1$ if $0 \leqq x \leqq 1$ and 0 otherwise, graphed over the xt-plane for $0 \leqq x \leqq 2$, $0.01 \leqq t \leqq 1.5$. Note that the curves of $u(x, t)$ for constant t resemble those in Fig. 299. ■

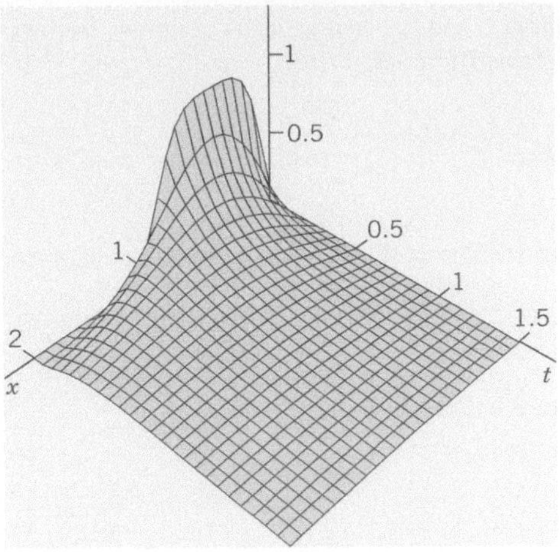

Fig. 300. Solution (20) in Example 4

PROBLEM SET 12.7

1. **CAS PROJECT. Heat Flow.** **(a)** Graph the basic Fig. 299.

   (b) In (a) apply animation to "see" the heat flow in terms of the decrease of temperature.

   (c) Graph $u(x, t)$ with $c = 1$ as a surface over a rectangle of the form $-a < x < a, \quad 0 < y < b$.

2–8 SOLUTION IN INTEGRAL FORM

Using (6), obtain the solution of (1) in integral form satisfying the initial condition $u(x, 0) = f(x)$, where

2. $f(x) = 1 \quad \text{if } |x| < a$ and 0 otherwise
3. $f(x) = 1/(1 + x^2)$.
   Hint. Use (15) in Sec. 11.7.
4. $f(x) = e^{-|x|}$
5. $f(x) = |x| \quad \text{if } |x| < 1$ and 0 otherwise
6. $f(x) = x \quad \text{if } |x| < 1$ and 0 otherwise
7. $f(x) = (\sin x)/x$.
   Hint. Use Prob. 4 in Sec. 11.7.
8. Verify that u in the solution of Prob. 7 satisfies the initial condition.

9–12 CAS PROJECT. Error Function.

$$\text{erf}\, x = \frac{2}{\sqrt{\pi}} \int_0^x e^{-w^2}\, dw \tag{21}$$

This function is important in applied mathematics and physics (probability theory and statistics, thermodynamics, etc.) and fits our present discussion. Regarding it as a typical case of a special function defined by an integral that cannot be evaluated as in elementary calculus, do the following.

9. Graph the **bell-shaped curve** [the curve of the integrand in (21)]. Show that erf x is odd. Show that

$$\int_a^b e^{-w^2}\, dw = \frac{\sqrt{\pi}}{2}(\text{erf}\, b - \text{erf}\, a).$$

$$\int_{-b}^b e^{-w^2}\, dw = \sqrt{\pi}\, \text{erf}\, b.$$

10. Obtain the Maclaurin series of erf x from that of the integrand. Use that series to compute a table of erf x for $x = 0(0.01)3$ (meaning $x = 0, 0.01, 0.02, \cdots, 3$).
11. Obtain the values required in Prob. 10 by an integration command of your CAS. Compare accuracy.
12. It can be shown that erf $(\infty) = 1$. Confirm this experimentally by computing erf x for large x.
13. Let $f(x) = 1$ when $x > 0$ and 0 when $x < 0$. Using erf $(\infty) = 1$, show that (12) then gives

$$u(x, t) = \frac{1}{\sqrt{\pi}} \int_{-x/(2c\sqrt{t})}^{\infty} e^{-z^2}\, dz$$
$$= \frac{1}{2} - \frac{1}{2}\, \text{erf}\left(-\frac{x}{2c\sqrt{t}}\right) \qquad (t > 0).$$

14. Express the temperature (13) in terms of the error function.
15. Show that $\Phi(x) = \dfrac{1}{\sqrt{2\pi}} \displaystyle\int_{-\infty}^{x} e^{-s^2/2}\, ds$

$$= \frac{1}{2} + \frac{1}{2}\, \text{erf}\left(\frac{x}{\sqrt{2}}\right).$$

Here, the integral is the definition of the "distribution function of the normal probability distribution" to be discussed in Sec. 24.8.

12.8 Modeling: Membrane, Two-Dimensional Wave Equation

Since the modeling here will be similar to that of Sec. 12.2, you may want to take another look at Sec. 12.2.

The vibrating string in Sec. 12.2 is a basic one-dimensional vibrational problem. Equally important is its two-dimensional analog, namely, the motion of an elastic membrane, such as a drumhead, that is stretched and then fixed along its edge. Indeed, setting up the model will proceed almost as in Sec. 12.2.

Physical Assumptions

1. The mass of the membrane per unit area is constant ("homogeneous membrane"). The membrane is perfectly flexible and offers no resistance to bending.
2. The membrane is stretched and then fixed along its entire boundary in the xy-plane. The tension per unit length T caused by stretching the membrane is the same at all points and in all directions and does not change during the motion.
3. The deflection $u(x, y, t)$ of the membrane during the motion is small compared to the size of the membrane, and all angles of inclination are small.

Although these assumptions cannot be realized exactly, they hold relatively accurately for small transverse vibrations of a thin elastic membrane, so that we shall obtain a good model, for instance, of a drumhead.

Derivation of the PDE of the Model ("Two-Dimensional Wave Equation") from Forces. As in Sec. 12.2 the model will consist of a PDE and additional conditions. The PDE will be obtained by the same method as in Sec. 12.2, namely, by considering the forces acting on a small portion of the physical system, the membrane in Fig. 301 on the next page, as it is moving up and down.

Since the deflections of the membrane and the angles of inclination are small, the sides of the portion are approximately equal to Δx and Δy. The tension T is the force per unit length. Hence the forces acting on the sides of the portion are approximately $T\Delta x$ and $T\Delta y$. Since the membrane is perfectly flexible, these forces are tangent to the moving membrane at every instant.

Horizontal Components of the Forces. We first consider the horizontal components of the forces. These components are obtained by multiplying the forces by the cosines of the angles of inclination. Since these angles are small, their cosines are close to 1. Hence the horizontal components of the forces at opposite sides are approximately equal. Therefore, the motion of the particles of the membrane in a horizontal direction will be negligibly small. From this we conclude that we may regard the motion of the membrane as transversal; that is, each particle moves vertically.

Vertical Components of the Forces. These components along the right side and the left side are (Fig. 301), respectively,

$$T\Delta y \sin\beta \qquad \text{and} \qquad -T\Delta y \sin\alpha.$$

Here α and β are the values of the angle of inclination (which varies slightly along the edges) in the middle of the edges, and the minus sign appears because the force on the

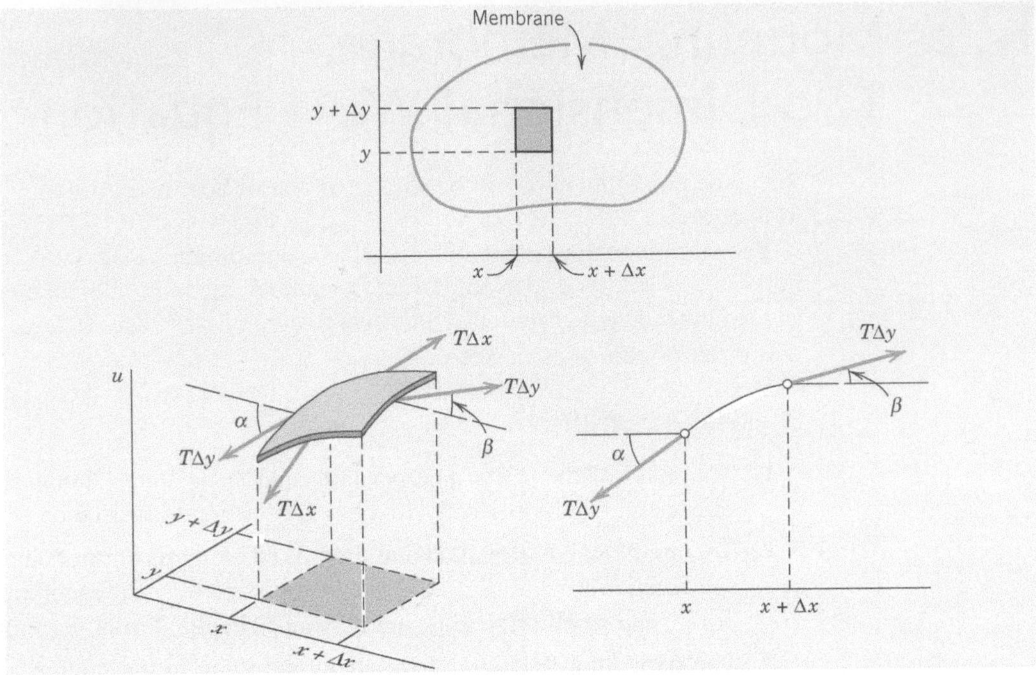

Fig. 301. Vibrating membrane

left side is directed downward. Since the angles are small, we may replace their sines by their tangents. Hence the resultant of those two vertical components is

$$(1)\qquad \begin{aligned} T\,\Delta y(\sin\beta - \sin\alpha) &\approx T\,\Delta y(\tan\beta - \tan\alpha) \\ &= T\,\Delta y[u_x(x+\Delta x, y_1) - u_x(x, y_2)] \end{aligned}$$

where subscripts x denote partial derivatives and y_1 and y_2 are values between y and $y + \Delta y$. Similarly, the resultant of the vertical components of the forces acting on the other two sides of the portion is

$$(2)\qquad T\,\Delta x[u_y(x_1, y+\Delta y) - u_y(x_2, y)]$$

where x_1 and x_2 are values between x and $x + \Delta x$.

Newton's Second Law Gives the PDE of the Model. By Newton's second law (see Sec. 2.4) the sum of the forces given by (1) and (2) is equal to the mass $\rho\,\Delta A$ of that small portion times the acceleration $\partial^2 u/\partial t^2$; here ρ is the mass of the undeflected membrane per unit area, and $\Delta A = \Delta x\,\Delta y$ is the area of that portion when it is undeflected. Thus

$$\begin{aligned} \rho\,\Delta x\,\Delta y\,\frac{\partial^2 u}{\partial t^2} &= T\,\Delta y[u_x(x+\Delta x, y_1) - u_x(x, y_2)] \\ &\quad + T\,\Delta x[u_y(x_1, y+\Delta y) - u_y(x_2, y)] \end{aligned}$$

where the derivative on the left is evaluated at some suitable point $(\tilde{x}, \tilde{y})$ corresponding to that portion. Division by $\rho\,\Delta x\,\Delta y$ gives

$$\frac{\partial^2 u}{\partial t^2} = \frac{T}{\rho}\left[\frac{u_x(x+\Delta x, y_1) - u_x(x, y_2)}{\Delta x} + \frac{u_y(x_1, y+\Delta y) - u_y(x_2, y)}{\Delta y}\right].$$

If we let Δx and Δy approach zero, we obtain the PDE of the model

$$\frac{\partial^2 u}{\partial t^2} = c^2\left(\frac{\partial^2 u}{\partial x^2} + \frac{\partial^2 u}{\partial y^2}\right) \qquad c^2 = \frac{T}{\rho}. \tag{3}$$

This PDE is called the **two-dimensional wave equation**. The expression in parentheses is the Laplacian $\Delta^2 u$ of u (Sec. 10.8). Hence (3) can be written

$$\frac{\partial^2 u}{\partial t^2} = c^2 \Delta^2 u. \tag{3'}$$

Solutions of the wave equation (3) will be obtained and discussed in the next section.

12.9 Rectangular Membrane. Double Fourier Series

Now we develop a solution for the PDE obtained in Sec. 12.8. Details are as follows.

The model of the vibrating membrane for obtaining the displacement $u(x, y, t)$ of a point (x, y) of the membrane from rest ($u = 0$) at time t is

$$\frac{\partial^2 u}{\partial t^2} = c^2\left(\frac{\partial^2 u}{\partial x^2} + \frac{\partial^2 u}{\partial y^2}\right) \tag{1}$$

$$u = 0 \text{ on the boundary} \tag{2}$$

$$u(x, y, 0) = f(x, y) \tag{3a}$$

$$u_t(x, y, 0) = g(x, y). \tag{3b}$$

Here (1) is the **two-dimensional wave equation** with $c^2 = T/\rho$ just derived, (2) is the **boundary condition** (membrane fixed along the boundary in the xy-plane for all times $t \geqq 0$), and (3) are the **initial conditions** at $t = 0$, consisting of the given *initial displacement* (initial shape) $f(x, y)$ and the given *initial velocity* $g(x, y)$, where $u_t = \partial u/\partial t$. We see that these conditions are quite similar to those for the string in Sec. 12.2.

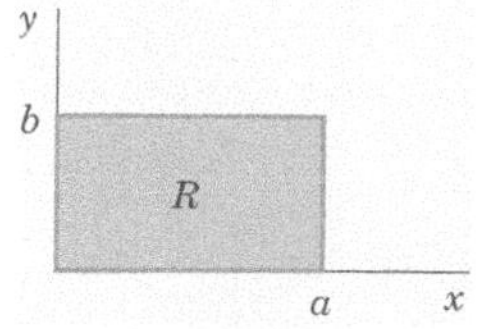

Fig. 302. Rectangular membrane

Let us consider the **rectangular membrane R** in Fig. 302. This is our first important model. It is much simpler than the circular drumhead, which will follow later. First we note that the boundary in equation (2) is the rectangle in Fig. 302. We shall solve this problem in three steps:

Step 1. By separating variables, first setting $u(x, y, t) = F(x, y)G(t)$ and later $F(x, y) = H(x)Q(y)$ we obtain from (1) an ODE (4) for G and later from a PDE (5) for F two ODEs (6) and (7) for H and Q.

Step 2. From the solutions of those ODEs we determine solutions (13) of (1) ("**eigenfunctions**" u_{mn}) that satisfy the boundary condition (2).

Step 3. We compose the u_{mn} into a double series (14) solving the whole model (1), (2), (3).

Step 1. Three ODEs From the Wave Equation (1)

To obtain ODEs from (1), we apply two successive separations of variables. In the first separation we set $u(x, y, t) = F(x, y)G(t)$. Substitution into (1) gives

$$F\ddot{G} = c^2(F_{xx}G + F_{yy}G)$$

where subscripts denote partial derivatives and dots denote derivatives with respect to t. To separate the variables, we divide both sides by c^2FG:

$$\frac{\ddot{G}}{c^2G} = \frac{1}{F}(F_{xx} + F_{yy}).$$

Since the left side depends only on t, whereas the right side is independent of t, both sides must equal a constant. By a simple investigation we see that only negative values of that constant will lead to solutions that satisfy (2) without being identically zero; this is similar to Sec. 12.3. Denoting that negative constant by $-\nu^2$, we have

$$\frac{\ddot{G}}{c^2G} = \frac{1}{F}(F_{xx} + F_{yy}) = -\nu^2.$$

This gives two equations: for the "**time function**" $G(t)$ we have the ODE

$$\ddot{G} + \lambda^2 G = 0 \qquad \text{where } \lambda = c\nu, \tag{4}$$

and for the "**amplitude function**" $F(x, y)$ a PDE, called the *two-dimensional* **Helmholtz**[3] **equation**

$$F_{xx} + F_{yy} + \nu^2 F = 0. \tag{5}$$

[3]HERMANN VON HELMHOLTZ (1821–1894), German physicist, known for his fundamental work in thermodynamics, fluid flow, and acoustics.

Separation of the Helmholtz equation is achieved if we set $F(x, y) = H(x)Q(y)$. By substitution of this into (5) we obtain

$$\frac{d^2H}{dx^2}Q = -\left(H\frac{d^2Q}{dy^2} + \nu^2 HQ\right).$$

To separate the variables, we divide both sides by HQ, finding

$$\frac{1}{H}\frac{d^2H}{dx^2} = -\frac{1}{Q}\left(\frac{d^2Q}{dy^2} + \nu^2 Q\right).$$

Both sides must equal a constant, by the usual argument. This constant must be negative, say, $-k^2$, because only negative values will lead to solutions that satisfy (2) without being identically zero. Thus

$$\frac{1}{H}\frac{d^2H}{dx^2} = -\frac{1}{Q}\left(\frac{d^2Q}{dy^2} + \nu^2 Q\right) = -k^2.$$

This yields two ODEs for H and Q, namely,

$$\frac{d^2H}{dx^2} + k^2 H = 0 \qquad (6)$$

and

$$\frac{d^2Q}{dy^2} + p^2 Q = 0 \qquad \text{where } p^2 = \nu^2 - k^2. \qquad (7)$$

Step 2. Satisfying the Boundary Condition

General solutions of (6) and (7) are

$$H(x) = A\cos kx + B\sin kx \qquad \text{and} \qquad Q(y) = C\cos py + D\sin py$$

with constant A, B, C, D. From $u = FG$ and (2) it follows that $F = HQ$ must be zero on the boundary, that is, on the edges $x = 0, x = a, y = 0, y = b$; see Fig. 302. This gives the conditions

$$H(0) = 0, \qquad H(a) = 0, \qquad Q(0) = 0, \qquad Q(b) = 0.$$

Hence $H(0) = A = 0$ and then $H(a) = B\sin ka = 0$. Here we must take $B \neq 0$ since otherwise $H(x) \equiv 0$ and $F(x, y) \equiv 0$. Hence $\sin ka = 0$ or $ka = m\pi$, that is,

$$k = \frac{m\pi}{a} \qquad (m \text{ integer}).$$

In precisely the same fashion we conclude that $C = 0$ and p must be restricted to the values $p = n\pi/b$ where n is an integer. We thus obtain the solutions $H = H_m$, $Q = Q_n$, where

$$H_m(x) = \sin\frac{m\pi x}{a} \quad \text{and} \quad Q_n(y) = \sin\frac{n\pi y}{b}, \qquad \begin{matrix} m = 1, 2, \cdots, \\ n = 1, 2, \cdots. \end{matrix}$$

As in the case of the vibrating string, it is not necessary to consider $m, n = -1, -2, \cdots$ since the corresponding solutions are essentially the same as for positive m and n, expect for a factor -1. Hence the functions

$$\text{(8)} \qquad F_{mn}(x, y) = H_m(x)Q_n(y) = \sin\frac{m\pi x}{a}\sin\frac{n\pi y}{b}, \qquad \begin{matrix} m = 1, 2, \cdots, \\ n = 1, 2, \cdots, \end{matrix}$$

are solutions of the Helmholtz equation (5) that are zero on the boundary of our membrane.

Eigenfunctions and Eigenvalues. Having taken care of (5), we turn to (4). Since $p^2 = \nu^2 - k^2$ in (7) and $\lambda = c\nu$ in (4), we have

$$\lambda = c\sqrt{k^2 + p^2}.$$

Hence to $k = m\pi/a$ and $p = n\pi/b$ there corresponds the value

$$\textbf{(9)} \qquad \lambda = \lambda_{mn} = c\pi\sqrt{\frac{m^2}{a^2} + \frac{n^2}{b^2}}, \qquad \begin{matrix} m = 1, 2, \cdots, \\ n = 1, 2, \cdots, \end{matrix}$$

in the ODE (4). A corresponding general solution of (4) is

$$G_{mn}(t) = B_{mn}\cos\lambda_{mn}t + B^*_{mn}\sin\lambda_{mn}t.$$

It follows that the functions $u_{mn}(x, y, t) = F_{mn}(x, y)G_{mn}(t)$, written out

$$\textbf{(10)} \qquad u_{mn}(x, y, t) = (B_{mn}\cos\lambda_{mn}t + B^*_{mn}\sin\lambda_{mn}t)\sin\frac{m\pi x}{a}\sin\frac{n\pi y}{b}$$

with λ_{mn} according to (9), are solutions of the wave equation (1) that are zero on the boundary of the rectangular membrane in Fig. 302. These functions are called the **eigenfunctions** or *characteristic functions,* and the numbers λ_{mn} are called the **eigenvalues** or *characteristic values* of the vibrating membrane. The frequency of u_{mn} is $\lambda_{mn}/2\pi$.

Discussion of Eigenfunctions. It is very interesting that, depending on a and b, several functions F_{mn} may correspond to the same eigenvalue. Physically this means that there may exists vibrations having the same frequency but entirely different **nodal lines** (curves of points on the membrane that do not move). Let us illustrate this with the following example.

EXAMPLE 1 Eigenvalues and Eigenfunctions of the Square Membrane

Consider the square membrane with $a = b = 1$. From (9) we obtain its eigenvalues

$$\lambda_{mn} = c\pi\sqrt{m^2 + n^2}. \tag{11}$$

Hence $\lambda_{mn} = \lambda_{nm}$, but for $m \neq n$ the corresponding functions

$$F_{mn} = \sin m\pi x \sin n\pi y \qquad \text{and} \qquad F_{nm} = \sin n\pi x \sin m\pi y$$

are certainly different. For example, to $\lambda_{12} = \lambda_{21} = c\pi\sqrt{5}$ there correspond the two functions

$$F_{12} = \sin \pi x \sin 2\pi y \qquad \text{and} \qquad F_{21} = \sin 2\pi x \sin \pi y.$$

Hence the corresponding solutions

$$u_{12} = (B_{12} \cos c\pi\sqrt{5}t + B_{12}^* \sin c\pi\sqrt{5}t)F_{12} \qquad \text{and} \qquad u_{21} = (B_{21} \cos c\pi\sqrt{5}t + B_{21}^* \sin c\pi\sqrt{5}t)F_{21}$$

have the nodal lines $y = \frac{1}{2}$ and $x = \frac{1}{2}$, respectively (see Fig. 303). Taking $B_{12} = 1$ and $B_{12}^* = B_{21}^* = 0$, we obtain

$$u_{12} + u_{21} = \cos c\pi\sqrt{5}t\,(F_{12} + B_{21}F_{21}) \tag{12}$$

which represents another vibration corresponding to the eigenvalue $c\pi\sqrt{5}$. The nodal line of this function is the solution of the equation

$$F_{12} + B_{21}F_{21} = \sin \pi x \sin 2\pi y + B_{21} \sin 2\pi x \sin \pi y = 0$$

or, since $\sin 2\alpha = 2 \sin \alpha \cos \alpha$,

$$\sin \pi x \sin \pi y(\cos \pi y + B_{21} \cos \pi x) = 0. \tag{13}$$

This solution depends on the value of B_{21} (see Fig. 304).

From (11) we see that even more than two functions may correspond to the same numerical value of λ_{mn}. For example, the four functions F_{18}, F_{81}, F_{47}, and F_{74} correspond to the value

$$\lambda_{18} = \lambda_{81} = \lambda_{47} = \lambda_{74} = c\pi\sqrt{65}, \qquad \text{because} \qquad 1^2 + 8^2 = 4^2 + 7^2 = 65.$$

This happens because 65 can be expressed as the sum of two squares of positive integers in several ways. According to a theorem by Gauss, this is the case for every sum of two squares among whose prime factors there are at least two different ones of the form $4n + 1$ where n is a positive integer. In our case we have $65 = 5 \cdot 13 = (4 + 1)(12 + 1)$. ■

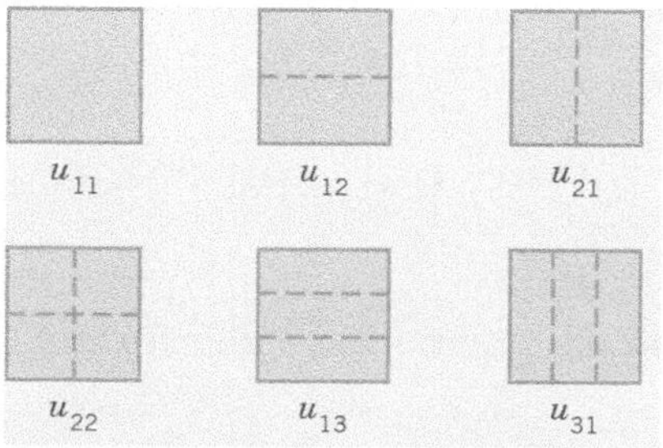

Fig. 303. Nodal lines of the solutions u_{11}, u_{12}, u_{21}, u_{22}, u_{13}, u_{31} in the case of the square membrane

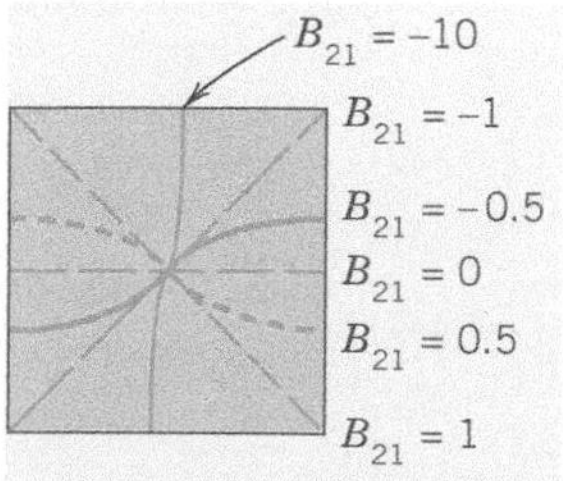

Fig. 304. Nodal lines of the solution (12) for some values of B_{21}

Step 3. Solution of the Model (1), (2), (3). Double Fourier Series

So far we have solutions (10) satisfying (1) and (2) only. To obtain the solutions that also satisfies (3), we proceed as in Sec. 12.3. We consider the double series

$$
\begin{aligned}
u(x, y, t) &= \sum_{m=1}^{\infty} \sum_{n=1}^{\infty} u_{mn}(x, y, t) \\
&= \sum_{m=1}^{\infty} \sum_{n=1}^{\infty} (B_{mn} \cos \lambda_{mn} t + B^*_{mn} \sin \lambda_{mn} t) \sin \frac{m\pi x}{a} \sin \frac{n\pi y}{b}
\end{aligned} \tag{14}
$$

(without discussing convergence and uniqueness). From (14) and (3a), setting $t = 0$, we have

$$
u(x, y, 0) = \sum_{m=1}^{\infty} \sum_{n=1}^{\infty} B_{mn} \sin \frac{m\pi x}{a} \sin \frac{n\pi y}{b} = f(x, y). \tag{15}
$$

Suppose that $f(x, y)$ can be represented by (15). (Sufficient for this is the continuity of $f, \partial f/\partial x, \partial f/\partial y, \partial^2 f/\partial x\, \partial y$ in R.) Then (15) is called the **double Fourier series** of $f(x, y)$. Its coefficients can be determined as follows. Setting

$$
K_m(y) = \sum_{n=1}^{\infty} B_{mn} \sin \frac{n\pi y}{b} \tag{16}
$$

we can write (15) in the form

$$
f(x, y) = \sum_{m=1}^{\infty} K_m(y) \sin \frac{m\pi x}{a}.
$$

For fixed y this is the Fourier sine series of $f(x, y)$, considered as a function of x. From (4) in Sec. 11.3 we see that the coefficients of this expansion are

$$
K_m(y) = \frac{2}{a} \int_0^a f(x, y) \sin \frac{m\pi x}{a}\, dx. \tag{17}
$$

Furthermore, (16) is the Fourier sine series of $K_m(y)$, and from (4) in Sec. 11.3 it follows that the coefficients are

$$
B_{mn} = \frac{2}{b} \int_0^b K_m(y) \sin \frac{n\pi y}{b}\, dy.
$$

From this and (17) we obtain the **generalized Euler formula**

$$
B_{mn} = \frac{4}{ab} \int_0^b \int_0^a f(x, y) \sin \frac{m\pi x}{a} \sin \frac{n\pi y}{b}\, dx\, dy \qquad \begin{matrix} m = 1, 2, \cdots \\ n = 1, 2, \cdots \end{matrix} \tag{18}
$$

for the **Fourier coefficients** of $f(x, y)$ in the double Fourier series (15).

The B_{mn} in (14) are now determined in terms of $f(x, y)$. To determine the B^*_{mn}, we differentiate (14) termwise with respect to t; using (3b), we obtain

$$\left.\frac{\partial u}{\partial t}\right|_{t=0} = \sum_{m=1}^{\infty} \sum_{n=1}^{\infty} B^*_{mn} \lambda_{mn} \sin \frac{m\pi x}{a} \sin \frac{n\pi y}{b} = g(x, y).$$

Suppose that $g(x, y)$ can be developed in this double Fourier series. Then, proceeding as before, we find that the coefficients are

$$\text{(19)} \qquad B^*_{mn} = \frac{4}{ab\lambda_{mn}} \int_0^b \int_0^a g(x, y) \sin \frac{m\pi x}{a} \sin \frac{n\pi y}{a} \, dx \, dy \qquad \begin{matrix} m = 1, 2, \cdots \\ n = 1, 2, \cdots . \end{matrix}$$

Result. *If f and g in* (3) *are such that u can be represented by* (14), *then* (14) *with coefficients* (18) *and* (19) *is the solution of the model* (1), (2), (3).

EXAMPLE 2 Vibration of a Rectangular Membrane

Find the vibrations of a rectangular membrane of sides $a = 4$ ft and $b = 2$ ft (Fig. 305) if the tension is 12.5 lb/ft, the density is 2.5 slugs/ft^2 (as for light rubber), the initial velocity is 0, and the initial displacement is

$$\text{(20)} \qquad f(x, y) = 0.1\,(4x - x^2)(2y - y^2) \text{ ft}.$$

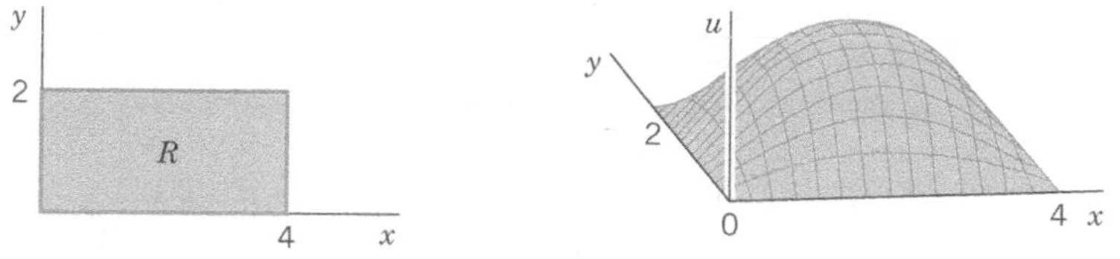

Membrane — Initial displacement

Fig. 305. Example 2

Solution. $c^2 = T/\rho = 12.5/2.5 = 5$ [ft^2/sec^2]. Also $B^*_{mn} = 0$ from (19). From (18) and (20),

$$B_{mn} = \frac{4}{4 \cdot 2} \int_0^2 \int_0^4 0.1(4x - x^2)(2y - y^2) \sin \frac{m\pi x}{4} \sin \frac{n\pi y}{2} \, dx \, dy$$

$$= \frac{1}{20} \int_0^4 (4x - x^2) \sin \frac{m\pi x}{4} \, dx \int_0^2 (2y - y^2) \sin \frac{n\pi y}{2} \, dy.$$

Two integrations by parts give for the first integral on the right

$$\frac{128}{m^3\pi^3}[1 - (-1)^m] = \frac{256}{m^3\pi^3} \qquad (m \text{ odd})$$

and for the second integral

$$\frac{16}{n^3\pi^3}[1 - (-1)^n] = \frac{32}{n^3\pi^3} \qquad (n \text{ odd}).$$

For even m or n we get 0. Together with the factor 1/20 we thus have $B_{mn} = 0$ if m or n is even and

$$B_{mn} = \frac{256 \cdot 32}{20 m^3 n^3 \pi^6} \approx \frac{0.426050}{m^3 n^3} \qquad (m \text{ and } n \text{ both odd}).$$

From this, (9), and (14) we obtain the answer

$$u(x, y, t) = 0.426050 \sum_{m,n \text{ odd}} \sum \frac{1}{m^3 n^3} \cos\left(\frac{\sqrt{5}\pi}{4}\sqrt{m^2 + 4n^2}\right) t \sin\frac{m\pi x}{4} \sin\frac{n\pi y}{2}$$

$$(21) \qquad = 0.426050\left(\cos\frac{\sqrt{5}\pi\sqrt{5}}{4} t \sin\frac{\pi x}{4}\sin\frac{\pi y}{2} + \frac{1}{27}\cos\frac{\sqrt{5}\pi\sqrt{37}}{4} t \sin\frac{\pi x}{4}\sin\frac{3\pi y}{2}\right.$$
$$\left. + \frac{1}{27}\cos\frac{\sqrt{5}\pi\sqrt{13}}{4} t \sin\frac{3\pi x}{4}\sin\frac{\pi y}{2} + \frac{1}{729}\cos\frac{\sqrt{5}\pi\sqrt{45}}{4} t \sin\frac{3\pi x}{4}\sin\frac{3\pi y}{2} + \cdots\right).$$

To discuss this solution, we note that the first term is very similar to the initial shape of the membrane, has no nodal lines, and is by far the dominating term because the coefficients of the next terms are much smaller. The second term has two horizontal nodal lines ($y = \frac{2}{3}, \frac{4}{3}$), the third term two vertical ones ($x = \frac{4}{3}, \frac{8}{3}$), the fourth term two horizontal and two vertical ones, and so on. ■

PROBLEM SET 12.9

1. Frequency. How does the frequency of the eigenfunctions of the rectangular membrane change (a) If we double the tension? (b) If we take a membrane of half the density of the original one? (c) If we double the sides of the membrane? Give reasons.

2. Assumptions. Which part of Assumption 2 cannot be satisfied exactly? Why did we also assume that the angles of inclination are small?

3. Determine and sketch the nodal lines of the square membrane for $m = 1, 2, 3, 4$ and $n = 1, 2, 3, 4$.

4–8 DOUBLE FOURIER SERIES

Represent $f(x, y)$ by a series (15), where

4. $f(x, y) = 1, \quad a = b = 1$

5. $f(x, y) = y, \quad a = b = 1$

6. $f(x, y) = x, \quad a = b = 1$

7. $f(x, y) = xy, \quad a$ and b arbitrary

8. $f(x, y) = xy(a - x)(b - y), \quad a$ and b arbitrary

9. CAS PROJECT. Double Fourier Series. (a) Write a program that gives and graphs partial sums of (15). Apply it to Probs. 5 and 6. Do the graphs show that those partial sums satisfy the boundary condition (3a)? Explain why. Why is the convergence rapid?

(b) Do the tasks in (a) for Prob. 4. Graph a portion, say, $0 < x < \frac{1}{2}, 0 < y < \frac{1}{2}$, of several partial sums on common axes, so that you can see how they differ. (See Fig. 306.)

(c) Do the tasks in (b) for functions of your choice.

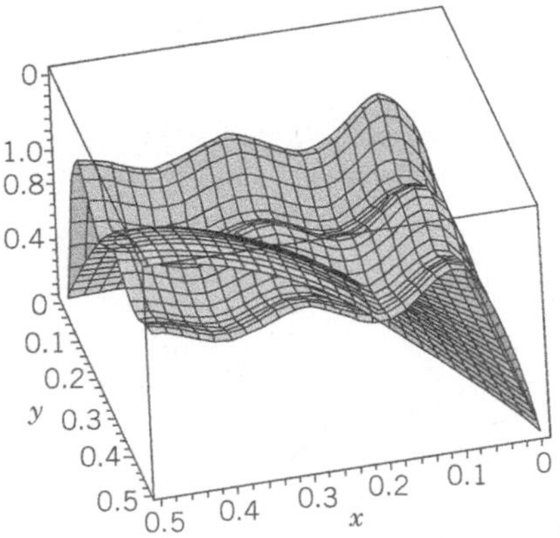

Fig. 306. Partial sums $S_{2,2}$ and $S_{10,10}$ in CAS Project 9b

10. CAS EXPERIMENT. Quadruples of F_{mn}. Write a program that gives you four numerically equal λ_{mn} in Example 1, so that four different F_{mn} correspond to it. Sketch the nodal lines of $F_{18}, F_{81}, F_{47}, F_{74}$ in Example 1 and similarly for further F_{mn} that you will find.

11–13 SQUARE MEMBRANE

Find the deflection $u(x, y, t)$ of the square membrane of side π and $c^2 = 1$ for initial velocity 0 and initial deflection

11. $0.1 \sin 2x \sin 4y$

12. $0.01 \sin x \sin y$

13. $0.1 xy(\pi - x)(\pi - y)$

14–19 RECTANGULAR MEMBRANE

14. Verify the discussion of (21) in Example 2.
15. Do Prob. 3 for the membrane with $a = 4$ and $b = 2$.
16. Verify B_{mn} in Example 2 by integration by parts.
17. Find eigenvalues of the rectangular membrane of sides $a = 2$ and $b = 1$ to which there correspond two or more different (independent) eigenfunctions.
18. **Minimum property.** Show that among all rectangular membranes of the same area $A = ab$ and the same c the square membrane is that for which u_{11} [see (10)] has the lowest frequency.
19. **Deflection.** Find the deflection of the membrane of sides a and b with $c^2 = 1$ for the initial deflection

$$f(x, y) = \sin \frac{6\pi x}{a} \sin \frac{2\pi y}{b} \quad \text{and initial velocity 0.}$$

20. **Forced vibrations.** Show that forced vibrations of a membrane are modeled by the PDE $u_{tt} = c^2\nabla^2 u + P/\rho$, where $P(x, y, t)$ is the external force per unit area acting perpendicular to the xy-plane.

12.10 Laplacian in Polar Coordinates. Circular Membrane. Fourier–Bessel Series

It is a ***general principle*** in boundary value problems for PDEs to *choose coordinates that make the formula for the boundary as simple as possible*. Here polar coordinates are used for this purpose as follows. Since we want to discuss circular membranes (drumheads), we first transform the Laplacian in the wave equation (1), Sec. 12.9,

$$u_{tt} = c^2\nabla^2 u = c^2(u_{xx} + u_{yy}) \tag{1}$$

(subscripts denoting partial derivatives) into **polar coordinates** r, θ defined by $x = r\cos\theta$, $y = r\sin\theta$; thus,

$$r = \sqrt{x^2 + y^2}, \qquad \tan\theta = \frac{y}{x}.$$

By the chain rule (Sec. 9.6) we obtain

$$u_x = u_r r_x + u_\theta \theta_x.$$

Differentiating once more with respect to x and using the product rule and then again the chain rule gives

$$\begin{aligned} u_{xx} &= (u_r r_x)_x + (u_\theta \theta_x)_x \\ &= (u_r)_x r_x + u_r r_{xx} + (u_\theta)_x \theta_x + u_\theta \theta_{xx} \\ &= (u_{rr} r_x + u_{r\theta}\theta_x) r_x + u_r r_{xx} + (u_{\theta r} r_x + u_{\theta\theta}\theta_x)\theta_x + u_\theta \theta_{xx}. \end{aligned} \tag{2}$$

Also, by differentiation of r and θ we find

$$r_x = \frac{x}{\sqrt{x^2 + y^2}} = \frac{x}{r}, \qquad \theta_x = \frac{1}{1 + (y/x)^2}\left(-\frac{y}{x^2}\right) = -\frac{y}{r^2}.$$

Differentiating these two formulas again, we obtain

$$r_{xx} = \frac{r - xr_x}{r^2} = \frac{1}{r} - \frac{x^2}{r^3} = \frac{y^2}{r^3}, \qquad \theta_{xx} = -y\left(-\frac{2}{r^3}\right)r_x = \frac{2xy}{r^4}.$$

We substitute all these expressions into (2). Assuming continuity of the first and second partial derivatives, we have $u_{r\theta} = u_{\theta r}$, and by simplifying,

$$\text{(3)} \qquad u_{xx} = \frac{x^2}{r^2}u_{rr} - 2\frac{xy}{r^3}u_{r\theta} + \frac{y^2}{r^4}u_{\theta\theta} + \frac{y^2}{r^3}u_r + 2\frac{xy}{r^4}u_\theta.$$

In a similar fashion it follows that

$$\text{(4)} \qquad u_{yy} = \frac{y^2}{r^2}u_{rr} + 2\frac{xy}{r^3}u_{r\theta} + \frac{x^2}{r^4}u_{\theta\theta} + \frac{x^2}{r^3}u_r - 2\frac{xy}{r^4}u_\theta.$$

By adding (3) and (4) we see that the **Laplacian of *u* in polar coordinates** is

$$\textbf{(5)} \qquad \nabla^2 u = \frac{\partial^2 u}{\partial r^2} + \frac{1}{r}\frac{\partial u}{\partial r} + \frac{1}{r^2}\frac{\partial^2 u}{\partial \theta^2}.$$

Circular Membrane

Circular membranes are important parts of drums, pumps, microphones, telephones, and other devices. This accounts for their great importance in engineering. Whenever a circular membrane is plane and its material is elastic, but offers no resistance to bending (this excludes thin metallic membranes!), its vibrations are modeled by the **two-dimensional wave equation in polar coordinates** obtained from (1) with $\nabla^2 u$ given by (5), that is,

$$\textbf{(6)} \qquad \frac{\partial^2 u}{\partial t^2} = c^2\left(\frac{\partial^2 u}{\partial r^2} + \frac{1}{r}\frac{\partial u}{\partial r} + \frac{1}{r^2}\frac{\partial^2 u}{\partial \theta^2}\right) \qquad c^2 = \frac{T}{\rho}.$$

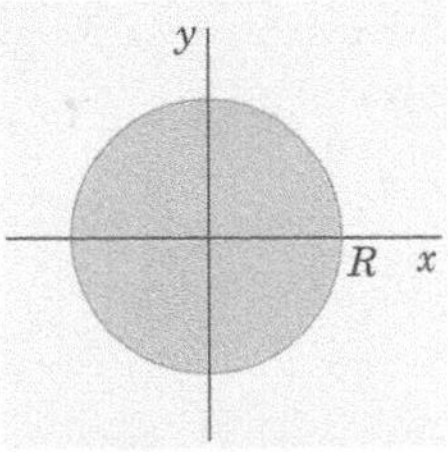

Fig. 307. Circular membrane

We shall consider a membrane of radius R (Fig. 307) and determine solutions $u(r, t)$ that are radially symmetric. (Solutions also depending on the angle θ will be discussed in the problem set.) Then $u_{\theta\theta} = 0$ in (6) and the model of the problem (the analog of (1), (2), (3) in Sec. 12.9) is

$$\textbf{(7)} \qquad \frac{\partial^2 u}{\partial t^2} = c^2\left(\frac{\partial^2 u}{\partial r^2} + \frac{1}{r}\frac{\partial u}{\partial r}\right)$$

$$\textbf{(8)} \qquad u(R, t) = 0 \text{ for all } t \geqq 0$$

$$\textbf{(9a)} \qquad u(r, 0) = f(r)$$

$$\textbf{(9b)} \qquad u_t(r, 0) = g(r).$$

Here (8) means that the membrane is fixed along the boundary circle $r = R$. The initial deflection $f(r)$ and the initial velocity $g(r)$ depend only on r, not on θ, so that we can expect radially symmetric solutions $u(r, t)$.

Step 1. Two ODEs From the Wave Equation (7). Bessel's Equation

Using the **method of separation of variables**, we first determine solutions $u(r, t) = W(r)G(t)$. (We write W, not F because W depends on r, whereas F, used before, depended on x.) Substituting $u = WG$ and its derivatives into (7) and dividing the result by c^2WG, we get

$$\frac{\ddot{G}}{c^2G} = \frac{1}{W}\left(W'' + \frac{1}{r}W'\right)$$

where dots denote derivatives with respect to t and primes denote derivatives with respect to r. The expressions on both sides must equal a constant. This constant must be negative, say, $-k^2$, in order to obtain solutions that satisfy the boundary condition without being identically zero. Thus,

$$\frac{\ddot{G}}{c^2G} = \frac{1}{W}\left(W'' + \frac{1}{r}W'\right) = -k^2.$$

This gives the two linear ODEs

$$\ddot{G} + \lambda^2 G = 0 \qquad \text{where } \lambda = ck \tag{10}$$

and

$$W'' + \frac{1}{r}W' + k^2W = 0. \tag{11}$$

We can reduce (11) to Bessel's equation (Sec. 5.4) if we set $s = kr$. Then $1/r = k/s$ and, retaining the notation W for simplicity, we obtain by the chain rule

$$W' = \frac{dW}{dr} = \frac{dW}{ds}\frac{ds}{dr} = \frac{dW}{ds}k \qquad \text{and} \qquad W'' = \frac{d^2W}{ds^2}k^2.$$

By substituting this into (11) and omitting the common factor k^2 we have

$$\frac{d^2W}{ds^2} + \frac{1}{s}\frac{dW}{ds} + W = 0. \tag{12}$$

This is **Bessel's equation** (1), Sec. 5.4, with parameter $\nu = 0$.

Step 2. Satisfying the Boundary Condition (8)

Solutions of (12) are the Bessel functions J_0 and Y_0 of the first and second kind (see Secs. 5.4, 5.5). But Y_0 becomes infinite at 0, so that we cannot use it because the deflection of the membrane must always remain finite. This leaves us with

$$W(r) = J_0(s) = J_0(kr) \qquad (s = kr). \tag{13}$$

On the boundary $r = R$ we get $W(R) = J_0(kR) = 0$ from (8) (because $G \equiv 0$ would imply $u \equiv 0$). We can satisfy this condition because J_0 has (infinitely many) positive zeros, $s = \alpha_1, \alpha_2, \cdots$ (see Fig. 308), with numerical values

$$\alpha_1 = 2.4048, \quad \alpha_2 = 5.5201, \quad \alpha_3 = 8.6537, \quad \alpha_4 = 11.7915, \quad \alpha_5 = 14.9309$$

and so on. (For further values, consult your CAS or Ref. [GenRef1] in App. 1.) These zeros are slightly irregularly spaced, as we see. Equation (13) now implies

$$kR = \alpha_m \qquad \text{thus} \qquad k = k_m = \frac{\alpha_m}{R}, \qquad m = 1, 2, \cdots. \tag{14}$$

Hence the functions

$$W_m(r) = J_0(k_m r) = J_0\left(\frac{\alpha_m}{R} r\right), \qquad m = 1, 2, \cdots \tag{15}$$

are solutions of (11) that are zero on the boundary circle $r = R$.

Eigenfunctions and Eigenvalues. For W_m in (15), a corresponding general solution of (10) with $\lambda = \lambda_m = ck_m = c\alpha_m/R$ is

$$G_m(t) = A_m \cos \lambda_m t + B_m \sin \lambda_m t.$$

Hence the functions

$$u_m(r, t) = W_m(r)G_m(t) = (A_m \cos \lambda_m t + B_m \sin \lambda_m t)J_0(k_m r) \tag{16}$$

with $m = 1, 2, \cdots$ are solutions of the wave equation (7) satisfying the boundary condition (8). These are the **eigenfunctions** of our problem. The corresponding **eigenvalues** are λ_m.

The vibration of the membrane corresponding to u_m is called the mth **normal mode**; it has the frequency $\lambda_m/2\pi$ cycles per unit time. Since the zeros of the Bessel function J_0 are not regularly spaced on the axis (in contrast to the zeros of the sine functions appearing in the case of the vibrating string), the sound of a drum is entirely different from that of a violin. The forms of the normal modes can easily be obtained from Fig. 308 and are shown in Fig. 309. For $m = 1$, all the points of the membrane move up (or down) at the same time. For $m = 2$, the situation is as follows. The function $W_2(r) = J_0(\alpha_2 r/R)$ is zero for $\alpha_2 r/R = \alpha_1$, thus $r = \alpha_1 R/\alpha_2$. The circle $r = \alpha_1 R/\alpha_2$ is, therefore, **nodal line**, and when at some instant the central part of the membrane moves up, the outer part ($r > \alpha_1 R/\alpha_2$) moves down, and conversely. The solution $u_m(r, t)$ has $m - 1$ nodal lines, which are circles (Fig. 309).

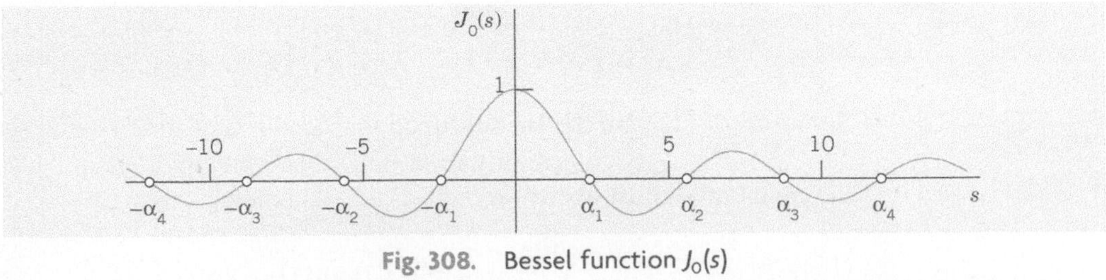

Fig. 308. Bessel function $J_0(s)$

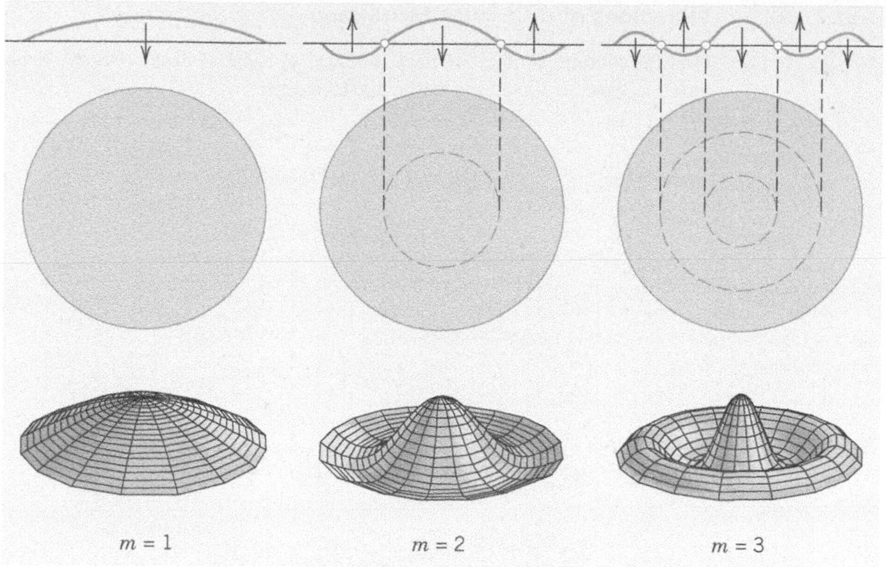

Fig. 309. Normal modes of the circular membrane in the case of vibrations independent of the angle

Step 3. Solution of the Entire Problem

To obtain a solution $u(r, t)$ that also satisfies the initial conditions (9), we may proceed as in the case of the string. That is, we consider the series

$$u(r,t) = \sum_{m=1}^{\infty} W_m(r)G_m(t) = \sum_{m=1}^{\infty} (A_m \cos \lambda_m t + B_m \sin \lambda_m t) J_0\left(\frac{\alpha_m}{R} r\right) \tag{17}$$

(leaving aside the problems of convergence and uniqueness). Setting $t = 0$ and using (9a), we obtain

$$u(r,0) = \sum_{m=1}^{\infty} A_m J_0\left(\frac{\alpha_m}{R} r\right) = f(r). \tag{18}$$

Thus for the series (17) to satisfy the condition (9a), the constants A_m must be the coefficients of the **Fourier–Bessel series** (18) that represents $f(r)$ in terms of $J_0(\alpha_m r/R)$; that is [see (9) in Sec. 11.6 with $n = 0$, $\alpha_{0,m} = \alpha_m$, and $x = r$],

$$A_m = \frac{2}{R^2 J_1^2(\alpha_m)} \int_0^R r f(r) J_0\left(\frac{\alpha_m}{R} r\right) dr \qquad (m = 1, 2, \cdots). \tag{19}$$

Differentiability of $f(r)$ in the interval $0 \leqq r \leqq R$ is sufficient for the existence of the development (18); see Ref. [A13]. The coefficients B_m in (17) can be determined from (9b) in a similar fashion. Numeric values of A_m and B_m may be obtained from a CAS or by a numeric integration method, using tables of J_0 and J_1. However, numeric integration can sometimes be ***avoided***, as the following example shows.

EXAMPLE 1 Vibrations of a Circular Membrane

Find the vibrations of a circular drumhead of radius 1 ft and density 2 slugs/ft^2 if the tension is 8 lb/ft, the initial velocity is 0, and the initial displacement is

$$f(r) = 1 - r^2 \text{ [ft]}.$$

Solution. $c^2 = T/\rho = \frac{8}{2} = 4$ [ft^2/sec^2]. Also $B_m = 0$, since the initial velocity is 0. From (10) in Sec. 11.6, since $R = 1$, we obtain

$$A_m = \frac{2}{J_1^2(\alpha_m)} \int_0^1 r(1 - r^2) J_0(\alpha_m r)\, dr$$
$$= \frac{4J_2(\alpha_m)}{\alpha_m^2 J_1^2(\alpha_m)}$$
$$= \frac{8}{\alpha_m^3 J_1(\alpha_m)}$$

where the last equality follows from (21c), Sec. 5.4, with $\nu = 1$, that is,

$$J_2(\alpha_m) = \frac{2}{\alpha_m} J_1(\alpha_m) - J_0(\alpha_m) = \frac{2}{\alpha_m} J_1(\alpha_m).$$

Table 9.5 on p. 409 of [GenRef1] gives α_m and $J_0'(\alpha_m)$. From this we get $J_1(\alpha_m) = -J_0'(\alpha_m)$ by (21b), Sec. 5.4, with $\nu = 0$, and compute the coefficients A_m:

m	α_m	$J_1(\alpha_m)$	$J_2(\alpha_m)$	A_m
1	2.40483	0.51915	0.43176	1.10801
2	5.52008	−0.34026	−0.12328	−0.13978
3	8.65373	0.27145	0.06274	0.04548
4	11.79153	−0.23246	−0.03943	−0.02099
5	14.93092	0.20655	0.02767	0.01164
6	18.07106	−0.18773	−0.02078	−0.00722
7	21.21164	0.17327	0.01634	0.00484
8	24.35247	−0.16170	−0.01328	−0.00343
9	27.49348	0.15218	0.01107	0.00253
10	30.63461	−0.14417	−0.00941	−0.00193

Thus

$$f(r) = 1.108J_0(2.4048r) - 0.140J_0(5.5201r) + 0.045J_0(8.6537r) - \cdots.$$

We see that the coefficients decrease relatively slowly. The sum of the explicitly given coefficients in the table is 0.99915. The sum of *all* the coefficients should be 1. (Why?) Hence by the Leibniz test in App. A3.3 the partial sum of those terms gives about three correct decimals of the amplitude $f(r)$.

Since

$$\lambda_m = ck_m = c\alpha_m/R = 2\alpha_m,$$

from (17) we thus obtain the solution (with r measured in feet and t in seconds)

$$u(r, t) = 1.108J_0(2.4048r) \cos 4.8097t - 0.140J_0(5.5201r) \cos 11.0402t + 0.045J_0(8.6537r) \cos 17.3075t - \cdots.$$

In Fig. 309, $m = 1$ gives an idea of the motion of the first term of our series, $m = 2$ of the second term, and $m = 3$ of the third term, so that we can "see" our result about as well as for a violin string in Sec. 12.3. ■

PROBLEM SET 12.10

1–3 RADIAL SYMMETRY

1. Why did we introduce polar coordinates in this section?

2. Radial symmetry reduces (5) to $\nabla^2 u = u_{rr} + u_r/r$. Derive this directly from $\nabla^2 u = u_{xx} + u_{yy}$. Show that the only solution of $\nabla^2 u = 0$ depending only on $r = \sqrt{x^2 + y^2}$ is $u = a \ln r + b$ with arbitrary constants a and b.

3. Alternative form of (5). Show that (5) can be written $\nabla^2 u = (ru_r)_r/r + u_{\theta\theta}/r^2$, a form that is often practical.

BOUNDARY VALUE PROBLEMS. SERIES

4. TEAM PROJECT. Series for Dirichlet and Neumann Problems

(a) Show that $u_n = r^n \cos n\theta$, $u_n = r^n \sin n\theta$, $n = 0, 1, \cdots$, are solutions of Laplace's equation $\nabla^2 u = 0$ with $\nabla^2 u$ given by (5). (What would u_n be in Cartesian coordinates? Experiment with small n.)

(b) Dirichlet problem (See Sec. 12.6) Assuming that termwise differentiation is permissible, show that a solution of the Laplace equation in the disk $r < R$ satisfying the boundary condition $u(R, \theta) = f(\theta)$ (R and f given) is

$$u(r,\theta) = a_0 + \sum_{n=1}^{\infty} \left[a_n \left(\frac{r}{R}\right)^n \cos n\theta + b_n \left(\frac{r}{R}\right)^n \sin n\theta \right] \tag{20}$$

where a_n, b_n are the Fourier coefficients of f (see Sec. 11.1).

(c) Dirichlet problem. Solve the Dirichlet problem using (20) if $R = 1$ and the boundary values are $u(\theta) = -100$ volts if $-\pi < \theta < 0$, $u(\theta) = 100$ volts if $0 < \theta < \pi$. (Sketch this disk, indicate the boundary values.)

(d) Neumann problem. Show that the solution of the Neumann problem $\nabla^2 u = 0$ if $r < R$, $u_N(R, \theta) = f(\theta)$ (where $u_N = \partial u/\partial N$ is the directional derivative in the direction of the outer normal) is

$$u(r, \theta) = A_0 + \sum_{n=1}^{\infty} r^n (A_n \cos n\theta + B_n \sin n\theta)$$

with arbitrary A_0 and

$$A_n = \frac{1}{\pi n R^{n-1}} \int_{-\pi}^{\pi} f(\theta) \cos n\theta \, d\theta,$$

$$B_n = \frac{1}{\pi n R^{n-1}} \int_{-\pi}^{\pi} f(\theta) \sin n\theta \, d\theta.$$

(e) Compatibility condition. Show that (9), Sec. 10.4, imposes on $f(\theta)$ in (d) the *"compatibility condition"*

$$\int_{-\pi}^{\pi} f(\theta) \, d\theta = 0.$$

(f) Neumann problem. Solve $\nabla^2 u = 0$ in the annulus $1 < r < 2$ if $u_r(1, \theta) = \sin \theta$, $u_r(2, \theta) = 0$.

5–8 ELECTROSTATIC POTENTIAL. STEADY-STATE HEAT PROBLEMS

The electrostatic potential satisfies Laplace's equation $\nabla^2 u = 0$ in any region free of charges. Also the heat equation $u_t = c^2 \nabla^2 u$ (Sec. 12.5) reduces to Laplace's equation if the temperature u is time-independent ("**steady-state case**"). Using (20), find the potential (equivalently: the steady-state temperature) in the disk $r < 1$ if the boundary values are (sketch them, to see what is going on).

5. $u(1, \theta) = 220$ if $-\frac{1}{2}\pi < \theta < \frac{1}{2}\pi$ and 0 otherwise

6. $u(1, \theta) = 400 \cos^3 \theta$

7. $u(1, \theta) = 110|\theta|$ if $-\pi < \theta < \pi$

8. $u(1, \theta) = \theta$ if $-\frac{1}{2}\pi < \theta < \frac{1}{2}\pi$ and 0 otherwise

9. CAS EXPERIMENT. Equipotential Lines. Guess what the equipotential lines $u(r, \theta) = \text{const}$ in Probs. 5 and 7 may look like. Then graph some of them, using partial sums of the series.

10. Semidisk. Find the electrostatic potential in the semidisk $r < 1$, $0 < \theta < \pi$ which equals $110\theta(\pi - \theta)$ on the semicircle $r = 1$ and 0 on the segment $-1 < x < 1$.

11. Semidisk. Find the steady-state temperature in a semicircular thin plate $r < a$, $0 < \theta < \pi$ with the semicircle $r = a$ kept at constant temperature u_0 and the segment $-a < x < a$ at 0.

CIRCULAR MEMBRANE

12. CAS PROJECT. Normal Modes. (a) Graph the normal modes u_4, u_5, u_6 as in Fig. 306.

(b) Write a program for calculating the A_m's in Example 1 and extend the table to $m = 15$. Verify numerically that $\alpha_m \approx (m - \frac{1}{4})\pi$ and compute the error for $m = 1, \cdots, 10$.

(c) Graph the initial deflection $f(r)$ in Example 1 as well as the first three partial sums of the series. Comment on accuracy.

(d) Compute the radii of the nodal lines of u_2, u_3, u_4 when $R = 1$. How do these values compare to those of the nodes of the vibrating string of length 1? Can you establish any empirical laws by experimentation with further u_m?

13. Frequency. What happens to the frequency of an eigenfunction of a drum if you double the tension?

14. Size of a drum. A small drum should have a higher fundamental frequency than a large one, tension and density being the same. How does this follow from our formulas?

15. Tension. Find a formula for the tension required to produce a desired fundamental frequency f_1 of a drum.

16. Why is $A_1 + A_2 + \cdots = 1$ in Example 1? Compute the first few partial sums until you get 3-digit accuracy. What does this problem mean in the field of music?

17. Nodal lines. Is it possible that for fixed c and R two or more u_m [see (16)] with different nodal lines correspond to the same eigenvalue? (Give a reason.)

18. Nonzero initial velocity is more of theoretical interest because it is difficult to obtain experimentally. Show that for (17) to satisfy (9b) we must have

$$B_m = K_m \int_0^R r g(r) J_0(\alpha_m r/R)\, dr \tag{21}$$

where $K_m = 2/(c\alpha_m R)J_1^2(\alpha_m)$.

VIBRATIONS OF A CIRCULAR MEMBRANE DEPENDING ON BOTH r AND θ

19. (Separations) Show that substitution of $u = F(r, \theta)G(t)$ into the wave equation (6), that is,

$$u_{tt} = c^2\left(u_{rr} + \frac{1}{r}\, u_r + \frac{1}{r^2}\, u_{\theta\theta}\right), \tag{22}$$

gives an ODE and a PDE

$$\ddot{G} + \lambda^2 G = 0, \qquad \text{where } \lambda = ck, \tag{23}$$

$$F_{rr} + \frac{1}{r}\, F_r + \frac{1}{r^2}\, F_{\theta\theta} + k^2 F = 0. \tag{24}$$

Show that the PDE can now be separated by substituting $F = W(r)Q(\theta)$, giving

$$Q'' + n^2 Q = 0, \tag{25}$$

$$r^2 W'' + rW' + (k^2 r^2 - n^2)W = 0. \tag{26}$$

20 Periodicity. Show that $Q(\theta)$ must be periodic with period 2π and, therefore, $n = 0, 1, 2, \cdots$ in (25) and (26). Show that this yields the solutions $Q_n = \cos n\theta$, $Q_n^* = \sin n\theta$, $W_n = J_n(kr)$, $n = 0, 1, \cdots$.

21. Boundary condition. Show that the boundary condition

$$u(R, \theta, t) = 0 \tag{27}$$

leads to $k = k_{mn} = \alpha_{mn}/R$, where $s = \alpha_{nm}$ is the mth positive zero of $J_n(s)$.

22. Solutions depending on both r and θ. Show that solutions of (22) satisfying (27) are (see Fig. 310)

$$\begin{aligned} u_{nm} &= (A_{nm} \cos ck_{nm}t + B_{nm} \sin ck_{nm}t) \\ &\quad \times J_n(k_{nm}r) \cos n\theta \\ u_{nm}^* &= (A_{nm}^* \cos ck_{nm}t + B_{nm}^* \sin ck_{nm}t) \\ &\quad \times J_n(k_{nm}r) \sin n\theta \end{aligned} \tag{28}$$

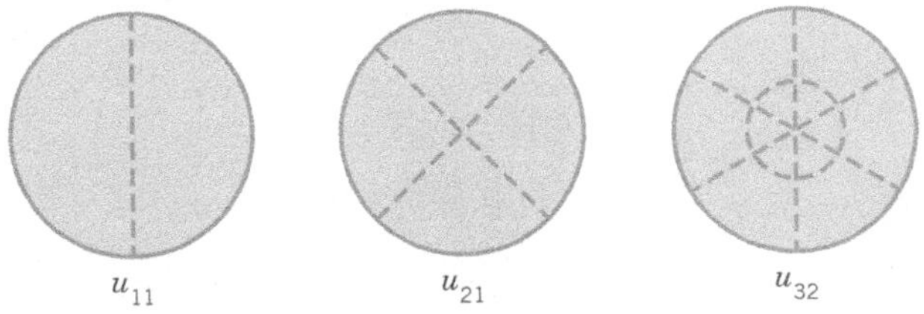

Fig. 310. Nodal lines of some of the solutions (28)

23. Initial condition. Show that $u_t(r, \theta, 0) = 0$ gives $B_{nm} = 0$, $B_{nm}^* = 0$ in (28).

24. Show that $u_{0m}^* = 0$ and u_{0m} is identical with (16) in this section.

25. Semicircular membrane. Show that u_{11} represents the fundamental mode of a semicircular membrane and find the corresponding frequency when $c^2 = 1$ and $R = 1$.

12.11 Laplace's Equation in Cylindrical and Spherical Coordinates. Potential

One of the most important PDEs in physics and engineering applications is **Laplace's equation**, given by

$$\nabla^2 u = u_{xx} + u_{yy} + u_{zz} = 0. \tag{1}$$

Here, x, y, z are Cartesian coordinates in space (Fig. 167 in Sec. 9.1), $u_{xx} = \partial^2 u/\partial x^2$, etc. The expression $\nabla^2 u$ is called the **Laplacian** of u. The theory of the solutions of (1) is called **potential theory**. Solutions of (1) that have ***continuous*** second partial derivatives are known as **harmonic functions**.

Laplace's equation occurs mainly in **gravitation, electrostatics** (see Theorem 3, Sec. 9.7), steady-state **heat flow** (Sec. 12.5), and **fluid flow** (to be discussed in Sec. 18.4).

Recall from Sec. 9.7 that the gravitational **potential** $u(x, y, z)$ at a point (x, y, z) resulting from a single mass located at a point (X, Y, Z) is

$$u(x, y, z) = \frac{c}{r} = \frac{c}{\sqrt{(x - X)^2 + (y - Y)^2 + (z - Z)^2}} \qquad (r > 0) \tag{2}$$

and u satisfies (1). Similarly, if mass is distributed in a region T in space with density $\rho(X, Y, Z)$, its potential at a point (x, y, z) not occupied by mass is

$$u(x, y, z) = k \iiint_T \frac{\rho(X, Y, Z)}{r} \, dX\, dY\, dZ. \tag{3}$$

It satisfies (1) because $\nabla^2(1/r) = 0$ (Sec. 9.7) and ρ is not a function of x, y, z.

Practical problems involving Laplace's equation are boundary value problems in a region T in space with boundary surface S. Such problems can be grouped into three types (see also Sec. 12.6 for the two-dimensional case):

(I) First boundary value problem or Dirichlet problem if u is prescribed on S.

(II) Second boundary value problem or **Neumann problem** if the normal derivative $u_n = \partial u/\partial n$ is prescribed on S.

(III) Third or **mixed boundary value problem** or **Robin problem** if u is prescribed on a portion of S and u_n on the remaining portion of S.

In general, when we want to solve a boundary value problem, we have to first select the appropriate coordinates in which the boundary surface S has a simple representation. Here are some examples followed by some applications.

Laplacian in Cylindrical Coordinates

The first step in solving a boundary value problem is generally the introduction of coordinates in which the boundary surface S has a simple representation. Cylindrical symmetry (a cylinder as a region T) calls for cylindrical coordinates r, θ, z related to x, y, z by

$$x = r\cos\theta, \qquad y = r\sin\theta, \qquad z = z \qquad \text{(Fig. 311).} \tag{4}$$

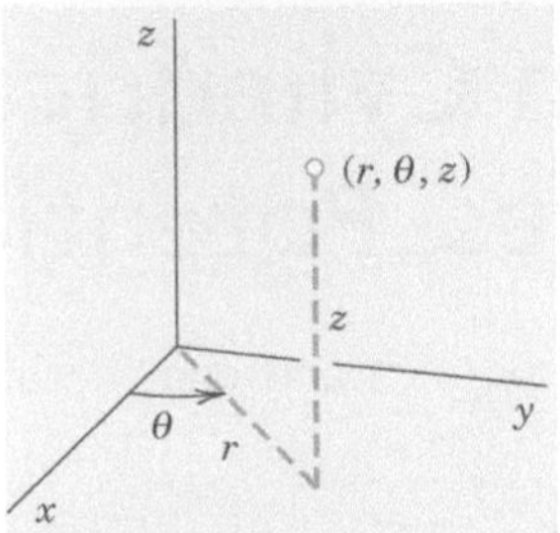

Fig. 311. Cylindrical coordinates ($r \geqq 0, 0 \leqq \theta \leqq 2\pi$)

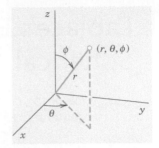

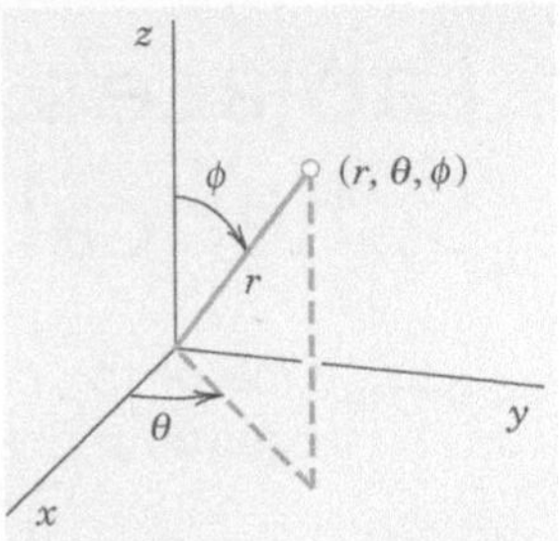

Fig. 312. Spherical coordinates ($r \geqq 0, 0 \leqq \theta \leqq 2\pi, 0 \leqq \varphi \leqq \pi$)

For these we get $\nabla^2 u$ immediately by adding u_{zz} to (5) in Sec. 12.10; thus,

$$\nabla^2 u = \frac{\partial^2 u}{\partial r^2} + \frac{1}{r}\frac{\partial u}{\partial r} + \frac{1}{r^2}\frac{\partial^2 u}{\partial \theta^2} + \frac{\partial^2 u}{\partial z^2}. \tag{5}$$

Laplacian in Spherical Coordinates

Spherical symmetry (a ball as region T bounded by a sphere S) requires **spherical coordinates** r, θ, ϕ related to x, y, z by

$$x = r\cos\theta\,\sin\phi, \qquad y = r\sin\theta\,\sin\phi, \qquad z = r\cos\phi \quad \text{(Fig. 312)}. \tag{6}$$

Using the chain rule (as in Sec. 12.10), we obtain $\nabla^2 u$ in spherical coordinates

$$\nabla^2 u = \frac{\partial^2 u}{\partial r^2} + \frac{2}{r}\frac{\partial u}{\partial r} + \frac{1}{r^2}\frac{\partial^2 u}{\partial \phi^2} + \frac{\cot\phi}{r^2}\frac{\partial u}{\partial \phi} + \frac{1}{r^2\sin^2\phi}\frac{\partial^2 u}{\partial \theta^2}. \tag{7}$$

We leave the details as an exercise. It is sometimes practical to write (7) in the form

$$\nabla^2 u = \frac{1}{r^2}\left[\frac{\partial}{\partial r}\left(r^2\frac{\partial u}{\partial r}\right) + \frac{1}{\sin\phi}\frac{\partial}{\partial \phi}\left(\sin\phi\frac{\partial u}{\partial \phi}\right) + \frac{1}{\sin^2\phi}\frac{\partial^2 u}{\partial \theta^2}\right]. \tag{7'}$$

Remark on Notation. Equation (6) is used in calculus and extends the familiar notation for polar coordinates. Unfortunately, some books use θ and ϕ interchanged, an extension of the notation $x = r\cos\phi$, $y = r\sin\phi$ for polar coordinates (used in some European countries).

Boundary Value Problem in Spherical Coordinates

We shall solve the following **Dirichlet problem** in spherical coordinates:

$$\nabla^2 u = \frac{1}{r^2}\left[\frac{\partial}{\partial r}\left(r^2\frac{\partial u}{\partial r}\right) + \frac{1}{\sin\phi}\frac{\partial}{\partial \phi}\left(\sin\phi\frac{\partial u}{\partial \phi}\right)\right] = 0. \tag{8}$$

$$u(R, \phi) = f(\phi) \tag{9}$$

$$\lim_{r\to\infty} u(r, \phi) = 0. \tag{10}$$

The PDE (8) follows from (7) or (7′) by assuming that the solution u will not depend on θ because the Dirichlet condition (9) is independent of θ. This may be an electrostatic potential (or a temperature) $f(\phi)$ at which the sphere S: $r = R$ is kept. Condition (10) means that the potential at infinity will be zero.

Separating Variables by substituting $u(r, \phi) = G(r)H(\phi)$ into (8). Multiplying (8) by r^2, making the substitution and then dividing by GH, we obtain

$$\frac{1}{G}\frac{d}{dr}\left(r^2\frac{dG}{dr}\right) = -\frac{1}{H\sin\phi}\frac{d}{d\phi}\left(\sin\phi\frac{dH}{d\phi}\right).$$

By the usual argument both sides must be equal to a constant k. Thus we get the two ODEs

$$\frac{1}{G}\frac{d}{dr}\left(r^2\frac{dG}{dr}\right) = k \qquad \text{or} \qquad r^2\frac{d^2G}{dr^2} + 2r\frac{dG}{dr} = kG \tag{11}$$

and

$$\frac{1}{\sin\phi}\frac{d}{d\phi}\left(\sin\phi\frac{dH}{d\phi}\right) + kH = 0. \tag{12}$$

The solutions of (11) will take a simple form if we set $k = n(n+1)$. Then, writing $G' = dG/dr$, etc., we obtain

$$r^2G'' + 2rG' - n(n+1)\,G = 0. \tag{13}$$

This is an **Euler–Cauchy equation**. From Sec. 2.5 we know that it has solutions $G = r^a$. Substituting this and dropping the common factor r^a gives

$$a(a-1) + 2a - n(n+1) = 0. \qquad \text{The roots are} \qquad a = n \quad \text{and} \quad -n-1.$$

Hence solutions are

$$G_n(r) = r^n \qquad \text{and} \qquad G_n^*(r) = \frac{1}{r^{n+1}}. \tag{14}$$

We now solve (12). Setting $\cos\phi = w$, we have $\sin^2\phi = 1 - w^2$ and

$$\frac{d}{d\phi} = \frac{d}{dw}\frac{dw}{d\phi} = -\sin\phi\frac{d}{dw}.$$

Consequently, (12) with $k = n(n+1)$ takes the form

$$\frac{d}{dw}\left[(1-w^2)\frac{dH}{dw}\right] + n(n+1)H = 0. \tag{15}$$

This is **Legendre's equation** (see Sec. 5.3), written out

(15′)
$$(1 - w^2)\frac{d^2H}{dw^2} - 2w\frac{dH}{dw} + n(n + 1)H = 0.$$

For integer $n = 0, 1, \cdots$ the Legendre polynomials

$$H = P_n(w) = P_n(\cos\phi) \qquad n = 0, 1, \cdots,$$

are solutions of Legendre's equation (15). We thus obtain the following two sequences of solution $u = GH$ of Laplace's equation (8), with constant A_n and B_n, where $n = 0, 1, \cdots,$

(16)
$$\text{(a)}\quad u_n(r, \phi) = A_n r^n P_n(\cos\phi), \qquad \text{(b)}\quad u_n^*(r, \phi) = \frac{B_n}{r^{n+1}} P_n(\cos\phi)$$

Use of Fourier–Legendre Series

Interior Problem: Potential Within the Sphere S. We consider a series of terms from (16a),

(17)
$$u(r, \phi) = \sum_{n=0}^{\infty} A_n r^n P_n(\cos\phi) \qquad (r \leqq R).$$

Since S is given by $r = R$, for (17) to satisfy the Dirichlet condition (9) on the sphere S, we must have

(18)
$$u(R, \phi) = \sum_{n=0}^{\infty} A_n R^n P_n(\cos\phi) = f(\phi);$$

that is, (18) must be the **Fourier–Legendre series** of $f(\phi)$. From (7) in Sec. 5.8 we get the coefficients

(19*)
$$A_n R^n = \frac{2n + 1}{2}\int_{-1}^{1} \tilde{f}(w)\, P_n(w)\, dw$$

where $\tilde{f}(w)$ denotes $f(\phi)$ as a function of $w = \cos\phi$. Since $dw = -\sin\phi\, d\phi$, and the limits of integration -1 and 1 correspond to $\phi = \pi$ and $\phi = 0$, respectively, we also obtain

(19)
$$A_n = \frac{2n + 1}{2R^n}\int_0^{\pi} f(\phi) P_n(\cos\phi)\sin\phi\, d\phi, \qquad n = 0, 1, \cdots.$$

If $f(\phi)$ and $f'(\phi)$ are piecewise continuous on the interval $0 \leqq \phi \leqq \pi$, then the series (17) with coefficients (19) solves our problem for points inside the sphere because it can be shown that under these continuity assumptions the series (17) with coefficients (19) gives the derivatives occurring in (8) by termwise differentiation, thus justifying our derivation.

Exterior Problem: Potential Outside the Sphere S. Outside the sphere we cannot use the functions u_n in (16a) because they do not satisfy (10). But we can use the u_n^* in (16b), which do satisfy (10) (but could not be used inside S; why?). Proceeding as before leads to the solution of the exterior problem

$$u(r, \phi) = \sum_{n=0}^{\infty} \frac{B_n}{r^{n+1}} P_n(\cos \phi) \qquad (r \geqq R) \tag{20}$$

satisfying (8), (9), (10), with coefficients

$$B_n = \frac{2n+1}{2} R^{n+1} \int_0^{\pi} f(\phi) P_n(\cos \phi) \sin \phi \, d\phi. \tag{21}$$

The next example illustrates all this for a sphere of radius 1 consisting of two hemispheres that are separated by a small strip of insulating material along the equator, so that these hemispheres can be kept at different potentials (110 V and 0 V).

EXAMPLE 1 Spherical Capacitor

Find the potential inside and outside a spherical capacitor consisting of two metallic hemispheres of radius 1 ft separated by a small slit for reasons of insulation, if the upper hemisphere is kept at 110 V and the lower is grounded (Fig. 313).

Solution. The given boundary condition is (recall Fig. 312)

$$f(\phi) = \begin{cases} 110 & \text{if} \quad 0 \leqq \phi < \pi/2 \\ 0 & \text{if} \quad \pi/2 < \phi \leqq \pi. \end{cases}$$

Since $R = 1$, we thus obtain from (19)

$$A_n = \frac{2n+1}{2} \cdot 110 \int_0^{\pi/2} P_n(\cos \phi) \sin \phi \, d\phi$$

$$= \frac{2n+1}{2} \cdot 110 \int_0^{1} P_n(w) \, dw$$

where $w = \cos \phi$. Hence $P_n(\cos \phi) \sin \phi \, d\phi = -P_n(w) \, dw$, we integrate from 1 to 0, and we finally get rid of the minus by integrating from 0 to 1. You can evaluate this integral by your CAS or continue by using (11) in Sec. 5.2, obtaining

$$A_n = 55(2n+1) \sum_{m=0}^{M} (-1)^m \frac{(2n-2m)!}{2^n m!(n-m)!(n-2m)!} \int_0^1 w^{n-2m} \, dw$$

where $M = n/2$ for even n and $M = (n-1)/2$ for odd n. The integral equals $1/(n - 2m + 1)$. Thus

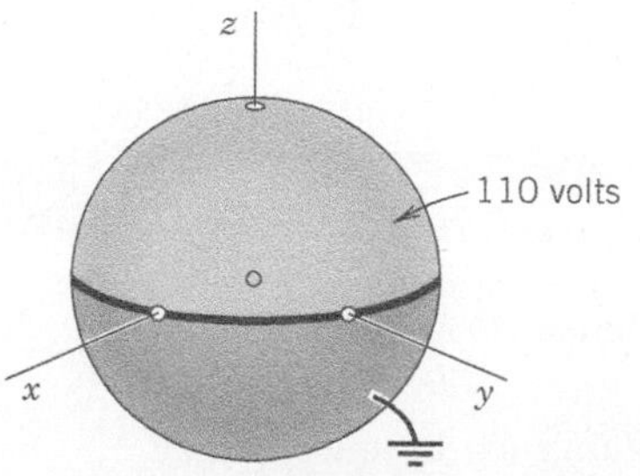

Fig. 313. Spherical capacitor in Example 1

$$(22) \qquad A_n = \frac{55(2n+1)}{2^n} \sum_{m=0}^{M} (-1)^m \frac{(2n-2m)!}{m!(n-m)!(n-2m+1)!}.$$

Taking $n = 0$, we get $A_0 = 55$ (since $0! = 1$). For $n = 1, 2, 3, \cdots$ we get

$$A_1 = \frac{165}{2} \cdot \frac{2!}{0!1!2!} = \frac{165}{2},$$

$$A_2 = \frac{275}{4}\left(\frac{4!}{0!2!3!} - \frac{2!}{1!1!1!}\right) = 0,$$

$$A_3 = \frac{385}{8}\left(\frac{6!}{0!3!4!} - \frac{4!}{1!2!2!}\right) = -\frac{385}{8}, \qquad \text{etc.}$$

Hence the ***potential* (17) *inside the sphere*** is (since $P_0 = 1$)

$$(23) \qquad u(r, \phi) = 55 + \frac{165}{2} r P_1(\cos\phi) - \frac{385}{8} r^3 P_3(\cos\phi) + \cdots \qquad \text{(Fig. 314)}$$

with $P_1, P_3, \cdots$ given by (11′), Sec. 5.21. Since $R = 1$, we see from (19) and (21) in this section that $B_n = A_n$, and (20) thus gives the ***potential outside the sphere***

$$(24) \qquad u(r, \phi) = \frac{55}{r} + \frac{165}{2r^2} P_1(\cos\phi) - \frac{385}{8r^4} P_3(\cos\phi) + \cdots.$$

Partial sums of these series can now be used for computing approximate values of the inner and outer potential. Also, it is interesting to see that far away from the sphere the potential is approximately that of a point charge, namely, $55/r$. (Compare with Theorem 3 in Sec. 9.7.) ■

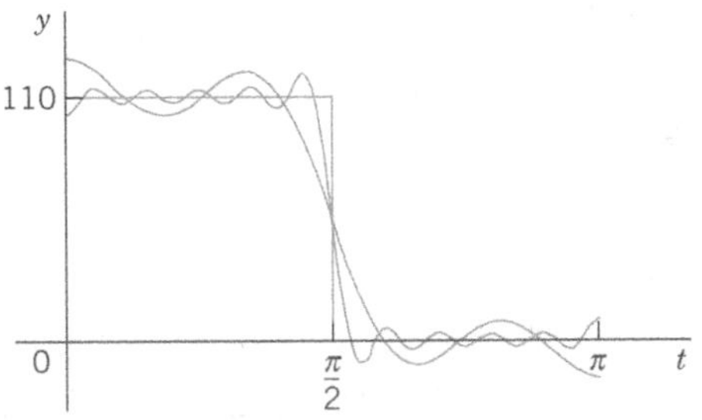

Fig. 314. Partial sums of the first 4, 6, and 11 nonzero terms of (23) for $r = R = 1$

EXAMPLE 2 Simpler Cases. Help with Problems

The technicalities encountered in cases that are similar to the one shown in Example 1 can often be avoided. For instance, find the potential inside the sphere S: $r = R = 1$ when S is kept at the potential $f(\phi) = \cos 2\phi$. (Can you see the potential on S? What is it at the North Pole? The equator? The South Pole?)

Solution. $w = \cos\phi$, $\cos 2\phi = 2\cos^2\phi - 1 = 2w^2 - 1 = \frac{4}{3}P_2(w) - \frac{1}{3} = \frac{4}{3}(\frac{3}{2}w^2 - \frac{1}{2}) - \frac{1}{3}$. Hence the potential in the interior of the sphere is

$$u = \tfrac{4}{3}r^2 P_2(w) - \tfrac{1}{3} = \tfrac{4}{3}r^2 P_2(\cos\phi) - \tfrac{1}{3} = \tfrac{2}{3}r^2(3\cos^2\phi - 1) - \tfrac{1}{3}.$$ ■

PROBLEM SET 12.11

1. **Spherical coordinates.** Derive (7) from $\nabla^2 u$ in spherical coordinates.
2. **Cylindrical coordinates.** Verify (5) by transforming $\nabla^2 u$ back into Cartesian coordinates.
3. Sketch $P_n(\cos\theta)$, $0 \leqq \theta \leqq 2\pi$, for $n = 0, 1, 2$. (Use (11′) in Sec. 5.2.)
4. **Zero surfaces.** Find the surfaces on which u_1, u_2, u_3 in (16) are **zero**.

5. CAS PROBLEM. Partial Sums. In Example 1 in the text verify the values of A_0, A_1, A_2, A_3 and compute $A_4, \cdots, A_{10}$. Try to find out graphically how well the corresponding partial sums of (23) approximate the given boundary function.

6. CAS EXPERIMENT. Gibbs Phenomenon. Study the Gibbs phenomenon in Example 1 (Fig. 314) graphically.

7. Verify that u_n and u_n^* in (16) are solutions of (8).

8–15 POTENTIALS DEPENDING ONLY ON *r*

8. Dimension 3. Verify that the potential $u = c/r$, $r = \sqrt{x^2 + y^2 + z^2}$ satisfies Laplace's equation in spherical coordinates.

9. Spherical symmetry. Show that the only solution of Laplace's equation depending only on $r = \sqrt{x^2 + y^2 + z^2}$ is $u = c/r + k$ with constant c and k.

10. Cylindrical symmetry. Show that the only solution of Laplace's equation depending only on $r = \sqrt{x^2 + y^2}$ is $u = c \ln r + k$.

11. Verification. Substituting $u(r)$ with r as in Prob. 9 into $u_{xx} + u_{yy} + u_{zz} = 0$, verify that $u'' + 2u'/r = 0$, in agreement with (7).

12. Dirichlet problem. Find the electrostatic potential between coaxial cylinders of radii $r_1 = 2$ cm and $r_2 = 4$ cm kept at the potentials $U_1 = 220$ V and $U_2 = 140$ V, respectively.

13. Dirichlet problem. Find the electrostatic potential between two concentric spheres of radii $r_1 = 2$ cm and $r_2 = 4$ cm kept at the potentials $U_1 = 220$V and $U_2 = 140$ V, respectively. Sketch and compare the equipotential lines in Probs. 12 and 13. Comment.

14. Heat problem. If the surface of the ball $r^2 = x^2 + y^2 + z^2 \leqq R^2$ is kept at temperature zero and the initial temperature in the ball is $f(r)$, show that the temperature $u(r, t)$ in the ball is a solution of $u_t = c^2(u_{rr} + 2u_r/r)$ satisfying the conditions $u(R, t) = 0$, $u(r, 0) = f(r)$. Show that setting $v = ru$ gives $v_t = c^2 v_{rr}$, $v(R, t) = 0$, $v(r, 0) = rf(r)$. Include the condition $v(0, t) = 0$ (which holds because u must be bounded at $r = 0$), and solve the resulting problem by separating variables.

15. What are the analogs of Probs. 12 and 13 in heat conduction?

16–20 BOUNDARY VALUE PROBLEMS IN SPHERICAL COORDINATES r, θ, ϕ

Find the potential in the interior of the sphere $r = R = 1$ if the interior is free of charges and the potential on the sphere is

16. $f(\phi) = \cos \phi$

17. $f(\phi) = 1$

18. $f(\phi) = 1 - \cos^2 \phi$

19. $f(\phi) = \cos 2\phi$

20. $f(\phi) = 10 \cos^3 \phi - 3 \cos^2 \phi - 5 \cos \phi - 1$

21. Point charge. Show that in Prob. 17 the potential exterior to the sphere is the same as that of a point charge at the origin.

22. Exterior potential. Find the potentials exterior to the sphere in Probs. 16 and 19.

23. Plane intersections. Sketch the intersections of the equipotential surfaces in Prob. 16 with xz-plane.

24. TEAM PROJECT. Transmission Line and Related PDEs. Consider a long cable or telephone wire (Fig. 315) that is imperfectly insulated, so that leaks occur along the entire length of the cable. The source S of the current $i(x, t)$ in the cable is at $x = 0$, the receiving end T at $x = l$. The current flows from S to T and through the load, and returns to the ground. Let the constants R, L, C, and G denote the resistance, inductance, capacitance to ground, and conductance to ground, respectively, of the cable per unit length.

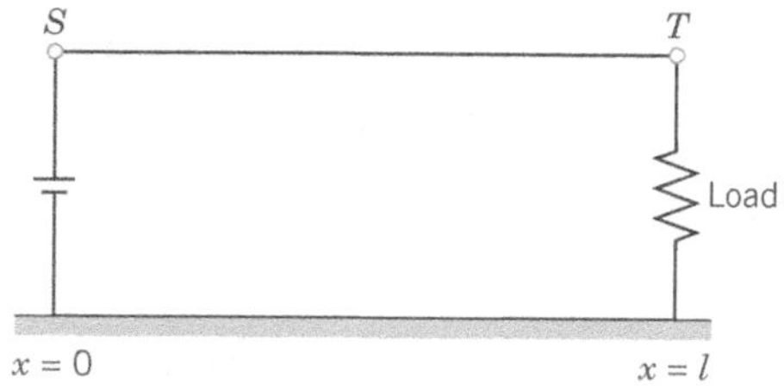

Fig. 315. Transmission line

(a) Show that (**"first transmission line equation"**)

$$-\frac{\partial u}{\partial x} = Ri + L\frac{\partial i}{\partial t}$$

where $u(x, t)$ is the potential in the cable. *Hint:* Apply Kirchhoff's voltage law to a small portion of the cable between x and $x + \Delta x$ (difference of the potentials at x and $x + \Delta x$ = resistive drop + inductive drop).

(b) Show that for the cable in (a) (**"second transmission line equation"**),

$$-\frac{\partial i}{\partial x} = Gu + C\frac{\partial u}{\partial t}.$$

Hint: Use Kirchhoff's current law (difference of the currents at x and $x + \Delta x$ = loss due to leakage to ground + capacitive loss).

(c) Second-order PDEs. Show that elimination of i or u from the transmission line equations leads to

$$u_{xx} = LCu_{tt} + (RC + GL)u_t + RGu,$$
$$i_{xx} = LCi_{tt} + (RC + GL)i_t + RGi.$$

(d) Telegraph equations. For a submarine cable, G is negligible and the frequencies are low. Show that this leads to the so-called *submarine cable equations* or **telegraph equations**

$$u_{xx} = RCu_t, \qquad i_{xx} = RCi_t.$$

Find the potential in a submarine cable with ends ($x = 0, x = l$) grounded and initial voltage distribution $U_0 = \text{const}$.

(e) High-frequency line equations. Show that in the case of alternating currents of high frequencies the equations in (c) can be approximated by the so-called **high-frequency line equations**

$$u_{xx} = LCu_{tt}, \qquad i_{xx} = LCi_{tt}.$$

Solve the first of them, assuming that the initial potential is

$$U_0 \sin(\pi x/l),$$

and $u_t(x, 0) = 0$ and $u = 0$ at the ends $x = 0$ and $x = l$ for all t.

25. Reflection in a sphere. Let r, θ, ϕ be spherical coordinates. If $u(r, \theta, \phi)$ satisfies $\nabla^2 u = 0$, show that $v(r, \theta, \phi) = u(1/r, \theta, \phi)/r$ satisfies $\nabla^2 v = 0$.

12.12 Solution of PDEs by Laplace Transforms

Readers familiar with Chap. 6 may wonder whether Laplace transforms can also be used for solving *partial* differential equations. The answer is yes, particularly if one of the independent variables ranges over the positive axis. The steps to obtain a solution are similar to those in Chap. 6. For a PDE in two variables they are as follows.

1. Take the Laplace transform with respect to one of the two variables, usually t. This gives an ***ODE for the transform*** of the unknown function. This is so since the derivatives of this function with respect to the other variable slip into the transformed equation. The latter also incorporates the given boundary and initial conditions.
2. Solving that ODE, obtain the transform of the unknown function.
3. Taking the inverse transform, obtain the solution of the given problem.

If the coefficients of the given equation do not depend on t, the use of Laplace transforms will simplify the problem.

We explain the method in terms of a typical example.

EXAMPLE 1 Semi-Infinite String

Find the displacement $w(x, t)$ of an elastic string subject to the following conditions. (We write w since we need u to denote the unit step function.)

(i) The string is initially at rest on the x-axis from $x = 0$ to ∞ (*"semi-infinite string"*).

(ii) For $t > 0$ the left end of the string ($x = 0$) is moved in a given fashion, namely, according to a single sine wave

$$w(0, t) = f(t) = \begin{cases} \sin t & \text{if } 0 \leqq t \leqq 2\pi \\ 0 & \text{otherwise} \end{cases} \qquad \text{(Fig. 316).}$$

(iii) Furthermore, $\lim_{x \to \infty} w(x, t) = 0$ for $t \geqq 0$.

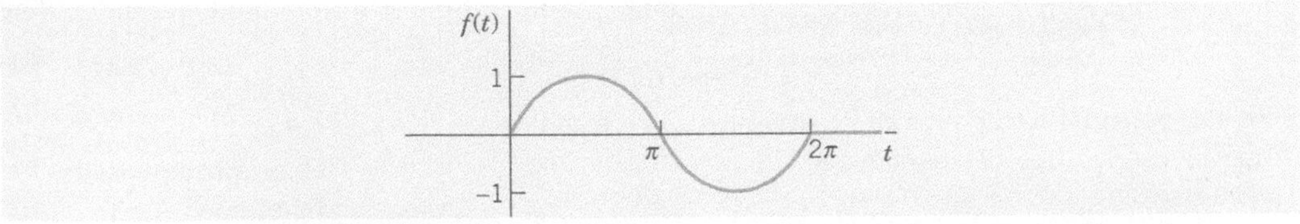

Fig. 316. Motion of the left end of the string in Example 1 as a function of time t

Of course there is no infinite string, but our model describes a long string or rope (of negligible weight) with its right end fixed far out on the x-axis.

Solution. We have to solve the wave equation (Sec. 12.2)

$$\frac{\partial^2 w}{\partial t^2} = c^2 \frac{\partial^2 w}{\partial x^2}, \qquad c^2 = \frac{T}{\rho} \tag{1}$$

for positive x and t, subject to the "boundary conditions"

$$w(0, t) = f(t), \qquad \lim_{x \to \infty} w(x, t) = 0 \qquad (t \geqq 0) \tag{2}$$

with f as given above, and the initial conditions

$$\text{(a)}\quad w(x, 0) = 0, \qquad \text{(b)}\quad w_t(x, 0) = 0. \tag{3}$$

We take the Laplace transform ***with respect to t***. By (2) in Sec. 6.2,

$$\mathcal{L}\left\{\frac{\partial^2 w}{\partial t^2}\right\} = s^2\mathcal{L}\{w\} - sw(x, 0) - w_t(x, 0) = c^2\mathcal{L}\left\{\frac{\partial^2 w}{\partial x^2}\right\}.$$

The expression $-sw(x, 0) - w_t(x, 0)$ drops out because of (3). On the right we assume that we may interchange integration and differentiation. Then

$$\mathcal{L}\left\{\frac{\partial^2 w}{\partial x^2}\right\} = \int_0^\infty e^{-st}\frac{\partial^2 w}{\partial x^2}\,dt = \frac{\partial^2}{\partial x^2}\int_0^\infty e^{-st}w(x, t)\,dt = \frac{\partial^2}{\partial x^2}\mathcal{L}\{w(x, t)\}.$$

Writing $W(x, s) = \mathcal{L}\{w(x, t)\}$, we thus obtain

$$s^2 W = c^2\frac{\partial^2 W}{\partial x^2}, \qquad \text{thus} \qquad \frac{\partial^2 W}{\partial x^2} - \frac{s^2}{c^2}W = 0.$$

Since this equation contains only a derivative with respect to x, it may be regarded as an ***ordinary differential equation*** for $W(x, s)$ considered as a function of x. A general solution is

$$W(x, s) = A(s)e^{sx/c} + B(s)e^{-sx/c}. \tag{4}$$

From (2) we obtain, writing $F(s) = \mathcal{L}\{f(t)\}$,

$$W(0, s) = \mathcal{L}\{w(0, t)\} = \mathcal{L}\{f(t)\} = F(s).$$

Assuming that we can interchange integration and taking the limit, we have

$$\lim_{x \to \infty} W(x, s) = \lim_{x \to \infty}\int_0^\infty e^{-st}w(x, t)\,dt = \int_0^\infty e^{-st}\lim_{x \to \infty} w(x, t)\,dt = 0.$$

This implies $A(s) = 0$ in (4) because $c > 0$, so that for every fixed positive s the function $e^{sx/c}$ increases as x increases. Note that we may assume $s > 0$ since a Laplace transform generally exists for *all* s greater than some fixed k (Sec. 6.2). Hence we have

$$W(0, s) = B(s) = F(s),$$

so that (4) becomes

$$W(x, s) = F(s)e^{-sx/c}.$$

From the second shifting theorem (Sec. 6.3) with $a = x/c$ we obtain the inverse transform

$$w(x, t) = f\left(t - \frac{x}{c}\right)u\left(t - \frac{x}{c}\right) \qquad \text{(Fig. 317)} \tag{5}$$

that is,

$$w(x, t) = \sin\left(t - \frac{x}{c}\right) \quad \text{if} \quad \frac{x}{c} < t < \frac{x}{c} + 2\pi \quad \text{or} \quad ct > x > (t - 2\pi)c$$

and zero otherwise. This is a single sine wave traveling to the right with speed c. Note that a point x remains at rest until $t = x/c$, the time needed to reach that x if one starts at $t = 0$ (start of the motion of the left end) and travels with speed c. The result agrees with our physical intuition. Since we proceeded formally, we must verify that (5) satisfies the given conditions. We leave this to the student. ■

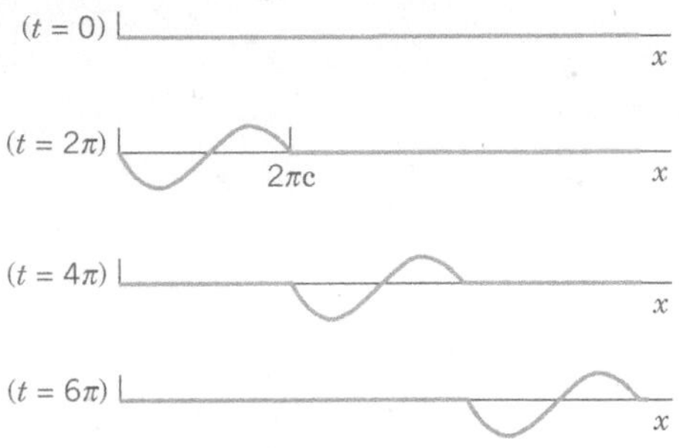

Fig. 317. Traveling wave in Example 1

We have reached the end of Chapter 12, in which we concentrated on the most important partial differential equations (PDEs) in physics and engineering. We have also reached the end of Part C on Fourier Analysis and PDEs.

Outlook

We have seen that PDEs underlie the modeling process of various important engineering application. Indeed, PDEs are the subject of many ongoing research projects.

Numerics for PDEs follows in Secs. 21.4–21.7, which, by design for greater flexibility in teaching, are independent of the other sections in Part E on numerics.

In the next part, that is, Part D on **complex analysis**, we turn to an area of a different nature that is also highly important to the engineer. The rich vein of examples and problems will signify this. It is of note that Part D includes another approach to the two-dimensional **Laplace equation** with applications, as shown in Chap. 18.

PROBLEM SET 12.12

1. Verify the solution in Example 1. What traveling wave do we obtain in Example 1 for a nonterminating sinusoidal motion of the left end starting at $t = 2\pi$?

2. Sketch a figure similar to Fig. 317 when $c = 1$ and $f(x)$ is "triangular," say, $f(x) = x$ if $0 < x < \frac{1}{2}$, $f(x) = 1 - x$ if $\frac{1}{2} < x < 1$ and 0 otherwise.

3. How does the speed of the wave in Example 1 of the text depend on the tension and on the mass of the string?

4–8 SOLVE BY LAPLACE TRANSFORMS

4. $\dfrac{\partial w}{\partial x} + x\dfrac{\partial w}{\partial t} = x, \quad w(x, 0) = 1, \quad w(0, t) = 1$

5. $x\dfrac{\partial w}{\partial x} + \dfrac{\partial w}{\partial t} = xt, \quad w(x, 0) = 0 \text{ if } x \geqq 0,$
$w(0, t) = 0 \text{ if } t \geqq 0$

6. $\dfrac{\partial w}{\partial x} + 2x\dfrac{\partial w}{\partial t} = 2x, \quad w(x, 0) = 1, \quad w(0, t) = 1$

7. Solve Prob. 5 by separating variables.

8. $\dfrac{\partial^2 w}{\partial x^2} = 100\dfrac{\partial^2 w}{\partial t^2} + 100\dfrac{\partial w}{\partial t} + 25w,$

$w(x, 0) = 0 \text{ if } x \geqq 0, \quad w_t(x, 0) = 0 \text{ if } t \geqq 0,$
$w(0, t) = \sin t \text{ if } t \geqq 0$

9–12 HEAT PROBLEM

Find the temperature $w(x, t)$ in a semi-infinite laterally insulated bar extending from $x = 0$ along the x-axis to infinity, assuming that the initial temperature is 0, $w(x, t) \to 0$ as $x \to \infty$ for every fixed $t \geqq 0$, and $w(0, t) = f(t)$. Proceed as follows.

9. Set up the model and show that the Laplace transform leads to

$$sW = c^2 \frac{\partial^2 W}{\partial x^2} \qquad (W = \mathcal{L}\{w\})$$

and

$$W = F(s)e^{-\sqrt{s}x/c} \qquad (F = \mathcal{L}\{f\}).$$

10. Applying the convolution theorem, show that in Prob. 9,

$$w(x, t) = \frac{x}{2c\sqrt{\pi}} \int_0^t f(t - \tau)\tau^{-3/2} e^{-x^2/(4c^2\tau)} d\tau.$$

11. Let $w(0, t) = f(t) = u(t)$ (Sec. 6.3). Denote the corresponding w, W, and F by w_0, W_0, and F_0. Show that then in Prob. 10,

$$w_0(x, t) = \frac{x}{2c\sqrt{\pi}} \int_0^t \tau^{-3/2} e^{-x^2/(4c^2\tau)} d\tau$$
$$= 1 - \operatorname{erf}\left(\frac{x}{2c\sqrt{t}}\right)$$

with the error function erf as defined in Problem Set 12.7.

12. Duhamel's formula.[4] Show that in Prob. 11,

$$W_0(x, s) = \frac{1}{s} e^{-\sqrt{s}x/c}$$

and the convolution theorem gives *Duhamel's formula*

$$W(x, t) = \int_0^t f(t - \tau) \frac{\partial w_0}{\partial \tau} d\tau.$$

CHAPTER 12 REVIEW QUESTIONS AND PROBLEMS

1. For what kinds of problems will modeling lead to an ODE? To a PDE?
2. Mention some of the basic physical principles or laws that will give a PDE in modeling.
3. State three or four of the most important PDEs and their main applications.
4. What is "separating variables" in a PDE? When did we apply it twice in succession?
5. What is d'Alembert's solution method? To what PDE does it apply?
6. What role did Fourier series play in this chapter? Fourier integrals?
7. When and why did Legendre's equation occur? Bessel's equation?
8. What are the eigenfunctions and their frequencies of the vibrating string? Of the vibrating membrane?
9. What do you remember about types of PDEs? Normal forms? Why is this important?
10. When did we use polar coordinates? Cylindrical coordinates? Spherical coordinates?
11. Explain mathematically (not physically) why we got exponential functions in separating the heat equation, but not for the wave equation.
12. Why and where did the error function occur?
13. How do problems for the wave equation and the heat equation differ regarding additional conditions?
14. Name and explain the three kinds of boundary conditions for Laplace's equation.
15. Explain how the Laplace transform applies to PDEs.

16–18 Solve for $u = u(x, y)$:

16. $u_{xx} + 25u = 0$
17. $u_{yy} + u_y - 6u = 18$
18. $u_{xx} + u_x = 0, \quad u(0, y) = f(y), \quad u_x(0, y) = g(y)$

19–21 NORMAL FORM

Transform to normal form and solve:

19. $u_{xy} = u_{yy}$
20. $u_{xx} + 6u_{xy} + 9u_{yy} = 0$
21. $u_{xx} - 4u_{yy} = 0$

22–24 VIBRATING STRING

Find and sketch or graph (as in Fig. 288 in Sec. 12.3) the deflection $u(x, t)$ of a vibrating string of length π, extending from $x = 0$ to $x = \pi$, and $c^2 = T/\rho = 4$ starting with velocity zero and deflection:

22. $\sin 4x$
23. $\sin^3 x$
24. $\frac{1}{2}\pi - |x - \frac{1}{2}\pi|$

[4] JEAN–MARIE CONSTANT DUHAMEL (1797–1872), French mathematician.

25–27 HEAT

Find the temperature distribution in a laterally insulated thin copper bar ($c^2 = K/(\sigma\rho) = 1.158\ \text{cm}^2/\text{sec}$) of length 100 cm and constant cross section with endpoints at $x = 0$ and 100 kept at 0°C and initial temperature:

25. $\sin 0.01\pi x$ **26.** $50 - |50 - x|$

27. $\sin^3 0.01\pi x$

28–30 ADIABATIC CONDITIONS

Find the temperature distribution in a laterally insulated bar of length π with $c^2 = 1$ for the adiabatic boundary condition (see Problem Set 12.6) and initial temperature:

28. $3x^2$ **29.** $100 \cos 2x$

30. $2\pi - 4|x - \frac{1}{2}\pi|$

31–32 TEMPERATURE IN A PLATE

31. Let $f(x, y) = u(x, y, 0)$ be the initial temperature in a thin square plate of side π with edges kept at 0°C and faces perfectly insulated. Separating variables, obtain from $u_t = c^2\nabla^2 u$ the solution

$$u(x, y, t) = \sum_{m=1}^{\infty}\sum_{n=1}^{\infty} B_{mn} \sin mx \sin ny\, e^{-c^2(m^2+n^2)t}$$

where

$$B_{mn} = \frac{4}{\pi^2}\int_0^{\pi}\int_0^{\pi} f(x, y) \sin mx \sin ny\, dx\, dy.$$

32. Find the temperature in Prob. 31 if $f(x, y) = x(\pi - x)y(\pi - y)$.

33–37 MEMBRANES

Show that the following membranes of area 1 with $c^2 = 1$ have the frequencies of the fundamental mode as given (4-decimal values). Compare.

33. Circle: $\alpha_1/(2\sqrt{\pi}) = 0.6784$

34. Square: $1/\sqrt{2} = 0.7071$

35. Rectangle with sides $1{:}2{:}\sqrt{5/8} = 0.7906$

36. Semicircle: $3.832/\sqrt{8\pi} = 0.7643$

37. **Quadrant** of circle: $\alpha_{21}/(4\sqrt{\pi}) = 0.7244$ ($\alpha_{21} = 5.13562 =$ first positive zero of J_2)

38–40 ELECTROSTATIC POTENTIAL

Find the potential in the following charge-free regions.

38. Between two concentric spheres of radii r_0 and r_1 kept at potentials u_0 and u_1, respectively.

39. Between two coaxial circular cylinders of radii r_0 and r_1 kept at the potentials u_0 and u_1, respectively. Compare with Prob. 38.

40. In the interior of a sphere of radius 1 kept at the potential $f(\phi) = \cos 3\phi + 3 \cos \phi$ (referred to our usual spherical coordinates).

SUMMARY OF CHAPTER 12
Partial Differential Equations (PDEs)

Whereas ODEs (Chaps. 1–6) serve as models of problems involving only *one* independent variable, problems involving *two or more* independent variables (space variables or time t and one or several space variables) lead to PDEs. This accounts for the enormous importance of PDEs to the engineer and physicist. Most important are:

(1) $u_{tt} = c^2 u_{xx}$ One-dimensional wave equation (Secs. 12.2–12.4)

(2) $u_{tt} = c^2(u_{xx} + u_{yy})$ Two-dimensional wave equation (Secs. 12.8–12.10)

(3) $u_t = c^2 u_{xx}$ One-dimensional heat equation (Secs. 12.5, 12.6, 12.7)

(4) $\nabla^2 u = u_{xx} + u_{yy} = 0$ Two-dimensional Laplace equation (Secs. 12.6, 12.10)

(5) $\nabla^2 u = u_{xx} + u_{yy} + u_{zz} = 0$ Three-dimensional Laplace equation (Sec. 12.11).

Equations (1) and (2) are hyperbolic, (3) is parabolic, (4) and (5) are elliptic.

In practice, one is interested in obtaining the solution of such an equation in a given region satisfying given additional conditions, such as **initial conditions** (conditions at time $t = 0$) or **boundary conditions** (prescribed values of the solution u or some of its derivatives on the boundary surface S, or boundary curve C, of the region) or both. For (1) and (2) one prescribes two initial conditions (initial displacement and initial velocity). For (3) one prescribes the initial temperature distribution. For (4) and (5) one prescribes a boundary condition and calls the resulting problem a (see Sec. 12.6)

Dirichlet problem if u is prescribed on S,
Neumann problem if $u_n = \partial u/\partial n$ is prescribed on S,
Mixed problem if u is prescribed on one part of S and u_n on the other.

A general method for solving such problems is the method of **separating variables** or **product method**, in which one assumes solutions in the form of products of functions each depending on one variable only. Thus equation (1) is solved by setting $u(x, t) = F(x)G(t)$; see Sec. 12.3; similarly for (3) (see Sec. 12.6). Substitution into the given equation yields ***ordinary*** differential equations for F and G, and from these one gets infinitely many solutions $F = F_n$ and $G = G_n$ such that the corresponding functions

$$u_n(x, t) = F_n(x)G_n(t)$$

are solutions of the PDE satisfying the given boundary conditions. These are the **eigenfunctions** of the problem, and the corresponding **eigenvalues** determine the frequency of the vibration (or the rapidity of the decrease of temperature in the case of the heat equation, etc.). To satisfy also the initial condition (or conditions), one must consider infinite series of the u_n, whose coefficients turn out to be the Fourier coefficients of the functions f and g representing the given initial conditions (Secs. 12.3, 12.6). Hence **Fourier series** (and *Fourier integrals*) are of basic importance here (Secs. 12.3, 12.6, 12.7, 12.9).

Steady-state problems are problems in which the solution does not depend on time t. For these, the heat equation $u_t = c^2\nabla^2 u$ becomes the Laplace equation.

Before solving an initial or boundary value problem, one often transforms the PDE into coordinates in which the boundary of the region considered is given by simple formulas. Thus in polar coordinates given by $x = r\cos\theta$, $y = r\sin\theta$, the **Laplacian** becomes (Sec. 12.11)

$$\nabla^2 u = u_{rr} + \frac{1}{r}u_r + \frac{1}{r^2}u_{\theta\theta}; \tag{6}$$

for spherical coordinates see Sec. 12.10. If one now separates the variables, one gets ***Bessel's equation*** from (2) and (6) (vibrating circular membrane, Sec. 12.10) and ***Legendre's equation*** from (5) transformed into spherical coordinates (Sec. 12.11).

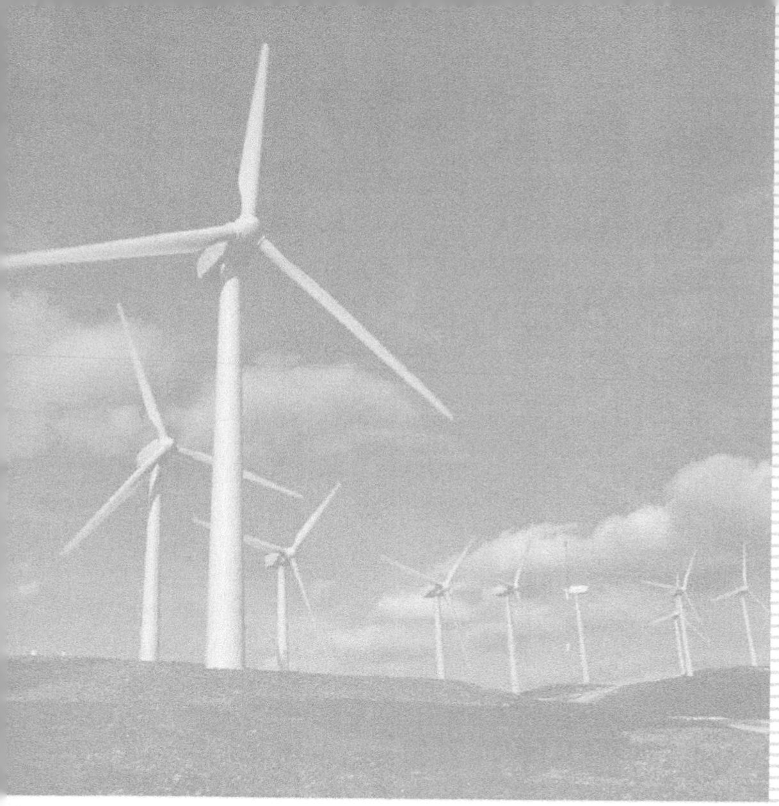

PART D

Complex Analysis

Complex analysis has many applications in heat conduction, fluid flow, electrostatics, and in other areas. It extends the familiar "real calculus" to "complex calculus" by introducing complex numbers and functions. While many ideas carry over from calculus to complex analysis, there is a marked difference between the two. For example, analytic functions, which are the "good functions" (differentiable in some domain) of complex analysis, have derivatives of all orders. This is in contrast to calculus, where real-valued functions of real variables may have derivatives only up to a certain order. Thus, in certain ways, problems that are difficult to solve in real calculus may be much easier to solve in complex analysis. Complex analysis is important in applied mathematics for three main reasons:

1. Two-dimensional potential problems can be modeled and solved by methods of analytic functions. This reason is the real and imaginary parts of analytic functions satisfy Laplace's equation in two real variables.

2. Many difficult integrals (real or complex) that appear in applications can be solved quite elegantly by complex integration.

3. Most functions in engineering mathematics are analytic functions, and their study as functions of a complex variable leads to a deeper understanding of their properties and to interrelations in complex that have no analog in real calculus.

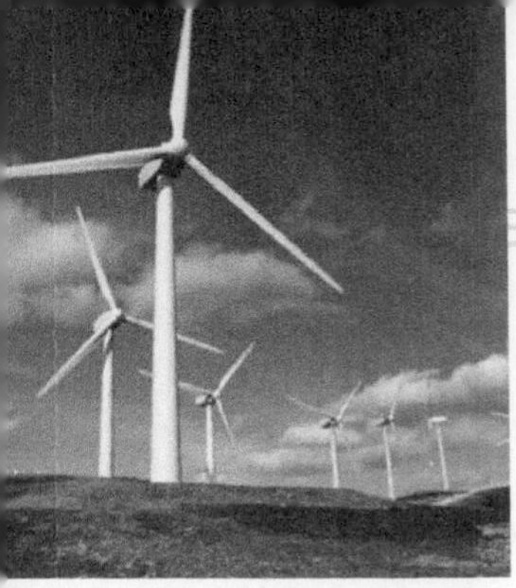

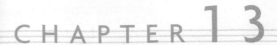

CHAPTER 13

Complex Numbers and Functions. Complex Differentiation

The transition from "real calculus" to "complex calculus" starts with a discussion of ***complex numbers*** and their geometric representation in the ***complex plane***. We then progress to ***analytic functions*** in Sec. 13.3. We desire functions to be analytic because these are the "useful functions" in the sense that they are differentiable in some domain and operations of complex analysis can be applied to them. The most important equations are therefore the Cauchy–Riemann equations in Sec. 13.4 because they allow a test of analyticity of such functions. Moreover, we show how the Cauchy–Riemann equations are related to the important ***Laplace equation***.

The remaining sections of the chapter are devoted to elementary complex functions (exponential, trigonometric, hyperbolic, and logarithmic functions). These generalize the familiar real functions of calculus. Detailed knowledge of them is an absolute necessity in practical work, just as that of their real counterparts is in calculus.

Prerequisite: Elementary calculus.
References and Answers to Problems: App. 1 Part D, App. 2.

13.1 Complex Numbers and Their Geometric Representation

The material in this section will most likely be familiar to the student and serve as a review.

Equations without *real* solutions, such as $x^2 = -1$ or $x^2 - 10x + 40 = 0$, were observed early in history and led to the introduction of complex numbers.[1] By definition, a **complex number** z is an ordered pair (x, y) of real numbers x and y, written

$$z = (x, y).$$

[1]First to use complex numbers for this purpose was the Italian mathematician GIROLAMO CARDANO (1501–1576), who found the formula for solving cubic equations. The term "complex number" was introduced by CARL FRIEDRICH GAUSS (see the footnote in Sec. 5.4), who also paved the way for a general use of complex numbers.

x is called the **real part** and y the **imaginary part** of z, written

$$x = \operatorname{Re} z, \qquad y = \operatorname{Im} z.$$

By definition, two complex numbers are **equal** if and only if their real parts are equal and their imaginary parts are equal.

(0, 1) is called the **imaginary unit** and is denoted by i,

$$i = (0, 1). \tag{1}$$

Addition, Multiplication. Notation $z = x + iy$

Addition of two complex numbers $z_1 = (x_1, y_1)$ and $z_2 = (x_2, y_2)$ is defined by

$$z_1 + z_2 = (x_1, y_1) + (x_2, y_2) = (x_1 + x_2, \quad y_1 + y_2). \tag{2}$$

Multiplication is defined by

$$z_1 z_2 = (x_1, y_1)(x_2, y_2) = (x_1 x_2 - y_1 y_2, \quad x_1 y_2 + x_2 y_1). \tag{3}$$

These two definitions imply that

$$(x_1, 0) + (x_2, 0) = (x_1 + x_2, 0)$$

and

$$(x_1, 0)(x_2, 0) = (x_1 x_2, 0)$$

as for real numbers x_1, x_2. Hence the complex numbers "***extend***" the real numbers. We can thus write

$$(x, 0) = x. \qquad \text{Similarly,} \qquad (0, y) = iy$$

because by (1), and the definition of multiplication, we have

$$iy = (0, 1)y = (0, 1)(y, 0) = (0 \cdot y - 1 \cdot 0, \quad 0 \cdot 0 + 1 \cdot y) = (0, y).$$

Together we have, by addition, $(x, y) = (x, 0) + (0, y) = x + iy$.

In practice, complex numbers $z = (x, y)$ are written

$$z = x + iy \tag{4}$$

or $z = x + yi$, e.g., $17 + 4i$ (instead of $i4$).

Electrical engineers often write j instead of i because they need i for the current.

If $x = 0$, then $z = iy$ and is called **pure imaginary**. Also, (1) and (3) give

$$i^2 = -1 \tag{5}$$

because, by the definition of multiplication, $i^2 = ii = (0, 1)(0, 1) = (-1, 0) = -1$.

For **addition** the standard notation (4) gives [see (2)]

$$(x_1 + iy_1) + (x_2 + iy_2) = (x_1 + x_2) + i(y_1 + y_2).$$

For **multiplication** the standard notation gives the following very simple recipe. Multiply each term by each other term and use $i^2 = -1$ when it occurs [see (3)]:

$$(x_1 + iy_1)(x_2 + iy_2) = x_1x_2 + ix_1y_2 + iy_1x_2 + i^2y_1y_2$$

$$= (x_1x_2 - y_1y_2) + i(x_1y_2 + x_2y_1).$$

This agrees with (3). And it shows that $x + iy$ is a more practical notation for complex numbers than (x, y).

If you know vectors, you see that (2) is vector addition, whereas the multiplication (3) has no counterpart in the usual vector algebra.

EXAMPLE 1   **Real Part, Imaginary Part, Sum and Product of Complex Numbers**

Let $z_1 = 8 + 3i$ and $z_2 = 9 - 2i$. Then $\operatorname{Re} z_1 = 8$, $\operatorname{Im} z_1 = 3$, $\operatorname{Re} z_2 = 9$, $\operatorname{Im} z_2 = -2$ and

$$z_1 + z_2 = (8 + 3i) + (9 - 2i) = 17 + i,$$

$$z_1z_2 = (8 + 3i)(9 - 2i) = 72 + 6 + i(-16 + 27) = 78 + 11i. \qquad \blacksquare$$

Subtraction, Division

Subtraction and **division** are defined as the inverse operations of addition and multiplication, respectively. Thus the **difference** $z = z_1 - z_2$ is the complex number z for which $z_1 = z + z_2$. Hence by (2),

(6)
$$z_1 - z_2 = (x_1 - x_2) + i(y_1 - y_2).$$

The **quotient** $z = z_1/z_2$ ($z_2 \neq 0$) is the complex number z for which $z_1 = zz_2$. If we equate the real and the imaginary parts on both sides of this equation, setting $z = x + iy$, we obtain $x_1 = x_2x - y_2y$, $y_1 = y_2x + x_2y$. The solution is

(7*)
$$z = \frac{z_1}{z_2} = x + iy, \qquad x = \frac{x_1x_2 + y_1y_2}{x_2^2 + y_2^2}, \qquad y = \frac{x_2y_1 - x_1y_2}{x_2^2 + y_2^2}.$$

The *practical rule* used to get this is by multiplying numerator and denominator of z_1/z_2 by $x_2 - iy_2$ and simplifying:

(7)
$$z = \frac{x_1 + iy_1}{x_2 + iy_2} = \frac{(x_1 + iy_1)(x_2 - iy_2)}{(x_2 + iy_2)(x_2 - iy_2)} = \frac{x_1x_2 + y_1y_2}{x_2^2 + y_2^2} + i\frac{x_2y_1 - x_1y_2}{x_2^2 + y_2^2}.$$

EXAMPLE 2   **Difference and Quotient of Complex Numbers**

For $z_1 = 8 + 3i$ and $z_2 = 9 - 2i$ we get $z_1 - z_2 = (8 + 3i) - (9 - 2i) = -1 + 5i$ and

$$\frac{z_1}{z_2} = \frac{8 + 3i}{9 - 2i} = \frac{(8 + 3i)(9 + 2i)}{(9 - 2i)(9 + 2i)} = \frac{66 + 43i}{81 + 4} = \frac{66}{85} + \frac{43}{85}i.$$

Check the division by multiplication to get $8 + 3i$. $\qquad \blacksquare$

Complex numbers satisfy the same commutative, associative, and distributive laws as real numbers (see the problem set).

Complex Plane

So far we discussed the algebraic manipulation of complex numbers. Consider the geometric representation of complex numbers, which is of great practical importance. We choose two perpendicular coordinate axes, the horizontal x-axis, called the **real axis**, and the vertical y-axis, called the **imaginary axis**. On both axes we choose the same unit of length (Fig. 318). This is called a **Cartesian coordinate system**.

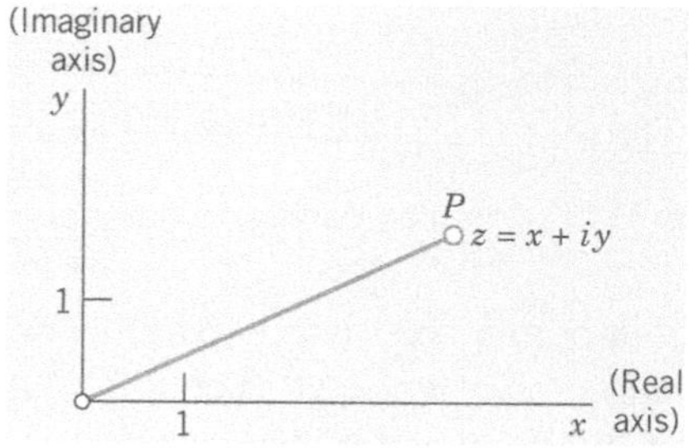

Fig. 318. The complex plane

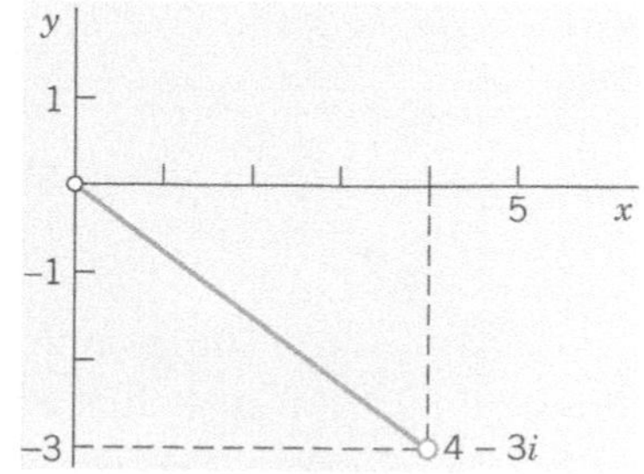

Fig. 319. The number $4 - 3i$ in the complex plane

We now plot a given complex number $z = (x, y) = x + iy$ as the point P with coordinates x, y. The xy-plane in which the complex numbers are represented in this way is called the **complex plane**.[2] Figure 319 shows an example.

Instead of saying "the point represented by z in the complex plane" we say briefly and simply "***the point*** z ***in the complex plane***." This will cause no misunderstanding.

Addition and subtraction can now be visualized as illustrated in Figs. 320 and 321.

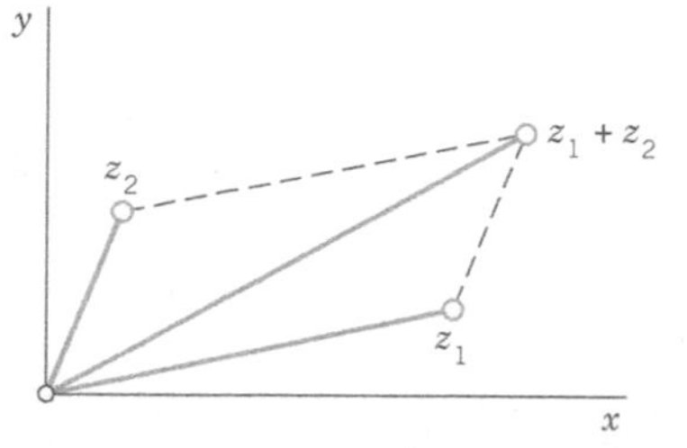

Fig. 320. Addition of complex numbers

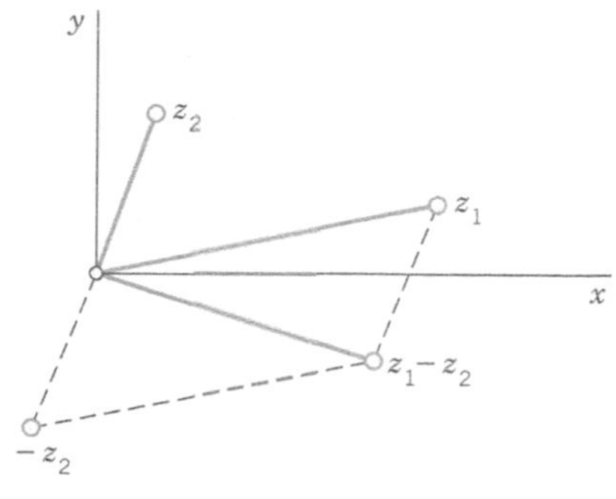

Fig. 321. Subtraction of complex numbers

[2]Sometimes called the **Argand diagram**, after the French mathematician JEAN ROBERT ARGAND (1768–1822), born in Geneva and later librarian in Paris. His paper on the complex plane appeared in 1806, nine years after a similar memoir by the Norwegian mathematician CASPAR WESSEL (1745–1818), a surveyor of the Danish Academy of Science.

Complex Conjugate Numbers

The complex conjugate $\bar{z}$ of a complex number $z = x + iy$ is defined by

$$\bar{z} = x - iy.$$

It is obtained geometrically by reflecting the point z in the real axis. Figure 322 shows this for $z = 5 + 2i$ and its conjugate $\bar{z} = 5 - 2i$.

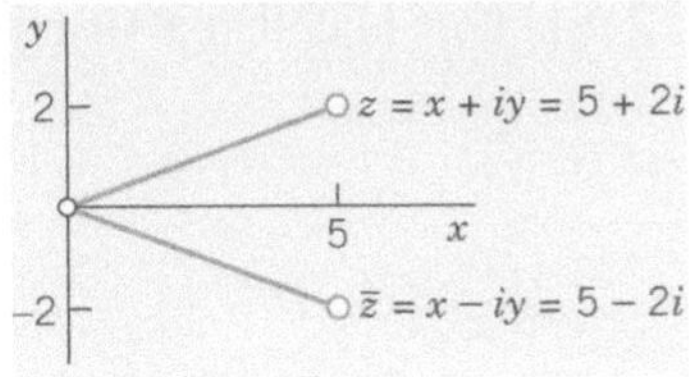

Fig. 322. Complex conjugate numbers

The complex conjugate is important because it permits us to switch from complex to real. Indeed, by multiplication, $z\bar{z} = x^2 + y^2$ (verify!). By addition and subtraction, $z + \bar{z} = 2x$, $z - \bar{z} = 2iy$. We thus obtain for the real part x and the imaginary part y (not iy!) of $z = x + iy$ the important formulas

(8)
$$\text{Re } z = x = \tfrac{1}{2}(z + \bar{z}), \qquad \text{Im } z = y = \frac{1}{2i}(z - \bar{z}).$$

If z is real, $z = x$, then $\bar{z} = z$ by the definition of $\bar{z}$, and conversely. Working with conjugates is easy, since we have

(9)
$$\overline{(z_1 + z_2)} = \bar{z}_1 + \bar{z}_2, \qquad \overline{(z_1 - z_2)} = \bar{z}_1 - \bar{z}_2,$$
$$\overline{(z_1 z_2)} = \bar{z}_1 \bar{z}_2, \qquad \overline{\left(\frac{z_1}{z_2}\right)} = \frac{\bar{z}_1}{\bar{z}_2}.$$

EXAMPLE 3   Illustration of (8) and (9)

Let $z_1 = 4 + 3i$ and $z_2 = 2 + 5i$. Then by (8),

$$\text{Im } z_1 = \frac{1}{2i}[(4 + 3i) - (4 - 3i)] = \frac{3i + 3i}{2i} = 3.$$

Also, the multiplication formula in (9) is verified by

$$\overline{(z_1 z_2)} = \overline{(4 + 3i)(2 + 5i)} = \overline{(-7 + 26i)} = -7 - 26i,$$
$$\bar{z}_1 \bar{z}_2 = (4 - 3i)(2 - 5i) = -7 - 26i. \qquad \blacksquare$$

PROBLEM SET 13.1

1. **Powers of i.** Show that $i^2 = -1$, $i^3 = -i$, $i^4 = 1$, $i^5 = i, \cdots$ and $1/i = -i$, $1/i^2 = -1$, $1/i^3 = i, \cdots$.

2. **Rotation.** Multiplication by i is geometrically a counterclockwise rotation through $\pi/2$ (90°). Verify this by graphing z and iz and the angle of rotation for $z = 1 + i$, $z = -1 + 2i$, $z = 4 - 3i$.

3. **Division.** Verify the calculation in (7). Apply (7) to $(26 - 18i)/(6 - 2i)$.

4. Law for conjugates. Verify (9) for $z_1 = -11 + 10i$, $z_2 = -1 + 4i$.

5. Pure imaginary number. Show that $z = x + iy$ is pure imaginary if and only if $\bar{z} = -z$.

6. Multiplication. If the product of two complex numbers is zero, show that at least one factor must be zero.

7. Laws of addition and multiplication. Derive the following laws for complex numbers from the corresponding laws for real numbers.

$$z_1 + z_2 = z_2 + z_1, \quad z_1 z_2 = z_2 z_1 \quad \textit{(Commutative laws)}$$

$$(z_1 + z_2) + z_3 = z_1 + (z_2 + z_3), \quad \textit{(Associative laws)}$$
$$(z_1 z_2) z_3 = z_1 (z_2 z_3)$$

$$z_1(z_2 + z_3) = z_1 z_2 + z_1 z_3 \quad \textit{(Distributive law)}$$

$$0 + z = z + 0 = z,$$

$$z + (-z) = (-z) + z = 0, \qquad z \cdot 1 = z.$$

8–15 COMPLEX ARITHMETIC

Let $z_1 = -2 + 11i$, $z_2 = 2 - i$. Showing the details of your work, find, in the form $x + iy$:

8. $z_1 z_2$, $\overline{(z_1 z_2)}$

9. $\text{Re}(z_1^2)$, $(\text{Re}\, z_1)^2$

10. $\text{Re}(1/z_2^2)$, $1/\text{Re}(z_2^2)$

11. $(z_1 - z_2)^2/16$, $(z_1/4 - z_2/4)^2$

12. z_1/z_2, z_2/z_1

13. $(z_1 + z_2)(z_1 - z_2)$, $z_1^2 - z_2^2$

14. $\bar{z}_1/\bar{z}_2$, $\overline{(z_1/z_2)}$

15. $4(z_1 + z_2)/(z_1 - z_2)$

16–20 Let $z = x + iy$. Showing details, find, in terms of x and y:

16. $\text{Im}(1/z)$, $\text{Im}(1/z^2)$

17. $\text{Re}\, z^4 - (\text{Re}\, z^2)^2$

18. $\text{Re}[(1 + i)^{16} z^2]$

19. $\text{Re}(z/\bar{z})$, $\text{Im}(z/\bar{z})$

20. $\text{Im}(1/\bar{z}^2)$

13.2 Polar Form of Complex Numbers. Powers and Roots

We gain further insight into the arithmetic operations of complex numbers if, in addition to the xy-coordinates in the complex plane, we also employ the usual polar coordinates r, θ defined by

$$x = r\cos\theta, \qquad y = r\sin\theta. \tag{1}$$

We see that then $z = x + iy$ takes the so-called **polar form**

$$z = r(\cos\theta + i\sin\theta). \tag{2}$$

r is called the **absolute value** or **modulus** of z and is denoted by $|z|$. Hence

$$|z| = r = \sqrt{x^2 + y^2} = \sqrt{z\bar{z}}. \tag{3}$$

Geometrically, $|z|$ is the distance of the point z from the origin (Fig. 323). Similarly, $|z_1 - z_2|$ is the distance between z_1 and z_2 (Fig. 324).

θ is called the **argument** of z and is denoted by arg z. Thus $\theta = \arg z$ and (Fig. 323)

$$\tan\theta = \frac{y}{x} \qquad (z \neq 0). \tag{4}$$

Geometrically, θ is the directed angle from the positive x-axis to OP in Fig. 323. Here, as in calculus, all ***angles are measured in radians and positive in the counterclockwise sense***.

For $z = 0$ this angle θ is undefined. (Why?) For a given $z \neq 0$ it is determined only up to integer multiples of 2π since cosine and sine are periodic with period 2π. But one often wants to specify a unique value of $\arg z$ of a given $z \neq 0$. For this reason one defines the **principal value** Arg z (with capital A!) of $\arg z$ by the double inequality

$$-\pi < \operatorname{Arg} z \leqq \pi. \tag{5}$$

Then we have Arg $z = 0$ for positive real $z = x$, which is practical, and Arg $z = \pi$ (not $-\pi$!) for negative real z, e.g., for $z = -4$. The principal value (5) will be important in connection with roots, the complex logarithm (Sec. 13.7), and certain integrals. Obviously, for a given $z \neq 0$, the other values of $\arg z$ are $\arg z = \operatorname{Arg} z \pm 2n\pi$ $(n = \pm 1, \pm 2, \cdots)$.

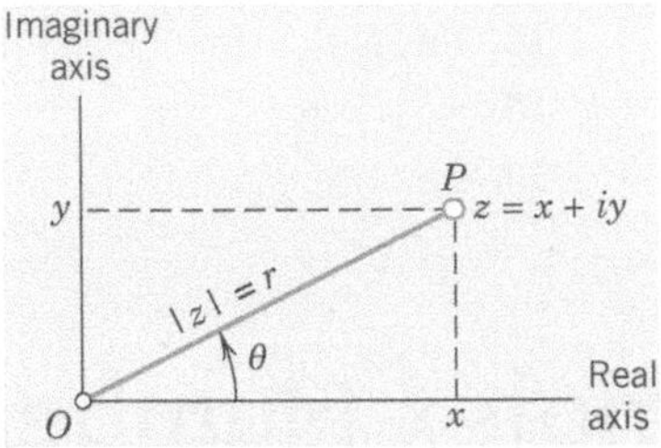

Fig. 323. Complex plane, polar form of a complex number

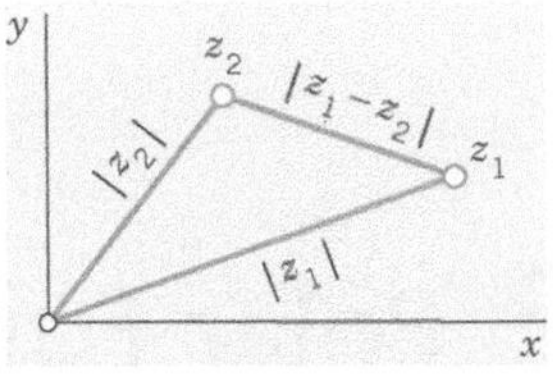

Fig. 324. Distance between two points in the complex plane

EXAMPLE 1 Polar Form of Complex Numbers. Principal Value Arg z

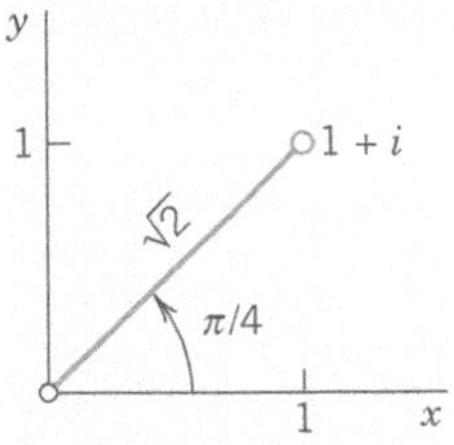

Fig. 325. Example 1

$z = 1 + i$ (Fig. 325) has the polar form $z = \sqrt{2}\,(\cos \frac{1}{4}\pi + i \sin \frac{1}{4}\pi)$. Hence we obtain

$$|z| = \sqrt{2}, \quad \arg z = \tfrac{1}{4}\pi \pm 2n\pi \ (n = 0, 1, \cdots), \qquad \text{and} \qquad \operatorname{Arg} z = \tfrac{1}{4}\pi \ \text{(the principal value)}.$$

Similarly, $z = 3 + 3\sqrt{3}i = 6\,(\cos \frac{1}{3}\pi + i \sin \frac{1}{3}\pi)$, $|z| = 6$, and Arg $z = \frac{1}{3}\pi$. ■

CAUTION! In using (4), we must pay attention to the quadrant in which z lies, since $\tan \theta$ has period π, so that the arguments of z and $-z$ have the same tangent. *Example:* for $\theta_1 = \arg(1 + i)$ and $\theta_2 = \arg(-1 - i)$ we have $\tan \theta_1 = \tan \theta_2 = 1$.

Triangle Inequality

Inequalities such as $x_1 < x_2$ make sense for *real* numbers, but not in complex because *there is no natural way of ordering complex numbers*. However, inequalities between absolute values (which are real!), such as $|z_1| < |z_2|$ (meaning that z_1 is closer to the origin than z_2) are of great importance. The daily bread of the complex analyst is the **triangle inequality**

$$|z_1 + z_2| \leqq |z_1| + |z_2| \tag{6}$$ (Fig. 326)

which we shall use quite frequently. This inequality follows by noting that the three points 0, z_1, and $z_1 + z_2$ are the vertices of a triangle (Fig. 326) with sides $|z_1|$, $|z_2|$, and $|z_1 + z_2|$, and one side cannot exceed the sum of the other two sides. A formal proof is left to the reader (Prob. 33). (The triangle degenerates if z_1 and z_2 lie on the same straight line through the origin.)

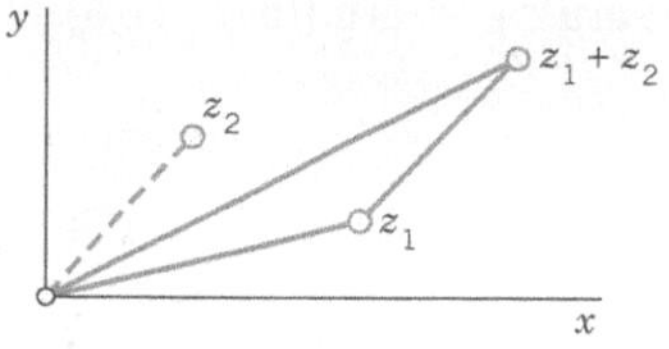

Fig. 326. Triangle inequality

By induction we obtain from (6) the **generalized triangle inequality**

$$|z_1 + z_2 + \cdots + z_n| \leqq |z_1| + |z_2| + \cdots + |z_n|; \tag{6*}$$

that is, *the absolute value of a sum cannot exceed the sum of the absolute values of the terms.*

EXAMPLE 2 Triangle Inequality

If $z_1 = 1 + i$ and $z_2 = -2 + 3i$, then (sketch a figure!)

$$|z_1 + z_2| = |-1 + 4i| = \sqrt{17} = 4.123 < \sqrt{2} + \sqrt{13} = 5.020. \qquad \blacksquare$$

Multiplication and Division in Polar Form

This will give us a "geometrical" understanding of multiplication and division. Let

$$z_1 = r_1(\cos\theta_1 + i\sin\theta_1) \qquad \text{and} \qquad z_2 = r_2(\cos\theta_2 + i\sin\theta_2).$$

Multiplication. By (3) in Sec. 13.1 the product is at first

$$z_1 z_2 = r_1 r_2[(\cos\theta_1\cos\theta_2 - \sin\theta_1\sin\theta_2) + i(\sin\theta_1\cos\theta_2 + \cos\theta_1\sin\theta_2)].$$

The addition rules for the sine and cosine [(6) in App. A3.1] now yield

$$z_1 z_2 = r_1 r_2[\cos(\theta_1 + \theta_2) + i\sin(\theta_1 + \theta_2)]. \tag{7}$$

Taking absolute values on both sides of (7), we see that *the absolute value of a product equals the* **product** *of the absolute values of the factors,*

$$|z_1 z_2| = |z_1||z_2|. \tag{8}$$

Taking arguments in (7) shows that *the argument of a product equals the* **sum** *of the arguments of the factors,*

$$\arg(z_1 z_2) = \arg z_1 + \arg z_2 \qquad \text{(up to multiples of } 2\pi\text{)}. \tag{9}$$

Division. We have $z_1 = (z_1/z_2)z_2$. Hence $|z_1| = |(z_1/z_2)z_2| = |z_1/z_2||z_2|$ and by division by $|z_2|$

$$\left|\frac{z_1}{z_2}\right| = \frac{|z_1|}{|z_2|} \qquad (z_2 \neq 0). \tag{10}$$

Similarly, $\arg z_1 = \arg [(z_1/z_2)z_2] = \arg (z_1/z_2) + \arg z_2$ and by subtraction of $\arg z_2$

(11) $$\arg \frac{z_1}{z_2} = \arg z_1 - \arg z_2 \qquad \text{(up to multiples of } 2\pi\text{)}.$$

Combining (10) and (11) we also have the analog of (7),

(12) $$\frac{z_1}{z_2} = \frac{r_1}{r_2} [\cos (\theta_1 - \theta_2) + i \sin (\theta_1 - \theta_2)].$$

To comprehend this formula, note that it is the polar form of a complex number of absolute value r_1/r_2 and argument $\theta_1 - \theta_2$. But these are the absolute value and argument of z_1/z_2, as we can see from (10), (11), and the polar forms of z_1 and z_2.

EXAMPLE 3 Illustration of Formulas (8)–(11)

Let $z_1 = -2 + 2i$ and $z_2 = 3i$. Then $z_1z_2 = -6 - 6i$, $z_1/z_2 = \frac{2}{3} + (\frac{2}{3})i$. Hence (make a sketch)

$$|z_1z_2| = 6\sqrt{2} = 3\sqrt{8} = |z_1||z_2|, \qquad |z_1/z_2| = 2\sqrt{2}/3 = |z_1|/|z_2|,$$

and for the arguments we obtain $\text{Arg } z_1 = 3\pi/4$, $\text{Arg } z_2 = \pi/2$,

$$\text{Arg } (z_1z_2) = -\frac{3\pi}{4} = \text{Arg } z_1 + \text{Arg } z_2 - 2\pi, \qquad \text{Arg}\left(\frac{z_1}{z_2}\right) = \frac{\pi}{4} = \text{Arg } z_1 - \text{Arg } z_2.$$ ■

EXAMPLE 4 Integer Powers of z. De Moivre's Formula

From (8) and (9) with $z_1 = z_2 = z$ we obtain by induction for $n = 0, 1, 2, \cdots$

(13) $$z^n = r^n (\cos n\theta + i \sin n\theta).$$

Similarly, (12) with $z_1 = 1$ and $z_2 = z^n$ gives (13) for $n = -1, -2, \cdots$. For $|z| = r = 1$, formula (13) becomes **De Moivre's formula**[3]

(13*) $$(\cos \theta + i \sin \theta)^n = \cos n\theta + i \sin n\theta.$$

We can use this to express $\cos n\theta$ and $\sin n\theta$ in terms of powers of $\cos \theta$ and $\sin \theta$. For instance, for $n = 2$ we have on the left $\cos^2 \theta + 2i \cos \theta \sin \theta - \sin^2 \theta$. Taking the real and imaginary parts on both sides of (13*) with $n = 2$ gives the familiar formulas

$$\cos 2\theta = \cos^2 \theta - \sin^2 \theta, \qquad \sin 2\theta = 2 \cos \theta \sin \theta.$$

This shows that *complex* methods often simplify the derivation of *real* formulas. Try $n = 3$. ■

Roots

If $z = w^n$ $(n = 1, 2, \cdots)$, then to each value of w there corresponds *one* value of z. We shall immediately see that, conversely, to a given $z \neq 0$ there correspond precisely n distinct values of w. Each of these values is called an ***n*th root** of z, and we write

[3] ABRAHAM DE MOIVRE (1667–1754), French mathematician, who pioneered the use of complex numbers in trigonometry and also contributed to probability theory (see Sec. 24.8).

$$w = \sqrt[n]{z}. \tag{14}$$

Hence this symbol is ***multivalued***, namely, *n-valued.* The n values of $\sqrt[n]{z}$ can be obtained as follows. We write z and w in polar form

$$z = r(\cos\theta + i\sin\theta) \qquad \text{and} \qquad w = R(\cos\phi + i\sin\phi).$$

Then the equation $w^n = z$ becomes, by De Moivre's formula (with ϕ instead of θ),

$$w^n = R^n(\cos n\phi + i\sin n\phi) = z = r(\cos\theta + i\sin\theta).$$

The absolute values on both sides must be equal; thus, $R^n = r$, so that $R = \sqrt[n]{r}$, where $\sqrt[n]{r}$ is positive real (an absolute value must be nonnegative!) and thus uniquely determined. Equating the arguments $n\phi$ and θ and recalling that θ is determined only up to integer multiples of 2π, we obtain

$$n\phi = \theta + 2k\pi, \qquad \text{thus} \qquad \phi = \frac{\theta}{n} + \frac{2k\pi}{n}$$

where k is an integer. For $k = 0, 1, \cdots, n-1$ we get n *distinct* values of w. Further integers of k would give values already obtained. For instance, $k = n$ gives $2k\pi/n = 2\pi$, hence the w corresponding to $k = 0$, etc. Consequently, $\sqrt[n]{z}$, for $z \neq 0$, has the n distinct values

$$\sqrt[n]{z} = \sqrt[n]{r}\left(\cos\frac{\theta + 2k\pi}{n} + i\sin\frac{\theta + 2k\pi}{n}\right) \tag{15}$$

where $k = 0, 1, \cdots, n-1$. These n values lie on a circle of radius $\sqrt[n]{r}$ with center at the origin and constitute the vertices of a regular polygon of n sides. The value of $\sqrt[n]{z}$ obtained by taking the principal value of arg z and $k = 0$ in (15) is called the **principal value** of $w = \sqrt[n]{z}$.

Taking $z = 1$ in (15), we have $|z| = r = 1$ and Arg $z = 0$. Then (15) gives

$$\sqrt[n]{1} = \cos\frac{2k\pi}{n} + i\sin\frac{2k\pi}{n}, \qquad k = 0, 1, \cdots, n-1. \tag{16}$$

These n values are called the ***n*th roots of unity**. They lie on the circle of radius 1 and center 0, briefly called the **unit circle** (and used quite frequently!). Figures 327–329 show $\sqrt[3]{1} = 1, -\frac{1}{2} \pm \frac{1}{2}\sqrt{3}i$, $\sqrt[4]{1} = \pm 1, \pm i$, and $\sqrt[5]{1}$.

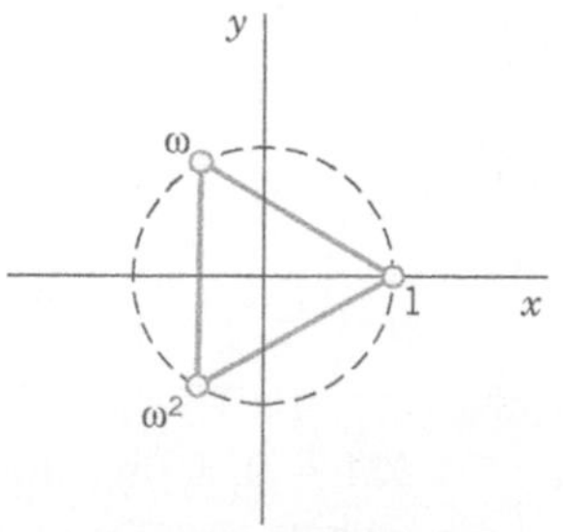

Fig. 327. $\sqrt[3]{1}$

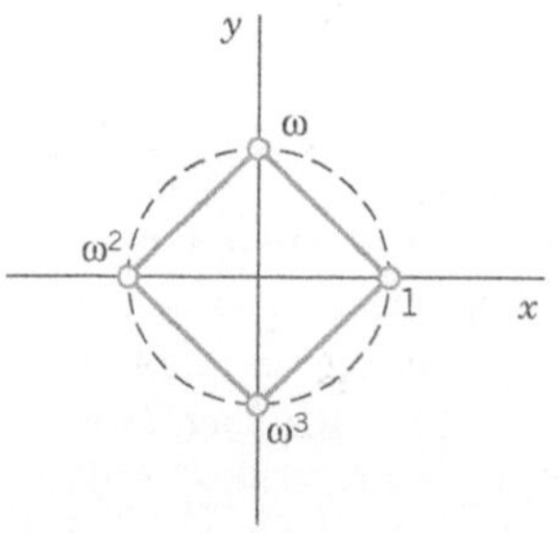

Fig. 328. $\sqrt[4]{1}$

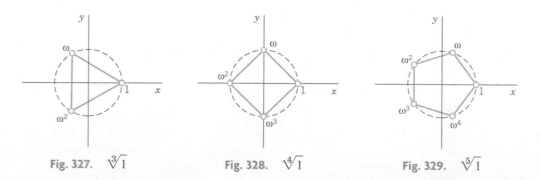

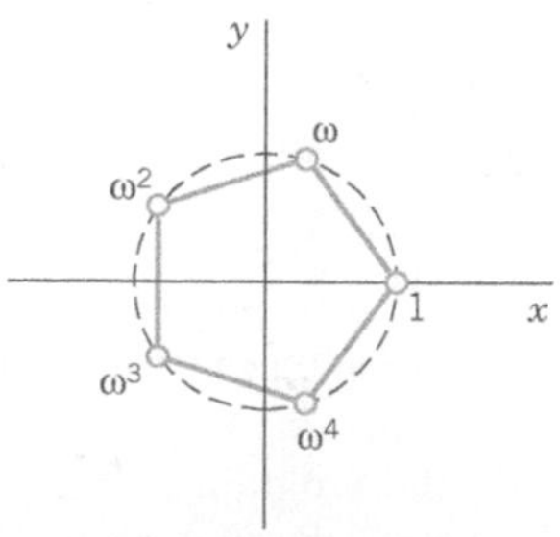

Fig. 329. $\sqrt[5]{1}$

If ω denotes the value corresponding to $k = 1$ in (16), then the n values of $\sqrt[n]{1}$ can be written as

$$1, \omega, \omega^2, \cdots, \omega^{n-1}.$$

More generally, if w_1 is any nth root of an arbitrary complex number $z\ (\neq 0)$, then the n values of $\sqrt[n]{z}$ in (15) are

$$w_1, \quad w_1\omega, \quad w_1\omega^2, \quad \cdots, \quad w_1\omega^{n-1} \tag{17}$$

because multiplying w_1 by ω^k corresponds to increasing the argument of w_1 by $2k\pi/n$. Formula (17) motivates the introduction of roots of unity and shows their usefulness.

PROBLEM SET 13.2

1–8 POLAR FORM

Represent in polar form and graph in the complex plane as in Fig. 325. Do these problems very carefully because polar forms will be needed frequently. Show the details.

1. $1 + i$
2. $-4 + 4i$
3. $2i, \quad -2i$
4. -5
5. $\dfrac{\sqrt{2} + i/3}{-\sqrt{8} - 2i/3}$
6. $\dfrac{\sqrt{3} - 10i}{-\frac{1}{2}\sqrt{3} + 5i}$
7. $1 + \frac{1}{2}\pi i$
8. $\dfrac{-4 + 19i}{2 + 5i}$

9–14 PRINCIPAL ARGUMENT

Determine the principal value of the argument and graph it as in Fig. 325.

9. $-1 + i$
10. $-5, \quad -5 - i, \quad -5 + i$
11. $3 \pm 4i$
12. $-\pi - \pi i$
13. $(1 + i)^{20}$
14. $-1 + 0.1i, \quad -1 - 0.1i$

15–18 CONVERSION TO $x + iy$

Graph in the complex plane and represent in the form $x + iy$:

15. $3 (\cos \frac{1}{2}\pi - i \sin \frac{1}{2}\pi)$
16. $6 (\cos \frac{1}{3}\pi + i \sin \frac{1}{3}\pi)$
17. $\sqrt{8} (\cos \frac{1}{4}\pi + i \sin \frac{1}{4}\pi)$
18. $\sqrt{50} (\cos \frac{3}{4}\pi + i \sin \frac{3}{4}\pi)$

ROOTS

19. **CAS PROJECT. Roots of Unity and Their Graphs.** Write a program for calculating these roots and for graphing them as points on the unit circle. Apply the program to $z^n = 1$ with $n = 2, 3, \cdots, 10$. Then extend the program to one for arbitrary roots, using an idea near the end of the text, and apply the program to examples of your choice.

20. **TEAM PROJECT. Square Root. (a)** Show that $w = \sqrt{z}$ has the values

$$\begin{aligned} w_1 &= \sqrt{r}\left[\cos \frac{\theta}{2} + i \sin \frac{\theta}{2}\right], \\ w_2 &= \sqrt{r}\left[\cos\left(\frac{\theta}{2} + \pi\right) + i \sin\left(\frac{\theta}{2} + \pi\right)\right] \\ &= -w_1. \end{aligned} \tag{18}$$

(b) Obtain from (18) the often more practical formula

$$\sqrt{z} = \pm[\sqrt{\tfrac{1}{2}(|z| + x)} + (\text{sign } y) i \sqrt{\tfrac{1}{2}(|z| + x)}] \tag{19}$$

where sign $y = 1$ if $y \geqq 0$, sign $y = -1$ if $y < 0$, and all square roots of positive numbers are taken with positive sign. *Hint:* Use (10) in App. A3.1 with $x = \theta/2$.

(c) Find the square roots of $-14i, -9 - 40i$, and $1 + \sqrt{48}i$ by both (18) and (19) and comment on the work involved.

(d) Do some further examples of your own and apply a method of checking your results.

21–27 ROOTS

Find and graph all roots in the complex plane.

21. $\sqrt[3]{1 + i}$
22. $\sqrt[3]{3 + 4i}$
23. $\sqrt[3]{216}$
24. $\sqrt[4]{-4}$
25. $\sqrt[4]{i}$
26. $\sqrt[8]{1}$
27. $\sqrt[5]{-1}$

28–31 EQUATIONS

Solve and graph the solutions. Show details.

28. $z^2 - (6 - 2i)z + 17 - 6i = 0$
29. $z^2 + z + 1 - i = 0$
30. $z^4 + 324 = 0$. Using the solutions, factor $z^4 + 324$ into quadratic factors with *real* coefficients.
31. $z^4 - 6iz^2 + 16 = 0$

32–35 **INEQUALITIES AND EQUALITY**

32. Triangle inequality. Verify (6) for $z_1 = 3 + i$, $z_2 = -2 + 4i$

33. Triangle inequality. Prove (6).

34. Re and Im. Prove $|\text{Re } z| \leqq |z|$, $|\text{Im } z| \leqq |z|$.

35. Parallelogram equality. Prove and explain the name

$$|z_1 + z_2|^2 + |z_1 - z_2|^2 = 2\,(|z_1|^2 + |z_2|^2).$$

13.3 Derivative. Analytic Function

Just as the study of calculus or real analysis required concepts such as domain, neighborhood, function, limit, continuity, derivative, etc., so does the study of complex analysis. Since the functions live in the complex plane, the concepts are slightly more difficult or *different* from those in real analysis. This section can be seen as a reference section where many of the concepts needed for the rest of Part D are introduced.

Circles and Disks. Half-Planes

The **unit circle** $|z| = 1$ (Fig. 330) has already occurred in Sec. 13.2. Figure 331 shows a general circle of radius ρ and center a. Its equation is

$$|z - a| = \rho$$

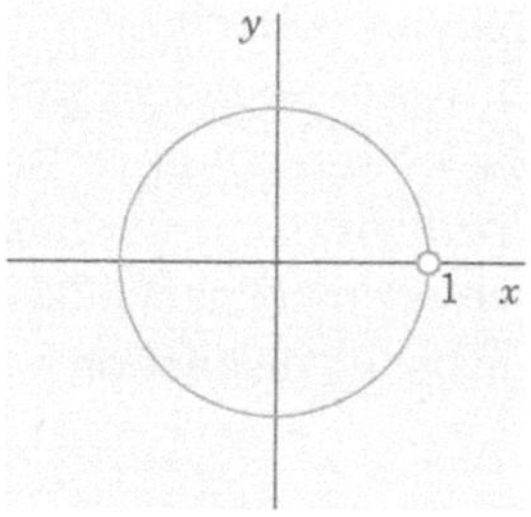

Fig. 330. Unit circle

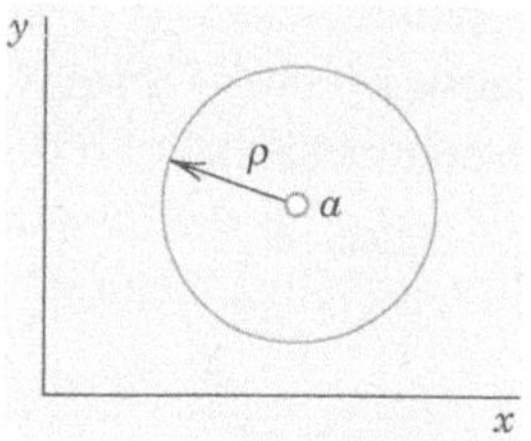

Fig. 331. Circle in the complex plane

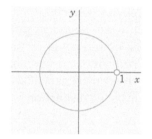

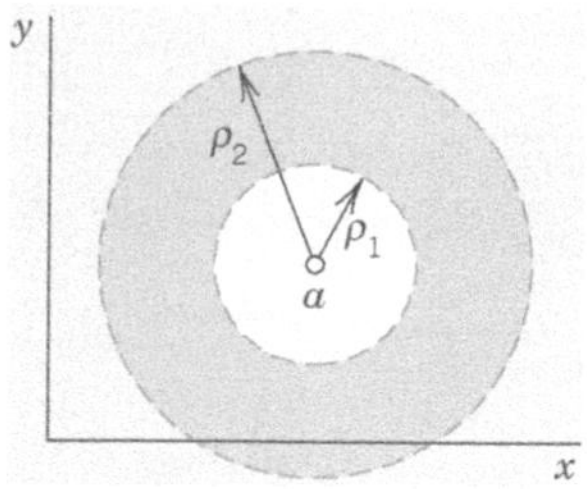

Fig. 332. Annulus in the complex plane

because it is the set of all z whose distance $|z - a|$ from the center a equals ρ. Accordingly, its interior ("**open circular disk**") is given by $|z - a| < \rho$, its interior plus the circle itself ("**closed circular disk**") by $|z - a| \leqq \rho$, and its exterior by $|z - a| > \rho$. As an example, sketch this for $a = 1 + i$ and $\rho = 2$, to make sure that you understand these inequalities.

An open circular disk $|z - a| < \rho$ is also called a **neighborhood** of a or, more precisely, a *ρ-neighborhood* of a. And a has infinitely many of them, one for each value of ρ (> 0), and a is a point of each of them, by definition!

In modern literature *any set* containing a ρ-neighborhood of a is also called a *neighborhood* of a.

Figure 332 shows an **open annulus** (circular ring) $\rho_1 < |z - a| < \rho_2$, which we shall need later. This is the set of all z whose distance $|z - a|$ from a is greater than ρ_1 but less than ρ_2. Similarly, the **closed annulus** $\rho_1 \leqq |z - a| \leqq \rho_2$ includes the two circles.

Half-Planes. By the (open) *upper* **half-plane** we mean the set of all points $z = x + iy$ such that $y > 0$. Similarly, the condition $y < 0$ defines the *lower half-plane*, $x > 0$ the *right half-plane*, and $x < 0$ the *left half-plane*.

For Reference: Concepts on Sets in the Complex Plane

To our discussion of special sets let us add some general concepts related to sets that we shall need throughout Chaps. 13–18; keep in mind that you can find them here.

By a **point set** in the complex plane we mean any sort of collection of finitely many or infinitely many points. Examples are the solutions of a quadratic equation, the points of a line, the points in the interior of a circle as well as the sets discussed just before.

A set S is called **open** if every point of S has a neighborhood consisting entirely of points that belong to S. For example, the points in the interior of a circle or a square form an open set, and so do the points of the right half-plane $\text{Re } z = x > 0$.

A set S is called **connected** if any two of its points can be joined by a chain of finitely many straight-line segments all of whose points belong to S. An open and connected set is called a **domain.** Thus an open disk and an open annulus are domains. An open square with a diagonal removed is not a domain since this set is not connected. (Why?)

The **complement** of a set S in the complex plane is the set of all points of the complex plane that ***do not belong*** to S. A set S is called **closed** if its complement is open. For example, the points on and inside the unit circle form a closed set ("closed unit disk") since its complement $|z| > 1$ is open.

A **boundary point** of a set S is a point every neighborhood of which contains both points that belong to S and points that do not belong to S. For example, the boundary points of an annulus are the points on the two bounding circles. Clearly, if a set S is open, then no boundary point belongs to S; if S is closed, then every boundary point belongs to S. The set of all boundary points of a set S is called the **boundary** of S.

A **region** is a set consisting of a domain plus, perhaps, some or all of its boundary points. WARNING! "Domain" is the *modern* term for an open connected set. Nevertheless, some authors still call a domain a "region" and others make no distinction between the two terms.

Complex Function

Complex analysis is concerned with complex functions that are differentiable in some domain. Hence we should first say what we mean by a complex function and then define the concepts of limit and derivative in complex. This discussion will be similar to that in calculus. Nevertheless it needs great attention because it will show interesting basic differences between real and complex calculus.

Recall from calculus that a *real* function f defined on a set S of real numbers (usually an interval) is a rule that assigns to every x in S a real number $f(x)$, called the *value* of f at x. Now in complex, S is a set of *complex* numbers. And a **function** f defined on S is a rule that assigns to every z in S a complex number w, called the *value* of f at z. We write

$$w = f(z).$$

Here z varies in S and is called a **complex variable**. The set S is called the *domain of definition* of f or, briefly, the **domain** of f. (In most cases S will be open and connected, thus a domain as defined just before.)

Example: $w = f(z) = z^2 + 3z$ is a complex function defined for all z; that is, its domain S is the whole complex plane.

The set of all values of a function f is called the **range** *of f.*

w is complex, and we write $w = u + iv$, where u and v are the real and imaginary parts, respectively. Now w depends on $z = x + iy$. Hence u becomes a real function of x and y, and so does v. We may thus write

$$w = f(z) = u(x, y) + iv(x, y).$$

This shows that a *complex* function $f(z)$ is equivalent to a *pair* of *real* functions $u(x, y)$ and $v(x, y)$, each depending on the two real variables x and y.

EXAMPLE 1 Function of a Complex Variable

Let $w = f(z) = z^2 + 3z$. Find u and v and calculate the value of f at $z = 1 + 3i$.

Solution. $u = \text{Re} f(z) = x^2 - y^2 + 3x$ and $v = 2xy + 3y$. Also,

$$f(1 + 3i) = (1 + 3i)^2 + 3(1 + 3i) = 1 - 9 + 6i + 3 + 9i = -5 + 15i.$$

This shows that $u(1, 3) = -5$ and $v(1, 3) = 15$. Check this by using the expressions for u and v. ■

EXAMPLE 2 Function of a Complex Variable

Let $w = f(z) = 2iz + 6\bar{z}$. Find u and v and the value of f at $z = \frac{1}{2} + 4i$.

Solution. $f(z) = 2i(x + iy) + 6(x - iy)$ gives $u(x, y) = 6x - 2y$ and $v(x, y) = 2x - 6y$. Also,

$$f(\tfrac{1}{2} + 4i) = 2i(\tfrac{1}{2} + 4i) + 6(\tfrac{1}{2} - 4i) = i - 8 + 3 - 24i = -5 - 23i.$$

Check this as in Example 1. ■

Remarks on Notation and Terminology

1. Strictly speaking, $f(z)$ denotes the value of f at z, but it is a convenient abuse of language to talk about *the function* $f(z)$ (instead of *the function* f), thereby exhibiting the notation for the independent variable.

2. We assume all functions to be ***single-valued relations***, as usual: to each z in S there corresponds but *one* value $w = f(z)$ (but, of course, several z may give the same value $w = f(z)$, just as in calculus). Accordingly, we shall *not use* the term "multivalued function" (used in some books on complex analysis) for a multivalued relation, in which to a z there corresponds more than one w.

Limit, Continuity

A function $f(z)$ is said to have the **limit** l as z approaches a point z_0, written

$$\lim_{z \to z_0} f(z) = l, \tag{1}$$

if f is defined in a neighborhood of z_0 (except perhaps at z_0 itself) and if the values of f are "close" to l for all z "close" to z_0; in precise terms, if for every positive real ϵ we can find a positive real δ such that for all $z \neq z_0$ in the disk $|z - z_0| < \delta$ (Fig. 333) we have

$$|f(z) - l| < \epsilon; \tag{2}$$

geometrically, if for every $z \neq z_0$ in that δ-disk the value of f lies in the disk (2).

Formally, this definition is similar to that in calculus, but there is a big difference. Whereas in the real case, x can approach an x_0 only along the real line, here, by definition,

z may approach z_0 ***from any direction*** in the complex plane. This will be quite essential in what follows.

If a limit exists, it is unique. (See Team Project 24.)

A function $f(z)$ is said to be **continuous** at $z = z_0$ if $f(z_0)$ is defined and

$$\lim_{z \to z_0} f(z) = f(z_0). \tag{3}$$

Note that by definition of a limit this implies that $f(z)$ is defined in some neighborhood of z_0.

$f(z)$ is said to be *continuous in a domain* if it is continuous at each point of this domain.

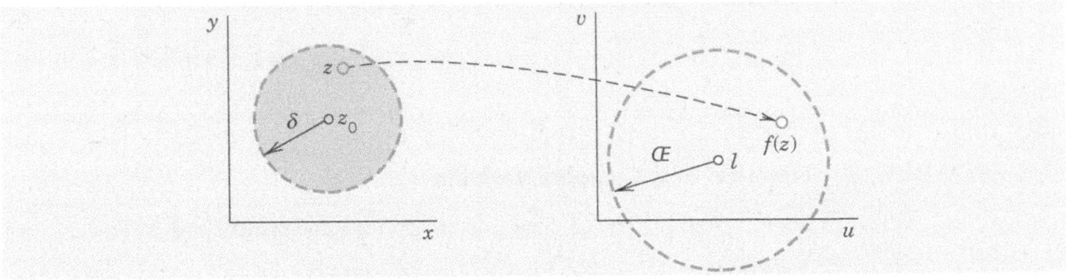

Fig. 333. Limit

Derivative

The **derivative** of a complex function f at a point z_0 is written $f'(z_0)$ and is defined by

$$f'(z_0) = \lim_{\Delta z \to 0} \frac{f(z_0 + \Delta z) - f(z_0)}{\Delta z} \tag{4}$$

provided this limit exists. Then f is said to be **differentiable** at z_0. If we write $\Delta z = z - z_0$, we have $z = z_0 + \Delta z$ and (4) takes the form

$$f'(z_0) = \lim_{z \to z_0} \frac{f(z) - f(z_0)}{z - z_0}. \tag{4'}$$

Now comes an ***important point***. Remember that, by the definition of limit, $f(z)$ is defined in a neighborhood of z_0 and z in (4′) may approach z_0 from any direction in the complex plane. Hence differentiability at z_0 means that, along whatever path z approaches z_0, the quotient in (4′) always approaches a certain value and all these values are equal. This is important and should be kept in mind.

EXAMPLE 3 Differentiability. Derivative

The function $f(z) = z^2$ is differentiable for all z and has the derivative $f'(z) = 2z$ because

$$f'(z) = \lim_{\Delta z \to 0} \frac{(z + \Delta z)^2 - z^2}{\Delta z} = \lim_{\Delta z \to 0} \frac{z^2 + 2z\,\Delta z + (\Delta z)^2 - z^2}{\Delta z} = \lim_{\Delta z \to 0} (2z + \Delta z) = 2z.$$ ■

The ***differentiation rules*** *are the same as in real calculus,* since their proofs are literally the same. Thus for any differentiable functions f and g and constant c we have

$$(cf)' = cf', \quad (f + g)' = f' + g', \quad (fg)' = f'g + fg', \quad \left(\frac{f}{g}\right)' = \frac{f'g - fg'}{g^2}$$

as well as the chain rule and the power rule $(z^n)' = nz^{n-1}$ (n integer).

Also, if $f(z)$ is differentiable at z_0, it is continuous at z_0. (See Team Project 24.)

EXAMPLE 4 **$\bar{z}$ not Differentiable**

It may come as a surprise that there are many complex functions that do not have a derivative at any point. For instance, $f(z) = \bar{z} = x - iy$ is such a function. To see this, we write $\Delta z = \Delta x + i\Delta y$ and obtain

$$\frac{f(z + \Delta z) - f(z)}{\Delta z} = \frac{\overline{(z + \Delta z)} - \bar{z}}{\Delta z} = \frac{\overline{\Delta z}}{\Delta z} = \frac{\Delta x - i\Delta y}{\Delta x + i\Delta y}. \tag{5}$$

If $\Delta y = 0$, this is $+1$. If $\Delta x = 0$, this is -1. Thus (5) approaches $+1$ along path I in Fig. 334 but -1 along path II. Hence, by definition, the limit of (5) as $\Delta z \to 0$ does not exist at any z. ■

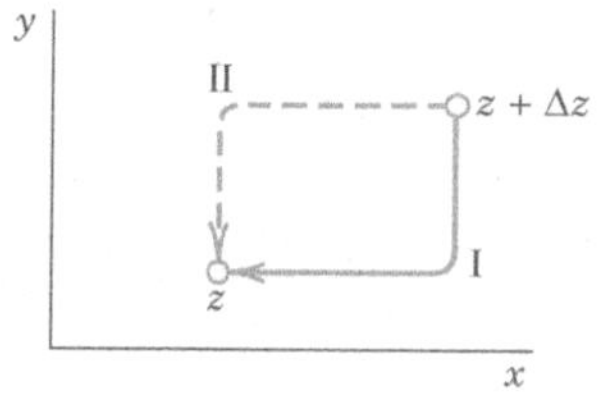

Fig. 334. Paths in (5)

Surprising as Example 4 may be, it merely illustrates that differentiability of a *complex* function is a rather severe requirement.

The idea of proof (approach of z from different directions) is basic and will be used again as the crucial argument in the next section.

Analytic Functions

Complex analysis is concerned with the theory and application of "analytic functions," that is, functions that are differentiable in some domain, so that we can do "calculus in complex." The definition is as follows.

DEFINITION

Analyticity

A function $f(z)$ is said to be *analytic in a domain D* if $f(z)$ is defined and differentiable at all points of D. The function $f(z)$ is said to be *analytic at a point $z = z_0$* in D if $f(z)$ is analytic in a neighborhood of z_0.

Also, by an **analytic function** we mean a function that is analytic in *some* domain.

Hence analyticity of $f(z)$ at z_0 means that $f(z)$ has a derivative at every point in some neighborhood of z_0 (including z_0 itself since, by definition, z_0 is a point of all its neighborhoods). This concept is *motivated* by the fact that it is of no practical interest if a function is differentiable merely at a single point z_0 but not throughout some neighborhood of z_0. Team Project 24 gives an example.

A more modern term for *analytic in D* is *holomorphic in D.*

EXAMPLE 5 Polynomials, Rational Functions

The nonnegative integer powers $1, z, z^2, \cdots$ are analytic in the entire complex plane, and so are **polynomials**, that is, functions of the form

$$f(z) = c_0 + c_1 z + c_2 z^2 + \cdots + c_n z^n$$

where $c_0, \cdots, c_n$ are complex constants.

The quotient of two polynomials $g(z)$ and $h(z)$,

$$f(z) = \frac{g(z)}{h(z)},$$

is called a **rational function**. This f is analytic except at the points where $h(z) = 0$; here we assume that common factors of g and h have been canceled.

Many further analytic functions will be considered in the next sections and chapters. ■

The concepts discussed in this section extend familiar concepts of calculus. Most important is the concept of an analytic function, the exclusive concern of complex analysis. Although many simple functions are not analytic, the large variety of remaining functions will yield a most beautiful branch of mathematics that is very useful in engineering and physics.

PROBLEM SET 13.3

1–8 REGIONS OF PRACTICAL INTEREST

Determine and sketch or graph the sets in the complex plane given by

1. $|z + 1 - 5i| \leqq \frac{3}{2}$
2. $0 < |z| < 1$
3. $\pi < |z - 4 + 2i| < 3\pi$
4. $-\pi < \operatorname{Im} z < \pi$
5. $|\arg z| < \frac{1}{4}\pi$
6. $\operatorname{Re}(1/z) < 1$
7. $\operatorname{Re} z \geqq -1$
8. $|z + i| \geqq |z - i|$
9. **WRITING PROJECT. Sets in the Complex Plane.** Write a report by formulating the corresponding portions of the text in your own words and illustrating them with examples of your own.

COMPLEX FUNCTIONS AND THEIR DERIVATIVES

10–12 Function Values. Find Re f, and Im f and their values at the given point z.

10. $f(z) = 5z^2 - 12z + 3 + 2i$ at $4 - 3i$
11. $f(z) = 1/(1 - z)$ at $1 - i$
12. $f(z) = (z - 2)/(z + 2)$ at $8i$
13. **CAS PROJECT. Graphing Functions.** Find and graph Re f, Im f, and $|f|$ as surfaces over the z-plane. Also graph the two families of curves $\operatorname{Re} f(z) = \text{const}$ and $\operatorname{Im} f(z) = \text{const}$ in the same figure, and the curves $|f(z)| = \text{const}$ in another figure, where **(a)** $f(z) = z^2$, **(b)** $f(z) = 1/z$, (c) $f(z) = z^4$.

14–17 Continuity. Find out, and give reason, whether $f(z)$ is continuous at $z = 0$ if $f(0) = 0$ and for $z \neq 0$ the function f is equal to:

14. $(\operatorname{Re} z^2)/|z|$
15. $|z|^2 \operatorname{Im}(1/z)$
16. $(\operatorname{Im} z^2)/|z|^2$
17. $(\operatorname{Re} z)/(1 - |z|)$

18–23 Differentiation. Find the value of the derivative of

18. $(z - i)/(z + i)$ at i
19. $(z - 4i)^8$ at $= 3 + 4i$
20. $(1.5z + 2i)/(3iz - 4)$ at any z. Explain the result.
21. $i(1 - z)^n$ at 0
22. $(iz^3 + 3z^2)^3$ at $2i$
23. $z^3/(z + i)^3$ at i
24. **TEAM PROJECT. Limit, Continuity, Derivative**

    (a) Limit. Prove that (1) is equivalent to the pair of relations

    $$\lim_{z \to z_0} \operatorname{Re} f(z) = \operatorname{Re} l, \qquad \lim_{z \to z_0} \operatorname{Im} f(z) = \operatorname{Im} l.$$

    (b) Limit. If $\lim_{z \to z_0} f(x)$ exists, show that this limit is unique.

    (c) Continuity. If $z_1, z_2, \cdots$ are complex numbers for which $\lim_{n\to\infty} z_n = a$, and if $f(z)$ is continuous at $z = a$, show that $\lim_{n\to\infty} f(z_n) = f(a)$.

(d) Continuity. If $f(z)$ is differentiable at z_0, show that $f(z)$ is continuous at z_0.

(e) Differentiability. Show that $f(z) = \text{Re } z = x$ is not differentiable at any z. Can you find other such functions?

(f) Differentiability. Show that $f(z) = |z|^2$ is differentiable only at $z = 0$; hence it is nowhere analytic.

25. WRITING PROJECT. Comparison with Calculus. Summarize the second part of this section beginning with *Complex Function*, and indicate what is conceptually analogous to calculus and what is not.

13.4 Cauchy–Riemann Equations. Laplace's Equation

As we saw in the last section, to do complex analysis (i.e., "calculus in the complex") on any complex function, we require that function to be *analytic on some domain* that is differentiable in that domain.

The Cauchy–Riemann equations are the most important equations in this chapter and one of the pillars on which complex analysis rests. They provide a criterion (a test) for the analyticity of a complex function

$$w = f(z) = u(x, y) + iv(x, y).$$

Roughly, f is analytic in a domain D if and only if the first partial derivatives of u and v satisfy the two **Cauchy–Riemann equations**[4]

$$u_x = v_y, \qquad u_y = -v_x \tag{1}$$

everywhere in D; here $u_x = \partial u/\partial x$ and $u_y = \partial u/\partial y$ (and similarly for v) are the usual notations for partial derivatives. The precise formulation of this statement is given in Theorems 1 and 2.

Example: $f(z) = z^2 = x^2 - y^2 + 2ixy$ is analytic for all z (see Example 3 in Sec. 13.3), and $u = x^2 - y^2$ and $v = 2xy$ satisfy (1), namely, $u_x = 2x = v_y$ as well as $u_y = -2y = -v_x$. More examples will follow.

THEOREM 1

Cauchy–Riemann Equations

Let $f(z) = u(x, y) + iv(x, y)$ be defined and continuous in some neighborhood of a point $z = x + iy$ and differentiable at z itself. Then, at that point, the first-order partial derivatives of u and v exist and satisfy the Cauchy–Riemann equations (1).

Hence, if $f(z)$ is analytic in a domain D, those partial derivatives exist and satisfy (1) *at all points of D.*

[4]The French mathematician AUGUSTIN-LOUIS CAUCHY (see Sec. 2.5) and the German mathematicians BERNHARD RIEMANN (1826–1866) and KARL WEIERSTRASS (1815–1897; see also Sec. 15.5) are the founders of complex analysis. Riemann received his Ph.D. (in 1851) under Gauss (Sec. 5.4) at Göttingen, where he also taught until he died, when he was only 39 years old. He introduced the concept of the integral as it is used in basic calculus courses, and made important contributions to differential equations, number theory, and mathematical physics. He also developed the so-called Riemannian geometry, which is the mathematical foundation of Einstein's theory of relativity; see Ref. [GenRef9] in App. 1.

PROOF By assumption, the derivative $f'(z)$ at z exists. It is given by

$$f'(z) = \lim_{\Delta z \to 0} \frac{f(z + \Delta z) - f(z)}{\Delta z}. \tag{2}$$

The idea of the proof is very simple. By the definition of a limit in complex (Sec. 13.3), we can let Δz approach zero along any path in a neighborhood of z. Thus we may choose the two paths I and II in Fig. 335 and equate the results. By comparing the real parts we shall obtain the first Cauchy–Riemann equation and by comparing the imaginary parts the second. The technical details are as follows.

We write $\Delta z = \Delta x + i\,\Delta y$. Then $z + \Delta z = x + \Delta x + i(y + \Delta y)$, and in terms of u and v the derivative in (2) becomes

$$f'(z) = \lim_{\Delta z \to 0} \frac{[u(x + \Delta x, y + \Delta y) + iv(x + \Delta x, y + \Delta y)] - [u(x, y) + iv(x, y)]}{\Delta x + i\,\Delta y}. \tag{3}$$

We first choose path I in Fig. 335. Thus we let $\Delta y \to 0$ first and then $\Delta x \to 0$. After Δy is zero, $\Delta z = \Delta x$. Then (3) becomes, if we first write the two u-terms and then the two v-terms,

$$f'(z) = \lim_{\Delta x \to 0} \frac{u(x + \Delta x, y) - u(x, y)}{\Delta x} + i \lim_{\Delta x \to 0} \frac{v(x + \Delta x, y) - v(x, y)}{\Delta x}.$$

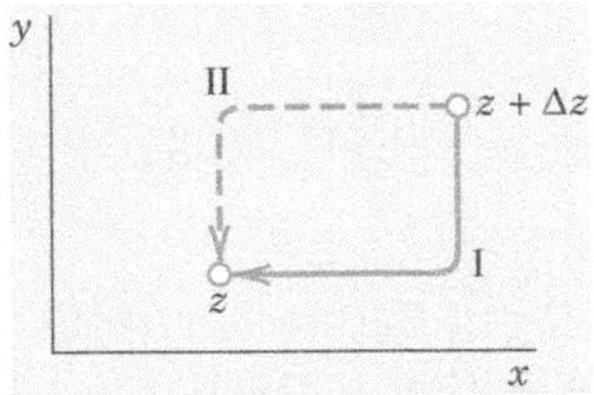

Fig. 335. Paths in (2)

Since $f'(z)$ exists, the two real limits on the right exist. By definition, they are the partial derivatives of u and v with respect to x. Hence the derivative $f'(z)$ of $f(z)$ can be written

$$f'(z) = u_x + iv_x. \tag{4}$$

Similarly, if we choose path II in Fig. 335, we let $\Delta x \to 0$ first and then $\Delta y \to 0$. After Δx is zero, $\Delta z = i\,\Delta y$, so that from (3) we now obtain

$$f'(z) = \lim_{\Delta y \to 0} \frac{u(x, y + \Delta y) - u(x, y)}{i\,\Delta y} + i \lim_{\Delta y \to 0} \frac{v(x, y + \Delta y) - v(x, y)}{i\,\Delta y}.$$

Since $f'(z)$ exists, the limits on the right exist and give the partial derivatives of u and v with respect to y; noting that $1/i = -i$, we thus obtain

$$f'(z) = -iu_y + v_y. \tag{5}$$

The existence of the derivative $f'(z)$ thus implies the existence of the four partial derivatives in (4) and (5). By equating the real parts u_x and v_y in (4) and (5) we obtain the first

Cauchy–Riemann equation (1). Equating the imaginary parts gives the other. This proves the first statement of the theorem and implies the second because of the definition of analyticity. ∎

Formulas (4) and (5) are also quite practical for calculating derivatives $f'(z)$, as we shall see.

EXAMPLE 1 **Cauchy–Riemann Equations**

$f(z) = z^2$ is analytic for all z. It follows that the Cauchy–Riemann equations must be satisfied (as we have verified above).

For $f(z) = \bar{z} = x - iy$ we have $u = x, v = -y$ and see that the second Cauchy–Riemann equation is satisfied, $u_y = -v_x = 0$, but the first is not: $u_x = 1 \neq v_y = -1$. We conclude that $f(z) = \bar{z}$ is not analytic, confirming Example 4 of Sec. 13.3. Note the savings in calculation! ∎

The Cauchy–Riemann equations are fundamental because they are not only necessary but also sufficient for a function to be analytic. More precisely, the following theorem holds.

THEOREM 2 **Cauchy–Riemann Equations**

If two real-valued continuous functions $u(x, y)$ and $v(x, y)$ of two real variables x and y have ***continuous*** *first partial derivatives that satisfy the Cauchy–Riemann equations in some domain D, then the complex function $f(z) = u(x, y) + iv(x, y)$ is analytic in D.*

The proof is more involved than that of Theorem 1 and we leave it optional (see App. 4).

Theorems 1 and 2 are of great practical importance, since, by using the Cauchy–Riemann equations, we can now easily find out whether or not a given complex function is analytic.

EXAMPLE 2 **Cauchy–Riemann Equations. Exponential Function**

Is $f(z) = u(x, y) + iv(x, y) = e^x(\cos y + i \sin y)$ analytic?

Solution. We have $u = e^x \cos y$, $v = e^x \sin y$ and by differentiation

$$u_x = e^x \cos y, \qquad v_y = e^x \cos y$$

$$u_y = -e^x \sin y, \qquad v_x = e^x \sin y.$$

We see that the Cauchy–Riemann equations are satisfied and conclude that $f(z)$ is analytic for all z. ($f(z)$ will be the complex analog of e^x known from calculus.) ∎

EXAMPLE 3 **An Analytic Function of Constant Absolute Value Is Constant**

The Cauchy–Riemann equations also help in deriving general properties of analytic functions.

For instance, show that if $f(z)$ is analytic in a domain D and $|f(z)| = k = \text{const}$ in D, then $f(z) = \text{const}$ in D. (We shall make crucial use of this in Sec. 18.6 in the proof of Theorem 3.)

Solution. By assumption, $|f|^2 = |u + iv|^2 = u^2 + v^2 = k^2$. By differentiation,

$$uu_x + vv_x = 0,$$

$$uu_y + vv_y = 0.$$

Now use $v_x = -u_y$ in the first equation and $v_y = u_x$ in the second, to get

$$\text{(6)} \qquad \begin{aligned} &\text{(a)} \quad uu_x - vu_y = 0, \\ &\text{(b)} \quad uu_y - vu_x = 0. \end{aligned}$$

To get rid of u_y, multiply (6a) by u and (6b) by v and add. Similarly, to eliminate u_x, multiply (6a) by $-v$ and (6b) by u and add. This yields

$$(u^2 + v^2)u_x = 0,$$
$$(u^2 + v^2)u_y = 0.$$

If $k^2 = u^2 + v^2 = 0$, then $u = v = 0$; hence $f = 0$. If $k^2 = u^2 + v^2 \neq 0$, then $u_x = u_y = 0$. Hence, by the Cauchy–Riemann equations, also $v_x = v_y = 0$. Together this implies $u = \text{const}$ and $v = \text{const}$; hence $f = \text{const}$. ■

We mention that, if we use the polar form $z = r(\cos\theta + i\sin\theta)$ and set $f(z) = u(r, \theta) + iv(r, \theta)$, then the **Cauchy–Riemann equations** are (Prob. 1)

$$\text{(7)} \qquad \begin{aligned} u_r &= \frac{1}{r} v_\theta, \\ v_r &= -\frac{1}{r} u_\theta \end{aligned} \qquad (r > 0).$$

Laplace's Equation. Harmonic Functions

The great importance of complex analysis in engineering mathematics results mainly from the fact that both the real part and the imaginary part of an analytic function satisfy Laplace's equation, the most important PDE of physics. It occurs in gravitation, electrostatics, fluid flow, heat conduction, and other applications (see Chaps. 12 and 18).

THEOREM 3

Laplace's Equation

If $f(z) = u(x, y) + iv(x, y)$ is analytic in a domain D, then both u and v satisfy **Laplace's equation**

$$\text{(8)} \qquad \nabla^2 u = u_{xx} + u_{yy} = 0$$

(∇^2 read "nabla squared") *and*

$$\text{(9)} \qquad \nabla^2 v = v_{xx} + v_{yy} = 0,$$

in D and have continuous second partial derivatives in D.

PROOF Differentiating $u_x = v_y$ with respect to x and $u_y = -v_x$ with respect to y, we have

$$\text{(10)} \qquad u_{xx} = v_{yx}, \qquad u_{yy} = -v_{xy}.$$

Now the derivative of an analytic function is itself analytic, as we shall prove later (in Sec. 14.4). This implies that u and v have continuous partial derivatives of all orders; in particular, the mixed second derivatives are equal: $v_{yx} = v_{xy}$. By adding (10) we thus obtain (8). Similarly, (9) is obtained by differentiating $u_x = v_y$ with respect to y and $u_y = -v_x$ with respect to x and subtracting, using $u_{xy} = u_{yx}$. ■

Solutions of Laplace's equation having ***continuous*** second-order partial derivatives are called **harmonic functions** and their theory is called **potential theory** (see also Sec. 12.11). Hence the real and imaginary parts of an analytic function are harmonic functions.

If two harmonic functions u and v satisfy the Cauchy–Riemann equations in a domain D, they are the real and imaginary parts of an analytic function f in D. Then v is said to be a **harmonic conjugate function** of u in D. (Of course, this has absolutely nothing to do with the use of "conjugate" for $\bar{z}$.)

EXAMPLE 4 How to Find a Harmonic Conjugate Function by the Cauchy–Riemann Equations

Verify that $u = x^2 - y^2 - y$ is harmonic in the whole complex plane and find a harmonic conjugate function v of u.

Solution. $\nabla^2 u = 0$ by direct calculation. Now $u_x = 2x$ and $u_y = -2y - 1$. Hence because of the Cauchy–Riemann equations a conjugate v of u must satisfy

$$v_y = u_x = 2x, \qquad v_x = -u_y = 2y + 1.$$

Integrating the first equation with respect to y and differentiating the result with respect to x, we obtain

$$v = 2xy + h(x), \qquad v_x = 2y + \frac{dh}{dx}.$$

A comparison with the second equation shows that $dh/dx = 1$. This gives $h(x) = x + c$. Hence $v = 2xy + x + c$ (c any real constant) is the most general harmonic conjugate of the given u. The corresponding analytic function is

$$f(z) = u + iv = x^2 - y^2 - y + i(2xy + x + c) = z^2 + iz + ic. \qquad \blacksquare$$

Example 4 illustrates that *a conjugate of a given harmonic function is uniquely determined up to an arbitrary real additive constant.*

The Cauchy–Riemann equations are the most important equations in this chapter. Their relation to Laplace's equation opens a wide range of engineering and physical applications, as shown in Chap. 18.

PROBLEM SET 13.4

1. **Cauchy–Riemann equations in polar form.** Derive (7) from (1).

2–11 CAUCHY–RIEMANN EQUATIONS

Are the following functions analytic? Use (1) or (7).

2. $f(z) = iz\bar{z}$
3. $f(z) = e^{-2x}(\cos 2y - i \sin 2y)$
4. $f(z) = e^x(\cos y - i \sin y)$
5. $f(z) = \text{Re}(z^2) - i\,\text{Im}(z^2)$
6. $f(z) = 1/(z - z^5)$
7. $f(z) = i/z^8$
8. $f(z) = \text{Arg}\, 2\pi z$
9. $f(z) = 3\pi^2/(z^3 + 4\pi^2 z)$
10. $f(z) = \ln |z| + i\,\text{Arg}\, z$
11. $f(z) = \cos x \cosh y - i \sin x \sinh y$

12–19 HARMONIC FUNCTIONS

Are the following functions harmonic? If your answer is yes, find a corresponding analytic function $f(z) = u(x, y) + iv(x, y)$.

12. $u = x^2 + y^2$
13. $u = xy$
14. $v = xy$
15. $u = x/(x^2 + y^2)$
16. $u = \sin x \cosh y$
17. $v = (2x + 1)y$
18. $u = x^3 - 3xy^2$
19. $v = e^x \sin 2y$
20. **Laplace's equation.** Give the details of the derivative of (9).

21–24 Determine a and b so that the given function is harmonic and find a harmonic conjugate.

21. $u = e^{\pi x} \cos av$
22. $u = \cos ax \cosh 2y$
23. $u = ax^3 + bxy$
24. $u = \cosh ax \cos y$
25. **CAS PROJECT. Equipotential Lines.** Write a program for graphing equipotential lines $u = \text{const}$ of a harmonic function u and of its conjugate v on the same axes. Apply the program to **(a)** $u = x^2 - y^2$, $v = 2xy$, **(b)** $u = x^3 - 3xy^2$, $v = 3x^2y - y^3$.
26. Apply the program in Prob. 25 to $u = e^x \cos y$, $v = e^x \sin y$ and to an example of your own.

27. Harmonic conjugate. Show that if u is harmonic and v is a harmonic conjugate of u, then u is a harmonic conjugate of $-v$.

28. Illustrate Prob. 27 by an example.

29. Two further formulas for the derivative. Formulas (4), (5), and (11) (below) are needed from time to time. Derive

$$(11) \qquad f'(z) = u_x - iu_y, \qquad f'(z) = v_y + iv_x.$$

30. TEAM PROJECT. Conditions for $f(z) =$ const. Let $f(z)$ be analytic. Prove that each of the following conditions is sufficient for $f(z) =$ const.

(a) $\operatorname{Re} f(z) =$ const

(b) $\operatorname{Im} f(z) =$ const

(c) $f'(z) = 0$

(d) $|f(z)| =$ const (see Example 3)

13.5 Exponential Function

In the remaining sections of this chapter we discuss the basic elementary complex functions, the exponential function, trigonometric functions, logarithm, and so on. They will be counterparts to the familiar functions of calculus, to which they reduce when $z = x$ is real. They are indispensable throughout applications, and some of them have interesting properties not shared by their real counterparts.

We begin with one of the most important analytic functions, the complex **exponential function**

$$e^z, \qquad \text{also written} \qquad \exp z.$$

The definition of e^z in terms of the real functions e^x, $\cos y$, and $\sin y$ is

$$(1) \qquad e^z = e^x(\cos y + i \sin y).$$

This definition is motivated by the fact the e^z ***extends*** the real exponential function e^x of calculus in a natural fashion. Namely:

(A) $e^z = e^x$ for real $z = x$ because $\cos y = 1$ and $\sin y = 0$ when $y = 0$.

(B) e^z is analytic for all z. (Proved in Example 2 of Sec. 13.4.)

(C) The derivative of e^z is e^z, that is,

$$(2) \qquad (e^z)' = e^z.$$

This follows from (4) in Sec. 13.4,

$$(e^z)' = (e^x \cos y)_x + i(e^x \sin y)_x = e^x \cos y + ie^x \sin y = e^z.$$

REMARK. This definition provides for a relatively simple discussion. We could define e^z by the familiar series $1 + x + x^2/2! + x^3/3! + \cdots$ with x replaced by z, but we would then have to discuss complex series at this very early stage. (We will show the connection in Sec. 15.4.)

Further Properties. A function $f(z)$ that is analytic for all z is called an **entire function**. Thus, e^z is entire. Just as in calculus the ***functional relation***

$$(3) \qquad e^{z_1+z_2} = e^{z_1}e^{z_2}$$

holds for any $z_1 = x_1 + iy_1$ and $z_2 = x_2 + iy_2$. Indeed, by (1),

$$e^{z_1}e^{z_2} = e^{x_1}(\cos y_1 + i \sin y_1)e^{x_2}(\cos y_2 + i \sin y_2).$$

Since $e^{x_1}e^{x_2} = e^{x_1+x_2}$ for these *real* functions, by an application of the addition formulas for the cosine and sine functions (similar to that in Sec. 13.2) we see that

$$e^{z_1}e^{z_2} = e^{x_1+x_2}[\cos(y_1 + y_2) + i \sin(y_1 + y_2)] = e^{z_1+z_2}$$

as asserted. An interesting special case of (3) is $z_1 = x$, $z_2 = iy$; then

$$e^z = e^x e^{iy}. \tag{4}$$

Furthermore, for $z = iy$ we have from (1) the so-called **Euler formula**

$$e^{iy} = \cos y + i \sin y. \tag{5}$$

Hence the **polar form** of a complex number, $z = r(\cos\theta + i \sin\theta)$, may now be written

$$z = re^{i\theta}. \tag{6}$$

From (5) we obtain

$$e^{2\pi i} = 1 \tag{7}$$

as well as the important formulas (verify!)

$$e^{\pi i/2} = i, \qquad e^{\pi i} = -1, \qquad e^{-\pi i/2} = -i, \qquad e^{-\pi i} = -1. \tag{8}$$

Another consequence of (5) is

$$|e^{iy}| = |\cos y + i \sin y| = \sqrt{\cos^2 y + \sin^2 y} = 1. \tag{9}$$

That is, for pure imaginary exponents, the exponential function has absolute value 1, a result you should remember. From (9) and (1),

$$|e^z| = e^x. \qquad \text{Hence} \qquad \arg e^z = y \pm 2n\pi \quad (n = 0, 1, 2, \cdots), \tag{10}$$

since $|e^z| = e^x$ shows that (1) is actually e^z in polar form.

From $|e^z| = e^x \neq 0$ in (10) we see that

$$e^x \neq 0 \qquad \text{for all } z. \tag{11}$$

So here we have an entire function that never vanishes, in contrast to (nonconstant) polynomials, which are also entire (Example 5 in Sec. 13.3) but always have a zero, as is proved in algebra.

Periodicity of e^x with period $2\pi i$,

$$e^{z+2\pi i} = e^z \qquad \text{for all } z \tag{12}$$

is a basic property that follows from (1) and the periodicity of $\cos y$ and $\sin y$. Hence all the values that $w = e^z$ can assume are already assumed in the horizontal strip of width 2π

$$-\pi < y \leqq \pi \qquad \text{(Fig. 336).} \tag{13}$$

This infinite strip is called a **fundamental region** of e^z.

EXAMPLE 1 Function Values. Solution of Equations

Computation of values from (1) provides no problem. For instance,

$$e^{1.4-0.6i} = e^{1.4}(\cos 0.6 - i \sin 0.6) = 4.055(0.8253 - 0.5646i) = 3.347 - 2.289i$$

$$|e^{1.4-1.6i}| = e^{1.4} = 4.055, \qquad \text{Arg}\, e^{1.4-0.6i} = -0.6.$$

To illustrate (3), take the product of

$$e^{2+i} = e^2(\cos 1 + i \sin 1) \qquad \text{and} \qquad e^{4-i} = e^4(\cos 1 - i \sin 1)$$

and verify that it equals $e^2e^4(\cos^2 1 + \sin^2 1) = e^6 = e^{(2+i)+(4-i)}$.

To solve the equation $e^z = 3 + 4i$, note first that $|e^z| = e^x = 5$, $x = \ln 5 = 1.609$ is the real part of all solutions. Now, since $e^x = 5$,

$$e^x \cos y = 3, \qquad e^x \sin y = 4, \qquad \cos y = 0.6, \qquad \sin y = 0.8, \qquad y = 0.927.$$

Ans. $z = 1.609 + 0.927i \pm 2n\pi i$ $(n = 0, 1, 2, \cdots)$. These are infinitely many solutions (due to the periodicity of e^z). They lie on the vertical line $x = 1.609$ at a distance 2π from their neighbors. ■

To summarize: many properties of $e^z = \exp z$ parallel those of e^x; an exception is the periodicity of e^z with $2\pi i$, which suggested the concept of a fundamental region. Keep in mind that e^z is an *entire function.* (Do you still remember what that means?)

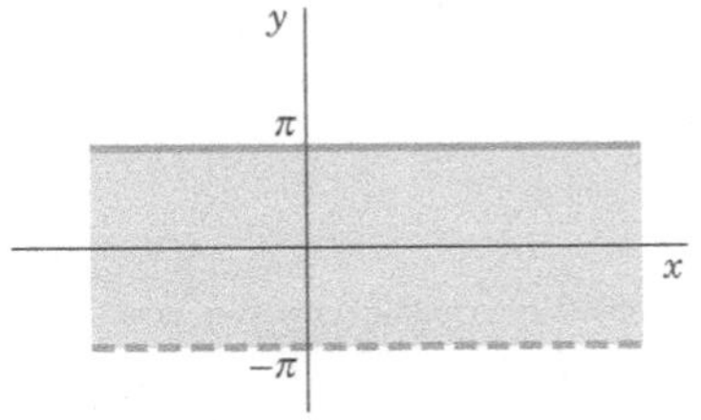

Fig. 336. Fundamental region of the exponential function e^z in the z-plane

PROBLEM SET 13.5

1. e^z is entire. Prove this.

2–7 **Function Values.** Find e^z in the form $u + iv$ and $|e^z|$ if z equals

2. $3 + 4i$

3. $2\pi i(1 + i)$

4. $0.6 - 1.8i$

5. $2 + 3\pi i$

6. $11\pi i/2$

7. $\sqrt{2} + \frac{1}{2}\pi i$

8–13 **Polar Form.** Write in exponential form (6):

8. $\sqrt[n]{z}$

9. $4 + 3i$

10. $\sqrt{i}$, $\sqrt{-i}$

11. -6.3

12. $1/(1 - z)$

13. $1 + i$

14–17 **Real and Imaginary Parts.** Find Re and Im of

14. $e^{-\pi z}$

15. $\exp(z^2)$

16. $e^{1/z}$

17. $\exp(z^3)$

18. TEAM PROJECT. Further Properties of the Exponential Function. (a) Analyticity. Show that e^z is entire. What about $e^{1/z}$? $e^{\bar{z}}$? $e^x(\cos ky + i \sin ky)$? (Use the Cauchy–Riemann equations.)

(b) Special values. Find all z such that (i) e^z is real, (ii) $|e^{-z}| < 1$, (iii) $e^{\bar{z}} = \overline{e^z}$.

(c) Harmonic function. Show that $u = e^{xy} \cos(x^2/2 - y^2/2)$ is harmonic and find a conjugate.

(d) Uniqueness. It is interesting that $f(z) = e^z$ is uniquely determined by the two properties $f(x + i0) = e^x$ and $f'(z) = f(z)$, where f is assumed to be entire. Prove this using the Cauchy–Riemann equations.

19–22 **Equations.** Find all solutions and graph some of them in the complex plane.

19. $e^z = 1$

20. $e^z = 4 + 3i$

21. $e^z = 0$

22. $e^z = -2$

13.6 Trigonometric and Hyperbolic Functions. Euler's Formula

Just as we extended the real e^x to the complex e^z in Sec. 13.5, we now want to extend the familiar *real* trigonometric functions to *complex trigonometric functions.* We can do this by the use of the Euler formulas (Sec. 13.5)

$$e^{ix} = \cos x + i \sin x, \qquad e^{-ix} = \cos x - i \sin x.$$

By addition and subtraction we obtain for the *real* cosine and sine

$$\cos x = \tfrac{1}{2}(e^{ix} + e^{-ix}), \qquad \sin x = \frac{1}{2i}(e^{ix} - e^{-ix}).$$

This suggests the following definitions for complex values $z = x + iy$:

$$\cos z = \tfrac{1}{2}(e^{iz} + e^{-iz}), \qquad \sin z = \frac{1}{2i}(e^{iz} - e^{-iz}). \tag{1}$$

It is quite remarkable that here in complex, functions come together that are unrelated in real. This is not an isolated incident but is typical of the general situation and shows the advantage of working in complex.

Furthermore, as in calculus we define

$$\tan z = \frac{\sin z}{\cos z}, \qquad \cot z = \frac{\cos z}{\sin z} \tag{2}$$

and

$$\sec z = \frac{1}{\cos z}, \qquad \csc z = \frac{1}{\sin z}. \tag{3}$$

Since e^z is entire, $\cos z$ and $\sin z$ are entire functions. $\tan z$ and $\sec z$ are not entire; they are analytic except at the points where $\cos z$ is zero; and $\cot z$ and $\csc z$ are analytic except where $\sin z$ is zero. Formulas for the derivatives follow readily from $(e^z)' = e^z$ and (1)–(3); as in calculus,

$$(4) \qquad (\cos z)' = -\sin z, \qquad (\sin z)' = \cos z, \qquad (\tan z)' = \sec^2 z,$$

etc. Equation (1) also shows that **Euler's formula *is valid in complex***:

$$\textbf{(5)} \qquad e^{iz} = \cos z + i \sin z \qquad \text{for all } z.$$

The real and imaginary parts of $\cos z$ and $\sin z$ are needed in computing values, and they also help in displaying properties of our functions. We illustrate this with a typical example.

EXAMPLE 1 Real and Imaginary Parts. Absolute Value. Periodicity

Show that

$$\textbf{(6)} \qquad \begin{aligned} &\textbf{(a)} \quad \cos z = \cos x \cosh y - i \sin x \sinh y \\ &\textbf{(b)} \quad \sin z = \sin x \cosh y + i \cos x \sinh y \end{aligned}$$

and

$$(7) \qquad \begin{aligned} &\textbf{(a)} \quad |\cos z|^2 = \cos^2 x + \sinh^2 y \\ &\textbf{(b)} \quad |\sin z|^2 = \sin^2 x + \sinh^2 y \end{aligned}$$

and give some applications of these formulas.

Solution. From (1),

$$\begin{aligned} \cos z &= \tfrac{1}{2}(e^{i(x+iy)} + e^{-i(x+iy)}) \\ &= \tfrac{1}{2}e^{-y}(\cos x + i \sin x) + \tfrac{1}{2}e^{y}(\cos x - i \sin x) \\ &= \tfrac{1}{2}(e^{y} + e^{-y}) \cos x - \tfrac{1}{2}i(e^{y} - e^{-y}) \sin x. \end{aligned}$$

This yields (6a) since, as is known from calculus,

$$(8) \qquad \cosh y = \tfrac{1}{2}(e^{y} + e^{-y}), \qquad \sinh y = \tfrac{1}{2}(e^{y} - e^{-y});$$

(6b) is obtained similarly. From (6a) and $\cosh^2 y = 1 + \sinh^2 y$ we obtain

$$|\cos z|^2 = (\cos^2 x)(1 + \sinh^2 y) + \sin^2 x \sinh^2 y.$$

Since $\sin^2 x + \cos^2 x = 1$, this gives (7a), and (7b) is obtained similarly.

For instance, $\cos(2 + 3i) = \cos 2 \cosh 3 - i \sin 2 \sinh 3 = -4.190 - 9.109i$.

From (6) we see that $\sin z$ and $\cos z$ are ***periodic with period*** 2π, just as in real. Periodicity of $\tan z$ and $\cot z$ with period π now follows.

Formula (7) points to an essential difference between the real and the complex cosine and sine; whereas $|\cos x| \leqq 1$ and $|\sin x| \leqq 1$, the complex cosine and sine functions are ***no longer bounded*** but approach infinity in absolute value as $y \to \infty$, since then $\sinh y \to \infty$ in (7). ■

EXAMPLE 2 Solutions of Equations. Zeros of cos z and sin z

Solve (a) $\cos z = 5$ (which has no real solution!), (b) $\cos z = 0$, (c) $\sin z = 0$.

Solution. (a) $e^{2iz} - 10e^{iz} + 1 = 0$ from (1) by multiplication by e^{iz}. This is a quadratic equation in e^{iz}, with solutions (rounded off to 3 decimals)

$$e^{iz} = e^{-y+ix} = 5 \pm \sqrt{25 - 1} = 9.899 \quad \text{and} \quad 0.101.$$

Thus $e^{-y} = 9.899$ or 0.101, $e^{ix} = 1$, $y = \pm 2.292$, $x = 2n\pi$. *Ans.* $z = \pm 2n\pi \pm 2.292i$ $(n = 0, 1, 2, \cdots)$.

Can you obtain this from (6a)?

(b) $\cos x = 0$, $\sinh y = 0$ by (7a), $y = 0$. *Ans.* $z = \pm\frac{1}{2}(2n + 1)\pi$ $(n = 0, 1, 2, \cdots)$.
(c) $\sin x = 0$, $\sinh y = 0$ by (7b), *Ans.* $z = \pm n\pi$ $(n = 0, 1, 2, \cdots)$.
Hence the only zeros of $\cos z$ and $\sin z$ are those of the real cosine and sine functions. ■

General formulas *for the real trigonometric functions continue to hold for complex values.* This follows immediately from the definitions. We mention in particular the addition rules

$$\begin{aligned} \cos(z_1 \pm z_2) &= \cos z_1 \cos z_2 \mp \sin z_1 \sin z_2 \\ \sin(z_1 \pm z_2) &= \sin z_1 \cos z_2 \pm \sin z_2 \cos z_1 \end{aligned} \tag{9}$$

and the formula

$$\cos^2 z + \sin^2 z = 1. \tag{10}$$

Some further useful formulas are included in the problem set.

Hyperbolic Functions

The complex **hyperbolic cosine** and **sine** are defined by the formulas

$$\cosh z = \tfrac{1}{2}(e^z + e^{-z}), \qquad \sinh z = \tfrac{1}{2}(e^z - e^{-z}). \tag{11}$$

This is suggested by the familiar definitions for a real variable [see (8)]. These functions are entire, with derivatives

$$(\cosh z)' = \sinh z, \qquad (\sinh z)' = \cosh z, \tag{12}$$

as in calculus. The other hyperbolic functions are defined by

$$\begin{aligned} \tanh z &= \frac{\sinh z}{\cosh z}, \qquad & \coth z &= \frac{\cosh z}{\sinh z}, \\ \operatorname{sech} z &= \frac{1}{\cosh z}, \qquad & \operatorname{csch} z &= \frac{1}{\sinh z}. \end{aligned} \tag{13}$$

Complex Trigonometric and Hyperbolic Functions Are Related. If in (11), we replace z by iz and then use (1), we obtain

$$\cosh iz = \cos z, \qquad \sinh iz = i \sin z. \tag{14}$$

Similarly, if in (1) we replace z by iz and then use (11), we obtain conversely

$$\cos iz = \cosh z, \qquad \sin iz = i \sinh z. \tag{15}$$

Here we have another case of *unrelated* real functions that have *related* complex analogs, pointing again to the advantage of working in complex in order to get both a more unified formalism and a deeper understanding of special functions. This is one of the main reasons for the importance of complex analysis to the engineer and physicist.

PROBLEM SET 13.6

1–4 FORMULAS FOR HYPERBOLIC FUNCTIONS

Show that

1. $\cosh z = \cosh x \cos y + i \sinh x \sin y$

$\sinh z = \sinh x \cos y + i \cosh x \sin y.$

2. $\cosh (z_1 + z_2) = \cosh z_1 \cosh z_2 + \sinh z_1 \sinh z_2$

$\sinh (z_1 + z_2) = \sinh z_1 \cosh z_2 + \cosh z_1 \sinh z_2.$

3. $\cosh^2 z - \sinh^2 z = 1, \quad \cosh^2 z + \sinh^2 z = \cosh 2z$

4. Entire Functions. Prove that $\cos z$, $\sin z$, $\cosh z$, and $\sinh z$ are entire.

5. Harmonic Functions. Verify by differentiation that $\operatorname{Im} \cos z$ and $\operatorname{Re} \sin z$ are harmonic.

6–12 Function Values. Find, in the form $u + iv$,

6. $\sin 2\pi i$

7. $\cos i, \quad \sin i$

8. $\cos \pi i, \quad \cosh \pi i$

9. $\cosh (-1 + 2i), \quad \cos (-2 - i)$

10. $\sinh (3 + 4i), \quad \cosh (3 + 4i)$

11. $\sin \pi i, \quad \cos (\frac{1}{2}\pi - \pi i)$

12. $\cos \frac{1}{2}\pi i, \quad \cos [\frac{1}{2}\pi(1 + i)]$

13–15 Equations and Inequalities. Using the definitions, prove:

13. $\cos z$ is even, $\cos(-z) = \cos z$, and $\sin z$ is odd, $\sin(-z) = -\sin z$.

14. $|\sinh y| \leqq |\cos z| \leqq \cosh y, |\sinh y| \leqq |\sin z| \leqq \cosh y$. Conclude that the complex cosine and sine are not bounded in the whole complex plane.

15. $\sin z_1 \cos z_2 = \frac{1}{2}[\sin(z_1 + z_2) + \sin(z_1 - z_2)]$

16–19 Equations. Find all solutions.

16. $\sin z = 100$

17. $\cosh z = 0$

18. $\cosh z = -1$

19. $\sinh z = 0$

20. Re tan z and Im tan z. Show that

$$\operatorname{Re} \tan z = \frac{\sin x \cos x}{\cos^2 x + \sinh^2 y},$$

$$\operatorname{Im} \tan z = \frac{\sinh y \cosh y}{\cos^2 x + \sinh^2 y}.$$

13.7 Logarithm. General Power. Principal Value

We finally introduce the *complex logarithm*, which is more complicated than the real logarithm (which it includes as a special case) and historically puzzled mathematicians for some time (so if you first get puzzled—which need not happen!—be patient and work through this section with extra care).

The **natural logarithm** of $z = x + iy$ is denoted by $\ln z$ (sometimes also by $\log z$) and is defined as the inverse of the exponential function; that is, $w = \ln z$ is defined for $z \neq 0$ by the relation

$$e^w = z.$$

(Note that $z = 0$ is impossible, since $e^w \neq 0$ for all w; see Sec. 13.5.) If we set $w = u + iv$ and $z = re^{i\theta}$, this becomes

$$e^w = e^{u+iv} = re^{i\theta}.$$

Now, from Sec. 13.5, we know that e^{u+iv} has the absolute value e^u and the argument v. These must be equal to the absolute value and argument on the right:

$$e^u = r, \qquad v = \theta.$$

$e^u = r$ gives $u = \ln r$, where $\ln r$ is the familiar *real* natural logarithm of the positive number $r = |z|$. Hence $w = u + iv = \ln z$ is given by

(1)
$$\ln z = \ln r + i\theta \qquad\qquad (r = |z| > 0, \quad \theta = \arg z).$$

Now comes an important point (without analog in real calculus). Since the argument of z is determined only up to integer multiples of 2π, *the complex natural logarithm* $\ln z \ (z \neq 0)$ *is infinitely many-valued.*

The value of $\ln z$ corresponding to the principal value Arg z (see Sec. 13.2) is denoted by Ln z (Ln with capital L) and is called the **principal value** of $\ln z$. Thus

(2)
$$\text{Ln } z = \ln |z| + i \text{ Arg } z \qquad\qquad (z \neq 0).$$

The uniqueness of Arg z for given $z \ (\neq 0)$ implies that Ln z is single-valued, that is, a function in the usual sense. Since the other values of arg z differ by integer multiples of 2π, the other values of $\ln z$ are given by

(3)
$$\ln z = \text{Ln } z \pm 2n\pi i \qquad\qquad (n = 1, 2, \cdots).$$

They all have the same real part, and their imaginary parts differ by integer multiples of 2π.

If z is positive real, then Arg $z = 0$, and Ln z becomes identical with the real natural logarithm known from calculus. If z is negative real (so that the natural logarithm of calculus is not defined!), then Arg $z = \pi$ and

$$\text{Ln } z = \ln |z| + \pi i \qquad\qquad (z \text{ negative real}).$$

From (1) and $e^{\ln r} = r$ for positive real r we obtain

(4a)
$$e^{\ln z} = z$$

as expected, but since arg $(e^z) = y \pm 2n\pi$ is multivalued, so is

(4b)
$$\ln (e^z) = z \pm 2n\pi i, \qquad\qquad n = 0, 1, \cdots.$$

EXAMPLE 1 **Natural Logarithm. Principal Value**

$$\ln 1 = 0, \pm 2\pi i, \pm 4\pi i, \cdots \qquad\qquad \text{Ln } 1 = 0$$

$$\ln 4 = 1.386294 \pm 2n\pi i \qquad\qquad \text{Ln } 4 = 1.386294$$

$$\ln (-1) = \pm \pi i, \pm 3\pi i, \pm 5\pi i, \cdots \qquad\qquad \text{Ln } (-1) = \pi i$$

$$\ln (-4) = 1.386294 \pm (2n + 1)\pi i \qquad\qquad \text{Ln } (-4) = 1.386294 + \pi i$$

$$\ln i = \pi i/2, -3\pi/2, 5\pi i/2, \cdots \qquad\qquad \text{Ln } i = \pi i/2$$

$$\ln 4i = 1.386294 + \pi i/2 \pm 2n\pi i \qquad\qquad \text{Ln } 4i = 1.386294 + \pi i/2$$

$$\ln (-4i) = 1.386294 - \pi i/2 \pm 2n\pi i \qquad\qquad \text{Ln } (-4i) = 1.386294 - \pi i/2$$

$$\ln (3 - 4i) = \ln 5 + i \arg (3 - 4i) \qquad\qquad \text{Ln } (3 - 4i) = 1.609438 - 0.927295i$$

$$= 1.609438 - 0.927295i \pm 2n\pi i \qquad\qquad (\text{Fig. 337}) \qquad\qquad \blacksquare$$

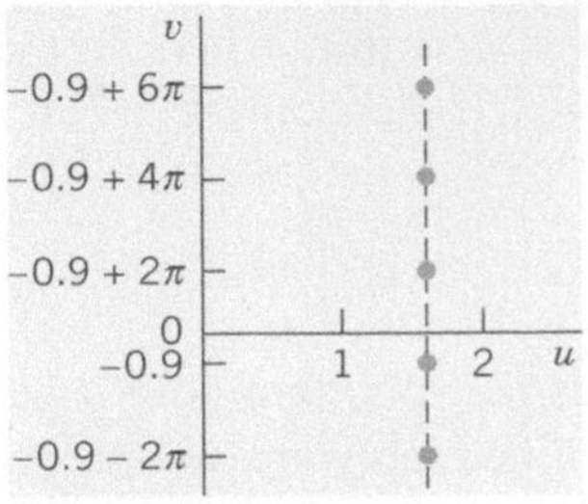

Fig. 337. Some values of ln (3 − 4*i*) in Example 1

The familiar relations for the natural logarithm continue to hold for complex values, that is,

$$\text{(5)} \qquad \text{(a)} \quad \ln(z_1 z_2) = \ln z_1 + \ln z_2, \qquad \text{(b)} \quad \ln(z_1/z_2) = \ln z_1 - \ln z_2$$

but these relations are to be understood in the sense that each value of one side is also contained among the values of the other side; see the next example.

EXAMPLE 2 **Illustration of the Functional Relation (5) in Complex**

Let

$$z_1 = z_2 = e^{\pi i} = -1.$$

If we take the principal values

$$\operatorname{Ln} z_1 = \operatorname{Ln} z_2 = \pi i,$$

then (5a) holds provided we write $\ln(z_1 z_2) = \ln 1 = 2\pi i$; however, it is not true for the principal value, $\operatorname{Ln}(z_1 z_2) = \operatorname{Ln} 1 = 0$. ■

THEOREM 1 **Analyticity of the Logarithm**

For every $n = 0, \pm 1, \pm 2, \cdots$ *formula* (3) *defines a function, which is analytic, except at* 0 *and on the negative real axis, and has the derivative*

$$\text{(6)} \qquad (\ln z)' = \frac{1}{z} \qquad (z \text{ not } 0 \text{ or negative real}).$$

PROOF We show that the Cauchy–Riemann equations are satisfied. From (1)–(3) we have

$$\ln z = \ln r + i(\theta + c) = \frac{1}{2}\ln(x^2 + y^2) + i\left(\arctan\frac{y}{x} + c\right)$$

where the constant c is a multiple of 2π. By differentiation,

$$u_x = \frac{x}{x^2 + y^2} = v_y = \frac{1}{1 + (y/x)^2} \cdot \frac{1}{x}$$

$$u_y = \frac{y}{x^2 + y^2} = -v_x = -\frac{1}{1 + (y/x)^2}\left(-\frac{y}{x^2}\right).$$

Hence the Cauchy–Riemann equations hold. [Confirm this by using these equations in polar form, which we did not use since we proved them only in the problems (to Sec. 13.4).] Formula (4) in Sec. 13.4 now gives (6),

$$(\ln z)' = u_x + iv_x = \frac{x}{x^2 + y^2} + i\,\frac{1}{1 + (y/x)^2}\left(-\frac{y}{x^2}\right) = \frac{x - iy}{x^2 + y^2} = \frac{1}{z}. \qquad \blacksquare$$

Each of the infinitely many functions in (3) is called a **branch** of the logarithm. The negative real axis is known as a **branch cut** and is usually graphed as shown in Fig. 338. The branch for $n = 0$ is called the **principal branch** of $\ln z$.

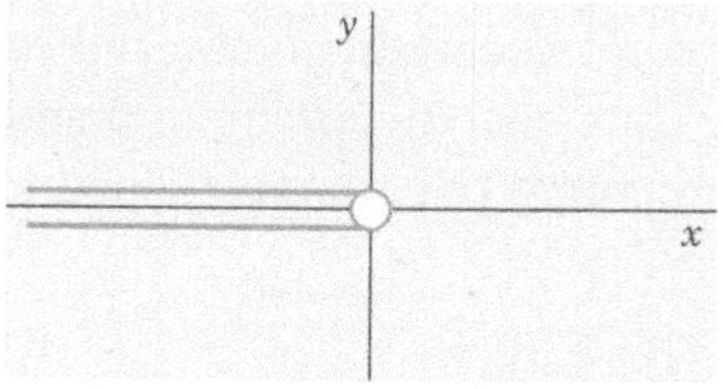

Fig. 338.   Branch cut for $\ln z$

General Powers

General powers of a complex number $z = x + iy$ are defined by the formula

(7) $$z^c = e^{c \ln z} \qquad\qquad (c \text{ complex, } z \neq 0).$$

Since $\ln z$ is infinitely many-valued, z^c will, in general, be multivalued. The particular value

$$z^c = e^{c\,\mathrm{Ln}\,z}$$

is called the **principal value** *of* z^c.

If $c = n = 1, 2, \cdots$, then z^n is single-valued and identical with the usual nth power of z. If $c = -1, -2, \cdots$, the situation is similar.

If $c = 1/n$, where $n = 2, 3, \cdots$, then

$$z^c = \sqrt[n]{z} = e^{(1/n)\ln z} \qquad\qquad (z \neq 0),$$

the exponent is determined up to multiples of $2\pi i/n$ and we obtain the n distinct values of the nth root, in agreement with the result in Sec. 13.2. If $c = p/q$, the quotient of two positive integers, the situation is similar, and z^c has only finitely many distinct values. However, if c is real irrational or genuinely complex, then z^c is infinitely many-valued.

EXAMPLE 3   **General Power**

$$i^i = e^{i \ln i} = \exp(i \ln i) = \exp\left[i\left(\frac{\pi}{2}i \pm 2n\pi i\right)\right] = e^{-(\pi/2)\mp 2n\pi}.$$

All these values are real, and the principal value ($n = 0$) is $e^{-\pi/2}$.

Similarly, by direct calculation and multiplying out in the exponent,

$$(1 + i)^{2-i} = \exp\left[(2 - i)\ln(1 + i)\right] = \exp\left[(2 - i)\{\ln \sqrt{2} + \tfrac{1}{4}\pi i \pm 2n\pi i\}\right]$$

$$= 2e^{\pi/4 \pm 2n\pi}\left[\sin(\tfrac{1}{2}\ln 2) + i\cos(\tfrac{1}{2}\ln 2)\right]. \qquad \blacksquare$$

It is a ***convention*** that for real positive $z = x$ the expression z^c means $e^{c \ln x}$ where $\ln x$ is the elementary real natural logarithm (that is, the principal value $\text{Ln } z$ $(z = x > 0)$ in the sense of our definition). Also, if $z = e$, the base of the natural logarithm, $z^c = e^c$ is *conventionally* regarded as the unique value obtained from (1) in Sec. 13.5.

From (7) we see that for any complex number a,

$$a^z = e^{z \ln a}. \tag{8}$$

We have now introduced the complex functions needed in practical work, some of them (e^z, $\cos z$, $\sin z$, $\cosh z$, $\sinh z$) entire (Sec. 13.5), some of them ($\tan z$, $\cot z$, $\tanh z$, $\coth z$) analytic except at certain points, and one of them ($\ln z$) splitting up into infinitely many functions, each analytic except at 0 and on the negative real axis.

For the **inverse trigonometric** and **hyperbolic functions** see the problem set.

PROBLEM SET 13.7

1–4 VERIFICATIONS IN THE TEXT

1. Verify the computations in Example 1.
2. Verify (5) for $z_1 = -i$ and $z_2 = -1$.
3. Prove analyticity of $\text{Ln } z$ by means of the Cauchy–Riemann equations in polar form (Sec. 13.4).
4. Prove (4a) and (4b).

COMPLEX NATURAL LOGARITHM *ln z*

5–11 Principal Value Ln z. Find $\text{Ln } z$ when z equals

5. -11
6. $4 + 4i$
7. $4 - 4i$
8. $1 \pm i$
9. $0.6 + 0.8i$
10. $-15 \pm 0.1i$
11. ei

12–16 All Values of ln z. Find all values and graph some of them in the complex plane.

12. $\ln e$
13. $\ln 1$
14. $\ln(-7)$
15. $\ln(e^i)$
16. $\ln(4 + 3i)$
17. Show that the set of values of $\ln(i^2)$ differs from the set of values of $2 \ln i$.

18–21 Equations. Solve for z.

18. $\ln z = -\pi i/2$
19. $\ln z = 4 - 3i$
20. $\ln z = e - \pi i$
21. $\ln z = 0.6 + 0.4i$

22–28 General Powers. Find the principal value. Show details.

22. $(2i)^{2i}$
23. $(1 + i)^{1-i}$
24. $(1 - i)^{1+i}$
25. $(-3)^{3-i}$
26. $(i)^{i/2}$
27. $(-1)^{2-i}$
28. $(3 + 4i)^{1/3}$
29. How can you find the answer to Prob. 24 from the answer to Prob. 23?
30. **TEAM PROJECT. Inverse Trigonometric and Hyperbolic Functions.** By definition, the **inverse sine** $w = \arcsin z$ is the relation such that $\sin w = z$. The **inverse cosine** $w = \arccos z$ is the relation such that $\cos w = z$. The **inverse tangent, inverse cotangent, inverse hyperbolic sine,** etc., are defined and denoted in a similar fashion. (Note that all these relations are ***multivalued***.) Using $\sin w = (e^{iw} - e^{-iw})/(2i)$ and similar representations of $\cos w$, etc., show that

    (a) $\arccos z = -i \ln(z + \sqrt{z^2 - 1})$

    (b) $\arcsin z = -i \ln(iz + \sqrt{1 - z^2})$

    (c) $\text{arccosh } z = \ln(z + \sqrt{z^2 - 1})$

    (d) $\text{arcsinh } z = \ln(z + \sqrt{z^2 + 1})$

    (e) $\arctan z = \dfrac{i}{2} \ln \dfrac{i + z}{i - z}$

    (f) $\text{arctanh } z = \dfrac{1}{2} \ln \dfrac{1 + z}{1 - z}$

    (g) Show that $w = \arcsin z$ is infinitely many-valued, and if w_1 is one of these values, the others are of the form $w_1 \pm 2n\pi$ and $\pi - w_1 \pm 2n\pi$, $n = 0, 1, \cdots$. (The *principal value of* $w = u + iv = \arcsin z$ is defined to be the value for which $-\pi/2 \leqq u \leqq \pi/2$ if $v \geqq 0$ and $-\pi/2 < u < \pi/2$ if $v < 0$.)

CHAPTER 13 REVIEW QUESTIONS AND PROBLEMS

1. Divide $15 + 23i$ by $-3 + 7i$. Check the result by multiplication.
2. What happens to a quotient if you take the complex conjugates of the two numbers? If you take the absolute values of the numbers?
3. Write the two numbers in Prob. 1 in polar form. Find the principal values of their arguments.
4. State the definition of the derivative from memory. Explain the big difference from that in calculus.
5. What is an analytic function of a complex variable?
6. Can a function be differentiable at a point without being analytic there? If yes, give an example.
7. State the Cauchy–Riemann equations. Why are they of basic importance?
8. Discuss how e^z, $\cos z$, $\sin z$, $\cosh z$, $\sinh z$ are related.
9. $\ln z$ is more complicated than $\ln x$. Explain. Give examples.
10. How are general powers defined? Give an example. Convert it to the form $x + iy$.

11–16 **Complex Numbers.** Find, in the form $x + iy$, showing details,

11. $(2 + 3i)^2$
12. $(1 - i)^{10}$
13. $1/(4 + 3i)$
14. $\sqrt{i}$
15. $(1 + i)/(1 - i)$
16. $e^{\pi i/2}$, $e^{-\pi i/2}$

17–20 **Polar Form.** Represent in polar form, with the principal argument.

17. $-4 - 4i$
18. $12 + i$, $12 - i$
19. $-15i$
20. $0.6 + 0.8i$

21–24 **Roots.** Find and graph all values of:

21. $\sqrt{81}$
22. $\sqrt{-32i}$
23. $\sqrt[4]{-1}$
24. $\sqrt[3]{1}$

25–30 **Analytic Functions.** Find $f(z) = u(x, y) + iv(x, y)$ with u or v as given. Check by the Cauchy–Riemann equations for analyticity.

25 $u = xy$
26. $v = y/(x^2 + y^2)$
27. $v = -e^{-2x} \sin 2y$
28. $u = \cos 3x \cosh 3y$
29. $u = \exp(-(x^2 - y^2)/2) \cos xy$
30. $v = \cos 2x \sinh 2y$

31–35 **Special Function Values.** Find the value of:

31. $\cos(3 - i)$
32. $\text{Ln}(0.6 + 0.8i)$
33. $\tan i$
34. $\sinh(1 + \pi i)$, $\sin(1 + \pi i)$
35. $\cosh(\pi + \pi i)$

SUMMARY OF CHAPTER 13
Complex Numbers and Functions. Complex Differentiation

For arithmetic operations with **complex numbers**

$$(1) \qquad z = x + iy = re^{i\theta} = r(\cos\theta + i\sin\theta),$$

$r = |z| = \sqrt{x^2 + y^2}$, $\theta = \arctan(y/x)$, and for their representation in the complex plane, see Secs. 13.1 and 13.2.

A complex function $f(z) = u(x, y) + iv(x, y)$ is **analytic** in a domain D if it has a **derivative** (Sec. 13.3)

$$(2) \qquad f'(z) = \lim_{\Delta z \to 0} \frac{f(z + \Delta z) - f(z)}{\Delta z}$$

everywhere in D. Also, $f(z)$ is *analytic at a point* $z = z_0$ if it has a derivative in a neighborhood of z_0 (not merely at z_0 itself).

If $f(z)$ is analytic in D, then $u(x, y)$ and $v(x, y)$ satisfy the (very important!) **Cauchy–Riemann equations** (Sec. 13.4)

$$\frac{\partial u}{\partial x} = \frac{\partial v}{\partial y}, \qquad \frac{\partial u}{\partial y} = -\frac{\partial v}{\partial x} \tag{3}$$

everywhere in D. Then u and v also satisfy **Laplace's equation**

$$u_{xx} + u_{yy} = 0, \qquad v_{xx} + v_{yy} = 0 \tag{4}$$

everywhere in D. If $u(x, y)$ and $v(x, y)$ are continuous and have *continuous* partial derivatives in D that satisfy (3) in D, then $f(z) = u(x, y) + iv(x, y)$ is analytic in D. See Sec. 13.4. (More on Laplace's equation and complex analysis follows in Chap. 18.)

The complex **exponential function** (Sec. 13.5)

$$e^z = \exp z = e^x (\cos y + i \sin y) \tag{5}$$

reduces to e^x if $z = x$ $(y = 0)$. It is periodic with $2\pi i$ and has the derivative e^z.

The **trigonometric functions** are (Sec. 13.6)

$$\begin{aligned} \cos z &= \tfrac{1}{2}(e^{iz} + e^{-iz}) = \cos x \cosh y - i \sin x \sinh y \\ \sin z &= \frac{1}{2i}(e^{iz} - e^{-iz}) = \sin x \cosh y + i \cos x \sinh y \end{aligned} \tag{6}$$

and, furthermore,

$$\tan z = (\sin z)/\cos z, \qquad \cot z = 1/\tan z, \quad \text{etc.}$$

The **hyperbolic functions** are (Sec. 13.6)

$$\cosh z = \tfrac{1}{2}(e^z + e^{-z}) = \cos iz, \qquad \sinh z = \tfrac{1}{2}(e^z - e^{-z}) = -i \sin iz \tag{7}$$

etc. The functions (5)–(7) are **entire**, that is, analytic everywhere in the complex plane.

The **natural logarithm** is (Sec. 13.7)

$$\ln z = \ln|z| + i \arg z = \ln|z| + i \operatorname{Arg} z \pm 2n\pi i \tag{8}$$

where $z \neq 0$ and $n = 0, 1, \cdots$. Arg z is the **principal value** of arg z, that is, $-\pi < \operatorname{Arg} z \leqq \pi$. We see that $\ln z$ is infinitely many-valued. Taking $n = 0$ gives the **principal value** Ln z of ln z; thus $\operatorname{Ln} z = \ln|z| + i \operatorname{Arg} z$.

General powers are defined by (Sec. 13.7)

$$z^c = e^{c \ln z} \qquad (c \text{ complex}, z \neq 0). \tag{9}$$

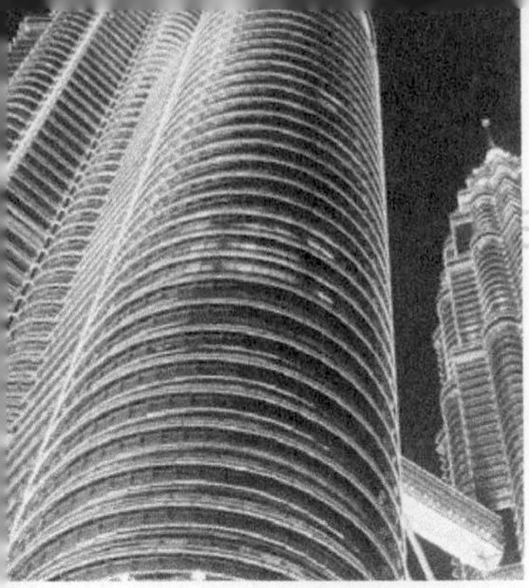

APPENDIX 1

References

Software *see* at the beginning of Chaps. 19 and 24.

General References

[GenRef1] Abramowitz, M. and I. A. Stegun (eds.), *Handbook of Mathematical Functions.* 10th printing, with corrections. Washington, DC: National Bureau of Standards. 1972 (also New York: Dover, 1965). See also [W1]

[GenRef2] Cajori, F., *History of Mathematics.* 5th ed. Reprinted. Providence, RI: American Mathematical Society, 2002.

[GenRef3] Courant, R. and D. Hilbert, *Methods of Mathematical Physics.* 2 vols. Hoboken, NJ: Wiley, 1989.

[GenRef4] Courant, R., *Differential and Integral Calculus.* 2 vols. Hoboken, NJ: Wiley, 1988.

[GenRef5] Graham, R. L. et al., *Concrete Mathematics.* 2nd ed. Reading, MA: Addison-Wesley, 1994.

[GenRef6] Ito, K. (ed.), *Encyclopedic Dictionary of Mathematics.* 4 vols. 2nd ed. Cambridge, MA: MIT Press, 1993.

[GenRef7] Kreyszig, E., *Introductory Functional Analysis with Applications.* New York: Wiley, 1989.

[GenRef8] Kreyszig, E., *Differential Geometry.* Mineola, NY: Dover, 1991.

[GenRef9] Kreyszig, E. *Introduction to Differential Geometry and Riemannian Geometry.* Toronto: University of Toronto Press, 1975.

[GenRef10] Szegö, G., *Orthogonal Polynomials.* 4th ed. Reprinted. New York: American Mathematical Society, 2003.

[GenRef11] Thomas, G. et al., *Thomas' Calculus, Early Transcendentals Update.* 10th ed. Reading, MA: Addison-Wesley, 2003.

Part A. Ordinary Differential Equations (ODEs) (Chaps. 1–6)

See also Part E: Numeric Analysis

[A1] Arnold, V. I., *Ordinary Differential Equations.* 3rd ed. New York: Springer, 2006.

[A2] Bhatia, N. P. and G. P. Szego, *Stability Theory of Dynamical Systems.* New York: Springer, 2002.

[A3] Birkhoff, G. and G.-C. Rota, *Ordinary Differential Equations.* 4th ed. New York: Wiley, 1989.

[A4] Brauer, F. and J. A. Nohel, *Qualitative Theory of Ordinary Differential Equations.* Mineola, NY: Dover, 1994.

[A5] Churchill, R. V., *Operational Mathematics.* 3rd ed. New York: McGraw-Hill, 1972.

[A6] Coddington, E. A. and R. Carlson, *Linear Ordinary Differential Equations.* Philadelphia: SIAM, 1997.

[A7] Coddington, E. A. and N. Levinson, *Theory of Ordinary Differential Equations.* Malabar, FL: Krieger, 1984.

[A8] Dong, T.-R. et al., *Qualitative Theory of Differential Equations.* Providence, RI: American Mathematical Society, 1992.

[A9] Erdélyi, A. et al., *Tables of Integral Transforms.* 2 vols. New York: McGraw-Hill, 1954.

[A10] Hartman, P., *Ordinary Differential Equations.* 2nd ed. Philadelphia: SIAM, 2002.

[A11] Ince, E. L., *Ordinary Differential Equations.* New York: Dover, 1956.

[A12] Schiff, J. L., *The Laplace Transform: Theory and Applications.* New York: Springer, 1999.

[A13] Watson, G. N., *A Treatise on the Theory of Bessel Functions.* 2nd ed. Reprinted. New York: Cambridge University Press, 1995.

[A14] Widder, D. V., *The Laplace Transform.* Princeton, NJ: Princeton University Press, 1941.

[A15] Zwillinger, D., *Handbook of Differential Equations.* 3rd ed. New York: Academic Press, 1998.

Part B. Linear Algebra, Vector Calculus (Chaps. 7–10)

For books on *numeric* linear algebra, *see also* Part E: Numeric Analysis.

[B1] Bellman, R., *Introduction to Matrix Analysis.* 2nd ed. Philadelphia: SIAM, 1997.

[B2] Chatelin, F., *Eigenvalues of Matrices.* New York: Wiley-Interscience, 1993.

[B3] Gantmacher, F. R., *The Theory of Matrices.* 2 vols. Providence, RI: American Mathematical Society, 2000.

[B4] Gohberg, I. P. et al., *Invariant Subspaces of Matrices with Applications.* New York: Wiley, 2006.

[B5] Greub, W. H., *Linear Algebra.* 4th ed. New York: Springer, 1975.

[B6] Herstein, I. N., *Abstract Algebra.* 3rd ed. New York: Wiley, 1996.

[B7] Joshi, A. W., *Matrices and Tensors in Physics.* 3rd ed. New York: Wiley, 1995.

[B8] Lang, S., *Linear Algebra.* 3rd ed. New York: Springer, 1996.

[B9] Nef, W., *Linear Algebra.* 2nd ed. New York: Dover, 1988.

[B10] Parlett, B., *The Symmetric Eigenvalue Problem.* Philadelphia: SIAM, 1998.

Part C. Fourier Analysis and PDEs (Chaps. 11–12)

For books on *numerics* for PDEs *see also* Part E: Numeric Analysis.

[C1] Antimirov, M. Ya., *Applied Integral Transforms.* Providence, RI: American Mathematical Society, 1993.

[C2] Bracewell, R., *The Fourier Transform and Its Applications.* 3rd ed. New York: McGraw-Hill, 2000.

[C3] Carslaw, H. S. and J. C. Jaeger, *Conduction of Heat in Solids.* 2nd ed. Reprinted. Oxford: Clarendon, 2000.

[C4] Churchill, R. V. and J. W. Brown, *Fourier Series and Boundary Value Problems.* 6th ed. New York: McGraw-Hill, 2006.

[C5] DuChateau, P. and D. Zachmann, *Applied Partial Differential Equations.* Mineola, NY: Dover, 2002.

[C6] Hanna, J. R. and J. H. Rowland, *Fourier Series, Transforms, and Boundary Value Problems.* 2nd ed. New York: Wiley, 2008.

[C7] Jerri, A. J., *The Gibbs Phenomenon in Fourier Analysis, Splines, and Wavelet Approximations.* Boston: Kluwer, 1998.

[C8] John, F., *Partial Differential Equations.* 4th edition New York: Springer, 1982.

[C9] Tolstov, G. P., *Fourier Series.* New York: Dover, 1976.

[C10] Widder, D. V., *The Heat Equation.* New York: Academic Press, 1975.

[C11] Zauderer, E., *Partial Differential Equations of Applied Mathematics.* 3rd ed. New York: Wiley, 2006.

[C12] Zygmund, A. and R. Fefferman, *Trigonometric Series.* 3rd ed. New York: Cambridge University Press, 2002.

Part D. Complex Analysis (Chaps. 13–18)

[D1] Ahlfors, L. V., *Complex Analysis.* 3rd ed. New York: McGraw-Hill, 1979.

[D2] Bieberbach, L., *Conformal Mapping.* Providence, RI: American Mathematical Society, 2000.

[D3] Henrici, P., *Applied and Computational Complex Analysis.* 3 vols. New York: Wiley, 1993.

[D4] Hille, E., *Analytic Function Theory.* 2 vols. 2nd ed. Providence, RI: American Mathematical Society, Reprint V1 1983, V2 2005.

[D5] Knopp, K., *Elements of the Theory of Functions.* New York: Dover, 1952.

[D6] Knopp, K., *Theory of Functions.* 2 parts. New York: Dover, Reprinted 1996.

[D7] Krantz, S. G., *Complex Analysis: The Geometric Viewpoint.* Washington, DC: The Mathematical Association of America, 1990.

[D8] Lang, S., *Complex Analysis.* 4th ed. New York: Springer, 1999.

[D9] Narasimhan, R., *Compact Riemann Surfaces.* New York: Springer, 1996.

[D10] Nehari, Z., *Conformal Mapping.* Mineola, NY: Dover, 1975.

[D11] Springer, G., *Introduction to Riemann Surfaces.* Providence, RI: American Mathematical Society, 2001.

Part E. Numeric Analysis (Chaps. 19–21)

[E1] Ames, W. F., *Numerical Methods for Partial Differential Equations.* 3rd ed. New York: Academic Press, 1992.

[E2] Anderson, E., et al., *LAPACK User's Guide.* 3rd ed. Philadelphia: SIAM, 1999.

[E3] Bank, R. E., *PLTMG. A Software Package for Solving Elliptic Partial Differential Equations: Users' Guide 8.0.* Philadelphia: SIAM, 1998.

[E4] Constanda, C., *Solution Techniques for Elementary Partial Differential Equations.* Boca Raton, FL: CRC Press, 2002.

[E5] Dahlquist, G. and A. Björck, *Numerical Methods.* Mineola, NY: Dover, 2003.

[E6] DeBoor, C., *A Practical Guide to Splines.* Reprinted. New York: Springer, 2001.

[E7] Dongarra, J. J. et al., *LINPACK Users Guide.* Philadelphia: SIAM, 1979. (See also at the beginning of Chap. 19.)

[E8] Garbow, B. S. et al., *Matrix Eigensystem Routines: EISPACK Guide Extension.* Reprinted. New York: Springer, 1990.

[E9] Golub, G. H. and C. F. Van Loan, *Matrix Computations.* 3rd ed. Baltimore, MD: Johns Hopkins University Press, 1996.

[E10] Higham, N. J., *Accuracy and Stability of Numerical Algorithms.* 2nd ed. Philadelphia: SIAM, 2002.

[E11] IMSL (International Mathematical and Statistical Libraries), *FORTRAN Numerical Library.* Houston, TX: Visual Numerics, 2002. (See also at the beginning of Chap. 19.)

[E12] IMSL, *IMSL for Java.* Houston, TX: Visual Numerics, 2002.

[E13] IMSL, *C Library.* Houston, TX: Visual Numerics, 2002.

[E14] Kelley, C. T., *Iterative Methods for Linear and Nonlinear Equations.* Philadelphia: SIAM, 1995.

[E15] Knabner, P. and L. Angerman, *Numerical Methods for Partial Differential Equations.* New York: Springer, 2003.

[E16] Knuth, D. E., *The Art of Computer Programming.* 3 vols. 3rd ed. Reading, MA: Addison-Wesley, 1997–2009.

[E17] Kreyszig, E., *Introductory Functional Analysis with Applications.* New York: Wiley, 1989.

[E18] Kreyszig, E., On methods of Fourier analysis in multigrid theory. *Lecture Notes in Pure and Applied Mathematics* 157. New York: Dekker, 1994, pp. 225–242.

[E19] Kreyszig, E., Basic ideas in modern numerical analysis and their origins. *Proceedings of the Annual Conference of the Canadian Society for the History and Philosophy of Mathematics.* 1997, pp. 34–45.

[E20] Kreyszig, E., and J. Todd, *QR* in two dimensions. *Elemente der Mathematik* 31 (1976), pp. 109–114.

[E21] Mortensen, M. E., *Geometric Modeling.* 2nd ed. New York: Wiley, 1997.

[E22] Morton, K. W., and D. F. Mayers, *Numerical Solution of Partial Differential Equations: An Introduction.* New York: Cambridge University Press, 1994.

[E23] Ortega, J. M., *Introduction to Parallel and Vector Solution of Linear Systems.* New York: Plenum Press, 1988.

[E24] Overton, M. L., *Numerical Computing with IEEE Floating Point Arithmetic.* Philadelphia: SIAM, 2004.

[E25] Press, W. H. et al., *Numerical Recipes in C: The Art of Scientific Computing.* 2nd ed. New York: Cambridge University Press, 1992.

[E26] Shampine, L. F., *Numerical Solutions of Ordinary Differential Equations.* New York: Chapman and Hall, 1994.

[E27] Varga, R. S., *Matrix Iterative Analysis.* 2nd ed. New York: Springer, 2000.

[E28] Varga, R. S., *Geršgorin and His Circles.* New York: Springer, 2004.

[E29] Wilkinson, J. H., *The Algebraic Eigenvalue Problem.* Oxford: Oxford University Press, 1988.

Part F. Optimization, Graphs (Chaps. 22–23)

[F1] Bondy, J. A. and U.S.R. Murty, *Graph Theory with Applications.* Hoboken, NJ: Wiley-Interscience, 1991.

[F2] Cook, W. J. et al., *Combinatorial Optimization.* New York: Wiley, 1997.

[F3] Diestel, R., *Graph Theory.* 4th ed. New York: Springer, 2006.

[F4] Diwekar, U. M., *Introduction to Applied Optimization.* 2nd ed. New York: Springer, 2008.

[F5] Gass, S. L., *Linear Programming. Method and Applications.* 3rd ed. New York: McGraw-Hill, 1969.

[F6] Gross, J. T. and J.Yellen (eds.), *Handbook of Graph Theory and Applications.* 2nd ed. Boca Raton, FL: CRC Press, 2006.

[F7] Goodrich, M. T., and R. Tamassia, *Algorithm Design: Foundations, Analysis, and Internet Examples.* Hoboken, NJ: Wiley, 2002.

[F8] Harary, F., *Graph Theory.* Reprinted. Reading, MA: Addison-Wesley, 2000.

[F9] Merris, R., *Graph Theory.* Hoboken, NJ: Wiley-Interscience, 2000.

[F10] Ralston, A., and P. Rabinowitz, *A First Course in Numerical Analysis.* 2nd ed. Mineola, NY: Dover, 2001.

[F11] Thulasiraman, K., and M. N. S. Swamy, *Graph Theory and Algorithms.* New York: Wiley-Interscience, 1992.

[F12] Tucker, A., *Applied Combinatorics.* 5th ed. Hoboken, NJ: Wiley, 2007.

Part G. Probability and Statistics (Chaps. 24–25)

[G1] American Society for Testing Materials, *Manual on Presentation of Data and Control Chart Analysis.* 7th ed. Philadelphia: ASTM, 2002.

[G2] Anderson, T. W., *An Introduction to Multivariate Statistical Analysis.* 3rd ed. Hoboken, NJ: Wiley, 2003.

[G3] Cramér, H., *Mathematical Methods of Statistics.* Reprinted. Princeton, NJ: Princeton University Press, 1999.

[G4] Dodge, Y., *The Oxford Dictionary of Statistical Terms.* 6th ed. Oxford: Oxford University Press, 2006.

[G5] Gibbons, J. D. and S. Chakraborti, *Nonparametric Statistical Inference.* 4th ed. New York: Dekker, 2003.

[G6] Grant, E. L. and R. S. Leavenworth, *Statistical Quality Control.* 7th ed. New York: McGraw-Hill, 1996.

[G7] IMSL, *Fortran Numerical Library.* Houston, TX: Visual Numerics, 2002.

[G8] Kreyszig, E., *Introductory Mathematical Statistics. Principles and Methods.* New York: Wiley, 1970.

[G9] O'Hagan, T. et al., *Kendall's Advanced Theory of Statistics 3-Volume Set.* Kent, U.K.: Hodder Arnold, 2004.

[G10] Rohatgi, V. K. and A. K. MD. E. Saleh, *An Introduction to Probability and Statistics.* 2nd ed. Hoboken, NJ: Wiley-Interscience, 2001.

Web References

[W1] upgraded version of [GenRef1] online at http://dlmf.nist.gov/. Hardcopy and CD-Rom: Oliver, W. J. et al. (eds.), *NIST Handbook of Mathematical Functions.* Cambridge; New York: Cambridge University Press, 2010.

[W2] O'Connor, J. and E. Robertson, MacTutor History of Mathematics Archive. St. Andrews, Scotland: University of St. Andrews, School of Mathematics and Statistics. Online at http://www-history.mcs.st-andrews.ac.uk. (Biographies of mathematicians, etc.).

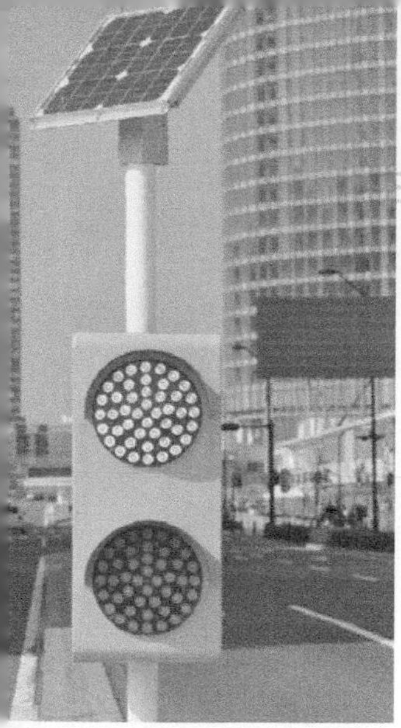

APPENDIX 2

Answers to Odd-Numbered Problems

Problem Set 1.1, page 8

1. $y = \frac{1}{\pi} \cos 2\pi x + c$

3. $y = ce^x$

5. $y = 2e^{-x}(\sin x - \cos x) + c$

7. $y = \frac{1}{5.13} \sinh 5.13x + c$

9. $y = 1.65e^{-4x} + 0.35$

11. $y = (x + \frac{1}{2})e^x$

13. $y = 1/(1 + 3e^{-x})$

15. $y = 0$ and $y = 1$ because $y' = 0$ for these y

17. $\exp(-1.4 \cdot 10^{-11}t) = \frac{1}{2}$, $t = 10^{11}(\ln 2)/1.4$ [sec]

19. Integrate $y'' = g$ twice, $y'(t) = gt + v_0$, $y'(0) = v_0 = 0$ (start from rest), then $y(t) = \frac{1}{2}gt^2 + y_0$, where $y(0) = y_0 = 0$

Problem Set 1.2, page 11

11. Straight lines parallel to the x-axis

13. $y = x$

15. $mv' = mg - bv^2$, $v' = 9.8 - v^2$, $v(0) = 10$, $v' = 0$ gives the limit$/9.8 = 3.1$ [meter/sec]

17. Errors of steps 1, 5, 10: 0.0052, 0.0382, 0.1245, approximately

19. $x_5 = 0.0286$ (error 0.0093), $x_{10} = 0.2196$ (error 0.0189)

Problem Set 1.3, page 18

1. If you add a constant later, you may not get a solution.
Example: $y' = y$, $\ln |y| = x + c$, $y = e^{x+c} = \tilde{c}e^x$ but not $e^x + c$ (with $c \neq 0$)

3. $\cos^2 y \, dy = dx$, $\frac{1}{2}y + \frac{1}{4}\sin 2y + c = x$

5. $y^2 + 36x^2 = c$, ellipses

7. $y = x \arctan (x^2 + c)$

9. $y = x/(c - x)$

11. $y = 24/x$, hyperbola

13. $dy/\sin^2 y = dx/\cosh^2 x$, $-\cot y = \tanh x + c$, $c = 0$, $y = -\text{arccot} (\tanh x)$

15. $y^2 + 4x^2 = c = 25$

17. $y = x \arctan (x^3 - 1)$

19. $y_0 e^{kt} = 2y_0$, $e^k = 2$ (1 week), $e^{2k} = 2^2$ (2 weeks), $e^{4k} = 2^4$

21. 69.6% of y_0

23. $PV = c =$ const

25. $T = 22 - 17e^{-0.5306t} = 21.9$ [°C] when $t = 9.68$ min

27. $e^{-k \cdot 10} = \frac{1}{2}$, $k = \frac{1}{10}$, $\ln \frac{1}{2}$, $e^{-kt_0} = 0.01$, $t = (\ln 100)/k = 66$ [min]

29. No. Use Newton's law of cooling.

31. $y = ax$, $y' = g(y/x) = a =$ const, independent of the point (x, y)

33. $\Delta S = 0.15S\Delta\phi$, $dS/d\phi = 0.15S$, $S = S_0 e^{0.15\phi} = 1000S_0$, $\phi = (1/0.15) \ln 1000 = 7.3 \cdot 2\pi$. Eight times.

Problem Set 1.4, page 26

1. Exact, $2x = 2x$, $x^2y = c$, $y = c/x^2$
3. Exact, $y = \arccos(c/\cos x)$
5. Not exact, $y = \sqrt{x^2 + cx}$
7. $F = e^{x^2}$, $e^{x^2} \tan y = c$
9. Exact, $u = e^{2x}\cos y + k(y)$, $u_y = -e^{2x}\sin y + k'$, $k' = 0$. *Ans.* $e^{2x}\cos y = 1$
11. $F = \sinh x$, $\sinh^2 x \cos y = c$
13. $u = e^x + k(y)$, $u_y = k' = -1 + e^y$, $k = -y + e^y$. *Ans.* $e^x - y + e^y = c$
15. $b = k$, $ax^2 + 2kxy + ly^2 = c$

Problem Set 1.5, page 34

3. $y = ce^x - 5.2$
5. $y = (x + c)e^{-kx}$
7. $y = x^2(c + e^x)$
9. $y = (x - 2.5/e)e^{\cos x}$
11. $y = 2 + c \sin x$
13. Separate. $y - 2.5 = c \cosh^4 1.5x$
15. $(y_1 + y_2)' + p(y_1 + y_2) = (y_1' + py_1) + (y_2' + py_2) = 0 + 0 = 0$
17. $(y_1 + y_2)' + p(y_1 + y_2) = (y_1' + py_1) + (y_2' + py_2) = r + 0 = r$
19. Solution of $cy_1' + pcy_1 = c(y_1' + py_1) = cr$
21. $y = uy^*$, $y' + py = u'y^* + uy^{*\prime} + puy^* = u'y^* + u(y^{*\prime} + py^*) = u'y^* + u \cdot 0$ $= r$, $u' = r/y^* = re^{\int p\,dx}$, $u = \int e^{\int p\,dx} r\,dx + c$. Thus, $y = uy_h$ gives (4). We shall see that this method extends to higher-order ODEs (Secs. 2.10 and 3.3).
23. $y^2 = 1 + 8e^{-x^2}$
25. $y = 1/u$, $u = ce^{-3.2x} + 10/3.2$
27. $dx/dy = 6e^y - 2x$, $x = ce^{-2y} + 2e^y$
31. $T = 240e^{kt} + 60$, $T(10) = 200$, $k = -0.0539$, $t = 102$ min
33. $y' = A - ky$, $y(0) = 0$, $y = A(1 - e^{-kt})/k$
35. $y' = 175(0.0001 - y/450)$, $y(0) = 450 \cdot 0.0004 = 0.18$,
$y = 0.135e^{-0.3889t} + 0.045 = 0.18/2$,
$e^{-0.3889t} = (0.09 - 0.045)/0.135 = 1/3$,
$t = (\ln 3)/0.3889 = 2.82$. *Ans.* About 3 years
37. $y' = y - y^2 - 0.2y$, $y = 1/(1.25 - 0.75e^{-0.8t})$, limit 0.8, limit 1
39. $y' = By^2 - Ay = By(y - A/B)$, $A > 0$, $B > 0$. Constant solutions $y = 0$, $y = A/B$, $y' > 0$ if $y > A/B$ (unlimited growth), $y' < 0$ if $0 < y < A/B$ (extinction). $y = A/(ce^{At} + B)$, $y(0) > A/B$ if $c < 0$, $y(0) < A/B$ if $c > 0$.

Problem Set 1.6, page 38

1. $x^2/(c^2 + 9) + y^2/c^2 - 1 = 0$
3. $y - \cosh(x - c) - c = 0$
5. $y/x = c$, $y'/x = y/x^2$, $y' = y/x$, $\tilde{y}' = -x/\tilde{y}$, $\tilde{y}^2 + x^2 = \tilde{c}$, circles
7. $2\tilde{y}^2 - x^2 = \tilde{c}$
9. $y' = -2xy$, $\tilde{y}' = 1/(2x\tilde{y})$, $x = \tilde{c}e^{\tilde{y}^2}$
11. $\tilde{y} = \tilde{c}x$
13. $y' = -4x/9y$. Trajectories $\tilde{y}' = 9\tilde{y}/4x$, $\tilde{y} = \tilde{c}x^{9/4}$ ($\tilde{c} > 0$). Sketch or graph these curves.
15. $u = c$, $u_x\,dx + u_y\,dy = 0$, $y' = -u_x/u_y$. Trajectories $\tilde{y}' = u_{\tilde{y}}/u_x$. Now $v = \tilde{c}$, $v_x\,dx + v_y\,dy = 0$, $y' = -v_x/v_y$. This agrees with the trajectory ODE in u if $u_x = v_y$ (equal denominators) and $u_y = -v_x$ (equal numerators). But these are just the Cauchy–Riemann equations.

Problem Set 1.7, page 42

1. $y' = f(x, y) = r(x) - p(x)y$; hence $\partial f/\partial y = -p(x)$ is continuous and is thus bounded in the closed interval $|x - x_0| \leqq a$.

3. In $|x - x_0| < a$; just take b in $\alpha = b/K$ large, namely, $b = \alpha K$.

5. R has sides $2a$ and $2b$ and center $(1, 1)$ since $y(1) = 1$. In R, $f = 2y^2 \leqq 2(b + 1)^2 = K$, $\alpha = b/K = b/(2(b + 1)^2)$, $d\alpha/db = 0$ gives $b = 1$, and $\alpha_{opt} = b/K = \frac{1}{8}$. Solution by $dy/y^2 = 2\,dx$, etc., $y = 1/(3 - 2x)$.

7. $|1 + y^2| \leqq K = 1 + b^2$, $\alpha = b/K$, $d\alpha/db = 0$, $b = 1$, $\alpha = \frac{1}{2}$.

9. No. At a common point (x_1, y_1) they would both satisfy the "initial condition" $y(x_1) = y_1$, violating uniqueness.

Chapter 1 Review Questions and Problems, page 43

11. $y = ce^{-2x}$

13. $y = 1/(ce^{-4x} + 4)$

15. $y = ce^{-x} + 0.01 \cos 10x + 0.1 \sin 10x$

17. $y = ce^{-2.5x} + 0.640x - 0.256$

19. $25y^2 - 4x^2 = c$

21. $F = x$, $x^3e^y + x^2y = c$

23. $y = \sin (x + \frac{1}{4}\pi)$

25. $3 \sin x + \frac{1}{3} \sin y = 0$

27. $e^k = 1.25$, $(\ln 2)/\ln 1.25 = 3.1$, $(\ln 3)/\ln 1.25 = 4.9$ [days]

29. $e^k = 0.9$, 6.6 days. 43.7 days from $e^{kt} = 0.5$, $e^{kt} = 0.01$

Problem Set 2.1, page 53

1. $F(x, z, z') = 0$

3. $y = c_1e^{-x} + c_2$

5. $y = (c_1x + c_2)^{-1/2}$

7. $(dz/dy)z = -z^3 \sin y$, $-1/z = -dx/dy = \cos y + \tilde{c}_1$, $x = -\sin y + c_1y + c_2$

9. $y_2 = x^3 \ln x$

11. $y = c_1e^{2x} + c_2$

13. $y(t) = c_1e^{-t} + kt + c_2$

15. $y = 3 \cos 2.5x - \sin 2.5x$

17. $y = -0.75x^{3/2} - 2.25x^{-1/2}$

19. $y = 15e^{-x} - \sin x$

Problem Set 2.2, page 59

1. $y = c_1e^{-2.5x} + c_2e^{2.5x}$

3. $y = c_1e^{-2.8x} + c_2e^{-3.2x}$

5. $y = (c_1 + c_2x)e^{-\pi x}$

7. $y = c_1 + c_2e^{-4.5x}$

9. $y = c_1e^{-2.6x} + c_2e^{0.8x}$

11. $y = c_1e^{-x/2} + c_2e^{3x/2}$

13. $y = (c_1 + c_2x)e^{5x/3}$

15. $y = e^{-0.27x}(A \cos (\sqrt{\pi}\,x) + B \sin (\sqrt{\pi}\,x))$

17. $y'' + 2\sqrt{5}y' + 5y = 0$

19. $y'' + 4y' + 5y = 0$

21. $y = 4.6 \cos 5x - 0.24 \sin 5x$

23. $y = 6e^{2x} + 4e^{-3x}$

25. $y = 2e^{-x}$

27. $y = (4.5 - x)e^{-\pi x}$

29. $y = \dfrac{1}{\sqrt{\pi}} e^{-0.27x} \sin (\sqrt{\pi}\,x)$

31. Independent

33. $c_1x^2 + c_2x^2 \ln x = 0$ with $x = 1$ gives $c_1 = 0$; then $c_2 = 0$ for $x = 2$, say. Hence independent

35. Dependent since $\sin 2x = 2 \sin x \cos x$

37. $y_1 = e^{-x}$, $y_2 = 0.001e^x + e^{-x}$

Problem Set 2.3, page 61

1. $4e^{2x}$, $-e^{-x} + 8e^{2x}$, $-\cos x - 2 \sin x$
3. 0, 0, $(D - 2I)(-4e^{-2x}) = 8e^{-2x} + 8e^{-2x}$
5. 0, $5e^{2x}$, 0
7. $(2D - I)(2D + I)$, $y = c_1e^{0.5x} + c_2e^{-0.5x}$
9. $(D - 2.1I)^2$, $y = (c_1 + c_2x)e^{2.1x}$
11. $(D - 1.6I)(D - 2.4I)$, $y = c_1e^{1.6x} + c_2e^{2.4x}$
15. Combine the two conditions to get $L(cy + kw) = L(cy) + L(kw) = cLy + kLw$. The converse is simple.

Problem Set 2.4, page 69

1. $y' = y_0 \cos \omega_0 t + (v_0/\omega_0) \sin \omega_0 t$. At integer t (if $\omega_0 = \pi$), because of periodicity.
3. (i) Lower by a factor $\sqrt{2}$, (ii) higher by $\sqrt{2}$
5. 0.3183, 0.4775, $\sqrt{(k_1 + k_2)/m}/(2\pi) = 0.5738$
7. $mL\theta'' = -mg \sin \theta \approx -mg\theta$ (tangential component of $W = mg$),
$\theta'' + \omega_0{}^2\theta = 0$, $\omega_0/(2\pi) = \sqrt{g/L}/(2\pi)$
9. $my'' = -\tilde{a}\gamma y$, where $m = 1$ kg, $ay = \pi \cdot 0.01^2 \cdot 2y$ meter^3 is the volume of the water that causes the restoring force $a\gamma y$ with $\gamma = 9800$ nt (= weight/meter^3).
$y'' + \omega_0{}^2 y = 0$, $\omega_0{}^2 = a\gamma/m = a\gamma = 0.000628\gamma$. Frequency $\omega_0/2\pi = 0.4$ [sec^{-1}].
13. $y = [y_0 + (v_0 + \alpha y_0)t]e^{-\alpha t}$, $y = [1 + (v_0 + 1)t]e^{-t}$;
(ii) $v_0 = -2, -\frac{3}{2}, -\frac{4}{3}, -\frac{5}{4}, -\frac{6}{5}$
15. $\omega^* = [\omega_0{}^2 - c^2/(4m^2)]^{1/2} = \omega_0[1 - c^2/(4mk)]^{1/2} \approx \omega_0(1 - c^2/8mk) = 2.9583$
17. The positive solutions of $\tan t = 1$, that is, $\pi/4$ (max), $5\pi/4$ (min). etc
19. $0.0231 = (\ln 2)/30$ [kg/sec] from $\exp(-10 \cdot 3c/2m) = \frac{1}{2}$.

Problem Set 2.5, page 73

3. $y = (c_1 + c_2 \ln x)x^{-1.8}$
5. $\sqrt{x}\,(c_1 \cos(\ln x) + c_2 \sin(\ln x))$
7. $y = c_1x^2 + c_2x^3$
9. $y = (c_1 + c_2 \ln x)x^{0.6}$
11. $y = x^2(c_1 \cos(\sqrt{6} \ln x) + c_2 \sin(\sqrt{6} \ln x))$
13. $y = x^{-3/2}$
15. $y = (3.6 + 4.0 \ln x)/x$
17. $y = \cos(\ln x) + \sin(\ln x)$
19. $y = -0.525x^5 + 0.625x^{-3}$

Problem Set 2.6, page 79

3. $W = -2.2e^{-3x}$
5. $W = -x^4$
7. $W = a$
9. $y'' + 25y = 0$, $W = 5$, $y = 3 \cos 5x - \sin 5x$
11. $y'' + 5y + 6.34 = 0$, $W = 0.3e^{-5x}$, $3e^{-2.5} \cos 0.3x$
13. $y'' + 2y' = 0$, $W = -2e^{-2x}$, $y = 0.5(1 + e^{-2x})$
15. $y'' - 3.24y = 0$, $W = 1.8$, $y = 14.2 \cosh 1.8x + 9.1 \sinh 1.8x$

Problem Set 2.7, page 84

1. $y = c_1e^{-x} + c_2e^{-4x} - 5e^{-3x}$
3. $y = c_1e^{-2x} + c_2e^{-x} + 6x^2 - 18x + 21$
5. $y = (c_1 + c_2x)e^{-2x} + \frac{1}{2}e^{-x} \sin x$
7. $y = c_1e^{-x/2} + c_2e^{-3x/2} + \frac{4}{5}e^x + 6x - 16$
9. $y = c_1e^{4x} + c_2e^{-4x} + 1.2xe^{4x} - 2e^x$
11. $y = \cos(\sqrt{3}x) + 6x^2 - 4$

13. $y = e^{x/4} - 2e^{x/2} + \frac{1}{5}e^{-x} + e^{x}$ **15.** $y = \ln x$

17. $y = e^{-0.1x}(1.5 \cos 0.5x - \sin 0.5x) + 2e^{0.5x}$

Problem Set 2.8, page 91

3. $y_p = 1.0625 \cos 2t + 3.1875 \sin 2t$

5. $y_p = -1.28 \cos 4.5t + 0.36 \sin 4.5t$

7. $y_p = 25 + \frac{4}{3} \cos 3t + \sin 3t$

9. $y = e^{-1.5t}(A \cos t + B \sin t) + 0.8 \cos t + 0.4 \sin t$

11. $y = A \cos \sqrt{2}t + B \sin \sqrt{2}t + t(\sin \sqrt{2}t - \cos \sqrt{2}t)/(2\sqrt{2})$

13. $y = A \cos t + B \sin t - (\cos \omega t)/(\omega^2 - 1)$

15. $y = e^{-2t}(A \cos 2t + B \sin 2t) + \frac{1}{4} \sin 2t$

17. $y = \frac{1}{3} \sin t - \frac{1}{15} \sin 3t - \frac{1}{105} \sin 5t$

19. $y = e^{-t}(0.4 \cos t + 0.8 \sin t) + e^{-t/2}(-0.4 \cos \frac{1}{2}t + 0.8 \sin \frac{1}{2}t)$

25. CAS Experiment. The choice of ω needs experimentation, inspection of the curves obtained, and then changes on a trail-and-error basis. It is interesting to see how in the case of beats the period gets increasingly longer and the maximum amplitude gets increasingly larger as $\omega/(2\pi)$ approaches the resonance frequency.

Problem Set 2.9, page 98

1. $RI' + I/C = 0, \quad I = ce^{-t/(RC)}$

3. $LI' + RI = E, \quad I = (E/R) + ce^{-Rt/L} = 4.8 + ce^{-40t}$

5. $I = 2(\cos t - \cos 20t)/399$

7. I_0 is maximum when $S = 0$; thus, $C = 1/(\omega^2 L)$.

9. $I = 0$ **11.** $I = 5.5 \cos 10t + 16.5 \sin 10t$ A

13. $I = e^{-5t}(A \cos 10t + B \sin 10t) - 400 \cos 25t + 200 \sin 25t$ A

15. $R > R_{crit} = 2\sqrt{L/C}$ is Case I, etc.

17. $E(0) = 600, I'(0) = 600, I = e^{-3t}(-100 \cos 4t + 75 \sin 4t) + 100 \cos t$

19. $R = 2\ \Omega, \quad L = 1\ \text{H}, \quad C = \frac{1}{12}\ \text{F}, \quad E = 4.4 \sin 10t\ \text{V}$

Problem Set 2.10, page 102

1. $y = A \cos 3x + B \sin 3x + \frac{1}{9}(\cos 3x) \ln |\cos 3x| + \frac{1}{3}x \sin 3x$

3. $y = c_1 x + c_2 x^2 - x \sin x$ **5.** $y = A \cos x + B \sin x + \frac{1}{2}x(\cos x + \sin x)$

7. $y = (c_1 + c_2 x)e^{2x} + x^{-2}e^{2x}$ **9.** $y = (c_1 + c_2 x)e^{x} + 4x^{7/2}e^{x}$

11. $y = c_1 x^2 + c_2 x^3 + 1/(2x^4)$ **13.** $y = c_1 x^{-3} + c_2 x^3 + 3x^5$

Chapter 2 Review Questions and Problems, page 102

7. $y = c_1 e^{-4.5x} + c_2 e^{-3.5x}$ **9.** $y = e^{-3x}(A \cos 5x + B \sin 5x)$

11. $y = (c_1 + c_2 x)e^{0.8x}$ **13.** $y = c_1 x^{-4} + c_2 x^3$

15. $y = c_1 e^{2x} + c_2 e^{-x/2} - 3x + x^2$ **17.** $y = (c_1 + c_2 x)e^{1.5x} + 0.25x^2 e^{1.5x}$

19. $y = 5 \cos 4x - \frac{3}{4} \sin 4x + e^{x}$ **21.** $y = -4x + 2x^3 + 1/x$

23. $I = -0.01093 \cos 415t + 0.05273 \sin 415t$ A

25. $I = \frac{1}{73}(50 \sin 4t - 110 \cos 4t)$ A
27. RLC-circuit with $R = 20\ \Omega$, $L = 4$ H, $C = 0.1$ F, $E = -25 \cos 4t$ V
29. $\omega = 3.1$ is close to $\omega_0 = \sqrt{k/m} = 3$, $y = 25(\cos 3t - \cos 3.1t)$.

Problem Set 3.1, page 111

9. Linearly independent
11. Linearly independent
13. Linearly independent
15. Linearly dependent

Problem Set 3.2, page 116

1. $y = c_1 + c_2 \cos 5x + c_3 \sin 5x$
3. $y = c_1 + c_2 x + c_3 \cos 2x + c_4 \sin 2x$
5. $y = A_1 \cos x + B_1 \sin x + A_2 \cos 3x + B_2 \sin 3x$
7. $y = 2.398 + e^{-1.6x}(1.002 \cos 1.5x - 1.998 \sin 1.5x)$
9. $y = 4e^{-x} + 5e^{-x/2} \cos 3x$
11. $y = \cosh 5x - \cos 4x$
13. $y = e^{0.25x} + 4.3e^{-0.7x} + 12.1 \cos 0.1x - 0.6 \sin 0.1x$

Problem Set 3.3, page 122

1. $y = (c_1 + c_2 x + c_3 x^2)e^{-x} + \frac{1}{8}e^{x} - x + 2$
3. $y = c_1 \cos x + c_2 \sin x + c_3 \cos 3x + c_4 \sin 3x + 0.1 \sinh 2x$
5. $y = c_1 x^2 + c_2 x + c_3 x^{-1} - \frac{1}{12} x^{-2}$
7. $y = (c_1 + c_2 x + c_3 x^2)e^{3x} - \frac{1}{4}(\cos 3x - \sin 3x)$
9. $y = \cos x + \frac{1}{2} \sin 4x$
11. $y = e^{-3x}(-1.4 \cos x - \sin x)$
13. $y = 2 - 2 \sin x + \cos x$

Chapter 3 Review Questions and Problems, page 122

7. $y = c_1 + e^{-2x}(A \cos 3x + B \sin 3x)$
9. $y = c_1 \cosh 2x + c_2 \sinh 2x + c_3 \cos 2x + c_4 \sin 2x + \cosh x$
11. $y = (c_1 + c_2 x + c_3 x^2)e^{-1.5x}$
13. $y = (c_1 + c_2 x + c_3 x^2)e^{-2x} + x^2 - 3x + 3$
15. $y = c_1 x + c_2 x^{1/2} + c_3 x^{3/2} - \frac{10}{3}$
17. $y = 2e^{-2x} \cos 4x + 0.05\, x - 0.06$
19. $y = 4e^{-4x} + 5e^{-5x}$

Problem Set 4.1, page 136

1. Yes
5. $y_1' = 0.02(-y_1 + y_2)$, $y_2' = 0.02(y_1 - 2y_2 + y_3)$, $y_3' = 0.02(y_2 - y_3)$
7. $c_1 = 1$, $c_2 = -5$
9. $c_1 = 10$, $c_2 = 5$
11. $y_1' = y_2$, $y_2' = y_1 + + \frac{15}{4} y_2$, $\mathbf{y} = c_1 [1 \quad 4]^{\mathsf{T}} e^{4t} + c_2 [1 \quad -\frac{1}{4}]^{\mathsf{T}} e^{-t/4}$
13. $y_1' = y_2$, $y_2' = 24y_1 - 2y_2$, $y_1 = c_1 e^{4t} + c_2 e^{-6t} = y$, $y_2 = y'$
15. **(a)** For example, $C = 1000$ gives -2.39993, -0.000167. **(b)** -2.4, 0.
(d) $a_{22} = -4 + 2\sqrt{6.4} = 1.05964$ gives the critical case. C about 0.18506.

Problem Set 4.3, page 147

1. $y_1 = c_1e^{-2t} + c_2e^{2t}, \quad y_2 = -3c_1e^{-2t} + c_2e^{2t}$
3. $y_1 = 2c_1e^{2t} + 2c_2, \quad y_2 = c_1e^{2t} - c_2$
5. $y_1 = 5c_1 + 2c_2e^{14.5t}$
$y_2 = -2c_1 + 5c_2e^{14.5t}$
7. $y_1 = -c_2 \cos \sqrt{2}t + c_3 \sin \sqrt{2}t + c_1$
$y_2 = c_2\sqrt{2} \sin \sqrt{2}t + c_3\sqrt{2} \cos \sqrt{2}t$
$y_3 = c_2 \cos \sqrt{2}t - c_3 \sin \sqrt{2}t + c_1$
9. $y_1 = \frac{1}{2}c_1e^{-18t} + 2c_2e^{9t} - c_3e^{18t}$
$y_2 = c_1e^{-18t} + c_2e^{9t} + c_3e^{18t}$
$y_3 = c_1e^{-18t} - 2c_2e^{9t} - \frac{1}{2}c_3e^{18t}$
11. $y_1 = -20e^t + 8e^{-t/2}$
$y_2 = 4e^t - 4e^{-t/2}$
13. $y_1 = 2 \sinh t, \quad y_2 = 2 \cosh t$
15. $y_1 = \frac{1}{2}e^t$
$y_2 = \frac{1}{2}e^t$
17. $y_2 = y_1' + y_1, \quad y_2' = y_1'' + y_1' = -y_1 - y_2 = -y_1 - (y_1' + y_1),$
$y_1'' + 2y_1' + 2y_1 = 0, \quad y_1 = e^{-t}(A \cos t + B \sin t),$
$y_2 = y_1' + y_1 = e^{-t}(B \cos t - A \sin t)$. Note that $r^2 = y_1^2 + y_2^2 = e^{-2t}(A^2 + B^2)$.
19. $I_1 = c_1e^{-t} + 3c_2e^{-3t}, I_2 = -3c_1e^{-t} - c_2e^{-3t}$

Problem Set 4.4, page 151

1. Unstable improper node, $y_1 = c_1e^t, \quad y_2 = c_2e^{2t}$
3. Center, always stable, $y_1 = A \cos 3t + B \sin 3t, \quad y_2 = 3B \cos 3t - 3A \sin 3t$
5. Stable spiral, $y_1 = e^{-2t}(A \cos 2t + B \sin 2t), \quad y_2 = e^{-2t}(B \cos 2t - A \sin 2t)$
7. Saddle point, always unstable, $y_1 = c_1e^{-t} + c_2e^{3t}, \quad y_2 = -c_1e^{-t} + c_2e^{3t}$
9. Unstable node, $y_1 = c_1e^{6t} + c_2e^{2t}, \quad y_2 = 2c_1e^{6t} - 2c_2e^{2t}$
11. $y = e^{-t}(A \cos t + B \sin t)$. Stable and attractive spirals
15. $p = 0.2 \neq 0$ (was 0), $\quad \Delta < 0$, spiral point, unstable.
17. For instance, **(a)** -2, **(b)** -1, **(c)** $= -\frac{1}{2}$, **(d)** $=1$, **(e)** 4.

Problem Set 4.5, page 159

5. Center at (0, 0). At (2, 0) set $y_1 = 2 + \tilde{y}_1$. Then $\tilde{y}_2' = \tilde{y}_1$. Saddle point at (2, 0).
7. (0, 0), $y_1' = -y_1 + y_2, \quad y_2' = -y_1 - y_2$, stable and attractive spiral point; $(-2, 2)$,
$y_1 = -2 + \tilde{y}_1, \quad y_2 = 2 + \tilde{y}_2, \quad \tilde{y}_1' = -\tilde{y}_1 - 3\tilde{y}_2, \quad \tilde{y}_2' = -\tilde{y}_1 - \tilde{y}_2$, saddle point
9. (0, 0) saddle point, $(-3, 0)$ and (3, 0) centers
11. $(\frac{1}{2}\pi \pm 2n\pi, 0)$ saddle points; $(-\frac{1}{2}\pi \pm 2n\pi, 0)$ centers.
Use $-\cos(\pm\frac{1}{2}\pi + \tilde{y}_1) = \sin(\pm\tilde{y}_1) \approx \pm\tilde{y}_1$.
13. $(\pm 2n\pi, 0)$ centers; $y_1 = (2n + 1)\pi + \tilde{y}_1', \quad (\pi \pm 2n\pi, 0)$ saddle points
15. By multiplication, $y_2y_2' = (4y_1 - y_1^3)y_1'$. By integration,
$y_2^2 = 4y_1^2 - \frac{1}{2}y_1^4 + c^* = \frac{1}{2}(c + 4 - y_1^2)(c - 4 + y_1^2)$, where $c^* = \frac{1}{2}c^2 - 8$.

Problem Set 4.6, page 163

3. $y_1 = c_1e^{-t} + c_2e^t, \quad y_2 = -c_1e^{-t} + c_2e^t - e^{3t}$
5. $y_1 = c_1e^{5t} + c_2e^{2t} - 0.43t - 0.24, \quad y_2 = c_1e^{5t} - 2c_2e^{2t} + 1.12t + 0.53$

7. $y_1 = c_1e^t + 4c_2e^{2t} - 3t - 4 - 2e^{-t}$, $y_2 = -c_1e^t - 5c_2e^{2t} + 5t + 7.5 + e^{-t}$

9. The formula for **v** shows that these various choices differ by multiples of the eigenvector for $\lambda = -2$, which can be absorbed into, or taken out of, c_1 in the general solution $y^{(h)}$.

11. $y_1 = -\frac{8}{3}\cosh t - \frac{4}{3}\sinh t + \frac{11}{3}e^{2t}$, $y_2 = -\frac{8}{3}\sinh t - \frac{4}{3}\cosh t + \frac{4}{3}e^{2t}$

13. $y_1 = \cos 2t + \sin 2t + 4\cos t$, $y_2 = 2\cos 2t - 2\sin 2t + \sin t$

15. $y_1 = 4e^{-t} - 4e^t + e^{2t}$, $y_2 = -4e^{-t} + t$

17. $I_1 = 2c_1e^{\lambda_1 t} + 2c_2e^{\lambda_2 t} + 100$,
$I_2 = (1.1 + \sqrt{0.41})c_1e^{\lambda_1 t} + (1.1 - \sqrt{0.41})c_2e^{\lambda_2 t}$,
$\lambda_1 = -0.9 + \sqrt{0.41}$, $\lambda_2 = -0.9 - \sqrt{0.41}$

19. $c_1 = 17.948$, $c_2 = -67.948$

Chapter 4 Review Questions and Problems, page 164

11. $y_1 = c_1e^{4t} + c_2e^{-4t}$, $y_2 = 2c_1e^{4t} - 2c_2e^{-4t}$. Saddle point

13. $y_1 = e^{-4t}(A\cos t + B\sin t)$, $y_2 = \frac{1}{5}e^{-4t}[(B - 2A)\cos t - (A + 2B)\sin t]$; asymptotically stable spiral point

15. $y_1 = c_1e^{-5t} + c_2e^{-t}$, $y_2 = c_1e^{-5t} - c_2e^{-t}$. Stable node

17. $y_1 = e^{-t}(A\cos 2t + B\sin 2t)$, $y_2 = e^{-t}(B\cos 2t - A\sin 2t)$. Stable and attractive spiral point

19. Unstable spiral point

21. $y_1 = c_1e^{-4t} + c_2e^{4t} - 1 - 8t^2$, $y_2 = -c_1e^{-4t} + c_2e^{4t} - 4t$

23. $y_1 = 2c_1e^{-t} + 2c_2e^{3t} + \cos t - \sin t$, $y_2 = -c_1e^{-t} + c_2e^{3t}$

25. $I_1' + 2.5(I_1 - I_2) = 169\sin t$, $2.5(I_2' - I_1') + 25I_2 = 0$,
$I_1 = (19 + 32.5t)e^{-5t} - 19\cos t + 62.5\sin t$,
$I_2 = (-6 - 32.5t)e^{-5t} + 6\cos t + 2.5\sin t$

27. $(0, 0)$ saddle point; $(-1, 0)$, $(1, 0)$ centers

29. $(n\pi, 0)$ center when n is even and saddle point when n is odd

Problem Set 5.1, page 174

3. $\sqrt{|k|}$

5. $\sqrt{3/2}$

7. $y = a_0(1 - x^2 + x^4/2! - x^6/3! + - \cdots) = a_0e^{-x^2}$

9. $y = a_0 + a_1x - \frac{1}{2}a_0x^2 - \frac{1}{6}a_1x^3 + \cdots = a_0\cos x + a_1\sin x$

11. $a_0(1 - \frac{1}{12}x^4 - \frac{1}{60}x^5 - \cdots) + a_1(x + \frac{1}{2}x^2 + \frac{1}{6}x^3 + \frac{1}{24}x^4 - \frac{1}{24}x^5 - \cdots)$

13. $a_0(1 - \frac{1}{2}x^2 - \frac{1}{24}x^4 + \frac{13}{720}x^6 + \cdots) + a_1(x - \frac{1}{6}x^3 - \frac{1}{24}x^5 + \frac{5}{1008}x^7 + \cdots)$

15. $\displaystyle\sum_{m=1}^{\infty} \frac{(m+1)(m+2)}{(m+1)^2+1}x^m$, $\displaystyle\sum_{m=5}^{\infty}\frac{(m-4)^2}{(m-3)!}x^m$

17. $s = 1 + x - x^2 - \frac{5}{6}x^3 + \frac{2}{3}x^4 + \frac{11}{24}x^5$, $s(\frac{1}{2}) = \frac{923}{768}$

19. $s = 4 - x^2 - \frac{1}{3}x^3 + \frac{1}{30}x^5$, $s(2) = -\frac{8}{5}$; but $x = 2$ is too large to give good values. Exact: $y = (x - 2)^2e^x$

Problem Set 5.2, page 179

5. $P_6(x) = \frac{1}{16}(231x^6 - 315x^4 + 105x^2 - 5)$,
$P_7(x) = \frac{1}{16}(429x^7 - 693x^5 + 315x^3 - 35x)$

11. Set $x = az$. $y = c_1 P_n(x/a) + c_2 Q_n(x/a)$

15. $P_1^1 = \sqrt{1 - x^2}$, $P_2^1 = 3x\sqrt{1 - x^2}$, $P_2^2 = 3(1 - x^2)$, $P_4^2 = (1 - x^2)(105x^2 - 15)/2$

Problem Set 5.3, page 186

3. $y_1 = 1 - \dfrac{x^2}{3!} + \dfrac{x^4}{5!} - + \cdots = \dfrac{\sin x}{x}$, $y_2 = \dfrac{1}{x} - \dfrac{x}{2!} + \dfrac{x^3}{4!} - + \cdots = \dfrac{\cos x}{x}$

5. $b_0 = 1$, $c_0 = 0$, $r^2 = 0$, $y_1 = e^{-x}$, $y_2 = e^{-x} \ln x$

7. $y_1 = 1 + \frac{1}{2}x^2 - \frac{1}{6}x^3 + \frac{1}{24}x^4 - \frac{1}{30}x^5 + \frac{1}{144}x^6 - \cdots$,
$y_2 = x + \frac{1}{6}x^3 - \frac{1}{12}x^4 + \frac{1}{120}x^5 - \frac{1}{120}x^6 + \cdots$

9. $y_1 \sqrt{x}$, $y_2 = 1 + x$

11. $y_1 = e^x$, $y_2 = e^x/x$

13. $y_1 = e^x$, $y_2 = e^x \ln x$

15. $y = AF(1, 1, -\frac{1}{2}; x) + Bx^{3/2}F(\frac{5}{2}, \frac{5}{2}, \frac{5}{2}; x)$

17. $y = A(1 - 8x + \frac{32}{5}x^2) + Bx^{3/4}F(\frac{7}{4}, -\frac{5}{4}, \frac{7}{4}; x)$

19. $y = c_1F(2, -2, -\frac{1}{2}; t - 2) + c_2(t - 2)^{3/2}F(\frac{7}{2}, -\frac{1}{2}, \frac{5}{2}; t - 2)$

Problem Set 5.4, page 195

3. $c_1 J_0(\sqrt{x})$

5. $c_1 J_\nu(\lambda x) + c_2 J_{-\nu}(\lambda x)$, $\nu \neq 0, \pm 1, \pm 2, \cdots$

7. $c_1 J_{1/2}(\frac{1}{2}x) + c_2 J_{-1/2}(\frac{1}{2}x) = x^{-1/2}(\tilde{c}_1 \sin \frac{1}{2}x + \tilde{c}_2 \cos \frac{1}{2}x)$

9. $x^{-\nu}(c_1 J_\nu(x) + c_2 J_{-\nu}(x))$, $\nu \neq 0, \pm 1, \pm 2, \cdots$

13. $J_n(x_1) = J_n(x_2) = 0$ implies $x_1^{-n} J_n(x_1) = x_2^{-n} J_n(x_2) = 0$ and $[x^{-n}J_n(x)]' = 0$ somewhere between x_1 and x_2 by Rolle's theorem. Now use (21b) to get $J_{n+1}(x) = 0$ there. Conversely, $J_{n+1}(x_3) = J_{n+1}(x_4) = 0$, thus $x_3^{n+1} J_{n+1}(x_3) = x_4^{n+1} J_{n+1}(x_4) = 0$ implies $J_n(x) = 0$ in between by Rolle's theorem and (21a) with $\nu = n + 1$.

15. By Rolle, $J_0' = 0$ at least once between two zeros of J_0. Use $J_0' = -J_1$ by (21b) with $\nu = 0$. Together $J_1 = 0$ at least once between two zeros of J_0. Also use $(xJ_1)' = xJ_0$ by (21a) with $\nu = 1$ and Rolle.

19. Use (21b) with $\nu = 0$, (21a) with $\nu = 1$, (21d) with $\nu = 2$, respectively.

21. Integrate (21a).

23. Use (21a) with $\nu = 1$, partial integration, (21b) with $\nu = 0$, partial integration.

25. Use (21d) to get

$$\begin{aligned}\int J_5(x)\,dx &= -2J_4(x) + \int J_3(x)\,dx = -2J_4(x) - 2J_2(x) + \int J_1(x)\,dx \\ &= -2J_4(x) - 2J_2(x) - J_0(x) + c.\end{aligned}$$

Problem Set 5.5, page 200

1. $c_1 J_4(x) + c_2 Y_4(x)$

3. $c_1 J_{2/3}(x^2) + c_2 Y_{2/3}(x^2)$

5. $c_1 J_0(\sqrt{x}) + c_2 Y_0(\sqrt{x})$

7. $\sqrt{x}\,(c_1 J_{1/4}(\tfrac{1}{2}kx^2) + c_2 Y_{1/4}(\tfrac{1}{2}kx^2))$
9. $x^3(c_1 J_3(x) + c_2 Y_3(x))$
11. Set $H^{(1)} = kH^{(2)}$ and use (10).
13. Use (20) in Sec. 5.4.

Chapter 5 Review Questions and Problems, page 200

11. $\cos 2x$, $\sin 2x$
13. $(x-1)^{-5}$, $(x-1)^7$; Euler–Cauchy with $x - 1$ instead of x
15. $J_{\sqrt{5}}(x)$, $J_{-\sqrt{5}}(x)$
17. e^x, $1 + x$
19. $\sqrt{x}\,J_1(\sqrt{x})$, $\sqrt{x}\,Y_1(\sqrt{x})$

Problem Set 6.1, page 210

1. $3/s^2 + 12/s$
3. $s/(s^2 + \pi^2)$
5. $1/((s-2)^2 - 1)$
7. $(\omega\cos\theta + s\sin\theta)/(s^2 + \omega^2)$
9. $\dfrac{1}{s} + \dfrac{e^{-s} - 1}{s^2}$
11. $\dfrac{1 - e^{-bs}}{s^2} - \dfrac{be^{-bs}}{s}$
13. $\dfrac{(1 - e^{-s})^2}{s}$
15. $\dfrac{e^{-s} - 1}{2s^2} - \dfrac{e^{-s}}{2s} + \dfrac{1}{s}$
19. Use $e^{at} = \cosh at + \sinh at$.
23. Set $ct = p$. Then $\mathcal{L}(f(ct)) = \displaystyle\int_0^\infty e^{-st} f(ct)\,dt = \int_0^\infty e^{-(s/c)p} f(p)\,dp/c = F(s/c)/c.$
25. $0.2\cos 1.8t + \sin 1.8t$
27. $\dfrac{1}{L^2}\cos\dfrac{n\pi t}{L}$
29. $2t^3 - 1.9t^5$
31. $\mathcal{L}^{-1}\left(\dfrac{4}{s-2} - \dfrac{3}{s+1}\right) = 4e^{2t} - 3e^{-t}$
33. $\dfrac{2}{(s+3)^3}$
35. $\dfrac{0.5 \cdot 2\pi}{(s+4.5)^2 + 4\pi^2}$
37. $\pi t e^{-\pi t}$
39. $\tfrac{7}{2} t^3 e^{-t\sqrt{2}}$
41. $e^{-5\pi t}\sinh \pi t$
43. $e^{3t}(2\cos 3t + \tfrac{5}{3}\sin 3t)$
45. $(k_0 + k_1 t)e^{-at}$

Problem Set 6.2, page 216

1. $y = 1.25e^{-5.2t} - 1.25\cos 2t + 3.25\sin 2t$
3. $(s-3)(s+2) = 11s + 28 - 11 = 11s + 17, \quad Y = 10/(s-3) + 1/(s+2),$
$y = 10e^{3t} + e^{-2t}$
5. $(s^2 - \tfrac{1}{4})Y = 12s, \quad y = 12\cosh\tfrac{1}{2}t$
7. $y = \tfrac{1}{2}e^{3t} + \tfrac{5}{2}e^{-4t} + \tfrac{1}{2}e^{-3t}$
9. $y = e^t - e^{3t} + 2t$
11. $(s + 1.5)^2 Y = s + 31.5 + 3 + 54/s^4 + 64/s,$
$Y = 1/(s+1.5) + 1/(s+1.5)^2 + 24/s^4 - 32/s^3 + 32/s^2,$
$y = (1+t)e^{-1.5t} + 4t^3 - 16t^2 + 32t$
13. $t = \tilde{t} - 1, \quad \tilde{Y} = 4/(s-6), \quad \tilde{y} = 4e^{6t}, \quad y = 4e^{6(t+1)}$

15. $t = \tilde{t} + 1.5, \quad (s - 1)(s + 4)\tilde{Y} = 4s + 17 + 6/(s - 2), \quad y = 3e^{t-1.5} + e^{2(t-1.5)}$

17. $\dfrac{1}{(s + a)^2}$ **19.** $\dfrac{2\omega^2}{s(s^2 + 4\omega^2)}$

21. $\mathcal{L}(f') = \mathcal{L}(\sinh 2t) = s\mathcal{L}(f) - 1.$ *Answer:* $(s^2 - 2)/(s^3 - 4s)$

23. $12(1 - e^{-t/4})$ **25.** $(1 - \cos \omega t)/\omega^2$

27. $\frac{1}{9}(1 + t - \cos 3t - \frac{1}{3}\sin 3t)$ **29.** $\dfrac{1}{a^2}(e^{-at} - 1) + \dfrac{t}{a}$

Problem Set 6.3, page 223

3. $\mathcal{L}((t - 2)u(t - 2)) = e^{-2s}/s^2$

5. $\left(e^t\left(1 - u\left(t - \frac{1}{2}\pi\right)\right)\right) = \dfrac{1}{s - 1}(1 - e^{-\pi s/2 + \pi/2})$

7. $\dfrac{1}{s + \pi}(e^{-2(s+\pi)} - e^{-4(s+\pi)})$

9. $e^{-3s/2}\left(\dfrac{2}{s^3} + \dfrac{3}{s^2} + \dfrac{\frac{9}{4}}{s}\right)$

11. $(se^{-\pi s/2} + e^{-\pi s})/(s^2 + 1)$ **13.** $2[1 + u(t - \pi)]\sin 3t$

15. $(t - 3)^3 u(t - 3)/6$ **17.** $e^{-t}\cos t \ (0 < t < 2\pi)$

19. $\frac{1}{3}(e^t - 1)^3 e^{-5t}$ **21.** $\sin 3t + \sin t \ (0 < t < \pi); \ \frac{4}{3}\sin 3t \ (t > \pi)$

23. $e^t - \sin t \ (0 < t < 2\pi), \quad e^t - \frac{1}{2}\sin 2t \ (t > 2\pi)$

25. $t - \sin t \ (0 < t < 1), \quad \cos(t - 1) + \sin(t - 1) - \sin t \ (t > 1)$

27. $t = 1 + \tilde{t}, \quad \tilde{y}'' + 4\tilde{y} = 8(1 + \tilde{t})^2(1 - u(\tilde{t} - 4)), \quad \cos 2t + 2t^2 - 1$ if $t < 5$,
$\cos 2t + 49\cos(2t - 10) + 10\sin(2t - 10)$ if $t > 5$

29. $0.1i' + 25i = 490e^{-5t}[1 - u(t - 1)],$
$i = 20(e^{-5t} - e^{-250t}) + 20u(t - 1)[-e^{-5t} + e^{-250t+245}]$

31. $Rq' + q/C = 0, \quad Q = \mathcal{L}(q), \quad q(0) = CV_0, \quad i = q'(t),$
$R(sQ - CV_0) + Q/C = 0, \quad q = CV_0 e^{-t/(RC)}$

33. $10I + \dfrac{100}{s}I = \dfrac{100}{s^2}e^{-2s}, \quad I = e^{-2s}\left(\dfrac{1}{s} - \dfrac{1}{s + 10}\right), \quad i = 0$ if $t < 2$ and
$1 - e^{-10(t-2)}$ if $t > 2$

35. $i = (10\sin 10t + 100\sin t)(u(t - \pi) - u(t - 3\pi))$

37. $(0.5s^2 + 20)I = 78s(1 + e^{-\pi s})/(s^2 + 1),$
$i = 4\cos t - 4\cos\sqrt{40}t - 4u(t - \pi)[\cos t + \cos(\sqrt{40}(t - \pi))]$

39. $i' + 2i + 2\displaystyle\int_0^t i(\tau)\,d\tau = 1000(1 - u(t - 2)), \quad I = 1000(1 - e^{-2s})/(s^2 + 2s + 2),$

$i = 1000e^{-t}\sin t - 1000u(t - 2)e^{-t+2}\sin(t - 2)$

Problem Set 6.4, page 230

3. $y = 8\cos 2t + \frac{1}{2}u(t - \pi)\sin 2t$

5. $\sin t \ (0 < t < \pi); \quad 0 \ (\pi < t < 2\pi); \quad -\sin t \ (t > 2\pi)$

7. $y = e^{-t} + 4e^{-3t}\sin\frac{1}{2}t + \frac{1}{2}u(t - \frac{1}{2})e^{-3(t-1/2)}\sin(\frac{1}{2}t - \frac{1}{4})$

9. $y = 0.1[e^t + e^{-2t}(-\cos t + 7\sin t)] + 0.1u(t - 10)[-e^{-t} +$
$e^{-2t+30}(\cos(t - 10) - 7\sin(t - 10))]$

11. $y = -e^{-3t} + e^{-2t} + \frac{1}{6}u(t-1)(1 - 3e^{-2(t-1)} + 2e^{-3(t-1)}) + u(t-2)(e^{-2(t-2)} - e^{-3(t-2)})$

15. $ke^{-ps}/(s - se^{-ps})\ (s > 0)$

Problem Set 6.5, page 237

1. t

3. $(e^t - e^{-t})/2 = \sinh t$

5. $\frac{1}{2}t \sin \omega t$

7. $e^t - t - 1$

9. $y - 1 * y = 1, \quad y = e^t$

11. $y = \cos t$

13. $y(t) + 2\int_0^t e^{t-\tau} y(\tau)\, d\tau = te^t, \quad y = \sinh t$

17. $e^{4t} - e^{-1.5t}$

19. $t \sin \pi t$

21. $(\omega t - \sin \omega t)/\omega^2$

23. $4.5(\cosh 3t - 1)$

25. $1.5t \sin 6t$

Problem Set 6.6, page 241

3. $\dfrac{\frac{1}{2}}{(s+3)^2}$

5. $\dfrac{s^2 - \omega^2}{(s^2 + \omega^2)^2}$

7. $\dfrac{2s^3 + 24s}{(s^2 - 4)^3}$

9. $\dfrac{\pi(3s^2 - \pi^2)}{(s^2 + \pi^2)^3}$

11. $\dfrac{4s^2 - \pi^2}{(s^2 + \frac{1}{4}\pi^2)^2}$

15. $F(s) = -\dfrac{1}{2}\left(\dfrac{1}{s^2 - 9}\right)', \quad f(t) = \frac{1}{6}t \sinh 3t$

17. $\ln s - \ln (s-1); \quad (-1 + e^t)/t$

19. $[\ln (s^2+1) - 2 \ln (s-1)]' = 2s/(s^2+1) - 2/(s-1); \quad 2(-\cos t + e^t)/t$

Problem Set 6.7, page 246

3. $y_1 = -e^{-5t} + 4e^{2t}, \quad y_2 = e^{-5t} + 3e^{2t}$

5. $y_1 = -\cos t + \sin t + 1 + u(t-1)[-1 + \cos (t-1) - \sin (t-1)]$
$y_2 = \cos t + \sin t - 1 + u(t-1)[1 - \cos (t-1) - \sin (t-1)]$

7. $y_1 = -e^{-2t} + 4e^t + \frac{1}{3}u(t-1)(-e^{3-2t} + e^t),$
$y_2 = -e^{-2t} + e^t + \frac{1}{3}u(t-1)(-e^{3-2t} + e^t)$

9. $y_1 = (3 + 4t)e^{3t}, \quad y_2 = (1 - 4t)e^{3t}$

11. $y_1 = e^t + e^{2t}, \quad y_2 = e^{2t}$

13. $y_1 = -4e^t + \sin 10t + 4 \cos t, \quad y_2 = 4e^t - \sin 10t + 4 \cos t$

15. $y_1 = e^t, \quad y_2 = e^{-t}, \quad y_3 = e^t - e^{-t}$

19. $4i_1 + 8(i_1 - i_2) + 2i_1' = 390 \cos t, \quad 8i_2 + 8(i_2 - i_1) + 4i_2' = 0,$
$i_1 = -26e^{-2t} - 16e^{-8t} + 42 \cos t + 15 \sin t,$
$i_2 = -26e^{-2t} + 8e^{-8t} + 18 \cos t + 12 \sin t$

Chapter 6 Review Questions and Problems, page 251

11. $\dfrac{5s}{s^2 - 4} - \dfrac{3}{s^2 - 1}$

13. $\frac{1}{2}(1 - \cos \pi t), \quad \pi^2/(2s^3 + 2\pi^2 s)$

15. $e^{-3s+3/2}/(s - \frac{1}{2})$

17. Sec. 6.6; $2s^2/(s^2+1)^2$

19. $12/(s^2(s+3))$

21. $tu(t-1)$

23. $\sin(\omega t + \theta)$

25. $3t^2 + t^3$

27. $e^{-2t}(3\cos t - 2\sin t)$

29. $y = e^{-2t}(13\cos t + 11\sin t) + 10t - 8$

31. $e^{-t} + u(t-\pi)[1.2\cos t - 3.6\sin t + 2e^{-t+\pi} - 0.8e^{2t-2\pi}]$

33. $0\ (0 \leqq t \leqq 2),\quad 1 - 2e^{-(t-2)} + e^{-2(t-2)}\quad (t > 2)$

35. $y_1 = 4e^t - e^{-2t},\quad y_2 = e^t - e^{-2t}$

37. $y_1 = \cos t - u(t-\pi)\sin t + 2u(t-2\pi)\sin^2 \tfrac{1}{2}t,$
$y_2 = -\sin t - 2u(t-\pi)\cos^2 \tfrac{1}{2}t + u(t-2\pi)\sin t$

39. $y_1 = (1/\sqrt{10})\sin\sqrt{10}t,\quad y_2 = -(1/\sqrt{10})\sin\sqrt{10}t$

41. $1 - e^{-t}\ (0 < t < 4),\quad (e^4 - 1)e^{-t}\quad (t > 4)$

43. $i(t) = e^{-4t}(\tfrac{3}{26}\cos 3t - \tfrac{10}{39}\sin 3t) - \tfrac{3}{26}\cos 10t + \tfrac{8}{65}\sin 10t$

45. $5i_1' + 20(i_1 - i_2) = 60,\quad 30i_2' + 20(i_2' - i_1') + 20i_2 = 0,$
$i_1 = -8e^{-2t} + 5e^{-0.8t} + 3,\quad i_2 = -4e^{-2t} + 4e^{-0.8t}$

Problem Set 7.1, page 261

3. $3 \times 3,\quad 3 \times 4,\quad 3 \times 6,\quad 2 \times 2,\quad 2 \times 3,\quad 3 \times 2$

5. $\mathbf{B} = \tfrac{1}{5}\mathbf{A},\quad \tfrac{1}{10}\mathbf{A}$

7. No, no, yes, no, no

9. $\begin{bmatrix} 0 & 6 & 12 \\ 18 & 15 & 15 \\ 3 & 0 & -9 \end{bmatrix},\quad \begin{bmatrix} 0 & 2.5 & 1 \\ 2.5 & 1.5 & 2 \\ -1 & 2 & -1 \end{bmatrix},\quad \begin{bmatrix} 0 & 8.5 & 13 \\ 20.5 & 16.5 & 17 \\ 2 & 2 & -10 \end{bmatrix},$ undefined

11. $\begin{bmatrix} 0 & 26 \\ 34 & 32 \\ 28 & -10 \end{bmatrix},$ same, $\begin{bmatrix} 5.4 & 0.6 \\ -4.2 & 2.4 \\ -0.6 & 0.6 \end{bmatrix},$ same

13. $\begin{bmatrix} 70 & 28 \\ -28 & 56 \\ 14 & 0 \end{bmatrix},$ same, $-\mathbf{D}$, undefined

15. $\begin{bmatrix} 5.5 \\ 33.0 \\ -11.0 \end{bmatrix},$ same, undefined, undefined

17. $\begin{bmatrix} -4.5 \\ -27.0 \\ 9.0 \end{bmatrix}$

Problem Set 7.2, page 270

5. 10, $n(n+1)/2$

7. $\mathbf{0},\quad \mathbf{I},\quad \begin{bmatrix} 1 & 0 \\ 0 & 0 \end{bmatrix},\quad \begin{bmatrix} 1 & 1 \\ 0 & 0 \end{bmatrix}$

11. $\begin{bmatrix} 10 & -14 & -6 \\ -5 & 7 & -12 \\ -5 & -1 & -4 \end{bmatrix}$, same, $\begin{bmatrix} 10 & -5 & -15 \\ -14 & 7 & -33 \\ -2 & -4 & -4 \end{bmatrix}$, same

13. $\begin{bmatrix} 1 & 2 & 0 \\ 2 & 13 & -6 \\ 0 & -6 & 4 \end{bmatrix}$, $\begin{bmatrix} -9 & -5 \\ 3 & -1 \\ 4 & 0 \end{bmatrix}$, undefined, $\begin{bmatrix} -9 & 3 & 4 \\ -5 & -1 & 0 \end{bmatrix}$

15. Undefined, $\begin{bmatrix} 8 \\ -4 \\ -3 \end{bmatrix}$, $[7 \quad -1 \quad 3]$, same

17. $\begin{bmatrix} -30 & -18 \\ 45 & 9 \\ 5 & -7 \end{bmatrix}$, undefined, $\begin{bmatrix} 22 \\ 4 \\ -12 \end{bmatrix}$, undefined

19. Undefined, $\begin{bmatrix} 10.5 \\ 0 \\ -3 \end{bmatrix}$, $\begin{bmatrix} 7 \\ -3 \\ 1 \end{bmatrix}$, same

25. **(d)** $\mathbf{AB} = (\mathbf{AB})^{\mathsf{T}} = \mathbf{B}^{\mathsf{T}}\mathbf{A}^{\mathsf{T}} = \mathbf{BA}$; etc.
(e) *Answer.* If $\mathbf{AB} = -\mathbf{BA}$.

29. $\mathbf{p} = [85 \quad 62 \quad 30]^{\mathsf{T}}$, $\mathbf{v} = [44{,}920 \quad 30{,}940]^{\mathsf{T}}$

Problem Set 7.3, page 280

1. $x = -2, \quad y = 0.5$
3. $x = 1, \quad y = 3, \quad z = -5$
5. $x = 6, \quad y = -7$
7. $x = -3t, \quad y = t$ arb., $\quad z = 2t$
9. $x = 3t - 1, \quad y = -t + 4, \quad z = t$ arb.
11. $w = 1, \quad x = t_1$ arb., $\quad y = 2t_2 - t_1, \quad z = t_2$ arb.
13. $w = 4, \quad x = 0, \quad y = 2, \quad z = 6$
17. $I_1 = 2, \quad I_2 = 6, \quad I_3 = 8$
19. $I_1 = (R_1 + R_2)E_0/(R_1R_2)$ A, $\quad I_2 = E_0/R_1$ A, $\quad I_3 = E_0/R_2$ A
21. $x_2 = 1600 - x_1, \quad x_3 = 600 + x_1, \quad x_4 = 1000 - x_1$. No
23. C: $3x_1 - x_3 = 0$, H: $8x_1 - 2x_4 = 0$, O: $2x_2 - 2x_3 - x_4 = 0$, thus $C_3H_8 + 5O_2 \rightarrow 3CO_2 + 4H_2O$

Problem Set 7.4, page 287

1. 1; $[2 \quad -1 \quad 3]$; $[2 \quad -1]^{\mathsf{T}}$
3. 3; $\{[3 \quad 5 \quad 0], \ [0 \quad 3 \quad 5], \ [0 \quad 0 \quad 1]\}$
5. 3; $\{[2 \quad -1 \quad 4], \ [0 \quad 1 \quad -46], \ [0 \quad 0 \quad 1]\}$; $\{[2 \quad 0 \quad 1], \ [0 \quad 3 \quad 23], \ [0 \quad 0 \quad 1]\}$

7. 2; [8 0 4 0], [0 2 0 4]; [8 0 4], [0 2 0]
9. 3; [9 0 1 0], [0 9 8 9], [0 0 1 0]
11. **(c)** 1
17. No
19. Yes
21. No
23. Yes
25. Yes
27. 2, [−2 0 1], [0 2 1]
29. No
31. No
33. 1, solution of the given system $c[1 \;\; \frac{10}{3} \;\; 3]$, basis $[1 \;\; \frac{10}{3} \;\; 3]$
35. 1, $[4 \;\; 2 \;\; \frac{4}{3} \;\; 1]$

Problem Set 7.7, page 300

7. $\cos(\alpha + \beta)$
9. 1
11. 40
13. 289
15. −64
17. 2
19. 2
21. $x = 3.5, \quad y = -1.0$
23. $x = 0, \quad y = 4, \quad z = -1$
25. $w = 3, \quad x = 0, \quad y = 2, \quad z = -2$

Problem Set 7.8, page 308

1. $\begin{bmatrix} 1.20 & 4.64 \\ 0.50 & 3.60 \end{bmatrix}$

3. $\begin{bmatrix} 54 & 0.9 & -3.4 \\ 2 & 0.2 & -0.2 \\ -30 & -0.5 & 2 \end{bmatrix}$

5. $\begin{bmatrix} 1 & 0 & 0 \\ -2 & 1 & 0 \\ 3 & -4 & 1 \end{bmatrix}$

7. $\mathbf{A}^{-1} = \mathbf{A}$

9. $\begin{bmatrix} 0 & 0 & \frac{1}{2} \\ \frac{1}{8} & 0 & 0 \\ 0 & \frac{1}{4} & 0 \end{bmatrix}$

11. $(\mathbf{A}^2)^{-1} = (\mathbf{A}^{-1})^2 = \begin{bmatrix} 3.760 & 22.272 \\ 2.400 & 15.280 \end{bmatrix}$

15. $\mathbf{AA}^{-1} = \mathbf{I}, \quad (\mathbf{AA}^{-1})^{-1} = (\mathbf{A}^{-1})^{-1}\mathbf{A}^{-1} = \mathbf{I}$. Multiply by $\mathbf{A}$ from the right.

Problem Set 7.9, page 318

1. $[1 \;\; 0]^\mathsf{T}, \; [0 \;\; 1]^\mathsf{T}; \; [1 \;\; 0]^\mathsf{T}, \; [0 \;\; -1]^\mathsf{T}; \; [1 \;\; 1]^\mathsf{T}, \; [-1 \;\; 1]^\mathsf{T}$
3. 1, $[1 \;\; 11 \;\; -7]^\mathsf{T}$
5. No
7. Dimension 2, basis xe^{-x}, e^{-x}
9. 3; basis $\begin{bmatrix} 1 & 0 \\ 0 & -1 \end{bmatrix}, \begin{bmatrix} 0 & 1 \\ 0 & 0 \end{bmatrix}, \begin{bmatrix} 0 & 0 \\ 1 & 0 \end{bmatrix}$
11. $x_1 = 5y_1 - y_2, \quad x_2 = 3y_1 - y_2$
13. $x_1 = 2y_1 - 3y_2, \quad x_2 = -10y_1 + 16y_2 + y_3, \quad x_3 = -7y_1 + 11y_2 + y_3$

15. $\sqrt{26}$

17. $\sqrt{5}$

19. 1

21. $k = -20$

23. $\mathbf{a} = [3 \quad 1 \quad -4]^T$, $\mathbf{b} = [-4 \quad 8 \quad -1]^T$, $\|\mathbf{a} + \mathbf{b}\| = \sqrt{107} \leqq 5.099 + 9$

25. $\mathbf{a} = [5 \quad 3 \quad 2]^T$, $\mathbf{b} = [3 \quad 2 \quad -1]^T$, $90 + 14 = 2(38 + 14)$

Chapter 7 Review Questions and Problems, page 318

11. $\begin{bmatrix} -1 & 6 & 1 \\ -18 & 8 & -7 \\ -13 & -2 & -7 \end{bmatrix}$, $\begin{bmatrix} 1 & 18 & 13 \\ -6 & -8 & 2 \\ -1 & 7 & 7 \end{bmatrix}$

13. $[21 \quad -8 \quad -31]^T$, $[21 \quad -8 \quad 31]$

15. 197, 0

17. -5, $\det \mathbf{A}^2 = (\det \mathbf{A})^2 = 25$, 0

19. $\begin{bmatrix} -2 & -12 & -12 \\ -12 & 16 & -9 \\ -12 & -9 & -14 \end{bmatrix}$

21. $x = 4$, $y = -2$, $z = 8$

23. $x = 6$, $y = 2t + 2$, $z = t$ arb.

25. $x = 0.4$, $y = -1.3$, $z = 1.7$

27. $x = 10$, $y = -2$

29. Ranks 2, 2, ∞

31. Ranks 2, 2, 1

33. $I_1 = 16.5$ A, $I_2 = 11$ A, $I_3 = 5.5$ A

35. $I_1 = 4$ A, $I_2 = 5$ A, $I_3 = 1$ A

Problem Set 8.1, page 329

1. 3, $[1 \quad 0]^T$; -0.6, $[0 \quad 1]^T$

3. -4, $[2 \quad 9]^T$; 3, $[1 \quad 1]^T$

5. $-3i$, $[1 \quad -i]$; $3i$, $[1 \quad i]$, $i = \sqrt{-1}$

7. $\lambda^2 = 0$, $[1 \quad 0]^T$

9. $0.8 + 0.6i$, $[1 \quad -i]^T$; $0.8 - 0.6i$, $[1 \quad i]^T$

11. $-(\lambda^3 - 18\lambda^2 + 99\lambda - 162)/(\lambda - 3) = -(\lambda^2 - 15\lambda + 54)$; 3, $[2 \quad -2 \quad 1]^T$; 6, $[1 \quad 2 \quad 2]^T$; 9, $[2 \quad 1 \quad -2]^T$

13. $-(\lambda - 9)^3$; 9, $[2 \quad -2 \quad 1]^T$, defect 2

15. $(\lambda + 1)^2(\lambda^2 + 2\lambda - 15)$; -1, $[1 \quad 0 \quad 0 \quad 0]^T$, $[0 \quad 1 \quad 0 \quad 0]^T$; -5, $[-3 \quad -3 \quad 1 \quad 1]^T$, 3, $[3 \quad -3 \quad 1 \quad -1]^T$

17. $\begin{bmatrix} 0 & -1 \\ 1 & 0 \end{bmatrix}$. Eigenvalues i, $-i$. Corresponding eigenvectors are complex, indicating that no direction is preserved under a rotation.

19. $\begin{bmatrix} 0 & 0 \\ 0 & 1 \end{bmatrix}$; 1, $\begin{bmatrix} 0 \\ 1 \end{bmatrix}$; 0, $\begin{bmatrix} 1 \\ 0 \end{bmatrix}$. A point onto the x_2-axis goes onto itself, a point on the x_1-axis onto the origin.

23. Use that real entries imply real coefficients of the characteristic polynomial.

Problem Set 8.2, page 333

1. 1.5, $[1 \quad -1]^\mathsf{T}$, $-45°$; 4.5, $[1 \quad 1]^\mathsf{T}$, 45°

3. 1, $[-1/\sqrt{6} \quad 1]^\mathsf{T}$, 112.2°; 8, $[1 \quad 1/\sqrt{6}]^\mathsf{T}$, 22.2°

5. 0.5, $[1 \quad -1]^\mathsf{T}$; 1.5, $[1 \quad 1]^\mathsf{T}$; directions $-45°$ and 45°

7. $[5 \quad 8]^\mathsf{T}$

9. $[11 \quad 12 \quad 16]^\mathsf{T}$

11. 1.8

13. $c[10 \quad 18 \quad 25]^\mathsf{T}$

15. $\mathbf{x} = (\mathbf{I} - \mathbf{A})^{-1}\mathbf{y} = [0.6747 \quad 0.7128 \quad 0.7543]^\mathsf{T}$

17. $\mathbf{A}\mathbf{x}_j = \lambda_j\mathbf{x}_j$ $(\mathbf{x}_j \neq 0)$, $(\mathbf{A} - k\mathbf{I})\mathbf{x}_j = \lambda_j\mathbf{x}_j - k\mathbf{x}_j = (\lambda_j - k)\mathbf{x}_j$.

19. From $\mathbf{A}\mathbf{x}_j = \lambda_j\mathbf{x}_j$ $(\mathbf{x}_j \neq \mathbf{0})$ and Prob. 18 follows $k_p\mathbf{A}^p\mathbf{x}_j = k_p\lambda_j^p\mathbf{x}_j$ and $k_q\mathbf{A}^q\mathbf{x}_j = k_q\lambda_j^q\mathbf{x}_j$ $(p \geqq 0, q \geqq 0$, integer). Adding on both sides, we see that $k_p\mathbf{A}^p + k_q\mathbf{A}^q$ has the eigenvalue $k_p\lambda_j^p + k_q\lambda_j^q$. From this the statement follows.

Problem Set 8.3, page 338

1. $0.8 \pm 0.6i$, $[1 \quad \pm i]^\mathsf{T}$; orthogonal

3. $2 \pm 0.8i$, $[1 \quad \pm i]$. Not skew-symmetric!

5. 1, $[0 \quad 2 \quad 1]^\mathsf{T}$; 6, $[1 \quad 0 \quad 0]^\mathsf{T}$, $[0 \quad 1 \quad -2]^\mathsf{T}$; symmetric

7. 0, $\pm 25i$, skew-symmetric

9. 1, $[0 \quad 1 \quad 0]^\mathsf{T}$; i, $[1 \quad 0 \quad i]^\mathsf{T}$; $-i$, $[1 \quad 0 \quad -i]^\mathsf{T}$, orthogonal

15. No

17. $\mathbf{A}^{-1} = (-\mathbf{A}^\mathsf{T})^{-1} = -(\mathbf{A}^{-1})^\mathsf{T}$

19. No since $\det \mathbf{A} = \det(\mathbf{A}^\mathsf{T}) = \det(-\mathbf{A}) = (-1)^3 \det(\mathbf{A}) = -\det(\mathbf{A}) = 0$.

Problem Set 8.4, page 345

1. $\begin{bmatrix} -25 & 12 \\ -50 & 25 \end{bmatrix}$, -5, $\begin{bmatrix} 3 \\ 5 \end{bmatrix}$; 5, $\begin{bmatrix} 2 \\ 5 \end{bmatrix}$; $\mathbf{x} = \begin{bmatrix} -2 \\ 4 \end{bmatrix}$, $\begin{bmatrix} 2 \\ 1 \end{bmatrix}$

3. $\begin{bmatrix} 3.008 & -0.544 \\ 5.456 & 6.992 \end{bmatrix}$, 4, $\begin{bmatrix} -17 \\ 31 \end{bmatrix}$; 6, $\begin{bmatrix} -2 \\ 11 \end{bmatrix}$; $\mathbf{x} = \begin{bmatrix} 25 \\ 25 \end{bmatrix}$, $\begin{bmatrix} 10 \\ 5 \end{bmatrix}$

5. $\begin{bmatrix} 4 & 3 & -9 \\ 0 & -5 & 15 \\ 0 & -5 & 15 \end{bmatrix}$, 0, $\begin{bmatrix} 0 \\ 3 \\ 1 \end{bmatrix}$; 4, $\begin{bmatrix} 1 \\ 0 \\ 0 \end{bmatrix}$; 10, $\begin{bmatrix} -1 \\ 1 \\ 1 \end{bmatrix}$; $\mathbf{x} = \begin{bmatrix} 3 \\ 0 \\ 1 \end{bmatrix}$, $\begin{bmatrix} 0 \\ 1 \\ 0 \end{bmatrix}$, $\begin{bmatrix} 1 \\ -1 \\ 1 \end{bmatrix}$

9. $\begin{bmatrix} \frac{1}{5} & \frac{2}{5} \\ -\frac{2}{5} & \frac{1}{5} \end{bmatrix} \mathbf{A} \begin{bmatrix} 1 & -2 \\ 2 & 1 \end{bmatrix} = \begin{bmatrix} 5 & 0 \\ 0 & 0 \end{bmatrix}$

11. $\begin{bmatrix} -2 & 1 \\ 3 & -1 \end{bmatrix} \mathbf{A} \begin{bmatrix} 1 & 1 \\ 3 & 2 \end{bmatrix} = \begin{bmatrix} 2 & 0 \\ 0 & -5 \end{bmatrix}$

$$\textbf{13.}\ \begin{bmatrix} 1 & 0 & 0 \\ -2 & 1 & 0 \\ 1 & -2 & 1 \end{bmatrix} \mathbf{A} \begin{bmatrix} 1 & 0 & 0 \\ 2 & 1 & 0 \\ 3 & 2 & 1 \end{bmatrix} = \begin{bmatrix} 4 & 0 & 0 \\ 0 & -2 & 0 \\ 0 & 0 & 1 \end{bmatrix}$$

$$\textbf{15.}\ \begin{bmatrix} \frac{1}{3} & \frac{1}{3} & \frac{1}{3} \\ -\frac{1}{3} & \frac{1}{6} & \frac{1}{6} \\ 0 & -\frac{1}{2} & \frac{1}{2} \end{bmatrix} \mathbf{A} \begin{bmatrix} 1 & -2 & 0 \\ 1 & 1 & -1 \\ 1 & 1 & 1 \end{bmatrix} = \begin{bmatrix} 10 & 0 & 0 \\ 0 & 1 & 0 \\ 0 & 0 & 5 \end{bmatrix}$$

$$\textbf{17.}\ \mathbf{C} = \begin{bmatrix} 7 & 3 \\ 3 & 7 \end{bmatrix}, \quad 4y_1^2 + 10y_2^2 = 200, \quad \mathbf{x} = \frac{1}{\sqrt{2}} \begin{bmatrix} 1 & 1 \\ -1 & 1 \end{bmatrix} \mathbf{y}, \quad \text{ellipse}$$

$$\textbf{19.}\ \mathbf{C} = \begin{bmatrix} 3 & 11 \\ 11 & 3 \end{bmatrix}, \quad 14y_1^2 - 8y_2^2 = 0, \quad \mathbf{x} = \frac{1}{\sqrt{2}} \begin{bmatrix} 1 & 1 \\ 1 & -1 \end{bmatrix} \mathbf{y}; \quad \text{pair of straight lines}$$

$$\textbf{21.}\ \mathbf{C} = \begin{bmatrix} 1 & -6 \\ -6 & 1 \end{bmatrix}, \quad 7y_1^2 - 5y_2^2 = 70, \quad \mathbf{x} = \frac{1}{\sqrt{2}} \begin{bmatrix} -1 & 1 \\ 1 & 1 \end{bmatrix} \mathbf{y}, \quad \text{hyperbola}$$

$$\textbf{23.}\ \mathbf{C} = \begin{bmatrix} -11 & 42 \\ 42 & 24 \end{bmatrix}, \quad 52y_1^2 - 39y_2^2 = 156, \quad \mathbf{x} = \frac{1}{\sqrt{13}} \begin{bmatrix} 2 & 3 \\ 3 & -2 \end{bmatrix} \mathbf{y}, \quad \text{hyperbola}$$

Problem Set 8.5, page 351

1. Hermitian, 5, $[-i \quad 1]^\mathsf{T}$, 7, $[i \quad 1]^\mathsf{T}$

3. Unitary, $(1 - i\sqrt{3})/2$, $[-1 \quad 1]^\mathsf{T}$; $(1 + i\sqrt{3})/2$, $[1 \quad 1]^\mathsf{T}$

5. Skew-Hermitian, unitary, $-i$, $[0 \quad -1 \quad 1]^\mathsf{T}$, i, $[1 \quad 0 \quad 0]^\mathsf{T}$, $[0 \quad 1 \quad 1]^\mathsf{T}$

7. Eigenvalues -1, 1; eigenvectors $[1 \quad -1]^\mathsf{T}$, $[1 \quad 1]^\mathsf{T}$; $[1 \quad -i]^\mathsf{T}$, $[1 \quad i]^\mathsf{T}$; $[0 \quad 1]^\mathsf{T}$, $[1 \quad 0]^\mathsf{T}$, resp.

9. Hermitian, 16

11. Skew-Hermitian, $-6i$

13. $\overline{(\mathbf{ABC})}^\mathsf{T} = \overline{\mathbf{C}}^\mathsf{T}\overline{\mathbf{B}}^\mathsf{T}\overline{\mathbf{A}}^\mathsf{T} = \mathbf{C}^{-1}(-\mathbf{B})\mathbf{A}$

15. $\mathbf{A} = \mathbf{H} + \mathbf{S}$, $\mathbf{H} = \frac{1}{2}(\mathbf{A} + \overline{\mathbf{A}}^\mathsf{T})$, $\mathbf{S} = \frac{1}{2}(\mathbf{A} - \overline{\mathbf{A}}^\mathsf{T})$ (**H** Hermitian, **S** skew-Hermitian)

19. $\mathbf{A}\overline{\mathbf{A}}^\mathsf{T} - \overline{\mathbf{A}}^\mathsf{T}\mathbf{A} = (\mathbf{H} + \mathbf{S})(\mathbf{H} - \mathbf{S}) - (\mathbf{H} - \mathbf{S})(\mathbf{H} + \mathbf{S}) = 2(-\mathbf{HS} + \mathbf{SH}) = \mathbf{0}$ if and only if $\mathbf{HS} = \mathbf{SH}$.

Chapter 8 Review Questions and Problems, page 352

11. 3, $[1 \quad 1]^\mathsf{T}$; 2, $[1 \quad -1]^\mathsf{T}$

13. 3, $[1 \quad 5]^\mathsf{T}$; 7, $[1 \quad 1]^\mathsf{T}$

15. 0, $[2 \quad -2 \quad 1]^\mathsf{T}$; $9i$, $[-1 + 3i \quad 1 + 3i \quad 4]^\mathsf{T}$; $-9i$, $[-1 - 3i \quad 1 - 3i \quad 4]^\mathsf{T}$

$$\textbf{17.}\ -1, 1; \quad \mathbf{A} = \frac{1}{16} \begin{bmatrix} 5 & -3 \\ -3 & 5 \end{bmatrix} \begin{bmatrix} 23 & 2 \\ 39 & 1 \end{bmatrix} = \frac{1}{8} \begin{bmatrix} -1 & 1 \\ 63 & 1 \end{bmatrix}$$

19. $\frac{1}{3}\begin{bmatrix} 2 & -1 \\ -1 & 2 \end{bmatrix} \mathbf{A} \begin{bmatrix} 2 & 1 \\ 1 & 2 \end{bmatrix} = \begin{bmatrix} -0.9 & 0 \\ 0 & 0.6 \end{bmatrix}$

21. $\frac{1}{3}\begin{bmatrix} 1 & 1 & -1 \\ 1 & -1 & 0 \\ 0 & 1 & 1 \end{bmatrix} \mathbf{A} \begin{bmatrix} 1 & 2 & 1 \\ 1 & -1 & 1 \\ -1 & 1 & 2 \end{bmatrix} = \begin{bmatrix} 4 & 0 & 0 \\ 0 & -20 & 0 \\ 0 & 0 & 22 \end{bmatrix}$

23. $\mathbf{C} = \begin{bmatrix} 4 & 12 \\ 12 & -14 \end{bmatrix}$, $10y_1^2 - 20y_2^2 = 20$, $\mathbf{x} = \frac{1}{\sqrt{5}}\begin{bmatrix} 2 & 1 \\ 1 & -2 \end{bmatrix}\mathbf{y}$, hyperbola

25. $\mathbf{C} = \begin{bmatrix} 3.7 & 1.6 \\ 1.6 & 1.3 \end{bmatrix}$, $4.5y_1^2 + 0.5y_2^2 = 4.5$, $\mathbf{x} = \frac{1}{\sqrt{5}}\begin{bmatrix} 2 & 1 \\ 1 & -2 \end{bmatrix}\mathbf{y}$, ellipse

Problem Set 9.1, page 360

1. 5, 1, 0; $\sqrt{26}$; $[5/\sqrt{26}, 1/\sqrt{26}, 0]$

3. 8.5, −4.0, 1.7; $\sqrt{91.14}$, $[0.890, -0.419, 0.178]$

5. 2, 1, −2; $\mathbf{u} = [\frac{2}{3}, \frac{1}{3}, -\frac{2}{3}]$, position vector of Q

7. Q: $(4, 0, \frac{1}{2})$, $|\mathbf{v}| = \sqrt{16.25}$

9. Q: $(0, 0, -8)$, $|\mathbf{v}| = 8$

11. $[6, 4, 0]$, $[\frac{3}{2}, 1, 0]$, $[-3, -2, 0]$

13. $[1, 5, 8]$

15. $7[9, -7, 8] = [63, -49, 56]$

17. $[12, 8, 0]$

21. $[4, 9, -3]$, $\sqrt{106}$

23. $[0, 0, 5]$, 5

25. $[6, 2, -14] = 2\mathbf{u}$, $\sqrt{236}$

27. $\mathbf{p} = [0, 0, -5]$

29. $\mathbf{v} = [v_1, v_2, 3]$, v_1, v_2 arbitrary

31. $k = 10$

33. $|\mathbf{p} + \mathbf{q} + \mathbf{u}| \leqq 18$. Nothing

35. $v_B - v_A = [-19, 0] - [22/\sqrt{2}, 22/\sqrt{2}] = [-19 - 22/\sqrt{2}, -22/\sqrt{2}]$

37. $\mathbf{u} + \mathbf{v} + \mathbf{p} = [-k, 0] + [l, l] + [0, -1000] = \mathbf{0}$, $-k + l + 0 = 0$,
$0 + l - 1000 = 0$, $l = 1000, k = 1000$

Problem Set 9.2, page 367

1. 44, 44, 0

3. $\sqrt{35}$, $\sqrt{320}$, $\sqrt{86}$

5. $|[2, 9, 9]| = \sqrt{166} = 12.88 < \sqrt{80} + \sqrt{86} = 18.22$

7. $|-24| = 24$, $|a||c| = \sqrt{35}\sqrt{86} = \sqrt{3010} = 54.86$; cf. (6)

9. 300; cf. (5a) and (5b)

13. Use (1) and $|\cos\gamma| \leqq 1$.

15. $|\mathbf{a} + \mathbf{b}|^2 + |\mathbf{a} - \mathbf{b}|^2 = \mathbf{a} \cdot \mathbf{a} + 2\mathbf{a} \cdot \mathbf{b} + \mathbf{b} \cdot \mathbf{b} + (\mathbf{a} \cdot \mathbf{a} - 2\mathbf{a} \cdot \mathbf{b} + \mathbf{b} \cdot \mathbf{b})$
$= 2|\mathbf{a}|^2 + 2|\mathbf{b}|^2$

17. $[2, 5, 0] \cdot [2, 2, 2] = 14$

19. $[0, 4, 3] \cdot [-3, -2, 1] = -5$ is negative! Why?

21. Yes, because $W = (\mathbf{p} + \mathbf{q}) \cdot \mathbf{d} = \mathbf{p} \cdot \mathbf{d} + \mathbf{q} \cdot \mathbf{d}$.

23. $\arccos 0.5976 = 53.3°$

27. $\beta - \alpha$ is the angle between the unit vectors **a** and **b**. Use (2).

29. $\gamma = \arccos(12/(6\sqrt{13})) = 0.9828 = 56.3°$ and $123.7°$

31. $a_1 = -\frac{28}{3}$

33. $\pm[\frac{3}{5}, -\frac{4}{5}]$

35. $(\mathbf{a} + \mathbf{b}) \cdot (\mathbf{a} - \mathbf{b}) = |\mathbf{a}|^2 - |\mathbf{b}|^2 = 0$, $|\mathbf{a}| = |\mathbf{b}|$. A square.

37. 0. Why?

39. If $|\mathbf{a}| = |\mathbf{b}|$ or if **a** and **b** are orthogonal.

Problem Set 9.3, page 374

5. $-\mathbf{m}$ instead of $\mathbf{m}$, tendency to rotate in the opposite sense.

7. $|\mathbf{v}| = |[0, 20, 0] \times [8, 6, 0]| = |[0, 0, -160]| = 160$

9. Zero volume in Fig. 191, which can happen in several ways.

11. $[0, 0, 7]$, $[0, 0, -7]$, -4

13. $[6, 2, 7]$, $[-6, -2, -7]$

15. $\mathbf{0}$

17. $[-32, -58, 34]$, $[-42, -63, 19]$

19. 1, -1

21. $[-48, -72, -168]$, $12\sqrt{248} = 189.0, 189.0$

23. 0, 0, 13

25. $\mathbf{m} = [-2, -2, 0] \times [2, 3, 0] = [0, 0, -10]$, $m = 10$ clockwise

27. $[6, 2, 0] \times [1, 2, 0] = [0, 0, 10]$

29. $\frac{1}{2}|[-12, 2, 6]| = \sqrt{46}$

31. $3x + 2y - z = 5$

33. $474/6 = 79$

Problem Set 9.4, page 380

1. Hyperbolas

3. Parallel straight lines (planes in space) $y = \frac{3}{4}x + c$

5. Circles, centers on the y-axis

7. Ellipses

9. Parallel planes

11. Elliptic cylinders

13. Paraboloids

Problem Set 9.5, page 390

1. Circle, center $(3, 0)$, radius 2

3. Cubic parabola $x = 0, z = y^3$

5. Ellipse

7. Helix

9. A "Lissajous curve"

11. $\mathbf{r} = [3 + \sqrt{13}\cos t, 2 + \sqrt{13}\sin t, 1]$

13. $\mathbf{r} = [2 + t, 1 + 2t, 3]$

15. $\mathbf{r} = [t, 4t - 1, 5t]$

17. $\mathbf{r} = [\sqrt{2}\cos t, \sin t, \sin t]$

19. $\mathbf{r} = [\cosh t, (\sqrt{3}/2)\sinh t, -2]$

21. Use $\sin(-\alpha) = -\sin\alpha$.

25. $\mathbf{u} = [-\sin t, 0, \cos t]$. At P, $\mathbf{r}' = [-8, 0, 6]$. $\mathbf{q}(w) = [6 - 8w, i, 8 + 6w]$.

27. $\mathbf{q}(w) = [2 + w, \frac{1}{2} - \frac{1}{4}w, 0]$

29. $\sqrt{\mathbf{r}' \cdot \mathbf{r}'} = \cosh t$, $l = \sinh l = 1.175$

31. $\sqrt{\mathbf{r}' \cdot \mathbf{r}'} = a$, $l = a\pi/2$

33. Start from $\mathbf{r}(t) = [t, f(t)]$.

35. $\mathbf{v} = \mathbf{r}' = [1, 2t, 0]$, $|\mathbf{v}| = \sqrt{1 + 4t^2}$, $\mathbf{a} = [0, 2, 0]$

37. $\mathbf{v}(0) = (\omega + 1)R\mathbf{i}$, $\mathbf{a}(0) = -\omega^2 R\mathbf{j}$

39. $\mathbf{v} = [-\sin t - 2\sin 2t, \cos t - 2\cos 2t]$, $|\mathbf{v}|^2 = 5 - 4\cos 3t$,

$\mathbf{a} = [-\cos t - 4\cos 2t, -\sin t + 4\sin 2t]$, and $\mathbf{a}_{\tan} = \dfrac{6\sin 3t}{5 - 4\cos 3t}\mathbf{v}$.

41. $\mathbf{v} = [-\sin t, 2\cos 2t, -2\sin 2t]$, $|\mathbf{v}|^2 = 4 + \sin^2 t$,

$\mathbf{a} = [-\cos t, -4\sin 2t, -4\cos 2t]$, and $\mathbf{a}_{\tan} = \dfrac{\frac{1}{2}\sin 2t}{4 + \sin^2 t}\mathbf{v}$.

43. 1 year $= 365 \cdot 86{,}400$ sec, $R = 30 \cdot 365 \cdot 86{,}400/2\pi = 151 \cdot 10^6$ [km], $|\mathbf{a}| = \omega^2 R = |\mathbf{v}|^2/R = 5.98 \cdot 10^{-6}$ [km/sec^2]

45. $R = 3960 + 80$ mi $= 2.133 \cdot 10^7$ ft, $g = |\mathbf{a}| = \omega^2 R = |\mathbf{v}|^2/R$, $|\mathbf{v}| = \sqrt{gR} = \sqrt{6.61 \cdot 10^8} = 25{,}700$ [ft/sec] $= 17{,}500$ [mph]

49. $\mathbf{r}(t) = [t, y(t), 0]$, $\mathbf{r}' = [1, y', 0]$ $\mathbf{r} \cdot \mathbf{r}' = 1 + y'^2$, etc.

51. $\dfrac{d\mathbf{r}}{ds} = \dfrac{d\mathbf{r}}{dt}\Big/\dfrac{ds}{dt}, \quad \dfrac{d^2\mathbf{r}}{ds^2} = \dfrac{d^2\mathbf{r}}{dt^2}\Big/\left(\dfrac{ds}{dt}\right)^2 + \cdots, \quad \dfrac{d^3\mathbf{r}}{ds^3} = \dfrac{d^3\mathbf{r}}{dt^3}\Big/\left(\dfrac{ds}{dt}\right)^3 + \cdots$

53. $3/(1 + 9t^2 + 9t^4)$

Problem Set 9.7, page 402

1. $[2y - 1, 2x + 2]$
3. $[-y/x^2, 1/x]$
5. $[4x^3, 4y^3]$
7. Use the chain rule.
9. Apply the quotient rule to each component and collect terms.
11. $[y, x]$, $[5, -4]$
13. $[2x/(x^2 + y^2), 2y/(x^2 + y^2)]$, $[0.16, 0.12]$
15. $[8x, 18y, 2z]$, $[40, -18, -22]$
17. For P on the x- and y-axes.
19. $[-1.25, 0]$
21. $[0, -e]$
23. Points with $y = 0, \pm\pi, \pm 2\pi, \cdots$.
25. $-\nabla T(P) = [0, 4, -1]$
31. $\nabla f = [32x, -2y]$, $\nabla f(P) = [160, -2]$
33. $[12x, 4y, 2z]$, $[60, 20, 10]$
35. $[-2x, -2y, 1]$, $[-6, -8, 1]$
37. $[2, 1] \bullet [1, -1]/\sqrt{5} = 1/\sqrt{5}$
39. $[1, 1, 1] \bullet [-3/125, 0, -4/125]/\sqrt{3} = -7/(125\sqrt{3})$
41. $\sqrt{8/3}$
43. $f = xyz$
45. $f = \int v_1\,dx + \int v_2\,dy + \int v_3\,dz$

Problem Set 9.8, page 405

1. $2x + 8y + 18z$; 7
3. 0, after simplification; solenoidal
5. $9x^2y^2z^2$; 1296
7. $-2e^x(\cos y)z$
9. **(b)** $(fv_1)_x + (fv_2)_y + (fv_3)_z = f[(v_1)_x + (v_2)_y + (v_3)_z] + f_xv_1 + f_yv_2 + f_zv_3$, etc.
11. $[v_1, v_2, v_3] = \mathbf{r}' = [x', y', z'] = [y, 0, 0]$, $z' = 0, z = c_3$, $y' = 0, y = c_2$, and $x' = y = c_2, x = c_2t + c_1$. Hence as t increases from 0 to 1, this "shear flow" transforms the cube into a parallelepiped of volume 1.
13. $\operatorname{div}(\mathbf{w} \times \mathbf{r}) = 0$ because v_1, v_2, v_3 do not depend on x, y, z, respectively.
15. $-2\cos 2x + 2\cos 2y$
17. 0
19. $2/(x^2 + y^2 + z^2)^2$

Problem Set 9.9, page 408

3. Use the definitions and direct calculation.
5. $[x(z^2 - y^2), y(x^2 - z^2), z(y^2 - x^2)]$
7. $e^{-x}[\cos y, \sin y, 0]$
9. $\operatorname{curl} \mathbf{v} = [-6z, 0, 0]$ incompressible, $\mathbf{v} = \mathbf{r}' = [x', y', z'] = [0, 3z^2, 0]$, $x = c_1$, $z = c_3$, $y' = 3z^2 = 3c_3^2$, $y = 3c_3^2t + c_2$
11. $\operatorname{curl} \mathbf{v} = [0, 0, -3]$, incompressible, $x' = y$, $y' = -2x$, $2xx' + yy' = 0$, $x^2 + \frac{1}{2}y^2 = c, z = c_3$
13. $\operatorname{curl} \mathbf{v} = 0$, irrotational, $\operatorname{div} \mathbf{v} = 1$, compressible, $\mathbf{r} = [c_1e^t, c_2e^t, c_3e^{-t}]$. Sketch it.
15. $[-1, -1, -1]$, same (why?)
17. $-yz - zx - xy$, 0 (why?), $-y - z - x$
19. $[-2z - y, -2x - z, -2y - x]$, same (why?)

Chapter 9 Review Questions and Problems, page 409

11. $-10, \quad 1080, \quad 1080, \quad 65$
13. $[-10, -30, 0], \quad [10, 30, 0], \quad \mathbf{0}, \quad 40$
15. $[-1260, -1830, -300], \quad [-210, 120, -540]$, undefined
17. $-125, \quad 125, \quad -125$
19. $[70, -40, -50], \quad 0, \quad \sqrt{35^2 + 20^2 + 25^2} = \sqrt{2250}$
21. $[-2, -6, -13]$
23. $\gamma_1 = \arccos(-10/\sqrt{65 \cdot 40}) = 1.7682 = -101.3°, \gamma_2 = 23.7°$
25. $[5, 2, 0] \bullet [4 - 1, 3 - 1, 0] = 19$
27. $\mathbf{v} \bullet \mathbf{w}/|\mathbf{w}| = 22/\sqrt{8} = 7.78$
29. $[0, 0, -14]$, tendency of clockwise rotation
31. 4
33. $1, \quad -2y$
35. 0, same (why?), $2(y^2 + x^2 - xz)$
37. $[0, -2, 0]$
39. $9/\sqrt{225} = \frac{3}{5}$

Problem Set 10.1, page 418

3. 4
5. $\mathbf{r} = [2 \cos t, \quad 2 \sin t], \quad 0 \leqq t \leqq \pi/2; \quad \frac{8}{5}$
7. "Exponential helix," $(e^{6\pi} - 1)/3$
9. $23.5, \quad 0$
11. $2e^{-t} + 2te^{-t^2}, \quad -2e^{-2} - e^{-4} + 3$
15. $18\pi, \quad \frac{4}{3}(4\pi)^3, \quad 18\pi$
17. $[4 \cos t, \quad + \sin t, \quad \sin t, \quad 4 \cos t], \quad [2, 2, 0]$
19. $144t^4, \quad 1843.2$

Problem Set 10.2, page 425

3. $\sin \frac{1}{2}x \cos 2y, \quad 1 - 1/\sqrt{2} = 0.293$
5. $e^{xy} \sin z, \quad e - 0$
7. $\cosh 1 - 2 = -0.457$
9. $e^x \cosh y + e^z \sinh y, \quad e - (\cosh 1 + \sinh 1) = 0$
13. $e^{a^2} \cos 2b$
15. Dependent, $x^2 \neq -4y^2$, etc.
17. Dependent, $4 \neq 0$, etc.
19. $\sin(a^2 + 2b^2 + c^2)$

Problem Set 10.3, page 432

3. $8y^3/3, \quad 54$
5. $\displaystyle\int_0^1 [x - x^3 - (x^2 - x^5)]\, dx = \frac{1}{12}$
7. $\cosh 2x - \cosh x, \quad \frac{1}{2}\sinh 4 - \sinh 2$
9. $36 + 27y^2, \quad 144$
11. $z = 1 - r^2, \quad dx\, dy = r\, dr\, d\theta$, *Answer*: $\pi/2$
13. $\bar{x} = 2b/3, \quad \bar{y} = h/3$
15. $\bar{x} = 0, \quad \bar{y} = 4r/3\pi$
17. $I_x = bh^3/12, \quad I_y = b^3h/4$
19. $I_x = (a + b)h^3/24, \quad I_y = h(a^4 - b^4)/(48(a - b))$

Problem Set 10.4, page 438

1. $(-1 - 1) \cdot \pi/4 = -\pi/2$
3. $9(e^2 - 1) - \frac{8}{3}(e^3 - 1)$
5. $2x - 2y, \quad 2x(1 - x^2) - (2 - x^2)^2 + 1, \quad x = -1 \cdots 1, \quad -\frac{56}{15}$
7. 0. Why?
9. $\frac{16}{5}$
13. $\nabla^2 w = \cosh x, \quad y = x/2 \cdots 2, \quad \frac{1}{2}\cosh 4 - \frac{1}{2}$

15. $\nabla^2 w = 6xy,\quad 3x(10 - x^2)^2 - 3x,\quad 486$
17. $\nabla^2 w = 6x - 6y,\quad -38.4$
19. $|\text{grad } w|^2 = e^{2x},\quad \frac{5}{2}(e^4 - 1)$

Problem Set 10.5, page 442

1. Straight lines, $\mathbf{k}$
3. $z = c\sqrt{x^2 + y^2}$, circles, straight lines, $[-cu \cos v,\quad -cu \sin v,\quad u]$
5. $z = x^2 + y^2$, circles, parabolas, $[-2u^2 \cos v,\quad -2u^2 \sin v,\quad u]$
7. $x^2/a^2 + y^2/b^2 + z^2/c^2 = 1,\quad [bc \cos^2 v \cos u,\quad ac \cos^2 v \sin u,\quad ab \sin v \cos v]$, ellipses
11. $[\tilde{u},\quad \tilde{v},\quad \tilde{u}^2, + \tilde{v}^2],\quad \tilde{\mathbf{N}} = [-2\tilde{u},\quad -2\tilde{v},\quad 1]$
13. Set $x = u$ and $y = v$.
15. $[2 + 5 \cos u,\quad -1 + 5 \sin u,\quad v],\quad [5 \cos u,\quad 5 \sin u,\quad 0]$
17. $[a \cos v \cos u,\quad -2.8 + a \cos v \sin u,\quad 3.2 + a \sin v],\quad a = 1.5;$ $[a^2 \cos^2 v \cos u,\quad a^2 \cos^2 v \sin u,\quad a^2 \cos v \sin v]$
19. $[\cosh u,\quad \sinh u,\quad v],\quad [\cosh u,\quad -\sinh u,\quad 0]$

Problem Set 10.6, page 450

1. $\mathbf{F}(\mathbf{r}) \cdot \mathbf{N} = [-u^2,\quad v^2,\quad 0] \cdot [-3,\quad 2,\quad 1] = 3u^2 + 2v^2,\quad 29.5$
3. $\mathbf{F}(\mathbf{r}) \cdot \mathbf{N} = \cos^3 v \cos u \sin u$ from (3), Sec. 10.5. *Answer:* $\frac{1}{3}$
5. $\mathbf{F}(\mathbf{r}) \cdot \mathbf{N} = -u^3, -128\pi$
7. $\mathbf{F} \cdot \mathbf{N} = [0,\quad \sin u,\quad \cos v] \cdot [1, -2u, 0],\quad 4 + (-2 + \pi^2/16 - \pi/2)\sqrt{2} = -0.1775$
9. $\mathbf{r} = [2 \cos u,\quad 2 \sin u,\quad v],\quad 0 \leqq u \leqq \pi/4,\quad 0 \leqq v \leqq 5$. Integrate $2 \sinh v \sin u$ to get $2(1 - 1/\sqrt{2})(\cosh 5 - 1) = 42.885$.
13. $7\pi^3/\sqrt{6} = 88.6$
15. $G(\mathbf{r}) = (1 + 9u^4)^{3/2},\quad |\mathbf{N}| = (1 + 9u^4)^{1/2}$. *Answer:* 54.4
21. $I_{x=y} = \iint_S [\frac{1}{2}(x - y)^2 + z^2]\, \sigma\, dA$
23. $[u \cos v,\quad u \sin v,\quad u],\quad \int_0^{2\pi}\int_0^h u^2 \cdot u\sqrt{2}\, du\, dv = \frac{\pi}{\sqrt{2}} h^4$
25. $[\cos u \cos v,\quad \cos u \sin v,\quad \sin u],\quad dA = (\cos u)\, du\, dv$, B the z-axis, $I_B = 8\pi/3$, $I_K = I_B + 1^2 \cdot 4\pi = 20.9$.

Problem Set 10.7, page 457

1. 224
3. $-e^{-1-z} + e^{-y-z},\quad -2e^{-1-z} + e^{-z},\quad 2e^{-3} - e^{-2} - 2e^{-1} + 1$
5. $\frac{1}{2}(\sin 2x)(1 - \cos 2x),\quad \frac{1}{8},\quad \frac{3}{4}$
7. $[r \cos u \cos v,\quad \cos u \sin v,\quad r \sin u],\quad dV = r^2 \cos u\, dr\, du\, dv,\quad \sigma = v,\quad 2\pi^2 a^3/3$
9. $\operatorname{div} \mathbf{F} = 2x + 2z$, 48
11. $12(e - 1/e) = 24 \sinh 1$
13. $\operatorname{div} \mathbf{F} = -\sin z$, 0
15. $1/\pi + \frac{5}{24} = 0.5266$
17. $h^4\pi/2$
19. $8abc(b^2 + c^2)/3$
21. $(a^4/4) \cdot 2\pi \cdot h = ha^4\pi/2$
23. $h^5\pi/10$
25. Do Prob. 20 as the last one.

Problem Set 10.8, page 462

1. $x = 0, y = 0, z = 0$, no contributions. $x = a$: $\partial f/\partial n = \partial f/\partial x = -2x = -2a$, etc. Integrals $x = a$: $(-2a)bc$, $y = b$: $(-2b)ac$, $z = c$: $(4c)\,ab$. Sum 0
3. The volume integral of $8y^2 + [0, 8y] \cdot [2x, 0] = 8y^2$ is $8y^3/3 = \frac{8}{3}$. The surface integral of $f \partial g/\partial n = f \cdot 2x = 2f = 8y^2$ over $x = 1$ is $8y^3/3 = \frac{8}{3}$. Others 0.
5. The volume integral of $6y^2 \cdot 4 - 2x^2 \cdot 12$ is 0; $8(x = 1)$, $-8(y = 1)$, others 0.
7. $\mathbf{F} = [x, \ 0, \ 0]$, $\operatorname{div} \mathbf{F} = 1$, use (2*), Sec. 10.7, etc.
9. $z = 0$ and $z = \sqrt{a^2 - x^2 - y^2} = \sqrt{a^2 - r^2}$, $dx\,dy = r\,dr\,d\theta$, $-2\pi \cdot \frac{1}{2}(a^2 - r^2)^{3/2} \cdot \frac{2}{3} \Big|_0^a = \frac{2}{3}\pi a^3$
11. $r = a$, $\phi = 0$, $\cos \phi = 1$, $v = \frac{1}{3} a \cdot (4\pi a^2)$

Problem Set 10.9, page 468

1. S: $z = y \ (0 \leqq x \leqq 1, 0 \leqq y \leqq 4)$, $[0, 2z, -2z] \bullet [0, -1, 1]$, ± 20
3. $[2e^{-z} \cos y, \ -e^{-z}, \ 0] \bullet [0, \ -y, \ 1] = ye^{-z}$, $\pm(2 - 2/\sqrt{e})$
5. $[0, 2z, \frac{3}{2}] \bullet [0, 0, 1] = \frac{3}{2}$, $\pm\frac{3}{2}a^2$
7. $[-e^z, \ -e^x, \ -e^y] \bullet [-2x, \ 0, \ 1]$, $\pm(e^4 - 2e + 1)$
9. The sides contribute a, $3a^2/2$, $-a$, 0.
11. -2π; curl $\mathbf{F} = \mathbf{0}$
13. $5\mathbf{k}$, 80π
15. $[0, -1, 2x - 2y] \bullet [0, 0, 1]$, $\frac{1}{3}$
17. $\mathbf{r} = [\cos u, \ \sin u, \ v]$, $[-3v^2, \ 0, \ 0] \bullet [\cos u, \ \sin u, \ 0]$, -1
19. $\mathbf{r} = [u \cos v, \ u \sin v, \ u]$, $0 \leqq u \leqq 1$, $0 \leqq v \leqq \pi/2$, $[-e^z, \ 1, \ 0] \bullet [-u \cos v, \ -u \sin v, \ u]$. *Answer*: 1/2

Chapter 10 Review Questions and Problems, page 469

11. $\mathbf{r} = [4 - 10t, \ 2 + 8t]$, $\mathbf{F}(\mathbf{r}) \bullet d\mathbf{r} = [2(4 - 10t)^2, \ -4(2t + 8t)^2] \bullet [-10, \ 8]\,dt$; $-4528/3$. Or using exactness.
13. Not exact, curl $\mathbf{F} = (5 \cos x)\mathbf{k}$, ± 10
15. 0 since curl $\mathbf{F} = \mathbf{0}$
17. By Stokes, $\pm 18\pi$
19. $\mathbf{F} = \operatorname{grad}(y^2 + xz)$, 2π
21. $M = 8$, $\bar{x} = \frac{8}{5}$, $\bar{y} = \frac{16}{5}$
23. $M = \frac{63}{20}$, $\bar{x} = \frac{8}{7} = 1.14$, $\bar{y} = \frac{118}{49} = 2.41$
25. $M = 4k/15$, $\bar{x} = \frac{5}{16}$, $\bar{y} = \frac{4}{7}$
27. $288(a + b + c)\pi$
29. $\operatorname{div} \mathbf{F} = 20 + 6z^2$. *Answer*: 21
31. $24 \sinh 1 = 28.205$
33. Direct integration, $\frac{224}{3}$
35. 72π

Problem Set 11.1, page 482

1. $2\pi, 2\pi, \pi, \pi, 1, 1, \frac{1}{2}, \frac{1}{2}$
5. There is no *smallest* $p > 0$.
13. $\dfrac{4}{\pi}\left(\cos x + \dfrac{1}{9}\cos 3x + \dfrac{1}{25}\cos 5x + \cdots\right) + 2\left(\sin x + \dfrac{1}{3}\sin 3x + \dfrac{1}{5}\sin 5x + \cdots\right)$
15. $\frac{4}{3}\pi^2 + 4(\cos x + \frac{1}{4}\cos 2x + \frac{1}{9}\cos 3x + \cdots) - 4\pi(\sin x + \frac{1}{2}\sin 2x + \frac{1}{3}\sin 3x + \cdots)$
17. $\dfrac{\pi}{2} + \dfrac{4}{\pi}\left(\cos x + \dfrac{1}{9}\cos 3x + \dfrac{1}{25}\cos 5x + \cdots\right)$

19. $\frac{\pi}{4} - \frac{2}{\pi}\left(\cos x + \frac{1}{9}\cos 3x + \frac{1}{25}\cos 5x + \cdots\right) + \sin x - \frac{1}{2}\sin 2x + \frac{1}{3}\sin 3x - + \cdots$

21. $2(\sin x + \frac{1}{2}\sin 2x + \frac{1}{3}\sin 3x + \frac{1}{4}\sin 4x + \frac{1}{5}\sin 5x + \cdots)$

Problem Set 11.2, page 490

1. Neither, even, odd, odd, neither **3.** Even **5.** Even

9. Odd, $L = 2$, $\frac{4}{\pi}\left(\sin\frac{\pi x}{2} + \frac{1}{3}\sin\frac{3\pi x}{2} + \frac{1}{5}\sin\frac{5\pi x}{2} + \cdots\right)$

11. Even, $L = 1$, $\frac{1}{3} - \frac{4}{\pi^2}\left(\cos \pi x - \frac{1}{4}\cos 2\pi x + \frac{1}{9}\cos 3\pi x - + \cdots\right)$

13. Rectifier, $L = \frac{1}{2}$, $\frac{1}{8} - \frac{1}{\pi^2}\left(\cos 2\pi x + \frac{1}{9}\cos 6\pi x + \frac{1}{25}\cos 10\pi x + \cdots\right) + \frac{1}{\pi}\left(\frac{1}{2}\sin 2\pi x - \frac{1}{4}\sin 4\pi x + \frac{1}{6}\sin 6\pi x - \frac{1}{8}\sin 8\pi x + - \cdots\right)$

15. Odd, $L = \pi$, $\frac{4}{\pi}\left(\sin x - \frac{1}{9}\sin 3x + \frac{1}{25}\sin 5x - + \cdots\right)$

17. Even, $L = 1$, $\frac{1}{2} + \frac{4}{\pi^2}\left(\cos \pi x + \frac{1}{9}\cos 3\pi x + \frac{1}{25}\cos 5\pi x + \cdots\right)$

19. $\frac{3}{8} + \frac{1}{2}\cos 2x + \frac{1}{8}\cos 4x$

23. $L = 4$, **(a)** 1, **(b)** $\frac{4}{\pi}\left(\sin\frac{\pi x}{4} + \frac{1}{3}\sin\frac{3\pi x}{4} + \frac{1}{5}\sin\frac{5\pi x}{4} + \cdots\right)$

25. $L = \pi$, **(a)** $\frac{\pi}{2} + \frac{4}{\pi}\left(\cos x + \frac{1}{9}\cos 3x + \frac{1}{25}\cos 5x + \cdots\right)$,
(b) $2(\sin x + \frac{1}{2}\sin 2x + \frac{1}{3}\sin 3x + \frac{1}{4}\sin 4x + \cdots)$

27. $L = \pi$, **(a)** $\frac{3\pi}{8} + \frac{2}{\pi}\left(\cos x - \frac{1}{2}\cos 2x + \frac{1}{9}\cos 3x + \frac{1}{25}\cos 5x - \frac{1}{18}\cos 6x + \frac{1}{49}\cos 7x + \frac{1}{81}\cos 9x - \frac{1}{50}\cos 10x + \frac{1}{121}\cos 11x + \cdots\right)$
(b) $\left(1 + \frac{2}{\pi}\right)\sin x + \frac{1}{2}\sin 2x + \left(\frac{1}{3} - \frac{2}{9\pi}\right)\sin 3x + \frac{1}{4}\sin 4x + \left(\frac{1}{5} + \frac{2}{25\pi}\right)\sin 5x + \frac{1}{6}\sin 6x + \cdots$

29. Rectifier, $L = \pi$,
(a) $\frac{2}{\pi} - \frac{4}{\pi}\left(\frac{1}{1\cdot 3}\cos x + \frac{1}{3\cdot 5}\cos 3x + \frac{1}{5\cdot 7}\cos 5x + \cdots\right)$, **(b)** $\sin x$

Problem Set 11.3, page 494

3. The output becomes a pure cosine series.
5. For A_n this is similar to Fig. 54 in Sec. 2.8, whereas for the phase shift B_n the sense is the same for all n.

7. $y = C_1 \cos \omega t + C_2 \sin \omega t + a(\omega) \sin t, \quad a(\omega) = 1/(\omega^2 - 1) = -1.33,$ $-5.26, 4.76, 0.8, 0.01$. Note the change of sign.

11. $y = C_1 \cos \omega t + C_2 \sin \omega t + \dfrac{4}{\pi}\left(\dfrac{1}{\omega^2 - 9} \sin t + \dfrac{1}{\omega^2 - 49} \sin 3t + \dfrac{1}{\omega^2 - 121} \sin 5t + \cdots\right)$

13. $y = \sum_{n=1}^{N} (A_n \cos nt + B_n \sin nt), \quad A_n = [(1 - n^2)a_n - nb_nc]/D_n,$ $B_n = [(1 - n^2)b_n + nca_n]/D_n, \quad D_n = (1 - n^2)^2 + n^2c^2$

15. $b_n = (-1)^{n+1} \cdot 12/n^3$ (n odd), $\quad y = \sum_{n=1}^{\infty} (A_n \cos nt + B_n \sin nt),$ $A_n = (-1)^n \cdot 12nc/n^3D_n, \quad B_n = (-1)^{n+1} \cdot 12(1 - n^2)/(n^3D_n)$ with D_n as in Prob. 13.

17. $I = 50 + A_1 \cos t + B_1 \sin t + A_3 \cos 3t + B_3 \sin 3t + \cdots, A_n = (10 - n^2)a_n/D_n,$ $B_n = 10na_n/D_n, \quad a_n = -400/(n^2\pi), D_n = (n^2 - 10)^2 + 100n^2$

19. $I(t) = \sum_{n=1}^{\infty} (A_n \cos nt + B_n \sin nt), \quad A_n = (-1)^{n+1} \dfrac{2400(10 - n^2)}{n^2D_n},$ $B_n = (-1)^{n+1} \dfrac{24{,}000}{nD_n}, \quad D_n = (10 - n^2)^2 + 100n^2$

Section 11.4, page 498

3. $F = \dfrac{\pi}{2} - \dfrac{4}{\pi}\left(\cos x + \dfrac{1}{9} \cos 3x + \dfrac{1}{25} \cos 5x + \cdots\right), E^* = 0.0748,$ $0.0748, 0.0119, 0.0119, 0.0037$

5. $F = \dfrac{4}{\pi}\left(\sin x + \dfrac{1}{3} \sin 3x + \dfrac{1}{5} \sin 5x + \cdots\right), E^* = 1.1902, 1.1902, 0.6243, 0.6243,$ $0.4206 \quad (0.1272$ when $N = 20)$

7. $F = 2[(\pi^2 - 6) \sin x - \frac{1}{8}(4\pi^2 - 6) \sin 2x + \frac{1}{27}(9\pi^2 - 6) \sin 3x - + \cdots];$ $E^* = 674.8, 454.7, 336.4, 265.6, 219.0$. Why is E^* so large?

Section 11.5, page 503

3. Set $x = ct + k$.

5. $x = \cos \theta, dx = -\sin \theta \, d\theta$, etc.

7. $\lambda_m = (m\pi/10)^2, m = 1, 2, \cdots; y_m = \sin (m\pi x/10)$

9. $\lambda = [(2m + 1)\pi/(2L)]^2, m = 0, 1, \cdots, y_m = \sin ((2m + 1)\pi x/(2L))$

11. $\lambda_m = m^2, m = 1, 2, \cdots, y_m = x \sin (m \ln |x|)$

13. $p = e^{8x}, q = 0, r = e^{8x}, \lambda_m = m^2, y_m = e^{-4x} \sin mx, m = 1, 2, \cdots$

Section 11.6, page 509

1. $8(P_1(x) - P_3(x) + P_5(x))$

3. $\frac{4}{5}P_0(x) - \frac{4}{7}P_2(x) - \frac{8}{35}P_4(x)$

9. $-0.4775P_1(x) - 0.6908P_3(x) + 1.844P_5(x) - 0.8236P_7(x) + 0.1658P_9(x) + \cdots,$ $m_0 = 9$. ***Rounding*** seems to have considerable influence in Probs. 8–13.

11. $0.7854P_0(x) - 0.3540P_2(x) + 0.0830P_4(x) - \cdots,\ m_0 = 4$
13. $0.1212P_0(x) - 0.7955P_2(x) + 0.9600P_4(x) - 0.3360P_6(x) + \cdots,\ m_0 = 8$
15. (c) $a_m = (2/J_1^2(\alpha_{0,m}))(J_1(\alpha_{0,m})/\alpha_{0,m}) = 2/(\alpha_{0,m}J_1(\alpha_{0,m}))$

Section 11.7, page 517

1. $f(x) = \pi e^{-x}(x > 0)$ gives $A = \displaystyle\int_0^\infty e^{-v}\cos wv\,dv = \frac{1}{1+w^2},\ B = \frac{w}{1+w^2}$ (see Example 3), etc.

3. Use (11); $B = \dfrac{2}{\pi}\displaystyle\int_0^\infty \frac{\pi}{2}\sin wv\,dv = \frac{1-\cos \pi w}{w}$

5. $B(w) = \dfrac{2}{\pi}\displaystyle\int_0^1 \frac{1}{2}\pi v\sin wv\,dv = \frac{\sin w - w\cos w}{w^2}$

7. $\dfrac{2}{\pi}\displaystyle\int_0^\infty \frac{\sin w\cos xw}{w}\,dw$

9. $A(w) = \dfrac{2}{\pi}\displaystyle\int_0^\infty \frac{\cos wv}{1+v^2}\,dv = e^{-w}\ (w > 0)$

11. $\dfrac{2}{\pi}\displaystyle\int_0^\infty \frac{\cos \pi w + 1}{1-w^2}\cos xw\,dw$

15. For $n = 1, 2, 11, 12, 31, 32, 49, 50$ the value of $\mathrm{Si}(n\pi) - \pi/2$ equals 0.28, −0.15, 0.029, −0.026, 0.0103, −0.0099, 0.0065, −0.0064 (rounded).

17. $\dfrac{2}{\pi}\displaystyle\int_0^\infty \frac{1-\cos w}{w}\sin xw\,dw$

19. $\dfrac{2}{\pi}\displaystyle\int_0^\infty \frac{w - e(w\cos w - \sin w)}{1+w^2}\sin xw\,dw$

Section 11.8, page 522

1. $\hat{f}_c(w) = \sqrt{(2/\pi)}\,(2\sin w - \sin 2w)/w$
3. $\hat{f}_c(w) = \sqrt{(2/\pi)}\,(\cos 2w + 2w\sin 2w - 1)/w^2$
5. $\hat{f}_c(w) = \sqrt{\dfrac{2}{\pi}}\,\dfrac{(w^2-2)\sin w + 2w\cos w}{w^3}$
7. Yes. No
9. $\sqrt{2/\pi}\,w/(a^2+w^2)$
11. $\sqrt{2/\pi}\,((2-w^2)\cos w + 2w\sin w - 2)/w^3$
13. $\mathcal{F}_s(e^{-x}) = \dfrac{1}{w}\left(-\mathcal{F}_c(e^{-x}) + \sqrt{\dfrac{2}{\pi}}\cdot 1\right) = \dfrac{1}{w}\left(\sqrt{\dfrac{2}{\pi}}\cdot\dfrac{1}{w^2+1} + \sqrt{\dfrac{2}{\pi}}\right) = \sqrt{\dfrac{2}{\pi}}\,\dfrac{w}{w^2+1}$

Problem Set 11.9, page 533

3. $i(e^{-ibw} - e^{-iaw})/(w\sqrt{2\pi})$ if $a < b$; 0 otherwise
5. $[e^{(1-iw)a} - e^{-(1-iw)a}]/(\sqrt{2\pi}(1-iw))$
7. $(e^{-iaw}(1+iaw) - 1)/(\sqrt{2\pi}w^2)$
9. $\sqrt{2/\pi}(\cos w + w\sin w - 1)/w^2$
11. $i\sqrt{2/\pi}\,(\cos w - 1)/w$
13. $e^{-w^2/2}$ by formula 9

17. No, the assumptions in Theorem 3 are not satisfied.

19. $[f_1 + f_2 + f_3 + f_4,\ \ f_1 - if_2 - f_3 + if_4,\ \ f_1 - f_2 + f_3 - f_4,\ \ f_1 + if_2 - f_3 - if_4]$

21. $\begin{bmatrix} 1 & 1 \\ 1 & -1 \end{bmatrix}\begin{bmatrix} f_1 \\ f_2 \end{bmatrix} = \begin{bmatrix} f_1 + f_2 \\ f_1 - f_2 \end{bmatrix}$

Chapter 11 Review Questions and Problems, page 537

11. $1 + \frac{4}{\pi}\left(\sin\frac{\pi x}{2} + \frac{1}{3}\sin\frac{3\pi x}{2} + \frac{1}{5}\sin\frac{5\pi x}{2} + \cdots\right)$

13. $\frac{1}{4} - \frac{2}{\pi^2}\left(\cos \pi x + \frac{1}{9}\cos 3\pi x + \frac{1}{25}\cos 5\pi x + \cdots\right) +$
$\frac{1}{\pi}\left(\sin \pi x - \frac{1}{2}\sin 2\pi x + \frac{1}{3}\sin 3\pi x - + \cdots\right)$

15. $\cosh x$, $\sinh x$ $(-5 < x < 5)$, respectively

17. Cf. Sec. 11.1.

19. $\frac{1}{2} - \frac{4}{\pi^2}\left(\cos \pi x + \frac{1}{9}\cos 3\pi x + \cdots\right)$, $\frac{2}{\pi}\left(\sin \pi x - \frac{1}{2}\sin 2\pi x + - \cdots\right)$

21. $y = C_1 \cos \omega t + C_2 \sin \omega t + \frac{\pi^2}{\omega^2} - 12\left(\frac{\cos t}{\omega^2 - 1} - \frac{1}{4}\cdot\frac{\cos 2t}{\omega^2 - 4} + \frac{1}{9}\cdot\frac{\cos 3t}{\omega^2 - 9}\right.$
$\left. - \frac{1}{16}\cdot\frac{\cos 4t}{\omega^2 - 16} + - \cdots\right)$

23. 0.82, 0.50, 0.36, 0.28, 0.23

25. 0.0076, 0.0076, 0.0012, 0.0012, 0.0004

27. $\frac{1}{\pi}\int_0^\infty \frac{(\cos w + w \sin w - 1)\cos wx + (\sin w - w \cos w)\sin wx}{w^2}\, dw$

29. $\sqrt{2/\pi}\,(\cos aw - \cos w + aw \sin aw - w \sin w)/w^2$

Problem Set 12.1, page 542

1. $L(c_1u_1 + c_2u_2) = c_1L(u_1) + c_2L(u_2) = c_1 \cdot 0 + c_2 \cdot 0 = 0$

3. $c = 2$

5. $c = a/b$

7. Any c and ω

9. $c = \pi/25$

15. $u = 110 - (110/\ln 100)\ln(x^2 + y^2)$

17. $u = a(y)\cos 4\pi x + b(y)\sin 4\pi x$

19. $u = c(x)\, e^{-y^3/3}$

21. $u = e^{-3y}(a(x)\cos 2y + b(x)\sin 2y) + 0.1e^{3y}$

23. $u = c_1(y)x + c_2(y)/x^2$ (Euler–Cauchy)

25. $u(x, y) = axy + bx + cy + k$; a, b, c, k arbitrary constants

Problem Set 12.3, page 551

5. $k \cos 3\pi t \sin 3\pi x$

7. $\frac{8k}{\pi^3}\left(\cos \pi t \sin \pi x + \frac{1}{27}\cos 3\pi t \sin 3\pi x + \frac{1}{125}\cos 5\pi t \sin 5\pi x + \cdots\right)$

9. $\frac{0.8}{\pi^2}\left(\cos \pi t \sin \pi x - \frac{1}{9}\cos 3\pi t \sin 3\pi x + \frac{1}{25}\cos 5\pi t \sin 5\pi x - + \cdots\right)$

11. $\frac{2}{\pi^2}\Big((2-\sqrt{2})\cos \pi t \sin \pi x - \frac{1}{9}(2+\sqrt{2})\cos 3\pi t \sin 3\pi x$

$+ \frac{1}{25}(2+\sqrt{2})\cos 5\pi t \sin 5\pi x - + \cdots\Big)$

13. $\frac{4}{\pi^3}\Big((4-\pi)\cos \pi t \sin \pi x + \cos 2\pi t \sin 2\pi x + \frac{4+3\pi}{27}\cos 3\pi t \sin 3\pi x$

$+ \frac{4-5\pi}{125}\cos 5\pi t \sin 5\pi x + \cdots\Big)$. No terms with $n = 4, 8, 12, \cdots$.

17. $u = \frac{8L^2}{\pi^3}\left(\cos\left[c\left(\frac{\pi}{L}\right)^2 t\right]\sin\frac{\pi x}{L} + \frac{1}{3^3}\cos\left[c\left(\frac{3\pi}{L}\right)^2 t\right]\sin\frac{3\pi x}{L} + \cdots\right)$

19. **(a)** $u(0, t) = 0$, **(b)** $u(L, t) = 0$, **(c)** $u_x(0, t) = 0$, **(d)** $u_x(L, t) = 0$. $C = -A$, $D = -B$ from (a), (c). Insert this. The coefficient determinant resulting from (b), (d) must be zero to have a nontrivial solution. This gives (22).

Problem Set 12.4, page 556

3. $c^2 = 300/[0.9/(2 \cdot 9.80)] = 80.83^2\ [\text{m}^2/\text{sec}^2]$

9. Elliptic, $u = f_1(y + 2ix) + f_2(y - 2ix)$

11. Parabolic, $u = xf_1(x - y) + f_2(x - y)$

13. Hyperbolic, $u = f_1(y - 4x) + f_2(y - x)$

15. Hyperbolic, $xy'^2 + yy' = 0$, $y = v$, $xy = w$, $u_w = z$, $u = \frac{1}{y}f_1(xy) + f_2(y)$

17. Elliptic, $u = f_1(y - (2 - i)x) + f_2(y - (2 + i)x)$. Real or imaginary parts of any function u of this form are solutions. Why?

Problem Set 12.6, page 566

3. $u_1 = \sin x\, e^{-t}$, $u_2 = \sin 2x\, e^{-4t}$, $u_3 = \sin 3x\, e^{-9t}$ differ in rapidity of decay.

5. $u = \sin 0.1\pi x\, e^{-1.752\pi^2 t/100}$

7. $u = \frac{800}{\pi^3}\left(\sin 0.1\pi x\, e^{-0.01752\pi^2 t} + \frac{1}{3^3}\sin 0.3\pi x\, e^{-0.01752(3\pi)^2 t} + \cdots\right)$

9. $u = u_I + u_{II}$, where $u_{II} = u - u_I$ satisfies the boundary conditions of the text,

so that $u_{II} = \sum_{n=1}^{\infty} B_n \sin\frac{n\pi x}{L}\, e^{-(cn\pi/L)^2 t}$, $B_n = \frac{2}{L}\int_0^L [f(x) - u_I(x)]\sin\frac{n\pi x}{L}\,dx$.

11. $F = A\cos px + B\sin px$, $F'(0) = Bp = 0$, $B = 0$, $F'(L) = -Ap\sin pL = 0$, $p = n\pi/L$, etc.

13. $u = 1$

15. $\frac{1}{2} + \frac{4}{\pi^2}\left(\cos x\, e^{-t} + \frac{1}{9}\cos 3x\, e^{-9t} + \frac{1}{25}\cos 5x\, e^{-25t} + \cdots\right)$

17. $-\frac{K\pi}{L}\sum_{n=1}^{\infty} nB_n e^{-\lambda_n^2 t}$

19. $u = 1000(\sin\frac{1}{2}\pi x \sinh\frac{1}{2}\pi y)/\sinh \pi$

21. $u = \frac{100}{\pi}\sum_{n=1}^{\infty}\frac{1}{(2n-1)\sinh(2n-1)\pi}\sin\frac{(2n-1)\pi x}{24}\sinh\frac{(2n-1)\pi y}{24}$

23. $u = A_0 x + \sum_{n=1}^{\infty} A_n \frac{\sinh(n\pi x/24)}{\sinh n\pi} \cos\frac{n\pi y}{24}$,

$A_0 = \frac{1}{24^2}\int_0^{24} f(y)\,dy, \quad A_n = \frac{1}{12}\int_0^{24} f(y)\cos\frac{n\pi y}{24}\,dy$

25. $\sum_{n=1}^{\infty} A_n \sin\frac{n\pi x}{a}\sinh\frac{n\pi(b-y)}{a}, \ A_n = \frac{2}{a\sinh(n\pi b/a)}\int_0^a f(x)\sin\frac{n\pi x}{a}\,dx$

Problem Set 12.7, page 574

3. $A = \frac{2}{\pi}\int_0^{\infty}\frac{\cos pv}{1+v^2}\,dv = \frac{2}{\pi}\cdot\frac{\pi}{2}e^{-p}, \ u = \int_0^{\infty} e^{-p-c^2p^2t}\cos px\,dp$

5. $A = \frac{2}{\pi}\int_0^1 v\cos pv\,dv = \frac{2}{\pi}\cdot\frac{\cos p + p\sin p - 1}{p^2}$, etc.

7. $A = \frac{2}{\pi}\int_0^{\infty}\frac{\sin v}{v}\cos pv\,dv = \frac{2}{\pi}\cdot\frac{\pi}{2} = 1$ if $0 < p < 1$ and 0 if $p > 1$,

$u = \int_0^1 \cos px\, e^{-c^2p^2t}\,dp$

9. Set $w = -v$ in (21) to get $\text{erf}(-x) = -\text{erf}\,x$.

13. In (12) the argument $x + 2cz\sqrt{t}$ is 0 (the point where f jumps) when $z = -x/(2c\sqrt{t})$. This gives the lower limit of integration.

15. Set $w = s/\sqrt{2}$ in (21).

Problem Set 12.9, page 584

1. **(a)**, **(b)** It is multiplied by $\sqrt{2}$. **(c)** Half

5. $B_{mn} = (-1)^{n+1}8/(mn\pi^2)$ if m odd, 0 if m even

7. $B_{mn} = (-1)^{m+n}4ab/(mn\pi^2)$

11. $u = 0.1\cos\sqrt{20}t\sin 2x\sin 4y$

13. $\frac{6.4}{\pi^2}\sum_{\substack{m=1\\ m,n\ \text{odd}}}^{\infty}\sum_{n=1}^{\infty}\frac{1}{m^3n^3}\cos(t\sqrt{m^2+n^2})\sin mx\sin ny$

17. $c\pi\sqrt{260}$ (corresponding eigenfunctions $F_{4,16}$ and $F_{16,14}$), etc.

19. $\cos\left(\pi t\sqrt{\frac{36}{a^2}+\frac{4}{b^2}}\right)\sin\frac{6\pi x}{a}\sin\frac{4\pi y}{b}$

Problem Set 12.10, page 591

5. $110 + \frac{440}{\pi}\left(r\cos\theta - \frac{1}{3}r^3\cos 3\theta + \frac{1}{5}r^5\cos 5\theta - + \cdots\right)$

7. $55\pi - \frac{440}{\pi}\left(r\cos\theta + \frac{1}{9}r^3\cos 3\theta + \frac{1}{25}r^5\cos 5\theta + \cdots\right)$

11. Solve the problem in the disk $r < a$ subject to u_0 (given) on the upper semicircle and $-u_0$ on the lower semicircle.

$$u = \frac{4u_0}{\pi}\left(\frac{r}{a}\sin\theta + \frac{1}{3a^3}r^3\sin 3\theta + \frac{1}{5a^5}r^5\sin 5\theta + \cdots\right)$$

13. Increase by a factor $\sqrt{2}$ **15.** $T = 6.826\rho R^2 f_1^2$

17. No **25.** $\alpha_{11}/(2\pi) = 0.6098$; See Table A1 in App. 5.

Problem Set 12.11, page 598

5. $A_4 = A_6 = A_8 = A_{10} = 0, \quad A_5 = 605/16, \quad A_7 = -4125/128, \quad A_9 = 7315/256$

9. $\nabla^2 u = u'' + 2u'/r = 0, \quad u''/u' = -2/r, \quad \ln|u'| = -2\ln|r| + c_1,$
$u' = \tilde{c}/r^2, u = c/r + k$

13. $u = 320/r + 60$ is smaller than the potential in Prob. 12 for $2 < r < 4$.

17. $u = 1$

19. $\cos 2\phi = 2\cos^2\phi - 1$, $2w^2 - 1 = \frac{4}{3}P_2(w) - \frac{1}{3}$, $u = \frac{4}{3}r^2 P_2(\cos\phi) - \frac{1}{3}$

25. Set $1/r = \rho$. Then $u(\rho, \theta, \phi) = rv(r, \theta, \phi)$, $u_\rho = (v + rv_r)(-1/\rho^2)$,
$u_{\rho\rho} = (2v_r + rv_{rr})(1/\rho^4) + (v + rv_r)(2/\rho^3)$, $u_{\rho\rho} + (2/\rho)u_\rho = r^5(v_{rr} + (2/r)v_r)$.
Substitute this and $u_{\phi\phi} = rv_{\phi\phi}$ etc. into (7) [written in terms of ρ] and divide by r^5.

Problem Set 12.12, page 602

5. $W = \dfrac{c(s)}{x^s} + \dfrac{x}{s^2(s + 1)}$, $W(0, s) = 0$, $c(s) = 0$, $w(x, t) = x(t - 1 + e^{-t})$

7. $w = f(x)g(t)$, $xf'g + f\dot{g} = xt$, take $f(x) = x$ to get $g = ce^{-t} + t - 1$ and $c = 1$ from $w(x, 0) = x(c - 1) = 0$.

11. Set $x^2/(4c^2\tau) = z^2$. Use z as a new variable of integration. Use $\operatorname{erf}(\infty) = 1$.

Chapter 12 Review Questions and Problems, page 603

17. $u = c_1(x)e^{-3y} + c_2(x)e^{2y} - 3$ **19.** Hyperbolic, $f_1(x) + f_2(y + x)$

21. Hyperbolic, $f_1(y + 2x) + f_2(y - 2x)$ **23.** $\frac{3}{4}\cos 2t \sin x - \frac{1}{4}\cos 6t \sin 3x$

25. $\sin 0.01\pi x\, e^{-0.001143t}$

27. $\frac{3}{4}\sin 0.01\pi x\, e^{-0.001143t} - \frac{1}{4}\sin 0.03\pi x\, e^{-0.01029t}$

29. $100\cos 2x\, e^{-4t}$

39. $u = (u_1 - u_0)(\ln r)/\ln(r_1/r_0) + (u_0 \ln r_1 - u_1 \ln r_0)/\ln(r_1/r_0)$

Problem Set 13.1, page 612

1. $1/i = i/i^2 = -i, \quad 1/i^3 = i/i^4 = i$ **3.** $4.8 - 1.4i$

5. $x - iy = -(x + iy), \quad x = 0$ **9.** $-117, \quad 4$

11. $-8 - 6i$ **13.** $-120 - 40i$

15. $3 - i$ **17.** $-4x^2y^2$

19. $(x^2 - y^2)/(x^2 + y^2), \quad 2xy/(x^2 + y^2)$

Problem Set 13.2, page 618

1. $\sqrt{2}\,(\cos\frac{1}{4}\pi + i\sin\frac{1}{4}\pi)$

3. $2(\cos\frac{1}{2}\pi + i\sin\frac{1}{2}\pi), \quad 2(\cos\frac{1}{2}\pi - i\sin\frac{1}{2}\pi)$

5. $\frac{1}{2}(\cos \pi + i \sin \pi)$

7. $\sqrt{1 + \frac{1}{4}\pi^2}\,(\cos \arctan \frac{1}{2}\pi + i \sin \arctan \frac{1}{2}\pi)$

9. $3\pi/4$

11. $\pm\arctan\left(\frac{4}{3}\right) = \pm 0.9273$

13. -1024. *Answer:* π

15. $-3i$

17. $2 + 2i$

21. $\sqrt[6]{2}\,(\cos \frac{1}{12}k\pi + i \sin \frac{1}{12}\pi), \quad k = 1, 9, 17$

23. $6, \quad -3 \pm 3\sqrt{3}i$

25. $\cos\left(\frac{1}{8}\pi + \frac{1}{2}k\pi\right) + i \sin\left(\frac{1}{8}\pi + \frac{1}{2}k\pi\right), \quad k = 0, 1, 2, 3$

27. $\cos \frac{1}{5}\pi \pm i \sin \frac{1}{5}\pi, \quad \cos \frac{3}{5}\pi \pm i \sin \frac{3}{5}\pi, \quad -1$

29. $i, \quad -1 - i$

31. $\pm(1 - i), \quad \pm(2 + 2i)$

33. $|z_1 + z_2|^2 = (z_1 + z_2)\overline{(z_1 + z_2)} = (z_1 + z_2)(\bar{z}_1 + \bar{z}_2)$. Multiply out and use $\text{Re}\, z_1\bar{z}_2 \leqq |z_1\bar{z}_2|$ (Prob. 34).
$z_1\bar{z}_1 + z_1\bar{z}_2 + z_2\bar{z}_1 + z_2\bar{z}_2 = |z_1|^2 + 2\,\text{Re}\, z_1\bar{z}_2 + |z_2|^2 \leqq |z_1|^2 + 2|z_1||z_2| + |z_2|^2 = (|z_1| + |z_2|)^2$. Hence $|z_1 + z_1|^2 \leqq (|z_1| + |z_2|)^2$. Taking square roots gives (6).

35. $[(x_1 + x_2)^2 + (y_1 + y_2)^2] + [(x_1 - x_2)^2 + (y_1 - y_2)^2] = 2(x_1^2 + y_1^2 + x_2^2 + y_2^2)$

Problem Set 13.3, page 624

1. Closed disk, center $-1 + 5i$, radius $\frac{3}{2}$

3. Annulus (circular ring), center $4 - 2i$, radii π and 3π

5. Domain between the bisecting straight lines of the first quadrant and the fourth quadrant.

7. Half-plane extending from the vertical straight line $x = -1$ to the right.

11. $u(x, y) = (1 - x)/((1 - x)^2 + y^2), \quad u(1, -1) = 0,$
$v(x, y) = y((1 - x)^2 + y^2), \quad v(1, -1) = -1$

15. Yes, since $\text{Im}\,(|z|^2/z) = \text{Im}\,(|z|^2\bar{z}/(z\bar{z})) = \text{Im}\,\bar{z} = -r \sin \theta \to 0$.

17. Yes, because $\text{Re}\, z = r \cos \theta \to 0$ and $1 - |z| \to 1$ as $r \to 0$.

19. $f'(z) = 8(z - 4i)^7$. Now $z - 4i = 3$, hence $f'(3 + 4i) = 8 \cdot 3^7 = 17{,}496$.

21. $n(1 - z)^{-n-1}i, \quad ni$

23. $3iz^2/(z + i)^4, \quad -3i/16$

Problem Set 13.4, page 629

1. $r_x = x/r = \cos \theta, \quad r_y = \sin \theta, \quad \theta_x = -(\sin \theta)/r, \quad \theta_y = (\cos \theta)/r$
(a) $0 = u_x - v_y = u_r \cos \theta + u_\theta(-\sin \theta)/r - v_r \sin \theta - v_\theta(\cos \theta)/r$
(b) $0 = u_y + v_x = u_r \sin \theta + u_\theta(\cos \theta)/r + v_r \cos \theta + v_\theta(-\sin \theta)/r$
Multiply (a) by $\cos \theta$, (b) by $\sin \theta$, and add. Etc.

3. Yes

5. No, $f(z) = \overline{(z^2)}$

7. Yes, when $z \neq 0$. Use (7).

9. Yes, when $z \neq 0, \quad -2\pi i, \quad 2\pi i$

11. Yes

13. $f(z) = -\frac{1}{2}i(z^2 + c)$, c real

15. $f(z) = 1/z + c$ (c real)

17. $f(z) = z^2 + z + c$ (c real)

19. No

21. $a = \pi, \quad v = e^{\pi x} \sin \pi y$

23. $a = 0, \quad v = \frac{1}{2}b(y^2 - x^2) + c$

27. $f = u + iv$ implies $if = -v + iu$.

29. Use (4), (5), and (1).

Problem Set 13.5, page 632

3. $e^{2\pi i}e^{-2\pi} = e^{-2\pi} = 0.001867$

5. $e^2(-1) = -7.389$

7. $e^{\sqrt{2}}i = 4.113i$

9. $5e^{i \arctan (3/4)} = 5e^{0.644i}$

11. $6.3e^{\pi i}$

13. $\sqrt{2}e^{\pi i/4}$

15. $\exp(x^2 - y^2)\cos 2xy, \quad \exp(x^2 - y^2)\sin 2xy$
17. $\text{Re}(\exp(z^3)) = \exp(x^3 - 3xy^2)\cos(3x^2y - y^3)$
19. $z = 2n\pi i, \quad n = 0, 1, \cdots$

Problem Set 13.6, page 636

1. Use (11), then (5) for e^{iy}, and simplify. **7.** $\cosh 1 = 1.543$, $i \sinh 1 = 1.175i$
9. Both $-0.642 - 1.069i$. Why? **11.** $i \sinh \pi = 11.55i$, both
15. Insert the definitions on the left, multiply out, and simplify.
17. $z = \pm(2n + 1)i/2$ **19.** $z = \pm n\pi i$

Problem Set 13.7, page 640

5. $\ln 11 + \pi i$ **7.** $\frac{1}{2}\ln 32 - \pi i/4 = 1.733 - 0.785i$
9. $i \arctan(0.8/0.6) = 0.927i$ **11.** $\ln e + \pi i/2 = 1 + \pi i/2$
13. $\pm 2n\pi i, \quad n = 0, 1, \cdots$

15. $\ln|e^i| + i \arctan \dfrac{\sin 1}{\cos 1} \pm 2n\pi i = 0 + i + 2n\pi i, \quad n = 0, 1, \cdots$

17. $\ln(i^2) = \ln(-1) = (1 \pm 2n)\pi i, \quad 2 \ln i = (1 \pm 4n)\pi i, n = 0, 1, \cdots$
19. $e^{4-3i} = e^4(\cos 3 - i \sin 3) = -54.05 - 7.70i$
21. $e^{0.6}e^{0.4i} = e^{0.6}(\cos 0.4 + i \sin 0.4) = 1.678 + 0.710i$
23. $e^{(1-i)\,\text{Ln}\,(1+i)} = e^{\ln\sqrt{2} + \pi i/4 - i\ln\sqrt{2} + \pi/4} = 2.8079 + 1.3179i$
25. $e^{(3-i)(\ln 3 + \pi i)} = 27e^{\pi}(\cos(3\pi - \ln 3) + i \sin(3\pi - \ln 3)) = -284.2 + 556.4i$
27. $e^{(2-i)\,\text{Ln}(-1)} = e^{(2-i)\pi i} = e^{\pi} = 23.14$

Chapter 13 Review Questions and Problems, page 641

1. $2 - 3i$ **3.** $27.46e^{0.9929i}, \quad 7.616e^{1.976i}$
11. $-5 + 12i$ **13.** $0.16 - 0.12i$
15. i **17.** $4\sqrt{2}e^{-3\pi i/4}$
19. $15e^{-\pi i/2}$ **21.** $\pm 3, \quad \pm 3i$
23. $(\pm 1 \pm i)/\sqrt{2}$ **25.** $f(z) = -iz^2/2$
27. $f(z) = e^{-2z}$ **29.** $f(z) = e^{-z^2/2}$
31. $\cos 3 \cosh 1 + i \sin 3 \sinh 1 = -1.528 + 0.166i$
33. $i \tanh 1 = 0.7616i$
35. $\cosh \pi \cos \pi + i \sinh \pi \sin \pi = -11.592$

Problem Set 14.1, page 651

1. Straight segment from (2, 1) to (5, 2.5).
3. Parabola $y = x^2$ from (1, 2) to (2, 8).
5. Circle through (0, 0), center (3, −1), radius $\sqrt{10}$, oriented clockwise.
7. Semicircle, center 2, radius 4.
9. Cubic parabola $y = x^3 \quad (-2 \leqq x \leqq 2)$
11. $z(t) = t + (2 + t)i \quad (-1 \leqq t \leqq 1)$
13. $z(t) = 2 - i + 2e^{it} \quad (0 \leqq t \leqq \pi)$

15. $z(t) = 2\cosh t + i\sinh t\ (-\infty < t < \infty)$
17. Circle $z(t) = -a - ib + re^{-it} \quad (0 \leqq t \leqq 2\pi)$
19. $z(t) = t + (1 - \frac{1}{4}t^2)i \quad (-2 \leqq t \leqq 2)$
21. $z(t) = (1 + i)t \quad (1 \leqq t \leqq 3), \quad \text{Re}\, z = t, \quad z'(t) = 1 + i$. *Answer:* $4 + 4i$
23. $e^{2\pi i} - e^{\pi i} = 1 - (-1) = 2$
25. $\frac{1}{2}\exp z^2|_1^i = \frac{1}{2}(e^{-1} - e^1) = -\sinh 1$
27. $\tan\frac{1}{4}\pi i - \tan\frac{1}{4} = i\tanh\frac{1}{4} - 1$
29. $\text{Im}\, z^2 = 2xy = 0$ on the axes. $z = 1 + (-1 + i)t \quad (0 \leqq t \leqq 1)$,
$(\text{Im}\, z^2)\,\dot{z} = 2(1 - t)y(-1 + i)$ integrated: $(-1 + i)/3$.
35. $|\text{Re}\, z| = |x| \leqq 3 = M$ on C, $L = \sqrt{8}$

Problem Set 14.2, page 659

1. Use (12), Sec. 14.1, with $m = 2$. **3.** Yes **5.** 5
7. **(a)** Yes. **(b)** No, we would have to move the contour across $\pm 2i$.
9. 0, yes
11. πi, no
13. 0, yes
15. $-\pi$, no
17. 0, no
19. 0, yes
21. $2\pi i$
23. $1/z + 1/(z - 1)$, hence $2\pi i + 2\pi i = 4\pi i$.
25. 0 (Why?)
27. 0 (Why?)
29. 0

Problem Set 14.3, page 663

1. $2\pi i z^2/(z - 1)|_{z=-1} = -\pi i$
3. 0
5. $2\pi i(\cos 3z)/6|_{z=0} = \pi i/3$
7. $2\pi i(i/2)^3/2 = \pi/8$
11. $2\pi i \cdot \left.\dfrac{1}{z + 2i}\right|_{z=2i} = \dfrac{\pi}{2}$
13. $2\pi i(z + 2)|_{z=2} = 8\pi i$
15. $2\pi i\cosh(-\pi^2 - \pi i) = -2\pi i\cosh\pi^2 = -60{,}739i$ since $\cosh\pi i = \cos\pi = -1$ and $\sinh\pi i = i\sin\pi = 0$.
17. $2\pi i \left.\dfrac{\text{Ln}\,(z + 1)}{z + i}\right|_{z=i} = 2\pi i\dfrac{\text{Ln}\,(1 + i)}{2i} = \pi(\ln\sqrt{2} + i\pi/4) = 1.089 + 2.467i$
19. $2\pi i e^{2i}/(2i) = \pi e^{2i}$

Problem Set 14.4, page 667

1. $(2\pi i/3!)(-\cos 0) = -\pi i/3$
3. $(2\pi i/(n - 1)!)e^0$
5. $\dfrac{2\pi i}{3!}(\cosh 2z)''' = \dfrac{\pi i}{3} \cdot 8\sinh 1 = 9.845i$
7. $(2\pi i/(2n)!)\,(\cos z)^{(2n)}|_{z=0} = (2\pi i/(2n)!)(-1)^n\cos 0 = (-1)^n 2\pi i/(2n)!$
9. $-2\pi i(\tan\pi z)'\Big|_{z=0} = \left.\dfrac{-2\pi i \cdot \pi}{\cos^2\pi z}\right|_{z=0} = -2\pi^2 i$
11. $\dfrac{2\pi i}{4}((1 + z)\sin z)'\Big|_{z=1/2} = \frac{1}{2}\pi i(\sin z + (1 + z)\cos z)|_{z=1/2}$
$= \frac{1}{2}\pi i(\sin\frac{1}{2} + \frac{3}{2}\cos\frac{1}{2})$
$= 2.821i$

13. $2\pi i \cdot \frac{1}{z}\Big|_{z=2} = \pi i$

15. 0. Why?

17. 0 by Cauchy's integral theorem for a doubly connected domain; see (6) in Sec. 14.2.

19. $(2\pi i/2!)4^{-3}(e^{3z})''|_{z=\pi i/4} = -9\pi(1 + i)/(64\sqrt{2})$

Chapter 14 Review Questions and Problems, page 668

21. $\frac{1}{2}\cosh(-\frac{1}{4}\pi^2) - \frac{1}{2} = 2.469$

23. $2\pi i(e^z)^{(4)}|_{z=0} = ie^z/12|_{z=0} = \pi i/12$ by Cauchy's integral formula.

25. $-2\pi i(\tan \pi z)'|_{z=1} = -2\pi^2 i/\cos^2 \pi z|_{z=1} = -2\pi^2 i$

27. 0 since $z^2 + \bar{z} - 2 = 2(x^2 - y^2)$ and $y = x$

29. $-4\pi i$

Problem Set 15.1, page 679

1. $z_n = (2i/2)^n$; bounded, divergent, ± 1, $\pm i$

3. $z_n = -\frac{1}{2}\pi i/(1 + 2/(ni))$ by algebra; convergent to $-\pi i/2$

5. Bounded, divergent, $\pm 1 + 10i$

7. Unbounded, hence divergent

9. Convergent to 0, hence bounded

17. Divergent; use $1/\ln n > 1/n$.

19. Convergent; use $\Sigma 1/n^2$.

21. Convergent

23. Convergent

25. Divergent

29. By absolute convergence and Cauchy's convergence principle, for given $\epsilon > 0$ we have for every $n > N(\epsilon)$ and $p = 1, 2, \cdots$

$$|z_{n+1}| + \cdots + |z_{n+p}| < \epsilon,$$

hence $|z_{n+1} + \cdots + z_{n+p}| < \epsilon$ by (6*), Sec. 13.2, hence convergence by Cauchy's principle.

Problem Set 15.2, page 684

1. No! Nonnegative integer powers of z (or $z - z_0$) only!

3. At the center, in a disk, in the whole plane

5. $\Sigma a_n z^{2n} = \Sigma a_n (z^2)^n, \quad |z^2| < R = \lim |a_n/a_{n+1}|$; hence $|z| < \sqrt{R}$.

7. $\pi/2, \infty$

9. $i, \sqrt{3}$

11. $0, \sqrt{\frac{26}{5}}$

13. $-i, \frac{1}{2}$

15. $2i$, 1

17. $1/\sqrt{2}$

Problem Set 15.3, page 689

3. $f = \sqrt[n]{n}$. Apply l'Hôpital's rule to $\ln f = (\ln n)/n$.

5. 2

7. $\sqrt{3}$

9. $1/\sqrt{2}$

11. $\sqrt{\frac{7}{3}}$

13. 1

15. $\frac{3}{4}$

Problem Set 15.4, page 697

3. $2z^2 - \dfrac{(2z^2)^3}{3!} + \cdots = 2z^2 - \dfrac{4}{3}z^6 + \dfrac{4}{15}z^{10} - + \cdots, \quad R = \infty$

5. $\frac{1}{2} - \frac{1}{4}z^4 + \frac{1}{8}z^8 - \frac{1}{16}z^{12} + \frac{1}{32}z^{16} - + \cdots, \quad R = \sqrt[4]{2}$

7. $\dfrac{1}{2} + \dfrac{1}{2}\cos z = 1 - \dfrac{1}{2 \cdot 2!}z^2 + \dfrac{1}{2 \cdot 4!}z^4 - \dfrac{1}{2 \cdot 6!}z^6 + - \cdots, \quad R = \infty$

9. $\displaystyle\int_0^z \left(1 - \frac{1}{2}t^2 + \frac{1}{8}t^4 - + \cdots\right) dt = z - \frac{1}{6}z^3 + \frac{1}{40}z^5 - + \cdots, \quad R = \infty$

11. $z^3/(1!3) - z^7/(3!7) + z^{11}/(5!11) - + \cdots, \quad R = \infty$

13. $(2/\sqrt{\pi})(z - z^3/3 + z^5/(2!5) - z^7/(3!7) + \cdots), \quad R = \infty$

17. Team Project. (a) $(\text{Ln}(1 + z))' = 1 - z + z^2 - + \cdots = 1/(1 + z)$.
(c) Use that the terms of $(\sin iy)/(iy)$ are all positive, so that the sum cannot be zero.

19. $\frac{1}{2} + \frac{1}{2}i + \frac{1}{2}i(z - i) + (-\frac{1}{4} + \frac{1}{4}i)(z - i)^2 - \frac{1}{4}(z - i)^3 + \cdots, \quad R = \sqrt{2}$

21. $1 - \dfrac{1}{2!}\left(z - \dfrac{1}{2}\pi\right)^2 + \dfrac{1}{4!}\left(z - \dfrac{1}{2}\pi\right)^4 - \dfrac{1}{6!}\left(z - \dfrac{1}{2}\pi\right)^6 + - \cdots, \quad R = \infty$

23. $-\frac{1}{4} - \frac{2}{8}i(z - i) + \frac{3}{16}(z - i)^2 + \frac{4}{32}i(z - i)^3 - \frac{5}{64}(z - i)^4 + \cdots, \quad R = 2$

25. $2\left(z - \dfrac{1}{2}i\right) + \dfrac{2^3}{3!}\left(z - \dfrac{1}{2}i\right)^3 + \dfrac{2^5}{5!}\left(z - \dfrac{1}{2}i\right)^5 + \cdots, \quad R = \infty$

Problem Set 15.5, page 704

3. $|z + i| \leqq \sqrt{3} - \delta, \quad \delta > 0$
5. $|z + \frac{1}{2}i| \leqq \frac{1}{4} - \delta, \quad \delta > 0$
7. Nowhere
9. $|z - 2i| \leqq 2 - \delta, \quad \delta > 0$
11. $|z^n| \leqq 1$ and $\Sigma 1/n^2$ converges. Use Theorem 5.
13. $|\sin^n |z|| \leqq 1$ for all z, and $\Sigma 1/n^2$ converges. Use Theorem 5.
15. $R = 4$ by Theorem 2 in Sec. 15.2; use Theorem 1.
17. $R = 1/\sqrt{\pi} > 0.56$; use Theorem 1.

Chapter 15 Review Questions and Problems, page 706

11. 1
13. 3
15. $\frac{1}{2}$
17. $\infty, \quad e^{2z}$
19. $\infty, \quad \cosh \sqrt{z}$

21. $\displaystyle\sum_{n=0}^{\infty} \frac{z^{4n}}{(2n + 1)!}, \quad R = \infty$

23. $\displaystyle\frac{1}{2} + \frac{1}{2}\cos 2z = 1 + \frac{1}{2}\sum_{n=1}^{\infty} \frac{(-1)^n}{(2n)!}(2z)^{2n}, \quad R = \infty$

25. $\displaystyle\sum_{n=1}^{\infty} \frac{(-1)^{n+1}}{n!} z^{2n-2}, \quad R = \infty$

27. $\cos[(z - \frac{1}{2}\pi) + \frac{1}{2}\pi] = -(z - \frac{1}{2}\pi) + \frac{1}{6}(z - \frac{1}{2}\pi)^3 - + \cdots = -\sin(z - \frac{1}{2}\pi)$

29. $\ln 3 + \dfrac{1}{3}(z - 3) - \dfrac{1}{2 \cdot 9}(z - 3)^2 + \dfrac{1}{3 \cdot 27}(z - 3)^3 - + \cdots, \quad R = 3$

Problem Set 16.1, page 714

1. $z^{-4} - \frac{1}{2}z^{-2} + \frac{1}{24} - \frac{1}{720}z^2 + - \cdots, \quad 0 < |z| < \infty$

3. $z^{-3} + z^{-1} + \frac{1}{2}z + \frac{1}{6}z^3 + \frac{1}{24}z^5 + \cdots, \quad 0 < |z| < \infty$

5. $z^{-2} + z^{-1} + 1 + z + z^2 + \cdots, \quad 0 < |z| < 1$

7. $z^3 + \frac{1}{2}z + \frac{1}{24}z^{-1} + \frac{1}{720}z^3 + \cdots, \quad 0 < |z| < \infty$

9. $\exp[1 + (z-1)](z-1)^{-2} = e \cdot [(z-1)^{-2} + (z-1)^{-1} + \frac{1}{2} + \frac{1}{6}(z-1) + \cdots],$
$0 < |z-1| < \infty$

11. $\dfrac{[\pi i + (z - \pi i)]^2}{(z - \pi i)^4} = \dfrac{(\pi i)^2}{(z - \pi i)^4} + \dfrac{2\pi i}{(z - \pi i)^3} + \dfrac{1}{(z - \pi i)^2}$

13. $i^{-3}\left(1 + \dfrac{z-i}{i}\right)^{-3}(z-i)^{-2} = \displaystyle\sum_{n=0}^{\infty}\binom{-3}{n} i^{-3-n}(z-i)^{n-2} = i(z-i)^{-2}$
$-3(z-i)^{-1} - 6i + 10(z-i) + \cdots, \quad 0 < |z-i| < 1$

15. $(-\cos(z-\pi))(z-\pi)^{-2} = -(z-\pi)^{-2} + \frac{1}{2} - \frac{1}{24}(z-\pi)^2 + - \cdots,$
$0 < |z-\pi| < \infty$

19. $\displaystyle\sum_{n=0}^{\infty} z^{2n}, \quad |z| < 1, \qquad -\sum_{n=0}^{\infty}\frac{1}{z^{2n+2}}, \quad |z| > 1$

21. $-(z + \frac{1}{2}\pi)^{-1}\cos(z + \frac{1}{2}\pi) = -(z + \frac{1}{2}\pi)^{-1} + \frac{1}{2}(z + \frac{1}{2}\pi) - \frac{1}{24}(z + \frac{1}{2}\pi)^3 + - \cdots,$
$|z + \frac{1}{2}\pi| > 0$

23. $z^8 + z^{12} + z^{16} + \cdots, \quad |z| < 1, \quad -z^4 - 1 - z^{-4} - z^{-8} - \cdots, \quad |z| > 1$

25. $\dfrac{i}{(z-i)^2} + \dfrac{1}{z-i} + i + (z-i)$

Section 16.2, page 719

1. $0 \pm 2\pi, \pm 4\pi, \cdots$, fourth order

3. $-81i$, fourth order

5. $\pm 1, \pm 2, \cdots$, second order

7. $\pm(2 + 2i), \pm i$, simple

9. $\frac{1}{2}\sin 4z$, $z = 0, \pm\pi/4, \pm\pi/2, \cdots$, simple

11. $f(z) = (z - z_0)^n g(z)$, $g(z_0) \neq 0$, hence $f^2(z) = (z - z_0)^{2n} g^2(z)$.

13. Second-order poles at i and $-2i$

15. Simple pole at ∞, essential singularity at $1 + i$

17. Fourth-order poles at $\pm n\pi i$, $n = 0, 1, \cdots$, essential singularity at ∞

19. $e^z(1 - e^z) = 0$, $e^z = 1$, $z = \pm 2n\pi i$ simple zeros. *Answer:* simple poles at $\pm 2n\pi i$, essential singularity at ∞

21. $1, \infty$ essential singularities, $\pm 2n\pi i$, $n = 0, 1, \cdots$, simple poles

Section 16.3, page 725

3. $\frac{4}{15}$ at 0

5. $\pm 4i$ at $\mp i$

7. $1/\pi$ at $0, \pm 1, \cdots$

9. -1 at $\pm 2n\pi i$

11. $(e^z)''/2!|_{z=\pi i} = -\frac{1}{2}$ at $z = \pi i$

15. Simple pole at $\frac{1}{4}$ inside C, residue $-1/(2\pi)$. *Answer:* $-i$

17. Simple poles at $\pi/2$, residue $e^{\pi/2}/(-\sin\pi/2)$, and at $-\pi/2$, residue $e^{-\pi/2}/\sin\pi/2 = e^{-\pi/2}$. *Answer:* $-4\pi i \sinh \pi/2$

19. $2\pi i(\sinh\frac{1}{2}i)/2 = -\pi\sin\frac{1}{2}$

21. $z^{-5}\cos\pi z = \cdots + \pi^4/(4!z) - + \cdots$. *Answer:* $2\pi^5 i/24$

23. Residues $\frac{1}{2}$ at $z = \frac{1}{2}$, 2 at $z = \frac{1}{3}$. *Answer:* $5\pi i$

25. Simple poles inside C at $2i, -2i, 3i, -3i$, residues $(2i \cosh 2i)/(4z^3 + 26z)|_{z=2i} = \frac{1}{10}, \frac{1}{10}, \frac{1}{10}, \frac{1}{10}$, respectively. *Answer:* $2\pi i \cdot \frac{4}{10}$

Problem Set 16.4, page 733

1. $2\pi/\sqrt{k^2 - 1}$

3. $\pi/\sqrt{2}$

5. $5\pi/12$

7. $2a\pi/\sqrt{a^2 - 1}$

9. 0. Why? (Make a sketch.)

11. $\pi/2$

13. 0. Why?

15. $\pi/3$

17. 0. Why?

19. Simple poles at $\pm 1, i$ (and $-i$); $2\pi i \cdot \frac{1}{4}i + \pi i(-\frac{1}{4} + \frac{1}{4}) = -\frac{1}{2}\pi$

21. Simple poles at 1 and $\pm 2\pi i$, residues i and $-i$. *Answer:* $\dfrac{\pi}{5}(\cos 1 - e^{-2})$

23. $-\pi/2$

25. 0

27. Let $q(z) = (z - a_1)(z - a_2) \cdots (z - a_k)$. Use (4) in Sec. 16.3 to form the sum of the residues $1/q'(a_1) + \cdots + 1/q'(a_k)$ and show that this sum is 0; here $k > 1$.

Chapter 16 Review Questions and Problems, page 733

11. $6\pi i$

13. $2\pi i(-10 - 10)$

15. $2\pi i(25z^2)'|_{z=5} = 500\pi i$

17. 0 (n even), $(-1)^{(n-1)/2}2\pi i/(n - 1)!$ (n odd)

19. $\pi/6$

21. $\pi/60$

23. 0. Why?

25. $\operatorname{Res}_{z=i} e^{iz}/(z^2 + 1) = 1/(2ie)$. *Answer:* π/e.

Problem Set 17.1, page 741

5. Only in size

7. $x = c, w = -y + ic$; $y = k, w = -k + ix$

9. Parallel displacement; each point is moved 2 to the right and 1 up.

11. $|w| \leqq \frac{1}{4}$, $-\pi/4 < \operatorname{Arg} w < \pi/4$

13. $-5 \leqq \operatorname{Re} z \leqq -2$

15. $u \geqq 1$

17. Annulus $\frac{1}{2} \leqq |w| \leqq 4$

19. $0 < u < \ln 4$, $\pi/4 < v \leqq 3\pi/4$

21. $z^3 + az^2 + bz + c$, $z = -\frac{1}{3}(a \pm \sqrt{a^2 - 3b})$

23. $z = (-1 \pm \sqrt{3})/2$

25. $\sinh z = 0$ at $z = 0$, $\pm\pi i, \pm 2\pi i, \cdots$

29. $M = |z| = 1$ on the unit circle, $J = |z|^2$

31. $|w'| = 1/|z|^2 = 1$ on the unit circle, $J = 1/|z|^4$

33. $M = e^x = 1$ for $x = 0$, the y-axis, $J = e^{2x}$

35. $M = 1/|z| = 1$ on the unit circle, $J = 1/|z|^2$

Problem Set 17.2, page 745

7. $z = \dfrac{w + i}{2w}$

9. $z = \dfrac{4w + i}{-3iw + 1}$

11. $z = 0$, $1/(a + ib)$

13. $z = 0$, $\pm\frac{1}{2}$, $\pm = \pm i/2$

15. $z = i, 2i$ **17.** $w = \dfrac{az}{cz + a}$ **19.** $w = \dfrac{az + b}{-bz + a}$

Problem Set 17.3, page 750

3. Apply the inverse g of f on both sides of $z_1 = f(z_1)$ to get $g(z_1) = g(f(z_1)) = z_1$.
9. $w = iz$, a rotation. Sketch to see.
11. $w = (z + i)/(z - i)$
13. $w = 1/z$, almost by inspection
15. $w = 1/z - 1$
17. $w = (2z - i)/(-iz - 2)$
19. $w = (z^4 - i)(-iz^4 + 1)$

Problem Set 17.4, page 754

1. Circle $|w| = e^c$
3. Annulus $1/\sqrt{e} \leqq |w| \leqq \sqrt{e}$
5. w-plane without $w = 0$
7. $1 < |w| < e, v > 0$
9. $\pm(2n + 1)\pi/2, \quad n = 0, 1, \cdots$
11. $u^2/\cosh^2 2 + v^2/\sinh^2 2 < 1, \quad u > 0, v > 0$
13. Elliptic annulus bounded by $u^2/\cosh^2 1 + v^2/\sinh^2 1 = 1$ and $u^2/\cosh^2 3 + v^2/\sinh^2 3 = 1$
15. $\cosh z = \cos iz = \sin(iz + \frac{1}{2}\pi)$
17. $0 < \text{Im}\, t < \pi$ is the image of R under $t = z^2/2$. *Answer:* $e^t = e^{z^2/2}$.
19. Hyperbolas $u^2/\cos^2 c - v^2/\sin^2 c = \cosh^2 c - \sinh^2 c = 1$ when $c \neq 0, \pi$, and $u = \pm\cosh y$ (thus $|u| \geqq 1$), $v = 0$ when $c = 0, \pi$.
21. Interior of $u^2/\cosh^2 2 + v^2/\sinh^2 2 = 1$ in the fourth quadrant, or map $\pi/2 < x < \pi, 0 < y < 2$ by $w = \sin z$ (why?).
23. $v < 0$
25. The images of the five points in the figure can be obtained directly from the function w.

Problem Set 17.5, page 756

1. w moves once around the circle $|w| = \frac{1}{2}$.
3. Four sheets, branch point at $z = -1$
5. $-i/4$, three sheets
7. z_0, n sheets
9. $\sqrt{z(z - i)(z + i)}$, 0, $\pm i$, two sheets

Chapter 17 Review Questions and Problems, page 756

11. $1 < |w| < 4$, $|\arg w| < \pi/4$
13. Horizontal strip $-8 < v < 8$
15. $u = 1 - \frac{1}{4}v^2$, same (why?)
17. $|w| > 1$
19. $\frac{1}{3} < |w| < \frac{1}{2}, \quad v < 0$
21. $w = 1 + iv, \quad v < 0$
23. $w = \dfrac{10z + 5i}{z + 2i}$
25. Rotation $w = iz$
27. $w = 1/z$
29. $z = 0$
31. $z = 2 \pm \sqrt{6}$
33. $z = 0, \pm i, \pm 3i$
35. $w = e^{4z}$
37. $w = iz^2 + 1$
39. $w = z^2/(2c)$

Problem Set 18.1, page 762

1. 2.5 mm = 0.25 cm; $\Phi = \text{Re}\, 110(1 + (\text{Ln}\, z)/\ln 4)$

3. $\Phi = \text{Re}\left(30 - \dfrac{20}{\ln 10}\text{Ln}\, z\right)$

5. $\Phi(x) = \text{Re}\,(375 + 25z)$

7. $\Phi(r) = \text{Re}\,(32 - z)$

13. Use Fig. 391 in Sec. 17.4 with the z- and w-planes interchanged and $\cos z = \sin(z + \frac{1}{2}\pi)$.

15. $\Phi = 220(x^3 - 3xy^2) = \text{Re}\,(220z^3)$

Problem Set 18.2, page 766

3. $w = iz^2$ maps R onto the strip $-2 \leqq u \leqq 0$; and $\Phi^* = U_2 + (U_1 - U_2)(1 + \frac{1}{2}u) = U_2 + (U_1 - U_2)(1 - xy)$.

5. (a) $\dfrac{(x-2)(2x-1) + 2y^2}{(x-2)^2 + y^2} = c$, **(b)** $x^2 - y^2 = c$, $xy = c$, $e^x \cos y = c$

7. See Fig. 392 in Sec. 17.4. $\Phi = \text{Re}\,(\sin^2 z)$, $\sin^2 x\ (y = 0)$, $\sin^2 x \cosh^2 1 - \cos^2 x \sinh^2 1\ (y = 1)$, $-\sinh^2 y\ (x = 0, \pi)$.

9. $\Phi(x, y) = \cos^2 x \cosh^2 y - \sin^2 x \sinh^2 y$; $\cosh^2 y\ (x = 0)$, $-\sinh y\ (x = \frac{\pi}{2})$, $\cos^2 x\ (y = 0)$, $\cos^2 x \cosh^2 1 - \sin^2 x \sinh^2 1\ (y = 1)$

13. Corresponding rays in the w-plane make equal angles, and the mapping is conformal.

15. Apply $w = z^2$.

17. $z = (2Z - i)/(-iZ - 2)$ by (3) in Sec. 17.3.

19. $\Phi = \dfrac{5}{\pi}\text{Arg}\,(z - 2)$, $F = -\dfrac{5i}{\pi}\text{Ln}\,(z - 2)$

Problem Set 18.3, page 769

1. $(80/d)y + 20$. Rotate through $\pi/2$.

5. $\dfrac{80}{\pi}\arctan\dfrac{y}{x} = \text{Re}\left(-\dfrac{80i}{\pi}\text{Ln}\, z\right)$

7. $T_1 + \dfrac{2}{\pi}(T_2 - T_1)\arctan\dfrac{y}{x} = \text{Re}\left(T_1 - \dfrac{2i}{\pi}(T_2 - T_1)\,\text{Ln}\, z\right)$

9. $\dfrac{T_1}{\pi}\left(\arctan\dfrac{y}{x-b} - \arctan\dfrac{y}{x-a}\right) = \text{Re}\left(\dfrac{iT_1}{\pi}\text{Ln}\dfrac{z-a}{z-b}\right)$

11. $\dfrac{100}{\pi}(\text{Arg}\,(z-1) - \text{Arg}\,(z+1)) = \text{Re}\left(\dfrac{100i}{\pi}\text{Ln}\dfrac{z+1}{z-1}\right)$

13. $\dfrac{100}{\pi}[\text{Arg}\,(z^2 - 1) - \text{Arg}\,(z^2 + 1)]$ from $w = z^2$ and Prob. 11.

15. $-20 + (320/\pi)\,\text{Arg}\, z = \text{Re}\left(-20 - \dfrac{320i}{\pi}\text{Ln}\, z\right)$

17. $\text{Re}\, F(z) = 100 + (200/\pi)\,\text{Re}\,(\arcsin z)$

Problem Set 18.4, page 776

1. $V(z)$ continuously differentiable.

3. $|F'(iy)| = 1 + 1/y^2$, $|y| \geqq 1$, is maximum at $y = \pm 1$, namely, 2.

5. Calculate or note that ∇^2 = div grad and curl grad is the zero vector; see Sec. 9.8 and Problem Set 9.7.
7. Horizontal parallel flow to the right.
9. $F(z) = z^4$
11. Uniform parallel flow upward, $V = \overline{F'} = iK$, $V_1 = 0$, $V_2 = K$
13. $F(z) = z^3$
15. $F(z) = z/r_0 + r_0/z$
17. Use that $w = \arccos z$ gives $z = \cos w$ and interchanging the roles of the z- and w-planes.
19. $y/(x^2 + y^2) = c$ or $x^2 + (y - k)^2 = k^2$

Problem Set 18.5, page 781

5. $\Phi = \frac{3}{2} r^3 \sin 3\theta$
7. $\Phi = \frac{1}{2}a + \frac{1}{2}ar^8 \cos 8\theta$
9. $\Phi = 3 - 4r^2 \cos 2\theta + r^4 \cos 4\theta$
11. $\Phi = \dfrac{2}{\pi}\left(r \sin \theta - \dfrac{1}{2}r^2 \sin 2\theta + \dfrac{1}{3}r^3 \sin 3\theta - + \cdots\right)$
13. $\Phi = \dfrac{2}{\pi} r \sin \theta + \dfrac{1}{2}r^2 \sin 2\theta - \dfrac{2}{9\pi}r^3 \sin 3\theta - \dfrac{1}{4}r^4 \sin 4\theta + + - \cdots$
15. $\Phi = \dfrac{1}{2} + \dfrac{2}{\pi}\left(r \cos \theta - \dfrac{1}{3}r^3 \cos 3\theta + \dfrac{1}{5}r^5 \cos 5\theta - + \cdots\right)$
17. $\Phi = \dfrac{1}{3} - \dfrac{4}{\pi^2}\left(r \cos \theta - \dfrac{1}{4}r^2 \cos 2\theta + \dfrac{1}{9}r^3 \cos 3\theta - + \cdots\right)$

Problem Set 18.6, page 784

1. Use (2). $F(z_0 + e^{i\alpha}) = (\frac{7}{2} + e^{i\alpha})^3$, etc. $F(\frac{5}{2}) = \frac{343}{8}$
3. Use (2). $F(z_0 + e^{i\alpha}) = (2 + 3e^{i\alpha})^2$, etc. $F(4) = 100$
5. No, because $|z|$ is not analytic.
7. $$\Phi(2, -2) = -3 = \frac{1}{\pi}\int_0^1\int_0^{2\pi} (1 + r\cos\alpha)(-3 + r\sin\alpha)r\,dr\,d\alpha$$
$$= \frac{1}{\pi}\int_0^1\int_0^{2\pi} (-3r + \cdots)\,dr\,d\alpha = \frac{1}{\pi}\left(-\frac{3}{2}\right)\cdot 2\pi$$
9. $$\Phi(1, 1) = 3 = \frac{1}{\pi}\int_0^1\int_0^{2\pi} (3 + r\cos\alpha + r\sin\alpha + r^2\cos\alpha\sin\alpha)r\,dr\,d\alpha$$
$$= \frac{1}{\pi}\cdot\frac{3}{2}\cdot 2\pi$$
13. $|F(z)| = [\cos^2 x + \sinh^2 y]^{1/2}$, $z = \pm i$, Max $= [1 + \sinh^2 1]^{1/2} = 1.543$
15. $|F(z)|^2 = \sinh^2 2x \cos^2 2y + \cosh^2 2x \sin^2 2y = \sinh^2 2x + 1 \cdot \sin^2 2y$, $z = 1$, Max $= \sinh 2 = 3.627$
17. $|F(z)|^2 = 4(2 - 2\cos 2\theta)$, $z = \pi/2,\ 3\pi/2$, Max $= 4$
19. No. Make up a counterexample.

Chapter 18 Review Questions and Problems, page 785

11. $\Phi = 10(1 - x + y), \quad F = 10 - 10(1 + i)z$

13. $\Phi = \text{Re}\,(220 - 95.54\,\text{Ln}\,z) = 220 - \dfrac{220}{\ln 10}\ln r = 220 - 95.54\ln r.$

17. $2(1 - (2/\pi)\,\text{Arg}\,z)$

19. $30(1 - (2/\pi)\,\text{Arg}\,(z - 1))$

21. $\Phi = x + y = \text{const}, \quad V = \overline{F'(z)} = 1 - i,$ parallel flow

23. $T(x, y) = x(2y + 1) = \text{const}$

25. $F'(z) = \bar{z} + 1 = x + 1 - iy$

Problem Set 19.1, page 796

1. $0.84175 \cdot 10^2, \quad -0.52868 \cdot 10^3, \quad 0.92414 \cdot 10^{-3}, \quad -0.36201 \cdot 10^6$

3. 6.3698, 6.794, 8.15, impossible

5. Add first, then round.

7. 29.9667, 0.0335; 29.9667, 0.0333704 (6S-exact)

9. 29.97, 0.035; 29.97, 0.03337; 30, 0.0; 30, 0.033

11. $|\epsilon| = |x + y - (\tilde{x} + \tilde{y})| = |(x - \tilde{x}) + (y - \tilde{y})| = |\epsilon_x + \epsilon_y|$
$\leqq |\epsilon_x| + |\epsilon_y| = \beta_x + \beta_y$

13. $\dfrac{a_1}{a_2} = \dfrac{\tilde{a}_1 + \epsilon_1}{\tilde{a}_2 + \epsilon_2} = \dfrac{\tilde{a}_1 + \epsilon_1}{\tilde{a}_2}\left(1 - \dfrac{\epsilon_2}{\tilde{a}_2} + \dfrac{\epsilon_2^2}{\tilde{a}_2^2} - + \cdots\right) \approx \dfrac{\tilde{a}_1}{\tilde{a}_2} + \dfrac{\epsilon_1}{\tilde{a}_2} - \dfrac{\epsilon_2}{\tilde{a}_2} \cdot \dfrac{\tilde{a}_1}{\tilde{a}_2},$

hence $\left|\left(\dfrac{a_1}{a_2} - \dfrac{\tilde{a}_1}{\tilde{a}_2}\right)\Big/\dfrac{a_1}{a_2}\right| \approx \left|\dfrac{\epsilon_1}{a_1} - \dfrac{\epsilon_2}{a_2}\right| \leqq |\epsilon_{r1}| + |\epsilon_{r2}| \leqq \beta_{r1} + \beta_{r2}$

15. **(a)** $1.38629 - 1.38604 = 0.00025$, **(b)** $\ln 1.00025 = 0.000249969$ is 6S-exact.

19. In the present case, (b) is slightly more accurate than (a) (which may produce nonsensical results; cf. Prob. 20).

21. $c_4 \cdot 2^4 + \cdots + c_0 \cdot 2^0 = (1\,0\,1\,1\,1.)_2$, NOT $(1\,1\,1\,0\,1.)_2$

23. The algorithm in Prob. 22 repeats 0011 infinitely often.

25. $n = 26$. The beginning is 0.09375 ($n = 1$).

27. $I_{14} = 0.1812$ (0.1705 4S-exact), $I_{13} = 0.1812$ (0.1820), $I_{12} = 0.1951$ (0.1951), $I_{11} = 0.2102$ (0.2103), etc.

29. $-0.126 \cdot 10^{-2}, -0.402 \cdot 10^{-3}; \quad -0.266 \cdot 10^{-6}, -0.847 \cdot 10^{-7}$

Problem Set 19.2, page 807

3. $g = 0.5\cos x, \quad x = 0.450184$ ($= x_{10}$, exact to 6S)

5. Convergence to 4.7 for all these starting values.

7. $x = x/(e^x \sin x)$; 0.5, 0.63256, $\cdots$ converges to 0.58853 (5S-exact) in 14 steps.

9. $x = x^4 - 0.12; \quad x_0 = 0, x_3 = -0.119794$ (6S-exact)

11. $g = 4/x + x^3/16 - x^5/576; \quad x_0 = 2, x_n = 2.39165$ ($n \geqq 6$), 2.405 4S-exact

13. This follows from the intermediate value theorem of calculus.

15. $x_3 = 0.450184$

17. Convergence to $x = 4.7, 4.7, 0.8, -0.5$, respectively. Reason seen easily from the graph of f.

19. 0.5, 0.375, 0.377968, 0.377964; (b) $1/\sqrt{7}$

21. 1.834243 ($= x_4$), 0.656620 ($= x_4$), −2.49086 ($= x_4$)

23. $x_0 = 4.5$, $x_4 = 4.73004$ (6S-exact)

25. **(a) ALGORITHM BISECT (f, a_0, b_0, ϵ, N) Bisection Method**

This algorithm computes the solution c of $f(x) = 0$ (f continuous) within the tolerance ϵ, given an initial interval $[a_0, b_0]$ such that $f(a_0)f(b_0) < 0$.

INPUT: Continuous function f, initial interval $[a_0, b_0]$, tolerance ϵ, maximum number of iterations N.

OUTPUT: A solution c (within the tolerance ϵ), or a message of failure.

For $n = 0, 1, \cdots, N - 1$ do:

$c = \frac{1}{2}(a_n + b_n)$

If $f(c) = 0$ then OUTPUT c Stop. [*Procedure completed*]

Else if $f(a_n)f(b_n) < 0$ then set $a_{n+1} = a_n$ and $b_{n+1} = c$.

Else set $a_{n+1} = c$, and $b_{n+1} = b_n$.

If $|a_{n+1} - b_{n+1}| < \epsilon|c|$ then OUTPUT c. Stop. [*Procedure completed*]

End

OUTPUT $[a_N, b_N]$ and a message "Failure". Stop.

[*Unsuccessful completion; N iterations did not give an interval of length not exceeding the tolerance.*]

End BISECT

Note that $[a_N, b_N]$ gives $(a_N + b_N)/2$ as an approximation of the zero and $(b_N - a_N)/2$ as a corresponding error bound.

(b) 0.739085; **(c)** 1.30980, 0.429494

27. $x_2 = 1.5$, $x_3 = 1.76471, \cdots$, $x_7 = 1.83424$ (6S-exact)

29. 0.904557 (6S-exact)

Problem Set 19.3, page 819

1. $L_0(x) = -2x + 19$, $L_1(x) = 2x - 18$, $p_1(9.3) = L_0(9.3) \cdot f_0 + L_1(9.3) \cdot f_1$ $= 0.1086 \cdot 9.3 + 1.230 = 2.2297$

3. $$p_2(x) = \frac{(x - 1.02)(x - 1.04)}{(-0.02)(-0.04)} \cdot 1.0000 + \frac{(x - 1)(x - 1.04)}{0.02(-0.02)} \cdot 0.9888 + \frac{(x - 1)(x - 1.02)}{0.04 \cdot 0.02} \cdot 0.9784 = x^2 - 2.580x + 2.580;$$ 0.9943, 0.9835

5. 0.8033 (error −0.0245), 0.4872 (error −0.0148); quadratic: 0.7839 (−0.0051), 0.4678 (0.0046)

7. $p_2(x) = 1.1640x - 0.3357x^2$; −0.5089 (error 0.1262), 0.4053 (−0.0226), 0.9053 (0.0186), 0.9911 (−0.0672)

9. $p_2(x) = -0.44304x^2 + 1.30896x - 0.023220$, $p_2(0.75) = 0.70929$ (5S-exact 0.71116)

11. $L_0 = -\frac{1}{6}(x - 1)(x - 2)(x - 3)$, $L_1 = \frac{1}{2}x(x - 2)(x - 3)$, $L_2 = -\frac{1}{2}x(x - 1)(x - 3)$, $L_3 = \frac{1}{6}x(x - 1)(x - 2)$; $p_3(x) = 1 + 0.039740x - 0.335187x^2 + 0.060645x^3$; $p_2(0.5) = 0.943654$, $p_3(1.5) = 0.510116$, $p_3(2.5) = -0.047991$

13. $2x^2 - 4x + 2$

15. $p_3(x) = 2.1972 + (x - 9) \cdot 0.1082 + (x - 9)(x - 9.5) \cdot 0.005235$

17. $r = -1.5$, $p_2(0.3) = 0.6039 + (-1.5) \cdot 0.1755 + \frac{1}{2}(-1.5)(-0.5) \cdot (-0.0302)$ $= 0.3293$

Problem Set 19.4, page 826

9. $[-1.39(x-5)^2 + 0.58(x-5)^3]'' = 0.004$ at $x = 5.8$ (due to roundoff; should be 0).

11. $1 - \frac{5}{4}x^2 + \frac{1}{4}x^4$

13. $1 - x^2$, $-2(x-1) - (x-1)^2 + 2(x-1)^3$, $-1 + 2(x-2) + 5(x-2)^2 - 6(x-2)^3$

15. $4 + x^2 - x^3$, $-8(x-2) - 5(x-2)^2 + 5(x-2)^3$, $4 + 32(x-4) + 25(x-4)^2 - 11(x-4)^3$

17. Use the fact that the third derivative of a cubic polynomial is constant, so that g''' is piecewise constant, hence constant throughout under the present assumption. Now integrate three times.

19. Curvature $f''/(1 + f'^2)^{3/2} \approx f''$ if $|f'|$ is small.

Problem Set 19.5, page 839

1. 0.747131, which is larger than 0.746824. Why?

3. 0.5, 0.375, 0.34375, 0.335 (exact)

5. $\epsilon_{0.5} \approx 0.03452$ ($\epsilon_{0.5} = 0.03307$), $\epsilon_{0.25} \approx 0.00829$ ($\epsilon_{0.25} = 0.00820$)

7. 0.693254 (6S-exact 0.693147)

9. 0.073930 (6S-exact 0.073928)

11. 0.785392 (6S-exact 0.785398)

13. $(0.785398126 - 0.785392156)/15 = 0.39792 \cdot 10^{-6}$

15. **(a)** $M_2 = 2$, $|KM_2| = 2/(12n^2) = 10^{-5}/2$, $n = 183$. **(b)** $f^{\text{iv}} = 24/x^5$, $M_4 = 24$, $|CM_4| = 24/(180 \cdot (2m)^4) = 10^{-5}/2$, $2m = 12.8$, hence 14.

17. 0.94614588, 0.94608693 (8S-exact 0.94608307)

19. 0.9460831 (7S-exact)

21. 0.9774586 (7S-exact 0.9774377)

23. Set $x = \frac{1}{2}(t+1)$, 0.2642411177 (10S-exact), $1 - 2/e$

25. $x = \frac{1}{2}(t+1)$, $dx = \frac{1}{2}dt$, 0.746824127 (9S-exact 0.746824133)

27. 0.08, 0.32, 0.176, 0.256 (exact)

29. $5(0.1040 - \frac{1}{2} \cdot 0.1760 + \frac{1}{3} \cdot 0.1344 - \frac{1}{4} \cdot 0.0384) = 0.256$

Chapter 19 Review Questions and Problems, page 841

17. 4.375, 4.50, 6.0, impossible

19. $44.885 \leqq s \leqq 44.995$

21. The same as that of $\tilde{a}$.

23. $x = 20 \pm \sqrt{398} = 20.00 \pm 19.95$, $x_1 = 39.95$, $x_2 = 0.05$, $x_2 = 2/39.95 = 0.05006$ (error less than 1 unit of the last digit)

25. $x = x^4 - 0.1$, -0.1, -0.999, -0.99900399

27. 0.824

29. $-x + x^3$, $2(x-1) + 3(x-1)^2 - (x-1)^3$

31. 0.26, $M_2 = 6$, $M_2^* = 0$, $-0.02 \leqq \epsilon \leqq 0$, 0.01

33. 0.90443, 0.90452 (5S-exact 0.90452)

35. (a) $(0.4^3 - 2 \cdot 0.2^3 + 0)/0.04 = 1.2$, (b) $(0.3^3 - 2 \cdot 0.2^3 + 0.1^3)/0.01 = 1.2$ (exact)

Problem Set 20.1, page 851

1. $x_1 = 7.3, \quad x_2 = -3.2$ **3.** No solution **5.** $x_1 = 2, \quad x_2 = 1$

7. $\begin{bmatrix} -3 & 6 & -9 & -46.725 \\ 0 & 9 & -13 & -51.223 \\ 0 & 0 & -2.88889 & -7.38689 \end{bmatrix}$
$x_1 = 3.908, \quad x_2 = -1.998, \quad x_3 = 2.557$

9. $\begin{bmatrix} 13 & -8 & 0 & 178.54 \\ 0 & 6 & 13 & 137.86 \\ 0 & 0 & -16 & -253.12 \end{bmatrix}$
$x_1 = 6.78, \quad x_2 = -11.3, \quad x_3 = 15.82$

11. $\begin{bmatrix} 3.4 & -6.12 & -2.72 & 0 \\ 0 & 0 & 4.32 & 0 \\ 0 & 0 & 0 & 0 \end{bmatrix}$
$x_1 = t_1$ arbitrary, $x_2 = (3.4/6.12)t_1, \quad x_3 = 0$

13. $\begin{bmatrix} 5 & 0 & 6 & -0.329193 \\ 0 & -4 & -3.6 & -2.143144 \\ 0 & 0 & 2.3 & -0.4 \end{bmatrix}$
$x_1 = 0.142856, \quad x_2 = 0.692307, \quad x_3 = -0.173912$

15. $\begin{bmatrix} -1 & -3.1 & 2.5 & 0 & -8.7 \\ 0 & 2.2 & 1.5 & -3.3 & -9.3 \\ 0 & 0 & -1.493182 & -0.825 & 1.03773 \\ 0 & 0 & 0 & 6.13826 & 12.2765 \end{bmatrix}$
$x_1 = 4.2, \quad x_2 = 0, \quad x_3 = -1.8, \quad x_4 = 2.0$

Problem Set 20.2, page 857

1. $\begin{bmatrix} 1 & 0 \\ 3 & 1 \end{bmatrix} \begin{bmatrix} 4 & 5 \\ 0 & -1 \end{bmatrix}, \quad \begin{matrix} x_1 = -4 \\ x_2 = 6 \end{matrix}$

3. $\begin{bmatrix} 1 & 0 & 0 \\ 2 & 1 & 0 \\ 2 & 5 & 1 \end{bmatrix} \begin{bmatrix} 5 & 4 & 1 \\ 0 & 1 & 2 \\ 0 & 0 & 3 \end{bmatrix}, \quad \begin{matrix} x_1 = 0.4 \\ x_2 = 0.8 \\ x_3 = 1.6 \end{matrix}$

5. $\begin{bmatrix} 1 & 0 & 0 \\ 6 & 1 & 0 \\ 3 & 9 & 1 \end{bmatrix} \begin{bmatrix} 3 & 9 & 6 \\ 0 & -6 & 3 \\ 0 & 0 & -3 \end{bmatrix}, \quad \begin{matrix} x_1 = -\frac{1}{15} \\ x_2 = \frac{4}{15} \\ x_3 = \frac{2}{5} \end{matrix}$

7. $\begin{bmatrix} 3 & 0 & 0 \\ 2 & 3 & 0 \\ 4 & 1 & 3 \end{bmatrix} \begin{bmatrix} 3 & 2 & 4 \\ 0 & 3 & 1 \\ 0 & 0 & 3 \end{bmatrix}$, $\begin{matrix} x_1 = 0.6 \\ x_2 = 1.2 \\ x_3 = 0.4 \end{matrix}$

9. $\begin{bmatrix} 0.1 & 0 & 0 \\ 0 & 0.4 & 0 \\ 0.3 & 0.2 & 0.1 \end{bmatrix} \begin{bmatrix} 0.1 & 0 & 0.3 \\ 0 & 0.4 & 0.2 \\ 0 & 0 & 0.1 \end{bmatrix}$, $\begin{matrix} x_1 = 2 \\ x_2 = -11 \\ x_3 = 4 \end{matrix}$

11. $\begin{bmatrix} 1 & 0 & 0 & 0 \\ -1 & 2 & 0 & 0 \\ 3 & -1 & 3 & 0 \\ 2 & 0 & -1 & 4 \end{bmatrix} \begin{bmatrix} 1 & -1 & 3 & 2 \\ 0 & 2 & -1 & 0 \\ 0 & 0 & 3 & -1 \\ 0 & 0 & 0 & 4 \end{bmatrix}$, $\begin{matrix} x_1 = 2 \\ x_2 = -3 \\ x_3 = 4 \\ x_4 = -1 \end{matrix}$

13. No, since $\mathbf{x}^{\mathsf{T}}(-\mathbf{A})\mathbf{x} = -\mathbf{x}^{\mathsf{T}}\mathbf{A}\mathbf{x} < 0$; yes; yes; no

15. $\begin{bmatrix} -3.5 & 1.25 \\ 3.0 & -1.0 \end{bmatrix}$

17. $\dfrac{1}{36}\begin{bmatrix} 584 & 104 & -66 \\ 104 & 20 & -12 \\ -66 & -12 & 9 \end{bmatrix}$

19. $\dfrac{1}{16}\begin{bmatrix} 21 & -6 & -14 & 6 \\ -6 & 36 & -12 & -4 \\ -14 & -12 & 20 & -4 \\ 6 & -4 & -4 & 4 \end{bmatrix}$

Problem Set 20.3, page 863

5. Exact 0.5, 0.5, 0.5

7. $x_1 = 2$, $x_2 = -4$, $x_3 = 8$

9. Exact 2, 1, 4

11. **(a)** $\mathbf{x}^{(3)\mathsf{T}} = [0.49983 \quad 0.50001 \quad 0.500017]$,
(b) $\mathbf{x}^{(3)\mathsf{T}} = [0.50333 \quad 0.49985 \quad 0.49968]$

13. 8, −16, 43, 86 steps; spectral radius 0.09, 0.35, 0.72, 0.85, approximately

15. $[1.99934 \quad 1.00043 \quad 3.99684]^{\mathsf{T}}$ (Jacobi, Step 5); $[2.00004 \quad 0.998059 \quad 4.00072]^{\mathsf{T}}$ (Gauss–Seidel)

19. $\sqrt{306} = 17.49$, 12, 12

Problem Set 20.4, page 871

1. 18, $\sqrt{110} = 10.49$, 8, $[0.125 \quad -0.375 \quad 1 \quad 0 \quad -0.75 \quad 0]$

3. 5.9, $\sqrt{13.81} = 3.716$, 3, $\frac{1}{3}[0.2 \quad 0.6 \quad -2.1 \quad 3.0]$

5. 5, $\sqrt{5}$, 1, $[1 \quad 1 \quad 1 \quad 1 \quad 1]$

7. $ab + bc + ca = 0$

9. $\kappa = 5 \cdot \frac{1}{2} = 2.5$ **11.** $\kappa = (5 + \sqrt{5})(1 + 1/\sqrt{5}) = 6 + 2\sqrt{5}$
13. $\kappa = 19 \cdot 13 = 247$; ill-conditioned
15. $\kappa = 20 \cdot 20 = 400$; ill-conditioned
17. $167 \leqq 21 \cdot 15 = 315$
19. $[-2 \quad 4]^{\mathsf{T}}$, $[-144.0 \quad 184.0]^{\mathsf{T}}$, $\kappa = 25{,}921$, extremely ill-conditioned
21. Small residual $[0.145 \quad 0.120]$, but large deviation of $\tilde{\mathbf{x}}$.
23. 27, 748, 28,375, 943,656, 29,070,279

Problem Set 20.5, page 875

1. $1.846 - 1.038x$ **3.** $1.48 + 0.09x$
5. $s = 90t - 675$, $v_{av} = 90$ km/hr **9.** $-11.36 + 5.45x - 0.589x^2$
11. $1.89 - 0.739x + 0.207x^2$
13. $2.552 + 16.23x$, $-4.114 + 13.73x + 2.500x^2$, $2.730 + 1.466x - 1.778x^2 + 2.852x^3$

Problem Set 20.7, page 884

1. 5, 0, 7; radii 6, 4, 6. Spectrum $\{-1, 4, 9\}$
3. Centers 0; radii 0.5, 0.7, 0.4. Skew-symmetric, hence $\lambda = i\mu$, $-0.7 \leqq \mu \leqq 0.7$.
5. 2, 3, 8; radii $1 + \sqrt{2}$, 1, $\sqrt{2}$; actually (4S) 1.163, 3.511, 8.326
7. $t_{11} = 100$, $t_{22} = t_{33} = 1$
9. They lie in the intervals with endpoints $a_{jj} \pm (n-1) \cdot 10^{-5}$. Why?
11. $\rho(\mathbf{A}) \leqq$ Row sum norm $\|\mathbf{A}\|_\infty = \max_j \sum_k |a_{jk}| = \max_j(|a_{jj}| + \text{Gerschgorin radius})$
13. $\sqrt{122} = 11.05$
15. $\sqrt{0.52} = 0.7211$
17. Show that $\mathbf{A}\bar{\mathbf{A}}^{\mathsf{T}} = \bar{\mathbf{A}}^{\mathsf{T}}\mathbf{A}$.
19. 0 lies in no Gerschgorin disk, by (3) with $>$; hence $\det \mathbf{A} = \lambda_1 \cdots \lambda_n \neq 0$.

Problem Set 20.8, page 887

1. $q = 10, 10.9908, 10.9999$; $|\epsilon| \leqq 3, 0.3028, 0.0275$
3. $q \pm \delta = 4 \pm 1.633$, 4.786 ± 0.619, 4.917 ± 0.398
5. Same answer as in Prob. 3, possibly except for small roundoff errors.
7. $q = 5.5, 5.5738, 5.6018$; $|\epsilon| \leqq 0.5, 0.3115, 0.1899$; eigenvalues (4S) 1.697, 3.382, 5.303, 5.618
9. $\mathbf{y} = \mathbf{A}\mathbf{x} = \lambda\mathbf{x}$, $\mathbf{y}^{\mathsf{T}}\mathbf{x} = \lambda\mathbf{x}^{\mathsf{T}}\mathbf{x}$, $\mathbf{y}^{\mathsf{T}}\mathbf{y} = \lambda^2\mathbf{x}^{\mathsf{T}}\mathbf{x}$,
$\epsilon^2 \leqq \mathbf{y}^{\mathsf{T}}\mathbf{y}/\mathbf{x}^{\mathsf{T}}\mathbf{x} - (\mathbf{y}^{\mathsf{T}}\mathbf{x}/\mathbf{x}^{\mathsf{T}}\mathbf{x})^2 = \lambda^2 - \lambda^2 = 0$
11. $q = 1, \cdots, -2.8993$ approximates -3 (0 of the given matrix), $|\epsilon| \leqq 1.633, \cdots, 0.7024$ (Step 8)

Problem Set 20.9, page 896

1. $$\begin{bmatrix} 0.98 & -0.4418 & 0 \\ -0.4418 & 0.8702 & 0.3718 \\ 0 & 0.3718 & 0.4898 \end{bmatrix}$$

3. $\begin{bmatrix} 7 & -3.6056 & 0 \\ -3.6056 & 13.462 & 3.6923 \\ 0 & 3.6923 & 3.5385 \end{bmatrix}$

5. $\begin{bmatrix} 3 & -67.59 & 0 & 0 \\ -67.59 & 143.5 & 45.35 & 0 \\ 0 & 45.35 & 23.34 & 3.126 \\ 0 & 0 & 3.126 & -33.87 \end{bmatrix}$

7. Eigenvalues 16, 6, 2

$$\begin{bmatrix} 11.2903 & -5.0173 & 0 \\ -5.0173 & 10.6144 & 0.7499 \\ 0 & 0.7499 & 2.0952 \end{bmatrix}, \begin{bmatrix} 14.9028 & -3.1265 & 0 \\ -3.1265 & 7.0883 & 0.1966 \\ 0 & 0.1966 & 2.0089 \end{bmatrix}, \begin{bmatrix} 15.8299 & -1.2932 & 0 \\ -1.2932 & 6.1692 & 0.0625 \\ 0 & 0.0625 & 2.0010 \end{bmatrix}$$

9. Eigenvalues (4S) 141.4, 68.64, −30.04

$$\begin{bmatrix} 141.1 & 4.926 & 0 \\ 4.926 & 68.97 & 0.8691 \\ 0 & 0.8691 & -30.03 \end{bmatrix}, \begin{bmatrix} 141.3 & 2.400 & 0 \\ 2.400 & 68.72 & 0.3797 \\ 0 & 0.3797 & -30.04 \end{bmatrix}, \begin{bmatrix} 141.4 & 1.166 & 0 \\ 1.166 & 68.66 & 0.1661 \\ 0 & 0.1661 & -30.04 \end{bmatrix}$$

Chapter 20 Review Questions and Problems, page 896

15. $[3.9 \quad 4.3 \quad 1.8]^\mathsf{T}$

17. $[-2 \quad 0 \quad 5]^\mathsf{T}$

19. $\begin{bmatrix} 0.28193 & -0.15904 & -0.00482 \\ -0.15904 & 0.12048 & -0.00241 \\ -0.00482 & -0.00241 & 0.01205 \end{bmatrix}$

21. $\begin{bmatrix} 5.750 \\ 3.600 \\ 0.838 \end{bmatrix}, \begin{bmatrix} 6.400 \\ 3.559 \\ 1.000 \end{bmatrix}, \begin{bmatrix} 6.390 \\ 3.600 \\ 0.997 \end{bmatrix}$
Exact: $[6.4 \quad 3.6 \quad 1.0]^\mathsf{T}$

23. $\begin{bmatrix} 1.700 \\ 1.180 \\ 4.043 \end{bmatrix}, \begin{bmatrix} 1.986 \\ 0.999 \\ 4.002 \end{bmatrix}, \begin{bmatrix} 2.000 \\ 1.000 \\ 4.000 \end{bmatrix}$
Exact: $[2 \quad 1 \quad 4]^\mathsf{T}$

25. 42, $\sqrt{674} = 25.96$, 21

27. 30

29. 5

31. $115 \cdot 0.4458 = 51.27$

33. $5 \cdot \frac{21}{63} = \frac{5}{3}$

35. $1.514 + 1.129x - 0.214x^2$

37. Centers 15, 35, 90; radii 30, 35, 25, respectively. Eigenvalues (3S) 2.63, 40.8, 96.6

39. Centers 0, −1, −4; radii 9, 6, 7, respectively; eigenvalues 0, 4.446, −9.446

Problem Set 21.1, page 910

1. $y = 5e^{-0.2x}$, 0.00458, 0.00830 (errors of y_5, y_{10})

3. $y = x - \tanh x$ (set $y - x = u$), 0.00929, 0.01885 (errors of y_5, y_{10})

5. $y = e^x$, 0.0013, 0.0042 (errors of y_5, y_{10})

7. $y = 1/(1 - x^2/2)$, 0.00029, 0.01187 (errors of y_5, y_{10})

9. Errors 0.03547 and 0.28715 of y_5 and y_{10} much larger

11. $y = 1/(1 - x^2/2)$; error -10^{-8}, $-4 \cdot 10^{-8}$, $\cdots$, $-6 \cdot 10^{-7}$, $+9 \cdot 10^{-6}$; $\epsilon = 0.0002/15 = 1.3 \cdot 10^{-5}$ (use RK with $h = 0.2$)

13. $y = \tan x$; error $0.83 \cdot 10^{-7}$, $0.16 \cdot 10^{-6}$, $\cdots$, $-0.56 \cdot 10^{-6}$, $+0.13 \cdot 10^{-5}$

15. $y = 3 \cos x - 2 \cos^2 x$; error $\cdot\ 10^7$: 0.18, 0.74, 1.73, 3.28, 5.59, 9.04, 14.3, 22.8, 36.8, 61.4

17. $y' = 1/(2 - x^4)$; error $\cdot\ 10^9$: 0.2, 3.1, 10.7, 23.2, 28.5, -32.3, -376, -1656, -3489, $+80444$

19. Errors for Euler–Cauchy 0.02002, 0.06286, 0.05074; for improved Euler–Cauchy -0.000455, 0.012086, 0.009601; for Runge–Kutta. 0.0000011, 0.000016, 0.000536

Problem Set 21.2, page 915

1. $y = e^x$, $y_5^* = 1.648717$, $y_5 = 1.648722$, $\epsilon_5 = -3.8 \cdot 10^{-8}$, $y_{10}^* = 2.718276$, $y_{10} = 2.718284$, $\epsilon_{10} = -1.8 \cdot 10^{-6}$

3. $y = \tan x$, $y_4, \cdots, y_{10}$ (error $\cdot\ 10^5$) 0.422798 (-0.49), 0.546315 (-1.2), 0.684161 (-2.4), 0.842332 (-4.4), 1.029714 (-7.5), 1.260288 (-13), 1.557626 (-22)

5. RK error smaller in absolute value, error $\cdot\ 10^5 = 0.4, 0.3, 0.2, 5.6$ (for $x = 0.4, 0.6, 0.8, 1.0$)

7. $y = 1/(4 + e^{-3x})$, $y_4, \cdots, y_{10}$ (error $\cdot\ 10^5$) 0.232490 (0.34), 0.236787 (0.44), 0.240075 (0.42), 0.242570 (0.35), 0.244453 (0.25), 0.245867 (0.16), 0.246926 (0.09)

9. $y = \exp(x^3) - 1$, $y_4, \cdots, y_{10}$ (error $\cdot\ 10^7$) 0.008032 (-4), 0.015749 (-10), 0.027370 (-17), 0.043810 (-26), 0.066096 (-39), 0.095411 (-54), 0.133156 (-74)

13. $y = \exp(x^2)$. Errors $\cdot\ 10^5$ from $x = 0.3$ to 0.7: -5, -11, -19, -31, -41

15. **(a)** 0, 0.02, 0.0884, 0.215848, $y_4 = 0.417818$, $y_5 = 0.708887$ (poor)
(b) By 30–50%

Problem Set 21.3, page 922

1. $y_1 = -e^{-2x} + 4e^x$, $y_2 = -e^{-2x} + e^x$; errors of y_1 (of y_2) from 0.002 to 0.5 (from -0.01 to 0.1), monotone

3. $y_1' = y_2$, $y_2' = -\frac{1}{4}y_1$, $y = y_1 = 1$, 0.99, 0.97, 0.94, 0.9005, error -0.005, -0.01, -0.015, -0.02, -0.0229; exact $y = \cos \frac{1}{2}x$

5. $y_1' = y_2$, $y_2' = y_1 + x$, $y_1(0) = 1$, $y_2(0) = -2$, $y = y_1 = e^{-x} - x$, $y = 0.8$ (error 0.005), 0.61 (0.01), 0.429 (0.012), 0.2561 (0.0142), 0.0905 (0.0160)

7. By about a factor 10^5. $\epsilon_n(y_1) \cdot 10^6 = -0.082, \cdots, -0.27$, $\epsilon_n(y_2) \cdot 10^6 = 0.08, \cdots, 0.27$

9. Errors of y_1 (of y_2) from $0.3 \cdot 10^{-5}$ to $1.3 \cdot 10^{-5}$ (from $0.3 \cdot 10^{-5}$ to $0.6 \cdot 10^{-5}$)

11. $(y_1, y_2) = (0, 1), (0.20, 0.98), (0.39, 0.92), \cdots, (-0.23, -0.97), (-0.42, -0.91), (-0.59), (-0.81)$; continuation will give an "ellipse."

Problem Set 21.4, page 930

3. $-3u_{11} + u_{12} = -200, \quad u_{11} - 3u_{12} = -100$

5. 105, 155, 105, 115; Step 5: 104.94, 154.97, 104.97, 114.98

7. 0, 0, 0, 0. All equipotential lines meet at the corners (why?). Step 5: 0.29298, 0.14649, 0.14649, 0.073245

9. 0.108253, 0.108253, 0.324760, 0.324760; Step 10: 0.108538, 0.108396, 0.324902, 0.324831

11. **(a)** $u_{11} = -u_{12} = -66$. **(b)** Reduce to 4 equations by symmetry. $u_{11} = u_{31} = -u_{15} = -u_{35} = -92.92$, $u_{21} = -u_{25} = -87.45$, $u_{12} = u_{32} = -u_{14} = -u_{34} = -64.22$, $u_{22} = -u_{24} = -53.98$, $u_{13} = u_{23} = u_{33} = 0$

13. $u_{12} = u_{32} = 31.25, \quad u_{21} = u_{23} = 18.75, \quad u_{jk} = 25$ at the others

15. $u_{21} = u_{23} = 0.25, \quad u_{12} = u_{32} = -0.25, \quad u_{jk} = 0$ otherwise

17. $\sqrt{3}$, $u_{11} = u_{21} = 0.0849$, $u_{12} = u_{22} = 0.3170$. (0.1083, 0.3248 are 4S-values of the solution of the linear system of the problem.)

Problem Set 21.5, page 935

5. $u_{11} = 0.766, \quad u_{21} = 1.109, \quad u_{12} = 1.957$, $u_{22} = 3.293$

7. **A**, as in Example 1, right sides $-220, -220, -220, -220$. Solution $u_{11} = u_{21} = 125.7$, $u_{21} = u_{22} = 157.1$

13. $-4u_{11} + u_{21} + u_{12} = -3$, $u_{11} - 4u_{21} + u_{22} = -12$, $u_{11} - 4u_{12} + u_{22} = 0$, $2u_{21} + 2u_{12} - 12u_{22} = -14$, $u_{11} = u_{22} = 2$, $u_{21} = 4$, $u_{12} = 1$. Here $-\frac{14}{3} = -\frac{4}{3}(1 + 2.5)$ with $\frac{4}{3}$ from the stencil.

15. $\mathbf{b} = [-200, -100, -100, 0]^\mathsf{T}$; $\quad u_{11} = 73.68$, $u_{21} = u_{12} = 47.37$, $u_{22} = 15.79$ (4S)

Problem Set 21.6, page 941

5. 0, 0.6625, 1.25, 1.7125, 2, 2.1, 2, 1.7125, 1.25, 0.6625, 0

7. Substantially less accurate, 0.15, 0.25 ($t = 0.04$), 0.100, 0.163 ($t = 0.08$)

9. Step 5 gives 0, 0.06279, 0.09336, 0.08364, 0.04707, 0.

11. Step 2: 0 (exact 0), 0.0453 (0.0422), 0.0672 (0.0658), 0.0671 (0.0628), 0.0394 (0.0373), 0 (0)

13. 0.3301, 0.5706, 0.4522, 0.2380 ($t = 0.04$), 0.06538, 0.10603, 0.10565, 0.6543 ($t = 0.20$)

15. 0.1018, 0.1673, 0.1673, 0.1018 ($t = 0.04$), 0.0219, 0.0355, $\cdots$ ($t = 0.20$)

Problem Set 21.7, page 944

1. $u(x, 1) = 0, -0.05, -0.10, -0.15, -0.20, 0$

3. For $x = 0.2, 0.4$ we obtain 0.24, 0.40 ($t = 0.2$), 0.08, 0.16 ($t = 0.4$), $-0.08, -0.16$ ($t = 0.6$), etc.

5. 0, 0.354, 0.766, 1.271, 1.679, 1.834, $\cdots$ ($t = 0.1$); 0, 0.575, 0.935, 1.135, 1.296, 1.357, $\cdots$ ($t = 0.2$)

7. 0.190, 0.308, 0.308, 0.190, (3S-exact: 0.178, 0.288, 0.288, 0.178)

Chapter 21 Review Questions and Problems, page 945

17. $y = e^x$, 0.038, 0.125 (errors of y_5 and y_{10})

19. $y = \tan x$; 0 (0), 0.10050 (−0.00017), 0.20304 (−0.00033), 0.30981 (−0.00048), 0.42341 (−0.00062), 0.54702 (−0.00072), 0.68490 (−0.00076), 0.84295 (−0.00066), 1.0299 (−0.0002), 1.2593 (0.0009), 1.5538 (0.0036)

21. 0.1003346 $(0.8 \cdot 10^{-7})$ 0.2027099 $(1.6 \cdot 10^{-7})$, 0.3093360 $(2.1 \cdot 10^{-7})$, 0.4227930 $(2.3 \cdot 10^{-7})$, 0.5463023 $(1.8 \cdot 10^{-7})$

23. $y = \sin x$, $y_{0.8} = 0.717366$, $y_{1.0} = 0.841496$ (errors $-1.0 \cdot 10^{-5}$, $-2.5 \cdot 10^{-5}$)

25. $y_1' = y_2$, $y_2' = x^2 y_1$, $y = y_1 = 1, 1, 1, 1.0001, 1.0006, 1.002$

27. $y_1' = y_2$, $y_2' = 2e^x - y_1$, $y = e^x - \cos x$, $y = y_1 = 0, 0.241, 0.571, \cdots$; errors between 10^{-6} and 10^{-5}

29. 3.93, 15.71, 58.93

31. 0, 0.04, 0.08, 0.12, 0.15, 0.16, 0.15, 0.12, 0.08, 0.04, 0 ($t = 0.3$. 3 time steps)

33. $u(P_{11}) = u(P_{31}) = 270$, $u(P_{21}) = u(P_{13}) = u(P_{23}) = u(P_{33}) = 30$, $u(P_{12}) = u(P_{32}) = 90$, $u(P_{22}) = 60$

35. 0.043330, 0.077321, 0.089952, 0.058488 ($t = 0.04$), 0.010956, 0.017720, 0.017747, 0.010964 ($t = 0.20$)

Problem Set 22.1, page 953

3. $f(\mathbf{x}) = 2(x_1 - 1)^2 + (x_2 + 2)^2 - 6$; Step 3: (1.037, −1.926), value −5.992

9. Step 5: (0.11247, −0.00012), value 0.000016

Problem Set 22.2, page 957

7. No

9. x_3, x_4 is the unused time on M_1, M_2, respectively.

11. $f(2.5, 2.5) = 100$

13. $f(-\frac{11}{3}, \frac{26}{3}) = 198\frac{1}{3}$

15. $f(9, 6) = 360$

17. $0.5x_1 + 0.75x_2 \leqq 45$ (copper), $0.5x_1 + 0.25x_2 \leqq 30, f = 120x_1 + 100x_2$, $f_{\max} = f(45, 30) = 8400$

19. $f = x_1 + x_2$, $2x_1 + 3x_2 \leqq 1200$, $4x_1 + 2x_2 \leqq 1600$, $f_{\max} = f(300, 200) = 500$

21. $x_1/3 + x_2/2 \leqq 100$, $x_1/3 + x_2/6 \leqq 80, f = 150x_1 + 100x_2, f_{\max} = f(210, 60) =$ 37,500

Problem Set 22.3, page 961

3. $f(120/11, 60/11) = 480/11$

5. Eliminate in Column 3, so that 20 goes. $f_{\min} = f(0, \frac{1}{2}) = -10$.

7. $f_{\max} = f(\frac{60}{21}, 0, \frac{1500}{105}, 0) = \frac{2200}{7}$

9. $f_{\max} = 6$ on the segment from (3, 0, 0) to (0, 0, 2)

11. We minimize! The augmented matrix is

$$\mathbf{T}_0 = \begin{bmatrix} 1 & 1.8 & 2.1 & 0 & 0 & 0 \\ 0 & 15 & 30 & 1 & 0 & 150 \\ 0 & 600 & 500 & 0 & 1 & 3900 \end{bmatrix}.$$

The pivot is 600. The calculation gives

$$\mathbf{T}_1 = \begin{bmatrix} 1 & 0 & \frac{6}{10} & 0 & -\frac{3}{1000} & -\frac{117}{10} \\ 0 & 0 & \frac{35}{2} & 1 & -\frac{1}{40} & \frac{105}{2} \\ 0 & 600 & 500 & 0 & 1 & 3900 \end{bmatrix} \quad \begin{matrix} \text{Row 1} - \frac{1.8}{600}\text{ Row 3} \\ \text{Row 2} - \frac{15}{600}\text{ Row 3} \\ \text{Row 3} \end{matrix}$$

The next pivot is $\frac{35}{2}$. The calculation gives

$$\mathbf{T}_2 = \begin{bmatrix} 1 & 0 & 0 & -\frac{6}{175} & -\frac{3}{1400} & -\frac{27}{2} \\ 0 & 0 & \frac{35}{2} & 1 & -\frac{1}{40} & \frac{105}{2} \\ 0 & 600 & 0 & -\frac{200}{7} & \frac{12}{7} & 2400 \end{bmatrix} \quad \begin{matrix} \text{Row 1} - \frac{1.2}{35}\text{ Row 2} \\ \text{Row 2} \\ \text{Row 3} - \frac{1000}{35}\text{ Row 2} \end{matrix}$$

Hence $-f$ has the maximum value -13.5, so that f has the minimum value 13.5, at the point

$$(x_1, x_2) = \left(\frac{2400}{600}, \frac{105/2}{35/2}\right) = (4, 3).$$

13. $f_{\max} = f(5, 4, 6) = 478$

Problem Set 22.4, page 968

1. $f(6, 3) = 84$
3. $f(20, 20) = 40$
5. $f(10, 5) = 5500$
7. $f(1, 1, 0) = 13$
9. $f(4, 0, \frac{1}{2}) = 9$

Chapter 22 Review Questions and Problems, page 968

9. Step 5: $[0.353 \quad -0.028]^{\mathsf{T}}$. Slower. Why?
11. Of course! Step 5: $[-1.003 \quad 1.897]^{\mathsf{T}}$
17. $f(2, 4) = 100$
19. $f(3, 6) = -54$

Problem Set 23.1, page 974

9. $\begin{bmatrix} 0 & 1 & 0 \\ 0 & 0 & 1 \\ 1 & 0 & 0 \end{bmatrix}$

11. $\begin{bmatrix} 0 & 1 & 1 & 1 \\ 0 & 0 & 0 & 0 \\ 1 & 0 & 0 & 0 \\ 0 & 0 & 0 & 0 \end{bmatrix}$

13. $\begin{bmatrix} 0 & 1 & 1 \\ 0 & 0 & 1 \\ 1 & 1 & 0 \end{bmatrix}$

15.
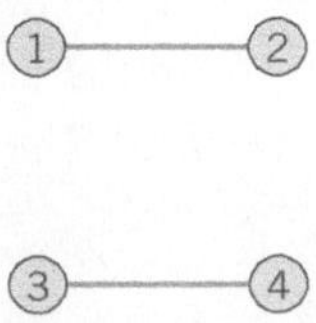

17. If G is complete.

19.

Edge

		e_1	e_2	e_3	e_4
Vertex	1	-1	-1	1	-1
	2	1	0	0	0
	3	0	1	-1	0
	4	0	0	0	1

Problem Set 23.2, page 979

1. 5   **3.** 4

5. The idea is to go backward. There is a v_{k-1} adjacent to v_k and labeled $k-1$, etc. Now the only vertex labeled 0 is s. Hence $\lambda(v_0) = 0$ implies $v_0 = s$, so that $v_0 - v_1 - \cdots - v_{k-1} - v_k$ is a path $s \rightarrow v_k$ that has length k.

15. Delete the edge $(2, 4)$.

17. No

Problem Set 23.3, page 983

1. $(1, 2), (2, 4), (4, 3);$ $L_2 = 12, L_3 = 36, L_4 = 28$

5. $(1, 2), (2, 4), (3, 4), (3, 5);$ $L_2 = 2, L_3 = 4, L_4 = 3, L_5 = 6$

7. $(1, 2), (2, 4), (3, 4);$ $L_2 = 10, L_3 = 15, L_4 = 13$

9. $(1, 5), (2, 3), (2, 6), (3, 4), (3, 5);$ $L_2 = 9, L_3 = 7, L_4 = 8, L_5 = 4, L_6 = 14$

Problem Set 23.4, page 987

1.
$$\begin{array}{c} 2 \\ \diagdown \\ 1 \end{array} 4 - 3 - 5 \quad L = 10$$

3. $5 - 3 - 6 \begin{array}{c} 1 \\ \diagup \\ \diagdown \\ 2 - 4 \end{array} \quad L = 17$

5. $1 \begin{array}{c} 2 \\ \diagup \\ \diagdown \\ 4 \diagup 3 \\ \diagdown \\ 5 \end{array} \quad L = 12$

9. Yes

11. $1 - 3 - 4 \begin{array}{c} 2 \\ \diagup \\ \diagdown \\ 5 - 6 \end{array} \quad L = 38$

13. New York–Washington–Chicago–Dalles–Denver–Los Angeles

15. G is connected. If G were not a tree, it would have a cycle, but this cycle would provide two paths between any pair of its vertices, contradicting the uniqueness.

19. If we add an edge (u, v) to T, then since T is connected, there is a path $u \to v$ in T which, together with (u, v), forms a cycle.

Problem Set 23.5, page 990

1. If G is a tree.
3. A shortest spanning tree of the largest connected graph that contains vertex 1.
7. (1, 4), (1, 3), (1, 2), (2, 6), (3, 5); $L = 32$
9. (1, 4), (4, 3), (4, 2), (3, 5); $L = 20$
11. (1, 4), (4, 3), (4, 5), (1, 2); $L = 12$

Problem Set 23.6, page 997

1. $\{3, 6\}$, $11 + 3 = 14$
3. $\{4, 5, 6\}$, $10 + 5 + 13 = 28$
5. $\{3, 6, 7\}$, $8 + 4 + 4 = 16$
7. $S = \{1, 4\}$, $8 + 6 = 14$
9. One is interested in flows *from s to t,* not in the opposite direction.
13. $\Delta_{12} = 5, \Delta_{24} = 8, \Delta_{45} = 2$; $\Delta_{12} = 5, \Delta_{25} = 3$; $\Delta_{13} = 4, \Delta_{35} = 9$
P_1: $1 - 2 - 4 - 5, \Delta f = 2$; P_2: $1 - 2 - 5, \Delta f = 3$; P_3: $1 - 3 - 5, \Delta f = 4$
15. $1 - 2 - 5, \Delta f = 2$; $1 - 4 - 2 - 5, \Delta f = 2$, etc.
17. $f_{13} = f_{35} = 8$, $f_{14} = f_{45} = 5$, $f_{12} = f_{24} = f_{46} = 4$, $f_{56} = 13$, $f = 4 + 13 = 17$, $f = 17$ is unique.
19. For instance, $f_{12} = 10$, $f_{24} = f_{45} = 7$, $f_{13} = f_{25} = 5$, $f_{35} = 3$, $f_{32} = 2$, $f = 3 + 5 + 7 = 15$, $f = 15$ is unique.

Problem Set 23.7, page 1000

3. (2, 3) and (5, 6)
5. By considering only edges with one labeled end and one unlabeled end
7. $1 - 2 - 5, \Delta_t = 2$; $1 - 4 - 2 - 5, \Delta_t = 1$; $f = 6 + 2 + 1 = 9$, where 6 is the given flow
9. $1 - 2 - 4 - 6, \Delta_t = 2$; $1 - 3 - 5 - 6, \Delta_t = 1$; $f = 4 + 2 + 1 = 7$, where 4 is the given flow
15. $S = \{1, 2, 4, 5\}$, $T = \{3, 6\}$, $\text{cap}(S, T) = 14$

Problem Set 23.8, page 1005

1. No
3. No
5. Yes, $S = \{1, 4, 5, 8\}$
7. Yes, $S = \{1, 3, 5\}$
11. $1 - 2 - 3 - 7 - 5 - 4$
13. $1 - 2 - 3 - 7 - 5 - 4$ is augmenting and gives $1 - 2 - 3 - 7 - 5 - 4$ and (1, 2), (3, 7), (5, 4) is of maximum cardinality.
15. $1 - 4 - 3 - 6 - 7 - 8$ is augmenting and gives $1 - 4 - 3 - 6 - 7 - 8$ and (1, 4), (3, 6), (7, 8) is of maximum cardinality.
19. 3
21. 2
23. 3
25. K_4

Chapter 23 Review Questions and Problems, page 1006

11. $\begin{bmatrix} 0 & 0 & 1 & 1 \\ 0 & 0 & 1 & 1 \\ 1 & 1 & 0 & 0 \\ 1 & 1 & 0 & 0 \end{bmatrix}$

13. To vertex 1 2 3 4, From vertex 1–4:

$\begin{bmatrix} 0 & 1 & 0 & 1 \\ 1 & 0 & 1 & 0 \\ 0 & 1 & 0 & 1 \\ 1 & 0 & 1 & 0 \end{bmatrix}$

15.

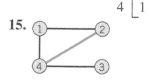

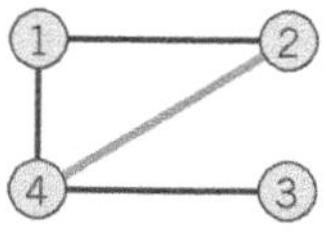

17.

Vertex	Incident Edges
1	(1, 2), (1, 4)
2	(2, 1), (2, 4)
3	(3, 4)
4	(4, 1), (4, 2), (4, 3)

19. (1, 2), (1, 4), (2, 3); $L_2 = 2, L_3 = 5, L_4 = 5$

23. (1, 6), (4, 5), (2, 3), (7, 8)

Problem Set 24.1, page 1015

1. $q_L = 19, q_M = 20, q_U = 20.5$ **3.** $q_L = 138, q_M = 144, q_U = 154$
5. $q_L = 199, q_M = 201, q_U = 201$ **7.** $q_L = 1.3, q_M = 1.4, q_U = 1.45$
9. $q_L = 89.9, q_M = 91.0, q_U = 91.8$ **11.** $\bar{x} = 19.875, s = 0.835$, IQR $= 1.5$
13. $\bar{x} = 144.67, s = 8.9735$, IQR $= 16$ **15.** $\bar{x} = 1.355, s = 0.136$, IQR $= 0.15$
17. 3.54, 1.29

Problem Set 24.2, page 1017

1. 2^3 outcomes: *RRR, RRL, RLR, LRR, RLL, LRL, LLR, LLL*
3. $6^2 = 36$ outcomes (1, 1), (1, 2), $\cdots$, (6, 6), first number (second number) referring to the first die (second die)
5. Infinitely many outcomes *H TH TTH TTTH* $\cdots$ (*H = Head, T = Tail*)
7. The space of ordered pairs of numbers
9. 10 outcomes: *D ND NND* $\cdots$ *NNNNNNNNND*
11. Yes
17. $A \cup B = B$ implies $A \subseteq B$ by the definition of union. Conversely. $A \subseteq B$ implies that $A \cup B = B$ because always $B \subseteq A \cup B$, and if $A \subseteq B$, we must have equality in the previous relation.

Problem Set 24.3, page 1024

1. $1 - 4/216 = 98.15\%$, by Theorem 1
3. (a) $0.9^3 = 72.9\%$, **(b)** $\frac{90}{100} \cdot \frac{89}{99} \cdot \frac{88}{98} = 72.65\%$
5. $\frac{8}{9}$
7. Small sample from a large population containing *many* items in each class we are interested in (defectives and nondefectives, etc.)
9. $\frac{498}{500} \cdot \frac{497}{499} \cdot \frac{496}{498} \cdot \frac{495}{497} \cdot \frac{494}{496} \approx 0.98008$
11. (a) $\frac{100}{200} \cdot \frac{99}{199} = 24.874\%$, **(b)** $\frac{100}{200} \cdot \frac{100}{199} + \frac{100}{200} \cdot \frac{100}{199} = 50.25\%$, **(c)** same as (a). (a) + (b) + (c) = 1. Why?
13. $1 - 0.96^3 = 11.5\%$
15. $1 - 0.875^4 = 0.4138 < 1 - 0.75^2 = 0.4375 < 0.5 \quad (c < b < a)$
17. $A = B \cup (A \cap B^c)$, hence $P(A) = P(B) + P(A \cap B^c) \geqq P(B)$ by disjointedness of B and $A \cap B^c$

Problem Set 24.4, page 1028

1. In $10! = 3{,}628{,}800$ ways
3. $\frac{2}{6} \cdot \frac{1}{5} \cdot \frac{4}{4} \cdot \frac{3}{3} \cdot \frac{2}{2} \cdot \frac{1}{1} = \frac{4}{6} \cdot \frac{3}{5} \cdot \frac{2}{4} \cdot \frac{1}{3} \cdot \frac{2}{2} \cdot \frac{1}{1} = \frac{4!2!}{6!} = \frac{2}{6} \cdot \frac{1}{5} = \frac{1}{15}$
5. $\binom{10}{3}\binom{5}{2}\binom{6}{2} = 18{,}000$
7. 210, 70, 112, 28
9. In $6!/6 = 120$ ways
11. $9 \cdot 8 = 72$
13. (b) $1/(12n)$
15. P (*No two people have a birthday in common*) $= 365 \cdot 364 \cdots 346/365^{20} = 0.59$. *Answer:* 41%, which is surprisingly large.

Problem Set 24.5, page 1034

1. $k = \frac{1}{55}$ by (6)
3. $k = \frac{1}{4}$ by (10), $P(0 \leqq X \leqq 2) = \frac{1}{2}$
5. No, because of (6)
7. $k = \frac{1}{100}$ because of (6) and $1 + 8 + 27 + 64 = 100$
9. $k = 5$; 50%
11. $0.5^3 = 12.5\%$
13. $F(x) = 0$ if $x < -1$, $F(x) = \frac{1}{2}(x + 1)^2$ if $-1 \leqq x < 0$
$F(x) = 1 - \frac{1}{2}(x - 1)^2$ if $0 \leqq x < 1$, $F(x) = 1$ if $x \leqq 1$
Answer: 500 cans, $P = 0.125, 0$
15. $X > b, X \geqq b, X < c, X \leqq c$, etc.

Problem Set 24.6, page 1038

1. $k = \frac{1}{2}, \mu = \frac{4}{3}, \sigma^2 = \frac{2}{9}$
3. $\mu = \pi, \sigma^2 = \pi^2/3$; cf. Example 2
5. $\mu = \frac{1}{4}, \sigma^2 = \frac{1}{16}$
7. $C = \frac{1}{2}, \mu = 2, \sigma^2 = 4$
9. 750, 1, 0.002
11. $c = 0.073$
13. \$643.50
15. $\frac{1}{2}, \frac{1}{20}, (X - \frac{1}{2})\sqrt{20}$
17. $X =$ *Product of the 2 numbers.* $E(X) = 12.25$, 12 cents
19. $(0 + 1 \cdot 3 + 3 \cdot 8 + 1 \cdot 27)/8 = 54/8 = 6 \cdot 75$

Problem Set 24.7, page 1044

3. 38%
5. $\binom{5}{x}$ 0.5^5, 0.03125, 0.15625, $1 - f(0) = 0.96875$, 0.96875
7. 0.265
9. $f(x) = 0.5^x e^{-0.5}/x!$, $f(0) + f(1) = e^{-0.5}(1.0 + 0.5) = 0.91$. *Answer:* 9%
11. $13\frac{1}{4}$ %
13. 42%, 47.2%, 10.5%, 0.3%
15. $1 - e^{-0.2} = 18\%$

Problem Set 24.8, page 1050

1. 0.1587, 0.5, 0.6915, 0.6247
3. 45.065, 56.978, 2.022
5. 15.9%
7. 31.1%, 95.4%
9. About 58%
11. $t = 1084$ hours
13. About 683 (Fig. 521a)

Problem Set 24.9, page 1059

1. $\frac{1}{8}, \frac{3}{16}, \frac{3}{8}$
3. $\frac{2}{9}, \frac{1}{9}, \frac{1}{2}$
5. $f_2(y) = 1/(\beta_2 - \alpha_2)$ if $\alpha_2 < y < \beta_2$
7. 27.45 mm, 0.38 mm
11. 25.26 cm, 0.0078 cm
13. 50%
15. The distributions in Prob. 17 and Example 1
17. No

Chapter 24 Review Questions and Problems, page 1060

11. $Q_L = 110$, $Q_M = 112$, $Q_U = 115$
13. $\bar{x} = 111.9$, $s = 4.0125$, $s^2 = 16.1$
21. $x_{\min} \leqq x_j \leqq x_{\max}$. Sum over j from 1.
17. $\bar{x} = 6$, $s = 3.65$
19. $f(x) = \binom{50}{x} 0.03^x 0.97^{50-x} \approx 1.5^x e^{-1.5}/x!$
21. $f(x) = 2^{-x}$, $x = 1, 2, \cdots$
23. $1, \frac{1}{2}$
25. 0.1587, 0.6306, 0.5, 0.4950

Problem Set 25.2, page 1067

1. In Example 1, $\mu = 0$ so $\sum_{j=1}^{n} x_j = 0$. $\partial \ln \ell/\partial \ell = 0$ and $\tilde{\sigma}^2$ is as before.
3. $\ell = e^{-n\mu}\mu^{(x_1+\cdots+x_n)}/(x_1!\cdots x_n!)$, $\partial \ln \ell/\partial\mu = -n + (x_1 + \cdots + x_n)/\mu = 0$, $n\hat{\mu} = n\bar{x}$, $\hat{\mu} = \bar{x} = 15.3$
5. $l = p^k(1 - p)^{n-k}$, $\hat{p} = k/n$, k = number of successes in n trails
7. 7/12
9. $l = f = p(1 - p)^{x-1}$, etc., $\hat{p} = 1/x$
11. $\hat{\theta} = n/\Sigma x_j = 1/\bar{x}$
13. $\hat{\theta} = 1$
15. Variability larger than perhaps expected

Problem Set 25.3, page 1077

3. Shorter by a factor $\sqrt{2}$ **5.** 4, 16

7. $c = 1.96, \bar{x} = 126, s^2 = 126 \cdot 674/800 = 106.155, k = cs/\sqrt{n} = 0.714$, $\text{CONF}_{0.95}\{125.3 \leqq \mu \leqq 126.7\}$, $\text{CONF}_{0.95}\{0.1566 \leqq p \leqq 0.1583\}$

9. $\text{CONF}_{0.99}\{63.72 \leqq \mu \leqq 66.28\}$

11. $n - 1 = 5, F(c) = 0.995, c = 4.03, \bar{x} = 9533.33, s^2 = 49{,}666.67$, $k = 366.66$ (Table 25.2), $\text{CONF}_{0.99}\{9166.7 \leqq \mu \leqq 9900\}$

13. $\text{CONF}_{0.95}\{0.023 \leqq \sigma^2 \leqq 0.085\}$

15. $n - 1 = 99$ degrees of freedom. $F(c_1) = 0.025$, $c_1 = 74.2$, $F(c_2) = 0.975$, $c_2 = 129.6$. Hence $k_1 = 12.41$, $k_2 = 7.10$. $\text{CONF}_{0.95}\{7.10 \leqq \sigma^2 \leqq 12.41\}$.

17. $\text{CONF}_{0.95}\{0.74 \leqq \sigma^2 \leqq 5.19\}$

19. $Z = X + Y$ is normal with mean 105 and variance 1.25. *Answer:* $P(104 \leqq Z \leqq 106) = 63\%$

Problem Set 25.4, page 1086

3. $t = (0.286 - 0)/(4.31/\sqrt{7}) = 0.18 < c = 1.94$; accept the hypothesis.

5. $c = 6090 > 6019$: do not reject the hypothesis.

7. $\sigma^2/n = 1.8$, $c = 57.8$, accept the hypothesis.

9. $\mu < 58.69$ or $\mu > 61.31$

11. Alternative $\mu \neq 5000$, $t = (4990 - 5000)/(20/\sqrt{50}) = -3.54 < c = -2.01$ (Table A9, Appendix 5). Reject the hypothesis $\mu = 5000$ g.

13. Two-sided. $t = (0.55 - 0)/\sqrt{0.546/8} = 2.11 < c = 2.37$ (Table A9, Appendix 5), no difference

15. $19 \cdot 1.0^2/0.8^2 = 29.69 < c = 30.14$ (Table A10. Appendix 5), accept the hypothesis

17. By (12), $t_0 = \sqrt{16}(20.2 - 19.6)/\sqrt{0.16 + 0.36} > c = 1.70$. Assert that B is better.

Problem Set 25.5, page 1091

1. $\text{LCL} = 1 - 2.58 \cdot 0.02/2 = 0.974$, $\text{UCL} = 1.026$

3. 27

5. Choose 4 times the original sample size

9. $2.58\sqrt{0.0004}/\sqrt{2} = 0.036$, $\text{LCL} = 3.464$, $\text{UCL} = 3.536$

11. $\text{LCL} = np - 3\sqrt{np(1-p)}$, $\text{CL} = np$, $\text{UCL} = np + 3\sqrt{np(1-p)}$

13. In about 30% (5%) of the cases

15. $\text{LCL} = \mu - 3\sqrt{\mu}$ is negative in (b) and we set $\text{LCL} = 0$, $\text{CL} = \mu = 3.6$, $\text{UCL} = \mu + 3\sqrt{\mu} = 9.3$.

Problem Set 25.6, page 1095

1. 0.9825, 0.9384, 0.4060

3. 0.8187, 0.6703, 0.1353

5. $e^{-25\theta}(1 + 25\theta)$, $P(A; 1.5) = 94.5$, $\alpha = 5.5\%$

7. 19.5%, 14.7%

9. $(1 - \theta)^n + n\theta(1 - \theta)^{n-1}$

11. $(1 - \frac{1}{2})^3 + 3 \cdot \frac{1}{2}(1 - \frac{1}{2})^2 = \frac{1}{2}$

13. $\sum_{x=0}^{9} \binom{100}{x} 0.12^x\, 0.88^{100-x} = 22\%$ (by the normal approximation)

15. $(1 - \theta)^5$, $[\theta(1 - \theta)^{5-1}]' = 0$, $\theta = \frac{1}{5}$, $\text{AOQL} = 6.7\%$

Problem Set 25.7, page 1099

3. $\chi_0^2 = (40 - 50)^2/50 + (60 - 50)^2/50 = 4 > c = 3.84$; no
5. $\chi_0^2 = \frac{16}{10} > 11.07$; yes
7. $\chi_0^2 = 10.264 < 11.07$; yes
9. 42 even digits, accept.
13. $\chi_0^2 = \dfrac{(355 - 358.5)^2}{358.5} + \dfrac{(123 - 119.5)^2}{119.5} = 0.137 < c = 3.84$ (1 degree of freedom, 95%)
15. Combining the last three nonzero values, we have $K - r - 1 = 9$ ($r = 1$ since we estimated the mean, $\frac{10{,}094}{2608} \approx 3.87$). $\chi_0^2 = 12.8 < c = 16.92$. Accept the hypothesis.

Problem Set 25.8, page 1102

3. $(\frac{1}{2})^8 + 8 \cdot (\frac{1}{2})^8 = 3.5\%$ is the probability that 7 cases in 8 trials favor A under the hypothesis that A and B are equally good. Reject.
5. $(\frac{1}{2})^{18}(1 + 18 + 153 + 816) = 0.0038$
7. $\bar{x} = 9.67$, $s = 11.87$. $t_0 = 9.67/(11.87/\sqrt{15}) = 3.16 > c = 1.76$ ($\alpha = 5\%$). Hypothesis rejected.
9. Hypothesis $\tilde{\mu} = 0$. Alternative $\tilde{\mu} > 0$, $\bar{x} = 1.58$, $t = \sqrt{10} \cdot 1.58/1.23 = 4.06 > c = 1.83$ ($\alpha = 5\%$). Hypothesis rejected.
11. Consider $y_j = x_j - \tilde{\mu}_0$.
13. $n = 8$; 4 transpositions, $P(T \leqq 4) = 0.007$. Assert that fertilizing increases yield.
15. $P(T \leqq 2) = 2.8\%$. Assert that there is an increase.

Problem Set 25.9, page 1111

1. $y = 0.98 + 0.495x$
3. $y = -11{,}457.9 + 43.2x$
5. $y = -10 + 0.55x$
7. $y = 0.5932 + 0.1138x$, $R = 1/0.1138$
9. $y = 0.32923 + 0.00032x$, $y(66) = 0.35035$
13. $c = 3.18$ (Table A9), $k_1 = 43.2$, $q_0 = 54{,}878$, $K = 1.502$, $\mathrm{CONF}_{0.95}\{41.7 \leqq \kappa_1 \leqq 44.7\}$.
15. $y - 1.875 = 0.067(x - 25)$, $3s_x^2 = 500$, $q_0 = 0.023$, $K = 0.021$, $\mathrm{CONF}_{0.95}\{0.046 \leqq \kappa_1 \leqq 0.088\}$

Chapter 25 Review Questions and Problems, page 1111

15. $\hat{\mu} = 20.325$, $\hat{\sigma}^2 = (\frac{7}{8})s^2 = 3.982$
17. $\mathrm{CONF}_{0.99}\{27.94 \leqq \mu \leqq 34.81\}$
19. $c = 14.74 > 14.5$, reject μ_0; $\Phi((14.74 - 14.50)/\sqrt{0.025}) = 0.9353$
21. $2.58 \cdot \sqrt{0.00024}/\sqrt{2} = 0.028$, LCL $= 2.722$, UCL $= 2.778$
23. $\alpha = 1 - (1 - \theta)^6 = 5.85\%$, when $\theta = 0.01$. For $\theta = 15\%$ we obtain $\beta = (1 - \theta)^6 = 37.7\%$. If n increases, so does α, whereas β decreases.
25. $y = 3.4 - 1.85x$

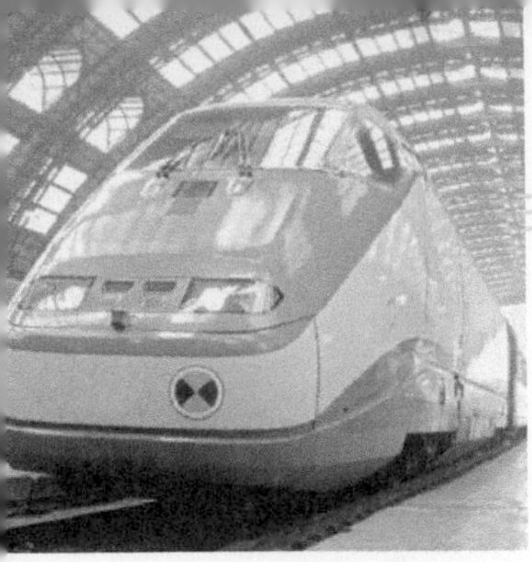

APPENDIX 3

Auxiliary Material

A3.1 Formulas for Special Functions

For tables of numeric values, see Appendix 5.

Exponential function e^x (Fig. 545)

$$e = 2.71828\ 18284\ 59045\ 23536\ 02874\ 71353$$

(1) $$e^x e^y = e^{x+y}, \qquad e^x/e^y = e^{x-y}, \qquad (e^x)^y = e^{xy}$$

Natural logarithm (Fig. 546)

(2) $$\ln(xy) = \ln x + \ln y, \qquad \ln(x/y) = \ln x - \ln y, \qquad \ln(x^a) = a \ln x$$

$\ln x$ is the inverse of e^x, and $e^{\ln x} = x$, $e^{-\ln x} = e^{\ln(1/x)} = 1/x$.

Logarithm of base ten $\log_{10} x$ or simply $\log x$

(3) $$\log x = M \ln x, \qquad M = \log e = 0.43429\ 44819\ 03251\ 82765\ 11289\ 18917$$

(4) $$\ln x = \frac{1}{M} \log x, \qquad \frac{1}{M} = \ln 10 = 2.30258\ 50929\ 94045\ 68401\ 79914\ 54684$$

$\log x$ is the inverse of 10^x, and $10^{\log x} = x$, $10^{-\log x} = 1/x$.

Sine and cosine functions (Figs. 547, 548). In calculus, angles are measured in radians, so that $\sin x$ and $\cos x$ have period 2π.

$\sin x$ is odd, $\sin(-x) = -\sin x$, and $\cos x$ is even, $\cos(-x) = \cos x$.

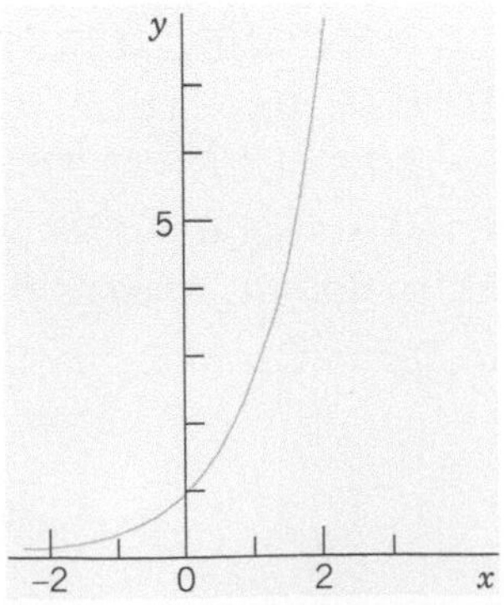

Fig. 545. Exponential function e^x

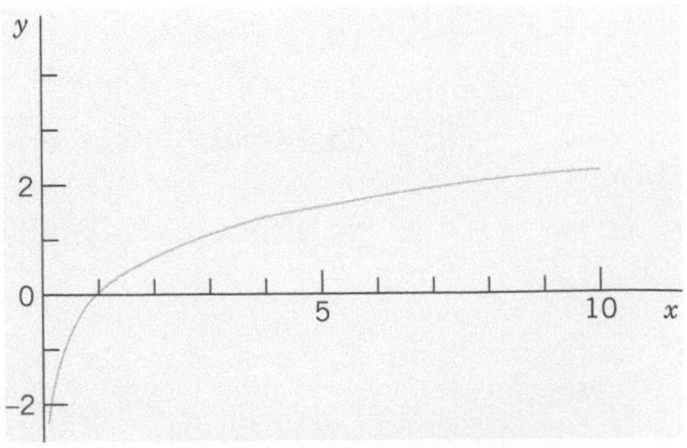

Fig. 546. Natural logarithm $\ln x$

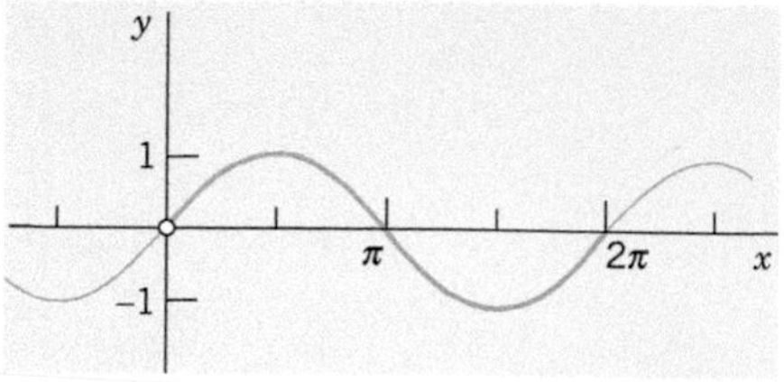

Fig. 547. sin x

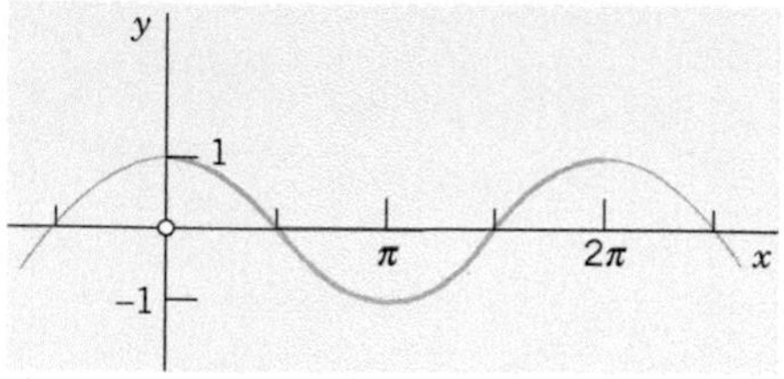

Fig. 548. cos x

$$1° = 0.01745\ 32925\ 19943 \text{ radian}$$

$$1 \text{ radian} = 57°\ 17'\ 44.80625''$$

$$= 57.29577\ 95131°$$

(5) $$\sin^2 x + \cos^2 x = 1$$

(6) $$\begin{cases} \sin(x+y) = \sin x \cos y + \cos x \sin y \\ \sin(x-y) = \sin x \cos y - \cos x \sin y \\ \cos(x+y) = \cos x \cos y - \sin x \sin y \\ \cos(x-y) = \cos x \cos y + \sin x \sin y \end{cases}$$

(7) $$\sin 2x = 2 \sin x \cos x, \qquad \cos 2x = \cos^2 x - \sin^2 x$$

(8) $$\begin{cases} \sin x = \cos\left(x - \frac{\pi}{2}\right) = \cos\left(\frac{\pi}{2} - x\right) \\ \cos x = \sin\left(x + \frac{\pi}{2}\right) = \sin\left(\frac{\pi}{2} - x\right) \end{cases}$$

(9) $$\sin(\pi - x) = \sin x, \qquad \cos(\pi - x) = -\cos x$$

(10) $$\cos^2 x = \tfrac{1}{2}(1 + \cos 2x), \qquad \sin^2 x = \tfrac{1}{2}(1 - \cos 2x)$$

(11) $$\begin{cases} \sin x \sin y = \tfrac{1}{2}[-\cos(x+y) + \cos(x-y)] \\ \cos x \cos y = \tfrac{1}{2}[\cos(x+y) + \cos(x-y)] \\ \sin x \cos y = \tfrac{1}{2}[\sin(x+y) + \sin(x-y)] \end{cases}$$

(12) $$\begin{cases} \sin u + \sin v = 2 \sin \frac{u+v}{2} \cos \frac{u-v}{2} \\ \cos u + \cos v = 2 \cos \frac{u+v}{2} \cos \frac{u-v}{2} \\ \cos v - \cos u = 2 \sin \frac{u+v}{2} \sin \frac{u-v}{2} \end{cases}$$

(13) $$A \cos x + B \sin x = \sqrt{A^2 + B^2} \cos(x \pm \delta), \qquad \tan \delta = \frac{\sin \delta}{\cos \delta} = \mp \frac{B}{A}$$

(14) $$A \cos x + B \sin x = \sqrt{A^2 + B^2} \sin(x \pm \delta), \qquad \tan \delta = \frac{\sin \delta}{\cos \delta} = \pm \frac{A}{B}$$

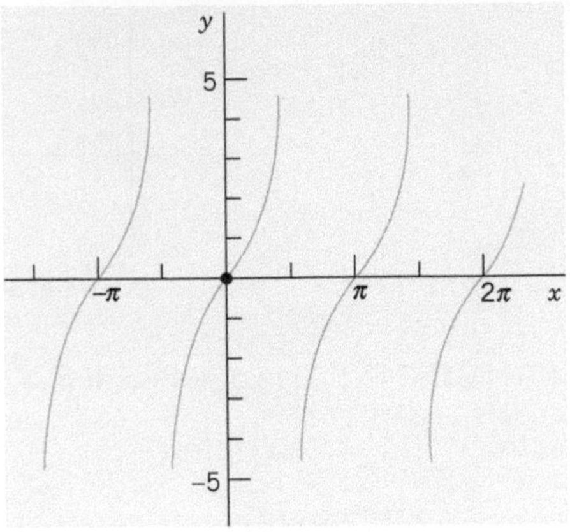

Fig. 549. tan x

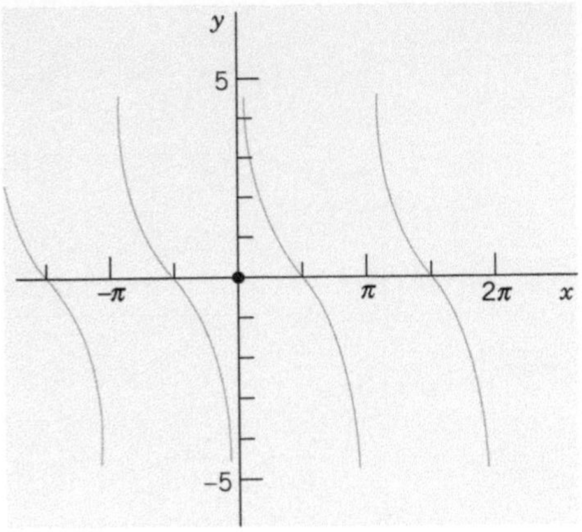

Fig. 550. cot x

Tangent, cotangent, secant, cosecant (Figs. 549, 550)

$$(15) \quad \tan x = \frac{\sin x}{\cos x}, \qquad \cot x = \frac{\cos x}{\sin x}, \qquad \sec x = \frac{1}{\cos x}, \qquad \csc x = \frac{1}{\sin x}$$

$$(16) \quad \tan(x + y) = \frac{\tan x + \tan y}{1 - \tan x \tan y}, \qquad \tan(x - y) = \frac{\tan x - \tan y}{1 + \tan x \tan y}$$

Hyperbolic functions (hyperbolic sine sinh x, etc.; Figs. 551, 552)

$$(17) \quad \sinh x = \tfrac{1}{2}(e^x - e^{-x}), \qquad \cosh x = \tfrac{1}{2}(e^x + e^{-x})$$

$$(18) \quad \tanh x = \frac{\sinh x}{\cosh x}, \qquad \coth x = \frac{\cosh x}{\sinh x}$$

$$(19) \quad \cosh x + \sinh x = e^x, \qquad \cosh x - \sinh x = e^{-x}$$

$$(20) \quad \cosh^2 x - \sinh^2 x = 1$$

$$(21) \quad \sinh^2 x = \tfrac{1}{2}(\cosh 2x - 1), \qquad \cosh^2 x = \tfrac{1}{2}(\cosh 2x + 1)$$

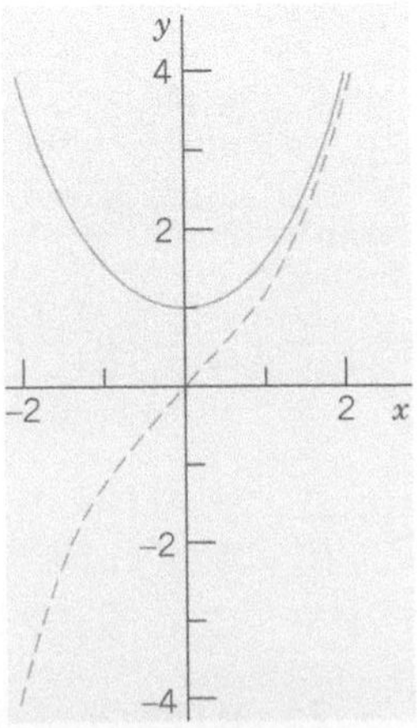

Fig. 551. sinh x (dashed) and cosh x

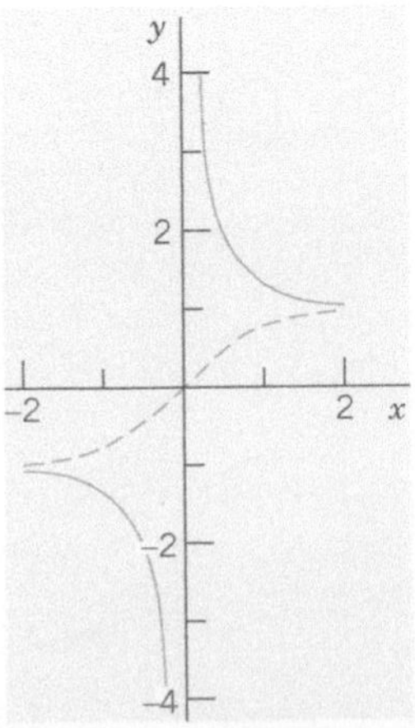

Fig. 552. tanh x (dashed) and coth x

$$\text{(22)} \qquad \begin{cases} \sinh(x \pm y) = \sinh x \cosh y \pm \cosh x \sinh y \\ \cosh(x \pm y) = \cosh x \cosh y \pm \sinh x \sinh y \end{cases}$$

$$\text{(23)} \qquad \tanh(x \pm y) = \frac{\tanh x \pm \tanh y}{1 \pm \tanh x \tanh y}$$

Gamma function (Fig. 553 and Table A2 in App. 5). The gamma function $\Gamma(\alpha)$ is defined by the integral

$$\text{(24)} \qquad \Gamma(\alpha) = \int_0^\infty e^{-t} t^{\alpha-1}\, dt \qquad (\alpha > 0),$$

which is meaningful only if $\alpha > 0$ (or, if we consider complex α, for those α whose real part is positive). Integration by parts gives the important *functional relation of the gamma function,*

$$\text{(25)} \qquad \Gamma(\alpha + 1) = \alpha\Gamma(\alpha).$$

From (24) we readily have $\Gamma(1) = 1$; hence if α is a positive integer, say k, then by repeated application of (25) we obtain

$$\text{(26)} \qquad \Gamma(k + 1) = k! \qquad (k = 0, 1, \cdots).$$

This shows that *the gamma function can be regarded as a generalization of the elementary factorial function.* [Sometimes the notation $(\alpha - 1)!$ is used for $\Gamma(\alpha)$, even for noninteger values of α, and the gamma function is also known as the **factorial function**.]

By repeated application of (25) we obtain

$$\Gamma(\alpha) = \frac{\Gamma(\alpha+1)}{\alpha} = \frac{\Gamma(\alpha+2)}{\alpha(\alpha+1)} = \cdots = \frac{\Gamma(\alpha+k+1)}{\alpha(\alpha+1)(\alpha+2)\cdots(\alpha+k)}$$

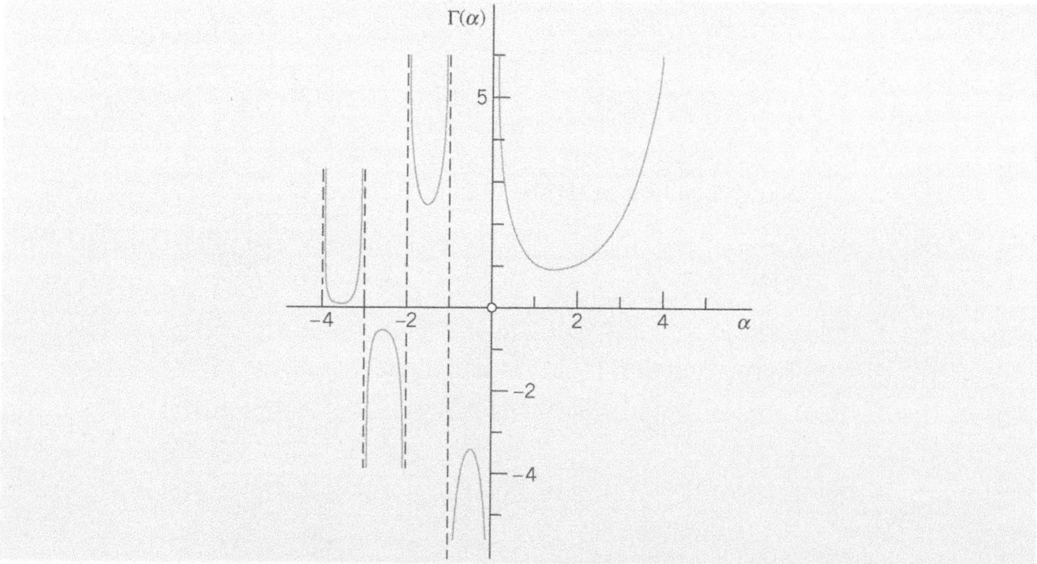

Fig. 553. Gamma function

and we may use this relation

$$\Gamma(\alpha) = \frac{\Gamma(\alpha + k + 1)}{\alpha(\alpha + 1) \cdots (\alpha + k)} \qquad (\alpha \neq 0, -1, -2, \cdots), \tag{27}$$

for defining the gamma function for negative α ($\neq -1, -2, \cdots$), choosing for k the smallest integer such that $\alpha + k + 1 > 0$. *Together with* (24), *this then gives a definition of* $\Gamma(\alpha)$ *for all* α *not equal to zero or a negative integer* (Fig. 553).

It can be shown that the gamma function may also be represented as the limit of a product, namely, by the formula

$$\Gamma(\alpha) = \lim_{n \to \infty} \frac{n!\, n^{\alpha}}{\alpha(\alpha + 1)(\alpha + 2) \cdots (\alpha + n)} \qquad (\alpha \neq 0, -1, \cdots). \tag{28}$$

From (27) or (28) we see that, for complex α, the gamma function $\Gamma(\alpha)$ is a meromorphic function with simple poles at $\alpha = 0, -1, -2, \cdots$.

An approximation of the gamma function for large positive α is given by the **Stirling formula**

$$\Gamma(\alpha + 1) \approx \sqrt{2\pi\alpha} \left(\frac{\alpha}{e}\right)^{\alpha} \tag{29}$$

where e is the base of the natural logarithm. We finally mention the special value

$$\Gamma(\tfrac{1}{2}) = \sqrt{\pi}. \tag{30}$$

Incomplete gamma functions

$$P(\alpha, x) = \int_0^x e^{-t} t^{\alpha-1}\, dt, \qquad Q(\alpha, x) = \int_x^{\infty} e^{-t} t^{\alpha-1}\, dt \qquad (\alpha > 0) \tag{31}$$

$$\Gamma(\alpha) = P(\alpha, x) + Q(\alpha, x) \tag{32}$$

Beta function

$$B(x, y) = \int_0^1 t^{x-1}(1 - t)^{y-1}\, dt \qquad (x > 0, y > 0) \tag{33}$$

Representation in terms of gamma functions:

$$B(x, y) = \frac{\Gamma(x)\Gamma(y)}{\Gamma(x + y)} \tag{34}$$

Error function (Fig. 554 and Table A4 in App. 5)

$$\operatorname{erf} x = \frac{2}{\sqrt{\pi}} \int_0^x e^{-t^2}\, dt \tag{35}$$

$$\operatorname{erf} x = \frac{2}{\sqrt{\pi}} \left(x - \frac{x^3}{1!3} + \frac{x^5}{2!5} - \frac{x^7}{3!7} + - \cdots\right) \tag{36}$$

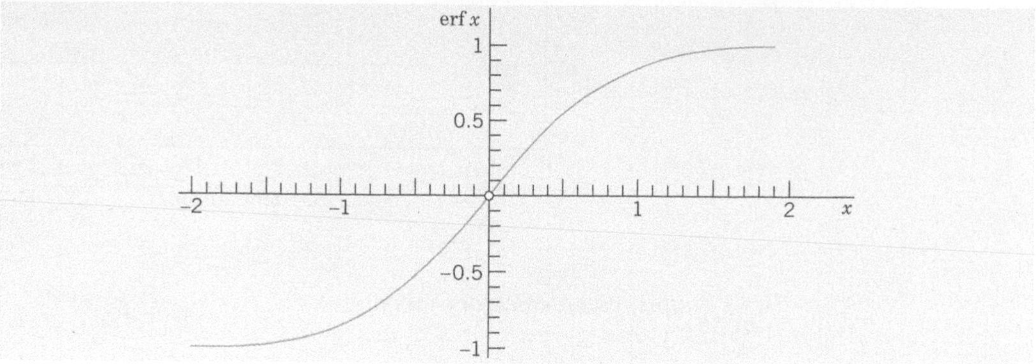

Fig. 554. Error function

erf $(\infty) = 1$, *complementary error function*

(37)
$$\operatorname{erfc} x = 1 - \operatorname{erf} x = \frac{2}{\sqrt{\pi}} \int_x^{\infty} e^{-t^2}\, dt$$

Fresnel integrals[1] (Fig. 555)

(38)
$$C(x) = \int_0^x \cos(t^2)\, dt, \qquad S(x) = \int_0^x \sin(t^2)\, dt$$

$C(\infty) = \sqrt{\pi/8}$, $S(\infty) = \sqrt{\pi/8}$, *complementary functions*

(39)
$$c(x) = \sqrt{\frac{\pi}{8}} - C(x) = \int_x^{\infty} \cos(t^2)\, dt$$
$$s(x) = \sqrt{\frac{\pi}{8}} - S(x) = \int_x^{\infty} \sin(t^2)\, dt$$

Sine integral (Fig. 556 and Table A4 in App. 5)

(40)
$$\operatorname{Si}(x) = \int_0^x \frac{\sin t}{t}\, dt$$

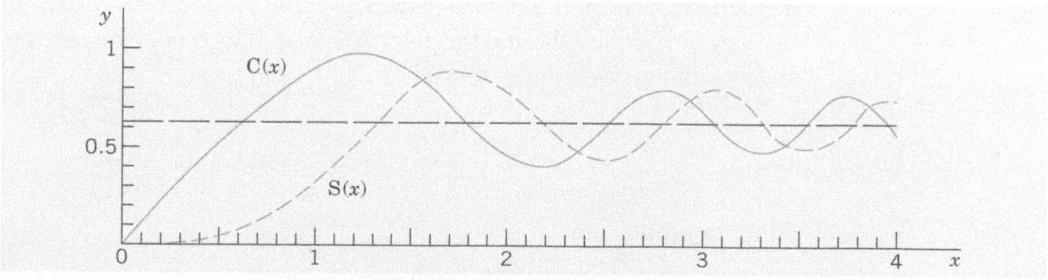

Fig. 555. Fresnel integrals

[1]AUGUSTIN FRESNEL (1788–1827), French physicist and mathematician. For tables see Ref. [GenRef1].

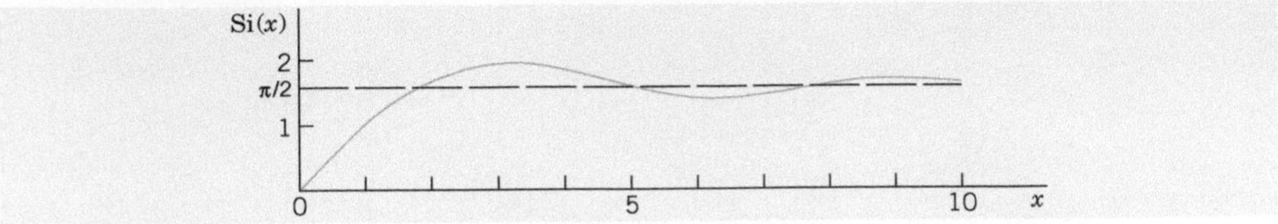

Fig. 556. Sine integral

$\mathrm{Si}(\infty) = \pi/2$, *complementary function*

(41)
$$\mathrm{si}(x) = \frac{\pi}{2} - \mathrm{Si}(x) = \int_x^{\infty} \frac{\sin t}{t}\, dt$$

Cosine integral (Table A4 in App. 5)

(42)
$$\mathrm{ci}(x) = \int_x^{\infty} \frac{\cos t}{t}\, dt \qquad (x > 0)$$

Exponential integral

(43)
$$\mathrm{Ei}(x) = \int_x^{\infty} \frac{e^{-t}}{t}\, dt \qquad (x > 0)$$

Logarithmic integral

(44)
$$\mathrm{li}(x) = \int_0^{x} \frac{dt}{\ln t}$$

A3.2 Partial Derivatives

For differentiation formulas, see inside of front cover.

Let $z = f(x, y)$ be a real function of two independent real variables, x and y. If we keep y constant, say, $y = y_1$, and think of x as a variable, then $f(x, y_1)$ depends on x alone. If the derivative of $f(x, y_1)$ with respect to x for a value $x = x_1$ exists, then the value of this derivative is called the **partial derivative** of $f(x, y)$ *with respect to x at the point* (x_1, y_1) and is denoted by

$$\left.\frac{\partial f}{\partial x}\right|_{(x_1,y_1)} \qquad \text{or by} \qquad \left.\frac{\partial z}{\partial x}\right|_{(x_1,y_1)}.$$

Other notations are

$$f_x(x_1, y_1) \qquad \text{and} \qquad z_x(x_1, y_1);$$

these may be used when subscripts are not used for another purpose and there is no danger of confusion.

We thus have, by the definition of the derivative,

$$\left.\frac{\partial f}{\partial x}\right|_{(x_1,y_1)} = \lim_{\Delta x \to 0} \frac{f(x_1 + \Delta x, y_1) - f(x_1, y_1)}{\Delta x}. \tag{1}$$

The partial derivative of $z = f(x, y)$ with respect to y is defined similarly; we now keep x constant, say, equal to x_1, and differentiate $f(x_1, y)$ with respect to y. Thus

$$\left.\frac{\partial f}{\partial y}\right|_{(x_1,y_1)} = \left.\frac{\partial z}{\partial y}\right|_{(x_1,y_1)} = \lim_{\Delta y \to 0} \frac{f(x_1, y_1 + \Delta y) - f(x_1, y_1)}{\Delta y}. \tag{2}$$

Other notations are $f_y(x_1, y_1)$ and $z_y(x_1, y_1)$.

It is clear that the values of those two partial derivatives will in general depend on the point (x_1, y_1). Hence the partial derivatives $\partial z/\partial x$ and $\partial z/\partial y$ at a variable point (x, y) are functions of x and y. The function $\partial z/\partial x$ is obtained as in ordinary calculus by differentiating $z = f(x, y)$ with respect to x, ***treating y as a constant***, and $\partial z/\partial y$ is obtained by differentiating z with respect to y, ***treating x as a constant***.

EXAMPLE 1 Let $z = f(x, y) = x^2y + x \sin y$. Then

$$\frac{\partial f}{\partial x} = 2xy + \sin y, \qquad \frac{\partial f}{\partial y} = x^2 + x \cos y.$$ ■

The partial derivatives $\partial z/\partial x$ and $\partial z/\partial y$ of a function $z = f(x, y)$ have a very simple **geometric interpretation**. The function $z = f(x, y)$ can be represented by a surface in space. The equation $y = y_1$ then represents a vertical plane intersecting the surface in a curve, and the partial derivative $\partial z/\partial x$ at a point (x_1, y_1) is the slope of the tangent (that is, $\tan \alpha$ where α is the angle shown in Fig. 557) to the curve. Similarly, the partial derivative $\partial z/\partial y$ at (x_1, y_1) is the slope of the tangent to the curve $x = x_1$ on the surface $z = f(x, y)$ at (x_1, y_1).

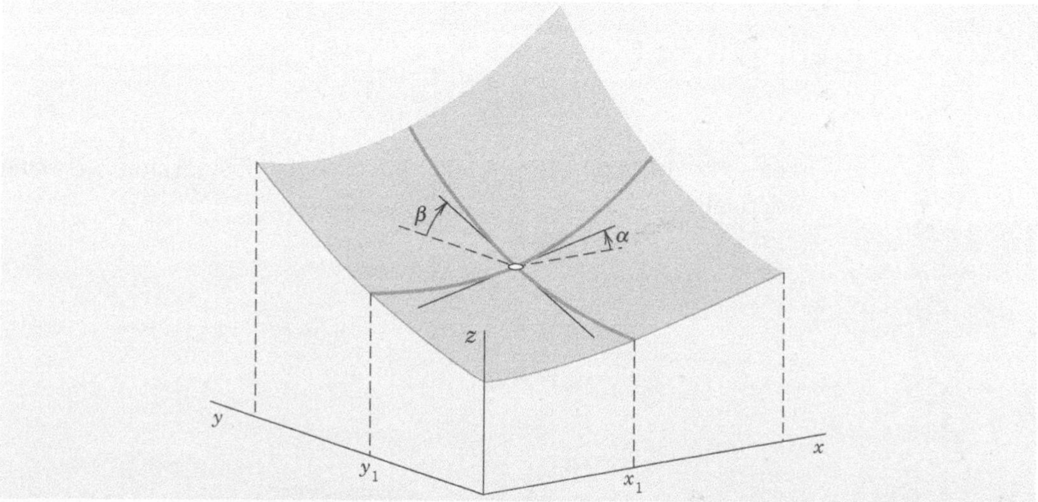

Fig. 557. Geometrical interpretation of first partial derivatives

The partial derivatives $\partial z/\partial x$ and $\partial z/\partial y$ are called *first partial derivatives* or *partial derivatives of first order*. By differentiating these derivatives once more, we obtain the four *second partial derivatives* (or *partial derivatives of second order*)[2]

$$
\begin{aligned}
\frac{\partial^2 f}{\partial x^2} &= \frac{\partial}{\partial x}\left(\frac{\partial f}{\partial x}\right) = f_{xx} \\
\frac{\partial^2 f}{\partial x\, \partial y} &= \frac{\partial}{\partial x}\left(\frac{\partial f}{\partial y}\right) = f_{yx} \\
\frac{\partial^2 f}{\partial y\, \partial x} &= \frac{\partial}{\partial y}\left(\frac{\partial f}{\partial x}\right) = f_{xy} \\
\frac{\partial^2 f}{\partial y^2} &= \frac{\partial}{\partial y}\left(\frac{\partial f}{\partial y}\right) = f_{yy}.
\end{aligned} \tag{3}
$$

It can be shown that if all the derivatives concerned are continuous, then the two mixed partial derivatives are equal, so that the order of differentiation does not matter (see Ref. [GenRef4] in App. 1), that is,

$$
\frac{\partial^2 z}{\partial x\, \partial y} = \frac{\partial^2 z}{\partial y\, \partial x}. \tag{4}
$$

EXAMPLE 2 For the function in Example 1.

$$
f_{xx} = 2y, \qquad f_{xy} = 2x + \cos y = f_{yx}, \qquad f_{yy} = -x \sin y.
$$

■

By differentiating the second partial derivatives again with respect to x and y, respectively, we obtain the *third partial derivatives* or *partial derivatives of the third order* of f, etc.

If we consider a function $f(x, y, z)$ of ***three independent variables***, then we have the three first partial derivatives $f_x(x, y, z)$, $f_y(x, y, z)$, and $f_z(x, y, z)$. Here f_x is obtained by differentiating f with respect to x, ***treating both y and z as constants***. Thus, analogous to (1), we now have

$$
\left.\frac{\partial f}{\partial x}\right|_{(x_1, y_1, z_1)} = \lim_{\Delta x \to 0} \frac{f(x_1 + \Delta x, y_1, z_1) - f(x_1, y_1, z_1)}{\Delta x},
$$

etc. By differentiating f_x, f_y, f_z again in this fashion we obtain the second partial derivatives of f, etc.

EXAMPLE 3 Let $f(x, y, z) = x^2 + y^2 + z^2 + xy\,e^z$. Then

$$
\begin{aligned}
f_x &= 2x + y\,e^z, & f_y &= 2y + x\,e^z, & f_z &= 2z + xy\,e^z, \\
f_{xx} &= 2, & f_{xy} &= f_{yx} = e^z, & f_{xz} &= f_{zx} = y\,e^z, \\
f_{yy} &= 2, & f_{yz} &= f_{zy} = x\,e^z, & f_{zz} &= 2 + xy\,e^z.
\end{aligned}
$$

■

[2] **CAUTION!** In the subscript notation, the subscripts are written in the order in which we differentiate, whereas in the "∂" notation the order is opposite.

A3.3 Sequences and Series

See also Chap. 15.

Monotone Real Sequences

We call a real sequence $x_1, x_2, \cdots, x_n, \cdots$ a **monotone sequence** if it is either **monotone increasing**, that is,

$$x_1 \leqq x_2 \leqq x_3 \leqq \cdots$$

or **monotone decreasing**, that is,

$$x_1 \geqq x_2 \geqq x_3 \geqq \cdots.$$

We call $x_1, x_2, \cdots$ a **bounded sequence** if there is a positive constant K such that $|x_n| < K$ for all n.

THEOREM 1

If a real sequence is bounded and monotone, it converges.

PROOF Let $x_1, x_2, \cdots$ be a bounded monotone increasing sequence. Then its terms are smaller than some number B and, since $x_1 \leqq x_n$ for all n, they lie in the interval $x_1 \leqq x_n \leqq B$, which will be denoted by I_0. We bisect I_0; that is, we subdivide it into two parts of equal length. If the right half (together with its endpoints) contains terms of the sequence, we denote it by I_1. If it does not contain terms of the sequence, then the left half of I_0 (together with its endpoints) is called I_1. This is the first step.

In the second step we bisect I_1, select one half by the same rule, and call it I_2, and so on (see Fig. 558).

In this way we obtain shorter and shorter intervals $I_0, I_1, I_2, \cdots$ with the following properties. Each I_m contains all I_n for $n > m$. No term of the sequence lies to the right of I_m, and, since the sequence is monotone increasing, all x_n with n greater than some number N lie in I_m; of course, N will depend on m, in general. The lengths of the I_m approach zero as m approaches infinity. Hence there is precisely one number, call it L, that lies in all those intervals,[3] and we may now easily prove that the sequence is convergent with the limit L.

In fact, given an $\epsilon > 0$, we choose an m such that the length of I_m is less than ϵ. Then L and all the x_n with $n > N(m)$ lie in I_m, and, therefore, $|x_n - L| < \epsilon$ for all those n. This completes the proof for an increasing sequence. For a decreasing sequence the proof is the same, except for a suitable interchange of "left" and "right" in the construction of those intervals. ■

[3]This statement seems to be obvious, but actually it is not; it may be regarded as an axiom of the real number system in the following form. Let $J_1, J_2, \cdots$ be closed intervals such that each J_m contains all J_n with $n > m$, and the lengths of the J_m approach zero as m approaches infinity. Then there is precisely one real number that is contained in all those intervals. This is the so-called **Cantor–Dedekind axiom**, named after the German mathematicians GEORG CANTOR (1845–1918), the creator of set theory, and RICHARD DEDEKIND (1831–1916), known for his fundamental work in number theory. For further details see Ref. [GenRef2] in App. 1. (An interval I is said to be **closed** if its two endpoints are regarded as points belonging to I. It is said to be **open** if the endpoints are not regarded as points of I.)

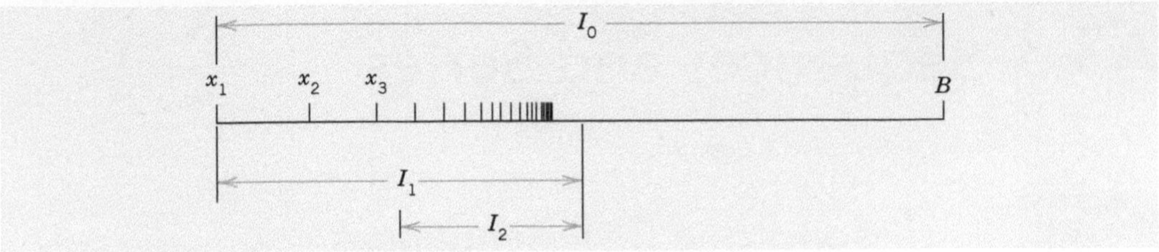

Fig. 558. Proof of Theorem 1

Real Series

THEOREM 2

Leibniz Test for Real Series

Let $x_1, x_2, \cdots$ *be real and monotone decreasing to zero, that is,*

$$(1) \qquad \text{(a)} \quad x_1 \geqq x_2 \geqq x_3 \geqq \cdots, \qquad \text{(b)} \quad \lim_{m \to \infty} x_m = 0.$$

Then the series with terms of alternating signs

$$x_1 - x_2 + x_3 - x_4 + - \cdots$$

converges, and for the remainder R_n *after the nth term we have the estimate*

$$(2) \qquad |R_n| \leqq x_{n+1}.$$

PROOF Let s_n be the nth partial sum of the series. Then, because of (1a),

$$s_1 = x_1, \qquad s_2 = x_1 - x_2 \leqq s_1,$$

$$s_3 = s_2 + x_3 \geqq s_2, \qquad s_3 = s_1 - (x_2 - x_3) \leqq s_1,$$

so that $s_2 \leqq s_3 \leqq s_1$. Proceeding in this fashion, we conclude that (Fig. 559)

$$(3) \qquad s_1 \geqq s_3 \geqq s_5 \geqq \cdots \geqq s_6 \geqq s_4 \geqq s_2$$

which shows that the odd partial sums form a bounded monotone sequence, and so do the even partial sums. Hence, by Theorem 1, both sequences converge, say,

$$\lim_{n \to \infty} s_{2n+1} = s, \qquad \lim_{n \to \infty} s_{2n} = s^*.$$

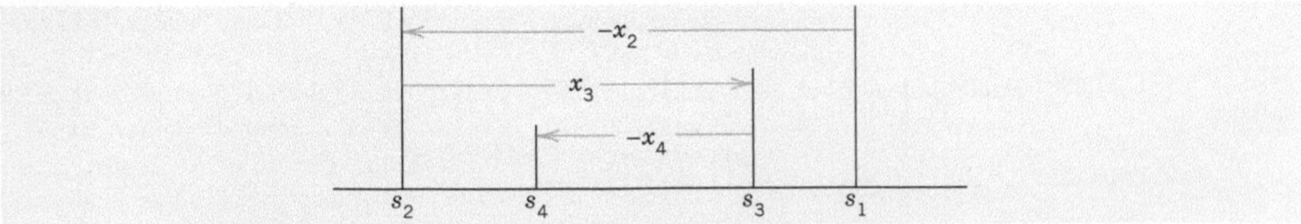

Fig. 559. Proof of the Leibniz test

Now, since $s_{2n+1} - s_{2n} = x_{2n+1}$, we readily see that (Ib) implies

$$s - s^* = \lim_{n\to\infty} s_{2n+1} - \lim_{n\to\infty} s_{2n} = \lim_{n\to\infty} (s_{2n+1} - s_{2n}) = \lim_{n\to\infty} x_{2n+1} = 0.$$

Hence $s^* = s$, and the series converges with the sum s.

We prove the estimate (2) for the remainder. Since $s_n \to s$, it follows from (3) that

$$s_{2n+1} \geqq s \geqq s_{2n} \qquad \text{and also} \qquad s_{2n-1} \geqq s \geqq s_{2n}.$$

By subtracting s_{2n} and s_{2n-1}, respectively, we obtain

$$s_{2n+1} - s_{2n} \geqq s - s_{2n} \geqq 0, \qquad 0 \geqq s - s_{2n-1} \geqq s_{2n} - s_{2n-1}.$$

In these inequalities, the first expression is equal to x_{2n+1}, the last is equal to $-x_{2n}$, and the expressions between the inequality signs are the remainders R_{2n} and R_{2n-1}. Thus the inequalities may be written

$$x_{2n+1} \geqq R_{2n} \geqq 0, \qquad 0 \geqq R_{2n-1} \geqq -x_{2n}$$

and we see that they imply (2). This completes the proof. ■

A3.4 Grad, Div, Curl, ∇^2 in Curvilinear Coordinates

To simplify formulas, we write Cartesian coordinates $x = x_1$, $y = x_2$, $z = x_3$. We denote curvilinear coordinates by q_1, q_2, q_3. Through each point P there pass three coordinate surfaces $q_1 = \text{const}$, $q_2 = \text{const}$, $q_3 = \text{const}$. They intersect along coordinate curves. We assume the three coordinate curves through P to be **orthogonal** (perpendicular to each other). We write coordinate transformations as

$$(1) \qquad x_1 = x_1(q_1, q_2, q_3), \qquad x_2 = x_2(q_1, q_2, q_3), \qquad x_3 = x_3(q_1, q_2, q_3).$$

Corresponding transformations of grad, div, curl, and ∇^2 can all be written by using

$$(2) \qquad h_j^2 = \sum_{k=1}^{3} \left(\frac{\partial x_k}{\partial q_j}\right)^2.$$

Next to Cartesian coordinates, most important are **cylindrical coordinates** $q_1 = r$, $q_2 = \theta$, $q_3 = z$ (Fig. 560a) defined by

$$(3) \quad x_1 = q_1 \cos q_2 = r \cos\theta, \qquad x_2 = q_1 \sin q_2 = r \sin\theta, \qquad x_3 = q_3 = z$$

and **spherical coordinates** $q_1 = r$, $q_2 = \theta$, $q_3 = \phi$ (Fig. 560b) defined by[4]

$$(4) \quad \begin{aligned} x_1 &= q_1 \cos q_2 \sin q_3 = r\cos\theta\sin\phi, \qquad x_2 = q_1 \sin q_2 \sin q_3 = r \sin\theta\sin\phi \\ x_3 &= q_1 \cos q_3 = r\cos\phi. \end{aligned}$$

[4]This is the notation used in calculus and in many other books. It is logical since in it, θ plays the same role as in polar coordinates. CAUTION! Some books interchange the roles of θ and ϕ.

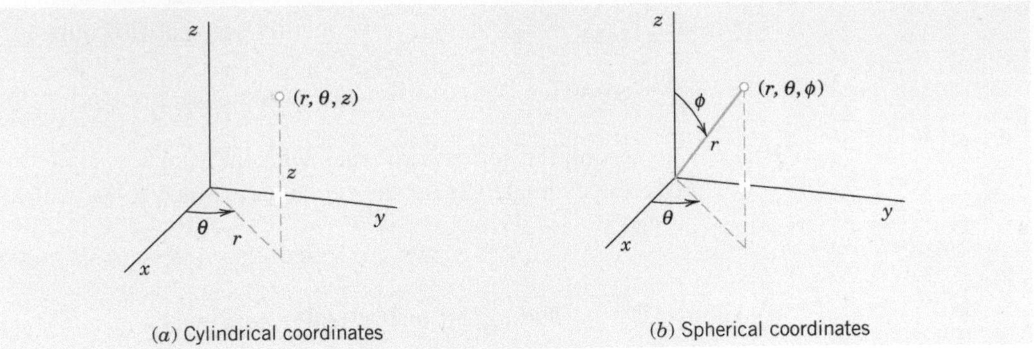

Fig. 560. Special curvilinear coordinates

In addition to the general formulas for any orthogonal coordinates q_1, q_2, q_3, we shall give additional formulas for these important special cases.

Linear Element *ds*. In Cartesian coordinates,

$$ds^2 = dx_1^2 + dx_2^2 + dx_3^2 \qquad \text{(Sec. 9.5).}$$

For the q-coordinates,

$$ds^2 = h_1^2\, dq_1^2 + h_2^2\, dq_2^2 + h_3^2\, dq_3^2. \tag{5}$$

$$ds^2 = dr^2 + r^2\, d\theta^2 + dz^2 \qquad \text{(Cylindrical coordinates).} \tag{5'}$$

For polar coordinates set $dz^2 = 0$.

$$ds^2 = dr^2 + r^2 \sin^2 \phi\, d\theta^2 + r^2\, d\phi^2 \qquad \text{(Spherical coordinates).} \tag{5''}$$

Gradient. grad $f = \nabla f = [f_{x_1}, \ f_{x_2}, \ f_{x_3}]$ (partial derivatives; Sec. 9.7). In the q-system, with **u, v, w** denoting unit vectors in the positive directions of the q_1, q_2, q_3 coordinate curves, respectively,

$$\text{grad} f = \nabla f = \frac{1}{h_1}\frac{\partial f}{\partial q_1}\mathbf{u} + \frac{1}{h_2}\frac{\partial f}{\partial q_2}\mathbf{v} + \frac{1}{h_3}\frac{\partial f}{\partial q_3}\mathbf{w} \tag{6}$$

$$\text{grad} f = \nabla f = \frac{\partial f}{\partial r}\mathbf{u} + \frac{1}{r}\frac{\partial f}{\partial \theta}\mathbf{v} + \frac{\partial f}{\partial z}\mathbf{w} \qquad \text{(Cylindrical coordinates)} \tag{6'}$$

$$\text{grad} f = \nabla f = \frac{\partial f}{\partial r}\mathbf{u} + \frac{1}{r \sin \phi}\frac{\partial f}{\partial \theta}\mathbf{v} + \frac{1}{r}\frac{\partial f}{\partial \phi}\mathbf{w} \qquad \text{(Spherical coordinates).} \tag{6''}$$

Divergence div $\mathbf{F} = \nabla \cdot \mathbf{F} = (F_1)_{x_1} + (F_2)_{x_2} + (F_3)_{x_3}$ ($\mathbf{F} = [F_1, F_2, F_3]$, Sec. 9.8);

$$\text{div}\, \mathbf{F} = \nabla \cdot \mathbf{F} = \frac{1}{h_1 h_2 h_3}\left[\frac{\partial}{\partial q_1}(h_2 h_3 F_1) + \frac{\partial}{\partial q_2}(h_3 h_1 F_2) + \frac{\partial}{\partial q_3}(h_1 h_2 F_3)\right] \tag{7}$$

$$\text{div}\, \mathbf{F} = \nabla \cdot \mathbf{F} = \frac{1}{r}\frac{\partial}{\partial r}(rF_1) + \frac{1}{r}\frac{\partial F_2}{\partial \theta} + \frac{\partial F_3}{\partial z} \qquad \text{(Cylindrical coordinates)} \tag{7'}$$

$$(7'') \qquad \operatorname{div} \mathbf{F} = \nabla \bullet \mathbf{F} = \frac{1}{r^2}\frac{\partial}{\partial r}(r^2 F_1) + \frac{1}{r \sin\phi}\frac{\partial F_2}{\partial \theta} + \frac{1}{r\sin\phi}\frac{\partial}{\partial\phi}(\sin\phi\, F_3)$$

(Spherical coordinates).

Laplacian $\nabla^2 f = \nabla \bullet \nabla f = \operatorname{div}(\operatorname{grad} f) = f_{x_1x_1} + f_{x_2x_2} + f_{x_3x_3}$ (Sec. 9.8):

$$(8) \quad \nabla^2 f = \frac{1}{h_1h_2h_3}\left[\frac{\partial}{\partial q_1}\left(\frac{h_2h_3}{h_1}\frac{\partial f}{\partial q_1}\right) + \frac{\partial}{\partial q_2}\left(\frac{h_3h_1}{h_2}\frac{\partial f}{\partial q_2}\right) + \frac{\partial}{\partial q_3}\left(\frac{h_1h_2}{h_3}\frac{\partial f}{\partial q_3}\right)\right]$$

$$(8') \qquad \nabla^2 f = \frac{\partial^2 f}{\partial r^2} + \frac{1}{r}\frac{\partial f}{\partial r} + \frac{1}{r^2}\frac{\partial^2 f}{\partial \theta^2} + \frac{\partial^2 f}{\partial z^2}$$

(Cylindrical coordinates)

$$(8'') \qquad \nabla^2 f = \frac{\partial^2 f}{\partial r^2} + \frac{2}{r}\frac{\partial f}{\partial r} + \frac{1}{r^2\sin^2\phi}\frac{\partial^2 f}{\partial\theta^2} + \frac{1}{r^2}\frac{\partial^2 f}{\partial\phi^2} + \frac{\cot\phi}{r^2}\frac{\partial f}{\partial\phi}$$

(Spherical coordinates).

Curl (Sec. 9.9):

$$(9) \qquad \operatorname{curl} \mathbf{F} = \nabla \times \mathbf{F} = \frac{1}{h_1h_2h_3}\begin{vmatrix} h_1\mathbf{u} & h_2\mathbf{v} & h_3\mathbf{w} \\ \dfrac{\partial}{\partial q_1} & \dfrac{\partial}{\partial q_2} & \dfrac{\partial}{\partial q_3} \\ h_1F_1 & h_2F_2 & h_3F_3 \end{vmatrix}.$$

For cylindrical coordinates we have in (9) (as in the previous formulas)

$$h_1 = h_r = 1, \qquad h_2 = h_\theta = q_1 = r, \qquad h_3 = h_z = 1$$

and for spherical coordinates we have

$$h_1 = h_r = 1, \qquad h_2 = h_\theta = q_1 \sin q_3 = r\sin\phi, \qquad h_3 = h_\phi = q_1 = r.$$

APPENDIX 4

Additional Proofs

Section 2.6, page 74

PROOF OF THEOREM 1 Uniqueness[1]

Assuming that the problem consisting of the ODE

$$y'' + p(x)y' + q(x)y = 0 \tag{1}$$

and the two initial conditions

$$y(x_0) = K_0, \qquad y'(x_0) = K_1 \tag{2}$$

has two solutions $y_1(x)$ and $y_2(x)$ on the interval I in the theorem, we show that their difference

$$y(x) = y_1(x) - y_2(x)$$

is identically zero on I; then $y_1 \equiv y_2$ on I, which implies uniqueness.

Since (1) is homogeneous and linear, y is a solution of that ODE on I, and since y_1 and y_2 satisfy the same initial conditions, y satisfies the conditions

$$y(x_0) = 0, \qquad y'(x_0) = 0. \tag{11}$$

We consider the function

$$z(x) = y(x)^2 + y'(x)^2$$

and its derivative

$$z' = 2yy' + 2y'y''.$$

From the ODE we have

$$y'' = -py' - qy.$$

By substituting this in the expression for z' we obtain

$$z' = 2yy' - 2py'^2 - 2qyy'. \tag{12}$$

Now, since y and y' are real,

$$(y \pm y')^2 = y^2 \pm 2yy' + y'^2 \geqq 0.$$

[1]This proof was suggested by my colleague, Prof. A. D. Ziebur. In this proof, we use some formula numbers that have not yet been used in Sec. 2.6.

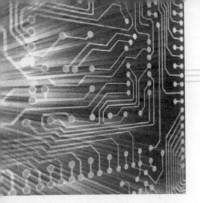

From this and the definition of z we obtain the two inequalities

$$(13) \qquad \text{(a)} \quad 2yy' \leqq y^2 + y'^2 = z, \qquad \text{(b)} \quad -2yy' \leqq y^2 + y'^2 = z.$$

From (13b) we have $2yy' \geqq -z$. Together, $|2yy'| \leqq z$. For the last term in (12) we now obtain

$$-2qyy' \leqq |-2qyy'| = |q||2yy'| \leqq |q|z.$$

Using this result as well as $-p \leqq |p|$ and applying (13a) to the term $2yy'$ in (12), we find

$$z' \leqq z + 2|p|y'^2 + |q|z.$$

Since $y'^2 \leqq y^2 + y'^2 = z$, from this we obtain

$$z' \leqq (1 + 2|p| + |q|)z$$

or, denoting the function in parentheses by h,

$$(14a) \qquad z' \leqq hz \qquad \text{for all } x \text{ on } I.$$

Similarly, from (12) and (13) it follows that

$$(14b) \qquad \begin{aligned} -z' &= -2yy' + 2py'^2 + 2qyy' \\ &\leqq z + 2|p|z + |q|z = hz. \end{aligned}$$

The inequalities (14a) and (14b) are equivalent to the inequalities

$$(15) \qquad z' - hz \leqq 0, \qquad z' + hz \geqq 0.$$

Integrating factors for the two expressions on the left are

$$F_1 = e^{-\int h(x)\,dx} \qquad \text{and} \qquad F_2 = e^{\int h(x)\,dx}.$$

The integrals in the exponents exist because h is continuous. Since F_1 and F_2 are positive, we thus have from (15)

$$F_1(z' - hz) = (F_1 z)' \leqq 0 \qquad \text{and} \qquad F_2(z' + hz) = (F_2 z)' \geqq 0.$$

This means that $F_1 z$ is nonincreasing and $F_2 z$ is nondecreasing on I. Since $z(x_0) = 0$ by (11), when $x \leqq x_0$ we thus obtain

$$F_1 z \geqq (F_1 z)_{x_0} = 0, \qquad F_2 z \leqq (F_2 z)_{x_0} = 0$$

and similarly, when $x \geqq x_0$,

$$F_1 z \leqq 0, \qquad F_2 z \geqq 0.$$

Dividing by F_1 and F_2 and noting that these functions are positive, we altogether have

$$z \leqq 0, \qquad z \geqq 0 \qquad \text{for all } x \text{ on } I.$$

This implies that $z = y^2 + y'^2 \equiv 0$ on I. Hence $y \equiv 0$ or $y_1 \equiv y_2$ on I. ■

Section 5.3, page 182

PROOF OF THEOREM 2 Frobenius Method. Basis of Solutions. Three Cases

The formula numbers in this proof are the same as in the text of Sec. 5.3. An additional formula not appearing in Sec. 5.3 will be called (A) (see below).

The ODE in Theorem 2 is

$$y'' + \frac{b(x)}{x}\, y' + \frac{c(x)}{x^2}\, y = 0, \tag{1}$$

where $b(x)$ and $c(x)$ are analytic functions. We can write it

$$x^2y'' + xb(x)y' + c(x)y = 0. \tag{1$'$}$$

The indicial equation of (1) is

$$r(r-1) + b_0 r + c_0 = 0. \tag{4}$$

The roots r_1, r_2 of this quadratic equation determine the general form of a basis of solutions of (1), and there are three possible cases as follows.

Case 1. Distinct Roots Not Differing by an Integer. A first solution of (1) is of the form

$$y_1(x) = x^{r_1}(a_0 + a_1x + a_2x^2 + \cdots) \tag{5}$$

and can be determined as in the power series method. For a proof that in this case, the ODE (1) has a second independent solution of the form

$$y_2(x) = x^{r_2}(A_0 + A_1x + A_2x^2 + \cdots), \tag{6}$$

see Ref. [A11] listed in App. 1.

Case 2. Double Root. The indicial equation (4) has a double root r if and only if $(b_0 - 1)^2 - 4c_0 = 0$, and then $r = \frac{1}{2}(1 - b_0)$. A first solution

$$y_1(x) = x^r(a_0 + a_1x + a_2x^2 + \cdots), \qquad r = \tfrac{1}{2}(1 - b_0), \tag{7}$$

can be determined as in Case 1. We show that a second independent solution is of the form

$$y_2(x) = y_1(x)\ln x + x^r(A_1x + A_2x^2 + \cdots) \qquad (x > 0). \tag{8}$$

We use the method of reduction of order (see Sec. 2.1), that is, we determine $u(x)$ such that $y_2(x) = u(x)y_1(x)$ is a solution of (1). By inserting this and the derivatives

$$y_2' = u'y_1 + uy_1', \qquad y_2'' = u''y_1 + 2u'y_1' + uy_1''$$

into the ODE (1$'$) we obtain

$$x^2(u''y_1 + 2u'y_1' + uy_1'') + xb(u'y_1 + uy_1') + cuy_1 = 0.$$

Since y_1 is a solution of (1′), the sum of the terms involving u is zero, and this equation reduces to

$$x^2 y_1 u'' + 2x^2 y_1' u' + xby_1 u' = 0.$$

By dividing by $x^2 y_1$ and inserting the power series for b we obtain

$$u'' + \left(2\frac{y_1'}{y_1} + \frac{b_0}{x} + \cdots\right) u' = 0.$$

Here, and in the following, the dots designate terms that are constant or involve positive powers of x. Now, from (7), it follows that

$$\frac{y_1'}{y_1} = \frac{x^{r-1}[ra_0 + (r+1)a_1 x + \cdots]}{x^r[a_0 + a_1 x + \cdots]}$$

$$= \frac{1}{x}\left(\frac{ra_0 + (r+1)a_1 x + \cdots}{a_0 + a_1 x + \cdots}\right) = \frac{r}{x} + \cdots.$$

Hence the previous equation can be written

$$\text{(A)} \qquad u'' + \left(\frac{2r + b_0}{x} + \cdots\right) u' = 0.$$

Since $r = (1 - b_0)/2$, the term $(2r + b_0)/x$ equals $1/x$, and by dividing by u' we thus have

$$\frac{u''}{u'} = -\frac{1}{x} + \cdots.$$

By integration we obtain $\ln u' = -\ln x + \cdots$, hence $u' = (1/x)e^{(\cdots)}$. Expanding the exponential function in powers of x and integrating once more, we see that u is of the form

$$u = \ln x + k_1 x + k_2 x^2 + \cdots.$$

Inserting this into $y_2 = uy_1$, we obtain for y_2 a representation of the form (8).

Case 3. Roots Differing by an Integer. We write $r_1 = r$ and $r_2 = r - p$ where p is a ***positive*** integer. A first solution

$$\text{(9)} \qquad y_1(x) = x^{r_1}(a_0 + a_1 x + a_2 x^2 + \cdots)$$

can be determined as in Cases 1 and 2. We show that a second independent solution is of the form

$$\text{(10)} \qquad y_2(x) = ky_1(x)\ln x + x^{r_2}(A_0 + A_1 x + A_2 x^2 + \cdots)$$

where we may have $k \neq 0$ or $k = 0$. As in Case 2 we set $y_2 = uy_1$. The first steps are literally as in Case 2 and give Eq. (A),

$$u'' + \left(\frac{2r + b_0}{x} + \cdots\right) u' = 0.$$

Now by elementary algebra, the coefficient $b_0 - 1$ of r in (4) equals minus the sum of the roots,

$$b_0 - 1 = -(r_1 + r_2) = -(r + r - p) = -2r + p.$$

Hence $2r + b_0 = p + 1$, and division by u' gives

$$\frac{u''}{u'} = -\left(\frac{p+1}{x} + \cdots\right).$$

The further steps are as in Case 2. Integrating, we find

$$\ln u' = -(p+1)\ln x + \cdots, \qquad \text{thus} \qquad u' = x^{-(p+1)}e^{(\cdots)}$$

where dots stand for some series of nonnegative integer powers of x. By expanding the exponential function as before we obtain a series of the form

$$u' = \frac{1}{x^{p+1}} + \frac{k_1}{x^p} + \cdots + \frac{k_{p-1}}{x^2} + \frac{k_p}{x} + k_{p+1} + k_{p+2}x + \cdots.$$

We integrate once more. Writing the resulting logarithmic term first, we get

$$u = k_p \ln x + \left(-\frac{1}{px^p} - \cdots - \frac{k_{p-1}}{x} + k_{p+1}x + \cdots\right).$$

Hence, by (9) we get for $y_2 = uy_1$ the formula

$$y_2 = k_p y_1 \ln x + x^{r_1 - p}\left(-\frac{1}{p} - \cdots - k_{p-1}x^{p-1} + \cdots\right)(a_0 + a_1 x + \cdots).$$

But this is of the form (10) with $k = k_p$ since $r_1 - p = r_2$ and the product of the two series involves nonnegative integer powers of x only. ■

Section 7.7, page 293

THEOREM

Determinants

The definition of a determinant

$$(7) \qquad D = \det \mathbf{A} = \begin{vmatrix} a_{11} & a_{12} & \cdots & a_{1n} \\ a_{21} & a_{22} & \cdots & a_{2n} \\ \cdot & \cdot & \cdots & \cdot \\ \cdot & \cdot & \cdots & \cdot \\ a_{n1} & a_{n2} & \cdots & a_{nn} \end{vmatrix}$$

as given in Sec. 7.7 is unambiguous, that is, it yields the same value of D no matter which rows or columns we choose in the development.

PROOF In this proof we shall use formula numbers not yet used in Sec. 7.7.

We shall prove first that *the same value is obtained no matter which* **row** *is chosen.*

The proof is by induction. The statement is true for a second-order determinant, for which the developments by the first row $a_{11}a_{22} + a_{12}(-a_{21})$ and by the second row $a_{21}(-a_{12}) + a_{22}a_{11}$ give the same value $a_{11}a_{22} - a_{12}a_{21}$. Assuming the statement to be true for an $(n-1)$st-order determinant, we prove that it is true for an nth-order determinant.

For this purpose we expand D in terms of each of two arbitrary rows, say, the ith and the jth, and compare the results. Without loss of generality let us assume $i < j$.

First Expansion. We expand D by the ith row. A typical term in this expansion is

$$a_{ik}C_{ik} = a_{ik} \cdot (-1)^{i+k}M_{ik}. \tag{19}$$

The minor M_{ik} of a_{ik} in D is an $(n-1)$st-order determinant. By the induction hypothesis we may expand it by any row. We expand it by the row corresponding to the jth row of D. This row contains the entries a_{jl} $(l \neq k)$. It is the $(j-1)$st row of M_{ik}, because M_{ik} does not contain entries of the ith row of D, and $i < j$. We have to distinguish between two cases as follows.

Case I. If $l < k$, then the entry a_{jl} belongs to the lth column of M_{ik} (see Fig. 561). Hence the term involving a_{jl} in this expansion is

$$a_{jl} \cdot (\text{cofactor of } a_{jl} \text{ in } M_{ik}) = a_{jl} \cdot (-1)^{(j-1)+l}M_{ikjl} \tag{20}$$

where M_{ikjl} is the minor of a_{jl} in M_{ik}. Since this minor is obtained from M_{ik} by deleting the row and column of a_{jl}, it is obtained from D by deleting the ith and jth rows and the kth and lth columns of D. We insert the expansions of the M_{ik} into that of D. Then it follows from (19) and (20) that the terms of the resulting representation of D are of the form

$$a_{ik}a_{jl} \cdot (-1)^{b}M_{ikjl} \qquad (l < k) \tag{21a}$$

where

$$b = i + k + j + l - 1.$$

Case II. If $l > k$, the only difference is that then a_{jl} belongs to the $(l-1)$st column of M_{ik}, because M_{ik} does not contain entries of the kth column of D, and $k < l$. This causes an additional minus sign in (20), and, instead of (21a), we therefore obtain

$$-a_{ik}a_{jl} \cdot (-1)^{b}M_{ikjl} \qquad (l > k) \tag{21b}$$

where b is the same as before.

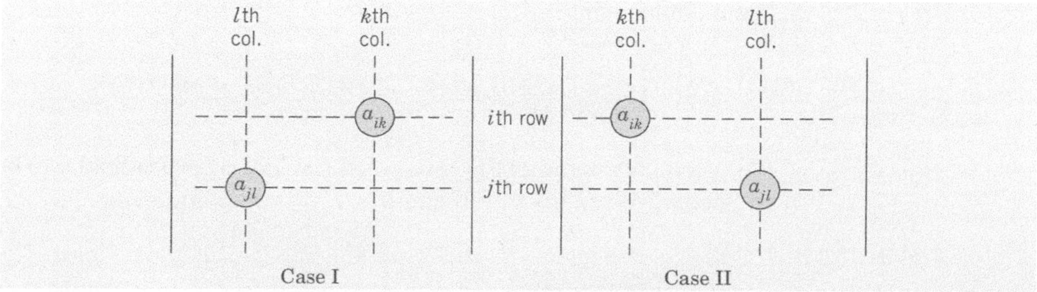

Fig. 561. Cases I and II of the two expansions of D

Second Expansion. We now expand D at first by the jth row. A typical term in this expansion is

$$a_{jl}C_{jl} = a_{jl} \cdot (-1)^{j+l} M_{jl}. \tag{22}$$

By the induction hypothesis we may expand the minor M_{jl} of a_{jl} in D by its ith row, which corresponds to the ith row of D, since $j > i$.

Case I. If $k > l$, the entry a_{ik} in that row belongs to the $(k - 1)$st column of M_{jl}, because M_{jl} does not contain entries of the lth column of D, and $l < k$ (see Fig. 561). Hence the term involving a_{ik} in this expansion is

$$a_{ik} \cdot (\text{cofactor of } a_{ik} \text{ in } M_{jl}) = a_{ik} \cdot (-1)^{i+(k-1)} M_{ikjl}, \tag{23}$$

where the minor M_{ikjl} of a_{ik} in M_{jl} is obtained by deleting the ith and jth rows and the kth and lth columns of D [and is, therefore, identical with M_{ikjl} in (20), so that our notation is consistent]. We insert the expansions of the M_{jl} into that of D. It follows from (22) and (23) that this yields a representation whose terms are identical with those given by (21a) when $l < k$.

Case II. If $k < l$, then a_{ik} belongs to the kth column of M_{jl}, we obtain an additional minus sign, and the result agrees with that characterized by (21b).

We have shown that the two expansions of D consist of the same terms, and this proves our statement concerning rows.

The proof of the statement concerning ***columns*** is quite similar; if we expand D in terms of two arbitrary columns, say, the kth and the lth, we find that the general term involving $a_{jl}a_{ik}$ is exactly the same as before. This proves that not only all column expansions of D yield the same value, but also that their common value is equal to the common value of the row expansions of D.

This completes the proof and shows that *our definition of an nth-order determinant is unambiguous.* ■

Section 9.3, page 368

PROOF OF FORMULA (2)

We prove that in right-handed Cartesian coordinates, the vector product

$$\mathbf{v} = \mathbf{a} \times \mathbf{b} = [a_1, \quad a_2, \quad a_3] \times [b_1, \quad b_2, \quad b_3]$$

has the components

$$v_1 = a_2b_3 - a_3b_2, \qquad v_2 = a_3b_1 - a_1b_3, \qquad v_3 = a_1b_2 - a_2b_1. \tag{2}$$

We need only consider the case $\mathbf{v} \neq \mathbf{0}$. Since $\mathbf{v}$ is perpendicular to both $\mathbf{a}$ and $\mathbf{b}$, Theorem 1 in Sec. 9.2 gives $\mathbf{a} \cdot \mathbf{v} = 0$ and $\mathbf{b} \cdot \mathbf{v} = 0$; in components [see (2), Sec. 9.2],

$$\begin{aligned} a_1v_1 + a_2v_2 + a_3v_3 &= 0 \\ b_1v_1 + b_2v_2 + b_3v_3 &= 0. \end{aligned} \tag{3}$$

Multiplying the first equation by b_3, the last by a_3, and subtracting, we obtain

$$(a_3b_1 - a_1b_3)v_1 = (a_2b_3 - a_3b_2)v_2.$$

Multiplying the first equation by b_1, the last by a_1, and subtracting, we obtain

$$(a_1b_2 - a_2b_1)v_2 = (a_3b_1 - a_1b_3)v_3.$$

We can easily verify that these two equations are satisfied by

$$(4) \qquad v_1 = c(a_2b_3 - a_3b_2), \qquad v_2 = c(a_3b_1 - a_1b_3), \qquad v_3 = c(a_1b_2 - a_2b_1)$$

where c is a constant. The reader may verify, by inserting, that (4) also satisfies (3). Now each of the equations in (3) represents a plane through the origin in $v_1v_2v_3$-space. The vectors **a** and **b** are normal vectors of these planes (see Example 6 in Sec. 9.2). Since $\mathbf{v} \neq \mathbf{0}$, these vectors are not parallel and the two planes do not coincide. Hence their intersection is a straight line L through the origin. Since (4) is a solution of (3) and, for varying c, represents a straight line, we conclude that (4) represents L, and every solution of (3) must be of the form (4). In particular, the components of **v** must be of this form, where c is to be determined. From (4) we obtain

$$|\mathbf{v}|^2 = v_1^2 + v_2^2 + v_3^2 = c^2\left[(a_2b_3 - a_3b_2)^2 + (a_3b_1 - a_1b_3)^2 + (a_1b_2 - a_2b_1)^2\right].$$

This can be written

$$|\mathbf{v}|^2 = c^2\left[(a_1^2 + a_2^2 + a_3^2)(b_1^2 + b_2^2 + b_3^2) - (a_1b_1 + a_2b_2 + a_3b_3)^2\right],$$

as can be verified by performing the indicated multiplications in both formulas and comparing. Using (2) in Sec. 9.2, we thus have

$$|\mathbf{v}|^2 = c^2\left[(\mathbf{a} \bullet \mathbf{a})(\mathbf{b} \bullet \mathbf{b}) - (\mathbf{a} \bullet \mathbf{b})^2\right].$$

By comparing this with formula (12) in Prob. 4 of Problem Set 9.3 we conclude that $c = \pm 1$.

We show that $c = +1$. This can be done as follows.

If we change the lengths and directions of **a** and **b** continuously and so that at the end $\mathbf{a} = \mathbf{i}$ and $\mathbf{b} = \mathbf{j}$ (Fig. 188a in Sec. 9.3), then **v** will change its length and direction continuously, and at the end, $\mathbf{v} = \mathbf{i} \times \mathbf{j} = \mathbf{k}$. Obviously we may effect the change so that both **a** and **b** remain different from the zero vector and are not parallel at any instant. Then **v** is never equal to the zero vector, and since the change is continuous and c can only assume the values $+1$ or -1, it follows that at the end c must have the same value as before. Now at the end $\mathbf{a} = \mathbf{i}$, $\mathbf{b} = \mathbf{j}$, $\mathbf{v} = \mathbf{k}$ and, therefore, $a_1 = 1$, $b_2 = 1$, $v_3 = 1$, and the other components in (4) are zero. Hence from (4) we see that $v_3 = c = +1$. This proves Theorem 1.

For a left-handed coordinate system, $\mathbf{i} \times \mathbf{j} = -\mathbf{k}$ (see Fig. 188b in Sec. 9.3), resulting in $c = -1$. This proves the statement right after formula (2). ■

Section 9.9, page 408

PROOF OF THE INVARIANCE OF THE CURL

This proof will follow from two theorems (A and B), which we prove first.

THEOREM A **Transformation Law for Vector Components**

For any vector $\mathbf{v}$ *the components* v_1, v_2, v_3 *and* v_1^*, v_2^*, v_3^* *in any two systems of Cartesian coordinates* x_1, x_2, x_3 *and* x_1^*, x_2^*, x_3^**, respectively, are related by*

$$
\begin{aligned}
v_1^* &= c_{11}v_1 + c_{12}v_2 + c_{13}v_3 \\
v_2^* &= c_{21}v_1 + c_{22}v_2 + c_{23}v_3 \\
v_3^* &= c_{31}v_1 + c_{32}v_2 + c_{33}v_3,
\end{aligned}
\tag{1}
$$

and conversely

$$
\begin{aligned}
v_1 &= c_{11}v_1^* + c_{21}v_2^* + c_{31}v_3^* \\
v_2 &= c_{12}v_1^* + c_{22}v_2^* + c_{32}v_3^* \\
v_3 &= c_{13}v_1^* + c_{23}v_2^* + c_{33}v_3^*
\end{aligned}
\tag{2}
$$

with coefficients

$$
\begin{aligned}
c_{11} &= \mathbf{i}^* \cdot \mathbf{i} & c_{12} &= \mathbf{i}^* \cdot \mathbf{j} & c_{13} &= \mathbf{i}^* \cdot \mathbf{k} \\
c_{21} &= \mathbf{j}^* \cdot \mathbf{i} & c_{22} &= \mathbf{j}^* \cdot \mathbf{j} & c_{23} &= \mathbf{j}^* \cdot \mathbf{k} \\
c_{31} &= \mathbf{k}^* \cdot \mathbf{i} & c_{32} &= \mathbf{k}^* \cdot \mathbf{j} & c_{33} &= \mathbf{k}^* \cdot \mathbf{k}
\end{aligned}
\tag{3}
$$

satisfying

$$
\sum_{j=1}^{3} c_{kj}c_{mj} = \delta_{km} \qquad (k, m = 1, 2, 3),
\tag{4}
$$

where the **Kronecker delta**[2] *is given by*

$$
\delta_{km} = \begin{cases} 0 & (k \neq m) \\ 1 & (k = m) \end{cases}
$$

and $\mathbf{i}, \mathbf{j}, \mathbf{k}$ *and* $\mathbf{i}^*, \mathbf{j}^*, \mathbf{k}^*$ *denote the unit vectors in the positive* x_1-, x_2-, x_3- *and* x_1^*-, x_2^*-, x_3^*-*directions, respectively.*

---

[2]LEOPOLD KRONECKER (1823–1891), German mathematician at Berlin, who made important contributions to algebra, group theory, and number theory.

We shall keep our discussion completely independent of Chap. 7, but readers familiar with matrices should recognize that we are dealing with **orthogonal transformations and matrices** and that our present theorem follows from Theorem 2 in Sec. 8.3.

PROOF The representation of $\mathbf{v}$ in the two systems are

$$\text{(5)} \qquad \text{(a)} \quad \mathbf{v} = v_1\mathbf{i} + v_2\mathbf{j} + v_3\mathbf{k} \qquad \text{(b)} \quad \mathbf{v} = v_1^*\mathbf{i}^* + v_2^*\mathbf{j}^* + v_3^*\mathbf{k}^*.$$

Since $\mathbf{i}^* \cdot \mathbf{i}^* = 1$, $\mathbf{i}^* \cdot \mathbf{j}^* = 0$, $\mathbf{i}^* \cdot \mathbf{k}^* = 0$, we get from (5b) simply $\mathbf{i}^* \cdot \mathbf{v} = v_1^*$ and from this and (5a)

$$v_1^* = \mathbf{i}^* \cdot \mathbf{v} = \mathbf{i}^* \cdot v_1\mathbf{i} + \mathbf{i}^* \cdot v_2\mathbf{j} + \mathbf{i}^* \cdot v_3\mathbf{k} = v_1\mathbf{i}^* \cdot \mathbf{i} + v_2\mathbf{i}^* \cdot \mathbf{j} + v_3\mathbf{i}^* \cdot \mathbf{k}.$$

Because of (3), this is the first formula in (1), and the other two formulas are obtained similarly, by considering $\mathbf{j}^* \cdot \mathbf{v}$ and then $\mathbf{k}^* \cdot \mathbf{v}$. Formula (2) follows by the same idea, taking $\mathbf{i} \cdot \mathbf{v} = v_1$ from (5a) and then from (5b) and (3)

$$v_1 = \mathbf{i} \cdot \mathbf{v} = v_1^*\mathbf{i} \cdot \mathbf{i}^* + v_2^*\mathbf{i} \cdot \mathbf{j}^* + v_3^*\mathbf{i} \cdot \mathbf{k}^* = c_{11}v_1^* + c_{21}v_2^* + c_{31}v_3^*,$$

and similarly for the other two components.

We prove (4). We can write (1) and (2) briefly as

$$\text{(6)} \qquad \text{(a)} \quad v_j = \sum_{m=1}^{3} c_{mj}v_m^*, \qquad \text{(b)} \quad v_k^* = \sum_{j=1}^{3} c_{kj}v_j.$$

Substituting v_j into v_k^*, we get

$$v_k^* = \sum_{j=1}^{3} c_{kj} \sum_{m=1}^{3} c_{mj}v_m^* = \sum_{m=1}^{3} v_m^* \left(\sum_{j=1}^{3} c_{kj}c_{mj} \right),$$

where $k = 1, 2, 3$. Taking $k = 1$, we have

$$v_1^* = v_1^* \left(\sum_{j=1}^{3} c_{1j}c_{1j} \right) + v_2^* \left(\sum_{j=1}^{3} c_{1j}c_{2j} \right) + v_3^* \left(\sum_{j=1}^{3} c_{1j}c_{3j} \right).$$

For this to hold for *every* vector $\mathbf{v}$, the first sum must be 1 and the other two sums 0. This proves (4) with $k = 1$ for $m = 1, 2, 3$. Taking $k = 2$ and then $k = 3$, we obtain (4) with $k = 2$ and 3, for $m = 1, 2, 3$. ■

THEOREM B **Transformation Law for Cartesian Coordinates**

The transformation of any Cartesian $x_1x_2x_3$-coordinate system into any other Cartesian $x_1^*x_2^*x_3^*$-coordinate system is of the form

$$\text{(7)} \qquad x_m^* = \sum_{j=1}^{3} c_{mj}x_j + b_m, \qquad m = 1, 2, 3,$$

with coefficients (3) *and constants* b_1, b_2, b_3; *conversely,*

$$\text{(8)} \qquad x_k = \sum_{n=1}^{3} c_{nk}x_n^* + \tilde{b}_k, \qquad k = 1, 2, 3.$$

Theorem B follows from Theorem A by noting that the most general transformation of a Cartesian coordinate system into another such system may be decomposed into a transformation of the type just considered and a translation; and under a translation, corresponding coordinates differ merely by a constant.

PROOF OF THE INVARIANCE OF THE CURL

We write again x_1, x_2, x_3 instead of x, y, z, and similarly x_1^*, x_2^*, x_3^* for other Cartesian coordinates, assuming that both systems are right-handed. Let a_1, a_2, a_3 denote the components of curl $\mathbf{v}$ in the $x_1x_2x_3$-coordinates, as given by (1), Sec. 9.9, with

$$x = x_1, \qquad y = x_2, \qquad z = x_3.$$

Similarly, let a_1^*, a_2^*, a_3^* denote the components of curl $\mathbf{v}$ in the $x_1^*x_2^*x_3^*$-coordinate system. We prove that the length and direction of curl $\mathbf{v}$ are independent of the particular choice of Cartesian coordinates, as asserted. We do this by showing that the components of curl $\mathbf{v}$ satisfy the transformation law (2), which is characteristic of vector components. We consider a_1. We use (6a), and then the chain rule for functions of several variables (Sec. 9.6). This gives

$$a_1 = \frac{\partial v_3}{\partial x_2} - \frac{\partial v_2}{\partial x_3} = \sum_{m=1}^{3} \left(c_{m3} \frac{\partial v_m^*}{\partial x_2} - c_{m2} \frac{\partial v_m^*}{\partial x_3} \right)$$

$$= \sum_{m=1}^{3} \sum_{j=1}^{3} \left(c_{m3} \frac{\partial v_m^*}{\partial x_j^*} \frac{\partial x_j^*}{\partial x_2} - c_{m2} \frac{\partial v_m^*}{\partial x_j^*} \frac{\partial x_j^*}{\partial x_3} \right).$$

From this and (7) we obtain

$$a_1 = \sum_{m=1}^{3} \sum_{j=1}^{3} (c_{m3}c_{j2} - c_{m2}c_{j3}) \frac{\partial v_m^*}{\partial x_j^*}$$

$$= (c_{33}c_{22} - c_{32}c_{23}) \left(\frac{\partial v_3^*}{\partial x_2^*} - \frac{\partial v_2^*}{\partial x_3^*} \right) + \cdots$$

$$= (c_{33}c_{22} - c_{32}c_{23})a_1^* + (c_{13}c_{32} - c_{12}c_{33})a_2^* + (c_{23}c_{12} - c_{22}c_{13})a_3^*.$$

Note what we did. The double sum had $3 \times 3 = 9$ terms, 3 of which were zero (when $m = j$), and the remaining 6 terms we combined in pairs as we needed them in getting a_1^*, a_2^*, a_3^*.

We now use (3), Lagrange's identity (see Formula (15) in Team Project 24 in Problem Set 9.3) and $\mathbf{k}^* \times \mathbf{j}^* = -\mathbf{i}^*$ and $\mathbf{k} \times \mathbf{j} = -\mathbf{i}$. Then

$$c_{33}c_{22} - c_{32}c_{23} = (\mathbf{k}^* \bullet \mathbf{k})(\mathbf{j}^* \bullet \mathbf{j}) - (\mathbf{k}^* \bullet \mathbf{j})(\mathbf{j}^* \bullet \mathbf{k})$$

$$= (\mathbf{k}^* \times \mathbf{j}^*) \bullet (\mathbf{k} \times \mathbf{j}) = \mathbf{i}^* \bullet \mathbf{i} = c_{11}, \qquad \text{etc.}$$

Hence $a_1 = c_{11}a_1^* + c_{21}a_2^* + c_{31}a_3^*$. This is of the form of the first formula in (2) in Theorem A, and the other two formulas of the form (2) are obtained similarly. This proves the theorem for right-handed systems. If the $x_1x_2x_3$-coordinates are left-handed, then $\mathbf{k} \times \mathbf{j} = +\mathbf{i}$, but then there is a minus sign in front of the determinant in (1), Sec. 9.9. ■

Section 10.2, page 420

PROOF OF THEOREM 1, PART (b) We prove that if

$$\int_C \mathbf{F}(\mathbf{r}) \cdot d\mathbf{r} = \int_C (F_1\,dx + F_2\,dy + F_3\,dz) \tag{1}$$

with continuous F_1, F_2, F_3 in a domain D is independent of path in D, then F = grad f in D for some f; in components

$$F_1 = \frac{\partial f}{\partial x}, \qquad F_2 = \frac{\partial f}{\partial y}, \qquad F_3 = \frac{\partial f}{\partial z}. \tag{2'}$$

We choose any fixed A: (x_0, y_0, z_0) in D and any B: (x, y, z) in D and define f by

$$f(x, y, z) = f_0 + \int_A^B (F_1\,dx^* + F_2\,dy^* + F_3\,dz^*) \tag{3}$$

with any constant f_0 and any path from A to B in D. Since A is fixed and we have independence of path, the integral depends only on the coordinates x, y, z, so that (3) defines a function $f(x, y, z)$ in D. We show that $\mathbf{F} = \text{grad}\, f$ with this f, beginning with the first of the three relations (2′). Because of independence of path we may integrate from A to B_1: (x_1, y, z) and then parallel to the x-axis along the segment B_1B in Fig. 562 with B_1 chosen so that the whole segment lies in D. Then

$$f(x, y, z) = f_0 + \int_A^{B_1} (F_1\,dx^* + F_2\,dy^* + F_3\,dz^*) + \int_{B_1}^{B} (F_1\,dx^* + F_2\,dy^* + F_3\,dz^*).$$

We now take the partial derivative with respect to x on both sides. On the left we get $\partial f/\partial x$. We show that on the right we get F_1. The derivative of the first integral is zero because A: (x_0, y_0, z_0) and B_1: (x_1, y, z) do not depend on x. We consider the second integral. Since on the segment B_1B, both y and z are constant, the terms $F_2\,dy^*$ and

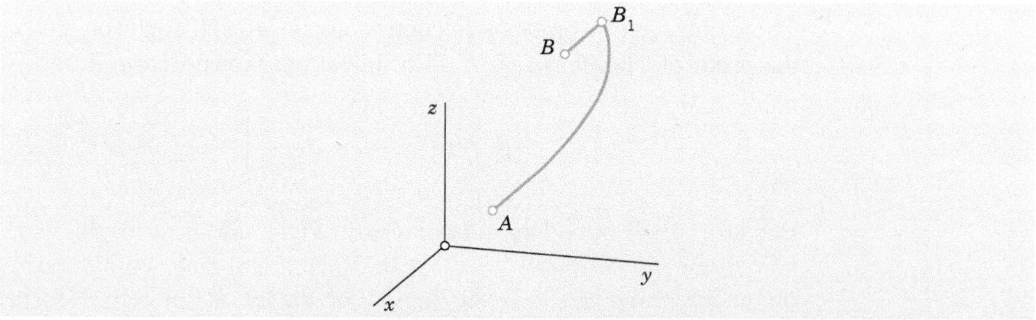

Fig. 562. Proof of Theorem 1

$F_3\,dz^*$ do not contribute to the derivative of the integral. The remaining part can be written as a definite integral,

$$\int_{B_1}^{B} F_1\,dx^* = \int_{x_1}^{x} F_1(x^*, y, z)\,dx^*.$$

Hence its partial derivative with respect to x is $F_1(x, y, z)$, and the first of the relations $(2')$ is proved. The other two formulas in $(2')$ follow by the same argument. ■

Section 11.5, page 500

THEOREM

Reality of Eigenvalues

If p, q, r, and p' in the Sturm–Liouville equation (1) *of Sec.* 11.5 *are real-valued and continuous on the interval $a \leqq x \leqq b$ and $r(x) > 0$ throughout that interval (or $r(x) < 0$ throughout that interval), then all the eigenvalues of the Sturm–Liouville problem* (1), (2), *Sec.* 11.5, *are real.*

PROOF Let $\lambda = \alpha + i\beta$ be an eigenvalue of the problem and let

$$y(x) = u(x) + iv(x)$$

be a corresponding eigenfunction; here α, β, u, and v are real. Substituting this into (1), Sec. 11.5, we have

$$(pu' + ipv')' + (q + \alpha r + i\beta r)(u + iv) = 0.$$

This complex equation is equivalent to the following pair of equations for the real and the imaginary parts:

$$(pu')' + (q + \alpha r)u - \beta rv = 0$$

$$(pv')' + (q + \alpha r)v + \beta ru = 0.$$

Multiplying the first equation by v, the second by $-u$ and adding, we get

$$-\beta(u^2 + v^2)r = u(pv')' - v(pu')'$$

$$= [(pv')u - (pu')v]'.$$

The expression in brackets is continuous on $a \leqq x \leqq b$, for reasons similar to those in the proof of Theorem 1, Sec. 11.5. Integrating over x from a to b, we thus obtain

$$-\beta \int_a^b (u^2 + v^2)r\,dx = \Big[p(uv' - u'v)\Big]_a^b.$$

Because of the boundary conditions, the right side is zero; this is as in that proof. Since y is an eigenfunction, $u^2 + v^2 \not\equiv 0$. Since y and r are continuous and $r > 0$ (or $r < 0$) on the interval $a \leqq x \leqq b$, the integral on the left is not zero. Hence, $\beta = 0$, which means that $\lambda = \alpha$ is real. This completes the proof. ■

Section 13.4, page 627

PROOF OF THEOREM 2   Cauchy–Riemann Equations

We prove that Cauchy–Riemann equations

(1)
$$u_x = v_y, \qquad u_y = -v_x$$

are sufficient for a complex function $f(z) = u(x, y) + iv(x, y)$ to be analytic; precisely, *if the real part u and the imaginary part v of f(z) satisfy* (1) *in a domain D in the complex plane and if the partial derivatives in* (1) *are* **continuous** *in D, then f(z) is analytic in D.*

In this proof we write $\Delta z = \Delta x + i\Delta y$ and $\Delta f = f(z + \Delta z) - f(z)$. The idea of proof is as follows.

(a) We express Δf in terms of first partial derivatives of u and v, by applying the mean value theorem of Sec. 9.6.

(b) We get rid of partial derivatives with respect to y by applying the Cauchy–Riemann equations.

(c) We let Δz approach zero and show that then $\Delta f/\Delta z$, as obtained, approaches a limit, which is equal to $u_x + iv_x$, the right side of (4) in Sec. 13.4, regardless of the way of approach to zero.

(a) Let $P: (x, y)$ be any fixed point in D. Since D is a domain, it contains a neighborhood of P. We can choose a point $Q: (x + \Delta x, y + \Delta y)$ in this neighborhood such that the straight-line segment PQ is in D. Because of our continuity assumptions we may apply the mean value theorem in Sec. 9.6. This yields

$$u(x + \Delta x, y + \Delta y) - u(x, y) = (\Delta x)u_x(M_1) + (\Delta y)u_y(M_1)$$

$$v(x + \Delta x, y + \Delta y) - v(x, y) = (\Delta x)v_x(M_2) + (\Delta y)v_y(M_2)$$

where M_1 and M_2 ($\neq M_1$ in general!) are suitable points on that segment. The first line is Re Δf and the second is Im Δf, so that

$$\Delta f = (\Delta x)u_x(M_1) + (\Delta y)u_y(M_1) + i[(\Delta x)v_x(M_2) + (\Delta y)v_y(M_2)].$$

(b) $u_y = -v_x$ and $v_y = u_x$ by the Cauchy–Riemann equations, so that

$$\Delta f = (\Delta x)u_x(M_1) - (\Delta y)v_x(M_1) + i[(\Delta x)v_x(M_2) + (\Delta y)u_x(M_2)].$$

Also $\Delta z = \Delta x + i\Delta y$, so that we can write $\Delta x = \Delta z - i\Delta y$ in the first term and $\Delta y = (\Delta z - \Delta x)/i = -i(\Delta z - \Delta x)$ in the second term. This gives

$$\Delta f = (\Delta z - i\Delta y)u_x(M_1) + i(\Delta z - \Delta x)v_x(M_1) + i[(\Delta x)v_x(M_2) + (\Delta y)u_x(M_2)].$$

By performing the multiplications and reordering we obtain

$$\Delta f = (\Delta z)u_x(M_1) - i\Delta y\{u_x(M_1) - u_x(M_2)\}$$
$$+ i[(\Delta z)v_x(M_1) - \Delta x\{v_x(M_1) - v_x(M_2)\}].$$

Division by Δz now yields

$$\text{(A)}\quad \frac{\Delta f}{\Delta z} = u_x(M_1) + iv_x(M_1) - \frac{i\Delta y}{\Delta z}\{u_x(M_1) - u_x(M_2)\} - \frac{i\Delta x}{\Delta z}\{v_x(M_1) - v_x(M_2)\}.$$

(c) We finally let Δz approach zero and note that $|\Delta y/\Delta z| \leqq 1$ and $|\Delta x/\Delta z| \leqq 1$ in (A). Then Q: $(x + \Delta x, y + \Delta y)$ approaches P: (x, y), so that M_1 and M_2 must approach P. Also, since the partial derivatives in (A) are assumed to be continuous, they approach their value at P. In particular, the differences in the braces $\{\cdots\}$ in (A) approach zero. Hence the limit of the right side of (A) exists and is independent of the path along which $\Delta z \to 0$. We see that this limit equals the right side of (4) in Sec. 13.4. This means that $f(z)$ is analytic at every point z in D, and the proof is complete. ■

Section 14.2, pages 653–654

GOURSAT'S PROOF OF CAUCHY'S INTEGRAL THEOREM Goursat proved Cauchy's integral theorem without assuming that $f'(z)$ is continuous, as follows.

We start with the case when C is the boundary of a triangle. We orient C counterclockwise. By joining the midpoints of the sides we subdivide the triangle into four congruent triangles (Fig. 563). Let C_{I}, C_{II}, C_{III}, C_{IV} denote their boundaries. We claim that (see Fig. 563).

$$\text{(1)}\qquad \oint_C f\,dz = \oint_{C_{\text{I}}} f\,dz + \oint_{C_{\text{II}}} f\,dz + \oint_{C_{\text{III}}} f\,dz + \oint_{C_{\text{IV}}} f\,dz.$$

Indeed, on the right we integrate along each of the three segments of subdivision in both possible directions (Fig. 563), so that the corresponding integrals cancel out in pairs, and the sum of the integrals on the right equals the integral on the left. We now pick an integral on the right that is biggest in absolute value and call its path C_1. Then, by the triangle inequality (Sec. 13.2),

$$\left|\oint_C f\,dz\right| \leqq \left|\oint_{C_{\text{I}}} f\,dz\right| + \left|\oint_{C_{\text{II}}} f\,dz\right| + \left|\oint_{C_{\text{III}}} f\,dz\right| + \left|\oint_{C_{\text{IV}}} f\,dz\right| \leqq 4\left|\oint_{C_1} f\,dz\right|.$$

We now subdivide the triangle bounded by C_1 as before and select a triangle of subdivision with boundary C_2 for which

$$\left|\oint_{C_1} f\,dz\right| \leqq 4\left|\oint_{C_2} f\,dz\right|. \qquad \text{Then} \qquad \left|\oint_C f\,dz\right| \leqq 4^2\left|\oint_{C_2} f\,dz\right|.$$

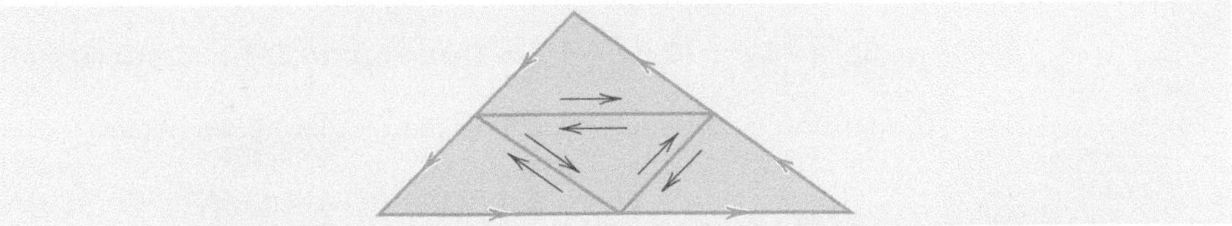

Fig. 563. Proof of Cauchy's integral theorem

Continuing in this fashion, we obtain a sequence of triangles $T_1, T_2, \cdots$ with boundaries $C_1, C_2, \cdots$ that are similar and such that T_n lies in T_m when $n > m$, and

(2)
$$\left|\oint_C f\,dz\right| \leqq 4^n \left|\oint_{C_n} f\,dz\right|, \qquad n = 1, 2, \cdots.$$

Let z_0 be the point that belongs to all these triangles. Since f is differentiable at $z = z_0$, the derivative $f'(z_0)$ exists. Let

(3)
$$h(z) = \frac{f(z) - f(z_0)}{z - z_0} - f'(z_0).$$

Solving this algebraically for $f(z)$ we have

$$f(z) = f(z_0) + (z - z_0)f'(z_0) + h(z)(z - z_0).$$

Integrating this over the boundary C_n of the triangle T_n gives

$$\oint_{C_n} f(z)\,dz = \oint_{C_n} f(z_0)\,dz + \oint_{C_n} (z - z_0)f'(z_0)\,dz + \oint_{C_n} h(z)(z - z_0)dz.$$

Since $f(z_0)$ and $f'(z_0)$ are constants and C_n is a closed path, the first two integrals on the right are zero, as follows from Cauchy's proof, which is applicable because the integrands do have continuous derivatives (0 and const, respectively). We thus have

$$\oint_{C_n} f(z)\,dz = \oint_{C_n} h(z)(z - z_0)\,dz.$$

Since $f'(z_0)$ is the limit of the difference quotient in (3), for given $\epsilon > 0$ we can find a $\delta > 0$ such that

(4)
$$|h(z)| < \epsilon \qquad \text{when} \qquad |z - z_0| < \delta.$$

We may now take n so large that the triangle T_n lies in the disk $|z - z_0| < \delta$. Let L_n be the length of C_n. Then $|z - z_0| < L_n$ for all z on C_n and z_0 in T_n. From this and (4) we have $|h(z)(z - z_0)| < \epsilon L_n$. The *ML*-inequality in Sec. 14.1 now gives

(5)
$$\left|\oint_{C_n} f(z)\,dz\right| = \left|\oint_{C_n} h(z)(z - z_0)\,dz\right| \leqq \epsilon L_n \cdot L_n = \epsilon L_n^2.$$

Now denote the length of C by L. Then the path C_1 has the length $L_1 = L/2$, the path C_2 has the length $L_2 = L_1/2 = L/4$, etc., and C_n has the length $L_n = L/2^n$. Hence $L_n^2 = L^2/4^n$. From (2) and (5) we thus obtain

$$\left|\oint_C f\,dz\right| \leqq 4^n \left|\oint_{C_n} f\,dz\right| \leqq 4^n \epsilon L_n^2 = 4^n \epsilon \frac{L^2}{4^n} = \epsilon L^2.$$

By choosing ϵ (> 0) sufficiently small we can make the expression on the right as small as we please, while the expression on the left is the definite value of an integral. Consequently, this value must be zero, and the proof is complete.

The proof for *the case in which C is the boundary of a polygon* follows from the previous proof by subdividing the polygon into triangles (Fig. 564). The integral corresponding to each such triangle is zero. The sum of these integrals is equal to the integral over C, because we integrate along each segment of subdivision in both directions, the corresponding integrals cancel out in pairs, and we are left with the integral over C.

The case of a general simple closed path C can be reduced to the preceding one by inscribing in C a closed polygon P of chords, which approximates C "sufficiently accurately," and it can be shown that there is a polygon P such that the integral over P differs from that over C by less than any preassigned positive real number $\tilde{\epsilon}$, no matter how small. The details of this proof are somewhat involved and can be found in Ref. [D6] listed in App. 1. ■

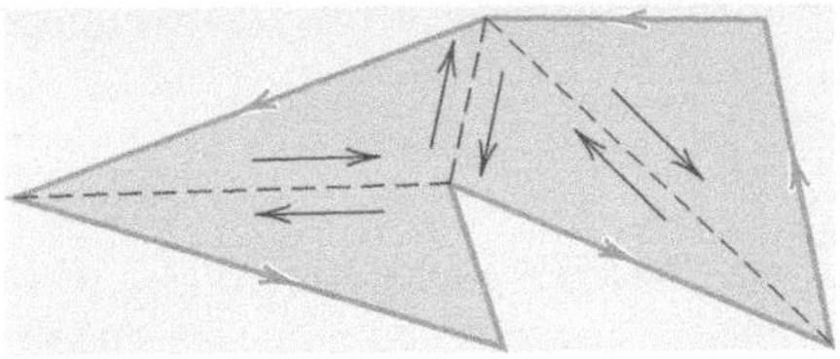

Fig. 564. Proof of Cauchy's integral theorem for a polygon

Section 15.1, page 674

PROOF OF THEOREM 4 Cauchy's Convergence Principle for Series

(a) In this proof we need two concepts and a theorem, which we list first.

1. A **bounded sequence** $s_1, s_2, \cdots$ is a sequence whose terms all lie in a disk of (sufficiently large, finite) radius K with center at the origin; thus $|s_n| < K$ for all n.

2. A **limit point** a of a sequence $s_1, s_2, \cdots$ is a point such that, given an $\epsilon > 0$, there are infinitely many terms satisfying $|s_n - a| < \epsilon$. (Note that this does *not* imply convergence, since there may still be infinitely many terms that do not lie within that circle of radius ϵ and center a.)

Example: $\frac{1}{4}, \frac{3}{4}, \frac{1}{8}, \frac{7}{8}, \frac{1}{16}, \frac{15}{16}, \cdots$ has the limit points 0 and 1 and diverges.

3. A bounded sequence in the complex plane has at least one limit point. (Bolzano–Weierstrass theorem; proof below. Recall that "sequence" always means *infinite* sequence.)

(b) We now turn to the actual proof that $z_1 + z_2 + \cdots$ converges if and only if, for every $\epsilon > 0$, we can find an N such that

$$|z_{n+1} + \cdots + z_{n+p}| < \epsilon \qquad \text{for every } n > N \text{ and } p = 1, 2, \cdots. \tag{1}$$

Here, by the definition of partial sums,

$$s_{n+p} - s_n = z_{n+1} + \cdots + z_{n+p}.$$

Writing $n + p = r$, we see from this that (1) is equivalent to

$$|s_r - s_n| < \epsilon \qquad \text{for all } r > N \text{ and } n > N. \tag{1*}$$

Suppose that $s_1, s_2, \cdots$ converges. Denote its limit by s. Then for a given $\epsilon > 0$ we can find an N such that

$$|s_n - s| < \frac{\epsilon}{2} \qquad \text{for every } n > N.$$

Hence, if $r > N$ and $n > N$, then by the triangle inequality (Sec. 13.2),

$$|s_r - s_n| = |(s_r - s) - (s_n - s)| \leqq |s_r - s| + |s_n - s| < \frac{\epsilon}{2} + \frac{\epsilon}{2} = \epsilon,$$

that is, (1*) holds.

(c) Conversely, assume that $s_1, s_2, \cdots$ satisfies (1*). We first prove that then the sequence must be bounded. Indeed, choose a fixed ϵ and a fixed $n = n_0 > N$ in (1*). Then (1*) implies that all s_r with $r > N$ lie in the disk of radius ϵ and center s_{n_0} and only *finitely many terms* $s_1, \cdots, s_N$ may not lie in this disk. Clearly, we can now find a circle so large that this disk and these finitely many terms all lie within this new circle. Hence the sequence is bounded. By the Bolzano–Weierstrass theorem, it has at least one limit point, call it s.

We now show that the sequence is convergent with the limit s. Let $\epsilon > 0$ be given. Then there is an N^* such that $|s_r - s_n| < \epsilon/2$ for all $r > N^*$ and $n > N^*$, by (1*). Also, by the definition of a limit point, $|s_n - s| < \epsilon/2$ for *infinitely many* n, so that we can find and fix an $n > N^*$ such that $|s_n - s| < \epsilon/2$. Together, for *every* $r > N^*$,

$$|s_r - s| = |(s_r - s_n) + (s_n - s)| \leqq |s_r - s_n| + |s_n - s| < \frac{\epsilon}{2} + \frac{\epsilon}{2} = \epsilon;$$

that is, the sequence $s_1, s_2, \cdots$ is convergent with the limit s.   ■

THEOREM

Bolzano–Weierstrass Theorem[3]

A bounded infinite sequence $z_1, z_2, z_3, \cdots$ in the complex plane has at least one limit point.

PROOF

It is obvious that we need both conditions: a finite sequence cannot have a limit point, and the sequence $1, 2, 3, \cdots$, which is infinite but not bounded, has no limit point. To prove the theorem, consider a bounded infinite sequence $z_1, z_2, \cdots$ and let K be such that $|z_n| < K$ for all n. If only finitely many values of the z_n are different, then, since the sequence is infinite, some number z must occur infinitely many times in the sequence, and, by definition, this number is a limit point of the sequence.

We may now turn to the case when the sequence contains infinitely many *different* terms. We draw a large square Q_0 that contains all z_n. We subdivide Q_0 into four congruent squares, which we number 1, 2, 3, 4. Clearly, at least one of these squares (each taken with its complete boundary) must contain infinitely many terms of the sequence. The square of this type with the lowest number (1, 2, 3, or 4) will be denoted by Q_1. This is

---

[3]BERNARD BOLZANO (1781–1848), Austrian mathematician and professor of religious studies, was a pioneer in the study of point sets, the foundation of analysis, and mathematical logic.
For Weierstrass, see Sec. 15.5.

the first step. In the next step we subdivide Q_1 into four congruent squares and select a square Q_2 by the same rule, and so on. This yields an infinite sequence of squares Q_0, $Q_1, Q_2, \cdots, Q_n, \cdots$ with the property that the side of Q_n approaches zero as n approaches infinity, and Q_m contains all Q_n with $n > m$. It is not difficult to see that the number which belongs to all these squares,[4] call it $z = a$, is a limit point of the sequence. In fact, given an $\epsilon > 0$, we can choose an N so large that the side of the square Q_N is less than ϵ and, since Q_N contains infinitely many z_n, we have $|z_n - a| < \epsilon$ for infinitely many n. This completes the proof. ■

Section 15.3, pages 688–689

PART (b) OF THE PROOF OF THEOREM 5

We have to show that

$$\sum_{n=2}^{\infty} a_n \left[\frac{(z + \Delta z)^n - z^n}{\Delta z} - nz^{n-1} \right]$$

$$= \sum_{n=2}^{\infty} a_n \, \Delta z[(z + \Delta z)^{n-2} + 2z(z + \Delta z)^{n-3} + \cdots + (n-1)z^{n-2}],$$

thus,

$$\frac{(z + \Delta z)^n - z^n}{\Delta z} - nz^{n-1}$$

$$= \Delta z[(z + \Delta z)^{n-2} + 2z(z + \Delta z)^{n-3} + \cdots + (n-1)z^{n-2}].$$

If we set $z + \Delta z = b$ and $z = a$, thus $\Delta z = b - a$, this becomes simply

$$\text{(7a)} \qquad \frac{b^n - a^n}{b - a} - na^{n-1} = (b - a)A_n \qquad (n = 2, 3, \cdots),$$

where A_n is the expression in the brackets on the right,

$$\text{(7b)} \qquad A_n = b^{n-2} + 2ab^{n-3} + 3a^2b^{n-4} + \cdots + (n-1)a^{n-2};$$

thus, $A_2 = 1$, $A_3 = b + 2a$, etc. We prove (7) by induction. When $n = 2$, then (7) holds, since then

$$\frac{b^2 - a^2}{b - a} - 2a = \frac{(b + a)(b - a)}{b - a} - 2a = b - a = (b - a)A_2.$$

Assuming that (7) holds for $n = k$, we show that it holds for $n = k + 1$. By adding and subtracting a term in the numerator and then dividing we first obtain

$$\frac{b^{k+1} - a^{k+1}}{b - a} = \frac{b^{k+1} - ba^k + ba^k - a^{k+1}}{b - a} = b\,\frac{b^k - a^k}{b - a} + a^k.$$

---

[4] The fact that such a unique number $z = a$ exists seems to be obvious, but it actually follows from an axiom of the real number system, the so-called *Cantor–Dedekind axiom:* see footnote 3 in App. A3.3.

By the induction hypothesis, the right side equals $b[(b-a)A_k + ka^{k-1}] + a^k$. Direct calculation shows that this is equal to

$$(b-a)\{bA_k + ka^{k-1}\} + aka^{k-1} + a^k.$$

From (7b) with $n = k$ we see that the expression in the braces $\{\cdots\}$ equals

$$b^{k-1} + 2ab^{k-2} + \cdots + (k-1)ba^{k-2} + ka^{k-1} = A_{k+1}.$$

Hence our result is

$$\frac{b^{k+1} - a^{k+1}}{b-a} = (b-a)A_{k+1} + (k+1)a^k.$$

Taking the last term to the left, we obtain (7) with $n = k + 1$. This proves (7) for any integer $n \geqq 2$ and completes the proof. ■

Section 18.2, page 763

ANOTHER PROOF OF THEOREM 1 ***without the use of a harmonic conjugate***

We show that if $w = u + iv = f(z)$ is analytic and maps a domain D conformally onto a domain D^* and $\Phi^*(u, v)$ is harmonic in D^*, then

$$\Phi(x, y) = \Phi^*(u(x, y), v(x, y)) \tag{1}$$

is harmonic in D, that is, $\nabla^2\Phi = 0$ in D. We make no use of a harmonic conjugate of Φ^*, but use straightforward differentiation. By the chain rule,

$$\Phi_x = \Phi^*_u u_x + \Phi^*_v v_x.$$

We apply the chain rule again, underscoring the terms that will drop out when we form $\nabla^2\Phi$:

$$\Phi_{xx} = \underline{\Phi^*_u u_{xx}} + (\Phi^*_{uu} u_x + \underline{\Phi^*_{uv} v_x}) u_x$$
$$+ \underline{\Phi^*_v v_{xx}} + (\underline{\Phi^*_{vu} u_x} + \Phi^*_{vv} v_x) v_x.$$

Φ_{yy} is the same with each x replaced by y. We form the sum $\nabla^2\Phi$. In it, $\Phi^*_{vu} = \Phi^*_{uv}$ is multiplied by

$$u_x v_x + u_y v_y$$

which is 0 by the Cauchy–Riemann equations. Also $\nabla^2 u = 0$ and $\nabla^2 v = 0$. There remains

$$\nabla^2\Phi = \Phi^*_{uu}(u_x^2 + u_y^2) + \Phi^*_{vv}(v_x^2 + v_y^2).$$

By the Cauchy–Riemann equations this becomes

$$\nabla^2\Phi = (\Phi^*_{uu} + \Phi^*_{vv})(u_x^2 + v_x^2)$$

and is 0 since Φ^* is harmonic. ■

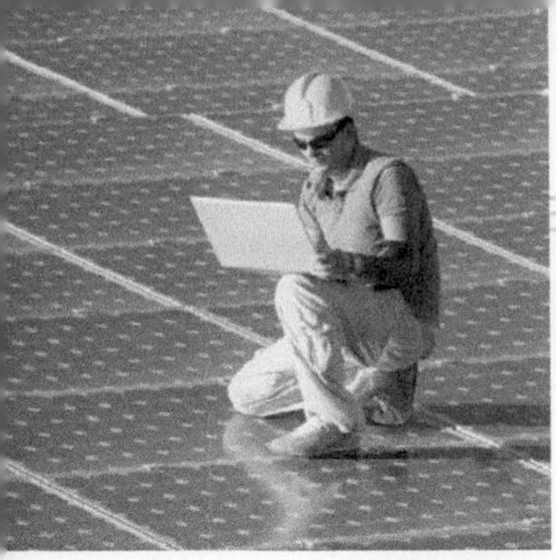

APPENDIX 5

Tables

For Tables of Laplace Transforms see Secs. 6.8 and 6.9.
For Tables of Fourier Transforms see Sec. 11.10.

If you have a Computer Algebra System (CAS), you may not need the present tables, but you may still find them convenient from time to time.

Table A1 Bessel Functions

For more extensive tables see Ref. [GenRef1] in App. 1.

x	$J_0(x)$	$J_1(x)$	x	$J_0(x)$	$J_1(x)$	x	$J_0(x)$	$J_1(x)$
0.0	1.0000	0.0000	3.0	−0.2601	0.3391	6.0	0.1506	−0.2767
0.1	0.9975	0.0499	3.1	−0.2921	0.3009	6.1	0.1773	−0.2559
0.2	0.9900	0.0995	3.2	−0.3202	0.2613	6.2	0.2017	−0.2329
0.3	0.9776	0.1483	3.3	−0.3443	0.2207	6.3	0.2238	−0.2081
0.4	0.9604	0.1960	3.4	−0.3643	0.1792	6.4	0.2433	−0.1816
0.5	0.9385	0.2423	3.5	−0.3801	0.1374	6.5	0.2601	−0.1538
0.6	0.9120	0.2867	3.6	−0.3918	0.0955	6.6	0.2740	−0.1250
0.7	0.8812	0.3290	3.7	−0.3992	0.0538	6.7	0.2851	−0.0953
0.8	0.8463	0.3688	3.8	−0.4026	0.0128	6.8	0.2931	−0.0652
0.9	0.8075	0.4059	3.9	−0.4018	−0.0272	6.9	0.2981	−0.0349
1.0	0.7652	0.4401	4.0	−0.3971	−0.0660	7.0	0.3001	−0.0047
1.1	0.7196	0.4709	4.1	−0.3887	−0.1033	7.1	0.2991	0.0252
1.2	0.6711	0.4983	4.2	−0.3766	−0.1386	7.2	0.2951	0.0543
1.3	0.6201	0.5220	4.3	−0.3610	−0.1719	7.3	0.2882	0.0826
1.4	0.5669	0.5419	4.4	−0.3423	−0.2028	7.4	0.2786	0.1096
1.5	0.5118	0.5579	4.5	−0.3205	−0.2311	7.5	0.2663	0.1352
1.6	0.4554	0.5699	4.6	−0.2961	−0.2566	7.6	0.2516	0.1592
1.7	0.3980	0.5778	4.7	−0.2693	−0.2791	7.7	0.2346	0.1813
1.8	0.3400	0.5815	4.8	−0.2404	−0.2985	7.8	0.2154	0.2014
1.9	0.2818	0.5812	4.9	−0.2097	−0.3147	7.9	0.1944	0.2192
2.0	0.2239	0.5767	5.0	−0.1776	−0.3276	8.0	0.1717	0.2346
2.1	0.1666	0.5683	5.1	−0.1443	−0.3371	8.1	0.1475	0.2476
2.2	0.1104	0.5560	5.2	−0.1103	−0.3432	8.2	0.1222	0.2580
2.3	0.0555	0.5399	5.3	−0.0758	−0.3460	8.3	0.0960	0.2657
2.4	0.0025	0.5202	5.4	−0.0412	−0.3453	8.4	0.0692	0.2708
2.5	−0.0484	0.4971	5.5	−0.0068	−0.3414	8.5	0.0419	0.2731
2.6	−0.0968	0.4708	5.6	0.0270	−0.3343	8.6	0.0146	0.2728
2.7	−0.1424	0.4416	5.7	0.0599	−0.3241	8.7	−0.0125	0.2697
2.8	−0.1850	0.4097	5.8	0.0917	−0.3110	8.8	−0.0392	0.2641
2.9	−0.2243	0.3754	5.9	0.1220	−0.2951	8.9	−0.0653	0.2559

$J_0(x) = 0$ for $x = 2.40483, 5.52008, 8.65373, 11.7915, 14.9309, 18.0711, 21.2116, 24.3525, 27.4935, 30.6346$

$J_1(x) = 0$ for $x = 3.83171, 7.01559, 10.1735, 13.3237, 16.4706, 19.6159, 22.7601, 25.9037, 29.0468, 32.1897$

Table A1 ***(continued)***

x	$Y_0(x)$	$Y_1(x)$	x	$Y_0(x)$	$Y_1(x)$	x	$Y_0(x)$	$Y_1(x)$
0.0	$(-\infty)$	$(-\infty)$	2.5	0.498	0.146	5.0	−0.309	0.148
0.5	−0.445	−1.471	3.0	0.377	0.325	5.5	−0.339	−0.024
1.0	0.088	−0.781	3.5	0.189	0.410	6.0	−0.288	−0.175
1.5	0.382	−0.412	4.0	−0.017	0.398	6.5	−0.173	−0.274
2.0	0.510	−0.107	4.5	−0.195	0.301	7.0	−0.026	−0.303

Table A2 Gamma Function [see (24) in App. A3.1]

α	$\Gamma(\alpha)$	α	$\Gamma(\alpha)$	α	$\Gamma(\alpha)$	α	$\Gamma(\alpha)$	α	$\Gamma(\alpha)$
1.00	1.000 000	1.20	0.918 169	1.40	0.887 264	1.60	0.893 515	1.80	0.931 384
1.02	0.988 844	1.22	0.913 106	1.42	0.886 356	1.62	0.895 924	1.82	0.936 845
1.04	0.978 438	1.24	0.908 521	1.44	0.885 805	1.64	0.898 642	1.84	0.942 612
1.06	0.968 744	1.26	0.904 397	1.46	0.885 604	1.66	0.901 668	1.86	0.948 687
1.08	0.959 725	1.28	0.900 718	1.48	0.885 747	1.68	0.905 001	1.88	0.955 071
1.10	0.951 351	1.30	0.897 471	1.50	0.886 227	1.70	0.908 639	1.90	0.961 766
1.12	0.943 590	1.32	0.894 640	1.52	0.887 039	1.72	0.912 581	1.92	0.968 774
1.14	0.936 416	1.34	0.892 216	1.54	0.888 178	1.74	0.916 826	1.94	0.976 099
1.16	0.929 803	1.36	0.890 185	1.56	0.889 639	1.76	0.921 375	1.96	0.983 743
1.18	0.923 728	1.38	0.888 537	1.58	0.891 420	1.78	0.926 227	1.98	0.991 708
1.20	0.918 169	1.40	0.887 264	1.60	0.893 515	1.80	0.931 384	2.00	1.000 000

Table A3 Factorial Function and Its Logarithm with Base 10

n	$n!$	log ($n!$)	n	$n!$	log ($n!$)	n	$n!$	log ($n!$)
1	1	0.000 000	6	720	2.857 332	11	39 916 800	7.601 156
2	2	0.301 030	7	5 040	3.702 431	12	479 001 600	8.680 337
3	6	0.778 151	8	40 320	4.605 521	13	6 227 020 800	9.794 280
4	24	1.380 211	9	362 880	5.559 763	14	87 178 291 200	10.940 408
5	120	2.079 181	10	3 628 800	6.559 763	15	1 307 674 368 000	12.116 500

Table A4 Error Function, Sine and Cosine Integrals [see (35), (40), (42) in App. A3.1]

x	erf x	Si(x)	ci(x)	x	erf x	Si(x)	ci(x)
0.0	0.0000	0.0000	∞	2.0	0.9953	1.6054	−0.4230
0.2	0.2227	0.1996	1.0422	2.2	0.9981	1.6876	−0.3751
0.4	0.4284	0.3965	0.3788	2.4	0.9993	1.7525	−0.3173
0.6	0.6039	0.5881	0.0223	2.6	0.9998	1.8004	−0.2533
0.8	0.7421	0.7721	−0.1983	2.8	0.9999	1.8321	−0.1865
1.0	0.8427	0.9461	−0.3374	3.0	1.0000	1.8487	−0.1196
1.2	0.9103	1.1080	−0.4205	3.2	1.0000	1.8514	−0.0553
1.4	0.9523	1.2562	−0.4620	3.4	1.0000	1.8419	0.0045
1.6	0.9763	1.3892	−0.4717	3.6	1.0000	1.8219	0.0580
1.8	0.9891	1.5058	−0.4568	3.8	1.0000	1.7934	0.1038
2.0	0.9953	1.6054	−0.4230	4.0	1.0000	1.7582	0.1410

Table A5 Binomial Distribution

Probability function $f(x)$ [see (2), Sec. 24.7] and distribution function $F(x)$

n	x	$p = 0.1$		$p = 0.2$		$p = 0.3$		$p = 0.4$		$p = 0.5$	
		$f(x)$	$F(x)$	$f(x)$	$F(x)$	$f(x)$	$F(x)$	$f(x)$	$F(x)$	$f(x)$	$F(x)$
		0.		0.		0.		0.		0.	
1	0	9000	0.9000	8000	0.8000	7000	0.7000	6000	0.6000	5000	0.5000
	1	1000	1.0000	2000	1.0000	3000	1.0000	4000	1.0000	5000	1.0000
	0	8100	0.8100	6400	0.6400	4900	0.4900	3600	0.3600	2500	0.2500
2	1	1800	0.9900	3200	0.9600	4200	0.9100	4800	0.8400	5000	0.7500
	2	0100	1.0000	0400	1.0000	0900	1.0000	1600	1.0000	2500	1.0000
	0	7290	0.7290	5120	0.5120	3430	0.3430	2160	0.2160	1250	0.1250
3	1	2430	0.9720	3840	0.8960	4410	0.7840	4320	0.6480	3750	0.5000
	2	0270	0.9990	0960	0.9920	1890	0.9730	2880	0.9360	3750	0.8750
	3	0010	1.0000	0080	1.0000	0270	1.0000	0640	1.0000	1250	1.0000
	0	6561	0.6561	4096	0.4096	2401	0.2401	1296	0.1296	0625	0.0625
	1	2916	0.9477	4096	0.8192	4116	0.6517	3456	0.4752	2500	0.3125
4	2	0486	0.9963	1536	0.9728	2646	0.9163	3456	0.8208	3750	0.6875
	3	0036	0.9999	0256	0.9984	0756	0.9919	1536	0.9744	2500	0.9375
	4	0001	1.0000	0016	1.0000	0081	1.0000	0256	1.0000	0625	1.0000
	0	5905	0.5905	3277	0.3277	1681	0.1681	0778	0.0778	0313	0.0313
	1	3281	0.9185	4096	0.7373	3602	0.5282	2592	0.3370	1563	0.1875
5	2	0729	0.9914	2048	0.9421	3087	0.8369	3456	0.6826	3125	0.5000
	3	0081	0.9995	0512	0.9933	1323	0.9692	2304	0.9130	3125	0.8125
	4	0005	1.0000	0064	0.9997	0284	0.9976	0768	0.9898	1563	0.9688
	5	0000	1.0000	0003	1.0000	0024	1.0000	0102	1.0000	0313	1.0000
	0	5314	0.5314	2621	0.2621	1176	0.1176	0467	0.0467	0156	0.0156
	1	3543	0.8857	3932	0.6554	3025	0.4202	1866	0.2333	0938	0.1094
	2	0984	0.9841	2458	0.9011	3241	0.7443	3110	0.5443	2344	0.3438
6	3	0146	0.9987	0819	0.9830	1852	0.9295	2765	0.8208	3125	0.6563
	4	0012	0.9999	0154	0.9984	0595	0.9891	1382	0.9590	2344	0.8906
	5	0001	1.0000	0015	0.9999	0102	0.9993	0369	0.9959	0938	0.9844
	6	0000	1.0000	0001	1.0000	0007	1.0000	0041	1.0000	0156	1.0000
	0	4783	0.4783	2097	0.2097	0824	0.0824	0280	0.0280	0078	0.0078
	1	3720	0.8503	3670	0.5767	2471	0.3294	1306	0.1586	0547	0.0625
	2	1240	0.9743	2753	0.8520	3177	0.6471	2613	0.4199	1641	0.2266
7	3	0230	0.9973	1147	0.9667	2269	0.8740	2903	0.7102	2734	0.5000
	4	0026	0.9998	0287	0.9953	0972	0.9712	1935	0.9037	2734	0.7734
	5	0002	1.0000	0043	0.9996	0250	0.9962	0774	0.9812	1641	0.9375
	6	0000	1.0000	0004	1.0000	0036	0.9998	0172	0.9984	0547	0.9922
	7	0000	1.0000	0000	1.0000	0002	1.0000	0016	1.0000	0078	1.0000
	0	4305	0.4305	1678	0.1678	0576	0.0576	0168	0.0168	0039	0.0039
	1	3826	0.8131	3355	0.5033	1977	0.2553	0896	0.1064	0313	0.0352
	2	1488	0.9619	2936	0.7969	2965	0.5518	2090	0.3154	1094	0.1445
	3	0331	0.9950	1468	0.9437	2541	0.8059	2787	0.5941	2188	0.3633
8	4	0046	0.9996	0459	0.9896	1361	0.9420	2322	0.8263	2734	0.6367
	5	0004	1.0000	0092	0.9988	0467	0.9887	1239	0.9502	2188	0.8555
	6	0000	1.0000	0011	0.9999	0100	0.9987	0413	0.9915	1094	0.9648
	7	0000	1.0000	0001	1.0000	0012	0.9999	0079	0.9993	0313	0.9961
	8	0000	1.0000	0000	1.0000	0001	1.0000	0007	1.0000	0039	1.0000

Table A6 Poisson Distribution

Probability function $f(x)$ [see (5), Sec. 24.7] and distribution function $F(x)$

	$\mu = 0.1$		$\mu = 0.2$		$\mu = 0.3$		$\mu = 0.4$		$\mu = 0.5$	
x	$f(x)$	$F(x)$	$f(x)$	$F(x)$	$f(x)$	$F(x)$	$f(x)$	$F(x)$	$f(x)$	$F(x)$
	0.		0.		0.		0.		0.	
0	9048	0.9048	8187	0.8187	7408	0.7408	6703	0.6703	6065	0.6065
1	0905	0.9953	1637	0.9825	2222	0.9631	2681	0.9384	3033	0.9098
2	0045	0.9998	0164	0.9989	0333	0.9964	0536	0.9921	0758	0.9856
3	0002	1.0000	0011	0.9999	0033	0.9997	0072	0.9992	0126	0.9982
4	0000	1.0000	0001	1.0000	0003	1.0000	0007	0.9999	0016	0.9998
5							0001	1.0000	0002	1.0000

	$\mu = 0.6$		$\mu = 0.7$		$\mu = 0.8$		$\mu = 0.9$		$\mu = 1$	
x	$f(x)$	$F(x)$	$f(x)$	$F(x)$	$f(x)$	$F(x)$	$f(x)$	$F(x)$	$f(x)$	$F(x)$
	0.		0.		0.		0.		0.	
0	5488	0.5488	4966	0.4966	4493	0.4493	4066	0.4066	3679	0.3679
1	3293	0.8781	3476	0.8442	3595	0.8088	3659	0.7725	3679	0.7358
2	0988	0.9769	1217	0.9659	1438	0.9526	1647	0.9371	1839	0.9197
3	0198	0.9966	0284	0.9942	0383	0.9909	0494	0.9865	0613	0.9810
4	0030	0.9996	0050	0.9992	0077	0.9986	0111	0.9977	0153	0.9963
5	0004	1.0000	0007	0.9999	0012	0.9998	0020	0.9997	0031	0.9994
6			0001	1.0000	0002	1.0000	0003	1.0000	0005	0.9999
7									0001	1.0000

	$\mu = 1.5$		$\mu = 2$		$\mu = 3$		$\mu = 4$		$\mu = 5$	
x	$f(x)$	$F(x)$	$f(x)$	$F(x)$	$f(x)$	$F(x)$	$f(x)$	$F(x)$	$f(x)$	$F(x)$
	0.		0.		0.		0.		0.	
0	2231	0.2231	1353	0.1353	0498	0.0498	0183	0.0183	0067	0.0067
1	3347	0.5578	2707	0.4060	1494	0.1991	0733	0.0916	0337	0.0404
2	2510	0.8088	2707	0.6767	2240	0.4232	1465	0.2381	0842	0.1247
3	1255	0.9344	1804	0.8571	2240	0.6472	1954	0.4335	1404	0.2650
4	0471	0.9814	0902	0.9473	1680	0.8153	1954	0.6288	1755	0.4405
5	0141	0.9955	0361	0.9834	1008	0.9161	1563	0.7851	1755	0.6160
6	0035	0.9991	0120	0.9955	0504	0.9665	1042	0.8893	1462	0.7622
7	0008	0.9998	0034	0.9989	0216	0.9881	0595	0.9489	1044	0.8666
8	0001	1.0000	0009	0.9998	0081	0.9962	0298	0.9786	0653	0.9319
9			0002	1.0000	0027	0.9989	0132	0.9919	0363	0.9682
10					0008	0.9997	0053	0.9972	0181	0.9863
11					0002	0.9999	0019	0.9991	0082	0.9945
12					0001	1.0000	0006	0.9997	0034	0.9980
13							0002	0.9999	0013	0.9993
14							0001	1.0000	0005	0.9998
15									0002	0.9999
16									0000	1.0000

Table A7 Normal Distribution

Values of the distribution function $\Phi(z)$ [see (3), Sec. 24.8]. $\Phi(-z) = 1 - \Phi(z)$

z	$\Phi(z)$	z	$\Phi(z)$	z	$\Phi(z)$	z	$\Phi(z)$	z	$\Phi(z)$	z	$\Phi(z)$
	0.		**0.**		**0.**		**0.**		**0.**		**0.**
0.01	5040	0.51	6950	1.01	8438	1.51	9345	2.01	9778	2.51	9940
0.02	5080	0.52	6985	1.02	8461	1.52	9357	2.02	9783	2.52	9941
0.03	5120	0.53	7019	1.03	8485	1.53	9370	2.03	9788	2.53	9943
0.04	5160	0.54	7054	1.04	8508	1.54	9382	2.04	9793	2.54	9945
0.05	5199	0.55	7088	1.05	8531	1.55	9394	2.05	9798	2.55	9946
0.06	5239	0.56	7123	1.06	8554	1.56	9406	2.06	9803	2.56	9948
0.07	5279	0.57	7157	1.07	8577	1.57	9418	2.07	9808	2.57	9949
0.08	5319	0.58	7190	1.08	8599	1.58	9429	2.08	9812	2.58	9951
0.09	5359	0.59	7224	1.09	8621	1.59	9441	2.09	9817	2.59	9952
0.10	5398	0.60	7257	1.10	8643	1.60	9452	2.10	9821	2.60	9953
0.11	5438	0.61	7291	1.11	8665	1.61	9463	2.11	9826	2.61	9955
0.12	5478	0.62	7324	1.12	8686	1.62	9474	2.12	9830	2.62	9956
0.13	5517	0.63	7357	1.13	8708	1.63	9484	2.13	9834	2.63	9957
0.14	5557	0.64	7389	1.14	8729	1.64	9495	2.14	9838	2.64	9959
0.15	5596	0.65	7422	1.15	8749	1.65	9505	2.15	9842	2.65	9960
0.16	5636	0.66	7454	1.16	8770	1.66	9515	2.16	9846	2.66	9961
0.17	5675	0.67	7486	1.17	8790	1.67	9525	2.17	9850	2.67	9962
0.18	5714	0.68	7517	1.18	8810	1.68	9535	2.18	9854	2.68	9963
0.19	5753	0.69	7549	1.19	8830	1.69	9545	2.19	9857	2.69	9964
0.20	5793	0.70	7580	1.20	8849	1.70	9554	2.20	9861	2.70	9965
0.21	5832	0.71	7611	1.21	8869	1.71	9564	2.21	9864	2.71	9966
0.22	5871	0.72	7642	1.22	8888	1.72	9573	2.22	9868	2.72	9967
0.23	5910	0.73	7673	1.23	8907	1.73	9582	2.23	9871	2.73	9968
0.24	5948	0.74	7704	1.24	8925	1.74	9591	2.24	9875	2.74	9969
0.25	5987	0.75	7734	1.25	8944	1.75	9599	2.25	9878	2.75	9970
0.26	6026	0.76	7764	1.26	8962	1.76	9608	2.26	9881	2.76	9971
0.27	6064	0.77	7794	1.27	8980	1.77	9616	2.27	9884	2.77	9972
0.28	6103	0.78	7823	1.28	8997	1.78	9625	2.28	9887	2.78	9973
0.29	6141	0.79	7852	1.29	9015	1.79	9633	2.29	9890	2.79	9974
0.30	6179	0.80	7881	1.30	9032	1.80	9641	2.30	9893	2.80	9974
0.31	6217	0.81	7910	1.31	9049	1.81	9649	2.31	9896	2.81	9975
0.32	6255	0.82	7939	1.32	9066	1.82	9656	2.32	9898	2.82	9976
0.33	6293	0.83	7967	1.33	9082	1.83	9664	2.33	9901	2.83	9977
0.34	6331	0.84	7995	1.34	9099	1.84	9671	2.34	9904	2.84	9977
0.35	6368	0.85	8023	1.35	9115	1.85	9678	2.35	9906	2.85	9978
0.36	6406	0.86	8051	1.36	9131	1.86	9686	2.36	9909	2.86	9979
0.37	6443	0.87	8078	1.37	9147	1.87	9693	2.37	9911	2.87	9979
0.38	6480	0.88	8106	1.38	9162	1.88	9699	2.38	9913	2.88	9980
0.39	6517	0.89	8133	1.39	9177	1.89	9706	2.39	9916	2.89	9981
0.40	6554	0.90	8159	1.40	9192	1.90	9713	2.40	9918	2.90	9981
0.41	6591	0.91	8186	1.41	9207	1.91	9719	2.41	9920	2.91	9982
0.42	6628	0.92	8212	1.42	9222	1.92	9726	2.42	9922	2.92	9982
0.43	6664	0.93	8238	1.43	9236	1.93	9732	2.43	9925	2.93	9983
0.44	6700	0.94	8264	1.44	9251	1.94	9738	2.44	9927	2.94	9984
0.45	6736	0.95	8289	1.45	9265	1.95	9744	2.45	9929	2.95	9984
0.46	6772	0.96	8315	1.46	9279	1.96	9750	2.46	9931	2.96	9985
0.47	6808	0.97	8340	1.47	9292	1.97	9756	2.47	9932	2.97	9985
0.48	6844	0.98	8365	1.48	9306	1.98	9761	2.48	9934	2.98	9986
0.49	6879	0.99	8389	1.49	9319	1.99	9767	2.49	9936	2.99	9986
0.50	6915	1.00	8413	1.50	9332	2.00	9772	2.50	9938	3.00	9987

Table A8 Normal Distribution

Values of z for given values of $\Phi(z)$ [see (3), Sec. 24.8] and $D(z) = \Phi(z) - \Phi(-z)$
Example: $z = 0.279$ if $\Phi(z) = 61\%$; $z = 0.860$ if $D(z) = 61\%$.

%	$z(\Phi)$	$z(D)$	%	$z(\Phi)$	$z(D)$	%	$z(\Phi)$	$z(D)$
1	−2.326	0.013	41	−0.228	0.539	81	0.878	1.311
2	−2.054	0.025	42	−0.202	0.553	82	0.915	1.341
3	−1.881	0.038	43	−0.176	0.568	83	0.954	1.372
4	−1.751	0.050	44	−0.151	0.583	84	0.994	1.405
5	−1.645	0.063	45	−0.126	0.598	85	1.036	1.440
6	−1.555	0.075	46	−0.100	0.613	86	1.080	1.476
7	−1.476	0.088	47	−0.075	0.628	87	1.126	1.514
8	−1.405	0.100	48	−0.050	0.643	88	1.175	1.555
9	−1.341	0.113	49	−0.025	0.659	89	1.227	1.598
10	−1.282	0.126	50	0.000	0.674	90	1.282	1.645
11	−1.227	0.138	51	0.025	0.690	91	1.341	1.695
12	−1.175	0.151	52	0.050	0.706	92	1.405	1.751
13	−1.126	0.164	53	0.075	0.722	93	1.476	1.812
14	−1.080	0.176	54	0.100	0.739	94	1.555	1.881
15	−1.036	0.189	55	0.126	0.755	95	1.645	1.960
16	−0.994	0.202	56	0.151	0.772	96	1.751	2.054
17	−0.954	0.215	57	0.176	0.789	97	1.881	2.170
18	−0.915	0.228	58	0.202	0.806	97.5	1.960	2.241
19	−0.878	0.240	59	0.228	0.824	98	2.054	2.326
20	−0.842	0.253	60	0.253	0.842	99	2.326	2.576
21	−0.806	0.266	61	0.279	0.860	99.1	2.366	2.612
22	−0.772	0.279	62	0.305	0.878	99.2	2.409	2.652
23	−0.739	0.292	63	0.332	0.896	99.3	2.457	2.697
24	−0.706	0.305	64	0.358	0.915	99.4	2.512	2.748
25	−0.674	0.319	65	0.385	0.935	99.5	2.576	2.807
26	−0.643	0.332	66	0.412	0.954	99.6	2.652	2.878
27	−0.613	0.345	67	0.440	0.974	99.7	2.748	2.968
28	−0.583	0.358	68	0.468	0.994	99.8	2.878	3.090
29	−0.553	0.372	69	0.496	1.015	99.9	3.090	3.291
30	−0.524	0.385	70	0.524	1.036			
31	−0.496	0.399	71	0.553	1.058	99.91	3.121	3.320
32	−0.468	0.412	72	0.583	1.080	99.92	3.156	3.353
33	−0.440	0.426	73	0.613	1.103	99.93	3.195	3.390
34	−0.412	0.440	74	0.643	1.126	99.94	3.239	3.432
35	−0.385	0.454	75	0.674	1.150	99.95	3.291	3.481
36	−0.358	0.468	76	0.706	1.175	99.96	3.353	3.540
37	−0.332	0.482	77	0.739	1.200	99.97	3.432	3.615
38	−0.305	0.496	78	0.772	1.227	99.98	3.540	3.719
39	−0.279	0.510	79	0.806	1.254	99.99	3.719	3.891
40	−0.253	0.524	80	0.842	1.282			

Table A9 *t*-Distribution

Values of z for given values of the distribution function $F(z)$ (see (8) in Sec. 25.3). Example: For 9 degrees of freedom, $z = 1.83$ when $F(z) = 0.95$.

$F(z)$	Number of Degrees of Freedom									
	1	2	3	4	5	6	7	8	9	10
0.5	0.00	0.00	0.00	0.00	0.00	0.00	0.00	0.00	0.00	0.00
0.6	0.32	0.29	0.28	0.27	0.27	0.26	0.26	0.26	0.26	0.26
0.7	0.73	0.62	0.58	0.57	0.56	0.55	0.55	0.55	0.54	0.54
0.8	1.38	1.06	0.98	0.94	0.92	0.91	0.90	0.89	0.88	0.88
0.9	3.08	1.89	1.64	1.53	1.48	1.44	1.41	1.40	1.38	1.37
0.95	6.31	2.92	2.35	2.13	2.02	1.94	1.89	1.86	1.83	1.81
0.975	12.7	4.30	3.18	2.78	2.57	2.45	2.36	2.31	2.26	2.23
0.99	31.8	6.96	4.54	3.75	3.36	3.14	3.00	2.90	2.82	2.76
0.995	63.7	9.92	5.84	4.60	4.03	3.71	3.50	3.36	3.25	3.17
0.999	318.3	22.3	10.2	7.17	5.89	5.21	4.79	4.50	4.30	4.14

$F(z)$	Number of Degrees of Freedom									
	11	12	13	14	15	16	17	18	19	20
0.5	0.00	0.00	0.00	0.00	0.00	0.00	0.00	0.00	0.00	0.00
0.6	0.26	0.26	0.26	0.26	0.26	0.26	0.26	0.26	0.26	0.26
0.7	0.54	0.54	0.54	0.54	0.54	0.54	0.53	0.53	0.53	0.53
0.8	0.88	0.87	0.87	0.87	0.87	0.86	0.86	0.86	0.86	0.86
0.9	1.36	1.36	1.35	1.35	1.34	1.34	1.33	1.33	1.33	1.33
0.95	1.80	1.78	1.77	1.76	1.75	1.75	1.74	1.73	1.73	1.72
0.975	2.20	2.18	2.16	2.14	2.13	2.12	2.11	2.10	2.09	2.09
0.99	2.72	2.68	2.65	2.62	2.60	2.58	2.57	2.55	2.54	2.53
0.995	3.11	3.05	3.01	2.98	2.95	2.92	2.90	2.88	2.86	2.85
0.999	4.02	3.93	3.85	3.79	3.73	3.69	3.65	3.61	3.58	3.55

$F(z)$	Number of Degrees of Freedom									
	22	24	26	28	30	40	50	100	200	∞
0.5	0.00	0.00	0.00	0.00	0.00	0.00	0.00	0.00	0.00	0.00
0.6	0.26	0.26	0.26	0.26	0.26	0.26	0.25	0.25	0.25	0.25
0.7	0.53	0.53	0.53	0.53	0.53	0.53	0.53	0.53	0.53	0.52
0.8	0.86	0.86	0.86	0.85	0.85	0.85	0.85	0.85	0.84	0.84
0.9	1.32	1.32	1.31	1.31	1.31	1.30	1.30	1.29	1.29	1.28
0.95	1.72	1.71	1.71	1.70	1.70	1.68	1.68	1.66	1.65	1.65
0.975	2.07	2.06	2.06	2.05	2.04	2.02	2.01	1.98	1.97	1.96
0.99	2.51	2.49	2.48	2.47	2.46	2.42	2.40	2.36	2.35	2.33
0.995	2.82	2.80	2.78	2.76	2.75	2.70	2.68	2.63	2.60	2.58
0.999	3.50	3.47	3.43	3.41	3.39	3.31	3.26	3.17	3.13	3.09

Table A10 Chi-square Distribution

Values of x for given values of the distribution function $F(z)$ (see Sec. 25.3 before (17)). Example: For 3 degrees of freedom, $z = 11.34$ when $F(z) = 0.99$.

$F(z)$	Number of Degrees of Freedom									
	1	2	3	4	5	6	7	8	9	10
0.005	0.00	0.01	0.07	0.21	0.41	0.68	0.99	1.34	1.73	2.16
0.01	0.00	0.02	0.11	0.30	0.55	0.87	1.24	1.65	2.09	2.56
0.025	0.00	0.05	0.22	0.48	0.83	1.24	1.69	2.18	2.70	3.25
0.05	0.00	0.10	0.35	0.71	1.15	1.64	2.17	2.73	3.33	3.94
0.95	3.84	5.99	7.81	9.49	11.07	12.59	14.07	15.51	16.92	18.31
0.975	5.02	7.38	9.35	11.14	12.83	14.45	16.01	17.53	19.02	20.48
0.99	6.63	9.21	11.34	13.28	15.09	16.81	18.48	20.09	21.67	23.21
0.995	7.88	10.60	12.84	14.86	16.75	18.55	20.28	21.95	23.59	25.19

$F(z)$	Number of Degrees of Freedom									
	11	12	13	14	15	16	17	18	19	20
0.005	2.60	3.07	3.57	4.07	4.60	5.14	5.70	6.26	6.84	7.43
0.01	3.05	3.57	4.11	4.66	5.23	5.81	6.41	7.01	7.63	8.26
0.025	3.82	4.40	5.01	5.63	6.26	6.91	7.56	8.23	8.91	9.59
0.05	4.57	5.23	5.89	6.57	7.26	7.96	8.67	9.39	10.12	10.85
0.95	19.68	21.03	22.36	23.68	25.00	26.30	27.59	28.87	30.14	31.41
0.975	21.92	23.34	24.74	26.12	27.49	28.85	30.19	31.53	32.85	34.17
0.99	24.72	26.22	27.69	29.14	30.58	32.00	33.41	34.81	36.19	37.57
0.995	26.76	28.30	29.82	31.32	32.80	34.27	35.72	37.16	38.58	40.00

$F(z)$	Number of Degrees of Freedom									
	21	22	23	24	25	26	27	28	29	30
0.005	8.0	8.6	9.3	9.9	10.5	11.2	11.8	12.5	13.1	13.8
0.01	8.9	9.5	10.2	10.9	11.5	12.2	12.9	13.6	14.3	15.0
0.025	10.3	11.0	11.7	12.4	13.1	13.8	14.6	15.3	16.0	16.8
0.05	11.6	12.3	13.1	13.8	14.6	15.4	16.2	16.9	17.7	18.5
0.95	32.7	33.9	35.2	36.4	37.7	38.9	40.1	41.3	42.6	43.8
0.975	35.5	36.8	38.1	39.4	40.6	41.9	43.2	44.5	45.7	47.0
0.99	38.9	40.3	41.6	43.0	44.3	45.6	47.0	48.3	49.6	50.9
0.995	41.4	42.8	44.2	45.6	46.9	48.3	49.6	51.0	52.3	53.7

$F(z)$	Number of Degrees of Freedom							
	40	50	60	70	80	90	100	> 100 (Approximation)
0.005	20.7	28.0	35.5	43.3	51.2	59.2	67.3	$\frac{1}{2}(h - 2.58)^2$
0.01	22.2	29.7	37.5	45.4	53.5	61.8	70.1	$\frac{1}{2}(h - 2.33)^2$
0.025	24.4	32.4	40.5	48.8	57.2	65.6	74.2	$\frac{1}{2}(h - 1.96)^2$
0.05	26.5	34.8	43.2	51.7	60.4	69.1	77.9	$\frac{1}{2}(h - 1.64)^2$
0.95	55.8	67.5	79.1	90.5	101.9	113.1	124.3	$\frac{1}{2}(h + 1.64)^2$
0.975	59.3	71.4	83.3	95.0	106.6	118.1	129.6	$\frac{1}{2}(h + 1.96)^2$
0.99	63.7	76.2	88.4	100.4	112.3	124.1	135.8	$\frac{1}{2}(h + 2.33)^2$
0.995	66.8	79.5	92.0	104.2	116.3	128.3	140.2	$\frac{1}{2}(h + 2.58)^2$

In the last column, $h = \sqrt{2m - 1}$, where m is the number of degrees of freedom.

Table A11 *F*-Distribution with (*m*, *n*) Degrees of Freedom

Values of z for which the distribution function $F(z)$ [see (13), Sec. 25.4] has the value **0.95**

Example: For (7, 4) d.f., $z = 6.09$ if $F(z) = 0.95$.

n	$m = 1$	$m = 2$	$m = 3$	$m = 4$	$m = 5$	$m = 6$	$m = 7$	$m = 8$	$m = 9$
1	161	200	216	225	230	234	237	239	241
2	18.5	19.0	19.2	19.2	19.3	19.3	19.4	19.4	19.4
3	10.1	9.55	9.28	9.12	9.01	8.94	8.89	8.85	8.81
4	7.71	6.94	6.59	6.39	6.26	6.16	6.09	6.04	6.00
5	6.61	5.79	5.41	5.19	5.05	4.95	4.88	4.82	4.77
6	5.99	5.14	4.76	4.53	4.39	4.28	4.21	4.15	4.10
7	5.59	4.74	4.35	4.12	3.97	3.87	3.79	3.73	3.68
8	5.32	4.46	4.07	3.84	3.69	3.58	3.50	3.44	3.39
9	5.12	4.26	3.86	3.63	3.48	3.37	3.29	3.23	3.18
10	4.96	4.10	3.71	3.48	3.33	3.22	3.14	3.07	3.02
11	4.84	3.98	3.59	3.36	3.20	3.09	3.01	2.95	2.90
12	4.75	3.89	3.49	3.26	3.11	3.00	2.91	2.85	2.80
13	4.67	3.81	3.41	3.18	3.03	2.92	2.83	2.77	2.71
14	4.60	3.74	3.34	3.11	2.96	2.85	2.76	2.70	2.65
15	4.54	3.68	3.29	3.06	2.90	2.79	2.71	2.64	2.59
16	4.49	3.63	3.24	3.01	2.85	2.74	2.66	2.59	2.54
17	4.45	3.59	3.20	2.96	2.81	2.70	2.61	2.55	2.49
18	4.41	3.55	3.16	2.93	2.77	2.66	2.58	2.51	2.46
19	4.38	3.52	3.13	2.90	2.74	2.63	2.54	2.48	2.42
20	4.35	3.49	3.10	2.87	2.71	2.60	2.51	2.45	2.39
22	4.30	3.44	3.05	2.82	2.66	2.55	2.46	2.40	2.34
24	4.26	3.40	3.01	2.78	2.62	2.51	2.42	2.36	2.30
26	4.23	3.37	2.98	2.74	2.59	2.47	2.39	2.32	2.27
28	4.20	3.34	2.95	2.71	2.56	2.45	2.36	2.29	2.24
30	4.17	3.32	2.92	2.69	2.53	2.42	2.33	2.27	2.21
32	4.15	3.29	2.90	2.67	2.51	2.40	2.31	2.24	2.19
34	4.13	3.28	2.88	2.65	2.49	2.38	2.29	2.23	2.17
36	4.11	3.26	2.87	2.63	2.48	2.36	2.28	2.21	2.15
38	4.10	3.24	2.85	2.62	2.46	2.35	2.26	2.19	2.14
40	4.08	3.23	2.84	2.61	2.45	2.34	2.25	2.18	2.12
50	4.03	3.18	2.79	2.56	2.40	2.29	2.20	2.13	2.07
60	4.00	3.15	2.76	2.53	2.37	2.25	2.17	2.10	2.04
70	3.98	3.13	2.74	2.50	2.35	2.23	2.14	2.07	2.02
80	3.96	3.11	2.72	2.49	2.33	2.21	2.13	2.06	2.00
90	3.95	3.10	2.71	2.47	2.32	2.20	2.11	2.04	1.99
100	3.94	3.09	2.70	2.46	2.31	2.19	2.10	2.03	1.97
150	3.90	3.06	2.66	2.43	2.27	2.16	2.07	2.00	1.94
200	3.89	3.04	2.65	2.42	2.26	2.14	2.06	1.98	1.93
1000	3.85	3.00	2.61	2.38	2.22	2.11	2.02	1.95	1.89
∞	3.84	3.00	2.60	2.37	2.21	2.10	2.01	1.94	1.88

Table A11 ***F*-Distribution with (*m*, *n*) Degrees of Freedom *(continued)***

Values of z for which the distribution function $F(z)$ [see (13), Sec. 25.4] has the value **0.95**

n	$m = 10$	$m = 15$	$m = 20$	$m = 30$	$m = 40$	$m = 50$	$m = 100$	∞
1	242	246	248	250	251	252	253	254
2	19.4	19.4	19.4	19.5	19.5	19.5	19.5	19.5
3	8.79	8.70	8.66	8.62	8.59	8.58	8.55	8.53
4	5.96	5.86	5.80	5.75	5.72	5.70	5.66	5.63
5	4.74	4.62	4.56	4.50	4.46	4.44	4.41	4.37
6	4.06	3.94	3.87	3.81	3.77	3.75	3.71	3.67
7	3.64	3.51	3.44	3.38	3.34	3.32	3.27	3.23
8	3.35	3.22	3.15	3.08	3.04	3.02	2.97	2.93
9	3.14	3.01	2.94	2.86	2.83	2.80	2.76	2.71
10	2.98	2.85	2.77	2.70	2.66	2.64	2.59	2.54
11	2.85	2.72	2.65	2.57	2.53	2.51	2.46	2.40
12	2.75	2.62	2.54	2.47	2.43	2.40	2.35	2.30
13	2.67	2.53	2.46	2.38	2.34	2.31	2.26	2.21
14	2.60	2.46	2.39	2.31	2.27	2.24	2.19	2.13
15	2.54	2.40	2.33	2.25	2.20	2.18	2.12	2.07
16	2.49	2.35	2.28	2.19	2.15	2.12	2.07	2.01
17	2.45	2.31	2.23	2.15	2.10	2.08	2.02	1.96
18	2.41	2.27	2.19	2.11	2.06	2.04	1.98	1.92
19	2.38	2.23	2.16	2.07	2.03	2.00	1.94	1.88
20	2.35	2.20	2.12	2.04	1.99	1.97	1.91	1.84
22	2.30	2.15	2.07	1.98	1.94	1.91	1.85	1.78
24	2.25	2.11	2.03	1.94	1.89	1.86	1.80	1.73
26	2.22	2.07	1.99	1.90	1.85	1.82	1.76	1.69
28	2.19	2.04	1.96	1.87	1.82	1.79	1.73	1.65
30	2.16	2.01	1.93	1.84	1.79	1.76	1.70	1.62
32	2.14	1.99	1.91	1.82	1.77	1.74	1.67	1.59
34	2.12	1.97	1.89	1.80	1.75	1.71	1.65	1.57
36	2.11	1.95	1.87	1.78	1.73	1.69	1.62	1.55
38	2.09	1.94	1.85	1.76	1.71	1.68	1.61	1.53
40	2.08	1.92	1.84	1.74	1.69	1.66	1.59	1.51
50	2.03	1.87	1.78	1.69	1.63	1.60	1.52	1.44
60	1.99	1.84	1.75	1.65	1.59	1.56	1.48	1.39
70	1.97	1.81	1.72	1.62	1.57	1.53	1.45	1.35
80	1.95	1.79	1.70	1.60	1.54	1.51	1.43	1.32
90	1.94	1.78	1.69	1.59	1.53	1.49	1.41	1.30
100	1.93	1.77	1.68	1.57	1.52	1.48	1.39	1.28
150	1.89	1.73	1.64	1.54	1.48	1.44	1.34	1.22
200	1.88	1.72	1.62	1.52	1.46	1.41	1.32	1.19
1000	1.84	1.68	1.58	1.47	1.41	1.36	1.26	1.08
∞	1.83	1.67	1.57	1.46	1.39	1.35	1.24	1.00

Table A11 *F*-Distribution with (*m*, *n*) Degrees of Freedom *(continued)*

Values of z for which the distribution function $F(z)$ [see (13), Sec. 25.4] has the value **0.99**

n	$m = 1$	$m = 2$	$m = 3$	$m = 4$	$m = 5$	$m = 6$	$m = 7$	$m = 8$	$m = 9$
1	4052	4999	5403	5625	5764	5859	5928	5981	6022
2	98.5	99.0	99.2	99.2	99.3	99.3	99.4	99.4	99.4
3	34.1	30.8	29.5	28.7	28.2	27.9	27.7	27.5	27.3
4	21.2	18.0	16.7	16.0	15.5	15.2	15.0	14.8	14.7
5	16.3	13.3	12.1	11.4	11.0	10.7	10.5	10.3	10.2
6	13.7	10.9	9.78	9.15	8.75	8.47	8.26	8.10	7.98
7	12.2	9.55	8.45	7.85	7.46	7.19	6.99	6.84	6.72
8	11.3	8.65	7.59	7.01	6.63	6.37	6.18	6.03	5.91
9	10.6	8.02	6.99	6.42	6.06	5.80	5.61	5.47	5.35
10	10.0	7.56	6.55	5.99	5.64	5.39	5.20	5.06	4.94
11	9.65	7.21	6.22	5.67	5.32	5.07	4.89	4.74	4.63
12	9.33	6.93	5.95	5.41	5.06	4.82	4.64	4.50	4.39
13	9.07	6.70	5.74	5.21	4.86	4.62	4.44	4.30	4.19
14	8.86	6.51	5.56	5.04	4.69	4.46	4.28	4.14	4.03
15	8.68	6.36	5.42	4.89	4.56	4.32	4.14	4.00	3.89
16	8.53	6.23	5.29	4.77	4.44	4.20	4.03	3.89	3.78
17	8.40	6.11	5.18	4.67	4.34	4.10	3.93	3.79	3.68
18	8.29	6.01	5.09	4.58	4.25	4.01	3.84	3.71	3.60
19	8.18	5.93	5.01	4.50	4.17	3.94	3.77	3.63	3.52
20	8.10	5.85	4.94	4.43	4.10	3.87	3.70	3.56	3.46
22	7.95	5.72	4.82	4.31	3.99	3.76	3.59	3.45	3.35
24	7.82	5.61	4.72	4.22	3.90	3.67	3.50	3.36	3.26
26	7.72	5.53	4.64	4.14	3.82	3.59	3.42	3.29	3.18
28	7.64	5.45	4.57	4.07	3.75	3.53	3.36	3.23	3.12
30	7.56	5.39	4.51	4.02	3.70	3.47	3.30	3.17	3.07
32	7.50	5.34	4.46	3.97	3.65	3.43	3.26	3.13	3.02
34	7.44	5.29	4.42	3.93	3.61	3.39	3.22	3.09	2.98
36	7.40	5.25	4.38	3.89	3.57	3.35	3.18	3.05	2.95
38	7.35	5.21	4.34	3.86	3.54	3.32	3.15	3.02	2.92
40	7.31	5.18	4.31	3.83	3.51	3.29	3.12	2.99	2.89
50	7.17	5.06	4.20	3.72	3.41	3.19	3.02	2.89	2.78
60	7.08	4.98	4.13	3.65	3.34	3.12	2.95	2.82	2.72
70	7.01	4.92	4.07	3.60	3.29	3.07	2.91	2.78	2.67
80	6.96	4.88	4.04	3.56	3.26	3.04	2.87	2.74	2.64
90	6.93	4.85	4.01	3.54	3.23	3.01	2.84	2.72	2.61
100	6.90	4.82	3.98	3.51	3.21	2.99	2.82	2.69	2.59
150	6.81	4.75	3.91	3.45	3.14	2.92	2.76	2.63	2.53
200	6.76	4.71	3.88	3.41	3.11	2.89	2.73	2.60	2.50
1000	6.66	4.63	3.80	3.34	3.04	2.82	2.66	2.53	2.43
∞	6.63	4.61	3.78	3.32	3.02	2.80	2.64	2.51	2.41

Table A11 *F*-Distribution with *(m, n)* Degrees of Freedom *(continued)*

Values of z for which the distribution function $F(z)$ [see (13), Sec. 25.4] has the value **0.99**

n	$m = 10$	$m = 15$	$m = 20$	$m = 30$	$m = 40$	$m = 50$	$m = 100$	∞
1	6056	6157	6209	6261	6287	6303	6334	6366
2	99.4	99.4	99.4	99.5	99.5	99.5	99.5	99.5
3	27.2	26.9	26.7	26.5	26.4	26.4	26.2	26.1
4	14.5	14.2	14.0	13.8	13.7	13.7	13.6	13.5
5	10.1	9.72	9.55	9.38	9.29	9.24	9.13	9.02
6	7.87	7.56	7.40	7.23	7.14	7.09	6.99	6.88
7	6.62	6.31	6.16	5.99	5.91	5.86	5.75	5.65
8	5.81	5.52	5.36	5.20	5.12	5.07	4.96	4.86
9	5.26	4.96	4.81	4.65	4.57	4.52	4.42	4.31
10	4.85	4.56	4.41	4.25	4.17	4.12	4.01	3.91
11	4.54	4.25	4.10	3.94	3.86	3.81	3.71	3.60
12	4.30	4.01	3.86	3.70	3.62	3.57	3.47	3.36
13	4.10	3.82	3.66	3.51	3.43	3.38	3.27	3.17
14	3.94	3.66	3.51	3.35	3.27	3.22	3.11	3.00
15	3.80	3.52	3.37	3.21	3.13	3.08	2.98	2.87
16	3.69	3.41	3.26	3.10	3.02	2.97	2.86	2.75
17	3.59	3.31	3.16	3.00	2.92	2.87	2.76	2.65
18	3.51	3.23	3.08	2.92	2.84	2.78	2.68	2.57
19	3.43	3.15	3.00	2.84	2.76	2.71	2.60	2.49
20	3.37	3.09	2.94	2.78	2.69	2.64	2.54	2.42
22	3.26	2.98	2.83	2.67	2.58	2.53	2.42	2.31
24	3.17	2.89	2.74	2.58	2.49	2.44	2.33	2.21
26	3.09	2.81	2.66	2.50	2.42	2.36	2.25	2.13
28	3.03	2.75	2.60	2.44	2.35	2.30	2.19	2.06
30	2.98	2.70	2.55	2.39	2.30	2.25	2.13	2.01
32	2.93	2.65	2.50	2.34	2.25	2.20	2.08	1.96
34	2.89	2.61	2.46	2.30	2.21	2.16	2.04	1.91
36	2.86	2.58	2.43	2.26	2.18	2.12	2.00	1.87
38	2.83	2.55	2.40	2.23	2.14	2.09	1.97	1.84
40	2.80	2.52	2.37	2.20	2.11	2.06	1.94	1.80
50	2.70	2.42	2.27	2.10	2.01	1.95	1.82	1.68
60	2.63	2.35	2.20	2.03	1.94	1.88	1.75	1.60
70	2.59	2.31	2.15	1.98	1.89	1.83	1.70	1.54
80	2.55	2.27	2.12	1.94	1.85	1.79	1.65	1.49
90	2.52	2.24	2.09	1.92	1.82	1.76	1.62	1.46
100	2.50	2.22	2.07	1.89	1.80	1.74	1.60	1.43
150	2.44	2.16	2.00	1.83	1.73	1.66	1.52	1.33
200	2.41	2.13	1.97	1.79	1.69	1.63	1.48	1.28
1000	2.34	2.06	1.90	1.72	1.61	1.54	1.38	1.11
∞	2.32	2.04	1.88	1.70	1.59	1.52	1.36	1.00

Table A12 Distribution Function $F(x) = P(T \leqq x)$ of the Random Variable T in Section 25.8

x	$n = 3$
	0.
0	167
1	500

x	$n = 4$
	0.
0	042
1	167
2	375

x	$n = 5$
	0.
0	008
1	042
2	117
3	242
4	408

x	$n = 6$
	0.
0	001
1	008
2	028
3	068
4	136
5	235
6	360
7	500

x	$n = 7$
	0.
1	001
2	005
3	015
4	035
5	068
6	119
7	191
8	281
9	386
10	500

x	$n = 8$
	0.
2	001
3	003
4	007
5	016
6	031
7	054
8	089
9	138
10	199
11	274
12	360
13	452

x	$n = 9$
	0.
4	001
5	003
6	006
7	012
8	022
9	038
10	060
11	090
12	130
13	179
14	238
15	306
16	381
17	460

x	$n = 10$
	0.
6	001
7	002
8	005
9	008
10	014
11	023
12	036
13	054
14	078
15	108
16	146
17	190
18	242
19	300
20	364
21	431
22	500

x	$n = 11$
	0.
8	001
9	002
10	003
11	005
12	008
13	013
14	020
15	030
16	043
17	060
18	082
19	109
20	141
21	179
22	223
23	271
24	324
25	381
26	440
27	500

x	$n = 12$
	0.
11	001
12	002
13	003
14	004
15	007
16	010
17	016
18	022
19	031
20	043
21	058
22	076
23	098
24	125
25	155
26	190
27	230
28	273
29	319
30	369
31	420
32	473

x	$n = 13$
	0.
14	001
15	001
16	002
17	003
18	005
19	007
20	011
21	015
22	021
23	029
24	038
25	050
26	064
27	082
28	102
29	126
30	153
31	184
32	218
33	255
34	295
35	338
36	383
37	429
38	476

x	$n = 14$
	0.
18	001
19	002
20	002
21	003
22	005
23	007
24	010
25	013
26	018
27	024
28	031
29	040
30	051
31	063
32	079
33	096
34	117
35	140
36	165
37	194
38	225
39	259
40	295
41	334
42	374
43	415
44	457
45	500

x	$n = 15$
	0.
23	001
24	002
25	003
26	004
27	006
28	008
29	010
30	014
31	018
32	023
33	029
34	037
35	046
36	057
37	070
38	084
39	101
40	120
41	141
42	164
43	190
44	218
45	248
46	279
47	313
48	349
49	385
50	423
51	461
52	500

x	$n = 16$
	0.
27	001
28	002
29	002
30	003
31	004
32	006
33	008
34	010
35	013
36	016
37	021
38	026
39	032
40	039
41	048
42	058
43	070
44	083
45	097
46	114
47	133
48	153
49	175
50	199
51	225
52	253
53	282
54	313
55	345
56	378
57	412
58	447
59	482

x	$n = 17$
	0.
32	001
33	002
34	002
35	003
36	004
37	005
38	007
39	009
40	011
41	014
42	017
43	021
44	026
45	032
46	038
47	046
48	054
49	064
50	076
51	088
52	102
53	118
54	135
55	154
56	174
57	196
58	220
59	245
60	271
61	299
62	328
63	358
64	388
65	420
66	452
67	484

x	$n = 18$
	0.
38	001
39	002
40	003
41	003
42	004
43	005
44	007
45	009
46	011
47	013
48	016
49	020
50	024
51	029
52	034
53	041
54	048
55	056
56	066
57	076
58	088
59	100
60	115
61	130
62	147
63	165
64	184
65	205
66	227
67	250
68	275
69	300
70	327
71	354
72	383
73	411
74	441
75	470
76	500

x	$n = 19$
	0.
43	001
44	002
45	002
46	003
47	003
48	004
49	005
50	006
51	008
52	010
53	012
54	014
55	017
56	021
57	025
58	029
59	034
60	040
61	047
62	054
63	062
64	072
65	082
66	093
67	105
68	119
69	133
70	149
71	166
72	184
73	203
74	223
75	245
76	267
77	290
78	314
79	339
80	365
81	391
82	418
83	445
84	473
85	500

x	$n = 20$
	0.
50	001
51	002
52	002
53	003
54	004
55	005
56	006
57	007
58	008
59	010
60	012
61	014
62	017
63	020
64	023
65	027
66	032
67	037
68	043
69	049
70	056
71	064
72	073
73	082
74	093
75	104
76	117
77	130
78	144
79	159
80	176
81	193
82	211
83	230
84	250
85	271
86	293
87	315
88	339
89	362
90	387
91	411
92	436
93	462
94	487

INDEX

I1

Some Constants

$$e = 2.71828\ 18284\ 59045\ 23536$$
$$\sqrt{e} = 1.64872\ 12707\ 00128\ 14685$$
$$e^2 = 7.38905\ 60989\ 30650\ 22723$$

$$\pi = 3.14159\ 26535\ 89793\ 23846$$
$$\pi^2 = 9.86960\ 44010\ 89358\ 61883$$
$$\sqrt{\pi} = 1.77245\ 38509\ 05516\ 02730$$

$$\log_{10} \pi = 0.49714\ 98726\ 94133\ 85435$$
$$\ln \pi = 1.14472\ 98858\ 49400\ 17414$$
$$\log_{10} e = 0.43429\ 44819\ 03251\ 82765$$
$$\ln 10 = 2.30258\ 50929\ 94045\ 68402$$

$$\sqrt{2} = 1.41421\ 35623\ 73095\ 04880$$
$$\sqrt[3]{2} = 1.25992\ 10498\ 94873\ 16477$$
$$\sqrt{3} = 1.73205\ 08075\ 68877\ 29353$$
$$\sqrt[3]{3} = 1.44224\ 95703\ 07408\ 38232$$
$$\ln 2 = 0.69314\ 71805\ 59945\ 30942$$
$$\ln 3 = 1.09861\ 22886\ 68109\ 69140$$

$$\gamma = 0.57721\ 56649\ 01532\ 86061$$
$$\ln \gamma = -0.54953\ 93129\ 81644\ 82234$$
(see Sec. 5.6)
$$1° = 0.01745\ 32925\ 19943\ 29577 \text{ rad}$$
$$1 \text{ rad} = 57.29577\ 95130\ 82320\ 87680° = 57°17'44.806''$$

Polar Coordinates

$$x = r\cos\theta \qquad y = r\sin\theta$$
$$r = \sqrt{x^2 + y^2} \qquad \tan\theta = \frac{y}{x}$$
$$dx\,dy = r\,dr\,d\theta$$

Series

$$\frac{1}{1-x} = \sum_{m=0}^{\infty} x^m \qquad (|x| < 1)$$
$$e^x = \sum_{m=0}^{\infty} \frac{x^m}{m!}$$
$$\sin x = \sum_{m=0}^{\infty} \frac{(-1)^m x^{2m+1}}{(2m+1)!}$$
$$\cos x = \sum_{m=0}^{\infty} \frac{(-1)^m x^{2m}}{(2m)!}$$
$$\ln(1-x) = -\sum_{m=1}^{\infty} \frac{x^m}{m} \qquad (|x| < 1)$$
$$\arctan x = \sum_{m=0}^{\infty} \frac{(-1)^m x^{2m+1}}{2m+1} \qquad (|x| < 1)$$

Greek Alphabet

α	Alpha	ν	Nu
β	Beta	ξ	Xi
γ, Γ	Gamma	o	Omicron
δ, Δ	Delta	π	Pi
ϵ, ε	Epsilon	ρ	Rho
ζ	Zeta	σ, Σ	Sigma
η	Eta	τ	Tau
$\theta, \vartheta, \Theta$	Theta	υ, Υ	Upsilon
ι	Iota	ϕ, φ, Φ	Phi
κ	Kappa	χ	Chi
λ, Λ	Lambda	ψ, Ψ	Psi
μ	Mu	ω, Ω	Omega

Vectors

$$\mathbf{a} \cdot \mathbf{b} = a_1 b_1 + a_2 b_2 + a_3 b_3$$
$$\mathbf{a} \times \mathbf{b} = \begin{vmatrix} \mathbf{i} & \mathbf{j} & \mathbf{k} \\ a_1 & a_2 & a_3 \\ b_1 & b_2 & b_3 \end{vmatrix}$$
$$\text{grad } f = \nabla f = \frac{\partial f}{\partial x}\mathbf{i} + \frac{\partial f}{\partial y}\mathbf{j} + \frac{\partial f}{\partial z}\mathbf{k}$$
$$\text{div } \mathbf{v} = \nabla \cdot \mathbf{v} = \frac{\partial v_1}{\partial x} + \frac{\partial v_2}{\partial y} + \frac{\partial v_3}{\partial z}$$
$$\text{curl } \mathbf{v} = \nabla \times \mathbf{v} = \begin{vmatrix} \mathbf{i} & \mathbf{j} & \mathbf{k} \\ \frac{\partial}{\partial x} & \frac{\partial}{\partial y} & \frac{\partial}{\partial z} \\ v_1 & v_2 & v_3 \end{vmatrix}$$

PHOTO CREDITS

Part A Opener: © Denis Jr. Tangney/iStockphoto
Part B Opener: © Jill Fromer/iStockphoto
Part C Opener: © Science Photo Library/Photo Researchers, Inc
Part D Opener: © Rafa Irusta/iStockphoto
Part E Opener: © Alberto Pomares/iStockphoto
Chapter 19, Figure 437: © Eddie Gerald/Alamy
Part F Opener: © Rainer Plendl/iStockphoto
Part G Opener: © Sean Locke/iStockphoto
Appendix 1 Opener: © Ricardo De Mattos/iStockphoto
Appendix 2 Opener: © joel-t/iStockphoto
Appendix 3 Opener: © Luke Daniek/iStockphoto
Appendix 4 Opener: © Andrey Prokhorov/iStockphoto
Appendix 5 Opener: © Pedro Castellano/iStockphoto